T0140339

Advances in Intelligent Systems and Computing

Volume 725

Series editor

Janusz Kacprzyk, Polish Academy of Sciences, Warsaw, Poland
e-mail: kacprzyk@ibspan.waw.pl

About this Series

The series "Advances in Intelligent Systems and Computing" contains publications on theory, applications, and design methods of Intelligent Systems and Intelligent Computing. Virtually all disciplines such as engineering, natural sciences, computer and information science, ICT, economics, business, e-commerce, environment, healthcare, life science are covered. The list of topics spans all the areas of modern intelligent systems and computing.

The publications within "Advances in Intelligent Systems and Computing" are primarily textbooks and proceedings of important conferences, symposia and congresses. They cover significant recent developments in the field, both of a foundational and applicable character. An important characteristic feature of the series is the short publication time and world-wide distribution. This permits a rapid and broad dissemination of research results.

More information about this series at http://www.springer.com/series/11156

Michael E. Auer
Thrasyvoulos Tsiatsos
Editors

Interactive Mobile Communication Technologies and Learning

Proceedings of the 11th IMCL Conference

Editors
Michael E. Auer
Carinthia University of Applied Sciences
Villach
Austria

Thrasyvoulos Tsiatsos
Department of Informatics
Aristotle University of Thessaloniki
Thessaloniki
Greece

ISSN 2194-5357 ISSN 2194-5365 (electronic)
Advances in Intelligent Systems and Computing
ISBN 978-3-319-75174-0 ISBN 978-3-319-75175-7 (eBook)
https://doi.org/10.1007/978-3-319-75175-7

Printed on acid-free paper

This Springer imprint is published by the registered company Springer International Publishing AG part of Springer Nature
The registered company address is: Gewerbestrasse 11, 6330 Cham, Switzerland

Preface

IMCL2017 was the 11th edition of the International Conference on Interactive Mobile Communication Technologies and Learning.

This interdisciplinary conference is part of an international initiative to promote technology-enhanced learning and online engineering worldwide. The IMCL2017 covered all aspects of mobile learning as well as the emergence of mobile communication technologies, infrastructures and services and their implications for education, business, governments and society.

The IMCL conference series actually aims to promote the development of mobile learning, to provide a forum for education and knowledge transfer, to expose students to latest ICTs and to encourage the study and implementation of mobile applications in teaching and learning. The conference was also platform for critical debates on theories, approaches, principles and applications of mobile learning among educators, developers, researchers, practitioners and policy makers.

IMCL2017 has been organized by Aristotle University of Thessaloniki, Greece, from 30 November to 01 December 2017.

This year's theme of the conference was "Concepts, Infrastructures and Applications".

Again, outstanding scientists from around the world accepted the invitation for keynote speeches:

- Tassos Anastasios Mikropoulos, University of Ioannina, Greece: Pedagogical Reasoning in Mobile Technologies
- Ana Pavani, Pontifical Catholic University of Rio de Janeiro, Brazil: Different problems, different solutions and one goal

Furthermore, two very interesting tutorials have been organized:

- "Smart Cities Energy Management: A Cooperative Perspectives" by Dr. John Vardakas, Senior Researcher, Iquadrat Informatica S.L. Barcelona, Spain & Dr. Christos Verikoukis, Fellow Researcher, Telecommunications Technological Centre of Catalonia, Spain

- “Design Guidelines for Location-based Mobile Games for Learning”, N. Avouris, N.Yiannoutsou & C. Sintoris, University of Patras, Greece

Since its beginning, this conference is devoted to new approaches in learning with a focus on mobile learning, mobile communication, mobile technologies and engineering education.

We are currently witnessing a significant transformation in the development of working and learning environments with a focus on mobile online communication.

Therefore, the following main topics have been discussed during the conference in detail:

- Future Trends and Emerging Mobile Technologies
- Design and Development Mobile Learning Apps and Content
- Mobile Games, Gamification and Mobile Learning
- Adaptive Mobile Environments
- Augmented Reality and Immersive Applications
- Tangible, Embedded and Embodied Interaction
- Interactive Collaborative and Blended Learning
- Digital Technology in Sports
- Mobile Health Care and Training
- Multimedia Learning in Music Education
- 5G Network Infrastructure
- Case Studies
- Real-World Experiences

The following Special Sessions have been organized:

- 5G Wireless and Optical Technologies for Mobile Communication Systems (5GWOTforMCS)
- Augmented Reality and Immersive Applications (ARIA)
- Digital Technology in Sports (DiTeS)
- Game-Based Learning (GBL)
- Mobile Health Care, new Trends and Technologies (MHCTT)
- Multimedia Learning in Music Education (mLME)
- 2nd IMCL Student International Competition for Mobile Apps

The conference accepted as submission types the following:

- Full Paper, Short Paper
- Work in Progress, Poster
- Special Sessions
- Roundtable Discussions, Workshops, Tutorials, Doctoral Consortium, Students’ Competition

All contributions were subject to a double-blind review. The review process was very competitive. We had to review about 200 submissions. A team of about 250 reviewers did this terrific job. Our thanks go to all of them.

Due to the time and conference schedule restrictions, we could finally accept only the best 99 submissions for presentation.

Our conference had again more than 180 participants from 28 countries.

IMCL2018 will be held at McMaster University Hamilton, Canada.

Our special thanks go to Sebastian Schreiter as the conference's Webmaster and technical editor of this proceedings.

Michael E. Auer
IMCL Steering Committee Chair
Thrasyvoulos Tsiatsos
IMCL General Chair

Committees

Steering Committee Chair

Michael E. Auer	CTI, Villach, Austria

General Conference Chair

Thrasyvoulos Tsiatsos	Aristotle University of Thessaloniki, Greece

International Chairs

Samir A. El-Seoud	The British University in Egypt (Africa)
Kumiko Aoki	Open University, Japan (Asia)
Alexander Kist	University of Southern Queensland, Australia (Australia/Oceania)
Doru Ursutiu	University Transylvania Brasov, Romania (Europe)
Arthur Edwards	Universidad de Colima, Mexico (Latin America)
David Guralnick	Kaleidoscope Learning New York, USA (North America)

Technical Programme Chairs

Stavros Demetriadis	Aristotle University of Thessaloniki, Greece
Sebastian Schreiter	IAOE, France

IEEE Liaison

Russ Meier	IEEE Education Society Meetings Chair

Workshop, Tutorial and Special Sessions Chair

Andreas Pester — Carinthia University of Applied Sciences Villach, Austria

Publication Chair

Sebastian Schreiter — IAOE, France

Local Organization Chair

Stella Douka — Aristotle University of Thessaloniki, Greece

Local Organization Committee Members

Christos Temertzoglou — Aristotle University of Thessaloniki, Greece
Vasiliki Peana — Aristotle University of Thessaloniki, Greece
Stavros Stavroulakis — Aristotle University of Thessaloniki, Greece

Programme Committee Members

Abul Azad — Northern Illinois University, USA
Achilles Kameas — Hellenic Open University, Greece
Agisilaos Konidaris — Technological Educational Institute of Ionian Islands, Greece
Alexander Chatzigeorgiou — University of Macedonia, Greece
Anastasios Economides — University of Macedonia, Greece
Anastasios Karakostas — Information Technologies Institute, Greece
Anastasios Mikropoulos — University of Ioannina, Greece
Andreas Veglis — Aristotle University of Thessaloniki, Greece
Apostolos Gkamas — University Ecclesiastical Academy of Vella of Ioannina, Greece
Barbara Kerr — Ottawa University, Canada
Carlos Travieso-González — Universidad de Las Palmas de Gran Canaria, Spain
Charalampos Karagiannidis — University of Thessaly, Greece
Christos Bouras — University of Patras, Greece
Christos Douligeris — University of Piraeus, Greece
Christos Georgiadis — University of Macedonia, Greece
Christos Panagiotakopoulos — University of Patras, Greece

Panagiotis Bamidis	Aristotle University of Thessaloniki, Greece
Panagiotis Politis	University of Thessaly, Greece
Petros Lameras	The Serious Games Institute, UK
Petros Nicopolitidis	Aristotle University of Thessaloniki, Greece
Rhena Delport	University of Pretoria, South Africa
Santi Caballé	Open University of Catalonia, Spain
Stelios Xinogalos	University of Macedonia, Greece
Symeon Retalis	University of Piraeus, Greece
Tharenos Bratitsis	University of Western Macedonia, Greece
Ting-Ting Wu	National Yunlin University of Science and Technology, Taiwan
Vassilis Komis	University of Patras, Greece

Special Session "IMCL2015 Doctoral Consortium" Programme Committee

Chairs

Stavros Demetriadis	Aristotle University of Thessaloniki, Greece
Thrasyvoulos Tsiatsos	Aristotle University of Thessaloniki, Greece

Members

Anastasios Economides	University of Macedonia, Greece
Iraklis Paraskakis	South East European Research Centre, Greece
Charalampos Karagiannidis	Department of Special Education, University of Thessaly, Greece
George Palaigeorgiou	University of Western Macedonia, Greece
Tharenos Bratitsis	University of Western Macedonia, Greece

2nd IMCL Student International Competition for Mobile Apps

Chairs

Andreas Pester	Carinthia University of Applied Sciences, Austria
Ioannis Stamelos	Aristotle University of Thessaloniki, Greece

Judges

Petros Nikopolitidis	Aristotle University of Thessaloniki, Greece
Teresa Restivo	University of Porto, Portugal

Ilias Trohidis	Tero Consulting, Greece
Athena Vakali	Aristotle University of Thessaloniki, Greece
Panagiotis Siozos	Learnworlds, UK

Contents

Future Trends and Emerging Mobile Technologies

Digital Technology in Sports

Mobile Health Care and Training

5G Network Infrastructure

Case Studies

Future Trends and Emerging Mobile Technologies

Towards the Use of Cognitive Radio to Solve Cellular Network Challenges

Amine Hamdouchi[1(✉)], Aouatef El Biari[1], Badr Benmammar[2], and Youness Tabii[1]

[1] National School of Applied Sciences, Tetouan, Morocco
aminehamdouchi254@gmail.com, elbiariaouatef@gmail.com, youness.tabii@gmail.com
[2] LTT Laboratory, University of Tlemcen, Tlemcen, Algeria
badr.benmammar@gmail.com

Abstract. Programming and behavioral autonomy constitutes the main features of any cognitive system. In this regard, Cognitive Radio (CR) uses the concept of understanding-by-building in order to achieve two important objectives. These objectives are namely, establishing a long-lasting reliable communication, and allowing the most efficient use of the spectrum resources. Achieving said objectives is possible by supporting secondary users to access to the licensed spectrum in an opportunistic fashion, after vacant spectrum holes are detected once a primary user exits.

In this paper, we will look to enhance the cellular networks by integrating a cognitive radio to have the so-called cognitive radio cellular network (CRCN). We aim to overcome the current issues and difficulties facing the integration of CR in cellular networks. Overcoming these obstacles comes to improving the handover process in primary network and managing the secondary network using, respectively, the reinforcement learning (RL) and the k-nearest neighbors algorithm (k-NN). We also will come with a new architecture describing our vision for every case scenario in the two types of network.

Keywords: Cognitive radio · Spectrum management · Machine learning Emerging infrastructure

1 Introduction

The continuous progress of communication technologies and the rapid proliferation of radio communication standards and services have induced a spectrum overload, resulting in frequency scarcity.

Two separate studies[1] have revealed that spectrum usage varies between 15% and 85% and have shown that the spectrum is sporadically exploited, meaning frequency bands are exploited in particular locations and at particular times. These considerations have motivated the scientific community to develop a new radio technology aiming to maximize the exploitation rate of spectrum and so allow meeting the needs in terms of

[1] Studies in the Federal Communications Commission in the United States (FCC) and the Federal Office of Communications in the United Kingdom (OFCOM).

M. E. Auer and T. Tsiatsos (Eds.): IMCL 2017, AISC 725, pp. 3–10, 2018.
https://doi.org/10.1007/978-3-319-75175-7_1

reliability and quality of services. This new paradigm called Cognitive Radio (CR) was formally presented by Joseph Mitola III at a seminar in KTH, the Royal Institute of Technology, in 1998 [1]. It is a key technology that allows an opportunistic access to spectrum, with the aim of solving the problem of frequency scarcity. A CR system involves a licensed user called primary, and an unlicensed user called secondary or cognitive that monitors the spectral bands to detect the spectral holes. These are frequency bands which are not used by licensed users. The process optimizes the use of the available frequencies of the radio spectrum while minimizing interference with other users.

The principle of cognitive radio relies on understanding the temporal organization and the control states. It follows a series of steps starting with the observation and orientation of the environment, followed by the creation of the plans and then finally the decision making and an action taking. The main functions of cognitive radio are: Detection, sharing, decision, and spectral mobility [2].

Our current research aims to study the effects of implementing CR in cellular networks. The application of CR in these networks went far beyond what have previously been formalized. Unlike the case in the IEEE 802.22 standard where primary users occupy fixed locations and have deterministic or periodic temporal behavior. This standard is primarily designed to use cognitive radio in rural broadband wireless access.

This research describes a small step in the current vision of cognitive radio. CR is expected to go beyond what IEEE 802.22 allowed [3]. New solutions will be developed to allow access to different types of spectrum. In this regard, it is necessary to develop spectrum allocation and planning rules that encourage innovation and efficiency in spectrum management.

The implementation of cognitive radio in cellular networks is a very delicate task due to the spatiotemporal fluctuations of the primary user, in contrast to the IEEE 802.22 standard, where the behavior of said user is periodic and thus easily modeled. We deem necessary to include artificial intelligence into the process of the cognitive cycle in order to maintain an effective implementation of CR in cellular networks. This will allow the full advantages of the algorithmic progress that this domain has seen to apply it on cellular networks via cognitive radio while respecting the characteristics and constraints specific to cellular networks.

We suggest the application of reinforcement learning to handle primary users' handover which is a technique of unsupervised and online artificial intelligence. The technique's goal is to improve the system using simple modeling. The k-nearest neighbor algorithm will be helpful in determining the parameters affecting the secondary network's performance.

In this paper, we will discuss the progress of current researches in the application of cognitive radio in cellular network. Then we will propose an architecture that handles our problem, and will explain the way in which we want to apply machine learning in the different components of our architecture. We will proceed by analyzing the effectiveness of the presented solutions, and conclude by our perspectives and our vision for the cognitive radio cellular network (CRCN).

2 Application of Cognitive Radio in Cellular Networks

Several approaches to integrating cognitive radio in cellular networks have been suggested. They were a result of several measurement campaigns carried out all over the world, showing the extent of the spectral opportunities within this communication system intended for mobile devices.

The more densely populated urban environments are the more impactful the resulting opportunities offered by improving spectrum under-utilization are, as illustrated in [4] by Carolan et al. Knowing that the majority of the traffic in this type of cell is mainly handover type, the dynamic management of this type of cell becomes necessary, in order to benefit the secondary networks. The complex nature of spectral holes detection in cellular networks motivated the scientific community to integrate the methods of prediction in the different phases of the cognitive cycle. The main source of complexity being the spatiotemporal fluctuations in the behavior of the primary user and the sporadic availability of the spectrum.

The use of such methods opens the possibilities to both effectively implement CR and efficiently exploit the spectral holes.

Conventional approaches relying on instant information can not reflect the actual state of the environment accurately. The main responsible is the imperfection of the detection phase which provides false information leading to erroneous decision making. The type of situation that causes the dynamic sharing of inaccurate frequencies, and renders any analysis of the environmental state invalid, thus reducing the efficiency of the system. In order to equip the cognitive radio with intelligent processes and to deepen the level of abstraction, ways of predicting the main user's spatiotemporal behavior have been developed. Thus, the cognitive user will react in a proactive and anticipatory manner.

On the basis of historical data, spectral prediction on the temporal domain can spread over the four functions of the cognition cycle: spectral detection, spectral decision and spectral mobility or even spectral sharing as [5] dissected them.

In the space domain, a spectral map (REM: Radio Environment Map) was used, based on mathematical interpolation methods (Shepard method, Kriging, moving medium, etc.…) that differ according to the formula and the distance used to extract the unknown from the known. The aim is to estimate the field in any point of space [6, 7]. However, all these methods are based on measurement campaigns that have attempted to model the behavior of the primary user by laws of known probability in order to understand the spatial and temporal behavior of the primary users [8–16]. We note that there is no consensus among the scientific community over what a law describing the behavior of the primary user. Even after laborious measurement campaigns, the results widely vary, but we accept within reason that the Poisson point process can model the behavior of primary and secondary users.

In addition, the nature of the spectrum sharing problem has been widely addressed by the scientific community in a multitude of fields (game theory, auction theory, multi agent system and Markov chains) [17]. Using the Markov approach, Yao et al. in [18] integrated cognitive radio into cellular network. They showed that end-of-service throughput and average waiting time increase with the queue length and decrease with the augmentation of the number of buffer slots used in the secondary users handoff. But we note that the approach used a static definition of the various parameters appearing in the algorithmic implementation.

Artificial intelligence, through reinforcement learning algorithms, is also an alternative to solve the problem of sharing and choosing the appropriate channels for the secondary user [19]. These algorithms reduce the time of spectral detection and increase the spectrum usage rate.

3 Objective and Contribution

3.1 Primary Network

Managing the handover process in cellular networks translates to reserving spectral bands for the handover of the primary user. The main advantage of this static method of reservation resides in its simplicity. The exchange of control information between the different base stations is not required. However, this method is not suited for traffic management, which means it cannot be adapted to real traffic conditions. Also, the main goal of static methods is finding the optimal number of reservation channels, since this number has a huge impact on the performance of wireless cellular networks. The fact that the number of reserved channels is an integer makes identifying the optimal solution harder. Since this approach affects the primary user's performances it will automatically affect cellular cognitive radio. In other terms, statically reserving a G number of spectral bands will result in the secondary user being able to use only C minus G channels (C given number of channels).

The objective of this paper is to find the way to improve spectrum usage rate by the dynamic reservation of the spectral bands for the handover, and by the dynamic management of the secondary users' queue (Fig. 1).

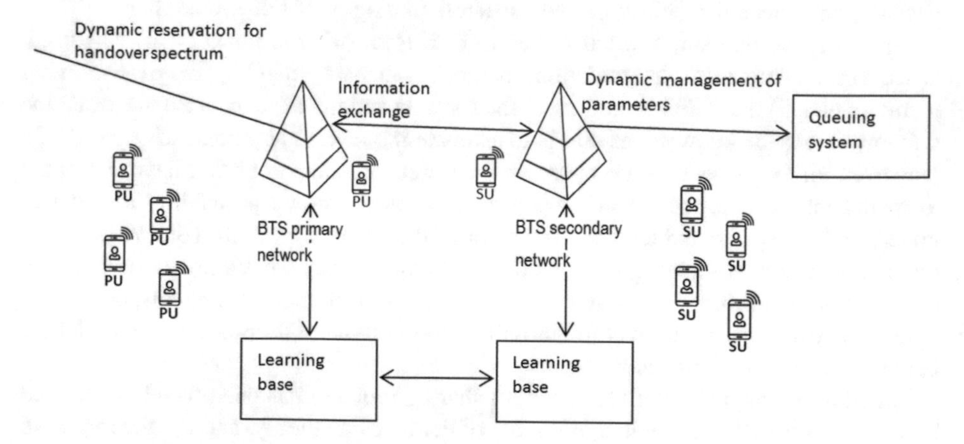

Fig. 1. Proposition of architecture to cognitive radio cellular network

The suggested approach, as described above, will ensure a permanent exchange of information between the primary and secondary networks. Thus, we can see the architecture of the network as a decision-making element, instead of being only a relay of information. The architecture will also reduce spectral detection time and increase the efficiency of the detection itself, which is essential to any implementation of CR. Thanks to this the secondary networks will be able to receive the information of the environment in real time and in a permanent manner.

In order to solve the problem of dynamic spectrum allocation (for handover), we choose to resort to the Q-learning algorithm due to its simplicity compared to others reinforcement learning algorithms. The later does not need any information about the transition's probability between states. The intelligent agent only goal is to learn an action-value function Q (a, s) which is the sum of discounted rewards, given the current action and forever afterwards using a certain policy. This helps in simplifying the implementation of this algorithm and making it easier in addition to increasing its problem solving capacity.

Thus, the Q-learning algorithm needs only a:

- Definition of the state space and the set of actions
- Determination of the value of the reward function.

Let L be the total length of the spectrum, H is the part reserved for spectral handover, we will then:

1. **Determine the states and actions:**
   In our problematic, the definition of a new status is subject to two cases:

   – A new primary Handover user will trigger a new state; therefore the learning agent (base station) must identify 3 parameters: the incoming user, those who have left the initial or handover call, and the length of H, in order to build the current state. Because we will have several scenarios as shown in the Table 1.

Table 1. Rewards according to actions and states

State	Actions						
	Increase H with			Reduce H with			Sleep
	1	2	3	1	2	3	
Arrival of a PU, 0 PU leaves, Length of H	**15**	10	5	−5	−5	−5	−5
Arrival of a PU, 1 PU leaves, Length of H	10	5	0	−5	−5	−5	**15**
Arrival of a PU, 2 PU leave, Length of H	5	0	0	**15**	−5	−5	10
Arrival of a PU, 3 PU leave, Length of H	0	0	0	10	**15**	−5	−5
Arrival of a PU, 4 PU leave, Length of H	0	0	0	5	10	**15**	0

   – After a 'μ' time (Mean arrival rate of inter-handoff PU calls to cell), if no primary user in handover is detected in the cell, the intelligent agent updates its learning base to release the bands reserved to Handover in the case of the exit of the primary users of the cell.

2. **Define the reward function:**
   The role of the reward function is the orientation of the intelligent agent to choose the action that maximizes the Q (a, s) according to the formula:

$$Q(s_n, a_n) \leftarrow Q(s_n, a_n) + \alpha[r_n + \gamma max_a Q(s_{n+1}, a) - Q(s_n, a_n)]$$

$$r(s,a) = \begin{cases} -20 \; \textit{if blocking initial call} > \textit{threshold to be fixed} \\ \qquad\qquad \textit{see the table } 1 \textit{ if else} \end{cases}$$

3.2 Secondary Network

The queuing method is the most effective way to achieve the goals of cognitive radio. The challenge is to reach the balance point between waiting time and service completion.

We need to find the most adequate H and Q pair, where H is the maximum queue size and Q is the number of reserved buffers slots for inter hand-off SU calls (Q < H). The results in [16] indicate that the service completion throughput and the average waiting time increase with the queue length and decrease with the number of reserved buffer slots for inter hand-off SU calls.

Determining these parameters must be made dynamically taking into account the specifics of each cell (population density, traffic load, weekday and hour of the day). The learning algorithm must solve two optimization problems:

- Finding the minimal waiting time.
- Reaching the optimal Completion Service Rate.

We will solve this problem using the k-nearest neighbors algorithm. This algorithm stores all available cases and classifies new cases based on a similarity measure.

It is a novel approach to dealing with this classification problem with the help of measurement campaigns.

The H parameter (number of reserved buffer slots for inter-handover) is a percentage of the Q parameter (queue size). We suggest defining 8 classes. A = class 10%, B = class 20%, C = class 30%, D = class 40%, E = class 50%, F = class 60%, G = class 70%, H = class 80%. (The justification of this number of classes will be presented in our next work, once we gather historical data from telecom operators.)

The type of handover used is not in the scope of this research (vertical/horizon-tal), we will limit ourselves to using the basic definition of this process (Table 2).

The expert will assign a class to each vector (population, area, spectrum length, day, hour, traffic load) while taking into account the constraints described previously.

Table 2. Classification procedure

Cell	Characteristics							
	Population	Area	Spectrum length	Day	Hour PM	Traffic load from		Class
						Initial call	Handover	
Name of the cell				X	00			A
					01			B
					02			C
					03			D
					04			E
					05			F
					06			G
					07			H
					08			H
					09			H
					10			H
					11			H

4 Conclusion

The way the spectral handover is processed in the primary network affects the overall performance of the cognitive radio cellular network architecture. We suggested applying reinforcement learning via the Q-Learning algorithm to improve the usage rate of the spectrum in the primary network, and by doing so offering more opportunities to the secondary network. For the latter, we have developed a classification approach in order to overcome the inadequacy of static reservations.

In our next work, we aim to introduce the mathematical model managing the classification in the secondary network based on measurement campaigns. Also we will simulate the exchange of information between the primary network and the secondary network in order to observe its effect on the overall network.

The main objective of our work is to demonstrate the possibility of effectively implementing CR in cellular networks, while taking into account the complexity of the environment. Efforts in this direction can greatly benefit cellular networks efficiency, proving that learning algorithms can considerably accelerate the effective deployment of cognitive radio.

References

1. Mitola, J., Maguire, G.Q.: Cognitive radio: making software radios more personal. IEEE Pers. Commun. **6**(4), 13–18 (1999)
2. Akyildiz, I.F., Lee, W.-Y., Chowdhury, K.R.: CRAHNs: cognitive radio ad hoc networks. Ad Hoc Netw. **7**(5), 810–836 (2009)

3. Kasbekar, G.S., Sarkar, S.: Spectrum auction framework for access allocation in cognitive radio networks. IEEE/ACM Trans. Netw. (TON) **18**(6), 1841–1854 (2010)
4. Carolan, E., McLoone, S.C., Farrell, R.: A predictive model for minimising power usage in radio access networks. In: International Conference on Mobile Networks and Management. Springer, Cham (2015)
5. Xing, X., et al.: Spectrum prediction in cognitive radio networks. IEEE Wirel. Commun. **20** (2), 90–96 (2013)
6. Debroy, S., Bhattacharjee, S., Chatterjee, M.: Spectrum map and its application in resource management in cognitive radio networks. IEEE Trans. Cogn. Commun. Netw. **1**(4), 406–419 (2015)
7. Vizziello, A., et al.: Cognitive radio resource management exploiting heterogeneous primary users and a radio environment map database. Wirel. Netw. **19**(6), 1203–1216 (2013)
8. Willkomm, D., et al.: Primary user behavior in cellular networks and implications for dynamic spectrum access. IEEE Commun. Mag. **47**(3), 88–95 (2009)
9. Wellens, M., Riihijärvi, J., Mähönen, P.: Empirical time and frequency domain models of spectrum usage. Phys. Commun. **2**(1), 10–32 (2009)
10. Geirhofer, S., Tong, L., Salder, B.M.: Dynamic spectrum access in the time: modeling and exploiting white space. IEEE Commun. Mag. **45**(5), 66–72 (2007)
11. Paul, U., et al.: Understanding traffic dynamics in cellular data networks. In: 2011 IEEE Proceedings of INFOCOM. IEEE (2011)
12. Sung, K.W., Kim, S.-L., Zander, J.: Temporal spectrum sharing based on primary user activity prediction. IEEE Trans. Wirel. Commun. **9**(12), 3848–3855 (2010)
13. Chen, D., et al.: Mining spectrum usage data: a large-scale spectrum measurement study. In: Proceedings of 15th Annual International Conference on Mobile Computing and Networking. ACM (2009)
14. Riihijärvi, J., Nasreddine, J., Mähönen, P.: Impact of primary user activity patterns on spatial spectrum reuse opportunities. In: 2010 European Wireless Conference (EW). IEEE (2010)
15. Esenogho, E., Walingo, T.: Primary users ON/OFF behavior models in cognitive radio networks. In: International Conference on Wireless and Mobile Communication Systems (WMCS 2014), Lisbon, Portugal (2014)
16. Bütün, İ., et al.: Impact of mobility prediction on the performance of cognitive radio networks. In: 2010 Wireless Telecommunications Symposium (WTS). IEEE (2010)
17. Benmammar, B., Amraoui, A., Krief, F.: A survey on dynamic spectrum access techniques in cognitive radio networks. Int. J. Commun. Netw. Inf. Secur. **5**(2), 68 (2013)
18. Yao, Y., Popescu, A., Popescu, A.: On prioritized opportunistic spectrum access in cognitive radio cellular networks. Trans. Emerg. Telecommun. Technol. **27**(2), 294–310 (2016)
19. Yau, K.L.A., Poh, G.S., Chien, S.F., Al-Rawi, H.A.: Application of rein-forcement learning in cognitive radio networks: models and algorithms. Sci. World J. **2014**, 23 p. (2014). Article ID 209810

Designing an Augmented Reality Smartphone Application for the Enhancement of Asthma Care Education

Chris Janes[1(✉)], Tom Andrews[2], and Mohamed Adbel-Maguid[1]

[1] University of Suffolk, Ipswich, Suffolk, UK
c.janes@uos.ac.uk
[2] Orbital Media, Stowmarket, Suffolk, UK

Abstract. Uncontrolled asthma puts a significant financial burden on health care providers and is the highest cause of asthma related hospitalisation and mortality. Maintaining asthma control requires that patients follow several processes, including; correct usage of inhalers, identification and avoidance of environmental triggers, and the development of an asthma action plan. MySpira was developed as an approach to supplement existing asthma care educational materials for young people (aged between 7 & 13) that utilises mobile augmented reality technology and gamification techniques to improve knowledge retention and user engagement.

1 Introduction

Asthma is a chronic respiratory disease that affects 5.4 million people in the UK, 1.1 million of whom are children [1]. Asthma care costs the NHS £1.1 billion annually, 10% of which comes from trips to A&E and hospitalisation [2]. It is estimated that 75% of asthma related hospital visits and 90% of asthma related mortalities are preventable [3,4].

Incorrect or sub-optimal inhaler usage leads to a higher rate of uncontrolled asthma and increased treatment costs [5], studies have demonstrated that there has been no meaningful improvement in the rate of incorrect inhaler usage for a significant period of time [6]. The purpose of this project was to produce a mobile application to work as an aid to self-management, as part of a well-rounded asthma education programme [7].

Globally, smartphone usage amongst young people has grown explosively [8], in the UK 66% of households have at least one smartphone device available [9]. The application is aimed at young patients (between 7 and 13 years of age) who are early in their treatment cycle and who may be less engaged by traditional educational material. This approach allows patients to continue their education in their own time at home.

With the recent advances in mobile Augmented Reality (AR) technology, the popularity of applications such as Pokémon Go [10] and the growing recognition of AR as an effective educational platform [11,12], the team decided to make use

M. E. Auer and T. Tsiatsos (Eds.): IMCL 2017, AISC 725, pp. 11–17, 2018.
https://doi.org/10.1007/978-3-319-75175-7_2

of augmented reality as an approach to improving user immersion [13]. Mobile AR allows for any patient with access to a smartphone to make use of the application in their own time, outside of the guidance offered by healthcare professionals.

This project was undertaken as part of a Knowledge Transfer Partnership between the University of Suffolk and Orbital Media.

2 MySpira Application

The team selected to develop the application with Unity 3D [14], as it offers a rapid development environment and allowed for the maximum utilization of the team skills and expertise. An analysis of available AR plugins for Unity was undertaken and after development of several prototypes the Vuforia [15] plugin was selected for the implementation of the proof of concept application. During development, support for Apples ARKit and ARCore from Google was added to Unity, allowing for a version of the application that worked without registration markers to be produced.

The team planned the important educational elements that had to be included through meetings with local asthma specialists. Once these elements were decided upon, the team worked through how the application would be gamified to provide reinforcement for the key learning points.

The key messages are focused around preparing the inhaler and completing the inhalation process for the medicine to reach the affected areas of the lungs, as these were identified as the issues that most impacted patient health. The application takes the user through seven learning experiences, each designed to be engaging and fun to use. In each of the stages the user is educated about different aspects of asthma and its care.

In the early stages of the project development the team investigated what helped children to become engaged in other educational applications [16–18].

Because the key objective of the application is the communication of all the educational content without overwhelming the user it was important to ensure that all the gamified elements aided engagement but did not require significant time to learn [19]. Due to this, great care was taken in designing simple and intuitive rules for the gamified elements, this included implementing subtle visual cues to improve understanding (Fig. 1).

Using gamification techniques such as collection mechanics and awarding badges stimulates the user into working for more achievements [20]. In MySpira, we award badges as a way of rewarding the user for completing the learning activities.

Each stage is linked by a simple narrative, driven by an animated 2D character that speaks to the user to provide guidance and information. The entire process is designed to avoid referencing the user as a patient, instead a second character, Spira, is introduced and all actions regarding inhaler use are demonstrated by them in 3D via the AR viewport (Fig. 2).

Fig. 1. In the letter jumble game, correct positions are highlighted with animations

Fig. 2. Spira demonstrating the use of an inhaler device

The first stage introduces the narrator character, Alex and builds the narrative by having Alex request the user help him gather his lost notes. Alex is represented in 2D via a small cut-in that appears in the top left corner of the application. This leads to a simple letter jumble game that features tactile interaction and several small reward effects such as particles and sound effects. Similar effects are used throughout the project to make the user feel good about their actions and achievements.

The second and third stages introduce the user to the differences between the two most common inhaler types (preventer and reliever), during which the child is asked to collect all the pencils that they can see on screen, for which they are rewarded with colorful comics that make use of visual storytelling to show the importance of keeping a reliever inhaler to hand at all times.

The fourth stage has the user work through the process of preparing the inhaler. The child must complete the preparation by taking the cap off, shaking the inhaler and attaching the spacer. With tactile feedback and visual stimulation, it is intended to make preparing the inhaler engaging and fun to complete.

In the fifth stage, we introduce Spira as a fun, dopey character who needs to take his inhaler. The user is given responsibility in helping Alex show him the correct inhale procedure. The user is awarded a card after each step of the procedure which allows them to absorb new information at their own pace. As part of this process, Spira takes two doses. Between each dose animation, the user plays a small matching memory game that teaches the symptoms and triggers of asthma through cartoon infographics.

In the sixth stage, the user is shown two videos that explain how the lungs work and how asthma can cause difficulty breathing. With these visual descriptions, the child can understand why symptoms occur with certain triggers and how both inhalers treat the condition.

The seventh and final stage is a multiple-choice quiz that tests the users knowledge at the end of their run through the application. The quiz is built in such a way that they cannot complete a question until they have answered it correctly. This guarantees that the correct answers are reinforced and we minimize any negative impact by using positive language even when the user selects the incorrect answer.

Throughout the stages of the application, the intention is to fully engage the user and to encourage a sense of empathy towards the virtual characters as the user works through the application and helps the characters complete their tasks. This emotional engagement is something that is highly difficult to achieve with traditional approaches to asthma education and is something that the application can make use of to ensure the users complete the entire process.

3 First Round of Feedback

The team have undertaken a small-scale test with eighteen children, aged between 5 and 15 and of mixed gender. The results of this test have concentrated on the built-in quiz and these initial results show that users are engaged and answering the questions correctly, with an 85% accuracy rate among the children who took part.

The most challenging questions that got the most incorrect answers were related to the color of the inhaler and their function; "What colour is the reliever inhaler?", "Which inhaler is used for rapid treatment?" and "What colour is the preventer inhaler?". This suggests that children do not fully understand what each inhaler is intended to do. This is the fundamental reason why patients are not recognizing the importance of accurate inhaler use and provides us with a focus for further development of the educational aspect of the application.

4 Conclusion

Improving the level of available asthma care by providing enhancements to the educational material available is an attempt to resolve one of the major problems facing asthma sufferers and those who care for them.

In this study, we presented an overview of the MySpira mobile application, detailing how we are targeting a younger market while gamifying the experience to improve user engagement and retention of the educational material. The team focused on educating users with regards to the types and use of inhalers as these were discovered to be issues that had a major impact on asthma care.

The application has been shown to several healthcare professionals at each stage of the development process to ensure that all educational content was correct and suitably targeted. The initial feedback gained during user testing has demonstrated that while generally successful, there is scope for improvements around helping the users identify which inhaler performs which function.

5 Future Work

The next steps of the project are to continue the evaluation and testing process, making iterative changes to the content to resolve any issues high-lighted from the results. There are plans for a larger scale user test that will involve a direct comparison between the content of the application and more traditional educational material.

References

1. Asthma UK: Asthma UK—Asthma Facts and Statistics (2015). https://www.asthma.org.uk/about/media/facts-and-statistics/. Visited 26 Apr 2017
2. Anderson, H.R., Gupta, R., Strachan, D.P., Limb, E.S.: 50 years of asthma: UK trends from 1955 to 2004. Thorax **62**(1), 85–90 (2007). ISSN 0040-6376. https://doi.org/10.1136/thx.2006.066407. http://thorax.bmj.com/cgi/doi/10.1136/thx.2006.066407

3. National Institute for Health Care Excellence: Implementation Programme NICE support for commissioners and others using the quality standard on nutrition support in adults. Technical report, pp. 1–24. National Institute for Health Care Excellence, November 2012. https://www.nice.org.uk/guidance/qs25/resources/support-for-commissioners-and-others-using-the-quality-standard-for-asthma-252373933, https://www.nice.org.uk/guidance/qs24/resources/support-for-commissioners-and-others-using-the-quality-standard-on-nutriti
4. Anagnostou, K., Harrison, B., Iles, R., Nasser, S., Sullivan, T., Neukirch, F.: Risk factors for childhood asthma deaths from the UK Eastern Region Confidential Enquiry 2001–2006. Prim. Care Respir. J. **21**(1), 71–77 (2012). ISSN 1471-4418. https://doi.org/10.4104/pcrj.2011.00097, http://www.nature.com/articles/pcrj201197
5. Price, D., Bosnic-Anticevich, S., Briggs, A., Chrystyn, H., Rand, C., Scheuch, G., Bousquet, J.: Inhaler competence in asthma: common errors, barriers to use and recommended solutions. Respir. Med. **107**(1), 37–46 (2013). ISSN 0954-6111. https://doi.org/10.1016/j.rmed.2012.09.017, http://linkinghub.elsevier.com/retrieve/pii/S0954611112003587
6. Sanchis, J., Gich, I., Pedersen, S., Systematic review of errors in inhaler use: has patient technique improved over time? Chest **150**(2), 394–406 (2016). ISSN 1931-3543. https://doi.org/10.1016/j.chest.2016.03.041, http://linkinghub.elsevier.com/retrieve/pii/S0012369216475719
7. Boulet, L.-P.: Asthma education: an essential component in asthma management. Eur. Respir. J. **46**(5), 1262–1264 (2015). ISSN 0903-1936. https://doi.org/10.1183/13993003.01303-2015. http://erj.ersjournals.com/content/46/5/1262, http://erj.ersjournals.com/lookup/doi/10.1183/13993003.01303-2015
8. Roberts, D.F., Foehr, U.G., Rideout, V.: A Kaiser Family Foundation Study. https://kaiserfamilyfoundation.files.wordpress.com/2013/01/generation-m-media-in-the-lives-of-8-18-year-olds-report.pdf
9. Ofcom: The Communications Market Report 2014, September 2014. ISSN 1098-6596. https://doi.org/10.1017/CBO9781107415324.004, eprint: arXiv:1011.1669v3, https://www.ofcom.org.uk/__data/assets/pdf_file/0022/20668/cmr_uk_2015.pdf
10. SurveyMonkey: Pokémon GO is now the biggest mobile game in U.S. history (2016). https://www.cnbc.com/2016/07/13/pokemon-go-now-the-biggest-mobile-game-in-us-history.html, https://www.surveymonkey.com/business/intelligence/pokemon-go-biggest-mobile-game-ever/. Visited 04 Sept 2017
11. Dunleavy, M., Dede, C., Mitchell, R., Affordances and limitations of immersive participatory augmented reality simulations for teaching and learning. J. Sci. Educ. Technol. **18**(1), 7–22 (2009). ISSN 1059-0145. https://doi.org/10.1007/s10956-008-9119-1, arXiv: arXiv:1002.2562v1, http://link.springer.com/10.1007/s10956-008-9119-1
12. Wu, H.K., Lee, S.W.Y., Chang, H.Y., Liang, J.C.: Current status, opportunities and challenges of augmented reality in education. Comput. Educ. **62**, 41–49 (2013). ISSN 0360-1315. https://doi.org/10.1016/j.compedu.2012.10.024, arXiv:1204.1594, http://linkinghub.elsevier.com/retrieve/pii/S0360131512002527
13. Sekhavat, Y.A., Zarei, H.: Enhancing the sense of immersion and quality of experience in mobile games using augmented reality. J. Comput. Secur. **3**(1), 53–62 (2016). ISSN 2383-0417. http://www.jcomsec.org/index.php/JCS/article/view/298
14. Unity: Unity - Game Engine (2017). https://unity3d.com/. Visited 05 Sept 2017
15. Qualcomm: Vuforia—Augmented Reality (2013). https://www.vuforia.com/. Visited 05 Sept 2017

16. Lamberty, K.K.: Getting and keeping children engaged with a constructionist design tool for craft and math. Ph.D. thesis, January 2007. https://smartech.gatech.edu/handle/1853/14589
17. Aoki, N., Ohta, S., Masuda, H., Naito, T., Sawai, T., Nishida, K., Okada, T., Oishi, M., Iwasawa, Y., Toyomasu, K., Hira, K., Fukui, T.: Edutainment tools for initial education of type-1 diabetes mellitus: initial diabetes education with fun. Stud. Health Technol. Inform. **107**(Pt 2), 855–859 (2004). ISSN 0926-9630. https://doi.org/10.3233/978-1-60750-949-3-855, http://www.ncbi.nlm.nih.gov/pubmed/15360933
18. Lahm, E.A.: Software that engages young children with disabilities. Focus Autism Other Dev. Disabil. **11**(2), 115–124 (1996). ISSN 1088-3576. https://doi.org/10.1177/108835769601100207, http://journals.sagepub.com/doi/10.1177/108835769601100207
19. Falloon, G.: Young students using iPads: app design and content influences on their learning pathways. Comput. Educ. **68**, 505–521 (2013). ISSN 0360-1315. https://doi.org/10.1016/j.compedu.2013.06.006, http://linkinghub.elsevier.com/retrieve/pii/S0360131513001577
20. Kiryakova, G., Angelova, N., Yordanova, L.: Gamification in education (2014). http://dspace.uni-sz.bg/handle/123456789/12

The Impact of Background Music on an Active Video Game

Aikaterini Ganiti, Nikolaos Politopoulos(✉),
and Thrasyvoulos Tsiatsos

Department of Informatics, Aristotle University of Thessaloniki,
Thessaloniki, Greece
katganiti@gmail.com,
{npolitop, tsiatsos}@csd.auth.gr

Abstract. Video games are the most accepted form of entertainment. However, there are games not designed primarily for entertainment purposes and these are called serious games. In particular, developers have created video games that promote exercise and healthy lifestyles, called Active Video Games or "exergames". After 1970's developers added music to their games. The goal was to create an emotional engagement and give players a better experience. Background music, proved to be one of the most important features of a game, because it causes emotional responses. The element of music attributed to increased stimulation and mood is the tempo (fast or slow). The fast tempo affects positive space-related abilities compared to the slowest pace.

Keywords: Active Video Games · Background music · Emotions

1 Introduction

The present research has been designed with a view to research the use of background music in an active video game to improve reaction time, interested in seeing what influence, it will have, on players. Additionally, it is important to see what effect background will have on emotions and, as a result, to the performance, while playing the game.

The music background is likely to make it easier or harder for players to play at the time of play. We are interested in the influences on players of the changes in music (rhythm and scale) and the game variables (period and radius of the ball). Music is mainly used without vocals to avoid overloading the cognitive load from unnecessary information (Levy 2015). Previous research has shown that players prefer dynamic music that follows the intensity of the game (Naushad and Tufail 2013).

For sports games, we are interested in high song rates (94 bpm) (Levy 2015). The low-tempo music background, likely to cause boredom (Thompson et al. 2001) and not to improve arousal theory (Levy 2015). If the pace of the song is coordinated with the ball's pitch, we estimate that the players will be tuned at this pace resulting in better performance (Bacon et al. 2012). Otherwise, there will be disintegration while at the same time players make unnecessary moves influenced by the rhythm of the song. In general, it is accepted that the sooner the balls appear, the more difficult it is to perform as well.

M. E. Auer and T. Tsiatsos (Eds.): IMCL 2017, AISC 725, pp. 18–28, 2018.
https://doi.org/10.1007/978-3-319-75175-7_3

Scale is the one that gives the feeling that music is joyful (great) or melancholy (small). Scale mainly affects mood and feelings. In order to draw conclusions about the influence of the scale on stimulation and, by extension, on performance, players' musical preferences should be considered, since from preference theory, players are positively affected only by integrating music of their preferences.

Music has proven to be one of the most important features of the game that causes emotional responses (Young 2012). Therefore, it's important to see what effect the emotion is playing on the music's background when playing the game. Music influences mood and emotion, so it is useful to see how we can use it for our benefit (Eladhari et al. 2006).

In summary, this paper aims to answer the following questions:

1. If and to what extent, background music will create positive emotions to players,
2. If and whether there is an improvement in reaction time between players as a result of background music,
3. If and to what extent, the change of music's tempo will affect them,
4. Seeing the psychological flow of players towards the game.

2 Theoretical Background

2.1 Emotions

Scherer (2001) argues that emotions are the results of correlated and synchronized changes in the state of one of the organization's five subsystems in response to internal or external events. Personality, feelings, mood, and emotions are all different kinds of interaction between them. Lazarus (1991) argues that emotion has a mutual influence with mood, temperament, personality and motivation.

2.2 Adaptive Game

The goal of each game creator is to create a highly interactive and immersive environment (Craig 2016). Throughout a video game, the user interacts with a complex set of cause and effect scenarios, influencing the course of events. Because of this, game experiences inherently contain the elements of volatility and unpredictability. Classifying games in the different categories is a subjective issue. While the boundaries of a species are clear, many games cannot be entirely attributed to one species as Zehnder and Lipscomb (2006) claim. A serious game is a game designed primarily for entertainment purposes and finds application in health and sports.

2.3 Background Music

Video games' sound can increase the sense of presence (feeling of being part of the world), engagement (a psychological state of absorption) (Lombard and Ditton 1997), and deepening (a feeling that is completely surrounded by another reality) (Nacke et al. 2010). These types show significant feedback to the player and affect overall performance. An interesting feature of music is that it can be used to direct attention to important screen features to make mood available (Pignatiello et al. 1986). Therefore, adding musical background to a game requires a harmonious collaboration with all its features (Mullan 2010).

It has been proven that sound and music are the most important elements of the game in causing emotional reactions from users (Carlile 2011). Many researches have examined whether certain music is causing certain feelings. Willimek and Willimek (2013) certify that specific harmonic structures cause the same emotional responses as the participants gave identical answers, confirming that the link can be categorized systematically.

Studies have shown that music can exert a positive and significant impact on human performance. Dynamic muscle strength was found to be higher in participants who had warmed up listening to music (Jarraya et al. 2012). Synchronizing the movements of athletes with music while cycling can improve their endurance and exercise efficiency (Bacon et al. 2012). Participants who listened to music before practicing on a static bicycle spent more miles (Becker et al. 1994). Likewise, the music during the warm-up of volleyball players helped them to increase heart rate more, and they also exhibited maximum anaerobic strength (Eliakim et al. 2007).

In the mood hypothesis, positive moods increase excitation and hence cause increased performance in spatial work. However, negative moods while causing stimulation do not cause increased performance. This paradox explains the preference for users to be influenced by the species they prefer. The elements of music attributed to increased stimulation and mood are the rate (fast or slow) (Parsons et al. 1999). The fast pace affects positive space-related abilities compared to the slowest pace. In addition, the stimulation is affected by the rhythm changes.

In some cases, music may also have an adverse effect on cognitive functions due to the cognitive problem (Van Merrienboer and Sweller 2005). Through this hypothesis, a person's limited cognitive resources are unduly overloaded by the presence of music and the need to process it. In fact, Konecni (1982) formulated the theory that music processing exhausts cognitive resources with a resultant reduction in performance.

The presence of music can affect the performance of a person in typing (Jensen 1931). People who hear fast music drinking water will drink faster (McElrea and Standing 1992). Participants in a simulator drive faster and use more steering movements (Konz and Mcdougal 1968). Listening to music, an individual while conducting repetitive work can also increase alert levels (Fox 1971). As described in sports literature, people often synchronize their physical movements with the rhythm of music (Karageorghis et al. 2010). Sound is a vital tool for the designer to share information with the player (Collins 2008). However, it remains unknown how music influences player's performance, behavior and experience.

3 Method

3.1 Participants

In order to test the research hypothesis of this paper the sample consisted of 18 University students, 3 of which were female and 15 were male). Their age was between 20 and 28 years old with average 25,4 years.

3.2 Instruments

The questionnaires of the experiment were created with Google Forms. The standard questionnaires included are as follows. Attitudes Towards Computer Questionaire (ATCQ), translated and adapted for video games, on a Likert five-grade scale. The Positive and Negative Affect Schedule (PANAS), translated and weighted by B. Daskalos and E. Sagolitos, on a Likert scale of five grades. The Flow Scale, translated and weighted by Doganis et al. (2000), on a Likert scale of five grades. The questionnaire was closed with Short Test of Music Preferences and a few general questions.

3.3 Materials

For research, a piece of music was used on the Purple Planet Music page and modified for the needs of the Audacity Audio Editor. The choice of music had to meet some specific criteria such as: lack of vocals, belonging to the class of electronic music and an initial rate of more than 120 bpm. The game used for the first part of the experiment was Tennis Attack, created on the Unity platform and using Kinect technology.

4 Procedure

As we have already mentioned, this paper investigates whether and how to affects players emotions during their involvement with the game and the study of their psychological flow towards it. For research needs we used and modified a game developed to improve simple reaction time (one stimulus - one response) using the natural user interface Microsoft Kinect (Politopoulos et al. 2015). Game's theme is tennis. The athlete stands opposite the screen and repel the balls on the screen. Additionally, we modified the song in ten different tempos. Each time a player repels 3 consecutive balls the tempo changes and becomes faster. The experiment lasted for two weeks.

Every player had to answer a small inventory about his attitude for games, preferences for music and personal date. Then he had a 5 min session in front of the game. This simulates his experience in real training, where he would have to defend one hundred fifty real serves. At the end, he had to answer another inventory about his psychological flow and his emotions. Then we tried to find any variations between the internal groups of the sample (gender, video games experience, etc.).

5 Data Analysis

5.1 Attitude Towards Video Games

The subclasses of the ATCQ questionnaire are shown in Table 1 below. Generally, for their attitude we have mean = 75,28, while median = 77. Since the mean value is less than the median, we have negative asymmetry with the observations being on the right of the distribution. So, players have a positive attitude towards video games.

Table 1. Statistics

		ATCQ Comfort	ATCQ Interest	ATCQ Dehumanizing	ATCQ Efficiancy	ATCQ Utility	Attitude in General
N	Valid	18	18	18	18	18	18
	Missing	0	0	0	0	0	0
Mean		17.33	18.50	11.72	14.61	13.11	75.28
Median		17.50	18.50	11.50	15.00	14.50	77.00
Std. Deviation		4.015	3.502	2.866	2.404	2.698	9.099
Variance		16.118	12.265	8.212	5.781	7.281	82.801

5.2 Feelings

After applying the Paired Samples T test for the two sub scales of the PANAS questionnaire, it turns out that they have a negative correlation, with Mean = 10,278, SD = 8,757, t = 4.980, and because sig (2-tailed) = 0.000 we have an absolute statistically significant difference between positive and negative emotion (Table 2).

Table 2. Paired samples test

		Paired Differences							
					95% Confidence Interval of the Difference				
		Mean	Std. Deviation	Std. Error Mean	Lower	Upper	t	df	Sig. (2-tailed)
Pair 1	Positive Emotion Negative Emotion	10.278	8.757	2.064	5.923	14.632	4.980	17	.000

5.3 Performance

Game performance is a set of many variables: the average response time, the fastest ball, the total balls, the highest number of consecutive longest streaks and finally one Standard scoring calculation. The goal was to hit the balls as quickly as possible so as not to lose it and to have many successive balls in succession, which in turn leads to a larger number of overall balls and higher scores.

The rebound time can be influenced by the musical background, but it depends on the music preferences of electronic music. But there was no statistically significant difference after applying the Independent Samples T test between the two categories because sig (2-tailed) = 0.238 > 0.05, with t = 1.248 (Table 3). Therefore, electronic music did not affect players differently as anticipated by the theory of preference.

Table 3. Independent samples test

		Levene's Test for Equality of Variances		t-test for Equality of Means						
									95% Confidence Interval of the Difference	
		F	Sig.	t	df	Sig. (2-tailed)	Mean Difference	Std. Error Difference	Lower	Upper
Response Time	Equal variances assumed	4.808	.043	1.248	16	.230	.056444778	.045216051	-.039408967	.152298523
	Equal variances not assumed			1.248	11.088	.238	.056444778	.045216051	-.042978320	.155867876

It was also examined whether there is a link between the time of the strike and the general attitude of the players towards video games. From Table 4 we seem to have a statistically significant difference because sig (2-tailed) = 0.00 < 0.05 with t = −34.688 thus the reaction time is affected by the attitude in video games.

Table 4. Paired samples test

		Paired Differences					t	df	Sig. (2-tailed)
		Mean	Std. Deviation	Std. Error Mean	95% Confidence Interval of the Difference Lower	Upper			
Pair 1	Response Time Attitude in General	-74.3235236	9.090393291	2.142626247	-78.8440698	-69.8029774	-34.688	17	.000

In the performance results the player could see only the greatest number of successive successful balls. However, for the needs of research, we kept data of every attempt the player made. These data were used to conclude such as: how many times he lost, at what level of music he lost most often, or what level he went through more easily without losing. We again felt that music can affect the player and lose, so we need to check if there is a correlation between these variables. For this reason, the sample was divided into two groups, those who listen to electronic music and those who do not listen. However, there was no statistically significant difference between the groups because sig (2-tailed) = 0.063 > 0.05 and t = 2.078 (Table 5).

Table 5. Independent samples test

		Levene's Test for Equality of Variances		t-test for Equality of Means						
		F	Sig.	t	df	Sig. (2-tailed)	Mean Difference	Std. Error Difference	95% Confidence Interval of the Difference Lower	Upper
Loses	Equal variances assumed	5.787	.029	2.078	16	.054	9.111	4.385	-.185	18.407
	Equal variances not assumed			2.078	10.428	.063	9.111	4.385	-.605	18.827

Then, attempts were made to compare each player's failures within the different levels of music he listened to. At this point, it should be stated that no player has reached the seventh level of the ten, which will be compared to level seven, and the other levels will yield the same results. In addition, the results associated with level 7 are not interesting as they cannot give us information about the effect of this particular rhythm of music.

If we check every next level with the rest, we will see that the regularity gradually decreases. This is mainly due to the small percentage of players reaching higher levels, with the result that there is not enough data. However, each successive pair has a statistically significant difference between its levels. For the couples following the normal distribution, Malfunction T Test was followed, while for the other non-parametric control of Wilcoxon to see if there was a statistically significant difference.

5.4 Flow

In Table 6 we see the averages, the minimum and maximum values of each sub-scale of the psychological flow questionnaire. Generally, the psychological flow has an average value of 122.67 above the bisectual value of 108. This tells us that in general the players experienced restraint, immersion and enjoyment.

Table 6. Descriptive statistics

	N	Minimum	Maximum	Mean	Std. Deviation	Variance	Skewness		Kurtosis	
	Statistic	Statistic	Statistic	Statistic	Statistic	Statistic	Statistic	Std. Error	Statistic	Std. Error
Flow Action-awareness merging	18	6	18	14.50	2.595	6.735	-2.169	.536	6.565	1.038
Flow Challenge-skill balance	18	8	16	12.72	3.232	10.448	-.660	.536	-1.264	1.038
Flow Sense of control	18	4	17	11.56	4.314	18.614	-.414	.536	-1.041	1.038
Flow Clear goals	18	11	19	15.61	2.304	5.310	-.533	.536	-.292	1.038
Flow Autotelic experience	18	6	17	11.78	3.246	10.536	-.368	.536	-.754	1.038
Flow Concentration on task at hand	18	9	20	14.78	3.135	9.830	-.117	.536	-.689	1.038
Flow Loss of self-consciousness	18	5	20	14.22	3.813	14.536	-.357	.536	.794	1.038
Flow Transformation of time	18	8	17	13.50	2.093	4.382	-.801	.536	1.661	1.038
Flow Unambiguous feedback	18	8	19	14.00	3.413	11.647	-.479	.536	-.761	1.038
Flow In General	18	88	141	122.67	15.915	253.294	-1.010	.536	-.045	1.038
Valid N (listwise)	18									

Among the factors influencing the flow is the positive or negative attitude and past experience in the field. For this reason, a correlation will be made between them if a statistically significant difference is found.

Initially, there was a correlation between the psychological flow and the attitude towards video games. When testing with a Paired Sample T Test, it emerged that since the variable sig = 0.00 < 0.05 with t = −0.922, we have a statistically significant difference between the factors (Table 7).

Table 7. Paired samples test

		Paired Differences					t	df	Sig. (2-tailed)
					95% Confidence Interval of the Difference				
		Mean	Std. Deviation	Std. Error Mean	Lower	Upper			
Pair 1	Attitude in General - Flow in General	-47.389	21.807	5.140	-58.233	-36.545	-9.220	17	.000

Then, it was checked whether there was a correlation between the psychological flow and the older experience of the players in the game. In Table 8 we can see that sig (2-tailed) = 0.549 > 0.05 $, so there is no statistically significant difference in the psychological flow between those who have played the game again and those who did not re-play with t = 0.613.

Table 8. Independent samples test

		Levene's Test for Equality of Variances		t-test for Equality of Means						
									95% Confidence Interval of the Difference	
		F	Sig.	t	df	Sig. (2-tailed)	Mean Difference	Std. Error Difference	Lower	Upper
Flow in General	Equal variances assumed	.704	.414	.576	16	.573	4.519	7.851	-12.124	21.163
	Equal variances not assumed			.613	15.327	.549	4.519	7.373	-11.166	20.205

6 Conclusions

The results of emotions showed that there is an absolute statistically significant difference between positive and negative emotions with players experiencing positive feelings as they played the game. Asked if they played without music, only 22.22% answered "Yes", with the others supporting their opinions with arguments such as "becoming more interesting or enjoyable", "making me mood" and "raising adrenaline". Therefore, we come to the conclusion that music really had a positive effect on the players' feelings.

In addition, it was found to have a positive attitude towards video games. This result was expected as there was a large percentage of videogamers in the sample. Positive attitude and positive emotion help us to determine if there are aberrant behaviors. If the players had a negative attitude or experienced negative emotions at that time from the game or from internal factors, then we would expect the results of the survey to be objective. This would be because these two factors greatly influence the perception of things and it would be likely that the players would not enjoy the game as they would be influenced by their feelings.

As far as performance is concerned, we have seen that reaction time is not affected by the separation of players in those who listen to electronic music and those who do not. This may not be the same as the results we expected based on the preference theory, but electronic music is widely known, widely sounded and widely accepted. Even those who said they did not listen to this kind of music chose other similar items. Besides from the reviewed literature, we know that sports video games are the most appropriate. If we found a difference between the teams, we should review this idea and look for other types of music.

Reaction time also appeared to be influenced by the attitude of the sample towards video games. This could be explained by the fact that players who have a less positive attitude towards video games are also those who do not play in their free time. So, they are not familiar with the logic of the games, nor with the technology resulting in the same performance in the game. For the second performance variable, the number of consecutive balls, based on the times each player lost on each of the levels of music they listened to while playing, we did not have a statistically significant difference between those players who listen to electronic music and To those who do not listen. So, again, we have a paradoxical basis of preference theory, but as explained above, if there was a difference, we should revise the idea that this kind is unsuitable for such games.

But we had very large statistically significant differences between the times each player lost per level of music. By the level of music, we mean the bubble level, with level 1 = 122 bpm and level 10 = 140 bpm. This means that, as the rhythm of music grew, players were getting more and more difficult to lose ball. This leads to the conclusion of the research question whether the rapid increase in music ultimately makes it hard for the players to follow the sudden change of music and lose.

However, while there was a statistically significant difference between music levels, this cannot be considered as accurate. There were other factors that affected the game's performance in a negative way. The most basic thing was the difficulty of synchronizing the hand with the screen racket. During the experiment, as the players played, the synchronization was often lost and the handling of the racket was impossible, with the result that many balls were lost and the player's play had to be interrupted, which also affected the flow and performance But also the willingness of the participant. The bad synchronization of Kinect was also the main observation of the participants at the end of the questionnaire.

The question of whether players will synchronize to the rhythm of the music, resulting in improved performance, we have answered above, stating that the rapid increase in music probably had a negative influence on performance. In addition, the main problem was that players did not manage to reach the highest levels of music to synchronize at a higher rate than the one they hit the balls. For the same reason that they did not reach all levels of music, it is impossible to answer the question of which rhythm is the best because we do not have data for all levels.

Finally, it was found that the players had a good psychological flow in the set with the most subclasses exceeding the average. Below the bisectual value were found the sub-scales of independent experience and control. Low control rates can be explained by the feeling that players cannot control the virtual racket with their hand movements. Experience is also affected by this factor. Participants said they would prefer to choose those music they want to listen to, the intensity they want and could play for longer. For the submarine of clear feedback, the players said that their music helped them because they understood that the pace of music was changing. Likewise, to concentrate on the work they performed, they stated that they were holding them together, not Helped them find the rhythm. For the clarity of the goals they said they were helped by the instructions given to them while the game was relatively simple.

A significant statistical difference was found in the correlation of the flow with the attitude towards video games. From this we conclude that those who have a positive attitude are better absorbed in the work, in our case in the game, since they consider it more enjoyable. The flow also correlates with the experience factor. However, there is no statistically significant difference between those who have played the game and those who have not played it again. This result can be explained by the factor that may have played the game, may have been a while and, moreover, a few times. Based on this, they may have an earlier experience but they are not large enough to differentiate one group from another.

7 Further Research

Proposals for future research are first to involve more sample in order to achieve greater uniformity and to achieve more results. In addition, any future research should treat Kinect as a factor. For this reason, it is very important to get the hands on either change or technology or to change the game that was used. In addition, because the way we used the experiment to find the best possible rhythm did not yield results, we could follow a different method. For example, the original sample could be divided into uniform groups where the individual will play the games by listening to a different rhythm.

For performance, a different scenario that could yield is whether players play the game with two different versions. In the first version they will choose the music of their preference and in the second they will not be able to choose. This will allow comparison within the same player's performances. Finally, the ideal scenario of all, is to create and produce special music that meets all the needs of the game.

References

Bacon, C., Myers, T., Karageorghis, C.: Effect of music-movement synchrony on exercise oxygen consumption. J. Sports Med. Phys. Fitness **52**, 359–365 (2012)

Becker, N., Brett, S., Chambliss, C., Growers, K., Haring, P., Marsh, C., Montemayor, R.: Mellow and frenetic antecedent music during athletic performance of children, adults, and seniors. Percept. Mot. Skills **79**, 1043–1046 (1994)

Carlile, S.: The psychophysics of immersion and presence. In: 2011 Game Developers Conference. Moscone Center, San Francisco, 1 March 2011. Presentation

Collins, K.: Game Sound: An Introduction to the History, Theory, and Practice of Video Game Music and Sound Design. MIT Press, Cambridge (2008)

Craig, J.: Adaptive audio engine for EEG-based horror game. Thesis, New York University (2016)

Eladhari, M., Nieuwdorp, R., Fridenfalk, M.: The soundtrack of your mind: mind music adaptive audio for game characters. In: Proceedings of the 2006 ACM SIGCHI International Conference on Advances in Computer Entertainment Technology (2006)

Ekman, I.: Meaningful noise: understanding sound effects in computer games. In: Digital Arts and Culture (2005)

Eliakim, M., Meckel, Y., Nemet, D., Eliakim, A.: The effect of music during warm-up on consecutive anaerobic performance in elite adolescent volleyball players. Int. J. Sports Med. **28**, 321–325 (2007)

Fox, J.: Background music and industrial efficiency? A review. Appl. Ergon. **2**, 70–73 (1971)

Jarraya, M., Chtourou, H., Aloui, A., Hammouda, O., Chamari, K., Chaouachi, A., Souissi, N.: The effects of music on high-intensity short-term exercise in well trained athletes. Asian J. Sports Med. **3**, 233–238 (2012)

Jensen, M.B.: The influence of jazz and dirge music upon speed and accuracy of typing. J. Educ. Psychol. **22**, 458–462 (1931)

Karageorghis, C., Priest, D., Williams, L., Hirani, R., Lannon, K., Bates, B.: Ergogenic and psychological effects of synchronous music during circuit-type exercise. Psychol. Sport Exerc. **11**, 551–559 (2010)

Konecni, V.J.: Social interaction and musical preference. In: The Psychology of Music (1982)

Konz, S., Mcdougal, D.: The effect of background music on the control activity of an automobile driver. Hum. Factors: J. Hum. Factors Ergon. Soc. **10**, 233–243 (1968)
Lazarus, R.S.: Emotion and Adaptation. Oxford University Press, New York (1991)
Levy, L.: The effects of background music on video game play performance, behavior, and experience in extraverts and introverts. Thesis, Georgia Institute of Technology (2015)
Lombard, M., Ditton, T.: At the heart of it all: the concept of presence. J. Comput. Med. Commun. (1997)
McElrea, H., Standing, L.: Fast music causes fast drinking. Percept. Mot. Skills **75**, 362 (1992)
Moffat, B.: Creating personalities for synthetic actors. In: Personality Parameters and Programs. LNAI, vol. 1195. Springer-Verlag (1997)
Mullan, E.: Game Sound Technology and Player Interaction (2010)
Nacke, L.E., Grimshaw, M.N., Lindley, C.A.: More than a feeling: measurement of sonic user experience and psychophysiology in a first-person shooter game. Interact. Comput. **22**, 336–343 (2010)
Naushad, A., Muhammad, T.: Condition driven adaptive music generation for computer games. Int. J. Comput. Appl. (2013)
Parsons, L., Martizen, M., Delosh, E., Halpern, A., Thaut, M.: Musical and Visual Priming of Visualization and Mental Rotations Tasks: Experiment 1. University of Texas, San Antonio (1999)
Pignatiello, M., Camp, C., Rasar, L.: Musical mood induction: an alternative to the Velten technique. J. Abnorm. Pycholo. **95**, 295–297 (1986)
Politopoulos, N., Tsiatsos, T., Grouios, G., Ziagkas, E.: Implementation and evaluation of a game using natural user interfaces in order to improve response time. In: 2015 International Conference on Interactive Mobile Communication Technologies and Learning (IMCL), pp. 69–72. IEEE, November 2015
Scherer, K.R.: Appraisal considered as a process of multilevel sequential checking. In: Schorr, A., Johnstone, T. (eds.) Appraisal Processes in Emotion: Theory, Methods, Research. Oxford University Press, Oxford (2001)
Thompson, W.F., Schellenberg, E.G., Husain, G.: Arousal, mood, and the Mozart effect. Psychol. Sci. **12**, 248–251 (2001)
Van Merrienboer, J.J., Sweller, J.: Cognitive load theory and complex learning: recent developments and future directions. Educ. Psychol. Rev. **17**, 147–177 (2005)
Willimek, D., Willimek, B.: Music and Emotions (2013)
Young, D.: Adaptive game music: the evolution and future of dynamic music systems in video games. Thesis, Ohio University (2012)
Zehnder, S.M., Lipscomb, S.D.: The role of music in video games. In: Playing Video Games: Motives, Responses and Consequences (2006)
Doganis, G., Iosifidou, P., Vlachopoulos, S.: Factor structure and internal consistency of the Greek version of the flow state scale. Percept. Mot. Skills **91**, 1231–1240 (2000)

Teacher-Oriented Data Services for Learning Analytics

Dafinka Miteva[(✉)], Aleksandar Dimov, and Eliza Stefanova

Faculty of Mathematics and Informatics,
Sofia University "St. Kliment Ohridski", Sofia, Bulgaria
{dafinca, aldi, eliza}@fmi.uni-sofia.bg

Abstract. Modern education is benefitted from using Learning Management Systems (LMS) enabling automatic collection of large amount of data about students' activities, progress and results. However, currently there is a lack of enough means for teachers to effectively analyze and assess results from such data. This paper presents a research aiming to design and develop a new Learning Analytics (LA) solution which uses the existing LMS as distributed systems and storages of data through user-oriented services. It integrates the best features from LA and data visualization applications, from service composition platforms empower users to build complex business processes. Using teacher-oriented services data is retrieved directly from the relevant LMS as a response to educator's request and are visualized in appropriate way for fast and easy adoption. The paper describes the new approach and functionalities of this solution and ends with some challenges and conclusions.

Keywords: Learning Analytics · Service composition · Teacher-oriented

1 Introduction

One of the fastest growing IT areas in the last decade is Learning Analytics (LA). During the first conference Learning Analytics and Knowledge (LAK'11) it is defined as "the measurement, collection, analysis and reporting of data about learners and their contexts, for purposes of understanding and optimizing learning and the environments in which it occurs" [1]. For the most part modern education takes place in online environments based on advanced IT technologies, including mobile devices. All learners' activities are tracked by e-systems supporting training and a lot of student-related data is gathered and stored in a database. In particular, Learning Management Systems (LMS) like Moodle, Blackboard etc. assist teaching and learning by managing courses, but in addition they record details about each resource or activity review, submission or post in big log files. All this data could be analyzed, models for students' behavior and achievements could be extracted and taken into account to increase future success and to improve the environment where the learning occurs. This data need to be presented in a proper way because visualization is a key factor for data perception.

Data collected by LMS tend to be distributed in more than one e-system. In order to make such data useful for improvement of education, all of them should be merged

M. E. Auer and T. Tsiatsos (Eds.): IMCL 2017, AISC 725, pp. 29–35, 2018.
https://doi.org/10.1007/978-3-319-75175-7_4

together to give a more precise view about individual student. Integration of these data from existing databases may appear a tedious task, especially when it comes to big data.

In this article is presented a new approach to aggregating data from several separated LMS through teacher-oriented services which provides more comprehensive statistics, more accurate decisions, better training and outcomes. The paper starts with a short ground description of the research, followed by a review of related works, the approach and the proposed solution are described, faced challenges and conclusion are shared at the end.

2 Ground

The ultimate goal of each teacher is to give the learners knowledge and experience as much as possible. To what extend and how this goal is fulfilled it can be seen from students' achievement, from their attitude toward the course, from feedback polls. If instructor needs to be assessed based on students' surveys, the assessment will not be relevant if only one course learners' views are taken into account.

A worrisome problem is the ever higher percentage of students who drop out of training even though they have started with high activity and good results. There are also learners who demonstrate low efficiency and poor final results. In this case, to find out the reason, more data about student's background, activities and courses history is needed. Therefore one course LA does not reflect the overall student's attitude to learning, his/her priorities, strengths and weakness.

Unfortunately, LMS are course-oriented and LA reports are usually limited to data concerning a particular course. Teacher may track activities of each learner like number of visits to a resource, achievements on the respective topic, etc. and can analyze his/her behavior within a course but cannot compare student's performance in different classes of the curriculum and makes inferences in general.

For example in Moodle we can have several different instances of a course guided by one instructor at different times. In this case we might want statistics based on multiple editions of the same course as it is shown in Fig. 1.

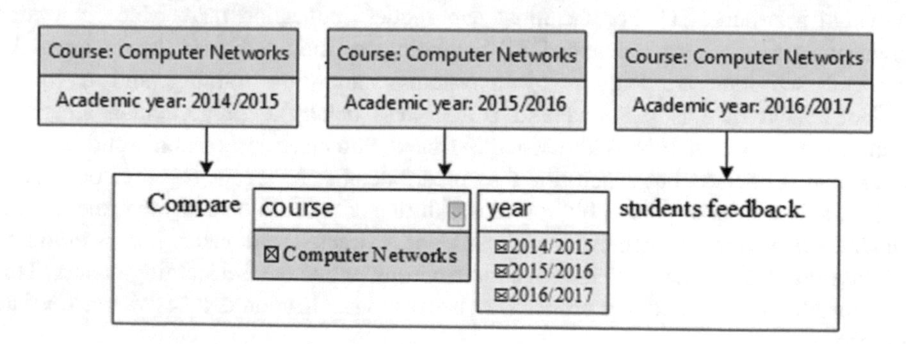

Fig. 1. Several instances of one course

On the other hand, various courses could be run by the same instructor whether at the same time or at different times. In this case we might want to make comparisons based on differences in learning content as it could be seen from Fig. 2.

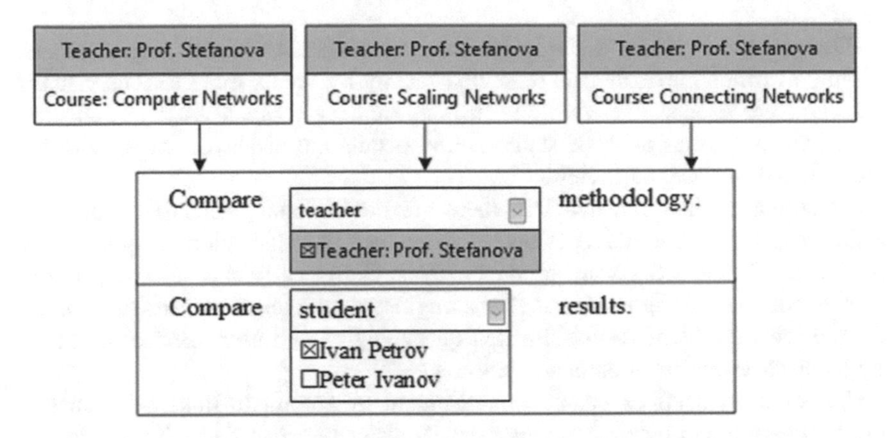

Fig. 2. Several courses of one teacher

This statistics and subsequent findings can be used to improve the quality of education by empowering the teacher to make self-analysis, by searching the reasons for the contrast in student's results from different editions of thc same course or from different courses, by evaluating the impact of teaching methods on final results.

Course analytics sections usually consist of a set of predefined reports and retrieving statistics beyond them may be a too complicated task for the average teacher. For example Moodle learning platform allows read-only access for instructor to write SQL queries directly to the database and to request details about badges, cohorts, courses and users. These kind of reports however require a very high computer literacy from the trainer.

From all the reasons mentioned above we can conclude that it is necessary to design and developed a new techniques, in terms of approaches, methodologies and software tools that are able to visualize the complex interconnection between individual learners, teachers and courses. Such tool could be a built-in unit of any LMS but retrieving cross-course statistics opposes to the course-oriented ideology of LMS, so we recommend a stand-alone application.

3 Related Work

We can distinguish three major directions that have direct relation to our work: learning analytics, data visualization and user oriented service compositions. Currently there are a lot of efforts to develop specific approaches, tools and environments in all of the above three areas. They offer a wide range of features and are very useful for the purposes they have been developed. However, none of them integrates the qualities of the other two categories of applications. In this section we briefly present each of these areas.

With respect to LA, beyond the built-in LMSs features, there is a number of plugins that provide additional options for data analysis and visualization. Such tools are Gismo [2] and SmartKlass [3]. Both of them use the LMS database and visualize statistics about one course [4] or about students and courses within the entire institution (SmartKlass).

When it comes to massive open online courses (MOOC) involving thousands of students, in which curriculum is presented mainly by video, more serious visual analytics tools are needed like VisMooc [5]. Considering content-based and clickstream data on video lectures analysis statistics and details are displayed graphically in list, content-based or dashboard view.

According to data visualization, there are professional interactive solutions and dashboards, which are not specifically developed for educational purposes. Good example are Tableau Software products [6] available as online, desktop, mobile or server version and designed for analysts, engineers, marketers and researchers. These software tools represent "a visualization query engine and user interface that make it easy to discover and communicate with data" [7].

User-oriented services enable establishment of automatic links between two or more software environments allowing data transfer between them. There are several service composition platforms that enable user to define such services like IFTTT [8], Zapier [9], Wufoo [10], eDream [11] which have already been analyzed and compared in [12].

For example IFTTT is a web platform or mobile application allowing user to compose interactive connections between favorite apps and devices. More than 450 ready-to-use recipes or applets are suggested for relating popular environments like Google, Microsoft, iOS, Android, topics like photography, social communications, weather et etc. The user is also given the opportunity to define his/her own channels. The pattern of service is described by the name "if *this* then *that*" which means if *this* condition happens an event is triggered and *that* action is executed as a result. For using this kind of pipes, services should be configured in advance with appropriate user's accounts and data.

The presented in this article solution brings together the advantages of LA tools and professional visualization software with user-oriented services to improve the LMS analytics.

4 Approach

As a stand-alone environment the newly designed LA tool works constantly communicating with all e-systems which are installed.

Data used by application is retrieved directly from the corresponding LMS instead of copying and duplicating the entire databases of all connected systems into one big data storage, no matter how actively this data is used.

Based on the notion of service oriented architecture [13] user-oriented services build a network of live links between data.

Each node of this net is an LMS which is joined in advance, shared data are defined and described by metadata. During the process of LMS association user account for authentication is required. In the case of Moodle this could be a teacher account which will restrict data access to the only courses the teacher has access to or an account with specifically defined Moodle role Analyst, which will provide read-only access to all courses, students' activities and results. Depending on the user account credentials potentially accessible categories of data are recommended to be described and available for service composition. Records can be limited by specific criteria.

To get statistics for learners or courses teacher needs to define a custom query or to adapt an existing one. Requests follow the structure "Get *data* for *whom/what* from *database* where …" where *data*, *whom/what, database,* and *dots* are configurable parameters. *Data* could be "grades", "competences, etc., *whom/what* is", a students' or courses' list, for now *database* is the Moodle instance database and optional *dots* replace one or more conditions the resulting data should perform. For example

"Get *grades* for *student Ivan Petrov* from *Moodle*
where *course is "Computer networks, winter, 2016/2017"*.

User's GET request forces service execution. It is sent to LMS, data is retrieved from the correspondent database and returned as a result of service to be visualized or further processed by another service.

Reports could be saved and stored in teacher's dashboard for reuse. Last results could be stored as long, as teacher decide.

5 Solution

The new solution is designed as a contemporary web application with user-friendly front-end interface predisposing even non-IT teacher to define their own questions with ease. There are five main sections corresponding to the categories of functionalities:

- Data description is the base of all available items for request compositions. During the LMS connection installation all available data types for students, courses, activities, competences etc. are recommended to be added here presented as pieces with data easy to drag-and-drop them in a query.
- Predefined reports is a list of the most frequently searched statistics ready for use or for further customization. In this section could be stored and share all newly defined quires by instructor.
- Request composition is the interactive visual area where complex database queries are designed in a simple way – just filling the pattern which data need to be displayed.
- Results visualization is the most attractive area where final results are presented graphically by charts. The proper types of diagrams are recommended and the user has the option to select.
- Recent reports is the history storage of recently used data results with an opportunity to set how long to keep them.

In the final version, these functionalities should be available as software services, which would be included into a user oriented service platform that we are currently developing [12].

6 Challenges and Conclusions

There are a lot of LA tools developed as a part of e-system gathering data and reporting statistics. Our solution presents two additional benefits: (1) more flexibility regarding data access and data collection and (2) data retrieving from more than one system giving a wider view of information as a result of collective intelligence of the e-systems.

The development of such solution faces a range of challenges to solve. For example making decision which metadata will best describe the LMS data types, how to synchronize connections between different LMS, how to discover the coincidences of the same student in different systems.

Supporting student-centered education implies educator to have sufficient knowledge about student's ability and skills. Giving a chance to ask the exact question to all relevant systems provides teacher with increased possibilities to get the right information at the right moment. This helps him/her to make the most accurate decisions. In addition the visual environment for query composition makes the tool widely available for the average teacher. The tool is being developed within a research project and is supposed to be actively tested over the next academic year.

Acknowledgment. The research is done with financial support of DFNI I02-2/2014 (ДФНИ И02-2/2014) project, funded by the National Science Fund, Ministry of Education and Science in Bulgaria and Sofia University "St. Kliment Ohridski" research science fund project N80-10-217/24.04.2017 "Inquiry-based learning in the field of high technologies as an application of modern information technologies".

References

1. 1st International Conference on Learning Analytics and Knowledge (2011). https://tekri.athabascau.ca/analytics. Accessed 26 April 2017
2. Blocks: GISMO. https://moodle.org/plugins/block_gismo. Accessed Sept. 2017
3. SmartKlass™: The Learning Analytics Plugin. http://klassdata.com/smartklass-learning-analytics-plugin/. Accessed Sept. 2017
4. Mazza, R., Milani, C.: GISMO: A graphical interactive student monitoring tool for course management systems. T.E.L.04 Technology Enhanced Learning 04 International Conference, Milan, Italy (2004)
5. Shi, C., Fu, S., Chen, Q.: VisMOOC: Visualizing video clickstream data from Massive Open Online Courses. In: IEEE Pacific Visualization Symposium (PacificVis), Hangzhou, China, pp. 159–166 (2015)
6. Tableau Software. https://www.tableau.com/products. Accessed Sept. 2017
7. Jones, B.: Communicating Data with Tableau: Designing, Developing, and Delivering Data Visualizations, p. 334. O'Reilly Media, Sebastopol (2014)
8. IFTTT. https://ifttt.com/. Accessed Sept. 2017
9. Zapier. https://zapier.com/app/explore. Accessed Sept. 2017
10. Wufoo. https://www.wufoo.com/. Accessed Sept. 2017

11. eDreams. https://www.edreams.com/
12. Dimov, A., Peltekova, E., Stefanova, E., Miteva, D.: User-oriented service composition platform. In: 2015 Proceeding of IMCL, Thessaloniki, Greece (2015)
13. Douglas, B.K.: Web Services, Service-Oriented Architectures, and Cloud Computing: The Savvy Manager's Guide, p. 248. Morgan Kaufmann Publishers, San Francisco (2012)

Improvement of Students' Achievement via VR Technology

Elitsa Peltekova(✉), Aleksandar Dimov, and Eliza Stefanova

Faculty of Mathematics and Informatics,
Sofia University "St. Kliment Ohridski", Sofia, Bulgaria
{epeltekova, aldi, eliza}@fmi.uni-sofia.bg

Abstract. Virtual Reality (VR) technology is developing very actively last few years. VR devices are more and more accessible, affordable and recognizable by youngsters in school and university. On one hand, areas of science, technology, engineering and mathematics (STEM) are some of the most rapidly developing disciplines. On the other hand students show unsatisfactory results and low interest in STEM subjects. This means it is needed something to be done students to level up their interest in STEM, so VR could be helpful and it could contribute for improving their achievements in STEM subjects. However, VR technology integration in educational process expects teachers to be aware of VR technology, to have appropriate educational VR scenarios and to be equipped with needed devices. The objective of our study is to propose a methodology for VR application in education, to provide teachers with VR scenarios which could improve students' achievements. In order to prepare VR scenarios, here in this paper we have started with formulating VR scenario's criteria which criteria are based on the answers from the interviews recently conducted with STEM teachers.

Keywords: Virtual Reality · Scenarios · Technology-enhanced learning STEM

1 Introduction

STEM stands for science, technology, engineering and mathematics and this will be what we mean in our paper also. Although definitions of STEM subjects can vary widely from country to country. 'Core' STEM subjects typically include: Mathematics; Chemistry; Computer Science; Biology; Physics; Architecture; and, General, Civil, Electrical, Electronics, Communications, Mechanical, and Chemical Engineering.

According to a study called "Encouraging STEM studies", the decline of pupils' interest in STEM subjects is particularly noticeable at the secondary school level (Caprile and Plamen 2015). Also according Organisation for Economic Co-operation and Development (OECD) Programme for International Students Assessment (PISA) results from 2015 - student achievements in STEM (with focus in Southeast Europe, where Bulgaria is) shows:

- The performance in science, reading and mathematics are with values below the OECD average;

M. E. Auer and T. Tsiatsos (Eds.): IMCL 2017, AISC 725, pp. 36–43, 2018.
https://doi.org/10.1007/978-3-319-75175-7_5

- Students' science beliefs, engagement and motivation are with a share of low achievers above the OECD average (Organisation for Economic Co-operation and Development 2015).

Analysis by CEDEFOP (ICF and Cedefop for the European Commission 2014) shows that employment of STEM professionals and associate professionals in the European Union has increased since 2000 in spite of the economic crisis and demand is expected to grow until 2025. In our digital era, development of STEM human capital is considered a critical factor for economic development and growth (Mcneely and Hahm 2012). There are three main policy approaches related to encouraging STEM studies and careers in Europe (Caprile 2015):

- curricular and teaching methods;
- teacher professional development;
- guiding young people to STEM.

It has been also noted that the way in which science subjects are taught has a great influence on students' attitudes towards science and on their motivation to study and, consequently, their achievement. Therefore the main goal of the paper is to formulate initial criteria for educational VR scenarios which will have a potential to enrich school and university STEM curricular and improve teaching methods, to bring teacher professional development and to guiding young people to STEM in new engaging way.

Our paper is divided into three main sections and conclusion. First section "Background" define what VR is and shows some statistics about VR devices and VR users worldwide. Second section "The Study" briefly presents interviewees, study's research method and describes the interview itself. Third section "Results" shows some interview's answers and outline VR criteria for the educational VR scenarios, and ideas for education VR scenarios.

2 Background

In this section we introduce how we understand VR and generally what kind of VR devices we plan to use in the future VR scenarios. VR definition are two in Fuchs's book "Virtual Reality Headsets – A Theoretical and Pragmatic Approach":

- Technical VR definition: Virtual reality is a scientific and technical domain that uses computer science (1) and behavioural interfaces (2) to simulate in a virtual world (3) the behaviour of 3D entities, which interact in real time (4) with each other and with one or more users in pseudo-natural immersion (5) via sensorimotor channels (Fuchs et al. 2017);
- Functional VR definition: Virtual reality will help user to come out of the physical reality to virtually change time, place and (or) the type of interaction: interaction with an environment simulating the reality or interaction with an imaginary or symbolic world (Fuchs et al. 2017).

The purpose of virtual reality is to make possible a sensorimotor and cognitive activity for a person (or persons) in a digitally created artificial world, which can be imaginary, symbolic or a simulation of certain aspects of the real world (Fuchs et al. 2017).

- “Behavioural interface” term to “user interface” term. They are made of “sensorial interfaces”, “motor interfaces” and “sensorimotor interfaces”:
- In sensorial interfaces (visual interface, tactile interface, audio interface, etc.), the user is informed about the development of the virtual world through his senses. A visual interface is always used: VR headset, CAVE1 or screen.
- Motor interfaces inform the computer about man’s motor actions on the virtual world (joystick, data glove …).
- Sensorimotor interfaces work in both directions (force feedback interface2) (Fuchs et al. 2017).

There are many different VR devices, so called VR viewers and head-mounted displays (HDM) for virtual reality experience and they come in a variety of forms, from single eye information displays to fully occluding stereoscopic headsets. We plan to use binocular HDM or viewer, where each eye receives its own separate viewing channel with slightly offset viewpoints mimicking the human visual system to create a stereoscopic view. (Aukstakalnis 2017). For example more affordable viewers as Google Cardboard (Google 2017) or similar to Samsung Gear VR viewers (Oculus VR 2017).

Use of flexible learning environments is a newer trend in schools, in teaching at all. VR fits this flexible learning model well in that it can easily satisfy the technology-driven portion of the lesson. More ideal would be availability of STEM resource and computer lab outfitted with VR to support an entire class simultaneously. The benefit with this scenario is exponential: while students work on their lesson in VR, the teacher is freed up to use his expertise and training in a more targeted manner with individual pupils as a facilitator, tutor, counselor, mentor, planner, or evaluator. This flexibility is priceless, as schools already recognize that many of the biggest gains in students’ progress are based on customizable, student-centered learning (Perkins n.d.).

Number of mobile VR users worldwide from 2015 to 2020 is expected highly to increase. The statistic shows the number of mobile VR users worldwide from 2015 to 2020 is expected to rise significantly. In 2015, there were 2 million mobile VR users, now in 2017 there are 53 million, and in 2020–135 million mobile VR users are expected (Statista Inc. 2017). In 2020, the number of AR/VR/MR devices (where AR stands for Augmented Reality and MR – Mixed Reality) shipped worldwide is expected to increase to 110.5 million units, now in 2017 shipped devices are 22.9 million units compared with 10.1 million in 2016, and 0.7 million in 2015. These numbers highly motivates us in conducting our study.

3 The Study

Along with the predicted statistics about high increase in VR devices usage and VR users above, here VR Intelligence (VR Intelligence 2017) share VRX Industry Survey results from industry’s views, opinions, as VR goes mainstream. Answer of the question: In which industries will VR have the most impact in the next 3 years?, unsurprisingly, most agree that gaming is the industry where VR will have the biggest impact. But as there will inevitably be those who have better knowledge of some industries than others, there is optimism across the board, particularly in healthcare and education. This is another reason motivating us to conduct our study.

3.1 Interviewees

The population for this study was a purposive sample of 6 teachers – 3 university teachers and 3 school teachers. The purpose of our study is to retrieve a qualitative data for writing criteria for development of VR scenarios. We claim that for this initial phase of our study, the number of 6 interviewees is enough. In our future research we will include more participants in the interview.

Participants were chosen because all they teach STEM subjects and all they have at least 5 years STEM teaching experience, and some of them even more than 10–20 years STEM teaching experience.

3.2 Research Method

In-depth interview was decided to be used as a main method to collect data for this stage of our study. In-depth interviewing, also known as unstructured interviewing, is a type of interview which researchers use to elicit information in order to achieve a holistic understanding of the interviewee's point of view or situation; it can also be used to explore interesting areas for further investigation. This type of interview involves asking interviewees open-ended questions, and probing wherever necessary to obtain data deemed useful by the researcher (Berry 1999).

Interviews were conducted in different ways with each interviewee – in person, through phone, and through Skype (Microsoft 2017). Interviews have been held in Bulgarian. Researcher have been taking notes during the whole interview process. After all the interviews have been conducted, all the information was combined in a common text file. Later on the information from the text document was translated in English and was transferred in table. It was used trail version of NVivo software (QSR International 2017), which is suitable and can help analyzing unstructured data like interviews as well.

As we already mentioned, in our study informants are Bulgarian teachers – both university and school teachers. Our interviews aim to prove hypothesis "teachers consider VR as helpful and useful, and they would rather implement VR technology usage in class".

3.3 The Interview

Below are listed all 11 interview questions:

1. Sex
2. Age
3. Level of education
4. Where do you teach (name of the school/the university)?
5. Which classes/courses do you teach?

These first five questions are rather socio-demographic compared to the next six questions which are specifically oriented to our study dedicated to VR.

6. Which disciplines do you teach?
7. What kind of educational content do you use in the teaching and learning process?
8. Is there something missing (any specific learning resource)?
9. What do you know about VR?
10. Do you have an observation about your students' VR awareness?
11. Does VR have the ability to bring pedagogical added value?

4 Results

First we start with presentation of information about our respondents. In Table 1 is presented the data about interviewees' sex, age, level of education, and classes/courses they teach.

Table 1. Generalized information about interviewees in our study

Quick Start Steps | NVivo-IMCL2017Analysis | Survey Respondent

	A : Sex	B : Age	C : Level of education	D : Which classes courses do you teach
1 : School Teacher 1	female	between 25 and 54	higher education	8-12 grade
2 : School Teacher 2	female	between 25 and 54	higher education	5-8 grade
3 : School teacher 3	female	between 25 and 54	higher education	5-11 grade
4 : University Teacher 1	female	between 25 and 54	higher education	Master students
5 : University Teacher 2	male	between 55 and 64	higher education	Bachelor and Master students
6 : University Teacher 3	male	between 55 and 64	higher education	Bachelor and Master students

Three of the questions and some of their answers are discussed below, we put our attention to the most VR-related once.

Question: *What do you know about VR?*

All the respondents have an idea about VR, one of them even teach their students about VR, he said: "In one of my courses, I have lectures on VR topic." Many of the interviewees have seen VR content with VR viewers. Most of them know VR technology as an expensive technology, not yet affordable for the schools or universities.

Question: *Do you have an observation about your students' awareness for VR?*

"Few of my students know about VR technology - 1 in 100."

"I do not know, honestly. Students use a variety of software. Many of the games they play are possibly VR games."

"I suppose, the students I teach have recently visited popular VR places in the shopping malls. Also, at recent conference I have attended in Vratsa, Bulgaria, students presented a VR classroom project."

"All they have seen are demo videos, but they do not use VR on a regular basis."

"I have been asking my students about VR, but I have not got any feedback. It is needed more information about VR among students."

"They are not very familiar with VR."

All these answers makes us clear that neither students in school, neither students in university are quite aware with VR technology according to their teachers, and current VR usage is connected mostly with gaming.

Question: *Does VR have the ability to bring pedagogical added value?*

This question categorically was answered with "Yes" by the teachers. This means VR have the ability to bring pedagogical added value. All the interviewees have the same opinion and even share ideas for VR scenarios they would be happy to have.

30 most frequently occurring words in our interviews were visualized with NVivo (Fig. 1). Some of the words – mathematics, computer, simulation, biology, informatics, class, materials, use, technologies, programming.

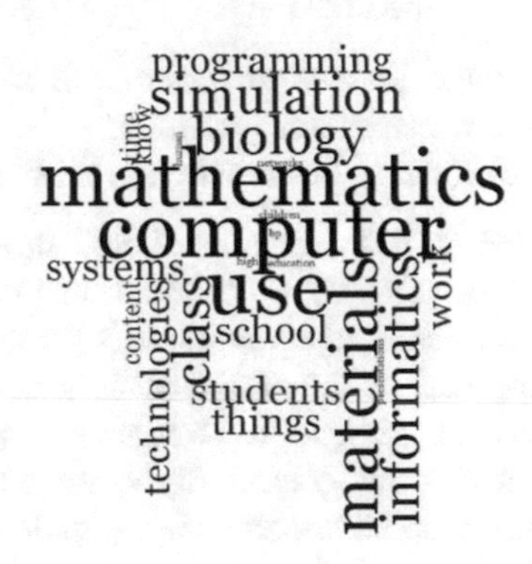

Fig. 1. NVivo's word cloud – most used words in the interviewees' answers

Below are listed criteria suitable for VR educational scenario, which have been proposed by the interviewees:

- Interactivity;
- Short VR content – not longer than 7–10 min;
- VR equipment - VR devices (viewers and/or VR HMD), mobile phones;
- Stable Internet connection - important prerequisite for VR usage because still most of VR content is available online.

Example scenarios proposed by informants:

- Demonstrations of:
  - dangerous situations;
  - experiments, when there is absence of chemical reagents, for example, in chemical laboratories or the experiment is too risky to be done by unexperienced people;
  - work with a fire extinguisher, since existing trainings only tells what should be done, but it is not done because if it would be done then this fire extinguisher becomes unusable and should be refilled.

- Trainings teaching how to act if we are in an accidents, if there is an earthquake, or if we experience any other disaster generally;
- Overcoming fears, narrow spaces, heights, and so on.

5 Conclusion and Future Work

The purpose of our study is creation and testing of STEM scenarios which use VR technology. In this stage of the study we understood teachers' awareness of VR, and their observation what students know about VR. Also it has been learned teachers' view for future VR application in class. But still much need to be done – interview with student, writing VR scenario together with teachers and testing scenario without students (initially) and with students. As part of the future work, we consider the following main issues briefly described below:

- A technology support should be created in terms of software tools that facilitate teacher in creation of VR scenarios for education.
- Scenario including VR element to be tested with teachers.

Technological support of VR STEM scenarios should provide tutors with means to easily browse within a set of existing VR tools and, select the ones that best suits their scenarios and further – to seamlessly integrate these VR tools within the scenario. For this purpose a promising approach is the notion of Service Oriented Architecture (Erl 2008). This is a well-known concept in software engineering, which provides tools and techniques for development and integration of heterogeneous distributed systems, using reusable and uniformly accessible components, called services.

We are currently developing a platform for autonomous composition of software services (Dimov et al. 2015), which should enable the integration of VR tools. For this purpose they should be wrapped as services, and registered within the platform. This way tutors would be facilitated in search and selection of appropriate VR tools.

Scenario for teacher training was created in the frame of ELITe project (Enhancing Learning In Teaching via e-inquiries, EC-Erasmus+project 2016-1-EL01-KA201-023647). Scenario design includes VR technology usage. Two of the main training objectives are (1) developing of "non-traditional" design training and (2) develop teacher's competences as knowing and using new technologies and apply them in class – to conduct technology-enhanced learning. So we will have an opportunity to collect data about teacher's perception about VR usage.

Acknowledgement. The work presented in this paper was partially supported by:

- Sofia University "St. Kliment Ohridski" research science fund project N80-10-217/24.04.2017 "Inquiry-based learning in the field of high technologies as an application of modern information technologies";
- DFNI I02-2/2014 (ДФНИ И02-2/2014) project, funded by the National Science Fund, Ministry of Education and Science in Bulgaria;
- ELITe project (Enhancing Learning In Teaching via e-inquiries, EC-Erasmus+project 2016-1-EL01-KA201-023647).

References

Aukstakalnis, S.: Practical Augmented Reality. Addison-Wesley, Boston (2017). ISBN 978-0-13-409423-6

Berry, R.S.: Collecting data by in-depth interviewing. In: British Educational Research Association Annual Conference, Brighton (1999). http://www.leeds.ac.uk/educol/documents/000001172.htm

Caprile, M., Plamen, R.: Encouraging STEM studies for the labour market (2015). www.europarl.europa.eu. http://www.europarl.europa.eu/RegData/etudes/STUD/2015/542199/IPOL_STU(2015)542199_EN.pdf

Dimov, A., Peltekova, E., Stefanova, E., Miteva, D.: User-oriented service composition platform. IEEE (2015)

Erl, T.: SOA: Principles of Service Design. Pearson, Boston (2008)

Fuchs, P., Guez, J., Hugues, O., Jégo, J.-F., Kemeny, A., Mestre, D.: Virtual Reality Headsets – A Theoretical and Pragmatic Approach. CRC Press/Balkema, London (2017)

Google (2017). Google Cardboard. https://vr.google.com/cardboard/

ICF and Cedefop for the European Commission: EU Skills Panorama (2014) STEM skills Analytical Highlight (2014)

Mcneely, C.L., Hahm, J.-O.: The Global Chase for Innovation: is STEM education the catalyst? Glob. Stud. Rev. **8**(1) (2012). http://www.globality-gmu.net/archives/2972

Microsoft (2017). Skype. https://www.skype.com/en/

Oculus VR (2017). Oculus VR. https://www.oculus.com/gear-vr/

Organisation for Economic Co-operation and Development: PISA 2015, Results in Focus (2015). https://www.oecd.org/pisa/pisa-2015-results-in-focus.pdf

Perkins, D.: 6 considerations for adapting virtual reality in education (n.d.). teachthought.com. https://teachthought.com/technology/6-considerations-quality-virtual-reality-education/

QSR International: Using NVivo for qualitative research (2017). NVivo 11 for Windows. http://help-nv11.qsrinternational.com/desktop/concepts/using_nvivo_for_qualitative_research.htm

Statista Inc.: Number of mobile virtual reality (VR) users worldwide from 2015 to 2020 (in millions) (2017). Statista. https://www.statista.com/statistics/650834/mobile-vr-users-worldwide/. Accessed 16 Sept 2017

VR Intelligence: VR Intelligence (2017). http://vr-intelligence.com/vrx/docs/VRX-2017-Survey.pdf. Accessed 16 Sept 2017

Contextualisation of eLearning Systems in Higher Education Institutions

Gerald Gwamba[1], Jaco Renken[1], Dianah Nampijja[2,3], Godfrey Mayende[3,4](✉), and Paul Birevu Muyinda[3]

[1] University of Manchester, Manchester, UK
[2] University of Agder, Kristiansand, Norway
[3] Makerere University, Kampala, Uganda
[4] University of Agder, Grimstad, Norway
godfrey.mayende@uia.no

Abstract. The proliferation of digital technologies, and the emergence of global lifelong learning has steered the transformation of education from the predominant classroom based learning to more flexible technology enhanced learning. However, realizing technology enhanced learning's much anticipated benefits towards improving the educational potential of Higher Educational Institutions (HEIs) in Developing Countries (DCs) is still a challenge because of the high information systems failure rate. Research suggests potentially substantial misalignment between DC HEIs' E-Learning Management Systems (ELMS) and their institutional contexts. To explore this gap, this research employs a qualitative approach based on a case study of the Makerere University Electronic Learning Environment (MUELE). We employ the Aparicio eLearning systems theoretical framework to explore the extent to which DC HEIs are aligning their ELMS within their institutional contexts. Results indicate that, ELMS are not aligned to the context of DC HEIs implying that stringent measures need to be taken to close misalignment gaps.

Keywords: E-learning · Learning Management Systems
Higher education institutions · Developing country context

1 Introduction

The proliferation of educational and mobile technologies, and the emergence of global lifelong learning has steered the transformation of education. From predominant classroom based learning to independent technology enhanced learning. Research suggests that there has been significant growth in eLearning across all products and services globally. Mobile learning is among the fastest growing subsectors within the online learning mainstream because of the improved access to affordable mobile technologies and the affordances that it brings among African citizens.

The rapid proliferation of eLearning within Higher Educational Institutions (HEIs) has been a result of social-economic drivers such as increased competitiveness (Sekiwu and Naluwemba 2014), cost effectiveness and reach (Lee-post 2009), emergence of a mobile citizenry (Hossein 2015; Vaiva et al. 2014), among others. In Uganda particularly, research suggests that government efforts to increase the literacy rates through

M. E. Auer and T. Tsiatsos (Eds.): IMCL 2017, AISC 725, pp. 44–55, 2018.
https://doi.org/10.1007/978-3-319-75175-7_6

Universal Primary Education (UPE) and Universal Secondary Education (USE) have spurred a significant increase in student enrolments. This has mounted pressure on educational institutions already limited by space and other resources. In a bid to meet this demand particularly in HEIs, the Ugandan government is encouraging investment in eLearning as a suitable alternative (Mayoka and Kyeyune 2012).

The emergence of digital mobile technologies has steered the incorporation of mobile devices, services and platforms into learning environments with increasing adaptation to suit the mobile delivery. This has propelled major changes in the design of today's Learning Management Systems (LMS) with mobile supported features allowing even further flexibility for eLearning delivery and accessibility.

However, realizing technology enhanced learning's much anticipated benefits towards improving the educational potential of HEIs in DCs is still a challenge. This is because most information system implementations fail in some way (Masiero 2016; Heeks 2003). Research suggests that there could exist substantial mismatches between the alignments of Developing Country (DC) HEIs' E-Learning Management Systems (ELMS) and their institutional contexts (Gwamba et al. 2017).

Moreover, some research focused on the development of information systems in DCs reveals that most IS projects fail to realise their anticipated benefits, and majority are externally funded especially when project funding is exhausted (Ssekakubo et al. 2011; Kinyua 2015). As a corrective, there is need for eLearning policy frameworks among African countries. The policy could allow for shared eLearning experiences of others through government led efforts providing centrally coordinated eLearning strategies that align with national goals, educational reforms and technology. Proposed eLearning strategies thus should lay out a road map for the eLearning architecture, resolve curriculum issues, guide on capacity development, and management of systems, content development and infrastructure among others (Manji et al. 2015).

Similar strategies have further focused on developing well-structured ideal solutions that all players should follow to achieve more successful eLearning developments (Manji et al. 2015; Mtebe 2013; Ssekakubo et al. 2011). This is however hard to achieve since the question of contextual alignment of eLearning takes precedence. To be specific, a one-size-fit-all approach is undesirable because of the variance in contextual factors. Some universities have introduced eLearning for administrative, management and registration context which is far from ELMS core educational functions. Given that institutions operate in different contexts, many factors shape eLearning development ranging from technology, to social and pedagogical aspects among others (Aparicio et al. 2016). Moreover, whilst government led policies where educational opportunities are identifiable; institutional realities within DCs present unique challenges most of which have not yet been adequately addressed.

This research employs an exploratory approach seeking to better understand the influence of the social and technical contexts of HEIs in DCs on their LMS implementations. We thus seek to answer the research question: To what extent are DC HEIs' E-LMS aligned within their institutional contexts? In this paper, we attempt to answer this question by using the holistic eLearning systems theoretical framework to analyse how the institutional context aligns with the LMS. In the next sections of this paper, we present the approaches and methods used in Sect. 2, followed by research findings in Sect. 3. Then discussions in Sect. 4. We then draw conclusions of the outcomes and limitations of the research in Sect. 5.

2 Research Approach and Methods

2.1 Research Framework

Aparicio et al. (2016) proposed a holistic E-Learning Systems Theoretical Framework (E-LSTF) which classifies key inter-connected factors that operate within an eLearning system domain that could lead to "successful" eLearning systems development. These factors are characterised in terms of people, technology and services dimensions.

In the framework, people interact with the systems and the technologies facilitate the direct and indirect interactions of different stakeholder groups, and further enable integration of eLearning content and collaborative tools (Aparicio et al. 2016). The services on the other hand integrate all the activities related to both instructional strategies and pedagogical models used within learning spaces. Therefore, the eLearning system provides the environment for complex interactions between various dimensions. In analysing the E-LSTF, we will critically assess the development of the different dimensions to gauge the extent of alignment of DC HEIs' ELMSes within their institutional contexts. The institutional context may focus on the environment which is constituent of a system of geographical, political, social-cultural and other factors as described in Heeks (2006, pp. 4–5).

Whereas other models may tend to focus on the extremes of social or technical aspects - a socio-technical approach like the e-government analogy in Heeks (2006); the E-LSTF combines both the soft social and hard technical extremes. The socio-technical/hybrid balance spans not only the system design side but also the implementation side. Accordingly, striking a social-technical balance could drive further alignment of IS within the institutional environment. In this regard, models that explore the nature of environmental forces whilst mapping them on technical aspects could be more ideal in informing IS in DC institutions, which resonates with the E-LSTF. It is however worth acknowledging that the E-LSTF is a newly published theory (2016) and has thus not yet gained a lot of validation putting our case among those to first validate this theory. The E-LSTF is well suited in the contextual alignment of ELMSes in DC HEIs since its constructs are focused on people, technology and activities, all of which are dimensions of the hybrid IS perspective (Aparicio et al. 2016; Heeks 2006). Investigating the extent of alignment of ELMS within the institutional context based on the proposed model seeks to pre-empt a discussion surrounding how well information systems fit within the micro/macro environment of DC HEIs.

2.2 Research Methods

Our study employed qualitative methods using guided interviews and observations. Interviews with 8 key respondents were based on a set of semi-structured questions derived from the E-LSTF conceptual framework; interviews lasted 30–45 min. All interviewed respondents were selected based on their positions as experienced eLearning practitioners at Makerere University. Data collection was further guided by themes including "context of DC HEI" and "LMS alignment" derived from the research question. We used the case study research strategy characterised with a rich background on IS adoption in a DC HEI to allow for further discussions. To freely allow respondents to further express themselves, emails questionnaires were in addition to face interviews sent to respondents. This further provided free expression especially on controversial topics. The interviews were transcribed, and inter-rater reliability was employed. Reference to institutional literature was vital in enhancing data triangulation. The data was analysed through thematic analysis based on ELMS system theoretical framework.

2.3 The Development of ELMS in Makerere University

Figure 1 shows the development of computer assisted learning dating back to 2001, ten (10) years since the inception of distance learning.

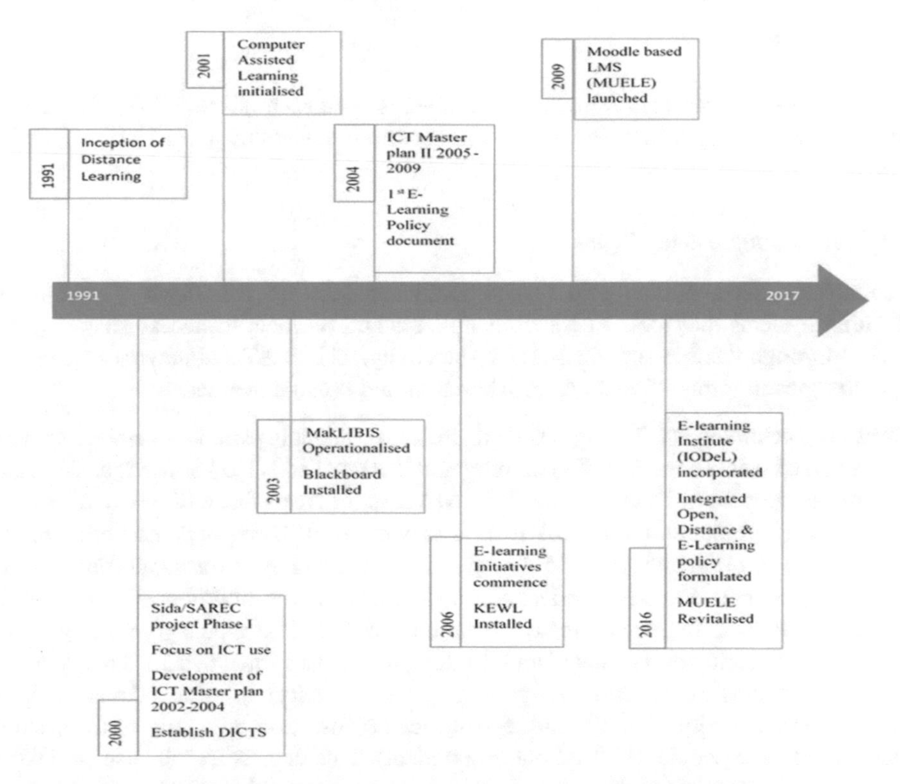

Fig. 1. Overview of the development of ELMS at Makerere University

Makerere University (MAK), through its present Institute of Open Distance and eLearning (ODeL), introduced first generation distance learning which later developed to second generation mode as referred to by (Taylor 2001). The university is gradually adopting the use of ICTs for teaching, learning, research and knowledge dissemination. In recognising the need to improve efficiency in service delivery for academic, library, financial and human resources, MAK embarked on a drive to integrate IS within its core business operations (Tusubira et al. 2008).

The eLearning environment at MAK is however characterised by poor student support systems. Justifiably, present eLearning is developed at a slow pace through adopting a blended approach to supplement the traditional face to face mode of instruction. According to Fig. 1, the first ELMS was introduced around 2003 in response to the ICT Master plan 2002–2004 (Greenberg and Versluis 2005). With support from Tufts University, MAK later introduced Blackboard LMS in 2003 but due to financial constraints, the replacement with an open source LMS – KEWL – in 2006 provided a sustainable alternative (Tusubira et al. 2008). KEWL's limited stability and errors at the time saw MAK opt for a more stable choice of Moodle which is better suited to handle a larger student population estimated at more than 40,000 students as of 2013/14 academic year.

3 Findings

Our findings are presented following a detailed look at each of the ELSTF framework dimensions offering valuable aspects to inform the contextual map of ELMS within DC HEIs.

3.1 eLearning Technologies

Technology plays a vital role in the implementation of collaborative web-based learning systems and tools that enable students and teachers to access shared workspaces through flexible and persistent connectivity. The ELST Framework categorises technologies in terms of content, collaboration and communication.

Content Technologies. Taking a keen interest on content is paramount since it is what delivers the business benefits of eLearning and it is what adds the knowledge, skills and capabilities to people. Data gathered from MAK staff reveals that a diversity of content technologies are supported on MUELE. Accounts from 8 respondents points to the parallel use of both analogue and digital content technologies for supporting institutional programmes. From respondent accounts, IODeL is currently focusing on training university teachers to develop instructional content based on e-pedagogy designs and content restructuring as a mechanism for interactive instructional design. This is one of the primary aims of the currently running projects. 'Most of what people do is to upload content, majorly knowledge repositories because from my audit of the system, that's what I see people do' (said one respondent). This depicts that the use context of the system is mainly for data storage as opposed to a learning platform which contradicts the use of the system by design. Conversely, 2 of the 8 respondents revealed that they had used commercial content development tools beyond Moodle with enhanced content development capabilities not yet affordable to MAK. Overall, content technologies identified were knowledge repositories, assessments, documents, text based authoring tools and limited digital audio/video. Whereas the above list of content technologies is dominant in MUELE, their overall alignment within the institutional context largely remains volatile, and partly inclined to social and institutionalization aspects.

Communication Technologies. One of the core functionalities of ELMS is the ability to offer flexible communication across diverse groups. Contextually, MAK offers traditional face to face programmes by establishment which affects the choice of communication technology to integrate in MUELE. Ideally the most popular communication technologies acknowledged are knowledge discussion forums, social networks and email. Data gathered suggests that videoconferencing for example was not planned during the university's establishment of eLearning. Based on staff accounts, some respondents argued that other institutions like the University of Nambia preferred videoconferencing due to harsh climatic conditions which hampered frequent human mobilities. However, the misalignment could emanate from the fact that MAK did not plan for such technologies within its establishments. Interestingly, from some respondents, 'the distance learning centres that were established as extramural centres were later transformed into population catchment centres - taking the university to the people'. Whereas by design, videoconferencing capabilities do exist for Moodle, they are highly resource dependent and would not be an ideal option for MAK given its financial constraints (Schenkel 2013; Ssekakubo et al. 2011; Tabaire and Okao 2010). For social networking, WhatsApp, Facebook and Twitter are outside the domain of MUELE. However, Moodle supports chat services with additional plugins for audio/video integration by design. These capabilities have not been utilised and according to some respondent accounts, Moodle inbuilt social networking tools have not been popularised, and rightly afforded within MUELE.

Collaborative Technologies. These offer unprecedented opportunities for environments that foster group related activities suitable for off campus learning geographically consolidating learners and teachers. Results indicate an individualized, and highly personal social cultural learning context with in eLearning at MAK. '…the pedagogical culture within MAK is more inclined towards individualised competitive teaching/learning as opposed to collaboration' (said one respondent). Such sentiments and more work against the culture of ELMS deterring group cooperation, which deters collaboration within MUELE. Through collaborative learning technologies, institutions and educators are in a better position to learn from each other, encouraging resource sharing to more openness that consolidate multidisciplinary user groups (The NMC Horizon Report 2017). Thus, sharing and learning in groups as opposed to individualised learning has the power to facilitate deeper learning to optimise the full versatility of the ELMS. Despite the availability of tools like forums, chats and multiuser areas like blogs that facilitate multidisciplinary sharing, evidence from 5 respondents indicates that the collaborative power of MUELE is yet to be utilized. Evidence of strong external collaborations beyond MUELE creates a competitive force emanating from social networking tools like Facebook, WhatsApp, among others. "Facebook, WhatsApp among other tools provide faster and more reliable connectivity compared to the platform characterised with frequent downtime," narrates respondent 7. Earlier research on major impediments on ICT developments in MAK confirms this sentiment (Mayoka and Kyeyune 2012; Opati 2013; Ssekakubo et al. 2011). However, all 8 respondents affirm that a clear roll plan for institutionalising eLearning has not yet been agreed upon causing serious misalignment issues. MUELE, on the one hand, provides several collaborative technologies, which we now found to be largely unused. On the other hand, collaborative technologies outside of MUELE are used by the university showing that the ELMS offering does not align with the collaboration requirements of the institution.

3.2 E-Learning Stakeholders

According to the Aparacio et al. (2016) ELMSF, stakeholders are all people within the wider institutional domain (internal and external) who make use of the IS including customers, suppliers, professional associations and board and others. Results indicate that students, academic and administrative staff are the principal stakeholders in MUELE. Whereas students and lectures are the main consumers, it would be pertinent for us to fully conceptualise the role that other groups play in shaping major ELMS change and development. However, the newly approved eLearning policy refers to eLearning stakeholders as "ODeL target clients" without clearly spelling out the scope of this domain but rather refers to other institutional policies (Makerere University 2015). This could create wider confusion for future application given the fact that further reference is made on an open admission policy that cartels many diverse groups of real, ideal or potential end-users. 'The current ELMS efforts are primarily targeting students alone and the system has not yet envisioned other groups within the university's establishment as potential customers' (said one respondent).

In the supplier categorisation, lecturers (facilitators) and MUELE technical team (administrators) emerged as the main other sub categories. No interests were noted on the remaining categorisations of professional associations, SIGs and Board & Shareholders. Exception however emerged when one respondent revealed the viability of donors as key stakeholders within the MUELE value chain. This argument certainly demonstrates the level of influence of donors in shaping major ELMS developments in DC echoed in earlier research findings (Faridha 2005; Mayoka and Kyeyune 2012; Moucheraud et al. 2017; Opati 2013; Ssekakubo et al. 2011; Tusubira et al. 2008). Some of these assertions depict how MUELE does not take on all stakeholders; yet a successful and sustainable eLearning system presuppose stakeholders' involvement. It is also equally important to recognise activities of all institutional stakeholders as vital within an LMS ecosystem to allow for LMS contextualization[1].

3.3 E-Learning Activities

E-learning activities integrate all the services delivered by the ELMS. MUELE makes use of some pedagogical models to practicalise eLearning within the institution. A total

[1] Contexualisation means using electronic learning management system, adding and replacing rich and interactive media content, and use of hyperlinks for non-linear navigation that befits individual courses.

of 7 respondents agreed that whereas MUELE offers collaborative learning groups, they have not engaged in any contributions online. However, results reveal that collaborative groups are open to both staff and students. Whereas MUELE offers this provision, the fact that none of the respondents was interested in participating in any collaborative group contributions for any reason demonstrates that adoption issues do exist for example negative user attitudes, poor user interfaces etc. Distributed learning is popularised in MAK although misalignments could be evidenced due to reliability issues. Results reveal that there exists limited participation in online knowledge building communities which resonates with earlier findings since MUELE is not institutionalised and current efforts to engage with students online are out of individual motivation alone. This lack of ELMS institutionalization is partially the cause of eLearning misalignments at MAK. MUELE has been customised to offer several instructional delivery strategies to supplement the face to face instructions. Only 1 respondent reported to have contextualised content to suit the course needs. All staff agreed that whereas there exist current efforts to contextualise courses to suit online delivery, many existing courses uploaded on the platform are not contextualised. And as such, the ELMS simply enable online replication of offline teaching and learning approaches.

4 Discussions

E-learning systems promote a culture of group based learning and thus could align well in societies that have a richer history of collaborative work principles. Our results indicated that it has been widely recognised that MAK community is more individualist as opposed to collectivist (Hofstede et al. 2010) and this wider recognition cuts across the administrative, academic and student domains. Collaborative learning promotes mutual assistance and sharing (Mayende et al. 2016), which is majorly not the reality in MAK given the high prevalence of individualised recognition from assessment, promotion and appraisal. Collaborative services like communities of practice are argued to offer wider opportunities for distance learners to connect with instructors and peers for constant feedback about a shared question or subject matter (The NMC Horizon Report 2017). Whilst some respondent showed use of other collaborative tools outside MUELE, issues of co-authorship and/or co-development of materials, reports and projects between different stakeholders were not well integrated; yet they are the backbone to collaborative eLearning systems. One respondent had this to say … "we are dealing with people in the same locality. eLearning is mainly for supporting face to face learners, yet we meet physically. Why go through the hustle of troublesome net that is off sometimes". This indicates mixed reactions on staff attitudes for supporting eLearning. Kisanga and Ireson (2015) acknowledges that formal stakeholder training in basic and professional ICT skills encourages a positive change in perception and attitude minimising resistance to change. Thus, misalignments could stem from the fact that while as MAK has deliberately decided to evolve its pedagogy to embrace collaborative learning activities, evidence from the ELMS analysis shows that this pedagogy has not yet diffused. The current eLearning system has not yet succeeded in enabling the pedagogical change and we therefore consider this an institutional misalignment. Other factors technological factors like limited internet could explain the limited use.

Beyond this, past institutional planning in comparison with current emerging eLearning developments could have taken the institution by surprise. MAK was originally founded as a technical school, and grew into a university with extramural centres. Results further revealed that MAK later opened regional population catchment centres for offering brick and mortar learning programmes in Lira, Jinja and Fort Portal. Since eLearning could shift the institution's traditional instructional mode, this could adversely affect the institution's rich traditional learning heritage and identity. The current blended approach at MAK provided a better contextual fit and alignment of ELMS within the existing traditional instructional framework.

The proliferation of other simpler collaborative technologies and the emergence of digital social networking tools have adversely affected the usability of MUELE. Further evidence beyond staff accounts suggests that many students spend a considerable amount of time on social media and enjoy more collaborative learning beyond MUELE (Opati 2013). Moreover, the portability, reliability and mobile accessibility of competing social media platforms like Facebook and WhatsApp perhaps makes them preferable to stakeholders.

The wider view of LMSes as repositories or archiving systems for educational materials by most stakeholders should not be down played. This wider application resonates well with the staff selection of knowledge repositories and content upload among the most popular content technologies. ELMSes as knowledge repositories would centralise educational content from various suppliers both internal and external to the institution for easy access. However, the lessening convergence of LMS to mere repositories of materials would render ELMSes library systems as opposed to learning platforms which contradicts the definition of LMSes (Oxagile 2016). On the other hand, absence of an institutionlised policy to guide the rollout of the ELMS at MAK could suggest that MUELE is still in its infancy. Relatedly, 'the current contestation of the IODeL policy by the MAK management is still far from resolution," (said a respondent). This lack of an operational institution policy is a major impediment to eLearning contexualsiation at MAK. MUELE like other information systems LMS has the power to integrate with other systems in what could be referred to as Enterprise Resource planning (ERP) creating a unified environment for centralised access and share point. This functionality is not yet utilised. 'One problem that we have is that MUELE and other systems within the university fraternity are not connected or integrated together' said a respondent. Integrating information systems has the potential to facilitate faster access and sharing of institutional resources, and to minimise data storage costs. Despite MAK's efforts to establish a road map to guide ICT institutionalisation, ELMS alignment within the university context remains a serious challenge. This is further reinforced by the fact that most ELMS development efforts are centered on satisfying donor interests as opposed to the larger institutional objectives. As such, the context of MAK may not be significantly different from other institutions in DCs (Ssekakubo et al. 2011) with similar conditions.

5 Conclusions

Our research revealed that DC HEIs have unique contexts as compared to developed countries from where ELMSes originate. DC contexts must put in consideration further continuity of ELMS initiatives beyond initial support. Contextual alignment of ELMS could be better referred to as a "hybrid" approach since it combines the extremes of social and technical aspects as a consolidation of "contextual aspects" and "eLearning systems dimensions". Outcomes indicate that to a certain extent, ELMSes are aligned to the context of DC HEIs in the following ways:

- The establishment of eLearning as an informal approach to supplementing the face to face education could account for its current development. eLearning has not yet been institutionalised, since there exist gaps in ELMS rollout.
- Elearning emanates from external funding whose target is primarily to fulfill donor objectives with the university only partially realising targets.

However, to a certain extent, DC HEI's ELMS are not aligned with their institutional contexts:

- There is a misalignment between culture and the ELMS design requirements. Our evidence highlighted this in terms of learning collaboration.
- Institutional establishment is misaligned with the strategic planning for eLearning.
- Resources disparities evidenced in MAK work against ELMS development and continuity.
- Misalignment in management support and strategic planning.
- Misalignment in stakeholder competences in content development, and ePedagogies.

Conclusively, MUELE is to a considerable extent, not aligned with the context of Makerere University. This may equally apply to other DC HEIs within similar contexts. This implies that ad-hoc actions need to be taken to close the misalignment gaps e.g. through hybrid approaches to assure sustainability and realization of potential benefits that can be derived from eLearning. Suggested approaches may include; practicalising ELMS institutionalisation, management commitment (for ownership, resource provision and leadership oversight), strengthening collaborative workplaces, ELMS strategic planning and forecasting, ELMS specific capacity development among others. In summary, our research is meant to raise awareness on the organisational environment in DC HEIs and analyse how foreign digital technologies align within DC institutions. Finally, our research reveals that, the organisational environment plays a significant role in shaping the alignment of information systems in general as evidenced from research results in resonance with the framework.

Acknowledgement. The work reported in this paper was financed by Equity & Merit Scholarship and the DELP project which is funded by NORAD. Acknowledgements also go to the University of Agder and Makerere University who are in research partnership.

References

Aparicio, M., Bacao, F., Oliveira, T.: An e-learning theoretical framework. Educ. Tech. Soc. **19** (1), 292–307 (2016)

Faridha, M.: Towards Enhancing Learning with Information and Communication Technology in Universities; A Framework for Adaptation of Online Learning. A Research Dissertation Submitted to the Graduate School in Partial Fulfillment of the Requirements for the award of t. Makerere University (2005)

Greenberg, A., Versluis, G.: Sida Supported ICT Project at Makerere University in Uganda. Evaluation (2005). http://www.sida.se/contentassets/48d6c85ee2b845fca035baa5758983ab/sida-supported-ict-project-at-makerere-university-in-uganda_2094.pdf. Accessed

Gwamba, G., Mayende, G., Isabwe, G.M.N., Muyinda, P.B.: Conceptualising design of learning management systems to address institutional realities. In: Paper presented at the International Conference on Interactive Collaborative Learning (2017)

Heeks, R.: Success and Failure Rates of eGovernment in Developing/Transitional Countries, eGov4Dev (2003)

Heeks, R.: Implementing and Managing eGovernment an International Text. SAGE, London (2006)

Hofstede, G., Hofstede, G.J., Minkov, M.: Cultures and Organizations: Software of the Mind. Revised and Expanded, 3rd ed. McGraw-Hill, New York (2010). http://geert-hofstede.com/books.html. Accessed

Hossein, M.: Investigating users' perspectives on e-learning an integration of TAM and IS success model. Comput. Hum. Behav. **45**, 359–374 (2015)

Kinyua, A.W.: Design reality gap framework for post implementation evaluation of the national ICT policy, July 2015

Kisanga, D., Ireson, G.: Barriers and strategies on adoption of e-learning in Tanzanian higher learning institutions: lessons for adopters. Int. J. Educ. Dev. Inf. Commun. Technol. **11**(2), 126 (2015)

Lee-post, A.: e-Learning success model: an information systems perspective. Electron. J. e-Learn. **7**(1), 61–70 (2009)

Makerere University: Open Distance and eLearning Policy of Makerere University (2015). https://policies.mak.ac.ug/sites/default/files/policies/Open-Distance-eLearning-Policy.pdf. Accessed 01 Feb 2018

Manji, F., Jal, E., Badisang, B., Opoku-mensah, A.: The Trajectory of change: Next steps for education. eLearning Africa Report (2015)

Masiero, S.: The origins of failure: seeking the causes of design–reality gaps. Inf. Technol. Dev. **22**(3), 487–502 (2016)

Mayende, G., Prinz, A., Isabwe, G.M.N., Muyinda, P.B.: Learning groups for MOOCs: lessons for online learning in higher education. Paper presented at the 19th International Conference on Interactive Collaborative Learning (ICL2016), Clayton Hotel, Belfast, UK, 21–23 September 2016

Mayoka, K., Kyeyune, R.: An analysis of E-learning information system adoption in Ugandan Universities case of Makerere University business school. Inf. Technol. Res. J. **2**(1), 1–7 (2012)

Moucheraud, C., Schwitters, A., Boudreaux, C., Giles, D., Kilmarx, P.H., Ntolo, N., Bossert, T.J., et al.: Sustainability of health information systems: a three-country qualitative study in Southern Africa. BMC Health Serv. Res. **17**, 1–12 (2017). https://doi.org/10.1186/s12913-016-1971-8

Mtebe, S.J.: Making Learning Management System Success for Blended Learning in Higher Education in sub-Saharan Africa (2013)

Opati, O.D.: The Use of ICT in Teaching and Learning at Makerere University The Case of College of Education and External Studies. University of Oslo, 102, 37 (2013)

Oxagile: History and Trends of Learning Management System (2016). http://www.oxagile.com/wp-content/uploads/2016/04/LMS-history01.png. Accessed

Schenkel, M.: The malaise of Makerere: underfunding, overcrowding (2013). http://gga.org/stories/editions/aif-15-off-the-mark/the-malaise-of-makerere-underfunding-overcrowding-strikes. Accessed

Sekiwu, D., Naluwemba, F.: E-learning for university effectiveness in the developing world. Glob. J. Hum.-Soc. Sci. **14**(3) (2014). http://www.socialscienceresearch.org/index.php/GJHSS/article/view/1162. Accessed

Ssekakubo, G., Suleman, H., Marsden, G.: Issues of adoption : have e-learning management systems fulfilled their potential in developing countries? p. 236 (2011)

Tabaire, B., Okao, J.: Reviving makerere university to a leading institution for academic excellence in Africa. In: Synthesis Report of the Proceedings of The 3rd State of the Nation Platform. Retrieved from http://www.acode-u.org/Files/Publications/PDS_8.pdf (2010)

Taylor, J.C.: Fifth Generation Distance Education (Higher Education Series No. 40) - taylor01.pdf (2001). http://www.c3l.uni-oldenburg.de/cde/media/readings/taylor01.pdf. Accessed

The NMC Horizon Report: NMC Horizon Report Preview 2017 Higher Education Edition (2017). http://cdn.nmc.org/media/2017-nmc-horizon-report-he-preview.pdf. Accessed

Tusubira, F., Mulira, N.K., Kahiingi, E.K., Kivunike, F.: Transforming Institutions through Information and Communication Technology: The Makerere University experience. Intersoft Business Services (2008)

Vaiva, Z., Edita, B., Daiva, V.-A., Vladislav, V.F., Kathy, K.-P.: E-Learning as a Socio-Cultural System: A Multidimensional Analysis, p. 259. IGI Global, Hershey (2014)

An Attempt for Critical Categorization of Android Applications Available for the Greek Kindergarten

Tharrenos Bratitsis(✉)

University of Western Macedonia, Florina, Greece
bratitsis@uowm.gr

Abstract. Nowadays, children are becoming familiarized with mobile devices from very young ages, having their first contact with them even under their 2nd year of age. Also, a lot of discussion is being made regarding the integration of mobile applications in educational settings, formal and informal. A very common concern of both the educators and the parents of children aged 4–6 years old is "what is out there for my children to use?". Although this is being dealt with for several years for older ages, there is not much work regarding this age-group. Based upon this fundamental question, this paper attempts to examine the availability of mobile applications which can be exploited by teachers in Early Childhood Education. Specifically, the focus of the paper is the Greek Kindergarten which is attended by children aged 4–6 six years old and Android based mobile devices. A thorough survey in Google Play was carried out in order to record all the corresponding results which were then categorized and discussed upon. Furthermore, the paper tries to shed light to the perspective of an inexperienced teacher who has the means and the will to integrate mobile devices in his/her classroom, by identifying the challenges that he/she will face during such an attempt. Thus, this paper aims at serving as a reference point for those who wish to initiate their trip in the mobile learning world by discussing all the aspects and the difficulties of such an attempt.

Keywords: Greek · Kindergarten · Mobile applications · Android

1 Introduction

The past decades, everyday life has changed, following economy growth and societal evolvement, as a result of the ubiquitous presence of technologies [1]. With the Internet as an initiation point, social media and other communication means have been developed, enabling people to stay connected to their friends and families in diverse locations [2]. Access to information and people is ubiquitous, easy and available to those who are fluent in using contemporary technologies [2]. Mobile devices, such as tablets and smartphones have become increasingly significant in technology based societies [3]. These changes are reflected even amongst youngsters, who are very insightfully described by Prensky [4] as "digital natives", due to their extensive familiarization with technologies and the corresponding artefacts. In fact, further extending this term, Marsh et al. [5] state that young children (up to 6 years of age) are

M. E. Auer and T. Tsiatsos (Eds.): IMCL 2017, AISC 725, pp. 56–68, 2018.
https://doi.org/10.1007/978-3-319-75175-7_7

"immersed in practices related to popular culture, media and new technologies from birth". As a result, they are very efficient in operating machines and, at the same time, develop skills, knowledge and understandings about the world they live in [2].

A report issued by the New Media Consortium Horizon Project in 2012 stated that tablet computing was already becoming a hot trend for technology adoption in schools in 2013 [6]. Based on the idea that tablet PCs and PDAs could hold significant educational value, early in the 2000s, the emergence of mobile devices (mainly tablets) lead the research strive to collect empirical evidence regarding their integration in schools [7]. Thus, a significant amount of research has been conducted towards that direction over the past 5–6 years, originating from the thrill that the first iPad brought in 2010. Several case studies or even larger scale studies have been carried out in all levels of education, although in the ICT-Education relationship research and emergent technologies primarily focus on tertiary or secondary education, before reaching, eventually, the Kindergarten level or even preschool education.

The focus of this paper is Kindergarten, specifically in a national context, that of Greece. Building upon the overall trend of integrating tablets in school classrooms and following the unofficial discussions with educators and parents who often wonder which are the appropriate mobile applications for their students and children, this paper attempts to shed light into this exact agony. A thorough search in the Google Play platform was carried out in order to see "what's out there" for these specific target groups, namely teachers and parents of children of 4 to 6 years of age. Then, categorization approaches where examined in order to reach some useful conclusions.

The rest of the paper is structured as follows: initially, a brief overview of the literature regarding mobile devices in Kindergarten and Early Childhood Education is presented. Then, some of the existing application categorization research papers, also focusing on these ages, are discussed upon. The last section is a concluding discussion which attempts to sum up the findings and explore the anticipated frustration of teachers (mainly) when they are involved in discussions regarding the topic under investigation.

2 Background of the Study

This section of the paper presents a brief overview of the literature, highlighting two main axes. The first is about tablets and mobile devices in preschool education, in general. The second regards mobile applications and corresponding categorization approaches. It is to be noted that not much work exists for the designated age group. Moreover, browsing through the literature using keywords like "tablets" or "mobile devices", one can find some work with the utilization of early Tablet PCs (running earlier Windows operating systems), iPods and PDAs, all which are more or less obsolete technology nowadays. This paper considers only works which focus on children under 6 years old and refer to tablets in their contemporary form.

2.1 Mobile Devices in Kindergarten

Mobile learning originally was described as any kind of learning which occurs in mobile settings, considering the mobility of technology, learners or even learning itself

[8]. With the deployment of tablets and smartphones, mobility was elevated to a whole new level and thus mainly when talking about mobile learning one refers to the utilization of these devices.

According to Terras and Ramsay [9], "Mobile devices have a number of unique characteristics such as portability, connectivity, convince, expediency, immediacy, accessibility, individuality and interactivity and hence offer the potential of educational applications above and beyond those of traditional information and communication technology". Newmann and Newmann [10] highlight the physical features of touch screen tablets which resemble writing pads in matters of size and aspect ratio while supporting multi point touch and gesture-based interaction. Furthermore, the integration of accelerometers allow interaction through physical movement and automated screen rotation, adding value to the interactivity of tablet [10].

It is a fact that such features allow infants, toddlers and preschoolers to begin the investigation of the digital world much earlier in their developmental process than ever before [11, 12]. Newmann and Newmann [10] argue upon the usefulness of tablets as literacy building media which allow children to progress from an immediate, concrete sensory experience to a more conceptual and abstract understanding, to independent operation of the device through the use of apps.

Focusing on formal education, Lindahl and Folkesson [13] claim that mobile device integration in preschool curricula aims at facilitating participation and collaboration among children. Examining technology adoption in school settings, an interesting observation is made by Egan and Hengst [14], who state that despite the availability of educational software for over 30 years, the Kindergarten sector initially resisted the integration of ICTs in teaching young children. For many years, educators and policy makers relied on the findings of developmental psychology in order to support the idea that only physical activity and interaction with tangible objects were necessary in order to construct an understanding of concepts and, eventually, the world. Nevertheless, over the past 15–20 years research has shown that computer mediated artifact manipulation can equally facilitate learning and therefore, ICTs hold position in Kindergarten classrooms [15]. But as a consequence, usually Kindergarten based research is posterior to the other levels of education. Thus, not much work is available in the literature.

In an indicative example, Sahin et al. [16] came up with only 18 research papers while searching in very well-known databases, using combinations of the following keywords: children, preschool education and tablet computers. Some of them were prior to the appearance of tablets as we know them today (2010 is considered as a milestone, as iPad was released then and then Android based devices were developed as well). According to their literature review there are four types of studies, those that focus on the teachers' perspective, the perspective of children, the software used and other related topics (such as the perspective of parents) [16].

Focusing on the exploitation of tablets in Kindergarten, the context becomes clearer. It is a fact that adults consider most mobile applications as games and are thus reluctant to engage children in using them. Kjällander and Moinan [17] studied the playful engagement of preschoolers with tablets, having in mind that digital games have often been criticized as hampering creativity. They showed that children can be very creative when using mobile applications and they actually make new meanings in a playful manner, thus learning by shifting from learning consumers to playing producers.

Case studies can be found in the literature regarding educational exploitation of tablets in Kindergarten, of which just a few indicative examples are presented hereinafter. Zaranis et al. [15] used various mobile applications for teaching realistic mathematics in Kindergarten, concluding that learning was enhanced, compared to traditional teaching approaches. Spencer [18] cultivated mathematical literacy through the use of iPads, which constituted mainly in number recognition, enumeration (up to 10) and guided number onscreen writing. The results favored the use of mobile devices.

Chiong and Shuler [19] used iPods for delivering specially designed audiovisual material to 3–7 year old children, observing that they drew significant gains in vocabulary and phonological awareness. Beschorner and Hutchinson [20] used tablets as literacy teaching tools in early childhood education settings, concluding that reading, writing, listening, speaking and expressing oneself within a predefined context, was facilitated significantly through the mobile devices and several applications.

Other educational practices involving tablets include their exploitation for means of collaboration among children while performing tasks or even completing game-like activities, in groups (e.g. [21]). But in all cases, there are several limitations. Up to date, no significant longitudinal studies have been carried out in order to examine the effects of utilizing tablets in Kindergarten, on the long term. To the author's knowledge, only one study exists, reporting on a three year investigation of iPads use in Australian schools of the Sydney region [3]. But nevertheless, it is more an exploratory research than a report of a curriculum-wide application of a pre-designed activities in order to examine the long term outcome of such an attempt.

Recently, interventions which involve tablet devices are appearing in the literature, exploiting applications which are not characterized as educational or directly aiming young children, but they are also in the form of small scale case studies. One example is the study conducted by Bratitsis et al. [22] in which augmented reality content creation applications were used in order to design teaching material for science education in early primary education (grade 2). This study aimed at facilitating the discovering of the Water Circle by the students through digital stories presented in the form of augmented reality content, attached to specific trigger images. Such applications seem to emerge in the current literature.

2.2 Types and Categories of Mobile Applications

In order to utilize tablet devices in classroom settings, appropriate applications have to be used. An application (or just app) is a mini-program which needs to be downloaded and installed in the mobile device. Some apps are free of charge and some require a small amount to be paid. There are cases of apps which provide a free of charge version with limited features. Examining the literature, very few studies can be found, addressing a significant problem which arises; which are the appropriate apps for a teacher to integrate in his/her teaching practices?

In a longitudinal study conducted by Goodwin [3], quite a few apps were used within the three years of the study's lifespan. They found out that 43% of them were of an instructive nature, 26% were manipulable and 31% were constructive. Instructive apps were those which match most of the cases discussed in the previous section of this paper, including those which show numbers and help children learn how to pronounce them, how to count from 1 to 10, etc. Also, this type of apps uses flashcards for vocabulary acquisition and literacy, consisting mainly drill & practice or presentation apps category which requires minimal cognitive investment by the learners. The manipulable category includes game-like apps, such as finding pairs, matching numbers with groups of elements, matching images with letters based on the first letter of the corresponding word, puzzles, etc. These apps require more cognitive involvement than the instructive but less that the constructive apps, as they are context-limited. The constructive apps are of a more free form of use, including for example drawing or musical applications [3].

An interesting result which derived from this study was that 53% of the students preferred constructive apps, whereas 36% preferred to use instructive and 11% manipulable apps. Thus, children seemed to not be bothered by the increased cognitive investment which was necessary. On the contrary, they enjoyed the opportunity to be creative and act as content creators, for example by producing drawings or composing their own music. Another significant outcome of the study was that there is a need for the creation of an evaluation rubric for mobile apps in the literature and the author suggested that an online database with apps information and classification is required to assist educators who are willing to incorporate mobile devices in their teaching.

Also, the study reported than in earlier studies less than half of the examined applications which were characterized by their creators as educational were actually appropriate for education. This is a very significant realization which further highlights the need for a proper categorization and evaluation scheme, which educators can consult while wishing to incorporate tablet devices in their teaching practices.

Ebner [23] in a study regarding mobile apps for math education, provided a different categorization, focusing on iPhone apps. Those were: stand-alone learning apps, game-based learning apps, collaborative apps and learning analytics apps. This proposal considered mainly technical aspects of the apps and further proposed a correlation of teaching strategies and app categories. The strategies were: self-directed learning, incidental learning, collaborative learning and learning analytics.

In the literature there are a few more studies, similar to the two aforementioned ones. Overall, there is no widely recognized and adopted classification scheme for mobile apps which are appropriate for educational settings and young children. To a point that is quite expected, as most of the studies are of an exploratory nature and in a rather small scale. There is additional reasoning for this, which is discussed in the next section.

In the examples discussed in the previous and this section of this paper, mainly iPad devices were used. It is a fact that less studies can be found in the literature which utilize Android based devices. One explanation for this is that mainly the existing studies have been conducted in countries in which organized attempts of incorporating tablet devices in schools, even on a national level, exist. These attempts include extensive funding schemes, provided by regional or national authorities or the private sector and they mainly target iPads. Thus, the need to shed some more light on the current status regarding Android based devices seems to be high.

3 The Study: Searching for Apps

Taking into account the existing literature, as discussed in the previous section, the author conducted a small research in order to examine the available applications for Android based devices, appropriate for educational purposes. For that matter, an extensive search was carried out in Google Play, the official platform for acquiring android applications, using the keywords “kindergarten”, “educational”, “games”, “applications”, “Greek” and “free”. These keywords were written in the Greek language and in all possible combinations. The search returned a total of 248 apps when using the keyword “Kindergarten” and subsets of this result list when combinations of keywords were used.

3.1 Aim

The core aim was to “see what is out there for Greek teachers and parents to use” and furthermore to attempt to categorize them. All the app data were stored in an excel file, including app name, acquisition URL, application type, PEGI level, disciplinary area, star based evaluation, average evaluation by users, number of evaluations, number of downloads and application language. These are features incorporated in Google Play and explained hereinafter.

Some comments were also added in the excel sheets, based on a draft evaluation of the end users’ comments for each application, an overview of the developer’s app description and an overview of the available screenshots. Furthermore, the possible disciplinary areas (if any) for which each app seemed appropriate for were marked.

All these information was analyzed in a quantitative manner in order to reach some preliminary conclusions.

3.2 Google Play

In this section, the features of Google Play and the problems one faces when searching for applications are discussed.

PEGI stands for “Pan-European Game Information” and it is a classification system which indicates the age groups for which a mobile app is appropriate for. The main aim for creating this system was to assist mainly parents to identify apps which are appropriate for their children. This system is used by 30 European countries. Developers choose the appropriate PEGI class of their applications which is further evaluated by a committee, which approves or not this choice. The classification is determined by a number which corresponds to the age of the end-user that the app is appropriate for. For example a PEGI 3 application is appropriate to be used by children of 3 years of age or more. Usually, the PEGI classifications are: 3, 7, 12, 16, 18 and “Parental guidance recommended”. Thus, in the case of Kindergarten, PEGI 3 applications were examined. Other countries, outside EU use similar classification systems which characterize the age appropriateness of the content.

The star evaluation system of Google Play is a user-based grading system. Each user can attribute an app with 1–5 five stars (including half numbers), following a likert-like grading system. Apps receiving a 5 are considered as excellent, whereas 1 is the worst possible grade. The indication in Google Play is the Average of the total grades submitted by the users, on a voluntary basis.

Another feature of Google Play is that the number of downloads for each app is displayed, thus indicating how famous they are. The number is not absolute, but a categorization is rather used. So an application falls in the "1–50", "50–100", "100–500", …, "10000–50000", "50000–100000" category and so on.

The languages in which an app is available is usually included in the short description provided by the developer.

One would think that these features can facilitate the app search by any user. Nevertheless, this extensive search revealed significant problems which mainly exist because of the architecture of the Google Play platform. These are thoroughly discussed in the Findings section, thus highlighting the problems that an educator, a parent or a researcher can face in similar attempts.

4 Findings

This section is divided into two parts. In the first part, the problems that were revealed during the Google Play search are explained and discussed upon, contributing to the categorization attempt which initiated this paper. The second part contains data analysis results and a position statement regarding the categorization of the apps which the search returned. Findings are generalized in the discussion section.

Regarding the Google Play structure, there is a search facility and the results appear as tiles. Each tile contains some of the information which were stored during this study. For one to reveal all the corresponding information, each tile has to be clicked and the app screen will come up. The search parameterization is rather limited, to resource type (application, movie, music and book), resource price (free or paid) and a choice between all resources or those graded with 4 stars and above. Consequently, the results need to be browsed manually, considering that there is no technical possibility to export or extract the result list for further analysis. There exist only two result storing tools which work only with plain Google search. Nothing can be found for the Google Play platform and thus for this study, all the results were copied manually, one by one.

Another significant issue with the Google Play platform is that there is no option to short the returned results in any kind of order (cost, star grading, popularity, etc.) and each time the search is conducted, the sequence of results is different in a manner which seems random. It is known that Google promotes apps, as it does plain search results, based on keywords, for a fee. Thus, there is a significant possibility that a good app will be so far down in the results' list that the seeker will never review it. Several test searches were conducted to reassure that at least the complete results' list is the same every time, which was the case.

Regarding the language, it seems that Google Play recognizes the locale setting of the browser and presents the results in that language. The main problem is that Google Translate is incorporated in the platform and thus, the results contain apps which do not actually support the language of search (in this case Greek) but the description of the application is automatically translated. In this study, this issue turned out to be very significant, as most of the results were not actually valid. Thus, the actual description for every application can be very misleading, as discussed further bellow.

Lastly, the ranking system of Google Play is strictly based on the end users' satisfaction and not on a scientifically acceptable system which makes it very subjective and increases the possibility of the ranking being misleading. When browsing through the comments which are uploaded by users who have graded an app, it is very common to see that many of them have based their evaluation on factors like "I enjoyed it as it reminded me of a game I used to play when I was young" or "my child spends a lot of time on it, being quiet". It is obvious that these assessments are not at all related to the educational value or potential of the corresponding app.

Proceeding to the analysis of the manually parsed data, as already stated a total of 248 applications were returned. Some of them (23) were assigned a PEGI greater than 3, marking them as not appropriate for the designated age group, although 4 of them were considered as appropriate after inspecting their content. On the contrary, 20 applications marked as PEGI 3 were not appropriate for Kindergarten (e.g. they included multiplications, spelling or other complex exercises).

Focusing on the number of downloads, most of the apps had been downloaded from 10000 to 5000000 times. Of course, although this classification is an indication of an apps popularity, it is not very accurate as the span of the categories is very wide. For example an app with 10,000,001 downloads will be in the same category with one having 49,999,999 downloads and also multiple downloads from the same user in various devices are not always counted as one. Shifting to the star rating of the apps, these spanned from 3 to 5. Less than half (121) were assigned a star rating over 4 and only 32 were assigned a 4.5 or greater. The number of evaluating users spanned from 0 to 69228 with an average of 2956 (SD: 8688). Also, in many cases, the less the number of evaluating users the higher the star rating was. This leads to the assumption that actually these numbers (downloads, star rating and number of evaluators) are not a safe indication of the most appropriate apps for educational purposes.

Examining the discipline of the recovered apps, 59 of them (23.79%) can be considered as related to Language learning, 64 (25.8%) related to Mathematics, 10 (4%) provided content related to Animals and 11 (4.4%) included content related to combinations of these disciplinary areas. Additionally, 7 apps dealt with Colors (identification and naming), 4 with music, 2 with Food, 1 with Sports (identifying sports and related terminology), 1 with Professions and 1 was about identifying the Hours on a clock. Overall that is a total of 158 apps, which is 63.7% of the results. The rest were strictly games which seemed to hold no actual educational value of any kind. Also, there were a few apps which allowed the design of a nursery room or a kindergarten classroom. The former is totally irrelevant, but it provided results as nursery in Greek means both Kindergarten (nursery school) and a child's bedroom.

Further examining the type of the apps, the decision was not based only on the developer's description. The indicative screenshots were examined, some of the users' evaluative comments and, in case of doubt, the app was installed and tested. The results reveal that 90 (36.3%) of them were games, 20 of which could be characterized as educational games as the children could actually learn something by fulfilling the goals of the game. For example, one of them included a puzzle-like game activity and when a puzzle with an image of an animal was completed, some information about that animal, including the way it would sound were presented. So, at extension, since the information were valid enough, the app could be characterized as an educational one. Also, 20 (8%) were just puzzle apps with no additional features, 13 (5.24%) included only some famous kids' songs and 4 were simple "coloring book" apps in which the child just touches an area of a picture in order to fill it with a selected color.

A total of 95 (38.3%) of the returned apps were characterized as educational, including various activities for various disciplinary areas, spanning from color recognition to enumeration, writing and reading letters and numbers. This number does not include the aforementioned 20 educational games.

Up to this point, the numbers indicate that despite the difficulties in searching throughout Google Play for educational apps appropriate for the 4–6 age group, considering that the keywords were input in the Greek Language, the actual results are rather different. A total of 33 (13.3%) apps were implemented in or supported the Greek language (among others). Only 1 app was actually in English, but it turns out that it can be used by all children as the language appears only in the initial screen and knowledge of English is not necessary when using it (it is about learning the numbers from 1 to 10 by matching the number with a group of items). The reason for that is, as already mentioned, the auto-translation feature of Google Play. Thus, the apps which do not correspond to the search criteria in this case were far more than the valid ones, which constitutes a problem of frustration for the end user.

Additionally, many of the apps which were characterized are educational, either have a very small number of integrated activities for free (and for the rest a fee is required, making them practically unusable for free) or the activities are very simple and not interesting. For example, when a child matches a voice with an animal, a number with a group of items or a letter/image with a word, the activities can be attractive at first but their actual educational value is rather limited, since not much can be done with them.

All these apps were characterized as educationally appropriate, since their content complies with what the official Kindergarten Curriculum in Greece dictates [24].

Finally, other apps that are not originally targeting the 4–6 age group can be actually exploited for teaching purposes, for this age-group, but a search approach such as the one described in this paper will not bring them up in the result list. For example, programming environments like ScratchJr or content creation application (e.g. Aurasma in the case of augmented reality approaches, as in the case of [22]).

Overall, the search which was conducted for the needs of this study revealed that the level of frustration for a teacher (mainly) who has the determination and the means to incorporate tablet devices in the classroom can be very high. In this case, a complete search through the result list took about a full month, working full time on the task and the results were not very intriguing. This leads to the rather solid conclusions that the average user will cease the attempts due to frustration.

5 Discussion

A complete report of the search approach in Google Play and the emergent results was described in the previous sections. So, the question remains for the average user; "What is there to use and how can I acquire and exploit it?".

Attempting to provide an answer, some reflective comments on the process followed in this paper are necessary. The first comment is that although the results' list at a glance seems promising, the fact is that very few apps can be actually used in the classroom and almost none "as it is". To elaborate on that, out of the 33 which support Greek, only 14 are "educational", 6 contain simple puzzles, 4 are games of finding pairs from flipped cards and the rest contain mainly flashcard based activities. Regarding the educational apps, almost all of them concern number, letter, color, shape and item recognition in various activities. Thus, they are game-like apps which can serve as drill & practice ones, being interesting and exploitable for a limited number of times (gradually becoming less interesting for the children) and mainly for individual use. Only 1 app was quite different, as it presented images which constituted a story, requiring from the child to put them in a correct chronological order. It is about language learning and its educational value relies on it operating as a trigger for argumentation and storytelling.

Also, many of the apps seemed to be good for ages younger than 4–6 years old. For instance, some of them simulated classic wooden puzzles for young children (mainly infants) in which a draft shaped piece with an image embedded on it (e.g. a car of a plane) has to be positioned to the corresponding slot. Thus, overall this attempt revealed how difficult it is for someone to find educationally exploitable apps, mainly due to the way Google Play is setup. Of course, this is the case in the context of a language specific search, as there are many apps which support the English language. But in any case, the app types are more or less the same.

Examining the latter and attempting to categorize the apps, several comments can be made, drawing also elements from the literature, in order to reach some conclusions. First of all, not taking into account the app's language the results' list shows that the available apps fall under the following categories: games, educational, puzzles, music/songs and coloring pages. The educational ones include activities based on flashcards, instructions for writing a letter or a number on screen, simple exercises (e.g. 2 dogs and 3 dogs are displayed in 2 groups and the child has to select 5 as the correct answer), matching exercises (pairs, a number with a group of items, an item with the starting letter, etc.) and recognition exercises (e.g. color blue is call out and the child has to select a blue item as the correct answer). All these apps can be used by an individual child or small groups of children, collaboratively selecting the correct actions. In the latter case, it is important to regulate turn taking or roles in order to avoid conflicts. Nevertheless, all these apps have limited value as standalone ones. Of course, the teacher may use some of them (e.g. flashcard based) as triggers for further discussions and activities in the classroom. Regarding the language, some of the apps which did not support Greek can be used. For example, a number recognition application can be used if the teacher makes the appropriate selections at the home screen or if the navigation requires a "next" or an "ok" button to be pressed. This is not applicable to language oriented applications, since the alphabet is totally different. But overall, the vast majority of the applications focused on Language and Mathematics (59 + 64 = 123 out of 248, excluding pure games).

Shifting back to the literature, similar problems were identified by [3] who identified 43% of the examined apps as game-like and instructive (similar to the ones described throughout this paper) which are more suitable for memorizing information. The same report refers to a study conducted by Watlington [25] who examined the top 100 selling Primary Education apps for IOS and found that only 48% were appropriate and Language was the main focus. Also, this further highlights the problem of many applications which come up when search for educational ones and it turns out that they cannot be characterized as such.

As mentioned in the Background section, Goodwin [3] suggests 3 categories of apps: Instructive, Manipulable and Constructive. This is based on the usage experience and not the disciplinary area or the content, as presented in this paper and thus these categories are supplemental. Ebner [23] proposes a categorization which is more generic, including: Standalone, Game-based, Collaborative and Learning Analytics apps. This approach is not very helpful since in the target age group of this study most of the apps are also standalone, game-based and potentially collaborative (considering that collaboration occurs among children who try to use 1 application together). Learning analytics, as described in [23] refers to apps not appropriate for Kindergarten.

Furthermore, Goodwin refers to the educators' preference who stated that constructive apps (75% of them) are more appropriate for young children as they allow them to be more creative, matching the students' preferences who chose the same category (53% of them) [3]. This make apparent that most of the instructive apps found in this study are not preferred by the children and the teachers, further strengthening the claim made at the beginning of this section that they are of limited interest and educational value.

This rather "chaotic" situation adds value to Goodwin's [3] and Sahin et al. [16] claims that principles for selecting appropriate apps and rubrics for evaluating them are necessary. This, in correlation with Watlington [25] finding that there is a lack of research in younger ages, complies with the argument upon which this study relied in the first place. Furthermore, it seems that children's and teacher's preferences in app types should also be considered, as also stated by [3, 16, 26]. Thus the need to conduct long-term and extensive evaluations by gathering information also from various case studies and implementing more seems necessary. The fact is that with keyword search, such as in the case of the current study doesn't reveal all the conducted studies (e.g. [22]) which apply in Kindergarten but with apps which are more generic.

Concluding, there is still not a solid framework for categorizing and evaluating mobile apps for educational use, in the literature. A way to proceed is to combine the categories mentioned in this paper with the categorization proposed by [3], thus incorporating both the content, use and the interaction method of the applications.

What seems necessary and is possibly more important though is the creation of some short of database, which teachers and maybe parents can consult when searching for applications. This is also highlighted by [3] but is still far from being implemented. Towards this direction, collaboration among researchers seems necessary by sharing information.

Thus, a next step, following this study would be the construction of a more solid categorization framework and the initiation of a data collection process from case studies, worldwide, in order to record the apps used in every case, along with the main educational outcomes. The "bigger picture" is the construction of a database which will be open and will support keyword based search to facilitate searches in the context of the one conducted in this study. The way the application platforms (Google Play and iTunes/AppStore) are operating at the moment increases the difficulty of such an attempt, but also highlights the need for it to be realized.

References

1. Yelland, N.J.: New technologies, playful experiences, and multimodal learning. In: Berson, I.R., Berson, M.J. (eds.) High Tech Tots: Childhood in a Digital World, pp. 5–22. Information Age, Charlotte (2010)
2. Yelland, N.J., Gilbert, C.L.: iPlay, iLearn, iGrow. A report for IBM. Victoria University, Melbourne (2013)
3. Goodwin, K.: Use of tablet technology in the classroom. Curriculum and Learning Innovation Centre, NSW Department of Education and Communities, Strathfield, NSW (2012)
4. Prensky, M.: Digital natives, digital immigrants. Horizon **9**(5), 1–6 (2001)
5. Marsh, J., Brooks, G., Hughes, J., Ritchie, L., Roberts, S., Wright, K.: Digital beginnings: young children's use of popular culture, media and new technologies. University of Sheffield, Sheffield (2005)
6. Johnson, L., Adams, S., Cummins, M.: The NMC Horizon Report: 2012 K-12 Edition. The New Media Consortium, Austin, Texas (2012)
7. Clark, W., Luckin, R.: iPads in the classroom. What the research says. London Knowledge Lab, Institute of Education, University of London (2013)
8. Sharples, M., Taylor, J., Vavoula, G.: A theory of learning for the mobile age. In: Andrews, R., Haythornthwaite, C. (eds.) The Sage Handbook of E-learning Research, pp. 221–247. Sage, London (2007)
9. Terras, M., Ramsay, J.: The five central psychological challenges facing effective mobile learning. Br. J. Edu. Technol. **43**(5), 820–832 (2012)
10. Newmann, M., Newmann, D.: Touch screen tablets and emergent literacy. Early Child. Educ. J. **42**, 231–239 (2014)
11. Orlando, J.: How young is too young? Mobile technologies and young children August 21, 2011 Posted by Editor21C in Directions in Education, Early Childhood Education, Engaging Learning Environments. University of Western Sydney (2011). http://learning21c.wordpress.com/2011/08
12. Plowman, L., Stevenson, O., Stephen, C., McPake, J.: Preschool children's learning with technology at home. Comput. Educ. **59**, 30–37 (2012)
13. Lindahl, G., Folkesson, A.: ICT in preschool: friend or foe? The significance of norms in a changing practice. Int. J. Early Years Educ. **20**, 422–436 (2012)
14. Egan, M., Hengst, R.: Software on demand: an early child-hood numeracy partnership. Contemp. Issues Technol. Teacher Educ. **12**, 328–342 (2012)
15. Zaranis, N., Kalogiannakis, M., Papadakis, S.: Using mobile devices for teaching realistic mathematics in Kindergarten education. Creat. Educ. **4**(7A1), 1–10 (2013). SciRes
16. Sahin, M.C., Tas, I., Ogul, I.G., Cilingir, E., Keles, O.: Literature review on the use of tablet computers in preschool education. Eur. J. Res. Soc. Stud. **1**(1), 80–83 (2014). IASSR

17. Kjällander, S., Moinan, F.: Digital tablets and applications in preschool – preschoolers creative transformation of didactic design. Des. Learn. **7**(1), 10–33 (2014)
18. Spencer, P.: iPads: improving numeracy learning in the early years. In: Steinle, V., Ball, L., Bardini, C. (eds.) Mathematics Education: Yesterday, Today and Tomorrow (Proceedings of the 36th Annual Conference of the Mathematics Education Research Group of Australasia), MERGA, Melbourne, VIC (2013)
19. Chiong, C., Shuler, C.: Learning: is there an app for that? Investigations of young children's usage and learning with mobile devices and apps. In: The Joan Ganz Cooney Center at Sesame Workshop, New York (2010)
20. Beschorner, B., Hutchison, A.: iPads as a literacy teaching tool in early childhood. Int. J. Educ. Math. Sci. Technol. **1**(1), 16–24 (2013)
21. Khoo, E., Merry, R., Nguyen, N.H., Bennett, T., MacMillan, N.: iPads and opportunities for teaching and learning for young children (iPads n kids). Wilf Malcolm Institute of Educational Research, Hamilton, New Zealand (2015)
22. Bratitsis, T., Bardanika, P., Ioannou, M.: Science education and augmented reality content: the case of the water circle. In: 17th IEEE International Conference on Advanced Learning Technologies - ICALT 2017, pp. 311–312. IEEE (2017)
23. Ebner, M.: Mobile applications for math education – how should they be done? In: Crompton, H., Traxler, J. (eds.) Mobile Learning and Mathematics. Foundations, Design, and Case Studies, pp. 20–32. Routledge, New York (2015)
24. CTCF: Cross thematic curriculum framework. Published by the Greek Pedagogical Institute on the Government Gazette, Issue B (2003)
25. Watlington, D.: Using iPod touch and iPad educational apps in the classroom. In: Koehler, M., Mishra, P. (eds.) Proceedings of Society for Information Technology & Teacher Education International Conference 2011, AACE, Chesapeake, VA, pp. 3112–3114 (2011)
26. Clarke, L., Abbot, L.: Young pupils', their teacher's and classroom assistants' experiences of iPads in a Northern Ireland school: "four and five years old, who would have thought they could do that?". Br. J. Edu. Technol. **47**(6), 1051–1064 (2015)

Proposing New Mobile Learning (M-Learning) Adoption Model for Higher Education Providers

Mohamed Sarrab[1(✉)], Hafedh Al-Shihi[2], Zuhoor Al-Khanjari[3], and Hadj Bourdoucen[4]

[1] Communication and Information Research Center, Sultan Qaboos University, Muscat, Oman
sarrab@squ.edu.om
[2] Department of Information System, Sultan Qaboos University, Muscat, Oman
hafedh@squ.edu.om
[3] Department of Computer Science, College of Science, Sultan Qaboos University, Muscat, Oman
zuhoor@squ.edu.om
[4] Department of Electrical and Computer Engineering, Sultan Qaboos University, Muscat, Oman
hadj@squ.edu.om

Abstract. Mobile learning remains a new frontier for many instructors and students. When compared to the traditional and E-learning systems, mobile learning requires the use of different learning and teaching strategies to be adopted and accepted in higher education. This article presents mobile learning concepts ecosystem with more focus on the mobile learning adoption. The article describes the results of research investigation to design a model of mobile learning adoption for higher education providers. The model main components were identified as: user initiation, user willingness and acceptance with different influencing factors including demographics, external variables (performance expectancy, effort expectancy, social influence, facilitating conditions), enjoyment, efficiency, sociability, flexibility, suitability and economic. The provided model is part of an Omani-funded research project investigating the development, adoption and dissemination of M-learning in Oman.

Keywords: Mobile learning · M-learning · Adoption · Higher education Ecosystem

1 Introduction

The rapid development in internet and mobile technologies had an influence on learning and education systems and led to a new approach of learning called M-learning [1]. M-learning is emerging as a promising market for learning, training and education industries which provides knowledge across multiple contexts, through material and social interactions, using mobile hardware such as PDAs, Tables, Smartphones or other mobile device technologies [2–4]. The mobility of technology and learning contents are

M. E. Auer and T. Tsiatsos (Eds.): IMCL 2017, AISC 725, pp. 69–76, 2018.
https://doi.org/10.1007/978-3-319-75175-7_8

equally important in the learning process to deliver adequate instructional contents as well as the mobility of learners. The number of potential M-learning users is increasing as mobile devices offer functional opportunities by providing users with learning resources that can be accessed anytime and anywhere [5–8].

In UK, a series of learning software tools for mobile users have already been developed [9]. Also, in China, a huge number of educational electronic handheld devices fully designed for mobile learning have already been developed and used by learners since 2006 [10], as well as in other countries e.g. US, Canada, Singapore and Japan. The M-learning market (products and services) will continue to prosper in the recent years all over the world. In fact, there is a growing interest in academic and business communities in M-learning but there are still several unresolved issues on how to promote users' adoption of M-learning. For instance, the mobile devices availability for learners does not guarantee their accessibility for educational purpose.

There appears to be an urgent need to explore the factors influencing user's behavioral intention and need to be considered while developing M-learning applications in order to provide acceptable services [11]. In organizational context, the technology acceptance has been widely studied. However, there is a need to explore the potentiality of learner's willingness and acceptance in a social context alike. Therefore, this research attempts to bridge the gap in the literature by deepening the understanding related to willingness and acceptance toward M-learning adoption.

This paper is organized as follows; Sect. 2 reviews the M-learning approaches in higher education providers and discusses the driving factors for M-learning adoption. Section 3 provides an overview of the proposed research model and discusses the proposed model components including external variables, demographics, willingness towards M-learning adoption and user acceptance. Finally, conclusion and future work are briefly reported.

2 Literature Review

It became compulsory for instructors to use modern information technology tools as learners have become more ICT savvy through what is called electronic learning to facilitate the instructors' tasks and achieve the objective of the learning process using different learning technologies. M-learning is a new form of electronic learning that utilizes the innovative technologies offered by recent mobile and network technologies [12–15]. The idea behind M-learning approach is the availability and user privacy to offer learning opportunity anywhere and anytime using various portable devices [16–18]. M-learning approach in higher education providers is still in the initial phases of development [19, 20].

With this rapid distribution and increased usage of mobile devices in all aspects of our life especially the education sector, there is a huge demand for researchers to study the factors that drive the acceptance and adoption of M-learning [21]. In this light, Carlsson et al. have discussed why and how learners adopt or do not adopt mobile services in 2006 [22]. Previous adoption and acceptance models are called to understand learners, acceptance process towards mobile services with appropriate extensions or modifications on their original structure.

Carlsson et al. in 2006 has reported that the technology acceptance model (TAM) and UTAUT were designed to examine the adoption of information technologies in organization "but the mobile technology adoption is more individual, more personalized and focused on the services made available by the technology" [22]. Sarrab et al. in 2016 discussed the driving factors for M-learning adoption [23]. They focused on the need to provide guidelines and develop strategic plans considering learners willingness and acceptance of M-learning in order to consider all driving factors for the sustainable M-learning deployment.

Among all acceptance and adoption models, TAM appears as the most widely applied models. Which did Davis develop in 1989 that considered two particular constructs of perceived ease of use and perceived usefulness as drivers of technology acceptance [24]. Moreover, compelled by the rapid development of different mobile technologies, M-learning has grown extensively as an accepted learning approach in educational environments [25]. Learner's and instructor's acceptance is an essential part in the M-learning solution deployment [26] as it is possible for learner's and instructor's to customize their learning contents according to their needs.

3 Proposed Research Model

Huge number of research studies that used UTAUT model as adaptation model have modified or extended the model by reducing existing variables or adding new ones to suit a specific study context. That is due to the difference between mobile devices penetration rate and M-learning in all countries [27].

Likewise, the provided study modified the model to suit the context of M-learning acceptance in Saltunate of Oman. Hence, the four core constructs of UTAUT have been modified and three core components have been considered as the main constructs, namely user attitude, willingness towards M-learning adoption and user acceptance as shown in Fig. 1. The main component and its subcomponent for the structure of our research model are detailed as follows:

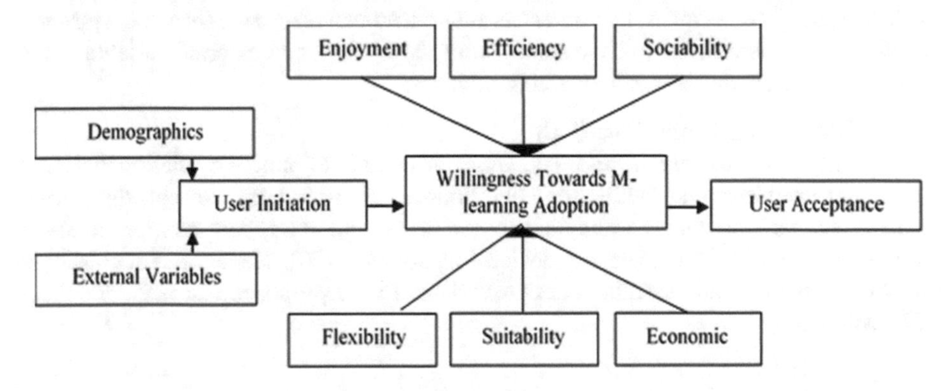

Fig. 1. Research model for adoption of M-learning.

3.1 User Initiation

User initiation consists of two main components, demographics and external variables:

3.1.1 External Variables

The proposed research model considers the following five external variables:

3.1.1.1 Performance Expectancy (PE)

Performance expectancy is closely related to user believes on the use of a particular system that will help attain the expected job performance [28]. In the context of education technology acceptance, a number of researches have already empirically studied and supported a positive relationship between willingness of adoption, behavioral intention and performance expectancy (e.g. Chiu and Wang 2008 [29]). Hence, in terms of M-learning adoption model, it is reasonable to consider performance expectancy as one of the proposed model component.

3.1.1.2 Effort Expectancy (EE)

Effort expectancy is concerned with the level of ease associated with the use of specific information system. Effort expectancy is closely related to complexity and ease of use of particular M-learning system or application. To some extent, effort expectancy leads to better performance, whereas, effort expectancy should have a direct influence on willingness and acceptance to use as well as performance expectancy.

Chiu and Wang in 2008 reported that in E-learning context an effort expectancy was positively related to behavioral willingness and intention, and performance expectancy [29]. Thus, the proposed model considers the same relationships would be found in mobile learning context.

3.1.1.3 Social Influence (SI)

Social influence is referred to the change in an individual behavior that someone causes in other person intentionally or unintentionally on believe that he or she should use the new developed system. UTAUT is also included social influence as a determinant of behavioral intention. Social influence has been suggested by previous studies as strong predictor of behavioral intention to use (e.g. Venkatesh and Davis 2000 [30]). In fact, other actors, such as peer instructor or other learners also influence the learner's decision, thus it is rational to have social influence as one of an external variable in the proposed research model for M-learning adoption.

3.1.1.4 Facilitating Conditions (FC)

In the context of mobile technology, there is a kind of positive relation between behavioral intentions and facilitating conditions as in M-learning concept, the learner decision and satisfaction are influenced by the perception of support from providers of learning material and the learners' device functionality [30]. Hence, facilitating conditions seems to be an essential external variable in the proposed research model for M-learning adoption.

3.2 Demographics

This study conducted an empirical study and visited higher education institutions to survey the willingness and acceptance toward the adoption of M-learning. The analysis stage provided a systematic insight on participants' willingness and acceptance of M-learning adoption into Omani higher education by empirically analyzing participants' demographic background. Key demographic factors were chosen, namely gender, age, and education. About 60% of the total 806 participants were found to be Females, about 60% of them were young learners aged 18 to 23, and the majority (23%) holds diploma and bachelor degrees.

3.3 Willingness Towards M-Learning Adoption

The research team in a previous study reported that the following factors of perceived innovative characteristics have more impact on learners' acceptance to adopt of M-learning [31–33].

3.3.1 Enjoyment

Perceived enjoyment refers to the enjoy-ability of an activity in its own right. Usually the learning process is associated with the sense of pressure; therefore, it is necessary to make more enjoy-able learning activities as it will promote learner's willingness and acceptance of M-learning. Hence, the conclusion of learner's about the acceptance of M-learning can be influenced by perceived enjoyment [31].

3.3.2 Efficiency

Perceived efficiency refers to the quick and efficient implementation of learning activities using mobile devices. M-learning features should manage all aspects related to learners and instructors including grading and assessment management. M-learning designed tool should be able to accommodate both novel and experienced users. Errors should be minimized and when they occur, M-learning users should be provided with descriptive messages before they commit to an action to improve M-learning efficiency [34].

3.3.3 Sociability

The design of M-learning should consider the importance of sociability factor in the adoption. An interactive discussion boards should be provided to offer more robust M-learning environment that supports real-time discussion and an effective communication between M-learning learners and instructors. This will facilitate both private and public interaction between learners and encourage effective collaboration. Accordingly, the choice of learners regarding the willingness and acceptance of M-learning can be influenced by the sociability of the learning approach [31, 34].

3.3.4 Flexibility

Flexibility in M-learning refers to the ability of accessing learning contents anytime and anywhere as well as easy to customize, modify, and learn how to use the adopted M-learning system. It is necessary for the adopted system to offer learners and instructors kind of freedom to start at their preferred time, learn at their own convenience, connect to a variety of learning styles and follow their progression [34].

3.3.5 Suitability

Suitability refers to learning goals; context and characteristics as well as high priority should be given to quick information access for the sake of learning experience enhancement. M-learning should be suitable for different type of users irrespective of their location and other physical barriers. Depending on suitability, course material and user needs, M-learning can be designed to include animation, graphics, video, presentations or podcasts [35].

3.3.6 Economic

The cost of learning activities is very essential factor in learners, instructors and organizations decision to adopt M-learning. It is very important to allow unlimited number of learners and instructors to create accounts and use M-learning services as much as they like at no cost [34, 35].

3.4 User Acceptance

User acceptance of M-learning approach demonstrates the willingness to employ mobile technologies for the learning tasks they are designed for. User acceptance of M-learning is more about the understanding of the influencing factors of M-learning adoption as planned by organization or individual user who has some degree of choice. According to Innovation Diffusion Theory (IDT) provided by Rogers in 1995, there are five different characteristics to determine the technology acceptance including complexity, observability relative advantage and compatibility [35].

4 Conclusion and Future Work

This research study constructed mobile learning adoption model for higher education providers. The goal was to identify the adoption main influencing factors. The mobile learning adoption model was constructed upon the three main identified components; user initiation, willingness and acceptance with different influencing factors including demographics, external variables (performance expectancy, effort expectancy, social influence, facilitating conditions), enjoyment, efficiency, sociability, flexibility, suitability and economic.

For future work, the developed model needs to be validated for applicability using different case studies in higher education providers. To show that well implementation of the proposed model can help improve M- learning adoption in Higher education. The researchers also intend to use this model as a base to propose a framework for assessing the success of mobile learning systems.

Acknowledgment. This article is based upon research work funded by The Research Council (TRC) of the Sultanate of Oman, under Grant No: ORG/SQU/ICT/13/006, (www.trc.gov.om).

References

1. Alzahrani, A., Alalwan, N., Sarrab, M.: Mobile cloud computing: advantage, disadvantage and open challenge. In: Proceedings of the 7th Euro American on Telematics and Information Systems, (EATIS 2014), Article No. 20. ACM, New York (2014)
2. Sarrab, M., Elgamel, L., Aldabbas, H.: Mobile learning (M-Learning) and educational environments. Int. J. Distrib. Parallel Syst. (IJDPS) **3**(4), 31–38 (2012)
3. Sarrab, M.: Mobile Learning (M-learning) Concepts, Characteristics, Methods, Components. Platforms and Frameworks. Nova Science Publishers, New York (2015). ISBN 978-1-63463-342-0
4. Al-Kindi, E., Al-Kindi, Z., Al-Khanjari, Z.: Development of post result mobile application using SOA for SQU students. In: International Conference on Innovation in Engineering and Management; Oman Vision 2020: Opportunities and Challenges (ICIEM-2016). Waljat College of Applied Sciences in Academic Partnership with Birla Institution of Technology, Ranchi, India, Muscat, Oman, pp. 439–444, 24–25 Feb 2016
5. Al-Khanjari, H., Al-Khanjari, Z., Sarrab, M.: Proposing a new design approach for m-learning applications. Int. J. Softw. Eng. Appl. (IJSEIA) **9**(11), 11–24 (2015)
6. Sarrab, M., Al-Shihi, H., Al-Manthari, B.: System quality characteristics for selecting mobile learning applications. Turkish Online J. Distance Educ. (TOJDE) **16**(4), 18–27 (2015). Article 2
7. Sarrab, M., Alzahrani, A., Alalwan, A., Alfarraj, O.: An empirical study on cloud computing requirements for better mobile learning services. Int. J. Mobile Learn. Organ. **09**(01), 1–20 (2015)
8. Al-Khanjari, Z., Al-Kindi, Z., Al-Kindi, E., Kraiem, N.: Developing educational mobile application architecture using SOA. Int. J. Multimed. Ubiquitous Eng. (IJMUE) **10**(9), 247–254 (2015)
9. Stead, G.: Moving mobile into the mainstream. In: mLearn 2005 Conference, Cape Town, South Africa (2005)
10. SINO: Market Analysis Report on China educational electronic devices in (2006). http://www.cciddata.com/2006_11/smxxj_ly.htm. Accessed 5 Feb 2008
11. Khan, A.I.: A., Al-Shihi, H., Al-khanjari, Z., Sarrab, M.: Mobile learning (M-Learning) adoption in the middle east: lessons learned from the educationally advanced countries. Telematics Inform. **32**(4), 909–920 (2015)
12. Ahmed, N., Al-Khanjari, Z.: Effective m-learning on instruction through a learning management system. In: International Conference on Innovation in Engineering and Management; Oman Vision 2020: Opportunities and Challenges (ICIEM-2016). Waljat College of Applied Sciences in Academic Partnership with Birla Institution of Technology, Ranchi, India, pp. 38–42, Muscat, Oman, 24–25 February 2016
13. Ciampa, K.: Learning in a mobile age: An investigation of student motivation. J. Comput. Assist. Learn. **30**(1), 82–96 (2014)
14. Isaiah, T., Martin, C.: Using management procedure gaps to enhance e-learning implementation in Africa. Comput. Educ. **90**, 64–79 (2015)
15. Liu, Y.: An adoption model for mobile learning. In: Proceeding for the IADIS International Conference e-Commerce, Amsterdam, The Netherlands (2008)
16. Al-Khanjari, Z., Al-Kindi, K., Al-Zidi, A., Baghdadi, Y.: M-Learning: the new horizon of learning in SQU. J. Eng. Res. (TJER) **11**(2), 14–25 (2014)
17. Moore, J., Camille, D., Krista, G.: e-Learning, online learning, and distance learning environments: Are they the same? Internet High. Educ. **14**, 129–135 (2011)

18. Park, Y.: A pedagogical framework for mobile learning: Categorizing educational applications of mobile technologies into four types. Int. Rev. Res. Open Distance Learn. **12**(2), 78–102 (2011)
19. Sarrab, M., Alzahrani, A., Alalwan, A., Alfarraj, O.: An empirical study on cloud computing requirements for better mobile learning services. Int. J. Mobile Learn. Organ. **9**(1), 1–20 (2015)
20. Al-Saifi, N., Al-Khanjari, Z., Sarrab, M.: An Evaluation of Mobile Learning (M-Learning) Design Approach Using ISO/IEC 12207:2008 (2017)
21. Uzunboylu, H., Ozdamli, F.: Teacher perception for m-learning: Scale development and teachers' perceptions. J. Comput. Assist. Learn. **27**(6), 544–556 (2011)
22. Carlsson, C., Carlsson, J., Hyvonen, K., Puhakainen, J., Walden, P.: Adoption of mobile devices/services searching for answers with the UTAUT. In: Proceedings of the 39th Annual Hawaii International Conference on System Sciences (HICSS 2006), Hawaii, USA, p. 132a (2006)
23. Davis, F.D.: Perceived usefulness, perceived ease of use, and user acceptance of information technology. MIS Q. **13**(3), 319–339 (1989)
24. Sarrab, M., Al Shibli, I., Badursha, N.: An empirical study of factors driving the adoption of mobile learning in Omani higher education. Int. Rev. Res. Open Distrib. Learn. (IRRODL) publication requirements **17**(4), 331–349 (2016)
25. Sarrab, M., Al-Shih, H., Rehman, O.: Exploring major challenges and benefits of m-learning adoption. British J. Appl. Sci. Technol. **3**(4), 826–839 (2013)
26. Liu, Y., Han, S., Li, H.: Understanding the factors driving m-learning adoption: A literature review. Campus-Wide Inf. Syst. **27**(4), 210–226 (2011)
27. Nassuora, A.: Students acceptance of mobile learning for higher education in Saudi Arabia. Int. J. Learn. Manage. Syst. **4**(2), 1–9 (2013)
28. Venkatesh, V., Michael, G., Gordon, B., Fred, D.: User acceptance of information technology: toward a unified view. MIS Q. **27**(3), 425–478 (2003)
29. Chiu, C., Wang, E.: Understanding web-based learning continuance intention: The role of subjective task value. Inf. Manag. **45**(3), 194–201 (2008)
30. Venkatesh, V., Davis, F.: A theoretical extension of the technology acceptance model: four longitudinal field studies. Manage. Sci. **45**(2), 186–204 (2000)
31. Sarrab, M., Elbasir, M.: Mobile learning (M-learning): A state-of-the-art review survey and analysis. Int. J. Innov. Learn. (IJIL) **20**(4), 347–383 (2016)
32. Khan, A., Al-Khanjari, Z., Sarrab, M.: Crowd sourced testing through end users for mobile learning application in the context of bring your own device. In: The Proceeding of the 7th IEEE Annual Information Technology, Electronics and Mobile Communication (IEEE IEMCON 2016). University of British Columbia, Vancouver, Canada, 13–15 October 2016
33. Khan, A., Al-Khanjari, Z., Sarrab, M.: Crowd sourced evaluation process for mobile learning application quality. In: The Proceeding of the 2017 2nd International Conference on Information Systems Engineering (ICISE2017). College of Charleston, South Carolina, USA 1–3 April 2017
34. Sarrab, M., Elbasir, M., Alnaeli, S.: Towards a quality model of technical aspects for mobile learning services: an empirical investigation. Comput. Hum. Behav. **55**, 100–112 (2016)
35. Rogers, E.: Diffusion of Innovations. Free Press, New York (1995)

On the Integration of Wearable Sensors in IoT Enabled mHealth and Quantified-self Applications

Andreas Menychtas[1], Dennis Papadimatos[1], Panayiotis Tsanakas[2], and Ilias Maglogiannis[1](✉)

[1] University of Piraeus, Piraeus, Greece
{amenychtas, dpapadimatos, imaglo}@unipi.gr
[2] Greek Research and Technology Network (GRNET), Athens, Greece
tsanakas@grnet.gr, panag@cs.ntua.gr

Abstract. The regular use of activity trackers and biosignal sensors from non-professional users is constantly increasing, while for patients with chronic diseases has already become part of their daily routine for a continuous assessment of their health condition. However, the integration of such devices into larger systems such as health monitoring stations or assisted living applications is not straightforward due to a series of business, operational and technical issues which introduce extreme complexity for end-users and developers to properly support their diverse functionally and features. This work presents a thorough analysis of the aforementioned issues, proposes a generic integration solution for Bluetooth devices and evaluates the respective prototype implementations for Android and Linux systems.

Keywords: mHealth · Biosignal sensors · Activity trackers · Bluetooth Integration · Android · Linux · IoT · Quantified-self

1 Introduction

The use of biosignals sensors and activity trackers is not something new. There is a large number of device models available in the market which cover a variety of usages and target different user segments. The features and capabilities of the devices vary as well based on its intended use (professional/medical vs. home use), the nature of the device (e.g. biosignal type that the device measures, the implementation of the measurement process, etc.) and of course, the business and marketing approach that is followed by the device manufacturer. An important fact in this area the last few years is that these devices are not only available to healthcare professionals and their cost is constantly decreasing, making them affordable for individuals and patients willing to monitor their physical activity and biosignals, while in parallel, their features and capabilities are constantly enriched and improved. Characteristics [1, 2] that were available only to professional devices in past, are now transparently exploited from individuals as part of the daily monitoring routine. The number of biosignals that can be measured is also increasing, as well as the technical and technological approaches

M. E. Auer and T. Tsiatsos (Eds.): IMCL 2017, AISC 725, pp. 77–88, 2018.
https://doi.org/10.1007/978-3-319-75175-7_9

that are used in the devices that realize modern scientific methodologies for more accurate and less intrusive measurement procedures.

One of the most important characteristics for the modern sensing and tracking devices is the communication with external systems for offloading and synchronization the biosignal and activity data. The advent of *BLE (Bluetooth Low Energy)* [3] and its adoption in all modern smartphones, tables and personal computing systems, revolutionized the communication capabilities of these sensors and eliminated the need for additional wireless communication hardware or complex wired communication setups. Unfortunately, the adoption of BLE in sensors does not ensure out-of-the-box, flawless integration with external systems due several issues described and analyzed in the next section. Our focus in this paper is the in-depth analysis of these issues, their categorization and the proposal of integration approaches for modern systems (i.e. Android and Linux) to overcome them. The use of mobile healthcare technologies and sensors can considerably extent the functionality of a wide range of applications and services, including telemedicine, assistive care, location-based medical services, emergency response, independent living and IoT. In this context, the reliable acquisition, aggregation and secure transmission of the biosignals or any kind of medical data to local or remote computing infrastructures are the major challenges during the establishment of point-of care systems [4].

2 Medical Sensors and Activity Trackers

Application developers and researches working with biosignal sensors and activity trackers face several challenges and difficulties for their effective integration and operation within their applications and projects. The types of the devices that this work focuses on are (a) *medical sensors* for biosignals such as heart-rate, oxygen saturation and weight, (b) *trackers* measuring activity and steps, and (c) *smartwatches* that also incorporate biosignal or activity measurement functionality. The challenges and integration issues the researchers and developers may face have many dependencies to the type and nature of each device, and an operational feature or requirement of a device, such as the need for initialization before each measurement, may have a substantial effect on the technical requirements and implementations so as to properly support it. Therefore, whatever approach is followed for the analysis and categorization for the sensors' characteristics and the related issues, they all result to technical implications and challenges which sometimes require tremendous effort, especially if multiple sensors need to be supported simultaneously or a sufficient level of usability is required for the non-technology savvy users. The results of the current analysis have been organized in following categories: (a) business, (b) operational and (c) technical.

2.1 Business Issues

Availability of Devices. The first step when developing an mHealth application with biosignal monitoring capabilities, either for end-users or for research purposes, is the market analysis in order to find appropriate devices. The abundance of devices makes this process extremely complex, while several aspects should be taken into consideration such as the maturity of the solution, the quality of the results, the battery life, and other characteristics, which are analyzed in the sections below. This market is highly competitive, with devices manufactured from big enterprises, SMEs and start-ups. In addition, the maturity and the purpose of the devices varies as well and devices are available for fully commercial uses or research and prototyping, and for several types of devices, such as the activity trackers, OEM versions can be found. Therefore, since the integration of such devices requires considerable effort, the organizations that use them as part of their applications should ensure availability of the devices from manufacturers for - at least - a mid-term period. There several examples of products that are discontinued immediately after the SME that was producing the device were acquired by a big enterprise. Also, the future versions of device, even with small changes in their functionality or firmware, may break the compatibility with the integrated systems, causing problems to developers and end-users.

Vendor Lock-In. The vendors have also different business and market strategies, and in many cases these strategies are reflected in the devices themselves in technical and operational level. The open source designs for such devices that can be used for producing or communicating with them are uncommon. The hardware and firmware of the majority of the devices is proprietary and closed source, hence establishing communication with the device as a third-party required guidance from the vendors either by providing access to the communication protocols or through libraries/APIs.

Open Specifications. The last few years we have seen a huge effort for specifications regarding the BLE integration of medical devices. For the BLE based communication, the development of Health Profiles (ISO/IEEE 11073-20601) [5] as part of the Bluetooth GATT specifications family [6] was the first step towards a generic data exchange specification for medical devices and data. The initiative of Personal Connected Health Alliance (former Continua Alliance) [7] from Healthcare Information and Management Systems Society and several enterprises and organizations in the area of device manufacturing, IT and healthcare, was very important to the definition of the Health Profiles and other standards. Besides their contributions to standards, they provide certification for the devices that members implement adopting the endorsed specifications and guidelines. It should be noted though that this certification does not always guarantee the full compatibility of a certified device. Given that the specifications are not always complete, covering for instance only the aspects of data exchange, while other processes, such as the device initialization, are independently defined from vendors, disallows implementations of holistic integration approaches.

Proprietary Specifications. Even though the considerable effort for open standards and specifications, the market is still dominated by proprietary solutions so as to support the business proposition of each vendor and to ensure lock-in. Therefore, the characteristics of the available devices vary in all steps of the device operation lifecycle. Here we could identify three main approaches to grant access to application developers for the acquisition of the monitoring data and for the management of the devices. *Communication Protocols,* allow direct communication with the devices. These protocols include information for establishing the communication and interpretation of the messages to and from the device. Typically, this information refers to the low level BLE messages exchanged such as the services and characteristics to be used, and the mapping of each message bytes to meaningful high-level data. In addition, depending on the device type include the required processes that should be followed (as messages to be sent to the device) for its initialization or before/after each measurement. *SDKs and APIs,* typically supporting only popular platforms such as Android and iOS are available to the developers for integrating such devices. The SDK libraries are written in the programming language of each platform therefore their use is restricted to the particular use only, which means that only a limited set of platforms is supported. In most of the cases, only Android and iOS are supported, as the platforms with the vast majority of end-users, obstructing their use in applications running on Linux or Windows environments. As with the communication protocols, these libraries provide the required functionality for initializing and managing the devices, as well as for synchronizing the measurement data however in this case, the communication protocol is already implemented in the library and is therefore transparent to the developer. An interesting finding in this analysis was that several SDKs include additional restrictions from the vendors, allowing them to *lock remotely the SDK* and thus the respective functionality of application that uses it. This is implemented with the use of access tokens provided to the developers in advance, which are validated at run time. During the initialization of the SDK from the application, the SDK validates the token calling the corresponding vendor services over Internet, and a failure on this process blocks its use, and as a result the communication with the device. A technical implication with these SDKs is that they cannot be used from applications that do not foresee Internet access, while in business level there is a high risk that the vendors can halt the use of the API at any time. *Cloud-based Communication* is also very common among the popular vendors and devices. In these cases, no protocol or SDK is provided to the developers, and the devices are controlled by companion applications of the vendors, which are available for a limited number of platforms (typically Android and iOS). The devices communicate with the companion applications and synchronize the data through them with the vendors' platforms. Other applications are allowed (even though this is not always supported) to acquire the measurement and activity data from APIs of the cloud platforms as with any other public API. This approach is mainly seen in trackers and smartwatches.

2.2 Operational Issues

The operations issues are related with the nature of the devices, their design and the specified use. The landscape is quite diverse, from medical sensors, which are used when a biosignal should be measured, to wearables that implement an always-on operational mode, monitoring continuously the users' biosignals and activity. These different operational modes may affect considerable the process of integrating the devices into third party systems such as mobile applications. An aspect related to this (will be analyzed in the technical issues section) which poses additional technical complexities is whether these devices include *internal memory* so as to temporary store the acquired data allowing the devices to operate offline, with the connectivity with external systems to be required only for data synchronization.

Measurement Operations. For the measurement operations, even though a categorization into biosignal sensors and wearables seems rational, it does not highlight the related particularities. The proposed categorization is the following:

- *Single Measurement:* The single measurement devices are only used the moment the user wants to measure the value of a biosignal (or values of distinct biosignals in a single measurement) such as blood pressure monitors and glucometers. Typically, these devices may operate offline and after a measurement, they connect to the system and send the measurement or if requested they synchronize all data.
- *Multiple Measurements:* This mode is followed from devices such as oximeters and ECGs, which monitor a single or a set of biosignals for a particular time frame (from seconds to hours) producing time series of monitoring data. Connectivity is usually required in such cases for pushing the raw data to the system in real-time, especially since the majority of the devices have limited internal storage capabilities.
- *Continuous Monitoring:* This category includes the wearables (activity trackers and smartwatches) as well as sensors that monitor continuously various biosignals. Such sensors often act also as actuators for regulating the medication intake for users with chronic diseases like the insulin dose for patients with diabetes. In this category, we see both devices that require connectivity for their operation, and devices that operate offline and periodically connect with an external system for synchronization.

Sensor Registration and Initialization. A step in the measurement process using wearables and biosignal sensors is their registration with the connected system and their proper initialization. The analysis results prove that vendors follow quite diverse approaches. Part of the Bluetooth and BLE specifications is the pairing or bonding of the sensor with the system that will communicate with. There are sensors available that communicate directly with a system (e.g. Android and Linux) as soon as they are discovered with no other restrictions. For other devices, bonding is necessary so that the user confirm their intention to allow the two endpoints to communicate. We have noticed also differentiations in the process of BLE bonding across devices and vendors. There examples where no pin is required so the users only need to acknowledge the connection, while in other cases a numerical pin is needed which may be the same for all devices of the same model, or unique for every single device. An interesting finding, is that some devices allow simultaneous bonding with different systems but for others, change of bonding from one system to the other requires hard reset of the device. It is evident that this affects the usability of the overall integrated system since the users may face difficulties to register devices that follow different approaches, but also lead to increased complexity in technical level for the implementation of the solution.

Sensor Management. Following the *registration* of a sensor to the system, its initialization is also mandatory in several cases based on their design and indented use. It is very common that sensors require initialization after their connection to the system. Through this process, a set of different parameters is set on the sensor such as the time/date or the memory that should be used. It should be noted that this *initialization* process might take place either once, after the sensor is first connected to the system, or before (and rarely after) the measurement process. This is prerequisite not only for the correctness of the measurement metadata (e.g. date) but also for the measurement itself since as part of this process, the *characteristics of the user profile* (gender, age, weight, etc.) are communicated to the sensor and are taken into consideration for the final measurement results or for a personalized assessment of the quality of the measurement. There are also sensors that *do not include any controls or screens* on them; therefore the communication with an external system is a prerequisite for the use and management. Part of the management process could be considered the functionality offered in protocols, SDKs or companion apps for *updating the firmware* of the devices.

2.3 Technical Issues

As already mentioned, the business decisions of the vendors as well as the design and specifications of the various devices regarding their communication and management also introduce technical difficulties on the development of generic solutions that integrate biosignal sensors and process their data. These issues – analyzed below – could be categorized in three main sets: (a) communication, (b) sensor control and (c) data management.

Communication. The first step for communicating with a BLE device is its discovery so as to ensure that it is in range and in turn, initiates the procedures for connecting with it. Core part in this process is the identification of the device, i.e. its type and the capabilities that it has, in order to check whether it is supported and how the communication with it can be established. Typically, the BLE devices advertise themselves so that they can detected from other systems. The advertisement packets include the name of the device, its mac address and the UUIDs of the main GATT services that the device supports (e.g. heart rate), which sometimes are aligned with open specifications, while in proprietary offerings they are custom. Therefore, a unified solution that supports different types of devices should be able to recognize from the three parameters above, the exact type of the device and trigger the proper workflow.

Sensor Control. The discovery process may complicate though the integration and operation of the devices as part of third-party system. There are sensors that broadcast only when they are powered on, while others broadcast continuously making difficult for the system to assess the sensor's state - ready for measurement/synchronization - requiring additional effort for its integration as well as the implementation of the respective functionality for its control. In some rare cases, the discovery process can be only performed through the proprietary SDKs, which cannot be replaced from the system native discovery mechanisms. The challenge in this case is then, how to simultaneously support different discovery operations using different mechanisms under the same system. The different operation lifecycle of each sensor also complicates its control from another system. Accordingly, it is essential to support and implement the post-measurement messages and commands to be send to the sensor, and at the same time set the system in a state which does block the discovery and future measurement. For example, there are sensors that (a) automatically power off after synchronization, (b) require special commands so as to power off, (c) power off after a timeout and (d) never power off, and obviously the effort to support the different cases above is substantial.

Data Management. In addition to the different communication specifications each manufacturer follows, the fact that they use different, and often non-standardized, data formats, introduces immense complexity to the third-party tools and applications intending communication and use of these sensors. Furthermore, the diverse types of biosignals and the different measuring techniques raise additional challenges. One of the main concerns during the design and implementation of the sensors integration is the use of a single format and model for all biosignals and activity data in order to simplify all operations of the system components that store, manage and analyze the data. Standards such as Open mHealth [10] and FHIR [11] should be exploited mapping the acquired data to their elements. One interesting finding from this analysis is related with the metadata of each measurement. There are devices such as blood pressure monitors that also assess some quality characteristics for the measurement that should be taken into consideration in the data model that will be implemented.

3 Implementation of a Holistic Approach

The businesses, operational and technical issues described in the previous sections complicate the design and implementation of a system that will effectively address them without decreasing the required levels of usability and security. There two approaches proposed in this work, which cover different technological areas and usage environments. We present the methodology that has been followed in an Android app that has been implemented focusing on the end-users as well as, an open solution based on Linux and Raspberry Pi targeting developers and rapid prototyping usages.

3.1 The Android OS Case

The Android implementation concentrated on all the issues highlighted in the previous section, addressing them effectively the technical implications. It should be noted that the usability was considered as the top requirement in this work so as to allow the use of the system and the sensors from non-technical savvy users such as elders or patients with chronic diseases. To this direction, we aimed for a non-intrusive application, which would disrupt their habits or the process that they follow already to measure their biosignals. Conceptually, for the design of the system architecture the problem was separated into two distinct - to a certain extent - problems, (a) *the registration and initialization of devices* and (b) the *measurement process and the data synchronization*. Therefore, the design of the system includes two separate subsystems dealing correspondingly with the two aforementioned challenges. Core element in this approach, was the implementation of operations that should be executed in each step of the workflow specifically for each device. In terms of implementation, we have named the respective code snippets "drivers". Several drivers have been implementing allowing the system-users and the application developers to send generic commands, which are in sequel translated in low level commands to be send to the device.

Sensor Drivers. The sensor drivers operate as the bridge between the operations in the registration of a device and the operations before/during/after the measurement process. Practically, and depending on the device type, the drivers either implement in high level the communication protocol and the processes that should be followed for the communication with the device, or in cases that SDKs are available, perform the transformation of application commands to the methods to be executed by the SDK. The UML class diagram in Fig. 1 depicts the interfaces specification that will be implemented in the various drivers, and a callback that is called when messages are send from the device to the system through the drivers. The *GenericDriver* interface includes the common methods that are required from all drivers during their registration and measurement operations, while the *ProtocolDriver* and *SdkDriver* interfaces include the specific operations of the respective device families (Fig. 2).

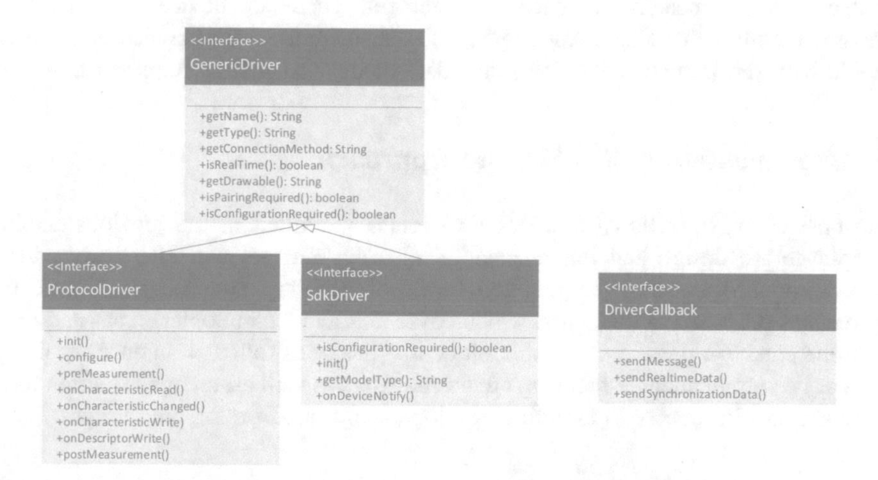

Fig. 1. Android drivers implementation

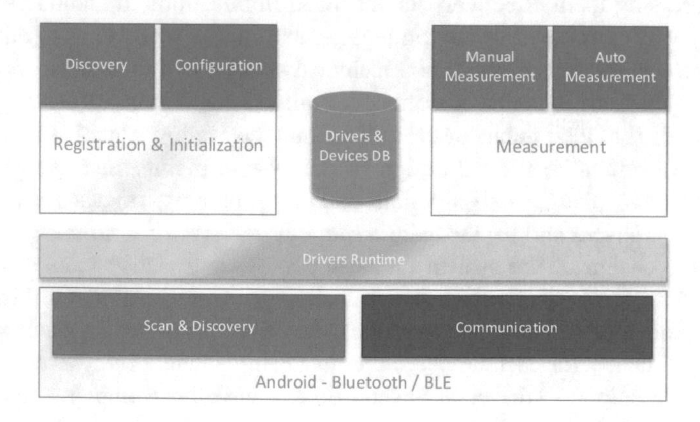

Fig. 2. Android system architecture

System Architecture. The architecture of the system by exploiting the drivers approach, becomes modular. The operations are separated into two subsystems. *The Registration and Initialization Operations* and the components that realize them are responsible for the discovery of devices and their proper identification. As soon as a device is discovered, the system searches in the local drivers & devices database to see if the device can be supported and if it is not already registered to start its registration. The registration process loads the appropriate driver and based on its type and the device requirements the appropriate initialization workflow is triggered from the *Drivers Runtime*. The initialization includes both the pairing/bonding of the device as well as it configuration if required (e.g. set the time, define the user profile etc.). In order, a device to be user for *Measurement Operations*, the device should be already registered as described above. The measurement may either take place automatically or initiated by the user. In the first case, the system scans continuously for registered devices and as soon as one becomes available (e.g. a user started using a pulse oximeter) the system loads the respective driver and starts the measurement process based on the methodology defined from the vendor. Based on the device type, different callbacks are used to inform the system for single measurements or for batch synchronization of data. The drivers also take care of the pre/post measurement processes and ensure that both the system and the driver will return to a normal operational state.

3.2 The Linux OS/Raspberry Pi Case

The approach that has been followed for the Linux/Raspberry Pi implementation is based on the AGILE-IoT Software [8]. The AGILE-IoT Software Architecture is a flexible software architecture aimed for single board computers (SBCs) running as an edge gateway (AGILE Gateway). The AGILE Gateway running the AGILE Software stack can currently use various Linux-based operating systems supported by the SBCs and use Docker [9] containers. Communication is mainly achieved via REST services exposed between the containers or Linux specific communication protocols to access the underlying hardware. Depending in the short distance networking abilities such as Bluetooth or Zigbee, the gateway can interface with various sensor devices and will allow the user to access data from local sensors, view and store the data on the gateway device. The AGILE Software Stack is composed of various modules implemented in docker containers which handle specific parts of the implemented functionality. The implementation of a sensor gateway application involves the extending the agile core functionality for the access to the new sensors and the application which uses the sensor data via REST calls. The application can be implemented as a separate docker container and can access Agile functions via the Agile SDK library.

The architecture of the AGILE Software stack allows the integration of new sensors which increase the usefulness of the Agile Gateway in many applications. In order to ease access to sensors data the Agile – IoT project has defined a set of REST commands which homogenize access to sensors. Unfortunately, due to the great variance of features and access protocols between devices, great effort is needed to integrate them via the implementation of corresponding device drivers. The Agile Software stack allows two methods of implementation of device drivers depending on the required level of integration:

- *Protocol based implementation:* Implementation of functionality which converts device communication to Agile compliant protocol primitives.
- *Device based implementation:* Implementation of device communication using existing AGILE PAN networking protocols. Bluetooth Low Energy and Zigbee are currently supported.

The selection of the appropriate method depends on the complexity of the sensor communication protocol. In order to integrate the sensors, the system we used the *Device based implementation* method. The *Device based implementation* uses a registration/publish/subscribe device model. Each device needs to be registered before use and the application can subscribe to the sensor data endpoints in order for published data to be streamed to the user. This model works well with BLE devices, which use GATT notifications [6] to stream data to the host. The Bluetooth GATT contains many standardized profiles and services but many device manufacturers have proprietary communication protocols mapped onto GATT services and characteristics. These include:

- *Single value characteristics:* these characteristics contain one sensor parameter. Driver needs to enable notification on the characteristic to retrieve the parameter. Driver may need to write to the characteristic to enable data flow.
- *Multiple value characteristics:* these characteristics contain multiple sensor parameters. Driver must filter the sensor value and route it to the correct endpoint.
- *Multi-characteristic control:* these sensors use a UART-type communication where there are different characteristics utilized for read and write.

Some sensor values need a sequence of accesses to return the correct value. Also, each sensor may have an initialization sequence which needs to be handled. The device driver needs to handle the intermittent connectivity of BLE sensors. It must keep the implementation up-to-date with the connection state of the sensors (Fig. 3).

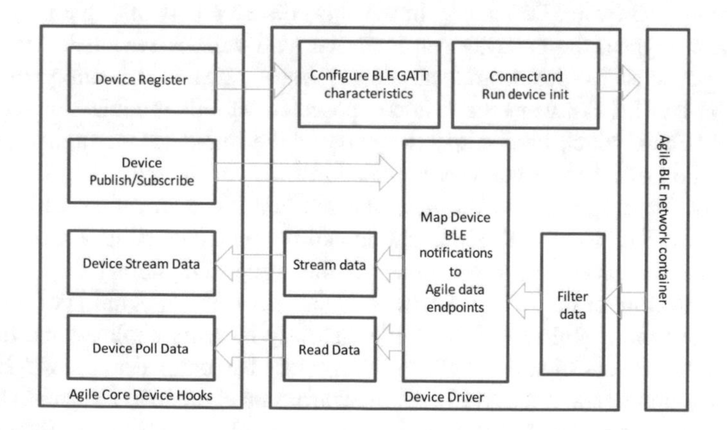

Fig. 3. Agile device driver architecture

4 Discussion and Conclusions

The utilization of captured biosignals is considered essential for understanding the health condition of an individuals while at the same time the trend of using quantified-self solutions approaches is increasing. The added value of adopting such solutions (in home and professional use) includes among others the following:

- Allows users to monitor their physical, physiological and emotional status
- Enhances their well-being
- Allows detection and management of potentially dangerous situations
- Motivates them to engage in social, physical and self-caring activities
- Models their behavior to improve self-care

In order for these solutions to be competitive though, they need to provide high levels of usability and robustness for all provided features, from the communication with sensors for self-tracking and biosignal monitoring, to the visualization and storage of the sensor data. The resolution of the issues related with the integration of sensors solutions for smartphones of personal computing systems is crucial for their success and wide adoption, since this is an aspect that affects the extent of their applicability and usability, enhance their flexibility to support additional sensors and features, and simplify their operation and maintenance.

The proposed solutions, following the in-depth analysis of different devices' operations and technical characteristics, can be used as the foundation for the sensor management subsystem for Android and Linux-based applications which make-up the vast majority of mHealth offerings. Both Android and Linux approaches follow a modular design and are therefore highly extensible allowing the integration of additional devices in the future, which may require different operational workflows and communication patterns. Besides the befits for usability for end-users and the simplicity for application developers, they also add value to business proposition of applications providers. These approaches enable the simultaneous integration and operation of multiple sensors extending the applications' functionality for monitoring additional biosignals in a single endpoint and minimize the vendor lock-in.

For the analysis and the proposed implementation, several biosignal sensors, wearables and smartwatches have been studied, a process that still is ongoing in order to better support the existing devices, as the used software and hardware evolves, but most importantly in order to enable the integration of new devices that are becoming available. In addition, prototypes for other systems like Windows and iOS are under development towards an ultimate goal for a holistic solution for communication and management of biosignal sensors and activity trackers promoting mHealth and self-care.

Acknowledgement. This work has been partly supported by the University of Piraeus Research Center.

Dr. A. Menychtas acknowledges the Greek State Scholarship Foundation (IKY). This research was implemented with a Scholarship from IKY and was funded from the action "Reinforcement of postdoctoral researchers" of the programme "Development of Human Resources, Education and Lifelong Learning", with priority axes 6, 8, 9 and it was co-financed by the European Social Fund-ESF and the Greek State.

References

1. Pantelopoulos, A., Bourbakis, N.G.: A survey on wearable sensor-based systems for health monitoring and prognosis. IEEE Trans. Syst. Man Cybern. Part C (Appl. Rev.) **40**(1), 1–12 (2010)
2. Kaniusas, E.: Biomedical Signals and Sensors I: Linking Physiological Phenomena and Biosignals. Springer Science & Business Media, Heidelberg (2012)
3. Mackensen, E., Lai, M., Wendt, T.M.: Bluetooth low energy (BLE) based wireless sensors. In: 2012 IEEE Sensors. IEEE (2012)
4. Menychtas, A., Tsanakas, P., Maglogiannis, I.: Automated integration of wireless biosignal collection devices for patient-centred decision-making in point-of-care systems. Healthc. Technol. Lett. **3**(1), 34–40 (2016)
5. Bluetooth - Health Device Profile. https://www.bluetooth.com/specifications/assigned-numbers/health-device-profile
6. Bluetooth GATT Specifications. https://www.bluetooth.com/specifications/gatt
7. Personal Connected Health Alliance. http://www.pchalliance.org
8. Menychtas, A., et al.: A versatile architecture for building IoT quantified-self applications. In: Proceedings of 30th IEEE International Symposium on Computer-Based Medical Systems - IEEE CBMS (2017)
9. Merkel, D.: Docker: lightweight Linux containers for consistent development and deployment. Linux J. **2014**(239), 2 (2014)
10. Open mHealth. http://www.openmhealth.org/organization/about/
11. FHIR (Fast Healthcare Interoperability Resources). https://www.hl7.org/fhir/overview.html

"Developing Interdisciplinary Instructional Design Through Creative Problem-Solving by the Pillars of STEAM Methodology"

Neofotistou Eleni[1(✉)] and Paraskeva Fotini[2]

[1] Department of Digital Systems, HAEF-Athens College Elementary School, University of Piraeus, Piraeus, Greece
eleni_neof@yahoo.gr

[2] Learning Psychology with Technology in the Department of Digital Systems, University of Piraeus, Piraeus, Greece
fparaske@unipi.gr

Abstract. STEAM methodology proposes a holistic view of educational topics and provides the tools and structures so as to combine them in everyday school practice. Science, Technology, Engineering, Arts and Mathematics are the five silos, through which each topic is explored. Arts, as the new extension, is considered to be the key in bringing technological topics closer to social and humanity sciences, in order to design real inter-disciplinary teaching and learning experiences. STEAMapT^{2}theGalaxy course aims to provide a proper e-learning environment so as to enable participants to manage to use STEAM activities in practice through problem-solving procedures that they design by themselves.

Keywords: STEAM · Creative Problem Solving · Six Thinking Hats Creativity · Inter-disciplinary

1 Introduction

E-learning is a necessity that education coordinators need to embrace in order to achieve better learning results and more importantly to facilitate learning opportunities that expand over the traditional classroom's walls. In addition, the ever-changing learning environment that modern societies tend to face, makes the skills of adaptation and problem-solving come to the foreground.

1.1 The Research Background and Motivation

The detailed consideration of the current research proposals, concluded to the two major factors that issued the research gap - the need of altering the way science is being approached in education today and the challenge of developing problem-solving skills throughout education. The previous statements are explained in detail below:

It is common sense that rational and divergent thinking are not to be separated in the fields of education. Opposite to this admission, STEM subjects tend to be combined to the convergent thinking and that is why the initial STEM methodology is criticized

M. E. Auer and T. Tsiatsos (Eds.): IMCL 2017, AISC 725, pp. 89–97, 2018.
https://doi.org/10.1007/978-3-319-75175-7_10

of being a technocratic teaching approach. On the other hand, research through the ages, beginning from the important influential personalities, such as Leonardo da Vinci, to the more recent findings [1, 2] proposes that teaching should aim into developing both creativity and rationality. Technology, science, social studies and arts are deeply connected. This relationship can clearly be identified in STEAM instructional approaches.

STEAM approach is considered a strong motivation for the learners involved and is related to better learning results, even in more technocratic fields, commonly known as "STEM disciplines", as described in [2, 3].

Future citizens need to familiarize themselves with change and learn how to react in different circumstances, in their personal and professional life. For this to be achieved, they need to learn how to combine different elements together and manage to adapt. STEM elements could be merged through art integration and provide a holistic learning experience [3].

Considering the educational reality nowadays, it is clearly assumed that the way science is being taught is not compatible with the students' and modern world's needs in any educational grade. Experimental and "try and error" procedures tend to be replaced by the "one and only" scientific truth, even if researchers and educators try to accomplish the opposite [4]. Creativity and personal contribution are excluded, as they are based on multiple perspectives and approaches [5].

International common cores tend to underline the significant role that creativity and problem-solving skills should hold in education today, due to the challenges future world proposes. Creativity and problem-solving are, in any case, the more mentioned and requested skills among the 21st century skills' framework [6]. So, in the years to come, stakeholders should invest on these skills along with STEM subjects, or just develop STEAM instructional frameworks [1].

According to further research, there is a prominent need to explore practical interdisciplinary implementations of STEAM methodology, so as to scaffold 21st century skills, such as problem-solving, scientific literacy and creativity [7].

This paper proposes an e-course, STEAMapT^{2}theGalaxy, designed by the principles of problem-based learning, purposing in a meaningful familiarization with the STEAM methodology, by its application in subjects that appeal to each one of the participants' interests.

1.2 The Purposes of the Research

The instructional design developed in this e-course and delivered through STEAMapT^{2}theGalaxy site, aims to:

- familiarize participants with STEAM methodology.
- help participants identify the educational potential of including STEAM approaches in their teaching practice and be able to apply them in real circumstances.
- develop problem-solving skills.
- enhance creative thinking.

2 Theoretical Framework

The e-course is based on the phases of "Creative Problem Solving" (CPS) model, proposed by Treffinger [8], which is combined to the thinking dispositions indicated by "The 6 Thinking Hats" learning and problem exploration strategy.

2.1 Problem-Based Learning Principles and the "Creative Problem Solving" Model

Problem-based learning refers to the teaching procedure that is based on real-life problems and problematic situations, which students explore within inter-disciplinary learning environments, usually working in groups, by the principles of active and collaborative learning [9].

The CPS model includes the initial problem solving steps, along with the parameter of inventing multiple innovative solutions in problems of gradual difficulty [8].

The steps of the CPS model are sorted in three larger dispositional groups, according to the attitude towards the "problem" that should be adopted in each one of them:

- Understanding the problem
- Generating ideas
- Planning for action

Each of the general groups named above, contains specific steps to be followed during the procedure. The activities included are executed in groups or individually, depending on the circumstances and the goals that have been set in the beginning of the process:

- Understanding the problem
  - Mess Finding (ph_1)
    In this step, the participant is informed about the subject of the e-course and tries to form an initial opinion towards it. That is the introductory problem-statement phase.
  - Data Finding (ph_2)
    Next, he is asked to collect information about the subject and clearly state the problem he is about to explore. Gathering data should be a careful and detailed process, as it is crucial for the steps to follow.
  - Problem Finding (ph_3)
    After information gathering and organizing, student is ready to clearly state the problem and decide the specific aspect of it that needs to be improved. Creative Problem Solving, either way, is about not only dealing with fully defined problems, but also figuring out ways to better manage a well-known situation.
- Generating ideas
  - Idea Finding (ph_4)
    Elaborative and flexible thinking is the key to this step, as it is important to come up with many ideas and possible solutions to the problem stated. The "one and only" solution is not accepted by the presenters of the model.

- Planning for action
  - Solution Finding (ph_4)
    As long as all the possible solutions are collected, selecting the one that best fits the given situation is feasible. This procedure includes ranking the options and identifying the advantages, disadvantages and the potential of each possible idea.
  - Acceptance Finding (ph_5)
    The final step is about planning the actual implementation of the solution chosen and predict all the assisters and resisters that may affect the whole attempt.

It is clearly stated that the steps numbered above can be used in the order that better facilitates the whole problem-solving process, without identifying wrong and right sequences.

2.2 "The 6 Thinking Hats" Strategy

"The 6 Thinking Hats" strategy [10] offers the framework for exploring the problematic situation stated in a creative and holistic way. Each one of the hats is related to a different thinking disposition, so as to cover all the aspects of the subject/problem discussed. Moreover, it has been chosen so as to present the steps that should be followed in an understandable and more descriptive way.

In contrast to the theoretical model, the strategy proposes a basic hat sequence depending on the circumstances under which is being used and the goals that have been set. For the "problem-solving" the most common sequence is the one that follows:

- Blue hat – controller of the whole process
- White hat – information about the subject
- Red hat – emotions and personal beliefs
- Black hat – possible dangers and obstacles
- Yellow hat – positive aspect and hope
- Green hat – creativity and possible solutions' generation
- (Blue hat – selecting the best solution and sums up the procedure).

2.3 Web-Based Educational Framework

As it has been mentioned before, the e-course was designed as an instructor-led and facilitated e-learning program that was delivered through an educational site hosted by Weebly. Building a site was the best option for the following reasons:

- The flexibility in design procedure [11].
- Instructor-led paths that could exceed the options given by a standard learning management system.
- Compatibility of multi-media tools.
- Wide variety of collaborative and scaffolding tools which could be embedded in the site.

3 Research Methodology

3.1 The Research Questions

The theories explained above were orchestrated properly in a web-based learning environment, called STEAMapT^{2}theGalaxy, in order to answer the research questions below:

- Can the workflow that is delivered through STEAMapT^{2}theGalaxy site be an effective instructional design by the principles of Creative Problem Solving in a STEAM course?
- What are the educational potential and affordances of STEAMapT^{2}theGalaxy which can contribute to the students' engagement and positively impact their learning outcomes?

In order to answer the questions above, a research methodology framework was designed.

3.2 The Instructional Design Framework

The phases of the "Creative Problem Solving" model where combined to the hats of the strategy explained above, creating a strong problem-solving framework. Participants had the chance to elaborate on a problem of their own choice, related to STEAM methodology, after they were introduced to the general concern about the STEM projects that are so popular in nowadays.

The phases were combined in order to better respond to the goals that have been set and the needs of the trainees' group. The "Idea Finding" and "Solution Finding" phases were grouped into one, on the grounds that in this way the creative process is fully approached and not disturbed. Furthermore, the "Problem Finding" phase was connected to three hats, as it is about defining the problem and predicting any clue that could affect its exploration (Fig. 1).

In each one of the steps, proper activities were included so as to help students reach the desired learning results. The activities in the first steps were individual, while in the middle the students worked in pairs and then in larger groups.

3.3 Key Performance Indicators and Assessment

Assessment was prominent in all phases, as an individual or group procedure. Individual assessment was included in every single phase as multiple choice questions, referring to the Key Performance Indicators (KPIs) that are examined in this educational scenario. The KPIs were grouped as follows:

- STEAM oriented
- e-learning system oriented
- creativity oriented

Moreover, students completed questionnaires about their prior knowledge (ph_0) and creative attitude (ph_4). In pairs, they assessed the lesson plans that had been

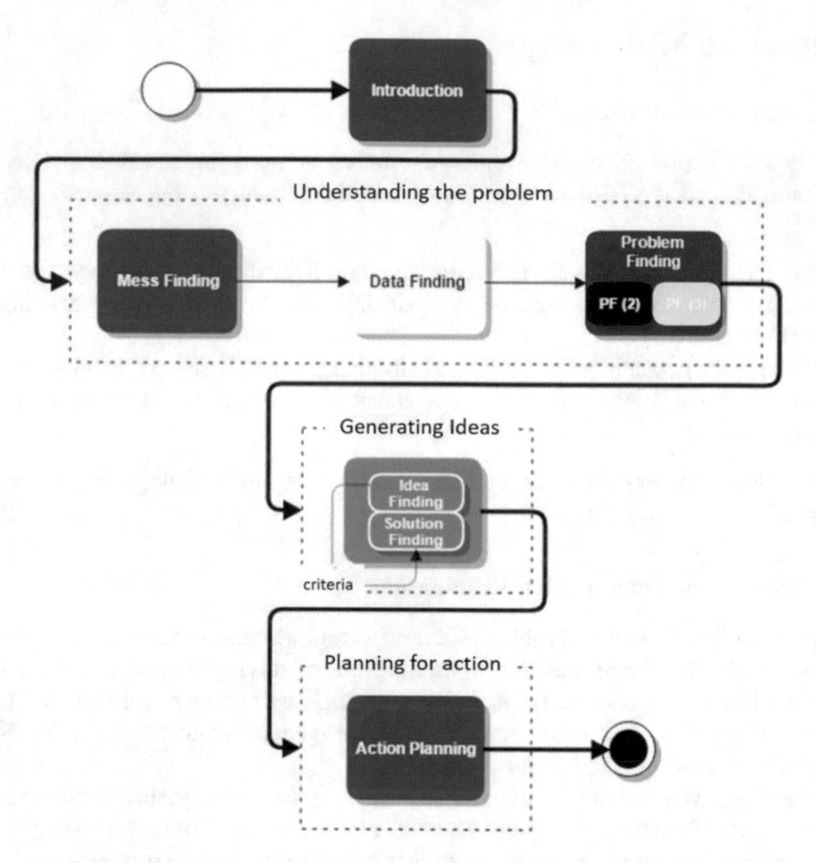

Fig. 1. The instructional design framework, by the principles of "Creative Problem Solving" and "The 6 Thinking Hats"

designed by their colleagues, in pairs as well, by providing detailed and constructive feedback. Ultimately, working individually, they assessed the course in general. In this way, assessment was an ongoing process so as to get specific and trustworthy results.

3.4 Tools Orchestration and Site Management

In the site, each phase of the educational scenario was presented in a separate page, while the navigation was linear. Any additional information and source was included as a hyperlink in "help" pages that were designed for every unit. In addition, communication forms enabled the communication loop between the instructor and the students and provided a strong basis for collaboration.

The case that was explored was delivered to the students as a story about traveling from the well-known STEM planet, to the unknown and full of secrets STEAM galaxy, after suspecting that the way STEM planet (projects, tools etc.) is being exploited today is not the proper one. Graphics were included in each page to correspond to this story, as well as make the browsing experience more understandable and fun. Finally, a blog was designed on Weebly host, in order to enable students to share their opinions and express themselves. Blog posting is also necessary in several activities included in the e-course. Collaboration was supported by embedding Padlet, in problem stating phase (ph_3) and feedback providing process (ph_5) as it makes it easier to post, view and respond in each one's posts in real time. In both communication procedures described above, instructor was present to help and monitor students' interaction.

4 Findings and Discussion

4.1 STEAM Oriented Results

The first research question, which was about the instructional effectiveness of the workflow by the principles of Creative Problem Solving, delivered by the STEA-MapT^{2}theGalaxy site, was answered by reviewing the students' responds in the rubrics that were related to STEAM methodology.

In the statistical analysis that was conducted, it was shown that the five aspects of STEAM methodology are supported equally and effectively with material and activities in the lesson plans that students design. Also, the students were able to conduct a careful research in terms of STEAM methodology and share their results, while they were successful in presenting and supporting methodology's advantages (new teaching practices, support different learning types, enhancing collaboration, enrich learning environment, include art in teaching etc.) by their work on them (Table 1).

Table 1. T-test analysis on the data collected towards STEAM basics supported in participants' lesson plans, that makes it possible to expand the conclusion to the relevant population

Test value = 4						
	t	df	Sig. (2-tailed)	Mean difference	95% confidence interval of the difference	
					Lower	Upper
Exploration	8,507	84	.000	.506	.39	.62
Learning styles	7,447	84	.000	.459	.34	.58
Teaching approaches	10,101	84	.000	.612	.49	.73
Learning environment	11,581	84	.000	.659	.55	.77
Collaboration	8,091	84	.000	.518	.39	.64
Understanding	6,507	84	.000	.412	.29	.54
Arts	5,191	84	.000	.400	.25	.55

After implementing T-test analysis for these specific research individuals, it is possible to expand the results in the related population. Therefore, the e-course proposed in general, was proved to be suitable for familiarization with the STEAM methodology.

4.2 System Oriented Results

The second research question was related to the web-based system that was used to deliver the educational program. Specifically, it was about the educational potential of STEAMapT^{2}theGalaxy site, which could contribute to the students' engagement and the accomplishment of better learning results.

After the statistical analysis, it was clearly indicated that the educational web-based environment was suitable to the students' needs and managed to fulfill their expectations towards the supportive material provided, the interface quality and the communication paths that were included (Table 2).

Table 2. Correlation between the factors that are proved to define the success of the e-learning tool designed and used

	Supportive material	Ease of navigation	Communication with the instructor	Communication with colleagues
Supportive material	1			
Ease of navigation	.594**	1		
Communication with the instructor	.619**	.493**	1	
Communication with colleagues	.456**	.500**	.426**	1

**.Correlation is significant at the 0.01 level (2-tailed).

In the table presented above, it is proved that the variables examined are strongly connected. Therefore, it is important for all of them to be taken into serious consideration when designing e-learning programs. Students who had experience in e-learning, underlined the importance of these aspects as well, while agreeing (by 50%) that the ease of navigation is the one that matters the most.

5 Conclusion

STEAM is a promising teaching and learning methodology that could lead to better learning results for teachers and students and enhance their motivation to a high degree. E-learning is a proper area to develop such training and educational programs, as it is well related to 21st century skills and needs and it is compatible to STEAM methodology in general.

STEAMapT^{2}theGalaxy has proven to be a proper web-based approach for familiarizing with the methodology and experimenting in designing lessons according to it.

The way in which the framework presented in this article could be implemented in different learning environments or possibly be enriched by adding learning materials that could expand the subjects examined, remains to be seen.

Acknowledgement. This work has been partly supported by the University of Piraeus Research Center.

REFERENCES

1. Daugherty, M.K.: The prospect of an "A" in STEM education. J. STEM Educ.: Innov. Res. **14**(2), 10–15 (2013). http://search.ebscohost.com/login.aspx?direct=true&db=eric&AN=EJ1006879&site=ehost-live, http://ojs.jstem.org/index.php?journal=JSTEM&page=article&op=view&path[]=1744&path[]=1520
2. Williams, P.J.: STEM education: proceed with caution. Des. Technol. Educ.: Int. J. **16**(1), 26–35 (2009)
3. Henriksen, D.: Full STEAM ahead: creativity in excellent STEM teaching practices. STEAM J. **1**(2), 15 (2014). https://doi.org/10.5642/steam.20140102.15
4. Tytler, R., Osborne, J., Wiiliams, G., Tytler, K., Cripps Clark, J.: Opening up pathways: engagement in STEM across the primary-secondary school transition (2008)
5. Sternberg, R.J.: The nature of creativity. Creat. Res. J. **18**(1), 87–98 (2006). https://doi.org/10.1207/s15326934crj1801
6. The Partnership for 21st Century Learning. P21 Framework Definitions, pp. 1–9 (2015)
7. Ge, X., Ifenthaler, D., Spector, J.M.: Moving forward with STEAM education research. In: Emerging Technologies for STEAM Education, pp. 383–395 (2015). http://doi.org/10.1007/978-3-319-02573-5
8. Treffinger, D.J.: Creative problem solving: overview and educational implications. Educ. Psychol. Rev. **7**(3), 301–312 (1995)
9. Wood, J.C., Mack, L.G.: Problem-based learning and interdisciplinary instruction (2001)
10. de Bono, E.: Six Thinking Hats. Educ. Psychol. Pract. **4**(4), 208–215 (1989). https://doi.org/10.1080/0266736890040408
11. E-Learning for Teacher Training: From Design to Implementation Handbook for Practitioners. European Training Foundation, Italy (2009)

Design and Development Mobile Learning Apps and Content

Design of Lifelong Learning Content Using the Mobile E-Time Capsule System

Kazuya Takemata[1(✉)], Akiyuki Minamide[1], Shintarou Wakayama[2], and Takumi Nishiyama[1]

[1] Kanazawa Technical College, Kanazawa, Japan
{takemata,minamide}@neptune.kanazawa-it.ac.jp,
takuman.sotuken@gmail.com
[2] Kurashiki Printing Co., Ltd., Tokyo, Japan
s.wakayama@kp-print.co.jp

Abstract. In this study, we develop a system called an "e-time capsule," where a "time capsule" event for students to put their mementos into a container, bury it in the school yard, and dig it out in a certain time later and open it is achieved on the cloud. The students' image data of the daily scenes in their classrooms and their products in classes, which are collected using a mobile device, are stored in a server located in the classrooms. The schedule of opening the time capsule is determined based on an agreement by all students of the class at the time when the capsule is buried. In Japan, a time capsule is often opened at the time of coming-of-age ceremonies when the participants attend a coming-of-age ceremony at the age of 20, also as a way of a reunion. In the system which is developed through this study, the opening event is achieved by distributing the data to the smart phones of the students who come to the reunion venue. This paper addresses the data collection and storage using mobile devices of the students.

Keywords: Time capsule · Portfolio · Learning activity

1 Introduction

One of the social issues of Japan is the ever-declining birthrate due to population decrease. As the number of children shrinks, the integration and abolition of elementary and junior high schools, which are part of compulsory education systems, progress. The abolition of schools results in the outflow of families with children from these affected areas, which will further facilitate the under-population of the areas [1]. In principle, there should be at least 6 classes (1 class per grade) for elementary schools and 3 classes (1 class per grade) for junior high schools. Due to population decline, if the schools cannot maintain 6 classes for elementary schools or 3 classes for junior high schools, the integration of nearby schools is discussed. In Japan, some say the "appropriate size for group learning" is 2 to 3 classes per grade, which enables the change of students' classes every year. If it is difficult to integrate schools due to geographic reasons, joint classes and joint events with other schools are offered using Information Communication Technology (ICT). However, local communities lose

M. E. Auer and T. Tsiatsos (Eds.): IMCL 2017, AISC 725, pp. 101–106, 2018.
https://doi.org/10.1007/978-3-319-75175-7_11

livelihood due to the integration and abolishment of schools, local community people often want to avoid the integration and abolishment of schools.

This study relates maintaining an event called time capsule with "the ever-declining birthrate due to population decline and under-population in local cities due to population concentration in the metropolitan area [2]." Sometimes, elementary or junior high schools disappear due to the integration and abolishment, as a result, the time capsule opening ceremonies must be given up before the time of coming-of-age ceremonies. Some schools with a small school yard are already running out of spaces to bury time capsules, and the time capsule burying ceremonies cannot be implemented. This study addresses this issue and proposes "e-time capsule" as a solution [3, 4].

This paper describes an experiment at an elementary school to test the beta version of the "e-time capsule" system where time capsules are created on the cloud and creating a mobile version of the system based on the experiment results. Joint leaning using ICT is promoted among the schools that have geographical difficulties for integration. The "e-time capsules" of this study is a system that can also contribute to the healthy class management of such schools.

2 Why is a Time Capsule Event Important in Japan?

Some elementary and junior high schools hold an event of putting students' products in classes and mementos in a special container and burying it in a corner of the schoolyard (a time capsule sealing ceremony). Especially, students are looking forward to the time capsule sealing ceremony when they are twelve years of age at the time of their graduation from elementary school. The schedule of the event to bury out the time capsule (a time capsule opening ceremony) is determined based on an agreement by all students of the class at the time when the capsule is buried. In Japan, a time capsule is often dug up and opened at the time of coming-of-age ceremonies when participants attend a coming-of-age ceremony at the age of 20, also as a way of a reunion. Seeing the contents of the time capsule, they relive their childhood and recall their feelings and friendship with their classmates at the time of burying it. The time capsule opening ceremony is used as a "common purpose" for the classmates who have grown up to get together. It is considered as a precious fun beyond generation in Japan.

In Japan, elementary and junior high schools play a central role of the local community they are located, and they have a potential function that connects the local communities and the young people who leave the communities. The time capsule events contribute to maintaining a connection between the local communities and young people, and inheriting these events as a form of "e-time capsule" supports the local communities. This is the background of the time-capsule events in Japan.

3 Experiment of an E-Time Capsule

We examined if using ICT for a time capsule event at an elementary school would not become a burden for the students, using the beta-version of the e-time capsule system. The beta-version system consists of 1 laptop computer and a few digital cameras. Due to security concerns, outsiders cannot connect an external system to the school's network within the Internet environment of elementary schools in Japan. Therefore, we instructed students to take pictures of memorable scenes with a digital camera, and store their data in an SD card, for using the beta version. The data was stored in a specified space of a laptop computer via an SD card used in the camera. We proposed that the collected data should be saved in a DVD, and stored at a school with their yearbook.

The experiment of trying the beta system was conducted at Kanazawa City Miwa Elementary School, which is located in the suburbs of Kanazawa City and has 18 ordinary classrooms and 2 special classrooms with a total of 541 students (as of fiscal 2015, in which the experiment was carried out). This is the average scale of elementary schools in Kanazawa City. In this elementary school, the parents' association is active, and the base for local communities is healthy in this region. Accordingly, the principal and guardians of the elementary school understood this study well, and accepted the trial operation of this system flexibly. In detail, 32 students of Grade 6, Class 1 were instructed to use the beta version of this system for two months from late January to early March 2016 (the continuation ceremony). They used the system smoothly, thanks to the constraint that only the pictures taken with a digital camera can be stored in an e-time capsule. The beta version system has the function to write the data students want to store in a time capsule into a laptop computer via an SD card. In the display of a PC, a character representing a student moves on a campus map to find a place for burying a capsule. This part has a game element, making the beta version attractive to students.

Figure 1 shows students registering data in the beta version of this system. The students are operating this system freely without any support from teachers. This scene indicates that the computer skills of 6th graders in Japan would suffice to operate the beta version. We were not able to conduct a questionnaire survey about the beta version targeted at the students of this class after the trial use, but the then principal Hideaki Kosaka, who directly instructed the students during the experiment, commented, "Students were using the system smoothly." Accordingly, it was considered that there would be no problems with practical use.

Fig. 1. Scene of students registering data in this system

4 Making E-Time Capsules Mobile

In the experiment of trying the beta version of this system, pupils were instructed to store memorable data taken with a digital camera into a specified space in a laptop computer via an SD card. There are two problems. The first is that when the data of an SD card is stored into a laptop computer, the computer is occupied by a single student. The second is that since a digital camera is shared, it is difficult to manage personal data inside an SD card. In order to solve the two problems, mobile terminals were used instead of digital cameras, for making e-time capsules mobile (hereinafter called "this system"). For overcoming the first problem, the communications function of mobile terminals is used. For solving the second problem, the login function of mobile terminals is utilized.

Figure 2 shows the application window of a mobile terminal in this system. (1) A pupil activates an application, selects his/her name from his/her class, and logs in to the system. In the figure, 6-1, 6-2, and 6-3 represent Classes 1, 2, and 3 in the sixth grade, respectively (Fig. 2a). After login, (2) the pupil saves data by utilizing the camera function (photo or video) of a mobile terminal (Fig. 2b). (3) A desirable place inside a campus is selected, and the collected data is buried there as shown in Fig. 3 (Here, data is transferred from a personal folder in a mobile terminal to a folder in a server). (4) After all students complete the task, all pieces of data in the server are transferred to the recording space in the cloud and external recording media.

a: To select your name from your class

b: To collect data you want to save in a mobile terminal

Fig. 2. Data registration function of the e-time capsule 1/2

To search for a place to bury the capsule

Fig. 3. Data registration function of the e-time capsule 2/2

5 Conclusions

This system is required to store the data of students during a period from the time capsule sealing ceremony to the time capsule opening ceremony. For safety, their data is stored in the data storage services in the cloud and external recording media. However, there remain some problems. The first is that there is a possibility that the external recording media will not be compatible with next-generation operating systems and it will be impossible to read the data. The second is that it is necessary to determine who should be responsible for returning (delivery) data to owners at the time capsule opening ceremony. As for the first problem with this system, it is necessary to predict the trend of IT devices about a decade from the saving of data. As for the second problem about data management and return, local communities can deal with it. Especially, the Ministry of Education, Culture, Sports, Science and Technology suggested the "promotion of development of an independent and cooperative society through learning" mainly at elementary school, in the report [5]. Based on this proposal, it is possible to position "e-time capsules" as a portfolio for children in local communities and maintain and manage them in local communities.

In this paper, we discussed the development of "e-time capsules" with mobile functions, based on the trial use of the beta version of the e-time capsule system. The trial use at an elementary school clarified the problems to be solved for utilizing the mobile functions for this system, and the authors were able to develop an appropriate system. The remaining issue is that it is necessary to conduct the experiment of trying this system again. This can be solved in cooperation with local communities, mainly elementary schools. To do so, the authors will cement the cooperation with local communities, and improve this system.

Acknowledgment. I would like to thank Mr. Hideaki Kosaka at the ex-principal of Kanazawa City Miwa Elementary School for his advice when writing this paper.

References

1. Ministry of Education, Culture, Sports, Science and Technology, Japan: Explanatory material of the regarding the rationalization of school scale, appropriate locations (2015). http://www.mext.go.jp/component/a_menu/education/micro_detail/__icsFiles/afieldfile/2015/01/29/1354768_3.pdf. Accessed 6 Sept 2017
2. Maxim, G.W.: Time capsules: tools of the classroom historian. Soc. Stud. **88**(5), 227–232 (1997). https://doi.org/10.1080/00377999709603784
3. Wade, A., Abrami, P., Sclater, J.: An electronic portfolio to support learning. Can. J. Learn. Technol. **31**(3) (2005). https://doi.org/10.21432/t2h30p
4. Takemata, K., Nishiyama, T., Minamide, A., Wakayama, S.: Design and trial use of an e-time capsule system. In: 2017 IEEE 17th International Conference on Advanced Learning Technologies, pp. 194–195 (2017)
5. Ministry of Education, Culture, Sports, Science and Technology, Japan: Material submitted by the Ministry of Education, Culture, Sports, Science and Technology: MEXT's efforts for regional vitalization: for developing lively communities (2014). http://www.kantei.go.jp/jp/singi/tiiki/platform/kakuryo/dai1/siryo3.pdf. Accessed 6 Sept 2017

Céos: A Collaborative Web-Based Application for Improving Teaching-Learning Strategies

Marcos Mincov Tenorio[1(✉)], Francisco Reinaldo[1], Rauany Jorge Esperandim[1], Rui Pedro Lopes[2], Lourival Gois[1], and Guataçara dos Santos Junior[1]

[1] Universidade Tecnológica Federal do Paraná, Curitiba, Brazil
{marcostenorio,reinaldo,gois,guata}@utfpr.edu.br, raulesperandim@gmail.com

[2] Instituto Politécnico de Bragança, Bragança, Portugal
rlopes@ipb.pt

Abstract. This paper reports the Ceos system to support and promote collaborative learning by gamification strategies and mobile technologies. Céos was developed for basic smartphones to support classroom exercises and to encourage students out of classrooms in poor communities. The machinery of Céos has built-in teaching-learning strategies, such as gamification and mobile learning. To evidence a new kind of active learning process, Céos can track and suggest tips and rewards, guiding students in their decisions during the collective process of knowledge acquisition. The student knowledge acquisition also happens via Céos during contribution of teachers, other students, and teammates. Considering the Declaration of Helsinki, an experimental research was carried out with students of a high school in the southwest of Paraná/Brazil. We produced an experimental study to observe students social interaction behavior during a survey test using Céos. The results indicated positive behavioral patterns, guiding them to a forum to complete the exercises. Céos provided a friendly educational environment and automatically promoted the collaborative process of knowledge construction by the use of gamification and decision making. This collaboration between Céos and students offered another point of view of using mobile technologies in teaching classrooms.

Keywords: Mobile learning · Collaborative work · Mobile applications

1 Introduction

Researchers from the Regional Center for Studies on the Development of the Information Society [17] pointed out that teenagers choose to communicate via virtual social media, i.e., Facebook chat, instead of traditional media - make a phone call or send an SMS. This choice is the result of many research about the collaborative benefits of quickly solving a given issue, all the readiness of a native

M. E. Auer and T. Tsiatsos (Eds.): IMCL 2017, AISC 725, pp. 107–114, 2018.
https://doi.org/10.1007/978-3-319-75175-7_12

WiFi connection in public places or services, the low cost of mobile broadband and friendly interfaces that invite them to use smartphone all the time [8].

Although significant social interactions emerge from this connectivity, there is a gap of tools and resources to monitor and intervene in the educational process of these adolescents [17]. Some research attempt to intervene in this scenario by developing Virtual Learning Environments (VLE) in desktop computers [6]. Nevertheless, several researchers noted that students did not use VLE because of its user interface complexity and the lack of expertise in computer systems [12,15].

In this scenario, smartphones suggest a clear interface, a high acceptance, low-cost and high computational power, but pedagogical practices using mobile devices were not implemented. Keengwe and Bhargava [10] shown that elements related to mobility and accessibility allow reducing the transactional distance. In this manner, traditional methods push students to social tools as a refuge. For Benson [1], therefore, smartphones allow greater interactions between participants and then gamification can be employed in this process of knowledge production.

This paper aims to present Céos, a web-based collaborative application developed for basic smartphones in order to facilitate mobile learning. Céos provides the collaborative learning inside and outside the classroom by the use of Mobile Learning (m-learning) to link physical and virtual spaces. More specifically, Céos has built-in gamification and artificial intelligence capabilities suggested in [4,9,14] to interact among groups of students to solve an educational problem.

So an experimental analyze was performed to find which effects Céos produced in students during solving a homework exercise (real life problems). The results demonstrated students were able to interact with the tool, obtaining positive advices during test sessions.

2 Céos

Céos was developed using an iterative and incremental modeling approach based on RUP (Rational Unified Process) [11]. RUP offers low maintenance costs and high process control and establishes a parallel communication channel among modules to be flexible during the framework development process. Pressman suggests that this approach is a successful alternative to the cascade model in Software Engineering [13].

The software has three main modules, such as Céos-mentor that provides a web-based application for teachers, Céos-app that supports a students' mobile app or mobile interface and Céos-root that bridges the final gap as a web-service to link Céos-mentor and Céos-app. Additionally, the clean layout of Céos-app offers a simplistic design and high user acceptance with a concise, familiar, responsive, consistent, attractive and efficient interface. These features were based on some characteristics of successful user interfaces, by Fadeyev [7].

Céos-mentor offers teachers a real-time support to the management of classrooms, organize groups of studies, enroll students, and create/publish students' activities. Other features are the student's enrollment to a team, support for the in-progress development of exercises and report of students achievement about their individual/collective activities.

In another vein, Céos-app offers students, Individually or in groups, a space to interact with teammates to perform their homework exercises. The most other common features are classroom enrollment, blackboard to teach and write draft solutions, and forum to chat with other students and Céos.

Céos implements a new kind of active learning process and knowledge acquisition. It is based on a mixed architecture that contains the best practices from Piaget's behavior-based observations, from Vygotsky's sociocultural theory [5], from Reinaldo's Hybrid Architecture of Behavioral Layer to Distance Learning [14] and studies involving mobile and social learning by Boticki [3].

The architectural modules were arranged in a computational layered architecture, following the bottom-up technique for behavior actions (taxies and instinct) and top-down (cognitive and social) for conflict resolution, as proposed by Reinaldo [14]. Dillenbourg also suggests embedding roles within tasks to have paired knowledge and solve conflicting viewpoints [5]. Therefore, the Céos learning machinery can interact with students and solve conflicts by a socio-constructivist approach, suggesting new solutions during the broadcast of messages for decision making. The application also allows teachers to track and guide students during the collective process of building knowledge. Finally, time constraint, limited resources, and challenge are the gamification challenges that Céos deals with as an engagement resource.

3 Materials and Methods

A supervised exploratory approach started a test trial to identify if Céos teaching-learning strategies affect the student's behavior and talk.

3.1 Participants

The school's inclusion criteria were to be a public school in the southwestern region of Paraná, with teachers having computer access at home and school, with lectures directed to teach workshops and students being part of the socioeconomic B and C classes. As a result, the SESI high school (Social Service of Industry) in the Francisco Beltrão city - Paraná, Brazil - was the only one to join the academic research.

The SESI high school has an educational system based on research for society. The teacher is the facilitator of knowledge. The school presents the vocational orientation of engineering field to the students by proposing workshops such as "Engineering Tracks." It also draws a connection between theory and practices via engineering tasks in real situations.

The student's inclusion criteria were to be an active and non-repeating student, with parents working in the industry, living at urban area, with Internet access and a basic smartphone. From 76 eligible students, we randomly sampled 31 participants with ages between 17 and 26 years old. The selected group were participants of "Engineering Tracks" workshop.

3.2 Procedure

To collect the data, teachers lecturing Chemistry and Mathematics developed a problem that contains two tasks. These tasks are in a survey format that contains structured questions obeying the 5-point Likert-Thurstone mixed scale.

First, teachers receive support to work with Céos-mentor. Next, students receive basic support to install and use the Céos-app. Following, the students were arranged into teams, and the problem was made available. For seven days, students used the Céos-app to solve the exercises and they kept in contact with teammates via forum.

3.3 Ethics

This study was conducted by the Declaration of Helsinki, which promotes respect for all participants by noninvasive, free of coercion, no affecting psychological and personal integrity and establishing the participant's right to privacy.

All participants were a volunteer and had no financial incentives. However, every participation was attested by a certificate of appreciation.

3.4 Statistical Analysis

Following a dual validation procedure, we initially checked if every student had prior knowledge when using mobile technology. After that, we apply inferential statistics to investigate if there were behavioral correlations among students.

4 Results

During the problem-solving process, Céos-mentor's intelligent machinery detected some typical students mistakes. To solve this, Céos-mentor guided students to communicate with their pairs via a forum and suggested some extra explanations. Céos-mentor also invited teachers to monitor the preliminary student's results for further understanding. Through, it was possible to measure the particular and collective student behavior during the task development.

4.1 Gamificating and Advising

To support students during finding the answers, Céos guide teams via a forum. Each student, who has achieved a certain threshold of wrong answers, received some advice from Céos expert system and their teammates. There was no direct teachers interference to help students, but students interact among them by chat.

Céos-app rewarded successful attempts with visual feedback. The search for public recognition of their work provided by a correct answer enables others to participate in the experiment enthusiastically. Students reported that when they were engaged in solving the problem, they spent more time studying with Céos-app. These comments evidence that students established a strong social connection between them and Céos, which promote engagement in pedagogical activities as suggested in [2,9].

Céos-mentor also provides effective negotiations using rewards and advice to help students attain the level, and not to give up after facing an error during a knowledge construction. Although some students caused unfavorable advising, it was detected early on. Hence, every students could establish a positive behavioral pattern on time. Thus, a hierarchical pattern between the students had made.

4.2 Correlation Between Messages and Right Answers

In addition to the qualitative results collected in the forum, Céos could extract some quantitative results regarding interaction (number of messages) in the two practices applied. Thus, we relate them to the number of right answers performed by every student in their teams.

The Fig. 1 shows the average and variance of individual messages among teams.

The Fig. 2 shows the average and variance of particular right answers grouped by teams in the two proposed tasks.

The Fig. 3 shows the correlation between interaction and right answers for every student reported engaging in the suggested exercises.

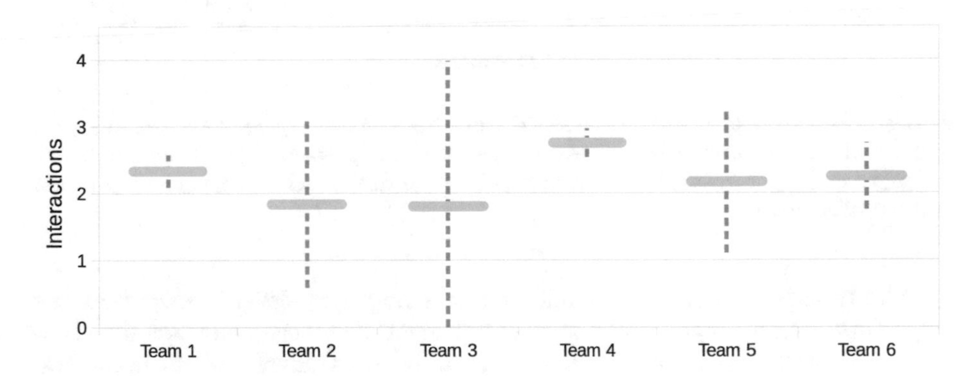

Fig. 1. The average of the interactions from every student of a team after completed the proposed exercise. The average message length (horizontal bar) is shown over the variance values (vertical dash).

5 Discussions

During the experiment, it was possible to obtain informal feedback from the students regarding Céos usability and functionality. Students reported us how easy is to use the Céos-app. They also reported the app has a fast responsive interface, and task flow is basically similar to the notion of common environments, in which they use for every day in social interactions.

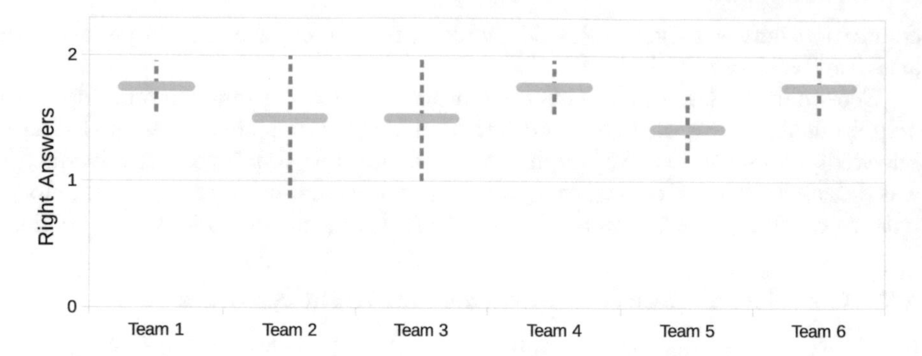

Fig. 2. The average of right answers from every student in a team after completed the proposed exercise. The average of right answers (horizontal bar) is shown with variance values (vertical dash).

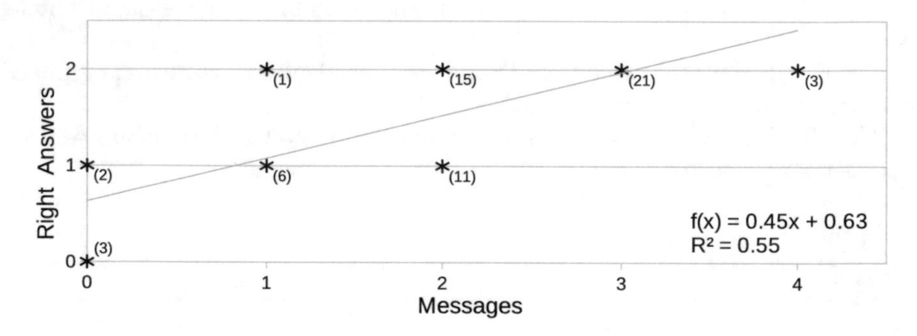

Fig. 3. The correlation of messages and right answers. Every student was analyzed individually in both exercises. In many cases students presented the same number of messages and right answers, therefore besides each dot, a value corresponding to the number of occurrences was shown.

The reward provided by gamification was another relevant factor that has helped students on engagement, and commitment. This technique uses time constraint, limited resources, and challenges to act on students' intrinsic motivations. Moreover, the game-like design has stimulated students, providing them emotional engagement during the forum activities. In the forum, we detected altruism, competition, cooperation and fellowship, which behavior were predicted by [16] in gamification contexts.

It is a worth noting point that in the behavior analysis, Céos reported in Fig. 1 that teams 4, 1 and 6 respectively presented the highest interaction average in a small variation (more homogeneous participation). In contrast, team 3 presented the lowest average of messages and a wide variation (more heterogeneous participation). Another measured variable was right answers in both exercises (Fig. 2). We noted that teams 4, 1 and 6 presented the higher right answers rate and low variation, in this order. These are the same teams with a higher average of messages, indicating a possible positive correlation between these factors. The result also suggests that if a team has homogeneous participation, in this case it tends to achieve success. This result reinforces the importance of tools that support and stimulate collaboration among students.

The Fig. 3 presents the correlation between interaction and right answers. This result indicates a positive relationship between these factors, and the R^2 outcome evidences it. This result may also support teachers to track students during the learning process, identifying some behaviors that tend to have bad results and remediate them.

6 Conclusions

In this paper, teachers and educational tutors used Céos as a resource to advise, monitor and analyze the collective and individual student's behavior, and his/her intellectual level viva a transactional real educational scenario. For students, Céos presented to be a promising tool inside and outside the classroom for collaborative learning. Céos resources have favored the interaction among students groups to solve educational problems.

The Céos cognitive maps and knowledge rules also suggested a change in the student's behavior, motivating and engaging it to participate outside the classroom. Additionally, who participated and interacted with colleagues tend to present good results in exercises.

In this scenario, teachers may use the Céos resources and track students behavior who presented low interaction with colleagues, then treat it separately, favoring the knowledge construction to every student.

References

1. Benson, P.: Learner autonomy. TESOL Q. **47**(4), 839–843 (2013)
2. Bista, S.K., Nepal, S., Colineau, N., Paris, C.: Using gamification in an online community. In: 2012 8th International Conference on Collaborative Computing: Networking, Applications and Worksharing (CollaborateCom), pp. 611–618. IEEE (2012)
3. Boticki, I., Baksa, J., Seow, P., Looi, C.K.: Usage of a mobile social learning platform with virtual badges in a primary school. Comput. Educ. **86**, 120–136 (2015). http://www.sciencedirect.com/science/article/pii/S0360131515000688
4. Deterding, S., Khaled, R., Nacke, L., Dixon, D.: Gamification: toward a definition. In: CHI 2011, pp. 12–15 (2011). http://gamification-research.org/wp-content/uploads/2011/04/02-Deterding-Khaled-Nacke-Dixon.pdf
5. Dillenbourg, P., Järvelä, S., Fischer, F.: The evolution of research on computer-supported collaborative learning. In: Technology-Enhanced Learning, pp. 3–19 (2009)
6. Dougiamas, M., Taylor, P.C.: Moodle: using learning communities to create an open source course management system. In: Dougiamas. Honolulu, Hawaii (2003). https://dougiamas.com/archives/edmedia2003/

7. Fadeyev, D.: Thoughts on design and user experience: 8 characteristics of successful user interfaces. The Usability Post web page, 15 April 2009. usabilitypost.com/2009/04/15/8-characteristics-of-successful-user-interfaces/. Accessed 16 Aug 2017
8. Ha, Y.W., Kim, J., Libaque-Saenz, C.F., Chang, Y., Park, M.C.: Use and gratifications of mobile SNSs: Facebook and KakaoTalk in Korea. Telemat. Inform. **32**(3), 425–438 (2015). http://www.sciencedirect.com/science/article/pii/S0736585314000744
9. Ibanez, M.B., Di-Serio, A., Delgado-Kloos, C.: Gamification for engaging computer science students in learning activities: a case study. IEEE Trans. Learn. Technol. **7**(3), 291–301 (2014)
10. Keengwe, J., Bhargava, M.: Mobile learning and integration of mobile technologies in education. Educ. Inf. Technol. **19**(4), 737–746 (2013)
11. Kruchten, P.: The Rational Unified Process: An Introduction, 3rd edn. Addison-Wesley Longman Publishing Co., Inc., Boston (2003)
12. Park, Y., Tech, V.: A pedagogical framework for mobile learning: categorizing educational applications of mobile technologies into four types. Int. Rev. Res. Open Distance Learn. **12**(2), 78–102 (2011)
13. Pressman, R.S.: Software Engineering a Practitioner's Approach, 7th edn. Roger S. Pressman, Boca Raton (2009)
14. Reinaldo, F., Camacho, R., Reis, L.P.: Arquitetura Híbrida das Teoria de Aprendizagem em Camadas Comportamentais para Ensino a Distância. In: 40th IGIP International Symposium on Engineering Education, vol. 1, pp. 820–824 (2010). ISBN 978-85-89120-87-6
15. Santana, M.A., Neto, B.S., Costa, E.B., Silva, I.C.L.: Avaliando o Uso das Ferramentas Educacionais no Ambiente Virtual de Aprendizagem Moodle. SBIE - Simpósio Brasileiro de Informática na Educação (Cbie), pp. 278–287 (2014)
16. Shneiderman, B.: Designing for fun: how can we design user interfaces to be more fun? Interactions **11**, 48–50 (2004)
17. UNESCO: Marco referencial metodológico para la medición del acceso y uso de las tecnologías de la información y la comunicación (tic) en educación. Technical report, UNESCO - Centro Regional de Estudios para el Desarrollo de la Sociedad de la Información (Cetic.br) (2017). http://cetic.br/media/docs/publicacoes/8/marco-referencial-metodologico-para-la-medicion-del-acceso-y-uso-de-las-tecnologias-de-la-informacion-y-la-comunicacion-en-educacion.pdf

An iOS Knowledge App to Support the Course "Developing Applications in a Programming Environment" of the Greek Lyceum

Eleni Seralidou(✉), Dimitrios Theodoropoulos, and Christos Douligeris

Department of Informatics, University of Piraeus, Piraeus, Greece
{eseralid, cdoulig}@unipi.gr,
theodoropoulos4@gmail.com

Abstract. There is no doubt that nowadays new technologies affect people's lives to a great extent. Mobile devices can be found everywhere, modifying the way of communication and interaction between people. This phenomenon expands in education as well, altering the teaching and learning methods. In this paper, the design and the implementation of an educational application for iOS mobiles devices is presented. The application's context covers the course of the Greek Lyceum curriculum "Developing Applications in a Programming Environment". The findings of the application's evaluation show us that this application is a useful tool, which combines the gaming experience with the solution of simple as well as of more complex problems.

Keywords: Mobile devices · Programming · Education
Smartphone applications

1 Introduction

The continuous technological development offers new opportunities in creating innovative teaching and learning environments and sets new foundations in the educational process. Since the 1990s, technology has entered people's everyday lives due to the simultaneous development of various new technologies and the world wide web [1]. Nowadays, all kinds of information can be accessed through various devices and many collaborative learning environments are used to make teaching customized to individual learners' needs [2].

Learning beyond traditional teaching methods and without limitations in space and time is a challenge for mobile application designers and for teachers in order to understand and apply a better method of supporting teaching [3]. Mobile devices nowadays have become so widespread that many countries have adopted several forms of e-learning and m-learning in their curriculum by recognizing the real value of using mobile devices in the field of education [4]. In addition, digital game-based learning is a particular research field within the wider context of education that attracts the keen interest of the educational and the scientific societies [5].

The algorithmic design of problem solving plays a key role in the cognitive development of students, as the algorithmic approach to any problem focuses on how

M. E. Auer and T. Tsiatsos (Eds.): IMCL 2017, AISC 725, pp. 115–126, 2018.
https://doi.org/10.1007/978-3-319-75175-7_13

to solve this problem, regardless of the origin of the problem [6]. In Greece, over the last two decades, the course "Developing Applications in a Programming Environment" (D.A.P.E.) is taught as an algorithm course, which aims at structuring the students' thinking and the teaching of problem-solving techniques. Specifically, it aims at developing methodological skills in students, such as design, development and control of algorithms, so that the students can solve problems regardless of the programming language they will use [7]. However, learning without the existence of a specific programming environment results in difficulties in assimilation of basic as well as of higher-level algorithmic and programming concepts. Considering the above discussion, important questions arise. Can an application for mobile devices with gaming characteristics work to support a particular course and raise students' interest? Can it enhance the learning process by combining the above mentioned concepts?

In this paper, an educational application for mobile devices based on the iOS platform, under the name AEPP Genius, is presented. This educational application fills a gap in the literature and it works as a supportive tool for the better understanding of the lesson's content, by providing theoretical and practical examples in the form of multiple choice questions. The application can be used by students and teachers and it aims at learning basic concepts of computer science and at acquiring methodological skills by fully complying to the objectives of the course and by providing students who are preparing for the pan-Hellenic exams for entering in University studies with a useful tool which combines the gaming experience with the repetition of the course's curriculum through the solving of various types of problems.

2 Mobile Learning

2.1 Mobile Devices in Education

Mobile devices, and especially smartphones, are a significant part in everyone's life, especially in the lives of young ones. Learning by using mobile devices is considered to be particularly effective [8]. Smartphone ownership has surpassed the ownership of desktop and laptop computers since 2009 [9]. The ease-of-use and the connectivity provided by these wireless portable devices works as a catalyst for pedagogical change by enabling the sharing of digital material of any kind based on the needs of each user [10]. More generally, the use of mobile devices allows the provision of instant feedback and makes the exchange of knowledge and information possible on a wide range of topics that are constantly enriched, while the exchange of school work is done directly with the electronic transfer of files and data [11]. Many universities and business entities have recognized the dynamics of mobile learning, due to its significant impact on teaching and research in education [12].

Through mobile applications, learners are allowed to create, share and participate online in authentic learning situations [13]. It has been estimated that smartphone users download about 50 to 80 applications on their mobile phones, while students who use mobile apps for educational purposes often use quizzes for assessment of knowledge and are more likely to monitor their progress than students who do not use them [14]. In terms of mobile devices use in education, according to recent studies, millions of i-Pads and i-Phone devices are already used in classrooms [12]. Considering the above, we decided to choose i-Pad and i-Phone devices for developing our suggested educational application, under the name AEPP Genius.

2.2 M-Learning Framework

The development of educational software is based on several learning theories, i.e. in behaviorism, cognitive psychology, constructivism and socio-cultural theories. Many of these theories have had to adapt to the new data brought by the knowledge society, and, in particular, to the new internet-based learning environments [15]. In developing educational applications, the challenge is no longer to find better ways to provide knowledge but to design, develop and implement interactive learning experiences that will inspire students to participate in the learning process and to eventually learn [16].

Recognizing the fact that an educational application should be based on an educational framework, the design and creation of AEPP Genius was carried out after carefully studying and exploiting the following three educational frameworks: the "conversational framework for the effective use of learning technologies" by Laurillard [17], the "pedagogical framework for mobile learning" by Park [1] and the "model for framing mobile learning" by Koole [18].

After researching the scope of the above frameworks in the creation of educational applications, we decided to use Diana Laurillard's "Conversational Model" for the design of the layout and the functions of AEPP Genius. This model consists of twelve steps (Fig. 1) which provide the framework for designing the environment of an educational application.

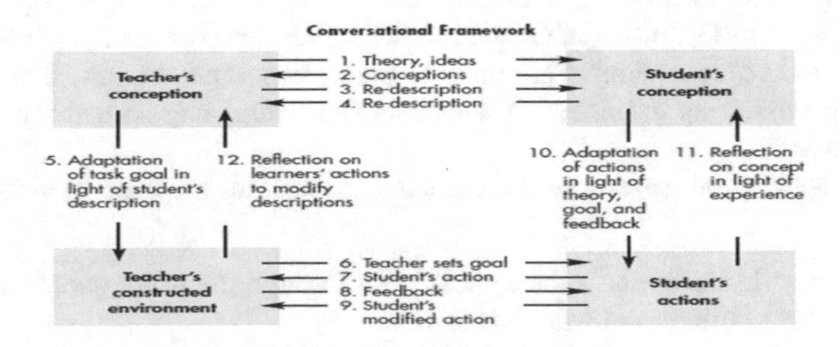

Fig. 1. Diana Laurillard's "Conversational Model"

According to the conversational model's characteristics, the interaction between the student and the educational application must be continuous until the student fully understands the course's content. New knowledge through this interaction is produced according to the user's perception capabilities and actions, while at the same time constant feedback is given until the concept that is taught is fully understood. Through questions the student comprehends the course's content and the evaluation process of the curriculums comprehension (Fig. 2).

In more detail, the questions of an application's module describe the theory or idea behind the particular learning objective (first step) and, thus, the learner's initial

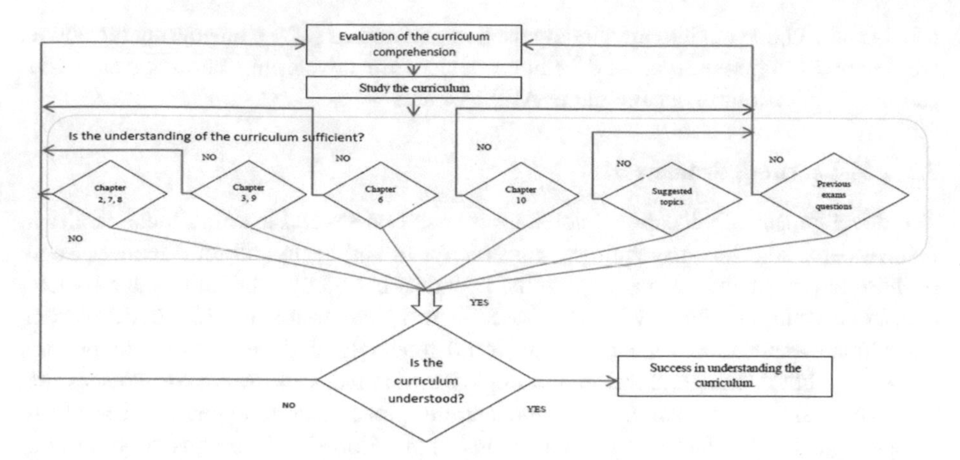

Fig. 2. Modified "Conversational Model" diagram, which responds to the learning of D.A.P.E. through mobile devices

perception of the subject (second step) is created. This perception is returned to the application via the user's response (third step). If the user's response is incorrect, the user should be allowed to repeat the same activity. The learner redefines his/hers original perception (fourth step) based on the corrections that were made in the previous step. The above results in a repeatable process where the learner takes action based on the given feedback (eighth step), while the cognitive gaps occurred from the previous steps determine the next learning objectives (fifth and sixth step). The learner adapts his actions based on the knowledge that emerged from the first steps (Theory - Ideas), to the sixth (Definition of Goals) and the eighth step (Feedback), while he is in a process of reflection regarding his original perceptions and the new knowledge he acquired by answering a number of questions and completing an activity (tenth and eleventh steps).

In this application, several of the framework's features are used in order to make students:

- Have access to the lesson's theory and ideas through the game questions.
- Express their opinion through their answers.
- Reach to a better understanding of the lesson by adjusting their actions.
- Make continuous practice, using the feedback that allows them to improve their performance.
- Have control over the learning goals so that they improve their learning capabilities.

3 Programming in Greek Secondary Education

3.1 Existing Curriculum

Programming is a powerful tool for the development of high-level skills. Programming is integrated into the curriculum of the Lyceum as an autonomous cognitive subject

[19]. The course D.A.P.E. of the 3rd grade of the Greek General Lyceum, is a basic course that prepares students for higher education, not only in IT departments but also in many other areas requiring a basic knowledge of algorithmic design. It is characterized as a course that is taught in classrooms and in laboratories and is examined in writing at a national level through the pan-Hellenic examinations. Emphasis is put on the “freedom” of algorithmic design combined with the strict syntax and vocabulary of a programming language [20]. The course is not intended to teach the use of a specific programming environment but to structure the students’ way of thinking and to teach problem solving techniques [7].

However, research has shown that the students experience a number of difficulties in learning basic concepts of programming, such as the concept of a variable, due to the lack of appropriate skills on the part of students, the difficulty in understanding specific programming skills, and the lack of logical-mathematical knowledge [21, 22]. These difficulties reflect the need to create new teaching methods and new programming environments. Moreover, understanding and developing algorithms is a milestone in computer science, so the development of innovative methods for teaching them is an important field of research for computer science education [23].

3.2 Games as Learning Tools

Digital games is a rapidly expanding field as they are the most popular technology in youth entertainment [24] and a source of motivation for users to test, develop and apply their knowledge while having fun [25]. Since the first researches made on the use of games in education, they were found to attract students and seemed to encourage them in a more constructive way than conventional education has done so far [26].

Learning through digital games can coexist with other forms of learning at all levels of education and in a variety of thematic areas in order to stimulate students and improve the educational process [23]. Prensky defined the following components of digital games: rules, goals, narrative-representation, conflict and competition, results and feedback, as well as interaction with the game or with other players [26]. These components should be included in an educational digital game in order for the game to be attractive.

Since the recent technological advances in mobile devices have led to the widespread use of smartphones and tablets with millions of users installing digital games on their devices, “mobile” games can provide opportunities for continuous, self-directed learning, anywhere, anytime, contributing in enhancing formal learning at school and informal learning in the student’s free time [27]. Through using applications and interactive multimedia environments in general, learners are offered with alternative and more tempting ways of learning, with greater cognitive transfer and a stronger learning motivation [23]. This educational use of mobile and wireless technologies and devices to create interactive learning experiences has led to the “Mobile Assisted Learning” [10, 13].

4 The AEPP Genius Application

4.1 Description

The design of applications for mobile learning cannot be achieved without first researching a set of parameters. So initially we conducted a research regarding existing applications that are in used for the support of D.A.P.E. The results have shown that there are no applications for the iOS operating system and the applications that exist for the android operating system do not meet the objectives of the course after the recent change of the Greek educational system [30, 31]. After that the AEPP Genius educational application was designed and developed in Greek based on the course's new curriculum.

The basic differentiation of this application, in relation to other pre-existing ones, is the fact that the writing of the questions was made from scratch in order to avoid content mistakes. The total number of questions is 600, including theoretical and practical examples, covering the entire range of the curriculum (Fig. 3). Since the application will be implemented on mobile devices, the amount of information to be provided should not be too large. Questions of different types, such as text input or code correction questions, would be hard to answer on a mobile device. The small screen combined with the large size of practical examples would make the application less useful. For these reasons, we decided to develop the application in the form of quiz questions with multiple responses, aiming at their direct understanding by students. AEPP Genius is addressed to Lyceum students and teachers and provides the learning of basic Information Technology (IT) concepts and of methodological skills acquisition.

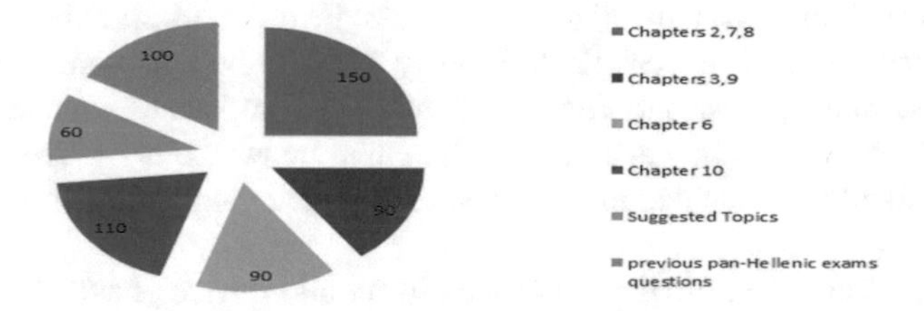

Fig. 3. Number of questions per module

The aim is to support the learning process without the application itself being an obstacle with complex user requirements or with a poorly developed interaction. For this reason, AEPP Genius has been designed so that each user can launch a new game on the desired module of the curriculum, making the minimum moves possible, and without customizing additional settings as they are all available at any time.

In addition, the use of pleasing colors, where each color corresponds to a module of the curriculum, in relation with the animation implementation in most user interactions improves the level of the user experience.

4.2 The Application

When initializing the AEPP Genius application, the list of available modules and content is displayed on the device's screen (Fig. 4).

Fig. 4. AEPP Genius application (in Greek)

The user can select the module of his interest and start a new game (Fig. 5). In addition, the user can change the number of questions he wants to play with (there is a minimum of four questions), the amount of time he wants to have in order to complete the answers and he can adjust the correct answer's appearance, where the user can witness in real time the result of his choice in each question of the game.

Before the new game starts, the applications must customize the user's options and take the following actions:

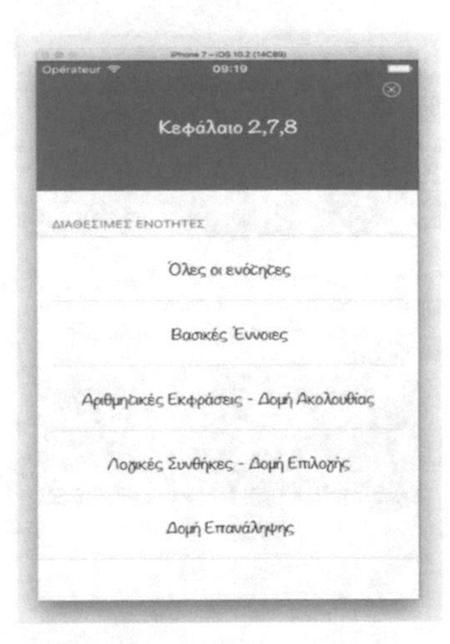

Fig. 5. Selected module (in Greek)

- Load the selected module's questions from the database according to the user's criteria.
- Initialize time and its counter.
- Prepare the user interface.

After completing the above the new game starts (Fig. 6).

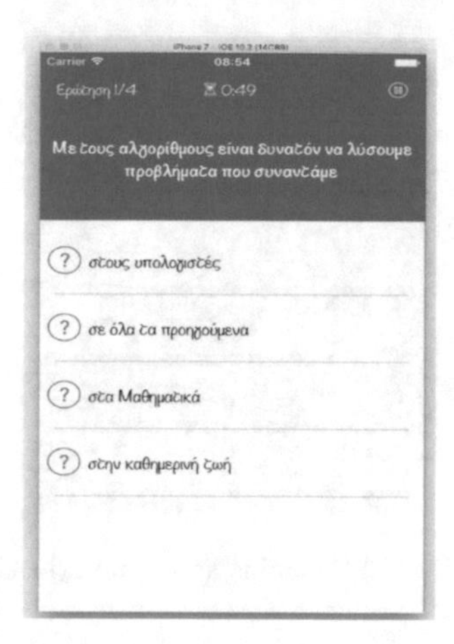

Fig. 6. New game – multiple choice question (in Greek)

It is noteworthy that in case the game is repeated, the answers appear each time in random order. While playing the game, the user can choose a response by tapping. If the answer is correct then it is marked in green and if it is wrong in red. If the user enables the display of the right answer in a real time selection in the game settings, he will simultaneously see the correct and the wrong answer, in case he gave the wrong answer to the question (Fig. 7).

Fig. 7. Display of the question's correct and wrong answer (in Greek)

A game can end in three ways, by answering all the questions, by the expiration of the given time in a question, or by pausing the game.

5 Evaluation

5.1 Methodology

In the evaluation research that was carried out in the context of this study, the questionnaire method [28] was used as a useful tool for the direct and detailed recording of the participant's opinions. The participants were thirty students of the third grade of public Greek schools, between the ages 16 to 18. The research was conducted in March 2017, while the anonymity of the participants was maintained. An important point is that the research took place at a time when students had completed the lesson's content, so it was possible to play with all the modules of the application.

The participants were asked to respond on the basis of a Likert numerical scale to state whether they are in favor or against a view and the degree of acceptance of that view. The total number of questions in the questionnaire was 10. The questions were separated in five main categories. The first had questions regarding the context of the application (questions 1 to 5), the second had questions regarding the usability and the easiness in the use of the application (questions 6 to 11), the third consisted of questions regarding the learning enhancement (questions 12 to 14) and the fourth (questions 15 to 17) focused on the uscrs interest. Finally, the fifth category regarded the application's technical specifications (questions 18 to 20). Some of the questions included are: Is it possible to select a specific learning target? The number of questions is limited? Do I feel confident about my acquired knowledge by playing with AEPP Genius?

The students used the application for a month and then they proceeded in answering the questionnaire.

5.2 Results

The results of the questionnaire indicated the following data: Students agree on the statement that the game's questions cover the lesson's curriculum by a percentage of 96.6, recognizing the existence of theoretical and practical examples (86.6%). In addition, they consider that the number of questions to be answered is satisfactory (90%). The students also responded that they agree or fully agree that it is easy to navigate through the game and that the interaction with the program is simple and easy (80%). They believe that simple and understandable language is used (73.3%) while agreeing to the statement that the user can re-answer the questions he failed (96.6%). In the statement that the application is complex, the majority of students disagrees completely or rather disagrees, reaching the percentage of 86.6.

Additionally, the students agree that immediate feedback is provided when playing with the game (90%), given that they see in real time if they answered correctly, and respond positively by 86.6% in the question "Do I feel confident about the acquired knowledge when playing with the game". An important data which needs to be mentioned is the fact that most of the participants state that they would use the application in the future in order to learn (93.3%).

In the question regarding the students' feelings almost all the students state that they feel excited while playing with the application (96.6%). Regarding the statement "The game is not attractive as graphics are limited," an important finding is the fact that most students declare disagreement (19 out of 30).

The last part of the questionnaire investigates the students' opinion about the technical specifications of the application. All the students say that the application did not stop or interrupted unexpectedly, while the majority of them agree with the statement that the response time for a new game is short (90%).

6 Conclusions and Future Steps

The development of the above discussed educational application, which was linked to the pedagogical approaches proposed by Diana Laurillard's "Conversational Model", can be introduced as a pedagogical tool that can be used creatively to support teaching. This application motivates and entertains users by providing incentives to achieve the desired goals. The students, as shown by the application's evaluation, feel excited by playing with the game and declare that they would often use it in the future in order to learn. AEPP Genius is easy to navigate without requiring additional assistance. The students can distinguish successfully from their failed actions in a plain and comprehensible language, while the number of theoretical and practical examples is considered adequate. The response time for creating a new game is satisfactory and the application has not displayed any errors while executing the game and while displaying the results.

Expanding the learner's cognitive level becomes more effective through the application by exploiting those features of digital games that are thought to be capable of motivating the learner, making the educational process more effective. It appears that this application can be a useful educational tool for the deeper understanding and learning for both educators and students.

In the future, we aim to extend the current research by addressing it to specialists, and more specifically to teachers who teach the D.A.P.E. lesson in Greek schools. Furthermore, a more extensive evaluation by students at a national level would result in additional evidence and significant conclusions. This evaluation will be held by distributing the application freely to be used in public schools in Greece. In addition, research could be made as to whether the learning of the students who have used the application is better supported than the learning of students who have not used the application. Also, there may be further extensions that can expand the application's specifications, such as its translation in other languages, extra graphics and multimedia, additional learning material, multiplayer experience, and more. Last but not least, the application can be also expanded in order to support more courses in various fields.

Acknowledgment. We thank the 30 students who offered their support through the use of the application and the completion of the questionnaire.

References

1. Park, Y.: A pedagogical framework for mobile learning: categorizing educational applications of mobile technologies into four types. Int. Rev. Res. Open Distance Learn. **2**(12), 1–25 (2011)
2. Garcia-Cabot, A., de-Marcos, L., Garcia-Lopez, E.: An empirical study on m-learning adaptation: learning performance and learning contexts. Comput. Educ. **82**, 450–459 (2015)
3. Eimler, S., Hoppe, H.U., Bollen, L.: SMS-based discussions-technology enhanced collaboration for a literature course. In: Proceedings of 2nd IEEE International Workshop Wireless and Mobile Technologies in Education, pp. 209–210 (2004)
4. Bryanm, A.: Going nomadic: mobile learning in higher education. Educause Rev. **39**, 29–35 (2004)
5. Bomsdorf, B.: Adaptation of Learning Spaces: Supporting Ubiquitous Learning in Higher Distance Education - Mobile Computing and Ambient Intelligence: The Challenge of Multimedia. Internationales Begegnungs, Dagstuhl (2005). (Ed. by, T. Kirste, H.S.N. Davies)
6. Douligeris, C., Kavounidis, T., Koilias, C., Pedros, A., Doukakis, S.: Introduction to the Principles of Computer Science, 1st edn. New Technologies Publishing Ed., Athens (2014). (in Greek, Institute of Technology of Computers and Publications "Diofantos")
7. Giannopoulos, H., Ioannidis, N., Kilias, X., Malamas, K., Manolopoulos, I., Politis, P., Vakali, A.: Developing Applications in a Programming Environment. New Technologies Publishing Ed., Athens (1999). (in Greek, Teaching Publications Organization, Schoolbook of General Lyceum - Pedagogical Institute)
8. Lonsdale, P., Baber, C., Sharples, M.: A context awareness architecture for facilitating mobile learning. In: Attewell, J., Savill-Smith, C. (eds.) Learning with Mobile Devices: Research and Development, pp. 79–85. Learning and Skills Development Agency, London (2004)
9. Baran, E.: A Review of Research on Mobile Learning in Teacher Education. Educ. Technol. Soc. **4**(17), 17–32 (2014). (Department of Educational Sciences, Faculty of Education, Middle East Technical University, Ankara, Turkey)
10. Roschelle, M., Sharples, J.: Guest editorial: special issue on mobile and ubiquitous technologies for learning. IEEE Trans. Learn. Technol. **3**(1), 4–5 (2010)
11. Levine, L., Smith, A., Stone, R., Johnson, S.: The 2010 Horizon Report - The New Media Consortium. Texas, Austin, USA (2010)
12. Advertising and Marketing Blog. Kid Tech According to Apple (2017). http://www.mdgadvertising.com/blog/kid-tech-according-to-apple-infographic/
13. Kukulska-Hulme, A.: Mobile usability in educational contexts: what have we learnt? Int. Rev. Res. Open Distrib. Learn. **8**(2), (2007). http://oro.open.ac.uk/8134/1/356–3034-1-PB.pdf
14. Garske, L.: Rogerian argument: maybe there is middle ground (2012). http://logang-c0300.weebly.com/rogerian-argument.html
15. Rikala, J.: Designing a Mobile Learning Framework for a Formal Educational Context, 1st edn. Department of Mathematical Information Technology: Publishing Unit, University Library of Jyvaskyla, Jyvaskyla (2015). (Ed. by, T. Mannikko)
16. Alrasheedi, L.F., Capretz, M.: Learner perceptions of a successful mobile learning platform: a systematic empirical. In: Proceedings of World Congress on Engineering and Computer Science (WCECS 2014), vol. 1, pp. 1–5 (2014)

17. Laurillard, D.: Pedagogical forms for mobile learning: framing research questions (Chap. 6). In: Pachler, N. (ed.) Mobile Learning: Towards a Research Agenda, pp. 153–175. London Knowledge Lab Institute of Education, London (2007)
18. Koole, M.L.: A model for framing mobile learning (Chap. 2). In: Mobile Learning: Transforming the Delivery of Education and Training, pp. 25–47. AU Press, Athabasca University, Athabasca (2009)
19. Komis, B.: Introduction to Computer Science, 1st edn. Kleidarithmos, Athens (2008). (in Greek)
20. Kanakis, I.N.: Teaching and Learning Using Modern Communication: From Facial Expression to Computers, Athens, Greece (1999). (in Greek)
21. Tzimogiannis, A.: The teaching of programming and algorithmic problem solving in the general Lyceum (2008). (in Greek). http://blogs.sch.gr/atsiozos/files/2008/06/ebookb4-programming.pdf
22. Brusilovsky, P.: Adaptive and intelligent technologies for web-based education. Special Issue on Intell. Syst. Teleteach. **4**, 19–25 (1999)
23. Hatzopoulos, A., Basdekidis, C.: Learning theories and education techniques. Applying an interactive multimedia environment for teaching, self-learning, programming and learning - assessment of pedagogical characteristics. In: 1st Educational Conference "Integration and Use of ICT in the Educational Process", Volos, Greece (2009). (in Greek)
24. Oikonomou, D., Papamagana, I., Zozas, I., Barbatsis, K.: Electronic games as educational tools. In: 2nd Panhellenic Educational Conference of Imathia, Veria, Greece (2010). (in Greek)
25. Karagiannidis, G., Vavoula, C.: Designing mobile learning experiences. In: Advances in Informatics, vol. 3746, pp. 534–544. Springer, Heidelberg (2005)
26. Prensky, M.: The motivation of gameplay or, the real 21st century learning revolution. Horizon **10**(1), 5–11 (2002)
27. Papastergiou, M., Zourbanos, N., Siakvaravas, I.: Mobile games in the teaching and learning of informatics (2012). (in Greek). http://www.etpe.gr/custom/pdf/etpe2387.pdf
28. Javeau, C.: Survey through questionnaire. In: The Handbook of Good Researcher, 3rd edn. TYPOTHITO/DARDANOS, Greece (2000). (in Greek)
29. Lu, J., Zhang, J.: Using mobile serious games for learning programming. In: Proceedings of 4th International Conference on Advanced Communications and Computation (INFOCOMP 2014), Paris, pp. 24–29 (2014)
30. Test4U – The official Test4U blog. AEPP Test4U (2015), http://blog.test4u.eu/tag/Aepp
31. LySTe – Lyceum Tests. LySTe Lyceum sos tests (2017). http://mstamos.gr/content/mstamos-quiz-app

WIP: Design, Development and Implementation of a "Web Technologies" Android Application for Higher Education

Eleni Seralidou(✉) and Christos Douligeris

Department of Informatics, University of Piraeus, Piraeus, Greece
{eseralid, cdoulig}@unipi.gr

Abstract. New technologies have been developing rapidly during the last few years affecting all aspects of everyday life. Mobile devices have been spread among higher education students altering the nature of communication and interaction between them. In this work in progress, we design and propose the implementation of an educational application for the android operating system. The application's context will be created according to the content of the course "Web Technologies", of the first year of studies, in the Department of Informatics of the University of Piraeus. After this application is completed, it will be evaluated and through this process we aim to find out whether such a mobile application can be intriguing for students and in addition whether it can enhance their learning experience.

Keywords: Education · Smartphones · Android applications

1 Introduction

The integration of "smart" devices in everyday life is paramount, leading to the appearance of numerous applications for educational and entertainment purposes. Moreover, the increased ubiquity of mobile computing devices in college campuses has the potential to create new options for higher education students, since mobile devices provide learners with opportunities to collaborate, discuss content with classmates and instructors, and create new meaning and understanding [1]. It is well-documented that mobile device applications can be used to support the students study, anytime and anywhere [3]. In parallel, the integration of mobile devices in education also dovetails with the broad goals of STEM (science, technology, engineering, mathematics) education and the more recent STEAM education, which includes the visual and performing arts [2]. The proper design of the technologies leads to a more effective m-Learning (mobile learning) and to the transformation of knowledge in the modern world [6]. Nowadays, smartphones may be used for texting, web surfing, emailing, downloading and listening to music, playing games and engaging in social networking. It is obvious that if students have clear and concise directions on how to use smartphones for educational purposes, this technology can reinforce the learning objectives and it can work as an instructional aide inside and outside of the classroom [7].

M. E. Auer and T. Tsiatsos (Eds.): IMCL 2017, AISC 725, pp. 127–134, 2018.
https://doi.org/10.1007/978-3-319-75175-7_14

Even though mobile technologies have seen a high penetration in all aspects of people's lives, their usage as an educational platform has been very slow. The development of m-Learning in higher education, in particular, is still in very early stages [4]. There are still several barriers to the adoption of an m-Learning platform, especially by institutions of higher learning [9]. So far, many universities provide free applications, but their content is primarily non-instructional [5].

In this paper, we design and propose the implementation of an educational application for android mobile platforms in higher education, under the name "WebTech app". The aim is to introduce university students in web technologies through a more interactive learning experience which exploits the benefits of mobile technologies and comes to full terms with the course "Web Technologies" of the first year of studies in the Department of Informatics of the University of Piraeus. By this way, we focus in the instructional way of use of an application and of smartphone devices within and also outside of a course, which is an area that needs further exploration.

In order to achieve the above, initially we researched for existing applications for the android operating system regarding the subject "web technologies". This step was essential in order to make decisions for the design and development of the mentioned application. Also, the final decisions were influenced by the selection of a proper pedagogical model for supporting and framing educational applications. So, after fully developing the application, the next step will include its evaluation by the students. Through the evaluation process we expect to find out whether mobile applications for educational purposes can be intriguing and interesting for students and whether they can actually enhance learning. This research also attempts to promote a different approach in the nature of teaching in universities.

2 Theoretical Foundations

2.1 Mobile Devices in Higher Education

Mobile devices, and especially smartphones, have become so widespread affecting the field of education as learning by using mobile devices is considered to be particularly effective [8]. By being wireless and portable they enable users to communicate while on the move. The popularity of these devices is therefore a consequence of their ability to function at multiple levels [6]. While m-Learning has the potential to support all forms of education, higher education is a particularly appropriate venue for the integration of student-centered m-Learning because mobile devices have become ubiquitous in college campuses [5]. So, mobile apps are very popular in higher education [14], while serious games are generally considered to induce positive effects in the areas of learning motivation and learning gains [13].

But the creation of new didactic sequences and educational activities that can be used to connect formal and informal learning settings into a congruent whole is required. As a result one major challenge for mobile research in higher education is to combine the teaching of university courses with the use of mobile devices [10].

2.2 The m-Learning Framework

By definition, m-Learning is learning through wireless devices that can be used wherever the learner's device can receive unbroken transmission signals [12]. The idea of m-Learning, a relatively new concept, became interesting by the way it blends the notion of mobility into the already popular electronic learning context [9]. The theory which seems to be more suitable for m-Learning is the Activity theory [15]. Its key point is the concept of activity which consists of a subject and an object mediated by a tool [16]. The activity theory offers an appropriate framework for designing and implementing software applications for mobile learning with emphasis to its context.

A framework that is also relevant, and is designed to describe the minimal requirements for supporting learning in formal education, is the "conversational framework" by Laurillard [19]. According to this framework, through questions the theory of the learning objective is described and the learner forms an initial perception. If the learner answers incorrectly new knowledge is produced through the repetition of the process. So, feedback is given until the concept is fully understood and the learner during the entire process revises his knowledge by answering the questions of an activity. This leads to a continuous interaction between the student and the educational application.

Additionally, learning must be continuous and it must come as a natural activity. Informal learning provides this type of learning, having a self-motivated learner "under the radar" of a tutor, individually or in a group, intentionally or tacitly, in response to an immediate or recent situation, perceived need, or serendipitously with the learner mostly being (meta-cognitively) unaware of what is being learnt [17]. What is interesting about this type of learning is that it is possible to continue happening inside and outside formal education settings. The ubiquity of mobile devices supports this type of learning, as well.

So interest rises in the potential of mobile learning to bridge pedagogically designed learning contexts, and to facilitate learner generated contexts, and content (both personal and collaborative), while providing personalization and ubiquitous social connectedness, that sets it apart from more traditional learning environments [8]. So why remain traditionalists where we can be revolutionary? Traditional teaching is the past and m-Learning is the future.

In this context, the differences in the learning approach, through our suggested application, and the innovations that derive from the implementation of m-Learning in the teaching process are:

- Using mobile phones in formal and informal educational settings, promoting continuous learning.
- Exploiting devices that the students already know how to use, by being their owners.
- Approaching a course's curriculum in a different and more playful character.

3 Suggested Application

3.1 Web Technologies Course Curriculum

The course "Web Technologies" is being taught in the first year of studies at the Department of Informatics of the University of Piraeus. The course's curriculum focuses on introductory topics of Internet Technologies such as: basic principles and Internet functions, Internet services, the World Wide Web and the Client - Server model, programming on the Internet from the client and from the server side, design and implementation of web applications, interactive websites using scripting languages and advanced application development [18]. It lasts one semester and includes four hours of theory and a two hour laboratory work per week. The laboratories focus on using HTML, CSS, JavaScript and PHP technologies for webpage development, and support the practical application of the theory taught.

The application we present in this work so far covers the four above mentioned technologies; in the form of four different multiple questions quiz games that focus on each module of the workshops parts of the curriculum.

3.2 The WebTech Application

The development of an educational application can provide a strong and creative pedagogical tool. In this study, an educational application is designed as a supporting tool for better understanding and teaching of the course "Web Technologies", under the name "WebTech App". The final software is addressed to university students, in their first year of studies, and aims at providing a useful supporting tool for learning and understanding the course's context.

Given that there are no pre-existing applications in the University of Piraeus for the support of the "Web Technologies" lesson, we decided to create one that will fully meet the objectives of the course. Before proceeding to the application's design, we performed a research on the existing applications, under the subject of "web technologies". The findings offered significant help in the design of our application [20, 21]. The initial development of the application comes in the Greek language and for the android operating system. The application is in the form of a questions quiz which supports four deferent categories, following the workshops curriculum of the lesson "Web Technologies", so far. The context of the application is not fully developed yet. We are at the process of writing and it is going to consist of original questions, which will be written from scratch in order to avoid mistakes and make sure that they follow the course's context.

So far, the basic functions of the application have been developed. So, when initializing the "WebTech" app a menu is available to the user, in order to select the category of the curriculum he/she desires (Fig. 1).

Fig. 1. WebTech game menu (in Greek)

After selecting a specific category, the application loads the selected module's questions from the database, prepares the user interface and the questions game starts (Fig. 2). The idea is for the user to be able to answer as many questions possible during a specific set of time.

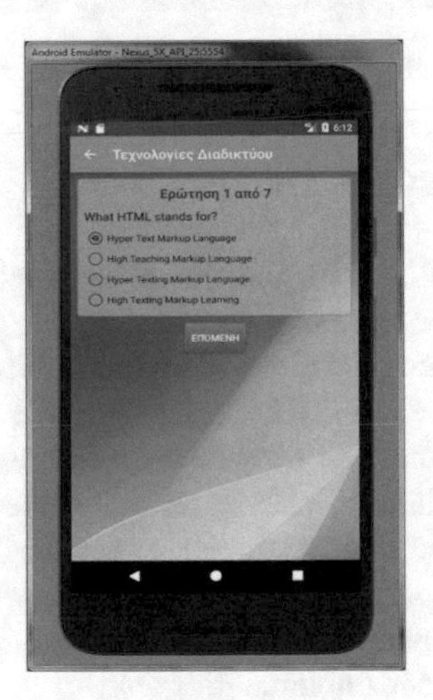

Fig. 2. WebTech game - multiple choice question

After completing the set of questions, the user can witness the results of his efforts by receiving a total rank (Fig. 3).

The user also has the ability to see, for each question, his answer in red letters and the correct one in green letters (Fig. 4).

From the results screen the user has the ability to start a new game, selecting a different module from the initial screen or proceed in the game's termination.

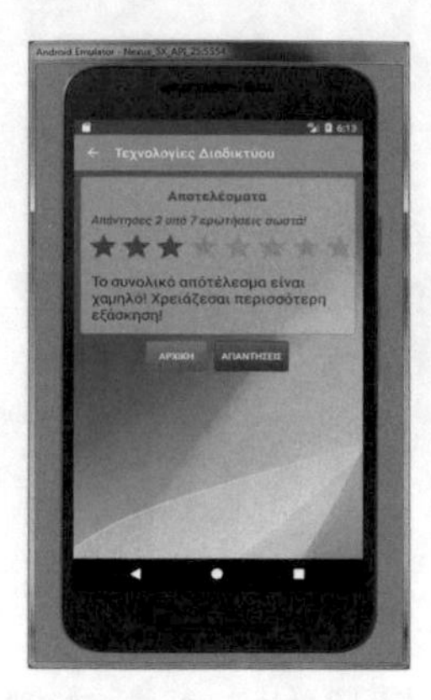

Fig. 3. WebTech results screen (in Greek)

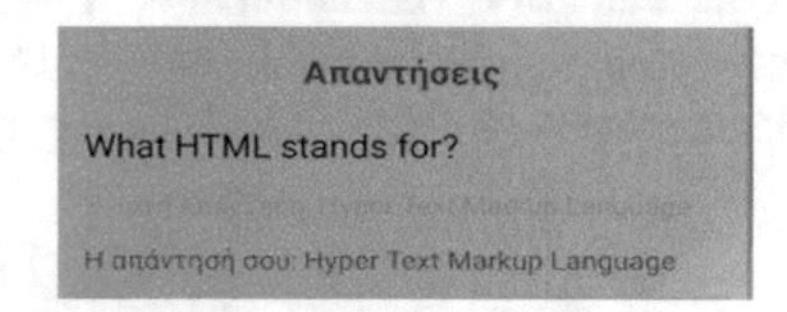

Fig. 4. WebTech single question result

4 Conclusions and Future Steps

Nowadays students are familiar with the use of smartphones and own, at least, one mobile device. While mobile technologies are not yet widely and routinely used in education, they have the potential to be used in a multitude of pedagogical and other contexts in higher education [11]. By implementing the adequate mobile device software, learning evolves in something new through a more flexible and easy process.

Although our work is still at an early stage we argue that it is really important for m-Learning implementation in higher education, because it can be used as a beneficial way to support teaching and learning. At the time of the writing we continue to work on fully developing the mobile application technically and add more specifications. Some of the features that we consider implementing are:

- Initialization of timer for each module.
- Randomly selected questions and answers.
- Multilanguage support.

Some other features that can be implemented in the future are multiplayer experience, extra graphics, multimedia etc. Also, an expansion of the application in order to support different lessons and curriculums is something that we will take under consideration.

Of course all the above won't be important if the context is not properly adjusted and implemented. So, the originality and difficulty of this application lays in the proper delivery of the lesson's content in order for it to work as a supporting tool for enhancing the learning process. As this application becomes available it will be tested and evaluated by students and teachers in higher education in Greece. The upcoming benefits, advantages or disadvantages of the suggested application are of great scientific interest.

Acknowledgment. This work has been partly supported by the European Project Mitigate and the University of Piraeus Research Center.

References

1. Gikas, J., Grant, M.M.: Mobile computing devices in higher education: student perspectives on learning with cellphones, smartphones & social media. Internet High. Educ. **19**, 18–26 (2013)
2. Ostler, E.: 21st century STEM education: a tactical model for long-range success. Int. J. Appl. Sci. Technol. **2**(1), 28–33 (2012). (University of N. at O.) in M.M. Grant "Using Mobile Devices to Support Formal, Informal and Semi-formal Learning". In: Ge, X., et al. (ed.) Emerging Technologies for STEAM Education, Educational Communications and Technology: Issues and Innovations, pp. 157–177. Springer International Publishing, Switzerland (2012)
3. Young, J.R.: Smartphones on campus: the search for "killer" apps. The chronicle of higher education, B6-B8 (2011). http://www.chronicle.com/article/Smartphones-on-Campus-the/127397
4. Park, Y.: A pedagogical framework for mobile learning: categorizing educational applications of mobile technologies into four types. Int. Rev. Res. Open Distance Learn. **12**(2), 78–102 (2011)
5. Cheon, J., Sangno, L., Crooks, S.M., Song, J.: An investigation of mobile learning readiness in higher education based on the theory of planned behaviour. Comput. Educ. **59**, 1054–1064 (2012)
6. Osman, M., El-Hussein, M., Cronje, J.C.: Defining mobile learning in the higher education landscape. Educ. Technol. Soc. **13**(3), 12–21 (2010)
7. Buck, J.L., McInnis, E., Randolph, C.: The new frontier of education: the impact of smartphone technology in the classroom. In: 2013 ASEE Southeast Section Conference on American Society for Engineering Education (2013). http://se.asee.org/proceedings/ASEE2013/Papers2013/177.PDF
8. Cochrane, T., Baterman, R.: Smartphones give you wings: pedagogical affordances of mobile Web 2.0. Aust. J. Educ. Technol. **26**(1), 1–14 (2010)
9. Alrasheedi, M., Capretz, L.F., Raza, A.: A systematic review of the critical factors for success of mobile learning in higher education (university students perspective), electrical and computer engineering publications. paper 67 (2015). http://ir.lib.uwo.ca/electricalpub/67
10. Vazquez-Cano, E.: Mobile distance learning with smartphones and apps in higher education. Educ. Sci. Theor. Pract. **14**(4), 1505–1520 (2014)
11. Herrington, A.: Using a smartphone to create digital teaching episodes as resources in adult education. In: Herrington, J., Herrington, A., Mantei, J., Olney, I., Ferry, B. (eds.) New Technologies, New Pedagogies: Mobile Learning in Higher Education, 138p. Faculty of Education, University of Wollongong, Wollongong (2009)
12. Yu, F.A.: Mobile/smartphone use in higher education. In: Proceedings of Southwest Decision Sciences Institute (SWDSI), pp. 831–839 (2012)

13. Iten, N., Petko, D.: Learning with serious games: is fun playing the game a predictor of learning success? Br. J. Educ. Technol. **47**(1), 151–163 (2016)
14. NMC Horizon report. Higher education edition (2014). http://cdn.nmc.org/media/2014-nmc-horizon-report-he-EN-SC.pdf
15. Impedovo, M.A.: Mobile learning and activity theory. J. e-learning Knowl. Soc. **7**(2), 103–109 (2011). English edition
16. Uden, L.: Activity theory for designing mobile learning. Int. J. Mob. Learn. Organ. **1**(1), 81–102 (2007)
17. Cook, J., Pachler, N.: Bridging the gap? Mobile phones at the interface between informal and formal learning. J. Res. Cent. Educ. Technol. **4**(1), 3–18 (2008)
18. Douligeris, C., Mavropodi, R., Kopanaki, E.: Web technologies. In: Operations Principles and Programming of Applications, 2nd edn. New Technologies (2004). (in Greek)
19. Laurillard, D.: Pedagogical forms for mobile learning: framing research questions. In: Pachler, N. (ed.) Mobile Learning: Towards a Research Agenda, pp. 153–175. London Knowledge Lab Institute of Education, London (2007). Chapter 6
20. Google Play Applications. Web Development (HTML, CSS, JS) (2017). https://play.google.com/store/apps/details?id=everyneedz.com.webdevelopment
21. Google Play Applications. SoloLearn: Learn to code for free (2017). https://play.google.com/store/apps/details?id=com.sololearn

Design of Communications System for Idea Creation in Team Activity

Akiyuki Minamide(✉), Kazuya Tekemata, and Arihiro Kodaka

Kanazawa Technical College, Kanazawa, Japan
{minamide, takemata, kodaka}@neptune.kanazawa-it.ac.jp

Abstract. In engineering education, implementation of Project-Based Learning (PBL) is increasing as problem solving class. The purpose of this study is to inform the team members in a timely manner the findings of the team activities and suggested ideas and to develop a communication system that can efficiently and effectively implement PBL. As a result of conducting PBL using the development system, student activities were more aggressive and communication between students was smooth. It became clear that the development communication system is more likely to be used for PBL and its preliminary lesson.

Keywords: Communications system · PBL · Team activity

1 Introduction

In engineering education, implementation of Project-Based Learning (PBL) [1] is increasing as problem solving class. The PBL finds various problems of the community by the students themselves and derives solutions. [2] Therefore, the student goes to the site by spending free time to do various investigations, and puts out the idea. After that, members of the team gather together and report the findings of each investigation and summarize ideas. Due to the schedule, discussions between members often have days since the survey, and in many cases the discussion will not be activated. If we can communicate the findings and ideas of each person to the members of the team in a timely manner using the mobile communication tool, the discussion will be more active and effective. The purpose of this study is to inform the team members in a timely manner the findings of the team activities and suggested ideas and to develop a communication system that can efficiently and effectively implement PBL.

In this paper, the prototype system and the outline of the PBL implemented using it are described.

2 Development of Communications System for PBL

First of all, the communications system that became the center of this education was designed, and the system was made for trial purposes with the following design concepts:

M. E. Auer and T. Tsiatsos (Eds.): IMCL 2017, AISC 725, pp. 135–141, 2018.
https://doi.org/10.1007/978-3-319-75175-7_15

1. It is possible to compete for the idea by dividing into the team so that the student may improve the learning effect.
2. The number of teams shall be two teams so that the judgment result of the competition is easy to understand.
3. Generally, the tag used well for the idea putting out in PBL can be done on this system.
4. The investigation data and the idea can be up-loaded from the investigation site.
5. Because it is the best that the specialist judges the competition, it is possible to judge it from the remote place where the specialist exists. The system was made for trial purposes with these design concepts.

Figure 1 shows a construction of the Communication System for ideas creation in a team activity. The system is composed of two or more smart phones for data input, a communications table that built in two touch panel displays, two personal computers and a router.

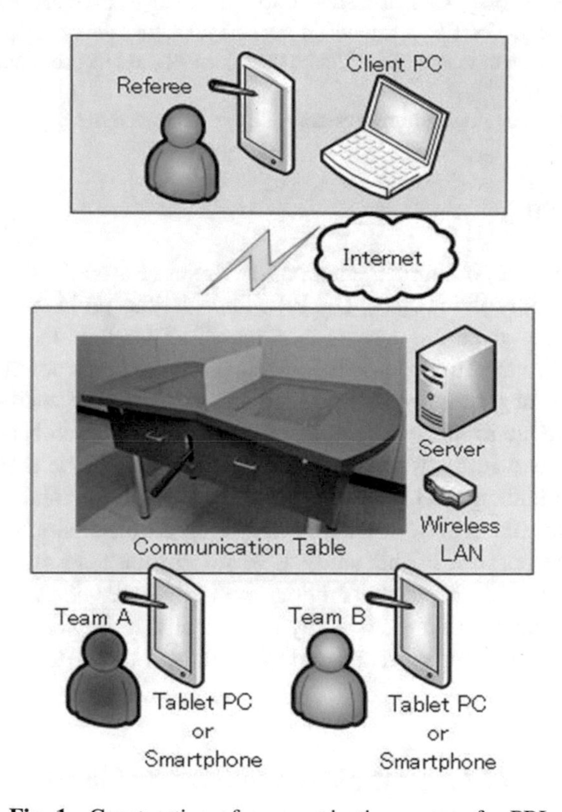

Fig. 1. Construction of communication system for PBL

Figure 2 is in the communications table. Two personal computers were built in inside the communications table, and the server that brought them together was set up outside of the table. Ideas from each team are created by students' tablet PCs, smartphones, etc., and the ideas are sequentially displayed on the display of their team using wireless LAN. As long as it is an input device connected to the Internet, data can be sent anytime from anywhere, so it can be sent even at the site investigation. The displayed idea becomes a display like a tag on the touch panel display, and it is sequentially displayed after the data transmission. These idea tags can be grouped, rearranged, enlarged or reduced using the touch panel.

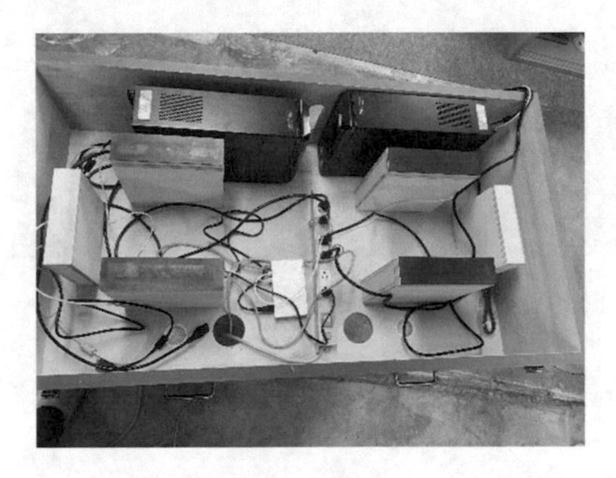

Fig. 2. Communications table inside

3 Execution of PBL by Communications System

3.1 Implementation of a Competition

In order to verify the prototype system, two teams conducted a competition with the theme “effectively communicating disaster countermeasures to students”. The students who participated in the competition were six of the 4th and 5th graders of Kanazawa Technical College (KTC) and the referee was the safety chairman of KTC. Prior to starting the competition, the referee explained the contents of the PBL theme and how to use the new communication system. Prior to starting the competition, the contents of the PBL theme and how to use the new communication system were explained from the referee, and the students deeply understood the contents of the theme.

Figure 3 shows the students suggesting ideas from smartphones. Many ideas were issued from students, and they seemed to be sufficiently used as a substitute for brainstorming using conventional tags. Because students are using digital equipment on a daily basis, it seems that they were able to respond immediately to the new system. After all the ideas came out, the students gathered around the communication table and grouped ideas and proposed new ideas. The day to announce the final idea was set after two weeks, allowing each team to freely make ideas and prototypes.

Fig. 3. Idea generation from smartphone to communication system.

3.2 Proposed Final Idea

Idea proposed by Team A

"Disaster Countermeasure Smartphone Application Proposal".

The problem of the Web page of the school safety committee shown in Fig. 4 and its remediation were examined.

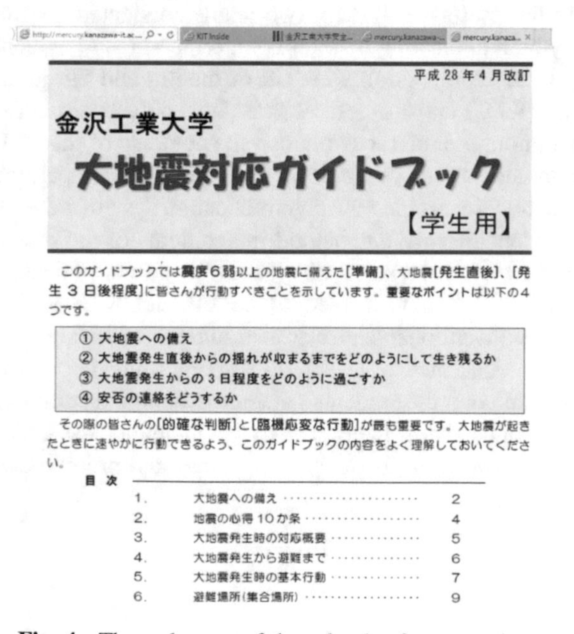

平成 28 年 4 月改訂

金沢工業大学

大地震対応ガイドブック

【学生用】

このガイドブックでは震度６弱以上の地震に備えた[準備]、大地震[発生直後]、[発生 3 日後程度]に皆さんが行動すべきことを示しています。重要なポイントは以下の４つです。

① 大地震への備え
② 大地震発生直後からの揺れが収まるまでをどのようにして生き残るか
③ 大地震発生からの 3 日程度をどのように過ごすか
④ 安否の連絡をどうするか

その際の皆さんの[的確な判断]と[臨機応変な行動]が最も重要です。大地震が起きたときに速やかに行動できるよう、このガイドブックの内容をよく理解しておいてください。

目 次

Fig. 4. The web page of the school safety committee.

The following two problems were pointed out by the students: since this web page is a PDF file, it is difficult for students to see what they want to see instantaneously, and if there is no internet environment it is not possible to see the file. As a countermeasure, it was suggested to create applications for smartphones that students carry around on a daily basis. Since the application has a search function, there is an advantage that the information which he/she wants to search can be accessed immediately. Furthermore, since it is a smartphone application, if installed on a smartphone it will be usable even in an environment without Wi-Fi.

Idea proposed by Team B

"Mobile Disaster Countermeasure Manual Proposal".

Team B started with examining the problem of the school safety committee's Web page just like Team A. Team B gave the following problems:

1. It can not be used without a Wi-Fi environment.
2. Handling of PDF files is troublesome.
3. When printing a PDF file, characters can not be read unless it is B5 size or larger, and it is difficult to carry around.
4. In disasters it is difficult to secure power supply and it is possible that digital equipment such as smartphone can not be used.
5. In disasters, belongings often get wet and it is difficult to use smartphones.

The above No. 1 to No. 3 are the same as Team A, but No. 4 and No. 5 were new findings. Based on the above study results, a portable disaster countermeasure manual that can be carried from Team B was proposed.

Figure 5 is a prototype manual. The contents of the manual are narrowed down to the minimum information necessary for disasters and are explained very concisely. Because it becomes business card size when you fold it so you can carry it, you can put it in a wallet etc. (Fig. 6). Furthermore, it was suggested to use Stone Paper [3] which is excellent in water resistance and durability without using general paper so that it will be fine even if wet.

The front side The back side

Fig. 5. Prototype manual

Fig. 6. Folded prototype manual

3.3 Comparison of Ideas

The referee judged the ideas of both teams and made it a victory for Team B. The reason is that although there was little difference in content between both ideas, it depends on the difference between the following two points.

1. Consideration of power loss situation at the time of disaster is taken into consideration.
2. Conditions with high possibility of occurring at the time of disaster are taken into consideration, and it is proposed to use stone paper which is okay if it gets wet.

4 Summary

In general, the PBL that is done is finally ended with a simple presentation. Because various themes are announced at the same time, it is difficult to make each team superior or inferior. In that respect, competition by two teams is easy to attach idea superiority, and it is clear that students' understanding and satisfaction is increased by receiving expert's explanation immediately. In addition, students seemed to be working more seriously with the aim of victory because they put merit on ideas. Based on the above results, it was clear that competition by the two teams was very meaningful, and that expert judges were effective as well. For the prototype communication system, the following was clarified.

1. Because students are good at using digital devices, ideas can be laid out smoothly using smartphones.
2. Since it is possible to upload ideas from a remote place and save ideas as digital data, it is easy to store and read data.
3. Experts who are in remote areas can also be refereed.

As a result, it became clear that the prototype communication system is likely to be used for PBL and its preliminary lesson.

Acknowledgment. This work was partially supported by a Grant-in-Aid for Scientific Research from the Ministry of Education, Culture, Sports, Science, and Technology.

References

1. Greeno, J.G.: Learning in activity. In: Sawyer, R.K. (ed.) The Cambridge Handbook of the Learning, pp. 79–90. Cambridge University Press, New York (2006)
2. Lee, H.-J., Lim, C.: Peer evaluation in blended team project-based learning: what do students find important? Educ. Technol. Soc. **15**, 214–224 (2012)
3. http://www.stone-paper.nl/what-is-stone-paper

Employing Theatrical Interactions and Audience Engagement to Enable Creative Learning Experiences in Formal and Informal Learning

Enriching Social and Community Theatre Practices with Digital Technologies

Nektarios Moumoutzis(✉), Nektarios Gioldasis, George Anestis, Marios Christoulakis, George Stylianakis, and Stavros Christodoulakis

Laboratory of Distributed Multimedia Information Systems and Applications, School of Electrical and Computer Engineering, Technical University of Crete, Chania, Greece
{nektar,nektarios,ganest,marios, gstylianakis,stavros}@ced.tuc.gr

Abstract. Social and Community Theatre (SCT) is a methodology promoting active participation in cultural events with the aim to enable cultural awareness and transformative learning. This paper presents how SCT is enriched with the use of digital technologies to promote active participation of audiences before, during and after cultural interventions that are organized within the context of the Caravan Next project (http://caravanext.eu). We present a mobile app (cNext App - http://cnext.tuc.gr/app) that enables audiences to get informed about cultural interventions, participate, share and reflect on their experiences when attending such interventions, thus promoting transformative learning in an informal inclusive learning framework. Furthermore, we present how special software employing innovative input devices enables the use of digital marionettes in schools thus promoting drama-based digital storytelling learning scenarios.

Keywords: Social community theatre · Audience engagement
Digital storytelling · Digital marionettes

1 Introduction

Extensive work from scholars and practitioners searching for a new "ritualization" of society has documented the fundamental role and heritage of theatre in political action, democratic "negotiation" of the social norms and rules, cultural dialogue and even social therapy. Social and Community Theatre (SCT) [1–3] builds on this work to create an inclusive environment for active participation of the society in cultural events with the explicit aim to promote cultural awareness and transformative learning. In this

M. E. Auer and T. Tsiatsos (Eds.): IMCL 2017, AISC 725, pp. 142–154, 2018.
https://doi.org/10.1007/978-3-319-75175-7_16

paper we describe in detail how SCT can be enriched with the use of digital technologies to promote active participation of audiences before, during and after cultural interventions that are organized within the context of the Caravan Next project (http://caravanext.eu).

The integration of SCT practices with digital technologies presents a challenge that could be framed within a wider research agenda on social sciences and humanities that follows a school of thought claiming that human behavior can be understood and analyzed by assuming that all human practices are *performed* so that actions can be seen as a public presentation of *self*. This is the conceptual basis of the methodological breakthrough titled *the performative turn* [4] that entered in cultural studies, social sciences and humanities in late 20th century and influenced disciplines like ethnology, anthropology, sociology, etc. The term *turn* indicates an alternative way to look at how members of groups and society at large interact and share knowledge within the context of groups and societies [5]. The major claim is that people create and recreate meaning and knowledge in social settings through performance. And even more: *The social reality itself is created through the actions of its members*. Thus, the focus is redirected to "the active social construction of reality rather than its representation" [6].

The roots of this approach go back to the 1940s or even earlier and can be attributed to the need to move beyond the prevailing focus on texts or symbolic representations to capture meaning. Performance is, above all, a meaning making bodily practice. Consequently, it is related to the theatre, rituals and other forms of spectacles and social practices. More than that, performance can be related to lifeless mediating objects, such as architectural objects or, in modern days, the digital systems that constitute our hyper-connected societies [7].

Beyond the main premises and the theoretical justification of the validity of performativity, one could attribute the significance of this paradigm to an *inherent* dramatic quality of human experience. Furthermore, it seems to be closely related to the capacity of digital technologies to extend our agencies, thus providing new ground for dramatic interaction (i.e. meaningful bodily and symbolic actions). By attributing to our daily lives a performative quality, the close relationship between drama as an art and drama as a social process is evident. The work reported in this paper explores this relationship, underlines the importance of learning in the interplay between dramatic arts and social reality and points out the similarities with modern learning theories.

The structure of the rest of this paper is the following: Sect. 2 presents the interplay between social drama and stage drama as a unified learning process. Section 3 presents how this unified learning process is supported with the use of digital technologies in the Caravan Next project. Section 3 presents the domain model of the cNext App and its framework for location-based services. Section 4 elaborates on issues related to the active participation of schools within the overall framework of the project through the use of special digital storytelling software that enables students create their own stories inspired by traditional storytelling techniques with puppets. Section 5 concludes and presents plans for future work.

2 Social Drama and Stage Drama as a Unified Learning Process

Beeman [8] offers a very interesting comparison and in-depth analysis of the relation between theatre and other performative genres: Revolutions, public demonstrations, campaigns, strikes, and other forms of participatory public action all have performative dimensions sharing certain features with the fundamental ritual processes. Such *social dramas* involve a break with normal structures of ongoing life, the entrance of groups of individuals into liminal transitory states, and the reincorporation of the liminalized individuals into a reconstituted social order. Beeman [8] goes on to analyze the interrelationship of stage drama, as a generalization of theatre, and social drama, as an inclusive term to describe all performative genres that aim at changing actual reality, employing a scheme initially proposed by Turner [9]. This scheme is depicted below:

Above the horizontal line Fig. 1 represents what is actual, visible and public while below the horizontal line what is hidden and virtual, i.e. implicit and internal. On the left of the vertical line *social drama* is represented, i.e. all performative genres related to social life while on the right any genre of cultural performance (*aesthetic or stage drama*). The arrows represent a circular process with a continuous feedback loop with four directions:

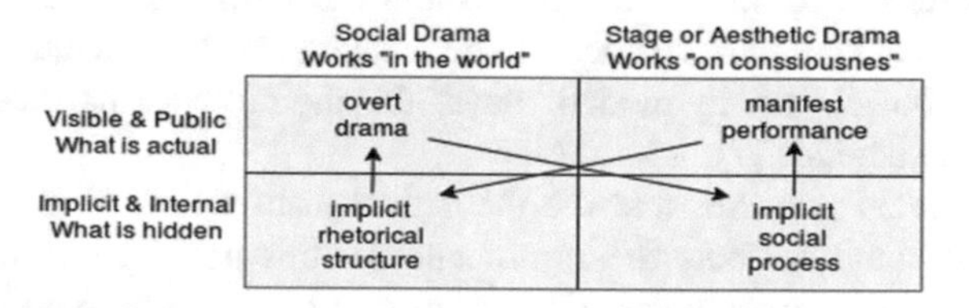

Fig. 1. The interrelationship between social drama and stage or aesthetic drama. Concepts depicted following the ideas of Turner [9].

1. Manifest social drama (i.e. visible social and political action) feeds into the hidden space of aesthetic drama influencing both form and content of the latter.
2. The latent space of stage drama feeds into manifest performance. This way, stage drama operates as an active or "magic" mirror meant to do more than entertain being a metacommentary on the major social dramas within the wider sociocultural context such as wars, revolutions, scandals, institutional changes etc.
3. Stage performance, within its own turn, feeds into the latent realm of social drama with its message and its rhetoric and partly account for its ritualization.
4. Finally, life itself stands as a mirror of art, of the stage drama, and the living perform their lives in a way that the protagonists of life are equipped with salient opinions, imageries and ideological perspectives created in stage drama.

The above feedback loop continues not as a cycle but rather as a helix: At each exchange new elements are added and other elements are left behind (forgotten or discarded). Beeman [8] underlines that human beings learn through experience, though all too often they repress painful experience. The deepest experience, he argues, is through drama; not through social drama, or stage drama (or its equivalent) alone, but in the circulatory or oscillatory process of their mutual and incessant modification. Philosophers feed their work into the spiraling process; poets feed poems into it; politicians feed their acts into it; and so on. Thus the result is not an endless cyclical repetitive pattern or a stable cosmology. The cosmology, he underlines, has always been destabilized, and society has always had to make efforts, through both social dramas and aesthetic dramas, to restabilize and actually produce cosmos.

It is interesting to see how this conception of reflective social process through which society looks at itself, learns from its experiences and continuously reconstructs or reinvents itself, resembles one of the most widely used models of learning: the learning cycle introduced by Kolb and Fry [10] and further elaborated by Honey and Mumford [11]. This model distinguishes four phases in the learning process of an individual that proceed iteratively as depicted below (Fig. 2):

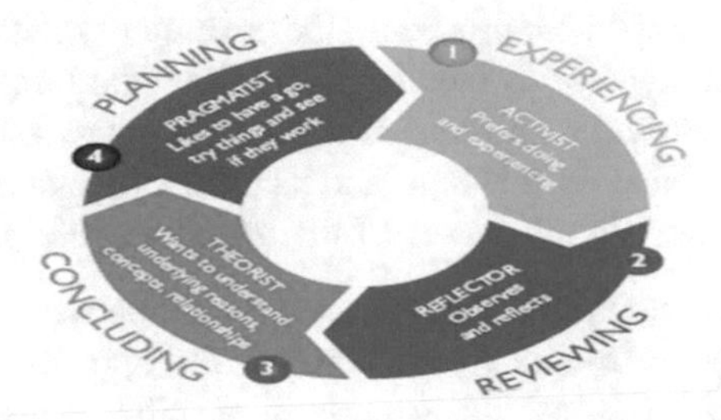

Fig. 2. The four phases of learning according Honey and Mumford [11]. Diagram available online at: https://www.talentlens.co.uk/develop/peter-honey-learning-style-series

In detail, the four learning phases (with their drama counterparts) proceed as follows:

1. The process starts from *experiencing reality*, an activity that is preferable by *activist* learners that try to actually do things and have concrete experiences. This is analogous to overt social drama discussed already.
2. The next learning phase is *reviewing and reflecting on the concrete experience*, the preferred mode of learning for *reflectors* that observe (their own or other peoples') actions. This is analogous to the latent realm of stage drama where social experiences are elaborated and give rise to art manifestations.
3. The third phase is *concluding from the experience* providing the means that will subsequently orient the individual in life. This is the preferred mode for *theorists*, i.e. people that build explanatory frameworks trying to find casual relationships and links to previous established norms and concepts in a way that resembles what is happening during the preparation and staging of drama manifestations.
4. Finally, the last phase is to *plan the next step* that will feed a new iteration. This is the preferred learning mode for *pragmatists* that try to exploit the knowledge accumulated in order to act in real life in an informed and purposeful manner. This is related to the latent realm of social drama where the art-refined social experience gets back into the social stage to enrich it with new concepts, plans and intentions.

3 Audience Engagement Using Mobile Technologies

The unified learning process based on the close interplay between social drama and stage drama is the conceptual baseline of the work done within the context of the Caravan Next project to enrich SCT with digital technologies for the creation of community-shared knowledge through audience active participation and energetic engagement. Audience' active participation in the reformation and the enrichment of the existing information is ideally achieved via a community-based organization and offers significant added value. This is due to the fact that each person acting as a potential participant and proportionally to his interests desires to participate and/or be informed about cultural events through other people that share the same interests with him. In particular he/she is interested to listen to their opinions and suggestions, take them into consideration, leave his comments and remarks and if possible participate in the creation of the knowledge and sharing experiences before, during and after an event. During this highly sociable procedure the creation of relationships among people is promoted, the communication is facilitated and the static information provided from official organizations (termed *cold data*) is augmented by *warm data* provided by the community itself. The cNext mobile app (and its back-end data & administration tools) supports SCT stakeholders in all phases of the learning cycle presented in Sect. 2:

1. *Experiencing:* cNext app promotes active audience participation in cultural events either by following instructions sent, through the app, from event organizers or specialized personnel (i.e. asking to perform specific task or provide some feed-back) or to play games employing location-based services provided by the app.
2. *Reviewing and reflecting:* Next, the app allows event participants to share their memories, exchange ideas, discuss and elaborate on them, thus reviewing what they have experienced during social cultural events.
3. *Concluding:* The warm data collected by the app are valuable to professionals (organizers, actors, scientists, theorists, etc.) of SCT in studying, analyzing and extracting conclusions about social interactions or event in evaluating the impact and the reactions caused by specific theatrical actions.
4. *Planning:* Finally, the cNext app supports the planning phase, not only for SCT professionals, (e.g. cultural event organizers) to plan, announce, and promote cultural activities, but also the community itself by informing it about upcoming or happening events in a location aware manner thus allowing individuals to plan their attendance or even to participate in their preparation.

To effectively support all the aforementioned phases an important initial step is the domain modeling as a way to describe and model real world entities and their relation-ships. From a system-level requirements point of view, identifying domain entities and their relationships provides an effective basis for understanding and helps practitioners design systems for maintainability, testability, and incremental development. In what follows we present the main aspects of the domain model developed in the Caravan Next project as the core data infrastructure that supports not only the mobile app, but also other web sites or applications used mainly in planning and promoting cultural events (e.g. the project's web site), or in supporting organizers and their specialized personnel to communicate with the audience or check their feedback.

Figure 3 shows the main entities that model the cold data of the Caravan Next application domain. Central role in this model has the notion of the event. An Event is a programmed cultural activity that takes place in specific Venue during a particular interval of time. Venues can be either specific indoor places, or outdoor public locations (like a square, a harbor, a street, etc.). Events belong to Categories (e.g. performance, parade, conference, meeting, etc.) and they are typically organized by some Organization (or co-organized by more than one) while can be sponsored by others. Events are described among others with name, description, start/end time, etc. They can also be complemented with multimedia: photos and/or videos that are provided either as promotional or as commemorative material by the event organizer.

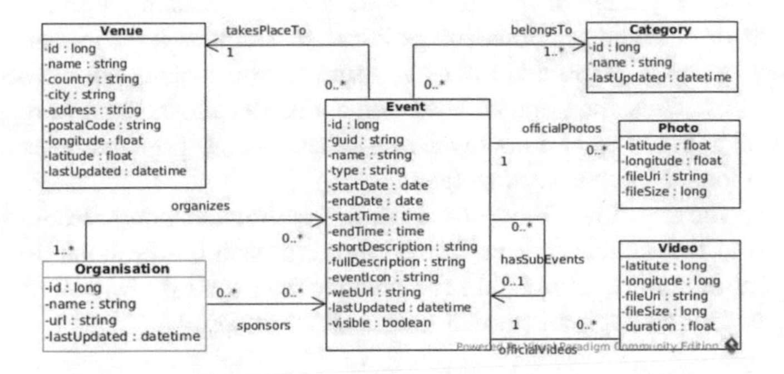

Fig. 3. *Cold data* of the Caravan Next application model.

Events in the Caravan Next application model can be quite complex in structure since a specific event can accommodate many sub-events. This allows the description of event series; a typical occasion in social community theater where cultural events last for several days and include activities in several venues or even cities.

Figure 4 illustrates the main concepts used to model the warm data of the Caravan Next application domain. The notion of User represents any person that uses the app or

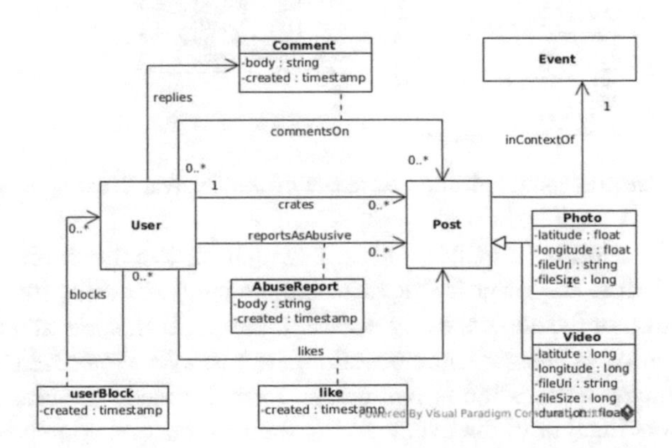

Fig. 4. *Warm data* of the Caravan Next application model.

other tools of the Caravan Next infrastructure. Users can be registered directly to the Caravan Next infrastructure, or use some other authentication provider (e.g. Facebook). User-generated content (i.e. warm data) is captured in the form posts, comments, and likes. A Post is a user message published in an online forum or virtual community called the Caravan Next "Wall". Posts can be simple text messages, videos, photos, or mixed. Posts are about encouraging engagement, action and active participation of the users to cultural events. To this end, posts are always published in the context of some particular event. Social interactions are supported by allowing users to react on posts made by other users. Thus users can start discussions by commenting on posts and/or replying to comments. Finally the notion of Like is used to allow users to express a quick positive reaction to a post.

In social networking where users with different social, cultural and ethical back-grounds or different religious and political sensibilities are communicating, pre-cautions have to be taken so that offensive users and/or content can be identified and blocked by users. Thus, the Caravan Next domain model allows users to report abusive posts, or event to block other users in the sense that content (posts/comments) from this users will no longer be presented to them.

Figure 5 shows how the downstream communication (from organizers/ professionals to the audience) is performed through push notifications. In the context of some event organizers or specialized personnel can send push notifications to user (audience) devices through a dedicated administration app.

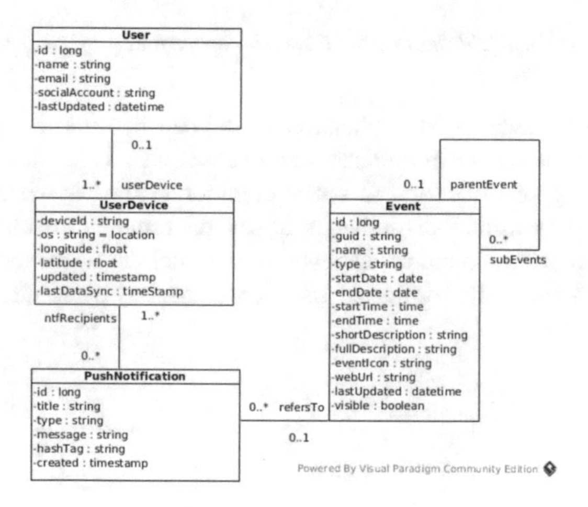

Fig. 5. Downstream communication elements of the Caravan Next application model.

A push notification is actually a message sent from event organizers/actors to the audience either before the event (to increase engagement) or during the event to enable active participation of the audience (e.g. by asking to perform some action or to interact with stage actors, or to answer some question, etc.) to event realization. The group of recipients to which a notification is sent can be formed based on current user locations (e.g. only to users that are in the event place), their interest on some specific event, or by selecting specific users.

Finally, Fig. 6 shows how location-based services are offered with the use of the cNext App. A location-based service in the context of the Caravan Next contains one or more event-driven use cases (i.e. functionality offered to the user) and are achieved with the deployment of a number of beacon devices [12]. When a Service is active, the cNext app is setting the user's device to scan for and track specific beacons that have been deployed in locations near to the event venue (e.g. in a room, in a city square, in the venue entrance etc.) or at specific spots. When a user device finds a beacon device and starts tracking it, three different (detection) events can be fired: (a) IN_RANGE: the user device came near to the beacon location, (b) OUT_OF_RANGE: the user device moved away from the beacon location, and (c) CLOSE: the user device is very close (less than a meter) to the beacon location. When such an event is fired, the cNext app knows that the user entered or exited some area (like the venue that an event is taking place), or even that is very close to some specific location spot (e.g. in front of some object monument) that is important to know. A location-based service may contain more than one use cases each one triggered by a specific detection event (IN_RANGE, CLOSE, or OUT_OF_RANGE) of one or more beacon devices. This framework allows the implementation of quite complex services with just a few beacon devices deployed. To ensure privacy and user control, a use case of a location-based service can be configured to require user confirmation. In such a case, when a use case is triggered, the very first step is to (describe what the use case is and) ask the user for confirmation who may reject or accept the use case execution. A use case consists of one or more functions that are executed according to the use case scenario. Such a function can be whatever the service designer wants or just a communication message to the user as a push notification.

The aforementioned framework supports a variety of usage scenarios. Two of them are presented next considering their relevance in promoting audience engagement and active participation in cultural events:

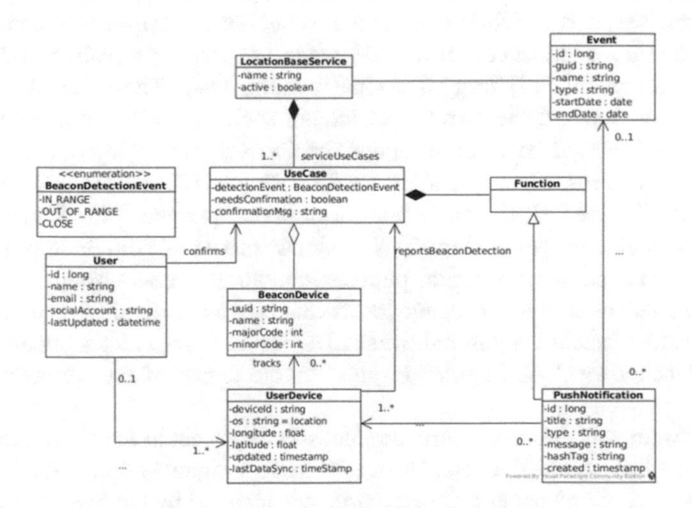

Fig. 6. Modeling of location-based services.

- *The Cultural Passport service* is about stimulating audience to increase the participation in cultural events. This is achieved by maintaining for each user a "cultural passport" that gets a stamp every time the user participates in some cultural event. For each cultural event, the service consists of two use cases: (a) check-in (entering the event location) and (b) check-out (leaving the event location). In particular, when entering the event location, the user is asked to confirm his check-in to that event. If the user accepts, his cultural passport will get automatically "stamped" (i.e. filled in with event data, title and date) and emailed to her/his mailbox, and a "certificate of attendance" will be generated and automatically posted to the cNext app Wall to let other people know that s/he is attending that event. On the other hand, when exiting an event location, a good bye message is sent inviting the user to upcoming events.
- *The Treasure Hunt service* is a technology-driven version of traditional treasure hunt. In the context of a social community theatre event, participants are encouraged to play a treasure hunt game using their mobile phones to follow certain paths. Such a service, requires the deployment of a number of beacon devices at specific location spots and the configuration of appropriate use cases that are triggered when a user comes close to those spots. The game starts by sending to all participants a quiz message that its solution may guide them to a specific location. Those who solve the quiz and get close to the right spot, are automatically detected and proceed to the next level where a new quiz or further directions are sent to them. Participants may be asked to perform physical (e.g. open the door in your left and enter room X) or virtual actions (e.g. upload a photo of the hidden object that you discovered).

4 Engaging Students with Digital Marionettes

Drama-based artistic forms can be effectively used in formal education to promote learner engagement and creativity. In particular, artistic forms inspired by traditional theatrical practices and enriched with digital technologies have been found to promote engaging learning experiences in a wide range of learning situations ranging from compulsory education [13] to professional training [14]. They can also effectively address learning beyond the typical curriculum such as during school festivals and exhibitions [15]. Based on these findings, the Caravan Next project implemented an extension of eShadow, already widely used in schools [13–15], interfacing it with a LEAP motion device [16] to enable control of digital puppets with fingers, thus simulating traditional marionettes (Fig. 3). The aim was to enable drama improvisations in real-time by interacting with digital puppets in cultural events organized by the Caravan Next project as well as to engage local schools in specially designed activities that take place before scheduled cultural events. During these activities students are invited to develop their own digital stories inspired by the theme of the forthcoming events (Fig. 7).

The software was evaluated during a pilot school project in the 5th Grade of the 19th Primary School of Chania in Crete, Greece, to create a digital story (available at: https://youtu.be/GqiOaCejFxs) using a digital marionette inspired by the myth of Europe. This

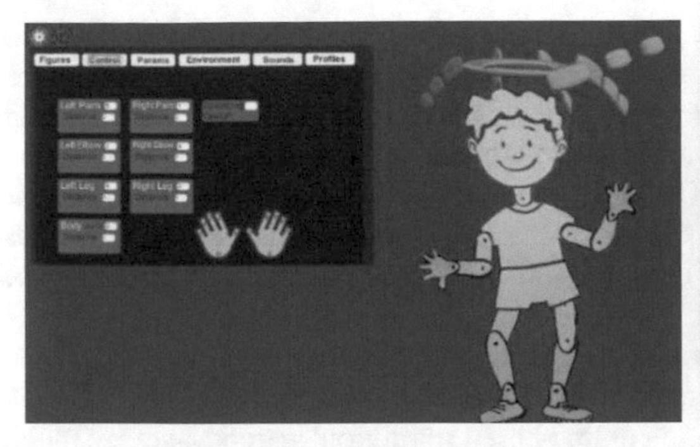

Fig. 7. Special software to support marionette-like interactions. Digital puppets are controlled with users' fingers via a Leap motion device [16]. Controlling fingers are configurable through the panel shown in the screenshot on the left. Several types of digital puppets are possible.

was directly related to the theme of the Caravan Next micro event organized in Heraklion, Crete, Greece, in September 2016. The students were invited to research on the myths of EU countries and prepare a scenario showing mythical Europe traveling in these countries and learning about their corresponding myths. After finalizing their scenarios, the students were asked to prepare drawings on the myths they have selected. The drawings were subsequently scanned and used with the digital marionette software as background images. Finally, the students used the software to animate their story. After finishing their project, the students and their teacher (21 participants in total) where asked to fill System Usability Scale (SUS) questionnaires [17].

The analysis of the questionnaires revealed that the average rating by the participants was high (78.55%). However, in terms of learnability, the score was rather moderate. In particular, this refers to the results recorded in SUS questionnaire questions that directly address the learnability of the system. To further investigate the learnability issue, a second experiment was designed and conducted to investigate the source of this lower performance in terms of learnability. The hypothesis was that the poor performance was related to the inherent difficulty in controlling the specific digital marionette (depicted on the right of Fig. 3). During this second experiment students were invited to use a different digital marionette (the one depicted on the left of Fig. 3) that is easier to control. The results of this second experimental round confirmed the overall high score with respect to the usability of the software (total score 78.62%) showing an improvement of learnability score by 11.48% (from a sum of 4.53 to a sum of 5.05 for the two questionnaire questions measuring learnability).

5 Conclusions and Future Work

In this paper we describe and provide details on the use of a mobile app (cNext App) that enables audiences to search for information for cultural interventions, sharing and reflecting on their experiences when participating in such interventions, thus promoting transformative learning in an informal learning setting. We present how the app supports SCT stakeholders and audiences before, during, and after cultural events. We discuss in detail the domain modeling aspects of the back-end Knowledge Base (KB) which models an information space consisting of events, categories, organizers, places, people, posts and comments, interaction devices and notifications and highlight how the developed model is exploited and supported by the app. The cNext app is available on both Android and iOS app stores [18] and the main provided services include: browsing/navigation of events by category and/or filtering by time (past, today, upcoming events), creation of multimedia posts and sharing in social media such as FB, association of posts with events, commenting, representation and access of events on interactive map, multilingual GUI, location-based services employing beacon devices for various interaction scenarios, etc. Furthermore, we present the use of digital marionettes in schools thus promoting drama-based learning scenarios. Both the mobile app and the digital marionette software are enriching the traditional SCT intervention methodologies in a way that promotes deeper engagement in cultural events of the broad public and give new expressive means that promote creativity and social interactions.

The major conclusion of this work is that digital technologies renovates theatre tradition by promoting active engagement in cultural events and creative learning in formal learning settings. It enables people to engage deeper with their cultural traditions and raise awareness about other cultures in a way that combines ICT in a creative and educationally significant framework. The use of digital technologies in school inspired by theatrical tradition, as it was observed, motivates students and promotes important social skills. The usability evaluation of the software demonstrated high usability scores. Further investigations are being done within the context of Erasmus + projects e-ARTinED (http://www.e-artined.eu/) and MultiLib (http://www.multilibproject.eu/) to validate the potential of digital arts in formal and informal learning settings [19].

Beyond schools, it should be noted that new learning opportunities are created within informal learning settings as well in the case of cultural events via the use of the cNext App. It has been extensively used in the context of the Caravan Next project where more than 250 event organizers have used the mobile app in more than 250 cultural events in order to promote those events, to develop (and communicate with) their audience, as well as to stimulate the participation of local communities in cultural activities. With the use of the app, event organizers were able to provide location-based services to their audience, as well as to communicate with it, before, during, and after the event. On the other hand, end-users were using the mobile app, to get informed about cultural events, to get help with the various organizational aspects of these events (time, place, directions, etc.), and exchange memories beliefs, and opinions on the themes of such events thus promoting learning through self-reflection and collaboration.

Although further analysis is needed with a focus on the mobile app usage in cultural interventions mainly with respect to its usability and learnability, there is qualitative evidence from its actual use in several events that it promotes and enhances audience active participation and opens up new learning opportunities within an informal learning framework. This is closely related to activities employing the location-based services of the platform as well and can extend previous research results [20] to blend SCT activities with learning activities in museums, archaeological sites, nature parks etc.

Acknowledgments. The work reported in this paper is implemented in the Caravan Next (Creative Europe 559286) project. Further use of the digital marionette software in schools is implemented in the e-ARTinED (Erasmus + ID 2015-1-SE01-KA201-012267) and the MultiLib (Erasmus + ID 2016-1-SE01-KA201-022101) projects.

The authors thank Nikos Blazakis for supporting this work by developing the original version of the marionette depicting the myth of Europe. They also thank Mrs Aliki Vitoriou, teachers in the 19th Primary School of Chania, Crete, Greece and her students for participating in the evaluation of the digital marionette software.

References

1. Jones, M., Kimberlee, R., Deave, T., Evans, S.: The role of community centre-based arts, leisure and social activities in promoting adult well-being and healthy lifestyles. Int. J. Environ. Res. Public Health **10**(5), 1948–1962 (2013)
2. Schinina, G.: Here we are. Social theatre and some open questions about its development. Drama Rev. **48**(T183), 17–31 (2004)
3. Bernardi, C.: On the dramaturgy of communities. In: Jennings, S. (ed.) Dramatherapy and Social Theatre: Necessary Dialogues. Routledge, New York (2009)
4. Dudina, V.: Performative turn and epistemological reconfiguration of social knowledge. In: XVIII ISA World Congress of Sociology, 13–19 July 2014. Isaconf (2014)
5. Cabitza, F., Simone, C.: Building socially embedded technologies: implications about design. In: Designing Socially Embedded Technologies in the Real-World, pp. 217–270. Springer, London (2015)
6. Dirksmeier, P., Helbrecht, I.: Time, non-representational theory and the 'Performative Turn' – towards a new methodology in qualitative social research. Forum Qual. Soc. Res. **9**(2), Art. 55 (2008)
7. Ganascia, J.-G.: Views and examples on hyper-connectivity. In: Floridi, L. (ed.) The Onlife Manifesto: Being Human in a Hyperconnected Era. Springer, Cham (2015)
8. Beeman, W.O.: The anthropology of theater and spectacle. Annu. Rev. Anthropol. **22**(1), 369–393 (1993)
9. Turner, V.: Are there universals of performance in myth, ritual and drama? In: Schechner, R., Appel, W. (eds.) By Means of Performance. Cambridge University Press, Cambridge (1990)
10. Kolb, D.A., Fry, R.E.: Toward an Applied Theory of Experiential Learning. MIT Alfred P. Sloan School of Management, Cambridge (1974)
11. Honey, P., Mumford, A.: Manual of Learning Styles. P. Honey, London (1982)
12. iBeacon. https://en.wikipedia.org/wiki/IBeacon
13. Moraiti, A., Moumoutzis, N., Christoulakis, M., Pitsiladis, A., Stylianakis, G., Sifakis, Y., Maragoudakis, I., Christodoulakis, S.: Playful creation of digital stories with eShadow. In: 11th International Workshop on Semantic and Social Media Adaptation and Personalization, SMAP 2016, pp. 139–144 (2016)
14. Moumoutzis, N., Christoulakis, M., Pitsiladis, A., Maragoudakis, I., Christodoulakis, S., Menioudakis, M., Koutsabesi, J., Tzoganidis, M.: Using new media arts to enable project-based learning in technological education. In: 2017 IEEE Global Engineering Education Conference (EDUCON), pp. 287–296 (2017). ISSN 2165-9567
15. Christoulakis, M., Pitsiladis, A., Moraiti, A., Moumoutzis, N., Christodoulakis, S.: eShadow: a tool for digital storytelling based on traditional Greek shadow theatre. In: Proceedings of the 8th International Conference on the Foundations of Digital Games (2013)

16. Leap Motion. https://en.wikipedia.org/wiki/Leap_Motion
17. Brooke, J.: SUS-a quick and dirty usability scale. Usability Eval. Ind. **189**(194), 4–7 (1996)
18. cNext App. http://cnext.tuc.gr/app
19. Moumoutzis, N., Christoulakis, M., Pitsiladis, A., Sifakis, G., Maragkoudakis, G., Christodoulakis, S.: The ALICE experience: a learning framework to promote gaming literacy for educators and its refinement. In: International Conference on Interactive Mobile Communication Technologies and Learning (IMCL), pp. 257–261 (2014)
20. Makris, D., Makris, K., Arapi, P., Christodoulakis, S.: PlayLearn: a platform for the development and management of learning experiences in location-based mobile games. In: 8th International Conference on Mobile, Hybrid, and On-line Learning (eLmL 2016), pp. 43–48 (2016)

Digital Mobile-Based Behaviour Change Interventions to Assess and Promote Critical Thinking and Research Skills Among Undergraduate Students

Yousef Asiri(✉), David Millard, and Mark Weal

University of Southampton, Southampton, UK
yaa1e15@soton.ac.uk, {dem,mjw}@ecs.soton.ac.uk

Abstract. University students' attempts at critical thinking in research projects frequently require supervisor interventions in the form of advice giving, feedback and supportive information. Providing such interventions to each student through classroom teaching or conventional meetings is restrictive as to timing and place. By using a Digital Behaviours Change Intervention (DBCI) technique, students can get continual assistance from supervisors in their research work through the web or mobile platforms. This research sought to understand students' perceptions of using digital mobile-based behaviour change interventions to improve their critical thinking and research skills. A survey instrument inspired by the Paul-Elder Critical Thinking Framework was designed to measure student self-perceived critical thinking abilities before and after an experiment. Five supervisors were interviewed to validate the instrument and the behaviour change intervention content. An experimental study was conducted to explore how students interact with a mobile app-based behaviour change intervention, which was developed using the *LifeGuide Toolbox* platform, to support critical thinking over three months of a research project. The results showed a significant improvement in students' critical thinking skills with respect to five intellectual standards (clarity, precision, relevance, logic, and fairness) after using the system.

Keywords: Critical thinking · Research skills
Digital Behaviour Change Intervention (DBCI)
Assessment and evaluation · Mobile learning

1 Introduction

Critical thinking is an essential activity in conducting research projects, and mobile technology, which now influences almost every other aspect of our lives, is being applied to aid in practicing and improving it also [1]. In particular, such new technologies as mobile platforms and Web 2.0 can be adapted for this purpose [2]. For example, online discussion forums allow educators and students to debate and discuss ideas, thereby honing their critical thinking skills

M. E. Auer and T. Tsiatsos (Eds.): IMCL 2017, AISC 725, pp. 155–166, 2018.
https://doi.org/10.1007/978-3-319-75175-7_17

through collaborative effort [3]. On such forums, students can openly discuss topics, brainstorm ideas, share thoughts, interact with others having differing views, and evaluate each other's viewpoints, thus providing the opportunity for learners from various majors and with different goals to interact and learn from one another [4].

Critical thinking can be viewed as a reflection of psychological and behavioural effects [5]. This definition best fit our study, and so the Paul-Elder's definition of critical thinking was adopted for it because of its focus on the behavioural aspects of critical thinking [6,7]. Paul and Elder described critical thinking as follows: "A systematic way to form and shape one's thinking. It functions purposefully and exactingly. It is thought that is disciplined, comprehensive, based on intellectual and behavioural standards, and, as a result, well-reasoned." [8]. Because critical thinking is an essential activity that is practiced on a daily basis, we chose mobile technology as the platform for our intervention tool so as to seamlessly integrate user behaviours. According to Alnuaim et al. [5], certain unique features of mobile technology, i.e., connectedness, portability, real-time interaction, and personalisation, could lead to significant results when this technology is used as an intervention tool to foster critical thinking and evaluation.

Digital behaviour change intervention is a model supporting both an intervention's creator and its user in developing a system to modify undesirable behaviours [10]. The Internet is becoming increasingly viable as a platform for delivering behavioural change interventions on a large scale [11]. With widespread and low-cost access to the Internet, behaviours change interventions are being increasingly adapted for online usage, thereby addressing many of the drawbacks of offline and conventional BCIs. In addition to personalisation and adaptation possibilities, cost-effectiveness of behavioural interventions, anonymity, confidentiality and accessibility are among other benefits of Internet-based systems [13]. A growing number of Internet-based and app-based interventions providing further flexibility are successfully providing end-users with these benefits along with on-the-go access and easier self-monitoring capabilities [12]. Delivering critical thinking interventions through mobile devices could help users, specifically students, to monitor decision making, change behaviours, acquire intervention support, and track thinking patterns through the set of critical thinking activities, diaries, reminders, and goals as well as receive notifications at any time or place.

The main goal of this research is to study how a digital mobile-based behaviour change intervention has the potential to measure and enhance students' critical thinking skills in their research work. To achieve this goal, the following set of research questions was formulated. Firstly, how can critical thinking skills of university students be measured in the DBCI context? Answering this question will aid in developing an instrument to assess students' critical thinking skills, which will require validation by supervisors. Secondly, what are supervisor expectations with respect to mobile behaviour change intervention in student critical thinking and research skills? This question aimed to identify supervisor requirements for a DBCI to promote students' critical thinking skills. Thirdly, how can we best design and implement a digital mobile-based behaviour change intervention to promote critical thinking skills in students? These questions highlight the areas requiring attention to improve future effectiveness of DBCI use for similar critical thinking situations.

2 Background

2.1 Definition of Critical Thinking

Critical thinking is defined as "a reasoned, purposive, and introspective approach to solving problems or addressing questions with incomplete evidence and information and for which an incontrovertible solution is unlikely" [14]. According to Garrison et al., Dewey [15] developed another significant definition of critical thinking which referred to reflective thinking. Due to the importance of critical thinking skills, scholars in several fields have conducted studies aimed at assessing and promoting critical thinking concepts: philosophy [16], education [15], psychology [17], and behaviourism [6,7]. According to Ennis [19], including critical thinking courses in curricula was the first technical step in adopting critical thinking in general. Since then, other strategies have gradually been adopted to improve such skills, involving such areas as collaboration [18,20], puzzle and problem solving [21], scaffolding [22], peer assessment [23], and online gaming [24], which employ critical thinking skills. In the fields of psychology and philosophy, studies have viewed critical thinking as a habit of mind that could be trained and practiced [6,16]. Reflection as a way to think critically is concerned with consciously looking at and thinking about his/her experiences, actions, feelings, behaviours, and responses and then interpreting or analysing them in order to learn from them [25]. This combination of different approaches, with the help of mobile technology, might lead to useful results in terms of promoting critical thinking [5]. In fact, experimental and theoretical research, which has applied various technologies to promote critical thinking, has generally yielded significant results in promoting students' critical thinking skills and will be discussed in Sect. 2.3.

2.2 The Paul-Elder Critical Thinking Framework

The Paul-Elder Critical Thinking Framework, which provides a comprehensive and broad definition of critical thinking [7,26], sets forth nine essential intellectual standards by which critical thinking can be assessed: clarity, accuracy, precision, significance, relevance, breadth, depth, logic, and fairness. The Paul-Elder framework indicates that these standards should be adopted by the systematic research steps called the Elements of Thought: purpose, questions, information, inference, concepts, assumptions, implications, and point of view. Indeed, Celuch and Slama [7], who used the Paul-Elder Framework in their study to examine the elements of the Theory of Planned Behaviour, found that doing so significantly improved students' attitudes toward critical thinking, self-efficacy for critical thinking, and self-identity as a critical thinker. In fact, they found that the Paul-Elder framework was effective in dealing with measuring critical thinking in relation to behavioural intentions, attitudes, social influences, perceived control, self-efficacy, and self-image. In a recent study by Ralston and Bays [26], the Paul-Elder Framework was used because of its "comprehensiveness, discipline neutral terminology and extensive high-quality resources.", they conducted to enhance critical thinking across the undergraduate experience for engineering students. The findings were significant, encouraging use of the framework in the efficacy of critical thinking in other areas.

2.3 Mobile Learning and Critical Thinking

For most of us, including students, mobile technology has become an important aspect of modern life [28]. In mobile learning, different educational environments no longer limited to specific times or places are possible for such activities as note taking, sharing ideas, watching lectures, making voice recordings, creating videos, accessing information and data on the Internet, engaging in group discussion, organising personal plans, and communicating with educators and other students [30]. The potential to accomplish all this in any online environment can help students to modify their daily behaviours for the better [27]. In fact, these educational activities can be implemented in promoting critical thinking skills. Getting in touch with educators, experts, and researchers at any time and from any location is a great advantage for students, allowing them to practice high level thinking with respect to tasks related to advanced research [29]. The advanced features of mobile technology, including individualised interfaces, real-time access to information, context sensitivity, quick communication, and instant feedback, have great potential to foster positive learning behaviour in users [30].

Many attempts have been made to use mobile devices in enhancing critical and creative thinking skills. A study by Boyinbode and Ng'ambi [31] resulted in the design of *MOBILect*, an interactive mobile lecturing tool that helps students and professors in higher education to overcome the challenges of podcasting and video-casting to promote deep thinking and learning. Another study by Wong Kung Fong [32] assessed the extent to which a mobile app called *CritlQ* could enhance students' ability to learn and communicate. Specifically, it is a mobile critique platform that enables and motivates undergraduate communication design learners to co-design artifacts with each other and criticise each other's work. A recent study by Alnuaim et al. [5] used the location-based app called *sLearn* to support human-computer interactions between students in their critical thinking and in their attempts to understand the context of the design work. The attempts described above revealed a significant improvement in students' ability to think critically and to make decisions. They incorporated a positive attitude toward using mobile devices to foster students' creativity and collaboration in design work. However, these studies focused on simple educational tasks which were given to students in certain situations or locations. In fact, critical thinking can require time [14] and can involve several stages and standards that should be tested separately. The studies described above included no accurate assessments of the students' progress with respect to critical thinking. In fact, these studies lacked continual interactive interventions between students and educators.

3 Methodology

Qualitative data were gathered by interviewing five supervisors in the areas of Computer Science, eLearning, and Human-Computer Interaction from the University of Southampton. Supervisors were approached by emails; describing the purpose of the study and the interview questions, alongside with the instrument statements and the intervention components. The aim of the interviews was to validate: (1) the instrument designed to measure critical thinking in the context of DBCI, (2) the intervention components with its content. In addition, the interviews identified supervisors' expectations and requirements regarding the use of DBCIs in critical thinking and research work. Interviews were audio-recorded and transcribed verbatim. Inductive thematic analysis was used to identify recurring patterns relevant to understanding supervisors' experiences of using mobile technology in critical thinking.

After validating the instrument and the intervention components, a total number of 30 undergraduate students in their third year (the final year of the undergraduate degree) in the Electronics and Computer Science Department of the University of Southampton participated individually in the mobile application experiment, which involved using the mobile application while carrying out their third-year research projects over a three-month period (Oct 2016–Dec 2016). The instrument (see Sect. 4), which was inspired by the Paul-Elder Critical Thinking Framework, was given as a pre- and post-experiment survey for students to self-report the perceived development in critical thinking skills before and after the experiment, respectively.

Students were recruited for the study by sending invitations through the third year students' email list. Based on the instrument developed, pre- and post-experiment online surveys, in Likert scale (Not at all = 1, Sometimes = 2, Not sure = 3, Usually = 4, Always = 5), were conducted to measure the self-perceived change in critical thinking skills of the students between the beginning and ending of the experiment due to their use of the mobile application. A paired sample t-test was used to investigate if there is a difference in average values between pre and post surveys, testing the perceived improvement in the critical thinking intellectual standards for the same respondents before and after using the system. The surveys consisted of a set of statements designed to gather quantitative data about students' reflections on their critical thinking skills. The surveys asked the students about the potential success of using mobile app-based behaviour change intervention to support their critical thinking. The study investigated students' engagement with the mobile application and the usefulness of using DBCIs in critical thinking.

4 An Instrument to Measure Critical Thinking Skills in a DBCI Context

The survey instrument in this study (Table 1), which was inspired by the Paul-Elder Critical Thinking Framework, was designed to measure university students' critical thinking in the context of a research project. The instrument statements were designed to reflect the nine intellectual standards of critical thinking: clarity, accuracy, precision, significance, relevance, depth, breadth, logic, and fairness.

Table 1. Instrument administered as an online survey to measure critical thinking skills of students in their research projects before and after the study's experiment

Intellectual standards	Critical thinking instrument statements
Clarity	My writing is clear and the meaning is obvious
	I think about the different levels of listeners and the readers when I speak or write
Accuracy	I think about evidence before I present my ideas
	When I examine an idea, I tend to falsify it in order to prove it holds
Precision	I am aware of the different contexts when I think about any situation
	I am able to be precise in every aspect in my life
Significance	In my analysis, I always consider both positive and negative results
	I worry about insignificant results
Relevance	I give relevant examples when I think about a certain problem
	I am mostly able to see the relevance between the different parts of my research work
Depth	I think deeply to understand any problem by focusing on the concepts
	I follow the same previous valid strategies that I normally do to solve any problem
Breadth	I tend to find new solutions with new ways when I think
	I criticise myself to improve my way of doing things in my daily life
Logic	I give reasonable explanations when I present my ideas
	I know what logical fallacies are
Fairness	Everything must have a meaning
	I am aware in my work that I may have hidden biases

The statements in the instrument were formed based on the definition of critical thinking from Paul-Elder Critical Thinking Framework [8]. To achieve *clarity*, clear answers are needed, including further elaboration that is required, examples, and meaningful illustration. Similarly, *accuracy* requires a testable idea whose truth or falsity can be determined. *Precision* of any solution requires that exact details to be verified. Information must be *relevant* to the problem and be linked with the studied issue. Critical thinking skills should cover the *depth* and *breadth* of the problem by asking what factors make the problem difficult and more complicated. To accomplish this latter, another perspective or point of view is needed. Moreover, having clarity, accuracy, and relevance along with a perspective exhibiting depth and breadth still does not apply all the metrics related to critical thinking. For example, *logical* evidence is also necessary when a solution is being considered. Results must be *significant* by focusing on the most important aspects of the problem. Lastly, *fairness* is one of the most neglected standards when critical thinking is required [7]. The main issues that can cause unfairness are dealing sympathetically with the problem and ignoring hidden biases.

5 Intervention Components and Design Principles for Implementing DBCIs in Critical Thinking and Research Skills

An existing DBCI platform, the *LifeGuide Toolbox*, was used to construct an intervention for fostering students' critical thinking. It was used to generate Android and iOS mobile applications for this study, named the *CriticalThinking* mobile application (Fig. 1).

Fig. 1. The *CriticalThinking* mobile application: homepage and questionnaire pages

The *LifeGuide Toolbox* is an open-source framework that enables intervention creators to design and generate digital behaviour change interventions in the form of web and mobile applications [13]. It helps experts with no programming skills to generate web and mobile applications for various purposes. It is part of the Lifeguide[1] programme of research which aims to promote cross-disciplinary research through the investigation of the use of web-based and app-based platforms in DBCIs. The mobile intervention was created using the accompanying authoring tool available with the *LifeGuide* framework. The aim of the mobile app is to investigate how mobile behaviour change intervention might improve students' critical thinking skills during their research projects. The mobile app was developed to provide several intervention components, such as critical thinking activities, short questionnaires about intellectual standards, goals and plans setting, writing diaries, and notifications. This mobile application's activities involve information about critical thinking and tasks where participants use the nine critical thinking standards to integrate them with the first stages of a research project: finding a research problem, searching the literature for pertinent sources, and writing the introduction. Email notifications were used to keep participants engaged and to support their critical thinking skills based on their performance and reflections in the mobile application. This was required to study the users' behaviours, performance and their activities in the mobile application.

Two types of recorded data in *LifeGuide Toolbox* digital interventions were used in this study: (1) Intervention data: responses to surveys, diary entries, planner entries. (2) Usage data: logs when users access different activities or receive notifications. According to Osmond et al. [9], there are many factors that make *LifeGuide Toolbox* unique and different than other tools, such as HTML editors, which can be used to design internet-based intervention: *LifeGuide Toolbox* is designed for making complex interventions and recording data without browsing server logs. It offers randomisation and stratification of participants. It is designed with a framework for running trials and usually allow for repeat visits. It contains a logic editor and graphical interface to easily design suitable interventions.

6 Results

6.1 Supervisors' Requirements and Validation

Supervisors validated the instrument for measuring perceived critical thinking skills. Supervisors also confirmed the intervention components used in the designed mobile application to be appropriate for students; such as critical thinking tasks, questionnaires to track students' critical thinking progress, notifications to get them engaged in their work through the app, setting goals and plans to help students to get a sense of achievement regarding their critical thinking and research skills:

> "*Besides the well-designed instrument and the app activities, you ask students questions, you get them thinking. Many kinds of questions can be asked, for example, after they have done part of their project, ask them*

[1] LifeGuide Website: https://www.lifeguideonline.org/.

> *[students]: have you looked back to the literature review, and what could you do differently? That's the types of questions they need at the end to change critical thinking behaviours through the web or the mobile app.*" [P2]

Based on the supervisors' views in the interviews, university students in general lack critical thinking skills in almost every step of their university research projects. In the interviews, supervisors stated that students demand technological support to help them in their critical thinking and research skills and encouraged mobile behaviour change intervention to be used and tested:

> "*What you can do is scaffold the learning by giving them clues as to have them represent it and media it. I might do this as a paper exercise. You can transfer this paper exercise to computers.*" [P1]

In the supervisors' opinions, DBCI could provide a constructive and useful tool for both students and their supervisors. These supervisors indicated that DBCIs would help them create relevant content as to what critical thinking is and what skills students need to learn to be critical thinkers in their research work:

> "*DBCI with its intervention components might actually have a strong effect to change students' behaviour in critical thinking specially in their research projects.*" [P3]

6.2 Findings from Mobile App Experiment

Students' experiences regarding using mobile technology in critical thinking were analysed and addressed. In the results, (94%) of students find mobile technology a helpful tool to communicate effectively and quickly in terms of their research work and to change their critical thinking behaviours because it is more flexible to carry all the time and everywhere. (84%) of students think it is always difficult to get quick and useful feedback from their peers or supervisors regarding their research inquiries in the classrooms or traditional meetings. In fact, (91%) of students were in demand for instant help during the research projects. Students' ability to engage and reflect on their critical thinking skills, through the mobile app and email notifications have significantly enhanced after using the DBCI technique with (76%) which was only (36%) in the beginning of the experiment. The results in (Fig. 2) show that the *significance* standard has the lowest average values in critical thinking standards for students, whereas the *breadth* standard has the highest average values critical thinking standard. The results also show that there is statistically a significant perceived improvement (at $p < 0.5$) in five critical thinking intellectual standards (*clarity* = *.010, precision* = *.019, relevance* = *.012, logic* = *.024, and fairness* = *.018*) after using the system. These critical thinking standards have improved because the study mainly focused on the first stages of the students' research projects which may affected their performance on the other standards. The standards that have not improved needed more time to apply in the last stages of the students' research projects such

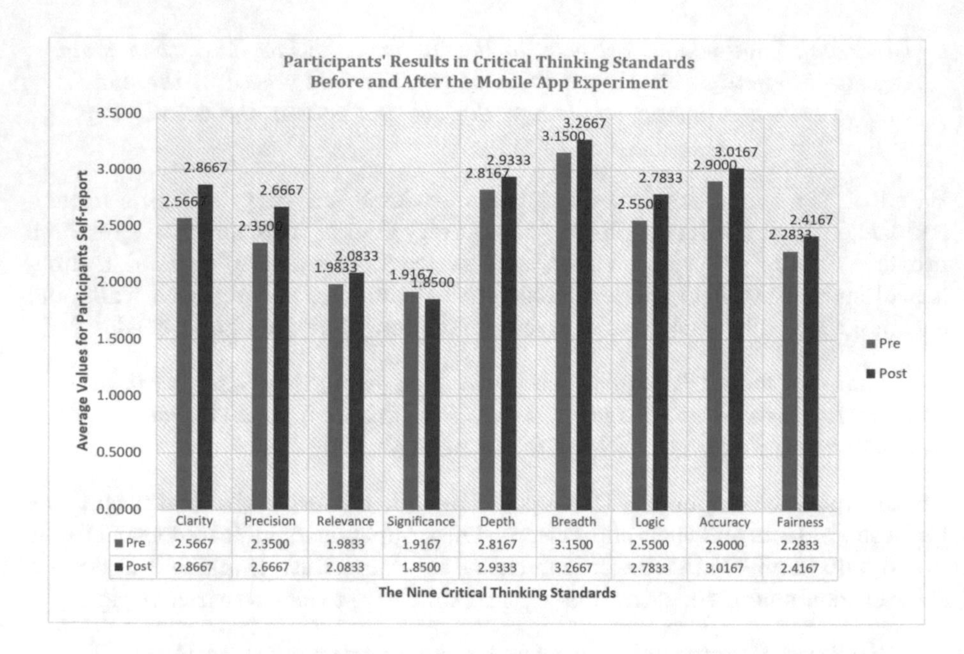

	Clarity	Precision	Relevance	Significance	Depth	Breadth	Logic	Accuracy	Fairness
Pre	2.5667	2.3500	1.9833	1.9167	2.8167	3.1500	2.5500	2.9000	2.2833
Post	2.8667	2.6667	2.0833	1.8500	2.9333	3.2667	2.7833	3.0167	2.4167

Fig. 2. Mean values for each critical thinking standard in pre- and post- surveys

as the analysis and results phases. The results were generally encouraging using mobile behaviour change intervention to keep students attached to their projects and to be aware of the critical thinking skills while working in the research. The investigation of this study led to explore the challenges of using mobile technology in critical thinking and research context. It highlights the major issues of behaviour learning when critical thinking is required and addressed the key components of a successful behavioural learning experience in critical thinking skills with university students.

7 Conclusion and Future Work

This study represented an instrument to measure perceived critical thinking in the context of DBCI which was validated by five supervisors in the interviews. This study demonstrates the design of the mobile intervention components and its implementation. In conclusion, using mobile behaviour change intervention supported students in improving their critical thinking skills in general. In fact, supervisors's views about critical thinking showed that it is not only a skill but it is also a behaviour that needs to be practised and monitored over time. The evaluation of students' performance and their reflections indicates that attaching smartphones as devices that are available most of the time for students during the day has a potential success to deliver interventions to change how they think about their work. In the future work, to reduce the effect of external factors on the study results, a control group will be included. Moreover, the System Usability Scale (SUS) evaluation test for the prototype of mobile behaviour change intervention technique will be used to assess the usability and utility of the tools. Semi-structured interviews for participants will be conducted after the experiment to understand why students may or may not struggle with certain critical thinking standards. This will help us to determine whether the use of DBCIs in critical thinking increases the success rate in achieving behaviour change and makes these changes more sustainable and positive. The analysis of the data will help to build a complete model for DBCIs which can be used when combining digital behavioural interventions and critical thinking.

References

1. Zheng, L., Chen, N.S., Li, X., Huang, R.: The impact of a two-round, mobile peer assessment on learning achievements, critical thinking skills, and meta-cognitive awareness. Int. J. Mob. Learn. Organ. **10**(4), 292306 (2016)
2. Cicchino, M.I.: Using game-based learning to foster critical thinking in student discourse. Interdisc. J. Prob.-Based Learn. **9**(2), 4 (2015)
3. Salleh, S.M., Tasir, Z., Shukor, N.A.: Web-based simulation learning framework to enhance students' critical thinking skills. Procedia-Soc. Behav. Sci. **64**, 372–381 (2012)
4. Yilmaz, F.G.K., Keser, H.: The impact of reflective thinking activities in e-learning: a critical review of the empirical research. Comput. Educ. **95**, 163–173 (2016)
5. Alnuaim, A., Caleb-Solly, P., Perry, C.: Evaluating the effectiveness of a mobile location-based intervention for improving human-computer interaction students' understanding of context for design. Int. J. Mob. Hum. Comput. Interact. (IJMHCI) **6**(3), 16–31 (2014)
6. Goldberg, I., Kingsbury, J., Bowell, T., Howard, D.: Measuring critical thinking about deeply held beliefs: can the california critical thinking dispositions inventory help? Inq.: Crit. Think. Across Discip. **30**(1), 40–50 (2015)
7. Celuch, K., Slama, M.: Promoting critical thinking and life-long learning: an experiment with the theory of planned behaviour. Mark. Educ. Rev. **12**(2), 13–22 (2002)
8. Paul, R., Elder, L.: Critical Thinking: Tools for Taking Charge of Your Professional and Personal Life. Pearson Education, London (2013)
9. Osmond, A., Hare, J., Price, J., Smith, A., Weal, M., Wills, G., Yang, Y., Yardley, L., De Roure, D.: Designing authoring tools for the creation of on-line behavioural interventions. In: Proceedings 5th International Conference on e-Social Science, Germany (2009)
10. Yardley, L., Spring, B.J., Riper, H., Morrison, L.G., Crane, D.H., Curtis, K., Merchant, G.C., Naughton, F., Blanford, A.: Understanding and promoting effective engagement with digital behavior change interventions. Am. J. Prev. Med. **51**(5), 833–842 (2016)
11. McCully, S.N., Don, B.P., Updegraff, J.A.: Using the Internet to help with diet, weight, and physical activity: results from the Health Information National Trends Survey (HINTS). J. Med. Internet Res. **15**(8), e148 (2013)
12. Kraft, P., Drozd, F., Olsen, E.: Digital therapy: addressing willpower as part of the cognitive affective processing system in the service of habit change. In: International Conference on Persuasive Technology, pp. 177–188. Springer, Heidelberg, June 2008

13. Yardley, L., Osmond, A., Hare, J., Wills, G., Weal, M., De Roure, D., Michie, S.: Introduction to the LifeGuide: software facilitating the development of interactive behaviour change internet interventions. In: Persuasive Technology and Digital Behaviour Intervention Symposium (AISB 2009), Edinburgh, GB, 06–09 April 2009, 4 p. (2009)
14. Baker, M., Rudd, R., Pomeroy, C.: Relationships between critical and creative thinking. J. South. Agric. Educ. Res. **51**(1), 173–188 (2001)
15. Garrison, D.R., Anderson, T., Archer, W.: Critical thinking, cognitive presence, and computer conferencing in distance education. Am. J. Distance Educ. **15**(1), 7–23 (2001)
16. Daniel, M., Auriac, E.: Philosophy, critical thinking and philosophy for children. Educ. Philos. Theory **43**(5), 415–435 (2011)
17. Toplak, M.E., West, R.F., Stanovich, K.E.: Rational thinking and cognitive sophistication: development, cognitive abilities, and thinking dispositions. Dev. Psychol. **50**(4), 1037 (2014)
18. Lima, M., Jouini, N., Namaci, L., Fabiani, T.: Social media as a learning resource for business students of the 'Net Generation': using active learning principles to empower creative and critical thinking. Int. J. Qual. High. Educ. Inst. **1**(1), 24–40 (2014)
19. Ennis, R.H.: Critical thinking assessment. Theory Pract. **32**(3), 179–186 (1993)
20. MacKnight, C.B.: Teaching critical thinking through online discussions. Educ. Q. **23**(4), 38–41 (2000)
21. Choi, E., Lindquist, R., Song, Y.: Effects of problem-based learning vs. traditional lecture on Korean nursing students' critical thinking, problem-solving, and self-directed learning. Nurse Educ. Today **34**(1), 52–56 (2014)
22. Wass, R., Harland, T., Mercer, A.: Scaffolding critical thinking in the zone of proximal development. High. Educ. Res. Dev. **30**(3), 317–328 (2011)
23. Hwang, G.J., Hung, C.M., Chen, N.S.: Improving learning achievements, motivations and problem-solving skills through a peer assessment-based game development approach. Educ. Tech. Res. Dev. **62**(2), 129–145 (2014)
24. Yang, Y.T.C., Chang, C.H.: Empowering students through digital game authorship: enhancing concentration, critical thinking, and academic achievement. Comput. Educ. **68**, 334–344 (2013)
25. Boud, D., Keogh, R., Walker, D.: Reflection: Turning Experience into Learning. Kogan Page, London (1994)
26. Ralston, P.A., Bays, C.L.: Enhancing critical thinking across the undergraduate experience: an exemplar from engineering. Am. J. Eng. Educ. **4**(2), 119 (2013)
27. McCann, S., Camp Pendleton, C.A.: Higher order mLearning: critical thinking in mobile learning. In: MODSIM World (2015)
28. Alhassan, R.: Mobile learning as a method of ubiquitous learning: students' attitudes, readiness, and possible barriers to implementation in higher education. J. Educ. Learn. **5**(1), 176 (2016)
29. Heflin, H., Shewmaker, J., Nguyen, J.: Impact of mobile technology on student attitudes, engagement, and learning. Comput. Educ. **107**, 91–99 (2017)
30. Hsu, Y.C., Ching, Y.H.: A review of models and frameworks for designing mobile learning experiences and environments. Can. J. Learn. Technol. **41**(3), 1–25 (2015)
31. Boyinbode, O., Ng'ambi, D.: MOBILect: an interactive mobile lecturing tool for fostering deep learning. Int. J. Mob. Learn. Organ. **9**(2), 182–200 (2015)
32. Wong Kung Fong, M.: CritIQ: a mobile critique app for undergraduate communication design learners. Des. Cult. **5**(3), 313–332 (2013)

The Design of a Mobile System for Voice e-Assessment and Vocal Hygiene e-Training

Eugenia I. Toki[1](✉), Konstantinos Plachouras[1], Georgios Tatsis[2], Spyridon K. Chronopoulos[3], Dionysios Tafiadis[2], Nausica Ziavra[1], and Vassiliki Siafaka[1]

[1] Laboratory of Audiology, Neurotology and Neurosciences, Department of Speech and Language Therapy, School of Health and Welfare, Technological Educational Institute (TEI) of Epirus, Ioannina, Greece
{toki, kpla, nziavra, vsiafaka}@ioa.teiep.gr

[2] Department of Speech and Language Therapy, School of Health and Welfare, Technological Educational Institute (TEI) of Epirus, Ioannina, Greece
{g.tatsis, d.tafiadis}@ioa.teiep.gr

[3] Department of Computer Engineering, Faculty of Applied Technology, Technological Educational Institute (TEI) of Epirus, Arta, Greece
spychro@teiep.gr

Abstract. Voice assessment is a complicated process and requires high level of expertise. Any attempt of designing and developing a mobile system dealing with voice evaluation and vocal hygiene e-training has to overcome various difficulties. The aim of this study, is to investigate the design of a mobile system with the knowledge on early voice evaluation and vocal hygiene. The user will have feedback on his/her voice profile and hygiene and if necessary will be pointed to consult an appropriate clinician (i.e. ear nose throat physician, gastroenterologist, speech therapist, and psychologist). The methodology tool is based on expert systems. The system's knowledge acquisition on voice assessment and hygiene is acquired by a combination of the literature, empirical data and interdisciplinary experts' collaboration. The system includes voice processing and user's voice self-perception providing findings and suggestions towards a better well-being.

Keywords: Mobile system · Voice assessment · Vocal hygiene
Online expert system

1 Introduction

Voice as an inseparable part of speech and communication, reveals the thoughts and to some point personal characteristics of a person [1, 2]. Voice characteristics (loudness, intensity and pitch) enable the partial recognition of a profession (e.g. the voice of a vendor). They may also, allow the recognition of a person's depressive symptoms via his/her voice production. Markel et al. [3, 4] have shown that various combinations of loudness, intensity and voice pitch correspond to different types of personality and is

M. E. Auer and T. Tsiatsos (Eds.): IMCL 2017, AISC 725, pp. 167–174, 2018.
https://doi.org/10.1007/978-3-319-75175-7_18

related to other standardized personality tests [5]. Besides the above, changes in voice characteristics may be the indicators of voice disorders [6].

It is estimated that 3–9% of the US population has a voice disorder, of which 25% are professional voice users [7]. There has also been extensive research into speech characteristics and disorders in teachers [8], students [9, 10] and other professional groups [6]. Voice disturbances can be detected in both adult [11] and pediatric populations [12, 13].

Various risk factors have been identified in the aforementioned populations [8, 14]. Among those are voice misuse [11], medication side effects [15], smoking [16], use and/or abuse of alcohol [17], hormonal factors [18], mental/emotional disturbances [19, 20], personality disorders [21], as well as work environment conditions (noise, poor acoustics, exposure to dust, smoke and/or chemical substances) [22]. Also, a major risk factor for voice disorders is gastroesophageal (GERD) and laryngopharyngeal (LPR) reflux [23–25]. The most common manifestation of LPR is reflux laryngitis, with or without granulomas [26–28]. In addition, reflux has been reported to be associated with subclinical stenosis, laryngeal carcinomas, polypoidal degeneration, laryngospasm, paradoxical movement of vocal cords and nodules [26–32]. In addition, LPR is correlated on developing a voice disorder in the pediatric population [33–35]. It has also been estimated that up to half of patients with laryngeal and voice disorders have a history of reflux [36, 37].

Voice problems have a negative impact on quality of life (QOL) [38] and on social interaction, resulting in low self-esteem and emotional disturbances such as depression [39].

Studies related to the traditional means of diagnosing voice disorders usually includes the examination of voice characteristics through speech samples, perceptual-acoustic measurements voice quality characteristics, objective measurements (otorhinolaryngological endoscopies), and self-perceived questionnaires (i.e. Voice Handicap Index) [37, 40–42]. Self-perceived questionnaires record how the subject understands his/her voice and/or a current voice problem.

On the other hand, rapid advances on digital technologies offer new opportunities to enhance aspects in healthcare. Healthcare professionals are increasingly turning to the use of mobile devices transforming various aspects of clinical practice, leading to a rapid growth in the development of medical software applications [43]. Traditional voice assessment can also benefit accordingly [44].

Various applications with electronic decision support [45] can serve as standalone [46] or online systems for adults [47, 48] and children [49–51]. Precisely, Maier et al. [44] described the Program for Evaluation and Analysis of all Kinds of Speech disorders (PEAKS), as a recording and analysis environment for the automatic or manual evaluation of voice and speech disorders. The system is adjusted to German language and can automatically rate the user's voice reporting on any voice and speech disorder. The automatic evaluation can be used by independent users or even a speech therapist to consider a second opinion during diagnosis. Evaluation of PEAKS reported results on voice and speech disorders on patients who underwent total laryngectomy, due to laryngeal cancer, and on children with cleft lip and palate. A panel of experts performed a perceptual rating to compare to this evaluation. Guiberson et al. [52] reported on a telehealth language screening system for use with Spanish-speaking children. Screening measures included a processing efficiency measure (Spanish nonword repetition [NWR]), language sampling, and a developmental language questionnaire. They reported for 82 preschoolers and their parents (34 with language impairment (LI), and 48 with typical language development), pointing towards the effectiveness of the system in screening childrens' language development.

Furthermore, an increased interest of interdisciplinary studying of voice disorders has been observed [17]. Also research findings on voice acoustic analysis in Greece have been presented [37, 40, 41, 53, 54]. Intergrading the aforementioned, in the design of a software system, early detection of vocal symptoms and vocal training can be thoroughly examined contributing towards the improvement of QOL.

Therefore, this study aimed to report on the design of an innovative mobile system for voice e-assessment via interdisciplinary approaches adjusted to Greek language and individualized vocal hygiene training.

2 The Design of the Expert Mobile System

Any Knowledge Base system can be used either by any non-expert user or by an expert user [55] (i.e. a speech pathologist). The system can provide solutions to the non-expert user to simple problems and/or to make valid decisions on behalf of the expert [51]. Thus, a non-expert user can use such a system to evaluate his/her voice for voice disorders and to train on vocal hygiene. Especially, the professional voice user can benefit when vocal training occurs at an early stage in his/her vocal demands [8, 39] and a mobile expert system can help towards that direction. On the other hand, it can be used as an advisory tool by a clinician (expert) to assist his/her final clinical decision. Thus, voice experts can improve their clinical effectiveness.

2.1 Ways of Knowledge Acquisition

Knowledge acquisition is considered to be one of the most difficult and time-consuming processes in the development of an expert system. For the expert knowledge extraction, literature as well as collaboration with experts is needed to establish the definition of factors of uncertainty in the diagnostic process of voice disorders [55].

Roy et al. [56] examined the recommended guidelines of voice disorders' classification for clinical voice evaluation. They found that acoustic, laryngeal imaging–based, and auditory–perceptual methods are among the most frequently used methods. Other common methods include case history, aerodynamic measures, functional measures, physical exams, and electroglottography [56].

Voice disorders reflect a range of different conditions as voice production combines functions among the laryngeal, respiratory, oral-pharyngeal, auditory, and somatosensory perception [57]. Therefore, the Classification Manual for Voice Disorders-I, suggests flexibility using a large variety of conditions affecting voice [57].

The state of the art on voice assessment is described by the ASHA [58]. Typical components comprising comprehensive assessment for voice disorders include, case history, self-assessment, respiratory assessment, auditory-perceptual assessment. In detail [58]:

- case history includes medical history/previous treatment, the individual's description on voice problem, onset and variability of symptoms, and vocal daily habits,
- self-assessment includes individual's assessment on how voice problems affects the ability to communicate and the emotions and self-image,
- oral-peripheral (orofacial) examination is a clinical examination administrated by the speech pathologist of the structural and functional integrity,
- assessment of respiration includes respiratory pattern, coordination with phonation, maximum phonation time, and glottal insufficiency assessment,
- auditory-perceptual assessment concerns vocal quality (consensus and additional perceptual features, like roughness, breathiness, strain, pitch deviations/instability, loudness deviations, diplophonia, tremor), resonance, phonation and rate,
- instrumental assessment may include laryngeal imaging, acoustic and aerodynamic assessment.

The accuracy of knowledge acquisition of the system was established by the new evaluation procedure, in collaboration with a team of three experienced clinicians.

2.2 Knowledge Representation

After establishing voice assessment procedures that a clinician has to follow manually, it is important to design how this knowledge can be presented in software systems.

Demirci states, *"generally a fuzzy system is a static mapping between its inputs and outputs. For a fuzzy system the mapping of the inputs to the outputs is characterized by a set of condition–action rules or, in modus pones (if–then) form, 'if premise then consequent'. Generally, the inputs of the fuzzy system are associated with the premise, and the outputs are associated with the consequences. These if–then rules can be represented in many forms. Multi-input multi-output (MIMO) and multi-input single output (MISO) are some of the standard forms"* [59].

The proposed mobile system is a MIMO system. Precisely, it is a complex system that can be considered as a composition of other simple subsystems. These subsystems are composed parts of MIMO subsystems. Figure 1, describes the design of the voice assessment expert system and how the various subsystems function in the Voice Evaluation Control Unit, in accordance to MIMO systems [60]. This system gathers first the data of the user (history symptoms and self-perception on voice aspects). Then the expert system with a range of subsystems i.e. Medical History Symptoms subsystem, Self-Assessment Symptoms subsystem, Respiration Symptoms subsystem, etc. applies sets of "if-then" rules to the data in order to produce Voice Quality results, Voice Acoustic results and Respiratory results. When all subsystems complete evaluation, the results are shown in the last face of the system that reports possible vocal deviations and suggested vocal Hygiene strategies (Fig. 1).

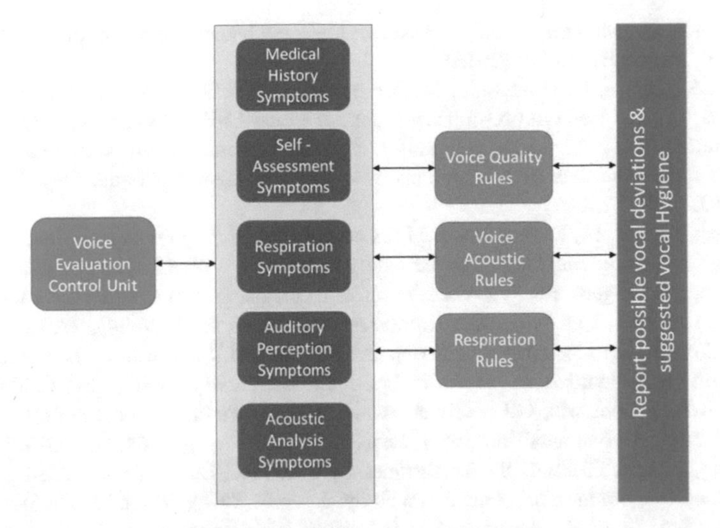

Fig. 1. The design of an expert system for voice e-assessment and vocal hygiene e-training

3 Conclusion

The proposed mobile system design is an advanced and innovative expert system indented to be used for voice assessment and vocal hygiene advice. It utilizes a series of voice assessment methods including the well-established procedures as well as the extraction and analysis of voice quality characteristics. The designed system can contribute towards the advances of new technologies in the healthcare setting in Greece. It enhances the presentation of possible solutions in terms of managing the diverse parameters of such a difficult problem in diagnosis. Additionally, it has the potential to diagnose voice disorders at an early stage and therefore may offer the chance of a successful vocal hygiene training that can help individuals in everyday activities.

References

1. Rosen, D.C., Sataloff, R.T.: Psychology of Voice Disorders. Singular, San Diego (1997)
2. Bruckert, L., Liénard, J.S., Lacroix, A., Kreutzer, M., Leboucher, G.: Women use voice parameters to assess men's characteristics. Proc. R. Soc. Lond. B: Biol. Sci. **273**(1582), 83–89 (2006)
3. Markel, N.N., Meisels, M., Houck, J.E.: Judging personality from voice quality. J. Abnorm. Soc. Psychol. **69**(4), 458 (1964)
4. Markel, N.N., Bein, M.F., Phillis, J.A.: The relationship between words and tone-of-voice. Lang. Speech **16**(1), 15–21 (1973)
5. Gawda, B.: Neuroticism, extraversion, and paralinguistic expression. Psychol. Rep. **100**(3), 721–726 (2007)
6. Roy, N., Stemple, J., Merrill, R.M., Thomas, L.: Epidemiology of voice disorders in the elderly: preliminary findings. Laryngoscope **117**(4), 628–633 (2007)

7. Shipley, K.G., McAfee, J.G.: Assessment in Speech-Language Pathology: A Resource Manual. Nelson Education (2015)
8. Titze, I.R., Lemke, J., Montequin, D.: Populations in the US workforce who rely on voice as a primary tool of trade: a preliminary report. J. Voice **11**(3), 254–259 (1997)
9. McKinnon, D.H., McLeod, S., Reilly, S.: The prevalence of stuttering, voice, and speech-sound disorders in primary school students in Australia. Lang. Speech Hear. Serv. Sch. **38**(1), 5–15 (2007)
10. Simberg, S., Sala, E., Rönnemaa, A.M.: A comparison of the prevalence of vocal symptoms among teacher students and other university students. J. Voice **18**(3), 363–368 (2004)
11. Schneider, B., Bigenzahn, W.: Vocal risk factors for occupational voice disorders in female teaching students. Eur. Arch. Oto-Rhino-Laryngol. Head Neck **262**(4), 272–276 (2005)
12. Angelillo, I.F., Di Costanzo, B., Costa, G., Barillari, M.R., Barillari, U.: Epidemiological study on vocal disorders in paediatric age. J. Prev. Med. Hyg. **49**(1), 1–5 (2008)
13. Shah, R.K., Woodnorth, G.H., Glynn, A., Nuss, R.C.: Pediatric vocal nodules: correlation with perceptual voice analysis. Int. J. Pediatr. Otorhinolaryngol. **69**(7), 903–909 (2005)
14. Carding, P.N., Roulstone, S., Northstone, K., ALSPAC Study Team: The prevalence of childhood dysphonia: a cross-sectional study. J. Voice **20**(4), 623–630 (2006)
15. Abaza, M.M., Levy, S., Hawkshaw, M.J., Sataloff, R.T.: Effects of medications on the voice. Otolaryngol. Clin. North Am. **40**(5), 1081–1090 (2007)
16. Byeon, H.: Relationships among smoking, organic, and functional voice disorders in Korean general population. J. Voice **29**(3), 312–316 (2015)
17. Colton, R.H., Casper, J.K., Leonard, R.: Understanding Voice Problems: A Physiological Perspective for Diagnosis and Treatment. Lippincott Williams & Wilkins, Philadelphia (2011)
18. Amir, O., Ashkenazi, O., Leibovitzh, T., Michael, O., Tavor, Y., Wolf, M.: Applying the Voice Handicap Index (VHI) to dysphonic and nondysphonic Hebrew speakers. J. Voice **20** (2), 318–324 (2006)
19. Cannizzaro, M., Harel, B., Reilly, N., Chappell, P., Snyder, P.J.: Voice acoustical measurement of the severity of major depression. Brain Cogn. **56**(1), 30–35 (2004)
20. Hashim, N.W., Wilkes, M., Salomon, R., Meggs, J., France, D.J.: Evaluation of voice acoustics as predictors of clinical depression scores. J. Voice **31**(2), 256-e1 (2016)
21. Roy, N., Bless, D.M., Heisey, D.: Personality and voice disorders: a multitrait-multidisorder analysis. J. Voice **14**(4), 521–548 (2000)
22. Beavan, V., Read, J., Cartwright, C.: The prevalence of voice-hearers in the general population: a literature review. J. Ment. Health **20**(3), 281–292 (2011)
23. Al-Sabbagh, G., Wo, J.M.: Supraesophageal manifestations of gastroesophageal reflux disease. In: Seminars in Gastrointestinal Disease, vol. 10, no. 3, pp. 113–119, July 1999
24. Toohill, R.J., Kuhn, J.C.: Role of refluxed acid in pathogenesis of laryngeal disorders. Am. J. Med. **103**(5), 100S–106S (1997)
25. Koufman, J.A., Wiener, G.J., Wu, W.C., Castell, D.O.: Reflux laryngitis and its sequelae: the diagnostic role of ambulatory 24-hour pH monitoring. J. Voice **2**(1), 78–89 (1988)
26. Ulualp, S.O., Toohill, R.J.: Laryngopharyngeal reflux: state of the art diagnosis and treatment. Otolaryngol. Clin. North Am. **33**(4), 785–801 (2000)
27. Havas, T.E., Priestley, J., Lowinger, D.S.: A management strategy for vocal process granulomas. Laryngoscope **109**(2), 301–306 (1999)
28. Koufman, J.A.: The otolaryngologic manifestations of gastroesophageal reflux disease (GERD): a clinical investigation of 225 patients using ambulatory 24-hour pH monitoring and an experimental investigation of the role of acid and pepsin in the development of laryngeal injury. Laryngoscope **101**(4 Pt 2 Suppl. 53), 1–78 (1991)

29. Kuhn, J., Toohill, R.J., Ulualp, S.O., Kulpa, J., Hofmann, C., Arndorfer, R., Shaker, R.: Pharyngeal acid reflux events in patients with vocal cord nodules. Laryngoscope **108**(8), 1146–1149 (1998)
30. Ross, J.A., Noordzji, J.P., Woo, P.: Voice disorders in patients with suspected laryngo-pharyngeal reflux disease. J. Voice **12**(1), 84–88 (1998)
31. Ward, P.H., Hanson, D.G.: Reflux as an etiological factor of carcinoma of the laryngopharynx. Laryngoscope **98**(11), 1195–1199 (1988)
32. Loughlin, C.J., Koufman, J.A.: Paroxysmal laryngospasm secondary to gastroesophageal reflux. Laryngoscope **106**(12), 1502–1505 (1996)
33. Tasker, A., Dettmar, P.W., Panetti, M., Koufman, J.A., Birchall, J.P., Pearson, J.P.: Reflux of gastric juice and glue ear in children. Lancet **359**(9305), 493 (2002)
34. Halstead, L.A.: Role of gastroesophageal reflux in pediatric upper airway disorders. Otolaryngol.-Head Neck Surg. **120**(2), 208–214 (1999)
35. Little, J.P., Matthews, B.L., Glock, M.S., Koufman, J.A., Reboussin, D.M., Loughlin, C.J., McGuirt Jr., W.F.: Extraesophageal pediatric reflux: 24-hour double-probe pH monitoring of 222 children. Ann. Otol. Rhinol. Laryngol. Suppl. **169**, 1–16 (1997)
36. Koufman, J.A., Amin, M.R., Panetti, M.: Prevalence of reflux in 113 consecutive patients with laryngeal and voice disorders. Otolaryngol. Head Neck Surg. **123**(4), 385–388 (2000)
37. Spantideas, N., Drosou, E., Karatsis, A., Assimakopoulos, D.: Voice disorders in the general Greek population and in patients with laryngopharyngeal reflux: prevalence and risk factors. J. Voice **29**(3), 389-e27 (2015)
38. Cohen, S.M., Dupont, W.D., Courey, M.S.: Quality-of-life impact of non-neoplastic voice disorders: a meta-analysis. Ann. Otol. Rhinol. Laryngol. **115**(2), 128–134 (2006)
39. Smith, E., Verdolini, K., Gray, S., Nichols, S., Lemke, J., Barkmeier-Kraemer, J.M., Heather, D., Hoffman, H.: Effect of voice disorders on quality of life. J. Med. Speech-Lang. Pathol. **4**(4), 223–244 (1996)
40. Helidoni, M.E., Murry, T., Moschandreas, J., Lionis, C., Printza, A., Velegrakis, G.A.: Cross-cultural adaptation and validation of the voice handicap index into Greek. J. Voice **24** (2), 221–227 (2010)
41. Tafiadis, D., Chronopoulos, S.K., Siafaka, V., Drosos, K., Kosma, E.I., Toki, E.I., Ziavra, N.: Comparison of voice handicap index scores between female students of speech therapy and other health professions. J. Voice **31**(5), 583–588 (2017)
42. Spantideas, N., Drosou, E., Bougea, A., Assimakopoulos, D.: Laryngopharyngeal reflux disease in the Greek general population, prevalence and risk factors. BMC Ear Nose Throat Disord. **15**(1), 7 (2015)
43. American Speech Hearing Association. Voice Evaluation Template. http://www.asha.org/uploadedFiles/slp/healthcare/AATVoiceEvaluation.pdf. Accessed 14 Feb 2015
44. Maier, A., Haderlein, T., Eysholdt, U., Rosanowski, F., Batliner, A., Schuster, M., Nöth, E.: PEAKS–a system for the automatic evaluation of voice and speech disorders. Speech Commun. **51**(5), 425–437 (2009)
45. Castaneda, C., Nalley, K., Mannion, C., Bhattacharyya, P., Blake, P., Pecora, A., Goy, A., Suh, K.S.: Clinical decision support systems for improving diagnostic accuracy and achieving precision medicine. J. Clin. Bioinform. **5**(1), 4 (2015)
46. Marchi, E., Ringeval, F., Schuller, B.: Voice-enabled assistive robots for handling autism spectrum conditions: an examination of the role of prosody. In: Neustein, A. (ed.) Speech and Automata in the Health Care, pp. 207–236. Walter de Gruyter GmbH & Co. KG (2014)
47. Maier, A., Haderlein, T., Stelzle, F., Nöth, E., Nkenke, E., Rosanowski, F., Schützenberger, A., Schuster, M.: Automatic speech recognition systems for the evaluation of voice and speech disorders in head and neck cancer. EURASIP J. Audio Speech Music Process. **2010** (1), 926951 (2009)

48. Martens, H., Van Nuffelen, G., De Bodt, M., Dekens, T., Latacz, L., Verhelst, W.: Automated assessment and treatment of speech rate and intonation in dysarthria. In: 2013 7th International Conference on Pervasive Computing Technologies for Healthcare (PervasiveHealth), pp. 382–384. IEEE, May 2013
49. Gerosa, M., Giuliani, D., Narayanan, S., Potamianos, A.: A review of ASR technologies for children's speech. In: Proceedings of the 2nd Workshop on Child, Computer and Interaction, p. 7. ACM, November 2009
50. Toki, E.I., Pange, J., Mikropoulos, T.A.: An online expert system for diagnostic assessment procedures on young children's oral speech and language. Procedia Comput. Sci. **14**, 428–437 (2012). https://doi.org/10.1016/j.procs.2012.10.049
51. Toki, E.I., Pange, J.: The design of an expert system for the e-assessment and treatment plan of preschoolers' speech and language disorders. Procedia – Soc. Behav. Sci. **9**, 815–819 (2010). https://doi.org/10.1016/j.sbspro.2010.12.240. Elsevier
52. Guiberson, M., Rodríguez, B.L., Zajacova, A.: Telemedicine and e-health. **21**(9), 714–720 (2015). https://doi.org/10.1089/tmj.2014.0190
53. Tafiadis, D., Tatsis, G., Ziavra, N., Toki, E.I.: Voice data on female smokers: coherence between the voice handicap index and acoustic voice parameters. AIMS Med. Sci. **4**(2), 151–163 (2017)
54. Tafiadis, D., Toki, E.I., Ziavra, N.: Deviations of Voice Characteristics in Female Speech Therapy Students that Smoke Using Dr. Speech. J. Commun. Disord. Assist. Technol. **1**, 1–18 (2017)
55. Vlahavas, I., Kefalas, P., Bassiliades, N., Kokkoras, F., Sakellariou, I.: Artificial Intelligence, 3rd edn. [in Greek: Τεχνητή Νοημοσύνη]. University of Macedonia Press, Thessaloniki (2011)
56. Roy, N., Barkmeier-Kraemer, J., Eadie, T., Sivasankar, M.P., Mehta, D., Paul, D., Hillman, R.: Evidence-based clinical voice assessment: a systematic review. Am. J. Speech-Lang. Pathol. **22**(2), 212–226 (2013)
57. Verdolini, K., Rosen, C.A., Rosen, C.A., Branski, R.C. (eds.): Classification Manual for Voice Disorders-I. Psychology Press (2014)
58. American Speech Hearing Association. Voice Disorders (2016). http://www.asha.org/PRPSpecificTopic.aspx?folderid=8589942600§ion=Assessment. Accessed 11 Sep 2017
59. Demirci, R.: Fuzzy adaptive anisotropic filter for medical images. Expert Syst. **27**, 219–229 (2010). https://doi.org/10.1111/j.1468-0394.2010.00525.x
60. Pange, J., Makris, P.: Informatics for Preschool Teachers. Publication of University of Ioannina, Greece (2000)

Mobilizing the Semantic Interpreter *Pythia* – Teaching Engineering Students to Integrate GIS and Soil Data During In Situ Measurements

Maria A. Papadopoulou[1,2(✉)] and George S. Ioannidis[3]

[1] Department of Civil Engineering, University of Applied Sciences of Thessaly, Trikala, Greece
m.papadopoulou@thessaly.gov.gr, dm.papadopoulou@gmail.com

[2] Thessaly Region Local Government, Larissa, Greece

[3] The Science Laboratory, University of Patras, Patras, Greece
gsioanni@upatras.gr, ioannidis_gs@hotmail.com

Abstract. A novel mobile Semantic Interpreter *Pythia* (SI) has been created by codifying an equally novel model-driven expert knowledge algorithm. Integrating data from databases and automating soil profiling, it offers consolidated knowledge and provides much-needed help to engineers in situ, operating either on-or off-line. Thus, preparatory work volume is reduced, and so is uncertainty. *Pythia* implements an ICT-based GIS-driven model focusing on the geotechnical aspect of soil conditions, while the mobile component undertakes in situ operations under GIS-guidance, by constantly exchanging and updating data. As the first SI *Pythia* trials have already been successfully completed, the need for appropriate education and training of future engineers looms large.

Keywords: Geotechnical engineering · Engineering education GIS · Semantic interpreter

1 Introduction

Engineering education is currently undergoing a phase of rapid change. Background theoretical engineering knowledge is constantly being augmented, widening the scope of engineering studies. However, it is perhaps in the field of student training to the understanding and use of the numerous novel ICT-based devices and applications where the biggest challenge lies, as the speed of their development is ever increasing. Such ICT applications include: digital orthophotography usage by any of the standard services, remote surveying and creation of topographic maps, geolocation and time information by the use of remote sensing applications and positioning system receivers of one of the known global navigation satellite systems (GPS, Galileo, etc.). Others are the development of hybrid mashup web-application systems using content from more than one earth mapping services into a single graphical interface (e.g. Google Earth), and internet-related services and products for engineers (e.g. soil data from GIS data

M. E. Auer and T. Tsiatsos (Eds.): IMCL 2017, AISC 725, pp. 175–186, 2018.
https://doi.org/10.1007/978-3-319-75175-7_19

sources). Educational challenges to engineering students include the development of familiarity with spatial data and the GIS-aided databases, familiarity with the Information and Communication Technology (ICT) and relevant computer applications, mobile applications, and all such. Their practical importance is immense as both remote sensing technology and data communication greatly facilitate engineers capture topography and extract the necessary field measurements.

Coming as a help to the aforementioned training task, the introduction of the novel GIS-driven expert knowledge algorithm called "Semantic Interpreter *Pythia*" (thereafter SI), the function of which is expanded in a recent thesis [1], can also be used as a teaching machine utilising a suitable mobile device. It is not just a matter of training in using a powerful ICT-based tool, but also education at a deeper level involved. Indeed, SI *Pythia* could be used as an education guide, to demonstrate the correct geotechnical methodologies and teach the accepted techniques by which engineering students can process precision-data and produce unified information to be used to resolve various geotechnical issues.

2 Background Information About Geotechnical Procedures

A typical civil engineering project commences with an extensive survey [2] of the area (or site) on which the engineering operations will take place (i.e. land size, topography, adjacent features, retaining walls, etc.). This stage aims to provide a rough knowledge of the possible problems to be encountered during construction. When necessary, a full topographic survey follows, aiming to identify and draw ground contours and existing features, located either on or slightly below the earth's surface. Such would be perimeters, distances between ends/vertices, elevations, inclination of sides, building blocks, streets and building lines, lines of easements on or crossing the property, perimeter of buildings, location of manholes, of utility poles, etc.).

The natural next step is the geotechnical modelling of soil conditions [2] something related to the surficial geological layers. Together with data collection and soil profiling, this is the essential preparatory step before any further data analysis [2] proceeds. Such analysis would determine how deposits would interact with the planed man-made constructions. Examples of such analyses would include one-dimensional seismic ground response under various load conditions, liquefaction potential, soil-foundation interactions, etc. Soil modelling, as presently performed is achieved by examining "site-specific soil columns" and "soil site conditions". To increase the precision of such modelling, one aims to further subdivide largish soil deposits into conceivably useful sublayers. This process begins by defining a number of representative types of soil profiles, including the stratigraphy and properties of layers, groundwater levels, local site effects, and discontinuities. Such "fixed type" of deposits could then be utilised during modelling by further specifying the positioning, and accurately determine dimensions and attributes of each such sublayer.

As the simulation of real world in the form of soil profiles is the basis of any geotechnical study, only the sufficiency of soil data [2] can ensure the proper design of suitably reliable simulations. The best way to gain detailed depth-related information of soil data (i.e. the physical properties and spatial distribution of soil and rock underlying and sometimes adjacent to the point of interest), is to acquire or collect such data from in situ geotechnical investigations. The disadvantages are that such investigations require hard work, incur high costs, and are subject to time delays. To avoid redundant repeated investigations superfluously duplicating already existing data, there exists a ubiquitous and growing effort for the development of a global spatial data infrastructure, using open data sharing. Such information would be open to scrutiny (thereby guarantying consistency and reliability), and shall be freely shared through the web, as promoted by international scientific organizations for the past two decades (see Sect. 3.6). Anyway, sufficiency of reliable soil data demands less tedious preparatory work for geotechnical engineers or engineers geologists, and implies less uncertainly and guesswork. This way, the role of civil engineers is directed towards its most essential aspects [3], which is to formulate realistic and consistent information utilizing all available data about soil conditions.

Acquisition of knowledge implies proper processing of all available soil data, followed by subsequent analysis, thus yielding a realistic model simulating soil conditions at any relevant depth. The automated acquisition and processing of previously available or newly acquired soil data, as promoted and facilitated by the SI, leads to the efficient and timely computation of the appropriate information, suitable for any individual engineering task. As demonstrated [1], this task is best achieved using a specially designed expert system like SI *Pythia* which contains in its code all the relevant methodology.

3 The Novel Semantic Interpreter *Pythia*

3.1 Semantic Interpreters Structure

SI *Pythia* has been carefully designed and developed in accordance to its novel concept [1]. In terms of structure, the underlying algorithm can be adjusted to interface with any of the widely used DBMSs (Data Base Management Systems) to consolidate the data contained therein, as well as to manage communication with other GIS-based systems.

Although the degree of interoperability should be more precisely adjusted when referring to specific cases, technical interoperability is composed of two dimensions: (a) the source algorithm, which interacts with data to form a data-process (and analysis) application, together with (b) its own core-database. All software applications developed use open standards to ensure compatibility and quality in their implementation, and long-term viability.

By utilizing ICT, the algorithm first attempts to extract all available multi-thematic data from various reputable sources, concentrating them, and digitally re-registering them into the homogenised format of the aforementioned core database. This is first attempted using web data sharing techniques, or (in cases of genuine lack of relevant data) by asking for totally new up-to-date measurements to be taken and manually typing those, as new input in the core database.

Then, the algorithm attempts to semantically integrate this wealth of multi-thematic raw data, consolidating them to provide unified geotechnical information, fully resolved and meaningful. This process of data-integration ensures semantic interoperability (see Sect. 3.3).

3.2 Multi-thematic Data and Soil Layer Subdivision Structure

SI *Pythia* is a model-driven algorithm that is primarily intended to help civil engineers cover the gap between: (a) "common knowledge" concerning soil properties, and (b) the real geometry of soil sublayers, as these have been gradually altered in shape and attributes over time, due to the overburden weight at ever-increasing depth. Referring to its specific method of operation, the algorithm relationally links multi-thematic data from various tables of SPT (Standard Penetration Test), CPT (Cone Pressure Test), geophysical measurements, and laboratory measurements of soil samplings, together with geological descriptions, data from earth observation, topographical models, engineering geology, and hydrological models and homogenizes them with the aforementioned standardized information components.

A stratum, amongst geologically visible layering of sedimentary rock or soil, is defined as a unit having internally consistent characteristics that distinguish it from others. To improve the accuracy of computations, every geology layer of a soil deposit is further subdivided into sublayers. Although it is sometimes called "layer" in general, or even "bed", a sublayer is the smallest division of a geologic formation or stratigraphic rock series distinguishing it from other layers and differentiated in various ways, including the deposition and the consolidation due to the influence of time.

By correlating raw-data samplings collected at various depths from a specific site, *Pythia* proceeds to utilise the aforementioned "thematic tables" of relevant multi-thematic data. This correlation aims to achieve a subdivision of the strata into virtual sublayers, the geometrical shape of which would greatly facilitate further processing, and behaviour forecasting. The geotechnical properties of each sublayer having being thus fixed in terms of soil behaviour, proceeds allowing the SI to continue with computing detailed soil conditions like stratigraphy, groundwater levels, local site effects and discontinuities, to conclude with the final design of representative soil profiles. Those are presented in a form appropriate to be semantically interoperable with other such geology results (see Sect. 3.3).

3.3 Semantic Interoperability During Data Integration Following DIKW

Proceeding with data processing, *Pythia* is simulating each sublayer in shape and dimensions, determining depth, and assigning the correct geotechnical properties to each sublayer. To this effect, the algorithm has to overcome any heterogeneity issues arising during the integration of data from various sources, while consolidating the increasingly refined information produced. Semantic interoperability is achieved during the data integration procedures by honouring the concept of DIKW pyramid, as depicted in simplified form in Fig. 1.

The base level of this pyramid consists of multi-thematic data. Each superior level is comprised of information, which is increasingly refined (and generalised) as one progresses upwards. This is because the information that comes as output from previous (i.e. inferior) level, acts as data-input to the adjacent level for the subsequent computations. However, the semantic content of requests between each hierarchical level needs to be unambiguously defined: what is sent is identical to what is understood by the receiving layer. The definition of the content of each layer needs to be

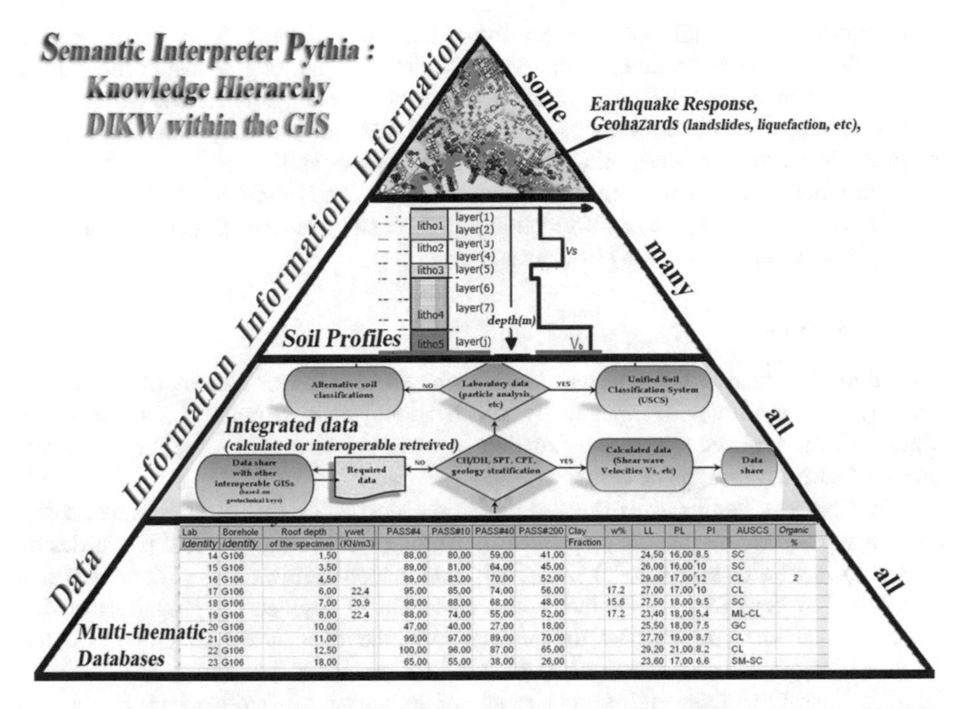

Fig. 1. DIKW knowledge hierarchy of SI *Pythia*: integrates raw, calculated, and shared data formulating composable levels of increasingly refined information; available to satisfy a variety of geotechnical demands.

unambiguous too. For example, spatial analysis is classified at a higher level than soil profiles, while soil layer characterization is set at a higher level than the values of soil properties, and so on. The classification of these levels inherent in the algorithm is original, and can be proposed as a future standard model, to achieve semantic interoperability in geoscience related computations.

Semantic interoperability concerns, in this case, the ability to automatically interpret the exchanged information in a meaningful and accurate manner so as the higher level can work flawlessly, producing useful information. As the level of usefulness of any final information produced is judged by the end users, the deliberate choice of the present design to achieve this goal was to formulate information hierarchically so as it passes gradually from one level to the one directly higher for further processing, as depicted in Fig. 1. However, intermediate level information also holds intrinsic practical value to engineers. The organizing of the intermediate results into this hierarchical structure is, therefore, an advantageous concept that allows users to choose readily available information from any level deemed appropriate, depending on the purpose of the engineering study, or even depending on the stage of the calculation.

In case data are somehow incomplete, the algorithm attempts recover and add those in its core database using data-sharing techniques. In case derivation of one or more missing values proves an impossible task, thereby leaving an overall ambiguity, the SI proceeds using alternative techniques (e.g. spatial interpolation, or other accepted computations that could even involve fuzzy logic techniques). All those adhere to standards accepted by various Geotechnical societies. This long semantic integration processing concludes with customised output results, and graphical visualisation of (vertical) subterranean soil sections offering complete soil profiling with detailed soil geometry computation, integrating physical properties of soil.

Furthermore, the algorithm organises, identifies, and relationally links data to provide consistent and precise information of a form suitable to be directly uploaded to appropriate data-bases used by engineers.

3.4 Soil Profiling

The algorithm searches, categorises, and proposes a series of similar (hence representative) soil profiles, accompanied by their relational attribute tables. Using those, the algorithm can proceed with the estimation of representative values for the soil properties of each stratum, as described previously.

This process begins with the 3-dimensional design of soil profiles (having the aforementioned representative properties), including the stratigraphy and properties of layers, groundwater levels, local site effects, and discontinuities. Then, it assesses soil conditions by depicting the geometry and geotechnical engineering properties of soil deposits in a form appropriate to compute how the deposit will interact with the intended man-made constructions (i.e. structural response to earthquake motions, bearing capacity of foundations, consistency of embankments and excavations against landslides, etc.). A number of techniques are used to quantitatively determine the representative geometry and properties of every layer, such as the interpolated construction of hypothetical data points, and the evaluation of mean property values. Both techniques work well within the range of a discrete set of known data points of adjacent site-specific soil columns. A successful soil profiling finally runs about 50–150 m vertically downwards into the ground. Each such section cuts through various layers of soil and rock types that overlay the quasi-geology bedrock (i.e. the unweathered parent rock material), or any alternative layer type compacted and cemented as it would be by the weight of the overlying horizontal layer.

3.5 ICT and Mobile SI Version

Though engineering structures are stable, humans move and so are the GIS data broadcasted by their laptops and mobile phones. Wireless ICT can access geotechnical databases to acquire information or contribute new data helpful to other engineers.

Figure 2 is depicting the mobile concept, which aims to be utilised by mobile phones or handy tablets during in situ engineering measurements, to obtain fully consolidated soil profiles accompanied by digital attribute tables of GIS-updated soil data. It inherently combines the display of technical information, syntax translation, and semantic interoperability functions to allow such exchanges between a range of computing systems, exchanging geotechnical data from several different data vendors, or even between past and future revisions of the same software product.

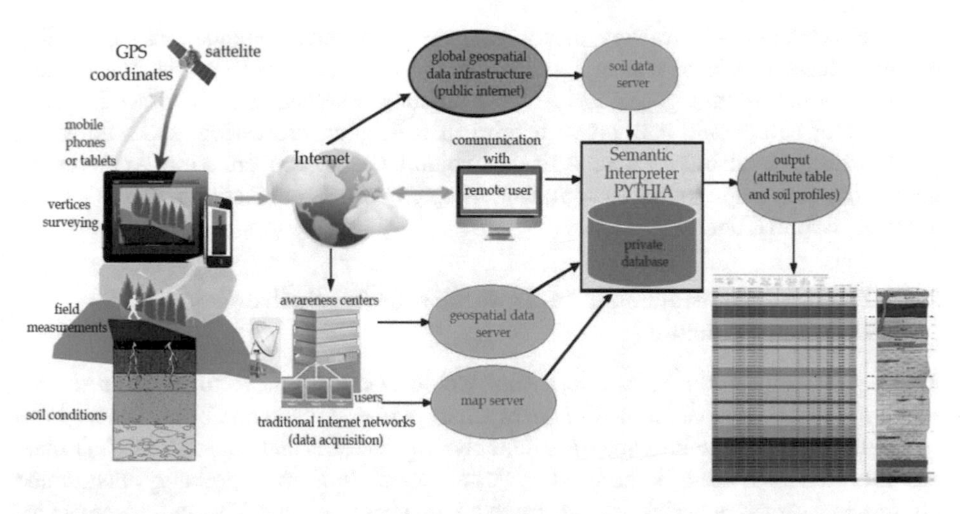

Fig. 2. Concept of the way SI *Pythia* can be called from mobile phones or tablets during in situ engineering measurements to automatically obtain full soil profiles accompanied with digital attribute tables of GIS-updated soil data.

Naturally, the next step in the evolution of this particular intelligent system would be to embed it into a suitably small Internet-connected device. Thus, it will gain the capacity to gather and analyse data in situ in direct communication with other systems, as in Fig. 3 is shown. The possibility of mobilizing the SI is the object of a current research study.

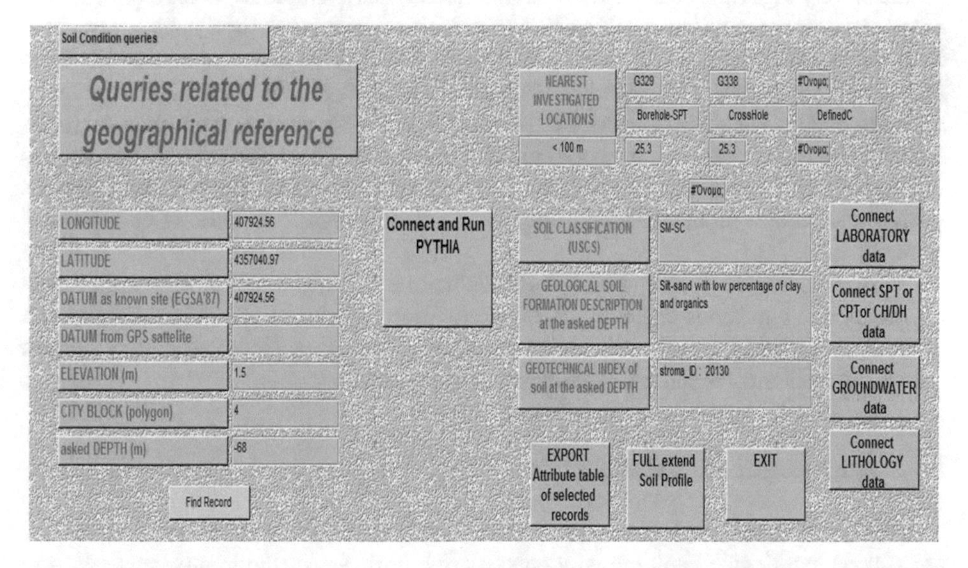

Fig. 3. User interface of the mobile SI *Pythia* for queries about the geotechnical soil condition of any geographical referenced site.

Such system could be very useful in many engineering applications demanding instant access to the SI's operation, while the engineer's position is dynamically updated by transmitting geolocation data to a database server, as its interface in Fig. 3 shows. Not only would they prove helpful in collecting, processing, and automating soil profiling, but also these measurements would be easily converted into a convenient geoengineering map-plotting system, increasing engineer's productivity, and providing accurate information.

3.6 Projected Contribution of SI Towards a Global Spatial Geotechnical Data Infrastructure

To be practically useful, a mobile *Pythia* would eventually have to work on a wide range of hand-held devices. Most modern smart-phones come with GPS or other global navigation features, while some offer suitably large screens and appropriate CPU main memory and GPU ready to be used as "thin clients" to a (most probably) distributed application server, working in tandem with database servers. The large number of different devices, screen sizes, firmware, and operating systems escalate this to a daunting task.

However, this projected mobile evolution of SI *Pythia* would prove invaluable to geotechnical data gathering, and eventually to the effort of creating (through data-sharing) a global infrastructure of geographically based engineering databases. The role of SI is to assist in acquiring, processing, merging, and to operationally offer these various data streams to all.

3.7 Semantic Interpreter *Pythia* as a Teaching Machine

Pythia's very algorithm can be used as an educational example to teach engineers the proper sequence for utilisation of available data to achieve the desired result. It can also be used to train engineers to use compatible mobile ICT devices for in situ measurements and geotechnical problem solving. In addition, it can be used educationally to account, spatially analyse, and award the most appropriate parameters to best simulate soil conditions, something that would include both geometry and mechanical properties for each of the multi-layered formations. The engineering students can then process precision-data and produce unified information corresponding to various geotechnical issues.

However, the use of SI *Pythia* transcends mere training and ventures into education (a much more ambitious objective in itself). As aforementioned, the mode of operation of SI *Pythia* could be used as guide to demonstrate what geotechnical methodologies and techniques are accepted as standard practice, and why.

4 Educational Example

The following is intended as a representative example of the way SI *Pythia* can be used as a teaching device, guiding students by its intrinsic methodology towards soil characterization (see Figs. 4 and 5) and soil profiling (see Table 1).

Given the sieve analysis, the Atterberg limits, and the SPT number of blows measurements of the 2-layered soil deposit shown in Fig. 4, layers A and B are examined up to a base-depth of 8.00 m and of 14.00 m, respectively. Formulate information provided, by answering the following questions in accordance with the Unified Soil Classification System (USCS) [4] and by adhering to all relevant consensus technical standards [5]. SI *Pythia* can be used to examine the reliability of geotechnical calculations.

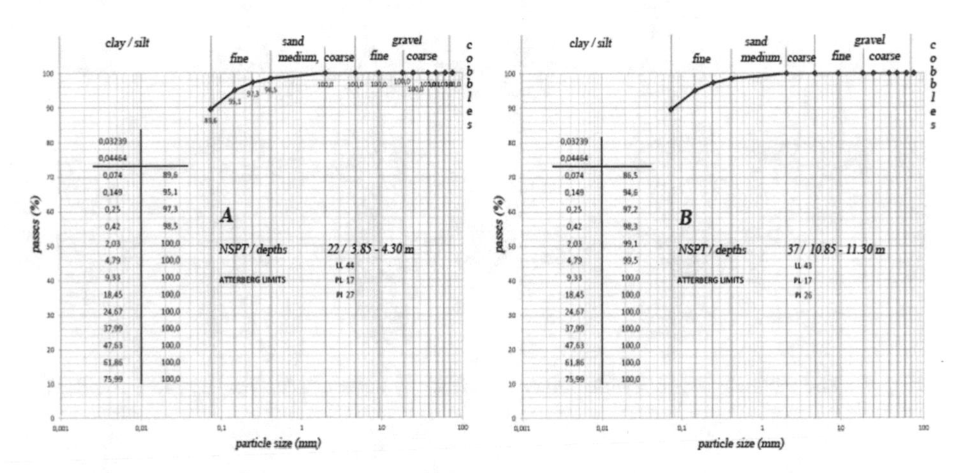

Fig. 4. Lab-tests (sieve analysis and Atterberg limits), and SPT in situ tests of the layers, A (left), and B (right) of a soil deposit (M. Papadopoulou, 2015: road-bridge construction study, Nikaia, Larissa, Greece).

Example of test-questions:

1. How is soil gradation methodologically determined.
2. What parameter (i.e. percent of soil passing a sieve) should an engineer obtain first, in order to determine the soil gradation.
3. Extract the appropriate input values from the attached diagram in order to compute and answer questions 4 and 6.
4. Characterize each layer based on the fine or coarse materials that dominate their content.
5. Characterize each layer on the basis of the particle sizes that dominate their content, as: uniformly-graded, or well-graded materials.
6. Characterize each layer after taking into consideration the plasticity index, based on the above results.
7. If in the content of each layer dominates the coarse graded grains, then characterize the corresponding layer based on the soil plasticity.
8. Attempt a proper soil classification for each layer in accordance with the AUSCS as guided by the corresponding methodology of SI *Pythia* in Fig. 5.
9. Characterize each sampling, based on the NSPT blows and EC8 soil categories.
10. Attempt the soil profile, guided by the SI *Pythia* algorithm.

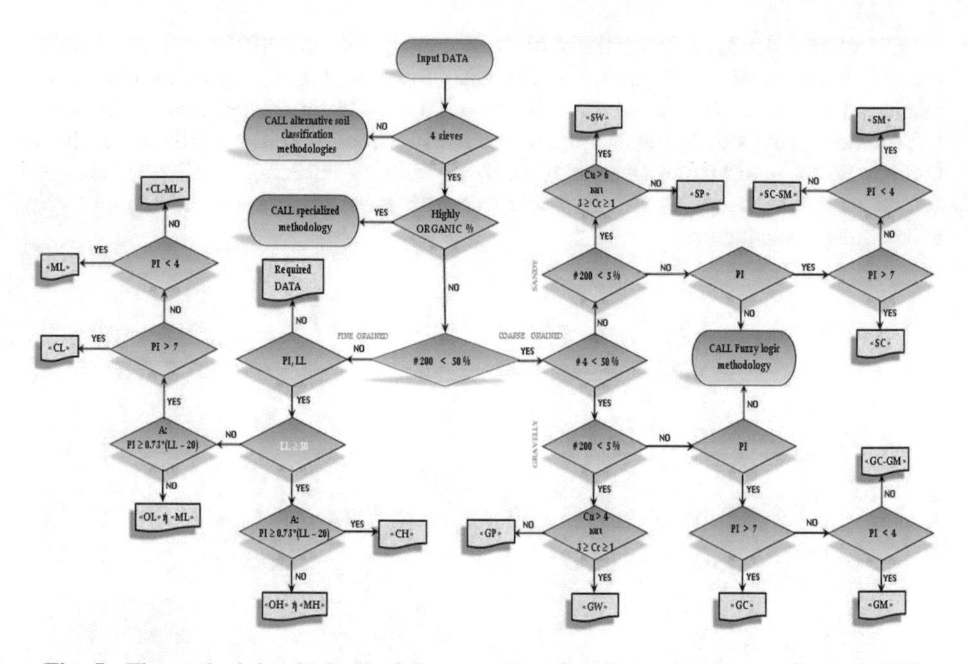

Fig. 5. The method that SI *Pythia* follows, to classify soils in accordance with the USCS.

Table 1. Indicative fields of the soil profile resulted by SI *Pythia* for the soil deposit of Fig. 5 (the results in *italic*, *: *recommended acquisition of new up-to-date measurements).*

Geology layer	NSPT (middle depth in m)	Lab sampling (middle depth in m)	Sub-layer	Thickness (m)	(Sub)layer depth (m)
1	*(*)*	*(*)*	*1*	*1*	*1.00*
1	*(*)*	*(*)*	2	*1*	*2.00*
1	4.15	*(*)*	*3*	*1*	*3.00*
1	4.15	*(*)*	*4*	*1*	*4.00*
1	4.15	*(*)*	*5*	*1*	*5.00*
1	4.15	6.00	*6*	*1*	*6.00*
1	*(*)*	6.00	*7*	*1*	*7.00*
1	*(*)*	6.00	*8*	*1*	*8.00*
2	*(*)*	*(*)*	*9*	*1*	*9.00*
2	11.15	*(*)*	*10*	*1*	*10.00*
2	11.15	12.5	*11*	*1*	*11.00*
2	11.15	12.5	*12*	*1*	*12.00*
2	11.15	12.5	*13*	*1*	*13.00*
2	11.15	12.5	*14*	*1*	*14.00*

5 Expected Outcome from the Educational Trial

One educational objective for engineering students is to train in the use of the novel software, and operate it in accordance to commonly accepted geotechnical practice. Such tasks comprise soil grading, field data taking concerning soil (and rock), calculation of additional parameters (coefficients of uniformity, of curvature, etc.) and so on, based on the geotechnical criteria and the consensus for technical standards. Using the present version of SI *Pythia*, such training includes: (a) To classify the soil and account the essential parameters for each of the multi-layered formations. (b) To spatially analyse, and award the most appropriate parameters to best simulate soil conditions. (c) To practise the method of subdividing any soil deposits into sublayers. As a corollary, find the way to assign geometrical and mechanical properties to each sublayer. (d) To practise on how to extract/select the necessary input data using the accompanying diagrams. (e) Also how to use the relationships and the empirical criteria for these classifications, in accordance to all formal rules of geotechnical engineering and empirically accepted criteria.

The educational objective, deeper as it really is from mere training, involves the use of the underlying structure of the SI to teach the interrelation of the various geotechnical parameters in engineering practice. However, good intentions alone do not ensure proper training. The results of the educational trial are essential to judge the level of its effectiveness.

6 Discussion

An imminent educational trial would prove doubly beneficial. Firstly, while training future engineers to the joys of handling what will effectively become a predominant tool of their future careers, any anticipated (yet unspecified) ergonomic and practical shortcomings would surely emerge, initiating a product improvement process. Furthermore, real educational benefits do arise from the expert guidance in applying trusted geotechnical methods to adapt soil data elevating them to composable levels suitable for further processing. Students would learn from the experience, as in addition to 4G mobile or Wi-Fi connectivity issues, they will be exposed to security related ones, all the while gaining remote monitoring and management experience, when achieving the integration of geographical and soil conditions into a single graphical interface. As intelligent systems comprise not just intelligent devices but also sets of interconnected such devices, often involving much more powerful computers (or clusters thereof), student learning involves the remote handling of sophisticated AI-based software systems (e.g. expert systems with chatbot interfaces). Furthermore, it is also expected (and envisaged) that skilled Engineer training and GIS technology will evolve in parallel in the near future, as system quality and reliability evolves, and new challenges emerge.

The benefit of the aforementioned technologies arises from their ability to extract and exchange relevant information utilising web-GIS, therebynot only solving individual locally-defined engineering problems, but also gradually creating a global infrastructure of spatially defined open data-sharing databases. A mobile web-connected GIS system is the ideal platform for such complex data processing and analyses, and is adaptable to various relevant yet diverse tasks. Such mobile devices may be assuming the role of "thin client" in interconnected IoT systems, and be either tailor-made units, or suitably customised general-purpose smart mobiles. As processing and displaying power of such intelligent terminals increases, they might undertake an increasingly larger share of the tasks for completion locally. The need to train future engineers becomes apparent, and so is the need to fine-tune the system interfaces (including visualisation models) to suit the multitude of tasks undertaken by engineers. These two tasks being interconnected, it is the aim of the authors to tackle them in parallel.

References

1. Papadopoulou, M.A.: Automated methodology for seismic hazard microzonation studies of interoperable geographic information systems – the case study of a Hellenic city. Doctoral/Ph. D. thesis, Department of Civil Engineering, University of Thessaly, Greece (2017). (in Greek)
2. Buck, P.: Soil mechanics. Department of the Navy, Naval Facilities Engineering Command, Section I: Civil Works Projects (1992)
3. Barnes, G.: Soil Mechanics: Principles and Practice (2011)
4. ASTM D2487-11, Standard Practice for Classification of Soils for Engineering Purposes (Unified Soil Classification System), ASTM International. https://doi.org/10.1520/d2487-11
5. ASTM D422-63(2007)e2, Standard Test Method for Particle-Size Analysis of Soils (2007–2016), ASTM International. https://doi.org/10.1520/d0422-63r07e02

Mobile Games – Gamification and Mobile Learning

A Game-Based Learning Platform for Vocational Education and Training

Dimitrios Kotsifakos(✉), Xenofon Zinoviou, Stefanos Monachos, and Christos Douligeris

Department of Informatics, University of Piraeus, Piraeus, Greece
dimkots@sch.gr, xzinoviou@gmail.com,
stefanosmonachos@hotmail.com, cdoulig@unipi.gr

Abstract. This article records our contribution to the international debate on the use of modern technologies in the learning process. Our theme is the development of online games that support the learning processes of Vocational Education and Training. The medium we used for this implementation concerns an online platform that includes polymorphic games. The learning content of the platform refers to topics related to web-based algorithms, techniques and data flow methods, such as the Bellman-Ford, Dijkstra, Floyd, and Johnson algorithms. Each digital scenario has a specific goal. The completion of each scenario will familiarize the student/user with the process of the particular algorithm and how each algorithm is implemented in a networking environment. Our main goal of developing the platform is to use all the available modern technologies to create a strong overall learning experience through gaming that will also broaden the students' knowledge of network algorithms.

Keywords: Game-based-learning · Vocational education training
Network flow algorithms · Education · Web technologies

1 Conceptual Framework

1.1 Game-Based Learning for VET

In an effort to predict the future, Santor, Turin's servant and mentor, used to talk about the evanescence of things: "Everything is temporary. I suppose, the joy of creation is the solitary purpose of time" (Tolkien 2007: 27). In this article, we avail ourselves of modern technologies combining the learning process with the joy of creating. What we seek through this work is to understand the connection between the time and the effort needed by the student to master the knowledge provided through the use of online media, such as online platforms. More specifically, we are interested in how game-based learning is evaluated in the area of Vocational Educational Training (VET). Our implementation is based on a website where a student-user can participate in a web-based game with the subject: "Algorithms". The game is divided into stages, where each stage has a cognitive base which is a particular network algorithm. The learning goal of this project is that the students-users, after finishing the game, will apply the acquired knowledge on the subject of algorithms and their applications by themselves.

M. E. Auer and T. Tsiatsos (Eds.): IMCL 2017, AISC 725, pp. 189–200, 2018.
https://doi.org/10.1007/978-3-319-75175-7_20

It is important to stress that research in the area of game-based learning in VET remains minimal. The preparation of VETs' students for the future working environment (Hillier 2009) faces some serious challenges. The nature of the knowledge that must be instilled demands the use and the design of new tools along with the optimization of the ones that are already available (Mavrikios et al. 2013). The scope of VET is to teach, educate and train. In what has been said, we need to add the necessity and the responsibility for the modernization of all the inventions and the applications of the VET curriculum that have taken place in VET specialty (Braha and Maimon 1997). These inventions and procedures are known with the term "technique" (Hubka and Eder 2012). A "technique" is based on knowledge-productive application sectors such as radar stations, tools and communication protocols, high frequency production mechanisms, and information and data transmission in technical and industrial production (Fujigaki 1998; Bourdieu 1975). Among these, we chose to focus our work on the algorithms that are used in the internet core and on data flow topics, techniques and methods.

This paper introduces an innovative, educational learning platform in the field of network algorithms according to both Greek and international educational standards. We establish in network course and Internet algorithm. The Internet algorithms are necessary for the proper functioning of the computer networks. Such chapters for chapters with corresponding content for the lesson of network are included in the curriculum of several classes of Vocational Education Training (Billet 2003). We acknowledge the fact that introductory courses in algorithms have already been successfully introduced not only Secondary and Post-Secondary General High Schools (GEL in Greece) but also in Vocational High Schools (EPAL in Greece) (Doukakis et al. 2014) and their significance and value have been thoroughly presented in the literature (Kotsifakos and Douligeris 2015).

According to Retalis et al. (2005), modern web technologies have introduced significant changes in the current educational map. It is important to emphasize the significance of using modern technologies in education, as this practice enhances the learning process. Depending on the educational philosophy each teacher espouses invaluable technological tools which can facilitate the learning experience of students (Douligeris et al. 2013). Furthermore, it should be highlighted that VET students as well as students who experience learning difficulties can greatly benefit from modern technologies (Kotsifakos et al. 2016). However, if we want to maximize the potential of modern technologies in education, there still exist major obstacles to overcome, such as the various school regulations and the often strict education policies in Vocational High Schools and Laboratory Centers (LC). For example, regulations about the use of cellular phones and other electronic devices (tablets, laptops) in Secondary Schools should become less rigid, since the digital skills are most likely to be enhanced only in a modern-classic education method. Several countries have already set national strategies to foster and develop digital skills in VET (People, machines, robots and skills 2017).

1.2 General View on Educational Issues

Stand up and tell me the 'This is the lazy man's voice' "said the griffin. "Oh these creatures! They like to command and make you memorize the courses!" Alice thought,

"Same thing with school" (Carroll 1865: 118). Two and a half centuries after these lines were written by L. Carroll, education does not seem to have undergone much progress in terms of teaching practices (Kontaxis 2017). Many parents as well as most students tend to consider that education remains dull and ineffective. It is true that in order to introduce post-modern web technologies into education (like for example the Web2.0 technologies) new and innovative educational approaches should be sought. Even today, modern schools seem to be preoccupied with the same student motivation issues, whilst the student's engagement in the school subjects remains a moving target (Lee and Hammer 2011). The use of educational games as learning tools in education seems very promising, especially with the abundance of modern technologies that can be easily adapted and used in the classroom (Zichermann and Cunnigham 2011; Deterding et al. 2011; Whitton and Moseley 2012). This is the reason why over the past few years, an international debate over the issue of incorporating gaming in school courses is taking place, with an ultimate goal to reap the maximum possible benefits of such a practice (Sheldon 2011; Sheldon 2014). In our work, we used all the existing knowledge in order to define and re-orientate the educational expectations through our platform. More specifically, we expect to have significant findings in the students' progress after a student has interacted with the platform for a certain period, whilst we expect interesting findings regarding the way teaching staff benefits as well.

In the following chapters, the technologies used in the development of the platform will be presented. Indicative examples of the features of the platform will also be illustrated. We will express our personal expectations for this project along with conclusions and data we have gathered so far. In the closing paragraph, our future work plans, future platform features and extensions will also be delineated.

2 Technical Structure - Design and Implementation

Our main goal was that the platform would be developed combining three significant factors: cost-efficiency, resource-efficient implementation and cross-browser compatibility. Based on these pillars, the platform was designed using the following technologies:

HTML5: HTML5 (Hyper Text Mark-Up Language Standard, 5th version) is used for designing, structuring and developing the platform content. Its main advantage is that it can be rendered by all modern browsers.

CSS3: CSS3 (Cascading Style Sheets, 3rd Version) is used for styling the content of the platform in a very efficient way, allowing the developer to make swift changes whenever needed without having any other dependencies. Tagging the elements of the html code with classes and ids makes it very easy to alter the layout of the content efficiently and, most importantly, quickly.

PHP: ((Personal HomePage Tools), Hypertext Preprocessor, 7.1 version) is used as the server-side scripting programming language. Thanks to PHP, we can access the platform's database and manipulate, save or insert data in it.

Javascript: Javascript is used as the client-side scripting programming language. Javascript's main feature is that it is loaded in all the modern browsers, so we use it to build the user-platform interaction. Maximizing the user experience (UE) is seriously taken under consideration.

Bootstrap 3: Bootstrap 3 is a framework that enables us to have a responsive design in the browser content, regardless of the medium it is browsed on. Technically, we use it to make our content adapt perfectly to every screen format we have without sacrificing any UE quality.

JQuery: JQuery is one of the most well-used Javascript libraries enabling the platform developer to be more efficient, having the same results while writing less code.

Ajax: Ajax (Asynchronous Javascript Xml) is a technology based on the server-client architecture. Ajax is used to render specific parts of the content, which means that when a request is sent, the server will respond but only to that part of the content that concerns the request. Thus, Ajax is used to reduce the response time since only a specific part of the content is refreshed.

MySQL: MySQL is the open source database management system used in our platform for storing, updating and retrieving user data. One of its strong points is that it has no platform dependencies. It is also user-friendly and can be easily managed through the phpMyAdmin interface (Fig. 1).

The Relational Model schema of the Database used in the project is presented in Fig. 1. Four tables have been used to help keep track of user info, statistics, high scores, questions and answers (both right and wrong). The "users" table is used for storing user data and each user is uniquely identified system-wise by his name. The "high_scores" table is used for storing high scores related to each user. The "courses" table is used for storing all the information regarding the title of a course, the author and the course itself. Last but not least, the table "questions" is used for storing information for each question; data that is valuable for the statistics.

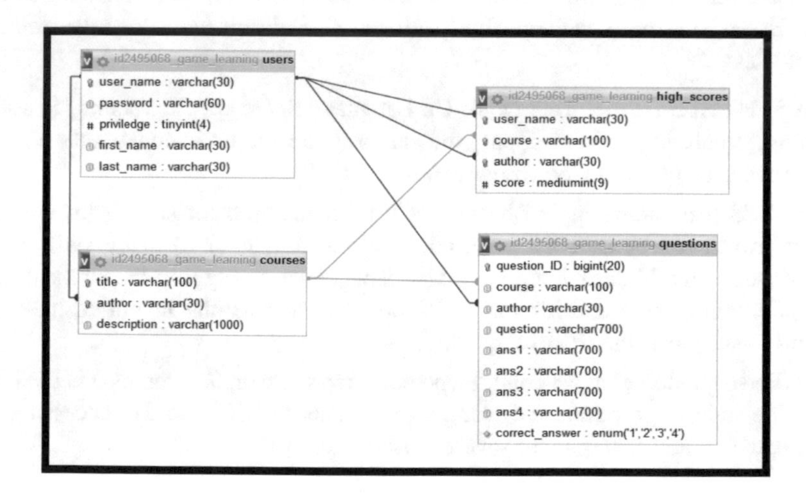

Fig. 1. Schema of the database used in the platform.

3 Case Study - Learning Objects

Each student-user will initially create a password-protected customized profile account which will enable him or her to use the features of the platform. All the students-users' data are saved in the database of the platform. Each student-user can progress through stages and can have access to invaluable personal behavioral information, such as scoring, competition, and successful tries. During their interaction with the game, students will be presented with educational information concerning the algorithm and its function in the digital playground of the game. Students can select any of the available subject learning areas in the platform and enhance their cognitive background. They are free to repeat each stage as many times as they like. For the development of this specific platform, we have considered all the potential factors so that the educational process can benefit from the use of modern web technologies. Therefore, students progress through stages enriching their knowledge and skills in the field of network algorithms. The development of this cognitive framework is based on the structural elements and procedures in the field of communications that take place at both the national and the international levels. We aim to contribute to an innovative and futuristic way of thinking regarding the skills and the philosophy of future engineers and employees. Specifically, European VET trends indicate that digital skills are included in syllabi for both students and teaching staff. This is one of the fundamental reasons why it is vital that the knowledge of digital skills be accessible on a larger scale in Europe. Not only will high-industrialized and high-income countries benefit from this accessibility, but also countries with a weaker economy. A competent employee can contribute substantially to a company's growth whist the latter can become more competitive and profitable if it ensures that its employees undergo continuous training.

Students that will use this platform should have a basic knowledge of the courses' content. The courses are designed in a multiple question format so that the student will experience a question-game participation feeling. The platform can also be used by teaching staff, who will be able to create new courses with questions on any subject of their preference. Regarding the user-side learning objects, the Student Users will be able to learn and enhance their theoretical background through interaction with the platform, in specific course subjects set by the teaching staff. Regarding the learning objects on the teaching staff-side, the Teachers will be able to learn how fast a user can enhance his theoretical background in a not-previously known subject. As an example from the use of platform, we choose the Bellman-Ford Algorithm. Bellman Ford is an algorithm with which the shortest path from a node to all the other nodes in a weighted graph can be found. The users will be able, without having any previous knowledge on the specific algorithm, to learn what this algorithm does and how it works. In addition, they will learn the circumstances under which it is best to use this algorithm.

4 Platform Roles - Handling

4.1 Platform Access - Platform Roles - Segmentation

To have the platform fully operational and supported, discrete roles with access permissions have been granted to each user segment. Five Role segments form the complete platform access management model. The roles are divided into 5 segments:

Architect, Developer, Administrator, Teaching Staff – Teacher, Student (Platform Access – Roles, Table 1). This role segmentation has been carefully decided, as the most functional regarding the platform, the platform safety, the operating procedures and the required technical support.

Architect Role: The Architect organizes the conception of the platform model and is responsible for designing and supervising the implementation. Therefore, unrestricted access to the Back-End of the platform has been granted to this role.

Developer Role: The Developer is responsible for the implementation of the Architect's concept. His role is to build the platform and to maintain it, in other words to support it. Unrestricted Back-End access has been granted to this role as well.

Administrator: The Administrator is responsible for managing requests from teaching staff and students. Such requests confirm a new student account, confirm a new member of the teaching staff account, and confirm a new teaching staff request for a new course creation or for the deletion of platform. Since there is no need to access the full featured Back-End of the platform, restricted Back-End access has been granted.

Teaching Staff – Teacher: The Teaching staff or the teacher will be responsible for creating new courses in the platform. In these courses, the teaching staff must have access to enter the questions and save them. For such actions, they are granted with unrestricted Front-End platform access.

Student: The Students are the ones with the least platform access privileges. They only need to be able to sign up and play. No special access is needed and no special requests can be made.

Table 1. Platform access - roles

Roles	Platform level access			
	Back-End (Unrestricted)	Back-End (Restricted)	Platform interface (Unrestricted)	Platform interface (Restricted)
Architect	•			
Developer	•			
Administrator		•		
Teaching staff			•	
Student				•

4.2 Platform Handling

When students use the platform, they will encounter the platform's first screen, the Login screen (Fig. 2).

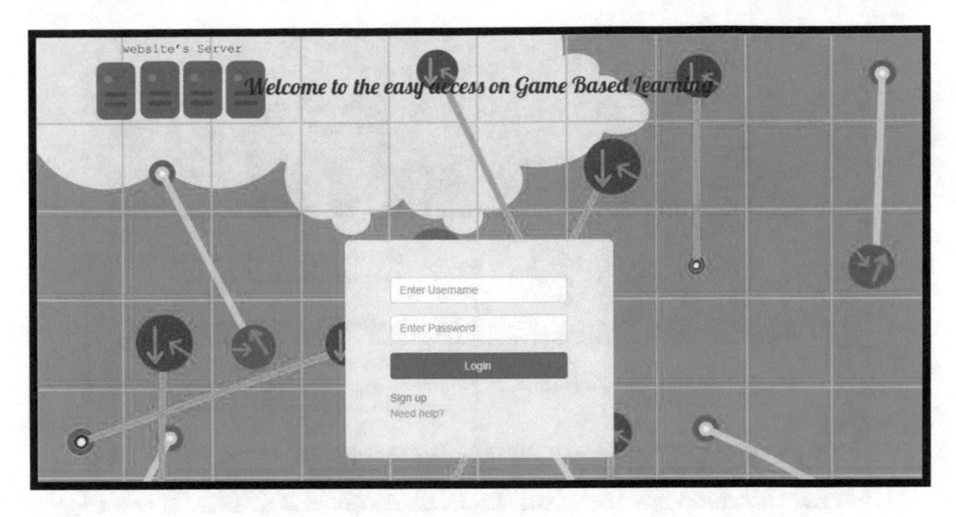

Fig. 2. Login screen

If a student doesn't have an account he must create one by entering his credentials. It is the only way he can use the platform (Fig. 3).

After creating an account, the student will be transferred to the course selection segment, where he can choose which course he wants to open. As mentioned above, the course is in a multiple question format, where only one can be the right answer (Fig. 4). The url of the game is http://monzin.000webhostapp.com/login.html (Fig. 5).

Teachers have access to create, edit and delete courses, as shown in Fig. 6.

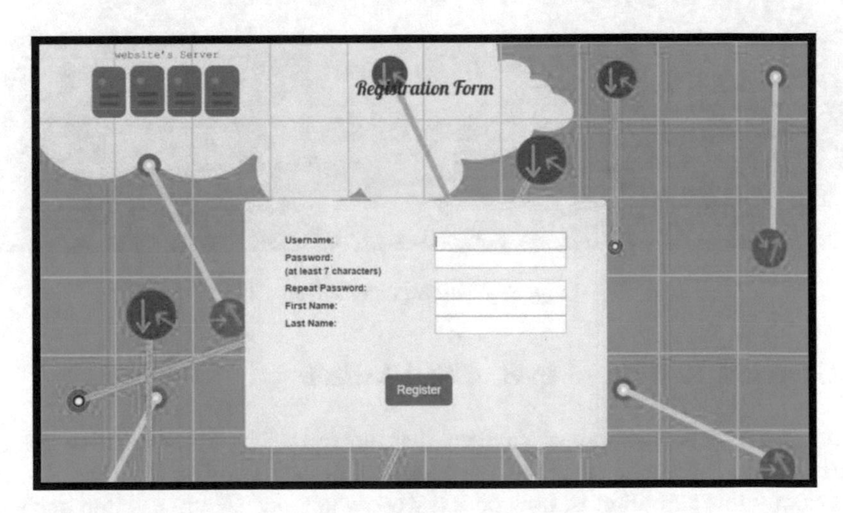

Fig. 3. Registration form

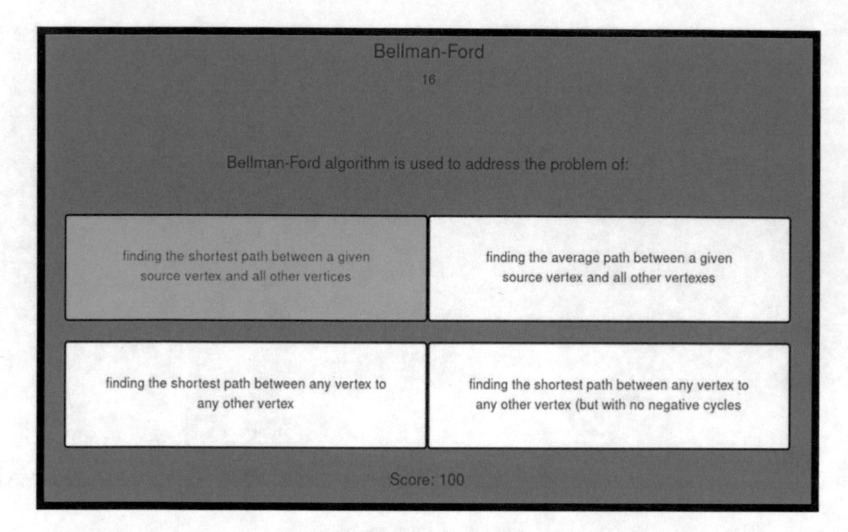

Fig. 4. Indicative Bellman-Ford algorithm questions

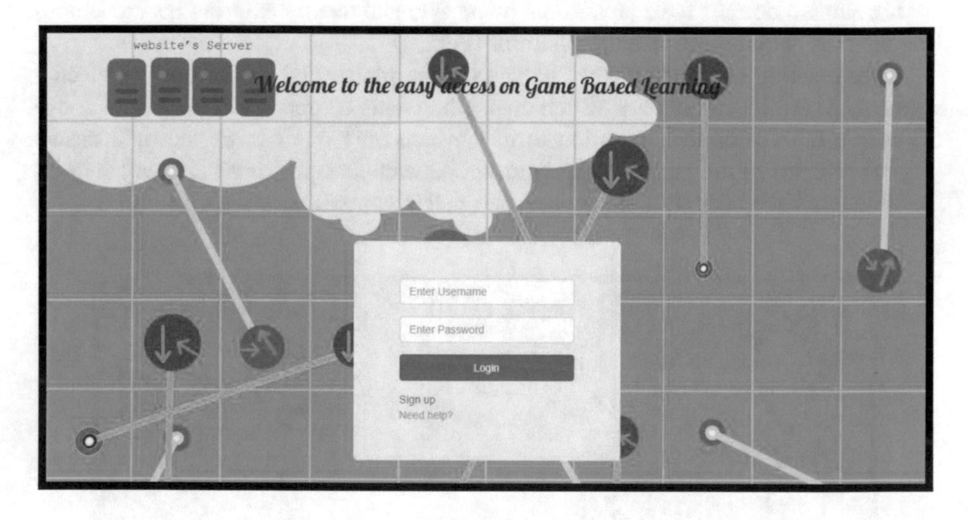

Fig. 5. Platform entry screen

5 Expected Results - Level of Satisfaction

On the student-side, we expect students that have previous knowledge on what an algorithm is, to fully understand the transitions and properties off an algorithm: to know what exactly this algorithm is, how does it work, when one should use it to one's best interest and which are its most identifying characteristics. We expect to see students having enhanced their academic background in a way they haven't thought as a learning procedure. On the teaching staff side, we expect to see the teaching staff interact with the platform and have a better overall understanding of the students'

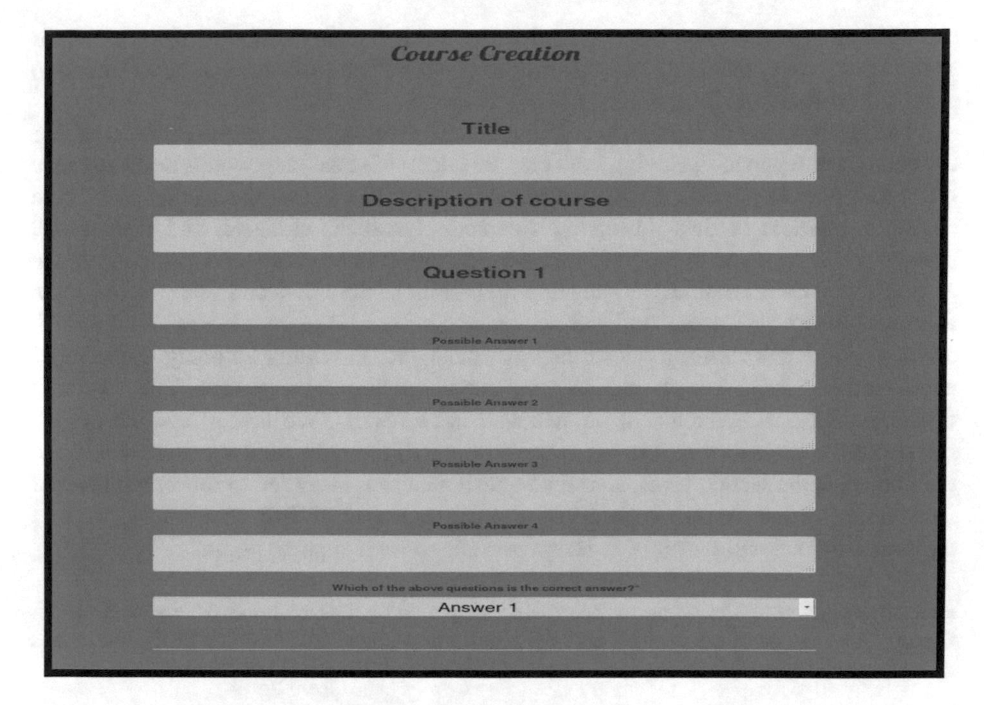

Fig. 6. Course creation form

theoretical background. Based on the students scoring, the teaching staff should be able to extract useful data concerning their initial status and how they have progressed, as they interact with the platform. From our point of view, as developers of this project, we expect to extract significant data regarding the platform interaction frequency on students and how they enhanced their theoretical background, judging by their scoring progress. The more the interaction with the platform increases, the better and more precise the data that will be collected.

6 Conclusions and Future Work

Regarding the design of the platform, we took under consideration all the difficulties that have been identified regarding the teaching of algorithms in Education (Tzimogiannis and Georgiou 1999). We also examined the evaluation criteria in the teaching material and the educational software used in all the VET's majors and in all the course programs (Gasouka et al. 2011). Relying on our observations, we tried to surpass the traditional teaching methods by designing a process which spurs game-based learning (Lin 2011; Gallear et al. 2014). Hence, we applied the fundamental principles of a learning-centered web game structure, using a set of modern web programming techniques (Pivec et al. 2003; Schell 2008; Ferdig 2009; Jaffal 2014; Bogost 2016). By using the platform, the students strengthen their knowledge about the issues of using and applying algorithms in the field of communications and on the web. The strengthening of this knowledge is achieved with the levels that the platform provides to the user - easy, moderate and difficult user levels. The platform can be extended to other algorithms besides the ones described above.

The user results collection is continuous and automatically stored in the database for each user separately. Having collected this data, the next step would be to evaluate them and plan improvements for the platform. The idea is that the platform will first collect a plethora of data during the interaction with the students, and then it will present them in a more sophisticated way to the teaching staff and possibly to the students as well (as a motive to develop their skills). More data can be collected, such as timestamps from starting and ending a course, or how long it took a student to finish a course, how many wrong questions in a course etc. Providing an analytical view of each student interaction should be very useful to the teaching staff. The thorough evaluation of our research will be achieved in two ways. We first record the performance of the students who use the platform through the individual scores stored in the database. On the second level, an online questionnaire is designed to reflect the level of satisfaction of the users involved in the process. Based on this assessment, we will expand, improve and possibly redesign various student support points.

Acknowledgments. This work has been partly supported by the University of Piraeus Research Center. The picture of the first entry of the platform was designed by Dimitrios Tzilivakis, who is an undergraduate student at the Department of Informatics, University of Piraeus.

References

Tolkien, J.R.R.: The Children of Hurin. Harper Collins Publisher (2007). Tolkien, C. (ed.) For Greece Publishing Ed. Aiolos, Athens

Hillier, Y.: Innovation in Teaching and Learning in Vocational Education and Training: International Perspectives. Research Overview. National Centre for Vocational Education Research (NCVER) (2009)

Mavrikios, D., Papakostas, N., Mourtzis, D., Chryssolouris, G.: On industrial learning and training for the factories of the future: a conceptual, cognitive and technology framework. J. Intell. Manuf. **24**(3), 473–485 (2013)

Braha, D., Maimon, O.: The design process: properties, paradigms, and structure. IEEE Trans. Syst. Man Cybern.-Part A: Syst. Hum. **27**(2), 146–166 (1997)

Hubka, V., Eder, W.E.: Theory of Technical Systems: A Total Concept Theory for Engineering Design. Springer Science & Business Media, Heidelberg (2012)

Fujigaki, Y.: Filling the gap between discussions on science and scientists' everyday activities: applying the autopoiesis system theory to scientific knowledge. Soc. Sci. Inf. **37**(1), 5–22 (1998)

Bourdieu, P.: The specificity of the scientific field and the social conditions of the progress of reason. Soc. Sci. Inf. **14**(6), 19–47 (1975)

Billett, S.: Vocational curriculum and pedagogy: an activity theory perspective. Eur. Educ. Res. J. **2**(1), 6–21 (2003)

Kotsifakos, D., Douligeris, C.: The Influence of Algorithmic Thinking on the Course of Computer Networks. In: 32th Congress of the Mathematical Society, Kastoria, Greece (2015)

Doukakis, S., Douligeris, C., Karvounidis, T., Koilias, C., Perdos, A.: Introduction to the Principles of Computer Science. Institute of Computer Technology and Publications – Diofantos, Greece (2014)

Tzimogiannis, A., Georgiou, B.: The difficulties of secondary school students in implementing the control structure for algorithms development. A case study. In: Proceedings of the Panhellenic Conference "Informatics and Education", Greece, pp. 183–192 (1999)

Gaskouka, M., Fokialis, P., Chionidou, M., Vassiliadis, A., Efthimiou, H., Doukakis, S., Siomadis, B.: Criteria for Evaluation of Educational Material for Classes-TEE Specialties, Greece (2011). http://repository.edulll.gr/edulll/handle/10795/990

Lin, Y.S.: Fostering creativity through education–a conceptual framework of creative pedagogy. Creative Educ. **2**(03), 149 (2011). http://www.scirp.org/Journal/PaperDownload.aspx?paperID=6710

Gallear, W., Lameras, P., Stewart, C.: Serendipitous learning & serious games: a pilot study. In: 2014 International Conference on Interactive Mobile Communication Technologies and Learning (IMCL), pp. 247–251. IEEE, November 2014

Pivec, M., Dziabenko, O., Schinnerl, I.: Aspects of game-based learning. In: Proceedings of I-KNOW 2003, Graz, Austria, 2–4 July 2003, pp. 216–225 (2003)

Schell, J.: The Art of Game Design: A Book of Lenses. Morgan Kaufmann Publishers is an imprint of Elsevier (2008)

Ferdig, R.E. (ed.): Handbook of Research on Effective Electronic Gaming in Education. Published in the United States of America by Information Science Reference (an imprint of IGI Global) and in the United Kingdom by Information Science Reference (an imprint of IGI Global) (2009). ISBN 978-1-59904-808-6 (hardcover) – ISBN 978-1-59904-811-6 (e-book)

Jaffal, Y.: A Practicl Introduction to 3D Game Development, 1st edn. (2014). bookboon.com. ISBN 978-87-403-0786-3

Bogost, I.: Play Anything. The Pleasure of Limits, the Uses of Boredom and the Secret of Games. Basic Civitas Books, New York (2016)

Retalis, S., Avouris, N., Anastasiadis, P., et al.: Advanced Internet Technologies for Learning. A. A. Kastaniotis, Greece (2005). ISBN: 960-03-3983-X Eudox: 16993

Douligeris, C., Mavropodi, P., Kopanaki, E.: Internet Technologies: Operating Principles & Programming of Internet Applications, Athens, Greece. Publications New Technologies (2013). ISBN 978-960-6759-90-1

Kotsifakos, D., Adamopoulos, P., Douligeris, C.: Design and development of a learning management system for vocational education. In: Proceedings of the SouthEast European Design Automation, Computer Engineering, Computer Networks and Social Media Conference, pp. 110–117. ACM, New York (2016). http://doi.acm.org/10.1145/2984393.2984413

People, Machines, Robots and Skills (Cedefop), 2017 Briefing note – 9121 EN Cat. No: TI-BB-17-003-EN-N ISBN 978-92-896-2316-2. https://doi.org/10.2801/057353. European Centre for the Development of Vocational Training. http://www.cedefop.europa.eu/en/publications-and-resources/publications/9121

Carroll, L.: Alice's Adventures in Wonderland (1865). Reprint Ed. Papadopulos, Athens (2006). ISBN 960-412-6024

Kontaxis, A.: How happy are our students and what can we do?, Greece (2017). https://www.esos.gr/arthra/50564/poso-eytyhismenoi-einai-oi-mathites-mas-kai-ti-mporoyme-na-kanoyme

Lee, J.J., Hammer, J.: Gamification in education: what, how, why bother. Acad. Exch. Q. **15**(2), 146 (2011). https://www.uwstout.edu/soe/profdev/resources/upload/Lee-Hammer-AEQ-2011.pdf

Zichermann, G., Cunnigham, C.: Gamification by Design: Implementing Game Mechanics in Web and Mobile Apps. O'Reilly Media (2011)

Deterding, S., Dixon, D., Khaled, R., Nacke, L.: From game design elements to gamefulness: defining "Gamification". In: Proceedings of MindTrek 2011, Finland (2011)
Whitton, N., Moseley, A. (eds.): Using Games to Enhance Learning and Teaching: A Beginner's Guide. Routledge (2012)
Sheldon, L.: The Multiplayer Classroom: Designing Coursework as a Game. Cengage Learning (2011)
Sheldon, L.: Character Development and Storytelling for Games. Nelson Education (2014)

An Overview of Location-Based Game Authoring Tools for Education

Spyridon Xanthopoulos(✉) and Stelios Xinogalos

University of Macedonia, Thessaloniki, Greece
{spyridon.xanthopoulos, stelios}@uom.edu.gr

Abstract. Nowadays, computer literacy is common at a very early age. As mobile devices and new social communication trends are increasingly integrating in the modern digital lifestyle new opportunities arise for harnessing this potential for innovative serious purposes. Location-based games that layer educational activities with game play can help players gain a new perspective through active engagement. In this paper, a range of key issues for implementing location based games for educational purposes using the two most commonly used open source authoring tools ARIS and TaleBlazer are presented. Although the reviewed authoring tools vary in functionality, they share a fairly common conceptual model and in both cases the software architecture is comprised of the same components. Both platforms provide the required game mechanics to create sophisticated gameplay interactions and game logic extension. However, a number of issues should be carefully considered. The conceptual models are analyzed with a critical view, pointing out limitations and challenges that can inform the design and implementation of next generation authoring tools.

Keywords: Location-based games · Serious games · Mobile learning

1 Introduction

Nowadays, computer literacy is common at a very early age as young people have spent almost their entire life interacting with diverse forms of computers and mobile devices. Over the last years technology has become predominantly mobile and ubiquitous; according to Gartner, global sales of smartphones to end users totaled 344 million units in the second quarter of 2016, a 4.3% increase over the same period in 2015. Furthermore, electronic games have become increasingly popular especially among young people; according to Google more than 1.5 billion apps and games are downloaded each month. As mobile devices and new social communication trends are increasingly integrating in the modern digital lifestyle, new opportunities arise for harnessing this potential for innovative serious purposes.

There is a considerable interest from researchers in exploiting the unique capabilities of mobile technologies to enhance innovative and engaging forms of learning, such as location-based games for educational purposes. Such games form a subcategory of location-based serious games (intended to support serious activities in a playful way) and the action is taking place in the real world enhanced with a virtual space where the player's location obtains a central role in configuring game play. Although

M. E. Auer and T. Tsiatsos (Eds.): IMCL 2017, AISC 725, pp. 201–212, 2018.
https://doi.org/10.1007/978-3-319-75175-7_21

there is evidence in the literature reporting increased intrinsic motivation through engaging and enjoyable learning experiences, the use of pedagogical games still is controversial and educational practitioners decline to use them. Literature mentions several reasons that can pose barriers, such as missing teacher efficacy in using appropriate software [11, 12, 29], the association of games with extra workload and unclear learning results, and logistical issues such as lack of technical support and equipment [29]. Moreover, remarkably little research exists that directly addresses educational practitioners [12].

This paper aims to explore location-based games implemented for educational purposes from a technical perspective and how free and open source authoring tools can support the implementation of such games. The specific aims of this study are: to review the different types of location-based games described in literature that are applicable to learning; identify common technical requirements for authoring tools; evaluate free and open source authoring tools used in research with a critical view pointing out problems and challenges as well as presenting key issues and guidelines to inform the design of such tools.

The rest of the paper is organized as follows. In Sect. 2, a review of the existing literature focusing on the area of practices, requirements and directives for tools used for authoring location-based experiences and games is presented. In Sect. 3, typical paradigms of location-based games for educational purposes are also presented and a number of common requirements for authoring tools in order to support effectively the creation of games by non-programmers are identified in three groups: authoring functionality; admin tools; and end-user client app functionality. In Sect. 4, the conceptual model adopted by the authoring tools ARIS and TaleBlazer is presented and analyzed. Finally, a comparative analysis of the tools is conducted, pointing out limitations and challenges that can inform the design and implementation of next generation authoring tools.

2 Related Work

Although there is a plethora of authoring tools promising to ease the development of locative interacting experiences there is a limited number of surveys focusing on creative practices, requirements and directives for tools to author location-based experiences and games to inform next generation authoring tools.

Paelke et al. [21] provided an overview of the area of mobile location-based gaming and its relation to maps and presented key aspects of exemplary commercial and research location-based games and authoring tools of the past (Mediascape, Caerus). They concluded that in order to be effective, these tools should support an appropriate conceptual model and should be designed so as to operate within a structured process that also integrates with other tools for media production and content management.

Winter [28] conducted a literature review in order to inform the development of a location-based games authoring tool for secondary school children and their teachers. Regarding educational aspects of location based games authoring the research question was how can location-based games authoring by students support their learning and how can it be integrated in educational practices (e.g. authoring in the context of ICT education emphasizes technical aspects of game development, requiring the use of various technologies to collect, share and manage resources). The discussed functionality and recommendations were synthesized into several generic design guidelines such as support for visual authoring and simulation mode.

In their survey Fidas et al. [8] presented five authoring tools for implementing cultural heritage experiences from a conceptual and architectural design (Hoppala, ARIS, TaggingCreaditor, LoCloud and CHEF). They concluded that a definition of a set of primitive elements is needed that can be used by cultural heritage experts as abstract building blocks without the help of professional developers and identified the need to streamline efforts with the aim of creating meta-models of such primitive elements.

Brundell et al. [5] focused upon the working practices of individuals, small independent artists and researchers, rather than designers or developers in large commercial companies or small to medium sized enterprises. A qualitative analysis of findings from the study was conducted to inform user requirements and design of next-generation authoring tools through a process of co-design.

3 Location Based Games for Education

Location-based games for education form a subcategory of location-based serious games intended to support serious activities in a playful way. Several location-based games within the genre of serious games have been developed mostly to support: *cultural heritage* purposes, such as supporting historical teaching and learning [2, 4, 16]; *museum* visits, experiencing the history of sites and in some cases participate in events and rituals that took place in ancient times [18, 22, 25, 26]; *simulations*, such as flood preparedness and fire evaluation scenarios [14, 19, 20]; *transportation*, exploring the symbiotic potential of social games in combination with location-based games and *tourism*, helping visitors engage with history and culture of their destination [10, 24].

Typical paradigms of location-based games for educational purposes have been developed mostly to support *situational language learning*. Learners are assigned missions (such as finding a specific book in the library) through location based game interaction activities. The learning goal of the games is to enhance English learning (listening and speaking), interest and motivation in a real environment. The conducted activities involve use of Bluetooth, indoor positioning, camera, QR codes attached to walls (revealing web links associated with relevant learning) with the aim to construct mixed reality game learning environments capable of integrating virtual objects with real scenes [6, 9, 13, 15, 17].

Other research efforts focused on engaging in *mathematical activities* and developing conceptual understanding on several topics. In the first case learners are expected to deepen their experiential knowledge of geometrical concepts related to shapes and orientation/navigation. In the exemplary game MobileMath [27] researchers observed the team activities during the game by accompanying a team outside as a participating researcher or by watching all teams in the game in real time on the website. The game was played by at least two teams with each team creating geometrical shapes on a previously defined playing field using a mobile phone with GPS functionality and an on-screen map. In the second case researchers focused on developing conceptual understanding on topics such as science, problem-solving and collaboration. In the exemplary game Outbreak [23] students used GPS handhelds to work together in groups of three in order to find the antidote to a pretend disease. As the students walk around their school's campus, each person took on the role of a different scientist in order to collect and share data with each other.

The proposed systems are based on the *situational learning approach* [6]. In the general case the system provides clues that learners have to identify (e.g. via listening in the case of language learning) where to go next in order to proceed in the game. The player's location is identified with the use of indoor and outdoor positioning techniques, activating game agents and other game mechanics.

A number of common requirements for authoring tools in order to support effectively the creation of games by non-programmers can be identified in three groups:

Authoring functionality – support for non-linear stories, organizing tasks as questions and answers with conditional branching, visual authoring, re-use of games, customization and personalization of the user interface to individual needs [28]; support for map-based authoring of game content considering also the alternative of on-site authoring within the physical environment, bridging the gap between desktop authoring and outdoor [21]; support for visual programming.

Admin tools – support for simulation mode to test mobile content and game flow; support for customized game analytics for assessing learning performance (e.g. results about a summative test with multiple choice questions [6]).

End-user client app functionality – support for communication (such as messaging or chat), support for multiple media formats (e.g. text, video, images, audio, HTML) [28]; support for the major mobile operating systems Android and iOS; support for augmented reality; support for QR Codes.

In the following section the conceptual model adopted by the most common open source authoring tools is presented and analyzed with a critical view and a comparative analysis is contacted with regard to the synthesized list of requirements.

4 Authoring Location-Based Games

Various research attempts tried to develop tools and technologies in order to encourage the creation of location-based experiences by non-technical users. The platforms COLLAGE[1], Games Atelier[2], ROAR[3] and TOTEM[4] were developed in the context of research projects in the past and successfully contributed towards the creation of innovative approaches of location-based playing and learning experiences. However, the aforementioned platforms are no longer available.

---

[1] http://www.celekt.info/projects/show/14.

[2] http://waag.org/en/project/games-atelier.

[3] http://gameslab.radford.edu/ROAR/.

[4] http://www.totem-games.org/.

MAGELLAN[5] is a promising 4 year project (2013–2017) aiming at researching and implementing an authoring and gaming platform based on visual authoring principles. The main objective of the platform is to enable non-programmers to cost-effectively author and publish multi-participant location-based experiences, as well as to support the browsing and execution of a massive number of such experiences [1, 7]. Currently, the platform is implemented as a proof-of concept [3]; however, access to documentation and platform functionality is not fully open to the general public.

7Scenes[6] is another platform for creating location-based experiences used in pedagogical settings mentioned in literature that started as a spinoff commercial product incubated at the Waag Society of the Netherlands. The community platform is targeting sightseeing tours, providing simplistic game like elements (time limit and a point system based on multiple choice questions). Finally, popular tools such as App Inventor[7] can support the implementation of multi-purpose mobile apps; however, no off-the-shelf authoring functionality (or other form of community extension) is provided for implementing location-based experiences or games.

Currently, commonly used authoring tools that actively support location-based software development providing open source code and free of charge services are ARIS[8] and TaleBlazer[9]. In the following sections, the conceptual model adopted by ARIS and TaleBlazer is presented and analyzed. ARIS is being developed by Wisconsin-Madison University, and TaleBlazer is being developed by MIT Scheller Teacher Education Program lab. Both platforms provide a way for creating mobile interactive stories and augmented location-based games.

4.1 ARIS

ARIS is an open source, free, cloud-based platform for creating and playing location-based augmented reality experiences on iOS devices. The conceptual model comprises of objects, triggers and scenes.

Objects – are containers for the content that players can see and interact with. Game objects provide support for starting conversations, inspecting items, viewing plaques and visiting web pages.

Triggers – provide a way of connecting actions in the physical world to ARIS objects. There are currently five types of triggers: *location triggers*, fired when the player is close enough to the geospatial coordinates of the trigger; *QR Code triggers*, fired when the players scan the image or enter the corresponding string code; *timer triggers*, fired when a particular period of time elapses; *beacon triggers*, fired according to the player's proximity to a Bluetooth beacon (iBeacon[10]); *AR View triggers*, fired when

[5] http://www.magellanproject.eu/.

[6] http://7scenes.com.

[7] http://appinventor.mit.edu/explore.

[8] http://www.arisgames.org.

[9] http://taleblazer.org.

[10] https://developer.apple.com/ibeacon.

specific views or images of the real world are recognized and trigger game objects (like images or video) that replace targets atop of the camera feed.

Scenes – are containers for collection of game objects. Scenes provide support for designing stories and help authors think about the progression of the game. They act as organizational units and represent different parts of the game.

ARIS has a story-based structure. The ARIS editor is browser-based with no local installation required. The platform allows authors to build *conversations* between the player and visual characters with the use of diagrams that help authors to visualize the flow (Fig. 1). The author defines the *lines* (text spoken by the character), and the *choices* for the player. Each *line* can also modify the player's environment by taking or giving items or setting the value of custom *attributes* defined by the author. ARIS allows authors to further customize game functionality (e.g. playing an audio clip in the background) based on the use of a JavaScript API. The custom code is executed when the platform renders HTML texts either entered or displayed to the player (e.g. presenting conversation lines).

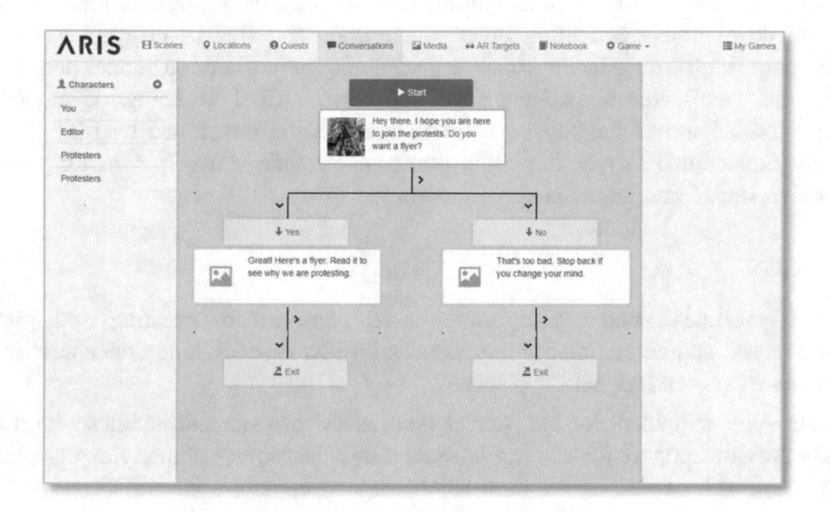

Fig. 1. ARIS conversations editor

4.2 TaleBlazer

TaleBlazer is an open-source, free, cloud-based platform for creating and playing location-based experiences on Android or iOS devices. The conceptual model comprises of agents, regions, scenarios and roles.

Agents – represent characters or objects that the player can interact with. The behavior of the agents is configured through *actions* that appear as buttons on the agent's dashboard. There are currently four types of actions that become activated when the player bumps into the agent: *text*, in the form of rich text format; *video*, playing an uploaded or YouTube video; *built-in functionality,* pickup and drop actions; *script*, executing a custom blocks-based script.

Regions – are real-world locations where the game takes place (Fig. 2).

Scenarios – are different versions of the same game that the players can pick from when they start the game (e.g. easy or hard). The game designer can use the player's choice to control the game logic with the use of block-scripting. The game designer can configure also multiple **roles** for the game in order to specify different interactions for players playing different roles through the use of block-scripting. The player picks the role when starting the game in order to access the role specific functionality.

TaleBlazer has a role-based structure. The TaleBlazer editor is browser-based with no local installation required (Fig. 2). The platform allows authors to define custom *traits* (for agents, roles or the game) which can be modified through block-based scripting. The editor uses a visual blocks-based scripting language that allows authors to create rich interactivity while helping authors avoid syntax errors. The interaction with the player is configured through *actions* that appear as buttons on the agent's dashboard. The platforms allow authors to modify the visibility of an action during game play through the use of blocks-based scripting. The game can be tested out with the use of an HTML emulator (Fig. 3) decreasing drastically debugging time.

Fig. 2. TaleBlazer editor

Fig. 3. TaleBlazer emulator

5 Comparative Analysis

In comparison to ARIS, TaleBlazer has a role-based structure. The use of traits provides an intuitive means to extend the data model with custom properties that can be easily modified through the use of blocks-based code. Furthermore, the visual blocks-based scripting language allows authors to create rich interactivity while helping users avoid syntax errors. On the other hand, ARIS has a story-based structure that helps authors to visualize game flow and provides convenient ways to segment the game.

In Table 1, key issues are synthesized in three groups: admin tools, authoring functionality and end-user client app features.

Table 1. ARIS and TaleBlazer features

Feature	ARIS	TaleBlazer
Authoring functionality		
Re-use games	Yes	Yes
Web visual authoring	Yes	Yes
Map-based authoring	Yes	Yes
Visual programming	No	Yes
Customization and extension	JavaScript API	Blocks–based scripting
In-situ authoring	No	No
Personalization	No	No
Non-linear stories	Yes	Yes
Admin tools		
Simulator	No	Yes
Analytics	No	Yes, only to officially featured organizations
End-user client app		
Mobile OS	iOS	Android, iOS
Connection needed	Yes	No
Augmented reality features	Yes	Yes
Multiplayer	Yes	No
Indoor navigation	Yes	Yes
QR Code	Yes	No
Tap to bump mode	Yes	Yes
On-screen map navigation	Yes	Yes
Communication	No	No
Multiple multimedia formats	Yes	Yes

5.1 Authoring Functionality

Both platforms provide authors with visually authoring map-based editors to create and re-use games. Although new games can be created as duplicates of existing ones (TaleBlazer *remixes* games, ARIS *duplicates* games), there seems to be no functionality to create libraries as collections of common scripts (or other resources) that can be imported in new games.

The extension of the game model is allowed by both tools through the use of custom properties which can be modified during game play providing also a way to control game flow (e.g. enabling or disabling actions according to the value of a property). The TaleBlazer visual blocks-based scripting language allows authors to create rich interactivity between game entities while helping authors avoid syntax errors. Although ARIS has no blocks-based scripting functionality, it provides sophisticated interaction management and various functions without the need for coding. However, the use of a JavaScript API for further customizing game functionality requires technical experience and effort to master.

In-situ authoring is not supported, neither personalization of both editor environments.

5.2 Admin Tools

TaleBlazer provides a mobile emulator to test the mobile content quickly and error check functionality to evaluate warnings and errors caused by invalid blocks-based code. TaleBlazer provides also a predefined set of analytics (such as total gameplays, duration times and average game play) and custom analytics events only to officially featured organizations. TaleBlazer *analytics events* can be used to collect specific data such as which choices players make or how many points they gained in the game.

5.3 End-User Client App

Both platforms rely on an end-user client-app for playing the games; ARIS supports only the iOS operating system, while TaleBlazer runs on either iOS or Android devices. Augmented reality features are implemented by the platforms in different ways. TaleBlazer provides a "heads up" functionality to show nearby agents as markers overlaid onto the video camera display based on the compass and GPS readings. On the other hand, ARIS uses "AR targets" created by Vuforia[11]. The player chooses views or images in the real world, recognized by the platform through the device's camera. The target views are replaced by the platform with custom media like images of video on the screen atop the existing camera feed.

Multiplayer and QR scanning functionality are available only to ARIS platform while both platforms provide indoor navigation, tap to bump and on-screen map navigation functionality.

6 Conclusions

Learning through mobile devices has been an active research area in recent years. Location-based games are estimated as a promising practice and have become an important strand to interact with the real world and use on learning environments. Various games have been developed in recent years and used to enhance the education and training efficiency of students in several subjects such as situational language learning, mathematical concepts, and conceptual understanding on topics such as science, problem-solving and collaboration. The potential of currently available authoring

[11] https://vuforia.com/, https://fielddaylab.wisc.edu/courses/aris-ar.

tools needs careful consideration in order to address effectively the diverse design and technical requirements posed by location-based games for educational purposes.

In this paper, a range of key issues for implementing location based games for educational purposes using two publicly available open source authoring tools, ARIS and TaleBlazer are presented. Although the reviewed authoring tools vary in functionality they share a fairly common conceptual model and in both cases the software architecture is comprised of the same components: a browser-based editor to author the game structure and mechanics, a back-end system to support concurrency and persistence and an end-user client-app for playing the games. Although both platforms provide the required game mechanics to create sophisticated gameplay interactions and game logic extension, a number of issues should be carefully considered.

Cross-platform development – in order to target as many possible mobile operating systems, the development of the end-user client app could be based on cross-platform development frameworks. Frameworks such as Ionic[12] provide free and open source mobile SDK, mobile components, typography, interactive paradigms and an extensible base theme for building and customizing apps for Android or iOS [30].

Interoperability and standardization – the fact that interoperability between authoring tools is not feasible, stresses the need for standardization for packages, libraries, game templates and integration with external systems (e.g. many universities are using single sign on systems for the authentication process).

Extension and customization – although both platforms provide sophisticated interaction management and default functionality, several means for extending the game logic should be provided by the authoring platform with the use of high-level scripting tools, templates and themes, and the provision of interfaces in order to customize and further extend the platform's core functionality and look & feel.

Testing – the integration testing process should provide the required means to evaluate the correct wiring of game container contexts and flow. TaleBlazer provides simulation functionality and evaluation of the custom blocks-based code, but further utilities are needed to support automated runtime unit tests for the entire application.

Analytics and reporting – assessing learning performance should be supported by customized analytics (e.g. results about summative tests in the game), logging and feedback (e.g. about exception or breakpoint conditions).

In conclusion, although several design and technical aspects of the location-based games were considered in this research, further consideration with regard to e-learning standardization and effective ways for embedding learning content must be considered. Future research should also focus on game design patterns specifically for location-based games and how they relate to learning outcomes. Games and tools need to be analyzed with a critical view, pointing out limitations, problems and challenges that can inform the design and implementation of next generation authoring tools.

[12] https://ionicframework.com/.

References

1. Anastasiadou, D., Lameras, P.: Identifying and classifying learning entities for designing location-based serious games. In: 2016 11th International Workshop on Semantic and Social Media Adaptation and Personalization (SMAP). IEEE (2016)
2. Anderson, E.F., McLoughlin, L., Liarokapis, F., Peters, C., Petridis, P., De Freitas, S.: Developing serious games for cultural heritage: a state-of-the-art review. Virtual Reality **14** (4), 255–275 (2010)
3. Balet, O., Koleva, B., Grubert, J., Yi, K.M., Gunia, M., Katsis, A., Castet, J.: Authoring and Living Next-Generation Location-Based Experiences. arXiv preprint arXiv:1709.01293 (2017)
4. Bellotti, F., Berta, R., De Gloria, A., D'ursi, A., Fiore, V.: A serious game model for cultural heritage. J. Comput. Cult. Herit. (JOCCH) **5**(4), 17 (2012)
5. Brundell, P., Koleva, B., Wetzel, R.: Supporting the design of location-based experiences by creative individuals. In: 2016 11th International Workshop on Semantic and Social Media Adaptation and Personalization (SMAP), pp. 112–116. IEEE, October 2016
6. Chen, C.M., Tsai, Y.N.: Interactive location-based game for supporting effective English learning. In: International Conference on Environmental Science and Information Application Technology, ESIAT 2009, vol. 3, pp. 523–526. IEEE, July 2009
7. Clarke, S., Lameras, P., Dunwell, I., Balet, O., Prados, T., Avantangelou, E.: A training framework for the creation of location-based experiences using a game authoring environment. In: European Conference on Games Based Learning, p. 125. Academic Conferences International Limited, October 2015
8. Fidas, C., Sintoris, C., Yiannoutsou, N., Avouris, N.: A survey on tools for end user authoring of mobile applications for cultural heritage. In: 2015 6th International Conference on Information, Intelligence, Systems and Applications (IISA), pp. 1–5. IEEE, July 2015
9. Godwin-Jones, R.: Emerging technologies games in language learning: opportunities and challenges. Lang. Learn. Technol. **18**(2), 9–19 (2014)
10. Grüntjens, D., Groß, S., Arndt, D., Müller, S.: Fast authoring for mobile game-based city tours. Procedia Comput. Sci. **25**, 41–51 (2013)
11. Kebritchi, M.: Factors affecting teachers' adoption of educational computer games: a case study. Br. J. Edu. Technol. **41**(2), 256–270 (2010)
12. Ketelhut, D.J., Schifter, C.C.: Teachers and game-based learning: improving understanding of how to increase efficacy of adoption. Comput. Educ. **56**(2), 539–546 (2011)
13. Liu, T.Y., Chu, Y.L.: Using ubiquitous games in an English listening and speaking course: impact on learning outcomes and motivation. Comput. Educ. **55**(2), 630–643 (2010)
14. Mannsverk, S.J.: Flooded - A Location-Based Game for Promoting Citizens' Flood Preparedness (2013)
15. Meyer, B., Sørensen, B.H.: Designing serious games for computer assisted language learning–a framework for development and analysis. In: Design and Use of Serious Games, pp. 69–82. Springer, Dordrecht (2009)
16. Mortara, M., Catalano, C.E., Fiucci, G., Derntl, M.: Evaluating the effectiveness of serious games for cultural awareness: the Icura user study. In: International Conference on Games and Learning Alliance, pp. 276–289. Springer International Publishing, Cham, October 2013
17. Mota, J.M., Ruiz-Rube, I., Dodero, J.M., Figueiredo, M.: Visual Environment for Designing Interactive Learning Scenarios with Augmented Reality. International Association for Development of the Information Society (2016)

18. Nilsson, T., Blackwell, A., Hogsden, C., Scruton, D.: Ghosts! A Location-Based Bluetooth LE Mobile Game for Museum Exploration. arXiv preprint arXiv:1607.05654 (2016)
19. Oliveira, M., Pereira, N., Oliveira, E., Almeida, J.E., Rossetti, R.J.: A multi-player approach in serious games: testing pedestrian fire evacuation scenarios. In: Oporto, DSIE 2015, January 2015
20. Onencan, A., Kortmann, R., Kulei, F., Enserin, B.: MAFURIKO: design of Nzoia basin location based flood game. Procedia Eng. **159**, 133–140 (2016)
21. Paelke, V., Oppermann, L., Reimann, C.: Mobile location-based gaming. In: Map-Based Mobile Services, pp. 310–334 (2008)
22. Peng, S.T., Hsu, S.Y., Hsieh, K.C.: An interactive immersive serious game application for Kunyu Quantu world map. ISPRS Ann. Photogramm. Remote Sens. Spat. Inf. Sci. **2**(5), 221 (2015)
23. Rosenbaum, E., Klopfer, E., Perry, J.: On location learning: authentic applied science with networked augmented realities. J. Sci. Educ. Technol. **16**(1), 31–45 (2007)
24. Rossetti, R.J., Almeida, J.E., Kokkinogenis, Z., Gonçalves, J.: Playing transportation seriously: applications of serious games to artificial transportation systems. IEEE Intell. Syst. **28**(4), 107–112 (2013)
25. Rubino, I., Barberis, C., Xhembulla, J., Malnati, G.: Integrating a location-based mobile game in the museum visit: evaluating visitors' behaviour and learning. J. Comput. Cult. Herit. (JOCCH) **8**(3), 15 (2015)
26. Sanchez, E., Pierroux, P.: Gamifying the museum: teaching for games based informal learning. In: European Conference on Games Based Learning, p. 471. Academic Conferences International Limited, October 2015
27. Wijers, M., Jonker, V., Drijvers, P.: MobileMath: exploring mathematics outside the classroom. ZDM **42**(7), 789–799 (2010)
28. Winter, M.: Location based games authoring. literature review to inform the development of a location based games authoring tool for secondary school children (2009). http://www.cem.brighton.ac.uk/staff/mw159/winter_2009_location_based_games_authoring.pdf. Accessed 29 Aug 2017
29. Williamson, B.: Computer games, schools, and young people: a report for educators on using games for learning. Futurelab, Bristol (2009). https://www.nfer.ac.uk/publications/FUTL27/FUTL27.pdf. Accessed 29 Aug 2017
30. Xanthopoulos, S., Xinogalos, S.: A comparative analysis of cross-platform development approaches for mobile applications. In: Proceedings of the 6th Balkan Conference in Informatics, pp. 213–220. ACM, September 2013

Conceptual Framework of Microlearning-Based Training Mobile Application for Improving Programming Skills

Ján Skalka and Martin Drlík(✉)

Constantine the Philosopher University in Nitra,
Tr. A. Hlinku 1, 94974 Nitra, Slovakia
{jskalka,mdrlik}@ukf.sk

Abstract. Application of microlearning-based training in higher education raises several new questions about the didactical design of learning activities, effective methodologies as well as technical implementation issues. The paper describes a framework of microlearning-based training mobile application for improving programming skills, which will integrate the advantages of the web-based learning environments and benefits of microlearning approach into the existing virtual learning environment infrastructure of the educational organization. The proposed mobile application framework implements innovative methods based on microlearning approach, learning anytime and anywhere requirements with immediate feedback, creating the learning community, and applying gamification features. These innovations are designed to address the interests, learning and communication habits of young people, with the goals to develop their internal motivation and sustain their interest in acquiring required programming skills.

Keywords: Mobile learning · Interactive learning environment
Programming languages

1 Introduction

According to several sources, there is an increased demand for IT professionals in the job market today [1–3]. The educational system is not able to satisfy the job market requirements in a sufficient rate. An established computer science educational concept at the universities, yet still often based on the combination of lectures and labs, is outdated and inefficient. The knowledge acquisition and IT skills improvement of students are questionable and less effective [4].

Young people are not nowadays interested in spending time by listening to the lectures. They prefer immediate use or application of the obtained knowledge and skills. They require the options to learn anytime and anywhere, not only in the schools [5]. It is necessary therefore to try to understand that the contemporary young people are more active than their parents in general, and they select the knowledge and skills, which usefulness they can imagine or prove in a short time.

M. E. Auer and T. Tsiatsos (Eds.): IMCL 2017, AISC 725, pp. 213–224, 2018.
https://doi.org/10.1007/978-3-319-75175-7_22

The article presents a conceptual model of the system for improving programming skills based on the microlearning theory and mobile learning environment. The first sections of paper explain the definition and most important principles of microlearning and their connections to the pedagogical theories and deal with challenges of teaching programming languages. Next part presents the description of the framework, which covers these ideas and its features. The last section deals with the potential of presented concept and scretch of future research.

2 Microlearning

Microlearning is rather a new term which has been in use through many aspects of learning, didactics, and education. It is related to the relatively short efforts and low degrees of time consumption. It deals with small or very small content units and rather narrow topics [6]. It offers a new way of designing and organizing learning, like learning in small steps and small units of content, with structure and classification created by the learner [7].

By Dash in [8], microlearning consists of micro-content and micro-activities. Micro-content is information published in a short form. By [9] microlearning encourages learners to become active co-producers of content through active social participation. The modern definition of microlearning based on [7, 10] says that microlearning is an action oriented approach offering bite-sized learning that gets learners to learn, act, and practice. It provides the educational content in small well-planed units/nuggets mostly through the mobile applications, which do not require the long attention of the students. It can provide better educational results in comparison with the classical approaches if it is suitably combined and sufficiently interactive [11, 12].

Current view of the microlearning presents pedagogical approaches, technological models and wide experience gained over the past years in the e-learning. It can be seen as a simple version of e-learning with all its positives and negatives contained in pedagogical models. Microlearning refers to any pedagogy that encourages learning in short segments [13]. Teachers in many universities use microlearning in their classes because it engages students with the subject matter and results in deeper learning, encouraging them to connect the subject matter with their everyday lives and the world around them [14]. A lot of organizations offer microlearning as an efficient tool for varied training needs including, e.g., building soft skills/behavioural change, compliance, obtaining professional skills, learning languages, etc. [14, 15].

Baumgartner defined three models of education related to the pedagogical theories connected to the behaviourism, cognitivism and constructivism [16]. According to [17] each model demands different levels of guidance and requires the learning system to play a different role.

Microlearning is an innovative way of transfer of skills and knowledge, but there are some disadvantages too:

- Each piece of microlearning content must create a separate online training unit [18].
- Brief modules and online training activities are not typically the best choice for more complicated tasks, skills, or processes [19].

- Microlearning should be part of a larger online training strategy, but it should not be primary online training method [19].

According to [18], the course creators need to define and set only a single teaching goal for one lesson. It is necessary to come up with a well-thought concept, and eliminate the extra content.

There are three groups of tools used in relation to microlearning. The first group of tools and platforms were not prepared for microlearning, but they are used very often: YouTube, Twitter, Instagram, SlideShare. The second group is represented by LMSs, which were developed for building e-learning content in general, but contains useful tools for microlearning lesson creation. The special tools for microcontent creation represent the last group, e.g. Coursmos, Grovo, Yammer [20, 21].

Ubiquitous communication and computing technologies enable smartphones to be widely used to acquire instant knowledge anywhere and anytime [12]. Smartphones provide a one-to-one relationship with its owner, what is the basic assumption of education content personalization [22].

3 Teaching Programming Languages

The motivation in learning is considered an evitable part of the education process, especially in teaching programming. An intrinsic motivation represents the most important element of the educational process [23].

Teaching programming languages has several specific features [23, 24]. It offers a wide range of educational resources, approaches, and methods suitable for the development of critical and innovative thinking. The typical training activities are analyzing a problem, design data structure, programming code, and searching logical mistakes in code. Some of this activities are possible to design as microlearning units, but a lot of complex problems are out of microlearning time interval [25, 26].

Nowadays, there are available several well-known interactive solutions for teaching programming (e.g., Hackerrank, CodeWars, CodeHunt, CodingBat, etc.). They are usually integrated into the web portals, and content is often created by the software experts or engineers without adequate knowledge of the didactical approaches and principles. It is the reason why they are useful for exercising but not suitable for individual organized knowledge building.

The automated evaluation of the programming code written in different program-ming languages is a common feature of these environments. All of them contain a particular level of the gamification, provide an opportunity to create a community of learners and support different forms of competitiveness.

Since the majority of these successful solutions have web-based oriented architecture, it is hard to extend them with the elements, which would support the microlearning approach, connect to the existing virtual learning environments and which are typically easier provided by the mobile applications.

An application called Sololearn (www.sololearn.com) can be considered one of the most popular mobile applications for learning programming languages, which utilizes the elements of microlearning approach. This application has refined the concept of gamification, development of learning community, the common creation of the educational content. However, it provides only simple interactive elements without the possibility to write programs with automated testing and code evaluation as well as without the possibility to interconnect it with other educational environments.

4 Framework

The main aim of the paper is to describe a framework of microlearning-based training mobile application for improving programming skills, which will integrate the advantages of the web-based learning environments, benefits of the suitable combination of microlearning and mobile learning approaches.

The framework is based on the suitable combination of content and interactive microlearning objects. The content unit represents an elementary unit of information mostly in the form of HTML document, video or image. Its extent should not be larger as the screen display. Contrary, it should contain only sole information. Interactive objects follow each content unit and verify the students' understanding of the presented educational content.

The contemporary LMS provides many interactive activities, which can be effectively used. For example, the most frequently used types of questions like multiple choice, short answer, ordering words, represent the core part of the proposed framework from its initial phases. Other types of questions are possible to add to the system.

The lessons, which represent prepared educational content divided into several didactically evaluated topics, create the starting point of the study. Each lesson is defined as a sequence of static educational content and interactive activities (question or creating/adding parts of a programming code). The quiz follows each sequence of the educational units (lessons). It is focused on summarization and evaluation of the obtained knowledge. Problematic parts are identified in the case of failing the quiz. At the same time, the student is returned to the educational unit, while the interactive activities are personalized and not repeated.

A successful passing the lesson unlocks a complimentary set of programming assignments. The students can choose any of them. The correctness and functionality of the student's solution are automatically tested. Considering the differences in initial students' knowledge, there is also an option to attempt the quiz without the reading the lessons.

A proposed framework represents a conceptual and technological basis, which existence is irrelevant without the content itself. The content development represents a time consuming task, which is easier implementable by engaging a community. Learning in the community brings not only the possibility to create a richer content but also a deeper understanding of the topic by the students [27–29]. Moreover, it allows a better understanding of the content and the relations between educational units [16] (Fig. 1).

From a student's point of view, education can be divided into several layers, and besides the first layer, the student also creates new educational content. The layers allow:

- Learning based on the didactically ordered content and interactive units – obtaining basic knowledge and skills.

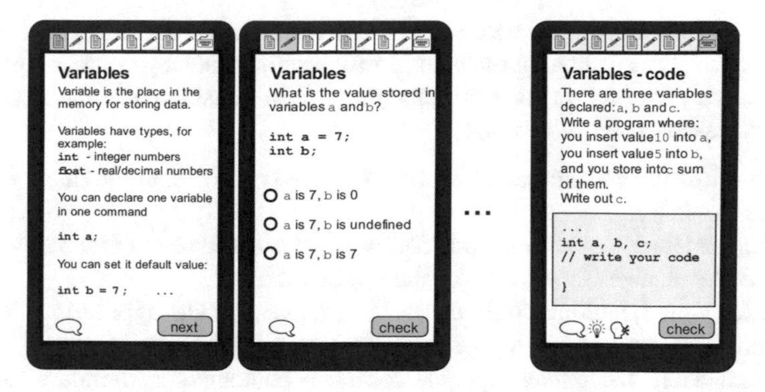

Fig. 1. The combination of content-units and interactive activities in the lesson.

- Involving students in the discussions, problem solving, peer-to-peer programming code evaluating, discussion about the effectiveness and programming code correctness.
- Creating of new questions and assignments, which extend provided educational content.
- Developing lessons for other programming languages and frameworks.
- Developing of new types of interactive activities, an alternative design of mobile and web applications. This action has great potential in the case of university study programs. The students obtain knowledge and practical skills because they develop real applications with immediate feedback of the users.

The quality of the educational content will be guaranteed by the authors in the beginning. Subsequently, the community will not only evaluate each microlearning object but also will be able to start a discussion.

Moreover, elements of gamification will be used to sustain the intrinsic motivation. According to the Review of Empirical Studies on Gamification [30] a majority of the reviewed studies, gamification does produce positive effects and benefits. Students who are intrinsically motivated are more engaged, retain information better, and are generally happier. However, on the other hand, educators should beware when thinking about implementing gaming into their classrooms because it can backfire on them.

Contrary, according to the results of the study [31] the combination of leaders boards, badges, and competition mechanics do not improve motivation, empowerment, or satisfactions, it actually harms them. The research of the overall contribution of the gamification will be therefore one of the aims of the proposed framework. Data generated by the students' activity will be stored in the suitable structure. Several reports and data visualizations will be created:

- Displaying grades from the attended activities divided into several categories based on the activity type (quizzes, discussions, etc.),
- Joining the categories and collecting badges with the aim to profiling student,
- Ranking the students considering different kinds of their activity
- Controlling the student's progress and comparison with others.

The possibility to contend with an arbitrary or random student in given conditions or the possibility to attend a tournament in programming belongs to the important part of gamification. The following requirements will be considered in the design of the framework based on microlearning:

- Definition of the content and structure of the lessons without the direct changes to the source code.
- Maximal interactivity with the possibility to add new types of interactive activities without the changes in the core of the application.
- Modularity and multiplatform character (native mobile application, web-based interface, desktop application) of individual parts of the system.
- Easily readable and interchangeable format – contemporary trends in data transformation are based on JSON, which is easily transformable into the text form and usable on both sides of the communication channel
- API definition for communication with external systems and applications, which will provide the module independence and their integration to the existing infrastructure of the university information systems and LMS.
- Communication with external systems, which will ensure the source code evaluation and its effectivity.

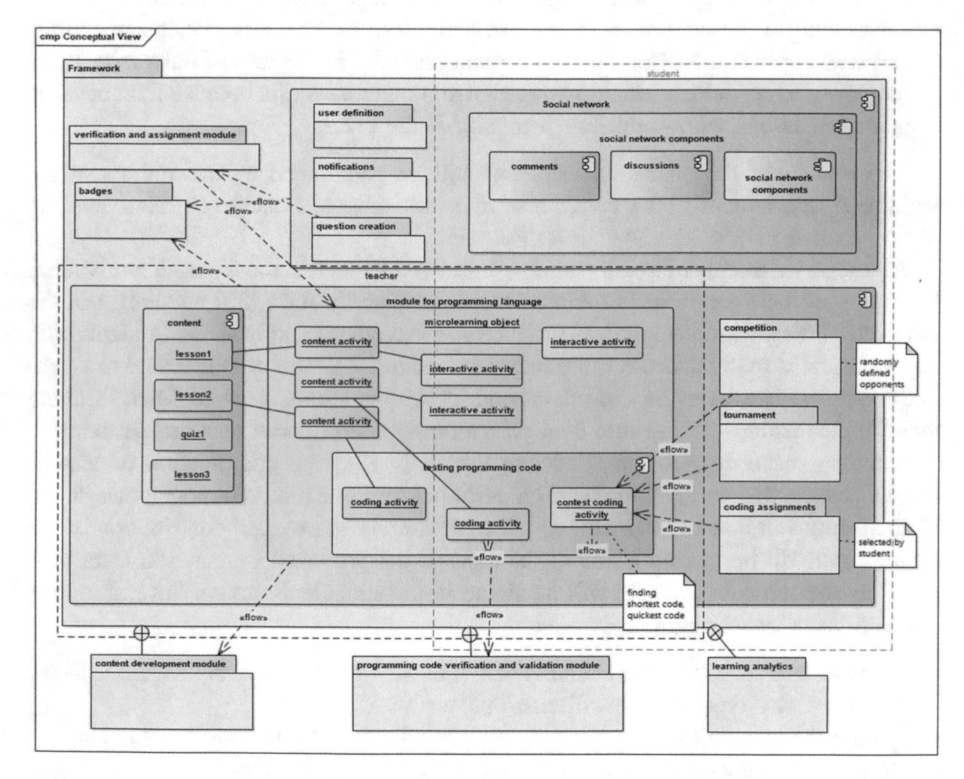

Fig. 2. The proposal of the microlearning framework for teaching programming languages.

It is necessary to design basic elements of the framework during the application design with an emphasis on its simple extensibility and modularity, which will allow building several more complex structures. The following units will be considered the basic microlearning objects (Fig. 2):

- content activity – elemental unit, which provides the content based on the particular type of the educational source (HTML document, picture, video etc.). This unit describes the content and allows starting a discussion regarding this content.
- interactive activity – contains a question, task or assignment, which requires the activity/reaction of the student. Several different kinds of interactive activities are based on the same structure. Their behaviour depends on the situation in which they are used. This activity will also support off-line mode.
- code activity – type of activity, which sends a programming code created by the students to the server and obtains the evaluation its correctness as a result (it will require an on-line connection).

These units create the following more complex logical units:

- lesson – the sequence of the activities, which combines educational content and interactive activities (quiz, code activity),
- quiz – the sequence of the activities, which contains different kinds of interactive activities. It can be used:
- for repeating in the context of the lesson,
- for skipping some lessons, if the student thought she already had obtained the presented knowledge.
- for attending the quiz as a feedback during the lesson or seminars.
- competition – the sequence of the interactive activities generated on the basis of the algorithms, which emphasis the correctness and speed of the solution,
- coding activity – elementary activity, which allows writing and checking the source code of the given assignment written by the student. The student chooses the assignment from several categories, which are open after she successfully finishes the lesson.
- tournament – single assignment or sequence of code activities, in which the correctness, speed as well as effectivity of the written source codes are rated.

These logical units have the form of container. They allow using arbitrary types of activities, even the activities, which were not created in the initial phases of the application development. This approach is possible due to the selected architecture of the application. This architecture uses JSON objects for the communication between the individual container and activity. Simultaneously, it retains the realization of all operations on the activity. Logical units are joined into the modules, which provide an educational content of the given programming language. Consequently, depending on the overall time and content difficulty, one or more modules can be created for one programming language.

The content part of the framework is closely connected with the social networks. Therefore it is easy for all stakeholders to evaluate or comment each microlearning object directly in the social network, which is very appreciated by the contemporary students. Moreover, additionally to these activities, the FAQ section creates an integral part of the proposed framework. It provides a platform for solving repeating problems or improving the learning process in general. The following algorithms will build a core of the proposed framework:

- Algorithms, which pass through the curriculum of individual programming languages with the elements of the personalization and evaluation of the quiz questions.
- Algorithms for testing the source code patterns or snippets, which represent the results of the practical assignments integrated into the curriculum.
- Algorithms, which define the structure of the competitions, challenges, battles, their evaluation and displaying the answers of the participants.
- Algorithms for ranking the students, creating top lists, displaying the main characteristics of the students based on the realised activities, badges and other outcomes of their activity in the environment.
- Algorithms for synchronization of the local data with the central data storage based on available APIs.
- Social network services (discussion, communication, evaluation, observation, scheduling).

5 Framework Features

A continual development of the content is considered one of the main assumptions of the sustainability of the proposed system. The community can fulfil this task most effectively because the community can use different social network channels and tools for intermediate feedback about the quality of individual microlearning objects, requirements to the new educational content, its modification. Moreover, this community generates huge data about the learners, which can be used in the process of system improvement or personalization.

The members of the community can participate in creating the new tasks after reaching a particular level. They can use predefined types of activities like questions or programming assignments. Subsequently, created tasks are validated by other members of the community or by the system developers. Finally, they are available in the corresponding lessons or in the particular content units in the lesson. The assignment can be done by the user with special privileges. After gathering enough data, it is possible to automate this process using machine learning approach. The community members can later get authorization to creating own quizzes, badges, as well as modules of programming languages. They all could encourage these students to specialize in given direction. All mentioned activities are independent of the activities of the framework development team. The community can realize these activities without any external intervention.

Automatized modifications of the content based on the students' behaviour as well as personalization belong to the requirements of modern learning environments [32–35]. The automatized modification of the content, like interactive object in a lesson in the case of the proposed framework, is based on:

- Students' evaluation – bad evaluation of the activity leads to its replacement by the activity with better results, which also belongs to the given content unit. The weakly evaluated content unit will be replaced by a new content unit with modified content.
- The success of the answers – It is necessary to implement the effective design of calculation of educational results [36, 37]. The rate of success should reach 80–95% in the initial phase. It will be lately improved considering further research. Activities will be replaced in the case of insufficient success of the results.

The personalization of the student will be partially based on the mathematic model of the student and partially on the content adaptation [38, 39]. The following main principles will be taken into account:

- Preference these types of tasks, which received a higher positive evaluation by the students.
- Recommendation for participating in other similar activities in the case of the lower rate of success.
- Selection of programming assignments based on the level of the students' knowledge and preferences.
- Identification of the problems with the understanding of content by students and recommendation to use slower and more detailed steps with repetition.

Automated programming code testing represents nowadays a tool, which provides immediate feedback to the student about the programming code she was written. There are several solutions, how to implement this tool into the framework [40]. I/O-based Assessment [41] is the most suitable for testing outputs of the program for given inputs. The programming code testing requires a sole server using the following steps:

- Mobile application sends programming code to the server as a text.
- The server creates the file with source code and checks it if it does not contain malicious code.
- If the compilation is required, the file will be compiled. The error message will be sent to the mobile application in the case of error, and the process will be stopped.
- Considering the type of the task the file will be launched with several inputs, and the outputs will be compared with the expected one, or the tests will be launched.
- The final protocol will be sent back to the mobile application. If some of the outputs will be not correct, the user receives a detailed explanation.

A framework should be designed in a way to be able to display the educational content not only in one language. It is necessary to create different language mutations of the educational content and quizzes in the native language of the students. A set of web-based tools connected with the backend learning management systems will be available and will provide effective and parallel development of the content in selected languages.

6 Conclusion

According to [7–9] microlearning raises new questions of the didactical design of learning activities based on micro-content and resulting in micro-content. Although there is plenty of micro-content everywhere on the Internet, there is a lack of research in microlearning activities that can utilize it in a microlearning educational context. There is still a need for research to develop and apply new innovative microlearning strategies and study their efficiency in education.

The aim of the presented framework is to implement innovative methods, new educational content and methodology to develop students' analytical and creative thinking through algorithms and programming in line with labour market requirements. These innovations must be designed to address the interests, learning and communication habits of young people, with goals to develop an internal motivation and sustain their interest. It can be assumed that the application of the conceptual framework can result in improving methods based on learning-by-doing, microlearning approach, learning anytime and anywhere with immediate feedback, creating the learning community, and applying gamification using standard features like points, rewards, badges, ranking, and comparison with other members, etc.

This model is designed not only for the university students but the general public and acquiring knowledge of professionals too. The university is only a starting point with a huge potential to validate the concept, to create content and to prepare students for one of the future forms of lifelong learning.

Although the concept will be verified on the platform of programming languages, it is usable for many areas of teaching and skills building. It is prepared for building new types of activities (e.g., parameterized able exercises of math, physics, languages, working with pictures/maps, etc.) and implementation in different areas. The parallel goals of research based on presented work are collection and evaluation of educational data, and text mining oriented to the preparation of micro-content from long texts.

References

1. Hüsing, T., Korte, W.B., Dashja, E.: e-Skills in Europe Trends and Forecasts for the European ICT Professional and Digital Leadership Labour Markets (2015–2020). http://eskills-lead.eu/fileadmin/lead/working_paper_-_supply_demand_forecast_2015_a.pdf
2. Eurostat: Europe 2020 Indicators – Employment. http://ec.europa.eu/eurostat/statistics-explained/index.php/Europe_2020_indicators_-_employment. Accessed 30 Aug 2017
3. World Economic Forum: The Future of Jobs Employment, Skills and Workforce Strategy. http://www3.weforum.org/docs/WEF_FOJ_Executive_Summary_Jobs.pdf. Accessed 30 Aug 2017
4. Guzman, J.L., Dormido, S., Berenguel, M.: Interactivity in education: an experience in the automatic control field. Comput. Appl. Eng. Educ. **21**(2), 360–371 (2013)
5. Alamri, S., Meccawy, M., Khoja, B., Bakhribah, H.: Learning anytime, anywhere: benefits and challenges of m-learning through the cloud. Int. Adv. Res. J. Sci. Eng. Technol. **4**(5), 37–41. https://doi.org/10.17148/IARJSET.2017.4508
6. Hug, T., Friesen, N.: Outline of a Microlearning Agenda. eLearning Papers (2009). (www.elearningpapers.eu). No 16. September 2009. ISSN 1887–1542

7. Kamilali, D., Sofianopoulou, C.: Lifelong Learning and Web 2.0: Microlearning and Self Directed Learning. In: Proceedings of EDULEARN 2013, Barcelona, pp. 0361–0366 (2013)
8. Lindner, M.: Use these tools, your mind will follow. Learning in immersive micromedia and microknowledge environments. In: The Next Generation: Research Proceedings of the 13th ALT-C Conference, pp. 41–49 (2006)
9. Kerres, M.: Microlearning as a challenge for instructional design. In: Didactics of Microlearning: Concepts, Discourses and Examples, pp. 98–109. Waxmann, Münster (2007)
10. Hug, T.: Micro learning and narration: exploring possibilities of utilization of narrations and storytelling for the design of "micro units" and didactical micro-learning arrangements. In: Proceedings of Media in Transition (2005)
11. Buchem, I., Hamelmann, H.: Microlearning: a strategy for ongoing professional development. eLearning Pap. **21**(7), 1–15 (2010). https://www.openeducationeuropa.eu/sites/default/files/old/media23707.pdf
12. Kovachev, D., et al.: Learn-as-you-go: new ways of cloud-based micro-learning for the mobile web. In: Advances in Web-Based Learning-ICWL 2011, pp. 51–61. Springer, Heidelberg (2011)
13. Trowbridge, S., Waterbury, C., Sudbury, L.: Learning in Bursts: Microlearning with Social Media. EDUCAUSE Review (2017). http://er.educause.edu/articles/2017/4/learning-in-bursts-microlearning-with-social-media. Accessed 30 Aug 2017
14. Pandey, A.: Micro Learning: 5 Killer Examples: How to Use Microlearning-Based Training Effectively (2016). https://elearningindustry.com/5-killer-examples-use-microlearning-based-training-effectively. Accessed 30 Aug 2017
15. Marks, S.: Get Inspired: Five Examples of Good Microlearning Design (2015). https://ilite.wordpress.com/2015/09/28/get-inspired-five-examples-of-good-microlearning-design/. Accessed 30 Aug 2017
16. Baumgartner, P.: Educational dimensions of microlearning–towards a taxonomy for microlearning. In: Designing Microlearning Experiences–Building up Knowledge in Organisations and Companies. Innsbruck University Press, Innsbruck (2013)
17. Göschlberger, B.: A platform for social microlearning. In: Verbert, K., Sharples, M., Klobučar, T. (eds.) Adaptive and Adaptable Learning: 11th European Conference on Technology Enhanced Learning, EC-TEL 2016, Lyon, France, 13–16 September 2016, pp. 513–516. Springer International Publishing, Cham
18. Zufic, J., Jurcan, B.: Micro learning and EduPsy LMS. In: Central European Conference on Information and Intelligent Systems. Faculty of Organization and Informatics Varazdin, p. 115 (2015)
19. Pappas, C.: Microlearning In Online Training: 5 Advantages and 3 Disadvantages. https://elearningindustry.com/microlearning-in-online-training-5-advantages-and-3-disadvantages. Accessed 30 Aug 2017
20. Giurgiu, L.: Microlearning an evolving elearning trend. Sci. Bull. **22**(1) (2017). https://doi.org/10.1515/bsaft-2017-0003. Accessed 30 Aug 2017. ISSN 2451-3148
21. Mykhalevych, N.: Microlearning In The Workplace: 7 Learning Tools And 4 Tips To Make Microlearning Exciting (2017). https://elearningindustry.com/microlearning-in-the-workplace-7-learning-tools-4-tips-make-microlearning-exciting. Accessed 30 Aug 2017
22. Wong, L.-H.: A learner-centric view of mobile seamless learning. Br. J. Educ. Technol. **43** (1), E19–E23 (2012). http://onlinelibrary.wiley.com/doi/10.1111/j.1467-8535.2011.01245.x/abstract#
23. Capay, M., Skalka, J., Drlik, M.: Computer science learning activities based on experience. In: The IEEE Global Engineering Education Conference (EDUCON) 2017: Conference Proceedings, 25–28 April 2017, Athens, CD-ROM, pp. 1367–1376. IEEE (2017). ISBN 978-1-5090-5466-4

24. Capay, M.: Algorithmic thinking observation: how students of applied informatics break the mystery of black box applications. In: World Engineering Education Forum (WEEF 2014): Interactive Collaborative Learning (ICL 2014), International Forum, 3rd–6th December 2014, Dubai, pp. 535–540. IEEE (2014). ISBN 978-1-4799-4438-5
25. Robins, A., Rountree, J., Rountree, N.: Learning and teaching programming: a review and discussion. Comput. Sci. Educ. **13**(2), 137–172 (2003)
26. Kalelioğlu, F.: A new way of teaching programming skills to K-12 students: Code.org. Comput. Hum. Behav. **52**, 200–210 (2015)
27. Hord, S.M.: Professional learning communities. J. Staff Dev. **30**(1), 40–43 (2009)
28. Palloff, M., Pratt, K.: Collaborating Online: Learning Together in Community. Wiley, Hoboken (2010)
29. Wenger, E.: Communities of practice and social learning systems: the career of a concept. Soc. Learn. Syst. Commun. Pract. **3**, 179–198 (2010)
30. Hamari, J., Kovisto, J., Sarsa, H.: Does gamification work?–a literature review of empirical studies on gamification. In: 2014 47th Hawaii International Conference on System Sciences (HICSS), pp. 3025–3034. IEEE (2014)
31. Hanus, M.D., Fox, J.: Assessing the effects of gamification in the classroom: a longitudinal study on intrinsic motivation, social comparison, satisfaction, effort, and academic performance. Comput. Educ. **80**, 152–161 (2015)
32. Brusilovky, P., Millán, E.: User models for adaptive hypermedia and adaptive educational systems. In: The Adaptive Web, pp. 3–53. Springer, Heidelberg (2007)
33. Chen, C.-M.: Personalized E-learning system with self-regulated learning assisted mechanisms for promoting learning performance. Expert Syst. Appl. **36**(5), 8816–8829 (2009)
34. Toktarova, I., et al.: Design and implementation of mobile learning tools and resources in the modern educational environment of university. Rev. Eur. Stud. **7**(8), 318 (2015)
35. Masthoff, J., Grasso, F., Ham, J.: Preface to the special issue on personalization and behavior change. User Model. User-Adap. Inter. **24**(5), 345–350 (2014)
36. Vasconcelos, L.A., Crilly, N.: Inspiration and fixation: questions, methods, findings, and challenges. Des. Stud. **42**, 1–32 (2016)
37. Crilly, N.: Fixation and creativity in concept development: the attitudes and practices of expert designers. Des. Stud. **38**, 54–91 (2015). ISSN 0142-694X
38. Graf, S., Kinshuk: An approach for dynamic student modelling of learning styles. In: Proceedings of the International Conference on Exploratory Learning in Digital Age (CELDA 2009), pp. 462–465. IADIS Press (2009)
39. Kinshuk, M.C., Graf, S., Yang, G.: Adaptivity and personalization in mobile learning. Technol. Instr. Cogn. Learn. **8**, 163–174 (2009). Paper presented at the Annual Meeting of the American Educational Research Association, San Diego, CA, 13–17 April 2009 (2009)
40. Staubitz, T., Klement, H., Renz, J., Teusner, R., Meinel, C.: Towards practical programming exercises and automated assessment in massive open online courses. In: Proceedings of 2015 IEEE International Conference on Teaching, Assessment and Learning for Engineering, TALE 2015, art. no. 7386010, pp. 23–30 (2015)
41. Pieterse, V.: Automated assessment of programming assignments. In: Proceedings of the 3rd Computer Science Education Research Conference on Computer Science Education Research, pp. 45–56 (2013)

The Game as a Way to Train the Mind Games of Reasoning

Marisol Elorriaga(✉), Mario Edelmiro Antúnez(✉), and Miguel Sandro Nicolino(✉)

Desarrollo de Entornos Virtuales de Enseñanza y Aprendizaje, UTN – Facultad Regional Delta - Campana, Buenos Aires, Argentina
{melorriaga, antunezm}@frd.utn.edu.ar, sandronicolino@gmail.com

Abstract. This work aims to contribute in a simple and positive way with everybody directly and indirectly is involved in an educational process by, offering notions about the importance of playing accompanied by methodological concepts, advantages and features for the teaching and the implementation of games with educational content; besides the interest of keeping them for their formative process as powerful elements in intellectual, ethical and physical education.

Keywords: Games · Reason · Destructuration · Gamification Mobile App

1 Introduction

"Gamification has been defined as the "process of using game thinking and mechanics to engage audiences and solve problems" [1].

WHY TO PLAY?

- Relaxes
- Not generates fear to the unknown
- Improving attention
- Developing skills and abilities
- It promotes the motivation
- Improving intellectual faculties
- It has not time limit.

For the process teaching learning we need:

Motivation, Attention, Ability to reason, Effort in a relaxed environmental and THIS IT IS FOUND IN THE GAMES!

The use of mechanical gaming environments and not playful applications in order to improve motivation, concentration and effort is called "Gamification".

This was the reason for using the games in the initial courses in our University.

M. E. Auer and T. Tsiatsos (Eds.): IMCL 2017, AISC 725, pp. 225–231, 2018.
https://doi.org/10.1007/978-3-319-75175-7_23

2 Objectives

- **To use the game for the teaching-learning process.**
- **To create a set of activities of reasoning on the University's platform.**
- **To create a mobile phone app. about reasoning games.**

To achieve the objectives, we started answering the following questions:

How do we relate the attentional processes with neuroplasticity?

For this, we develop games of reasoning, which can be solved at any time; it is a game, the students can play many times per day, if they are excited. As the time moves, the neurons solve the activities more quickly, they are acquiring more speed. These games destructuring the mind.

This proposal is based on several initial assumptions:

- All the people have minimum logical-analytical skills
- It is not true that some people may be unfit for science
- The student who is finishing high school, and is starting university or a higher course, represents malleable raw material in which the prejudices he brings can be changed
- The teacher is who has the ability to teach, to handle the necessary tools to achieve the objectives.

Why do we say destructuring?

Because all the exercises are mixed: calculations, reading and comprehension, exercises where just need to read, to write, and the four fundamental operations.

They are very simple exercises and this is the good news, because all the exercises of reasoning that we usually find in books or magazines are very difficult, ordinary people are not able to solve them, because they are for geniuses, and why the geniuses want to learn to reason?

The games and the reasoning activities are increasing their difficulty, but experience shows that, they progress fast.

3 Methodology

In the world of educational applications there are two clearly differentiated approaches, more serious applications where the objective is to learn a subject (either a language or any other type of knowledge), or more playful applications, where the only goal is fun through of general questions.

Let's look at the two highest exponents of both types of applications:

(a) **Duolingo**

It is a multiplatform language learning application (Web/Mobile/) in which the unique objective is to learn the language in question although with a less traditional approach. It uses test type questions, allocutions and images to not become boring. In addition the application is Gamificated (use of experience, lives, etc.) for a more pleasant and addictive game experience.

(b) **Triviados**

It's a multi-platform (Web/Mobile/Tablet) quiz based on the Trivial of life, where test questions on various topics are an excuse to play in multiplayer mode with your friends.

Pleasant appearance, multiplayer and ease of use and access are the strengths of this entertainment application.

This project is placed in an intermediate degree of these two points of view, mixing the intentionality of the user to learn with the fun and the emotion that provides the multiplayer mode.

These tools are applied in the initial courses, for all engineering specialties.

The pilot application was developed as follows:

- Simple problematic situations
- Multiple choice
- Texts lagoon
- Calculations
- Riddles
- Quiz

The games are in the educational virtual campus's platform, which was adapted for this purpose, in which the following considerations that are normally present in the gamification were taken into account.

These problematic situations, consist of reasoning activities.

The students who play, develops the following skills:

- Mental quickness.
- Abilities in the use of tools.

It is divided by categories and it is possible to compete for time, right questions and with other participants in the web.

The basis of the game is:

- Mechanics: Joining the game levels or badges.
- Aesthetics: The use of rewarding images to the player's view.
- Connection player-game: both are looking for a compromise between the player and the game.
- Motivation: The psychological predisposition of the person to participate in the game is certainly a trigger. And as time goes by, people learn from repetition, the challenges have to gradually increase to keep up with their growing skills.
- Promote learning: the gamification incorporates psychology techniques to encourage learning through play. Techniques such as assigning points and corrective feedback.
- Troubleshooting: It can be understood as the ultimate goal of the player, that is, reach the goal, solve the problem, cancel your enemy in combat, overcome obstacles, etc.,

It is an efficient training tool because it incorporates gaming elements challenges, fantasy, motivation, easy measuring (level, ranking, score), as well as satisfaction by the achievement of goal.

As a result of the experience and enthusiasm of students that is reflected in the increase of participants in the platform it was decided to develop an application for mobile phones in a joint project among teachers who use these tools and the Department of Engineering in Information Systems where teachers and students participate. Thus, we are integrating the whole educational community in a project that is meant to continue playing and enjoying continuous learning.

4 Project Overview

The project consists of the development of two applications:

(a) "*Brainify*", *a game for the Android platform*: The goal of the game is to allow the users to answer general knowledge questions and compete with their friends while doing so. Each question gives the user experience which can be accumulated allowing him/her to unlock new, more complex questions.
(b) "*Questions Manager*", *a Web application:* It is a web application with two objectives; it allows the loading of questions for each game level and provides questions to the Brainify application.
(c) The system will be developed in 7 stages, the first two, will be purely for testing and the last one the final candidate of the application.

Stage	Description
1	A complete prototype of the Brainify application will be made. The prototype will not be functional. The prototype will seek to illustrate the content of the application, the different buttons, design and configuration. The objective of this stage is to validate the aesthetics of the Brainfy application
2	The following modules of the Question Manager application will be developed: • User authentication • Level management • Questions and answers management • REST service of questions per level The following Brainify application modules will be developed: • Solitary mode • Current level question rounds • Question experience The objective of this was to validate the functioning of the round of questions
3	The following modules of the Question Manager application will be developed: User management The following Brainify application modules well be developed: Solitary mode Previous level question rounds Level upgrade

(*continued*)

(*continued*)

Stage	Description
	The objective of this stage was to finalize the operation of the round of questions in the Solitaire mode
4	The following modules of the Question Manager application will be developed: • Add players by Facebook or Google The following Brainify application modules will be developed: • Integrate social networks as a method of accessing the game. The objective of this step is to manage several users within the Brainify application
5	The following modules of the Question Manager application will be developed: • Relationship between players The following Brainify application modules will be developed: • It will allow you to view and list social networking friends within the game to invite them to the game or rounds of questions The aim of this stage is to manage the link between the user and his acquaintances in social networks
6	The following modules of the Question Manager application will be developed: • Competitive mode The following Brainify application modules will be developed: • Competitive modality It will allow you to invite other users to participate in a round of questions It will evaluate the possible combinations of questions and answers for both players It will calculate the experience obtained by the new modality The aim of this stage is to test the correct functioning of the competitive modality of the game
7	The development stage is completed. All stages are linked and the first final version is built. Seeks to validate the correct operation of the application

The App. starts with a first stage that is the development of skills such as: mental agility, logical thinking development, mental openness and lateral thinking.

The second part of the game is for students who have reached a certain basic level, now they will be able to select the subjects they want to reinforce, such as mathematics, physics, chemistry, etc.

This section of the App will be based on conceptual questions and reaffirmation of pre-acquired knowledge, playing we are constantly remembering what we have learned. The application of these activities is being carried out in the present elective cycle with the first year students of the Engineering degree.

This type of skills has already been evaluated on the platform with the students who are taking the introductory course to our university and the respective results were presented at IMCL 2016 [9].

Students who are currently in their first year of the Engineering degree have already had the possibility to use the reasoning games during the entrance course and are actively participating in higher level reasoning games, where challenges are made on the subjects learned.

– Students are more fluent in the use of subject-specific terms.
– Simple basic concept retention capability.

- The handling of dimensions and change of units is more fluid and presents fewer errors.
- The relationship between theoretical concepts and everyday life can be asociate them more clearly.

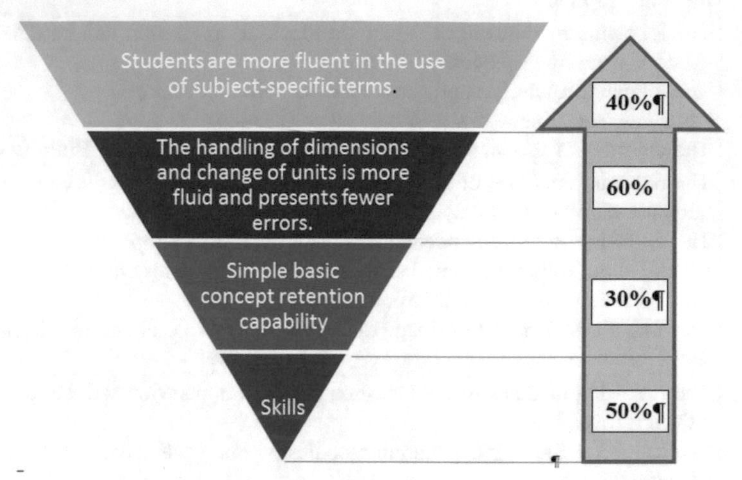

It can be clearly seen as brain training not only increases the speed of response, which does not only imply velocity but also reading comprehension, mental ordering, strategic thinking, detection data, incognita and necessary tools for resolution of problems.

New times require new strategies and the recent discoveries, given to us by cognitive neuroscience reveal that education today requires a profound restructuring, that does not prevent him to stay out of phase with the recent technological avalanche. Though we assume that education is not restricted to the school environment, school and teachers have to prepare future citizens for a changing world.

To do this, we have to eradicate teaching focused on the transmission of a series of abstract and decontextualized concepts that have no practical application. Our students have to learn to learn, facilitating, procurement of a series of useful skills that allow them to solve problems encountered in our daily life**: learning for life**. And for this primarily socioemotional intelligence is required.

Learning is optimized when the student is an active protagonist of that process, that is, you learn by acting and this is facilitated when it is a pleasant activity and occurs in a positive emotional climate.

References

1. Zichermann, G., Linder, J.: Game-Based Marketing: Inspire Customer Loyalty Through Rewards, Challenges, and Contest. Wiley, Hoboken (2010)
2. Kapp, K.: The Gamification of Learning and Instruction: Game Based Methods and Strategies for Training and Education. Wiley, San Francisco (2012)

3. Zichermann, G., Cunningham, C.: Gamification by Design: Implementing Game Mechanics in Web and Mobile Apps. O'Reilly Media, Cambridge (2011)
4. Guillen, J.: Escuela con Cerebro. https://escuelaconcerebro.wordpress.com/
5. Flores, A., Guillen, J.: Neuromitos en educación. Plataforma, New York (2015)
6. Manes, F., Niro, M.: Usar el Cerebro. Planeta, Madrid (2015)
7. Bachrach, E.: Agilmente. Conecta, Barcelona (2012)
8. Demeade, N.: Gammification with Moodle. Pack Publishing, Birmingham (2015)
9. Elorriaga, M., Antunez, M., Nicolino, M.: The game as a way to destructuration: Learning to reason through game. In: IMCL 2016 (2016)

Pattern-Based Game Apps for Collaborative Learning About Sustainable Management of Public Space

Panagiotis Tragazikis(✉) and Dimitris Gouscos

Department of Communication and Media Studies, National and Kapodistrian University of Athens, Athens, Greece
{ptragaz,gouscos}@media.uoa.gr

Abstract. The paper will introduce a research plan to establish dynamic learning experiences about public space management and the strategies selected. More specifically, we are concerned with an educational design approach for K-12 learners, focusing on sustainable use of public space, game-based learning with mobile devices, and integration of participants' viewpoints into building key elements of the digital environment that they will use to achieve specific learning goals. The research project aims to measure the concepts perception, related to public space use and management by participants through the proposed methodology. Additionally it aims, through that methodology, to highlight the conditions by which, a combination of digital game design for mobile devices based on basic design and narrative patterns, can be shown to be effective in approaching difficult concepts in a classroom setting.

Keywords: Game apps · Game design · Design patterns · Story patterns Collaborative learning · Public space · Sustainability

1 Context

There is an extensive research for public space use in a way that public space is a sweeping array of settings, including urban streets, plazas and squares, malls, parks, and other locations, and natural settings such as aquatic environments, national parks and forests, and wilderness areas that demand the attention of many disciplines and researchers, designers, managers and policy makers [1]. Additionally, the "use" could be related to the duration and the number of functions which took place supporting the general activity of public space, guiding design approaches to improve senses and communication [2], reshaping, taking into consideration machine communication, information, administration and the "pristine state of man" [3], or combining planning and architectural pedagogy, theory and every day practice, with domains related to art, activism and alternative planning and design praxis [4] and counting factors like economy, social inclusion, cultural diversity, environmental care and governance [5]. Furthermore, by empowering dynamics of public space with smart technologies, it seems that even physical activity is the basis for public space interaction, and cyber activity could be beneficial from the point of view of an individual human being [6].

M. E. Auer and T. Tsiatsos (Eds.): IMCL 2017, AISC 725, pp. 232–239, 2018.
https://doi.org/10.1007/978-3-319-75175-7_24

Moreover research; about use and management of public space, we count on historical, cultural, natural dimensions, in interactive dynamics [7], as a means to empower civility [8]. Converging all the above to humanistic design principles inspires designers and we can say that public spaces exist, modified or created but in any case, a critical discourse configured around them leading to evaluation models developing, such as remote assessment methods [9] or assessment model based on facts, related to cultural and historical reality of public space [10].

Based on the aforementioned, a research about public space use and management is conducted based on sustainable use of public space. Due to vast and multidisciplinary concepts, a simplification is applied in order to make the content suitable for K-12 students. Additionally, the overview of public space dimensions which are presented above, lead us to the thought that pubic space assessment consists a secure context for developing educational activities focusing on sustainability.

2 Purpose

In the above described context, the paper introduces a research plan to establish dynamic learning experiences about public space management and the strategies selected. More specifically, we are concerned with an educational design approach for K-12 learners, focusing on sustainable use of public space, game-based learning with mobile devices, and integration of participants' viewpoints into building key elements of the digital environment that they will use to achieve specific learning goals. The necessity to adopt sustainable attitudes about public space follows a number of facts: public space is constantly changing; active citizens need to develop methods and strategies to participate in its management; sustainable management conditions need to be applied. The learning goals to be achieved include understanding that the use of public space involves many groups, developing a management strategy, reflecting on the strategy selected, realizing that each choice is leading to specific results, distinguishing the more democratic strategy, acquiring the ability to develop an action plan based on the strategy chosen, understanding that each option may need to be redefined due to changes that had not been taken into account at the time of decision making. These goals derive from a spectrum of public space management matters, drawn from literature on education for sustainability, as a changing procedure which enlarges the public sphere, in which citizenship is conceived and practiced to include the environment. Furthermore, it is connected with public space as an effort towards regulation, equality and democracy [11], integration into formal education and non-formal and informal [12] and supporting the possible need to engage new players and prepared to slide down, before we can go forward to create a more sustainable future [13], which makes public space an educational matter, a fertile place to develop educational plans.

3 Approach

Our approach is based on Digital Game Based Learning which is precisely about fun and engagement, and the conjunction of serious learning and interactive entertainment into an emerging and highly exciting medium, Digital Learning Games [14]. It also makes kids able to do and understand so many complex things, achieving reasoning that the curriculum they are given appeared to be deteriorating their minds, [15]. Serious players of video games get their glory largely from being the first on the block to master the game that just came out, and this means that kids have a powerful incentive to get good at learning well and quickly [16]. Another dimension is mobile technologies, where mobile learning can be spontaneous, portable, personal, situational; it can be informal, unobtrusive, ubiquitous and disruptive [17]. Furthermore, many empirical studies have found mobile learning has positive influence on learning performance [18, 19] and it is suitable for communication establishment [20]. Taking into consideration the previous mentioned advantages and particularities of mobile learning, an approach was formatted to gain both from the mobility and digital games. In that line of thought, we consider that the creation of mini games, games with basic characteristics to support basic skills like a gaming environment promoting fast calculation [21], or conceptual mini-game focusing on a concrete concept to be taught, in order to transform the game into a learning object [22] can prove to be very effective. Scaffolding to learning experiences could also be the use of mobile devices in order to raise awareness of important social issues [23]. Moreover, in the field of social change, Villanueva [24] argues that mobile games in classroom allow players to experiment with different solutions to a problem, aiding in broadcasting their knowledge of the subject matter. Furthermore, a combination of collaborative learning [25, 26] and cooperative learning [27, 28], will be used depending on the way students react to their opinions' adoptions and modifications. Experience is going to be derived through applied cases in academic field of collaborative game based learning (CGBL) such as classroom actions with games [29], serious games building [30], serious games for collaborative learning [31] or collaborative models based on CBGL and motivation development. Critically approaching games narration as a considerable factor in game design, in order to show off its significant role, mini games will be connected to the same storyline, following Propp's model for storytelling [32]. Furthermore, these games will follow a game design model based on design patterns [33] as well as on learning patterns [34]. The design patterns to use can be negotiated with learners in order to accommodate the interests of the latter, whereas educational patterns will be selected by the teacher as a result of prior interaction with learners. We prefer the term educational because we consider that it contains all types of patterns which will be used in order to develop an educational context. In all cases, the mini games to be developed are based on the concept of "functional pattern". The latter is described as a game part which can be played as an autonomous game. A functional pattern, therefore, includes all necessary game elements like sounds, background, objects and heroes, together with their interactions, as well as tasks that should be accomplished. In this respect, a functional pattern looks like a game scene that could be played as a stand-alone game; still, due to its limited complexity, a functional pattern allows to integrate learning objectives and accommodate players' remarks more easily than in higher-complexity fully-fledged games. Connecting the term functional pattern with mini game, a mini game is composed of one or more functional patterns. With respect to research, an effort appears to understand the patterns in relation to the game mechanics, in a context that takes into account the design motifs, the framework, the mechanics and the code in relation to those in the processes involved that is to say players, designers, developers and researchers [35]. This approach highlights the context where a game develops. Additionally, it creates a guideline for games targeting educational domain and specific educational objectives and our approach is considering simplifying a multidimensional space of relations. Narration operates as a grammar for story generation [36], functions like a connective element between different mini games and maintaining linear relations between games providing the grounds for a creative wondering experience among them. Alongside this approach, a digital storytelling tool could be used for editing, saving and modifying the story that will liaise with the mini games. This approach based on the idea that digital game design patterns are created on a case-by-case basis, is based on a narrative event that can be described as "Slice of Life" (SoL), [37]. Furthermore, narrative scaffolding supports a fundamental procedure in games problem solving. In that line of thought, narrative amended to game either with backstory in order to provide the dramatic context for the game [38] or with cut scenes in order to further the story line, reinforces the mood and tone of the game and provides multiple information such as new elements or information that should be decoded by the player in order to select the appropriate strategy [39]. In that way Propp's narrative model facilitates game evolution on a pattern base line. Each mini game has a thematic core based on the "star model" for public space assessment proposed by Varna [10]. Varna's model considers space to be "less" or "more" public according to the responses that can be given to issues of ownership, control, physical configuration, animation and civility. Each category is approached under the concept of sustainability and in this way the overall educational design approach follows basic game design concepts [40, 41] and cross-references these with Varna's star model for the assessment of space publicity. For each domain of public space assessment according to this model, a mini game is designed. All mini games encompass story, action and trial. Furthermore, mini games follow a model of players' enjoinment [42] based on a core of basic design patterns [43, 44]. The enjoinment model describes elements such as challenge, abilities, tasks that should be accomplished, feedback, social interactions, design patterns and narration motif. According to the above, each mini game is designed and created with the use of a platform for mobile apps, and then introduced to learners as a game app running on a mobile device. Each final game app may comprise one functional pattern or a limited combination of functional patterns. In all cases, for any given game app, learners will be able to propose, within the lines de-limited by the underlying functional pattern(s), modifications that can be effected easily, such as transformations of the hero or amendments to the environment space. This kind of learner participation permits modifications that can make the same game app look like a different game. This can meet the interest of participants and lead to game apps that acquire personal characteristics. Based on previous references, design process is presented as follows:

Initially, a story is developed with participants, based on Propp's model with thematic content "Public space". Working in teams a short story under the thematic content mentioned before is created. The different short stories will be adapted creatively into one basic story.

The story provides the main hero; therefore designing could proceed to the settings such as icon, characteristics, abilities and moves. Additionally, for each part of the story a space will be designed. The gaming experience will be under improvement proposals by the participants, when played with mobile devices, respecting the story core and the functional patterns are used.

The following goal is to refine the educational orientation of mini game. In that domain, the work designed to be done contains only integration patterns and engagement patterns. Project focuses on finalization of these types of patterns in order to clarify that the educational goal is effectively perceived by the participants. The proposed design procedure considers that all the other types of learning patterns will be part of the whole effort and are not necessarily embedded in game design. That means that cognition patterns, social interaction patterns and presentation patterns will be developed during the procedure and participants could have an external assessment in order to evaluate the design process.

A monitoring process follows, in order to guide game design in sustainable concept. According to sustainable use of public space and the creation of such a context that could be embedded learning goals in educational patterns, as it was presented by education for sustainability references; matters for student development are critical thinking, awareness of complexity and active citizenship [45] in an uncertainty, ambiguity and complexity world [46]. Moreover, a method of digital games evaluation as a tool for sustainable development will be taken into serious [47]. But in our approach, it is expected that not all the sixteen evaluation criteria will be supported by the game itself but by the process that will be followed. Furthermore, each category of the public space assessment model based on "publicness", will be connected with functional patterns within a number of actions that could be performed. Therefore, the player should get decisions to provide those actions, modifying a given space more or less according to the characteristics of the model. In that line of thought, design will use the model in order to create a number of actions in mini games that they support more or less in each category. The above presented research plan will be conducted under an action research basis with K-12 participants. Due to difficulty of a serious number of concepts, and multidisciplinary approach, the research results will be deteriorating on the learning experiences that would be constructed based on K-12 dynamic participation as well as on the age restrictions on concept perceptions, which is measured.

4 Anticipated Outcomes

According to the above approach, we expect to come up with a series of mobile game apps for learning concepts and skills for public space sustainable management. These game apps are able to accommodate the following types of learning goals: Story-embedded goals which are similar to open questions and their purpose are to make the players aware and engage them in a critical discourse about issues of public space sustainable management. Goals of this type are initially set by the educator and cannot be modified; still, they could be placed at different parts of the story, if discussion with learners shows that this is advisable. Pattern-embedded goals, those goals are related to main areas of public space, sustainable management concepts and skills. They are set by the educator, and placed within certain parts of a functional pattern. Still, educators can propose different sequences along which pattern-embedded goals may appear during the course of a game, allowing in this way the game to better fit personal preferences of learners. This is particularly important with reference to different skills of different learners in the game tasks that need to be accomplished. Meta-game goals are set as a result of social interaction between learners, under educator guidance. They are based on the cognitive reflection of learners during game app design, during play, and past play. The purpose of meta-game goals is to evaluate, in a public space setting that calls for sustainable management, all possible actions that could be taken at some point together with their consequences. The response of learners towards meta-game goals will be based on what they have learnt from game design and game play, as well as from their ability to transliterate this learning in a real world public space setting. The latter could be a real physical public space, augmented with information about sustainable management issues, or a simulated public space with realistic sustainability issues. Additionally, the concepts perception which are dealt with, apart from the other parameters mentioned above, constitute the core measurement object of the research. In other words, a combination of narration building techniques, with basic structural elements of digital mobile games in order to achieve complex concepts and decisions selections constitute the overall matter they deal with.

5 Summary

Summing up, this paper elaborates on a research plan based on the above and presents a method of implementing such a research action. In the long term, this research effort aims at delivering an innovative proposal towards ways for embedding learning goals in game design to support topics like the sustainable use of public space, which are too broad and loosely structured to be effectively communicated through conventional mini games for learning. In this line of thought, a continuous action research endeavor is needed to reveal difficulties, barriers as well as hidden opportunities for the effective design of educational games that allow collaborative and blended learning as well as concepts perception.

References

1. Atman, I., Zube, E.H.: Public Places and Spaces. Plenum Press, New York (1989)
2. Gehl, J.: Life Between Buildings: Using Public Space. Island Press, Washington (2011)
3. Corbusier, Le: Concerning Town Planning. Yale University Press, New Haven (1948)
4. Tornaghi, C., Knierbein, S.: Public Space and Relational Perspectives: New Challenges for Architecture and Planning. Routledge, New York, London (2014)
5. Madanipour, A., Knierbein, S., Degros, A.: Public Space and the Challenges of Urban Transformation in Europe. Routledge, New York, London (2014)
6. Aurigi, A.: Making the Digital City: The Early Shaping of Urban Internet Space. Ashgate Publishing Company, Farnborough (2005)
7. Huat, C.B., Edwards, N.: Public Space: Design, Use and Management. Singapore University Press, Singapore (1992)
8. Dahnke, C., Spath, T.: Reclaiming Civility in the Public Square: 10 Rules That Work. WingSpan Press, Livemore (2007)

9. Taylor, B., et al.: Measuring the quality of public open space using Google Earth. Am. J. Prev. Med. **40**(2), 105–112 (2011)
10. Varna, G.: Measuring Public Space: The Star Model. Series: Design and the Built Environment. Ashgate, Farnham (2014)
11. Huckle, J., Stephen, S.: Education for Sustainability. Earthscan Publications, London (2014)
12. UNESCO: Shaping the Future We Want - UN Decade of Education for Sustainable Development, Final report (2014)
13. Tilbury, D.: Education for sustainability: a snakes and ladders game? Foro de Educación **13**(19), 7–10 (2015)
14. Prensky, M.: Digital Game-Based Learning. McGraw-Hill, New York (2001)
15. Prensky, M.: Don't Bother Me Mom, I'm Learning!. Paragon House, Saint Paul (2006)
16. Papert, S.: Does easy do it? Children, games, and learning. Game Developer Magazine. "Soapbox" section (1998)
17. Kukulska-Hulme, A., Traxler, J.: Mobile Learning: A Handbook for Educators and Trainers. Routledge, London (2004)
18. Hwang, G.J., Chang, H.F.: A formative assessment-based mobile learning approach to improving the learning attitudes and achievements of students. Comput. Educ. **56**, 1023–1031 (2011)
19. Bredl, K., Bösche, W.: Serious Games and Virtual Worlds in Education, Professional Development, and Healthcare. IGI Global, Hershey (2013)
20. de Vries, I.O.: Mobile telephony: realizing the dream of ideal communication? In: Hamill, L., Lasen, A. (eds.) Mobile World: Past, Present and Future, pp. 11–28. Springer, London (2005)
21. Panagiotakopoulos, C.: Applying a conceptual mini game for supporting simple mathematical calculation skills: students' perceptions and considerations. World J. Educ. **1**(1), 3–14 (2011). Google Scholar
22. Illanas, I., Galleg, F., Satorre, R., Llorens, F.: Conceptual mini-games for learning. In: IATED International Technology Education and Development Conference, Spain, Valencia (2011). http://rua.ua.es/dspace/bitstream/10045/8495/1/illanas08conceptual.pdf
23. Schreiner, K.: Digital Games Target Social Change. IEEE Comput. Graph. Appl. **28**(1), 12–17 (2008)
24. Villanueva, K., Vaidya, J.: Transforming Learning with Mobile Games: Learning with Mobile Games. Handbook of Research on Mobile Learning in Contemporary Classrooms. IGI Global, Hershey (2016)
25. Dillenbourg, P.: What do you mean by collaborative learning? In: Dillenbourg, P. (ed.) Collaborative-Learning: Cognitive and Computational Approaches, pp. 1–19. Elsevier, Oxford (1999)
26. Johnson, D.W.: Circles of learning: cooperation in the classroom. In: VA Association for Supervision and Curriculum Development, Alexandria (1984)
27. Panitz, T.: Collaborative versus cooperative learning: a comparison of the two concepts which will help us understand the underlying nature of interactive learning. Coop. Learn. Coll. Teach. **8**(2), 1–13 (1997)
28. James, S.: Revisiting an old friend: the practice and promise of cooperative learning for the twenty-first century. Soc. Stud. **102**(2), 88–93 (2016)
29. Squire, K.: Video Games and Learning: Teaching and Participatory Culture in the Digital Age. Teachers College Press, New York (2011)
30. Oksanen, K.: Serious game design: supporting collaborative learning and investigating learners' experiences. Finnish Institute for Educational Research (2014)

31. Romero, M.: Supporting collaborative game based learning knowledge construction through the use of knowledge group awareness. NoE Games and Learning Alliance, Lecture at the GaLa 1st Alignment School, 20 June, Edinburgh (2011)
32. Propp, V.: Morphology of the Folktale. Trans. Laurence Scott. Print. Trans. of Morfológi-jaskázki published in 1928, 2000, U of Texas Press (1928)
33. Bjork, S., Holopainen, J.: Patterns in Game Design (Game Development Series). Charles River Media Inc, Rockland (2004)
34. Kiili, K.: Call for learning-game design patterns. In: Educational Games: Design, Learning, and Applications, Nova Publishers, New York (2010)
35. Olsson, C.M., Björk, S., Dahlskog, S.: The conceptual relationship model: understanding patterns and mechanics. In: Game Design 2014, Proceedings of the 2014 DiGRA International Conference, Visby, Sweden (2014)
36. Gervás, P.: Propp's morphology of the folk tale as a grammar for generation. In: Finlayson, M., Fisseni, B., Löwe, B., Meister, J.C. (eds.) Workshop on Computational Models of Narrative 2013, Hamburg, Germany, pp. 106–122 (2013)
37. Maciuszek, D., Martens, A.: Patterns for the design of educational games. In: Edvardsen, F., Halsten, K. (eds.) Educational Games: Design, Learning and Applications. Nova Publishers, New York (2010)
38. Crawford, C.: Chris Crawford on Game Design. New Riders Publishing, Indianapolis (2003)
39. Dickey, M.: Appropriating adventure game design narrative devices and techniques for the design of interactive learning environments. In: Educational Technology Research and Development, vol. 54, no. 3, June 2006, pp. 245–263 (2006)
40. Salen, K., Zimmerman, E.: Rules of Play - Game Design Fundamentals. The MIT Press Cambridge, London (2004)
41. Schell, J.: The Art of Game Design: A Book of Lenses. CRC Press, Boca Raton (2014)
42. Sweetscr, P., Wyeth, P.: GameFlow: a model for evaluating player enjoyment in games. In: Computers in Entertainment, vol. 3, no. 3, pp. 1–24. ACM (2005)
43. Järvinen, A.: Games without Frontiers: Theories and Methods for Game Studies and Design. Tampere University Press, Tampere (2008)
44. Moore, M.: Basics of Game Design. Taylor & Francis Group, Boca Raton (2011)
45. Liarakou, G., Flogaitis, E.: From Environmental Education to Education for Sustainable Development. Nissos, Athens (2007)
46. Taubriz, R.: A pedagogy for uncertain times. In: Lambrechts, W., Hindson, J. (eds.) Development Exploring Collaborative Networks, Critical Characteristics and Evaluation Practices. Environment and School Initiatives – ENSI, Vienna, Austria (2016)
47. Liarakou, G., Sakka, E., Gavrilakis, C., Tsolakidis, C.: Evaluation of serious games, as a tool for education for sustainable development. Eur. J. Open, Distance E-learning, Spec. Issue, Best EDEN **2011**, 96–110 (2012)

Serious Games and Motivation

Jenny Pange[1], Aspa Lekka[1], and Sotiria Katsigianni[2(✉)]

[1] University of Ioannina, Ioannina, Greece
jennypagge@yahoo.gr, lekka.aspa@gmail.com
[2] Primary Education, Filippiada, Greece
geo_sotk@hotmail.com

Abstract. Serious games are gaining a lot of interest in education because they are effective learning tools that engage and motivate students. However, serious games are not ad-hoc motivational but they must follow specific elements to assure motivation. There are several types of motivation that reinforce the effectiveness of serious games and engage students to the task. When these types are combined they provide high learning outcomes. The preliminary study examines how primary school teachers apprehend serious games and how their concepts are related with intrinsic motivation. The findings of the current study provide clear evidence that teachers believe that serious games intrinsically motivate students to learn by developing significant learning skills.

Keywords: Serious games · Effectiveness · Intrinsic motivation

1 Introduction

In recent years, education faces new challenges that include crucial changes in the content of education and the type of learning. Nowadays, students are digital natives and they perceive information in different ways, hence educators have to adopt new teaching methods that will actively involve students to the teaching process [1–5].

School teachers consider that knowledge construction is supported in learning environments based in ICTs and serious games [6–8]. Serious games are the educational tools that respond better to all these emerging educational challenges as they provide knowledge and entertainment at the same time. The use of serious games in school education has been increased due to their positive outcomes in learning [9]. In addition, serious games draw the attention of researchers because they are considered to be helpful in the development of several skills in students by combining different learning objects [10–12]. It is also argued that they are useful tools not only for students' education but also for teachers' training [13, 14] as they engage and motivate learners.

The learning method that is based on the use of digital games is known as "Digital Game-Based Learning" (DGBL). This model incorporates related learning theories in the learning content in order to keep students engaged [15, 16]. Furthermore, DGBL provides to the students the opportunity to deal with the enriched technological environment of the 21th century. DGBL is divided into three categories. The first one is related to the adaptation of cognition, the second one is related to skills acquisition and the last one is in relation to the change of attitudes and behavior of students [17]. In

M. E. Auer and T. Tsiatsos (Eds.): IMCL 2017, AISC 725, pp. 240–246, 2018.
https://doi.org/10.1007/978-3-319-75175-7_25

another classification, DGBL is also grouped in other three categories, where the first category is related to the student's engagement, the second examines the learning process and the third one reflects on the learning outcomes [18]. In a digital game-based environment a student can experience engagement, participation, involvement, presence, motivation and flow [19].

2 Motivation in Serious Games

Motivation has to be a "necessary prerequisite" in serious games in order them to be effective. Additionally, a serious game has to be appropriately designed to motivate students [17, 18] and it must not be considered as ad-hoc motivational just because it is a game. A learning environment based on a well-designed serious game has better learning outcomes compared to a non-DGBL environment [20–22]. In addition, the selection of the suitable game is another factor that reinforces motivation and ensures effectiveness in DGBL. Consequently, the educator has to select the game that responds better to the student's special characteristics like the gender, the age and the prior game-experience etc. [23]. Specifically, gender seems to be a significant factor of the effectiveness of a game as males and females have different attitudes towards serious games [9, 24, 25].

2.1 Types of Motivation

Many researchers declare that there are two kinds of motivation: "intrinsic" and "extrinsic" [17, 18, 26, 27].

Extrinsic motivation is related with prizes, achievements and the player's desire to win [27]. According to Ref. [18], extrinsic motivation is correlated with the less important kinds of learning and it may undermine the intrinsic one. Nevertheless, autonomy in extrinsic motivation delineate the situation in which activities are conducted out of feelings of guilt, obligation or need for something to be proved. An important characteristic of games that ensures extrinsic motivation is the existence of "achievements" as well. Achievements are separated in expected and unexpected and their mixture enlarges motivation. Achievements motivate players and increase the playtime and the efforts [18, 28].

On the other hand, an activity is intrinsically motivational if it engages students regardless of achievements and rewards [29]. Intrinsic motivation is related to higher levels of enjoyment, interest, performance, higher learning levels and increased self-esteem. Furthermore, imagination (endogenous and exogenous) is acclaimed to be an important element that makes games internally motivational. Ref. [29] argued that intrinsic motivation is related to challenge, fantasy and curiosity. Challenge is associated with goals and uncertain outcomes, fantasy is correlated with mental images of objects or situations that are not present, and curiosity is related with the balance between complexity and learner's existing knowledge. The more an activity follows the foresaid elements, the more motivational it is.

A combination of intrinsic and extrinsic motivation reinforces "flow", namely the situation in which a person is so absorbed in the goal-driven activity he is interested in nothing else. Additionally, flow furnishes motivation and helps students to maintain their attention for a long time [19, 30].

So, motivation is a significant factor of a game's effectiveness and there are several types of motivation that can make a serious game not only entertaining but also engaging.

The aim of this preliminary study is to investigate how elementary school teachers appreciate the concept of serious game and how their ideas are related to the digital games' motivation.

3 Research

3.1 Materials and Methods

In order to investigate the way that primary school teachers perceive motivation in serious games, we used the selective sampling technique for a current pilot research. Hence, we selected 21 primary school teachers from urban area, Greece, with previous knowledge of serious games.

The research was guided by two research questions:

1. How do primary school teachers apprehend serious games?
2. How the definition that teachers give is related with intrinsic motivation?

In order to address the research questions, it was given a questionnaire to 21 participants. The first part of questions was dealing with the definition of 'serious games' and aimed to collect data for the first research question. Primary school teachers were free to give their own definition and select more than one answer from a list about the characteristics of serious games. In the second part of the questionnaire primary school teachers had to declare their views about serious games, students' engagement and intrinsic motivation in order to reply to the second research question.

3.2 Results

The sample consisted by 21 primary school teachers, from Ioannina, Greece. The teachers that participated to the research were all females with mean age 26,14 years (SD ± 2,85) and median age 26 years (Range 22–30). They had recently graduated from the University (i.e. the last 8 years). In addition, they answered that they all used serious games at school and had internet connection at their schools. So, we expect that most of these teachers were digital natives.

As about the first research question and the way that teachers in primary education apprehend serious games, most teachers considered that serious games are commonly entertaining (85.7%, N = 18), and others believed that serious games are only educational activities (66.6%, N = 14). Only one third of the primary school teachers knew that serious games have rules (33.3%, N = 7) (Table 1).

In the teaching process, 19.0% (N = 4) of our teachers believed that serious games develop mental skills and 14.2% (N = 3) of them believed that they help students to cooperate. Only one teacher believed that serious games help students to express themselves.

Table 1.

Definition of serious games	N	Percent (%)
Serious games are entertaining	18	85.7
They are educational activities	14	66.6
They have rules	7	33.3
They develop mental skills	4	19.0
They help students to interact with others	3	14.2
They help students to express themselves	1	4.7

In the second part of questionnaire it was attempted the main elements of motivation to be approached in order to answer to the second research question. Hence, teachers had to agree or disagree on specific aspects about serious games, related to motivation. As we can see in Table 2, teachers agreed that serious games help students to select their own learning strategies (100%, N = 21) and make decisions (86.71%) during the game. Furthermore, teachers believed that in games students play roles (95.24%, N = 20) that help them emotionally, as they keep their interest in the learning process. So, students had better enjoyment and engagement in the learning process.

Table 2.

Teachers' aspects about serious games	N	Percent (%)
Serious games need to have defined purpose	17	80.95
Students learned how to concentrate their mind	16	76.19
Students make decisions in problems easily	18	86.71
Students learn how to choose in between different strategies in a game	21	100.00
Students play roles in serious games	20	95.24
Students are able to evaluate data	12	57.14

Moreover, most of the teachers (80.95%, N = 17) had the point of view that serious games need to have defined purpose. This is a significant finding because it shows that teachers appreciate the importance of goals in intrinsically motivating environments.

Primary school teachers also believed that during the game, students learn how to concentrate their mind in the course material (76.16%, N = 16). Mind concentration is related with engagement and flow in the teaching process.

In addition, teachers had the aspect that while playing a game, students were able to calculate records and evaluate data (57.14%, N = 12), so they increased self-esteem.

3.3 Discussion and Conclusions

According to the results of this study primary school teachers defined serious games as entertaining educational activities where students follow rules. This finding is linked with the first research question about how teachers apprehend serious games.

The definition that primary school teachers gave, is also linked with "entertainment" (enjoyment) and "education" (learning), two basic elements of intrinsic motivation. Analyzing the second research question through the answers of the second questionnaire it is concluded that teachers believed that serious games help students learn how to choose strategies and make decisions, as also they believe that games have to have defined purpose. All three features are related with challenge in a game. The fact that students have the choice to make decisions increases their motivation according to Ref. [31]. In addition, teachers had the point of view that students play roles in games, a fact that immerses students in virtual reality and promote fantasy. Challenge and fantasy are important elements of intrinsic motivation according to Ref. [29]. According to the findings of this study students also learn how to concentrate their mind and evaluate data during the game. These are high mental skills related to intrinsic motivation. Additionally, teachers believe that serious games promote interaction between students, as other researchers have found [21] a fact that reinforces interpersonal motivation [29]. From the findings of the study we can support that teachers have the aspect that serious games have elements that make them engaging and intrinsic motivational. This result is in accordance with other researches [33, 34].

In conclusion, the primary school teachers in our pilot study believed that serious games are entertaining activities that intrinsically motivate students to acquire important skills. However, the selective sample that took part in the current research cannot be representative but it can lead researchers to carry out further research in the aspects of teachers about motivation in serious games.

References

1. Guillen-Neto, V., Aleson-Carbonell, M.: Serious games and learning effectiveness: the case of it's a deal. Comput. Educ. **58**, 435–448 (2012)
2. Mikropoulos, T.: Computer as a Cognitive Tool. Ellinika Grammata, Athens (2006). (in greek)
3. Pange, J.: Educational Technology and Web Applications. Disigma, Thessaloniki (2016). (in greek)
4. Prensky, M.: Digital natives, digital immigrants part 1. Horizon **9**(5), 1–6 (2001)
5. Toki, E.I., Pange, J.: Traditional and computer-based evaluation of preschoolers' oral language in Greek – a review of the literature. Sino-US English Teach. **9**(1), 840–845 (2012)
6. Hoyles, C.: Illuminations and reflections – teachers, methodologies and mathematics. In: 16th Conference: The Psychology of Mathematics Education, New Hampshire, pp. 263–283 (1992)
7. Noss, R.: Computers as commodities. In: Di Sessa, A., Hoyles, C. (eds.) Computers and Exploratory Learning, pp. 363–381. Springer, Heidelberg (1995)
8. Papert, S.: Mindstorms. Basic Books, Inc., Publishers, New York (1980)
9. Stege, L., Lankveld, G., Spronck, P.: Serious games in education. Int. J. Comput. Sci. Sport **10**(1), 1–9 (2011)
10. De Grove, F., Bourgonjon, J., Van Looy, J.: Digital games in the classroom? A contextual approach to teachers' adoption intention of digital games in formal education. Comput. Hum. Behav. **28**(6), 2023–2033 (2012)

11. Lekka, A., Sakellariou, M.: Computer games and ethical issues-a literature review. In: Presented at 2014 International Conference on Interactive Mobile Communication Technologies and Learning, Thessaloniki, Greece (2014)
12. Iten, N., Petko, D.: Learning with serious games: is fun playing the game a predictor of learning success? Br. J. Educ. Technol. **47**(1), 151–163 (2014)
13. Stavroulia, K., Botsari, E., Kekkeris, G., Psycharis, S.: Educating with the use of games. In: Proceedings of EEEP-DTPE "Education in the age of ICTs", Athens (2013). (in greek)
14. Stavroulia, K., Makri-Botsari, E., Psycharis, S., Kekkeris, G.: Emotional experiences in simulated classroom training environments. Int. J. Inf. Learn. Technol. **33**(3), 172–185 (2016)
15. Duplaa, E., Shirmohammadi, S.: Video Games in the classroom (2010). http://www.edu.gov.on.ca/eng/literacynumeracy/inspire/research/WW_Video_Games.pdf
16. Coffey, H.: Digital game-based learning. Digital game-based learning (2016). http://www.learnnc.org/lp/pages/4970
17. All, A., Nunez Castellar, E., Van Looy, J.: Measuring effectiveness in digital game-based learning: a methodological review. Int. J Ser. Games **1**(2) (2014)
18. Westera, W.: Games are motivating, aren't they? Disputing the arguments for digital game-based learning. Int. J. Ser. Games **2**(2) (2015)
19. Kiili, K., Perttula, A., Lindstedt, A., Arnab, S., Suominen, M.: Flow experience as a quality measure in evaluating physically activating collaborative serious games. Int. J. Ser. Games **1** (3) (2014)
20. Costabile, M.F., De Angeli, A., Roselli, T., Lanzilotti, R., Plantamura, P.: Evaluating the educational impact of tutorning hypermedia for children. Inf. Technol. Childhood Educ. Ann. 289–308 (2003)
21. Tobias, S., Fletcher, J.D., Wind, A.: Game-based learning. In: Spector, M., Merill, D., Elen, J., Bishop, M.J. (eds.) Handbook of Research on Educational Communications and Technology. Springer Science+Business Media, New York (2014)
22. Clark, D.B., Tanner-Smith, E.E., Killingsworth, S.S.: Digital games, design and learning, a systematic review and meta-analysis. Rev. Educ. Res. **86**(1) (2016)
23. Deubel, P.: Game on! T. H. E. (Technological Horizons in Education) J. **33**(6), 30–35 (2006)
24. Paliokas, I., Kekkeris, G., Georgiadou, K.: Study of users' behaviour in virtual reality environments. Int. J. Technol. Knowl. Soc. **4**, 121–132 (2008)
25. Papafilippou, N., Tsiatsos, T., Manousou, E., Lionarakis, A.: Investigation of complementary distance learning in the context of mathematics support teaching with the use of educational software. Open Educ. – J. Open Distance Educ. Educ. Technol. Special Edition One School Distance Educ. **12**, 73–89 (2016). (in greek)
26. Van Eck, R.: Digital game-based learning: it's not just the digital natives who are restless. Educause Rev. **2**(41) (2006). https://www.researchgate.net/profile/Richard_Van_Eck/publication/242513283_Digital_Game_Based_LEARNING_It's_Not_Just_the_Digital_Natives_Who_Are_Restless/links/0a85e53cd61cf43e29000000.pdf
27. Schrader, C.: Understanding the role of achievements in game-based learning. Interact. Des. Archit. J. **19**, 38–46 (2013)
28. Blair, L., Bowers, C., Cannon-Bowers, J., Gonzalez-Holland, E.: Understanding the role of achievements in game-based learning. Int. J. Ser. Games **3**(4) (2016)
29. Malone, T.: Toward a theory of intrinsically motivating instruction*. Cogn. Sci. **5**(4), 333–369 (1981)
30. Perttula, A., Kiili, K., Lindstedt, A., Tuomi, P.: Flow experience in game based learning – a systematic literature review. Int. J. Ser. Games **4**(1) (2017)

31. Zimbardo, P.G.: The human choice: individuation, reason, and order versus deindividuation, impulse, and chaos. In: Arnold, W.D., Levine, D. (eds.) Nebraska Symposium on Motivation, University of Nebraska, Lincoln, pp. 237–307 (1969)
32. Sanchez, R., Brown, E., Kocher, K., DeRosier, M.: Improving children's mental health with a digital social skills development game: a randomized controlled efficacy trial of adventures aboard the S.S. GRIN. Games Health J. **6**(1), 19–27 (2017)
33. Owston, R., Wideman, H., Ronda, N., Brown, C.: Computer game development as a literacy activity. Comput. Educ. **53**(3), 977–989 (2009)
34. Woo, J.C.: Digital game-based learning supports student motivation, cognitive success, and performance outcomes. Educ. Technol. Soc. **17**(3), 291–307 (2014)

Trials of the Acropolis: Teaching Greek Mythology Using Virtual Reality and Game Based Learning

Pantelis Chintiadis[1], Ioannis Kazanidis[2](✉), and Avgoustos Tsinakos[2]

[1] Computer and Informatics Engineering Department, Eastern Macedonia and Thrace Institute of Technology, Agios Loukas, Kavala, Greece
pantchin@teiemt.gr

[2] Advanced Educational Technologies and Mobile Applications Lab, Eastern Macedonia and Thrace Institute of Technology, Agios Loukas, Kavala, Greece
{kazanidis,tsinakos}@teiemt.gr

Abstract. The evolution of learning environments is huge in recent years in the fields of mobile, blended learning and telecommunication technologies. Many courses are using VR tools to recreate historic and natural sites, while teachers use VR, guiding students to never before seen historic places.

This paper presents a VR game for the instruction of the 3rd grade Greek history, and in particular the Greek mythology. The main objective of this paper is a brief description of the technologies used, the design and implementation procedure, the presentation of the adopted educational scenario and the preliminary results of formative evaluation.

Keywords: Virtual reality · Mobile learning · Game based learning Greek mythology · Primary education

1 Introduction

Mobile devices and technologies are evolving at frenzy rates, enhancing our access to every part of the surrounding world. New ways to learn, interact, publish and explore our everyday living, have been invented and established already.

Virtual Reality (VR) stands among these technologies that thrive in nowadays, causing a paradigm shift in human machine interaction field. Since the 1950s virtual reality has been hovering through the noir themed streets, struggling to be the biggest invention the world has ever seen, but at some pointed failed to achieve commercial adoption. Fast forward today reports say that until 2019 VR will grow to $15.9 billion industry [1]. With this powerful new technology, learning and entertainment can be experienced from a different angle and can help all levels of education sector guide students to new pools of information.

With so much attention in the last few years for virtual reality, the developers have been constantly thinking new ways to discover uses of this promising technology in our world. Today VR applications are very diversive, however they serve the same goal

M. E. Auer and T. Tsiatsos (Eds.): IMCL 2017, AISC 725, pp. 247–257, 2018.
https://doi.org/10.1007/978-3-319-75175-7_26

which apparently is life-long learning. Some of the most popular applications categories are outlined below [2, 3]:

Scientific and Architectural data visualization
Medicine engineering and scientific fields year by year try to expand and the push their boundaries in order to make the human world better. Using appropriate input devices like data gloves and head mounted displays, scientists and engineers can create and prototype whatever new technology they design and test it in a virtual environment. On the architectural side historians and archaeologists can make their imaginations reality, by walking and interacting in long lost ruins and temples of the ancient world.

Simulation and Training
One of the most common simulation targets is the training category. Establishing a safe virtual world, that can emulate a real and dangerous environment, can be a pretty big plus for training organizations or companies that wish to give their employees more experience in real world situations. Vehicle simulation, battlefield visualization, flight simulation and medical procedures [4] are some of the categories that benefit using virtual reality equipment and suitable software.

Entertainment
The most expansive and ever growing area that takes advantage of VR is entertainment. From videogames, to movies and videos, users have a plethora of choices to select and entertain themselves. All kinds of videogame genres are available for every kind of customer and their graphics are constantly getting better and better, with wand-like devices and different platforms the diversity is huge.

Virtual reality's hardware is mainly split into two main categories: Tethered and Mobile. Tethered like the Oculus Rift, Playstation VR, and HTC Vive are connected with a PC, except PS VR which connects with a Playstation 4. The headset has a built-in LCD display, motion sensors, a camera tracker and with a combination of a powerful desktop computer capable of running virtual reality apps the immersion is huge [5]. Mobile headsets like Google Daydream and Gear VR on the other hand are shells with lenses, where a smartphone is placed. These kind of devices can be used anywhere since there is no need for a cable or PCs. However, since all the processing is done by the phone, mobile headsets suffer on performance in comparison with tethered devices.

In this paper we are going to explain the followed process, the tools used to develop the proposed educational VR videogame, the adopted game theory and introduce the main game scenario and game environment.

The main objective of this paper is to present the approach we followed and the developed VR game which aims to help the students of the 3rd grade, and anyone who wants to learn and play simultaneously, learn about the ancient Greek myths and tails with an immersive learning experience. To achieve this goal we developed Trials of the Acropolis a VR game that takes place in Ancient Greece and tells its own story, using a collection of myths that are included in the History book of the 3rd grade.

2 VR in Education

Research has shown that the average person only remembers 20% of what they hear and 30% of what they see, but up to 90% of what they personally experience [6]. As a result, VR educational materials provide the scenario needed to build the attention gap, helping students to become more attentive during lessons. A meta-analysis on K-12 and higher education students has concluded that students learn better when immersed in virtual worlds [7]. Another analysis [8] of 54 studies on the use of VR in education, has discovered that user engagement and participation in VR worlds is contributing positively to learners achievements.

There is a broad range of VR apps and games developed with the education element in mind and the majority of them are completely free. Acropolis VR 3D [9] is a great example of such an app, available for a lot of different languages. Developed by Mozaik education and available for Google Cardboard, this educational app introduces Acropolis of Athens with a lot of history information and some quiz games. It's an architecture walkthrough where the user can walk in all of Acropolis in VR mode and when disabling this mode s/he can play a mini quiz game and read some actual facts. A similar VR approach is the Acropolis experience [10] available from the Unimersiv app for Gear VR devices, Google Cardboard and other VR devices. It has slightly better graphics in comparison with the Acropolis VR 3D and has a guide tour, where there is an autowalk script along with a narrated voice presenting the history of Acropolis, and a free tour where the user can walk across the environment.

3 The Case of Trials of the Acropolis

Trials of the Acropolis provides a game based learning approach to give education more ways to distribute knowledge. It offers puzzle solving quests, multiple choice quiz games, interaction with the game characters (heroes from the Greek mythology), provided in English and Greek audio, immersive original soundtrack and background music, and a unique story ready to be explored. The key element features of the produced game are presented in Fig. 1.

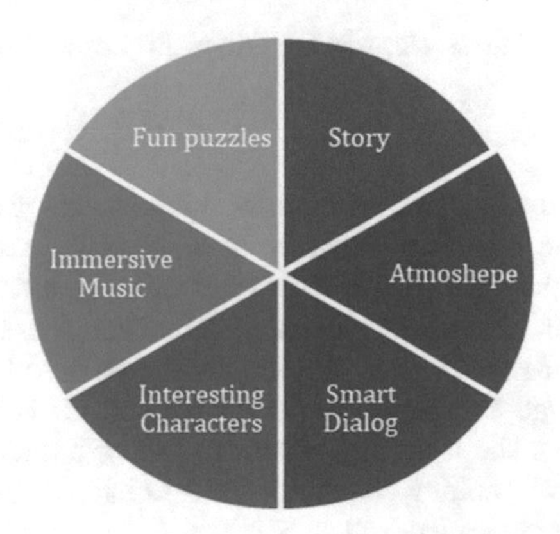

Fig. 1. Diagram showing the primary elements that Trials of the Acropolis is focused

Storytelling is a characteristic that a lot of games ignore and instead emphasize on gameplay. However, a good narration, great atmosphere and a mystery element raise the overall quality of a title and guarantee better interest levels [11]. Creating believable characters with interesting personalities and explaining their motives and their real world problems, makes the player more immersed in the story, rather than the gameplay.

Game like activities have positive impact on students motivation, engagement and performance [12–15] triggering also, as a side effect, other skills such as imagination, storytelling, challenge and cooperation. These characteristics are in line with the Constructivist theory where fun, enjoyment and reward are essential pieces for the students' motivation [16].

Figure 2 depicts the approach for the game development, where at the first step the story, the game scenario and the dialogues of the characters are prepared. At a second step, the graphics and the sound of the game have to be implemented. Finally all the assets of the game are composed using Unity 3D game engine. The game was designed for Google Cardboard and Samsung Gear VR.

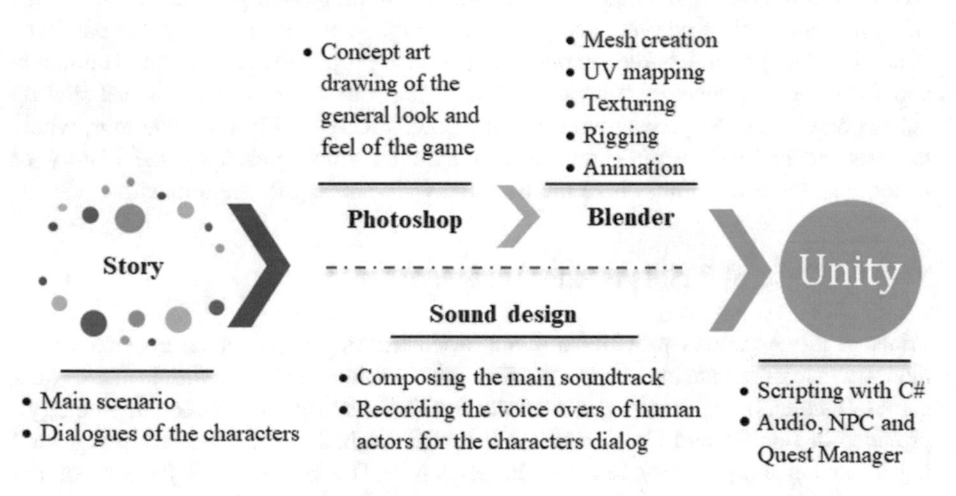

Fig. 2. Diagram of the adopted design and development of Trials of the Acropolis

A detailed analysis of these steps is described in the next sections.

3.1 Game Scenario

Trials of the Acropolis uses a game scenario where it challenges the player to fulfill an ancient myth, which no one in the history of Ancient Greece has ever completed. According to the myth there are scattered six trials, across the Acropolis of Athens, that test the wisdom, memory and concentration of a human. These ancient trials – myths were created by the gods themselves and the person who will complete all the trials with success, will emerge as the true leader of Ancient Greece and help the country during the difficult times she is facing. Each of the first 5 ancient trials-myths of the game, is based on a corresponding chapter of the 3rd Grade history book of Greece and involves two phases, as presented in Table 1.

Table 1. Trials characteristics

No	Trial name	Subject	Phase 1	Phase 2	Book section
1	A new hope	Gods and Titans	Puzzle	Quiz	1
2	Rise of the hero	Hercules	Puzzle	Quiz	2
3	Labyrinth madness	Theseus	Puzzle	Quiz	3
4	The mythic treasure	Jason and the Argonauts	Puzzle	Quiz	4
5	A melody from the stars	Odyssey	Music game	Quiz	6
6	The olympian wisdom	Revision of the above	Puzzle	Quiz	–

When the game initiates, the player is placed at the entrance of Propylaea, visiting the Acropolis for the first time and without knowing anything about the myth which will be confronted. The player is supposed to interact with the avatar/game narrator, sitting at the stairs and looking very disappointed (Fig. 3). When the player approaches, the narrator provides some input to the player, explaining that he has travelled a long way to come here and complete the trials/myths, in order to gain wealth and build a better future for his family. Though, despite his efforts, he failed to complete his task and therefore he is preparing himself emotionally, for the long way back to home. The narrator provides further information to the player related to the myth, on how the gods fought the titans, the general structure of the Olympic pantheon and how the trials were formed. He encourages the player to try and solve the trials and wishes him luck.

Fig. 3. A concept art depicting Propylaea with the wandering stranger just before the assignment of the first task.

Once, the introductory part is complete, the game starts, and the movement across the environment of Acropolis is done with predefined waypoints, guiding the player to the right destination. The first trial and generally all the trials consist of two phases: the puzzle phase and the quiz game phase.

After the first trial the player meets Megara the wife of Hercules. She tells the myth of how her husband was born, and a background of some of the labors he pursued, while also giving directions for the second trial.

The third trial is given by Theseus the legendary hero who killed the Minotaur. As all the NPC's (Non playable characters) of the game tell a different myth so does Theseus, giving the player important details of information about his past adventures. Theseus further explains that he didn't want to participate in these trials, because his name has already been written in thc history and it would be better for the new generation getting more confidence and experience by pursuing such challenges.

By successfully completing the third trial, the player needs to speak with the Temple Guardian of the Parthenon. His family and ancestors were guarding the temple for centuries, passing the torch to their strong children. When he was a little kid, temple guardian met Jason in Iolkos during a trip with his family. Jason knew that the guardians of Parthenon were very respectful and strong people, so he gave them a single golden hair from the legendary Golden Fleece as an act of honor. For that reason, the guardian tells the story of Jason and the Argonauts and also pointing that the fourth trial is located inside the Parthenon but warns the player, Acropolis is going to close so he has to hurry.

The penultimate trial is given by Orpheus near the Erechtheion. The part of story that Orpheus narrates, is based on the Odyssey and in particular, how the cunning Odysseus returned back to Ithaca. Only a small portion of people have come so far to complete the fifth trial, but all have failed. This trial has a different type of first phase. The player has to solve a memory game where Orpheus plays some melodies with his harp, which corresponds to specific buttons that are shown on screen. The player will then have to remember the pattern and press the buttons in the right order, so as to accomplish this part of the trial and proceed to the quiz phase.

The final trial, takes place in front of Zeus temple, behind the Parthenon. The player so far has collected all the five stones from the five trials and he has to place them in the correct order in front of the temple to invoke Zeus himself. Zeus is heard talking from the skies, but he cannot be seen. Zeus congratulates the player for making it this far, but warning him that he has to complete eight difficult questions in order to fulfill the myth. Completing correctly all the questions, the player emerges as the true leader of Greece and finishing the Trials of the Acropolis videogame.

3.2 Implementation

For the development of Trials of the Acropolis, mostly, free professional game development tools was used. Software like Autodesk 3ds Max, Maya, Cinema 4D are the top software for professional production, however, Blender 3d does almost the same job and its totally free. Similar at the game engine selection, we were pretty much close to choose the Unreal Engine 4, however, at the end, Unity was chosen since it offered a very flexible development process and a large community supporting, with a wide range of video tutorials and books. Therefore the primary tools used for the game development were Adobe Photoshop, Blender 3D [17] modelling software, Unity [18] game engine and Reaper [19] digital audio workstation.

Blender is free modelling software that packs all the essential tools needed for professional game model production, at zero cost. All the model meshes, textures, rigs and animations were done in Blender 3D. Blender is packed with intuitive functions like the modifiers category, where every single modifier does a useful and time saving task. For example the array modifier, can make copies of the selected object and by incrementing a particular value of the x, y and z axis the objects copies are placed along the selected axis. The array modifier was used on Trials of the Acropolis, during the modelling process of the temple columns, and saved a lot of development time. The characters models during the applied processing steps in Blender are presented in Fig. 4.

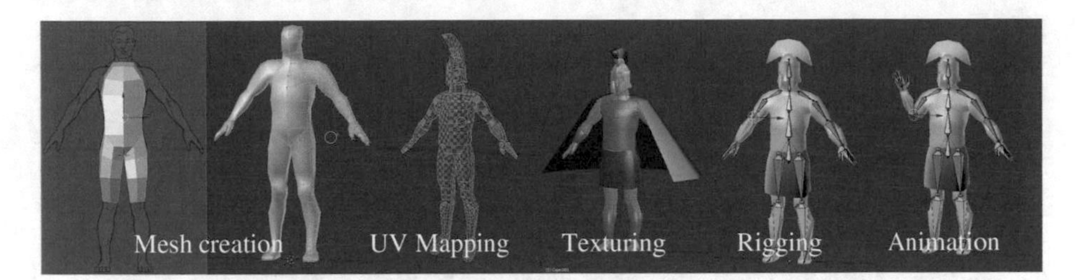

Fig. 4. Various models of game characters during the Blender processing steps

During the development of the objects meshes, the most common used functions were "loop cut and slide", subdivide, and extrude. Loop cut splits a loop of faces by inserting a new edge loop, intersecting the chosen edge. Subdividing splits selected edges and faces, by cutting them in half and finally extrusion tools duplicate vertices, while keeping the new geometry connected with the original vertices. Vertices are turned into edges and edges will form faces. After creating the model meshes, UV mapping takes place. UV is the 3D modeling process of projecting a 2D image to a 3D model's surface, for texture mapping. By selecting the mesh, pressing Cntrl+E and choosing mark seam, the selected area is cut, making the object flat and the mesh better for texturing (Fig. 5).

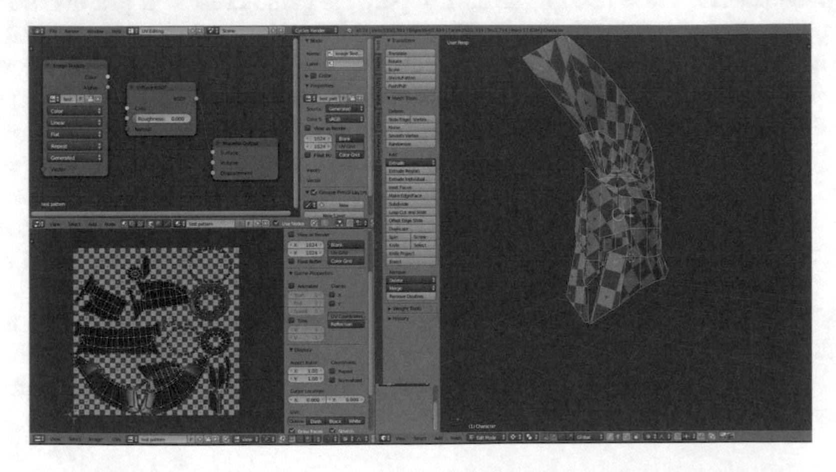

Fig. 5. UV mapping procedure of the Wandering stranger helmet.

After the UV mapping, the texture coloring is following. Blender has a lot of useful tools when coloring an object, like the fill function. For mobile VR games, taking into consideration the small screen size, and their processors it is proposed to decrease texture resolution.

When a character is fully textured, rigging process follows. At this stage the bones are inserted to the character. Rigging is necessary in order to animate the produced characters.

The final process of the character development cycle is the animation stage. Animation is making an object move or change shape over time and objects can be animated in many ways: (a) Moving as a whole object, changing parameters like position, orientation or scale in time, (b) Deforming objects, animating vertices, edges and faces and (c) Inherited animation, causing the object to move base on movements by another objects animation.

Completing all the games characters and the Acropolis environment, the objects are ready to be exported to Unity, using the fbx file format in the Blender export window.

The final stage of development is then passed to Unity game engine. Unity is an all-purpose game engine that supports 2D and 3D graphics, drag and drop functionality and scripting through C#. Unity today is the most used game engine making rapid prototypes of any videogame and speeding up the development process of all the computer game companies. At this development stage the quest, audio and NPC management takes place. All the scripting managers perform various tasks, like storing game data and audio initialization. Before the implementation of the game, some settings in the Unity interface needed to be adjusted, so as to have a steady 60 frames per second performance during the game, otherwise, the user will have motion sickness. Things like realistic shadows, ambient occlusion, real-time global illumination, are highly proposed to be disabled in order to achieve that.

3.3 Gameplay

One of the problems that every VR game has to deal with, is the user interaction with the game objects. We wanted to be able to play the game even with just the use of a smartphone and a Google Cardboard. Therefore the user interaction is based on mobile sensors instead of a joystick. Therefore we decided to allow user interact with the game objects, by looking at an object for three seconds. For example as we previously stated, traveling inside the game is accomplished with predefined waypoints. The player simply has to look at the next waypoint for three seconds and then a walk script initializes, moving the player to the marked destination. When a waypoint is reached, the player cannot go back to his previous destination. We made the decision to make a simple user interface, that the player can interact with all the game objects in the same way and avoid possible confusion.

NPC's interact with the player when s/he approaches their trigger zone, which is a collider that starts a specific function when the player object touches it. When the collider is touched, a specific NPC script is played starting the voiceovers and animations of the particular character (Fig. 6).

Fig. 6. Temple guardian in front of Parthenon during the 4th trial

3.4 Formative Evaluation and Preliminary Results

In order to proceed with the pilot use of the developed game, a formative evaluation with a small group of five teachers took place. The goal of the formative evaluation was to discover potential bugs of the game, to find out if the dialogues and game scenario is appropriate for the 3rd grade students (according to their teachers) and take a preliminary feedback about the perceived easy of use and usefulness of the game.

Our research preliminary results show that the five teachers that tested the game, were pretty satisfied and pointed out their strong believe that this game will help their students to experience the world of ancient Greece in a fun and educational way. The main observation that they made was about improving the length and context of the myths that the NPCs tell to the player. In addition they pointed some minor modifications on the dialogs and propose to be more concrete with the game objectives.

4 Discussion and Conclusion

The present study presented a VR game, along with the tools, techniques and the scenario which were used. The Trials of the Acropolis can contribute greatly to the school and education environment by providing an extra path, which uses the VR capabilities, supporting students and teachers. Our research concludes that game based learning applications, can be a useful tool for teachers and students, in this type of knowledge domain. It can greatly improve learning experience and help teachers engage their students during the course by planning which areas of the game they will discover for a particular lesson.

The preliminary results were very positive. However, there are some notable limitations that should be considered for the current research and especially the fact that no summative evaluation took place. The formative evaluation had a sample size of just five teachers and no students' opinion was discovered. However the main aim of the paper is to point the tools, the scenario and the idea behind the produced game.

It remains unclear how this game-based VR application can impact students' motivation and performance in the applied course. We are planning to use the Trials of the Acropolis during the next school year, in 3rd grade students, in order to confirm the preliminary results of this paper and study on its impact in student's motivation, engagement, satisfaction and performance.

References

1. Ezawa, k.: Virtual and augmented reality: are you sure it isn't real? Citi GPS: Global Perspectives & Solutions, October 2016
2. Mazuryk, T., Gervautz, M.: Virtual reality history, applications, technology and future. Institute of Computer Graphics, Vienna University of Technology, Austria (1996)
3. Giraldi, G.A., Silva, R., Oliveira, J.C.: Introduction to Virtual Reality, LNCC Research Report #06/2003, National Laboratory for Scientific Computation (2003). ISSN 0101 6113
4. LaValle, M.: Virtual Reality. University of Illinois, Cambridge University Press (2017)
5. Dredge, S.: The complete guide to virtual reality: everything you need to get started (2016). https://www.theguardian.com/technology/2016/nov/10/virtual-reality-guide-headsets-apps-games-vr
6. Davis, B., Summers, M.: Applying Dale's Cone of experience to increase learning and retention: a study of student learning in a foundational leadership course. In: QScience Proceedings: Engineering Leaders Conference (2014)
7. Merchant, Z., Goetz, E.T., Cifuentes, L., Keeney-Kennicutt, W., Davis, T.J.: Effectiveness of virtual reality-based instruction on students' learning outcomes in K-12 and higher education: a meta-analysis. Comput. Educ. **70**, 29–40 (2014)
8. Pellas, N., Kazanidis, I., Konstantinou, N., Georgiou, G.: Exploring the educational potential of three-dimensional multi-user virtual worlds for STEM education: a mixed-method systematic literature review. Educ. Inf. Technol. **22**(5), 2235–2279 (2017)
9. Mozaik Education. Acropolis VR 3D (2017). https://play.google.com/store/apps/details?id=com.rendernet.acropolis
10. Unimersiv. Acropolis experience (2017). https://unimersiv.com/just-released-acropolis-experience-unimersiv-app/
11. Markouzis, D., Fessakis, G.: Interactive storytelling and mobile augmented reality applications for learning and entertainment – a rapid prototyping perspective. Learning Technology & Educational Engineering Lab, University of the Aegean, Rhodes, Greece (2015)
12. Connolly, T.M., Boyle, E.A., MacArthur, E., Hainey, T., Boyle, J.M.: A systematic literature review of empirical evidence on computer games and serious games. Comput. Educ. **59**(2), 661–686 (2012)
13. Akl, E.A., Pretorius, R.W., Sackett, K., Erdley, W.S., Bhoopathi, P.S., Alfarah, Z., Schünemann, H.J.: The effect of educational games on medical students' learning outcomes: a systematic review: BEME Guide No 14. Med. Teach. **32**(1), 16–27 (2010)
14. Sitzmann, T.: A meta-analytic examination of the instructional effectiveness of computer-based simulation games. Personnel Psychol. **64**(2), 489–528 (2011)
15. Yien, J.M., Hung, C.M., Hwang, G.J., Lin, Y.C.: A game-based learning approach to improving students' learning achievements in a nutrition course. Turkish Online J. Educ. Technol. **10**(2), 1–10 (2011)

16. Nino, M., Evans, M.: Lessons learned using video games in the constructivist undergraduate engineering classroom. In: Proceedings of Twelfth LACCEI Latin American and Caribbean Conference for Engineering and Technology, LACCEI 2014, Guayaquil, Ecuador (2014)
17. Blender (2017). https://www.blender.org/
18. Unity (2017). https://unity3d.com/
19. Reaper (2017). https://www.reaper.fm/

Designing and Developing an Educational Game for Leadership Assessment and Soft Skill Optimization

Nikolaos Chatziantoniou(✉), Nikolaos Politopoulos, and Panagiotis Stylianidis

Department of Informatics, Aristotle University of Thessaloniki, Thessaloniki, Greece
{chafilnik, npolitop, pastylia}@csd.auth.gr

Abstract. It is known that education and the development of leadership skills can maximize productivity, build a positive culture, and promote harmony. But apart from their importance to organizations, one might consider them particularly useful in their everyday life. Enhancing and improving soft skills in leadership is a promising field that becomes considerably important. The aim of this work is to develop a game capable of simulating realistic situations for the development of skills useful in the professional field and in particular leadership skills. The principles of gaming-based learning and the technological tools for game support and development, which are widely disseminated in support or replacement of traditional teaching methods, can be used to achieve this goal. A significant part of the time was devoted to reviewing available technologies and specific applications that can serve the purposes of the system and to assess the strengths and weaknesses of each solution. The engine chosen to develop the game is Unity 3D. The additional Fungus visual novel and storytelling tool was used. Through this process a two-dimensional game has been implemented that allows the user to play and through it develop leadership-oriented skills that will help him in the professional field.

Keywords: Game based learning · Mobile learning · Soft skills Leadership · Self-assessment

1 Introduction

In the training industry we can find two main modules, soft skills training and hard skills training. Serious games are especially interesting for soft skills as they make the training of the theoretical contents possible, which is essential for efficient skills learning. For professionals, soft skills are increasingly important for their development. Using classic formats for soft skills training is proving ineffective, while serious games have become an essential tool for corporate training.

What makes soft skills that important is the current labor market, which is becoming more and more competitive in every field. In order to be successful in such a competitive environment, candidates should have, in addition to the knowledge and skills of their subject, skills and qualifications that make them stand out. Soft skills are those skills that

M. E. Auer and T. Tsiatsos (Eds.): IMCL 2017, AISC 725, pp. 258–265, 2018.
https://doi.org/10.1007/978-3-319-75175-7_27

shape the individual's personality and what ultimately will make a difference when it comes down to the level of knowledge and skills of the subject being the same.

This research describes the design and implementation of a game for optimizing the player's soft skills in leadership, with the use of the Unity platform and the additional Fungus tool. In the following sections, the theoretical backgrounds for the Game-based learning, the soft skills and the five leadership practices are described, followed by the analysis of the various game engines, the game design and development process and lastly the conclusions.

2 Theoretical Background

2.1 Game Based Learning

Game based learning (GBL) is a gameplay type that has defined numerous learning outcomes as GBL is designed to balance learning content with gameplay and the player's ability to retain and apply said content in real world applications.

GBL describes an approach to teaching, where students explore relevant aspects of games in a learning context designed by teachers. Teachers and students collaborate in order to add depth and perspective into the game experience.

Well designed and developed game-based learning applications can immerse users into virtual environments that have a familiar and relevant look and feel. In an effective game-based learning environment, the main goal is to experience the consequences of the actions/choices you make along the way. Making mistakes in a risk-free setting, through experimentation, enables active learning and practicing proper manners for each situation. This promotes high engagement in practicing behaviors that can easily be transferred from the simulated environment into real life.

2.2 Serious Games

Today, the term "serious games" is very popular. A Google search on "serious games" renders about 39,500,000 hits [17-10-2017]. The term itself is nowadays established, but there is no current singleton definition of the concept. According to Corti (2006, p. 1) game-based learning/serious games "is all about leveraging the power of computer games to captivate and engage end-users for a specific purpose, such as to develop new knowledge and skills". When searching the web, a number of different definitions are available. The number of hits when explicitly searching for definitions of serious games amounts to 1.5 million hits [17-10-2017]. Most web-pages, however, either do not define the concept or describe it vaguely. Serious games have been defined as entertaining games with non-entertainment goals (Raybourn 2006). Serious games are games that educate, train and inform (Michael and Chen 2006). Various authors anticipate the great opportunities of games (and simulations) in education, because of their positive effects on learning outcomes (e.g. Amory 2007; Prensky 2006). Games have been demonstrated to provoke active learner involvement through exploration, experimentation, competition and co-operation. They support learning because of increased visualization and challenged creativity. They also address the changing competences needed in the information age: self-regulation, information skills, networked cooperation, and problem solving strategies, critical thinking and creativity. Hence, games are an effective tool for mediating learning. Serious games not only convey hard skills such as the understanding of how complex systems operate, production networks being one of them, but also mediate soft skills like collaboration and communication (Scholz-Reiter et al. 2002).

2.3 Storytelling Games/Visual Novels

Visual novels are distinguished from other game types by minimal gameplay in general. Typically the majority of player interaction is limited to repetitive mouse clicks for keeping text, graphics and audio in motion (many recent games offer "play" or "fast-forward" commands that make this unnecessary), while making narrative choices along the way.

2.4 Soft Skills

The question "What exactly are soft skills" is not easy to answer, because the perception of what is a soft skill varies from field to field. A skill can be considered "soft" in a particular area and can be considered "hard" in another. Even internationally recognized encyclopedias have little to say about soft skills.

According to Wikipedia (https://en.wikipedia.org/wiki/Soft_skills), Soft Skills are a combination of people skills, social skills, communication skills, character traits, attitudes, career attributes, social intelligence and emotional intelligence quotients among others that enable people to effectively navigate their environment, work well with others, perform well, and achieve their goals with complementing hard skills. Harper Collins (Collins English dictionary, 1994) defines the term "soft skills" as "desirable qualities for certain forms of employment that do not depend on acquired knowledge: they include common sense, the ability to deal with people, and a positive flexible attitude.

The Five Practices of Exemplary Leadership

Created by James M. Kouzes and Barry Z. Posner in the early 1980s and recognized primarily in their internationally best-selling book "The Leadership Challenge, The Five Practices of Exemplary Leadership" approaches leadership as a measurable, learnable, and teachable set of behaviors. After conducting hundreds of interviews, reviewing thousands of case studies, and analyzing more than two million survey questionnaires to understand those times when leaders performed at their personal best, five practices common to making extraordinary things happen, emerged. These five practices are:

- **Model the Way** by finding your voice and affirming shared values
- **Inspire a Shared Vision** by envisioning the future and enlisting others in a common vision
- **Challenge the Process** by searching for opportunities and by experimenting, taking risks, and learning from mistakes
- **Enable Others to Act** by fostering collaboration and strengthening others
- **Encourage the Heart** by recognizing contributions and celebrating values and victories

3 Technological Analysis

3.1 Game Engines

In order to evaluate the existing game engines suitable to create these types of games, an evaluation table was created and importance points were assigned to features that are needed on the platform. Importance scale is 1 to 3, 1 slightly important, 2 important and 3 very important. If a game engine qualifies for a feature it was assigned an X on the table. The values are summarized at the bottom of the Table 1.

Table 1. Game engine comparison

Features	Degree of significance	Unity+ Fungus	GameMaker	Ren'Py	AGS	eAdventure
Dialog tools	3	X	X	X	X	X
Graphic tools	3	X	X	X	X	X
Audio tools	2	X	X	X	X	X
Variables	3	X	X	X	X	X
Free export to WEB	3	X	–	–	–	–
Free export to MOBILE	3	X	–	–	–	–
Script editor	3	X	X	X	X	X
VLE integration	1	–	–	–	–	X
Free cross-platform export	3	X	–	–	–	–
Total		23	14	14	14	15

Based on the review of available game engines, the table and related literature, it was decided to use the Unity3D engine along with the additional "Fungus" tool to develop the game. Unity3D is a cross-platform game engine developed by Unity Technologies, which is primarily used to develop video games and simulations for computers, consoles and mobile devices. Fungus is an extension tool that allows easy character insertion and the creation of story-driven games. It allows interactive story-telling to be added to games created through Unity3D.

4 Game Design and Development

The main idea was to create an interactive educational storytelling game that will allow the user to get a profile based on the five exemplary leadership practices, which he will be able to improve by developing and enhancing his "soft skills" in leadership. More detailed requirements of the game are:

- to implement a realistic and interesting scenario
- to save users and results on a database
- to identify the user and adapt the gameplay to him

- to provide assistance and suggestions to points where weaknesses have been observed.
- to provide a graphical presentation of the results for displaying the progress

A scenario was created in order to obtain the answers for the 30 questions of the LPI questionnaire: Leadership Practices Inventory of James M. Kouzes and Barry Z. Posnero. The user will assume the role of a department manager of a multinational company and will face situations in a working environment within a supposed one-week period. Based on the player's decisions and choices, specific values based on a grading model will be assigned to the questions in a similar scale with the questionnaire. Initially, the user will complete the game once to produce his profile based on the five practices. For every new session of the game, hints and tips will be triggered for each question that was poorly scored, in order for the user to improve. An account is necessary to access the game and for the system to store data such as player information and scores. The following UML sequence diagram (Fig. 1) was then created to determine the necessary procedures needed in a game session.

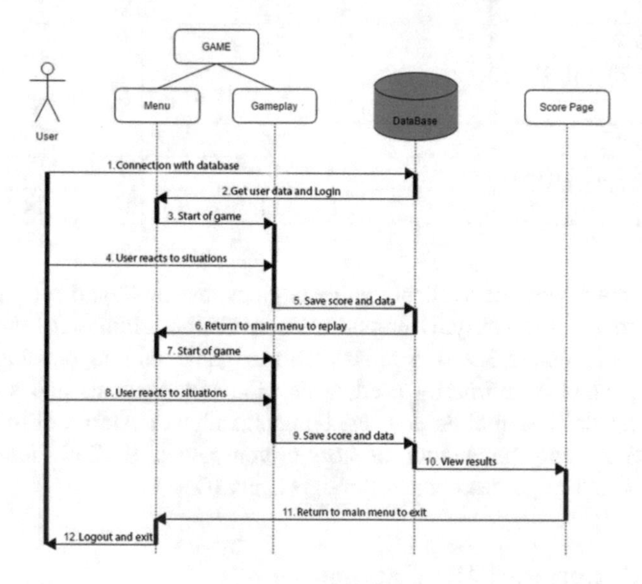

Fig. 1. UML sequence diagram of the game procedures

4.1 Script and Grading Model

The story of the game was derived by the thirty questions of the LPI. For each one of them a scenario was created with expert guidance where a situation is introduced to the player. In order to assess the player's choices a grading model was made (Fig. 2), based on the 10-scaled model used in the inventory. After the situation is introduced, three options are presented. Each option leads to two or four graded options depending on what the player chose.

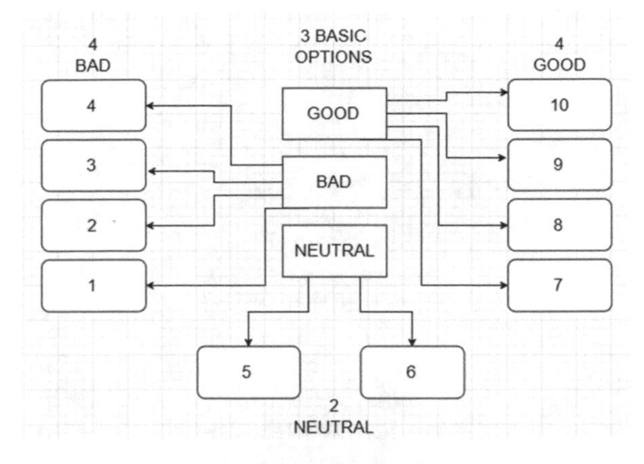

Fig. 2. Grading model diagram

4.2 Main Game Screen

The user controls the flow of the game and interacts with the options that will appear to him at various stages of the script. The characters will be displayed in the middle of the screen, the dialog messages at the bottom while the options at the top center of the screen. A time bar will also appear in some options. If the user has already played once and a profile has already been created, he is given the option to toggle the display of progress bars for the five practices, which are dynamically increasing, as well as the option to toggle hints and tips. The bars are placed in the top left of the screen. Also on the top right there is a button that shows/hides two control buttons. One is to toggle the sound and the other to reset the game and return the user to the login screen (Fig. 3).

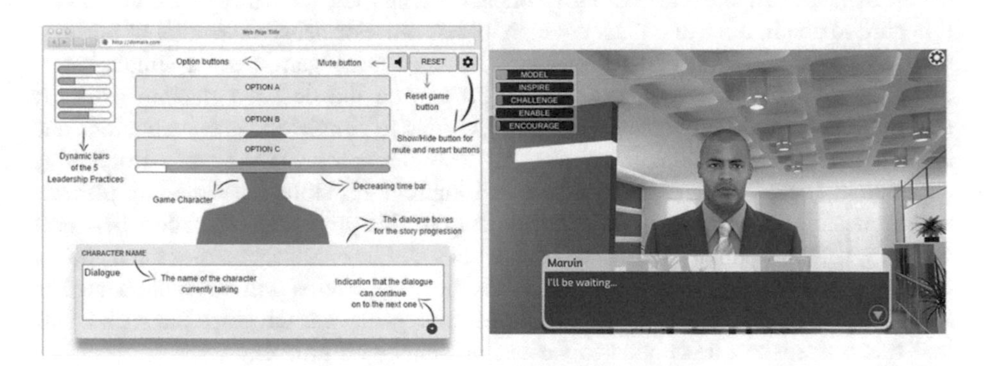

Fig. 3. (a) Mockup of main screen and (b) final main screen

4.3 Results Screen

In the results screen, the results of each attempt of the game are displayed to the user, as well as the average of every practice produced by all the players in the game (all the records stored in the base so far). The use of interactive graphs such as bar charts and a radar chart with the help of Chart.js was implemented (Fig. 4).

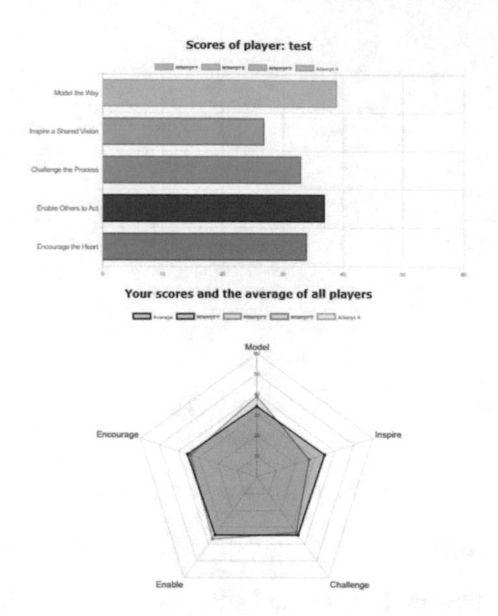

Fig. 4. The results page with the interactive graphs

5 Conclusions

It is well known that people growing up tend to focus more on gaining knowledge and acquiring hard skills required for their work. Most of the educational institutions provide the necessary theoretical background and practical skills. But what is often required in most of the 21st century jobs, especially in the leadership domain, are soft skills. Because soft skills are strongly connected with the personality of the individual, it is hard to obtain them and thus, are high in demand. For many years a lot of business focused games were created for training. Most of these games are a simulation of managing a store or a company, focusing more on the decision making skills by removing the human interaction part. LASSO tries to simulate realistic situations and encounters in the workplace, in an everyday level through a visual novel – storytelling experience. It can help the player assess and improve his skills overtime in a pleasant and interesting environment. The next big step of this project is evaluation. The user interface will be evaluated and some case studies will be conducted. After the feedback and the improvement of the application, additional scenarios will be implemented to keep the interest of players in high levels. Also the game was developed in such a way that this concept can be applied to various domains with little effort.

References

Amory, A.: Game object model version II: a theoretical framework for educational game development. Educ. Technol. Res. Dev. **55**(1), 51–77 (2007)

Corti, K.: Games-based Learning; a serious business application. PIXELearning Limited (2006). www.pixelearning.com/docs/games_basedlearning_pixelearning.pdf

Göbel, S., de Carvalho Rodrigues, A., Mehm, F., Steinmetz, R.: Narrative game-based learning objects for story-based digital educational games. Narrative **14**, 16 (2009)

Kouzes, J.M., Posner, B.Z.: Leadership is Everyone's Business. Jossey-Bass, San Francisco (2007)

Kouzes, J.M., Posner, B.Z.: The Leadership Challenge, vol. 3. Wiley, Hoboken (2006)

Kouzes, J.M., Posner, B.Z.: The Leadership Practices Inventory (LPI): Participant's Workbook, vol. 47. Wiley, Hoboken (2003)

Michael, D., Chen, S.: Serious Games: Games that Educate, Train, and Inform. Thomson Course Technology, Boston (2006)

Prensky, M.: Don't Bother Me, Mom, I'm Learning!: How Computer and Video Games are Preparing Your Kids for 21st Century Success and How You can Help! Paragon House, St. Paul (2006)

Romero, M., Usart, M., Ott, M.: Can serious games contribute to developing and sustaining 21st century skills? Games Culture **10**(2), 148–177 (2015)

Raybourn, E.M.: Applying simulation experience design methods to creating serious game-based adaptive training systems. Interact. Comput. **19**(2), 206–214 (2006)

Scholz-Reiter, B., Gavirey, S., Echelmeyer, W., Hamann, T., Doberenz, R.: Developing a virtual tutorial system for online simulation games. In: Proceedings of the 30th SEFI Annual Conference, Firenze, Italy (2002)

The Role of Adults in Giving and Receiving Feedback for Game Design Sessions with Students of the Early Childhood

George Kalmpourtzis(✉), Lazaros Vrysis, and George Ketsiakidis

Aristotle University of Thessaloniki, Thessaloniki, Greece
gkalmp@nured.auth.gr, lvrysis@auth.gr,
gketsiakidis@playcompass.com

Abstract. Recent technologic advances and the continuous increase of software and hardware integration in people's daily lives has extended research interest for the field of human computer interaction. User involvement in the design of software, among which games consist a big part, has also been on researchers' spotlight, even from the early childhood. The aim of the present study is to explore and identify the impact of adult feedback on early childhood students' design of games. The results presented are part of an ongoing and larger study that took place for three months and involved a team of eighteen kindergarten students, participating in game design sessions for a period of three months. The game design sessions were proposed, based on participatory design techniques and consisted of cross-generational teams of students and one adult. This paper studies how adult feedback influenced students' decisions and work while designing their games, offering empirical evidence on the field of participatory and game design for the work with children of this age.

Keywords: Participatory design · Game design · Game design education
Early childhood · Cross-generation design

1 Introduction

Rapid technologic advances have brought games in the forefront of software design, while their use in various learning fields has been subject of academic research [1–5]. This use of games in learning contexts shows researchers' and educators' interest about the impact of educational games on students' cognitive processes [6]. This is the reason why researchers from various fields like human computer interaction, learning and design have showed interest for the study of game design and its education [7–11].

Game design and its education have been studied from a variety of different perspectives, like participatory design [12, 13], education [14, 15], and game design [16, 17]. According to [18], games can be viewed as problems, presented in a playful attitude. Based on this perspective, this study viewed designing games as problem posing activities. A series of game design sessions, based on participatory design techniques and existing problem posing theory were proposed.

M. E. Auer and T. Tsiatsos (Eds.): IMCL 2017, AISC 725, pp. 266–275, 2018.
https://doi.org/10.1007/978-3-319-75175-7_28

This study aims at providing answers on the impact of feedback, provided by adults, during those game design sessions and the evolution of students' game design choices based on this feedback.

2 Theoretical Background

Research on students as game designers is not new. The development of game design skills, related to narratives, game mechanics, creativity and cognitive development has been a subject of several previous studies [19–21]. The involvement of students in the design of games that they will later play is consistent with the aims and premise of participatory design and has hence been the subject of previous research [22–24]. Children as design partners have been proposed by [25], who identified four roles of children as design partners: user, tester, informant and design partner. This interest gave also birth to the notion of cooperative inquiry, an approach addressing mainly young design partners, through a series of activities and approaches, providing students with necessary stimuli and connection with new technologies, as well as coordinating and facilitating the design process in cross-generational teams [26]. Communication among cross-generational teams and management of feedback has been considered as an important aspect of the design process [12, 27]. Participation in game design education activities has been suggested to have a positive impact on students' understanding of game related concepts [28] and knowledge transfer [29].

The analysis, study and, therefore, presentation and teaching of game design led to the proposal of several game analysis frameworks and models. Some of them target the field of game design from an educational aspect [6, 30, 31], others from a flow facilitation perspective [15] and others from a game mechanics perspective [32, 33]. Games have also been considered as problems, consisting of smaller problems, the solution of which, leads to gameplay [34]. Connection between creating games and problems has been identified in previous studies mainly for games' ability to present problem posing situations, which are hard and not motivating to students in other contexts [35]. This is also an argument of the use of games in activities about problem posing [14]. Therefore, engaging in playing games has similarities to solving problems. So, respectively designing games could be considered similar to problem posing, in the sense of problem construction.

In an attempt to organize the design of problem posing activities, Stoyanova and Ellerton [38] categorized them in three axes: structured, semi-structured and free of structure. This categorization seems to be relevant to game design situations in teaching settings, as it could classify and explain the initial support, concerning this design, that the teacher offer to students in the proposed situations.

This study presents a part of a larger, ongoing study where mixed method analysis has been applied. The larger study examined the development of early childhood students' development of game design skills and strategies as a result of participation in game design sessions, based on participatory design techniques.

3 Setting

3.1 Didactic Approach and Environment

See Fig. 1.

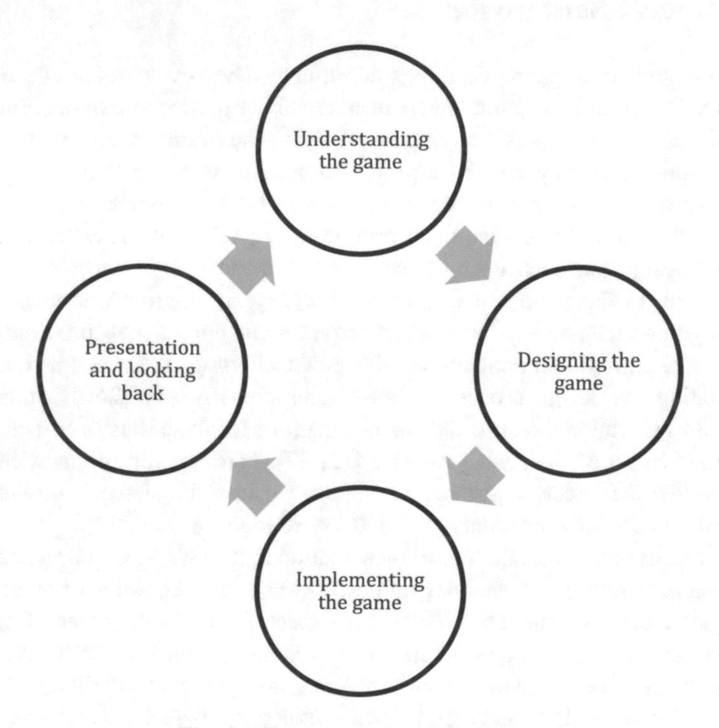

Fig. 1. The proposed design process

I. Design process

In order to facilitate the game design sessions with students of the early childhood, a design process was put in place. The design process was based on Polya's problem solving analysis, adapted for game design [36] and a scaffolding process, where external support is gradually removed when designers develop game design skills [37]. The approach that was adopted, consisted of a four-step iterative process:

- **Understanding the game**: consisting of processes like identifying existing resources, technical or problem restrictions posed by the researcher and deciding on what game students want to design.
- **Designing the game**: consisting of brainstorming collaborative session, where students discuss, analyze and propose the components of their games, along with their interaction and mechanics of the game.
- **Implementing the game**: describing the phase where designers actually design their games, based on their designs.

- **Presentation and looking back**: describing the phase where students present their games and others play them. After the presentation, students discuss their decisions and reflect upon their designs.

Those steps were not always sequential and there were times that those phases overlapped during the design process.

II. Design tools

In order to facilitate the design process, a set of design tools were proposed. The tools aimed at helping students facilitate their design process and organizing the delivery and reception of feedback among different peers of the team. The proposed tools were:

- **Guiding questions**: Consisting of a list of design related questions, proposed with the intention of facilitating student workflow. The questions targeted game components, the selection and analysis of which was based on the analysis of game analysis frameworks, presented in the theoretic review of the study. Guiding questions targeted consisted of the following format: "Where is the game played?", "What happens if..?", "How many players?". The proposal and posing of appropriate questions in problem posing situations can have a greater impact of relevant students' skills [38].
- **Material overload!**: The main utility of this tool is to present students with a diverse set of images and experiences using recent technologic feats of hardwarc and software. According to [39], young students' familiarity with new technologies can have a positive impact in their participation in participatory design activities. By using this tool, students were presented with a diverse set of devices and games, like AR/VR head mounts, tablets, computers, interactive whiteboards and tabletops. Additionally, they were presented with a diverse set of materials, among which they could choose which ones to use in order to construct their own low-tech prototypes. The materials consisted of pens, pencils, crayons, clay, LEGO blocks and other types of stationery.
- **Presentation and manipulation sessions**: During those sessions, students were are asked to present their games. After doing so, the team was asked to modify at least one of its components. The modification could be on any components and could be related to addition, removal or expansion of existing materials, rules, symbolic representations or content.

III. Game design situations

Based on the work of Stoyanova [40] on creating problem posing situations, a game design approach, using three game design situations was proposed. The three situations were: structured, semi-structured and free of structure ones. In structured situations, complete and well-known games were presented to the students, who were being fully aware of all their existing components and were asked to alter and extend them. At the beginning of semi-structured situations, half-made games or sets of game resources with a specific design were presented to the students, who were later encouraged to construct their own proposals, taking into consideration the initial game provided by the activity. During free structured situations, students were not provided with any existing game or other assistance and were asked to come up with their own suggestions. Their game could be of any type, genre or material.

IV. Selection of games

Based on previously proposed game analysis frameworks and models [15, 30–32], a list of games that would act as bases for the structured and semi-structured situations was proposed. The games were selected in a way that would present students with a diverse set of gaming components and perspectives throughout the duration of the three-month study. Games included digital and physical games from a variety of genres and incorporating different materials.

4 Methodology and Results

4.1 Sample

The study took place in a kindergarten classroom in Greece, involving 18 students, eight of which were girls and ten were boys. The students were of the age of 5 and 6. For a period of three months, the students participated in game design sessions. The sessions lasted for one hour. Two game design sessions were organized in different days every week during school hours.

4.2 Procedure

During each game design session, students were divided into working groups. Each group consisted of three to four students and one adult. Each working group was voice recorded, while the adult was also asked to take notes on aspects that could not be recorded only by voice. Student creations were also kept at the end of the study in order to be analyzed along with the working group recordings. Every session consisted of the procedure presented below and used the tools that were also presented below.

The role of adults was to facilitate the design process. Adults were free to intervene whenever they deemed necessary, especially in situations that working groups had difficulty advancing on their designs. This could be for communication or implementation purposes or whenever the team had disputes and conflict that could not be overcome by the kids themselves.

4.3 Analysis

For the purposes of the study, a qualitative research was selected, in the form of a collective case study [41]. The central phenomenon of the qualitative research was the impact of adult feedback on students' game design choices and creations [42]. The advice of Strauss and Corbin, saying that "qualitative methods can be used to obtain the intricate details about phenomena such as feelings, thought processes, and emotions that are difficult to extract or learn about through more conventional methods" [43] was taken into account for the conducting of the study (Fig. 2).

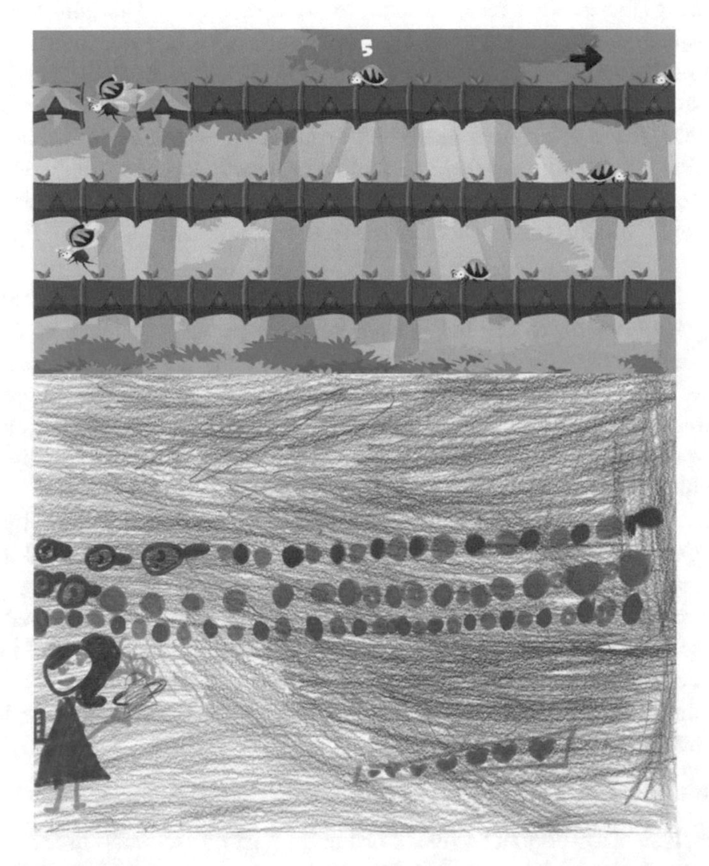

Fig. 2 A prototype created by working groups on a structured game design situation, based on Maskott's Forest Maths

Voice recordings of each working group were organized, transcribed and coded [44]. A line-by-line coding process was selected for the whole number of transcriptions. The analysis took into consideration adult participants' notes and the creations of the groups during the game design sessions. The codes were continuously reviewed, revised and, as a result, themes did emerge, leading to the later development of the research.

The analysis coming from the transcribed discussions of the working groups, focused on their reception of feedback and its impact on their creations as well as their delivery on behalf of the adults. From this analysis, two main thematic categories did emerge: (a) the role of adults in the design process and (b) the impact of feedback on early childhood students' decision-making process and game creations. The categories are analyzed below:

I. The ROLE of ADULTS in the DESIGN PROCESS

From the analysis of data, three roles of adult partners were identified: the non-interfering, the conflict resolver and the dominant ones. The non-interfering role describes situations, where adults rarely provided feedback, even in situations where teams had difficulty moving forward in the design process. In such occasions, adults would interfere only after considerable amount of time in order to provide students with questions that would offer them incentives or ideas on how they could move forward with their designs. Situations where this role was identified were usually characterized by either hesitation of adults to interfere in the design process or their belief that students would finally reach a conclusion or solution without their feedback and support.

The conflict resolver role describes situations where adults maintained an active role during the design process but did not offer proposals and ideas unless that was necessary for the evolution of the design process. Usually the reason why adults would intervene would be in order to resolve conflict among members of the team and provide solutions to situations that could not be handled by the students. In situations like this, adults were coordinators and facilitators of the process and their role was to provide students with feedback when they considered that this was necessary.

The dominant role describes adults that dominated the design process by providing feedback constantly. In those situations, adults would try to coordinate and take control of the design process from the beginning. They would assign roles to students and would ask for concrete deliverables from each one of them. Decision making might happen in common but cases where adults had a dominant role, they would undermine the process of decision making by indirectly presenting their opinion. The majority of cases that this role appeared was because adults did not have confidence in working groups' capabilities.

II. The IMPACT of FEEDBACK on STUDENTS' DECISION MAKING and CREATIONS

Data analysis showed that feedback has an impact on students' process of making decisions and, consequently, on their final game creations. This impact is connected with adults' role during the design process. Non-interfering adults had little impact on students' process of decision making and even less impact on students' final products. Even when adults with a non-interfering role would come up with questions, students would not change their process of thinking and discussing in order to provide answer to their posed questions, showing that they did not consider adults actually part of the team. Conflict resolvers, whose role was more active in the design process had a greater impact in the process of making decision and the final products of the working groups. Students were more responsive to their questions and tried several times to find answers to their questions. The questions presented by conflict resolvers affected the final games, where the impact of adults' feedback was obvious on the final designs. However, most of the decisions were made by students in this case.

When adults assumed dominant roles though, the final outcome was strongly connected with their feedback. The process of students' decision making appears to be less strong in those occasions and even if students were asked to make decisions, they were strongly influenced by adults' feedback, leading them to adopt their beliefs and perspective. This reception of feedback had an impact on the final game creations, which were influenced by adults' suggestions.

5 Conclusions and Discussion

The present study raises interesting points and lays the basis for further research on the field of game design education, its design, organizing and facilitation. The analysis that was presented is based previous study and analysis of game design frameworks, participatory design and mathematics and offers insights on the combination of those fields in this interdisciplinary subject. The study presents restrictions, notably for the number of students taking part in the qualitative study but also on the number of games that were selected for the limited three month duration of the game design sessions.

The data analysis shows that giving and receiving feedback in game design sessions of cross-generational teams has a strong impact on students' decision making process and the final outcomes of the produced games. From one side, different roles of adults as feedback providers were proposed, based on the depth, frequency and content of their feedback and the impact of those roles was studied later on based on the transcription and game creations analysis, leading to the conclusion that no or little feedback on behalf of adults leads to a small or non-observable impact on students' designs while adults with a dominant feedback providing role impact the process so much where at the end, students do not engage in decision making but end up following instructions and agreeing to the proposals and suggestions of adults.

Further analysis and exploration on this matter is considered necessary in order to study the relationship between giving and receiving impact in cross-generational teams in such game design situations.

References

1. Kalmpourtzis, G., Vrysis, L., Veglis, A.: Teaching game design to students of the early childhood through Forest Maths a pilot study. In: 2016 11th International Workshop Semantic and Social Media Adaptation and Personalization (SMAP), pp. 123–127 (2016)
2. Calvo-Ferrer, J.R.: Educational games as stand-alone learning tools and their motivational effect on L2 vocabulary acquisition and perceived learning gains. Br. J. Educ. Technol. **48** (2), 264–278 (2017)
3. Garneli, V., Giannakos, M., Chorianopoulos, K.: Serious games as a malleable learning medium: The effects of narrative, gameplay, and making on students' performance and attitudes. Br. J. Educ. Technol. **48**(3), 842–859 (2017)
4. Braghirolli, L.F., Ribeiro, J.L.D., Weise, A.D., Pizzolato, M.: Benefits of educational games as an introductory activity in industrial engineering education. Comput. Hum. Behav. **58**, 315–324 (2016)
5. Rodríguez Corral, J.M., Civit Balcells, A., Morgado Estévez, A., Jiménez Moreno, G., Ferreiro Ramos, M.J.: A game-based approach to the teaching of object-oriented programming languages. Comput. Educ. **73**, 83–92 (2014)
6. De Freitas, S., Oliver, M.: How can exploratory learning with games and simulations within the curriculum be most effectively evaluated? Comput. Educ. **46**(3), 249–264 (2006)

7. Winn, B., et al.: Learning from serious games? Arguments, evidence, and research suggestions. **12**(1) (2007)
8. Salen, K., Zimmerman, E.: Rules of play: game design fundamentals. Nihon Ronen Igakkai Zasshi, 672 (2004)
9. Schaefer, S., Warren, J.: Teaching computer game design and construction. Comput. Des. **36** (14), 1501–1510 (2004)
10. Baytak, A., Land, S.M.: A case study of educational game design by kids and for kids. Procedia - Soc. Behav. Sci. **2**(2), 5242–5246 (2010)
11. Games, D., Design, G., Ermi, L.: Player-centred game design: experiences in using scenario study to inform mobile game design introduction: players' role in game design the aims of the research. Game Stud. **5**, 1–12 (2005)
12. Tan, J.L., Goh, D.H.L., Ang, R.P., Huan, V.S.: Participatory evaluation of an educational game for social skills acquisition. Comput. Educ. **64**, 70–80 (2013)
13. Li, Q.: Digital game building: learning in a participatory culture. Educ. Res. **52**(4), 427–443 (2010)
14. Umetsu, T., Hirashima, T., Takeuchi, A.: Fusion method for designing computer-based learning game. In: Proceedings of International Conference Computers Education, vol. 1, pp. 124–128 (2002)
15. Kiili, K.: Educational game design : experiential gaming model revised. Building (2005)
16. Walsh, G., et al.: Layered elaboration: a new technique for co-design with children. In: Conference Human Factors Computing Systems, pp. 1237–1240 (2010)
17. Boyle, E.A., et al.: A narrative literature review of games, animations and simulations to teach research methods and statistics. Comput. Educ. **74**, 1–14 (2014)
18. Schell, J.: The Art of Game Design: A Book of Lenses. CRC Press, Boca Raton (2014)
19. Kafai, Y.: Making game artifacts to facilitate rich and meaningful learning. In: Annual Meeting of the American Educational Research Association (1995)
20. Moumoutzis, N., Christoulakis, M., Pitsiladis, A., Sifakis, G., Maragkoudakis, G., Christodoulakis, S.: The ALICE experience: a learning framework to promote gaming literacy for educators and its refinement. In: Proceedings of 2014 International Conference on Interactive Mobile Communication Technologies and Learning, IMCL 2014, pp. 257–261 (2015)
21. Vos, N., Van Der Meijden, H., Denessen, E.: Effects of constructing versus playing an educational game on student motivation and deep learning strategy use. Comput. Educ. **56** (1), 127–137 (2011)
22. Könings, K.D., Brand-Gruwel, S., van Merriënboer, J.J.G.: Participatory instructional redesign by students and teachers in secondary education: effects on perceptions of instruction. Instr. Sci. **39**(5), 737–762 (2011)
23. Nitsche, M., et al.: Designing procedural game spaces: a case study. In: Proceedings of Futureplay, pp. 10–12 (2006)
24. Cao, X., Kurniawan, S.H.: Designing mobile phone interface with children. In: Proceedings of ACM CHI 2007 Conference Human Factors Computing Systems, vol. 2, pp. 2309–2314 (2007)
25. Druin, A., Bederson, B., Boltman, A., Miura, A., Knotts-Callahan, D., Platt, M.: Children as our technology design partners. Des. Child. Technol. no. Age 8, 51–60 (1998)
26. Druin, A.: Cooperative inquiry: developing new technologies for children with children. Hum. Factors Comput. Syst. **14**(99), 592–599 (1999)
27. Reynolds, R., Caperton, I.H.: Contrasts in student engagement, meaning-making, dislikes, and challenges in a discovery-based program of game design learning. Educ. Technol. Res. Dev. **59**(2), 267–289 (2011)

28. Bermingham, S., et al.: Approaches to collaborative game making for fostering 21st century skills. In: Proceedings of 7th European Conference Games-Based Learning, pp. 45–52 (2013)
29. Habgood, M.P.J., Ainsworth, S., Benford, S.: Intrinsic fantasy: motivation and affect in educational games made by children. Learning **36**(4), 483–498 (2005)
30. Amory, A.: Game object model version II: a theoretical framework for educational game development. Educ. Technol. Res. Dev. **55**(1), 51–77 (2007)
31. Arnab, S., et al.: Mapping learning and game mechanics for serious games analysis. Br. J. Educ. Technol. **46**(2), 391–411 (2015)
32. Brousseau, G.: Theory of Didactical Situations in Mathematics (2002)
33. Hunicke, R., LeBlanc, M., Zubek, R.: MDA: a formal approach to game design and game research. In: Work. Challenges Game AI, pp. 1–4 (2004)
34. Kiili, K.: Digital game-based learning: towards an experiential gaming model. Internet High. Educ. **8**(1), 13–24 (2005)
35. Chang, K.E., Wu, L.J., Weng, S.E., Sung, Y.T.: Embedding game-based problem-solving phase into problem-posing system for mathematics learning. Comput. Educ. **58**(2), 775–786 (2012)
36. Polya, G.: Polya's problem solving techniques. In: How To Solve It, pp. 1–4 (1945)
37. Lajoie, S.P.: Extending the scaffolding metaphor. Instr. Sci. **33**(5–6), 541–557 (2005)
38. Stoyanova, E., Ellerton, N.: A framework for research into students' problem posing in school mathematics. In: Technology in Mathematics Education Proceedings of 19th Annual Conference Mathematics Education Research Group Australas, (MERGA), 30 June–3 July 1996, University of Melbourne (1996). (ISBN 0959684468), no. 1989, 1993
39. Druin, A.: The role of children in the design of new technology. Behav. Inf. Technol. **21**(1), 1–25 (2002)
40. Stoyanova, E.: Extending and exploring students' problem solving via problem posing (1997)
41. Yin, R.K.: Case Study Research : Design and Methods (2009)
42. Creswell, J.W.: Educational Research: Planning, Conducting, and Evaluating Quantitative and Qualitative Research, vol. 4 (2012)
43. Strauss, A., Corbin, J.: Basics of qualitative research: grounded theory procedure and techniques. Qual. Sociol. **13**(1), 3–21 (1990)
44. Creswell, J.W.: Research Design: Qualitative, Quantitative and Mixed Approaches, 3rd edn (2009)

Design and Evaluation of a Virtual-Reality Braille Writer-Simulator

Paul D. Hatzigiannakoglou(✉) and Evangelia Chytopoulou

Department of Educational and Social Policy, University of Macedonia, Thessaloniki, Greece
pxatzi@uom.gr, espl6179@uom.edu.gr

Abstract. This project presents and evaluates a novel virtual-reality Braille writer-simulator, which was developed for sighted tutors to learn and practice writing Braille [1]. Its novelty lies in the fact that it offers Braille trainers the opportunity to learn and practice the Braille writing system, by using a 3D virtual Braille writer [1]. Although Braille training software is available for computer users, we argue that learning Braille through the particular virtual device will be much more effective [2].

Keywords: Braille · Simulation · Virtual reality · Mobile learning

1 Introduction

Knowledge of the Braille code can improve the life of people with visual impairment in multiple ways. The quality of training of the visually-impaired on the Braille code largely depends on the skills of their instructors [3]. A common problem for Braille trainers is that they cannot afford to buy their own Braille writer to practice, owing to the high cost of traditional machines (current price $775). The aim of the present project is to offer a cost-free tool to Braille instructors for learning and practicing the Braille writing system. The usability and educational value of such a virtual machine "see Fig. 2" was evaluated through a research experiment conducted among sighted Braille instructors.

2 Related Work

A literature search regarding applications for practicing the Braille code was conducted from 30 October to 6 November 2017. The following databases were used: Scopus, PubMed, Eric. The keywords used were: Braille, mobile, virtual reality, applications, learning. We searched for such applications in all of the above databases. However, we failed to find any. Then, we used the Google search engine. Three results appeared from such a search, as explained below.

The first application is Visual Brailler App for iPad®, which is a Braille editor developed for iPad, using the six-dot Braille system. This application enables the user to type a text in Braille code. It was created by the American Printing House for the Blind, Inc. [4] and it runs IOS software.

M. E. Auer and T. Tsiatsos (Eds.): IMCL 2017, AISC 725, pp. 276–280, 2018.
https://doi.org/10.1007/978-3-319-75175-7_29

The second application is Braille Tutor-free, which consists of a series of lessons for practicing the Braille code. It is intended for both sighted and blind learners. It is a self-voiced application, and it is supported free of charge on IOS devices [5].

The third software is called Perky Duck and it is the most popular among sighted and blind learners. The program, which only runs on a conventional computer, enables the user to practice the Braille code and to create texts. It is a cost-free tool and it simulates the Perkins Brailler machine [6].

Compared to these three applications and their positive attributes, our VR application offers an additional advantage to the user and a certain novelty.

3 Method

To design and develop the project we used the "SketchUp" and "Unity 3D" software. The Perkins Braille embosser "see Fig. 1" served as a model for the 3D design of the machine [1]. Moreover, the Perkins Brailler is the most common device for manual Braille typing; it is used by the Panhellenic Association of the Blind-Regional Union of Central Macedonia, for educational purposes, and because, being resistant, it requires less maintenance compared to other Braille writing machines commercially available. Initially, we measured the dimensions of the moving and non-moving parts of the machine. Then we designed the 3D model engine, aiming to make it look like the actual machine, such that the adopted proportions and the tutors would get the optimal user experience, just like the real experience. Using the Unity 3D engine, the code was developed in C# [1].

Fig. 1. Perkins Braille embosser

The VR Brailler simulates, besides typing keys, all the control functions, such as character erasing, line change, and moving the Carriage Lever. Additionally, the corresponding sounds are produced, so that the user would experience the actual sounds as well [1]. When the user finishes typing the text, one can see a virtual page with one's text in Braille. The dimensions of the 3D model may be adjusted by the user "see Fig. 3" according to the dimensions of the tablet's display. The user may also rotate the machine to find the most convenient position.

Fig. 2. Virtual-reality Braille writer-simulator

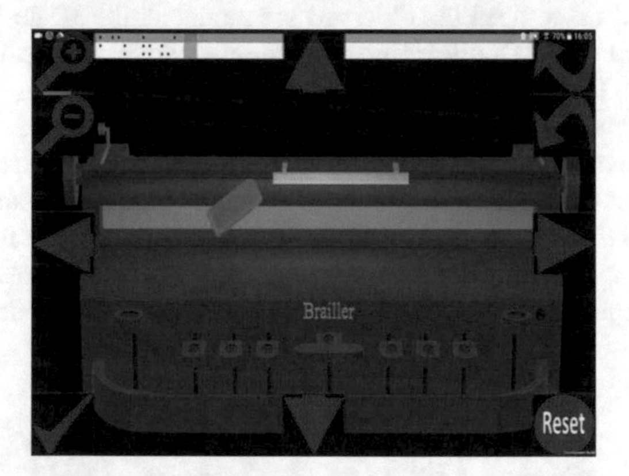

Fig. 3. Setup screen

When the user finishes typing the text, one can see a virtual page with one's text in Braille "see Fig. 4." The size of the virtual paper is similar to that of a real Braille writer, that is, the simulator can emboss 25 lines with 42 cells on a letter-size (8½ × 11″) sheet of paper.

Fig. 4. Virtual page

4 Results

After developing the VR Brailler Writer-Simulator, we carried out a short evaluation test. We visited the Panhellenic Association of the Blind-Regional Union of Central Macedonia, and asked 10 Braille instructors to use and evaluate our application. To this end, we created a survey diagram using a standard 5-point Likert scale. The outcome of this pilot research suggests that using the Brailler Writer-Simulator could prove of great value in the acquisition of Braille knowledge "see Table 1."

Table 1. Frequency of Braille instructors' evaluation of the application

	Strongly agree	Agree	Neutral	Disagree	Strongly disagree
The application is an innovative and creative tool for practicing	8	2	0	0	0
Usage of the application from the students will contribute to Braille code learning	7	3	0	0	0
Usage of the application will motivate the student, in order to study more about Braille code	6	2	2	0	0

4.1 Some Observations from Braille Instructors

The respondents made the following written or oral comments during the use or the VR device:

- "It would be a good idea to have the option to adjust the colors of the application, so that low-sighted users could use it."
- "It would be useful to provide an option for printing documents in Braille format."
- "If I could set the key response time myself, I would have reduced any unnecessary delay" [This happens because the VR Brailler detects the combinations of simultaneous touches within a fixed time range (150 ms). In fact, this time range should be inversely proportional to the degree of the user's experience. On the other hand, key response time in the Perkins Braille machine depends on the mechanical pressure of the buttons.]
- "If I had the chance to select between the Perkins Duck and the VR Brailler for my practice, I would definitely choose the VR Brailler."
- "While typing on the VR Brailler, I would like to sense vibrations in a similar way to the real Braille machine."

5 Future Work

In future versions, we will take into consideration the instructors' observations and will try to update it considerably. Tests for self-assessment will also be included. For example, a character will appear in a Braille cell, and the user will have to reproduce it.

As mentioned above, our sample consisted of 10 Braille instructors who expressed their opinion, evaluated whether our software can be compared to the Perkins Brailler machine, and assessed the extent to which it may assist the learning process of the Braille code. We intend to follow up our pilot study by using a larger sample of participants, toward the goal of consolidating the usefulness of the application.

6 Conclusion

We created a VR Braille learning machine, which was evaluated by 10 sighted Braille instructors. Our aim is to provide a cost-free tool for practicing this useful code of communication, the importance of which is immense for visually-impaired or low-sighted individuals. So, in search of the "optimal" possible educational result, we chose our virtual-reality machine-VR Braille to resemble the real Perkins Braille machine and to function in a similar way, albeit at no cost. Every detail of the Perkins Braille machine was recreated in our project, giving the user the perception of the real machine, but in a virtual reality environment. Based on our pilot data, we conclude that this new tool can prove to be helpful in the effective training of future Braille instructors. Compared to the applications currently available for learning and practicing the Braille code, ours is based on VR technology, which can facilitate the training process substantially in comparison with conventional software.

References

1. Hatzigiannakoglou, P.: VR-Brailler: a virtual-reality Braille writer-simulator. In: INTED2017 Proceedings, p. 934 (2017)
2. Alexander, A.L., Brunyé, T., Sidman, J., Weil, S.A.: From gaming to training: a review of studies on fidelity, immersion, presence, and buy-in and their effects on transfer in pc-based simulations and games. DARWARS Training Impact Group **5**, 1–14 (2005)
3. Hatzigiannakoglou, P.D., Kampouraki, M.T.: "Learn Braille": a serious game mobile app for sighted Braille learners. J. Eng. Sci. Technol. Rev. **9**(1), 174–176 (2016)
4. Visual Brailler on the App Store – iTunes – Apple. https://itunes.apple.com/us/app/visual-brailler/id888739587?mt=8
5. Braille Tutor. http://ienabletechnology.com/braille-tutor/
6. Free Download of Perky Duck. http://www.duxburysystems.com/perky.asp

ARTé Mecenas: In the Shoes of a Medici

Michalis Matthaios Lygkiaris[1(✉)], Fragiskos Gerasimos Bersimis[2], and André Thomas[3]

[1] Ionian University, Corfu, Greece
m.lygkiaris@yahoo.gr
[2] Harokopio University, Athens, Greece
fbersim@hua.gr
[3] Texas A&M University, College Station, USA

Abstract. Educational games aimed at teaching history or other humanities are often structured as quiz games, not differing much from a typical examination. However, games have many elements that could possibly contribute in increasing user/student motivation, such as their mechanics, interactive environments, character portrayal and story. This paper explores the concepts mentioned above, as applied in the educational game "ARTé: Mecenas" by Triseum, as well as their effects in a small sample of people who played the game. The conducted pilot study was focused on the avatar identification process, which was found affecting positively the effects of engagement and knowledge improvement.

Keywords: Game-Based Learning · Motivation · Engagement Identification

1 Introduction

The evolution of digital technology affected education in a more profound way than ever before. The Internet has made information readily available, including, not only text, but audiovisual resources as well. All these data are now extremely easy and fast to access and reproduce with minimum cost, providing students with an almost infinite library to expose themselves to. Furthermore, digital technology offers even more opportunities that could possibly change the actual form of education.

Game-Based Learning employs game mechanics and aesthetics to educate or train the user. It is not hard to understand that the use of interactivity in the teaching process is fundamentally different from lecture-based education. Games are fun, engaging and losing in a game is definitely less damaging, in all sorts of ways, than failing an exam.

However, all of the above are simply the minimum requirements for a typical educational game. Videogames are capable of incorporating quite a number of different aspects in them, which can contribute in a multivariate User Experience. Correspondingly, an educational game can utilize all these tools to create an interactive teaching environment that extends beyond information understanding and towards a learning experience. In this scenario, the user is not only accessing information and answering questions on the teaching subject, but rather making conscious choices that are interrelated with it. This could possibly lead to a deeper understanding of the subject, as the student has actually participated and affected the outcome of this experience.

M. E. Auer and T. Tsiatsos (Eds.): IMCL 2017, AISC 725, pp. 281–293, 2018.
https://doi.org/10.1007/978-3-319-75175-7_30

Such seems to be the case with ARTé: Mecenas, an educational game that puts the student in the place of a member of the Medici family in the Italian Renaissance. By playing the game, the student has the chance to encounter the opportunities and threats that the Medici were facing back then. Furthermore, in order to advance in the game, the player must find a way to balance wealth, reputation and relationship with the city guilds and the Catholic Church and at the same time sponsor the great artists of the time, just as the Medici did [1]. Since the student is now the one making these choices and forming, or rather reproducing, the story, they can better understand how the Medici came into power and influenced History.

ARTé: Mecenas is a turn-based point & click browser game which utilizes resource management and trading at its core mechanic, intermixing it with narrative decisions addressed to the user in the form of letters and text notes. The main interface depicts the Medici desk filled with interactive objects. By clicking on them the user may manage economic, diplomatic, politic and art-related issues. For example, a weighing scale opens up the market pop-up window where the player may buy or sell trading goods, while a feathered pen opens up the mail correspondence.

By having the student take on the Medici role, the content of the teaching subject can be presented in a way that promotes the development of interdisciplinary skills, such as strategic and critical thinking. Every decision the user makes, has an immediate effect on his Avatar's social and/or economical status. Thus, he can better understand the dynamics prevailing in the Italian Renaissance and use this knowledge to acknowledge the true opportunities and form a strategy to advance in the game.

Many aspects of ARTé: Mecenas are focused on creating an immersive environment and a link between the user and the Avatar. This study examines whether an immersive environment and user Identification with the Avatar, are affecting student engagement and have a positive impact on the learning outcome.

2 Previous Studies

It has been over ten years since Kiili [2] proposed an experiential gaming model for education. Even though, the progression in the field has been obvious, it seems that the medium is only just getting mature enough to support his proposal. Soutter and Hitchens [3] have recently shown that Identification[1] is strongly and positively related to Flow[2], which in turn leads to greater engagement. Their research however was targeted at recreational games and not educational ones. Still, other researchers, such as Bowman et al. [4] and Birk et al. [5] strengthen this hypothesis.

Bachen et al. [6] have shown that Identification and Presence (Spatial Immersion)[3] are significantly related, with Presence having a strong positive influence on Identification. They found however that Identification played only a lesser role in the learning outcome, when compared with Presence. It must be noted though that the game they

[1] The sense of connection with the Avatar.

[2] The user's absorption and enjoyment resulting from the activity.

[3] The sense of being in the game's environment.

used on their research was themed around a Haitian Earthquake and the participants were undergraduates in the Californian Universities. Identification of the users was mostly achieved with a specific game character from a related cultural background (an American Journalist) while the users developed mostly empathy towards Haitian game characters.

3 Aim - Materials and Methods: Study Design - Data Collection

The project's main target is to explore whether ARTé: Mecenas game fulfills basic aspects of learning and gaming. A pilot research was conducted by using a questionnaire (Appendix A) with questions regarding several attributes of Game Based Learning (GBL) during August of 2017. The pilot research has been designed so that data collection was completed by using simple random sampling (SRS) [7] from 22 individuals characterized as gamers (percentage of 55.6%) or not with art history knowledge (percentage of 38.9%) or not, among students of various university departments. All students participating were informed about the aims of the project, gave their written consent and have been involved in the game for at least 4 h, in two separate sessions. Briefly, the questionnaire (Appendix A) records the GBL structure of ARTé: Mecenas game by gathering information for basic educational gaming elements such as "Concentration", "Immersion", "Knowledge Improvement", "Engagement", "Clarity of Goals", "Autonomy"[4], "Identification" and "Challenge". Additional information obtained from this questionnaire includes whether the participants are familiar with gaming and art history. Questionnaire's questions/variables used the Likert scale [8] with seven partitions with appropriate coding of reverse questions, where -3 expresses the maximum degree of disagreement towards an attitude and 3 expresses the maximum degree of agreement towards an attitude. Secondary variables were generated expressing the average value of the corresponding blocks' questions. Therefore, these variables, known as components hereafter, express the above-mentioned concepts of "Concentration", "Immersion", "Knowledge Improvement", "Engagement", "Clarity of Goals", "Autonomy", "Identification" and "Challenge". The questionnaire was tested and validated before use as all the items used were adapted from previous studies [9].

4 Statistical Analysis

Comparisons between gamers and non gamers, as well as, between history art connoisseur and non history art connoisseur were conducted by calculating descriptive measures of central tendency and dispersion, such as the mean and the standard deviation, for the secondary variables. In order to test the statistical significance of all participants, parametric and non parametric tests were performed. One-Sample Kolmogorov-Smirnov (K-S) test [10] was performed for exploring whether factors

[4] The user's sense of immediate feedback and control on interactive media.

are distributed normally. In addition, one sample t-test [11] was conducted and one sample Wilcoxon signed-rank test [12] were used for testing the equality of components' means (medians) in relation to zero that corresponds to a neutral attitude (positive average value corresponds to a positive attitude towards the ARTé: Mecenas game). The non parametric correlation coefficient or Spearman's rho was used for exploring the degree of relation between "Identification", "Engagement", "Improving Knowledge". In addition, p-values less than 0.01 were considered statistically significant and 95% confidence intervals (CI) are given. Data analysis was performed using statistical software of IBM SPSS (Version 21).

5 Results

Table 1 lists the descriptive statistics of factors "Concentration", "Clarity of Goals", "Challenge", "Autonomy", "Immersion", "Identification", "Engagement", "Improving Knowledge" for the whole sample, and also for categories "gamer with art history knowledge", "gamer without art history knowledge", "non gamer with art history knowledge" and "non gamer without art history knowledge".

Table 1. Descriptive statistics of "Concentration", "Clarity of Goals", "Challenge", "Autonomy", "Immersion", 'Identification", "Engagement", "Improving Knowledge"

Factor	Total Sample Mean (Stdev) (22)	Gamers with art history knowledge (6)	Gamers without art history knowledge (6)	Non Gamers with art history knowledge (3)	Non Gamers without art history knowledge (7)
Concentration	1.46 (0.80)	1.25 (0.53)	1.75 (1.03)	0.63 (0.18)	1.67 (0.80)
Clarity of Goals	1.13 (0.76)	1.20 (0.82)	1.20 (0.65)	1.00 (0.26)	1.04 (1.04)
Challenge	1.53 (0.86)	1.55 (0.93)	1.60 (1.28)	1.50 (0.35)	1.46 (0.70)
Autonomy	0.69 (0.67)	0.48 (0.52)	1.16 (0.55)	0.70 (0.14)	0.47 (0.86)
Immersion	0.68 (0.96)	0.40 (1.01)	1.05 (0.95)	0.13 (0.18)	0.79 (1.10)
Identification	0.87 (0.80)	0.84 (0.43)	1.04 (1.00)	0.40 (0.85)	0.90 (0.99)
Engagement	0.67 (1.08)	0.20 (1.47)	1.08 (1.06)	0.30 (0.71)	0.83 (0.86)
ImprovingKnowledge	1.61 (0.83)	2.08 (0.46)	1.88 (0.67)	0.70 (0.14)	1.30 (1.03)

It is rather impressive that the Gamers without art history knowledge group presents the highest average value for factors "Concentration", "Clarity of Goals", "Challenge", "Autonomy", "Immersion", "Identification" and " Engagement", while the "Gamers with art history knowledge" group has the highest average value for "Improving Knowledge".

Analytically, in Fig. 1, the variables that contribute to the concept of "Concentration" are presented, where the highest average value is achieved by attitude *"Generally speaking, I can remain concentrated in the game"* position and the lowest average value is in the attitude *"Most of the gaming activities are related to the learning task"*. The negative value of the 4th attitude of the factor is justified because has a negative (inverse) concept against the factor "Concentration". In Fig. 2, variables contributing to the concept of "Clarity of Goals" are presented, where the highest average value is achieved by variable *"Intermediate game goals were presented in the beginning of each level"* and the lowest average is achieved by variable *"Overall, game goals were presented in the beginning of the game"*.

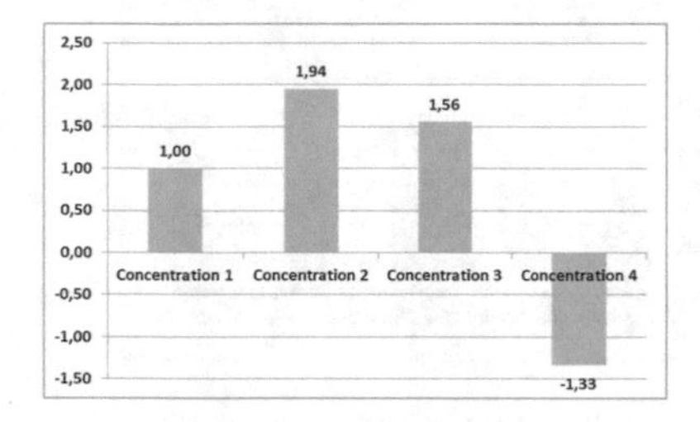

Fig. 1. Bar chart of variables average values of factor "Concentration"

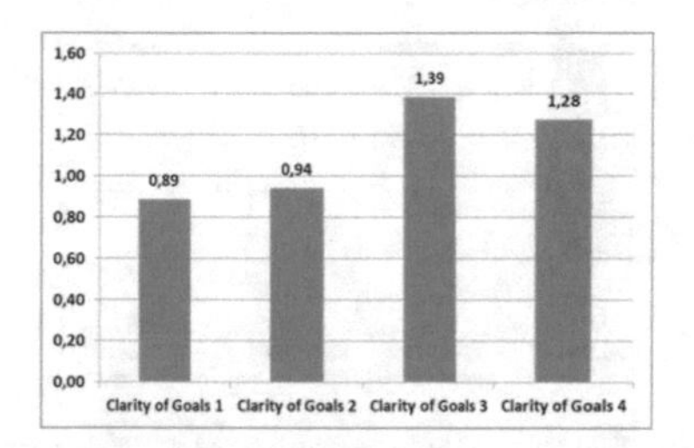

Fig. 2. Bar chart of variables average values of factor "Clarity of Goals"

In Fig. 3, variables contributing to the concept of "Challenge" are presented, where the highest average value is taken by the attitude *"My skill gradually improves through the course of making decisions and seeing the consequences"* and the lowest average is taken by the attitude *"The game provides 'hints' in text that help me make decisions"*.

In Fig. 4, variables contributing to the concept of "Autonomy" are presented, where the highest average value is achieved by variable *"I feel a sense of control over objects in the game"* and the lowest mean value is taken by variable *"I know the next step in the game"*. The negative value of second variable is justified because it has a negative (inverse) concept against the Autonomy factor.

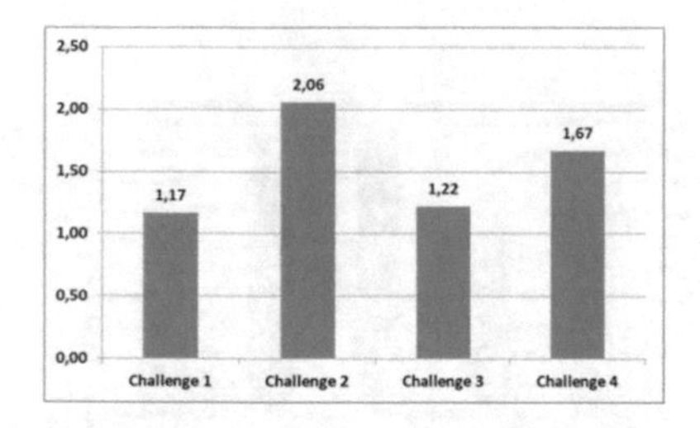

Fig. 3. Bar chart of variables average values of factor "Challenge"

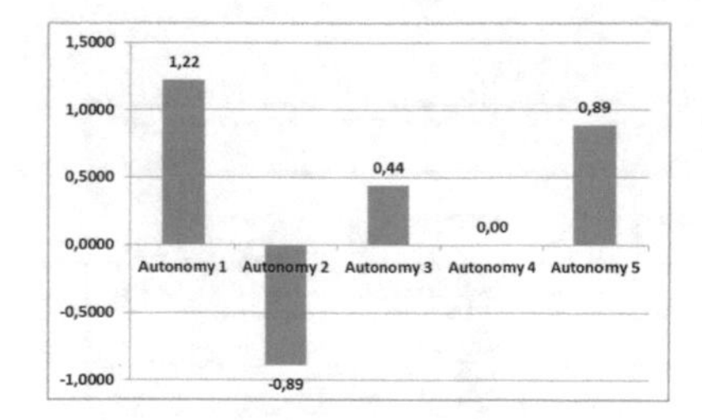

Fig. 4. Bar chart of variables average values of factor "Autonomy"

In Fig. 5, variables contributing to the concept "Immersion" are presented, where the highest average value is achieved by attitude *"I forget about time passing while playing the game"* and the lowest average value is achieved by attitude *"I feel viscerally involved in the game"*. In Fig. 6, variables contributing to the concept "Identification" are presented, where the highest average value is achieved by attitude *"I feel as if I am influencing the story or that I am a part of it"* position and the lowest average value is achieved by attitude *"My goals are the same with those of the avatar (Medici)".*

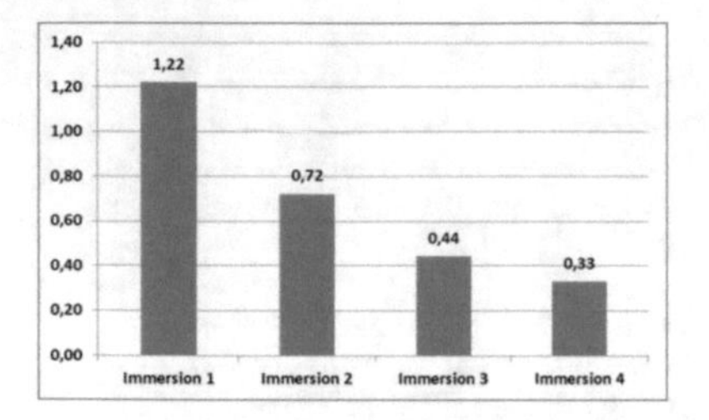

Fig. 5. Bar chart of variables average values of factor "Immersion"

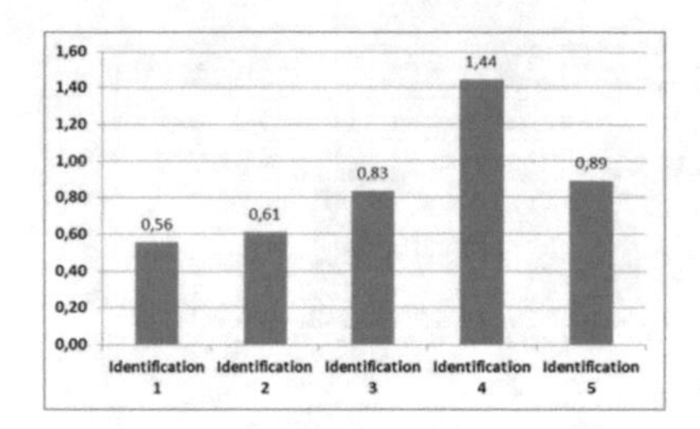

Fig. 6. Bar chart of variables average values of factor "Identification"

In Fig. 7, variables contributing to the concept "Engagement" are presented, where the highest average absolute value is taken by attitude *"The game tires me"* and the lowest average value is taken by attitude *"I make decisions intuitively or automatically"*. The negative value of Stage 2 is justified because it has a negative (inverse) concept against the "Engagement" factor. In Fig. 8, variables contributing to the concept "Improving Knowledge" are presented, where the highest average value is achieved by attitude *"I try to apply my knowledge in the game"* and the lowest average value is achieved by attitude *"I want to know more about the knowledge taught"*. The normality test of K-S showed that the assumption of normal; distribution was not violated for all variables/factors "Concentration", "Clarity of Goals", "Challenge", "Autonomy", "Immersion", "Identification", "Engagement", and "Improving Knowledge".

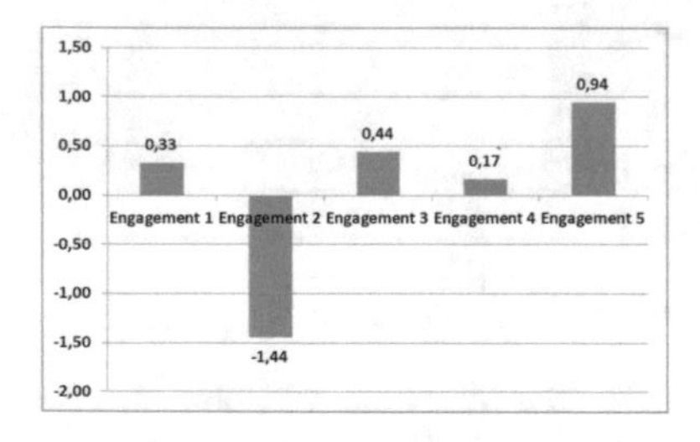

Fig. 7. Bar chart of variables average values of factor "Engagement"

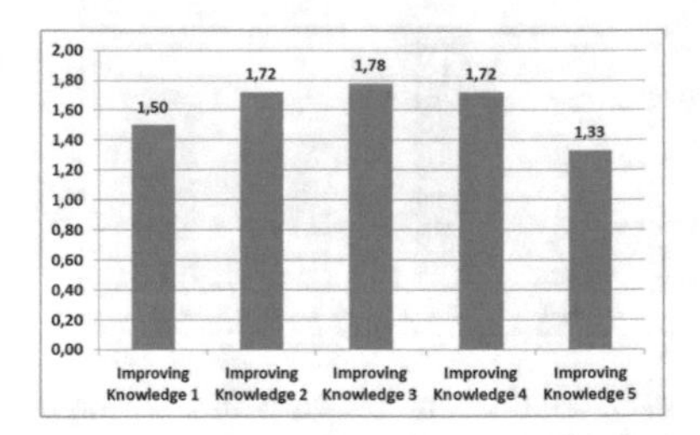

Fig. 8. Bar chart of variables average values of factor "Improving Knowledge"

Table 2 lists the results of one sample t-test and one-sample Wilcoxon signed-rank test, indicating that all the aforementioned variables/factors have a statistically significant higher mean (Median) value in relation to 0, which indicates the neutral position.

Table 2. One sample t-test - one-sample Wilcoxon signed-rank test results

	One sample t-test		One-sample Wilcoxon signed-rank test	
Factor	t	p-value	z	p-value
Concentration	7.725	<0.01	-3.628	<0.01
Clarity of Goals	6.253	<0.01	-3.644	<0.01
Challenge	7.726	<0.01	-3.627	<0.01
Autonomy	4.343	<0.01	-3.109	0.002<0.01
Immersion	3.014	0.008<0.01	-3.772	0.006<0.01
Identification	4.579	<0.01	-3.434	0.001<0.01
Engagement	2.623	0.018<0.05	-2.381	0.017<0.05
ImprovingKnowledge	8.233	<0.01	-3.627	<0.01

Therefore, results indicate that a statistically significant positive attitude of the participants in attitudes "Concentration", "Clarity of Goals", "Challenge", "Autonomy", "Immersion", "Identification", "Engagement", "Improving Knowledge" towards ARTé: Mecenas game exists.

Table 3 lists the results of Spearman's rho Correlation between Identification-Engagement and Identification-Improving Knowledge. The results are also presented in graph form in Figs. 9 and 10 correspondingly. Positive, moderate to strong, correlation was observed for both pairs of characteristics Identification-Engagement ($p_v = 0.013 < 0.05$) and Identification-Improving Knowledge ($p_v = 0.002 < 0.01$).

Table 3. Spearman's rho correlation

		Engagement	Improving Knowledge
Identification	Correlation Coefficient	0.571	0.670
	Sig. (2-tailed)	0.013<0.05	0.002<0.01

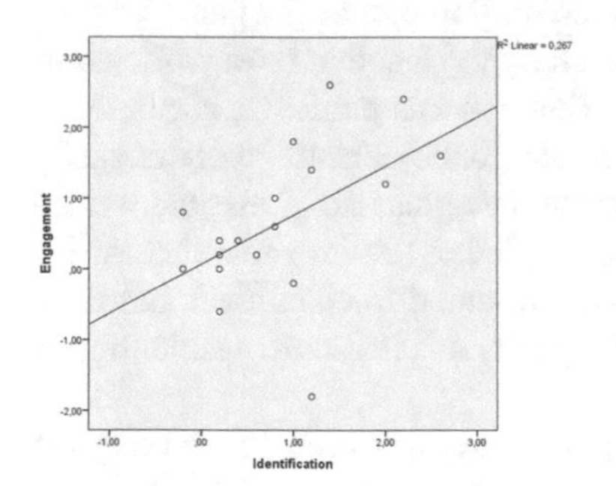

Fig. 9. Scatterplot of Identification vs Engagement

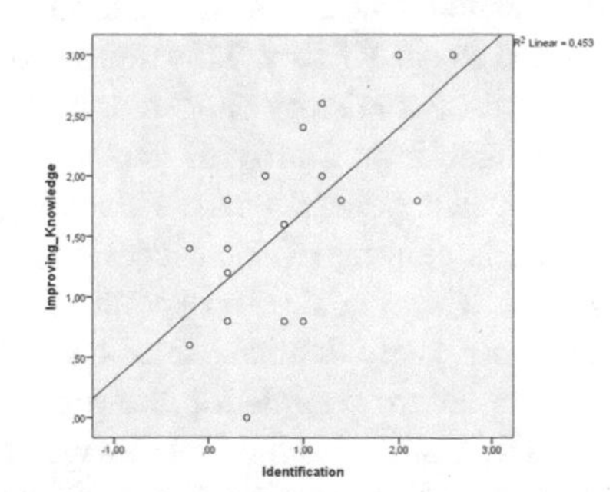

Fig. 10. Scatterplot of Identification vs Improving Knowledge

6 Discussion

The data collected in this research are rather positive for ARTé: Mecenas and the User Experience it offers to students. All participants, whether gamers or non gamers, with art history knowledge or not, seem to hold a positive attitude towards the game. However, gamers without art history knowledge are the ones who had the highest average value on every factor, except “Improving Knowledge”. This observation should be mainly attributed to the two following facts. First, a gamer can easily interact with gaming mechanics and interfaces, while a non gamer has to overcome the lack of familiarity with games and, second, their limited knowledge in the teaching subject, probably enhances their experience, since the game keeps them interested in its whole duration. Thus, participants belonging to this category are more likely to focus on the factors recorded in this research and, due to the exploration of unfamiliar grounds, maintain this interest long enough to feel immersed, engaged and challenged.

On the other hand, gamers who have also studied art history have a highest average value in “Improving Knowledge” than those who have not. This seems rather counterintuitive, since the latter should have more things to learn than the former. However, this could be attributed to the process of active learning. The way the game improves knowledge in art history is experiential and in a broader sense than that of a lecture. The student has the chance to experience the difficult choices a Medici was making in the Renaissance. An art history connoisseur could probably be more excited by this prospect and acknowledge the alternative viewpoint that the game offers, which could push higher attitude ratings. Since the player is actually making these choices, they can better understand what led the Medici act this way, and, because of his former knowledge, value this kind of knowledge improvement.

An examination of the “Concentration” factor graph (Fig. 1) shows that participants remain generally concentrated in the game, since the corresponding attitude has the highest value. The weaker positive value in *“Most of the gaming activities are related to the learning task”* could probably be attributed to the unusual and more global way of the game’s approach in teaching art history. Most participants would probably expect more Art related “questions” and thought that the economic/trading part of the game was not related to Art History. This trend is probably enhanced by the fact that most participants were not Art History educated, pulling the average down in this particular attribute.

Regarding “Clarity of Goals” (Fig. 2), intermediate goals seem to be significantly clearer than general goals. This is logical since intermediate goals are presented written in the beginning of each level and also appear in the Codex (an info interface in the game). These goals correspond to the specific years presented in each level and thus, are more easily defined. Since each level has unique goals, it is harder to see the bigger picture before completing the game.

The results for the “Challenge” factor (Fig. 3) are extremely useful in linking the game to active learning. The variable *“My skill gradually improves through the course of making decisions and seeing the consequences”*, which has the highest average than all others, is definitely the one more closely related to active learning, while the variable *“The game provides ‘hints’ in text that help me make decisions”*, which is the lowest, refers to reading; the same way a student learns from a book.

The active learning approach is obviously facilitated by the fact that the users found the interface rather intuitive as shown in the "Autonomy" graph (Fig. 4) and the *"I feel a sense of control over objects in the game"* variable, which scored higher than all others. The neutral result in the *"I know the next step in the game"* is probably due to the nature of the gameplay of ARTé: Mecenas. The game depends a lot on, unpredictable by the user, events, invoking a sense of uncertainty in accordance with its Renaissance theme.

Regarding "Immersion" (Fig. 5), the highest value in the variable corresponding to time perception in addition with the lowest value in the attitude *"I feel viscerally involved in the game"*, probably show that the game is more closely related to Flow concept than it is to Immersion. Further research could clarify this.

The process of identifying with the Avatar has not been explored enough in Educational Games, as it is not essential in teaching. It could however prove to be a valuable tool for engagement, as is the case with other media, such as movies and books. In Fig. 6, the attitudes *"I feel as if I am influencing the story or that I am a part of it"*, which has the highest average, and *"I feel as if I am actually experiencing the story of the game"* must probably be valued higher than the others, since they are the ones not mentioning the Avatar itself. As explained above, participants without former Art History knowledge outnumbered the ones who were familiar with the subject. Their limited knowledge on the Medici family could pull the average down on attitudes referencing them.

Engagement is an important factor for Game-Based Learning. Figure 7 shows that the variables immediately referring the time spent playing the game (*"I play more than I initially intend to"*) and the will to keep on playing (*"The game tires me"*) had significantly higher absolute average values, than all the others. Even though the time devoted to learning is the goal of engagement, the weak positive values in the rest of the variables show that ARTé: Mecenas could probably benefit from some improvements. In-game information (i.e. inside the Codex) could keep the students more focused in the game (*"While playing, I do not respond to external stimuli"*). Better controls (*"Controls in the game feel intuitive or automatic"*) and clearer goals (*"I make decisions intuitively or automatically"*) could also increase the value of the "Engagement" Factor.

The "Improving Knowledge" Factor had the highest mean (Table 1) than all of the other Factors, and all of its variables had average values above 1 (Fig. 5). The above observations prove the educational aspect of the game. The variable *"I want to know more about the knowledge taught"* has the lowest average value, probably due to participants already educated in Art History. As mentioned above however, the average value is still high, since the game offers a wider and more active training in the subject.

The moderate positive correlation of "Engagement" and "Knowledge Improvement" with the factor of "Identification" is a rather positive outcome. Even though only a moderate correlation was found between the factors, the approach of putting the user in the shoes of a Medici, as taken by the designers, seems to pay off. The results of this pilot study cannot be generalized, of course. However, it seems probable that game-based learning could notably benefit from incorporating this kind of narrative techniques to engage and draw students, turning the learning process into a richer user experience.

7 Conclusion

The results of this pilot research support not only the educational nature of ARTé: Mecenas, but its immersive and engaging characteristics as well. The hybrid form of the game combines elements of trading/economic or even diplomacy games with the typical questions of quiz educational games. This approach is not far from commercial entertaining games, which employ different mechanics or mini-games to keep the player engaged, challenged and interested. In addition, the variation and intermixing of these mechanics, when attuned to the game's theme and story, provide a deeper understanding of its characters and narrative, as well as a more immersive user experience.

It would not be farfetched to assume that utilizing the aforementioned elements in an educational game, would also lead to a deeper understanding and an immersive experience. This could be the case for ARTé: Mecenas, where the different mechanics used in the game, reflect different parts of a Medici life. The player/student experiences and learns about all these different aspects of the Art Patrons in the Renaissance. Identification with the Avatar is definitely facilitated by combining this dynamic learning experience with the game's story and by linking the challenges of the Medici with the choices of the user.

Immersion and Identification however are aspects that are mainly sought in entertainment games, while Knowledge Improvement is a necessary element in an educational game. Engagement on the other hand seems to be equally important for learning as is for an exciting interactive experience. This paper has helped in acknowledging all the above characteristics in ARTé: Mecenas, as well as finding some correlation between them. It must be noted however that, since the sample was small (N = 22) and this was only a pilot research, a further study should take place with a larger number of participants, in order to validate the results.

Appendix A

Questionaire Regarding User Experience from Playing «ARTé: Mecenas»
Rate each of the following statements according to your degree of agreement/disagreement.

Concentration	Strongly Disagree	Disagree	Somewhat Disagree	Neither Agree Nor Disagree	Somewhat Agree	Agree	Strongly Agree
Most of the gaming activities are related to the learning task							
Generally speaking, I can remain concentrated in the game.							
I am not distracted from tasks that the player should concentrate on within the game.							
I am burdened with tasks in the game that seem unrelated							

Clarity of Goals	Strongly Disagree	Disagree	Somewhat Disagree	Neither Agree Nor Disagree	Somewhat Agree	Agree	Strongly Agree
Overall, game goals were presented in the beginning of the game.							
Overall, game goals were presented clearly.							
Intermediate goals were presented in the beginning of each level.							
Intermediate goals were presented clearly.							

Challenge	Strongly Disagree	Disagree	Somewhat Disagree	Neither Agree Nor Disagree	Somewhat Agree	Agree	Strongly Agree
The game provides "hints" in text that help me make decisions.							
My skill gradually improves through the course of making decisions and seeing the consequences.							
The difficulty of the decisions increases as my skills improve.							
The game provides new decisions with an appropriate pacing.							

Autonomy	Strongly Disagree	Disagree	Somewhat Disagree	Neither Agree Nor Disagree	Somewhat Agree	Agree	Strongly Agree
I feel a sense of control over objects in the game.							
When I make errors, I cannot progress in the game.							
The game supports my recovery from errors.							
I know the next step in the game.							
I feel a sense of control over the game.							

Immersion	Strongly Disagree	Disagree	Somewhat Disagree	Neither Agree Nor Disagree	Somewhat Agree	Agree	Strongly Agree
I forget about time passing while playing the game.							
I become unaware of my surroundings while playing the game.							
I feel emotionally involved in the game.							
I feel viscerally involved in the game.							

Identification	Strongly Disagree	Disagree	Somewhat Disagree	Neither Agree Nor Disagree	Somewhat Agree	Agree	Strongly Agree
My goals are the same with those of the avatar (Medici).							
The choices I make correspond to those of a Medici.							
I feel as if I am actually experiencing the story of the game.							
I feel as if I am influencing the story or that I am a part of it.							
While playing the game, I feel as if I am a Medici.							

Engagement	Strongly Disagree	Disagree	Somewhat Disagree	Neither Agree Nor Disagree	Somewhat Agree	Agree	Strongly Agree
While playing, I do not respond to external stimuli.							
The game tires me.							
Controls in the game feel intuitive or automatic.							
I make decisions intuitively or automatically.							
I play more than I initially intend to.							

Knowledge Improvement	Strongly Disagree	Disagree	Somewhat Disagree	Neither Agree Nor Disagree	Somewhat Agree	Agree	Strongly Agree
The game increases my knowledge.							
I catch the basic ideas of the knowledge taught in the game.							
I try to apply my knowledge in the game.							
The game motivates the player to integrate the knowledge taught.							
I want to know more about the knowledge taught.							

References

1. Gombrich, E.H.: The Story of Art. Phaidon Press Limited, London (1995)
2. Kiili, K.: Digital game-based learning: towards an experiential gaming model. Internet High. Educ. **8**, 13–24 (2005)
3. Soutter, A.R.B., Hitchens, M.: The relationship between character identification and flow state within video games. Comput. Hum. Behav. **55**, 1030–1038 (2016)
4. Bowman, N.D., Oliver, M.B., Rogers, R., Sherrick, B., Woolley, J., Chung, M.Y.: In control or in their shoes? How character attachment differentially influences video game enjoyment and appreciation. J. Gaming Virtual Worlds **8**, 83–99 (2016)
5. Birk, M.V., Atkins, C., Bowey, C.T., Mandryk, R.L.: Fostering intrinsic motivation through avatar identification in digital games. In: Proceedings of the SIGCHI Conference on Human Factors in Computing Systems (CHI 2016), pp. 2982–2995 (2016)
6. Bachen, C.M., Hernandez-Ramos, P., Raphael, C., Waldron, A.: How do presence, flow, and character identification affect players' empathy and interest in learning from a serious computer game? Comput. Hum. Behav. **64**, 77–87 (2016)
7. Yates, D., Moore, D.S., Starnes, D.S.: The Practice of Statistics: TI-83/84/89 Graphing Calculator Enhanced. Macmillan Higher Education, New York (2007)
8. Likert, R.A.: A technique for the development of attitude scales. Educ. Psychol. Measur. **12**, 313–315 (1952)
9. Triseum: Educators and Administrators ARTé (2016). https://triseum.com/art-history/arte/mecenas/educators-administrators/
10. Stephens, M.A.: EDF statistics for goodness of fit and some comparisons. J. Am. Stat. Assoc. **69**(347), 730–737 (1974)
11. O'Connor, J.J., Robertson, E.F.: MacTutor history of mathematics archive. University of St Andrews, Scotland (2015)
12. Wilcoxon, F.: Individual comparisons by ranking methods. Biom. Bull. **1**(6), 80–83 (1945)

Game-Based Learning for IoT: The Tiles Inventor Toolkit

Anna Mavroudi(✉), Monica Divitini, Simone Mora, and Francesco Gianni

Norwegian University of Science and Technology, Trondheim, Norway
anna.mavroudi@idi.ntnu.no

Abstract. This paper presents a Game-Based Learning Design Pattern for designing Internet of Things (IoT) applications, as well as an instance of this particular pattern, namely the 'Tiles inventor game'. This educational game has been validated in various educational contexts aiming to understand IoT fundamentals. Due to lack of topics on IoT in the STEM curricula today as well as due to the lack of game-based learning design patterns, we propose these two artifacts, the pattern of designing IoT applications and the description of the Tiles inventor game, as a means of communicating best practice and contributing to bridging the gap between educational theory and informal practice in such a niche domain.

Keywords: Internet of Things · STEM innovative scenario
Game-Based Learning · Prototyping · Design · Programming

1 Introduction

Game-Based Learning (GBL) has a long tradition in education, although it was introduced to the Technology Enhanced Learning community at the beginning of the second millennium by Prensky (2001). GBL is part of entertainment education, which refers to any attempt to make learning enjoyable (Breuer and Bente 2010). It can be media-based, mediated or classroom-based and can include the use of any type of games for educational purposes (ibid): board games, card games, digital games, exergames and so on. The purpose of the Tiles inventor scenario presented in this paper is to engage upper high school students in designing a technology-augmented solution that relates to Internet of Things (IoT) through GBL. The development of the solution exposes students to aspects of four different scientific areas: human-device interaction, Internet of Things, design and programming.

We describe the Tiles game-based learning application as a case of a Game-Based Learning Design Pattern, as an 'intermediary form of knowledge'. Such forms of knowledge which are more concrete than generalized theories, but more general than a single instance (Prieto et al. 2017) are much needed to bridge the gap between theory and practice. Design patterns typically consist of (Huynh-Kim-Bang et al. 2010): the name/title, the context, the problem, conflicting interests (called "forces") that intervene in the problem, and a generic or canonical solution. Design Patterns are embraced by various domains, including education. Examples include:

M. E. Auer and T. Tsiatsos (Eds.): IMCL 2017, AISC 725, pp. 294–305, 2018.
https://doi.org/10.1007/978-3-319-75175-7_31

- the Design Principles Database (DPD, http://www.edu-design-principles.org/dp/designPatterns.php), an online repository of design principles and design patterns on the use of technologies for education (Kali 2006), and
- the "Empowering the School infrastructures for the Implementation of Sustainable Instructional Patterns" project (eSIT4SIP, http://www.esit4sip.eu/), which has provided a knowledge base of instructional patterns that consider the existing ICT school infrastructure and other local specificities of the school reality (Mavroudi et al. 2016). Game-based Learning Design Patterns (GBLDPs) is a valid way of collecting, documenting and disseminating best practice in educational games and in GBL (Ecker et al. 2011). In Sect. 3, we provide the GBLDP of the Tiles inventor game.

2 The Tiles Game-Based Learning Scenario

The Tiles scenario consists of four main phases, the first one being an introduction both to the design challenge and to the programming and prototyping challenge:

- the introductory activities
- the design challenge (i.e. create an IoT-infused design solution)
- the transition from design to programming/prototyping
- the programing/prototyping challenge (i.e. implement a design solution).

During the design challenge, students work collaboratively in small groups to create an IoT-infused design solution. As a result, this phase generates a number of product ideas of IoT ecologies for a specific domain. In the design challenge, students do not use any technological tools, but instead they make use of design storyboards and cards, while the guidance they receive from the facilitator(s) is minimal. The students initiate the ideation phase using two boards, namely, "users" and "contexts traces", that help them to frame their idea by selecting target audience and set the context for their design solution. That is, the domain and the context of the problem are given to the participant students as a set of different pre-defined options about scenarios and personas. Next, students are supported by a set of 110 ideation cards (Mora et al. 2017) used for the design thinking activities such as idea storyboarding and idea pitching. The cards (Fig. 1) display elements from both the physical and the digital domains, including everyday things, user interface metaphors, and online services. Important to this phase are also the criteria cards. Those constitute a set of critical lenses that enable participants to reflect on the ideas generated at the end of the design phase. An example of a criteria card is "Utility: How useful and practical to use is the IoT concept? Does the concept solve a real problem for their users? Can you see it being used every day?". The criteria cards add constraints to the ideation process that might foster creativity. At the end of this phase, they pitch their idea while presenting their design solution to their peers and the facilitator(s)/tutor(s).

Crucial to the success of the Tiles workshop is the intermediate (third) phase that facilitates the transition from design to programming. In this phase, the themes cards are important. They describe provocative design missions, centered on human behaviors and desires. Example of themes are: "Sixth-Sense: Create an IoT concept that

Fig. 1. Tiles ideation cards used in workshop setting

gives its owner some kind of superpower, like new types of senses, perceiving new information, etc." and "Trojan Horse: Create an IoT concept that seemingly does one thing, but where the intention is to produce another, deeper effect."

Prototyping complements idea generation with a set of technical tasks that allow participants to hands-on experience the ideas generated, by using making as a process to both evaluate ideas feasibility and to find input to iterate on concepts. Prototyping is supported by electronic bricks (Fig. 2) that connect to online data sources and third-party IoT devices; they can be programmed with a simple textual language via a companion software toolkit described in (Mora et al. 2016). Prototyping allow participants to play with code and electronics in order to implement creatively the idea generated in hybrid physical/digital artefacts.

To that end, the electronics bricks (Fig. 2) provided by Tiles allow participants to focus on exploring the combination of simple pre-implemented functionalities with a simple JavaScript-based language. Especially for the programming part of the prototyping phase, the workshop facilitator supports students while they make use of the Tiles toolkit (explained in the next section), physical artifacts and javascript templates to program and test the design solution. Usually, the same IoT-infused design solution is exploited in the design challenge phase, but it could also be a pre-defined mission from the themes cards.

Resources and Materials Needed for the Tiles Game-Based Learning Scenario
The completion of the scenario requires core (i.e. obligatory) and secondary (i.e. subsidiary) hardware. The former involves a PC connected to the internet, a projector, and IoT squares. Examples of IoT squares are the TILES squares (http://tilestoolkit.io/), and micro:bit (http://www.microbit.org/). Tiles is an inventor toolbox that supports the

Fig. 2. Tiles prototyping tools

design and making of interactive objects for learning and playing. It allows non-experts to create complex and distributed physical interfaces. Similarly, BBC micro:bit is a pocket-sized computer that users can interact with, code, customise and control to bring their ideas and applications to life. With respect to secondary hardware, it depends on the specific ready-made IoT scenarios that the instructor wishes to demonstrate or that the students wish to attempt. Examples include a 3D printer, a laser cutter, drones, and so on. Regarding software, the scenario requires a programming environment that supports the development of the javascript code snippets. In addition, in order to play the game, the Tiles generator idea board is needed, since it supports the ideation and design process step-by-step. The Tiles primitive design cards and the Tiles generator idea board used to play the game and overcome the design challenge are available online under Creative commons license in the 'Tiles IoT cards' repository (https://github.com/simonem/tiles-IoTcards).

3 Tiles Inventor Game and Skills Cultivation

This section outlines the student competences that can be developed with the Tiles inventor game by suggesting conceptual mappings with the game-based learning skill framework that emerged from a comprehensive review conducted by Mishra and Foster (2007). The definitions of the skills are in line with this framework. In doing that, we explain from the game designer viewpoint the intended skills cultivation, focusing on how and why it happens using the Tiles inventor game. Also, we explain how students can be assessed with respect to these skills in the context of the game.

Practical skills refer to learning which aims at the cultivation of skills that are applicable to the real world or to authentic settings. Good games afford expertise development, innovativeness, creativity and other skills needed for the 21st century workforce. From this theme, we identified the following skills:

- Digital/technological literacy – this involves the IoT knowledge acquired, as well as the prototyping and programming skills cultivation.
- Innovative/creativity/design skills – intended as sparking design thinking and idea generation. In a previous empirical evaluation conducted by the authors (Mora et al. 2017), most of the student participants stated that they had put forward ideas they

would not have had without the cards and that the themes cards helped them to be creative. A collaboration rubric, like the rubric suggested in (Mokhtar et al. 2010), may be used for self-assessment or for peer feedback. It assesses domain-general views of creativity across three axes: novelty, resolution and style. The whole Tiles scenario focuses on design, which involves how students "make use of the materials and resources that are available to them at a particular moment to create their representation" (Sanders and Albers 2010, p. 8).

- Multirepresentional understanding and multimodal literacy/processing – this involves engaging students with multimodal representations: auditory, textual, visual and interactive. For example: the introductory PowerPoint presentation, the boards and the cards in the design phase in order to overcome the design challenge, the code snippets in the programming phase in order to create a prototype, the storyboard creation, the idea pitch, and so on.

Cognitive skills in most game environments are cultivated by a learning-by-doing approach, where knowledge and hands-on experiences are inextricably linked. The review reports that the proponents of GBL report immediate feedback, cooperation and collaboration, scaffolds, problem solving, exploration and curiosity in a risk-free environment and transfer of knowledge to new situations as the main game affordances. From this theme, we identified the following skills:

- Deductive/inductive reasoning: while themes cards foster divergent thinking, criteria support refining ideas toward converging on a concept that satisfies one or more criterion.
- Systemic thinking: the Tiles scenario provides a 'lightweight' simulation environment of IoT ecosystems where students can inquire, problem solve, test hypotheses and apply their ideas.
- Causal/complex/iterative relations: the students use the themes cards to find ways to challenge one's idea, then go back and refine the storyboard contents. This provides triggers to diverge by iteratively modifying and expanding previous ideas.

Social skills can be cultivated when players collaborate with each other or when they learn to work with others in gaming situations. This theme focuses on the development of interpersonal skills, cooperation and the development of identities. From this theme, the Tiles scenario focuses on the cultivation of communication and collaboration skills. During the Tiles scenario, the student groups work together towards the creation of their IoT-infused scenario of use. Students are using all the materials ('users' board, 'context traces' board, 'Tiles idea generator' board, Tiles primitive design cards) and co-operate in order to design their own scenario of use. The criteria embedded in the game design act as triggers for collaborative reflection. During the programming phase, each student group also co-operates with their instructor in order to program the behavior of the physical artifacts by writing the java script code. To assess student communication and collaboration, we can consider assessment of collaborative projects with collective outcomes using formative assessment mechanisms. As an example, the collaboration rubric mentioned above, originally suggested by Mokhtar et al. (2010), can be used in conjunction with students' peer feedback after the ideas pitch.

Motivation pertains to the affordances of game environments to cultivate the intrinsic learner motivation. In turn, this revolves around the motivational design principles (e.g. grant power, autonomy, and adjusted challenge levels). From this theme, we identified the following skills:

- Immediate feedback/scaffolds- the materials are intended to guide students to succeed in the design and the prototyping challenge and the instructors act as facilitators throughout the learning process. Their role is to scaffold the learning experience by using different techniques, like explaining, providing examples, demonstrating, discussing, using metaphors, and encouraging the students. At the end of the design phase the student assessment on subject matter aspects can take place via a knowledge quiz. The questions can be integrated in a classroom response system, like kahoot!, which gives automated feedback to all participants after the time limit in each of the multiple questions. This can facilitate immediate feedback and reflection, student monitoring, and in-class discussion in tandem with game-like features (like competition, rewards and so on). The Kahoot! knowledge quiz is available online at https://play.kahoot.it/#/k/36df1589-fcff-4c5f-a03c-8d8190bfeecc.
- Control/choice/autonomy/clear goals- at the beginning of the scenario, the students go through the Tiles cards in order to check that they have understood their meaning and purpose of use in the game. Then, the instructor facilitates the process by providing feedback on an as-needed basis; in this case explaining to the students the meaning of those cards that they do not know or understand (probably due to lack of background knowledge). The problem is open ended, and the students are bounded only by the procedural rules of the game, there is no expert solution.

4 The Tiles Game-Based Learning Design Pattern

This section presents the design part of the Tiles Inventor game using the generic pattern template and the specific case study template, which are recommended by Ecker et al. (2011) particularly for GBLDPs. That is, the Tiles Inventor Game is manifested as a case study of the generic underlying GBLDP proposed. The basic structural elements of the pattern template are:

1. Formal aspects, which capture information about the pattern's history, contact details of the pattern creator(s), and so on; all this information aim to help readers, among others, advancing the pattern in a responsible way
2. Aspects of content, which involve crucial characteristics the learning situation
3. Conceptual aspects, which aim to externalize important background information of the learning experience
4. Examples and references to better understand the pattern.

GBLDP: design your IoT solution	
Formal aspects	
Author(s)	Mora S., Divitini M. and Gianni F. (2016)
Version	1.0
Date	12 September 2017
Aspects of content	
Basic problem	It is an open ended design problem on IoT applications. The pattern is: (1) Initiate the game, (2) Ideate your IoT solution, (3) Design your IoT solution iteratively
Approach	It covers the first two steps of the IoT application development process: (1) idea generation, (2) design, (3) prototyping, and (4) 3D print and assembly
General description	Game–based learning and learning by design
Conceptual aspects	
Educational concept	The pattern includes a series of design thinking and reflection activities performed collaboratively in a workshop setting. It aims at co-creating IoT–related concepts that are validated against several criteria, like creativity and feasibility, while a number of learning outcomes among workshop participants are facilitated
Learning objectives	The students will be able to: (a) define in their own words the concept of IoT (b) recognise IoT applications (c) summarise the basic elements associated with the concept of IoT (d) design an IoT-infused solution (e) judge the efficiency of their solution based on specified criteria (f) reflect on their progress
Activated cognitive skills	This pattern links with the Revised Bloom Taxonomy (Anderson et al. 2001): understand factual knowledge, create procedural knowledge, evaluate procedural knowledge, and evaluate metacognitive knowledge
Requirements	It requires an introductory activity to IoT and to the design challenge
Potential problems in using	Students might have misconceptions about IoT
Advise on the application	For the students' group formation, it is advisable that the students within one group have complementary skills
Examples and References	
Related patterns	'Create an artifact', 'Collaborate', 'Reflect' from the DMP, http://www.edu-design-principles.org/dp/designPatterns.php
Application sample(s)	Makerfairs, Hackathons, Designathons etc.
References	Divitini et al. 2017

The ensuing case study which is based on the proposed pattern is shown below.

Case study: The Tiles IoT design Game	
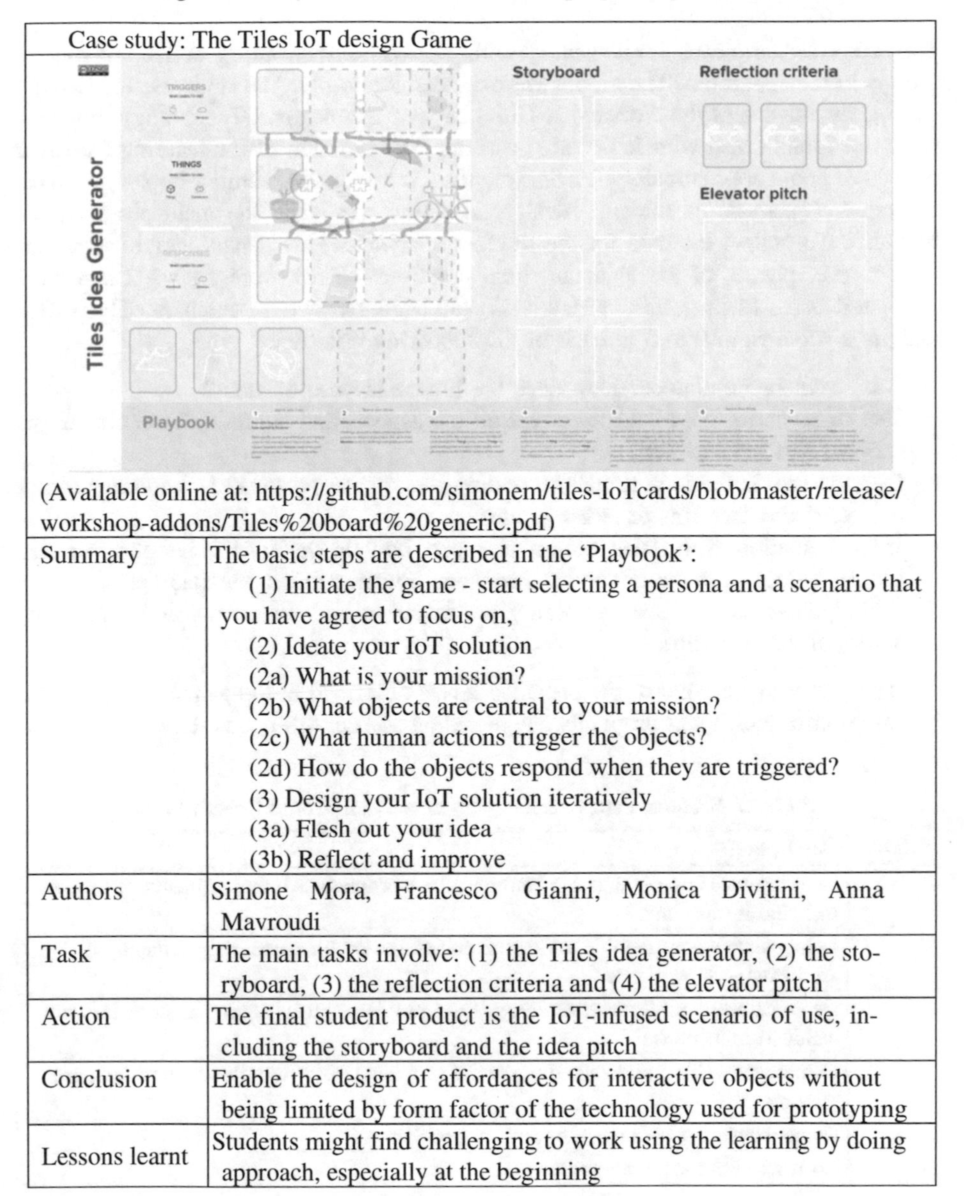 (Available online at: https://github.com/simonem/tiles-IoTcards/blob/master/release/workshop-addons/Tiles%20board%20generic.pdf)	
Summary	The basic steps are described in the ‘Playbook’: (1) Initiate the game - start selecting a persona and a scenario that you have agreed to focus on, (2) Ideate your IoT solution (2a) What is your mission? (2b) What objects are central to your mission? (2c) What human actions trigger the objects? (2d) How do the objects respond when they are triggered? (3) Design your IoT solution iteratively (3a) Flesh out your idea (3b) Reflect and improve
Authors	Simone Mora, Francesco Gianni, Monica Divitini, Anna Mavroudi
Task	The main tasks involve: (1) the Tiles idea generator, (2) the storyboard, (3) the reflection criteria and (4) the elevator pitch
Action	The final student product is the IoT-infused scenario of use, including the storyboard and the idea pitch
Conclusion	Enable the design of affordances for interactive objects without being limited by form factor of the technology used for prototyping
Lessons learnt	Students might find challenging to work using the learning by doing approach, especially at the beginning

5 Method

The Tiles scenario was implemented with 12 freshmen working collaboratively in groups in the classroom. The groups were created randomly. The students are studying Computer Science at the Norwegian University of Science and Technology, but they were not familiarized with IoT-related aspects. The duration of the learning endeavor was three hours approximately. To analyse the Tiles scenario learning impact, we used the artefact analysis of students' design solutions created in the main phase of the scenario. To collect the data for the artefact analysis we used photographical observations (i.e. photos of the students' final solutions). They were analysed against a dedicated assessment rubric, which was validated by two researchers. The rubric assigns a score from 0 to 5 in each of the following criterion:

1. Data exchange/online services: does the idea make use of them?
2. Use of scenario: does the idea cover all the aspects and address all the challenges proposed in the scenario?
3. Persona use: is the idea actually targeting the chosen persona? Is it too generic or too restrictive in terms of target group?
4. Implementation with Tiles design element: how is the idea using the required functionalities in terms of services, sensors, human actions, feedback etc.?
5. Participation: is the idea promoting participation and collaboration between the actors of the scenario?

The scores correspond to performance levels described in the generic template for holistic rubrics which was originally suggested by Mertler (2001), as shown in Table 1.

Table 1. Students performance levels description from Mertler (2001)

Score	Description
5	Demonstrates complete understanding of the problem. All requirements of task are included in response
4	Demonstrates considerable understanding of the problem. All requirements of task are included in response
3	Demonstrates partial understanding of the problem. Most requirements of task are included in response
2	Demonstrates little understanding of the problem. Many requirements of task are missing
1	Demonstrates no understanding of the problem
0	No response/task not attempted

6 Results

The idea of the first student group aims at connecting people and making new friends through social media. It describes an IoT-design solution that promotes co-located - yet distributed in the smart city- online communities by using proximity (and other) sensors in tandem with social media outdoors to connect with other people in the smart city. The second idea describes a smart watch that helps children adopt an eco-friendlier behavior indoors. For example, the smart watch vibrates when the persona, who is a six-year-old girl, forgets to turn off the lights whenever she leaves the room. The third idea is targeting Tom, a 43-year-old man who lives on a wheelchair, and it suggests the use of environmental sensors which adapt the wheelchair to the environment conditions. For example, the wheelchair has an autopilot system with GPS. Finally, the fourth student team designed a smart bin which is helping children to recycle better.

Concerning artefact analysis, Table 2 presents the scores of the student solutions against the predefined criteria of the assessment rubric mentioned above.

Table 2. The scores of the student solutions against the predefined performance indicators

Student group	Criterion 1	Criterion 2	Criterion 3	Criterion 4	Criterion 5	Total score
1	3	3	3	4	4	68%
2	5	4	4	4	2	76%
3	4	4	3	5	2	72%
4	5	4	5	3	2	76%

7 Conclusions

The paper suggests a GBL paradigm for learning about IoT based on previously validated templates and prior research work. The empirical results of this paper and the results mentioned in (Mora et al. 2017) have shown that the Tiles IoT inventor game can promote an effective teaching methodology for the basic design and development aspects of the IoT applications students. The scores of the artefact analysis, except for the scores of the fifth assessment criterion, are encouraging, considering that the students were not familiarized with IoT. The scores of the fifth criterion corresponding to the social aspect of the design solution of the students' artefacts suggest that it is more difficult to cultivate attitudes of IoT-related issues (such as collaboration and digital citizenship), than skills development (such as design skills) with the Tiles learning scenario.

The Tiles IoT inventor game incorporates attributes that have been identified as important pedagogical affordances of GBL (Kim et al. 2009; Pivec and Dziabenko 2004): (a) it encourages learners to combine knowledge from different areas in order to create their IoT-infused design solution or to make a decision at several points, (b) it promotes open-ended problem solving by enabling learners to test how the game outcome is dependent on their decisions and actions, and (c) it encourages learners within groups to discuss and negotiate subsequent steps and, in turn, cultivate their social skills. There is a growing body of educational environments that make use of Internet of Things, ubiquitous technologies or mobile computing, with very encouraging results. Luckily, the International Conference on Interactive Mobile Communication, Technologies and Learning (ICML) presents numerous such examples every year. Yet, this paper is advancing our views on the topic assuming that the citizens of tomorrow will not only have to make use of the technology, but they will actually design it (Resnick 2002).

The contribution of the paper is twofold. For the practitioners, it presents a generic LD pattern for GBL on IoT (Initiate the game, Ideate your IoT solution, and Design your IoT solution iteratively) that they could use in their respective contexts. In doing that, it identifies and explains the associated student competences, while creating conceptual mappings of students' competences with a dedicated framework for GBL and making use of previously validated templates. Also, practitioners could run the Tiles inventor game with their students by studying this paper and making use of the Tiles inventor game materials which are available online under creative commons license. The contribution of the paper for TEL research pertains to the creation of 'intermediate level design knowledge' on TEL which is much needed nowadays. It has been recently suggested that luckily the field of TEL has reached a maturation point where there exist many successful design instances and also enough theoretical frameworks that can guide our research, but not enough knowledge 'in the middle space', in this case a GBLD pattern (Prieto et al. 2017). This 'intermediate' type of TEL knowledge can evolve to a strong TEL concept provided testing in diverse contexts (Prieto et al. 2017; Sharma et al. 2017), which is in line with our future work.

Acknowledgements. The research work is funded by the EU project titled 'UMI-Sci-Ed - Exploiting Ubiquitous Computing, Mobile Computing and the Internet of Things to promote Science Education' (H2020-SEAC-2015-1).

References

Anderson, L.W., Krathwohl, D.R., Airasian, P., Cruikshank, K., Mayer, R., Pintrich, P., Raths, J., Wittrock, M.: A Taxonomy for Learning, Teaching and Assessing: A Revision of Bloom's Taxonomy. Longman Publishing, New York (2001). Artz, A.F., Armour-Thomas, E.: Development of a cognitive-metacognitive framework for protocol analysis of mathematical problem solving in small groups. Cogn. Instr. **9**(2), 137–175 (1992)

Breuer, J.S., Bente, G.: Why so serious? On the relation of serious games and learning. Eludamos: J. Comput. Game Cult. **4**(1), 7–24 (2010)

Divitini, M., Giannakos, M.N., Mora, S., Papavlasopoulou, S., Iversen, O.S.: Make2Learn with IoT: engaging children into joyful design and making of interactive connected objects. In: Proceedings of the 2017 Conference on Interaction Design and Children, pp. 757–760. ACM, June 2017

Ecker, M., Müller, W., Zylka, J.: Game-based learning design patterns: an approach to support the development. In: Handbook of Research on Improving Learning and Motivation through Educational Games: Multidisciplinary Approaches: Multidisciplinary Approaches, p. 137 (2011)

Huynh-Kim-Bang, B., Wisdom, J., Labat, J.M.: Design patterns in serious games: a blue print for combining fun and learning. Project SE-SG (2010). http://seriousgames.lip6.fr/DesignPatterns/designPatternsForSeriousGames.pdf

Hwang, G.J., Tsai, C.C.: Research trends in mobile and ubiquitous learning: a review of publications in selected journals from 2001 to 2010. Br. J. Educ. Technol. **42**(4) (2011)

Kali, Y.: Collaborative knowledge building using the design principles database. Int. J. Comput.-Support. Collab. Learn. **1**(2), 187–201 (2006)

Kim, B., Park, H., Baek, Y.: Not just fun, but serious strategies: using meta-cognitive strategies in game-based learning. Comput. Educ. **52**(4), 800–810 (2009)

Mavroudi, A., Miltiadous, M., Libbrecht, P., Müller, W., Hadzilacos, T., Otero, N., Barth, K., Georgiou, K.: Let me do it: towards the implementation of sustainable instructional patterns. In: 2016 IEEE 16th International Conference on Advanced Learning Technologies (ICALT), pp. 414–415. IEEE, July 2016

Mertler, C.A.: Designing scoring rubrics for your classroom. Pract. Assess., Res. Eval. **7**(25), 1–10 (2001)

Mishra, P., Foster, A.N.: The claims of games: a comprehensive review and directions for future research (2007, online submission)

Mokhtar, M.Z., Tarmizi, R.A., Ayub, A.M., Tarmizi, M.A.: Enhancing calculus learning engineering students through problem-based learning. WSEAS Trans. Adv. Eng. Educ. **7**(8), 255–264 (2010)

Mora, S., Gianni, F., Divitini, M.: Tiles: a card-based ideation toolkit for the internet of things. In: Proceedings of the 2017 Conference on Designing Interactive Systems, pp. 587–598. ACM, June 2017

Mora, S., Divitini, M., Gianni, F.: TILES: an inventor toolkit for interactive objects. In: Proceedings of the International Working Conference on Advanced Visual Interfaces, pp. 332–333. ACM, June 2016

Pivec, M., Dziabenko, O.: Game-based learning in universities and lifelong learning: UniGame: social skills and knowledge training game concept. J. Univ. Comput. Sci. **10**(1), 14–26 (2004). http://www.jucs.org/jucs_10_1/game_based_learning_in. Accessed 17 Sept 2017

Prensky, M.: Digital Game-Based Learning. McGraw-Hill, New York (2001)

Prieto, L.P., Alavi, H., Verma, H.: Strong technology-enhanced learning concepts. In: European Conference on Technology Enhanced Learning, pp. 454–459. Springer, Cham, September 2017

Resnick, M.: Rethinking learning in the digital age (2002). https://llk.media.mit.edu/papers/mres-wef.pdf. Accessed 8 May 2017

Sanders, J., Albers, P.: Multimodal literacies: an introduction. Literacies, the Arts and Multimodalities, pp. 1–43 (2010)

Sharma, K., Alavi, H.S., Jermann, P., Dillenbourg, P.: Looking THROUGH versus Looking AT: a strong concept in technology enhanced learning. In: European Conference on Technology Enhanced Learning, pp. 238–253. Springer, Cham, September 2017

Adaptive Mobile Environments

Ontological Support for Teaching the Blind Students Spatial Orientation Using Virtual Sound Reality

Dariusz Mikulowski(✉) and Marek Pilski(✉)

Faculty of Science, Siedlce University of Natural Sciences and Humanities, Siedlce, Poland
{dariusz.mikulowski,marek.pilski}@ii.uph.edu.pl

Abstract. The teaching spatial orientation of blind children currently takes place through traditional methods such as orientation courses in artificial environments with objects set in them. This paper proposes a virtual sound reality based on the use of semantic techniques such as specially created ontologies and relevant application machines that could considerably modernize these methods. Such virtual sound reality can be used to create systems for teaching the spatial orientation of the blind with appropriate software.

Keywords: Blind children · Spatial orientation · Sound reality Ontology

1 Introduction

One of the key skills that is necessary for independent functioning of a blind person in society is ability to exploration of an open environment on his own. In order to be able to move independently, the blind pupil has to master skills of recognizing obstacles, hearing their surroundings, recognizing dangerous elements of space such as moving cars etc. They usually do this using their senses i.e. hearing and touch and phenomenas like availability of hearing the sounds reflected from obstacles. They also benefit from the traditional teaching methods such as assistance of specially trained spatial orientation instructors who help them master these essential skills.

There are a few various electronic systems that support the orientation of such people, which are usually based on adapted GPS devices and their respective software [1]. However these systems help to move blind people in the real world, but they are suitable for people who have the experience and skills that are necessary to move independently. Therefore they are not sufficiently suited to apply them in process of teaching of beginner students who must master the basic skills of recognizing dangerous elements of space.

It seems that one of the ideas to solve this problem could be creating an artificial reality model available to the student in the form of sound and vibration signals. Continuing the research we had conducted in the framework of the

M. E. Auer and T. Tsiatsos (Eds.): IMCL 2017, AISC 725, pp. 309–316, 2018.
https://doi.org/10.1007/978-3-319-75175-7_32

project Blind-enT [2], we would like to propose the development of such sound model for the blind. We call it as a Virtual Sound Reality (shortly VRS). It should be smart enough, what suggests the use of artificial intelligence methods, such as semantic support by ontologies. Let us notice, that the construction of VRS should begin with the perception, that sighted people recognize their surroundings mainly through graphic information reaching them while blind people perceive reality through various sounds and acoustic phenomena associated with them. So to create this VRS, the development of three dedicated ontologies will be needed. The first one would be an ontology of graphical objects which is needed for creating the component of reality that is seen from perspective of sighted trainer. The second one would be an ontology of sound objects. It will serve creating a reality component perceived from a blind pupil perspective. The final important element that binds both ontologies mentioned above, will be the founding ontology that will collect general terms. It is needed for automatic inference a model described in graphical ontology to the adequate model expressed in terms of sound ontology.

2 The Related Works

Taking into account current research in terms of navigational aid for the blind it would seem that it could be divided into several groups. For example the first division may include systems operating in known or discovered environment. The next breakdown can be based on the type of devices used, e.g. GPS, odometry, RFID tags, camera, ultrasonic perception, etc. It turns out that most of the solutions are a combination of many of them.

Simoes and de Lucena [3] propose an indoor navigation system include audio guided built in a wearable device designed to work with a hybrid mapping. This mapping is based on two kinds of markers: radio frequency and visual markers located inside known environment. This system allows the blind users the safe guided navigation. The system works in two stages. In first stage the indoor mapping is made through construction of markers to generate a contextual database that increases the quality of location indication. The second stage, where the indoor navigation is performed, is based on the proximity method, visual pattern recognition, odometry, and ultrasonic perception of barriers.

Similarly Sammoud and Alrjoub [4] propose a mobile navigation intended for blind students, employees, or guests within King Saud University campus zone. The solution includes many components such as: the blind mobile device, RFID tags and Reader, GPS, text to speech, voice recognition, and Wi-Fi. The system detects the blind location using GPS, if internet connection is available, and uses RFID tags fixed outdoors and indoors on the building in the path, Wi-Fi routers are used indoors to detect the location. The system uses voice recognition and Synthetic speech to communicate with the blind user to give him the information about propper directions.

Another solution, proposed in [5] by Jelonkiewicz and Laskowski is a system allowing to assist the blind users in unknown environment by application stereovision based depth analysis module (camera), to recognize e.g. familiar persons and read texts on boards or signposts. All information gathered by stereo-vision module and processed by the system about obstacles in surrounding, result of face recognition, texts in visual field are passed by tactile interface and the voice system.

A similar to [5] project was presented in [6] by Kozik. Its system also uses the stereo camera and image processing algorithms to facilitate its user with object detection and recognition mechanisms. The system gives blind user the information about obstacles located in the discovered environment. The risk assessment based on ontology problem modeling allows to handle the risk, predict possible user's moves and pro-vide the user with appropriate set of suggestions that will eliminate or reduce the discovered risk.

Let us assume here, that most of the approaches mentioned above attempt to solve the problem of navigating a blind user in a real world environment. Almost none of the solutions address the problem of teaching a blind child to move separate so that it is easier for him to travel in the real world. An exception may be a project proposed by Maidenbaum and Amedi [7]. Authors describes several games which were developed in order to increase the traditional White-Cane with additional distance and angles. The role of the player is to solve simple virtual tasks such as finding the exit from a room or avoiding obstacles while wearing blindfolds.

We would like to combine these two recently mentioned elements and construct a technology that allows the teaching of blind children to orientation knowledge in real-world situations using a set of specially created ontologies.

3 The Ontological Support for Teaching Spatial Orientation

It is obvious that blind people perceive reality through their surrounding in another way then sighted one. They are able to recognize the different kinds of sounds, and the direction from which these sounds reach them. This type of sound is known as a bineural sound. Due to this, blind persons have an ability to revolt 3 dimensional model of a space fragment in their imagination. Another phenomenon is used by blind so that sound, published by user first, get on the obstacle and then return to the person due to the fact that the blind are able to detect large obstacles such as walls even from a distance of several meters. It seems that such phenomena could be also used in a computer simulation of artificial sound reality (VRS). Such sound artificial reality could be used to build systems that facilitate, for example, training of spatial orientation of blind children.

Let us notice, that construction of VRS should begin with the observation that sighted people perceive the world based mainly on graphical information, while blind people focus on different soundstages. For this reason, it seems that it is necessary to provide a way to automatically process a fragment of reality expressed graphically into its sound equivalent.

3.1 The Sound Reality Construction

To build VRS based system the following three components will be need:

1. A founding ontology, which collects basic data about the objects perceived by the user while browsing through the environment; This founding ontology is a model (a map), consisting of common objects, describing the user's environment.
2. Representation of objects from the founding ontology need to create a graphical model of a fragment of reality the sighted trainer can to make.
3. Representation of objects from the founding ontology need to create a sound model of fragment of reality the blind user can use to explore his environment.

Let us notice here, that every piece of graphical representation of virtual reality should to has an associated fragment of reality expressed in terms of sound objects. In addition, both representations use the same objects with the founding ontology and the corresponding inference machine which converts one representation to another one.

But how VRS can be used to teach a spatial orientation of the blind person?

- Let us imagine, that a blind student wants to learn how to travel from home to school on their own. The route connecting these two places passes through two crossroads and three fragments of the streets. So to pass the route the student must trail following steps:
  1. To leave the house, turn left and walk down the street to the nearest crossroad,
  2. To go through the pedestrian crossing and then turn left;
  3. To walk to the next crossroad and do not crossing the road then turn right;
  4. To pass the section of the street and after approximately 200 m way from the right side of the entrance to the school building.

  The route described above is presented schematically in Fig. 1.
- To learn this route, the student turns to his trainer of spatial orientation, which has an access to VRS based system.

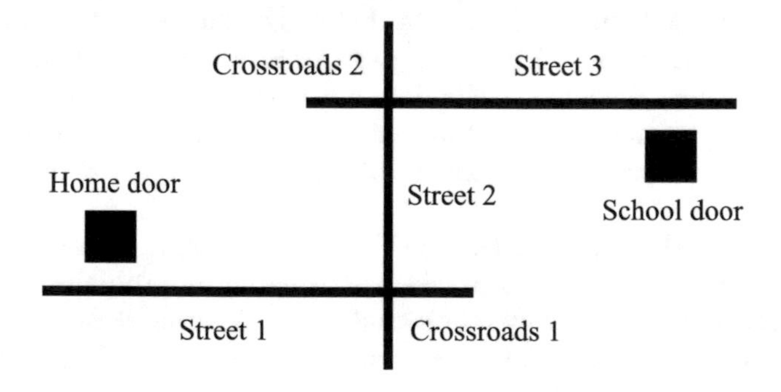

Fig. 1. Route from house to school

- The trainer creates a model of the route system by adding objects of streets, crossroads and doors of buildings;
- Then route model is converted from graph form in the form of sound in such a way that each graphical object is transferred to an analog signal. For example, a fragment of the street is converted into the sound of passing cars in the air in right channel of stereo and the sound of footsteps, broadcasting on both channels.
- The student is equipped with a special helmet that allows you to broadcast sounds bineuralnych and approximate white cane, which is able to radiate the energy of vibration as well as a mobile phone camera.
- Then the trainer starts the application that plays to the student's sounds, describing the route. In addition, the trainer explains to the student the individual sections of the route. In this way the student can pass through virtual route using its sound patterns and tips of the trainer. During this virtual journey student being watched through a webcam, allowing at any moment to stop or to repeat a section of the route. More over, the trainer has a view of students position on the route at any time he wants.
- Now, after exploration almost the entire route, the student can safe to go to her real introduction without the help of our system. However, the route is known to him and thanks to the use of the training he will be able without problems to overcome the actual route on their own.

3.2 A Founding Ontology for Virtual Reality

The basic element needed to build a VRS is a special founding ontology, gathering the concepts required to create a shared object map of the environment. This map should have a hierarhical structure in order to be able to easy handle it and automatically create the routes. On the upper level of this structure may be such facilities as a town or district, and below objects such as street or crossroad. On the lower level the objects, such as pavement, lawn, wall, building entrance, column or trash can be found. They are needed because they can be important milestones for navigation for a blind user.

These all kind of objects on the map must have different properties that describes their parameters, such as size or shape. The choice of these parameters needs to be implemented correctly, so that contained important information from the point of view of a blind user.

In addition to descriptions of objects in the founding ontology the corresponding relations should be passed into it. First of all it is a descendent relation that makes all hierarchical structure of a map. Other relations will be dependencies such as: *is_near_to*, *is_over*, *is_under*, *is_on_left*, *is_on_right* etc. They are important to determine the location of the object that represents the user in a particular location of the route. They are usefull also to track user during his navigation along the route.

Due to the structure described above, the founding ontology will allow to define routes between different objects, and will make the posibility to track the user during his movement along the route.

3.3 A Graphical Model for Virtual Reality

Such a basic ontology representing the map should be available to the spatial orientation trainer who should be able to add objects into map. He should be able also to edit existing objects in a route. Therefore, we need a graphical representation of the objects from the map and the appropriate software, which allow the trainer to manipulate of them. The most convenient way for the trainer would be probably to take pictures of part of the route, or to download them from web map, uploading these photos to the system based on VRS and creating thus a model of the route. But we understand well that the implementation of the application with such functionality is not a simple matter at least because of the difficulty in fairly accurate image recognition. So to make it easier we can accept that In a system based on VRS we apply a bit more simple solution. It can rely on the fact that we have a set of basic graphical objects (icons), such as: types of supports, roadway, wall, building entrances, cars and complicated objects, such as icons of the streets, crossroads, etc. Using a suitable GUI the trainer can easely compose these components into the hierarchical structure of a route. For example, if he wants to build the route shown in Fig. 1, he can connect 3 fragments of the streets and 2 the crossroad at the same hierarchy level. Then he can add snippets of streets, objects, walls, pillars, lawns, sidewalks and building entrances at a lower level. After creating the route the trainer can add relationships between objects, for example the relationship suggests that part of the street S1 has a direction East-West and a fragment of S2 has a street direction North-South. After reviewing all of the route the trainer can maintain its structure in the basic ontology of the object map.

3.4 A Sound Ontology for VRS

The structure of the route stored in the ontology map can be transformed into representation suitable for the perception of the blind user. As well as map representation for the trainer consists of graphical objects, so the equivalent model designed for the blind consists of specially developed sound objects.

But how a blind student can recognize the fragment of a route that he currently passing through? Let us imagine that user stopped at the crossroad before pedestrians crossing. When the student stands in front of the crossroad and there's light turns red, he hears a few sounds informing him about this. He hears the sounds of vehicles moving from left to right and moving from right to left in front of him. In addition, he also hears the signal of the light, but this sound comes to him from a great distance on the left hand side. At the moment when the light turns green, also the sounds coming towards him are changing. Instead of the passing cars sounds in front of him, the student hears the sounds of vehicles moving from back to front and moving front to back that comes from his left hand side. Similarly, instead of the distant sound of traffic the student begins to hear a louder signal coming from close range at his left side. When he begins to cross the road this sound goes away and the next sound of an impending traffic signal he is hearing. This signal comes to him from the right side with

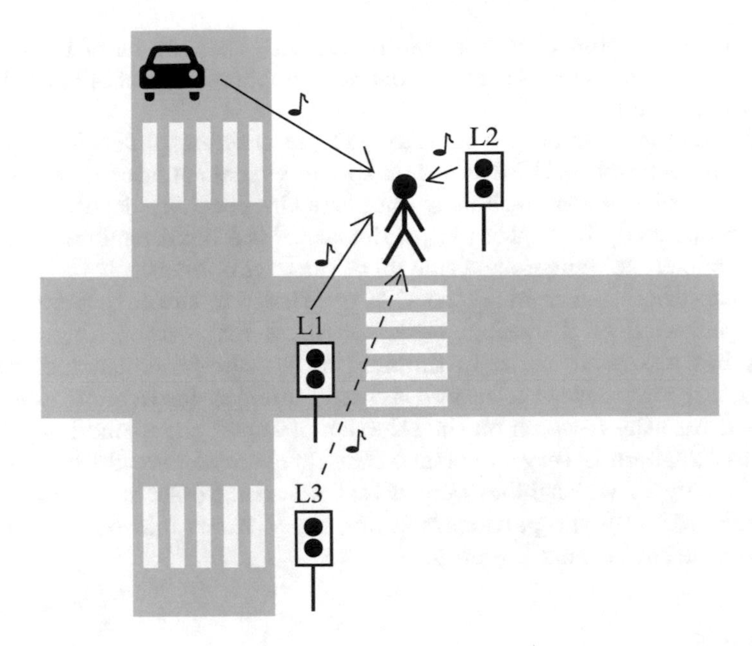

Fig. 2. Fragment of a route

close distance. It is the lights sound, located on the other side of the street, through which student actually passes through. The situation described above is schematically presented in Fig. 2.

As we can see, the environment that is described with a basic ontology should be presented to the blind through the appropriate sounds. For example, a graphic of a traffic light post is shown as the sound coming to user from the right side, distance and some time. At a similar way, this passing car object described in the example above, is presented to the user as the corresponding sound slighting from the left channel to the right. Such sound objects needs to be stored in the sound representation of the ontology for VRS. Each of these sound objects must be associated with its counterpart in the underlying ontology and in this way indirectly with his graphical representation. Also the relations between the individual objects and the relationship between the object representing the user and the objects included in the route are very important. During the movement of the user along the route, these relations must be changed, which caused changes in the sounds that come to him as we explained in above example.

4 Summary

The blind people face difficulties in independent movement, both in the outdoor environment and public buildings. Fortunately, they are compensated by abilities of the blind to recognize of various types of audio signals or hearing sounds that are reflected from obstacles. Moreover, with the advent of modern technologies (GPS devices or cameras) appeared an opportunity of enhanced spatial orientation of blind people.

In this article, a solution that can assist in training blind children spatial orientation was proposed. It is based on special virtual sound reality VRS which is designed with semantic technology in particular specially devoted ontologies. This approach could be helpfull in particular for the blind children before they travel to unlock the dangerous elements of the environment.

The ontologies used to build the VRS mentioned in this article are currently in development stage. The main work associated with establishing a founding ontology has just been made. In the near future, the other parts of VRS, i.e. the graphical representation as well as part intended for a blind user will be developed. Also the research on the selection of appropriate sound objects have been started. There is very important that these sounds should best to reflect the true reality in which blind person really works. So such studies have also been conducted with the participation of blind students and blind members of the team that implements the project.

References

1. Montanha, A., Escalon, M.J., Dominguez-Mayo, F.J., Polidorio, A.M.: A technological innovation to safely aid in the spatial orientation of blind people in a complex urban environment. In: International Conference on Image, Vision and Computing (ICIVC 2016) (2016)
2. Ambroszkiewicz, S., Bartyna, W., Mikulowski, D., et al.: Blind-enT: making objects visible for blind people. Nr 975 Institute of Computer Science Polish Academy of Sciences, Warszawa, Poland (2004)
3. Simoes, W.C.S.S., de Lucena Jr., V.F.: Hybrid indoor navigation assistant for visually impaired people based on fusion of proximity method and pattern recognition algorithm. In: IEEE 6th International Conference on Consumer Electronics - Berlin (ICCE-Berlin), 05–07 September 2016, Berlin, Germany (2016)
4. Sammouda, R., Alrjoub, A.: Mobile blind navigation system using RFID. In: 2015 Global Summit on Computer and Information Technology (GSCIT), 11–13 June 2015, Sousse, Tunisia (2015)
5. Jelonkiewicz, J., Laskowski, L.: System for independent living - new opportunity for visually impaired. In: Artificial Intelligence and Soft Computing, Part II, 11th International Conference on Artificial Intelligence and Soft Computing (ICAISC) 29 April–03 May 2012, Zakopane, Poland. Lecture Notes in Artificial Intelligence (2012)
6. Kozik, R., Burduk, R., Kurzynski, M., Wozniak, M., Zolnierek, A.: Stereovision system for visually impaired. In: Computer Recognition Sysyems 4. Advances in Intelligent and Soft Computing (2011)
7. Maidenbaum, S., Amedi, A.: Blind in a virtual world: mobility-training virtual reality games for users who are blind. In: IEEE Virtual Reality Conference (VR) Proceedings of the IEEE Virtual Reality Annual International Symposium, 23–27 March 2015, Arles, France (2015)

Participatory Design with Dyslectics: Design and Evaluation of an Enhancing Reading Skills Tool

Panagiota Vangeli(✉) and Jan Stage

Aalborg University of Computer Science, Aalborg, Denmark
panvangeli@gmail.com, jans@cs.aau.dk

Abstract. Participatory Design (PD) was used successfully in many projects but the question is how participatory design works with people with a cognitive disorder like dyslexia. In this study, we analyzed observations on PD sessions with dyslectic participants for developing designs of a reading software application by applying two participatory design methods: the IDEAS and CI methods. Furthermore, we conducted online surveys to gather information on dyslectics participants' and their special-education teachers' opinion regarding the participatory design process, methods and final designs. The results indicate that participatory design works effectively with dyslectic people provided the participation of Proxy Users to represent dyslectics, when it is necessary, the participation of an experienced on dyslexia facilitator who has the knowledge to address incidents caused due to dyslexia, and a proper allocation of the groups in proportion to the required tasks to prevent biases.

Keywords: Design Techniques · Cooperative design · Participatory design
Participatory Design (PD) methods · Dyslexia · Interactive design
Cooperative inquiry method · IDEAS method · Grounded Theory

1 Introduction

To begin with the PD (or alternatively cooperative-design, or brugerinddragende design), it has its origins in Scandinavia [1]. PD is an approach that requires the involvement of products' potential users as participants to the design process. It has been characterized as an innovative co-design methodology, since participants are focused on finding a solution through investigating a problem. The participants finally evaluate the solution in order to validate that the final solution meets their expectations [2]. PD methodology is being applied, to a large extent, in pedagogical experiments of Human-Computer Interaction (HCI) research [2]. Based on participatory design theory, this approach is an iterative process of 'making' things tangible by using verbal and visual tools, 'saying' (discussing) about existing related techniques-tools-practices, and 'acting' or alternatively 'developing through practicing' [3].

The usefulness of participatory design is that it increases the chance of the final design to meet the expectations of the product's users [4]. Moreover, by using participatory design method developers can avoid a common mistake of designing an

M. E. Auer and T. Tsiatsos (Eds.): IMCL 2017, AISC 725, pp. 317–330, 2018.
https://doi.org/10.1007/978-3-319-75175-7_33

application for themselves instead of designing it for the product's users [5]. Throughout participatory design, participants express their thoughts and perspectives on a design problem, and by cooperating with each other, they have the chance to build a design, which will finally correspond to their expectations and aims [6]. Dyslexia, as a learning disability, prevents dyslectics to read or write fluently. This situation can deeply affect dyslectics' life by making often their adoption to the community difficult; the result is for dyslectic to isolate themselves and lose the interest of living a normal life [7]. Studies have proved that assistive technology has helped dyslectics improve their reading skills [8, 9]. However, the majority of studies and experiments have been focused on children, even though cognitive impairments, like dyslexia, last for a lifetime and a high number of adults struggle with reading or writing due to dyslexia [10]. The above factors and the academic contribution of a research on participatory design with dyslectic participants were the mainspring of this empirical study. After a systematic research, we realized that studies on participatory design with dyslectic participants are very limited, even though dyslexia is a cognitive impairment, which strongly affects dyslectics' daily routine. With this study, we aim at contributing future studies on participatory design with dyslectic participants.

2 Related Work

For the purpose of this study, we have searched for methods that have been applied for participatory design involving people with other cognitive impairments than dyslexia, because of the fact that we were not able to find any method of participatory design with dyslectics. We selected two methods, namely IDEAS and CI methods, already used on participatory design for people with a certain cognitive impairment of Autism Spectrum Disorders (ASD). Both selected methods consist of specific features appropriate for people with cognitive impairments, such as ASD. At this point, it is important to make clear that there are many similarities between Dyslexia and Autism, since Dyslexia and Autism are relative cognitive impairments. That means that people suffered by autism often are diagnosed with dyslexia [11, 12]. To begin with, the two selected participatory design methods consist of the four steps of an Introduction, Discussion, Brainstorming and Design. The main difference between each other is that the IDEAS method makes use of High-tech materials, while the CI method is more traditional than the former and it focuses strictly on Low-tech materials. The presence of a facilitator is required in both methods. Below there is a thorough analysis of both participatory design methods, applied to autistic participants.

In the IDEAS (Interface Design Experience for the Autistic Spectrum) method [13], initially the facilitator makes an Introduction of the participatory design study to the participants by using High-tech materials, e.g. videos, to inform them about the concept of the study, its goals and the process. On the Discussion step based on the content of the Introduction, the participants have to discuss and criticize the existing technologies for users with ASD that they watched on the video and talk about the presented technologies' designs. Afterwards, in the Brainstorming step, the participants generate, write as many design ideas as possible for a new technology for them, which they grade then. The final step, based on the previous brainstorming step, is the Development of the top design idea, which has been rated by the participants as the most efficient design idea for the autistic participants. In this step, the participants can use High-tech materials, e.g. computers, internet, videos, smartphones etc. in order for them to be inspired to generate as much efficient paper designs as they can. For developing paper designs, facilitators provide the participants supportive Low-tech materials e.g. post-it notes, color pens, markers, and pre-printed plastic icons.

Regarding the CI (Cooperative Inquiry) participatory design method [14], the whole process is taking place without any use of High-tech materials. To be more precise, at the Introduction step the facilitator presents verbally one or two existing software applications for users with ASD. The Discussion step is this, where the participants discuss about the presented, by the facilitator, existing software applications for users with ASD. Then, at the Brainstorming step, they write their thoughts about the previously presented pieces of software for autistic users, and they notice likes, dislikes or possible improvements. The goal of this step is for the participants to generate a variety of design ideas through combining their likes, dislikes and proposals for improvements. After ending this step, participants write grade their ideas. On the final step of the Development of the design top rated design ideas, they create design paper prototypes for a new technology addressing to them. In this method, they are not allowed to use any High-tech material to get inspired. For developing paper designs, facilitators provide the participants supportive Low-tech materials e.g. post-it notes, color pens, markers, and pre-printed plastic icons. As we can see on the below table, the differences between the two participatory designs methods focus not only on the steps that they consist of, but also on the required Low- and High-tech materials on a case-by-case basis.

3 Empirical Study

3.1 Participants, Setting and Materials

Participants

Three AOF dyslexia organizations agreed to participate in the empirical study by allowing us to observe how the two selected participatory design methods work with dyslectic participants. On observations three groups participated, which consisted of dyslectic adults aged from 20 to 57 years old. It is important to point out that the participants of Group_1 have been diagnosed with a high-level of dyslexia, which prevents them to read and/or write in comparison to the participants of other two groups. Groups' special education teachers –Maria Hyttel and Rita Cassøe– played a crucial role, as people with knowledge of dyslexia impairment. They contributed the study by translating parts of the process both in Danish and in English, or even representing dyslectic participants, when it was necessary.

Setting

All the sessions took place at participants' schools of AOF-Vendsyssel of Hjallerup (Group1), AOF-Brønderslev (Group2), and AOF-Hjørring (Group3). AOF Danmark is a Danish educational organization, which provides free teaching to dyslectic adults to help them reduce their reading and/or writing difficulties, and improve their skills on reading-understanding, using and writing texts [15].

Materials

The materials used in participatory design process were of two kinds: (1) Low-tech materials, e.g. Post-it Notes, Plastic Icons, Highlighters, Colored Pens and Pre-Printed Graphical Design Elements, and (2) High-tech materials, e.g. a webcam software 'CyberLink YouCam 5' in order to record the development-sessions, laptops and mobile phones. The whole process took place by respecting the rules of the signed agreement between us and AOF organization regarding ethical considerations, in order for any participant not to be stigmatized by the process.

3.2 Procedure

The sessions were conducted twice per week for a period of four weeks with Group_1 and Group_3, and for a period of three weeks with Group_2. All the sessions focused on applying the selected participatory design methods and observing participants in order for us to find out, how participatory design works with the dyslectic participants through the selected PD methods. The whole process took place with the dyslectic participants to create a range of designs for a reading software application, with the goal for the generated designs that will finally correspond to the dyslectic participants' expectations. The sessions were divided into two categories of the IDEAS-method sessions and CI-method sessions, and each category was separated into two parts of the Development part and Evaluation part.

IDEAS-Method Sessions

The Development Part:

As it was mentioned earlier, the IDEAS participatory design method consists of the four steps of Introduction, Discussion, Brainstorming and Design. We applied the IDEAS participatory design method based on the method's guidelines [18]. In this participatory design method, all the three groups took part.

First step: At the Introduction step, the facilitator introduced the research to the participants by explaining them, in parallel, the process that they had to follow and the purpose of the research. A visual timetable presented the parts, rules and duration of each step. Regarding the video, we selected a video in Danish language, which video presented seven software applications for dyslectic users, as the most popular software applications tailor-made to dyslectic users' needs [19]. Regarding the video selection, we based on two main criteria of the language, which was the first and most serious criterion; the audience of the video consisted only of Danish people, who might face difficulties on understanding a foreign language. The second criterion was the descriptiveness of the video regarding the seven software applications.

Second step: At the Discussion step, the participants discussed with each other about what they liked or disliked on the software programs that they previously watched on the video. This discussion step was a preparation for the third step, since the discussion among the participants gave them the chance to think design ideas inspired by the previously presented software programs that they watched.

Third step: At the Brainstorming step, the participants received a pre-printed form (a Low-tech material), where they had to write a design 'Topic' for a reading software application. Then they should generate and write a number of eight design ideas related to the topic. This task was based on the previous 'Introduction' and 'Discussion' steps. Afterwards, the participants should rate their design ideas in order for them to find the special one that they should use on the Design step. All the 'Ideas Generation' pre-printed forms included rating boxes, where participants had to write a grade from 1 ('Terrible') to 10 ('Perfect'). The top rated design idea(s) should be built in the final Design step of the process. In case that someone of the participants faced difficulties in generating ideas, the facilitator was allowed to support them by giving them an extra pre-printed form (a Low-tech material) with a design topic and four, instead of eight, design ideas boxes. The pre-written design topic and the less number of design-ideas boxes, could facilitate and encourage dyslectic participants to generate design ideas. Below you can see a sample of the pre-printed form as an extra support. If the situation became more difficult and some of the participants were not able to think any idea, the facilitator was allowed to provide them a stronger support by giving more detailed pre-printed form (a Low-tech material). On this form, there were four pre-written design ideas of a specific design topic of a reading software application for dyslectic users. In this case, the participants had only to rate the design ideas, in order for them to draw it on the last step. Below you can see a sample of the extra supportive pre-printed form.

Fourth step: At the Design step, the participants received Low-tech materials, like Post-it Notes, Plastic Icons, Highlighters, Colored Pens and Pre-Printed Graphical Design Elements and a pre-printed Interface Design Template (See). With these Low-tech materials, they should build the best design idea(s) that came first among the total number of the design ideas after the rating process. At this step, it was allowed for the participants to use High-tech materials, e.g. internet, videos, mobile phones, etc. in order for them to be inspired to generate as much efficient design paper prototypes as they could.

CI-Method Sessions

The Development process:

Based on the literature, the CI (Cooperative Inquiry) participatory design method consists of the four parts of Introduction, Discussion, Brainstorming and Design. As this is a more traditional participatory design method, there was neither visual support to the participants, nor any type of help provided by the facilitator or any High-tech source [13, 14]. We applied the CI participatory design method based on the method's guideline [18]. In the development-part of the CI-method participated only Group_1 and Group_3.

First step: At the Introduction step, the facilitator presented verbally two software programs for dyslectic users: The 'IntoWords' [20] and 'CD-ORD' programs [21]. We chose these two programs based on two criteria from our perspective. The experience was the first criterion, since the participants were experienced with one of the programs, the CD-ORD, since it is systematically being used by the AOF organization. The fact, that the participants already knew and were familiar with the design of the CD-ORD program, would facilitate them on the process of criticizing and finding its pros and cons. Regarding the 'IntoWords' software program is similar to the first one, but it was available only to iPhones. It is a fancy program but with less functions in comparison to the 'CD-ORD'. In our perspective, this selection could make the comparison between the two selected programs and, in parallel, the inspiration of design ideas easier for our dyslectic participants.

Second step: At the Discussion step and before participants start discussing design ideas with each other, they received colored Post-it notes and a pre-printed form, namely 'Sticky-notes' (Low-tech materials). On the pre-printed forms and by using the Sticky-notes, the participants should individually write likes, dislikes and/or any possible improvements regarding the presented by the facilitator software programs/applications for dyslectic users. Below you can see one participant's Stick-notes form with his/her likes, dislikes and improvement ideas.

We decided to give an individual character to this task to 'protect' the participants from any influence by their fellow-participants. Once they had finished this first part of the Discussion step, we came back to the cooperative character of the process, and the discussion among participants started. Participants discussed about the two presented programs and they expressed their opinions on them. Below we can see the dyslectic participants of a group to have started discussing on the presented software applications. The purpose of this step was to prepare dyslectic participants for the next step of Brainstorming. When ending the Discussion step, the participants would have already in their mind a general picture about what they like, dislike or suggested improvements on the already presented software programs.

Third step: In the Brainstorming step, the participants received a pre-printed form (a Low-tech material) and colored pens. The purpose of this step was to make participants think and write design ideas about a design for a new software application based on data from the former step of the Discussion. Below we can see participants while thinking and writing design ideas on the given pre-printed form. Once participants completed the Brainstorming step, they rated their design ideas in order for them to find, which one or what kind of combinations among their design ideas could lead them to a final design of a reading software application for dyslectic users.

Fourth step: At the Development step, the participants received Low-tech materials, like Post-it Notes, Plastic Icons, Highlighters, Colored Pens and Pre-Printed Graphical Design materials and a pre-printed Interface Design Template, where and with which they should make a design paper-prototype based on the design idea that came first after the grading process. In both participatory design methods, once the participants completed their design-prototypes put their prototypes on a mobile paper-prototype in order for them to present their work to their fellow-participants and us. In this way, we would be informed about how their designs work, and what kind of help could a specific design offer to dyslectic users.

The Evaluation Part of IDEAS and CI methods:
Based on literature, once both IDEAS and CI participatory design processes were completed by all the groups, the final designs should be evaluated by the participating groups [13, 14, 18]. Specifically, each group should evaluate another group's generated designs. All the three groups took part in the evaluation process. Below you can see an

evaluation map for each group and a sample of the evaluation process. The average grade of each final design was on a scale from 1 to 10. These numbers reflect the level of the efficiency of the final designs for the participants. Based on results we cannot talk about one ideal final design, because the differences among the grades are barely noticeable, especially among the designs that got a high grade (from 7 'Very Good' to 10 'Perfect').

Based on the designs' grades and participants comments, the final designs being full of words and providing many choices got a low grade (from 5 'OK' to 6 'Good Idea'), since dyslectics evaluated them as impractical or hard to use. Additionally, any foreign language feature was a strong reason for a final design to be graded low, since for instance an English logo might make difficult for dyslectics to find a useful app.

Final designs' simplicity proved crucial for the dyslectic participants considering those final designs' grades with less functions, which got a grade from '8' ('Great Idea') to '9' ('Excellent Idea'). Final designs that explained distinctly their functions got a grade of '9' ('Excellent Idea'). This made us clear that descriptiveness of designs' functions was an important feature for the participants, since in other case the application may be distracting for them, as they said.

Final Designs' Evaluation by Greek participants:
Based on a previous literature research [22] the difficulty-level of a language depends also on the language-type. The difference between opaque and transparent types of languages affects the dyslectic peoples' reading performance [23, 24]. Therefore, we translated the Danish participants' final designs from the opaque Danish language into a transparent one (the Greek language in our case) and asked a number of nine dyslectic participants from Greece to evaluate them. Based on the evaluation we would be able to see, if there is any difference on the designs grading by participants taking a transparent language. The evaluation took place through online surveys. We cannot talk about one top-ideal final design, since the differences between the grades are barely noticeable, especially among the majority of the high graded final designs (from 7 'Very Good' to 10 'Perfect').

Designs with variable functions and colors got a higher grade in Greek language than in Danish one. The only design characterized as 'Complete' by the Greek dyslectic participants and taken the highest grade of '8.5' (higher than 'Great Idea' and lower than 'Excellent Idea') provided users variable functions and choices. On the other hand, Greek dyslectic participants did not show any preference on designs that were simple or with a small number of functions and colors. Conversely, a large number of them pointed out that such designs needed to become more attractive and they commented that 6 out of 8 designs need improvement. Similarly, to the Danish evaluation of the final designs, the designs explaining distinctly their functions got a better grade of '8' ('Great Idea'). This made us clear that descriptiveness of designs' functions was very crucial for the Greek participants –as it was for the Danish– since in other case the application may be distracting for them, as they claimed.

3.3 Data Collection

Throughout the empirical study, we collected data from sessions with the three groups of AOF regarding the collaboration of the dyslectic participants, problems that participants faced during the process, strengths of the empirical study during the process, and biases that may affect the evaluation of the final designs. We collected the participatory design data over a period of approximately one month. The data sources that we used were both qualitative (observations, video recordings, notes) and quantitative (online surveys). As we have already mentioned, the participatory sessions were divided into two main categories of (i) Development and (ii) Evaluation. All the groups took part in both categories of sessions except of the Group_2, which did not take part on the development session of the CI method due to the participants' dyslexia level. When possible, we collected data of the discussions and informal conversations among participants, facilitator and special education teachers though both video-records (in total 5 h 47 min) and notes (18 hand-written pages). We also gathered data through notes of participants' conversations on the final sessions for the design-evaluation and took photos of the final designs' rating. The table below is an overview of the total hours of the development process video records transcribed by evaluation session by group. Furthermore, we used two online surveys to gather information from both the participants and their special-education teachers. The online surveys were open-ended and they helped us gather information related to our areas of attention. Below is an overview of the three online surveys' focus: Online Survey 1: with a focus on the participants' opinion on the participatory design methods used and process. Online Survey 2: with a focus on the teachers' opinion on the participatory design, methods used and process.

4 Data Analysis

Video recordings' analysis was conducted in three phases inspired by the study of Hornbæk and Nørgaard [25]. Initially, we transcribed the video recordings and partially translated them from Danish to English. Then, we segmented the recordings based on four areas of attention: Collaboration, Problems, Strengths, and Biases. Finally, we analyzed the transcriptions by trying to find among them parts related to our areas of attention. We also kept notes and conducted surveys during the sessions, which we observed. The whole process of data analysis was based on the Grounded Theory and Chi's proposals on analyses of verbal data [26, 27]. On the Findings section, we describe data aspects, which are important for the future studies to know when working on participatory design with dyslectic participants.

5 Findings

As we mentioned on the previous section, we observed statements for the participants based on the four areas of attention: Collaboration, Problems, Strengths, and Biases. The areas of attention are described below with a focus on aspects that have been mentioned as important for the future studies on participatory design with dyslectic participants. Abbreviations have been used instead of the whole words in the following terms: Participant (P), Facilitator (F), and Teacher (T).

5.1 Collaboration

After observing the dyslectic participants' statements with a focus on the Collaboration area of attention, we found out that all the participants preferred to work as pairs or teams:

GROUP_1 – IDEAS:
P3: "Can we work together?" (P3 means as pairs)
F: "Of course you can. Why do you prefer to work together?"
P2: "Because it is easier for us when we work together."
GROUP_2 – IDEAS:
F: "Would it be possible for you to work as pairs?"
P10: "I'm not sure..."
T: "It is difficult for them to write anything."
F: "Then...we could do that as workshop. All together."
T: "Let's do it all together!"

To investigate the dyslectic participants' preference on collaboration we conducted the Online Survey 1, where all the participants answered that if they could choose how to work –with their fellow-participants or individually– they answered "with their fellow-participants". Furthermore, at the Online Survey 2 the special-education teachers corroborated the positive impact of collaboration for dyslectics. "They have been challenged to think creatively in terms of design. And they learned to collaborate and to accept a new way of collaboration." The results indicate that participatory design with dyslectics works more effectively, when participants collaborating between each other than working individually.

5.2 Problems

Throughout sessions, dyslectics seemed to enjoy the process to a large extent. However, after observing their statements, we found out that in some cases participatory design can become stressful for dyslectic participants or even make them have light tantrums. P3, P1 and P6 expressed the above statements under stress and even P6 had a light tantrum trying to 'translate' his/her design ideas into paper design prototype. These feelings may have been due to the difficulties facing dyslectics often on the writing conceptual area [22]. On the other hand, they may have been because of the lack of High-tech materials, which enhanced their creativity on the IDEAS developing process.

GROUP_1 – CI:
P3: "We have all the ideas in our mind but we cannot get it down on the paper!!!"
P1: "What we have to do is to draw the first page of an app, and write how it works and design it? How could we also use this to get help from dyslectics? I can't draw it!"
GROUP_3 – CI:
P6: "I think that it is more important to come up with ideas that could become real. I cannot draw them, but is more important for me to give you my ideas."

These results indicate that participatory design together with dyslectic people requires a facilitator experienced on dyslectics' emotional tantrums in order to conduct the process smoothly. In other case, dyslectics may feel frustrated and or even stop taking part to the process.

5.3 Strengths

On Strengths topic, which concerns the benefits of the participatory design for the participants, our findings focused on: Teachers as Proxy Users, and Facilitator experienced on dyslexia.

From the beginning of the sessions the special-education teachers had a vital role on the participatory design process as Proxy Users. Especially for the group where the participants had a high level of dyslexia the presence of their teacher as a Proxy User was necessary in order to represent them, when they were not able to express their thoughts.

> GROUP_2 – IDEAS:
> T: "First Idea: "One application with one function, text to speech and reader and a scanner that modify the text to speech", "Fourth Idea: "A translation function would be a nice idea", "Fifth Idea: Suggestions of words' dictation and grammatical marks like comma, full stop etc."

It was an advantage for the efficiency of the process the fact that teachers voluntarily took this role without even asking for this. The findings indicate that when dyslectic participants are diagnosed with a high-levelled dyslexia then participation of special-education teachers (or person from their close environment) as Proxy Users presence is deemed necessary, as they can contribute significantly the participatory design process. Similarly, it was advantageous for the smooth progress of the process the presence of an experienced facilitator, who was able to keep calm dyslectic people when being under stress, e.g. in a previous case, when participants got stressed or even they had light tantrums because they felt unable to carry through with design development.

An experienced facilitator has the knowledge to maintain a friendly and cozy atmosphere during the participatory design processes, and enhanced dyslectics' productivity by making them feel always an important and crucial part of the process:

> GROUP_1 – CI:
> F: "I don't want to make you feel stressed or what else. It is ok for us to have your ideas. We can draw them for you! We are here to enjoy the process!"
> GROUP_3 – IDEAS:
> F: "Don't worry, it's a game! Believe me! "Don't worry, it's a game! Believe me! You don't need to feel stressed. Just relax and enjoy the process. You just need to sketch and play with papers and pencils"

Finding indicates that there is a requirement of a highly experienced person to address potential incidents in a way that will boost the process by encouraging dyslectics.

5.4 Biases

On Biases area of attention, our findings focused on:

- Dyslectics' repetitive behaviors
- Same participants on bot design and evaluation parts

After observing the dyslectic participants' statements with a focus on the biases area of attention, we found that some participants' thoughts were affected by their fellow-participants' repetitive behavior:

GROUP_3 – IDEAS:
P6: "I think that this has to be very simple…", "I have written here simple. Because that is like the essence that it has to be simple.", "This is a very simple and actually it would help to be simple…" (P6's words during all the four steps of PD)

Repetitive behavior is a typical symptom that appears on dyslectics sometimes. Finding indicates that in such cases facilitators have to be prepared to address such repetitive behaviors in order to avoid biases. It is important to make dyslectic participants clear, when a participatory design step starts and ends, what includes each step and that in case that a step does not include a discussion then the discussion is not allowed in order for avoiding distractive factors. A strong bias caused by a 'bad' allocation of the tasks that each group should take. This 'bad' allocation biased the final designs' evaluation process, since some of the participants, who took part in the evaluation process, had already taken part on development process. Under such a circumstance, it was difficult for them to be fair and impartial. Findings indicate that regardless the limited number of dyslectic participants, there should be a clear allocation of the tasks that each group of participants takes, in order for the process to be unbiased.

6 Discussion

In this section, we compare the results to related work, in regards to how participatory design works with autistics and dyslectic. Furthermore, we discuss a dyslectics' preference on the IDEAS method. We also discuss the different expectations of dyslectics based on their language's type. Finally, we discuss the implication that might have the involvement of dyslectic participants in designing technologies for dyslectic users.

6.1 Comparison of Findings with Related Work

Participatory design proved as promising as it was with children with ASD [18]. However, in the case of participatory design with dyslectic people, the process showed a more smooth progress than in the case of participatory design with autistics, since in our case there was not any need for an extra support on the collaboration among dyslectic participants. Additionally, even though dyslectics struggled sometimes on generating design ideas, they did not make use of any extra pre-printed supportive form. Conversely, children with ASD received a support on both collaboration and generation of IDEAS method steps of participatory design [18].

6.2 Comparison Between IDEAS and CI Methods

Regarding the selected participatory design methods (IDEAS and CI), the IDEAS method seemed to be preferable to the dyslectic participants, since it worked better with them. Even though both participatory design methods gave the participants the chance to cooperate with each other, IDEAS method was proved that it enhanced dyslectics productiveness and creativity by providing them High-tech materials, in contrast to the CI method which made them struggle with generating ideas and developing of designs processes.

6.3 Participatory Design in Opaque and Transparent Languages

After comparing the final designs' evaluations' from Danish and Greek dyslectic participants, we realized that the Greek language, as a transparent and simpler than the Danish language, gives dyslectics the opportunity to use designs that are more complex. Conversely, opaque languages, like the Danish one, increase the level of the simplicity on designs, in order for them to be efficient and useful to Danish dyslectic users.

7 Implications

The current research indicates that involvement of dyslectic people in designing software applications results in a tailor-made design of software applications guided by dyslectics' expectations. Of course, participation of dyslectic users in participatory design may have some implications for it. Any potential problems on collaboration, emotional tantrums, difficulties on expressing thoughts, limited imagination, and maybe a distracting focus on unnecessary details, may have implications on participatory design.

8 Conclusion

In this paper, we have described an empirical study on participatory design with dyslectic participants. The main goal of this study was to examine how participatory design works with dyslectic participants. Overall and considering the findings, participatory design was beneficial for the dyslectic participants and worked effectively with them. Through participatory design, dyslectics had the chance to express their design ideas on a software application focused on dyslexia, and work on designing an application based on their expectations.

Based on findings participatory design can work efficiently with dyslectics only under certain conditions:

(i) collaboration among dyslectic participants is very important, since, in this way, dyslectics are more creative and productive,
(ii) a Proxy User participation is also important, since a Proxy User can represent participants with a high level of dyslexia, when it is necessary,
(iii) an experienced on dyslexia facilitator is necessary to take part to the process for avoiding incidences caused by dyslexia, and
(iv) a proper allocation of the groups by tasks is required for preventing biases.

Furthermore, based on dyslectics' special education teachers, through participatory design dyslectics managed to increase their attention on a task though cooperation and participating to this empirical study was a great motivation for collaboration among them.

In this study, one limitation that we faced was, in some cases, the language, which prevented us to communicate our thoughts efficiently and immediately to the participants. Language, in some cases, was a problem, since the majority of the Danish participants did not speak English and we were not native speakers of the Danish language. This limitation had as result the delay of the process sometimes. A second limitation was our skills on participatory design, since it was the first time that we worked on such a study. This limitation may have skewed to some extent some of the results.

Regarding the future work, our dyslectic participants showed an interest in a future participation on a real software application's development based on the current study, and the special-education teachers corroborated this suggestion as beneficial for their students. Based on this, future studies might focus on Participatory Development of the product with Dyslectics. Furthermore, it might be beneficial for a future research on participatory design with dyslectics to take into account factors like dyslectic participants' age groups, the participants' level of dyslexia, and the other four conceptual areas. In this way the results might generalized, something that is missed from this study.

Acknowledgements. We would like to thank the AOF-Vendsyssel of Hjallerup, Brønderslev and Hjørring for agreeing to take part in this empirical study, the special-education teachers for their time, and all the dyslectic participants for sharing their valuable ideas with us. Finally, I would personally like to extend my gratitude to my supervisor Jan Stage for his valuable feedback and guidance throughout the study period.

References

1. Stein, F.N.: Den Store Danske, 10 April 2017. http://denstoredanske.dk/index.php?sideId=255134
2. Spinuzzi, C.: The methodology of participatory design. Tech. Commun. **52**, 163–174 (2005)
3. Brandt, E., Binder, T., Sanders, E.B.-N.: A framework for organizing the tools and techniques (2010)
4. Ross, T., May, A., Sims, R., Parker, C., Mitchell, V.: Empirical investigation of the impact of using co-design methods when generating proposals for sustainable travel solutions. Int. J. CoCreat. Des. Arts (2015)
5. Anić, I.: UX Passion, Sunday November 2015. http://www.uxpassion.com/blog/participatory-design-what-makes-it-great/. Accessed Thursday Apr 2017

6. Foraker Labs: Usability First. Foraker Labs (2002–2015). http://www.usabilityfirst.com/usability-methods/participatory-design/. Accessed Thursday Apr 2017
7. National Joint Committee on Learning Disabilities: Learning disabilities: issues on definition. Asha **33**(Suppl. 5), 18–20 (1991)
8. Outhred, L.: Word processing: its impact on children's writing. J. Learn. Disabil. **22**, 262–264 (1989)
9. Wise, B.W., Olson, R.K.: Reading on the computer with orthographic and speech feedback: an overview of the Colorado remediation project. Read. Writ.: Interdisc. J. **4**, 107–144 (1992)
10. Lerner, J.: Learning Disabilities: Theories, Diagnosis, and Teaching Strategies, 7th edn. Houghton Mifflin Company, Boston (1997)
11. Ψυχολογοσ, Σ.N.D.K.: ΤΟ ΦΑΣΜΑ ΤΟΥ ΑΥΤΙΣΜΟΥ: ΔΙΑΧΥΤΕΣ ΑΝΑΠΤΥΞΙΑΚΕΣ ΔΙΑΤΑΡΑΧΕΣ ΕΝΑΣ ΟΔΗΓΟΣ ΓΙΑ ΤΗΝ ΟΙΚΟΓΕΝΕΙΑ. ΣΥΛΛΟΓΟΣ ΓΟΝΕΩΝ ΚΗΔΕΜΟΝΩΝ ΚΑΙ ΦΙΛΩΝ ΑΥΤΙΣΤΙΚΩΝ ΑΤΟΜΩΝ, ΛΑΡΙΣΑ (2006)
12. Tsermentseli, S., O'Brien, J.M., Spencer, J.V.: Comparison of form and motion coherence processing in autistic spectrum disorders and dyslexia. J. Autism Dev. Disord. **38**, 1201–1210 (2008)
13. Benton, L.: Participatory design and autism: supporting the participation, contribution and collaboration of children with ASD during the technology design process. University of Bath, Bath (2013)
14. Druin, A.: Cooperative inquiry: developing new technologies for children with children (1999)
15. AOF: AOF Danmark. https://aftenskole.aof.dk/temaer/ordblindeundervisning/
16. Muller, M.J.: PICTIVE-an exploration in participatory design, pp. 225–231 (1991)
17. CyberLink (2017). https://www.cyberlink.com/index_en_EU.html?r=1
18. Johnson, H., Ashwin, E., Brosnan, M., Grawemeyer, B., Benton, L.: Developing IDEAS: supporting children with autism within a participatory design team. In: CHI 2012, Texas (2012)
19. Svendsen, K.: Youtube - Top 7 apps til ordblinde (2012). https://www.youtube.com/watch?v=9fqf4MpyFj8
20. MV-NORDIC: MV-NORDIC. https://www.mv-nordic.com/dk/produkter/intowords
21. MV-NORDIC. https://www.mv-nordic.com/dk/produkter/cd-ord/vejledninger
22. Vangeli, P.: Literature research on interaction design of systems for dyslectic users. Aalborg University, Aalborg (2017)
23. Giovanna: Italo Bimbi. Italo Bimbi, 22 November 2011. http://www.italobimbi.it/en/blog/57-lingue-opache-e-lingue-trasparenti.html. Accessed 02 May 2017
24. Silva, C., Marcelino, L., Ferreira, P., Madeira, J.: Assistive mobile applications for dyslexia. Procedia Comput. Sci. **64**, 417–424 (2015)
25. Hornbæk, K., Nørgaard, M.: What do usability evaluators do in practice? An explorative study of think-aloud testing. In: Proceedings of the 6th Conference on Designing Interactive Systems, University Park, PA, USA (2006)
26. Pace, S.: A grounded theory of the flow experiences of web users. Int. J. Hum.-Comput. Stud. **60**, 347–363 (2004)
27. Chi, M.T.H.: Quantifying qualitative analyses of verbal data: a practical guide. J. Learn. Sci. **6**(3), 271–315 (1997)
28. Robertson, J., Good, J.: CARSS: a framework for learner-centred design with children (2006)
29. Nielsen, J.: Teaching experienced developers to design graphical user interfaces. In: CHI 1992, pp. 557–564 (1992)
30. Nesset, V., Beheshti, J., Bowler, L., Large, A.: Bonded design: a novel approach to intergenerational information technology design. Libr. Inf. Sci. Res. **28**, 64–82 (2006)

Literature Survey on Interaction Design and Existing Software Applications for Dyslectic Users

Panagiota Vangeli(✉) and Jan Stage

Aalborg University of Computer Science, Aalborg, Denmark
panvangeli@gmail.com, jans@cs.aau.dk

Abstract. The purpose of this study is a literature research on interaction design and existing software applications for dyslectic users. This literature research will contribute a future empirical study on how we could design a reading software application together with dyslectic users to enhance their reading skills. For the purpose of this research, we initially collected 175 studies, from which we selected, reviewed and made an overview table of 71 studies organized by areas of attention. The literature research on interaction design of systems for dyslectic users resulted in a presentation and comparison of interaction design (IxD) parameters. This process indicated common dimensions and elements among IxD parameters supporting users in improving their reading skills. Finally, reviewed studies on existing software applications resulted in a focus on improving dyslectics' reading performance. Our results showed that there is a trend on developing interaction designs focused on the reading conceptual area. We also discuss dyslectics users of existing software applications complaints, which resulted in a lack of existing software applications' design and system quality.

Keywords: Dyslexia · Causes of dyslexia · Dyslexia teaching strategies
Dyslexia treatments · Dyslexia theories · Dyslexia languages
Dyslexia technology tools · Mobile applications for dyslexia
IxD of SW apps for dyslexia · Design guidelines for dyslexia

1 Introduction

Dyslexia is a hidden learning disorder in reading, spelling and written language, and maybe in number work. It is a learning disability, which cannot be completely treated and has negative consequences for dyslectics' life by making it complicated [1]. Learning difficulties, caused by dyslexia, have often a negative impact on the way dyslectics are used to thinking, behaving and living. Statistics have shown that approximately 70–80% of people with reading problems are probably dyslectics, and one out of five students have a language-based learning disability [2]. Research have shown that dyslexia is a cognitive disorder, which affects deeply dyslectics' daily routine by isolating them often from the community. It is very usual for a dyslectic person to complain that (s)he is not able to be focused on a specific task, recall tasks, orders, messages, routes or even their daily schedule [3].

Furthermore, it is important to point out that research supports that there is a relation between dyslexia and the type of languages. A language can be either opaque

M. E. Auer and T. Tsiatsos (Eds.): IMCL 2017, AISC 725, pp. 331–344, 2018.
https://doi.org/10.1007/978-3-319-75175-7_34

(e.g. the English, Danish, French languages, etc.), or transparent (e.g. the Greek, Italian, Spanish languages, etc.), which difference affects the level of a language's complexity, and has an impact on dyslectics' reading and writing performance [4]. Studies have also proved that assistive technology contributes significantly the improvement of dyslectics' cognitive skills [4–6]. Technology is an alternative and modern way of helping people with dyslexia improve their skills on reading, writing, memory, organization or numeracy conceptual areas. Maybe technology is not able to treat dyslexia yet, but it is able to facilitate dyslectics by enhancing the motivation for improvement [7, 8]. Especially Human-Computer Interaction (HCI) field can enhance this trial through designing systems for building a dyslexia friendly environment [9].

After a systematic literature research on interaction design of systems and existing software applications supporting dyslectic users, we realized that related studies to the field of dyslexia are very limited, even though dyslexia is a cognitive disorder with strong impacts to dyslectics' life. With this study, our goal is to contribute future research on developing designs for software applications addressing to dyslectic users.

2 Related Work

In this section, two related works are presented regarding literature research on interaction design of systems for dyslectic users, and one related work on existing software applications that support dyslectic users to improve their cognitive skills.

2.1 Related Work on IxD Guidelines and Parameters

de Avelar et al. [10] investigated Web accessibility issues for users with dyslexia by involving in their study related literature studies on interaction design parameters. A number of related works on interaction design for dyslexia have been mentioned in their research. Some of them focused on functionality and some others on the user interface: From one hand, the studies of Freire et al. [11] and Al-Wabil et al. [12] focused on functionalities that could help dyslectic users improve their performance. In their studies, they refer to a number of 693 problems on accessibility and usability, which problems are related to difficulties in navigation, architecture of information, the form of texts, the organization of the content, the language and the amount of information that makes harder for dyslectics to scan a text. Because of the fact that such difficulties can be distracting for dyslectics, interaction design of systems for dyslexia has to be focused on fulfil these functionalities. On the other hand, the studies of Rello et al. [13–15], Santana et al. [16], Rello and Barbosa [17], and Rello and Baeza-Yates [18, 19] focused on user interface design-parameters. The recommended design-parameters allow users to highlight content of texts, adjust the size and type of fonts, the alignment of a text, the spacing of characters, the fore- and background colours, the length of texts, and its borders. Additionally, there are suggestions, which could improve dyslectics' reading skills: Rello and Baeza-Yates recommend Helvetica, Courier, Arial, Verdana and Computer Modern Unicode font types as the best font types for dyslectic users [18, 19]. McCarthy et al. [20, 21] included into their study a literature survey on interaction design for dyslectic users, which resulted in a number of parameters focused on the user interface as well. In this study, there have been mentioned features that allow dyslectic users to adjust the size of a text, and design parameters that refer to short sentences, use of pictures, dark background, and San Serif fonts of 12pt or larger. These recommendations are an overview of other researchers' studies [22, 23], which McCarthy provides us.

2.2 Related Work on Existing Software Applications

A literature survey on existing software applications for dyslectic users led us to the CALL Research and Development center of Scotland, which provides two 'App Wheels' that present a number of 180 different types of existing Android and iPad software applications categorized into five conceptual areas of Reading, Writing, Numeracy, Organization and Memory. A thorough research on the 'AppWheels' showed that there is a trend on the majority of the existing software applications for dyslectic users towards the reading conceptual area [24, 25].

3 Method

As we mentioned earlier, our literature research focused on interaction design of software applications for dyslectic users. We divided our literature research into four phases of the Review Focus, the Filtering, Categorization, and in-depth Analysis (Reading) of the literature. For reviewing the literature, we based on a combination of the Müller-Bloch and Kranz's literature review framework [20], and the Wolfswinkel's et al. [21]. Below we explain each research phase individually.

3.1 Phase 1: The Review Focus

For the research purpose, Google Scholar, Scopus and IEEE literature databases have been used, since they cover a variety of research fields. We did not limit our research on specific conferences, since the sources regarding dyslexia and interaction design are very limited. We included, for instance, in our research conferences such as ICCHP, ACM, WCE, ITiCSE, etc. Additionally we included published reports and books related to our focus area. Before starting, the literature research 6 keywords/key-phrases selected as leaders: dyslexia, causes of dyslexia, dyslexia theories, IxD of SW applications for dyslexia, dyslexia technology tools, mobile applications for dyslexia. Initially, the review of the literature was concerned with the keyword 'dyslexia'. This keyword led the survey to a large number literatures have been conducted with a more theoretical focus, which provided us general information on dyslexia. Therefore, the survey's focus particularized on 'dyslexia theories', and 'causes of dyslexia', which key-phrases enhanced our knowledge on dyslexia. Papers with a high focus on medical models of dyslexia were excluded. Moving on, the review concerned with the 'IxD of SW applications for dyslexia'. This key-phrase excluded papers suggesting methodologies of interaction design for developing software applications for users without dyslexia. Afterwards, the key-phrase of 'dyslexia technology tools' provided us various technological tools being used addressing to dyslectics, but due to the fact that our literature survey focus area was on IxD for mobile applications, we added the sub-keyword of 'mobile applications for dyslexia'. This sub-keyword excluded a number of literature works related to other types of assistive technologies, e.g. for desktops, developed for using by dyslectics.

3.2 Phase 2: Filtering Studies

In the filtering process, we should decide if the preselected literature should be included to the final selection. For the final selection of the studies, we added four questions as rules for the final selection of a study: "Is the study related to dyslexia?", "Is the study related to theories on dyslexia?", "Is the study related to mobile applications for dyslexia?", "Is the study related to interaction design for mobile technology?" If the answer on the question was 'Yes', the study was accepted, if the answer on the question was 'No', the study was rejected. The phase 1 resulted to a number of 175 studies before filtering: 63 about dyslexia, 33 about causes of dyslexia, 34 about dyslexia theories, 8 about IxD of SW apps for dyslexia, 17 about dyslexia technology tools, 20 about mobile applications for dyslexia.

Afterwards, we re-evaluated the found studies to be sure that they corresponded to our research. The final number of selected studies after filtering was 71 with 22 studies about dyslexia, 16 about causes of dyslexia, 18 about dyslexia theories, 9 about IxD of SW apps for dyslexia, 5 about dyslexia technology tools, and 1 about mobile applications for dyslexia. As we can see, the total number of studies was decreased significantly by reaching approximately the half number of the initially selected studies. Furthermore, in some cases the numbers reached only just one literature, e.g. on 'mobile applications for dyslexia', since the majority of literature was conducted either in a very theoretical background or it was referred very generally to the usefulness of the technology for dyslectic people.

3.3 Categorization of the Studies

In this phase, we categorized the selected studies based on their content. Specifically, we made a table consisting of four main areas of attention of 'Dyslexia Understanding' included dyslexia and causes of dyslexia keywords, 'Theories on Dyslexia' included dyslexia theories keyword, 'Technology on Dyslexia' included dyslexia technology tools and mobile applications for dyslexia keywords, and 'IxD and Dyslexia' included IxD of SW apps for dyslexia keyword. In these main categories (areas of attention), we classified the selected papers based on keywords/key-phrases. Categorization of papers facilitated us to the next phase of 'Reading of Literature', since it was easier for us to know beforehand, which papers –based on the related keywords/key-phrases- belong to which category (area of attention).

3.4 Reading of Literature

Through reading the selected studies, we found new pieces of information related to the mentioned areas of attention. After translating the new pieces of information into keywords/key-phrases, four keywords/key-phrases were added: dyslexia teaching strategies, dyslexia treatments (Dyslexia Understanding), dyslexia languages (Theories on Dyslexia), and design guidelines for dyslexia (IxD and Dyslexia). Finally, we had 10 keywords/key-phrases instead of 6. New keywords/key-phrases were added into three out of four areas of attention, which contributed highly our literature research, especially in the area of 'IxD and Dyslexia' attention.

4 Findings

This section presents the results we received from the literature review. Initially, an overview table illustrates the studies' number organized by areas of attention. Then interaction design guidelines and parameters are shown and compered with each other. The section ends looking at the existing software applications that support dyslectic users (Table 1).

Table 1. Literature references organized by areas of attention

Areas of attention	Keywords/key-phrases	Literature reference number
Dyslexia understanding	Dyslexia	1, 2, 3, 4, 9, 11, 12, 22, 25, 28, 30, 50, 51, 57, 63, 67, 69
	Dyslexia causes	31, 32, 33, 34, 36, 37, 39, 41, 42
	Dyslexia teaching strategies	5, 6, 8, 29
	Dyslexia treatments	7
Theories on dyslexia	Dyslexia theories	5, 10, 12, 13, 14, 23, 35, 36, 38, 40, 56
	Dyslexia languages	14, 15, 16, 18, 24, 52, 53, 11
Technology for dyslexia	Dyslexia technology tools	20, 21, 26, 27
	Dyslexia mobile applications	17, 29, 47, 48
IxD design and dyslexia	IxD dyslexia	19, 49, 65
	Design guidelines dyslexia	43, 44, 45, 46

At this point, there is an explanation of each area of attention. Additionally, we explain what kind of information each area of attention provided us:

Dyslexia Understanding: Studies of this area of attention provided us pieces of information enough to equip us with the necessary knowledge regarding dyslexia, its aspects, teaching strategies have been used until now and trials of treatment. Based on this area of attention studies, we managed to learn that dyslexia is a learning disability that belongs to cognitive disorders; dyslexia can be caused by phonological, magno-cellular, cerebellar, auditory and visual deficits; it is cognitive impairment that cannot be treated yet, but by using sophisticated teaching strategies, dyslectics' learning skills can be improved.

Theories on Dyslexia: Studies on this area of attention provided us pieces of information regarding theories related to the causes of dyslexia and the impact of the type of languages to the level of dyslexia. Five causal theories of dyslexia relate the difficulties facing dyslectic people to their causes. The phonological deficit theory relates dyslectics' difficulties in matching sounds to letters to phonological deficits. The magnocellular deficit theory relates dyslectics' visual, auditory and tactile difficulties to magnocellular deficits. The cerebellar deficit theory relates dyslectics' difficulties in automizing learn tasks to cerebellar deficits. The auditory deficit theory relates dyslectics' difficulties in perceiving rapid changes of sounds to auditory deficits. The visual deficit theory relates dyslectics' difficulties in reading to visual deficits.

Technology and Dyslexia: Studies on this area of attention provided us pieces of information regarding the contribution of technology to the improvement of dyslectics' performance. Assistive technology can be a useful tool for dyslectics. Especially mobile applications can help dyslectics improve their reading and writing skills and performance. In this area, there is also one study, which provided us an overview of a number of 180 existing software applications for dyslectic users. Based on this study the largest number of existing software applications aims at improving dyslectics' reading skills.

IxD and Dyslexia: Studies on this area of attention provided us pieces of information regarding the recommended interaction design guidelines and parameters focused on functionality and user interface. Based on studies of this area of attention, we learned that specific font size, colors and layouts facilitates dyslectics to improve their reading performance. Based on this area of attention studies, IxD guidelines and parameters focus on helping dyslectics improve their reading skills.

4.1 Interaction Design Guidelines and Parameters

Research on interaction design guidelines resulted in one design guideline with an emphasis on three design dimensions -Form, Content and Behavior- mentioned of high importance for software applications' design addressing to dyslectic users. To be more precise the interaction design guideline supports that these dimensions and their elements facilitate users, who address visual (the form dimension), or phonological deficits (the content and behavior dimensions) due to dyslexia [26–29]. Simple and clear layouts with font sizes from 12 to 14 and Sans Serif fonts, as well as features that allow dyslectic users to adjust the font size, the style, and colors, or specific combinations on colors and contrasts by avoiding bright colors, have been recommended as supportive to dyslectic users and able to improve their reading performance. Additionally, features that provide explanations, enrichment of texts with pictures and audio elements make reading tasks more accessible for users with dyslexia.

Moving forward, our literature analysis led us to Rello and Barbosa study on IxD parameters of software applications for dyslectic users. These interaction design parameters focus on the Form dimension, as visual deficits affect deeply dyslectic users' reading performance. This study recommends a number of layout-design parameters as appropriate to help dyslectic users improve their reading performance [17]. Specifically, *Font Type/Sizes:* Arial, Comic Sans Verdana, Century, Gothic, Trebuchet, Dan Sassoon Primary, Times New Roman, Courier, Dyslexie/12 or 14, and extra-large letter spacing, *Brightness-Colors:* Low Brightness & color differences among text and background, and Light grey as font color, *Space/Lines/Columns:* Lines of 60–70/Characters Clear Spacing between letter combinations/Line spacing: 1.3, 1.4, 1.5, 1.5–2/Narrow columns should be avoided [17].

Explaining the Rello and Barbosa text layout parameters, Sans Serif fonts of a size between 12 and 14, low brightness and light contrasts between background and fonts' colors have been recommended by their study. Furthermore, suggestions for lines of 60 to 70 characters maximum and clear spacing between letter combinations, as well as line spacing from 1.3 to 2, and avoidance of narrow columns have been recommended as supportive to dyslectic users and able to improve their reading performance [17]. Based on comparisons among the IxD guidelines/parameters there are many similarities on (i) the font type and size, (ii) the recommendations about avoiding bright colors and narrow columns, and (iii) the suggesting number of characters and line spacing (see Table 2).

Table 2. Comparisons among IxD guidelines/parameters

Design Guidelines' Synopsis (Table 4)			Rello and Barbosa's text layout parameters (Table 5)		
Fonts	**Colour**	**Layout**	**Fonts**	**Colour**	**Layout**
Font Type /Size	**Brightness/ Colours**	**Space/Lines/ Columns**	**Font Type /Size**	**Brightness/ Colours**	**Space/Lines/ Columns**
Font Type: Arial, Comic Sans Verdana, Century, Gothic, Trebuchet, Dan Sassoon **Font Size:** from **12 to 14** points or larger	Avoid bright green and red colours & **Suggested pale colour-codes:** #A4D5A6 #CCE685 #A8E685 #DED8E4 #87AA74 #9E9E7C #F19D3B	Lines of no more than **70 characters** Line **spacing:** 1.5 **Avoidance of narrow columns**	**Font Type:** Arial, Comic Sans Verdana, Century, Gothic, Trebuchet, Dan Sassoon **Font Size:** **12 or 14** extra-large letter spacing	**Low Brightness** & colour differences among text and background Light grey as **font colour**	Lines of 60-70 **characters** **Line spacing:** 1.3 1.4 **1.5** 1.5-2 **Narrow columns** should be **avoided**

Similarities on examined IxD guidelines/ parameters indicate that there is an agreement within the literature in interaction design guidelines for systems to support dyslectic users (See Table 3).

Table 3. Suggested IxD generated by comparisons

Suggested IxD		
Fonts	**Colour**	**Layout**
Font Type /Size	**Brightness/ Colours**	**Space/Lines/ Columns**
Font Type: Arial, Comic Sans Verdana, Century, Gothic, Trebuchet, Dan Sassoon **Font Size:** 12 - 14	Pale colours e.g. #A4D5A6 #CCE685 #A8E685 #DED8E4 #87AA74 #9E9E7C #F19D3B Avoidance of brightness	Lines of maximum 70 **characters** **Line spacing:** 1.5 Narrow **columns** should be avoided

As we can see on the table (See Table 3), design elements, such as Sans Serif fonts of 12 to 14 size, pale background colors, avoidance of brightness and narrow columns, and lines of maximum 70 characters with a line spacing of 1.5 have been recommended as the most appropriate for designs addressing to dyslectic users.

4.2 Existing SW Applications Supporting Dyslectic Users

As we mentioned earlier, the CALL study that we found on 'IxD and Dyslexia' area of attention provided us two 'AppWheels' presenting a number of 180 different types of Android and iPad software applications for dyslectic users, categorized into five conceptual areas of reading, writing, numeracy, memory and organization [24, 25] (See Fig. 1). Finally based on the 'AppWheels', we made an overview table presenting the total number and types of the existing software applications categorized by conceptual areas of reading, writing, numeracy, organization and memory. This table helped us learn that the majority of the existing software applications for dyslectic users focus mainly on improving the Reading conceptual area, even though there is no any clear and reliable scientific evidence on the reflection of those numbers (See Table 4).

As we can see on the table below, the majority of the existing software applications (70 out of 180) have been developed to help dyslectic users improve their performance on reading. A number of 64 software applications have been developed to help dyslectics improve their writing performance, and only a small total number of 41 applications have been developed to improve dyslectics' skills and these applications belong to the numeracy, organization and memory conceptual areas.

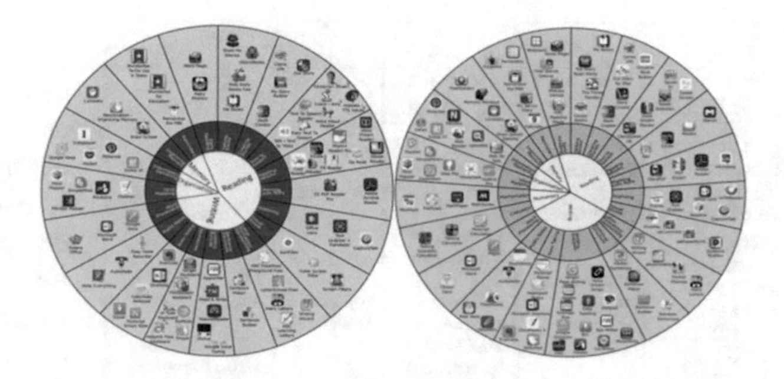

Fig. 1. Android and iPad apps for learners with dyslexia [24, 25]

Table 4. Overview of existing SW applications [24, 25]

	Categories				
Software Applications	**Reading**	**Writing**	**Numeracy**	**Organisation**	**Memory**
Early Reading	7				
Taking Books	8				
Creating Stories	9				
Text to Speech	12				
Reading eBooks	12				
Working with PDFs	6				
Scanning	9				
Visual Stress	7				
Letter Formation/Handwriting		9			
Sentence Structure		6			
Writing Support		18			
Keyboards		6			
Note Taking		11			
Audio Notes		8			
Word Processing		6			
Calculator			3		
Mathematics			4		
Mind Mapping				12	
Information Gathering				12	
Improving Memory					7
Reminders					8
Total	**70**	**64**	**7**	**24**	**15**

Unfortunately, there is not any clear and reliable scientific evidence of why the numbers are higher on the reading conceptual area. If we take into consideration the principle of dyslexia "Dyslexia is a hidden disorder in reading, spelling and written language" [30], these numbers may reflect that the reading one is this conceptual area, where dyslectics face the majority of their difficulties in comparison to the other four areas. Alternatively, the numbers of the existing software applications in this area are too high because this may be the conceptual area of developers' interest. Based on the overview table of the existing software applications (See Table 4), we divided the software applications into Android and iPad software applications, and made a with the purpose of finding, if the high numbers on the reading conceptual area are based on an increase of one OS's numbers, or if the numbers are high on both OS. The results indicate a trend on the software applications' development focused on improving dyslectic users' reading skills and performance. Because of the fact that there was not any clear scientific evidence to excuse the previously mentioned trend on the reading conceptual area, we decided to investigate it. For this purpose, we visited the Google Play Store [31] and collected dyslectic users' comments on the 29 first reading software applications addressing to dyslectic users. The total number of dyslectic users' complaints on reading software applications makes us clear that the main problem of users refers to applications' design and systems' quality, which make an application useless in the end, as they claim. This was partially surprising for us, since both related work and reviewed literature focused mainly on design, and there was not any suggestion focused on the systems' quality. Below we can see a sample of dyslectic users' complaints with the higher total numbers:

NOT WORKING

"It just doesn't work I speak into it and it comes out as a scramble of words."

"Amazing idea but failed to work. I am myself severely dyslexic and found the idea of this program amazing but sadly it fails to work every time. If the bugs can be worked out I would be very happy to re evaluate. Please do not abandon the attempt to get this program to work as I can see the potential in the idea."

"Does not work. All it keeps saying is that it could not extract text from image...even though very clear picture."

USELESS

"Without the option for font size, the app is useless if don't have a device it was made for. This font is too big and it's unreadable. Can't understand why there isn't an option for changing font size."

"Cool idea but doesn't give any controls to go back a few sentences or to navigate a table of contents. In the end, not quite usable. I look forward to future improvements."

BUGS/ BAD DESIGN

"Interface is also poor-need to keep pressing the record button for long text and keyboard editing capability (e.g. to correct errors) us very clunky."

"...the PDF to text translator doesn't seem to be able to handle the formatting. PDF documents written in LaTeX also fail. I'll keep it installed for the time being in hopes that this can be remedied. Thanks!"

The first indication generated by the above comments is that there is a need for developing system and design of high quality for software applications, especially because they address to users with dyslexia. Complaints like, "Interface is also poor", "It just doesn't work I speak into it and it comes out as a scramble of words", or "…I found the idea of this program amazing but sadly it fails to work every time", or "the PDF to text translator doesn't seem to be able to handle the formatting.", validate this indication. Dyslectic users can be patient or even contribute the improvement of promising software applications, if there are possibilities for future improvement. Comments, like "If the bugs can be worked out I would be very happy to re-evaluate.", or "I look forward to future improvements.", or "I'll keep it installed for the time being in hopes that this can be remedied.", validate this indication.

5 Discussion

In this section, we compare the findings to related work, in regards to the literature research on interaction design and existing software applications for dyslectic users. Specifically, we discuss the interaction design guidelines and parameters shown in the study and their relation to the related work. We also discuss possible reasons for the focus of the existing software applications' focus on the reading conceptual area. Finally, we discuss if there is any relation between the users' complaints on the existing software applications on Google Play Store and the results of the literature.

5.1 Interaction Design Guidelines and Parameters

Regarding the interaction design guidelines and parameters shown in the study, and their relation to the design parameters of the related works, there is a clear agreement among them. Both IxD guidelines/parameters of literature research and related work focused on user interface and functionalities that help dyslectic users improve their reading performance. In both related works and literature research's IxD guidelines, design parameters have been proposed for developing designs for software applications addressing to dyslectic users. Their aim? To facilitate and help dyslectic users improve their reading skills and performance.

5.2 Existing SW Applications Supporting Dyslectic Users

As for the existing software applications supporting dyslectic users and based on literature, the largest number of them have been developed with a focus on improving dyslectic users' reading skills and performance. Even though there is not any clear and reliable scientific evidence regarding this trend on the reading conceptual area, we assume that this trend may be based on three main factors: (i) The poor reading-performance is the first sign that a person has a cognitive disorder, like dyslexia [32]. (ii) As the IxD guidelines/parameters focus mainly on the reading conceptual area, it is reasonable for the developers to focus their work on this conceptual area. (iii) Developers may focus their interest on the reading conceptual area, as the better reading performance of dyslectic persons is, the better their daily routine becomes, since reading is a daily requirement for everybody either dyslectic or not.

5.3 Users' Complaints on Quality and Literature

Based on findings about users' complaints on reading software applications, the main problem was related to the designs and systems' quality of the existing software applications for dyslectic users. That was partially surprising for us. The elated work, the reviewed literature and the users comments have mentioned a focus on design parameters and guidelines for systems addressing to dyslectic users. On the contrary, there is not any mention on the quality of the systems.

5.4 Implications

Given the trend of developing reading software applications for dyslectic users and the users' comments about a need for improving the designs of applications for dyslectic users, the current research indicates a further investigation for appropriate designs to improve dyslectic users' reading performance. This indication may imply the involvement of dyslectic people to the developing process in order for the designs (but also the systems) to correspond to dyslectic users' expectations.

6 Conclusion

In this paper, we presented the results of a literature review on studies about interaction design of systems for dyslectic users. The studies have been selected without any limitation on specific conferences, since the sources regarding dyslexia and interaction design were very limited. The literature review resulted in a number of 71 studies related to interaction design and dyslexia. Each study was categorized by areas of attention including the related keywords. A table presented an overview of the reviewed studies references organized by areas of attention. Based on findings: (i) Interaction design guidelines and parameters focus mainly on reading conceptual area, as the most popular conceptual area among the writing, numeracy, memory and organization conceptual areas. (ii) the high numbers of existing Android and iPad reading software applications corroborates the former trend, and (iii) from the dyslectic users' perspective, there is a need for developing designs and systems of high quality. One limitation, which we faced throughout our literature survey, was regarding the filtering process. We tried not to miss important literatures and this resulted in a time consuming back and forth research until selecting the final number of the literature for reviewing. However, this was a study made by one person, so there is always the possibility of missing literatures. A second limitation was about the final selection of the studies. After collecting a number of 175 studies related to our study, we systematically applied four specific questions to all the 175 studies as rules for the final selection of the studies. Based on these questions, we came up with 71 studies, although there is always room for improvement of this method. Our literature study resulted in a trend on the Reading among five conceptual areas of Reading, Writing, Numeracy, Memory, and Organization. Considering the lack of finding any scientific evidence that excuse this trend, we believe that it would be beneficial a future research to finding why there is a focus on the reading conceptual area. Furthermore, Even though there is an agreement among the interaction design guidelines and parameters for developing designs of software applications for dyslectic users, then why did a number of dyslectic users express a need for higher quality of designs? This need indicates a further research on the efficiency of the existing interaction design parameters and guidelines. This indication can contribute a future research on this field with the aim of fulfilling at some point dyslectics' expectations. Based on our literature survey and on the existing software applications' users' complaints, this current literature survey will contribute a future empirical study of investigating how we could develop together with dyslectic users a design for a software application to enhance their reading skills. Furthermore, based on our literature survey's findings regarding the impact of the type of languages to the level of dyslexia, part of a future empirical study will be focused on investigating how the type of languages differentiates the efficiency-level of a software application's designs for dyslectic users.

Acknowledgements. I would like to thank our supervisor Jan Stage for his valuable feedback and guidance throughout the study period.

References

1. Hall, J., Tinklin, T.: Students First: The Experiences of Disabled Students in Higher Education. University of Glasgow, Glasgow (1998)
2. Swanson, H.L., Harris, K.R., Graham, S.: Language processes: keys to reading. In: Handbook of Learning Disabilities, pp. 213–277. The Guilford Press, New York (2013)
3. Theodora, M.: Nature of the problem and solution, Athens, pp. 61–66 (2004)
4. Madeira, J., Silva, C., Marcelino, L., Ferreira, P.: Assistive mobile applications for dyslexia. Procedia Comput. Sci. **64**, 417–424 (2015)
5. Lynne, O.: Word processing: its impact on children's writing. J. Learn. Disabil. **22**, 262–264 (1989)
6. Olson, R., Wise, B.W.: Reading on the computer with orthographic and speech feedback: an overview of the Colorado remediation project. Read. Writ.: Interdisc. J. **4**, 107–144 (1992)
7. Αθανάσιος, Μ.: ΟΙ ΝΕΕΣ ΤΕΧΝΟΛΟΓΙΕΣ ΣΤΗΝ ΕΙΔΙΚΗ ΑΓΩΓΗ. CVP ΠΑΙΔΑΓΩΓΙΚΗΣ & ΕΚΠΑΙΔΕΥΣΗΣ, 28 August 2015. ISSN: 2241-4665
8. Fragaki, M.: Η Τεχνολογία στην Ειδική Αγωγή: Ένα Εναλλακτικό Μέσο σε μια Πολυμορφική Εκπαίδευση. In: 6th International Conference in Open & Distance Learning, November 2011. Loutraci, Greece-Proceedings, Loutraki, Athens (2011)
9. Heim, A.: This young scientist is using technology to help dyslexics. TNW, 14 July 2013. https://thenextweb.com/insider/2013/07/14/this-young-scientist-is-using-technology-to-help-dyslexics/#.tnw_WoxOdZlY. Accessed 28 May 2017
10. de Avelar, L.O., Rezende, G.C., Friere, A.P.: WebHelpDyslexia: a browser extension to adapt web content for people with dyslexia. In: 6th International Conference on Software Development and Technologies for Enhancing Accessibility and Fighting Info-exclusion (DSAI 2015), Lavras, Brazil (2015)
11. Freire, A.P., Petrie, H., Power, C.: Empirical results from an evaluation of the accessibility of websites by dyslexic users. Human Computer Interaction Group, Department of Computer Science, University of York, Deramore Lane, York, UK (2011)

12. Al-Wabil, A., Zaphiris, P., Wilson, S.: Web navigation for individuals with dyslexia: an exploratory study. In: Proceedings of the 4th International Conference on Universal Access in Human Computer Interaction: Coping with Diversity. Springer, Heidelberg (2007)
13. Rello, L., Kanvinde, G., Baeza-Yates, R.: Layout guidelines for web text and a web service to improve accessibility for dyslexics. In: Proceedings of the International Cross-Disciplinary Conference on Web Accessibility, New York, NY, USA (2012)
14. Rello, L., Pielot, M., Marcos, M.C., Carlini, R.: Size matters (spacing not): 18 points for a dyslexic-friendly Wikipedia. In: Proceedings of the 10th International Cross-Disciplinary Conference on Web Accessibility, New York, NY, USA (2013)
15. Rello, L., Saggion, H., Baeza-Yates, R.: Keyword highlighting improves comprehension for people with dyslexia. In: Proceedings of the 3rd Workshop on Predicting and Improving Text Readability for Target Reader Populations (PITR), Gothenburg, Sweden (2014)
16. Santana, V.F., Oliveira, R., Almeida, L.D.A., Ito, M.: Firefixia: an accessibility web browser customization toolbar for people with dyslexia. In: Proceedings of the 10th International Cross-Disciplinary Conference on Web Accessibility, New York, NY, USA (2013)
17. Rello, L., Barbosa, S.D.J.: Do people with dyslexia need special reading software? In: Workshop on Rethinking Universal Accessibility: A Broader Approach Considering the Digital Gap, Cape Town, South Africa (2013)
18. Rello, L., Baeza-Yates, R., Bott, S., Saggion, H.: Simplify or help?: text simplification strategies for people with dyslexia. In: Proceedings of the 10th International Cross-Disciplinary Conference on Web Accessibility, New York, NY, USA (2013)
19. Rello, L., Baeza-Yates, R.: Good fonts for dyslexia. In: Proceedings of the 15th International ACM SIGACCESS Conference on Computers and Accessibility (2013)
20. McCarthy, J.: Dyslexia and accessibility/usability
21. McCarthy, J.E., Swierenga, S.J.: What wc know about dyslexia and Web accessibility: a rescarch review. Michigan State University, Michigan (2009)
22. Bradford, J.: Designing web pages for dyslexic users. Dyslexia Online Mag. (2005). http://www.dyslexia-parent.com/mag35.html
23. Vashti, Z.: Ten guidelines for improving accessibility for people with dyslexia. CETIS, University of Wales Bangor, Wales Bangor (2002)
24. The University of Edinburgh, CALL Scotland: Android Apps for Learners with Dyslexia/Reading and Writing Difficulties, Version 1.0. CALL Scotland, The University of Edinburgh, October 2015. http://www.callscotland.org.uk/downloads/posters-and-leaflets/android-apps-for-learners-with-dyslexia/. Accessed 11 May 2017
25. iPad Apps for Learners with Dyslexia/Reading and Writing Difficulties, Version 1.4. CALL Scotland, The University of Edinburgh, February 2016. http://www.callscotland.org.uk/downloads/posters-andleaflets/ipad-apps-for-learners-with-dyslexia/. Accessed 11 May 2017
26. Web Content Accessibility Guidelines 1.0. World Wide Web Consortium, 5 May 1999. https://www.w3.org/TR/WAI-WEBCONTENT/, https://www.w3.org/TR/WCAG10/. Accessed 10 May 2017
27. IMS Global Learning Consortium, July 2002. https://www.imsglobal.org/activity/accessibility, https://www.imsglobal.org/accessibility/accessiblevers/index.html
28. Benyon, D.: Designing Interactive Systems: A Comprehensive Guide to HCI. Centre for Interaction Design, Edinburgh Napier University (2013)
29. Husni, H., Aziz, F.A.: Interaction design for dyslexic children reading application: a guideline. In: Knowledge Management International Conference (KMICe), Johor Bahru, Malaysia (2012)
30. Fry, E.B., Kress, J.E.: The Reading Teachers Book of Lists. Jossey-Bass Teacher (2015). ISBN-13: 978-0787982577, ISBN-10: 0787982571
31. Google Play Store. Google. https://play.google.com/store

32. Stein, J., Walsh, V.: To see but not to read; the magnocellular theory of dyslexia. University of Oxford, Oxford (1997)
33. Bradley, L., Bryant, P.E.: Difficulties in auditory organisation as a possible cause of reading backwardness. Lett. Nat. **271**, 746–747 (1978)
34. Lovegrove, W.J., Bowling, A., Badcock, D., Blackwood, M.: Specific reading disability: differences in contrast sensitivity as a function of spatial frequency. Science **210**(4468), 439–440 (1982)
35. Nicolson, R., Fawcett, A., Berry, E., Jenkins, I., Dean, P., Brooks, D.: Association of abnormal cerebellar activation with motor learning difficulties in dyslexic adults. Lancet **353** (9165), 1662–1997 (1999)
36. Stein, J., Walsh, V.: To see but not to read; the magnocellular theory of dyslexia. Trends Neurosci. **20**(4), 147–152 (1997)
37. Tallal, P.: Auditory temporal perception, phonics, and reading disabilities in children. Brain Lang. **9**(2), 182–198 (1980)
38. The University of Edinburgh: Android Apps for Learners with Dyslexia/Reading and Writing Difficulties, Version 1.0. CALL Scotland, September 2015. http://www.callscotland.org.uk/downloads/posters-and-leaflets/android-apps-for-learners-with-dyslexia/. Accessed 10 May 2017
39. The University of Edinburgh: CALL Scotland. The University of Edinburgh, February 2016. http://www.callscotland.org.uk/downloads/posters-and-leaflets/ipad-apps-for-learners-with-dyslexia/. Accessed 10 May 2017
40. Olson, R., Foltz, G., Wise, B.: Reading instruction and remediation with the aid of computer speech. Behav. Res. Methods Instrum. Comput. **18**, 93–99 (1986)
41. Rello, L., Barbosa, S.D.J.: Do people with dyslexia need special reading software? Department of Information and Communication Technologies, University Pompeu Fabra, Barcelona (2013)
42. Kranz, J., Müller-Bloch, C.: A framework for rigorously identifying research gaps in qualitative literature reviews. In: International Conference on Information Systems (2015)
43. Wolfswinkel, J.F., Furtmueller, E., Wilderom, C.P.M.: Using grounded theory as a method for rigorously reviewing literature. Eur. J. Inf. Syst. **22**(1), 44–55 (2013)

Analytical Hierarchy Process and Human Plausible Reasoning for Providing Individualised Support in a Mobile Interface

Katerina Kabassi[1(✉)], Maria Virvou[2], and Efthimios Alepis[2]

[1] TEI of Ionian Islands, Zakynthos, Greece
kkabassi@teiion.gr
[2] University of Piraeus, Piraeus, Greece
{mvirvou, talepis}@unipi.gr

Abstract. This paper focuses on the combination of Human Plausible Reasoning (HPR) with Analytical Hierarchy Process (AHP) for the purposes of a system that provides individualized support to users of a mobile phone. The system is called MobIFM and uses a combination of a HPR with AHP to make hypothesis about users' goals and plans and provide individualized assistance in case a user action is considered as not intended.

Keywords: Cognitive theory · Analytical Hierarchy Process · Individualisation Intelligent help

1 Introduction

Developing software for handheld devices is still quite challenging since it requires special software development skills due to the huge diversity of mobile devices and also as a result of the large number of sensors that these devices incorporate in order to provide context awareness. A sophisticated UI should be dynamic and adapt its content in order to correspond to the user needs and at the same time handle specific device limitations. In view of the above, individualized assistance in mobile UIs could be of great value both for users and also for mobile software vendors, since it could facilitate user interaction and could result in improved user experience and less user frustration. Furthermore, for the provision of individualized assistance, it is rather helpful for a system to model the human experts' reasoning process when they are trying to help a user. Actually, the system will try to provide individualized help by reproducing the complex human reasoning. For this purpose cognitive theories seem rather appropriate. Multi-criteria decision making theories, on the other hand, can facilitate a system in reproducing experts' reasoning when selecting the item that seems to be most appropriate from a group of alternatives.

In view of the above, the paper focuses on a combination of a cognitive theory with a multi-criteria decision making theory for providing individualized support to users of a mobile phone. Human Plausible Reasoning has been combined before with a multi-criteria decision making theory in the domain of learning environments or graphical user interfaces (Virvou and Du Boulay 1999), (Virvou and Kabassi 2000,

M. E. Auer and T. Tsiatsos (Eds.): IMCL 2017, AISC 725, pp. 345–352, 2018.
https://doi.org/10.1007/978-3-319-75175-7_35

2002, 2004). However, the combination of Human Plausible Reasoning with Analytical Hierarchy Process has never been used before in any domain. The implementation of this novel combination for providing individualised support in a mobile interface seems rather promising as it can effectively simulate human reasoning.

The combination has been tested in system called MobIFM to make hypothesis about users' goals and plans and provide individualized assistance in case a user action is considered not intended. More specifically, HPR provides a unifying formal framework of inference patterns that is domain-independent. Additionally, the theory defines a set of criteria (certainty parameters) to be taken into account in order to select the best hypothesis. However, a main problem of the theory is that it does not specify precise mathematical formulas for combining these criteria. For this purpose the AHP is used. The main advantage of the latter theory is that it has formal way for calculating the weights of the criteria by making pair-wise comparisons of the criteria. Pair-wise comparisons are also made between the alternative actions generated by the system by applying the cognitive theory in order to sort them and find the most appropriate one with respect to the user's goals and plans.

2 Related Work

2.1 Human Plausible Reasoning

Human Plausible Reasoning (HPR) is a descriptive theory of human plausible inference which categorizes plausible inferences in terms of a set of frequently recurring inference patterns and a set of transformations on those patterns (Collins and Michalski 1989). The theory is used to formalize the plausible inferences that frequently occur in people's responses to questions for which they do not have ready answers. The theory is based on the analysis of people's answers to everyday questions about the world and consists of three parts:

1. a formal representation of plausible inference patterns; such as deductions, inductions, and analogies, that are frequently employed in answering everyday questions;
2. a set of parameters, such as conditional likelihood, typicality and similarity, that affect the certainty of people's answers to such questions; and
3. a system relating the different plausible inference patterns and the different certainty parameters.

HPR detects the relationship between a question and the knowledge retrieved from memory and drives the line (type) of inference. More specifically, HPR models the reasoning of people who have a patchy knowledge of certain domains such as geography. By patchy knowledge we mean partial knowledge of the facts and relations in the domain.

The theory assumes that a large part of human knowledge is represented in "dynamic hierarchies" that are always being updated, modified or expanded. Every time the user does not know the answer, several alternative statements are created using different relationships such as generalization (GEN), specialization (SPEC), similarity

(SIM) and dissimilarity (DIS) among concepts in hierarchies trying to assume what the user really intended to say.

The theory also introduces certainty parameters that affect the certainty of the inferences of these four kinds of expression. The certainty parameters are approximate numbers ranging between 0 and 1 that affect the certainty of different plausible inferences and are the following: Degree of similarity (σ), Typicality (τ), Frequency (ϕ), Dominance (δ), Degree of Certainty (γ). However, one problem that remained unsolved in HPR was the calculation of the certainty parameters and their combination. In the past, this problem has been solved by the application of Multi-Criteria Decision Making theories (Kabassi and Virvou 2006; 2015).

2.2 Analytical Hierarchy Process

AHP (Saaty 1980) is one of the most popular MCDM theories and consists of the following steps (Zhu and Buchman 2000):

1. **Developing a goal hierarchy:** The overall goal, criteria and decision alternatives are arranged in a hierarchical structure. After decomposing the problem into a hierarchy, alternatives at a given hierarchy level are compared in pairs to assess their relative preference with regard to each criterion at the higher level. A scale is needed to represent the varying degrees of preference. Saaty (1980) establishes a scale, where 9 is the upper limit and 1 is the lower limit and a unit difference between successive scale values is used. This scale is built based on psychological experiments, which have shown that individuals have difficulty to compare more than five to nine objects at one time.
2. **Setting up a pairwise comparison matrix of criteria:** A comparison is implemented among the elements that are on the same level of the goal hierarchy. In a comparison process, a V from the scale is assigned to the comparison result of two elements P and Q at first, then the value of comparison of Q and P is a reciprocal value of V, i.e. 1/V. The value of the comparison of P and P is 1. Following these rules a comparison matrix of criteria is built. The weights of the matrix attributes are calculated through finding the eigenvector associated with the maximal eigenvalue of this matrix.
3. **Ranking the relative importance between alternatives:** In this step, the relative importance between each pair of alternatives in terms of a criterion will be assessed. All matrices are normalized and the weight of each alternative is also derived.
4. **Calculating AHP values:** The AHP value is computed using the following formula:

$$AHP_i = \sum_{j=1}^{N} a_{ij} w_j, \text{for } i = 1, 2, 3, \ldots, M \tag{1}$$

where M is the number of alternatives and N is the number of criteria; a_{ij} denotes the score of the ith alternative related to the jth criterion; w_j denotes the weight of the jth criterion.

3 MobIFM: File Manipulation in Mobile Phones

Intelligent File Manipulator (MobIFM) is an interface for file manipulation in a mobile phone. The system reasons about every user's action and in case it diagnoses a problematic situation, it provides spontaneous advice. When MobIFM generates advice, it suggests to the user a command, other than the one issued, which was problematic. In this respect, MobIFM tries to find out what the error of the user has been and what his/her real intention was.

MobIFM collects information about the user and stores them in a user model. This information is used to make hypothesis about the user's goals and plans. The reasoning of MobIFM is similar to that of IFM for Personal Computers (Virvou and Kabassi 2004).

The algorithm used for the reasoning of user action is the following:

1. The user issues an action.
2. The system reasons about the action so as to categorize it in one of the four categories
   a. **Expected**: In this case the action is expected by the system in terms of the user's hypothesized goals.
   b. **Neutral**: In this case the action is neither expected nor contradictory to the user's hypothesized goals.
   c. **Suspect**: In this case the action contradicts the system's hypotheses about the user's goals.
   d. **Erroneous**: In this case the action is wrong and does not have a result.
3. If the action is categorized as "expected" or "neutral" it is executed.
4. If the action is categorized as "suspect" or "erroneous" then it is transformed based on an adaptation of HPR. The transformation of the given action is done so that similar alternatives can be found which would not be suspect or erroneous.
5. The system reasons about every alternative action so that it can categorize it in one of the four categories in a similar way as in step 2.
   a. If an alternative action is categorized as "suspect" or "erroneous" is ignored
   b. If an alternative action is categorized as "expected" or "neutral" is selected to the final set of alternatives.
6. For each alternative action of the set of alternative actions, the system applies the combination of HPR with AHP to calculate the degree of certainty for each alternative action.
7. If no better alternative can be found that could be compatible to users' goals and plans then the user's initial action is executed without the user realizing that the system was alerted.

4 HPR and AHP

If the user action is categorized by the reasoning mechanism of the system as "suspect" or "erroneous" then it is transformed using the statement transforms of HPR. MobIFM may not find any "expected" or "neutral" alternative action or it may find plenty. In that case the system may not know which one to propose while proposing several alternatives to a novice user may entangle him/her rather than help him/her. A solution to this problem is ordering the alternative actions in a way that the ones, which are most likely to have been intended by the user, come first. For this purpose, we have used five of the certainty parameters presented in HPR, which are considered the criteria for evaluating the different alternative actions generated by the system. The certainty parameters are:

- Degree of similarity (σ): The degree of similarity is used to calculate the resemblance of two commands or two objects.
- Typicality (τ): The typicality of a command is based on the estimated frequency of execution of the command by the particular user.
- Frequency (ϕ): The degree of frequency of a command represents how often a user has mistaken the execution of that command.
- Dominance (δ): The dominance of an error in the set of all users' errors as these have been recognised and stored in the user model reveals the weaknesses of a user (Virvou and Kabassi 2004).

Finally, all parameters are combined in order to calculate another certainty parameter, the degree of certainty (γ). Since, the exact way of calculation of its value, was not specified fully in HPR, we apply AHP to combine the above mentioned certainty parameters for the calculation of the degree of certainty (γ).

The first step for the application of AHP is the development of the goal hierarchy. The goal hierarchy consists of the overall goal, the set of criteria and the alternative action. The overall goal is to find the best advice to provide to the user, which means that one has to find the alternative action that the user probably meant to issue. The criteria are the certainty parameters of HPR. The alternative actions are the alternatives that have been produced by the reasoning mechanisms of the system after applying the statement transforms of HPR.

The second step of the application of AHP concerns the calculation of the weights of the criteria (certainty parameters). For this purpose a pair-wise comparison matrix is constructed. More specifically, a 4×4 table is formed and the group of decision makers has to complete it. The group of decision makers consists of 5 human experts in educating novice users in Computer Science were used to make the pair-wise comparisons of criteria.

The five matrixes (one from each expert acting as decision maker) that are collected and the value of each cell of the final matrix is calculated as a geometric mean of the corresponding cells of the other five matrixes collected by the human experts. As a result, the final matrix is built (Table 1).

After making pair-wise comparisons, estimations are made that result in the final set of weights of the criteria. In this step, the principal eigenvalue and the corresponding normalised right eigenvector of the comparison matrix give the relative importance of the various criteria being compared. The elements of the normalised eigenvector are the weights of criteria or sub-criteria. There are several methods for calculating the eigenvector. In terms of simplicity, we have used the 'Priority Estimation Tool' (PriEst) (Sirah et al. 2015), an open-source decision-making software that implements the AHP method, for making the calculations of AHP. The weights of the criteria were estimated

Table 1. Final matrix of the pair-wise comparison of certainty parameters

	σ	τ	φ	δ
σ	1.00	3.73	3.73	3.73
τ	0.27	1.00	0.28	2.17
φ	0.27	3.52	1.00	2.55
δ	0.27	0.46	0.39	1.00

to: $w_\sigma = 0.533$, $w_\tau = 0.127$, $w_\varphi = 0.247$, $w_\delta = 0.093$. Indeed, human experts thought that the most important criterion was considered to be the similarity. Second most important was degree of frequency of an error and other two criteria were considered less important.

The next step of the application of HPR is evaluating the different alternative actions in terms of the certainty parameters. More specifically, a comparison matrix is formed for each one of the certainty parameters of HPR, so that the alternative actions are pair-wise compared. As a result, four matrixes are generated. For the completion of these four tables, the system uses the information from the user model as well as the information of the knowledge representation component in order to complete the tables. The comparison matrix of alternative actions with respect the certainty parameter of similarity (σ) is completed taking into account the knowledge representation component of the system. The other three tables are completed using the information from the user model.

Finally, the degree of certainty is calculated as an AHP value

$$\gamma_i = \sum_{j=1}^{N} a_{ij} w_j, \text{for } i = 1, 2, 3, \ldots, M \quad (2)$$

where M is the number of alternatives and N is the number of criteria; a_{ij} denotes the score of the ith alternative related to the jth certainty parameter; w_j denotes the weight of the jth criterion.

5 Example of Operation

The user of MobIFM first creates a new folder called ‘NewPhone’. Then he copies a video file called ‘Sotiris’ and selects the folder ‘NewPhotos’. However, a file with that name already exists in the folder ‘NewPhotos’. The system finds the particular action as suspect and tries to generate alternative actions that are considered as expected or neutral. As a result, it generates the following alternatives applying the HPR transforms of HPR:

- Opt1: CopyTo(Photos)
- Opt2: CopyTo(NewPhone)
- Opt3: CopyTo (Download)

In order to select which one of the alternative actions the MobIFM is going to propose to the user, the system applies AHP to calculate the degree of certainty of each alternative. The following steps are implemented:

1. *Developing the goal hierarchy.* The overall goal is to find the most appropriate advice for the user, the criteria are the four certainty parameters and Opt1-Opt3 are the three alternatives.
2. *Calculating the weights of the criteria.* This step has already been done and there is no need of repeating it.
3. *Ranking the relative importance between alternatives.* For each one of the four certainty parameters a matrix 3 × 3 is created. The alternative actions are pair-wise compared. The values of the four matrixes are calculated using the values of the user model. All matrices are normalized and the weight of each alternative is also derived.
4. *Calculating AHP values.* The AHP value is used as the value of the degree of certainty. The three values are $\gamma_{opt2} = 0.485\, \gamma_{opt3} = 0.332\, \gamma_{opt1} = 0.185$.

Opt-2 was the most likely to have been intended by the user. However, opt3 also considered a possible alternative. Opt1 had the lower value of γ and, therefore, considered as not being likely to have been intended by the user.

6 Conclusions

The provision of individualised support in using a mobile interface is expected to help users that are not familiar with the technology of mobile phone. In this paper we have shown how a cognitive theory called HPR can provide individualised assistance to a user while s/he uses a mobile phone. The particular theory has been used in the past for providing individualised help in different interfaces (Virvou and Du Boulay 1999; Virvou and Kabassi 2004) for personal computers but never before in a mobile phone. A main drawback of the theory that was also noticed in previous applications of HPR was that the theory provides a definition of criteria (certainty parameters) that affect the certainty of the hypotheses made but it did not specify precise mathematical formulas for combining these criteria. Therefore, the appropriateness of a specific decision making theory, the AHP, was investigated as a means to complete HPR. HPR has been combined before with a decision making theory for the purposed of individualised help (Kabassi and Virvou 2015), however, AHP is quite different as it uses pair-wise comparisons to compare pairs of alternatives with respect to each certainty parameter. It is among our future plans to evaluate the performance of the combination of HRP with AHP.

References

Kabassi, K., Virvou, M.: Combining decision making theories with a cognitive theory for intelligent help: a comparison. IEEE Trans. Hum-. Mach. Syst. **45**(2), 176–186 (2015)

Saaty, T.: The Analytic Hierarchy Process. McGraw-Hill, New York (1980)

Virvou, M., Du Boulay, B.: Human plausible reasoning for intelligent help. User Model. User-Adapt. Interact. **9**, 321–375 (1999)

Virvou, M., Kabassi, K.: An intelligent learning environment for novice users of a GUI. In: Gauthier, G., Frasson, C., VanLehn, K. (eds.) Intelligent Tutoring Systems: Proceedings of the 5th International Conference on Intelligent Tutoring Systems-ITS 2000. Lecture Notes in Computer Science, vol. 1839, pp. 484–493. Springer, Berlin (2000)

Virvou, M., Kabassi, K.: Intelligent on-line training for a GUI. In: Fernstrom, K. (ed.) Proceedings of the International Conference on Information Communication Technologies in Education, National and Kapodistrian University of Athens, pp. 257–262 (2002). ISBN 960-8313-03-1, ISSN 1109-2084)

Virvou, M., Kabassi, K.: Adapting the human plausible reasoning theory to a graphical user interface. IEEE Trans. Syst. Man Cybern. Part A Syst. Hum. **34**(4), 546–563 (2004)

Personalized Museum Exploration by Mobile Devices

Efthimios Alepis[1], Katerina Kabassi[2(✉)], and Maria Virvou[1]

[1] University of Piraeus, Piraeus, Greece
{talepis,mvirvou}@unipi.gr
[2] TEI of Ionian Islands, Zakynthos, Greece
kkabassi@teiion.gr

Abstract. User interface (UI) and user experience (UX) design are considered as two of the most important factors in modern mobile app development. Both UI and UX design are very important not only in terms of what mobile users initially see, as well as in how the mobile UI behaves and corresponds to users actions. This paper focuses on the creation of a novel mobile app for museums, named MuseFy, that can provide personalized assistance to users through adaptive UIs. Such features add great value both for users and also for mobile software vendors, since they facilitate user interaction and result in improved user experience and less user frustration.

Keywords: Smartphone apps · Mobile development
Personalized user interaction · Multi-criteria decision making

1 Introduction

Both web traffic reports as well as market share statistics indicate that mobile technology is not only growing over the last decade, but is also dominating in most of the fields of human-machine interaction, including mobile educational systems. Despite their worldwide adoption by users, mobile devices have specific limitations regarding their size constraints as well as their resource usage that results in energy consumption. As a result, there are challenges to be faced, such as adaptive user interfaces and improved user experience. Indeed, UI and UX design are considered as two of the most important factors in mobile app development. Taking human-mobile interaction a step further, a sophisticated user interface should behave dynamically and adapt its content in order to correspond to user needs and at the same time handle all kinds of possible device limitations.

For the personalization of the user interaction, the system uses a double stereotype system for modeling the users interacting with the system. The information stored in the user model is used in combination with a multi-criteria decision making theory for finding the most appropriate way of presenting information about the exhibits to the user. More specifically, the system uses the

M. E. Auer and T. Tsiatsos (Eds.): IMCL 2017, AISC 725, pp. 353–360, 2018.
https://doi.org/10.1007/978-3-319-75175-7_36

Technique for Order Preference by Similarity to Ideal Solution (TOPSIS) [5]. The particular theory calculates the relative Euclidean distance of the alternative from a fictitious ideal alternative. The alternative closest to that ideal alternative and furthest from the negative-ideal alternative is chosen best.

The rest of this paper is organized as follows. In the next Sect. 2, we review the related work both in terms of personalized mobile applications and also in applications of user modeling. Then, in Sect. 3 we cover the developed application's main architecture and we also illustrate some examples of its usage. Section 4 discusses how the adaptation to each user's needs is achieved through the incorporation of the proposed user modeling approach. Finally, Sect. 5 provides some discussion and thoughts for future work.

2 Related Work

This section provides discussion in related work, both in the area of personalization in mobile applications and also in user modeling. The later is a field with a lot of research over the last decades, which nevertheless may also emerge in mobile software development. The following subsection illustrates on the one hand the significance of enabling personalization techniques in mobile environments and on the other, the novelty of our proposed system. After a thorough investigation in the related scientific literature we have come up to find that there is no mobile application that utilizes multi-criteria decision making techniques in order to provide personalization for educational purposes.

2.1 Personalized Mobile Applications

Personalization in mobile interfaces and more generally mobile user modeling is a promising field of research that gathers interest in an increasing pace. This is clearly stated in [7], where the challenge of how to provide users with personalized services anywhere and anytime, without requiring the user to bootstrap a user model from scratch every time is described. The authors of this paper try to address corresponding issues such as how can the environment or service provider access the UM and get the needed information from the model, and how can the UM respond to continuous requests for personal data required by services in ubiquitous computing.

The authors of [1] provide a very interesting review on mobile location-based games for learning. They introduce the theoretical and empirical considerations of mobile location-based games, and then discuss an analytical framework of their main characteristics through typical examples. More specifically, their paper focuses on the narrative structure of mobile location-based games, the interaction modes that they afford, their use of physical space as prop for action, the way this is linked to virtual space and the possible learning impact such game activities have.

The "GUIDE" system is presented in [2], which provides city visitors with an intelligent and context-aware tourist guide. This paper is also relevant to our work, since it focuses on the role of adaptive hypermedia within the developed system and the techniques used to tailor or adapt the presentation of web-based information. To this end, the context used by "GUIDE" includes the visitor's personal context, namely the visitor's current location and personal profile, and the environmental context, such as the opening times of the city's attractions.

Another relevant work is of [3], where the factors during design of personalized mobile applications in cultural heritage environments are investigated. This paper presents a formal description of these factors which allows both for a systematic survey of existing practice, and for supporting the design process of mobile cultural heritage applications in the future.

Another aspect of mobile personalization applications is described in [6], namely mobile health applications and remote health monitoring. Mobile user modeling is applied in order to provide remote health monitoring together with behavior change support features and persuasion strategies.

Finally, the authors of [9] examine also the cultural heritage domain and propose a framework on how to monitor visitor behavior on the go, in order to determine personality traits. As the authors state, this resultant knowledge can be then used along with context to give tailored advice, while methods of monitoring visitor behavior, converting that to traits and to personality types are also described. Different dimensions of how to give tailored advice based on personality are also discussed.

3 Architecture of MuseFy

The mobile educational app is expected to facilitate mobile users both during their navigation inside the premises of a museum by adapting its content and also outdoors in order to choose the best places of interests according to their specific user profile. To this end, a native smartphone application for the Android OS has been developed, supported by a Cloud-as-a-Service infrastructure. For the needs of the later, "Firebase" [4], has been used as it provides a wide range of services as a unified platform for mobile development. Including a realtime database and backend as a service, "Firebase" has provided us with an API that allows application data to be synchronized across clients and at the same time updated and stored on Firebase's cloud.

More specifically, the developed mobile application is capable of handling data deriving from two different domains. The first domain is related to the information provided by the museums and their exhibits. Such data include graphics, text, audio files and animations. All data is categorized and labeled in terms of provided information simplicity, language, educational background and age classification. The second domain is related to the created user models and consists of information related to user profiles, such as nationality, gender, age, educational background, familiarity with using mobile technology and available mobile device sensors. The information of the user model is used in combination with Multi-Criteria Decision Making for individualizing the experience of the user indoors or outdoors of the Museum. More specifically, the system uses the

Technique for Order Preference by Similarity to Ideal Solution (TOPSIS). The developed application's general architectural design is illustrated in Fig. 1, where the basic steps in the application's logic are also presented.

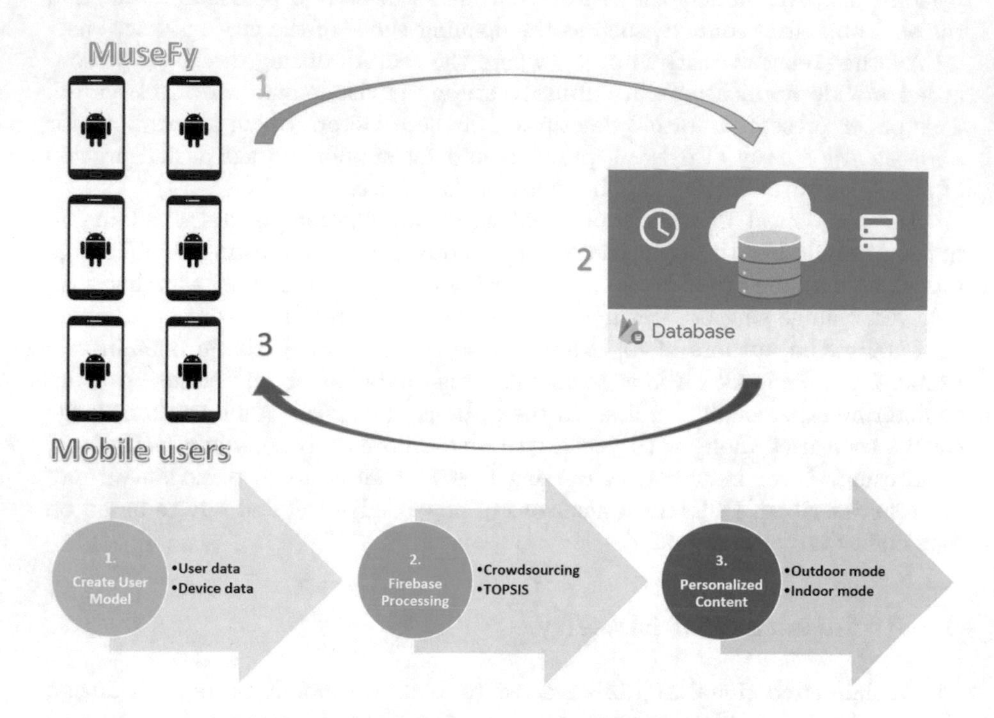

Fig. 1. MuseFy general architecture

The mobile application supports two basic modes of interaction with users, namely an outdoor mode and an indoor mode. More specifically, during the "outdoor" mode, mobile location services such as network and GPS are utilized in order to locate users and suggest a visiting "path" taking into consideration of users' available time, available Points Of Interest (POIs), traveling distance and of course the user profile, e.g. the user's interests. To achieve this, the mobile application also incorporates geo-fences handling mechanisms in combination with the location services. During the "indoor" mode, where a user has already visited one of the available museums, software mechanisms such as Near Field Communication (NFC) and QR code scanning are utilized in order to locate users and provide them with adaptable individualized content. User data and educational application data are stored both locally in the mobile device and also remotely to a Cloud as a Service infrastructure. This approach facilitates data synchronization and also enables the incorporation of user modelling software mechanisms in the system's back-end. The resulting mobile educational application is capable of dynamically adapting its content on real time according to user profiles and in combination with the available educational information.

MuseFy's snapshots in Fig. 2 illustrate two different personalized representations of the same domain, where the reconstruction of the mobile UI is based on the user model and the applied TOPSIS algorithmic approach. The app is illustrated operating in the "indoor" mode, while the available information is dynamically adapted in two different user profiles. Users' feedback while using the app is further utilized in order both to evaluate the actual outcomes of the incorporated model, and also to improve the interaction by fine-tuning the acquired user model.

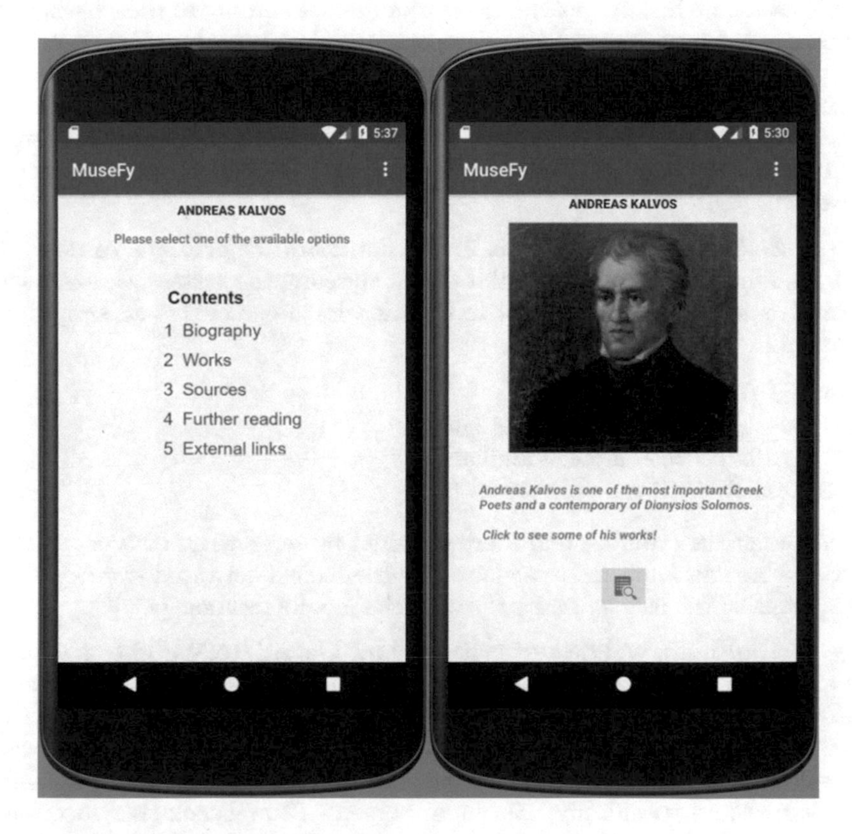

Fig. 2. Personalized UIs for the same museum exhibit

4 Back-End Personalization

In this paper, the main purpose for constructing personalized user models is in order to provide adaptation and customization towards museum touring educational purposes. In this sense, our proposed back-end service becomes more efficient and operates in order to best serve users' needs. As a result, the developed app builds user models by taking an initial input by the users and subsequently tracking user behavior while they use it. The available collected data is

anonymized and then securely send and stored to the on-line Cloud-as-a-Service Infrastructure for further processing as already mentioned. After incorporating user models into MuseFy, personalized data are displayed to each user, both in terms of UI and UX and also in terms of guidance and assistance.
The user interaction with the system and the personalization process can be summarized to the following steps:

Step 1: Activation of the double stereotype for the particular user
Every time a user is connected to the system a double- stereotype is activated for this user. The first stereotype categories users according to their background knowledge and interests and this categorization has been based on a previous study [8]. The resulting stereotypes are: (1) Tourist, (2) Archaeologist and (3) Conservator. The other stereotype categorize users according to the mobile's model and the users' familiarity with the mobile technology. Therefore, users are categorized into one of the following stereotypes: (1) Expert, (2) Intermediate (3) Novice.

Step 2: Selection of an exhibit and calculation of criteria values
Each time the user selects an exhibit of the museum, the system creates a set of alternative descriptions to present to the user. Each one of these descriptions is evaluated in terms of some criteria:

1. Interest (i)
2. Compatibility with background knowledge (k)
3. Compatibility with mobile familiarity (f)
4. Compatibility with mobile model (m).

The values of the criteria i and k are acquired by the first stereotype while the other two criteria f and m are acquired by the second activated stereotype that corresponds to the user's familiarity with the mobile technology.

Step 3: Acquiring weights of criteria and Calculate Weighted Ratings
The criteria are not equally important in the reasoning of the system, therefore, in order to find out which is the weight of importance of each criterion, a short experiment was conducted during the design phase of the system's life cycle. The values of the weights are static and used in all different cases of exhibits evaluation. More specifically, 10 human experts (2 archaeologists, 3 museum curators, 3 software engineers, 2 conservators) we asked to divide 10 points to rate the four criteria with respect to their importance while taking decisions on what to present to a user. For example, one expert gave 4 points to criterion i, 3 points to criterion k, 2 points to f and only 1 to m. The weights were calculated after summing up the rates given by all experts then divide them by 100 (10 points x 10 experts): $w_i = 0.33$, $w_k = 0.24$, $w_f = 0.26$, $w_m = 0.17$.
The weighted values for the j alternative are calculated as: $v_{ij} = w_i \cdot i_j$, $v_{kj} = w_k \cdot k_j$, $v_{fj} = w_f \cdot f_j$, $v_{mj} = w_m \cdot m_j$.

Step 4: Identify Positive-Ideal and Negative-Ideal Solutions
The positive ideal solution is the composite of all best criteria ratings attainable, and is denoted: $A^* = \{v_i^*, v_k^*, v_f^*, v_m^*\}$ where v_i^* is the best value for the criterion

i among all alternatives. The negative-ideal solution is the composite of all worst attribute ratings attainable, and is denoted: $A^- = \{v_i^-, v_k^-, v_f^-, v_m^-\}$ where v_i^- is the worst value for the criterion i among all alternatives.

Step 5: Calculate the separation measure from the positive-ideal and negative-ideal alternative
The separation of each alternative from the positive-ideal solution A^*, is given by the n-dimensional Euclidean distance: $S_j^* = \sqrt{(v_{ij} - v_i^*)^2 + (v_{kj} - v_k^*)^2 + (v_{fj} - v_f^*)^2 + (v_{mj} - v_m^*)^2}$, where j is the index related to the alternatives. Similarly, the separation from the negative-ideal solution A^- is given by $S_j^- = \sqrt{(v_{ij} - v_i^-)^2 + (v_{kj} - v_k^-)^2 + (v_{fj} - v_f^-)^2 + (v_{mj} - v_m^-)^2}$.

Step 6: *Calculate Similarity Indexes.* The similarity to positive-ideal solution, for alternative j, is finally given by $C_j^* = \frac{S_j^-}{S_j^* + S_j^-}$ with $0 \leq C_j^* \leq 1$. The alternatives are ranked according to C_j^* in descending order.

5 Conclusions

This paper focuses on presenting a novel museum exploration mobile application that utilizes user modeling and multi-criteria decision making in order to provide a personalized user experience in a domain that is highly challenging and popular. Our independent research has concluded that the incorporation of user modeling techniques in mobile development can be significantly beneficial for mobile end-users, improving user interaction and user experience. Positive feedback from a considerable number of app users could indicate that such an approach can be applied in a variety of interactive mobile applications.

It is in our future plans to evaluate the effectiveness and also the user friendliness of our proposed approach by distributing it through the Google Play app marketplace and more specifically by targeting its audience to users interested in using educational applications for museums. The evaluation feedback has been planned to be collected both from within the application, in terms of a short questionnaire after a specific period of usage and also through the Google Play's platform incorporated rating mechanism, where both scores and also user comments can be analyzed.

References

1. Avouris, N.M., Yiannoutsou, N.: A review of mobile location-based games for learning across physical and virtual spaces. J. UCS **18**(15), 2120–2142 (2012). https://doi.org/10.3217/jucs-018-15-2120
2. Cheverst, K., Davies, N., Mitchell, K., Smith, P.: Providing tailored (context-aware) information to city visitors. In: Adaptive Hypermedia and Adaptive Web-Based Systems, Proceedings of International Conference, AH 2000, Trento, Italy, 28–30 August 2000. LNCS, vol. 1892, pp. 73–85. Springer, Heidelberg (2000). https://doi.org/10.1007/3-540-44595-1_8

3. Fidas, C.A., Avouris, N.M.: Personalization of mobile applications in cultural heritage environments. In: 6th International Conference on Information, Intelligence, Systems and Applications, IISA 2015, Corfu, Greece, 6–8 July 2015, pp. 1–6 (2015). https://doi.org/10.1109/IISA.2015.7388114
4. Firebase: Firebase helps you build better mobile apps and grow your business (2017). https://firebase.google.com/
5. Hwang, C.L., Yoon, K.: Methods for multiple attribute decision making. Springer (1981)
6. Kirci, P., Ünal, P.: Personalization of mobile health applications for remote health monitoring. In: Proceedings of the International Workshop on Personalization in Persuasive Technology Co-located with the 11th International Conference on Persuasive Technology (PT 2016), Salzburg, Austria, 5 April 2016, pp. 120–125 (2016). http://ceur-ws.org/Vol-1582/10Kirci.pdf
7. Kuflik, T., Mumblat, Y., Dim, E.: Enabling mobile user modeling: infrastructure for personalization in ubiquitous computing. In: 2nd ACM International Conference on Mobile Software Engineering and Systems, MOBILESoft 2015, Florence, Italy, 16–17 May 2015, pp. 48–51 (2015). https://doi.org/10.1109/MobileSoft.2015.13
8. Ntelianidou, K.K.I.: Video personalisation in a folklore museum. Int. J. Digital Culture Electron. Tourism **2**(2), 155–169 (2017)
9. Wecker, A.J., Kuflik, T., Stock, O.: Dynamic personalization based on mobile behavior: from personality to personalization: a blueprint. In: Proceedings of the 18th International Conference on Human-Computer Interaction with Mobile Devices and Services Adjunct, MobileHCI 2016, pp. 978–983. ACM, New York (2016). https://doi.org/10.1145/2957265.2962645

Towards Technology for Supporting Effective Online Learning Groups

Godfrey Mayende[1,2](✉), Andreas Prinz[1], Paul Birevu Muyinda[2], and Ghislain Maurice Norbert Isabwe[1]

[1] Department of Information and Communication Technology, University of Agder, Grimstad, Norway
godfrey.mayende@uia.no
[2] Institute of Open, Distance and eLearning, Makerere University, Kampala, Uganda

Abstract. Group learning has been advocated for increasing active learning among distance learners. However, there is limited understanding on how to engage learners in online courses. Following the design science methodology, we iteratively developed guiding factors for supporting effective online learning groups. The factors for effective online learning groups cover five key dimensions, namely institutional policies, institutional technology, group activity, group composition, and facilitation. The factors are validated through repetitive evaluation using authentic online learning courses, as well as using a focus group discussion with experienced online facilitators. This way, the factors provide pedagogical and technological guidelines for introducing online course groups. Moreover, they give requirements for online learning systems supporting effective online learning groups.

Keywords: Online learning · Learning groups · Online learning systems

1 Introduction

Distance learning is a mode of study where students have minimal face-to-face contact with their facilitators; the learners learn on their own, away from the institutions, most of the time. Recently, distance learning has adopted the use of group assignments with the aim of encouraging students to work together to bridge the distance between the online students. Group work requires students coming together either physically or virtually through technology. A typical risk in group assignments is that a few students do the group assignments and just include other student's names. This causes high failure rates during summative assessment, since not all students engage with the course materials during the group assignment. Those students fail to harness the benefits of working in groups. On the positive side, group work leads to better and faster learning (Sven et al. 2015). To bring those benefits to online courses, effective ways of supporting online learning groups are essential for interactions. When there is interaction within online learning groups, meaningful learning is achieved. However, motivating and sustaining effective student interactions requires planning, coordination

M. E. Auer and T. Tsiatsos (Eds.): IMCL 2017, AISC 725, pp. 361–371, 2018.
https://doi.org/10.1007/978-3-319-75175-7_37

and implementation of curriculum, pedagogy and technology. Therefore, the creation of guidelines for introducing online learning groups can create possibilities of effective online learning.

The aim of this paper is to develop guidelines for introducing online course groups. The guidelines are informed by both e-pedagogy and online learning systems. They will help in ensuring that online learning groups are effectively supported within the online learning systems through answering the two research questions; What principles should guide the design of tools to support effective online learning groups? and What tools should be used for effective online learning groups?

The rest of this paper is organized in five sections. As a background, Sect. 2 provides an overview of collaborative learning. Section 3 explains our research methods and approaches. Section 4 presents the factors for effective online learning groups. In Sect. 5, the factors are discussed, and the paper is concluded in Sect. 6.

2 Collaborative Learning

Collaborative learning refers to instructional methods that encourage students to work together to find a common solution (Ayala and Castillo 2008). Ashley (2009) and Stahl et al. (2006) contend that collaborative learning involves joint intellectual effort by groups of students who are mutually searching for meanings, understanding or solutions through negotiation. This approach is learner-centered rather than teacher-centered; views knowledge as a social construct, facilitated by peer interaction, evaluation and cooperation; and learning as not only active but interactive (Vygotsky 1978). This interaction is in line with Anderson's online learning framework which argues that learning can be achieved through any of the following interactions: student-teacher, student-student, and student-content (Anderson 2003). This is also apt with Stahl et al. (2006) who asserts that learning takes place through student-student interactions, and it is in agreement with our own earlier studies (Mayende et al. 2014; Mayende et al. 2017). Ludvigsen and Mørch (2009) found out that students effectively develop deep learning when using computer supported collaborative learning. Therefore, careful integration of computer supported interaction can heavily increase learning in online learning systems.

Collaborative learning is based on consensus building through interaction by group members, in contrast to competition. This can be very helpful for distance learners, who are typically adults. Collaborative activities are essential to encourage information sharing, knowledge acquisition, and skill development (Collison et al. 2000). Different technological tools have been adopted for collaboration in distance learning.

Collaborative learning hinges on the belief that knowledge is socially constructed although each learner has control over his/her own learning. Online learning systems offer the possibility for these collaborations to be achieved through communication among learners. Collaborative learning (and also our study presented here) is underpinned by the social constructivist learning theory (Vygotsky 1978).

3 Approaches and Methods

The design science methodology was employed to find the factors. This methodology is aimed at iteratively coming up with an artefact, in this case the guidelines for the introduction of online learning groups. Figure 1 indicates the various stages in the design science methodology.

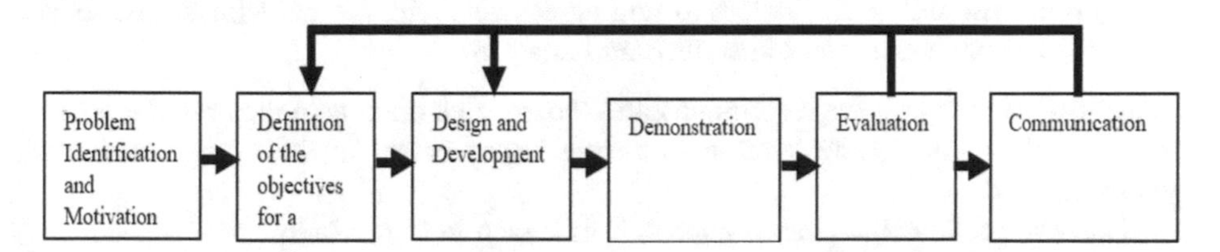

Fig. 1. Design science process (Peffers et al. 2007)

The distinct stages of the design science process as adopted from Peffers et al. (2007) are described below with corresponding methods used in each phase.

Problem identification and motivation. This stage defines the specific research problem and justifies the importance of a solution. The problem definition is later used to develop an artefact that can effectively provide a solution. Our problem emanates from thc need to support online learning groups and their importance for effective learning.

Define the objectives for a solution. This stage uses the problem definition and knowledge of what is possible and feasible to define the objectives. In this research study, we use research questions under three research directions, which are effectiveness of learning groups, processes to support effective online learning groups and tools to support online learning groups. Our overall aim is to determine solutions for supporting effective online learning groups.

Design and Development. This stage creates an artefact which is used in the study, based on the needs of the end users of the desired solution. In our study, the artefact is a set of factors that guide the introduction of online learning groups. We started the process by interviewing experienced online learning facilitators and looking at online learning interactions within the online learning systems. This input was transcribed and analysed and led to an initial set of factors, which was improved in the iterations of the study. This was done for two courses whenever the courses were run (in the demonstration stage). Figure 2 illustrates how the factors evolved through phases.

Demonstration. This stage demonstrates the use of the artefact. We used two online courses, one run in Norway and one run in Uganda. A MOOC course was run at the University of Agder (Mayende et al. 2016) and an undergraduate course was run at Makerere University (Mayende et al. 2015). Both courses were run in the real environment and used customised existing LMS's to verify and improve the factors.

Evaluation. This stage observes and measures how well the artefact provides a solution to the problem. It was during this stage that we used mixed methods in evaluating the online courses under demonstration. We iterated back to design and development to improve the artefact. Surveys were used in the online courses to understand the processes of online learning groups. In addition, we also observed the interaction logs in the online learning courses. With this data, we identified themes which informed the elements of the factors. The factors were then evaluated through focus group with online facilitators to find agreements with the guidelines. The focus group discussions were then transcribed and analysed.

Communication. This stage communicates the research outputs of the previous stages and Possibly starts a new iteration to ensure improvement in the artefact which is quality assured.

The study followed a phased approach as shown in Fig. 2 below.

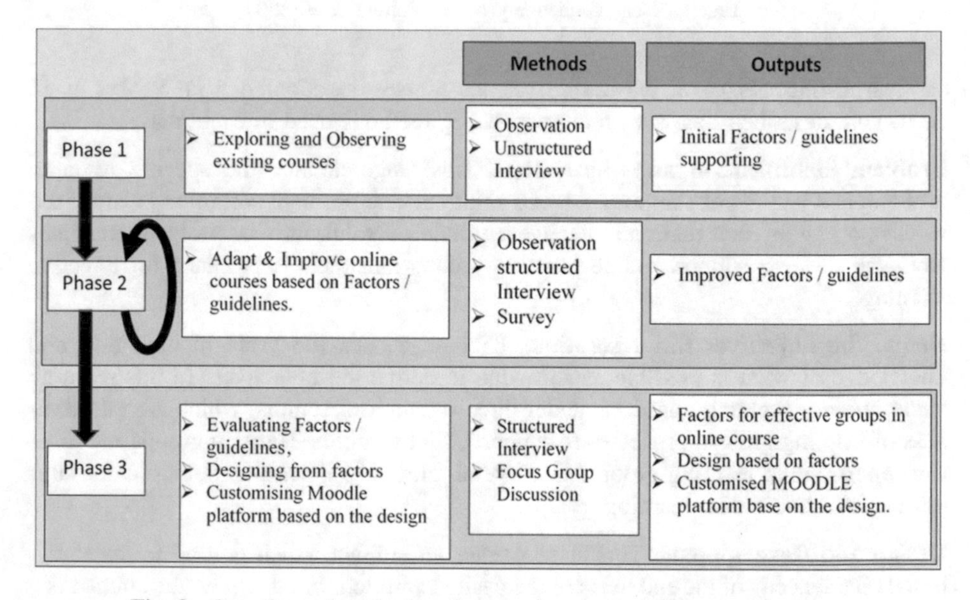

Fig. 2. Overview of methods and research outputs under research phases

In phase 1 the focus was on exploration and observation of existing online courses. Qualitative methods were used in the collection and analysis of data. Data logs were observed and analyzed. Unstructured interviews were used when interviewing experienced online facilitators. After the collection and analysis of the data we came up with the initial factors and guidelines. Then, in phase 2, we adapted and improved online courses based on the initial factors. Data to verify the factors was collected using observation of the interaction logs, structured interviews and surveys for large online classes (MOOCs). Finally, in phase 3, we adapted and improved the online courses based on the improved factors and evaluated the factors. This led to the final factors. Sections 4 and 5 elaborate the factors in detail.

4 Factors for Effective Groups in Online Courses

This section describes the factors for effective groups in online courses in the following dimensions: overview of the factors, supporting online learning group institutional policy, supporting online learning group institutional technology, quality of online learning group activity, quality of online learning group and quality of online learning group facilitation.

4.1 Overview of the Factors

Learning groups have been advocated for increasing interaction and learning. However, the use of learning groups has not been very easy in online learning systems. Therefore, this study provides guidelines for effective ways of using groups in online learning courses. These guidelines are given as factors in five dimensions as shown in Fig. 3.

Fig. 3. Factors for effective online learning groups

All the five dimensions contribute to the factors for effective online learning groups abbreviated as FEOLG. OLG stands for online learning groups. The dotted arrows indicate the order in which the dimension should be acted on, normally starting from supporting OLG institutional policy. The arrows indicate that the factors describe a continuous process that provides support to make online learning groups effective. The following sections describe the five dimensions in detail.

4.2 Supporting Online Learning Group Institutional Policy

The first-dimension concerns supporting online learning group institutional polices. Often online courses are run without having supporting policies to ensure their success. This can create problems in the running of the online course. Therefore, having the supporting institutional policies can create possibilities of groups in online learning courses. Under this dimension, the following key sub elements were identified through our iterative process.

Progressive group assessments policies are institutionalized. Respondents revealed the need of having an institutional policy that caters for the progressive group assessments. This was emphasized for helping in the reward of students during the online progressive group work. During the focus group discussion, the facilitators at Makerere University advised on the need for embedding such policies in the curriculum. When such policies are not available, administering group work online becomes difficult. The facilitators at the University of Agder emphasized the need of awarding between 40%–60% on progressive assessment. This helps the students to be rewarded given the amount of work involved in the online group activity.

Digital assessment for groups policies are institutionalized. The respondents emphasized the need for digital assessment for groups policies. One of the respondents said that "the policy should put emphasis on feedback for facilitators and peers". This is in line with the peer assessment based activity which revealed improved interaction among learners in groups (Mayende et al. 2015) and the individualized activity which also revealed improved interactions (Mayende et al. 2016). Digital assessment and feedback are key in online learning groups, and they need technological support.

Online facilitation and tutoring policies are institutionalized. Facilitation and tutoring must be scaled to enhance its effectiveness. With large online classes, there is a need for many online tutors to assist in scaffolding learners. It was revealed that facilitators with large classes at Makerere University are not assigned online tutors to help in the process. This is because of the costs involved in paying the online tutors. During the focus group discussion one of the respondents said that "lack of online tutors to help in the support for the online students create heavy information overload for the facilitators". The respondents revealed the importance of recruiting and remunerating online tutors to help in effective student support hence learning.

4.3 Supporting Online Learning Group Institutional Technology

Technology support for online learning groups is very important in enhancing effective online group work. Technology is key in supporting all the other dimensions. There are many online learning technologies available both commercial and open source and institutions should choose one institutional technology to use. This helps in having a single point of contact of the institution for the support and maintenance of the learning management system. To have good support for online learning groups, the technology should support the following elements.

Authentication. The technology should allow for users to login to access the platform. The users should be categorized differently to allow distinct access. The users may include the following facilitators/tutors, learners, eLearning administrators, and eLearning support team.

Systems administration. The technology should allow system administrators to administer the platform, including the possibility to add users and give them different access rights. Facilitators should have control of their online courses. Learners should have sufficient access rights to allow for interaction and submission.

Announcements. The technology should allow facilitators to send communication about the progress and course status. This triggers the learners to actively participate in the activities of the course. This could be implemented using the message boards which can be embedded in the user's home page.

Discussion. The technology should support users to interact with one another within groups, both synchronously and asynchronously. This can be implemented using forums. Discussion forums should be designed in such a way that students can discuss within their groups. Discussions can enable learning within the platform when learners are engaged and communicate through the platform.

Co-creation of artefacts. The technology should support learners to be able to create artefacts together in a group. Co-creation was emphasized because of its need for jointly creating knowledge together in a group. Examples are joint programming and writing a document together. This helps increasing learning through interaction, as emphasized in Mayende et al. (2017).

User support. The technology should support users (facilitators and learners) in the use of the system. This support can be embedded within each course such that learners can ask questions related to the technology. Technological experts should be available for each online course to allow for support within the course.

4.4 Quality of Online Learning Group Activity

A group activity is very important in ensuring that learners interact effectively within the groups. Activities with emphasis on interaction is important for online learning. In earlier papers we have suggested peer assessment based activity (Mayende et al. 2015) and individual based activity (Mayende et al. 2016) as a form of activity organization. Both increased interaction among learners within the online learning group. The following checklist can be used for ensuring effective group activity.

- The activity has a clear and relevant title.
- The activity is clearly marked as a group activity.
- The activity is connected to the course learning outcome.
- The purpose of the group activity is stated clearly and concisely.
- The activity has outlined the tasks that the groups will be required to do.
- The activity is simple enough to be completed with ease in the given time for most groups.
- The activity provides clear instructions.

- The activity identifies the tools that participants require performing the tasks.
- The activity clearly states the completion criteria of the task.
- The activity clearly states the time required for completion.
- The activity indicates the contribution to the final grade of the course.
- The activity has rewards.
- The activity is structured for peer feedback and assessment.
- The activity enables teacher assessment.

Following the above checklist will help in ensuring effective group interaction. Emphasis is put on the way these activities are structured to encourage interaction and feedback. Outcomes from the evaluation indicate that online facilitators agreed that the online group activity is central to the effectivity of online groups. The system should cater for structuring the online activity.

4.5 Quality of Online Learning Group

Group composition is also very important in ensuring effective interaction within the online environment. The following essential elements should be taken into consideration when creating groups: group size, diversity, unity and stability.

The group should be composed of between 2 and 7 members. The readings did not clearly indicate the exact number of students that are required for an effective learning group, although emphasis on small groups is indicated. During the demonstration, we used five members in the group in one course and in another course, we had seven members. Both showed effective interaction in the groups. Our indication of 2 to 7 members was not extensively empirically studied. More studies might be needed to establish the exact number of learners required in an online learning group.

The group composition should promote diversity. Our findings revealed the need for diversity in the groups (various levels of experience, diverse backgrounds, different age and gender). This helped to scaffold peer learning as indicated by Vygotsky (1978).

The group composition should promote unity. Unity was emphasized to allow for possibility of putting learners together to make it possible for physical meetings as well as finding a common base line for discussions.

The group members should be kept in the same group for a longer period. Preferably learners should be kept in a group for at least a semester or 6 months. This allows for better group dynamics and social connection. This can help a group to go through all the different stages of group development as illustrated by Tuckman and Jensen (1977). At University of Agder students were kept in the groups for the full semester and this improved group dynamics.

4.6 Quality of Online Learning Group Facilitation

Physical classroom teaching differs from online teaching. In both situations learners should be guided when interacting within a group. Physical groups allow to see what learners are doing in real time. This possibility gives facilitators the opportunity to identify learners with challenges and to assist them immediately. This can help learners to learn better through intervention and scaffolding of the students learning.

Also in online teaching, facilitators are encouraged to show their presence within the learning environment. When learners within the online system do not see and feel the presence of the teacher, their participation is discouraged. Therefore, it is important to have a manageable number of learners per facilitator. The system should also have means to detect problems and warn the facilitator for easier follow up. This can help the facilitator to intervene and offer solutions to learners who need help and guidance. Such intervention will help to increase motivation and group interaction which is a precursor for meaningful learning.

The findings also reveal the importance of online facilitation, which is different from traditional teaching. Facilitators play a leading role in motivating and sustaining learner interaction within the online learning groups. Interventions by facilitators can provoke the students to interact at higher levels of Bloom's taxonomy (Anderson et al. 2001). This can also be supported through automated intervention by checking the status of groups and the individual students in the groups and sending them emails in case of deviations.

5 Discussion

These factors are effective because they have been developed through an iterative process of design science. This has been done over three years period of the project. The study was done in phases as seen in Fig. 2. In the first phase, we started by exploring and observing the existing courses. Mainly online courses at the University of Agder were observed and experienced online facilitators were interviewed. This helped in coming up with the initial factors, which focused on the following important elements for effective online learning courses: courses design, trained online facilitators, motivation and sustaining interaction and peer assessment based activity (Mayende et al. 2015). This initial list was used in the demonstration and led to phase 2.

In phase 2, we adopted and improved online learning courses based on the initial factors. This was accomplished using different case studies. The case studies were from authentic online courses at the University of Agder and Makerere University in several different studies (Mayende et al. 2015; Mayende et al. 2017; Mayende et al. 2016; Mayende et al. 2017). In this phase, we used observation of online interactions, interviews of facilitators and learners and surveys. In the case study with peer assessment based activity we found enhanced engagement and interaction, and the quality of the peer feedback was improved (Mayende et al. 2015). This indicates the importance of the online learning group activity, in agreement with (Salmon et al. 2016). The second case study was a MOOC run at the University of Agder, which confirmed the importance of the online learning group activity in enhancing interaction. It also revealed the importance of facilitator feedback or interventions, the composition of a group and technology in enhancing interactions within the online learning groups (Mayende et al. 2016; Mayende et al. 2017). This is in line with Salmon et al. (2015); Salmon et al. (2016).

Finally phase 3 evaluates the factors using focus group discussion and interviews. This has been done in one case study and we are going to make more evaluations on another case. This was done in understanding best practices for online learning designs (Mayende et al. 2017).

6 Conclusions

This paper concludes with identifying five key elements for ensuring effective online learning groups. The five elements are supporting online learning group institutional policies, supporting online learning group institutional technology, quality online learning group activity, quality online learning group and quality online learning group facilitation. However, the main emphasis is put on the online group activity and its structure within the online learning systems to cater for effective interaction. Once the activity is well structured with interaction embedded in it, there is a good chance that the learners will actively interact within the group. This interaction should also be supported by well-trained online facilitators or tutors. The trained facilitator's intervention can help in motivating the learners and sustain the group interaction. For an effective support of the elements appropriate technology needs to be used. In addition to the design science process for developing these factors we are in the process of evaluating them on a case study and our developed online learning system that supports the factors.

Acknowledgement. The work reported in this paper was financed by the DELP project which is funded by NORAD. Acknowledgements also go to the University of Agder and Makerere University who are in research partnership.

References

Anderson, L.W., Krathwohl, D.R., Airasian, P.W., Cruikshank, K.A., Mayer, R.E., Pintrich, P. R., Raths, J., Wittrock, M.C.: A Taxonomy for Learning, Teaching, and Assessing: A Revision of Bloom's Taxonomy of Educational Objectives. Abridged edition. Longman, White Plains (2001)

Anderson, T.: Modes of interaction in distance education: recent developments and research questions. In: Moore, M., Anderson, G. (eds.) Handbook of Distance Education, pp. 129–144. Erlbaum, Mahwah (2003)

Ashley, D.: A Teaching with Technology White Paper. Collaborative Tools (2009). http://www.cmu.edu/teaching/technology/whitepapers/CollaborationTools_Jan09.pdf. Accessed 1 Nov 2014

Ayala, G., Castillo, S.: Towards computational models for mobile learning objects. Paper presented at the Fifth IEEE International Conference on Wireless, Mobile, and Ubiquitous Technology in Education, WMUTE 2008 (2008)

Collison, G., Elbaum, B., Haavind, S., Tinker, R.: Facilitating Online Learning: Effective Strategies for Moderators. ERIC (2000)

Ludvigsen, S., Mørch, A.: Computer-supported collaborative learning: basic concepts, multiple perspectives, and emerging trends. In: McGaw, B., Peterson, P., Baker, E. (eds.) The International Encyclopedia of Education, 3rd edn. Elsevier, Amsterdam (2009). (in press)

Mayende, G., Isabwe, G.M.N., Muyinda, P.B., Prinz, A.: Peer assessment based assignment to enhance interactions in online learning groups. Paper presented at the International Conference on Interactive Collaborative Learning (ICL), 20–24 September 2015, Florence, Italy (2015)

Mayende, G., Muyinda, P.B., Isabwe, G.M.N., Walimbwa, M., Siminyu, S.N.: Facebook mediated interaction and learning in distance learning at Makerere University. Paper presented at the 8th International Conference on e-Learning, 15–18 July, Lisbon, Portugal (2014)

Mayende, G., Prinz, A., Isabwe, G.M.N.: Improving communication in online learning systems. Paper presented at the Proceedings of the 9th International Conference on Computer Supported Education CSEDU, vol. 1, Porto, Portugal (2017)

Mayende, G., Prinz, A., Isabwe, G.M.N., Muyinda, P.B.: Supporting learning groups in online learning environment. Paper presented at the CSEDU 2015 - 7th International Conference on Computer Supported Education, Lisbon, Portugal (2015)

Mayende, G., Prinz, A., Isabwe, G.M.N., Muyinda, P.: Learning Groups for MOOCs: lessons for online learning in higher education. Paper presented at the 19th International Conference on Interactive Collaborative Learning (ICL2016), 21–23 September, Clayton Hotel, Belfast, UK (2016)

Mayende, G., Prinz, A., Isabwe, G.M.N., Muyinda, P.B.: Learning groups in MOOCs: lessons for online learning in higher education. Int. J. Eng. Pedagog., 109–124 (2017). https://doi.org/10.3991/ijep.v7i2.6925

Peffers, K., Tuunanen, T., Rothenberger, M.A., Chatterjee, S.: A design science research methodology for information systems research. J. Manag. Inf. Syst. **24**(3), 45–77 (2007)

Salmon, G., Gregory, J., Lokuge Dona, K., Ross, B.: Experiential online development for educators: the example of the Carpe Diem MOOC. Br. J. Educ. Technol. **46**(3), 542–556 (2015)

Salmon, G., Pechenkina, E., Chase, A.M., Ross, B.: Designing massive open online courses to take account of participant motivations and expectations. Br. J. Educ. Technol. **48**, 1284–1294 (2016)

Stahl, G., Koschmann, T., Suthers, D.: Computer-supported collaborative learning: an historical perspective. In: Cambridge handbook of the learning sciences (2006)

Sven, Å.B., Lazareva, A., Mayende, G., Nampijja, D., Isabwe, G.M.N.: Together we can. Team and online collaborative work (2015). http://grimstad.uia.no/puls/Groupwork/main.htm

Tuckman, B.W., Jensen, M.A.C.: Stages of small-group development revisited. Group Organ. Stud. **2**(4), 419–427 (1977)

Vygotsky, L.S.: Mind in society: the development of higher psychological processes. Harvard University Press, Cambridge (1978)

Human Factors Considerations in Mobile Learning Management Systems

Mohamed Sarrab[1(✉)], Zuhoor Al-Khanjari[2], Saleh Alnaeli[3], and Hadj Bourdoucen[4]

[1] Communication and Information Research Center, Sultan Qaboos University, Muscat, Oman
sarrab@squ.edu.om

[2] Computer Science Department, College of Science, Sultan Qaboos University, Muscat, Oman
zuhoor@squ.edu.om

[3] Computer Science Department, University of Wisconsin Colleges, UW-Fox Valley, Menasha, USA
saleh.alnaeli@uwc.edu

[4] Electrical and Computer Engineering Department, Sultan Qaboos University, Muscat, Oman
hadj@squ.edu.om

Abstract. The rapid development in Internet and mobile technologies had an influence on education and learning processes that led to emergence of mobile learning as potential part of learning management system. Many individual learners and instructors, in some cases, entire education providers are gravitating towards the adoption of mobile learning through comprehensive learning management systems (LMS) for managing courses contents and enhancing learners' education process. This research involves the development and validation of a survey that used to study 23 distinct Omani higher education providers empirically. A total sample of 806 university and higher college students and instructors participated in this study. A correlated six-factor Flexibility, Suitability, Enjoyment, Impact, Social and Economic were found to be as human influencing factors. Four different learning management systems Moodle, Blackboard, Schoology and Edmodo have been compared and evaluated in respect of the selected human factors. The effort is part of an Omani-funded research project investigating the development, adoption and dissemination of M-learning in Oman.

Keywords: Mobile learning · Human factors · Economic · Enjoyment
Impact · Flexibility · Social · Suitability

1 Introduction

Mobile devices have become ubiquitous companies' present discoveries and ideas in rapid succession, and as their cost has been declining overtime. This phenomenon had its impact on our everyday life, and the education is not an exception. With the availability of mobile devices with a relatively affordable cost, the interest in

M. E. Auer and T. Tsiatsos (Eds.): IMCL 2017, AISC 725, pp. 372–383, 2018.
https://doi.org/10.1007/978-3-319-75175-7_38

integrating this technology in the learning process has significantly increased as schools and educational institutions are trying endlessly to enhance the quality of their programs. This interest has led to emergence of M-learning as an essential part of their educational systems. Its introduction in the educational system satisfy the learning requirements including wider, fast access to learning materials and persistent needs for prompt communication [1]. M-learning is defined as "learning across multiple contexts, through social and content interactions, using handheld and mobile technologies such as personal digital assistants (PDAs), smartphones or other mobile devices. M-learning increases learning flexibility by adapting learning to be more personalized and learner-centered [2, 3]. The three elements that are interdependent and are equally important in making mobile devices viable for the delivery of higher education instructional contents are mobilities of technology, learners and learning. Exploiting mobile devices enhances the learning process by increasing the accessibility and the availability of the relevant resources and materials for all potential learners. That is, learners become able to access the learning resources with almost no time constrains and regardless their location. This flexibility is not always available for desktop users [4, 5]. With growing popularity of mobile devices, it is important that adequate learning and educational approaches are developed in order to utilize the benefits of technical devices effectively for an enhanced learning [6]. Acceptance by all stakeholders is the vital part in the development of a successful M-learning application. Therefore, there is a need to explore the factors that has to be considered while developing the M-learning application [7]. Well-implemented M-learning application can assist in the reduction of cognitive load by filtering available information based on all contextual factors. Therefore, a careful analysis of the factors that has to be considered while developing the M-learning application is essential [8]. The developers need to analyze their users and consider the device types used in order to meet the needs of their learners. Mobile device characteristics should be emphasized in design and development stage and the platform should support integration of various technologies [9, 10].

2 Importance of Human Factors

A successful development of any mobile and wireless technologies applications rely on different human factors consideration [11]. Human Factors are defined as knowledge about human abilities, human limitations, and other human characteristics that are relevant in development of the system. The application of human factors to the design of tools and systems can optimize the overall system performance [12]. The key factors that need to be considered are the physical attributes of devices, content and software applications, network speed and reliability, and the physical environment of use. Human factors focus on general behavior of users and their interaction with the system. The various human factors including functionality, performance, cost, reliability, maintenance and usability are equally important in the successful deployment of any system and a serious failure in any one of them can cause a failure of the entire system.

Human factors issues, arise at several stages and prototypes of the proposed system, can be tested long before the final system is in place; many levels of design iteration are now introduced into the development process [13]. These added methods enhance the software system and help ensure its acceptance, in addition to reducing the system maintenance. The direct benefits from adding the human factors includes a reduction in user learning times; a reduction in user errors; and a reduction in the cost of system maintenance [11, 14]. The costs of the decision to include human factors in the software are sensitive to a variety of phenomena. These include the type and number of system user, the complexity of the user interface being built and the amount and type of human factors stages that are included in the software development lifecycle [15, 16]. The human behavior can be divided into three levels: individual, group, and organization. The human factor analysis in software engineering should study the behavior traits, behavior model, categories of human errors [17]. The human factor failures frequently occur, suddenly, disorder, it is difficult to analyze their patterns; for the reason of systematic protection functions, many human factor failures only constitute potential failures, it brings about a large number of important human factor information lost [18]. However it is important to identify the set of human factors that can extend up to an adequate level of optimization and is compatible with given learning setting (including dynamic nature and behaviors of users), and how these factors are applied on the learning space during development stage in order to have the desired results and learning impact [19].

3 Identify Human Factors in M-learning Development

The utilization of human factors for constructing learning models, based on which M-learning application can be developed and offer more effective learning content that can satisfy user's needs by being effective, efficient, and consistent and also of high quality while targeting the major concerns of the user in most cases. Most researchers are emphasizing in the selection of human factors that concerns the learning/cognitive styles, visual and cognitive processing, and working memory span for building up more personalized and adapted M-learning environments. The presented study aimed to identify the human factors associated with the development of M-learning application. The study was carried out by conducting surveys, collecting data and analyzing the result for identifying the major human factors that influence the development of M-learning application [20].

3.1 Methodology (Survey)

The study was carried out by conducting a pilot survey in order to validate the accuracy and consistency of the human factors in M-learning development. A rigorous validation process was carried out for preparing the survey's questionnaire. The study encompasses participants selected from 23 distinct Omani higher education colleges. A questionnaire was prepared and distributed among students and faculties of various universities and higher colleges in Oman. A sample of 806 usable responses were obtained. Using analytic hierarchy process (AHP), 28 factors of perceived innovative characteristics have been analyzed to examine the relationship among the perceived innovative characteristics and willingness of M-learning adoption. The selected factors through the survey are tested. Five-point Likert scale was used for analyzing the questionnaire responses.

In this way, respondents expressed their views and thus obtained baseline data on which further analysis was performed by to determine the consistency ratio using the software AMOS 19.0. The demographic information indicates that 49% of respondents were males, and 51% were females. Majority of respondents (59%) belonged to the age group 18–23, followed by (21%) belonged to age group of 24–29 years and (20%) belonged to the age group of 29+ years. In addition, majority of respondents (50%) were undergraduate and diploma students, followed by (26%) bachelor students and (24%) students in other levels. The objective behind this survey was to understand the student's perception towards M-learning application development. This questionnaire is designed to undertake a survey on the students and instructors perceptions on the human factors considered while M-learning application development. The questionnaire consists of six sections. The first section deals with the demographic background of the respondents. The other section more specifically asked about the internet experience such as (how often they connect to internet through mobile device and the intention of their usage). The third section of the questionnaire focused on the E-learning experience, which includes (how long they use e-learning and the purpose of using e-learning). Fourth section deals with their experience of mobile device usage that involves (types of mobile devices in their opinion, frequently used software application in the mobile devicc, hours of usage and the purpose of using the mobile device). Fifth section asked about mobile learning experience such as (where they use mobile device for academic tasks, their motivation behind using it for learning and the intention of using M-learning). The last section of the questionnaire focused on their view point about mobile devices as a learning tool (in terms of flexibility, suitability, interactivity, enjoyment, impact of mobile devices use in learning and also the legal, economic and social feasibility of learning using mobile devices in Oman) [21].

3.2 Collected Data

The information and data collected from this survey is intended to help form the basis for making informed decisions before strategically designing, developing, implementing, and evaluating M-learning application. The data analysis is carried out using analytic hierarchy process (AHP) and 28 factors of perceived innovative characteristics have been obtained. These factors are further examined using confirmatory factor analysis (CFA). The mean values of all 28 factors ranged from 3.00 to 3.90 on the Likert scale of 5. The standard deviations ranged from 0.937 to 1.25 and the skew and kurtosis indices from −.93 to −.03 and −.279 to .779 respectively. Based on recommendations made in [22], the data in this research were considered univariate normal. A principal components analysis (PCA) with varimax rotation was employed on the 28 items to understand the underlying structure of the M-learning measures. The purpose of using principle component analysis was to identify the selected items group on one or more than one constructs. Factor loadings are considered very significant if these are more than 0.50 [23]. Appropriateness of factor analysis is tested using two important measures. The first measure is Kaiser-Meyer-Olkin (KMO) overall measure of sampling adequacy and its value was 0.913 which falls within the acceptable limit and was significant at 1% level of significance as $p < 0.001$. The other measure is Bartlett's test of sphericity and its value was 6445.107 and significant at 1% level of significance as $p < 0.001$. This measure indicates a highly significant correlation among the items of the constructs in the survey. The data validation process indicated low accuracy or consistency for more than half of the questions and factors. Consequently, only six human factors were chosen to be considered on the development of M-learning application including: Enjoyment felt using mobile devices for learning purposes. Flexibility of mobile devices in accessing learning contents anywhere and anytime. Suitability of using mobile devices in learning environment, Feasibility of using mobile devices in learning purposes related to costs. Usefulness felt using mobile devices in learning environment on students' productivity and consistency and impact of using mobile devices in learning on society in general and the relationship between instructors and learners in specific.

4 Comparing Different LMS in Respect of Human Factors

A Based on the selected six-factor (flexibility, suitability, enjoyment, impact, economic and social) an index system was developed that illustrates the main key human factors of M-learning development in Oman. The reliability measures in terms of Cronbach's alpha were above the recommended level of 0.70 as an indicator for adequate internal consistency [22, 23]. The principal component analysis with varimax rotation was used to find how and to what extent the items are associated with respective constructs. The number of components was retained on the recommendations of where Eigen value should be more than one. On this recommendation, six components were found suitable for retention with the total variance explained was 66.42%. The factor loadings of indicators in these components were greater than 0.70 as pre the recommendations made in [23]. The name of six extracted factors were flexibility (8 indicators), Suitability (4 indicators), Enjoyment (4 indicators), Impact (4 indicators), Economic feasibility (3 indicators), Social feasibility (5 indicators). The comparison of major M-learning applications like Moodle, Blackboard, Schoology and Edmodo were carried out by considering the six human factors obtained from the survey and analysis.

4.1 Flexibility

See Table 1

Table 1. Comparing different LMS approaches in respect of mobile learning flexibility.

Approach	Flexibility
Moodle	Moodle interface is complicated and 'clunky'. The users must be familiar with basic HTML, which makes it less flexible. Navigation through the system is frustrating and lack of support is another limitation of Moodle. Individual updating of every component of the system is essential when a new version is released. Learners may get distracted due to complexity of the LMS and unfriendly interface
Blackboard	Blackboard is highly complex and difficult to learn. Blackboard interface is "clunky" and unattractive. Instructors must have high technical knowledge in order to master its usage. Setting up of Blackboard is difficult. Navigation to the system is quite unclear and the contents are unorganized
Schoology	Schoology feature a Facebook-like interface that is clean, appealing to students and fairly easy to use. It uses simple student-self-sign-up-system. The navigation within the Google drive is creepy and hard to find folders or even access shared items. Instructor takes long time to learn and teach students. That is, it takes more time to get familiar with Schoology compared to the time needed to learn Edmodo. (due to many extended features)
Edmodo	Edmodo feature a Facebook-like interface that is clean, appealing to students and fairly easy to use with brighter interfaces and clean icons. No technical expertise is required to get started. Uploading files to the system takes fair amount of steps. Edmodo has minimal administrative features

4.2 Suitability

See Table 2

Table 2. Comparing different LMS approaches in respect of mobile learning suitability.

Approach	Suitability
Moodle	Moodle supports a diversity of content. It is challenging to export content in a format that is suitable for other LMS and file management is tough. It requires extensive customization and testing to turn specific features on and off. System upgradation can be drastic and complex
Blackboard	Blackboard options may be restricted to particular operating systems. The users with less technical experience find it difficult to work with the system. It limits creativity by confining instruction to a restricted format. Platform is relatively outdated with limited availability and functionality of tools, incompatibility of programs, and differences among browsers
Schoology	Schoology has well designed interface and best suited for Secondary students. Interface like Facebook presented the information in a format that was familiar to all age groups. It is not as full-featured like Moodle, and does not include private messaging between students. It requires more time to learn the system and more sophisticated and challenging than Edmodo
Edmodo	Edmodo has limited features and resources. It has no professional look with limited features for student activities, which makes it suitable for elementary schools. It is difficult for students to upload their work on the iPads – the app quite often force quits by itself. (Edmodo is less intuitive compared to other systems.)

4.3 Sociability

See Table 3

Table 3. Comparing different LMS approaches in respect of mobile learning sociability.

Approach	Sociability
Moodle	The learners must be connected online at all times in Moodle. The current Moodle interface and other course resources are effective for online learners with fast Internet connections. It is possible to send personal messages to users within Moodle. The students interact with their teachers and peers, as they desire
Blackboard	Blackboard lacks collaboration features and less expertise in email integration. Though discussion boards and chat rooms provide a wider range of interaction, interactive learning is still limited. Lessons also usually are broken with various parts of learning in various different files uploading and not in one comprehensive whole. Instructions provided online materials are difficult to follow, and being socially isolated when using Blackboard due to slow internet connectivity
Schoology	While Schoology works well for sharing information easily, it is in many ways a "closed system" that inhibits the degree of sharing that teachers might like i.e. Information is available only within the system. Schoology is designed like a digital classroom, with a teacher organizing information and moderating behavior. Schoology lacks the raw connectivity that social media offer
Edmodo	Edmodo interface is similar to Facebook and accessed using a web browser. It is difficult to organize and share materials with students and other instructors compare to other systems. On Edmodo, teachers are at the center of a powerful network that connects them to students, administrators, parents, and publishers. Only teachers and students can interact with one another, student-to-student messages are not allowed

4.4 Impact

See Table 4

Table 4. Comparing different LMS approaches in respect of mobile learning impact.

Approach	Impact
Moodle	Moodle relies on third-party add-ons to create functionality rather than including it as part of the core product. This can increase the workload for maintaining and updating the software as new versions are released. It also lacks a full-featured competency development and management toolset. It needs some technical expertise to get it up and running – and to maintain it. There is information overload and confusing navigation layout design, including navigation icons and symbols that are poor and inconsistent

(*continued*)

Table 4. (*continued*)

Approach	Impact
Blackboard	Users cannot access the underlying source code and thus cannot adjust the software, add features or correct bugs immediately. The Blackboard UI is unintuitive and redirects to different links instead of using hypertexts. It takes lots of time to sort the entire software out due to its complexity. The students face difficulties in logging on to the system
Schoology	Schoology is more intuitive and developed. While there is no limit to the number of files, the user can attach to materials in Schoology, but must upload each file individually. Schoology has a size limit of 512 MB per file. It can easily integrate new and existing content created in-house and it is easier to import and synchronize Schoology with existing systems
Edmodo	Edmodo features storage for text, images, video etc. in library but it is difficult to embed the content. Maximum file size is 100 megabytes while attachments are limited to 2 megabytes. Edmodo cluttered and its test feature has not advanced. Work always gets lost on the wall

4.5 Enjoyment

See Table 5

Table 5. Comparing different LMS approaches in respect of mobile learning enjoyment.

Approach	Enjoyment
Moodle	Confusing for the learner: spoils the main purpose of learning, which is ease and interactivity. Moodle have the "video conferencing" support. By the use of video conferencing tool, virtual class application can be performed by using tools such as; online chat, file transferring (.pdf, .swf, .doc, .docx, .xls, .xlsx, . ppt, and .pps), whiteboard application, two-side video and voice transfer on a specified date and time. Moodle owns debate forums, file transfer, e-mail, calendar, whiteboard, and real time chatting options
Blackboard	Blackboard presents a rigid hierarchy that users must follow. It does not include the visual representation of all available options on a single page. Instead, the users relies on drop down menus for the Add File and Add Content Link functions, which make it less user friendly and enjoyable
Schoology	Schoology's Resource Center and Groups allow users to collaborate with other instructors in a local-to-global community. They can share instructional methods, personal experiences, resources and more to help each other improve instruction. Schoology's platform allows for the creation of different classes, and even the ability to individually assign quizzes or assignments, but you cannot create small groups in Schoology
Edmodo	Edmodo has feature a Facebook-like interface that is clean, appealing to students and easy to use. Edmodo teacher dashboards and student interfaces are brighter and include clean icons. Includes chat and group creating tools. Edmodo allows students and teachers to connect their Google Drive accounts

4.6 Economic

See Table 6

Table 6. Comparing different LMS approaches in respect of mobile learning economic.

Approach	Economic
Moodle	Moodle does not charge for licensing being an open source system, but charges are involved in setting up the server and hosting the system. Customization of Moodle is costly complicated user interface, which needs administrative training to use. Additional costs may be involved to keep an administrator to manage the system. Separate costs for customization, may need to hire an outside vendor to do this
Blackboard	Blackboard is not free and can be quite costly depending on the amount of capabilities a user needs it to perform. It entails an annual license fee, which seems to keep rising. The cost of Blackboard depends on how many licenses are required. According to Blackboard executives, costs for their network environment products, including Blackboard Learning System, may start low but as subscribers integrate more functions into Blackboard, subscription licenses may be $200,000 to $400,000-a-year
Schoology	It's web-based and the basic package is free for instructors, with the option to upgrade to an Enterprise Package if you want specialized support or integration with your school's SIS platform. There are two versions of Schoology: free and paid. The paid version costs $10 per student. Schoology does charge for Enterprise support. It does not share the prices for the Enterprise Package on its website
Edmodo	Edmodo is currently a free service. The Edmodo founders have stated they wish to maintain all current services as free services. It allows unlimited number of teachers and students to create accounts and use their system at no cost. It is an ideal starting place for a low-risk, low-cost entry point

5 Discussion

The stage one of this research yielded in six mobile learning human factors that need to be considered in mobile learning management system. Based on the findings of this research, it was discovered that mobile learning system needs to be easy to use with simple step up, easy to customize, modify, and learn how to use. Moreover, LMS should got an inbuilt search option that helps users to find the exact information looking for, quickly and efficiently. Whereas, instructor can create courses, units and add quizzes in few simple steps followed by easy navigation to the course contents with easy course signing up. That gives learners more freedom to start at any time, follow their progression, and learn at their own convenience. The research findings in terms of suitability, it is possible to integrate and deliver content in different formats and ensures that the content is portable across different systems.

Depending on the suitability and needs of learners, course material can be designed to include graphics, animation, video, podcasts or presentations. Full video integration, share audio content, quiz upload and download should be made ease. From sociability perspective, mobile learning system should be designed in proper way that considers the system interface to be similar to social networking sites like Facebook. Including interactive discussion boards to help develop a more robust learning environment and live chat integration for real-time interaction for effective communication between students and instructors. To facilitates both public and private interaction among the users and encourage discussion and effective collaboration, users can view active online users and can initiate for a chat. Interactive questions, forums, chat and assessments options are provided within the course. In terms of impact, the designed mobile learning system should be designed to accommodate both novel and experienced users. Minimized page scrolling with display area that does not exceed the screen size of mobile devices. The frequency and number of errors are minimized. When error occurs, users are provided with clear and explanatory messages that presented with a confirmation option before they commit to an action. Contents are made adapt according to the device used. Specific menus are created for each possible operations for users to allow easy interaction with the system for improved efficiency. For the enjoyment, the layout of the application should give the impression of a social media platform with attractive and faster user interface. That, allow users to send friend requests to other users, share, and interact with them both privately and publicly similar to social media networking. The instructors should be able to create quizzes and other activities on general topics apart from course is to increase their interest to use the application and enhancing learning experience. Users can make the learning process more enjoyable via integrating various educational gaming tools to the system. In terms of economic, the study findings that it should be provided as free of license fee and other cost. It allows unlimited number of teachers and students to create accounts and use their system at no cost.

6 Conclusion

In this paper, a systematic empirical study approach was used to reveal find out most human influencing human factors within the contest of Mobile Learning. Based on A number of factors (e.g., such as user experiences, research background review, and different LMS quality standards), were considered to develop a number of criteria were developed that was used for scientifically conducting the empirical study procedure. The paper focused on human factors consideration in mobile learning management system including flexibility, suitability, social, enjoyment, impact, and economic. To show the feasibility of the proposed approach, four different learning management systems Moodle and Blackboard, Schoology and Edmodo have been compared and evaluated considering the selected human factors. Based on the findings, it was discovered that easy to use, customize, modify, and learn are key important criteria in the flexibility of mobile learning system. Moreover, depending on the suitability and needs of learners and instructors, learning material can be designed as well as the layout of the system should give the impression of a social media platform with attractive and faster user interface. Mobile learning management systems should be designed for a different level of users' experiences. Future research may be directed towards a comprehensive and better design of mobile learning management system. Another work direction may investigate the guidelines and policies that need to be in place to ensure the successful development and adoption of such system.

Acknowledgment. This article is based upon research work funded by The Research Council (TRC) of the Sultanate of Oman, under Grant No: ORG/SQU/ICT/13/006, (www.trc.gov.om).

References

1. Alzahrani, A., Alalwan, N., Sarrab, M.: Mobile cloud computing: advantage, disadvantage and open challenge. In: Proceedings of the 7th Euro American on Telematics and Information Systems, (EATIS 2014), article no. 20. ACM, New York (2014)
2. Sarrab, M., Elbasir, M., Alnaeli, S.: Towards a quality model of technical aspects for mobile learning services: an empirical investigation. Comput. Hum. Behav. **55**(Part A), 100–112 (2016)
3. Sarrab, M.: Mobile Learning (M-learning) Concepts, Characteristics, Methods, Components. Platforms and Frameworks. Nova Science Publishers, New York (2015). ISBN 978-1-63463-342-0
4. Al-Harrasi, H., Al-Khanjari, Z., Sarrab, M.: Proposing a new design approach for M-learning applications. Int. J. Softw. Eng. Appl. (IJSEIA) **9**(11), 11–24 (2015)
5. Sarrab, M., Al-Shihi, H., Al-Manthari, B.: System quality characteristics for selecting mobile learning applications. Turk. Online J. Distance Educ. (TOJDE) **16**(4), 18–27 (2015)
6. Sarrab, M., Alzahrani, A., Alalwan, N., Alfarraj, O.: An empirical study on cloud computing requirements for better mobile learning services. Int. J. Mob. Learn. Organ. **9**(1), 1–20 (2015)
7. Al Khan, A., Al-Shihi, H., Al-khanjari, Z., Sarrab, M.: Mobile learning adoption in the middle east: lessons learned from the educationally advanced countries. Telemat. Inform. **30**, 909–920 (2015)
8. Sarrab, M., Alzahrani, A., Alalwan, N., Alfarraj, O.: From T-learning into M-learning in education at the university level: undergraduate students perspective. Int. J. Mob. Learn. Organ. **8**(3/4), 167–186 (2014)
9. Sarrab, M., Al-Shihi, H., Rehman, O.: Exploring major challenges and benefits of M-learning adoption. Br. J. Appl. Sci. Technol. **3**(4), 826–839 (2013)
10. Sarrab, M., Elgamel, L.: Contextual M-learning system for higher education providers in Oman. World Appl. Sci. J. **22**(10), 1412–1419 (2013)
11. Sharples, M.: Mobile learning: research, practice and challenges. Distance Educ. China **3**(5), 5–11 (2013)
12. Chintan, A., Daneva, M., Damian, D.: Human factors in software development: on its underlying theories and the value of learning from related disciplines, a guest editorial introduction to the special issue. Inf. Softw. Technol. **56**, 1537–1542 (2014)
13. Marilyn, M., Toby, J.: Cost/benefit analysis for incorporating human factors in the software lifecycle. Commun. ACM **31**(4), 428–439 (1988)
14. Waard, I., Abajian, S., Gallagher, M., Hogue, R., Keskin, N., Koutropoulos, A., Rodriguez, O.C.: Using mLearning and MOOCs to understand chaos, emergence, and complexity in education. Int. Rev. Res. Open Distrib. Learn. **12**(7), 94–115 (2014)
15. Massimo, P.: Systems Lifecycle Cost-Effectiveness: The Commercial, Design and Human Factors of Systems Engineering. Gower Publishing Ltd., Aldershot (2014)

16. Angelia, S., Quesada, S., Andre, T.: Development of a human factors principles-based design tool for engineers. In: Proceedings of the Human Factors and Ergonomics Society Annual Meeting, vol. 58, no. 1. SAGE Publications (2014)
17. Neville, S., Salmon, P.: Human Factors Methods: A Practical Guide for Engineering and Design. Ashgate Publishing Ltd., Farnham (2012)
18. Mishra, D., Mishra, A.: Effective communication in collaboration and coordination in extreme programming: human-centric perspective in a small organization. Hum. Factors Ergon. Manuf. **19**(5), 438–456 (2009)
19. Mugambi, M.: Human coordination factors in software development. Int. J. Educ. Res. **2**(5) (2014)
20. Pınar, C., Kalıpsız, O.: Assessing the human factors in software development courses students project. In: International Conference on Education and Educational Technologies (2013)
21. Sarrab, M., Al-Shihi, H., AI Khan, A.: An empirical analysis of mobile learning (M-learning) awareness and acceptance in higher education. In: The 2015 International Conference on Computing and Network Communications (CoCoNet 2015), Symposium on Multimedia, Visualization and Human Computer Interaction (SMVH 2015), Trivandrum, Kerala, India (2015)
22. Hair, J., Black, W., Anderson, B., Tatham, R.: Multivariate Data Analysis, 7th edn. Prentice-Hall, Upper Saddle River (2010)
23. Nunnally, J.: Citation classic-psychometric theory. Curr. Contents/Soc. Behav. Sci. **22**, 12 (1979)

Augmented Reality and Immersive Applications

Augmented Reality Supporting Deaf Students in Mainstream Schools: Two Case Studies of Practical Utility of the Technology

Andri Ioannou(✉) and Vaso Constantinou

Cyprus Interaction Lab, Cyprus University of Technology, Limassol, Cyprus
{andri,vaso}@cyprusinteractionlab.com

Abstract. The study suggests a practical use of augmented reality (AR) in supporting Deaf and Hard of Hearing (DHH) students in mainstream schools. The paper presents two case studies. First, we used AR via wearable glasses to support the communication and feedback loop between the instructor and DHH learner during the lesson. Second, we used AR via a typical tablet to support the DHH students in acquiring vocabulary and subsequently, improving their reading-comprehension capabilities. In the first study, although the wearable AR glasses were not stylish enough to be attractive to the DHH adolescent, who felt uncomfortable wearing them in a real classroom setting, she was positive about the potential value of the technology in providing immediate feedback and communication with the educator. In the second study, the experience was positively endorsed by four participating six-graders and their special teacher who thought that AR technology can be a great supporting tool during reading-comprehension of difficult texts. The study demonstrates the utility of AR technology in real world settings, serving the needs of special education students, in this case, those with hearing loss, potentially contributing to more inclusive classroom environments in mainstream schools.

Keywords: Augmented reality · Wearable AR glasses · QR codes
Deaf · Hard of Hearing · Mainstream education · Special education
Inclusive education · Technology assisted learning

1 Introduction

Because of the advancement of medical technology (e.g., hearing aids, cochlear implants), the education of Deaf and Hard of Hearing (DHH) children has evolved into a change in speech therapy, educational audiology, and special education (Beal-Alvarez and Cannon 2014). Educational programs that serve DHH children have changed and keep changing, especially since most DHH children are now placed in mainstream schools rather than schools for the Deaf (Kelman and Branco 2009). DHH students, supported by medical technology, are now primarily educated in general/mainstream schools together with their hearing peers, whilst they benefit from special education classes providing additional support from special educators (Beal-Alvarez and Cannon 2014).

M. E. Auer and T. Tsiatsos (Eds.): IMCL 2017, AISC 725, pp. 387–396, 2018.
https://doi.org/10.1007/978-3-319-75175-7_39

Yet, mainstreaming DHH students holds challenges. It is often the case that the teachers and other student-peers have problems communicating with the DHH students or cannot understand them fully (Ditcharoen et al. 2010; DeWitt et al. 2015). Considering that interaction amongst students and their teacher is the most important aspect of a classroom environment, it becomes apparent that DHH students miss out on this learning opportunity in the mainstream classroom. Along these lines, in their survey of the educational needs of DHH students in mainstream education in New Zealand, Fitzgeralds and associates (as cited in Powell and Hyde 2014) elaborated on that mainstream schools do not always serve the needs of DHH students. They argued that there is a need for learning environments that make effective use of visual learning material, adaptive technology, speech-to-text captioning, or other technology to accommodate hearing loss. Building on these ideas, the present investigation suggests a practical use of AR technology to support DHH students in mainstream schools.

AR refers to "technologies that project digital materials onto real world objects" (Cuendet et al. 2013) and although several other definitions can be found in the literature, all of them point to the capability of AR to superimpose virtual information to physical objects. The number of published studies about AR in education has progressively increased during the last few years (Bacca et al. 2014), yet there seems to be limited work in the context of special education and DHH learners. This study presents two case studies of AR application and utility in supporting DHH students in mainstream schools. These studies are presented next along with prior relevant work, goals in each case, methodology, and findings from initial evaluation. A discussion and future possibilities conclude the article.

2 Study 1

The aim of this case study was to explore how the use of AR glasses can facilitate the communication and feedback loop between the DHH learner and instructor by (i) helping the DHH student check, in real-time, his/her understanding of the material and receive immediate feedback and (ii) helping the teacher assess the DHH student's status of understanding the lesson content. The concept was developed and refined in discussions with a local group of special educators who are part of a larger project aiming at inclusion and inclusive pedagogy in mainstream schools.

2.1 Relevant Work

A couple of previous works are relevant to the present investigation, although not in the context of special education. First, Zarraonandia et al. (2013) used AR technology to facilitate the feedback loop between students and instructor, on the premise of supporting students who fear of showing themselves up in front of their classmates. In this case, the instructor was equipped with a head mounted AR display and could see (augmented) symbols that student selected to represent their status of understanding the lecture content. A pilot evaluation in a university lecture demonstrated that AR technology could provide effective support to communication and interaction during lectures, especially as it matures and becomes less intrusive. A similar use of AR technology was previously noted by Ramsden (2008), who explained that the use of AR in education offers advantages over asking people to raise their hands as it can help to quantify the feedback and reply anonymously. For example, Ramsden (2008) used QR codes to gather formative feedback during a presentation. Two QR codes were used by the audience to scan/answer a yes/no question; answers were gathered on a server and presented to the instructor for subsequent action. With respect to using AR with DHH students, a relevant study was presented by Parton (2017) who reported on the so called, Glass Vision 3D project. Glass Vision 3D focused on developing and researching a Google Glass app that allowed DHH children to look at the QR code of an object in the classroom, which then overlaid a related video based on the American Sign Language. Pilot testing with four fifth-graders at a school for the Deaf demonstrated successful use of the Google Glass by all DHH participants along with enthusiastic engagement, although a lengthy learning curve and hardware issues distracting the learning process were noted. In the present investigation, we expand on the above-mentioned ideas to support real-time feedback loop and communication between the DHH student and the teacher in the mainstream classroom.

2.2 Methodology

AR glasses were deemed appropriate means for this work; compared to the use of a smart device (e.g., tablet), the wearable AR glasses would allow DHH students to look at the teacher and blackboard through them while taking notes (i.e., does not block sight or hands). We used Sony's SmartEyeglass because it was commercially available at the time of the study (unlike for example, Google Glass) and was affordable too. SmartEyeglass is lightweight, binocular eyewear with a wired controller (see Fig. 1); it has a 419 × 138 resolution monochrome display that is only able to display bright green (see Fig. 2).

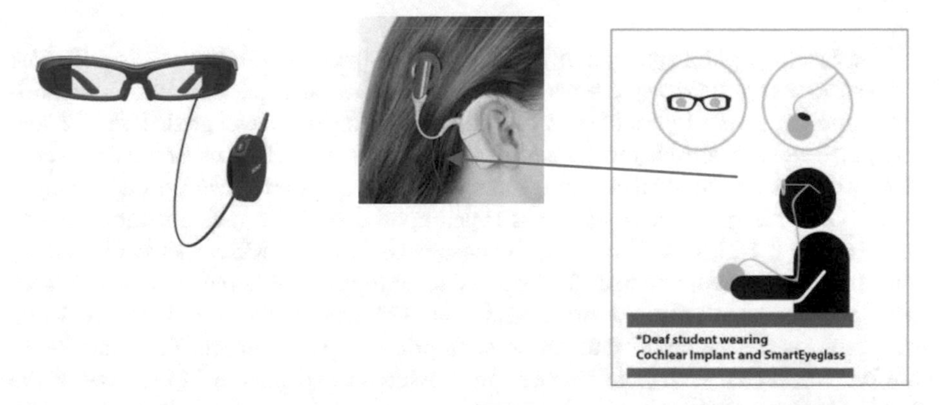

Fig. 1. SmartEyeglass by Sony (left), deaf students with cochlear implant wearing SmartEyeglass (right)

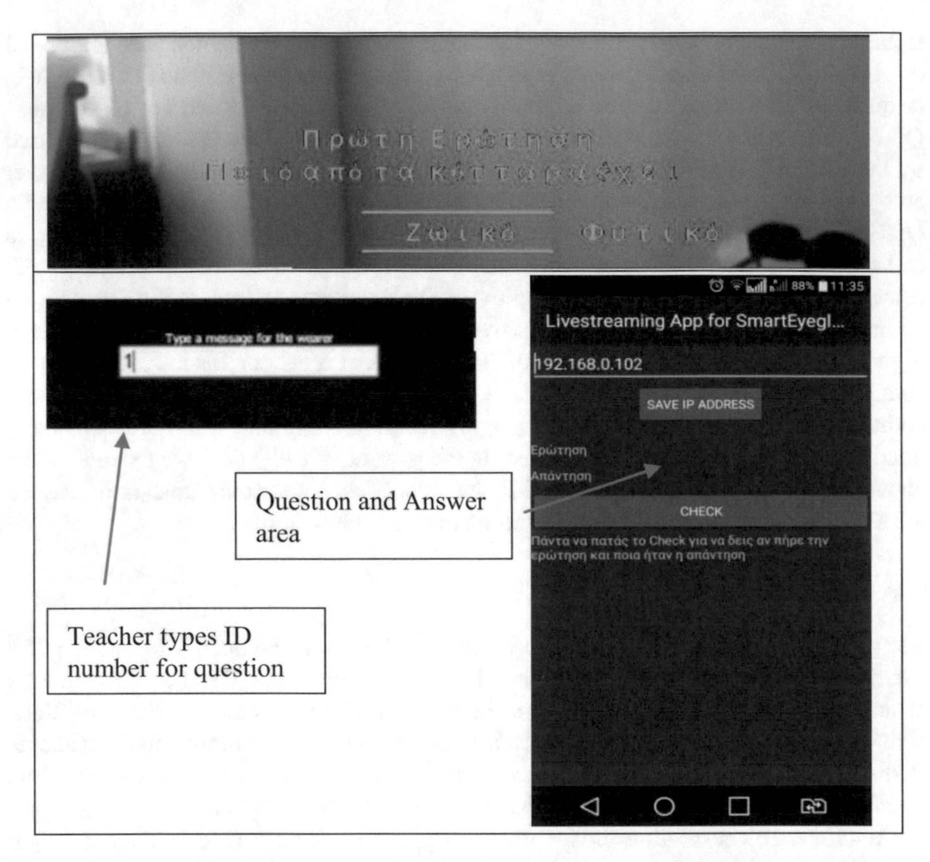

Fig. 2. Question with possible answers displayed through SmartEyeglass (top), teacher types the question to be displayed (bottom-left), teacher checks student's answer (bottom-right).

The SmartEyeglass application was designed to work as follows (see Fig. 2): The teacher prepares a list of questions before the lesson. Each question has an identification number, a set of possible answers and an explanatory message that will be sent to the student as feedback for the correct or wrong answer. The teacher sends predefined questions to the student through his computer or smart device, while giving a lesson. Once the question is sent, it is superimposed onto the user's natural field of view through the glasses. The student chooses one of the possible answers provided, using his controller. Immediate feedback is superimposed onto the student's field of view, as a message with a short explanation. The teacher can check the student's answer on his device and decide on an appropriate way to proceed. For example, in case one answer is wrong the teacher can provide an explanation in the flow of the ongoing lesson. In case of several consecutive answers being wrong, the teacher might wish to repeat aspects of the lesson or help the student upon the completion of the lesson. Aside from questions and answers, the SmartEyeglass application allows the teacher to write something for the student in real time, such as "SOS on what I will be presenting next" or "please see me after class about [topic]". Also, if there is a need to present a new (not predefined) question during the lesson, the teacher can write a true/false type of question by typing its text followed by "?" which automatically inserts the possible answers, true or false.

2.3 Evaluation and Results

The SmartEyeglass application and overall concept were evaluated with a 16-year old DHH girl. She is supported by a cochlear implant since very young age and has good hearing skills, approximately 80%–90% of what a hearing person can listen, based on her school's educational audiologist. She attends typical lessons in a mainstream high school, although she receives extra support by special educators in the school's special education unit. We began with in a laboratory setting aiming to follow up with a field study in the real classroom setting. In the laboratory setting, we simulated the experience of sitting in the classroom with other classmates and interacting with the instructor via the use of SmartEyeglass, as in Fig. 3.

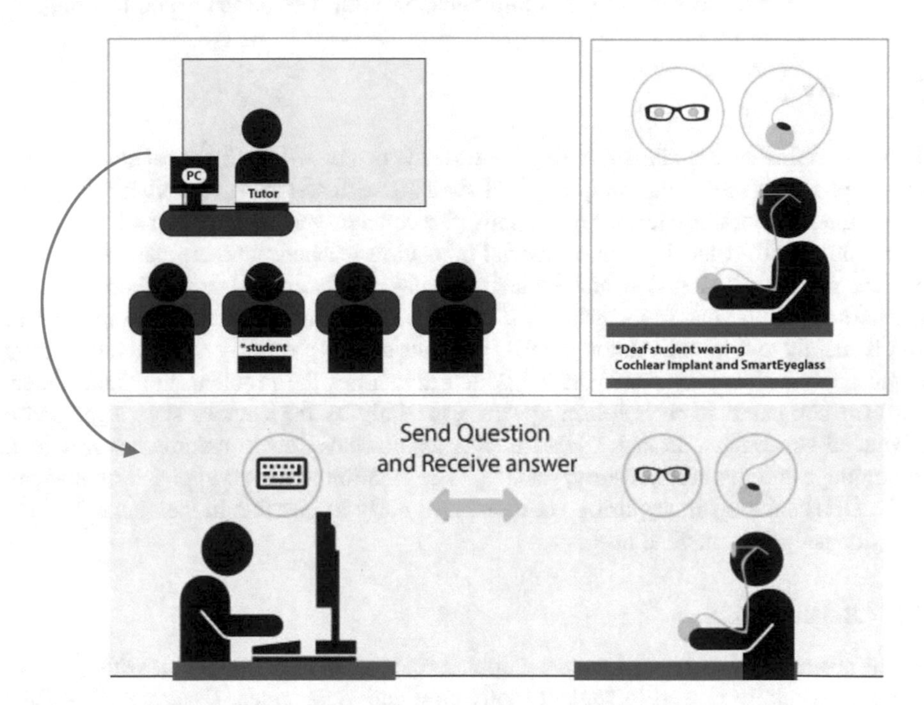

Fig. 3. Classroom setting simulated in the laboratory environment

Her first impression of the appearance and use of the SmartEyeglass was disappointing. When she tried to wear the SmartEyeglass her hearing aid (the extension of the cochlear implant that sits on top of the ear; see Fig. 1) got knocked off; it took a few minutes for her to place the SmartEyeglass in a fair position over her hearing aid and it was still not completely comfortable. Adding to the above, she found SmartEyeglass not stylistic enough for her to use; she said that she would not like to wear it in class since it would draw too much attention by her classmates who could then bully her about this.

Despite the initial reaction, we proceeded with the laboratory evaluation as planned. We let the student warm up with the use of the application, responding to sample questions and receiving feedback, for as long as she needed to get accustomed to the technology (in this case, 10 min of practice was enough). We then simulated a 15-min. lecture on a biology unit, during which a set of questions were sent to the student at various time-points as the lecture progressed. The student participated in the experience with interest, completed all the questions and received feedback without difficulties. At the end of the experience, the participant was engaged in a discussion with the researchers (i.e., unofficial interview). She elaborated on that the learning experience was positive and that the technology could be useful in class in supporting her understanding of the lecture. Yet, due to the appearance of the glasses, she did not consent for a follow-up investigation in her real classroom environment. She explicitly stated that if the AR glasses looked more like the regular glasses, she would indeed wear them in class to support her communication with the instructor and learning.

3 Study 2

The aim of this case study was to explore how the use of AR via a typical smart device (smart phone or tablet) can support DHH students with acquiring new vocabulary and understanding complex texts. Once again, the concept was developed and refined in discussions with a local group of special education teachers who are part of a larger project aiming at inclusion and inclusive pedagogy in mainstream schools. These teachers explained that in mainstream schools, the teaching material is the same for all and is mainly text-based. Although DHH students are supported by medical technology (e.g., hearing aids, cochlear implants) which moderates the problem, they still under-perform compared to their hearing peers, especially as the courses start using more advanced vocabulary. In fact, DHH students tend to have lots of unknown words (e.g. in language and literature, history, science). The creation of an AR application, that can assist DHH students in acquiring vocabulary, as early as possible in their school years, was deemed of immediate need.

3.1 Related Work

There are a lot of published studies about the use of AR in education with positive results, especially related to student motivation and engagement (Bacca et al. 2014). Most of these studies have used marker-based AR technology (Bacca et al. 2014); that is, a smart device acted as a QR code reader and by scanning a QR code or object the user's view was augmented with text, symbols, images and 3d graphics. In the context of special education and supporting DHH students, a couple of previous studies seem relevant to this work. Parton and Hancock (2011), aimed to augment oral or written English information for DHH children by providing videos in sign-language. They created a traditional storybook, called Lambert's Colorful World, which linked to multimedia components via radio frequency identification (RFID) tags. Using a RFID scanner, the children could scan the RFID tag and see a video in the American Sign Language (ASL) for that page of the story. Although effective and engaging, the specialized RFID reader and tags were negative factors. In a follow-up study, instead of using RFID reader and tags, Parton (2015) used photographed pages from the story book to augment with ASL videos. The technology was evaluated by a small group of kindergarten DHH children in a residential school for the Deaf, with positive results. Drifting away from using AR to overlay sign-language, the present study focused on assisting DHH students to improve their vocabulary and consequently their reading-comprehension capacity, via augmented images, 3D models, slideshows, videos and text-definitions.

3.2 Methodology

The AR application lets the user scan the unknown word from his textbook, using his/her personal smartphone or tablet and receive an appropriate explanation in the form of text, image, slideshow or video. In practice, the teacher prepares a list of anticipated unknown words and their explanations and works with the multimedia programmer to transform those into 3D models, slideshows, images with text etc. before the lesson. The DHH student installs the application on a smart device. When the app starts, the device's camera is initiated. The user places the device above each image (target) and the augmented information appears.

In this study, we worked closely with the special educator of a mainstream school, to select vocabulary from the six-grade's language and literature book, which DHH students would probably not be able to understand. Each unknown word together with a representative image was printed on an external bookmarker as in Figs. 4 and 5. The augmented explanation of the word could take the form of a slideshow, 3d model, video, or static image along with text definition, as decided by the educator. For example, when the mobile device is above the word "natives" a 3d model of an Indian girl appears along with a text definition, or when it is above the word "slaves" the user can see a relevant video (see Fig. 6). For the development of the application, Unity and Vuforia were used.

3.3 Evaluation and Results

The AR application was evaluated with four DHH students in sixth grade, who used it together with their special teacher, during several sessions in the special unit of the elementary school. Evaluation feedback from students indicated that they were very excited to use the application every time. As they explained, in addition to being fun to use, the technology helped them comprehend the meaning of the words and consequently the overall text within a page. Also, the augmented multimedia elements helped them remember the meaning of these words from session to session, as reported by the students and confirmed by the educator. Overall, both the students and the special educator argued that the technology can be a great supporting tool during reading-comprehension of difficult texts and she expressed interest in extending this work to more units and courses. As a minor drawback, the teacher noted that there must be plenty of light in the room for the application to scan the images and work properly.

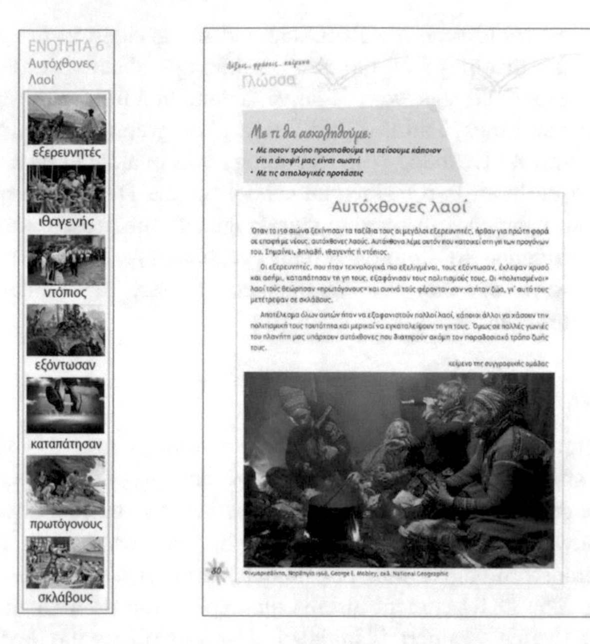

Fig. 4. A bookmark with images trigging AR videos, 3D models, images, slideshows and text definitions

Fig. 5. Sample bookmarks for AR experience for six-grade's language and literature book

Fig. 6. 3D models and definition for "natives" (top), video explanation for "slaves" (bottom)

4 Discussion and Conclusion

The number of published studies about AR in education has progressively in-creased during the last few years (Bacca et al. 2014), yet there seems to be limited work in the context of special education and DHH learners. The paper presented two case studies of the practical utility of AR technology in supporting DHH students in mainstream schools. First, we used AR via wearable glasses to support the communication and feedback loop between the instructor and a DHH learner during the lesson. Although the wearable AR glasses were not stylish enough to be attractive to the DHH adolescent, she was positive about the potential value of the technology in providing immediate feedback and communication with the educator. Second, we used AR via a typical tablet to support DHH students in acquiring vocabulary and subsequently, improving their reading-comprehension capabilities. The experience was positively endorsed by four participating six-graders, who responded with enthusiasm and demonstrated gains, consistent with previous work using AR with DHH children (i.e., Parton 2015, 2017). Along with their special teacher, they argued that AR technology can be a great supporting tool during reading-comprehension of difficult texts.

Overall, the study demonstrated how AR technology can serve the needs of DHH students, potentially contributing to more inclusive classroom environments in main-stream schools. The study contributes to the technology-enhanced learning and special education communities by presenting cases of real world impact of AR technology use. Our findings are promising and justify further investigation to fully exploit the possibilities of AR in this context, helping to build accommodative learning opportunities for children with special needs in general educational programs.

Acknowledgment. This project has received funding from the European Union's Horizon 2020 research and innovation programme under grant agreement No. 692058.

References

Bacca, J., Baldiris, S., Fabregat, R., Graf, S.: Augmented reality trends in education: a systematic review of research and applications. J. Educ. Technol. Soc. **17**(4), 133 (2014)

Beal-Alvarez, J., Cannon, J.E.: Technology intervention research with deaf and hard of hearing learners: levels of evidence. Am. Ann. Deaf **158**(5), 486–505 (2014)

Cuendet, S., Bonnard, Q., Do-Lenh, S., Dillenbourg, P.: Designing augmented reality for the classroom. Comput. Educ. **68**, 557–569 (2013). https://doi.org/10.1016/j.compedu.2013.02.015

DeWitt, D., Alias, N., Ibrahim, Z., Shing, N.K., Rashid, S.M.M.: Design of a learning module for the deaf in a higher education institution using padlet. Procedia-Soc. Behav. Sci. **176**, 220–226 (2015)

Ditcharoen, N., Naruedomkul, K., Cercone, N.: SignMT: an alternative language learning tool. Comput. Educ. **55**(1), 118–130 (2010)

Kelman, C.A., Branco, A.U.: (Meta) communication strategies in inclusive classes for deaf students. Am. Ann. Deaf **154**(4), 371–381 (2009). https://doi.org/10.1353/aad.0.0112

Parton, B., Hancock, R.: Interactive storybooks for deaf children. J. Technol. Integr. Classr. **3**(1) (2011). https://prezi.com/n3xwdj3m_fgf/creative-and-innovative-strategies-to-engage-students-in-literacy/

Parton, B.S.: Leveraging augmented reality apps to create enhanced learning environments for deaf students. Instr. Technol. **12**, 21 (2015)

Parton, B.S.: Glass vision 3D: digital discovery for the deaf. TechTrends **61**(2), 141–146 (2017)

Powell, D., Hyde, M.: Deaf education in New Zealand: where we have been and where we are going. Deafness Educ. Int. **16**(3), 129–145 (2014)

Ramsden, A.: The use of QR codes in education: a getting started guide for academics (2008). http://opus.bath.ac.uk/11408/. Accessed August 2017

Zainuddin, N.M.M., Badioze Zaman, H., Ahmad, A.: Learning science using AR-Book by blended learning strategies: a case study on preferred visual needs of deaf students. Malays. J. Educ. Technol. **9**(2), 5–20 (2009)

Zarraonandia, T., Aedo, I., Díaz, P., Montero, A.: An augmented lecture feedback system to support learner and teacher communication. Br. J. Educ. Technol. **44**(4), 616–628 (2013)

Enabling Social Exploration Through Virtual Guidance in Google Expeditions: An Exploratory Study

Antigoni Parmaxi(✉), Kostas Stylianou, and Panayiotis Zaphiris

Cyprus University of Technology, Limassol, Cyprus
{antigoni.parmaxi,kostas.stylianou,
panayiotis.zaphiris}@cut.ac.cy

Abstract. This paper reports on an exploratory study on the use of Google Expeditions in the context of an intensive 650-h Greek language course for specific academic purposes. Google Expeditions are collections of linked virtual reality (VR) content and supporting materials that can enable teachers to guide students through virtual trips to places throughout the world including museums, surgical processes, outer space, the ocean etc. Qualitative thematic analysis of instructors' field notes, students' reflections, interviews and focus group was employed aiming at identifying the potential of Google Expeditions as instructional tools that can extend the language course for specific academic purposes in topics related to Nursing. To triangulate the findings, the study also collected data by observing students' behavior in the use of Google Expeditions. The use of Google Expeditions enabled students to extend the borders of the classroom by making virtual walkthroughs in places that would normally be unreachable and trigger social exploration through inter- and extra-VR communication, sharing of ideas, concepts, experiences and artifacts. The outcomes have shown that actions taken by the students and instructor during a virtual trip in nursing-related places reveal results in favor of the use of Google Expeditions in the language classroom. Further implications for practitioners and researchers are also provided.

Keywords: Technology-Enhanced Learning · Google Expeditions
Constructionism · Social constructionism · Wearable · Virtual reality
VR · Low-cost VR

1 Introduction

1.1 Background

The onset of Virtual Reality (VR) technology can be traced to the 1960s in the entertainment industry. Since then, VR has been leveraged to meet the needs of training (e.g. with the use of flight simulators) and curricula in subjects such as mechanics, architecture, art, chemistry, medicine, science and language learning [1–6, 20, 21]. Several studies indicate that these technologies provide fertile ground for visualizing abstract concepts, enhance embodied learning but also give opportunities to visit and

M. E. Auer and T. Tsiatsos (Eds.): IMCL 2017, AISC 725, pp. 397–408, 2018.
https://doi.org/10.1007/978-3-319-75175-7_40

interact with places or object that time or spatial restrictions might limit [1–4], engaging learners in authentic, real-world situations and interactive communication [5, 6, 20].

Virtual Reality is a key concept to understand the effectiveness of experiential and experimental view of learning. According to Schwienhorst [7], as VR uses interface structures, it provides access to authentic resources and tools. VR is a highly interactive, computer-based, multimedia environment in which the user becomes a participant with the computer in a 'virtually real' world (Pantelides, 1993 in [7], p. 200). González et al. [8] defined virtual world as a persistent computer-simulated environment allowing large number of users, who are represented by avatars interacting in real-time with each other at the simulated environment. Learners engagement can be increased through the use of Virtual World, while it develops simulated activities which stimulate active participation [9]. Moreover, VR promisingly provides a wide range of opportunities especially when dealing with sensitive topics or concepts that cannot be easily approached through traditional audiovisual simulations. Ochs and Blache [10], for instance, refer to the European project TARDIS as an example of using virtual recruiters in order to train young people before attending their job interviews. Ochs and Blache [10] specifically examine the use of virtual patients as a training tool for doctors who are obliged to deliver bad news to patients or relatives. Such virtual characters (agents, recruiters, patients etc.) are designed after a deep research on the real behavior of humans in several social contexts and they should coordinate accordingly.

Immersive VR often requires an advanced technology that people hesitate or just do not have the necessary skills to use. As Yang et al. [11] state, nowadays learners have to wear devices such as Head-Mounted Displays (HDM) to experience a virtual world. These devices, are relatively expensive, heavy and easily damaged. On the other hand, desktop VR does not require any special device and regarding the price, it has the lowest cost among the applications used for VR. However, it offers a lower level of immersion compared to other solutions [11, p. 1346].

Even though several researchers have underlined the positive impact of VR in education [9, 11], there is also evidence which shows that teachers and trainers still hesitate to use it extensively. In fact, most teachers make a partial use of VR in classroom as part of a game-based activity or they hesitate to use it because it requires a high level of advanced technology knowledge. Moreover, the high cost of such devices -Yap [12] refers to Oculus VR, Google Glass and Samsung Gear VR- makes it inaccessible to schools, colleges or universities.

1.2 Theoretical Framework and Goals of the Current Research

This study incorporated Virtual Reality cardboard as a low-cost 3D viewer in conjunction with a smartphone in an intensive Greek Language Course for perspective Nursing students. VR Cardboard is a low-cost technology product compared to other VR devices discussed earlier. On 2016, via Google Expeditions, Google has launched a number of virtual environments using videos, 3D images and 360 panoramas aiming at giving teachers and students the opportunity to visit places or environments that are not able to visit ordinarily. Considering that nowadays most of the students own a smartphone, and also knowing that a VR cardboard can be purchased for less than 20 dollars, VR cardboard along with Google Expeditions is the cheapest solution in the market for creating VR environments. Google Expeditions give instructors the ability to control (using a tablet) what students are viewing and guide them to watch content that is related to the aim of the lesson. Moreover, it includes information and questions related to the 3D environment which can be used by teachers during the VR experience. The fact that all previous VR solutions required a desktop or a computer to view content is another advantage of Google Cardboard. Using a low-cost virtual reality headset in conjunction with a smartphone, Google Expeditions can simulate virtual objects and scenes and has a good record of educational applications in nursing, especially for visualisations of hospital departments, nursing processes and human anatomy.

Incorporating VR in the curricula of health-related professionals, such as nurses, we explored their exposure to a dynamic virtual environment with an eye to fostering social exploration. This study adopts the theoretical framework of constructionism and social constructionism. Constructionism is a theory of learning, teaching, and design, which can be summarized in the conviction that learning occurs more effectively when learners understand the world around them by creating meaningful artifacts that can be probed and shared [13–15]. Building on the notions of constructionism, social constructionism leverages the need for social interactions as well as the need for giving students opportunities to explore a specific topic or theme before proceeding to construction and evaluation of a shareable artifact [16–18].

To capture the use of Google Expeditions through the lens of social constructionism, the process that students adopted and the way technology and context fostered this process were analyzed. This research is unique in that it studies VR through the lens of social constructionism looking at the use of Google Expeditions as an instructional tool that can allow a small group of students to explore, experience and develop expertise in places and objects of their interest. Specifically, the research questions that guide this study are:

1. **How a low-cost VR kit in conjunction with Google Expeditions can facilitate social exploration of authentic, real-life nursing-related situations?** This question aims at studying students' and teachers' use of the VR in terms of their interactions with the content.
2. **What roles does the teacher and students adopt in such an environment?** This question aims at studying the role of students and teachers in a guided VR Google Expedition.

2 Setting

All data related to this study were collected at a public, Greek-speaking university in the Republic of Cyprus. The university accommodates approximately 2500 undergraduate and postgraduate students. The study took place in the context of a course related to Greek as a second language (L2), throughout October 2016 till June 2017. The class met face-to-face every day for five hours, for a total of 650 h. Activities were held face-to-face and online, whereas outdoor activities allowed students to practice the language in authentic, real-world situations. The course was particularly designed to meet the needs of university students who planned to study Nursing.

2.1 Students

The participants were three male students from Kenya, who came to Cyprus, for five years, on full scholarships. Students enrolled in the Greek course upon their arrival in Cyprus, had sessions every day for five hours. This study's horizon is to go in detail and in depth, despite the small sample, having participants work intensively with VR and collect data rich in detail about their use. The students' age ranged from 19–27 years. All students were fluent English speakers; none of them had any knowledge of Greek upon arrival in Cyprus. Their computer skills were in general at basic to intermediate level. Students had no knowledge of VR.

2.2 Instructors

The class was taught by two instructors with extensive experience in teaching Greek as an L2 (see Table 1). The instructors held weekly meetings for coordinating the progress of the course.

Table 1. Overview of the setting.

Students	Instructors	Scope of the course	Thematic units in Google Expeditions
3 male students Age: 19–27	1 female teacher with nine years of experience for teaching Greek as an L2 1 male teacher with eight years of experience for teaching Greek as an L2	Greek as a second language (L2) – 650 h	- A luxury house - Clinic admissions department - Surgical preparation - Human heart - Human respiratory system - Pregnancy development

2.3 Tools and Materials

Expeditions are group experiences with the instructor acting as a guide leading and the students following along (see Fig. 1). Moreover, it includes information and questions related to the 3D environment being explored. Students used a low-cost virtual reality viewer and Android mobile phones in which they downloaded Google Expeditions application.

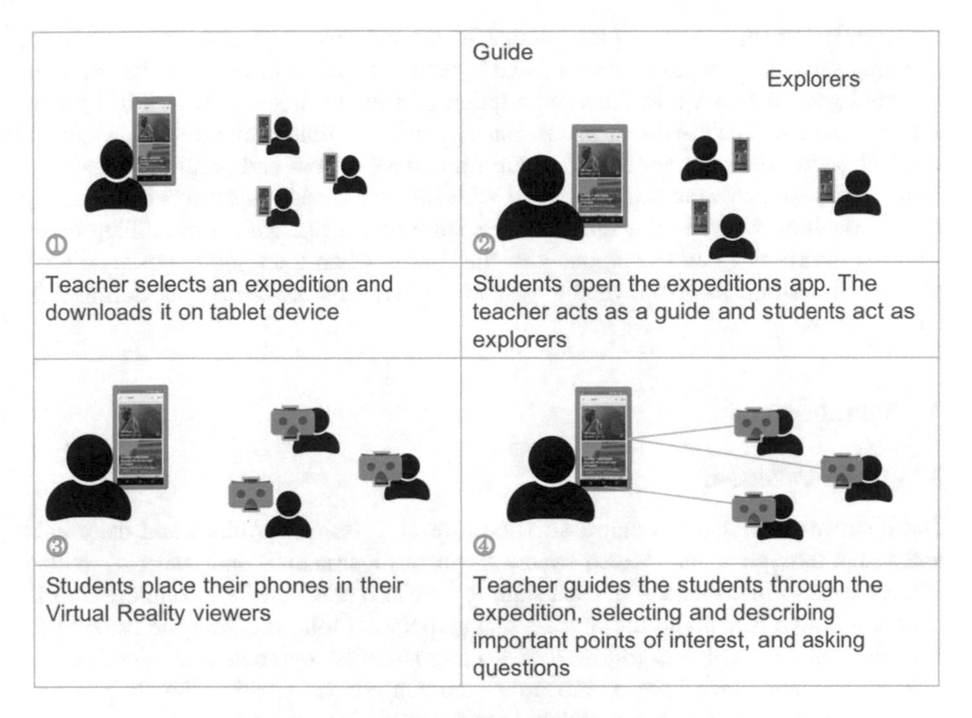

Fig. 1. Setting up Google Expeditions.

The instructors controlled and guided students through Google Expeditions from an Android tablet. Through the tablet the instructor could control what students were viewing and guide them to watch content that is related to the aim of the lesson (see Fig. 2).

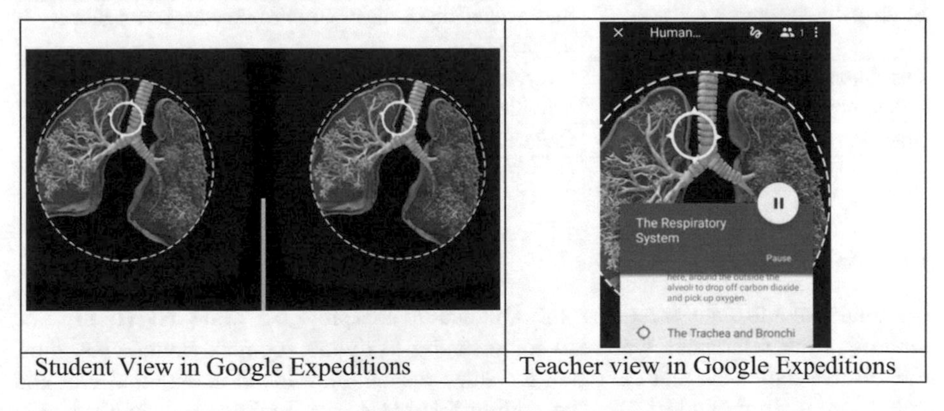

Fig. 2. Student and teacher view in Google Expeditions.

Google Expeditions was incorporated in the existing curriculum of the course, allowing students to explore topics related to their professional needs. Topics explored included general topics (such as a description of a luxury house), as well as specific topics related to health-related professionals (such as clinic admissions department, surgical preparation, human heart, human respiratory system and pregnancy development). These topics were closely linked with students' needs as perspective Nursing professionals and raised their interest in exploring the language for describing them. VR was incorporated as a component of the course, when the topic imposed so for a total of 30 h. Students also owned a wiki where they exposed their work during their Greek course.

3 Methodology

3.1 Data Collection

The data was collected through a questionnaire, in class observations and daily field notes kept throughout the course by the instructor, instructors' and learners' weekly reflective diary. Interviews with each student were also conducted which allowed us to elicit qualitative data about the process that participants followed with the use of VR. An interview protocol was followed to explore students' opinions and overall experiences. The interviews were tape recorded and transcribed verbatim. Finally, students participated in a focus group which lasted approximately 30 min, through which written notes were captured. Table 1 briefly describes the types of data collected (Table 2).

Table 2. Overview of data collected.

Data	Purpose
Questionnaire	Insight into students language and computer literacy
Students' reflections	Self-evaluation of their activities outcomes and process adopted
Instructors' field notes and reflections	Overview of the process adopted and reflections on activities held
Interviews	Reflection on activity process and outcomes
Focus group minutes	Overview of process adopted

3.2 Data Analysis

We analyzed the data set using the Qualitative Research Software Nvivo 11. The content of the utterances was read for meaning to define segment boundaries, thus, consecutive sentences that construct the same meaning are taken as one text unit and coded into a single code [19]. The coding focused on the actions that took place in order to socially explore a specific topic, as well as the affordances of the specific technology.

4 Findings and Discussion

The study identified that a low-cost VR kit in conjunction with Google Expeditions can allow a small group of students to explore, experience and develop expertise in authentic, real-life nursing-related situations, providing opportunities for communication, sharing of ideas, concepts, experiences and artifacts (see Fig. 3). The sections that follow summarize how a low-cost VR kit in conjunction with Google Expeditions can facilitate social exploration of authentic, real-life nursing related situations (RQ1), as well as the roles adopted by the teacher and students in such an environment (RQ2).

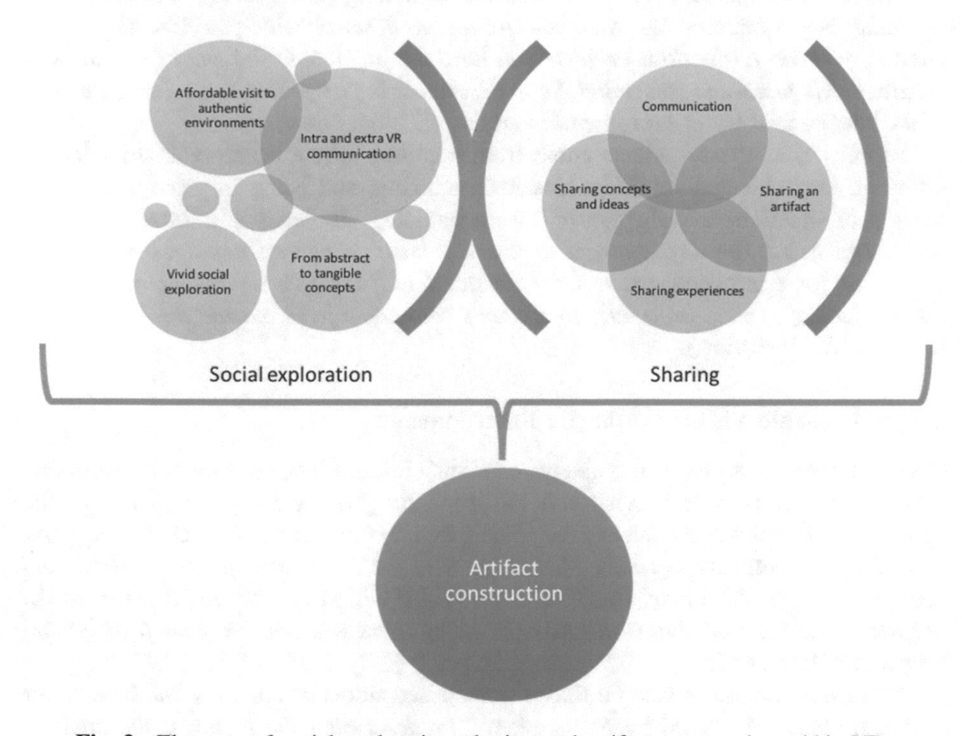

Fig. 3. Elements of social exploration, sharing and artifact construction within VR.

4.1 Intra- and Extra- VR Communication

Google Expeditions can facilitate intra- and extra- VR communication as it allows for (a) instructor to guide students and pose questions for specific important teaching moments given within the application, and (b) students to listen to their instructors' guidance and describe what they could see during the exploration and after the exploration. Intra- and extra- VR communication is particularly important for language learners, as it deals with two fundamental language skills: speaking and listening. Students voiced the importance of being guided by the instructor in practicing their speaking and listening skills, but also noted that being able to visualize information was of paramount importance for them: *[Not only producing and understanding a language is important] you have to see something… […] For example the heart… you see the heart. The first thing you see, then you have to think what is it* (S2, Interview).

4.2 From Abstract to Tangible Concepts

Guided VR tours within Google Expeditions allow students to explore together with their peers important points of interest. The Expeditions act as a common experience for the whole class making abstract concepts and ideas tangible and vivid. As noted by one student: *I was finding it hard to describe something [that I have not seen]. […] it can take time to imagine but when you are told to describe what you have seen in the glasses, you can easily do it because you have the picture. At the end of the day you create a vivid picture in your mind. So to describe it is just getting some things that are in the picture and try to bring words together* (S3, Interview).

Moreover, as these students come from a culture where health-related habits are different, they encountered difficulties in visualizing, and being in place to describe objects or situations that they have not encountered: *To imagine a student reading something he or she does not get a clue it's hard. How can you stick to reading something you don't know, you don't understand. […] I think it is very important for us for the idea of glasses especially to students who come from outside the Greek environment* (S3, Interview).

4.3 Affordable Visit to Authentic Environments

Google Expeditions provided a gateway for students and instructor to visit places that they would not be able to visit and consequently practice language related to the specific communicative context like '*being in the hospital*' (S2, focus group). As voiced by one student: *We have short time also for the lesson, for the course, and because we have short time, I want to save. […] it will take a day to go there [to the hospital], and the time that it will take for going there will limit the time for learning things* (S3, Interview).

Moreover, the use of Google Expeditions raised students' curiosity and interest for a specific topic. As noted by one student: "*by describing the pictures, the analysis makes learning interesting and sticks in the memory*" (S3, Reflections). Also, as students were exploring the development of pregnancy, they were triggered in proceeding to further research in order to find out more about the development of the child: *I had to go a little bit deeper, doing some kind of research, to get the facts. So I was talking about the curiosity, the curiosity from that made me go a little bit deeper, to collect the facts about what exactly happens in that trimester period* (S3, Interview).

4.4 Vivid Exploration Leading to Artifact Construction

Being guided to a specific environment, students were then tasked to construct an artifact related to the specific environment. More specifically, students were tasked to develop an artifact and share it with the rest of the group (e.g. describe the child development, based on the virtual trip they had experienced; identify the tasks of the admissions' officer etc.). To facilitate the completion of the task, students were instructed to switch their Google Expedition from explorers to guiders. The specific task allowed students to concretize their progress. As voiced by one student: *I remember when we used to be in school. In high school. Before the beginning of the lesson we were told "by the end of the lesson the learner should have learned a, b, c". So I found this was another type of not telling the student what he should have known at the end of the lesson, but driving the student through curiosity up to reaching a time, he or she will describe. Now it is tiresome to say "at the end of the lesson the learner should", "the learner should", "the learner should" for the whole page. Then the student can be astonished "how will I learn about all these things?". And it's like you are pressuring to know exactly because it is a must to know this and this. But with this [i.e. the glasses], I was driven from "at the end of the lesson I should do, this and this" by looking at it at the end of the day I describe, without that tiresome process of I should do this and this* (S3, Interview).

4.5 Role(s) Adopted by Instructor and Students

Instructor's role within a VR expedition can be marked as a guide and coach. The instructor guides students through a specific virtual environment, pointing out places of interest and addressing relevant questions to students. However, the role of the teacher as guider can deprive from the students the opportunity to construct knowledge. As noted by the instructor: *My main criticism towards this activity has to do with the role of the teacher. I felt like controlling the whole process and that was a disadvantage as students were expecting me to ask before they answer. I think that such activities would have better results if students are let free to present their experiences through the Virtual Environment provided* (Instructors' field notes).

Students act primarily as explorers, being guided by their instructor and describing the environment around them. At the same time, students also act as constructors of vivid artifacts closely related to their virtual experiences. Students are requested to act like "*professionals who are explaining to people what are they looking at*" (Instructors' reflections). Students, are being placed in an authentic environment in which are encouraged to use the language for describing the reality they have in front of them.

5 Conclusion and Future Work

Until recently, the use of VR in the classroom required a high-cost technology and specialized expertise. Currently, the use of Google Expeditions offers a low-cost, easy-to-use VR option, to be used in the classroom. This paper reported on an exploratory study on the use of a low-cost VR kit in conjunction with Google Expeditions in the context of a Greek language course for specific academic purposes. The use of Google Expeditions enabled students and instructors to extend the borders of the classroom by making virtual walkthroughs in places that would normally be unreachable, and trigger social exploration through inter- and extra-VR communication, sharing of ideas, concepts, experiences and artifacts. The outcomes have shown that actions taken by the students, as explorers and constructors, and instructors, as guiders and coaches, during a virtual trip reveal results in favor of the use of Google Expeditions as an instructional tool for allowing a small group of students to explore, experience and develop expertise in places and objects of their interest.

Traditional language classrooms invest a significant amount of time to bring to the classroom authentic situations that would boost students' need to use the target language. However, these attempts often fail as it is not feasible to bring in class all types of communicative situations that would encourage learners to use the target language. This is especially true when we refer to language teaching for specific academic purposes for perspective health-professionals. VR can facilitate social exploration in places related to health, giving opportunities for social learning. In addition, the classroom is no longer isolated to standardized activities and fill-in-the-gap exercises. Students and instructors can be linked to each other in an active, synchronous exploration. The experience obtained from inter-classroom communication and social exploration through Google Expeditions opens the door to new opportunities. It provides a strategy for connecting instructors and students, and for implementing new forms of engagement, interest and curiosity. Yet, further implications derive in terms of the instructional design of Expeditions. More specifically, more capabilities can be given in allowing students to be members of an Expedition or even in building an Expeditions' community that can provide media-rich constructionist environment for both students and instructors. Taking advantage of the extraordinary power of VR, its low-cost and easy-to-use software, Google Expeditions can support activities that were previously infeasible, making it better positioned to succeed than previous attempts to introduce VR in the classroom.

A low-cost VR kit can open new perspectives in teaching and learning, and its functionalities (in conjunction to Google Expeditions or other VR applications) can foster multiple uses for instructional designs, which remain unexploited until they are embodied and sustained in real-classroom environments. Google Expeditions are currently only available in English, and consequently limit or incommode their use to speakers of other languages. VR holds the potential to release teachers' load in introducing topics, concepts or ideas that are unknown to students. From this perspective, VR is a good mechanism for Technology-Enhanced Learning, especially for introducing ideas and concepts that students cannot understand or visualize from textual sources.

Expeditions are still in the early stage of their development, yet their popularity will most likely continue to increase as companies like Google invest resources into VR to engage and motivate students and teachers. A deeper understanding of the value of this technology can be obtained when different features of this low-cost VR will be incorporated in well-designed, theoretically grounded activities, aligned with the educational needs of the students. Drawing on new forms of teaching, learning and assessment for an interactive world, VR can make fundamental changes in the classroom ecology by leveraging new educational theories and practices such as productive failure, teachback and learning from the crowd [22]. Our working hypothesis is that, as students work closely with their instructors on personally meaningful projects such as virtual reality field trips, they can develop technological fluency, problem-solving skills, social interactions, teamwork skills, and self-confidence on a certain topic that serves them well in the wider spheres of their lives.

References

1. Bellan, J.M., Scheurman, G.: Actual and virtual reality: making the most of field trips. Soc. Educ. **62**(1), 35–40 (1998)
2. Merchant, Z., Goetz, E.T., Cifuentes, L., Keeney-Kennicutt, W., Davis, T.J.: Effectiveness of virtual reality-based instruction on students' learning outcomes in K-12 and higher education: a meta-analysis. Comput. Educ. **70**, 29–40 (2014)
3. Rasheed, F., Onkar, P., Narula, M.: Immersive virtual reality to enhance the spatial awareness of students. In: Proceedings of the 7th International Conference on HCI, IndiaHCI 2015, pp. 154–160. ACM, December 2015
4. Seo, J.H., Smith, B., Cook, M., Pine, M., Malone, E., Leal, S., Suh, J.: Anatomy builder VR: applying a constructive learning method in the virtual reality canine skeletal system. In: 2017 IEEE Virtual Reality (VR), pp. 399–400. IEEE, March 2017
5. Lin, C.-S., Kuo, M.S.: Adaptive networked learning environments using learning objects, learner profiles and inhabited virtual learning worlds. In: Proceedings of IEEE ICALT2005: the 5th International Conference on Advanced Learning Technology, pp. 116–118, Kaohsiung, Taiwan (2005)
6. Chen, Y.L.: The effects of virtual reality learning environment on student cognitive and linguistic development. Asia-Pac. Educ. Res. **25**(4), 637–646 (2016)
7. Schwienhorst, K.: Why virtual, why environments? Implementing virtual reality concepts in computer-assisted language learning. Simul. Gaming **33**(2), 196–209 (2002)
8. González, M.A., Santos, B.S.N., Vargas, A.R., Martín-Gutiérrez, J., Orihuela, A.R.: Virtual worlds. Opportunities and challenges in the 21st century. Procedia Comput. Sci. **25**, 330–337 (2013)
9. Loup, G., Serna, A., Iksal, S., George, S.: Immersion and persistence: improving learners' engagement in authentic learning situations. In: Verbert, K. (ed.) EC-TEL 2016, LNCS, vol. 9891, pp. 410–415. Springer International Publications, Cham (2016). https://doi.org/10.1007/978-3-319-45153-4_35
10. Ochs, M., Blache, P.: Virtual reality for training doctors to break bad news. In: Verbert, K. (ed.) EC-TEL 2016, LNCS, vol. 9891, pp. 466–471. Springer International Publications, Cham (2016). https://doi.org/10.1007/978-3-319-45153-4_44
11. Yang, J.C., Chen, C.H., Jeng, M.C.: Integrating video-capture virtual reality technology into physically interactive learning environment for English learning. Comput. Edu. **55**, 1346–1356 (2010). https://doi.org/10.1016/j.compedu.2010.06.005
12. Yap, M.: Google Cardboard for a K12 Social Studies Module (2016). http://hdl.handle.net/10125/40604
13. Papert, S.: Mindstorms: Children, Computers and Powerful Ideas. Basic Books, New York (1980)
14. Papert, S., Harel, I.: Situating constructionism. In: Constructionism, pp. 193–206. Ablex Publishing Corporation, Westport (1991)
15. Papert, S.: The Children's Machine: Rethinking School in the Age of the Computer. Basic Books, New York (1993)
16. Parmaxi, A., Zaphiris, P., Michailidou, E., Papadima-Sophocleous, S., Ioannou, A.: Introducing new perspectives in the use of social technologies in learning: social constructionism. In: Kotzé, P., et al. (eds.) Proceedings of INTERACT 2013, LNCS, vol. 8118, pp. 554–570. Springer, Heidelberg (2013)
17. Parmaxi, A., Zaphiris, P.: Developing a framework for social technologies in learning via design-based research. Educ. Med. Int. **52**(1), 33–46 (2015). https://doi.org/10.1080/09523987.2015.1005424

18. Parmaxi, A., Zaphiris, P., Ioannou, A.: Enacting artifact-based activities for social technologies in language learning using a design-based research approach. Comput. Hum. Behav. **63**, 556–567 (2016)
19. Chi, M.T.H.: Quantifying qualitative analyses of verbal data: a practical guide. J. Learn. Sci. **6**(3), 271–315 (1997)
20. Brown, A., Green, T.: Virtual reality: low-cost tools and resources for the classroom. TechTrends **60**(5), 517–519 (2016)
21. Lin, C.H., Hsu, P.H.: Integrating procedural modelling process and immersive VR environment for architectural design education. In: MATEC Web of Conferences, vol. 104, p. 03007. EDP Sciences (2017)
22. Sharples, M., de Roock, R., Ferguson, R., Gaved, M., Herodotou, C., Koh, E., Kukulska-Hulme, A., Looi, C.-K., McAndrew, P., Rienties, B., Weller, M., Wong, L.H.: Innovating pedagogy 2016: open university innovation report 5 (2016). https://iet.open.ac.uk/file/innovating_pedagogy_2016.pdf

Scoping the Window to the Universe; Design Considerations and Expert Evaluation of an Augmented Reality Mobile Application for Astronomy Education

Panagiotis E. Antoniou[1(✉)], Maria Mpaka[2], Ioanna Dratsiou[2], Katerina Aggeioplasti[3], Melpomeni Tsitouridou[2], and Panagiotis D. Bamidis[1]

[1] Medical Physics Laboratory, Medical School, Aristotle University of Thessaloniki, Thessaloniki, Greece
pantonio@otenet.gr, bamidis@med.auth.gr

[2] Education Science – Learning Technologies Post-graduate Program, Faculty of Education, Aristotle University of Thessaloniki, Thessaloniki, Greece
{mpaioamar, idratsiou, tsitouri}@nured.auth.gr

[3] Education for Environment and Sustainability Postgraduate Program, Department of Education Sciences in Early Childhood, Democritus University of Thrace, Alexandroupolis, Greece
kangeiop@psed.duth.gr

Abstract. Astronomy is a difficult to teach topic in primary education because its core concepts and topics have no relevance or fly contrary to pupils' limited observational perceptions. Augmented Reality (AR) experiential technologies offer a unique opportunity for enriching the observations of young pupils with engaging and scientifically valid educational content. This work describes the implementation details of a mobile AR application for astronomy education. A simple but pedagogically sound design and implementation process was followed and a detailed pupil evaluation strategy has been developed. The application has been presented to 15 primary school teachers and evaluated both on cognitive and affective axes. Cognitively, the difficulty of the content has been deemed adequate without challenge spikes, while affectively the application's only weakness has been identified as the lack of strong collaborative and personalization elements. These outcomes shall form the basis for future iterations of this work utilizing a participatory design strategy for a collaborative, personalized educational experience.

Keywords: Augmented Reality · Education technology · m-Education STEM · Astronomy

1 Introduction

Teaching Astronomy in primary school pupils is challenging because it is a field where personal experience goes contrary to contemporary scientific theory [1]. As it is established from cognitive theory children create their knowledge about the world and

M. E. Auer and T. Tsiatsos (Eds.): IMCL 2017, AISC 725, pp. 409–420, 2018.
https://doi.org/10.1007/978-3-319-75175-7_41

the universe based on two channels of information: personal observation and other people's explanations [2]. Detailing in this model a rigorous strand of research [2–6] has established three steps, three cognitive models for the establishment of scientifically based models of the universe and the world concepts. The first is the initial model which is based on children's own daily experiences (e.g., seeing the surface of the earth as a more or less flat place) which creates the first entrenchment of beliefs based on those early experiences. The second is the synthetic model where children reconcile their preconceptions with the scientific information as the children start receiving them from education and other contemporary information sources from adults. The third and final is the scientific model when this reconciliation is completed around adolescence where the scientific paradigm of world view becomes internalized and accepted as it emerges the most efficient way to interpret the young adult's more complex world interactions.

Especially in the previously described first stage, a source of serious problems can emerge if children do not get exposed early enough to the correct facts about the physical and astronomical phenomena through their daily life [7]. Children can then have so entrenched misconceptions that new scientific knowledge can get reinterpreted or skewed in accordance with preliminary models instead of the scientific ones [1, 6, 8]. It is exactly this experiential initial world view that can be assisted through the use of experiential means such as Augmented (AR) or Virtual Reality (VR).

AR utilizes advanced image recognition algorithms to track and superimposes over real objects virtual graphical objects through computer based visualizations, providing interactivity potential in real time for the user [9, 10]. Historically, the term AR has been coined by Boing in 1990, when, in a project, they devised this visualization technique to assist their workers in the assembly of their aircraft [11]. AR and VR [9], while utilizing both 3d content user immersion have significant differences. VR fully replaces the user's environmental perception with an artificial one, while AR only adds to the real world virtual objects. Thus AR can allow for outdoor activities as well as lower intensity (and thus overhead) experiential activities than VR [12, 13].

AR appears to be the correct mix of virtual and real field of simulations and feedback for teachers and learners [14]. It creates more authentic user experiences being able to tap into realistic, complex and tangible interplays between the virtual and the real [15].

"Tangible interface" is a term denoting a transparent UI design allowing the user to manipulate the real markers of AR and affect changes in the virtual superimposed content. This kind of interaction provides "sensorimotor feedback" [16]. This kind of manipulation alleviate the immediacy barrier between the user and the content that exists through standard mouse based interactions of virtual (computer generated) content [16, 17]. As has been postulated in the literature, AR interfaces radically change the way that educational content is understood through transparent sensory immersion and interaction while not altering its core [18]. AR, thus, is the correct mix of real and virtual, in order to offer maximum educational benefits through the support of transparent interaction between the real and the virtual environment, utilizing the tactile-tangible interface metaphor interactivity for a smooth transition between the virtual and the real [19].

Several applications of AR have been developed from its inception. For example in entertainment storytelling, applications like MagicBook [20] allow reading of a traditional book while a handheld display projects virtual objects. The users still can use the book without the technology, tacitly turning the pages, however the AR content intensifies the experience. Specifically for education AR has been implemented for immersive experiential education topics like medicine to physics [21, 22]. Specifically in the latter and in overall Science Technology Engineering and Mathematics (STEM) fields have had specific applications. For example, augmented Chemistry consisted of a workbench and a screen that allowed users to understand atom or molecule structure by displaying it in 3D. Use of booklets, cubes and other image registration objects has been used according to necessity to allow immediate and tactile interaction with atoms and molecules [23]. In other educational endeavors, apps like ‘The Book of Colours’ [24] used headgear to teach children, through immediate visual feedback, the basic theory of color.

In astronomy education several applications have been made, exploring AR in teaching concepts such as earth-sun relationship teaching solistices and axial tilt, as well as explore the whole of the solar system in book based educational tools [18, 25].

In this context the idea was born to create a card based, AR mobile app, that would allow learners to tacitly interact with the object of the solar system, while at the same time being able to explore information about each of them. This paper describes both the implementation details of this endeavor as well as the first evaluation results from educational experts (teachers) that would be called to utilize such a medium in experiential, collaborative, learner centric educational episodes.

2 Methods

Implementation

The “Window to the Universe I” application used in this study is an augmented reality application that uses appropriate surveillance cards to visually display information about the celestial bodies that consist the solar system. Using the mobile phone’s camera, the application recognizes each card and digitally overlays the three-dimensional model of each planet, along with encyclopedic knowledge for it (Fig. 1A).

Fig. 1. Augmented reality interactive card with A. overlaid 3D planet and B. overlaid infographic.

The user has the ability through a tactile interface with his mobile phone to interact with the three-dimensional content (by rotating the card to see each planet from another angle) and to obtain information about the planets through infographics suitably designed for each age group (Fig. 1B). The application starts with the user's choice of age and the way he wants to interact with the material. The application enables the student to use interactive cards both in the previously described form and as a means of comparing the sizes of the basic planets. Specifically, approaching the two interactive cards, the planets depicted take their comparative size scale, while their rotation speed is also adjusted in proportion to their real size. Thus, the student acquires a direct view of the sizes and movements of the planets of the solar system. The application was developed in Unity utilizing the Vuforia AR package for it. The digital content included commercially available assets of the solar system (3D models of the planets) overlaid in small handheld cards (one for each planet). This was a core design decision of the approach as to engage the user/learner through the tactility of "holding" a realistically looking planet in her hands, being able to manually rotate and "look closer" at it. Furthermore, in each card representing the planet, core knowledge was inserted as information buttons at the bottom of the cards area. These buttons, press-able in the mobile device changed the digital content of the card, from a tactile model planet to a dashboard that provides concise information in the form of stylized infographics). The user can scroll this infographic through her mobile device and explore the information on her own. The application has been developed with content for 3 age ranges (selectable at launch) with increasingly complex and detailed content according to it. The material that was included in the application has been co-authored by an astronomy expert in collaboration with an educational expert, in order to offer both topic and age specific content and phraseology.

Evaluation

Instructional scenario

For the evaluation of the app in the context of a specific educational episode, two versions of an instructional scenario were created in order to be distributed to students of a control and an experimental group respectively. The instructional scenario was designed for 2.5 instructional hours and it was compatible with the curriculum of the 6th grade of the Greek Primary School education, as its topic is included as a specific thematic unit in the subject of Geography. The two versions of the scenario were developed by the teachers that will act as the facilitators of the evaluation teaching episodes. For the control group, the facilitators created relevant printed material that contained the digital information presented in the app "Window to the Universe". For the experimental group no printed material was included, as all the information was contained in the app itself. The instructive-methodological approach that is aimed to be practiced in both groups' instructional scenarios is the collaborative exploratory method, which aims to actively involve students in the learning process through group based exploratory knowledge building. In that fashion, learner initiative will be fostered and the teacher's role shall be supportive and coordinating.

Learner's evaluation tools

Evaluation of the students' performance in both groups shall be conducted in two axes, one cognitive, regarding knowledge retention, and another affective, regarding the engagement of the students with the subject matter through the two (conventional and AR enabled) teaching episodes.

The cognitive part of the evaluation involves assessment sheets that shall be provided to students after the end of the teaching episodes in order to determine whether the learning objectives have been achieved and to determine the extent to which learning has been accomplished. The assessment sheet was structured on 6 subjects which include multiple-choice questions, "True-False" questions, matching questions, and open-ended questions with graded difficulty. It was considered appropriate to incorporate a combination of questions, both open and objective-closed ones, in the assessment test. Open-ended questions evaluate the creative and synthetic ability of the learner. On the other hand, objective questions, if properly formulated, require the learner to recall information and also other superior cognitive skills. Thus, it was considered preferable to use a combination of questions so that the pupils shall respond as well as possible according to their abilities. The subjects of the assessment sheet were divided according to the type of the question and its difficulty.

The affective part of the evaluation involves a validated instrument, namely the Engagement Versus Disaffection with Learning scale [26] for testing the experimental and the control classes for homogeneity regarding the mood and disposition towards educational engagement. It also utilizes the AffectLecture app for immediate self-reporting classroom affective analytics [27] of student mood before and after an educational episode using one from five emoticons that best represented their affective state (Fig. 2).

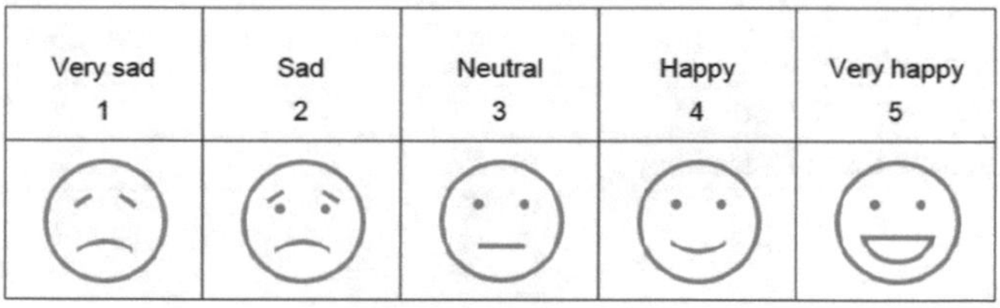

Very sad 1	Sad 2	Neutral 3	Happy 4	Very happy 5

Fig. 2. AffectLecture emoticon scale

Teacher's evaluation

With the full battery of both the experimental procedure and the evaluation procedure in place the actual evaluation is expected to take place at a time close to the students' actual involvement with the subject matter in the formal curriculum. However before deploying the educational episodes an expert evaluation was conducted amongst primary school teachers. This evaluation also included both a cognitive and an affective axis.

The cognitive axis was covered through an assessment of the completeness and suitability of the student cognitive assessment tool. Specifically, 15 primary school teachers were called to review the student assessment questionnaire and report their opinions in a teacher oriented survey. The evaluation was based on the following topics: the level of difficulty of the questions, the degree of overlapping between the application's content and the course material of Geography's unit about the Solar System, the appropriateness of the kind of questions and their structure and finally the completeness of the educational content. Difficulty was graded in a 1–10 scale for each assessment question (to match standard student assessment grading) while the other aspects were graded on a standard 5 points Likert scale with 1 being strong subject disagreement and 5 being strong subject agreement.

The affective axis of the evaluation consisted of a teachers' questionnaire exploring what factors would make the "Window to the Universe" app more engaging to the user. After teachers' interaction with it for a minimum of 30 min they were called to grade several aspects of it that would increase its affective impact and engagement potential. These items were organized in 10 categories (Table 1) with topics in each that included generics like graphics and audio, down to specifics like autonomous navigation and user centric design. All these items were graded on a scale of 1–10 with 1 being the least influential to the users' engagement and overall affective impact and 10 being the most influential.

Table 1. Mobile app implementation aspects affecting engagement potential and affective impact

Aesthetics
Subjective attractiveness
Novelty
Time endurability
User involvement
User attention focus
Perceived usability
User collaboration
User personalization
User feedback immediacy

3 Results

Cognitive Axis

The teachers' questionnaire was answered by 15 primary school teachers 11 women and 4 men. Teachers were asked to grade on a scale of 1 to 10 the level of difficulty of each subject separately as a reflection of the overall level of difficulty of the app's educational content. Most of them agreed that the 3rd subject, which was a matching type exercise, had a higher level of difficulty, as well as the first one that required the correct classification of the planets according to their distance from the Sun. They considered the 5th subject, which was an open-ended exercise, easier than the others because of the fact that students could give varying answers, many of which would be accepted as correct. The graph below (Fig. 3) presents the answers of teachers regarding the level of difficulty of each subject.

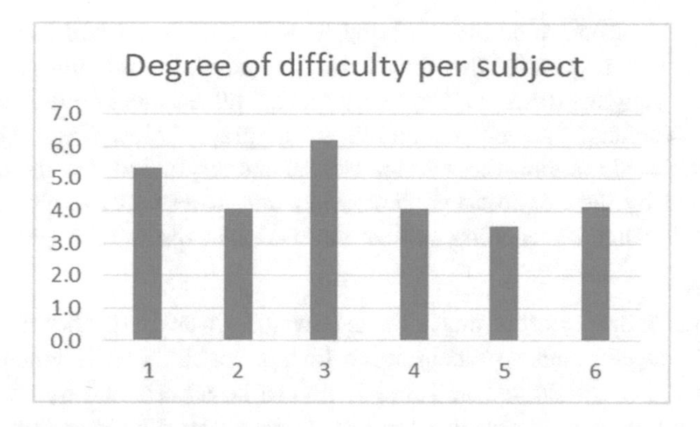

Fig. 3. Teachers' difficulty assessment for the app's topics as represented by the questionnaire.

The results from the specifics of the evaluation are summarized in Fig. 4. Specifically the content of the app, as reflected by the assessment sheet given to the students, overlapped appropriately, according to the teachers, the 6th grade Geography unit of the formal curriculum for the "Solar System" (Fig. 4A). Most teachers, as shown by the graph below (Fig. 4B), felt that the knowledge content was more than the content of the book's chapter, by approximately 41%–60%. Furthermore, most of teachers believed that the assessment methods that are going to be used for evaluation of the students' absorption of the app's material are very appropriate. In particular, responding to a five-point scale ranging from "no" to "very much," all teachers gave answers ranging from 3 to 5 (Fig. 4C).

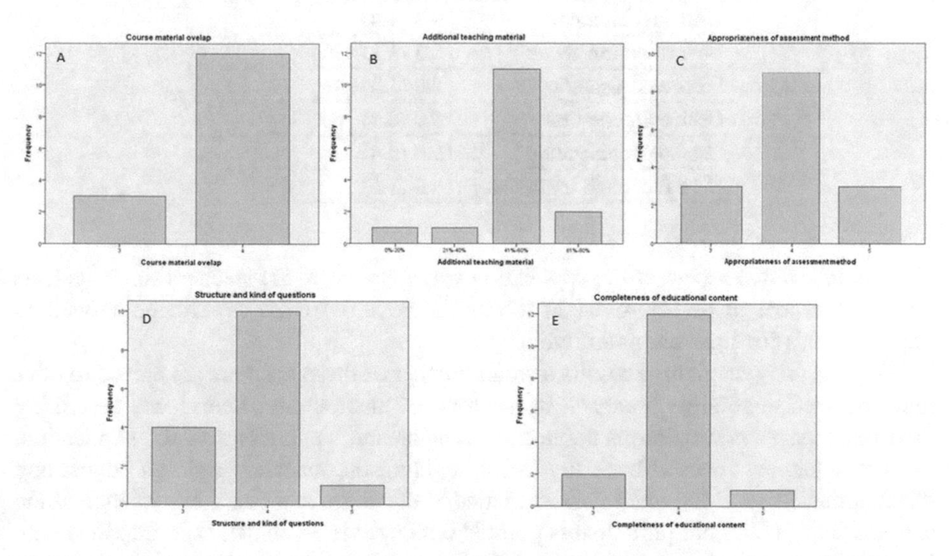

Fig. 4. Summary of teachers' assessment of app educational content and evaluation method as reflected by the cognitive content evaluation sheet.

Regarding the structure and type of questions that were selected on the assessment sheet, most teachers responded that they were quite good. In particular, responding to a five-point scale ranging from "no" to "very much," all teachers gave answers ranging from 3 to 5. Teachers' answers are shown in the graph below (Fig. 4D). Finally, in terms of educational content, the average of teachers responded that the content of the app as reflected by the cognitive content evaluation sheet was largely complete as it covered the full range of teaching subject and various aspects of it (Fig. 4E).

Affective axis

For this axis the same cohort of teacher's as previously was used. The teacher's grading of the various aspects that would improve the application's engagement and overall affective impact is summarized in Table 2. As can be ascertained by that, the aspects that received the highest attention from the teachers were the user collaboration and personalization capacity of the application. Specifically the teachers focused on even more customization, regarding not only the age range, but also the specific knowledges of the users of the application, as well as the inclusion of more collaborative tasks, in the application design.

Table 2. Teacher's perceptions regarding the augmenting of application's design aspects for increased engagement and affective impact

Affective impact categories	Teachers' score (STD)
Aesthetics	8.2 (1.8)
Subjective attractiveness	8.8 (1.3)
Novelty	7.5 (2.0)
Time endurability	8.2 (1.3)
User involvement	8.8 (1.4)
User attention focus	8.5 (1.7)
Perceived usability	8.8 (1.3)
User collaboration	9.9 (0.4)
User personalization	9.8 (0.4)
User feedback immediacy	8.2 (1.5)

The innovative aspect of the application was apparent to the teachers since the least important aspect to be improved in it was the novelty of the experience as well as improvements for time endurability.

Beyond this quantitative results a small number of these teachers (2) agreed to offer more detailed qualitative feedback in the form of unstructured interviews, describing their experiences and opinions about the strengths and weaknesses of the application.

The interview focused basically in highlighting the teachers' opinion concerning the functionality and the main characteristics of the app. As far as the aesthetics of the app are concerned, both participants pointed out that it is a sophisticated interface with impressive and interesting graphics, while the digital information that is presented, is legible and displayed in a linear and organized manner to the user. As for the functionality of the app, the participants seemed to agree that its use is simple by allowing the user to manage the material easily and quickly. In fact, one of them pointed out that one key factor of the app is that it allowed the user to manage the instructional information in the order and time he/she prefers and did not force her to study linearly the whole material. Concerning the content of the app, the teachers focused on the fact that it is covered in scope and depth with a critical scientific eye. They particularly emphasized that knowledge is being explored in a way that allows the learner to create his own learning path autonomously. Also, the approach of a complex and difficult subject such as that of the Solar System is reinforced by the app's diverse and playful concept, transforming the user's interaction with the content into an engaging and experiential activity. On the other hand, the coordinators stated certain weaknesses of the app agreeing that its load of data necessitates the installation of the app only on state-of-the-art mobile devices that meet high standards. Thus, the successful installation and use of the app may be difficult and the user has to possess the proper device and software. In addition, one of the coordinators said that in some cases the user is overloaded with the abundance of information with a risk of being dismantled. However, in general, the comments of both coordinators were positive, pointing out mightily that "Window to the Universe" it's about an innovative and interesting app that can engage the user in a multidimensional learning experience.

4 Discussion

Summarizing, regarding the cognitive teacher evaluation, we are coming to the following overall conclusions: teachers considered the level of difficulty of the evaluation sheet small to moderate. However, there were variations in individual subjects as they identified a greater degree of difficulty in close-ended questions such as matching and "True-False" questions, which required complex and abstract thinking, developed critical competence and co-thinking. On these topics students had to combine information from various parts of the app's content and to critically approach the questions.

Teachers felt that the types of questions that were chosen and their structure were very appropriate. In particular, different types of gradual difficulty questions were selected to challenge students' interest and help them all to respond as well as possible without exception, even those who may be experiencing special learning difficulties as well as those who learn at normal rates. In addition, they considered that the school content about the solar system was completely overlapped and outmatched by the applications content. In that context the participating teachers declared their extreme interest in discovering the results from the forthcoming learner evaluation.

Regarding the affective evaluation, the application has been evaluated very positively as a single-user engagement tool for involving the learner experientially into the subject matter. From the qualitative evaluation it became clear that the learner-centered, exploratory educational paradigm was very efficient and innovative as a means for a user to discover the solar system at her own pace. The need for collaborative content customized to each group according to their knowledges and their age was the core requirement for improving the application. While the technical overhead was mentioned as a weakness in the evaluation the rapid increase in processing power of mobile devices is going to alleviate very soon this weakness.

Successful educational episodes require a wide array of experts (technologists, teachers, students, artists, etc.) working in collaboration and are not straightforward regarding their deployment in the field [28]. Designing interfaces for children, transferring the material into effective AR experiences, adapting the experience for age and ability, designing for collaboration in the classroom are all topics open to research [7]. An approach that has been followed and that this work has also adhered to, was exploring the value that AR can offer to STEM education through the incorporation of "personalized lenses". This entails the inclusion of age- and ability-specific content on the target objects of the technological modality. Thus the experience becomes more accessible to various age groups and levels of familiarity with the subject, while allowing for an increased audience to participate. That way the road opens for educational collaboration potential in diverse learning environments, in and out of the classroom [29]. In that context, the idea of iterating with AR prototypes based on formal STEM curricula of varying "virtuality" for collaborative learning is not new. A step further, indeed, is the inclusion of participatory design processes with students and teachers [30]. This design paradigm has proven its merits in several occasions in the past. In educational interventions towards challenging target groups such as deaf students and people with dementia, participatory design at the early stages of it revealed unanticipated potential problems [31, 32]. In fact not following this methodology close enough at all design stages was something that was mentioned as a proven weakness in the course of the design process [32]. Participatory design processes in AR content for education has matured including checklists and rigorous predefined protocols in order to formally and efficiently support e.g. child oriented design [33] and definitive domain goals for the participation of both children and adults in these processes [30]. Furthermore, this design paradigm has been determined to be an excellent educational experience for the children included in it [34].

Given the previously outlined context, it must be noted that the next iteration of the application, that is going to be publicly released, shall integrate all the outcomes of this evaluation, as well as those from the forthcoming student evaluation. These outcomes form the basis for a next iteration of this application. This next iteration shall be the focus of a more massive, formative assessment of user preferences, as well as the appropriateness for various educational contexts by both teachers and pupils. In short this work is the first part of participatory content and user experience design process with the end goal of an effective personalized educational support tool for the experientially difficult topic of astronomy in the primary school age range.

References

1. Hannust, T., Kikas, E.: Young children's acquisition of knowledge about the Earth: a longitudinal study. J. Exp. Child Psychol. **107**, 164–180 (2010)
2. Vosniadou, S., Brewer, W.: Mental models of the earth: a study of conceptual change in childhood. Cogn. Psychol. **24**, 535–585 (1992)

3. Vosniadou, S.: Exploring the relationships between conceptual change and intentional learning. In: Intentional Conceptual Change (2003)
4. Vosniadou, S., Brewer, W.: Mental models of the day/night cycle. Cogn. Sci. **18**, 123–183 (1994)
5. Vosniadou, S., Ioannides, C., Dimitrakopoulou, A.: Designing learning environments to promote conceptual change in science. Learn. **11**, 381–419 (2001)
6. Vosniadou, S., Skopeliti, I., Ikospentaki, K.: Reconsidering the role of artifacts in reasoning: children's understanding of the globe as a model of the earth. Learn. Instr. **15**, 333–351 (2005)
7. Fleck, S., Simon, G.: An augmented reality environment for astronomy learning in elementary grades: an exploratory study. In: Proceedings of 25ième Conférence Francoph. l'Interaction Homme-Machine, vol. 13, pp. 14–22 (2013)
8. Hannust, T., Kikas, E.: Children's knowledge of astronomy and its change in the course of learning. Early Child. Res. Q. **22**, 89–104 (2007)
9. Azuma, R.: A survey of augmented reality. Presence Teleoperators Virtual Environ. **6**, 355–385 (1997)
10. Azuma, R., Baillot, Y., Behringer, R.: Recent advances in augmented reality. IEEE Comput. **21**, 34–37 (2001)
11. Barfield, W.: Fundamentals of Wearable Computers and Augmented Reality. CRC Press, Boca Raton (2015)
12. Ma, J., Choi, J.: The virtuality and reality of augmented reality. J. Multimed. **2**, 32–37 (2007)
13. Milgram, P., Takemura, H., Utsumi, A.: Augmented reality: a class of displays on the reality-virtuality continuum. In: Telemanipulator and Telepresence Technologies (1994)
14. Salmi, H., Kaasinen, A., Kallunki, V.: Towards an open learning environment via augmented reality (AR): visualising the invisible in science centres and schools for teacher education. Procedia-Soc. Behav. **45**, 284–295 (2012)
15. Arvanitis, T., Petrou, A., Knight, J., Savas, S.: Human factors and qualitative pedagogical evaluation of a mobile augmented reality system for science education used by learners with physical disabilities. Pers. Ubiquitous **13**, 243–250 (2009)
16. Shelton, B., Hedley, N.: Exploring a cognitive basis for learning spatial relationships with augmented reality. Instr. Cogn. Learn. **1**(4), 323 (2004)
17. Chen, Y.: A study of comparing the use of augmented reality and physical models in chemistry education. In: Proceedings of the 2006 ACM International Conference (2006)
18. Shelton, B.E., Hedley, N.R.: Using augmented reality for teaching Earth-Sun relationships to undergraduate geography students. In: ART 2002 - 1st IEEE International Augmented Reality Toolkit Workshop, Proceedings (2002)
19. Billinghurst, M.: Augmented reality in education. In: New Horizons for Learning (2002)
20. Billinghurst, M., Kato, H., Poupyrev, I.: The magicbook-moving seamlessly between reality and virtuality. IEEE Comput. Graph. **31**, 6–8 (2001)
21. Antoniou, P., Dafli, E., Arfaras, G., Bamidis, P.: Versatile mixed reality medical educational spaces; requirement analysis from expert users. Pers. Ubiquitous **21**, 1015–1024 (2017)
22. de Jong, T., Linn, M.C., Zacharia, Z.C.: Physical and virtual laboratories in science and engineering education. Science **340**(6130), 305–308 (2013)
23. Fjeld, M., Voegtli, B.: Augmented chemistry: an interactive educational workbench. In: Mixed and Augmented Reality 2002 (2002)
24. Ucelli, G., Conti, G., De Amicis, R., Servidio, R.: Learning using augmented reality technology: multiple means of interaction for teaching children the theory of colours. In: Intelligent Technologies (2005)
25. Sin, A., Zaman, H.: Tangible interaction in learning astronomy through augmented reality book-based educational tool. In: International Visual Informatics Conference (2009)

26. Skinner, E., Furrer, C., Marchand, G., Kindermann, T.: Engagement and disaffection in the classroom: part of a larger motivational dynamic? J. Educ. Psychol. **100**(4), 765–781 (2008)
27. Antoniou, P.E., Spachos, D., Kartsidis, P., Konstantinidis, E.I., Bamidis, P.D.: Towards classroom affective analytics. Validating an affective state self-reporting tool for the medical classroom. MedEdPublish **6**(3) (2017)
28. Vate-U-Lan, P.: Augmented reality 3D pop-up children book: instructional design for hybrid learning. In: e-Learning Industrial Electronics (ICELIE) (2011)
29. Kaufmann, H., Schmalstieg, D.: Mathematics and geometry education with collaborative augmented reality. Comput. Graph. **27**, 339–345 (2003)
30. Thompson, B., Leavy, L., Lambeth, A., Byrd, D., Alcaidinho, J., Radu, I., Gandy, M.: Participatory design of STEM education AR experiences for heterogeneous student groups: exploring dimensions of tangibility, simulation, and interaction. In: Adjunct Proceedings of the 2016 IEEE International Symposium on Mixed and Augmented Reality, ISMAR-Adjunct 2016, pp. 53–58 (2017)
31. Zainuddin, N., Zaman, H.: A participatory design in developing prototype an augmented reality book for deaf students. In: Computer Research (2010)
32. Slegers, K., Wilkinson, A., Hendriks, N.: Active collaboration in healthcare design: participatory design to develop a dementia care app. In: CHI 2013 Extended Abstracts (2013)
33. Van Mechelen, M., Sim, G., Zaman, B., Gregory, P.: Applying the CHECk tool to participatory design sessions with children. In: Proceedings of the 2014 Conference on Interaction Design and Children (2014)
34. Bower, M., Howe, C., McCredie, N.: Augmented reality in education–cases, places and potentials. Educ. Media **51**, 1–15 (2014)

Advanced Interactive Multimedia Delivery in 5G Networks

Radoslav Vargic(✉), Martin Medvecký, Juraj Londák, and Pavol Podhradský

Slovak University of Technology in Bratislava, Ilkovičova 3, 812 19 Bratislava, Slovakia
radoslav.vargic@stuba.sk

Abstract. In this paper, we present current state and aims of our ongoing research aimed at advanced interactive multimedia and mulsemedia delivery in 5G networks. We summarize the underlying and necessary properties of 5G networks, architecture of SDN/NFV and their usage in the multimedia and mulsemedia delivery. Advanced topics as content adaptation and user identification are discussed. Preliminary results regarding algorithms for content adaptation are presented. We show the advance of usage of AR/VR headsets as effective terminal for immersive application for this kind of delivery and discuss the advantages and disadvantages. We propose corresponding system architecture for advanced interactive multimedia delivery in 5G networks.

Keywords: Interactive multimedia · 5G networks · Content adaptation

1 Introduction

Interactive multimedia delivery is presently one of the challenging areas of the content delivery systems. It puts considerable requirement on the underlying technology. In this contribution, we concentrate on interactive multimedia delivery using 5G networks combined with Software Defined Networks (SDN) and Network Function Virtualization (NFV) technologies. Interactive multimedia content includes also serious games and gamification which are important elements which can be used e.g. in the Self-directed learning (SDL) process. We discuss how the 5G and SDN/NFV can be used and present some proposals how the delivery can be optimized and how the used methods can be reused in the wider context.

The paper is organized as follows. First, we introduce the 5G and SDN/NFV concept and system architecture in the Sects. 1.1 and 1.2. Based on this we describe the multimedia delivery, concentrating on the novel immersive aspects. The core part - the optimizations, i.e. content adaptation options are presented in the Sect. 2. The advanced optimization options are discussed in the Sect. 2.1. The whole proposed concept is synthetized in the Sect. 3, where the corresponding system architecture is proposed.

1.1 5G Networks

The 5G networks are the next generation of converged fixed/mobile networks. Although the standardization of 5G started by 3GPP only in 2nd quarter 2017, the basic

M. E. Auer and T. Tsiatsos (Eds.): IMCL 2017, AISC 725, pp. 421–430, 2018.
https://doi.org/10.1007/978-3-319-75175-7_42

performance criteria and features for 5G systems have been defined by ITU in [1]. International Mobile Telecommunications IMT-2020 key capabilities are shown in Table 1. The 5G networks have been consolidated to be used primary for Enhanced Mobile Broadband (eMBB), Massive Machine-type Communications (mMTC) and Ultra-reliable and Low-latency Communications (URLLC) [1], but the 5G networks are designed to be flexible to adapt to new use cases with a wide range of requirements.

Table 1. IMT-2020 key capabilities

Parameter	Value
Peak data rate	20 Gbit/s
User experienced data rate	100 Mbit/s
Area traffic capacity	10 Mbit/s/m^2
Network energy efficiency	100 x
Spectrum efficiency	3 x
Connection density	10^6 devices/km^2
Latency	1 ms
Mobility	500 km/h

The evolved 5G networks will be characterized by agile resilient converged fixed/mobile networks based on NFV and SDN technologies and capable of supporting network functions and applications encompassing different domains. To achieve these goals, network slicing, edge computing, security, reliability, and scalability should be taken into account in the context of virtualization. The main benefits of 5G are as follows:

- high throughput
- small latency
- network slicing support.

Network Slicing is the ability to create end-to-end slices (network-on-demand) on the same infrastructure for heterogeneous services. While legacy systems (e.g., 4G mobile networks) hosted multiple telco services (e.g. voice or SMS) on the same mobile network architecture (e.g., LTE/EPC), network slicing aims for building dedicated logical networks that exhibit functional architectures customized to the respective telco services. Moreover, legacy systems are characterized by monolithic network elements that have tightly coupled hardware, software, and functionality. In contrast, the 5G architecture decouples software-based network functions from the underlying infrastructure resources by means of utilizing different resource abstraction technologies.

The 5G architecture can be divided into different layers (see Fig. 1):

- **Service layer** - comprises Business Support Systems (BSSs) and business-level policy and decision functions as well as applications and services operated by the tenant. This includes the end-to-end orchestration system.
- **Management and Orchestration layer** - includes ETSI NFV MANO functions

- **Control layer** - accommodates the two main controllers (SDM-X and SDM-C) as well as other control applications.
- **Multi-Domain Network Operating System Facilities** - includes different adaptors and network abstractions above the networks and clouds heterogeneous fabrics.
- **Data layer** - comprises the VNFs (Virtual Network Functions) and PNFs (Physical Network Functions) needed to carry and process the user data traffic.

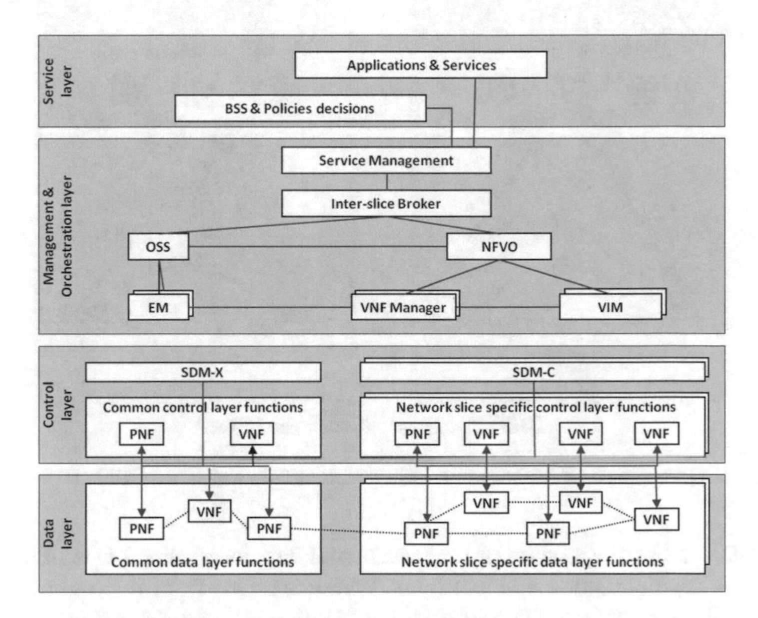

Fig. 1. 5G architecture functional layers

1.2 SDN/NFV

Software Defined Networking

Technical community has analyzed drawbacks of initial 70's design of TCP/IP and Internet architecture that is evolving over last 40 years. IETF has identified multiple aspects for potential improvements at early 90's [2]:

- Routing and Addressing limitations
- Support for Multi-Protocol Architecture
- Improve Security of Architecture
- E2E Traffic Control and State
- Ready for Advanced Applications.

Every current traditional packet network consists of nodes that integrate packet control logic and packet forwarding hardware in each node. In most of networking equipment there is specialized hardware for fast switching/routing of data between interfaces - Data Forwarding plane. The forwarding is managed by rules using routing algorithms, address translation and other higher functions executed by processor running operating system, - this is the Control plane. This concept limits introduction of new features. Growing complexity of control mechanisms generate more complex protocols which creates bottleneck at innovation of IP layer protocols because new features have to be distributed to all routing nodes. This traditional architecture is illustrated on the left side of Fig. 2.

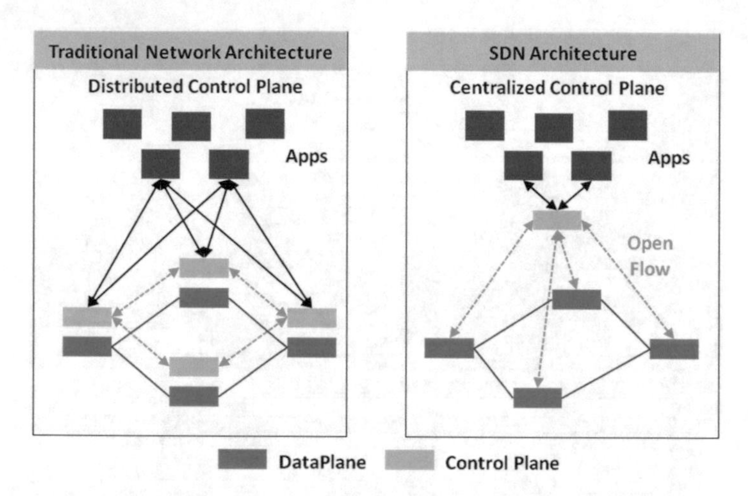

Fig. 2. Comparison of traditional network architecture and basic SDN architecture

The SDN defines separation of the control and data forwarding plane in network as depicted on the right side of Fig. 2 (called Network OS) from forwarding plane on network equipment. It is possible to centralize routing and switching decisions, as well as configuration of all network devices by separating control plane by the software to general purpose computer.

SDN based network can enable:

- Simpler architecture: Separation of control from data plane, Potentially integrated control plane of IP & Transport networks.
- Resilience and Automation: Self-healing mechanisms, scale in/out, higher level of automation, E2E network visibility.
- Time to market: Faster delivery of new service, hosting network functions in infrastructure cloud.

OpenFlow is an open standard originally developed at universities and currently maintained by Open Network Foundation (ONF), [3] – a non-profit consortium with mission to commercialize and promote OpenFlow based SDN. The OpenFlow specifications describe:

- the southbound protocol between OF controller and OF switches
- the operation of the OF switch.

Network Function Virtualisations

Network operators nowadays usually run network/service platforms and IT systems in separate datacenters and often on dedicated hardware. Trend of "Cloudification" creates motivation for re-platforming and move of applications to cloud based infrastructure and common x86 hardware. Virtualization and cloud orchestration helps share and better utilize the same cloud infrastructure for multiple network and IT applications. White paper that presented the first draft of NFV was published in October 2012, [4] driven mainly by service providers expectations. European Telecommunications Standards Institute (ETSI) manages requirements which are asked from the technology and describes the benefits that come with NFV technology. The architecture of NFV technology was designed, [5, 6] based on the following components:

- NFVI (NFV Infrastructure), [7] - provides virtual resources needed to support the implementation of virtualized network functions - commercial COTS hardware components for acceleration, layer of software that virtualizes and abstracts the underlying hardware.
- VNF (Virtualized Network Functions), [8] - software implementation of network functions that is able to run by NFVI and may be accompanied by EMS - Element Management System, which manages the VNF. VNF is an entity corresponding to today's network node, which is expected to be delivered as pure software independent of the hardware.
- NFV MANO (Management and Orchestration), [8] - covers orchestration and lifecycle management of physical and/or software tools that support the virtualization and infrastructure lifecycle management VNFs. NFV MANO focuses on virtualization management tasks, which is necessary for NFV framework. It also collaborates with external NFV OSS/BSS and enables integration NFV to existing networks.

1.3 Multimedia Delivery in 5G Networks Using SDN/NFV

Due to properties of 5G networks and SDN/NFV mentioned in previous text, they are extraordinary suitable for multimedia and mulsemedia delivery. The whole system provides the important scalability and flexibility for the content providers, while keeping the key parameters as latency and throughput. They enable immersive multimedia based services as:

- Video casting services (live/VOD) with immersive video options: 360°, 3D (stereoscopy), multiple viewpoints, free viewpoint television (FTV).
- Online gaming with immersive options: VR, real time interaction with objects/other players.

The 3D (stereoscopic) video and multiple viewpoints video transmission is standardized in H.264 (MPEG-4 AVC) compression standard, developed by MPEG and VCEG under the name Multiview Video coding (MVC). MVC uses 2D Plus Delta method which transmits properly channel for one eye and for second eye transmits the difference data, which are in the form of spatial stereo disparity, temporal predictive, bidirectional or optimized motion compensation. 3D video requires 30–60% more throughput as monoscopic video. 360° video is typically formatted into classical video stream using an equirectangular projection, is stitched from more cameras and is either monoscopic or stereoscopic. 360° video needs typically 3–5 more throughput as normal, single view video. Free viewpoint television enables arbitrary viewpoint for each watcher, generates individual video stream for them by combining the video data from more scene cameras.

As the most important aspect to create the above-mentioned services as immersive as possible we see the usage of VR/AR technology. Using AR/VR for the delivery of above mentioned services with interactivity, the MTP (motion to photon latency) below 20 ms is necessary to convince the user's mind that he is present in another place. Therefore, for the online/cloud based VR gaming and highly interactive VR communication is required the one-way time delay about 5 ms. The further crucial factors for AR/VR are the display resolution and necessary video data throughput. Required throughput for low resolution 360° VR experience is at least 25 Mbit/s. For VR experience, comparable to HD TV the required throughput jumps to 80–100 Mbit/s and comparable to 4k TV to about 600 Mbit/s. Here the 5G networks can help.

2 Content Adaptation Using 5G Networks

There are possible many types of multimedia content adaptation using 5G networks. We use the following taxonomy:

- static setup adaptation - this adaptation runs continuously as the content is being delivered, but adaptation changes its properties only occasionally, depending on end user device used, its abilities, device settings and user preferences (another device is used, user has changed his preferences, another user is logged in),
- dynamic adaptation - this adaptation has to adapt dynamically to varying conditions, e.g. varying transmission conditions, intended interactive user actions and behavior (e.g. movement of headset, change of camera field of view (FOV), free viewpoint, interaction with other users, objects, etc.).

Typical use case for the static adaptation for the 5G network is, that multiple users want to access 4k 360° video content, they have different devices, one has TV set with gesture control, another one the VR headset. The SDN/NFV 5G solution could be, that the 4K 360° data are down streamed to Dedicated service created in SDN/NFV environment (edge processing done in micro cloud) and then individually processed and delivered using low latency 5G network to the individual users. The TV user can control the displayed field of view using gestures and VR user by moving the head. The whole situation is depicted on Fig. 3.

There are many options how the video content will be adapted. On the device, there shall be always available margin data around the current FOV (Field of View) to be able to provide user with the data when moved the FOV until the low latency codec delivers chosen FOV. Further strategy may include reception of multiple FOV and final combination on the user device. The viewport adaptive system used in [9] favorites the current FOV and other parts of 360° video are transmitted as well but with decreasing quality as moving further from current FOV. After head movements, the quality

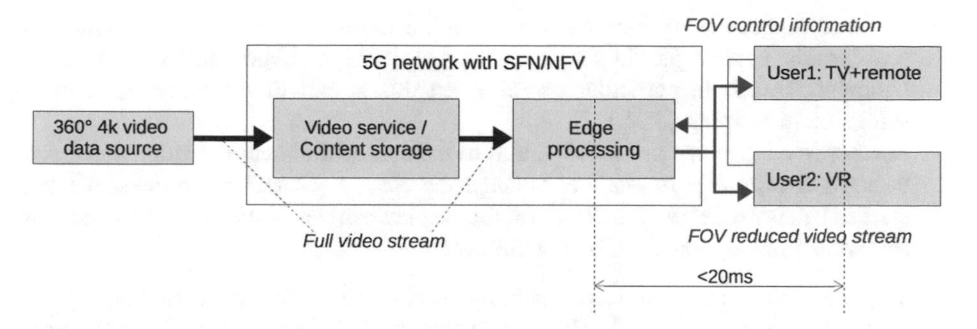

Fig. 3. Simplified architecture for multimedia content adaptation using SDN/NFV and 5G networks

degradation is only temporary and system recovers fast. The strategies how much of the data to order from edge processing, how to combine it on device and how to display can include novel decision information as eye tracker embedded into VR headset. There can be the strategies that include as parameter the intensity of user attention (measured e.g. using Brain Computer Interface). Both situations are discussed in next chapter.

2.1 Advanced Content Adaptation

It is known, the eye movements and the underlying user attention is very important source of the data that can be used for various purposes. These data can be used for content adaptation purposes in various ways. Some of them are summarized here:

- prediction of head movement - by using algorithms than provide better prediction of head movements the viewport adaptive systems can be used with better efficiency and better modelled quality decrease from the current FOV. From ongoing research on the area is known that head and eye movements are correlated and can be mutually modelled and predicted. There are coordinated eye and head movements during gaze shifts, [10, 11].
- usage of saliency information from eye fixations (fixation is then a gaze stops for at least 200–300 ms) as measure of user attention, [12]. This can information further enhance the viewport adaptive systems to focus quality not to whole FOV but mostly to salient area. This saliency information can also help to enhance the overall compression process as offline feedback (source video can be, based on feedback from several users re-encoded with special respect to salient areas in time). The similar principle is used e.g. in, [13].

The above-mentioned principles belong to content adaptation motivated by decreasing the data throughput while preserving the user experience. But the above-mentioned principles allow to use the content adaptation also for different purposes, e.g.

- measure of general or focused user attention during watching the content, this provides the content provider or even the user itself the important feedback about behavior. This is in particular useful when this is part of the teaching or even self-directed learning (SDL).
- interactive content can use the attention of the user as the input modality which can be used to various purposes, i.e. to adapt the content (characters on the scene can start to behave differently, lecturer on the display can notify the student to wake or say some joke to increase the attention, etc.).

In partial as source signal for adaptation based on attention can be used also Brain computer interface (BCI), [14]. BCI can measure (with certain limitation) the user attention, [15] and could provide independent verification model or the data can be fused with eye tracker data to obtain better reliability of the model. There exist BCI models that can be combined with VR headset.

In the Horizon 2020 N project we are focusing on two cases from the above-mentioned options:

- prediction of head motions to obtain better effectivity of the viewport adaptive system,
- tracking of user attention as important part of the SDL process and provide the feedback.

3 System Architecture Proposal

Based on content adaptation strategies described in previous chapters we propose generic architecture for two kinds of services. In the upper part of Fig. 4 there is depicted architecture for video delivery, in the lower part architecture for immersive

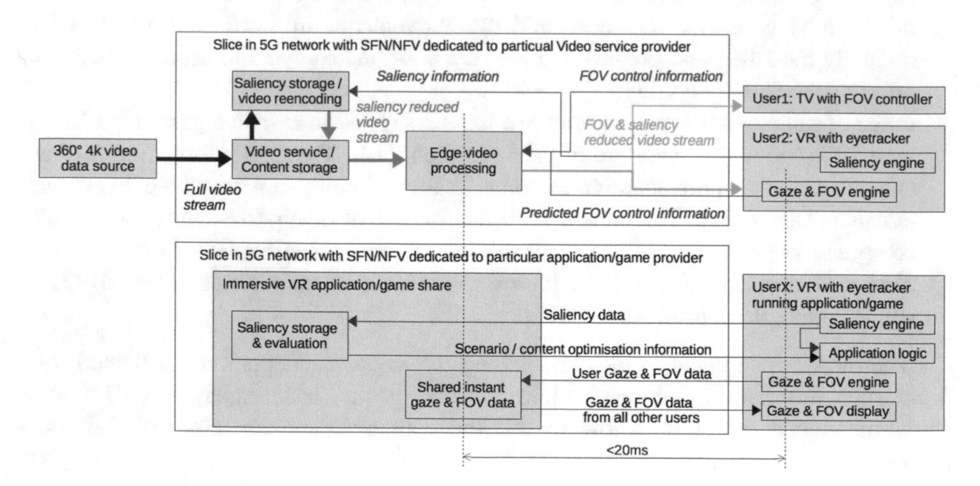

Fig. 4. Simplified architecture of proposed multimedia content adaptation system using SDN/NFV and 5G networks

applications with shared interactive content. Both cases are realized using dedicated slices in the 5G SDN/NFV part. Both cases have fast part, which requires quick response times from the network and slow part. In the video delivery service the slow optimization part consist of gathering of saliency information from all user in a period of time and then re-encoding the video based on this information and distributing this optimized video stream further for all users (also for non-VR ones). For VR users, the video stream can be further reduced by using predicted FOV information and thus minimize the side information necessary for random user head movements. The architecture for multiuser application contains fast data exchange for gaze & FOV data to be able to show to the other users of multiuser application where the other users look (especially informative e.g. for shooting games). The slower saliency information loop can be used to optimize scenario or content information in the application. Local loop to the application logic can be used in situations, where information from the same user is sufficient, e.g. when users falls asleep, the application/game scenario can wake him. These last scenarios can be make robust using additional BCI device, which can measure the user attention level and provide additional source of feedback. This could be especially suitable for SDL application/games.

All functional blocks of the proposed architecture shown in Fig. 4 can be realized as VNFs. In the adaptive video delivery scenario, the Video service/Content storage and Saliency storage/video re-encoding functions can be realized as a VNF cloud services in service provider's main cloud. The saliency reduced video stream will be distributed to an edge cloud located from the reason of low latency as close as possible to the subscriber. The FOV & saliency reduced video stream is subsequently distributed through 5G SDN RAN to the subscriber. In the second scenario proposed for immersive applications with shared interactive content the functions of Saliency storage & evaluation and Shared instant gaze & FOV data can be realized as VNFs on main and edge clouds. In this scenario, the content is distributed to the subscriber using a 5G RAN as well. In both scenarios RAN and core network can be implemented by SDN using VNF functional blocks proposed in, [16]. The VNF infrastructure is managed and orchestrated by Management & Orchestration layer as shown in Fig. 1.

4 Conclusions

In the presented contribution, we have introduced a novel concept of realization of multimedia delivery for streaming and application provisioning services using 5G networks, SDN NFV. The approach is based on adaptation mechanisms specific for advanced VR equipment that depend on gathering the saliency information and evaluating gaze and head movement data. We find the approach useful not only for minimizing the data throughput in 5G networks but also for optimizing the multimedia/application content. We find the approach useful for modern learning management systems (LMS) which provide lot of multimedia information, gather the user feedback and information for learning content optimization. In particular the self-directed learning, can profit from the feedback part.

Acknowledgment. The research described in the paper was financially supported by the H2020 project NEWTON, No. 688503 and VEGA project INOMET, No. 1/0800/16.

References

1. Recommendation ITU-R M.2083:2015 IMT Vision - "Framework and overall objectives of the future development of IMT for 2020 and beyond"
2. Clark, D., Chapin, L., Cerf, V., Braden, R., Hobby, R.: Future of Internet Architecture. In: IETF RFC 1287, December 1991
3. Open Networking Foundation. https://www.opennetworking.org
4. Chiosi, M., Clarke, D., Willis, P., Reid, A., Feger, J., Bugenhagen, M., Khan, W., et al.: Network Functions Virtualisation: An Introduction, Benefits, Enablers, Challenges and Call for Action. https://portal.etsi.org/nfv/nfv_white_paper.pdf
5. Chiosi, M., Wright, S., Clarke, D., Willis, P., et al.: Network Functions Virtualisation: Network Operator Perspectives on Industry Progress. http://portal.etsi.org/NFV/NFV_White_Paper2.pdf
6. ETSI GS NFV-SWA 001 V1.1.1, Network Functions Virtualisation: Virtual Network Functions Architecture, December 2014. http://www.etsi.org/deliver/etsi_gs/NFV-SWA/001_099/001/01.01.01_60/gs_NFV-SWA001v010101p.pdf
7. ETSI GS NFV-INF 001 V1.1.1, Network Functions Virtualisation: Infrastructure Overview. http://www.etsi.org/deliver/etsi_gs/NFV-INF/001_099/001/01.01.01_60/gs_NFV-INF001v010101p.pdf
8. ETSI GS NFV-MAN 001 V1.1.1, Network Functions Virtualisation: Management and Orchestration, December 2014. http://www.etsi.org/deliver/etsi_gs/NFV-MAN/001_099/001/01.01.01_60/gs_NFV-MAN001v010101p.pdf
9. Corbillon, X., Gwendal, S., Devlis, A., Chakareski, J.: Viewport-adaptive navigable 360-degree video delivery. In: Proceedings of 2017 IEEE International Conference on Communications (ICC), pp. 21–25, May 2017. ISSN 1938-1883
10. Hollands, M., Ziavra, N., Bronstein, A.: A new paradigm to investigate the roles of head and eye movements in the coordination of whole-body movements. Exp. Brain Res. **154**, 261 (2004). https://doi.org/10.1007/s00221-003-1718-8
11. Muñoz, M., Lee, J., Reimer, B., Mehler, B., Victor, T.: Analysis of drivers' head and eye movement correspondence: predicting drivers' glance location using head rotation data. In: Proceedings of the 8th International Driving Symposium on Human Factors in Driver Assessment, Training, and Vehicle Design, Snowbird, UT (2015)
12. Vargic, R., Kučerová, J., Polec, J.: Wavelet based image coding using saliency map. J. Electron. Imaging **25**(6), 061610 (2016). 10 pages
13. Gitman, Y., Erofeev, M., Vatolin, D., Bolshakov, A., Fedorov, A.: Semiautomatic visual-attention modeling and its application to video compression. In: 2014 IEEE International Conference on Image Processing (ICIP), pp. 1105–1109, October 2014
14. Vargic, R., Chlebo, M., Kačur J.: Human computer interaction using BCI based on sensorimotor rhythm. In: INES 2015: 19th International Conference on Intelligent Engineering Systems, Bratislava, Slovakia, 3–5 September 2015, pp. 91–95. IEEE, Danvers (2015). ISBN 978-1-4673-7939-7
15. Liu, N.-H., Chiang, C.-Y., Chu, H.-C.: Recognizing the degree of human attention using EEG signals from mobile sensors. Sensors (Basel, Switzerland) **13**(8), 10273–10286 (2013). https://doi.org/10.3390/s130810273
16. ETSI GS NFV 001 V1.1.1:2013, Network Functions Virtualisation (NFV); Use Cases. http://www.etsi.org/deliver/etsi_gs/NFV/001_099/001/01.01.01_60/gs_NFV001v010101p.pdf

ASAMPL: Programming Language for Mulsemedia Data Processing Based on Algebraic System of Aggregates

Yevgeniya Sulema(✉)

Igor Sikorsky Kyiv Polytechnic Institute, Kyiv, Ukraine
sulema@pzks.fpm.kpi.ua

Abstract. This paper presents the new programming language ASAMPL which is developed for effective processing of multimodal data. The language is based on the Algebraic System of Aggregates which is also presented in this paper. An aggregate is a set of tuples where every tuple represents data of certain modality (graphical, audio, video, olfactory, etc.). The proposed programming language is aimed at the development of mulsemedia applications for education, health care, industry, electronic commerce, entertainment, and other areas.

Keywords: Mulsemedia · Multimedia · Programming language
Algebraic system

1 Introduction

Nowadays, the technological level of computer systems hardware is high enough. Modern data storages and processing systems enable processing and storing big data volumes. By expert estimation [1], the storage devices of 100 TB will be available at the market till 2025. Essential growth is also expected in data processing productivity. First of all, it can be achieved by parallel computing [2–4]. At the same time, rapid development of new immersive technologies requires recording, processing, and storing reliable and detailed information about real word objects, process, and phenomena. Accumulation of knowledge and development of corresponding ontologies based on information presented by different modalities (graphical, audio, olfactory, tactile data, etc.) enable development of new generations of monitoring systems, knowledge management systems, and information retrieval systems. Multimodality will provide more detailed description of environments, objects, processes to be investigated. It will help to discover hidden resemblance, study behavior, fulfil scientific prognostication, etc.

Some modern trends in technology also require the introduction of new modalities. 3D printing [5, 6] and its further advancement enable custom production of 3D objects from a wide range of materials including eatable what fosters development of devices for food 3D printing [7, 8]. In its turn it produces new tasks for registration, storing, processing, and reproduction of such modality as taste. As further development of 3D printing, 4D printing [9] enables creation of 3D objects which later reshape themselves or self-assemble over time. By opinion of Gartner analysts [10], 4D printing is one of new emerging technologies which can be developed within next 10 years.

M. E. Auer and T. Tsiatsos (Eds.): IMCL 2017, AISC 725, pp. 431–442, 2018.
https://doi.org/10.1007/978-3-319-75175-7_43

Another promising trend which requires multimodal information is Digital Twin [11]. Digital Twin is a virtual model of a process, product or service. This pairing of the virtual and physical worlds allows analysis of data and monitoring of systems to head off problems before they even occur [12].

The global aim of multimodality or mulsemedia [13–15] is to define objects and processes in real (or in some cases virtual) world as precise as possible. Thus, there is a need in finding a proper way for complex representation of multiple data about a scene of real world, i.e. certain environment with objects in it. Multimodal information can be formalized and transformed into the data about a *multi-image* of the object or process which is stored as the whole but at the same time can be considered as set of separate elements which are data of different modalities.

Objects and processes of real world in most of cases can be described by multimodal data (signals). For example, heart beating of a patient can be described as a multimodal signal which includes the acoustic component (heart beating can be heard), tactile component (heart beating can be felt in tactile way if one puts a palm on a chest in the heart area) as well as it can be described by both electrical signal obtained from a cardiograph and visual signal obtained as a result of ultrasonic investigation or MRI. The complex representation of all these and/or some other additional data will allow a doctor to see the whole picture of the patient health status.

A concept of multi-image can be useful not only in health care but it is also promising for education, entertainment, e-commerce, military applications as well as scientific researches and some industries where surveillance of objects or processes is necessary.

2 Related Work

The concept of olfaction-enhanced multimedia applications is presented in [16]. The survey [17] explains relevant olfactory psychophysical terminology necessary for working with olfaction as a media component. In [18], the authors conclude that the presence of the audio has the ability to mask larger synchronization skews between the other media components in olfaction-enhanced multimedia presentations.

The analysis of the correlation between objective metrics such as both heart rate and electrodermal activity and user quality of experience (QoE) of immersive virtual reality environments is presented in [19]. The influence of users' age and gender on user QoE considering various scent is evaluated in [20]. The analysis of user QoE of olfaction-based mulsemedia when diverse scent types and video content were considered is reported in [21].

A time evaluation of the integration between a distributed mulsemedia platform called PlaySEM and an interactive application where users interact by gestures is presented in [22]. The framework for tactile and haptic interaction is presented as well as the corresponding ISO standards are outlined in [23]. The development of haptic applications based on 5DT Data Glove is presented in [24]. The authors of [25] report on the process of creating and integrating touchless feedback into short movie experiences.

3 Concept of Multi-image

The most detailed graphical information about a 3D object is represented by data structures based on a *voxel* (from *volume element*) which is a volume element of a unit size [26]. A voxel structure is the basis of voxel graphics. A voxel is broadening of the term *pixel* (from *picture element*) which is a unit element (dot) in raster graphics. Similar to a pixel, a voxel is characterized by color. Voxel graphics is used for visualization of medical information (e.g. volume visualization of results of MRI), visualization of scientific data, computer animation, etc. The further advancement and generalization of terms a 'pixel' and a 'voxel' is a *doxel* (from *dynamic element*) which is a voxel defined in time [27, 28]. The idea of keeping volume graphical information realized in both voxel and doxel graphics can be extended to keeping information of other modalities such as audio, olfactory, taste, tactile information, and information about physical features of either an object or environment (temperature, humidity, density). Thus, multimodal information about a scene of the real world can be kept by using volume element of the most generalized type that enables keeping volume information of every possible modality represented in terms of time. By analogy, let us name such an element *muxel* (from *multimodal element*).

For completed description of a state of a matter in a certain point of a 3D scene in a certain moment of time, the muxel data structure can include the following data:

- Time data slot, it stores a moment of time when multimodal information is captured;
- Graphical data slot, it sets color and its transparency is a certain point of a 3D scene;
- Audio data slot, it sets an amplitude value of acoustic signal in a certain time moment defined by time data slot;
- Olfactory data slot, it sets smell and its intensity in a certain point in a 3D scene;
- Taste data slot, it sets taste and its intensity in a certain point of a matter;
- Physical data slot, it sets both a type and a physical state of a matter in a certain point of a 3D scene; some examples of the matter type are air, water, wood, metal, glass; a physical state of the matter can be characterized by its temperature, pressure, humidity, density.

Then *multi-image* is described by a set of muxels. Every muxel can be described by a tuple of values which represent different modality of the object appearance in this point (muxel). However, in real situation there is a large set of muxels which describe the object as the whole. To process this set effectively, a special structure for complex data representation is necessary. Let us consider the algebraic system of such structures.

4 Algebraic System of Aggregates

Algebraic System of Aggregates (ASA) is an algebraic system with a carrier which is an arbitrary set of specific structures named *aggregates*.

Let us consider that an aggregate A is an ordered set of tuples:

$$<a_1, a_2, \ldots, a_l>, <b_1, b_2, \ldots, b_m>, <c_1, c_2, \ldots, c_n>, \ldots, <w_1, w_2, \ldots, w_q>$$

of arbitrary lengths *l*, *m*, *n*, …, *q*. Elements of these tuples belong to certain sets: $a_i \in M_1, b_j \in M_2, c_k \in M_3, \ldots, w_t \in M_N$. Thus, the aggregate is a tuple of arbitrary tuples.

A tuple can include strict values a_i and/or fuzzy values $\tilde{a}_i$.

Since the sequence order of sets is important for fulfilment of operations on aggregates, let us define the aggregate as follows:

$$A = [\![M_1, M_2, M_3, \ldots, M_N \mid <a_1, a_2, \ldots, a_l>, <b_1, b_2, \ldots, b_m>, <c_1, c_2, \ldots, c_n>, \ldots, <w_1, w_2, \ldots, w_q>]\!] \quad (1)$$

A tuple can be empty. Such tuple is marked as <∅>. The aggregate which includes no tuples is called *null-aggregate* $A_\varnothing = [\![\varnothing \mid <\varnothing>]\!]$. A null-aggregate plays a role of the neutral element.

A *length* |A| of the aggregate *A* is a quantity of tuples in it. The length of a null-aggregate is equal to zero: $|A_\varnothing| = 0$. A *cumulative length* ||A|| of the aggregate *A* is a sum of the lengths of the aggregate tuples. The cumulative length of a null-aggregate is equal to zero: $\|A_\varnothing\| = 0$.

The sequence order of sets and corresponding tuples is important. It defines how operations on aggregates will be fulfilled. If two aggregates have equal lengths and both the type and sequence order of these aggregates are the same, then let us consider these aggregates as *compatible*. If the type and sequence order of these aggregates coincide partly, then let us consider these aggregates as *quasi-compatible*. In this case their lengths can differ. In other cases, let us consider the aggregates as *incompatible*.

The basic operations on aggregates are: Union ∪, Intersection ∩, Difference /, and Symmetric Difference △. These operations are similar to their classical interpretation [29] with the difference caused by the fact that aggregates are ordered sets (tuples). It leads to certain specificity in the operation implementation. Thus, a *union* of two aggregates is an aggregate which includes tuple elements belonging to either the first or the second aggregate ordered according to the following rule:

1. If aggregates are compatible, then elements of *i*-tuple of the second aggregate are added into the end of *i*-tuple of the first aggregate:

$$A_1 = [\![M_1, M_2, \ldots, M_N \mid <a_1^1, a_2^1, \ldots, a_l^1>, <b_1^1, b_2^1, \ldots, b_m^1>, \ldots <w_1^1, w_2^1, \ldots, w_n^1>]\!] \quad (2)$$

$$A_2 = [\![M_1, M_2, \ldots, M_N \mid <a_1^2, a_2^2, \ldots, a_r^2>, <b_1^2, b_2^2, \ldots, b_q^2>, \ldots <w_1^2, w_2^2, \ldots, w_p^2>]\!] \quad (3)$$

$$A_1 \cup A_2 = [\![M_1, M_2, \ldots, M_N \mid <a_1^1, a_2^1, \ldots, a_l^1, a_1^2, a_2^2, \ldots, a_r^2>, <b_1^1, b_2^1, \ldots, b_m^1, b_1^2, b_2^2, \ldots, b_q^2>, \ldots, <w_1^1, w_2^1, \ldots, w_n^1, w_1^2, w_2^2, \ldots, w_p^2>]\!]; \quad (4)$$

2. If aggregates are incompatible, then tuples of the second aggregate are added into the end of the tuple of sets of the first aggregate:

$$A_1 = [\![M_1^1, M_2^1, \ldots, M_N^1 \mid <a_1, a_2, \ldots, a_l> , <b_1, b_2, \ldots, b_m> , \ldots, <w_1, w_2, \ldots, w_q>]\!] \tag{5}$$

$$A_2 = [\![M_1^2, M_2^2, \ldots, M_K^2 \mid <c_1, c_2, \ldots, c_r> , <d_1, d_2, \ldots, d_s> , \ldots, <z_1, z_2, \ldots, z_p>]\!] \tag{6}$$

$$A_1 \cup A_2 = [\![M_1^1, M_2^1, \ldots, M_N^1, M_1^2, M_2^2, \ldots, M_K^2 \mid <a_1, a_2, \ldots, a_l> , <b_1, b_2, \ldots, b_m> , \ldots, <w_1, w_2, \ldots, w_q> , <c_1, c_2, \ldots, c_r> , <d_1, d_2, \ldots, d_s> , \ldots, <z_1, z_2, \ldots, z_p>]\!]; \tag{7}$$

3. If aggregates are quasi-compatible, then elements of *i*-tuple of the second aggregate are added into the end of *i*-tuple of the first aggregate if their elements belong to the same set; otherwise the rule for incompatible aggregates is applied:

$$A_1 = [\![M_1^1, R, M_2^1, \ldots, M_N^1 \mid <a_1, a_2, \ldots, a_l> , <r_1^1, r_2^1, \ldots, r_s^1> , <b_1, b_2, \ldots, b_m> , \ldots, <w_1, w_2, \ldots, w_q>]\!] \tag{8}$$

$$A_2 = [\![M_1^2, R, M_2^2, \ldots, M_K^2 \mid <c_1, c_2, \ldots, c_r> , <r_1^2, r_2^2, \ldots, r_t^2> , <d_1, d_2, \ldots, d_m> , \ldots, <z_1, z_2, \ldots, z_p>]\!] \tag{9}$$

$$A_1 \cup A_2 = [\![M_1^1, R, M_2^1, \ldots, M_N^1, M_1^2, M_2^2, \ldots, M_K^2 \mid <a_1, a_2, \ldots, a_l> , <r_1^1, r_2^1, \ldots, r_s^1, r_1^2, r_2^2, \ldots, r_t^2> , <b_1, b_2, \ldots, b_m> , \ldots, <w_1, w_2, \ldots, w_q> , <c_1, c_2, \ldots, c_r> , <d_1, d_2, \ldots, d_m> , \ldots, <z_1, z_2, \ldots, z_p>]\!]. \tag{10}$$

The basic relations of aggregates are: Is Equivalent $\equiv$; Includes $\subset$; Is Included $\supset$; Precedes $\vdash$; Succeeds $\dashv$; Is Less $<$; Is Greater $>$. For example, if $B_1 = A_1 \cup A_2$ then $B_1 \subset A_2$ and $A_1 \equiv B_2 = B_1 \cap A_2$.

The Algebraic System of Aggregates has been developed as a basis for the new programming language focused on processing of multimodal data.

5 Programming Language ASAMPL

The programming language ASAMPL has got its name from ‘Algebraic System of Aggregates’ and ‘Mulsemedia data Processing Language’. This language is driven by data modality and time. It means that ASAMPL is a special-purpose programming language developed for easy and effective processing of mulsemedia data which, by definition, always exists in/belongs to certain time moments. The key concept in ASAMPL is the concept of a multi-image of a real-world object. To present such object in this programming language a special data structures – tuples and aggregates – are used. Processing of aggregates is fulfilled according to the ASA rules. The key features of ASAMPL are:

1. Orientation on external sources of multimodal data (data streams from remote sensors, media files in cloud storages).
2. Complex representation of multimodal data by using tuples and aggregates.
3. Linkage of data processing to a timeline.
4. Synchronization of data of different modalities.
5. Possibility to link external libraries, renderers, handlers for coding/decoding, processing and reproduction of data of a specific modality, use of specific file formats and specific devices.

In ASAMPL, any data can belong to one of three categories:

1. Element, it is a single data value.
2. Tuple, it is an ordered data set.
3. Aggregate, it is a tuple of tuples.

Data types defined in ASAMPL are:

1. Time, it is a time value, its format is <hours>:<minutes>:<seconds>:<milliseconds>, e.g. 10:30:24:01.
2. Date, it is a date value, its format is <year>/<month>/<day>, e.g. 2017/09/01.
3. Integer, it is an integer number, its format <integer value>, e.g. 104.
4. Real, it is a real number of a single precision, e.g. 10.4.
5. Double, it is a real number of a double precision, e.g. 1.04e−5.
6. Text, it is a textual string, e.g. "any text".
7. Link, it is a link or full file name related to certain data, e.g. 'http://cloud.tv/v1.mp4 '.

Mathematical operations (+, −, *, /) as well as relations and logical operations defined in the Algebraic System of Aggregates are used in ASAMPL.

The statements used in ASAMPL are:

1. Timeline (Timeline As) is a specific statement which enables application of a certain action to every element related to a certain time period. This statement can be considered as a kind of a repetitive statement. It has the following format:
   - Timeline <value1>:<step>:<value2> {<list of data processing actions>}
   - Timeline As <tuple> {<list of data processing actions>}.
2. If Then (If Then Else) is a branch statement. It has the same format as in other programming languages.
3. Case Of (Case Of Else) is a selection statement. It has the same format as in other programming languages.
4. Substitute For When is a specific statement which substitutes one data set (usually having lower resolution) for another one (usually having better resolution) if the data set, which has better quality, cannot be downloaded, uploaded or processed because of external conditions (mainly, low quality of network connection). This statement can be considered as a kind of a selection statement. It has the following format:
   - Substitute <variable1> For <variable2> When <condition>.

5. Download From (Download From With) is a specific statement which enables downloading data from a certain data source (remote or local, a file or a stream) and assignment of the data to a certain variable (element, tuple or aggregate). The conversion of the data from a file to a format of the variable is fulfilled by using a specified handler. This statement can be considered as a kind of an assignment statement. It has the following format:

   - Download <variable> From <source> With <handler>.

6. Upload To (Upload To With) is a specific statement which enables uploading data assigned to a certain variable (element, tuple or aggregate) to a certain resource (remote or local) for further storage as a file. The conversion of the data from a format of the variable to a file is fulfilled by using a specified handler. This statement can be considered as a kind of an assignment statement. It has the following format:

   - Upload <variable> To <target> With <handler>.

7. Is or =, it is an assignment statement. It has the following format:

   - <variable> Is <value>
   - <variable> = <value>.

8. Render With, it is a specific statement which enables reproduction of data by means of a certain rendering tool. It has the following format:

   - Render <variable> With <renderer>.

A program in ASAMPL consists of blocks which begin with correspondent keywords (Table 1).

Table 1. ASAMPL block titles

Keyword	Meaning
Aggregates	The beginning of a block which contains description of aggregates constructed from defined tuples
Handlers	The beginning of a block which contains descriptions of build-in and external modules for conversion of files with data of certain modality
Libraries	The beginning of a block which contains descriptions of build-in and external libraries of sub-programs
Actions	The beginning of a block which defines data processing actions
Program	The beginning of a program code
Renderers	The beginning of a block which contains descriptions of build-in and external modules for rendering of data of certain modality
Sets	The beginning of a block which describes data sets and their types
Sources	The beginning of a block which contains descriptions of data sources, i.e. resources where multimodal data sets are available
Tuples	The beginning of a block which contains description of tuples and their assessment to defined sets
Elements	The beginning of a block which contains description of variables and their assessment to defined sets

The program structure is the following:

```
Program <name> {
  Libraries {
    <library1> is <pathToLibray1>;
    <library2> is <pathToLibrary2>;
    ................................. }
  Handlers {
    <handler1> is <pathToHandler1>;
    <handler2> is <pathToHandler2>;
    ................................. }
  Renderers {
    <renderer1> is <pathToRenderer1>;
    <renderer2> is <pathToRenderer2>;
    ................................. }
  Sources {
    <source1> is <pathToSource1>;
    <source2> is <pathToSource2>;
    ................................. }
  Sets {
    <set1> is <type1>;
    <set2> is <type2>;
    ................................. }
  Elements {
    <element1> is <set>;
    <element2> = <value>;
    ................................. }
  Tuples {
    <tuple1> = <set1>;
    <tuple2> = <set2>;
    ................................. }
  Aggregates {
    <aggregate1> = [<tuple1>, …, <tupleM>];
    <aggregate2> = [<tupleK>, …, <tupleN>];
    ................................. }
  Actions {
    <list of statements>   }
}
```

A program code can include comments which begin with double slash in each line.

6 Discussion

Programming in ASAMPL is an easy process due to a number of build-in commands (statements) which allow a programmer to fulfil basic actions on multimodal data: download and upload data streams, synchronize data of different modalities, change their duration, and edit data.

Especial attention in the proposed language is devoted to relation of data to time. Time is a specific entity for multimodal data processing because all mulsemedia and multimedia data are developing in the course of time as well as every data delivering from a sensor or a device are registered (recorded) in certain time moments.

Possible applications of the proposed programming language relate to education and training, health care, telecommunications, industry, entertainment, etc.

Let us consider an example of the program representing mulsemedia content (multi-image) for a lesson in Geography.

```
Program GeographyLesson {
  Libraries {
    VisualLib Is 'D:\MultiImage\Library\vsl.lib';
    AudioLab Is 'D:\MultiImage\Library\aud.lib';
    OlfactLib Is 'D:\MultiImage\Library\olf.lib'}
  Handlers {
    MPEG2tuple Is 'D:\MultiImage\Handler\mpg2tup.exe';
    Tuple2MPEG Is 'D:\MultiImage\Handler\tup2mpg.exe';}
  Renderers {
    VisualRen Is 'C:\Renderer\vslren.exe';
    AudioRen Is 'C:\ Renderer\audren.exe';
    OlfactRen Is 'C:\Renderer\olfren.exe';}
  Sources {
    VideoFile Is 'http://edu.net/Forest&WildLife.mp4';
    VisualDataStream Is 'http:/webcam.edu.net/005201';
    OlfactoryFile Is 'D:\Lesson05\forest.dat';
    SceneFile Is 'D:\Lesson05\forest.agg';}
  Sets {
    Frame Is Integer(1920,1080);
    Audio Is Real;
    Scent Is Text;}
  Elements {
    duration is Time;
    time1 = 00:00:01;
    time2 = 00:15:00;
    step = 00:00:01;}
  Tuples {
    VisualDat Is Frame;
    VisualDat2 Is Frame;
    AudioDat Is Audio;
    OlfactoryDat Is Scent;}
  Aggregates {
    ForestScene = [VisualDat,AudioDat,OlfactoryDat];}
  Actions {
```

```
    Download AudioDat From VideoFile.audio
                      With MPEG2tuple;
    Download VisualDat2 From VideoFile.visual
                      With MPEG2tuple;
    Download OlfactoryDat From OlfactoryFile
                      With default.OlfactLib;
    Timeline time1 : step : time2 {
      Download VisualDat From VisualDataStream
                        With default.VisualLib;}
    Substitute VisualDat2 For VisualDat
             When VisualDataStream Is Equivalent Null;
    Upload ForestScene To SceneFile With default.all;
    Render ForestScene
         With [VisualRen, AudioRen, OlfactRen];}
}
```

This code enables creation of the multi-image of the scene in a forest. The scene is represented by the multi-image with three modalities: visual stream, audio stream, and scent. There are two possible sources of a visual stream: from the video recorded in advance and stored as a file and from a webcam installed in a real forest area and available via network. It is supposed that the webcam can be off for some reasons. For such case, the statement Substitute For When is used. It substitutes video from the file for the real visual data stream. The duration of the multi-image timeline is 15 min. During this time, the visual data stream is being downloaded and stored as a tuple. The multi-image is represented by one aggregate, stored as .agg file in a local disk, and reproduced by compound renderer including three renderers corresponding to data modalities used for the multi-image representation.

The practical realization of the multi-image reproduction can be achieved by using numerous devices available on a market [30, 31] as well as recently designed [32, 33].

7 Conclusion and Future Work

The proposed programming language ASAMPL (the name stands for ‘Algebraic System of Aggregates’ and ‘Mulsemedia data Processing Language’) enables development of software applications with multimodal content. It can be helpful in numerous use cases in education, health care, scientific researches, industry, and entertainment. The distinctive features of ASAMPL language are orientation on storing mulsemedia data in external storages (e.g. in cloud storages) and optimization of multimodal data processing. The latter can be achieved by using special data structures – aggregates which are the carrier at the Algebraic System of Aggregates.

The research presented in this paper is the ongoing one. The next stage of this research is focused on both the improvement of the compiler and the extension of the build-in libraries and tools for data processing and rendering. The special attention is focused on both parallel realization of build-in processing procedures and fuzzy procedures of multimodal data retrieval.

References

1. The site about storage devices. Disks of 100 TB capacity will be produced till 2025. http://vossozdat.com/k-2025-godu-budut-vypuskatsja-diski-v-100-tb/. Accessed 1 Feb 2017
2. Fuller, S.H., Millett, L.I.: The Future of Computing Performance: Game Over or Next Level?, p. 200. The National Academies Press, Washington (2011)
3. Future Directions for NSF Advanced Computing Infrastructure to Support U.S. Science and Engineering in 2017–2020: Interim Report. The National Academies Press, Washington, 34 p. (2014). ISBN 978-0-309-31379-7
4. Graham, S.L., Snir, M., Patterson, C.A.: Getting Up to Speed: The Future of Supercomputing, p. 308. The National Academies Press, Washington (2004)
5. Lipson, H., Kurman, M.: Fabricated: The New World of 3D Printing. Wiley, Indianapolis (2013)
6. Pearce, J., Blair, C., Laciak, K.J., Andrews, R., Nosrat, A.: 3-D printing of open source appropriate technologies for self-directed sustainable development. J. Sustain. Dev. **3**(4), 17–29 (2010)
7. 3D Food Printing. http://3dprinting.com/food/. Accessed 22 Oct 2017
8. 3D Printed Food Vending Machines on the Way. http://3dprintingindustry.com/2016/05/04/3d-printed-food-vending-machines-way/. Accessed 22 Oct 2017
9. The Next Wave: 4D Printing Programming the Material World. https://thebimhub.com/2015/05/13/the-next-wave-4d-printing-programming-the-material/. Accessed 22 Oct 2017
10. Gartner's Hypo Cycle. http://www.gartner.com/technology/home.jsp. Accessed 22 Oct 2017
11. Top Trends in the Gartner Hype Cycle for Emerging Technologies (2017). http://www.gartner.com/smarterwithgartner/top-trends-in-the-gartner-hype-cycle-for-emerging-technologies-2017/. Accessed 22 Oct 2017
12. What Is Digital Twin Technology - And Why Is It So Important? https://www.forbes.com/sites/bernardmarr/2017/03/06/what-is-digital-twin-technology-and-why-is-it-so-important/#573fc3c2e2a7. Accessed 22 Oct 2017
13. Ghinea, G., Timmerer, C., Lin, W., Gulliver, S.R.: Mulsemedia: state of the art, perspectives and challenges. ACM Trans. Multimed. Comput. Commun. Appl. **11**(1s) (2014). Article no. 17
14. Kannan, R., Andres, F.: Digital library for mulsemedia content management. In: International ACM Conference on Management of Emergent Digital EcoSystems (MEDES 2010), Bangkok, Thailand, pp. 275–276 (2010)
15. Yuan, Z., Chen, S., Ghinea, G., Muntean, G.M.: User quality of experience of mulsemedia applications. ACM Trans. Multimed. Comput. Commun. Appl. **11**(1s), 21 (2014)
16. Ghinea, G., Ademoye, O.A.: Olfaction-enhanced multimedia: perspectives and challenges. Multimed. Tools Appl. **55**, 601 (2011). https://doi.org/10.1007/s11042-010-0581-4
17. Murray, N., Lee, B., Qiao, Y., Muntean, G.M.: Olfaction-enhanced multimedia: a survey of application domains, displays, and research challenges. ACM Comput. Surv. (CSUR) **48**(4), 56 (2016)
18. Ademoye, O.A., Murray, N., Muntean, G.M., Ghinea, G.: Audio masking effect on inter-component skews in olfaction-enhanced multimedia presentations. ACM Trans. Multimed. Comput. Commun. Appl. **12**(4), 14 (2016). Article 51
19. Egan, D., Brennan, S., Barrett, J., Qiao, Y., Timmerer, C., Murray, N.: An evaluation of heart rate and electrodermal activity as an objective QoE evaluation method for immersive virtual reality environments. In: Proceedings of the 8th International Conference on Quality of Multimedia Experience (QoMEX) (2016)

20. Murray, N., Lee, B., Qiao, Y., Miro-Muntean, G.: The influence of human factors on olfaction based mulsemedia quality of experience. In: Proceedings of the 8th International Conference on Quality of Multimedia Experience (QoMEX) (2016)
21. Murray, N., Lee, B., Qiao, Y., Miro-Muntean, G.: The impact of scent type on olfaction-enhanced multimedia quality of experience. In: IEEE Transactions on Systems, Man, and Cybernetics: Systems, pp. 1–13 (2016)
22. Saleme, E.B., Celestrini, J.R., Santos, C.A.S.: Time evaluation for the integration of a gestural interactive application with a distributed mulsemedia platform. In: Proceedings of the 8th ACM on Multimedia Systems Conference, pp. 308–314 (2017)
23. van Erp, J.B.F., et al.: Setting the standards for haptic and tactile interactions: ISO's work. In: Lecture Notes in Computer Science, vol. 6192. Springer, Heidelberg (2010)
24. Sulema, Y.: Haptic interaction in educational applications. In: Proceedings of the 9th International Conference on Interactive Mobile Communication Technologies and Learning (IMCL 2015), Thessaloniki, Greece (2015)
25. Ablart, D., Velasco, C., Obrist, M.: Integrating mid-air haptics into movie experiences. In: Proceedings of the ACM International Conference on Interactive Experiences for TV and Online Video (TVX 2017), pp. 77–84. ACM, New York (2017)
26. Hill, D.L., Studholme, C., Hawkes, D.J.: Voxel similarity measures for automated image registration. In: SPIE Proceedings of Visualization in Biomedical Computing, vol. 2359, no. 205 (1994). https://doi.org/10.1117/12.185180
27. Carnero, J., Diaz-Pernil, D., Mari, J.L., Real, P.: Doxelo: towards a software for processing and visualizing topology computations in Doxel-based 3D+t images. In: Proceedings of the 16th International Conference on Applications of Computer Algebra ACA 2010 (2010)
28. Gonzalez–Diaz, R., Medrano, B., Real, P., Sánchez–Peláez, J.: Algebraic topological analysis of time-sequence of digital images. In: Lecture Notes in Computer Science, vol. 3718, pp. 208–219. Springer, Heidelberg (2005)
29. Acharjya, D.P.: Fundamental approach to discrete mathematics. New Age International, New Delhi (2008). ISBN 978-81-224-2607-6
30. Sulema, Y.: Mulsemedia vs. Multimedia: state of the art and future trends. In: Proceedings of the 23rd IEEE International Conference on Systems, Signals and Image Processing IWSSIP 2016, Bratislava, Slovakia, pp. 19–23 (2016)
31. Kovács, P.T., Murray, N., Rozinaj, G., Sulema, Y., Rybárová, R.: Application of immersive technologies for education: state of the art. In: Proceedings of the 9th International Conference on Interactive Mobile Communication Technologies and Learning (IMCL 2015), Thessaloniki, Greece (2015)
32. Ninu, A., Dosen, S., Farina, D., Rattay, F., Dietl, H.: A novel wearable vibro-tactile haptic device. In: IEEE International Conference on Consumer Electronics ICCE, pp. 51–52 (2013)
33. Obrist, M., Velasco, C., Vi, C.T., Ranasinghe, N., Israr, A., Cheok, A.D., Spence, C., Gopalakrishnakone, P.: Touch, taste, & smell user interfaces: the future of multisensory HCI. In: Proceedings of the 2016 CHI Conference Extended Abstracts on Human Factors in Computing Systems (CHI EA 2016), pp. 3285–3292. ACM, New York (2016)

Improving Physics Education Through Different Immersive and Engaging Laboratory Setups

Johanna Pirker[1(✉)], Michael Stefan Holly[1], Patrick Hipp[1], Christopher König[1], Dominik Jeitler[1], and Christian Gütl[1,2]

[1] Graz University of Technology, Graz, Austria
jpirker@iicm.edu, {michael.holly,patrick.hipp,christopher.koenig, dominik.jeitler}@student.tugraz.at, cguetl@iicm.edu
[2] Curtin University, Perth, WA, Australia

Abstract. Virtual and remote laboratories have been shown as valuable tools to support learners in understanding concepts. They provide an experimentation space in a safe, remote, and flexible way. However, the missing realism and lack of hands-on experience are often pointed out as a downside. This can also be described as a missing feeling of immersion and presence. Emerging virtual reality tools providing full-body tracking and even force-feedback when interacting with experiments support this sense of immersion and engage focused learning. Collaborative virtual setups additional add features to enable engaging discussions and social experiences. In this paper, we investigate learning experiences within a virtual laboratory environment in a room-scale virtual reality setup, a traditional screen-based solution, and mobile VR settings supporting multi-user setups. In two experiments setup as A/B studies, we investigate and compare the settings with a focus on comparing immersion, engagement, usability, and learning experience.

Keywords: Virtual reality · Immersion · Physics education

1 Introduction

In this paper, we want to investigate and compare learning experiences with the room-scale virtual reality setups of Maroon and compare it with a traditional screen-based version and introduce the concept of multi-user experiences in mobile VR settings. The focus is on identifying benefits and downsides, and realistic and interesting application scenarios for the different setups. Room-scale virtual reality setups can provide new and more immersive and realistic forms of interactions with experiments and can help students to stay focused on the learning experiences. In contrast, mobile VR solutions are described not as interactive and immersive but allow exciting live collaborations and in-class learning. Following, we compare such VR learning experiences with traditional

M. E. Auer and T. Tsiatsos (Eds.): IMCL 2017, AISC 725, pp. 443–454, 2018.
https://doi.org/10.1007/978-3-319-75175-7_44

screen-based solutions to explore and discuss potentials and downsides of both. This should contribute to a better understanding of the potential of the different technologies and should help to identify application scenarios in learning settings for these tools.

2 Related Work

Designing engaging STEM classes and learning experiences still pose a challenge. Active learning is a successful teaching method where students are directly involved in the learning process. This has been demonstrated to be an effective strategy for increasing the students' performance compared to traditional methods [1,9]. Simulations have been shown as valuable tools to support physics classes in active learning settings. They are used to visualize physical laws and let students experiment and understand physical concepts [5,8]. Such simulations are often more efficient, safe, cost-efficient, and take less preparation time compared to traditional experiments [13]. Other important tools in physics education are *virtual labs*. By using virtually simulated laboratories in a remote or an in-class setup, instructors can move students from mindless memorization to understanding and can make learning physics more effective, interesting and engaging [13]. In a large-scale study (306 participants) Corter et al. [3] compared the learning outcomes of different lab formats to measure their efficiency. They evaluated traditional hands-on labs, remotely operated labs, and simulations. After performing remote and simulated laboratories versus performing hands-on laboratories, they found that the learning outcomes are as high or higher than traditional hands-on laboratories. There are statistically significant differences between the students learning outcomes. These differences have consequences for the design of remote and virtual laboratory classes. Alternative access modes must be considered pedagogical options, rather than only logistical conveniences [7]. Emerging virtual reality (VR) technologies enable a new form of immersive virtual labs. However, virtual reality environments challenge designers and developers to make it an enjoyable experience for everyone. Wrong design decisions lead to nausea, disorientation, discomfort, and vomiting, called cybersickness [12]. Currently, traditional learning methods are leading in science education. However, game-based strategies and tools to engage learners are becoming more and more important. In [1] the authors have shown a 76% increase in learning outcomes by using a playful form of virtual laboratories compared to traditional methods. Another approach of interactive and engaging learning in virtual worlds is described by [11]. They implemented and evaluated a virtual world environment for physics education, where students can work together on experiments and discuss simulations. In a study, they found out that collaborative aspects are important, but engagement and immersion must be improved. One possible way to improve this is the use of game-specific elements. Motivational design aspects of such game-based educational tools are described by Pirker and Gütl [10] Immersion, engagement or even flow are often described as important factors for creating engaging and involving experiences. Flow is the feeling of a mental state of complete concentration and absorption in the process of an activity, where a person is fully immersed and involved. Based on qualitative interviews Csikszentmihalyi and Csikszentmihalyi [4] described various characteristics of the flow experience. Immersion is the feeling of being part of the digital experience. One way to achieve this is the use of virtual reality technologies (e.g., HTC Vive, Oculus Rift, Samsung Gear).

In this paper, we want to investigate further engagement through collaboration and immersion as tools to motivate learners in virtual labs.

3 Maroon - The Immersive Physics Laboratory

Maroon is an interactive three-dimensional virtual physics laboratory and experiment environment implemented in Unity. It was designed to simulate and illustrate various physical phenomena. The physics lab contains currently four electromagnetic and electrostatic experiments. (1) The first one demonstrates the electric field between a Van de Graff Generator and a grounding sphere. Users can change the distance between the grounding sphere and the generator to see how the frequency of the discharges changes. (2) The second experiment simulates the behavior of a balloon, which is placed between the grounding sphere and Van de Graaff Generator. To compare the learning effect between Desktop and VR, we developed two new experiments which provide much more user interactions. (3) The Falling Coil experiment demonstrates the dynamics of a conducting non-magnetic ring falling on the axis of a fixed magnet. (4) The Faraday's Law experiment illustrates the electromagnetic interaction between a conducting non-magnetic ring and a magnet; both constrained on the horizontal axis. Users can change the magnetic moment and the resistance of the coil to see how the magnetic flux and the induced current change. Also, the user can visualize the field lines, the vector field of the resulting field and the underlying magnet field using the iron filling visualization. To evaluate different aspects such as immersion, engagement, and different collaborative setups in VR and more traditional computer-based settings, we built different versions of Maroon supporting various degrees of immersion through VR and engagement through collaborative and social activities.

3.1 Maroon Desktop

Maroon Desktop was designed similar to classic computer games. The desktop environment includes different stations, which represent experiments or activities (see Fig. 1). The experience is controlled with keyboard and the mouse. The view direction determines the movement direction which can be changed freely in all directions. The arrow keys are used to control the player characters movement. Users can start the experiments or the activities if they are close enough to a checkpoint. In addition, the two other experiments with a magnet and a non-magnetic ring can be controlled by graphical control elements. Users can vary the ring resistance and the strength of the magnetic dipole moment to see how these parameters affect the dynamics of the ring by using the corresponding slider. The visualization of the magnetic field can also be modified by the user. The number of field lines and the vector field resolution can be changed with sliders, and a momentary view of the magnetic field can be displayed by clicking the iron filling button. The induced electric current in the ring over time is also displayed to the user as a graph (see Fig. 2).

Fig. 1. Desktop - lab overview

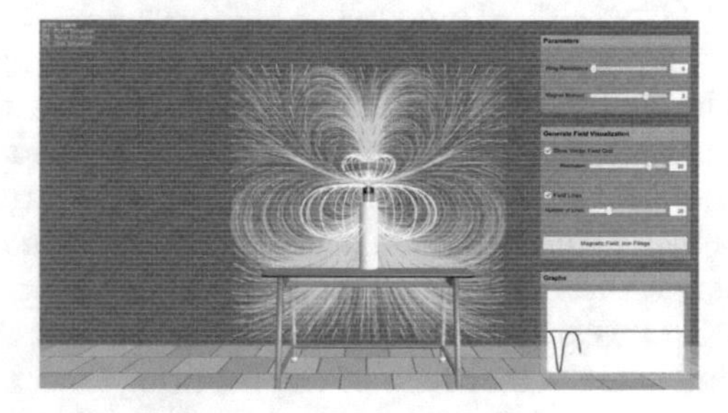

Fig. 2. Desktop - Falling Coil Experiment

3.2 Maroon VR

Maroon VR extends Marron with room-scale VR support for the HTC Vive. For this, the official SteamVR[1] plugin was used. The HTC Vice allows the user to move in 3D space and use a motion-tracked handled controller to interact with virtual objects and environments. Each hardware element in the Vive setup is tracked by two base stations named lighthouses. The base stations emit laser beams, which are recognized by photosensors on the headset and the controller. Based on the time difference between the impact of the laser beams on the respective sensors the exact position and orientation of the device can be calculated. The Vive controllers are specially designed for VR with intuitive control and

[1] http://store.steampowered.com/steamvr.

realistic haptic feedback. They include a highly-sensitive touchpad and individually programmable buttons for improved user interaction within virtual worlds [6]. Users can freely move in the environment and can additionally teleport themselves to other positions by pressing the touchpad on one of the controllers, which acts as a pointer to the preferred target and displays a precise colored beam for visual orientation. Experiments and Activities are started by entering a portal-like object through button press on the controller. The Van de Graaff Generator experiments are just controlled via the Vive controllers. Users can charge the Van de Graaff Generator and move the Grounder left and right by clicking a button on the controller. Charges, electric fields, and field lines are visualized immediately. In comparison, the other two electromagnetic experiments provide the possibility to change specific parameters and visualizations by interacting with a virtual control panel. An interactable object is highlighted when touched to signal that this object is usable. The ring resistance and the magnetic dipole moment are adjustable via sliders which can be moved with the controller. The induced electric current over time is also displayed to the user on a flat screen to keep them always in view (see Fig. 3).

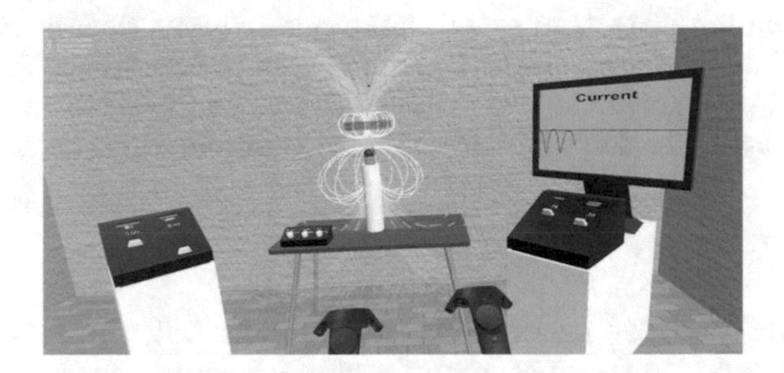

Fig. 3. VR - Falling Coil Experiment

3.3 Maroon Mobile VR

Maroon Mobile VR was developed for the Samsung Gear VR, a virtual reality headset released in 2015. It is compatible with Samsung Galaxy mobile devices which are inserted into the head-mounted gear. The Samsung Gear VR headset acts as a mount for the mobile device, rendering the VR applications, but also as a controller. The field of view is controlled through movement of the head. A combination of eye gaze, a virtual avatar and a touchpad on the side of the headset allows interactions with the virtual environment. The virtual avatar is controlled with a gaze point - the center of the screen - to move through the laboratory. Users can start experiments and interact with them by moving the gaze cursor to an interaction button.

3.4 Maroon Multi-user

Additionally, we designed a multi-user version for the computer-based version and the mobile virtual reality setup to support in-class collaboration scenarios. Students can conduct experiments together in the multi-user space. The multi-user variant is designed to enhance in-class experiences. Therefore, no chat functionalities are provided. While wearing the VR headset, students can talk to each other in the classroom. The mobile VR variant supports two scenarios. First, a free mode, which supports free interactions with the learning experiences. Second, a streaming mode, where one user (e.g., a teacher) controls the experiences while the other users (e.g., students) are watching.

4 Studies

To assess and compare the effect of immersion through virtual reality and engagement through collaboration, we performed two different A/B split user studies with a total of 40 participants. The aim was to get insights in advantages, disadvantages, and application scenarios of fully immersive room-scale virtual reality experiences compared to traditional desktop-based experiences and multi-user desktop-based experiences compared to multi-user immersive mobile virtual reality experiences. We chose the two different virtual reality settings to get insights in the different application scenarios of the two immersive settings.

5 Study 1

In the first user study, 20 participants evaluated the fully immersive room-scale virtual reality experience and Maroon Desktop with focus on evaluating engagement, immersion, learning experience, usability, and user experience. To compare the Desktop environment with HTC Vive variant, we made a multivariate AB/BA test. The first 11 users started with the HTC Vive (A) and continued with the Desktop version (B). The second 9 users started in the reverse order. In the first test environment, users had to answer physics questions after each experiment to measure the learning progress and compare the two environments.

5.1 Material and Setup

The Desktop setup consists of a classic workstation with keyboard and mouse. The setup for the HTC Vice contains the head-mounted display (HMD) itself, the two base stations and the two controllers. The room-scale experience was set up in an area of about 2 m × 2 m. For both environments, an Alienware AREA with two NVIDIA GeForce GTX 960 was used.

5.2 Method and Procedure

At the beginning of the study, every participant had to fill out a pre-questionnaire. The pre-questionnaire was used to gather demographic data and background information including experience with computers, VR technologies, and video games and their physics knowledge. They got an exact introduction into the test systems. We explained to them how to move in room-scale VR and how to interact with objects. Participants also got an introduction how they can interact with the desktop setting. After that, participants were given different task within the two environments. As the test was designed as an A/B study, users would either start with the VR experience or with the screen-based experience. The first task was to look around in the environment for two minutes to get an impression of the lab environment. The second task was to start the falling coil simulation and try to identify the relationship between the magnetic field and the electrical current. In the next task, participants had to start the Faradays Law experiment and move the magnet towards the coil to understand the force effects for a live conductor in a magnetic field. After each physic experiment participants had to answer physics questions to measure their learning progress. After finishing the two experiments, the participants could look at the rest of the environment. Also, they were asked to describe their impression of the environment in a short interview. After each iteration, users completed a corresponding post-questionnaire. It contains 10 open-ended questions about impression, 20 single-choice questions with ratings on a Likert-scale between 1 (fully disagree) and 7 (fully agree) and 19 standardized questions from the Game Engagement Questionnaire (GEQ) to measure the level of engagement on absorption, flow, presence, and immersion with ratings on a scale between 1 (not at all) and 5 (extremely). Finally, at the end of the test, each participant had to complete a common post-questionnaire with five open-ended questions about their experience in both environments.

5.3 Participants

20 participants (3 female) between 20 and 28 (AVG = 24.05; SD = 2.31) tested Maroon Desktop and Maroon VR. 18 of the participants were students. Most of them were in the field of computer science. 4 participants were from the fields industrial design, media design, mechanical engineering and business economics. 12 are very experienced in the use of computers (AVG = 4.4; SD = 0.82), 11 are experts in the usage of video-games (AVG = 4.2; SD = 1.06), and 18 like playing video games. In average the test group tended to play video games more often (AVG = 3.45; SD = 1.28). All of them rate themselves as not experienced in the usage of VR (AVG = 1.65; SD = 0.81). 18 had heard about VR devices, and 11 have already used such devices; two have used HTC Vive before. Only two had experienced cybersickness before. None of them rate themselves as very experienced in physics. In the following section, we discuss the results of the post-questionnaires and the interviews.

Table 1. Comparison of GEQ main elements between HTC Vive and a computer-based setup

	HTC Vive		Screen	
Category	AVG	SD	AVG	SD
Presence	3.0	1.4	2.2	1.4
Absorption	2.5	1.2	1.3	0.6
Flow	2.4	1.1	1.9	1.0
Immersion	3.7	1.1	2.2	1.1

5.4 Results

Experiencing Immersion and Engagement. As we can see in Table 1, immersion was perceived much higher in the virtual reality version compared to the computer-based version. One user even mentioned in the discussion that one would not lose track of time in the computer-based version compared to the VR version. The experience in the room-scale variant was described as more attractive (AVG = 6.0, SD = 1.3) compared to the computer-based variant (AVG = 5.3, SD = 1.3). One interesting point to highlight is that the room-scale variant was described as much more fun (AVG = 6.1, SD = 1.5) than the computer-based variant (AVG = 4.9, SD = 1.8). Participants described the VR experience as *"more cool and more fun, because one can touch everything"* and mentioned that they think that *"the learning experience is the same, but more motivating in VR"*. One user also indicated that he or she would prefer a step-by-step guided experience in the computer-based variant, whereas, in VR it is more interesting to try out different things in the form of a playground.

Usability and User Experience. As the laboratory was designed as large laboratory room and every experience was in this room, several users had issues with the scaling and the size of this room. Users have positively mentioned realistic elements, such as a practical and working clock. The computer-based variant was described to give a better overview and the controls (mouse and keyboard) are more familiar. However, the interaction with the lab in the VR variant more realistically and naturally was received by many users very positive. Therefore, many of the participants would prefer this interaction - even if it is not familiar - over the interaction through mouse and keyboard.

Limitations. This study was limited to a small number of participants, but already gives a good overview of potentials of the two different learning experiences. Many new research questions were opened up in this study. While the focus of this work is to learn more about motivational aspects such as immersion and engagement, we also identified several open questions looking at the participants' learning behavior in VR. Especially getting a deeper understanding of the differences in learning. Different kinds of learning concepts can be perceived differently in the two experiences. It would be important to understand how students learn, what concepts are suitable for what experience, and where (in the classroom or at home) students should learn these.

6 Experiment 2

6.1 The Setting

For this study, a simple reduced version of Maroon and Maroon Mobile VR were used. As introduced in Chap. 3, the experiments "Van de Graaff Generator" and "Balloon at Van de Graaff Generator" were the primary learning experiences for the setup.

6.2 Study Setup

We designed a qualitative A/B user study with 20 participants. Again, we focused on evaluating aspects supporting motivational environments such as immersion and engagement. Additionally, we wanted to get first insights into the learning experiences and the learning process as well. As it was designed as multi-user experience, the 20 participants were arranged in groups of two. They had to complete three tasks together in both setups.

Participants. In this study, 20 (four female) participants between 22 and 34 (AVG = 26.05; SD = 3.5) were asked to evaluate the two experiences. Most of the participant were students (15) and five employed. The fields of studies were mixed. The majority of the students were studying fields such as software engineering or electrical engineering. Others study architecture, chemistry, or account management. Their expertise with computer usage is also very mixed (AM = 3.6; SD = 1.34), as well as their experience with video games (AM = 3.4; SD = 1.7). Some play video games very often (five), some almost never (five), the arithmetic mean is 2.95 (SD = 1.54). The majority of them has no experience with VR, but 13 have already heard of the Google Cardboard or Samsung Gear VR; six have already tried the Gear VR, four the Google Cardboard. 16 have player online multi-user games before. They rate their physics knowledge in average rather low (AM = 2.4; S = 0.9).

Equipment and Setup. For the computer-based experience, two standard workstations were used. For the Mobile VR version, two Samsung Galaxy S6 together with two Samsung Gear VR were provided.

Method. Participants filled-out the pre-questionnaire about demographic information, previous experience with games, virtual reality, multi-user setups, and physics. The study was designed in an A/B format. Half of the users started the study with the VR setup; the other ten started with the computer-based experience. The study tasks were completed in pairs. In a first step, participants were introduced to the system. Then they had to complete three tasks together. After each iteration, they completed a small post-questionnaire with question about their impressions and 20 single-choice questions about their experience with answers as rating on a Likert scale between 1 (fully disagree) and 7 (fully agree) and the GEQ [2], rating aspects of immersion and engagement on a Likert scale between 1 (fully disagree) and 5 (fully agree). After completing both experiences including the two post-questionnaires, a post-questionnaire asking them to compare the two systems was given.

6.3 Findings

The setup of this study was very similar to the setup of analysis 4 to be able to compare results and systems.

Experiencing Immersion and Engagement. The differences between the two versions concerning immersion, flow, absorption, and presence are not significantly high. As illustrated in Table 2 The experience with Maroon Mobile VR was rated as a bit higher (AVG = 2.6) compared to the experience in the computer-based version (AVG = 2.2). Most elements of the GEQ are rated very similar when comparing the computer-based multi-user setup with the mobile VR multi-user setup. Experiences supporting flow such as *"I play without thinking about how to play"* are partly better rated in the computer-based version since users are more used to the controls and the setup. Elements, supporting immersion and absorption such as *"I feel spaced out"* are partly better supported by the VR setup.

Table 2. Comparison of GEQ main elements between the multi-user version of Mobile VR and the computer-based setup

	Screen		Mobile VR	
Category	AVG	SD	AVG	SD
Presence	2.2	1.4	2.3	1.2
Absorption	1.3	0.6	1.9	1.2
Flow	1.9	1.0	1.8	1.0
Immersion	2.2	1.1	2.6	1.0

Usability and User Experience. When asking the participants which device they would prefer 13 would prefer the VR version. Reasons for that were explained with "more fun", "easier to understand and follow the experiments", "new experience", and "more immersive". On the other hand, 7 participants would prefer the computer-based version of VR. Reasons were mainly the more natural controls, better resolution, and some participants experience dizziness

Table 3. Statements to different cooperative learning experiences on a Likert scale between 1 (not at all) and 7 (fully agree)

	AVG	SD
Interacting with other players helps me stay motivated	5.4	1.5
A teacher could use streaming mode for teaching students virtually	5.8	1.6
Mobile VR could be used as a virtual alternative for classrooms	5	1.6
I would learn content through a "virtual teacher" ("streaming mode")	5.6	1.2

in the mobile setup. Also, the learning experience was received a bit better in the VR version (Table 3).

Limitations. As mentioned earlier, this study is only designed as a first preliminary study of early prototypes presenting the multi-user capabilities. It was designed to get first insights and feedback to advance those capabilities further and evaluate design concepts and ideas such as the streaming mode.

7 Conclusion and Discussion

We can conclude, that the room-scale VR experience was received as more engaging and immersive when compared with a traditional computer-based variant. Participants mentioned that the interactive, realistic, and natural design in room-scale VR setups better support interactive and hands-on experiments. This also improved the feeling of immersion and the lose the track of time. When assessing their understanding of the experiments in a small quiz, the participants were able to explain the experiments after the VR experience better. VR also supports better concentration on learning tasks. However, it was also mentioned that it is easier to read and to take notes in the computer-based option. Participants would recommend the computer-based variant in the form of guided step-by-step experiences, whereas in VR, especially the exploratory approach was valued. Virtual reality experiences are described as more motivating, engaging, fun, and immersive. But also, in particular, the mobile setup, more complicated to learn and sometimes too complex for small experiments and learning concepts. In the multi-user setup, especially the streaming mode was received very well as learning structure. Also seeing other in the lab was mentioned as a positive experience when learning. Dizziness was mentioned as an issue in the mobile VR experience. We can conclude that especially short mobile virtual reality experience has the potential to change the way physics is taught in the classroom. Short learning experiences in a small physics lab are a valuable learning experience, and due to the short time in VR also cybersickness can be avoided. This scenario also allows collaborative learning scenarios through multi-user setups. A more immersive way of learning physics is provided through the room-scale setup. This setting is experienced as extremely immersive and more engaging compared to all other setups. However, the setup is more expensive and hard to install when compared to mobile VR. Thus, this scenario could be used as part of learning labs at school or learning institutions to allow concentrated and immersive learning experiences.

Acknowledgment. We would like to thank *John Winston Belcher* from the Department of Physics Massachusetts Institute of Technology. We thank all people who are and were involved in the development process. Details: gamelabgraz.com/maroon/.

References

1. Bonde, M.T., Makransky, G., Wandall, J., Larsen, M.V., Morsing, M., Jarmer, H., Sommer, M.O.: Improving biotech education through gamified laboratory simulations. Nat. Biotechnol. **32**(7), 694–697 (2014)
2. Brockmyer, J.H., Fox, C.M., Curtiss, K.A., McBroom, E., Burkhart, K.M., Pidruzny, J.N.: The development of the game engagement questionnaire: a measure of engagement in video game-playing. J. Exp. Soc. Psychol. **45**(4), 624–634 (2009)
3. Corter, J.E., Nickerson, J.V., Esche, S.K., Chassapis, C., Im, S., Ma, J.: Constructing reality: a study of remote, hands-on, and simulated laboratories. ACM Trans. Comput.-Hum. Interact. (TOCHI) **14**(2), 7 (2007)
4. Csikszentmihalyi, M., Csikszentmihalyi, I.: Optimal Experience: Psychological Studies of Flow in Consciousness. Cambridge University Press, Cambridge (1992). https://books.google.at/books?id=lNt6bdfoyxQC
5. Dori, Y.J., Hult, E., Breslow, L., Belcher, J.W.: How much have they retained? Making unseen concepts seen in a freshman electromagnetism course at MIT. J. Sci. Educ. Technol. **16**(4), 299–323 (2007)
6. HTC: Vive user guide (2017). http://dl4.htc.com/web_materials/Manual/Vive/Vive_User_Guide.pdf?_ga=2.82037588.318918390.1502103742-2091439008.1500636618
7. Lindsay, E., Good, M.: Virtual and distance experiments: pedagogical alternatives, not logistical alternatives. In: American Society for Engineering Education, pp. 19–21 (2006)
8. Lunce, L.M.: Simulations: bringing the benefits of situated learning to the traditional classroom. J. Appl. Educ. Technol. **3**(1), 37–45 (2006)
9. Olson, S., Riordan, D.G.: Engage to excel: producing one million additional college graduates with degrees in science, technology, engineering, and mathematics. Report to the President. Executive Office of the President (2012)
10. Pirker, J., Gütl, C.: Educational gamified science simulations. In: Gamification in Education and Business, pp. 253–275. Springer (2015)
11. Pirker, J., Gütl, C., Belcher, J.W., Bailey, P.H.: Design and evaluation of a learner-centric immersive virtual learning environment for physics education. In: Human Factors in Computing and Informatics, pp. 551–561. Springer (2013)
12. Settgast, V., Pirker, J., Lontschar, S., Maggale, S., Gütl, C.: Evaluating experiences in different virtual reality setups. In: International Conference on Entertainment Computing, pp. 115–125. Springer (2016)
13. Wieman, C., Perkins, K.: Transforming physics education. Phys. Today **58**(11), 36 (2005)

Tangible, Embedded and Embodied Interaction

Programming Human-Robot Interactions in Middle School: The Role of Mobile Input Modalities in Embodied Learning

Alexandros Merkouris(✉) and Konstantinos Chorianopoulos

Department of Informatics, Ionian University, Corfu, Greece
{cl4merk, choko}@ionio.gr

Abstract. Embodiment within robotics can serve as an innovative approach to attracting students to computer programming. Nevertheless, there is a limited number of empirical studies in authentic classroom environments to support this assumption. In this study, we explored the synergy between embodied learning and educational robotics through a series of programming activities. Thirty-six middle school students were asked to create applications for controlling a robot using diverse interaction modalities, such as touch, speech, hand and full body gestures. We measured students' preferences, views, and intentions. Furthermore, we evaluated students' interaction modalities selections during a semi-open problem-solving task. The results revealed that students felt more confident about their programming skills after the activities. Moreover, participants chose interfaces that were attractive to them and congruent to the programming tasks.

Keywords: Embodied learning · Educational robotics · Experiment
Children · Human-robot interaction

1 Introduction

Robotic computing has been proposed as an inspiring framework for getting students involved with STEM disciplines as well as with programming [4]. In most studies conducted on the use of educational robotics in schools, children are asked to enliven the robots by creating the appropriate computer programs [5]. The programmer has to think mainly about the goal of the robot and how the robot will interact with the environment. However, there is another crucial aspect that should also be considered, and this is if and how the user will interact with the robot. In particular, we are interested in the effects of programming human-robot interactions on learning performance and attitudes. Moreover, we are motivated by embodied learning findings that regard a broad spectrum of human motor-perceptual skills, which reach beyond the traditional desktop metaphor and keyboard-mouse as input devices.

Embodied cognition researchers argue that bodily experiences and physical interactions with the environment through sensorimotor modalities (touch, movement, speech, smell and vision) are considered essential factors in the learning process and the construction of knowledge [3, 22]. From a theoretical perspective, embodied learning is closely related to the principles of constructivist [20] and constructionist

M. E. Auer and T. Tsiatsos (Eds.): IMCL 2017, AISC 725, pp. 457–464, 2018.
https://doi.org/10.1007/978-3-319-75175-7_45

[18] learning theories. The core idea in Piaget's theory is that young learners construct knowledge and form the meaning of the world by interacting directly with physical objects [20]. Papert [18] believed that children are better learners when they construct knowledge voluntarily while playing with real-world metaphors or tangible objects, programming the turtle in the Logo environment or interactive robots.

The embodied approach is being widely used to cover the learning of abstract materials in a wide range of topics that extend from science, technology, engineering and mathematics (STEM) [10, 13–15] to computational thinking [6, 7, 19]. Specifically, concerning computational thinking, a practical learning approach is to have students physically enact the programming scripts through their bodies before creating the program [7]. Other scholars [6, 19] examined how embodied interaction in a virtual environment that processed students' dance movements can facilitate computational learning. Some educators and researchers believe that robotics education is a promising field for employing the embodied cognition view. Alimisis [1] points out that embodiment is an innovative approach for making robotic activities more attractive and meaningful to children. Lu et al. [16] examined how direct and surrogate bodily experiences in a robotic workshop can influence student's understanding of programming concepts. Similarly, Sung and colleagues [21] investigated how embodied experiences, with a different amount of embodiment [13] (full body and hand), can affect students' problem-solving skills. Having children enact [16, 21] or reenact the robots' moves through physical interaction seems a useful approach for learning abstract computational concepts.

This small sample of embodied research highlights the need to explore the positive learning effects of embodiment within robotics [1] in greater extent. Thus, the current study set out to investigate how various programming activities to control a robot using diverse interaction modalities, such as touch, speech, hand and full body gestures can affect students in exploring computational concepts. Allen-Conn's and Rose's work [2] for introducing powerful ideas (math and science) through programming with Squeak was the main inspiration for creating the intervention. Expanding their views "beyond the screen" by targeting a real robot, is one aspect of our study. The main contribution of our research is studying alternative types of human-robot interaction in the context of embodied learning. Our research questions centered on these major topics:

- Intention: Did the robotic workshop have any influence on students' attitudes towards computing?
- Interaction: What were students' interaction modalities selections for controlling the robot and what were the criteria for making such selections?

2 Methodology

2.1 Subjects

Thirty-six middle school students (17 girls, 19 boys), aged between fourteen and fifteen years, with little to no prior programming experience were recruited to participate in a seven-session robotic workshop. We randomly selected the participants from the third-level class of a middle school. The decision for selecting this specific age group was guided by the fact that none of the students had previously received teaching in computer programming as part of previous formal education. Students worked in pairs in each of the activities. Thus, fifteen same-gender and three mixed-gender pairs were created.

2.2 Activities

The workshop was divided into seven individual sessions. In the first session which served as the introductory activity, students were asked to assemble a three-wheel robot and create a simple mobile application for controlling the robot's arm with their mobile phone. In the second session, a remote control mobile app was developed by the students, and they controlled the movement of the robot by touching with their fingers the appropriate buttons on their phone's touchscreen. In the next session, students created a mobile app that engaged hand gesture movement for navigating the robot using the phone's orientation sensor. In the fourth session, they controlled the robot through speech commands by utilizing speech recognition technology. In the fifth session, students made use of computer vision technology by creating a program to control the robot through full body gestures. In the sixth session, students were asked to create a mobile app that integrated artificial intelligence to the robot so it could move autonomously on the track following a black line. Each of the above sessions followed a similar basic format: (1) Building the User Interface. A basic template application for each session was given to students, and they were given instructions to add the necessary UI elements, (2) Programming the application's behavior, and (3) Going further by enhancing the basic application with additional features such as variable speed. In the final session, a semi-open [23] problem-solving task was given to students. They were asked to create a program so that they could successfully navigate the robot on a fixed track and hit an object placed at a predefined spot with its robotic arm. No instructions were given to students on the final project, and they were prompted to choose any of the above interaction modalities they preferred. Moreover, they were allowed to reuse code from the previous sessions. Students attempted to solve the programming tasks by creating the following programming mechanisms: (1) robot navigation, (2) robotic arm control, and (3) power-speed control. The duration of each of the first six sessions was about 45 min while the Project activity lasted between 45–90 min.

2.3 Materials

App Inventor[1] [9] was employed as the development platform in the sessions that involved mobile technology and students used their own mobile phone devices in an attempt to reinforce the sense of ownership. For the session that involved full-body interaction, ScratchX[2] was employed as the development platform and was supported

[1] App Inventor: http://appinventor.mit.edu.

[2] ScratchX: http://scratchx.org/.

by the Kinect sensor for tracking the body [11]. With mobile technologies, as tablets or smart devices, the interaction space is expanded "to more physical and embodied modalities" [15] as touch screen, gyroscope based hand gestures, and speech interfaces can be used to interact with digital information [12]. Similarly, with the use of computer vision technologies full body interfaces can also be employed for interacting with information. The interaction modalities and the development platform employed in each of the activities can be found in Table 1.

Table 1. Overview of the interaction modalities and the development platforms for each session of the workshop.

Session	Activities	Interaction modalities	Development platform
1	Hello robot	Touch	App inventor
2	Remote control	Touch	App inventor
3	Remote sensor	Hand gestures	App inventor
4	Speak to robot	Speech	App inventor
5	Body control	Full body gestures	ScratchX
6	Line follow	Artificial intelligence	App inventor
7	Project	Student's selections	App inventor or ScratchX

The robots chosen for supporting the workshop were Lego Mindstorms[3] (NXT and EV3). Both App Inventor and ScratchX programming environments have the potential to be used for programming the Lego robots[4], and this was the main reason for their selection.

2.4 Measuring Instruments and Data Analysis

For the study, both qualitative and quantitative data were collected and analyzed. Concerning the quantitative data, the students filled out brief pre-test and post-test questionnaires. The pre-tests before the programming activities consisted of a five-level Likert questionnaire that recorded student's prior experience with programming, their views, and intentions towards computing, robotics, and mobile development. The post-tests after the programming activities included a five-level Likert questionnaire that recorded a change of students' views and intentions towards computing, robotics, and mobile development.

Regarding the qualitative data, student's projects in the final session were manually analyzed for investigating students' interaction modalities selections. We additionally employed a 30-min plus semi-structured interview that gave participants a chance to describe not only their projects but also their experiences. Finally, each of the students' workstation screens was recorded by Camtasia capture during the sessions. The

[3] Lego Mindstorms: https://www.lego.com/en-us/mindstorms.

[4] ScratchX extension for Ev3: http://kaspesla.github.io/ev3_scratch/.

qualitative data from the interviews and the Camtasia recordings are still being analyzed, so we intend to publish the results in a separate paper.

3 Findings

3.1 Students' Attitudes

Table 2 summarizes students' views and intentions towards computing before and after the workshop. We conducted six paired sampled t-tests were, to determine whether there were a significant change in students' views and intentions. The results indicated that participants reported having more programming skills after ($M = 2.86$, $SD = 0.899$) the workshop than before ($M = 2.25$, $SD = 0.77$). This difference, –0.61, BCa 95% CI $[-1.03, -0.19]$ was significant, $t(35) = -2.94$, $p = .006$ and represented a medium-sized effect, $d = 0.45$. The differences in the other cases were not significant.

Table 2. Views' and intentions' mean averages before and after the intervention.

Students' views and intentions	Pre-Test (N = 36)		Post-Test (N = 36)	
	M	SD	M	SD
How interested are you in computing education?	3.33	1.069	3.44	1.252
How difficult do you think computer programming is?	3.36	0.931	3.14	1.046
How many programming skills do you think you have?*	2.25	0.770	2.86	0.899
Would you like to learn programming in the future?	3.47	1.082	3.25	1.180
Would you like to create mobile applications in the future?	3.50	1.207	3.36	1.437
Would you like to build and program robots in the future?	3.22	1.333	3.19	1.191

3.2 Interaction Modalities

To complete the problem-solving task given to them in the project session, students had to create the appropriate programming mechanisms. First of all, they had to program the robot navigation mechanism so that the users of the application could move the robot on the track. Additionally, they had to program the robotic arm control mechanism for hitting the object with the robot's arm. Optionally, students could extend their application by adding a power control mechanism so that the robot could move with variable speed on the track. Figure 1 summarizes the interaction modalities that students selected while developing the programming mechanisms.

In total, eighteen projects were created in the final session as many as were the groups of students who participated in the workshop. In sum, all students were able to complete the main programming tasks by creating the robot navigation and the robotic arm control mechanisms, while ten groups extended their projects by adding the optional power control mechanism. Concerning the robot navigation mechanism, in most cases, full body gestures and touch sensorimotor were selected as the interaction modalities. For navigating the robot with accuracy on the track, the program must

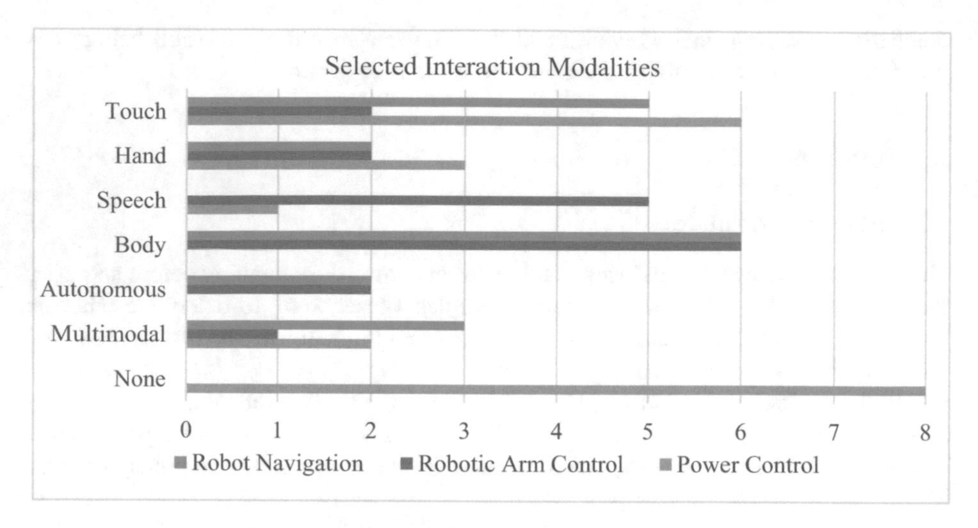

Fig. 1. Selected interaction modalities for each of the programming mechanisms.

respond immediately to the users' actions. For this reason, students avoided using speech commands for controlling the movement of the robot as there was a substantial delay in the speech recognition mechanism and in some cases failure to recognize the correct word. As for the robotic arm mechanism, participants showed a preference towards the full body and the speech interfaces. Students, in this case, used speech commands to trigger the movement of the robotic arm as any delay in speech recognition mechanism did not prevent them from hitting the object successfully. Finally, for the power control mechanism, most participants preferred to create a program that allowed users to change the speed of the robot with touch, by manipulating a power slider. None of the students created a body interface for controlling the speed of the robot even though the body interaction modality was the most popular in each of the main programming tasks.

4 Conclusion

Our results suggest that students felt more confident about their programming skills after the intervention. Moreover, students adopted various interaction modalities while developing the programming mechanisms in the problem-solving task. Body gestures were one of the most popular modalities used in the final session, as many groups selected them for navigating the robot and controlling the robotic arm. Surprisingly, none of the groups, which used the body interfaces implemented the power control mechanism. Students struggled to program a concurrent body gesture for controlling the speed of the robot, despite the fact that during the Body Control activity they were given instructions on how to create a mechanism for adjusting the speed of the robot depending on the distance between the users' knees. For the robot navigation, touch sensorimotor was also extensively used, as it allowed users to guide the robot more accurately. Although students did not use the speech interface for the navigation of the robot due to its affordance, they used it for triggering the robotic arm. In sum, it seems that the participants besides choosing interfaces that were attractive to them, they also chose interfaces that their affordances matched to the specific programming tasks [17].

One limitation that might influence students' modalities selections is the opportunity to use a new technology for controlling the robot, especially in the body control case. Moreover, it is possible that their choices might also be biased by other students' choices. Additionally, further analysis is needed to evaluate the learning outcomes of the current study. We intend to analyze students' final projects for assessing computational thinking. Finally, as a future investigation, it would be interesting to investigate whether students' choices are related to a particular learning style model [8].

The contribution of this paper is to provide additional insight on the synergy between embodied learning and educational robotics. Compared to previous studies, instead of exploring the learning outcomes by comparing a tangible interface to a digital one [24] we exposed students to a wide range of interactive possibilities and made an attempt to examine the problem-solving strategies that arose. We believe that the findings of our study might benefit teachers, assisting them in creating effective robotic interventions with an embodied learning perspective.

References

1. Alimisis, D.: Educational robotics: open questions and new challenges. Themes Sci. Technol. Educ. **6**(1), 63–71 (2013)
2. Allen-Conn, B.J., Rose, K.: Powerful ideas in the classroom using squeak to enhance math and science learning. Viewpoints Research Institute, Inc. (2003)
3. Barsalou, L.W.: Grounded cognition. Annu. Rev. Psychol. **59**, 617–645 (2008)
4. Benitti, F.B.V.: Exploring the educational potential of robotics in schools: a systematic review. Comput. Educ. **58**(3), 978–988 (2012)
5. Bers, M.U.: The TangibleK robotics program: applied computational thinking for young children. Early Child. Res. Pract. **12**(2), n2 (2010)
6. Daily, S.B., Leonard, A.E., Jörg, S., Babu, S., Gundersen, K.: Dancing alice: exploring embodied pedagogical strategies for learning computational thinking. In: Proceedings of the 45th ACM Technical Symposium on Computer Science Education, pp. 91–96. ACM, March 2014
7. Fadjo, C.L., Hallman Jr., G., Harris, R., Black, J.B.: Surrogate embodiment, mathematics instruction and video game programming. In: EdMedia: World Conference on Educational Media and Technology, pp. 2787–2792. Association for the Advancement of Computing in Education (AACE), June 2009
8. Fleming, N.D.: I'm different; not dumb. Modes of presentation (VARK) in the tertiary classroom. In: Research and Development in Higher Education, Proceedings of the 1995 Annual Conference of the Higher Education and Research Development Society of Australasia (HERDSA), vol. 18, pp. 308–313, July 1995
9. Grover, S., Pea, R.: Using a discourse-intensive pedagogy and android's app inventor for introducing computational concepts to middle school students. In: Proceeding of the 44th ACM Technical Symposium on Computer Science Education, pp. 723–728. ACM, March 2013

10. Han, I., Black, J.B.: Incorporating haptic feedback in simulation for learning physics. Comput. Educ. **57**(4), 2281–2290 (2011)
11. Howell, S.: Kinect2Scratch, Version 2.5 [Computer Software] (2012). http://scratch.saorog.com
12. Jacob, R.J., Girouard, A., Hirshfield, L.M., Horn, M.S., Shaer, O., Solovey, E.T., Zigelbaum, J.: Reality-based interaction: a framework for post-WIMP interfaces. In: Proceedings of the SIGCHI Conference on Human Factors in Computing Systems, pp. 201–210. ACM, April 2008
13. Johnson-Glenberg, M.C., Megowan-Romanowicz, C., Birchfield, D.A., Savio-Ramos, C.: Effects of embodied learning and digital platform on the retention of physics content: Centripetal force. Front. Psychol. **7**, 1819 (2016)
14. Kontra, C., Lyons, D.J., Fischer, S.M., Beilock, S.L.: Physical experience enhances science learning. Psychol. Sci. **26**(6), 737–749 (2015)
15. Lindgren, R., Tscholl, M., Wang, S., Johnson, E.: Enhancing learning and engagement through embodied interaction within a mixed reality simulation. Comput. Educ. **95**, 174–187 (2016)
16. Lu, C.M., Kang, S., Huang, S.C., Black, J.B.: Building student understanding and interest in science through embodied experiences with LEGO robotics. In: EdMedia: World Conference on Educational Media and Technology, pp. 2225–2232. Association for the Advancement of Computing in Education (AACE), June 2011
17. Oviatt, S., Cohen, A., Miller, A., Hodge, K., Mann, A.: The impact of interface affordances on human ideation, problem solving, and inferential reasoning. ACM Trans. Comput.-Hum. Interact. (TOCHI) **19**(3), 22 (2012)
18. Papert, S.: Mindstorms: Children, Computers, and Powerful Ideas. Basic Books Inc., New York (1980)
19. Parmar, D., Isaac, J., Babu, S.V., D'Souza, N., Leonard, A.E., Jörg, S., Gundersen, K., Daily, S.B.: Programming moves: design and evaluation of applying embodied interaction in virtual environments to enhance computational thinking in middle school students. In: Virtual Reality (VR), pp. 131–140. IEEE, March 2016
20. Piaget, J.: The Construction of Reality in the Child, vol. 82. Routledge, Abingdon (2013)
21. Sung, W., Ahn, J., Kai, S.M., Black, J.B.: Effective planning strategy in robotics education: an embodied approach. In: Society for Information Technology & Teacher Education International Conference, pp. 1065–1071. Association for the Advancement of Computing in Education (AACE), March 2017
22. Wilson, M.: Six views of embodied cognition. Psychon. Bull. Rev. **9**(4), 625–636 (2002)
23. Zhong, B., Wang, Q., Chen, J., Li, Y.: An exploration of three-dimensional integrated assessment for computational thinking. J. Educ. Comput. Res. **53**(4), 562–590 (2016)
24. Zhu, K., Ma, X., Wong, G.K.W., Huen, J.M.H.: How different input and output modalities support coding as a problem-solving process for children. In: Proceedings of the 15th International Conference on Interaction Design and Children, pp. 238–245. ACM, June 2016

FingerTrips on Tangible Augmented 3D Maps for Learning History

Iliana Triantafyllidou, Athina-Maria Chatzitsakiroglou, Stergiani Georgiadou, and George Palaigeorgiou(✉)

University of Western Macedonia, Florina, Greece
iliarozed@gmail.com, achatzitsakiroglou@gmail.com, stellageo123@gmail.com, gpalegeo@uowm.gr

Abstract. History education offers students the opportunity to learn about the past and make connections with the present. However, primary school students consider history lessons to be boring, dull and sterile. Integrating ICT in history teaching can enhance historical thinking and historical understanding, and may promote the exploration of the past with a critical approach rather than the passive accumulation of information. The objective of this study was to design and examine a low-cost augmented 3D tangible model of a historical site, in which students could interact with historical content through a virtual field trip by using their fingers. Twenty-six 6th grade students participated in a pilot study in order to evaluate the effectiveness and the efficiency of the proposed learning environment called FingerTrips. Participants played with the augmented model in 10 sessions and in groups of 2 or 3 members. Data were collected with an attitude questionnaire and semi-formal group interviews. Students' answers revealed that the FingerTrips environment enhanced their engagement and motivation in history learning, and made them feel as active participants in the historical event presented. Students considered their interactions as a real fieldtrip on the historical landscape model with the help of their fingers. Such an approach is closer to student's interactive experiences and expectations, gamifies learning, and exploits embodied learning affordances, in order to achieve efficient, effective, and enjoyable learning.

Keywords: Tangible interaction · Mixed reality interface · History learning FingerTrips · Virtual field trip

1 Introduction

History education gives students an opportunity to learn about the past and to make connections with the present, providing them with information that they can use to develop social understanding and make informed decisions about their social life [1–3]. In recent years, historical learning has moved from knowing facts, dates, and events, to the concept of "historical understanding", which in turn involves learning historical content, applying research methods, analyzing and evaluating resources, and reaching conclusions based on the given information [2, 3]. Students are encouraged not just to learn facts, but to familiarize themselves with the way historians work: they must

M. E. Auer and T. Tsiatsos (Eds.): IMCL 2017, AISC 725, pp. 465–476, 2018.
https://doi.org/10.1007/978-3-319-75175-7_46

collect sources, divide them into primary or secondary ones, analyze and interpret them, and construct their own, subjective, meaning about what the sources "are telling" to them [2, 3].

Although history is regarded as an important bridge between the past and the present, students usually consider history lessons to be boring, unattractive, "dull and sterile" [4]. Additionally, they have difficulties in historical understanding, since developing historical thinking requires the execution of complex cognitive tasks [5]. Students face challenges in reading primary sources, they are not familiar with historians' heuristics, they cannot advance their understanding beyond the facts presented, they confront difficulties in interpreting historical events, and they usually oversimplify them because of their limited or misapplied knowledge of the historical context in which these events had happened [5, 6]. Moreover, when describing historical changes, students confuse the concepts of time, change, and continuity [6, 7].

Researchers have thus shifted their focus of interest to a variety of teaching methods that aim at promoting historical thinking. For example, interdisciplinary lessons that combine History with other disciplines, such as Geography, Maths, Music, and Art, can prove to be ideal in promoting students' motivation and interest towards history thinking [8]. Field trips can also give the opportunity to both students and teachers to visit a historical site and become familiar with the area, make hypotheses about the relation between the place and its historical meaning and to evaluate their hypotheses, thus thinking and working like historians, in order to -ultimately- develop their historical thinking [9–13].

Several studies have suggested that the use of ICT may motivate students and help them develop historical thinking [1]. ICT seems to offer a lot of affordances that address history teaching needs, such as the use of video documentaries, exploiting the web for seeking historical sources and information, utilizing Web 2.0 technology for collaborative historical research, etc. [1, 2, 4]. Similarly, virtual field trips are ICT-supported field trips that have emerged as alternatives to traditional field trips. Virtual field trips range from simple teacher-directed class projects that involve navigating in online museum archives [14] to drone-based field trips [15], where students study an area of interest through the lens of a drone transmitting video in real time.

In this study, we propose that a virtual field trip into an interactive augmented tangible environment representing a historical area of interest may motivate students to be engaged with the critical study of the historical content. Such an experience can make learning enjoying and effective, while it also provides opportunities for embodied learning.

2 Literature Review

Several studies have underlined that the integration of ICT in history teaching can enhance historical understanding and may promote the exploration and critical approach of the past rather than the passive accumulation of information. Apps such as timelines and simulations of historical events allow participants to better understand the concept of time, the successions of historical events, and to capture how knowledge was discovered [7, 16]. Seeking for information, such as photographs, video and sound clips, or navigating through online newspapers' archives can help students to become familiar with the historical research and the historical thinking [17]. The use of participatory Web 2.0 tools such as blogs and wikis enable students to collaborate and gain much more autonomy in their historical learning process [2]. Moreover, the use of electronic games, i.e. for reconstructing ancient buildings, objects or even entire cities and civilizations, seems to increase students' engagement and interest in the History lesson [18]. However, teaching and learning history with ICT is an area that is still being developed [19], since history is usually thought to be in absolute contrast with technology and in that sense these two areas cannot exist together - although history is a vital subject for the modern world and technology is the main tool of our century.

Tangible User Interfaces (TUIs) for learning have received a lot of attention lately, with few however applications for history learning. TUIs "enable direct, hands-on interaction with physical objects" [20] and are considered a useful mean for enhancing students' engagement and motivation towards learning, as learners have the opportunity to use physical objects in a multisensory environment [20]. Many researchers have suggested that TUIs have a potential for supporting children's informal and formal learning and that they are highly suited to the design and development of learning activities because they leverage both familiar physical artifacts and digital computation. Tangibles can offer a natural and immediate form of interaction that is accessible to learners, promote active and hands-on engagement, allow for exploration, expression, discovery and reflection, provide learners with "tools to think with" and offer opportunities for collaborative activity among learners (e.g. [21–23]). TUIs have been used in several educational settings, as for example in museums [24] for science (i.e. [20, 24]), geography (i.e. [25]) and mathematics (i.e. [26]).

For creating a TUI for a historical place, we also focused on tangible user interfaces representing interactive landscapes. Continuous shape displays [27] in which a continuous physical model is coupled with a digital model are of special interest. For example, in Illuminating Clay [28], landscape models were constructed using clay support while the three-dimensional geometry was analyzed in real time using a laser scanner and it was easy to recognize changes such as shadow casting, land erosion, etc. TanGeoMS [29], is also an analogous geospatial modeling visualization system that combines a laser scanner, projector, and a flexible physical three-dimensional model with a geospatial information system to create a tangible user interface for terrain data. Similarly, the Augmented Reality Sandtable (ARES) [30] is a sand-based research testbed that uses commercial cheap prototyping tools to create a low-cost method of geospatial terrain visualization with a tangible user interface. Projection-Based City Atlas [31] is a low-cost exhibit of a projection-based city atlas. The exhibit could be replicated and reconfigured in accordance with the size of the mock-up and the spatial arrangement of the place where it will be located. Based on the description of events occurring over time, the interactive virtual tour aims to highlight how the urban fabric has changed throughout the historical and political periods that have affected the city. All those approaches take advantage of our natural ability to understand and manipulate physical forms while still harnessing the power of augmenting these forms with useful digital representations.

Moreover, these augmentations merge the digital with the physical and offer a vivid and immersive audiovisual interface for eliciting body activity. In essence, such approaches allow students to become part of the system they are trying to understand, giving them an insider perspective on the critical mechanisms and relationships that define the domain. New interaction technologies such as mixed reality environments can prove an excellent guide for students to perform physical actions that serve as "conceptual leverage" [32]. Under the umbrella of terms like embodied interaction, full-body interaction, motion-based interaction, gesture-based interaction, tangible interaction, bodily interaction, and kinesthetic interaction, several interactive learning environments based on novel interaction modalities have been developed. These interactive environments try to facilitate an embodied experience of a certain concept, to represent an abstract concept as a concrete instance or operationalize actions as means to express specific content or try to use space as a semiotic resource or even try to promote embodied metaphors. The new mediated environments seem to increase learner engagement since body-based experiences are more perceptually immersive and learners may feel that they are in a more authentic and meaningful educational space [33].

Being able to interact more naturally with digital enriched space enhances our spatial thinking, encouraging creativity, analytical exploration, and learning. The interaction in tangible interfaces is mostly done by touch. However, as Elo [34] supports, "insofar as digital interface design aims at haptic realism it conceives of the sense of touch in terms of narcissistic feedback and thus tends to conceal the pathic moment of touching." Elo [34] continues by indicating that the finger has been given the status of a switch but now it seems to be dragging the whole body along. Touch is not only a computational input device but also a human sense which is overly underestimated as a learning means.

2.1 FingerTrips for History Learning

In this study, we propose FingerTrips environment for history learning, in which an augmented interactive 3D model of a historical site aims at helping students relate to the historical content playfully and enable them to conduct interactive field trips over the landscape with their fingers. Our objective was to create a 3D augmented tangible map which would be of low-cost by using mainstream prototyping hardware and software, and which could be easily reconstructed by students and teachers, in order to be highly populated.

For assessing the idea of FingerTrips for history learning, we constructed a 3D model of the historical site of Fort Rupel which was built near the Greek-Bulgarian borders. Fort Rupel is situated on top of a hill and became famous for its defense during the German invasion of Greece in April 1941. The model was made with simple, recyclable materials: newspapers, cardboard, aluminum foils, a plastic box, and two pieces of polystyrene slabs, 120 $\times$ 60 cm each. The polystyrene pieces were used as the model's base. A plastic box and a carboard box were used as the main hill, while smaller pieces of newspaper were used to create paths leading to the top of the hill. Aluminum foil was also placed in different places of the model, in order to improve texture and give a better feeling of the mountains territory. Newspapers were put above it and was covered with paper tape. Finally, all the construction was painted white. In the following two images (Figs. 1 and 2), the augmented landscape is presented.

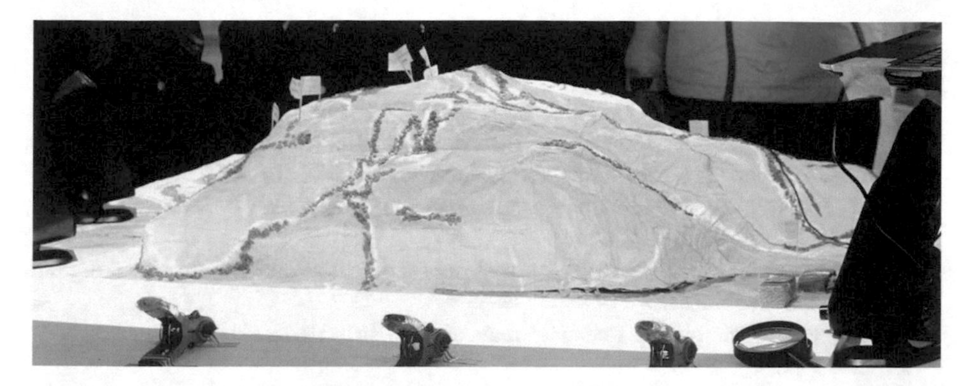

Fig. 1. The augmented 3D historical landscape of fort Rupel with various accessories.

Fig. 2. The back side of the augmented 3D model. Students interacted from all sides

In order to highlight the finger paths on the hill, lentils were glued on the paths. The lentils were also used to help the researchers hide the interaction points in the model, since students were called to explore the environment without knowing were the different events may unveil. With the use of a projector installed in the ceiling, images and animations were shown on the model, resembling the actual historical site.

2.2 Interacting with the Tangible Learning Environment

Learners are the protagonists of the game and role-play historians who have to assist Victor, the main character of the game plot. Victor asks students to help him learn the history of the place, by answering questions and walking together on the model. The students have to go through the paths by crossing bridges or climbing hills with their fingers, but also by reacting to unexpected events. In the course of their finger trip, different surprising events and activities were initiated since hidden interaction points were installed in between the lentils.

In order to promote historical thinking but also to provoke students' engagement, a variety of events and activities have been designed while students walk on the model's paths. For example:

- A song may start to play related to a historical event and students are called to listen it and infer answers to questions about that event or a bear producing sounds may appear running next to their finger, indicating that students need to pay attention to the paths they follow.
- Students may be asked to search for a clue or a place on the model by making and testing their hypotheses.
- A simple multiple-choice question may appear, in order to retain their focus on small details.
- Students may be asked to find rooms inside the fort, thus kneeling down and under the 3D model, in an attempt to create the feeling that the soldiers would have inside the Fort, or they may be called to surrender and find how to do it.

Hence, the journey is a constant swapping of activities, events and FingerTrips. Students interactions on the map were supported by two connected Makey Makey boards whose inputs were embedded into students' journey path and the interactions were programmed with MIT's Scratch. Both platforms are easy to use even for primary school students while the Makey Makey board is an affordable and powerful prototyping solution.

3 Methodology

3.1 Participants

In order to evaluate the FingerTrips approach to learning history, twenty-six (26) 6th grade students, 15 boys and 11 girls, participated in a study in the context of an interactive exhibition related to Tangible and Mixed Reality Interfaces for Elementary Schools. The participants played with the FingerTrips environment in ten (10) groups consisted of 2 or 3 students. Each session lasted about 30 min.

3.2 Procedures

At the beginning of the game brief instructions were given to each group, to help students become familiar with the concept of interacting with the 3D model before starting their FingerTrips game. The researchers offered guidance whenever the participants requested for. At the end of each session, students were asked to complete an online questionnaire about their experience. Seventeen (17) students afterwards, participated in brief group interview in a separate, quieter place.

3.3 Research Instrument

Data collection was based on a questionnaire and a semi-formal group interview. The questionnaire consisted of 22 5-point Likert questions and evaluated the tangible environment in regards to its usability and attractiveness. Some of the questionnaires' items were derived from AttrakDiff [35] and Flow State Scale [36]. The questionnaire consisted of the following variables:

Ease of Use (3 questions): Measure how easy to use the system is and its learnability;
Autotelic experience (3 questions): Measures the extent to which the system offers internal user satisfaction;
Perceived learning (3 questions): Measures students' perceptions on the educational value of the system;
User Focus (3 questions): Measures the concentration during the use of the system;
Pragmatic Quality (4 questions): Measures the extent to which the system allows a user to achieve his goals;
Hedonic Quality-Stimulation (3 questions): Measures the extent to which the system meets the user's need for innovation and whether it is of interest;
Hedonic Quality-Identity (3 questions): Measures the extent to which the system allows the user to identify with it.

All questions were adjusted to 5-point Likert scale answers, in order to become easier for the elementary students. All variables had satisfactory Cronbach's a (Table 1). The semi-formal interviews took place immediately after the end of each session and aimed at extracting the qualitative assessments of the students and at allowing them to describe in their own words their experience with the FingerTrips environment. The questions were focused on what students liked and disliked and their perceptions in regards to the learning effectiveness and efficiency of the environment. All audio-recorded interviews were transcribed and then encoded and compared within and between cases. Afterwards, the four researchers collaborated to reach consensus for the commonly identified issues.

Table 1. Statistics about the on-line questionnaire.

	Min	Max	Mean	SD	Cronbach's a
Easiness	3.33	5	4.28	0.61	0.69
Focus	1.67	5	4.20	0.77	0.82
Autotelic experience	1.67	5	4.37	0.69	0.78
Perceived learning	1.00	5	4.08	0.69	0.78
Pragmatic Quality	3.00	5	4.03	0.78	0.88
Hedonic Quality-Identity	2.00	5	4.41	0.58	0.76
Hed. Quality -Stimulation	3.00	5	4.40	0.54	0.74

4 Results

4.1 Quantitative Data: Questionnaire

As presented in Table 1, students claimed that they didn't have any difficulties in familiarizing with the FingerTrips style of interaction. They found the environment easy to use (M = 4.58, SD = .50) and they would like to interact with this environment often (M = 4.16, SD = .61). Moreover, even if they had been using the educational environment in the context of a noisy exhibition, they assessed that they maintained high levels of concentration on the task at hand. That's a proof of the engagement that similar activities and environments can provoke. The high score also on autotelic experience variable validates that the activity was marked by intrinsic pleasure, it was done by the students for its own sake, regardless of external reinforcements.

Students' answers show that FingerTrips environment can address the problem of engagement with the historical content. Moreover, students were also positive in regards to the learning efficiency of the environment and the possibility of exploiting it in the school. They claimed that they would learn faster (M = 4.35, SD = .98) than with current methods in school and would learn more and better (M = 4.15, SD = 1.19). As also shown in the minimum values of each variable (see Table 1), there were also a few students which were negative towards the FingerTrips environment. This is something that has to be investigated further. It is important to note that the interaction in FingerTrips is social, with students moving around the table, discussing alternatives, taking decisions and interacting with the landscape, and that can work positively for the learning results and negatively, in some cases, in how stressful the students consider the activity.

Students' answers in the mini AttrakDiff questionnaire validated that they considered the functions of the environments as appropriate to achieve the goal of understanding the history of Fort Rupel (pragmatic quality). Moreover, the variable hedonic quality, which is a measure of pleasure (fun, original, engaging) and avoidance of boredom and discomfort, had very high scores. Students' answers show that the environment made them identify themselves with it (Hedonic Quality-Identity) and believed that it offered inspiring and novel functions and interactions (Hedonic Quality-Stimulation).

4.2 Qualitative Data: Interviews

Students reinforced in the interviews the quantitative results already presented. The majority of the students stated that they enjoyed the FingerTrips environment and that they would like to work more often with similar environments. When asked to identify the characteristics that differentiate the specific approach for learning about history, they mentioned several themes.

The most repetitive comment was related to the authenticity and the realness of the learning environment. Students felt that they were living the historical events, that they shared the emotions of the protagonists, that they got a sense of the landscape and the way it formulated causes and effects. The 3D multimedia character of the presentation did manage to involve them more.

- "we thought that everything was real."
- "we saw airplanes, bears."
- "[you understand that] if you lived that, you would have been scared."
- "it was like the war was happening right that moment."
- "it was like reality, as we lived that."

Some students stated that they liked the environment just because it was 3D and that the details of the landscape representation were impressive.

- "I liked it because it was 3D and everything was alive."
- "The landscape had a lot of details."

Students also indicated that they enjoyed the proposed interaction style with the augmented 3D model. The continuous touching of the model, the transitions between activities on the mountain and the contextualization of the finger walking with unexpected events was intriguing and engaging.

- "I liked it a lot because we went on the mountain and we climbed and we had to search the right answer and touch it with our hand."
- "I liked that I had to do different things, to walk over the paths while planes and tanks appeared from nowhere."

Students also perceived FingerTrips learning environment as an effective one, they indicated that the required collaboration was appealing and that the game-based character of the interaction was fun.

- "It is easier to remember [with FingerTrips]."
- "It was big [refers to duration], but we liked the fact that we learned a lot."
- "I really liked our collaboration, the fact that we were trying to find the compasses and all these things together."

Regarding the difficulties students confronted, only a few issues were identified. One student said that he had difficulty in understanding the questions, while another found the paths on the model "a little bit confusing". Most of the participants had a problem hearing or understanding the story or the question presented, but they said they didn't face any major difficulty since they could also read the text of the question on the 3D-model's surface. Two of the participants said they didn't like the way they got feedback as "you had to listen to same things again, and it was a little boring". All these issues could be addressed better in a following version of the environment.

5 Discussion

In this pilot study, the FingerTrips environment succeeded in engaging the primary school students with historical events, made them feel as active participants in the event, and they also thought that they did make a fieldtrip on the historical landscape model with the help of their fingers. The environment provoked them to search, think, discover and study history as historians would. Students claimed that the environment was motivating and effective and that they would like to use more often similar approaches in their classroom since they would learn faster and more things.

Affordable augmented interactive 3D landscapes with FingerTrips can give life to history and offer a participatory experience for students, a much-needed quality in history learning. It is of equal importance that this approach can be followed by students and teachers by themselves since they can design, develop and build interactive landscapes for the historical events they prefer. The event-based programming of

Scratch environment, together with the ability of Makey Makey to make any conductive material to an interactive element, enable students and instructors to easily design and program FingerTrips with a variety of events and activities over an augmented map.

The preciseness of the 3D model is moderate since a single-layer image was projected into a 3D model while the lentils were positioned approximately over the mountain paths without investing a lot of time in the accuracy of the representation. However, that was exactly what was assessed in the study, that is whether these approximations which make FingerTrips an affordable and feasible solution, achieve satisfying effects on students' motivations and learning.

Our study has several limitations. The most important one is that we present the perceived learning evaluations of the students. Although their views are a good indicator for the acceptability of the proposed interface, they cannot offer definite answers for the learning effectiveness of the environment. Additionally, we do not analyze the underlying embodied mechanism for learning. Our hypothesis that touching the landscape may improve students understanding needs more evidence. This is the basic aim of our future research with FingerTrips for history learning.

References

1. Adesote, S.A., Fatoki, O.R.: The role of ICT in the teaching and learning of history in the 21st century. Educ. Res. Rev. **8**(21), 21–55 (2013)
2. Giannopoulos, D.: Italian presence in the dodecanese 1912–1943: teaching a history topic in weebly environment. Procedia Comput. Sci. **65**, 176–181 (2015)
3. Yilmaz, K.: A vision of history teaching and learning: thoughts on history education in secondary schools. High Sch. J. **92**(2), 37–46 (2008)
4. Boadu, G., Awuah, M., Ababio, A.M., Eduaquah, S.: An examination of the use of technology in the teaching of history: a study of selected senior high schools in the cape coast metropolis, Ghana. Int. J. Learn., Teach. Educ. Res. **8**(1), 187–214 (2014)
5. Nokes, J.D.: Recognizing and addressing the barriers to adolescent' "reading like historians". Hist. Teach. **44**(3), 379–404 (2011)
6. Van Drie, J., Van Boxtel, C.: Historical reasoning: towards a framework for analyzing students' reasoning about the past. Educ. Psychol. Rev. **20**(2), 87–110 (2008)
7. Galán, J.G.: Learning historical and chronological time practical applications. Eur. J. Sci. Theol. **12**(1), 5–16 (2016)
8. Bickford, J.H.: Initiating historical thinking in elementary schools. Soc. Stud. Res. Pract. **8** (3), 60–77 (2013)
9. Cengelci, T.: Social studies teachers' views on learning outside the classroom. Educ. Sci.: Theory Pract. **13**(3), 1836–1841 (2013)
10. Coughlin, P.K.: Making field trips count: collaborating for meaningful experiences. Soc. Stud. **101**(5), 200–210 (2010)
11. Stainfield, J., Fisher, P., Ford, B., Solem, M.: International virtual field trips: a new direction? J. Geogr. High. Educ. **24**(2), 255–262 (2000)
12. Strait, J.B., Fujimoto-Strait, A.R.: The mixed plate: a field experience on the cultural and environmental diversity of the big island of Hawai'i. Geogr. Teach. **14**(1), 5–24 (2017)
13. Stoddard, J.: Toward a virtual field trip model for the social studies. Contemp. Issues Technol. Teach. Educ. **9**(4), 412–438 (2009)

14. Lacina, J.G.: Technology in the classroom: designing a virtual field trip. Child. Educ. **80**(4), 221–222 (2004)
15. Palaigeorgiou, G., Malandrakis, G., Tsolopani, C.: Learning with drones: flying windows for classroom virtual field trips. In: 2017 IEEE 17th International Conference on Advanced Learning Technologies (ICALT), pp. 338–342. IEEE (2017)
16. Korallo, L., Foreman, N., Boyd-Davis, S., Moar, M., Coulson, M.: Do challenge, task experience or computer familiarity influence the learning of historical chronology from virtual environments in 8–9 year old children? Comput. Educ. **58**(4), 1106–1116 (2012)
17. Hillis, P.: Connecting authentic activities with multimedia to enhance teaching and learning, an exemplar from Scottish history. Aust. Educ. Comput. **24**(2), 21–27 (2010)
18. Bogdanovych, A., Ijaz, K., Simoff, S.: The city of Uruk: teaching ancient history in a virtual world. In: International Conference on Intelligent Virtual Agents, pp. 28–35. Springer, Heidelberg (2012)
19. Fisher, D.: History Teaching with ICT: the 21st century's' gift of Prometheus'? ACE Pap. **8** (7), 46–59 (2015)
20. Ma, J., Sindorf, L., Liao, I., Frazier, J.: Using a tangible versus a multi-touch graphical user interface to support data exploration at a museum exhibit. In: Proceedings of the Ninth International Conference on Tangible, Embedded, and Embodied Interaction, pp. 33–40. ACM (2015)
21. Antle, A.N., Wise, A.F.: Getting down to details: using theories of cognition and learning to inform tangible user interface design. Interact. Comput. **25**(1), 1–20 (2013)
22. Price, S., Jewitt, C.: A multimodal approach to examining embodiment in tangible learning environments. In: Proceedings of the 7th International Conference on Tangible, Embedded and Embodied Interaction, pp. 43–50. ACM (2013)
23. Resnick, M., Martin, F., Berg, R., Borovoy, R., Colella, V., Kramer, K., Silver-man, B.: Digital manipulatives: new toys to think with. In: Proceedings of the SIGCHI Conference on Human Factors in Computing Systems, pp. 281–287. ACM Press (1998)
24. Loparev, A., Westendorf, L., Flemings, M., Cho, J., Littrell, R., Scholze, A., Shaer, O.: BacPack: exploring the role of tangibles in a museum exhibit for bio-design. In: 2017 International Conference on Tangible, Embedded and Embodied Interactions, pp. 111–120. ACM Press (2017)
25. Palaigeorgiou, G., Karakostas, A., Skenderidou, K.: FingerTrips: learning geography through tangible finger trips into 3D augmented maps. In: 2017 IEEE 17th International Conference on Advanced Learning Technologies (ICALT), pp. 170–172. IEEE (2017)
26. Mpiladeri, M., Palaigeorgiou, G., & Lemonidis, C.: Fractangi: a tangible learning environment for learning about fractions with an interactive number line. In: Proceedings of 13th International Conference on Cognition and Exploratory Learning in Digital Age (CELDA 2016), pp. 157–164 (2016)
27. Petrasova, A., Harmon, B., Petras, V., Mitasova, H.: Tangible modeling with open source GIS. Springer International Publishing (2015)
28. Piper, B., Ratti, C., Ishii, H.: Illuminating clay: a 3-D tangible inter-face for landscape analysis. In: Proceedings of the SIGCHI Conference on Human Factors in Computing Systems, pp. 355–362. ACM (2002)
29. Tateosian, L., Mitasova, H., Harmon, B., Fogleman, B., Weaver, K., Harmon, R.: TanGeoMS: tangible geospatial modeling system. IEEE Trans. Vis. Comput. Graph. **16**(6), 1605–1612 (2010)
30. Amburn, C.R., Vey, N.L., Boyce, M.W., Mize, J.R.: The augmented reality sandtable (ARES). US Army Research Laboratory (2015)

31. Rossi, D., Petrucci, E., Olivieri, A.: Projection-based city atlas: an interactive, touchless, virtual tour of the urban fabric of Ascoli Piceno. In: 2014 International Conference on Virtual Systems & Multimedia (VSMM), pp. 310–317. IEEE (2014)
32. Lindgren, R., Tscholl, M., Wang, S., Johnson, E.: Enhancing learning and engagement through embodied interaction within a mixed reality simulation. Comput. Educ. **95**, 174–187 (2016)
33. Dede, C.: Immersive interfaces for engagement and learning. Science **323**(5910), 66–69 (2009)
34. Elo, M.: Digital finger: beyond phenomenological figures of touch. J. Aesthet. Cult. **4**(1), 14982 (2012)
35. Hassenzahl, M., Monk, A.: The inference of perceived usability from beauty. Hum.–Comput. Interact. **25**(3), 235–260 (2010)
36. Jackson, S.A., Marsh, H.W.: Development and validation of a scale to measure optimal experience: the flow state scale. J. Sport Exerc. Psychol. **18**(1), 17–35 (1996)

Embodied Learning About Time with Tangible Clocks

George Palaigeorgiou(✉), Dimitra Tsapkini, Tharrenos Bratitsis, and Stefanos Xefteris

University of Western Macedonia, Florina, Greece
{gpalegeo, bratitsis}@uowm.gr,
dimitratsapkini@hotmail.gr, xefteris@gmail.com

Abstract. Time is a complex concept to grasp for elementary students and time related competencies take years to fully develop. In this article, we present and evaluate an instructional approach for learning to read and write time through embodied interactions with tangible clocks. The instructional approach consists of four "time learning stations" that may facilitate groups of 12 students (separated in teams of 3) to learn about time. The "time stations" include (a) a game with a big tangible 3D clock, (b) a game with a miniature tangible clock, (c) two notebooks with learning games about time, (d) a set of typical hand-written worksheets about time. Each team explores each station for 10 min and afterwards students move in a circular pattern to the next station. In order to evaluate the instructional approach, 84 students participated in a pilot study forming 7 groups of 12 students that used the time stations for approximately 45 min. Focus groups were conducted after each round of runs. Students supported that the whole setting greatly helped them to get acquainted with time and clock reading. Students underlined that the big 3D tangible clock was the most useful and entertaining activity and pinpointed that the specific interface was more engaging, the interactions were more kinesthetic and unexpected while the learning representation was significantly different from any other that they have used in the past.

Keywords: Embodied learning · Learning about time · Clock reading Tangible interfaces

1 Introduction

Time is one of the most fundamental concepts that permeates all aspects of our lives. Yet, the way we count and record time lies more on a series of social conventions, rather than on concrete and objective factors. That is why time is a very complex concept for children to grasp and time related competencies take years to fully develop. Despite the difficulties in teaching time and all relevant concepts, it seems that this problem has not been tackled frequently [1, 2]. Researchers began to study time understanding more as developmental psychology began formulating new theories, emphasizing the children's natural development in the early 80's [3], but research from a pure educational perspective was started somewhat later [4, 5].

M. E. Auer and T. Tsiatsos (Eds.): IMCL 2017, AISC 725, pp. 477–486, 2018.
https://doi.org/10.1007/978-3-319-75175-7_47

The learning process for the concept of time spans all childhood years and ends when the student reaches adolescence. During elementary school, the pupil's understanding of time is confined to intuitable situations [6]. Children can use "time words", and learn to tell the time, but temporal operations are difficult. Mastering the concept of time and performing temporal operations is an iterative/evolving process, in the sense that children slowly improve their perception of time, proportionally to their development stage. Temporal concepts become more understandable and easily usable as children grow.

According to Piaget, young children can grasp time only through space, velocity and movement, as they cannot perceive "durations" independently [2, 5]. Burny et al. [6] underlined that time comprehension is not an isolated cognitive competency, but rather a skill that relies on a set of other developing competencies like literacy, numeracy, memory and spatial abilities [7]. This leads teachers to blend the development of time related concepts with other subjects, such as history (chronology), geography (deep time), mathematics (clocks and mechanical time) and literacy (time-related vocabulary).

Research in the field of time-related competences lack evidence and most times are rather driven by ideology, politics and marketing than by empirical evidence [8]. Even more the tools used in clock training have remained similar for many years while the technology supported interventions are few. For example, Wang et al. [9] used touchscreen tablets based on the embodied theory for learning to help children tell time with positive results.

In this paper, we propose an instructional approach which is also based on embodied learning and tangibles and consists of four "time learning stations" that may facilitate classrooms of 12 students to learn about time. The "time stations" include (a) a game with a big tangible 3D clock, (b) a game with a miniature tangible clock, (c) two notebooks with learning games about time, (d) a set of typical hand-written worksheets about time.

2 Embodied Interactive Learning

Embodied cognition theory and several relevant frameworks suggest that thinking and acting (or else mind and body) are intertwined in nature and that tangible engagement with objects or exploring spaces affects the way we think about them and vice-versa. Grounded Cognition describes that mental representations are grounded in motor areas of the cortex and that the perceptual and motor states acquired through experience are reactivated through simulation when knowledge is needed [10]. Similarly, the Embodied Metaphor theory suggests that abstract concepts and conceptual metaphors are based on image schemas that derive from physical actions [11]. And many more theoretical frameworks propose that full-body interaction has the potential to support learning by involving users at different levels such as sensorimotor experience, cognitive aspects and affective factors. The physical world seems to underpin one's internal mental representations [12]. The design rationale is that having learners act out and physicalize the systems processes, relationships, etc., will create conceptual anchors from which new knowledge can be built [13].

New interaction technologies can prove an excellent guide to perform physical actions that serve as "conceptual leverage" [13]. Under the umbrella of terms like embodied interaction, full-body interaction, motion-based interaction, gesture-based interaction, tangible interaction, bodily interaction, and kinesthetic interaction, several interactive learning environments based on novel interaction modalities have been developed. Following similar theoretical underpinnings, these interactive environments try to facilitate an embodied experience of a certain concept, to represent an abstract concept as a concrete instance or operationalize actions as means to express specific content, or try to use space as a semiotic resource or even try to become embodied metaphors. The new mediated environments seem to increase learner engagement since body-based experiences are more perceptually immersive and learners may feel that they are in a more authentic and meaningful educational space [14]. Following these theoretical underpinnings, tangibles are frequently used to teach children abstract concepts, in science and mathematics [15]. For example, Button Matrix [16] uses coupled tactile, vibration and visual feedback to highlight features of a physical experience with arithmetic concepts and cue reflection on the links between the physical experience and the mathematical symbols. Tangible Interactive Microbiology environment [17] also offers students an interface with microbiological living cells and tries to promote artistic expression and scientific exploration.

However, there is also another stream of research which indicates that "physicality is not important" and rather "their manipulability and meaningfulness make them [manipulatives] educationally effective". In many situations, children do not transfer performance with physical to symbolic representations of problems. Indeed, it has been suggested that previously identified virtues of physical manipulatives—learning through concrete and perceptually rich physical practices—are not the drivers of learning (e.g., [18]) and can even be detrimental to learning (e.g., [19]). However, a recent meta-analysis found that the use of physical manipulatives in math education tends to improve retention, problem solving, and transfer [20]. Additionally, the context of use seems also to have detrimental effects. For example, unconstrained physical manipulation has also been shown to be suboptimal for learning [21] or high interactivity can be overwhelming and may lead to a lower learning performance embedded learning, whereas self-guided problem-solving strategies can be effective, but seem to be moderated by the perception of possibilities for action on manipulatives (e.g., [15, 21]). Hence, the design of tangibles still holds great difficulty. Tangibles may differ in terms of the degree of metaphorical relationship between the physical and digital representation or may range from being completely analogous, to having no analogy at all and may also differ in terms of degree of 'embodiment'. Small representational differences may have great effect on performance differences [22].

In order to achieve the goal of designing efficient and effective learning tangibles, designers and researchers have to bring together specific knowledge about children's cognitive, physical, emotional, and social skills, the idiosyncratic characteristics and prior experience on each field domain and the opportunities provided of tangibles environments.

3 An Instructional Approach with Tangible Clocks

In this study, we propose an instructional approach for learning about reading and writing time which exploits the following design principles:

- Create new visual-spatial representations of an analogue clock since time-related competences are strongly connected with the related concrete representations.
- Exploit embodied interaction with the new tangible representations.
- Offer an authentic context of interacting with time-based activities since due to its abstract nature, time must be taught through realistic activities and authentic problem solving.
- Provide lots of activities with a constant switching between reading and writing time.
- Create a chain of activities which trigger different modalities and enable students to interact with several types of time representations.

The proposed approach consists of four "time learning stations" and targets classrooms of 12 students separated in four teams of 3. The four "time stations" are the following:

(a) *A tangible game with a big three-dimensional clock* (Fig. 1): In this game, one of the students becomes the hour indicator, the other one becomes the minutes indicator while the third one sits in the center of the clock and has to touch both of them and an orange (fruit) when they are ready to form a specific time in order for the program to evaluate their input. For the very first time, students become a part of a working clock and view an analogue clock from an unprecedented perspective. The program narrates different time activities and students have to form

Fig. 1. The big tangible clock with three students

the right combination of the two indicators (i.e. "the time is 12:00", or "let's say that 12 h and 10 min have passed"). If students answer wrongly, different layers of help are available. Students walk inside the clock, observe the position of their co-mates, discuss with them and try to answer to the presented challenges as fast as they can.

(b) *A tangible game with a miniature clock* (Fig. 2). The second learning station is the miniature of the first one with different however time-related activities such as calculating durations and time-intervals, pre-night and after midnight times, etc. Similarly, in this station, the one student is the hour indicator, the second one the minutes indicator and the third student is the interacting one. The activities relate to events of a girl's daily routine presented in a multimedia way. In this tangible clock, the difficulty is the exact opposite to the big clock since the hands of the students have to fit in a very limited space.

Fig. 2. The small tangible clock

(c) *Two notebooks with 3 learning games about time* (Fig. 3). The third learning station is based on common PC games about time which focus on converting digital to analogue time and vice versa. Multiple time representations coexist in the game screen and help students to make the required connections for the conversions.

(d) A desk with hand-written worksheets about time. The last learning station provides worksheets about time which have to be completed in a short period of time by the students. The worksheets include exercises addressing the objectives of the curriculum for 4th grade students and the related book.

Each team had to use each station for 10–12 min. After completing one station, students moved in a circular pattern to the next one.

Fig. 3. The other two clock stations

The two tangible clocks were developed having in mind simplicity, affordability, and ease of replication. Thus, the interfaces were created by exploiting 4 Makey Makey boards connected to two laptops running the web version of Scratch 2. Both platforms are easy to use even for primary school students while the Makey Makey board is an affordable and powerful prototyping solution. Such a setting can be reproduced and enhanced with new activities easily by teachers and students.

4 Methodology

4.1 Participants

The sample of the pilot study consisted of 84 students, from five 4th grade classes which were organized in 7 groups. The interventions were performed in a two-day activity in the context of an interactive exhibition related to Tangible and Mixed Reality Interfaces for Elementary Schools.

4.2 Data Collection

Each group of 12 students, after completing a full tour in all the learning stations lasting about 45 min, was gathered in a quitter place and a short focus group was conducted. The questions posed aimed at determining whether the proposed instructional intervention is effective and joyful way of learning, which of the four learning stations was the preferable and why and whether the tangible interfaces offered them new perspectives in time understanding, reading and writing. Exemplary questions of the focus group included

- Did these four games help you to become more familiar with clock reading?
- Which one did you enjoy most and why?
- Is there something new that you learned through this brief activity?
- What's new and useful in these games?

5 Results

According to students, the proposed instructional activities helped them to get more acquainted with time and clock reading. In their words

"We realized the exercises and thus learned to tell the time better."
"We exercised our time telling skills through games and the tangibles given to us."

The students learned about time in a pleasant and entertaining way, without getting bored or frustrated. When asked which learning station was the best in terms of learning efficiency, almost all students answered that they considered the big 3D tangible clock as the most helpful one. They identified the following advantages:

- The interface was more fun and engaging than the others learning stations. Students approached the big tangible clock as a joyful learning game.

"The big clock helped us more than the others, however both tangible games were fun."
"It's a more entertaining way of learning and we prefer this way since doing the same thing on paper is not fun."

- The interaction needed was kinesthetic and that was in opposition to the paper-based worksheets and the games on the computer. It is well documented that students want to abandon their classroom desks and learn in a more embodied way.

"The three-dimensional clock because we were moving."
"The big clock since we were running."
"I was running and laughing."
"The big clock because we were running and because we learned while having fun."

- It included several unexpected and fascinating for them interaction elements, such touching students, touching the numbers of the clock, touching an orange. Students were enthused by the workings of the tangible clocks, as they couldn't understand how touching each other became an input for the program. Natural interfaces seem a good fit for creating learning representations that intrigue the students.

"We liked the part where we touched each other and then pressed the orange to check the answer."
"The big clock because it was really cool to touch the orange."
"The big clock because there were wires and pressing the orange was fun...we had to ... touch our hands and then the orange."
"We liked the part where we touched each other and pressed the orange to check the answer."

- It offered a new learning representation significantly different from any other available representation at school. The big 3D interactive clock offered a novel visual-spatial perspective of analogue clocks, and an original role-playing game with hour and minute indicators which enabled them to re-visit and re-examine how to read time.

"The large clock because it is different from ordinary watches and we can learn more easily about time."
"It helped us because we were the clock indicators ourselves and we learned more."
"It helps someone to see the indicators [from a closer look] and understand time more."

"In typical watches that we wear in our hand we see the time, here we played with [emphasis] the time."

The big 3D clock was something completely new for the students, and that was noticeable from the beginning when they were awed and couldn't explain how such a game could evolve. The students' interest was heightened by the fact that they could move inside the space of the clock and that had to role-play the hour and minute indicators. There was a single outlier, a female student, who claimed that worksheets are a more effective way to learn and practice time telling skills.

Students presented some repeated difficulties in specific activities of the big clock. The more typical examples were:

- Activities like "quarter to five" in which the hour indicator usually touched number 4 and spoke loudly "4:45".
- Activities like "10 exactly" in which many students asked where is the zero number on the watch.
- Activities like "quarter past five" in which many children put the hour indicator in 5 but the minute indicator touched number 4.
- Activities relating to time calculations i.e. 5 h passed, while the time was half past nine. Students counted with fingers where to move the hour indicator.

Most of these problems come from the conversion of digital to analogue time. It seems that the tangible clocks may also be exploited as diagnostic tool for time misinterpretations or difficulties.

Finally, as to which were the new and useful things they encountered through the activity, students said:

"The new technologies",
"That was the first time we saw a clock made of Styrofoam which helped us learn more easily."
"Very useful and unusual. It is something new that does not happen in the classroom."

6 Discussion

The proposed setting addressed concurrently the needs of a big number of students and offered them multiple new perspectives in a short period of time by exploiting different learning media, from hand-written worksheets to tangible interfaces. Natural interfaces [23] and tangible interactions showed great impact on children's attitude towards learning about time and time-related calculations, difficult to grasp notions. This pilot study underlined that tangible clock representations in combination with a role-playing game can become a creative canvas for discussing time issues and for exercising typical time-related activities under a new perspective. Of course, teaching time to children demands a solid conceptual framework taking into account the children's evolving sense of the relevant concepts and the related math skills needed.

Low cost rapid prototyping hardware together with the uprising trend of arts and crafts fairs, tinkering and inventing have created a new trend of creating tangible technologies for learning. Researchers but also teachers and students can create or replicate tangible devices such as the ones described in this study with easiness. Hence, the focus now is on the effective exploitation of physical interfaces or on identifying adequate embodied metaphors and realizing them into interaction models [24].

References

1. Block, R.A.: Experiencing and remembering time: affordances, context, and cognition. In: Levin, I., Zakay, D. (eds.) Time and Human Cognition: A Life-Span Perspective, pp. 333–363. Amsterdam (1989). http://doi.org/10.1016/S0166-4115(08)61046-8
2. Piaget, J.: The Child's Conception of Time. Ballantine, New York (1969). Original Work Published 1946a (Trans. by, A.J. Pomerans)
3. Levin, I., Gilat, I.: A developmental analysis of early time concepts: the equivalence and additivity of the effect of interfering cues on duration comparisons of young children. Child Dev. 78–83 (1983)
4. Dawson, I.: Time for chronology? Ideas for developing chronological understanding. Teach. Hist. 14 (2004)
5. Hoodless, P.A.: An investigation into children's developing awareness of time and chronology in story. J. Curric. Stud. **34**(2), 173–200 (2002)
6. Burny, E., Valcke, M., Desoete, A.: Towards an agenda for studying learning and instruction focusing on time-related competences in children. Educ. Stud. **35**(5), 481–492 (2009). https://doi.org/10.1080/03055690902879093
7. Foreman, N., Boyd-Davis, S., Moar, M., Korallo, L., Chappell, E.: Can virtual environments enhance the learning of historical chronology? Instr. Sci. **36**(2), 155–173 (2008)
8. Slavin, R.E., Lake, C.: Effective programs in elementary mathematics: a best-evidence synthesis. Rev. Educ. Res. **78**(3), 427–515 (2008)
9. Wang, F., Xie, H., Wang, Y., Hao, Y., An, J.: Using touchscreen tablets to help young children learn to tell time. Front. Psychol. **7** (2016)
10. Barsalou, L.W.: Grounded cognition. Annu. Rev. Psychol. **59**, 617–645 (2008)
11. Antle, A.N.: The CTI framework: informing the design of tangible systems for children. In: Proceedings of the 1st International Conference on Tangible and Embedded Interaction, pp. 195–202. ACM (2007)
12. Malinverni, L., Pares, N.: Learning of abstract concepts through full-body interaction: a systematic review. Educ. Technol. Soc. **17**(4), 100–116 (2014). https://doi.org/10.2307/jeductechsoci.17.4.100
13. Lindgren, R., Tscholl, M., Wang, S., Johnson, E.: Enhancing learning and engagement through embodied interaction within a mixed reality simulation. Comput. Educ. **95**, 174–187 (2016)
14. Dede, C.: Immersive interfaces for engagement and learning. Science **323**(5910), 66–69 (2009)
15. Manches, A., O'Malley, C., Benford, S.: The role of physical representations in solving number problems: a comparison of young children's use of physical and virtual materials. Comput. Educ. **54**(3), 622–640 (2010)
16. Cramer, E.S., Antle, A.N.: Button matrix: how tangible interfaces can structure physical experiences for learning. In: Proceedings of the Ninth International Conference on Tangible, Embedded, and Embodied Interaction, pp. 301–304. ACM, January 2015
17. Lee, S.A., Chung, A.M., Cira, N., Riedel-Kruse, I.H.: Tangible interactive microbiology for informal science education. In: Proceedings of the Ninth International Conference on Tangible, Embedded, and Embodied Interaction, pp. 273–280. ACM, January 2015

18. Zacharia, Z.C., Olympiou, G.: Physical versus virtual manipulative experimentation in physics learning. Learn. Instr. **21**(3), 317–331 (2011)
19. Sloutsky, V.M., Kaminski, J.A., Heckler, A.F.: The advantage of simple symbols for learning and transfer. Psychon. Bull. Rev. **12**(3), 508–513 (2005)
20. Carbonneau, K.J., Marley, S.C., Selig, J.P.: A meta-analysis of the efficacy of teaching mathematics with concrete manipulatives. J. Educ. Psychol. **105**(2), 380 (2013)
21. Stull, A.T., Barrett, T., Hegarty, M.: Usability of concrete and virtual models in chemistry instruction. Comput. Hum. Behav. **29**(6), 2546–2556 (2013)
22. Goodman, S.G., Seymour, T.L., Anderson, B.R.: Achieving the performance benefits of hands-on experience when using digital devices: a representational approach. Comput. Hum. Behav. **59**, 58–66 (2016)
23. Steinberg, G.: Natural user interfaces. In: ACM SIGCHI Conference on Human Factors in Computing Systems. ACM (2012). https://www.cs.auckland.ac.nz/compsci705s1c/exams/SeminarReports/natural_user_interfaces_gste097.pdf
24. Mpiladeri, M., Palaigeorgiou, G., Lemonidis, C.: Fractangi: a tangible learning environment for learning about fractions with an interactive number line. Int. Assoc. Dev. Inf. Soc. (2016)

Android OS Mobile Technologies Meets Robotics for Expandable, Exchangeable, Reconfigurable, Educational, STEM-Enhancing, Socializing Robot

Nikolaos Fachantidis[1(✉)], Antonis G. Dimitriou[2], Sofia Pliasa[1], Vasileios Dagdilelis[1], Dimitris Pnevmatikos[3], Petros Perlantidis[4], and Alexis Papadimitriou[5]

[1] Department of Educational and Social Policy, University of Macedonia, Egnatia 156, 54636 Thessaloniki, Greece
nfachantidis@uom.edu.gr
[2] School of Electrical and Computer Engineering, Aristotle University of Thessaloniki, Egnatia, 54124 Thessaloniki, Greece
[3] Department of Primary Education, University of Western Macedonia, 3rd km Florina-Niki, 53100 Florina, Greece
[4] Department of Informatics and Telecommunications Engineering, University of Western Macedonia, Agios Dimitrios Park, 50100 Kozani, Greece
[5] MLS Multimedia SA, Technopolis, Pylaia, 55535 Thessaloniki, Greece

Abstract. In the context of the project "STIMEY" [1], we are constructing a socially assistive robotic artefact. The purpose of the robot is to represent a companion of the student, motivating and rewarding him for his "Science Technology Engineering and Math" (STEM) achievements. Furthermore, the robot should represent a mobile platform capable of deploying prototype STEM content; i.e. representing a powerful vehicle for the introduction of STEM activities in the classroom. Such robot should act as a socialization-means among the owners, thus socializing instead of secluding the user from his social environment, highlighting "STEM" as a fertile environment for social distinction. In this paper, we present existing prior-art on this field and demonstrate how an Android OS smartphone, representing the "head" of the robot, addresses the design requirements.

Keywords: STEM · Assistive robot · Educational robot

1 Introduction

As mobile devices are becoming richer in technological features and capabilities, their exploitation gets wider and many professional, entertainment and everyday applications are developing. For instance, health care professionals use medical devices and apps for many purposes: administration, health record maintenance and access, communications and consulting, reference and information gathering, and medical education [2]. In other applications, mobile system components with the ability to capture

M. E. Auer and T. Tsiatsos (Eds.): IMCL 2017, AISC 725, pp. 487–497, 2018.
https://doi.org/10.1007/978-3-319-75175-7_48

aspects of their environment, implement context-awareness and so the applications obtains and uses information on aspects of the user environment [3].

Education and learning are major areas of mobile devices exploitation. The applications of mobile learning range widely, from K–12 to higher education and corporate learning settings, from formal and informal learning to classroom learning, distance learning, and field study [4].

In the last years, many social robot manufacturers and researchers have integrated mobile devices into their creations. Social robots [5] become a tool in many cases, where we want to use the human-robot interaction as bridge in order to enhance humans' motivation. Robots and smartphones/tablets are currently adopted in education in a separate manner. Robots are used for developing engineering and programming skills in several platforms, like "Lego Mindstorms", where most of the time, the robot itself represents the subject or the tool of the course. Alternatively, they are also used as proxies for the teacher or the student, representing the telepresence-bridge in cases of distant learning, e.g. VGo, Giraff, Ohmni, AMY A1, PadBot P1. The latter are all equipped with a smartphone/tablet, which is only used as the visual representation of the missing teacher/student. Finally, robots have found their way into teaching students with Autism, where students tend to avoid interaction with others.

In the same time, tablets/smartphones are used in education, representing the most powerful vehicle of mobile learning, supporting the transmission and delivery of rich multimedia content or facilitating students according to the "Bring Your Own Device" approach. Several characteristics of mobile phones have been identified as positive factors for learning facilitation and social interaction enhancement in education [6].

Our goal is to study the deployment of an Android OS-Smartphone as the "brain" and the principal interaction environment of the robot and demonstrate how the "Smartphone + Robot" well-achieves the educational and technical goals of the project. For this reason, we have contacted a bibliography and market survey and in this paper, we are presenting the approaches in the architecture and visual representation of an educational or socializing robot, which we find in the bibliography or in the current market. We summarize below the physical, technical and educational characteristics and capabilities of the relative robots at the current market or bibliography.

2 Android OS Smartphone as the Robot's Brain

The android phone ensures great computational power, numerous sensors (camera, accelerometer, microphone, thermometer, etc.), while it provides several methods of interaction: audio recognition, visual interpretation (using the camera), and touching. Thanks to its small form factor, it fits well the requirement for a small desk-placed pet friend. It provides versatile communications options (Cellular, WiFi, Bluetooth, NFC, etc.) allowing for both web-based and peer to peer applications.

As an education tool, the Android OS ensures that rich content is already available to the teacher. A small list of examples includes: real-time retrieval of information from the internet, access to online frameworks for the support of classroom activities (project assignment, quizzes, auto correction, progress reporting etc.), screen-sharing among robots (hence also shared to the robot of the teacher which can be projected on the wall of the classroom), access to rich content with education material (through google play).

In addition to the aforementioned already-available material, the robot can also be used for customization and development of prototype STEM content; e.g. a programming interface, capable to create custom movements (like a dance), or be assigned specific tasks. Each student's achievements can be shared to the others (e.g. uploading the custom code), thus enhancing socialization among them. The smartphone itself can be extracted by the robot's body and exchanged among students, further enhancing socialization among them.

In addition, thanks to the rich communication potential of the smartphone, the robot can interact with sensors placed in the classroom, allowing exploration of the Internet of Things concept.

3 Smartphone-Based Social Robots

There is an increasing number of social robots that use a smartphone as the "brain" of the robot; i.e. the main processor that controls all actions of the robot. In the following paragraphs, we attempt to present the current state-of-the-art.

3.1 DragonBot, Tega, Huggable

"DragonBot" [7] (Fig. 1), is an android-smartphone based robot, developed by the Personal Robots Group at MIT Media Lab in 2011, under the guidance of Dr. Cynthia Breazeal. The robot runs on an Android phone, located on the head of the robot, which displays animated expressions. The phone provides sensory input (camera and microphone) and controls the actuation of the robot (motors and speakers). Internet access is essential for the robot to harness cloud-computing paradigms to learn from the collective interactions of multiple robots. The motors support 5 degrees of freedom, which allow the body of the robot to be expressive but the robot does not move from its position (by means of wheels or axes). The phone supports motor-control, realtime animation, computer vision, speech processing, data logging. DragonBot, can be tele-operated remotely (e.g. by a tablet). The purpose of DragonBot is to help with the process of learning. The platform was originally teleoperated by a human to test how children behaved to the robot in paradigms of story-telling and curiosity behavior.

"Tega" [8] (Fig. 2) is based on the DragonBot platform, with differences in the mechanical parts. Tega is designed for in-home interactions with children from vocabulary to storytelling. It has been used successfully in an experiment to teach Spanish words in pre-school (ages 3–5) children in [9]. Details on the Android-smartphone specifications are not included. The "Robot Operating System" (ROS) has been integrated into the smartphone. It handles all mechanical movements and has solved synchronization issues [10].

"Huggable" (Fig. 3) is another smartphone controlled social robot, based on the "DragonBot" platform. It hosts an HTC Vivid Android Smartphone, with a 1200 MHz dual core processor, 1 GB RAM, 16 GB memory and weighs 177 g. Huggable looks like a teddy bear. It is featured with a full body sensitive skin with over 1500 sensors, quiet back-drivable actuators, video-cameras in the eyes, microphones in the ears, an inertial measurement unit and a speaker. It represents a robotic companion for healthcare, education and social communication applications. An experiment in hospital pediatric care, where teleoperation of the robot took place, is presented in [11].

3.2 Romo

"Romo" (Fig. 4) started as a successfully funded kickstarter project. It is composed of two parts: (*a*) a robotic base with a docking station for (*b*) an IPhone 4 or IPhone 4S (1 GHz CPU). The user should purchase the base, install the appropriate software on his IOS and then control the assembled robot. The base is capable to tilt 5° forward or 45° backward to see where Romo is going, move, steer and rotate 360° around its vertical axis. Romo is using the IPhone's sensors (camera, microphone, audio, accelerometer and gyroscope) to interact with the user. The applications include 2 way video-audio streaming, remote control from another IOS device or from a web browser, facial tracking, "Romo training mode" where the user sets up desired expressions, taking pictures/video, playing music. Despite of its initial success, Romo was discontinued in 2016, probably because the company didn't adjust its platform to the annual IPhone (and IOS) updates.

Fig. 1. DragonBot.

Fig. 2. Tega.

Fig. 3. Huggable.

Fig. 4. Romo.

Fig. 5. RoboMe.

Fig. 6. Buddy.

Fig. 7. Synergy Swan.

Fig. 8. Jibo.

Fig. 9. Zenbo.

Fig. 10. Hub.

Fig. 11. AlphaEgg.

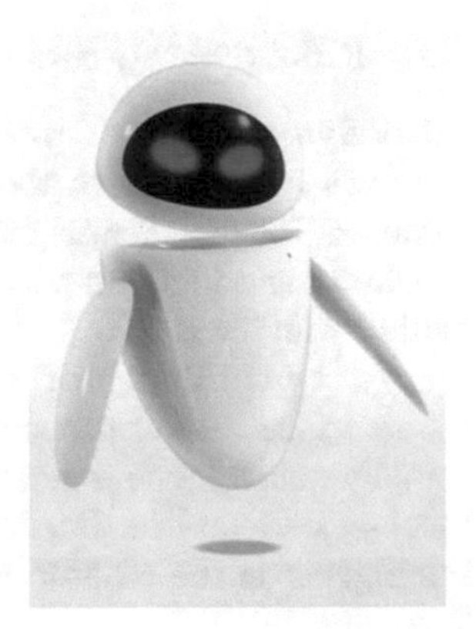

Fig. 12. Pixar's Eve.

3.3 WowWee RoboMe

"RoboMe" (Fig. 5) is a low-cost customizable companion robot toy (~40$). It has a docking station for an IPhone. The user can select the desired appearance and voice and associate actions with specific "triggers". These are 19 specific actions that the robot is pre-programmed to execute, like speaking specific voice-content, dancing, spinning, shaking hand, mainly controlled through voice orders. It includes an IR sensor to avoid obstacles and falling. It can be controlled by a remote control or another device, when the IPhone is connected. It also supports a telepresence mode.

3.4 "Buddy" from "Blue Frog" Robotics

"Buddy" (Fig. 6) is probably the most advanced and promising commercially available social assistant robot, costing around 600$. The range of applications include reminders, home protection (exploiting the camera), music and videos reproduction, interfacing and controlling compatible smart-home equipment, interaction with children (claiming to assist in the learning process), telepresence, making video/audio calls. Interaction with the user is accomplished by speech and touch. It supports real-time house mapping and localization, autonomous collision avoidance and can be remotely controlled.

The most exciting thing about Buddy is the potential for software and hardware evolution, suggested by the company. Developers can upgrade “Buddy” through the open source SDK accompanying the robot, supporting Unity 3D, Python, C++, C, C#, Java and Javascript. Furthermore, developers are free to build accessories for the robot.

From a technical point of view, Buddy hosts an 8” tablet, 4 range finder sensor (1 on the head and 3 on the body), 5 obstacle detection sensors, 5 ground detectors, 2 speakers and 4 motors, allowing head rotation along two axes and 360° base movement from two motor-controlled wheels. It supports WiFi and Bluetooth connectivity, moves at a maximum speed of 70 cm/s at a step of 1.5 cm (for obstacle detection). Its batteries last for 8–10 h, weighs above 5 kg with dimensions 56 cm × 35 cm × 35 cm.

3.5 R.Bot Synergy Swan

R.Bot Synergy Swan (Fig. 7) is a smartphone (or tablet) controlled telepresence robot. It serves as an alias of the user. It supports communication through speakers and cameras. The tablet/smartphone is not sold with the robot. It supports face-recognition capability and includes a human speech synthesizer. Only Android devices are compatible with the robot.

Apart from the above smartphone controlled robots, there are some new social robots in the market that include all parts of the smartphone architecture, but are not actually smartphone operated. Those later products, presented next, all share a touch-screen for the interaction with the user and aim at incorporating Artificial Intelligence in the robot to be able to interact with the user.

3.6 Jibo

“Jibo” (Fig. 8) targets to be a commercial product since 2014, when it was successfully funded through “Indiegogo” for 749$ per piece, raising 3,663,105$. However, three years later, it has not entered the market yet, even though initial Indiegogo funding has been raised by additional millions of dollars.

Jibo’s ambitious goal is to bring Artificial Intelligence in the house, representing a companion for the family, being able to understand and respond to natural language. It attempts to combine many AI technologies, like machine learning, natural language processing and computer vision to address its goals.

From a technical perspective, Jibo is 28 cm tall, weighs 2.7 kg, includes 360° sound localization, upper body touch sensors, 3 full revolute axes, an HD touchscreen, a high-end ARM-based mobile processor, WiFi and Bluetooth connectivity and runs on Linux kernel. Jibo’s body expressions are accomplished by rotating on inclined (instead of horizontal) planes.

3.7 Zenbo

Zenbo [12] (Fig. 9) is a companion-robot made by "Asus". It has not been commercially available yet. The following technical information is extracted from the FCC tests. It measures 37 cm by 37 cm by 62 cm, weighs 10 kg, has no limbs and fingers, moves on wheels, while a 10.1-inch WXGA touch screen display represents its face. It hosts a 4 GB RAM and will be available with at least 32 GB storage memory. It is powered by an Intel Atom Z850 processor. Its sensors include infrared, sonar and touch. It can process voice commands through 4 digital microphones, while it outputs audio by a 15 W speaker. It hosts a 13 megapixels camera for video and image capturing. The display changes into a typical tablet to display information, while it also operates as a touch-screen. The face of the robot is cartoon-like instead of human-like to avoid intimidating the user.

The purpose of the robot is to provide "assistance, entertainment and companionship" for all members of the family. The following applications are announced from the manufacturer: take photos, make videos, make video calls, speak reminders and support storytelling for children, hear and respond to requests and questions, reproduce music, connect and control smart-home devices, order items online, learn and adapt to one's preferences by proactive Artificial Intelligence (AI), express emotions with facial expressions.

Other major companies have presented prototypes of their upcoming products, like LG's "Hub" robot (Fig. 10) and Bosch's "Mykie" robot.

3.8 iFlytek

"iFlytek" is a Chinese company, specialized at voice recognition and speech synthesis [13]. Lately, the company has introduced its voice-expertise in developing commercial "toy-robots", capable to interact with children orally. Robot "AlphaEgg-TYR100" is a Child-education assistant, weighting 2.8 kg, with external dimensions of 25 cm 25 cm × 32 cm. It includes a single motor to rotate towards the child, a camera and a screen (representing the face of the robot) for animated expressions and reproduction of digital content, 20 W speakers, 2 GB RAM, 8 GB for storage, HDMI and USB connection-hubs at the rear of the body. It supports "games", like multiple-choice quizzes, "storytelling", reminders, video-calls, video/photo shooting. The GUI of the robot suggests that an Android-OS is installed (Fig. 11).

3.9 Summary

In summary, we can point out some common design specifications of social-robots, which use a smartphone or have similarities with smartrphone. They all share similar sizes, around 30 cm in the horizontal plane and 30 cm–50 cm height. With the exception of the "DragonBot" based robots, all others use a touchscreen (either as part of a smartphone or standalone device), where an animated character represents the face of the robot. This is an evident design approach to avoid the famous Mori's "Uncanny Valley" [14], where a non-perfect human-like face would greatly risk approval of the robot. Most commercial products demonstrate only the eyes (or one eye) on the screen and none shows an entire face (with nose, ears, etc.). All of them follow a simple clear-lines design approach with a big head and a body. In fact, most commercial products strongly resemble "Eve", an animated robotic-character that appeared in the film "Wall-E", by "Pixar Animation Studios", shown in Fig. 12. Almost all products lack hands; probably because performing delicate actions (like grabbing a pencil) is technically complex, while such exterior parts could lead to accidents (the hand could bump on an obstacle not "seen" by the body-sensor), frustrating the robot's owner.

The robots' processors clock between 1 GHz–1.5 GHz, including those robots that do not use a smartphone (e.g. Jibo uses an ARM processor designed for mobile devices and Zenbo the Atom processor). The screen varies from 5" to 8", depending on the design. With the exception of the 3 standing-still robots from Media Lab, all others deploy Infrared and Ultrasonic sensors for obstacle and fall avoidance, microphone(s) for audio input, speaker(s) for audio output, at least one camera and of course the face-display for reproducing image/video content. Some host touch sensors, light sensors and temperature sensors. Most robots don't use more than 4 motors (2 are used for movement and 2 for the body/head).

With the exception of Breazeal's group, no other robot aims in assisting in education. All commercial products seem to share similar applications-portfolio: oral and touchscreen interaction with the user, assistant-type services (phone-calls, reminders, video-calls), story-telling for children, surveillance (by teleoperating the robot and tele-viewing video from the camera).

Some products are advertised as being able to communicate with the user in everyday conversations (or at least the company is unclear on what they suggest by expressions like "responding to the user"). Such expectations are far from the reality of Human-Machine interaction and should be treated with skepticism. Understanding a native speaker falls in the field of Natural Language Processing, which deals with many sub-problems which are far from being solved. Such include "Natural language understanding"; i.e. interpreting the input from a microphone, under noise conditions, parsing it, segmenting it, identifying words in different languages and from different speakers, symbol grounding problem- interpreting the actual meaning of the sentence- and a whole lot of unsolved problems to an acceptable success-percentage for commercial deployment.

Probably the best AI engines are the "Google Assistant" or Amazon's "Alexa", which are still applications that handle specific orders (open an App, go to settings, perform a search in the internet etc.). Interestingly for smartphone-based robot developers, "Google Assistant" can be deployed in any android device since February 2017 and is also available in the IOS (all iphones). Furthermore, since April 2017, Google announced an SDK (Software Development Kit) for the Assistant, in order to be deployed in other devices (cars, smart homes etc.).

Bearing in mind the above technical analysis, modern Android OS smartphones in the price range of 100$–200$ outclass the minimum technical requirements of a smartphone-controlled social robot, in terms of CPU, RAM, Memory, camera-quality and wireless connectivity. All kinematics sensors, related to obstacle avoidance, interaction with the user, motion control, localization etc., need to be added. Furthermore, the range of "already available" applications is huge, including image recognition, voice recognition, text-to-speech engines, as well as rich content related to education. What is more important is that such android-applications are continuously improved by their developers, which is crucial for the improvement of the robot. For instance, "Google Assistant" will be continuously improved by "Google" thus enriching the AI capabilities of the robot.

4 Educational Aspects and Capabilities

Tablets are used to tele-control educational or interactive robots [15]. From an educational perspective, robots are used to support classroom activities: interact with the teacher's robot, receive project assignments, get rewards and motivation in many forms, implement learning activities through oral interaction [16]. The capability to program the robot actions through an app is given in some cases [17] and can be used for learning purposes. Thanks to the rich communication potential of the smartphone, the robots can interact and work as a pal or as a tutor [18]. Extended literature review in Mobile technologies and Learning [19] shows the capabilities of mobile devices to support a diversity of cognitive theories and educational cases.

Tablets and mobile devices with Android OS supports M-learning, which has been described as "a personal, unobtrusive, spontaneous, 'anytime, anywhere' way to learn and to access educational tools and material that enlarges access to education for all" [20]. Elias [21] extended and adapted the eight Universal Instructional Design (UID) principles, particularly useful in distance education, to the M-learning:

1. equitable use,
2. flexible use,
3. simple and intuitive,
4. perceptible information,
5. tolerance for error,
6. low physical and technical effort,
7. community of learners and support, and
8. instructional climate.

The relevance of almost all of these principles for designing inclusive online learning is further increased when designing inclusive m-learning.

In addition to the aforementioned material, the robot can also be used for implementation of STEM content activities. The students in some cases are equipped with a programming interface, capable to create custom movements (like a dance), or be assigned specific tasks. The smartphone itself can be detachable, which offers personalized capabilities.

5 Conclusions

In this paper, we studied through cases the context for the introduction of an Android OS smartphone in the creation of a social, educational robot. Many of the features of the current robots that host a smartphone, being based on special characteristics of the smartphones, which are being used today in teaching and learning and could enrich the capabilities of an educational robot with social features. This is particular interest for purposes like inclusive and accessible education, which should aspire to include all learners. The challenge in the our case is combine harmonically the features of the mobile devices, the capabilities that the m-learning offers and thw social character of the robot.

Acknowledgement. This project has received funding from the European Union's Horizon 2020 Research and Innovation Programme under Grant Agreement No 709515.

References

1. Assaad, M., Makio, J., Makela, T., Kankaanranta, M., Fachantidis, N., Dagdilelis, V., Piashkun, S.V.: Attracting European youths to STEM education and careers: a pedagogical approach to a hybrid learning environment. World academy of science, engineering and technology. Int. J. Educ. Pedag. Sci. **4**(10), 1730 (2017)
2. Ventola, C.L.: Mobile devices and apps for health care professionals: uses and benefits. Pharm. Ther. **39**(5), 356–364 (2014)
3. Akgul, F.O., Pahlavan, K.: Location awareness for everyday smart computing. In: International Conference on Telecommunications, Marrakech, Morocco (2009)
4. Yeonjeong, P.: A pedagogical framework for mobile learning: categorizing educational applications of mobile technologies into four types. Int. Rev. Res. Open Distrib. Learn. **12** (2), 78–102 (2011)
5. Hegel, F., Muhl, C., Wrede, B., Hielshcer-Fastabend, M., Sagerer, G.: Understanding social robots. In: 2009 Second International Conferences on Advances in Computer-Human Interactions (2009)
6. Cochrane, T., Roger, B.: Smartphones give you wings: pedagogical affordances of mobile Web 2.0. Australas. J. Educ. Technol. **26**(1) (2010)
7. Personal Robots Group, Media Lab. http://robotic.media.mit.edu/portfolio/dragonbot/. Accessed 27 Oct 2017
8. Westlund, J.K., Lee, J.J., Plummer, L., Faridi, F., Gray, J., Berlin, M., Quintus-Bosz, H., Hartmann, R., Hess, M., Dyer, S., Dos Santos, K., Adalgeirsson, S., Gordon, G., Spaulding, S., Martinez, M., Das, M., Archie, M., Jeong, S., Breazeal, C.: Tega: a social robot. In: Proceedings of the 11th ACM/IEEE International Conference on Human-Robot Interaction (2016)
9. Gordon, G., Spaulding, S., Westlund, J.K., Lee, J.J., Plummer, L., Martinez, M., Das, M., Breazeal, C.: Affective personalization of a social robot tutor for children's second language skill. In: Proceedings of the 30th AAAI Conference on Artificial Intelligence, Palo Alto, CA (2016)
10. Jeong, S., Logan, D., Goodwin, M., Graca, S., O'Connell, B., Goodenough, H., Anderson, L., Stenquist, N., Fitzpatrick, K., Zisook, M., Plummer, L., Breazeal, C., Weinstock, P.: A social robot to mitigate stress, anxiety, and pain in hospital pediatric care. In: Proceedings of the Tenth Annual ACM/IEEE International Conference on Human-Robot Interaction Extended Abstracts. ACM, New York (2015). https://doi.org/10.1145/2701973.2702028
11. Guizzo, E.: Cynthia Breazeal Unveils Jibo, a social robot for the home. In: IEEE Spectrum (2014)
12. ASUS – Zenbo Robot. https://zenbo.asus.com/. Accessed 27 Oct 2017
13. Liu, L.J., Ding, C., Jiang, Y., Zhou, M., Wei, S.: The IFLYTEK system for blizzard challenge 2017. In: The Blizzard Challenge 2017 Workshop, Stockholm (2017)

14. Mori, M., MacDorman, K.F., Kageki, N.: The uncanny valley. translation of original article included. IEEE Robot. Autom. Mag. **19**(2), 98–100 (2012). https://doi.org/10.1109/MRA.2012.2192811
15. Tanaka, F., Toshimitsu, T., Masahiko, M.: Tricycle-style operation interface for children to control a telepresence robot. Adv. Robot. **27**(17), 1375–1384 (2013)
16. Vogt, P., de Haas, M., de Jong, C., Baxter, P., Krahmer, E.: Child-robot interactions for second language tutoring to preschool children. Front. Hum. Neurosci. **11**, 73 (2017)
17. Redmer, P.: WowWee RoboMe review: full of character and ready to roll. http://www.robocommunity.com/article/20101/wowwee-robome-review-full-of-character-and-ready-to-roll. Accessed 27 Oct 2017
18. Gordon, G., Spaulding, S., Westlund, J.K., Lee, J.J., Plummer, L., Martinez, M., Das, M., Breazeal, C.: Affective personalization of a social robot tutor for children's second language skills. In: 13th AAAI Conference on Artificial Intelligence, Special Track: Integrated AI Capabilities, pp. 3951–3957 (2016)
19. Naismith, L., Sharples, M., Vavoula, G., Lonsdale, P.: Literature Review in Mobile Technologies and Learning. A NESTA Futurelab Series - report 11 (2004)
20. Traxler, J., Kukulska-Hulme, A.: Mobile learning in developing countries (2005). http://oro.open.ac.uk/49128/
21. Elias, T.: Universal instructional design principles for mobile learning. Int. Rev. Res. Open Distrib. Learn. **12**(2), 143–156 (2011)

Interactive Collaborative and Blended Learning

Developing Professional Skills Through Facilitated Dialogue

Pasi Juvonen(✉) and Anu Kurvinen(✉)

Faculty of Business Administration, Saimaa University of Applied Sciences, Lappeenranta, Finland
{pasi.juvonen, anu.kurvinen}@saimia.fi

Abstract. This article is a part of developing Experimental Development Ecosystem (EDE). EDE is a framework including researchers from University of Applied Sciences and a Technological University, students from both institutes, team entrepreneurs and lecturers, local enterprises and local cities. Learner-centric methods utilized with team entrepreneurs and practioners have now been tested in other contexts as well.

The article summarizes the experiences of piloting method of professional dialogue with (1) study course for professional MBA students in the international study programme (2) research project, and (3) long term coaching process.

In the study, survey data and field notes from observations of dialogues are used as main data sources. Ongoing discussions and feedback from stakeholder groups of the EDE have served as a secondary source of data. The data was analyzed with qualitative data analysis methods. Grounded theory is used as method for data analysis. The framework for developing the EDE has been action research.

The data shows that students are usually both motivated and excited about professional dialogue as a learning method. It was also revealed that making oneself familiar with professional dialogue requires repetition and reflection of what has been learned. The reflection needs to cover both content and process to be successful. As a consequence, this the method is also showing good results in learning as the students who graduate are having excellent competences for working life.

Local company representatives have also expressed their satisfaction to knowledge, skills, and character of the graduates. Therefore, the professional dialogue is currently spread to other student groups studying bachelor and master studies.

Keywords: Professional dialogue · Collaborative learning
Experimental Development Ecosystem
21st century education requirements · Learning

1 Background for the Study

Team learning and team entrepreneurship (TLE) in Bachelor level IT education and in Bachelor level Business and Administration education (specialization in marketing) has been utilized since 2009 at a Finnish University of Applied Sciences. It is based on experimental learning, action learning and methods of knowledge creation. The pedagogics needed in the school of future and entrepreneurship education have been

M. E. Auer and T. Tsiatsos (Eds.): IMCL 2017, AISC 725, pp. 501–511, 2018.
https://doi.org/10.1007/978-3-319-75175-7_49

utilized. In recent years, TLE has been expanded to Experimental Development Ecosystem (EDE) framework including researchers from university of applied sciences and a technological university, students from both institutes, team entrepreneurs and lecturers, local enterprises and local cities. These stakeholder groups have been cooperating in RDI projects for few years already. The experiences gained from learner-centric methods were now ready to be tested in other contexts as well. This article summarizes the recent experiences of deploying professional dialogue with three different contexts: (1) study course for professional MBA students (2) within research project, and (3) long term coaching process with team entrepreneurs.

The article is organised as follows. Chapters two and three present the background for the competence requirements for the future education and the organization of the current EDE. In chapter four data collection and analysis is discussed. Chapter five will present some observations on professional dialogue based on the data and Chapter six discusses and summarizes the conclusions.

2 Competence Requirements for Education

In the world of today, there is plenty of information available. The challenge lies in the fact that, one has to be able to think critically and have skills to synthesize and put the information into action in a wise way. According to Fadel et al. [1] in order to deepen the learning in the three essential dimensions – knowledge, skills and character qualities – there is an important dimension needed – meta-learning. Sometimes named also as 'learning to learn', meaning that there are some internal processes required for our learning, namely reflection and adaptation of our learning. Figure 1 presents the framework for 21st century learner and for the curriculum redesign.

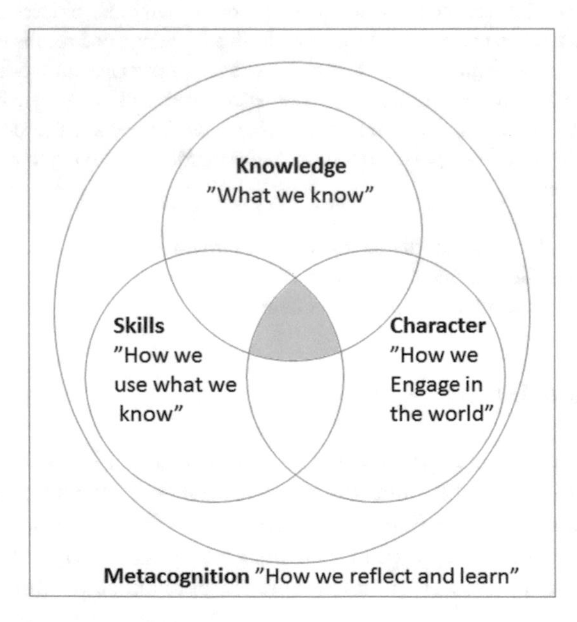

Fig. 1. The 21st century education framework presented by Center for Curriculum Redesign [1].

The Center for Curriculum Redesign (CCR) is trying to develop understanding to the question "What should students learn for the twenty-first century". According to Trilling and Fadel [2] there are four forces that are leading the learners towards new ways of learning for life in the 21st century. These forces are:

- Knowledge work – employees that are using brain power as well as digital tools for creating new solutions collaboratively in teams
- Thinking tools – knowledge workers use digital tools, devices and services
- Digital lifestyles – todays student generations are born in the digital society, and grown up with the digital devices
- Learning research – the research on learning done during the last 30 years has been revolutionary for our understanding on learning processes.

Key findings, also found in the learning research, that have proved to be successful in the EDE researched model, are; authentic learning, mental model building, internal motivation, multiple intelligences as well as social learning. The curriculum applied, as well as the piloted methodology applied in certain study courses, aims at developing following skills needed in the 21st century working life: team working skills, communal learning skills, problem solving skills, leadership and self-leadership skills, as well as innovativeness, shared expertise, and ability to reflect one's own values and attitude.

2.1 Learning for Life

Graduating students are expected to be ready for the working life needs [3–5]. E.g. Companies in the ICT field worldwide are in constant need of competent experts that are ready to adopt new tools and at the same time having an entrepreneur mindset [6] with understanding of current logic of the business world. The situation is very similar in Finland. A rough estimate of shortage of several hundreds of ICT developers in next few years has been stated for South Karelia region in Finland where our University of Applies Sciences is operating [7]. In order to meet these needs, the educators have to rethink the methods of learning and the curriculum. The key choice in the curriculum is, thus done between teaching skills vs. teaching knowledge. When considering teaching knowledge we have to notice, that a huge amount of new knowledge is discovered all the time. If we choose to concentrate to teach knowledge who has enough wisdom what knowledge to teach?

A growing amount of research from a wide range of fields has pointed to the need for students to balance content knowledge and understanding with skills that apply that knowledge to the real world [1, 2]. Also meta-level skills are nowadays required and emphasized. We have to decide whether we want our students to be passive receivers led by teachers or active agents supported by coaches and facilitators. A summary of the essential choices to be made is presented in Table 1.

Table 1. Teacher-directed and learner-centered approached compared [2]

Teacher-directed	Learner-centered
Direct instruction	Interactive exchange
Knowledge	Skills
Content	Process
Basic skills	Applied skills
Facts and principles	Questions and problems
Theory	Practice
Curriculum	Projects
Time-slotted	On-demand
One-size-fits-all	Personalized
Competitive	Collaborative
Classroom	Global community
Text-based	Web-based
Summative tests	Formative evaluations
Learning for school	Learning for life

2.2 Adopting the Skills

In constructivism, the learning of a human being is understood as a constant process where individuals are learning or creating their own understanding based on interaction between what they already know and believe, and ideas and knowledge with which they come into contact [8]. The constructivist learning thus involves at least the following five areas: (1) educator's attention to the learners, the students and their backgrounds (2) dialogue facilitation with the group with the purpose to creation of share understanding of a topic (3) planned or unplanned introduction of formal theory into the discussion (4) creating opportunities for the students to challenge or change the existing beliefs and conceptions by using tasks that are structured in a way that this is made possible, and (5) developing students own awareness of their own level of understanding and learning process [9].

In addition to the constructivism, we apply both dialogue [10–13] and the methods of team learning [14–16] and knowledge creation [17–19]. Open and honest dialogue is proven to be supporting the learning objectives. It is also a useful method for developing leadership competencies such as productive conflict resolution skills, perspective taking, finding creative solutions, expressing emotions, reaching out (from deadlock situations), reflective thinking and adaptive behavior [20].

3 The Experimental Development Ecosystem

The experimental development ecosystem (EDE) serves as a framework for developing learning methods and redesigning curricula for education, and for developing cooperation with stakeholder groups working in RDI projects. The EDE is based on Kolb's [21] theories of experiental learning, model of knowledge creation by Nonaka and Toyama [18], and thoughts about competence development and zone of proximal development presented by Vygotsky [22].

Two main types of professional dialogue are used with the EDE. A generative dialogue [23–25] and lateral thinking [26] mean different ideation techniques and other ways to help us to expand our thinking to gain non-obvious results. Common concept for this is also "out-of-the-box" thinking. A reflective dialogue is convergent and concluding [27–30]. In a safe presence of trusting peers it can bring to surface social, political and emotional data that arise from direct experience with others [30]. Principles for these different kind of dialogue sessions have been adapted from Von Krogh et al., and Nonaka's and Toyama's models of knowledge creation [17, 18, 31].

The team entrepreneurship curriculum is an essential part of the EDE. Figure 2 gives an overall picture of how team entrepreneur's studies are currently organized.

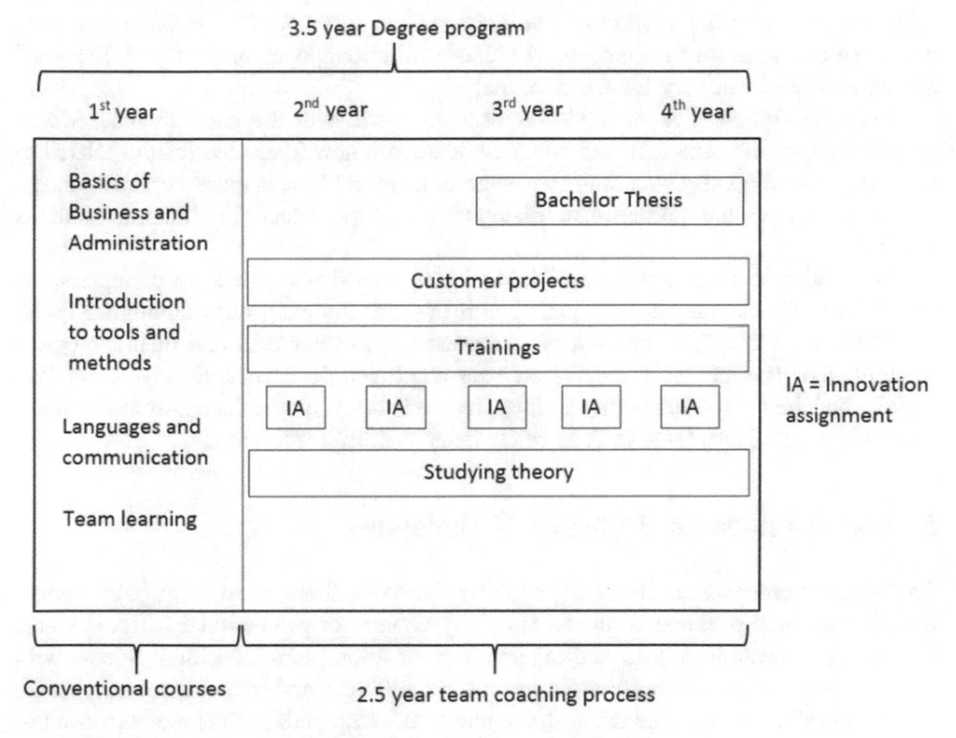

Fig. 2. Overview of curriculum of SUAS marketing students studying as team entrepreneurs [7].

In practice, the students who selected to specialize their studies in marketing establish a team enterprise (in a form of a cooperative) and run its operations together during their studies. As a method of learning the business theories, the students are reading books and articles, carrying out customer projects and trainings where a team coach coaches team members to develop their team and the team enterprise. During the studies, each team entrepreneur has to read 34–51 books depending on their scope. The literature is to be read from pre-defined fields of business expertise. Students are also free to suggest books. Besides reading, the students need to write book essays that are used for constructing reflection of the read theory or phenomena. In the book dialogues (twice a month) the main message of the book is also verbally communicated during the reflection the main insights of the books read are picked up.

4 Data Collection and Analysis

In the study, survey data and field notes from observations of dialogues served as data sources. The data was analysed with qualitative data analysis methods, open, axial and selective coding methods of grounded theory method [32–35] were utilized. In grounded theory a researcher lets "the data speak for itself" so understanding of "what is going on here?" is developed inductively. The research question: "*How professional dialogue as a learning method can be harnessed to serve the 21st century learner?*" was used as lenses for the analysis. "Evidence for strengths of professional dialogue" served as a seed category for the data analysis.

Constant comparative method was used to work with the data. In other words revisited the earlier data gathered when we learned a new interested feature related to the subject studied. By this way, we were able to achieve a more coherent understanding of how the professional dialogue was experienced in different contexts studied.

In addition to field notes, some theme-based interviews with team entrepreneurs, and short feedback sessions with participants of other student groups familiarized with dialogue as a method of learning were carried out to offer data and methodological triangulation. Researcher triangulation was employed in a natural way when two authors had their own viewpoint and interpretations based on the data. Discussing these viewpoints was fruitful and offered better inner validity for the study.

5 Observations on Professional Dialogue

This chapter presents observations made by the researchers in applying professional dialogue in three different contexts: (1) study course for professional MBA students (2) during a research project, and (3) as a part of a long term coaching process with team entrepreneurs. All these target groups are different and thus form a very interesting objective for the research with the aim to develop pedagogical methods used in the learning process. The main observations and comparison between the three groups are presented in Tables 2 and 3.

There are four practices present in dialogue; suspending, listening, respecting and voicing. Suspending assumptions and certainties means avoiding our natural tendency to jump into quick conclusions. Listening includes: following the disturbance and recognizing the resistance of hearing something and still continuing to listen. Respecting means respecting others' opinions as you were in a way getting to know a stranger. By respective listening you all are learning all the time compared to speaking when you usually are repeating something that you already know. Voicing means speaking your own voice. Listening what is inside you and bringing it out honestly.

Table 2. Observations on employing professional dialogue in three different contexts - key practices of dialogue

	RDI project (researchers and practitioners)	Long-term coaching process, after 6 months	International Group in Master education
Scope of practice	3 book dialogues via Skype for Business	Face-to-face book dialogue twice a month	Face-to-face book dialogue approx. 1/3 of the contact teaching time
Probing questions	Used mostly by researchers	Used by most participants, also by team coach if needed	Used by the lecturer and most participants
Roles (mover, follower, opposer, bystander)	Researchers served as movers	All roles represented within the group, few bystanders, opposers emerge when needed	All roles represented within the group depending on the students past experience or studies
Reflective dialogue	Reflection on what was learnt from books and how it is applicable to current business	Reflection of what has been learnt from the books and how it is applicable to customer projects	Reflection on what was learnt from books and how it is applicable to current business
Actions plans (how to utilize this in practice?)	Cautions plans to experiences something new with support of researchers	Are part of the coaching process – are made almost every time	Cautions plans to read more literature on some special area that is applicable in the own career or the employer's interests
Following actions	Some isolated small experiments	Several, continuous experimentation and reflection employed	Several new experimentation via the collective learning

This requires courage and needs to be practised. There may be a huge gap between what you need to say and what the organization or an individual can tolerate hearing. Here a very difficult question of values and authenticity comes in. Without making things visible, there is no possibility to develop them.

The different observation groups have clear similarities as shown in the table above. However, most of the data has been collected from young students studying ICT or marketing as team entrepreneurs. Learning motivation of students participating in dialogue sessions is on a relatively high level in each of the observation groups. Dialogue requires active participation and the more one tries to understand the topic, the more one can participate the dialogue as an equal part of the group. Dialogue as a method requires rehearsing and it shows from the observations. For team entrepreneur students who study a considerable part of their curriculum by reading books and using dialogue the method is more familiar and they are letting themselves more freely wondering new phenomena than the participants who already are having some professional identity formed.

Table 3. Summary of observations about four key practices of dialogue in three different contexts

	RDI project (researchers and practioners)	Long-term coaching process, after 6 months	International Group in Master education
Suspend assumptions and certainties	Specialist views dominate, now and then open ended questions were made	Open ended questions dominate, others' opinions are a matter of interest	Specialist views dominate, now and then open ended questions were made
Listening	Polite (partly because of Skype), some further questions were made	Lot of further questions and ideation (yes, and…) is made naturally	Some further questions were made
Respecting	Participants' opinions were respected as professionals of their own area	Some opinions are more respected than others. The team coach has to impact this now and then	Participants' opinions were respected as professionals of their own area
Voicing	Showing ignorance or wondering aloud something did not take place	Wondering aloud and showing ignorance (and promises of future learning) are common	Showing ignorance or wondering aloud something did not take place

The "RDI project group" and the "International Group in Master Education" had not tried this method before. In addition to this, these two groups are consisting of persons who already have certain types of professional identities formed and thus are more secure to present their ideas based on their working life experiences.

Dialogue skills develop through practice and reflection. Patience will be needed during the development process. Frustration may also occur due to slow development. With persistently employing the key practices of dialogue, one will develop. The survey data and field notes already show that students are usually both motivated and excited about professional dialogue as a learning method.

This the method has showed good results in learning as the students who graduate are having excellent competences for working life. Feedback from local companies about students in their working practice and also those starting their career after graduating has been strongly positive. Based on the feedback they have knowledge and skills, but above of that they have character. Most of them have inner motivation, curiosity and self-leadership skills needed in 2010's working life.

So far, we were able to answer to the research question: "*How professional dialogue as a learning method can be harnessed to serve the 21st century learner?*" partly. With the analysis of the current data, it is possible to argue that learning has been increased and that learners have been satisfied with the learning method.

6 Conclusions and Summary

At the moment, the use of professional dialogue seems to be effortless way for colleagues to employ new learning methods into their courses. Based on the previous feedback students are excited to use of professional dialogue as a learning method. Based on the feedback from company representatives the professional dialogue as a learning methods used has been appropriate to develop knowledge, skills and character of studets. The utilization of professional dialogue also seems to be the easiest way to spread the EDE.

The complex requirements for 21st century students' learning objectives require more collaboration - practices such as knowledge sharing and generative and reflective dialogue. We need to learn to cope with changes in education to avoid the risk of isolation from the real life. This has been noted by several authors [36–39].

Open dialogue can only exist if the educators, teachers or coaches themselves are ready to participate in the learning process together with the students. Studies of professional learning communities has been done to foster teacher collaboration and learning [40, 41]. The authors of this article are employing these same ideas with learning of several student groups. By admitting that we all have more to learn regardless our previous experience on the topic, it will be easier to start learning together. Professional dialogue is a powerful method of carrying this out.

Future work

Our aim is to spread the learning methods used with the EDE, especially professional dialogue to other degree programmes as well. The next steps are/will be:

- connect ICT bachelor students with marketing team entrepreneurs
- connect together students from different disciplines as well
- make professional dialogue as obligatory part of every degree program
- describe professional knowledge sharing practices in a way that any colleague is able to deploy them

The authors believe that learning environments can only be developed when the knowledge- and teacher- centered approaches are gradually transformed to learner-centered processes and methods with physical and mental environments, and curricula that support the idea of learning together. Professional dialogue, inspiration to collaborate and inner motivation are crucial. After enabling them, it will be fruitful to select appropriate low-tech and high-tech tools and systems for supporting the approach.

References

1. Fadel, C., Bialik, M., Trilling, B.: Four-Dimensional-Education – The Competencies Learners Need to Succeed. The Center for Curriculum Redesign, Boston (2015). 02130
2. Trilling, B., Fadel, C.: 21st Century Skills – Learning for Life in Our Times. Jossey-Bass, San Francisco (2009)
3. Wells, G., Claxton, G. (eds.): Learning for Life in the 21st Century. Blackwell Publishing, Hoboken (2002). Printed in Great Britain

4. OECD report: Moving up the value chain: staying competitive in the global economy (2007)
5. Meristö, T., Leppimäki, S., Laitinen, J., Tuohimaa, H.: The skill foresight of the finnish technology industries. (Tulevaisuuden osaamistarpeet teknologiateollisuudessa, yhteenvetoraportti toimialakohtaisista yrityskyselyistä). In: Teknologiateollisuus ry, Turku 2008 (2008). ISBN 978-952 12-1926-9. (in Finnish)
6. European Commission: Communication from the commission to the council, the European parliament, the European economic and social committee and the committee of the regions, implementing the community Lisbon programme: fostering entrepreneurial mindsets through education and learning (2006). http://eurlex.europa.eu/LexUriServ/site/en/com/2006/com2006_0033en01.pdf
7. Juvonen, P., Kurvinen, A.: Developing experimental development ecosystem to serve ICT education – a follow-up study of collaboration possibilities between stakeholder groups. In: EDUCON 2017, Presented in Athens, 27 April 2017 (2017)
8. Resnick, L.B.: Introduction. In: Resnick, L.B. (ed.) Knowing, Learning, and Instruction: Essays in Honor of Robert Glaser, pp. 1–24. Erlbaum, Hillsdale (1989)
9. Richardson, V.: Constructivist pedagogy. University of Michigan. Teachers College Record, vol. 105, No. 9, December 2003, p. 1626 (2003)
10. Bohm, D.: On Dialogue - Routledge Classics (1996)
11. Isaacs, W.: Taking flight: dialogue, collective thinking and organizational learning. Organ. Dyn. **22**(2), 24–40 (1993)
12. Jacobs, C.D., Heracleuos, L.T.: Answers for questions to come reflective dialogue as enabler of strategic innovation. J. Organ. Change Manag. **18**(4), 338–352 (2005)
13. Lefstein, A., Snell, J.: Better Than Best Practice. Developing Teaching and Learning Through Dialogue. Routledge, Abingdon (2014)
14. Katzenbach, J.R., Smith, D.K.: The Wisdom of Teams. Creating the High-Performance Organization. Harvard Business School Press, Brighton (1993)
15. Katzenbach, J.R., Smith, D.K.: The Discipline of Teams: A Mindbook-Workbook for Delivering Small Group Performance. Wiley, Hoboken (2001). Printed in the United States of America
16. de Vries, M.F.R.K.: The Hedgehog Effect. The Secrets of Building High Performance Teams. Jossey-Bass, San Francisco (2011). Printed in Great Britain
17. Von Krogh, G., Ichijo, K., Nonaka, I.: Enabling Knowledge Creation. How to Unlock the Mystery of Tacit Knowledge and Release the Power of Innovation. Oxford University Press, Oxford (2000)
18. Nonaka, I., Toyama, R.: The theory of the knowledge-creating firm: subjectivity, objectivity and synthesis. Ind. Corp. Change **14**(3), 419–436 (2005)
19. Sarder, R.: Building an Innovative Learning Organization: A Framework to Build a Smarter Workforce, Adapt to Change and Drive Growth. Wiley, Hoboken (2016). Printed in the United States of America
20. Beer, M.: High Commitment High Performance: How to Build a Resilient Organization for Sustained Advantage. Jossey-Bass, Hoboken (2009). Printed in the United States of America
21. Kolb, D.: Experiental Learning. Prentice-Hall, Englewood Cliffs (1984)
22. Vygotsky, L.S.: Mind in Society: The Development of Higher Psychological Processes. Harvard University Press, Cambridge (1978)
23. Isaacs, W.N.: Dialogic leadership. In: The Systems Thinker – Building Shared Understanding. vol. 10, No. 1, February 1999. Pegasus Communications Inc (1999)
24. Isaacs, W.: Dialogue: The Art of Thinking Together. Doubleday, Randomhouse Inc., New York (1999). Printed in the United States of America
25. Grill, C., Ahlborg, G., Lindgren, E.C.: Valuation and handling of dialogue in leadership. a grounded theory study in Swedish hospitals. J. Health Organ. Manag. **25**(1), 34–54 (2011)

26. De Bono, E.: Lateral Thinking. Penguin Books Ltd, London (1970)
27. Mezirow, J.D.: How critical reflection triggers transformative learning. In: Associates, J.M. A. (ed.) Fostering Critical Reflection in Adulthood: A Guide to Transformative and Emancipatory Learning. Jossey-Bass, San Francisco (1990)
28. Mezirow, J.D.: Transformative Dimensions of Adult Learning. Jossey-Bass, San Francisco (1991)
29. Boyd, D., Keogh, R., Walker, D.: Reflection: Turning Experience into Learning. Kogan Page, London (1985)
30. Boyd, D., Cressey, P., Docherty, P. (eds.): Productive Reflection at Work. Routledge, Abingdon (2006). Simultanously published in the USA and Canada
31. Wenger, E., McDermott, R., Snyder, W.M.: Cultivating Communities of Practice: A Guide to Managing Knowledge. Harvard Business School Press, Boston (2002)
32. Strauss, A., Corbin, J.: Grounded Theory in Practice. Sage Publications, Thousand Oaks (1997)
33. Glaser, B.G.: Doing Grounded Theory: Issues and Discussions. Sociology Press, Mill Valley (1998). Printed in United States of America
34. Corbin, J., Strauss, A.: Basics of Qualitative Research: Techniques and Procedures for Developing Grounded Theory, 3rd edn. Sage Publications, Thousand Oaks (2008)
35. Flick, U.: An Introduction to Qualitative Research, 4th edn. Sage, London (2009)
36. Engeström, Y.: Non scolae sed vitae discimus: toward overcoming the encapsulation of school learning. Learn. Instr.: Int. J. **1**, 243–259 (1991)
37. Michaelsen, L.K., Knight, A.B., Fink, L.D. (eds.): Team-Based Learning: A Transformative Use of Small Groups. Preager Publishers, Santa Barbara (2002). Printed in United States of America
38. Virkkunen, J., Ahonen, H., Schaupp, M., Lintula, L.: Toimintakonseptin yhteisen kehittämisen mahdollisuus. TYKES, raportteja 70 (2010). (in Finnish)
39. Kujala, T., Krause, C.M., Sajaniemi, N., Silvén, M., Jaakkola, T., Nyyssölä, K. (toim.): Aivot, oppimisen valmiudet ja koulunkäynti. Neuro- ja kognitiotieteellinen näkökulma. Tilannekatsaus tammikuu 2012. Opetushallitus (2012). (in Finnish)
40. Stoll, L., Bolam, R., McMahon, A., Wallace, M., Thomas. S.: Professional communities. A review of the literature. J. Educ. Change **7**, 221–258 (2006). https://doi.org/10.1007/s10833-006-0001-8. Springer
41. Vescio, V., Ross, D., Adams, A.: A review of research on the impact of professional learning communities on teaching practice and student learning. Teach. Teach. Educ. **24**, 80–91 (2006)

Evaluation of a Blog-Based Learning Analytics Tool: A Case Study Focusing on Teachers

Nikolaos Michailidis(✉), Panagiotis Kaiafas, and Thrasyvoulos Tsiatsos

Computer Science Department, Aristotle University of Thessaloniki, Thessaloniki, Greece
{nmicha, panakaia, tsiatsos}@csd.auth.gr

Abstract. The current paper presents a case study for evaluating an Interaction Analysis (IA) toolkit for blogs, called GIANT (Graphical Interaction Analysis Tool). The paper focuses on the perspective of the teachers and investigates the hypothesis that IA techniques could be implemented in educational group-blogging systems, as a tool for supporting teachers. The main objective of the study is to provide an indication that teachers could be facilitated, in the process of evaluating and assessing students' interaction and participation, through the real-time generated graphs from the GIANT toolkit. This paper also provides further evaluation evidence for the GIANT toolkit, from a teachers' point of view, in terms of overall teacher satisfaction, perceived usefulness and perceived ease of learning. Particularly, 32 secondary education teachers' used and evaluated GIANT during a collaborative blog-based activity with students from a literature course. Overall, teachers were to a large extent satisfied with their facilitation from the GIANT toolkit, during their blogging experience, rating the automatically generated IA graphs as useful, easy to understand and decode.

Keywords: Learning analytics · Regulation · Evaluation

1 Introduction

The role of the teacher is crucial in this case, as they are responsible for regulating and orchestrating collaborative activities. Such demanding tasks, though, require managing a huge amount of data from the teachers. Learning Analytics (LA) tools can assist in this directions, by recognizing, analyzing, collecting and interpreting educational data, aiming to help the teacher decide whether to utilize them or not. Moreover, cognitive procedures, such as observation, regulation and evaluation, are enhanced with the projection of these data. Finally, LA tools may support educational evaluation, a process that requires making decisions and reaching conclusions, as far as the performance of students is concerned.

The aim of this paper is to investigate whether an Interaction Analysis (IA) tool such as GIANT (Michailidis *et al.* 2013) can prove to be beneficial, and provide added value to collaborative activities conducted by teachers using blogs. The support that is to be provided is related to the process of regulation and intervention from the teachers'

M. E. Auer and T. Tsiatsos (Eds.): IMCL 2017, AISC 725, pp. 512–521, 2018.
https://doi.org/10.1007/978-3-319-75175-7_50

perspective throughout the activity, as well as the evaluation of the students' performance. Moreover, the perceived usefulness of the tool, how easy it is in terms of its use, and the overall satisfaction with it were also measured.

In the following section, we explore the theoretical background, and a brief description of the theoretical concepts upon which the paper was based is included. These are: Computer-Supported Collaborative Learning (CSCL), Blogs, Interaction Analysis (IA), Regulation, and Evaluation. In Sect. 2, we give information on the implementation of GIANT and its requirements, while in Sect. 3 the research method and the evaluation procedure adopted are analyzed. Finally, we reach the conclusion of the study, and discuss possible future improvements.

2 Theoretical Background

2.1 Computer-Supported Collaborative Learning

Nowadays, collaboration among students is considered to be an exceptionally efficient way to develop communicative skills, such as expressing one's self, and ex-changing ideas. The ever burgeoning use of Information and Communication Technologies (ICT) as an educational tool spawned the CSCL, which is based on the fact that ICT can support and facilitate collaborative procedures. In such cases, designing the learning process includes computer-supported tools, technologies and collaborative scripts that can enhance inter-action. In the words of Dillenbourg and Fischer (2007), technology should help collaboration to occur, when it is not possible otherwise. It is of vital importance, however, that collaborative learning, whether it is computer-supported or not, should be guided and structured, in order scaffold social interaction. Besides, it has been proven that self-guided learning is not beneficial, whether that is pursued individually or in groups (Dillenbourg and Fischer 2007).

2.2 Blogs

Blogs are a part of the Web 2.0, and are online journals that consist of hyperlinks, with posts being their main characteristic. They are not only maintained and updated by their owners, but by their users as well, who exchange opinions through posts. This stands as the major difference between blogs and ordinary websites, since they are an interactive medium, as opposed to the latter. In addition, blogs have been incorporated in the field of education, for their use in a subject could encourage discussion, collaboration, and exchanging ideas among groups. This, in turn, could help students who do not feel comfortable speaking to become more active by posting their thoughts online (Galanakis and Papadopoulos 2012). It is also worth mentioning that even students who serve as mere observers of the whole process can benefit (Lin et al. 2006). Blogs compared to other educational platforms are simpler, intuitive and easy to use (Kim 2008).

2.3 Interaction Analysis

In order to ascertain that technology-enhanced learning has positive learning results, it is necessary to record and analyze the learning experience of students while they interact with a digital educational environment (Chondrouli 2015). The data produced are vital to the guidance, self-regulation and evaluation of the performance of students (Theodoridis et al. 2013). Furthermore, these are presented to the students and/or the teacher, in order to enhance cognitive or meta-cognitive procedures, such as self-evaluation, observation and self-regulation (Bull and Kay 2008; Dimitracopoulou 2004). This is usually achieved through 'visualization', which is linked to the cognitive procedure that is facilitated by the use of visual representation (Bratitsis 2007). External representations aid memory, thought, and argumentation by creating internal representations. More specifically, visual representations should follow a set of rules, such as selecting the appropriate data for visualization, the appropriate medium, and the way data should be projected. Moreover, color code makes performance easy to compare and monitor. However, special attention should be paid so as for visual representation not to alter the meaning of the information presented.

2.4 Regulation

During CSCL activities, the role of the teacher is to support and encourage collaboration among students, and to regulate that when necessary (Kaendler *et al.* 2015). This means that a teacher has to be vigilant in order to help or intervene at any time, which is feasible by constantly observing students' actions. Thus, he/she can comprehend the current state of the activity, and act accordingly, based on the information he/she has collected (Van Leeuwen 2015). As a consequence, the teacher has to be able to manage the – sometimes – overwhelming amount of data, which can have a negative effect on his/her cognitive load (Van Diggelen *et al.* 2008), and decide whether it is best to intervene or let the students regulate themselves. But how can he/she rise to the challenge? The answer is through relying on Learning Analytics (LA) tools, which can visualize information in the form of graphs, therefore summing it up and making it easier for teachers to interpret, while lessening their workload (Spyrouklas 2016). These tools also assign a more active role to the teacher, and allow interaction with the students, making the collaboration more efficient. Lastly, time management improves, and students can be divided into groups according to their skills, abilities, and collaboration levels.

2.5 Educational Evaluation

Educational evaluation is the process of collecting data, making decisions, and reaching conclusions concerning the success and performance of students (Curtis 2010). The basic points are the following: observation, data, and making interventions, plus three major objectives which are the improvement of learning, instant guidance, and monitoring the system's performance. Teachers, in general, gather information that gives them the chance to evaluate individually or in groups, in real time, and not just after the activity has been completed, while securing the credibility of the whole process.

3 The Graphical Interaction ANalysis Toolkit (GIANT)

GIANT was developed by the multimedia lab of our university, and falls under the category of Interaction Analysis toolkits for blogs, designed for the popular WordPress blogging system. Its purpose is to analyze and graphically visualize in real time the social interactions that occur during blog-based collaborative activities. It collects data from interactions that take place in blogs, processes, and visualizes it in a way that is comprehensible by blog users. In a nutshell, GIANT creates graphs in real time, showing the individual and collective contribution of students during the activity. These graphs aim to support the students taking part in the activity by raising their awareness and self-regulation, so that they can reach the desired goals. The graphs are available through the blog's web site to the students to consult them whenever they log in the platform. Installing GIANT is a straightforward process similar to any other WordPress plugin installation. The main idea behind the prototype toolkit is to analyze and visualize blog-users' interactions, allowing them to constantly monitor and regulate those in real time, and assisting users in terms of awareness and metacognition, eventually leading to self-regulatory actions. Figure 1 illustrates the GIANTs' back-end main menu screen.

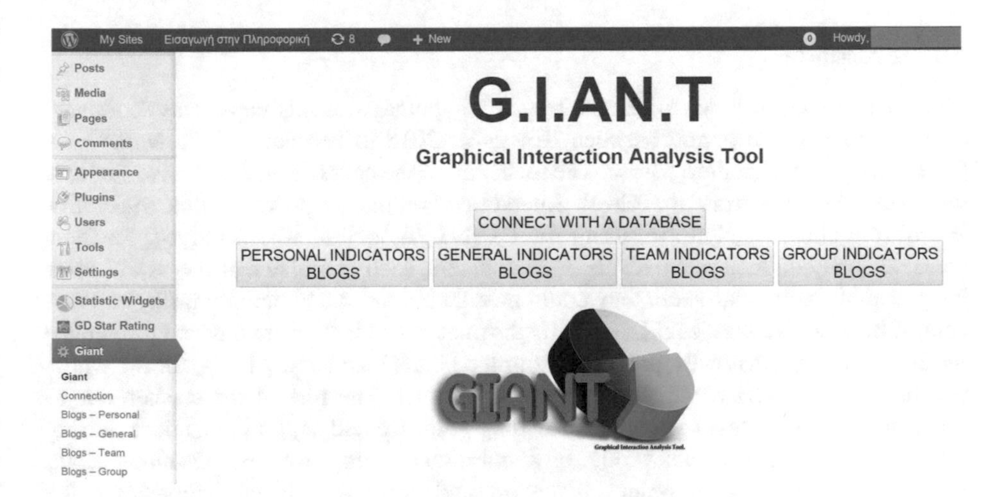

Fig. 1. GIANT's main menu screen

4 Method

The current paper describes a pilot study for evaluating the GIANT IA toolkit for blogs, mainly from a teacher's perspective. The main goal of the study is to investigate the hypothesis that teachers could be supported in improving their teaching process by monitoring their student's interaction and participation during a blog-based collaborative activity. Additionally, another goal of this paper is to advance the work of (Michailidis *et al.* 2013) providing further evaluation data regarding the strong and weak points of the GIANT toolkit for blogs, in terms of overall teacher satisfaction, usefulness and ease of use.

Towards the objectives of the case study, the authors applied a two-dimensional evaluation process. The first dimension focuses on usefulness, ease of use and overall teacher satisfaction from the GIANT's graphs, from the point of view of a teacher. The second dimension is related to the facilitation, if any, that GIANT can offer teachers in terms of regulation and evaluation of the overall blog-based activity.

4.1 Participants

A total of 32 secondary education Greek Literature teachers (14 females, and 18 males) and two assistant teachers participated in the study. All the teachers, were teaching the compulsory subject "New-Greek Literature Texts" on the 2nd grade of a selection of 23 Junior High Schools in the city of Thessaloniki, Greece. On average each teacher had about 26 participating students in the activity. Out of the 32 teachers only 30% had previous experience in blogging systems, while 50% of them had previous experience in other Web 2.0 environments. However, teachers had a satisfactory feeling of confidence in using blogs (68%), with 53.4% of them declaring that they like using blogs, 44.1% being neutral, and only 2.5% of them being negatively inclined to blogging

4.2 Procedure

The activity used in order to test the research hypothesis of this paper was conducted during a four-month period between November 2016 to February 2017, as part of a Greek Language Literature course. The main aim of the course is to familiarize students with selected texts from the Greek Literature. For the purposes of this research a WordPress blogging platform, using the GIANT IA toolkit, was deployed. Students were randomly divided into 6 teams of four student, each of which had access to a team blog from the network, where they could post their articles and activity tasks, and to a central blog, where they could read announcements. Students were advised to monitor the IA graphs automatically produced from the GIANT on a daily basis, for the whole duration of the activity, which was 3 weeks in total. The role of the teacher was to monitor all the available IA graphs, providing guidance and support to students during the blog activity, and answering any questions during weekly 45-minute long face-to-face meetings. In order to support and encourage social interactions, the learning activity was organized with a collaboration script of 3 distinct phases:

- Phase 1: Summary of the selected text, answers to comprehension questions and comments on co-students' posts.
- Phase 2: Group summary, group answer to a comprehension question and comments on co-students' posts.
- Phase 3: Final group answer to a comprehension question, which was revised after reading the other groups' suggestions.

Apart from the visualized IA indicators, the prototype GIANT toolkit attempts the implementation of an automated assessment algorithm on the blog activity data. Based on this predefined assessment algorithm the resulting table provides teacher with a set of proposed grades for each student, taking into account their participation and interaction during the educational group-blogging activity. The automated assessment algorithm was produced, from the manual assessment procedure followed by the teacher, on a pilot study prior the start of the activity.

4.3 Instruments

The experimental instruments employed in this study include: (a) the published content of the blogs (posts and comments), (b) an online questionnaire that all teachers answered after the activity, (c) a semi-structured interview with each participating teacher was conducted after the completion of the activity, and (d) the graphs produced by the GIANT toolkit for blogs.

In order to measure the usefulness and the ease of use of the toolkit, the Davis' (1989) "Perceived Usefulness and Ease of Use" questionnaire (PUEU) was employed. It consists of 12 questions grouped in two dimensions: (a) Usefulness and (b) Ease of use (6 in each category). Moreover, the questionnaire entitled "Teacher Satisfaction according to a blended e-learning system environment" by Wu et al. (2010) was used for evaluating the overall teacher satisfaction. It consists of 21 questions divided into 7 sub-categories: "Self-efficacy", "Performance Expectations", "Functionality", "Content Features", "Interaction", "Learning Climate" and "Satisfaction with Learning". Both questionnaires the questions' type is a 7-point Likert rating scale with the following anchors: "1" stands for "Completely Disagree" and "7" for "Completely Agree". At the end of the treatment period teachers, were asked to complete an online questionnaire, comprising of the above two questionnaires. The statistical analyses were performed using the SPSS 22 statistical package and the level of significance was set to 0.05.

Moreover, after the completion of the activity the qualitative method of a semi-structured interview was performed. The interview covered the aspects of regulation and evaluation of the students by the teacher. The interview comprised eighteen (18) open-ended questions, to which the teachers could answer without any restrictions whatsoever. The reasons for this choice are as follows: (a) The participants are able to express their opinions using more words, which adds a certain clarity to their answers (Creswell 2002) and (b) the main purpose was to investigate the subjective teachers' opinion.

5 Results

5.1 Usefulness, Ease of Use, and Level of Satisfaction

Descriptive measures of central tendency, such as mean and median, as well as frequencies, were estimated for each Likert-type item. Moreover, descriptive measures of dispersion were also investigated for each item, making the assumption that Likert-type variables can have internal order of numerical importance. The central tendency and dispersion measures estimated for the teachers' answers are being presented in Table 1.

As it can be seen, the teachers' answers are quite satisfactory. The GIANT's usability has also been evaluated in total, based on the three aforementioned

Table 1. Descriptive resuls for usability variables

Category	Minimum	Maximum	Std. Deviation	Mean
Self-efficacy	3.00	6.00	0.77	4.76
Performance Expectations	3.66	6.66	0.91	5.17
Functionality	3.66	6.66	0.80	5.09
Content Features	4.0	7.0	0.74	5.50
Interaction	3.33	7.0	0.86	5.30
Learning Climate	3.66	7.0	0.85	5.47
Satisfaction with Learning	4.0	7.0	0.76	5.78
Total Satisfaction	*3.61*	*6.76*	*0.81*	*5.29*
Perceived Usefulness	*3.66*	*7.0*	*0.82*	*5.23*
Perceived Ease of Use	*3.83*	*6.83*	*0.79*	*5.35*

dimensions, by averaging the responses of teachers on the individual questions which correspond to each dimension. As indicated by the results of the analysis, the teacher' overall satisfaction was positive, since the mean value (M) for the "Total Satisfaction" dimension is 5.29, with standard deviation (SD) = 0.81. Moreover, it could be argued that the teachers did not encounter any major difficulties in comprehending and using the graphs, since the mean value for the "Perceived Ease of Use" dimension is equal to 5.35, with SD = 0.79, while on the same time, teachers perceived the GIANT toolkit as a useful tool, with M = 5.23 and SD = 0.82, for the dimension "Perceived Usefulness". Additionally, students were also asked to write down the three most positive, as well as the three most negative aspects of the G.I.AN.T.'s graphs. Regarding the positive elements of the graphs, they were considered simple and interesting. Furthermore, according to the students' comments, the information presented could help them improve, as well as help their classmates. Lastly, concerning any negatives, these are pertinent to the lack of any interpretation instructions for the IA graph information. Their comments were then taken into consideration when improving the graphs and suggesting extensions.

5.2 Semi-structured Interviews Results

Proceeding with the analysis from the semi-structured interview, 87.5% of the target group stated that the tool helped them significantly in the way they accessed data, and that they used the tool to regulate the activity. Additionally, having access to these data helped them decide whether to intervene or not to the activity. In their own words: *"The categorization of the data lessened my workload towards drawing conclusions on the students' contribution."*, *"The data and especially graphs depicting quantitatively the students' actions, lead to effective decision making."* As a matter of fact, the sample in its entire-ty agreed that the presence of graphs was beneficial in orchestrating the activity, as *"they show exactly the level of the students' participation, both individually and group-wise,"* and that is why they said that their first thing to do after logging in the blog was to check the graphs. Subsequently, 68.7% reported that they became more aware of the progress of their students, which helped them make sure that they stayed on the right track, and only 21.8% said that they encountered difficulties in interpreting the awareness graphs, owing to the "*quantitative nature of the graphs, which is very clear and accurate*".

In contrast, the striking majority of the sample (71.8%) mentioned that the use of the GIANT toolkit was time-consuming, since it was mandatory for them to constantly monitor the blog, and every comment or post had to be assessed in a qualitative way. "*The benefits, however, are significantly more compared to the amount of work that needs to be done, making the time that is needed worth it in the long run*," argued one of the teachers.

To continue, 75% believed that GIANT was beneficial in attaining the learning goals set by the teachers themselves, because of the "*collaborative nature of the activity and the fact that the spotlight was on the students and not the teacher.*" They were also amazed to see that students who did not usually participate in the classroom took upon a more active role in this activity. Moreover, there was a positive stance with regard to the evaluation of students. More specifically, 90.6% mentioned that they took the automated evaluation feature of the toolkit into account, citing the "*accurate evaluation of every student's activity, based on the criteria and the final goal we set from the very beginning.*" In general, this feature contributed to measuring learning aspects which could not be measured otherwise ("*e.g. the level of collaboration among students*"), but once again the need for qualitative evaluation was underlined, since the teacher had to assess all the posts and comments one by one, "*by reading all the students' actions thoroughly*." This affects the credibility of the automated evaluation feature too in that the results "*were reliable only in terms of participation, while performance was measured strictly by the teacher*."

The majority of the participants praised the GIANT toolkit in the sense of "*enhancing collaborative learning, as the students were acting as part of a group, where exchanging ideas, receiving feedback, mutual trust, and working together towards a common goal were enhanced*." Last but not least, 71.8% of the interviewees who were asked whether managing the submitted work was easier gave a neutral answer. They said that, on the one hand, it was easier to find all the individual/group work "*just by pressing a button*", but, on the other hand, they had to *"read as many as approximately 100 posts and comments, which was tiresome and required a lot of time."*

6 Results

6.1 Discussion

According to the results of the PUEU questionnaire and those of the semi-structured interview, we can conclude that GIANT is useful to the teachers' work, and increase their efficiency and performance. In addition, it is useful for data collection relevant to the activity, which gives teachers the opportunity to decide whether and how they should intervene and support their students. As far as the new feature of the automated evaluation is concerned, the answers that the teachers provided to the interview suggested that it is useful only in a quantitative way, and not in a qualitative one. Hence, a teacher using it can see the level of participation and interest of students, but not the quality of their work. For the latter, it still seems necessary for the teacher to correct answers in the conventional way. Moving on to the overall usefulness of the IA toolkit, the findings of the PUEU questionnaire indicated a positive opinion, as the majority thought of GIANT as an easy to use tool, which allows for clear interaction with its user. Given that only 30% of the sample had previous experience using blogs, the results are quite encouraging. Finally, the "Teacher Satisfaction" questionnaire showed positive attitudes from the sample, as all sub-categories (besides "Effectiveness of the System") scored above the positive threshold. As a number of answers in the interview suggested, GIANT is very comprehensible in its data presentation, creates a pleasant learning atmosphere and stimulates interest and interaction among participants.

To sum up, the findings of the questionnaires and interview indicated that teachers are generally willing to adopt this kind of technology within in order to support the learning activities in theoretical courses, since GIANT provided them with information that would not be accessible in a traditional classroom. Not only was GIANT deemed satisfactory in terms of its use and ease to control, but it also proved to be useful in regulating the activity, making it easier for the teacher to decide if he/she should intervene or not, and in evaluating the work of students at the end of the activity. The automated grading algorithm implemented in the toolkit, in particular, facilitated the teachers in terms of assessing students' overall participation. Nevertheless, the fact that it adds to the already heavy workload of teachers should not be ignored, nor should the need for a qualitative kind of assessment to evaluate students' level of performance be discarded as less significant.

6.2 Future Outlook

In the future, the addition of qualitative analysis, such as the analysis of certain words or phrases, would be really useful to the teachers, since it would save them a great deal of work time. Alternatively, administrator rights could be given to them so that they can approve only the constructive posts and comments, reducing, thus, the amount of submitted work to be evaluated in the end. Finally, new case studies should be conducted in different educational contexts, with a larger sample, in order to further validate the findings of this paper.

References

Bratitsis, T.: Development of flexible support tools, asynchronous discussions, through analysis of interactions between participants, for technologically supported learning. Doctoral thesis, TEEPES, University of the Aegean (2007)

Bull, S., Kay, J.: Metacognition and open learner models. In: The 3rd Workshop on Meta-Cognition and Self-Regulated Learning in Educational Technologies, ITS2008, pp. 7–20 (2008)

Chondrouli, V.: Interaction analysis to support self-regulated and collaborative learning, using computers. Diploma thesis, Aristotle University of Thessaloniki, Department of Informatics (2015)

Creswell, J.W.: Educational Research: Planning, Conducting, and Evaluating Quantitative and Qualitative Research. Prentice Hall, Upper Saddle River (2002)

Curtis, D.D.: Defining, assessing and measuring generic competences. Doctoral dissertation, Flinders University of South Australia (2010)
Dillenbourg, P., Fischer, F.: Basics of computer-supported collaborative learning. Zeitschrift für Berufs- und Wirtschaftspädagogik. **21**, 111–130 (2007)
Dimitracopoulou, A.: State of the art on interaction analysis: "Interaction analysis indicators", ICALTS JEIRP Deliverable D.26.1. Kaleidoscope network of excellence (2004)
Galanakis, G., Papadopoulos, S.: "Blogs", Project in the course "Social Networking Technologies". Aristotle University of Thessaloniki (2012). https://learn20.wikispaces.com/Blogging
Kaendler, C., Wiedmann, M., Rummel, N., Spada, H.: Teacher competencies for the implementation of collaborative learning in the classroom: a framework and research review. Educ. Psychol. Rev. **27**(3), 505–536 (2015)
Kim, H.N.: The phenomenon of blogs and theoretical model of blog use in educational contexts. Comput. Educ. **51**, 1342–1352 (2008)
Lin, W.-J., Yueh, H.-P., Liu, Y.-L., Murakami, M., Kakusho, K., Minoh, M.: Blog as a tool to develop e-learning experience in an international distance course. In: Proceedings of the Sixth IEEE International Conference on Advanced Learning Technologies (ICALT 2006) (2006)
Michailidis, N., Nathanailidis, C., Papadopoulos, I., Tsiatsos, T.: Teachers' support using interaction analysis and visualization tools in educational group blogging. In: Proceedings of 2013 IEEE Global Engineering Education Conference (EDUCON 2013). IEEE Computer Society, pp. 790–797 (2013)
Spyrouklas, A.: Supporting teachers in computer-aided collaborative systems through Learning Analytics tools and improving the tool G.I.AN.T. Diploma thesis, Aristotle University of Thessaloniki, Department of Informatics (2016)
Theodoridis, D., Triantafyllidis, T., Tsiridis, P.: Developing a tool for analyzing co-interactions in blogs-wordpress. Diploma thesis, Aristotle University of Thessaloniki (2013)
Van Diggelen, W., Janssen, J., Overdijk, M.: Analysing and presenting interaction data: a teacher, student and researcher perspective. In: ICLS 2008 Proceedings of the 8th International Conference on International Conference for the Learning Sciences - Volume 3 (2008)
Van Leeuwen, A.: Teacher regulation of CSCL: exploring the complexity of teacher regulation and the supporting role of learning analytics. Interuniversity Center for Educational Research (2015)
Wu, J.H., Tennyson, R.D., Hsia, T.L.: A study of student satisfaction in a blended e-learning system environment. Comput. Educ. **55**(1), 155–164 (2010)

Lessons Learned from Synchronous Distance Learning in University Level at Congo

Apostolos Gkamas(✉) and Maria Rapti

University Ecclesiastical Academy of Vella,
P.O. Box 1144, 45001 Ioannina, Greece
gkamas@aeavellas.gr, m_rapti@yahoo.gr

Abstract. Multimedia and hypermedia had and still have a tremendous impact on the evolution of educational software. Internet based systems have been shown to be useful tools for supporting educational communication for teachers and students. In this paper, we analyze and present in detail the lessons learned during Synchronous Distance Learning in University level at Congo. The Synchronous Distance Learning took place between Greece (where the teacher located) and Congo (where the students located). In addition, we present the pedagogical and technical challenges during the above Distance Learning sessions between a European county and an African country.

Keywords: Synchronous Distance Learning · eLearning · eLearning in Africa

1 Introduction

The higher education sector in Africa faces challenges [15] related to critical shortage of quality faculty, limited capacity of management, limited financial support and limited infrastructures. In addition, there are limited capacity of research, knowledge generation and adaptation capabilities and problems in meeting increasing demand for equitable access.

Across Africa and disciplines, on average, only 70% of the required faculty positions are filled, and in some departments, this is only about 30–40%. Not less than 40% of the faculty in many universities in Africa is near retirement age, and over 30% of faculty sent overseas for training fail to return.

Today we can register a growing public interest in the Internet and especially in the World Wide Web. At the same time, the computer networks (which are becoming increasingly fast with the use of new technologies e.g. as telecommunications, computers, broadband networks, mobile networks, smartphones, tablets etc.) have enabled academic institutions all over the world to provide a flexible and more open learning environment for students. The above have as result a tremendous expansion in use of distant learning. In addition, the subject of distant learning and collaboration has engaged researches all over the world.

Higher education in sub-Sahara Africa [2, 14] is facing a critical challenge to meet new demands for the new century. The people who seek access to education at all levels - primary, secondary, and tertiary - will increase. Especially in tertiary – university level studies, Africa needs an educational environment that would be more

M. E. Auer and T. Tsiatsos (Eds.): IMCL 2017, AISC 725, pp. 522–530, 2018.
https://doi.org/10.1007/978-3-319-75175-7_51

responsive to challenges confronting the continent. An appealing approach to provide tertiary education is Distance Education. Distance Education makes it possible for students anywhere who have Internet and Web connections to enroll in online courses. Especially in sub-Sahara Africa the potential of Distance Education in tertiary education is great but many challenges must be meted in order to provide Distance Education.

Even though the application and use of information technology in education in sub-Saharan Africa has been severely underutilized, over the past few years there has been tremendous growth in the use of information technology. There have been pioneering efforts in Botswana, Madagascar, Namibia, South Africa, Tanzania, and Zimbabwe to apply information technology to higher education. Countries such as Cote d'Ivoire, Togo, and Congo are joining the Distance Education bandwagon by establishing pilot virtual programs.

Distance Education offers several advantages over the traditional educational system, including:

- Virtual access to faculty in higher institutions around the world
- Introduction of new interactive pedagogical techniques (e.g., more hands-on learning opportunities, independent research, less reliance on rote memorization)
- Creation of virtual institutions and linkages where resources could be shared by people and organization in physically unconnected places.

Distance Education requires teams of people performing different tasks and working at different levels to accomplish a common goal. Some of the staff might operate at the national level while others will be working at the local and institutional levels. The processes of recruitment, training and retention of staff should be well articulated and coordinated to ensure maximum efficiency and effectiveness. Some elements of staff development should include training for succession, sustainability and renewal mechanisms. Many of the Distance Education projects in Africa have inadvertently missed some of these elements.

Especially for the Republic of Congo faces a serious lack of telecommunication infrastructure [9], which hinders the development of several social sectors such as education.

This paper presents the pedagogical and technical challenges during the teaching of "Christian Catechism and Christian Pedagogics" course using Synchronous Distance Learning technologies. The teacher location was in University Ecclesiastical Academy of Vella in Ioannina, Greece and the student location was in Orthodox University of Congo in Kinshasa, Congo. The language of the course was French.

This paper has the following structure: The next section presents our motivation. The following section presents the various challenges we identified. In the next section, we present the used approach and after that we present the results evaluation. The last section concludes our paper.

2 Motivation

Distance Learning could be regarded as the process of learning with the use of Telematics that is the combination of telecommunication, information and multimedia technology and its services [5]. In such a scenario:

- All the interactions among students, teachers and instructional material, which are essential for the instructional process, can be implemented
- The information and the knowledge, which are essential for the instructional process are accessible and readable
- The place, time and the pace of learning are flexible.

Distance Learning has as target the development and promotion of special methods and techniques for the increase of the quality, the effectiveness and the suppleness of the learning. The Distance Learning has two main results:

1. The educational: The improvement of the existing learning methods and the development of new learning methods
2. The technological: The provision with new Distance Learning methods with the use of Information and Communication Technologies (ICT).

Distance Learning has the following main goals:

- The development of learning environments and methods suitable for the use of information technology to different learning environments
- The improvement of the organization environment, in which these new methods are applied, and the quality and manageability of the multimedia applications and the real-time services
- The encouragement achieved is recognizing the quality characteristics obtained through teaching with the use of new Distance Learning technologies and services
- The encouragement of the recognition of the quality characteristics which are gained from learning with the use of new Distance Learning technologies and services.

The last years we notice a shift in the training delivery. This gives rise to the need to implement tools that support Distance Learning (asynchronous learning, synchronous learning and Computer Support Collaborative Work for Learning - CSCW/L). In the aSynchronous Distance Learning the student selects the time, the duration and the pace of the lesson. During the synchronous lesson, there is live interaction between the participants (the teacher and the students). The CSCW/L functionalities include application sharing, bulleting boards, chat, e-mail and sharing workspace.

A Distance Learning environment combines various instructional scenarios such as collaborative learning and education with or without the live presentation of the Professor.

A Distance Learning environment provide a common environment for the implementation of all the above scenarios and the way to success in specific educational targets such as:

- The renewal of the pedagogical methods and environments in the Educational Institutes
- The incentives for the diffusion of information among the Educational Institutes in the World
- The encouragement for the collaboration
- The motivation of the students with the use of effective and modern equipment for the lesson
- The effective transmission and distribution of the instructional material to the students.

Besides the benefits of Distance Education, Distance Educations demands a great deal of personal sacrifice from the students. It requires students to have effective self-motivation, good learning skills, and the capability to learner autonomous [13]. There are incidents reported in several research studies of students dropping out Distance Education course due to the above demands and the other difficulties of every day live and the many roles may have in their live especially for developing counties like Congo. These roles bring with them a lot of challenges that will negatively affect students' studies. Faced with inadequate or no counselling students may fail to cope with studies resulting in low motivation to learn. They may therefore select to dropout or suspend studies. However, dropping out or suspension of studies has negatives social effects. They may be regarded as failures by the society in which they live. It is clear that, when counselling is not provided or is inadequate, student learning is curtailed resulting in low motivation to learn. As result the appropriate counselling is very important for the support of the Distance Education students and the successful completion of their studies.

3 Challenges

This paragraph presents the pedagogical and technical challenges during the teaching of "Christian Catechism and Christian Pedagogics" course using Synchronous Distance Learning technologies. The teacher location was in University Ecclesiastical Academy of Vella in Ioannina, Greece and the student location was in Orthodox University of Congo in Kinshasa, Congo. The language of the course was French.

Distance Education systems demand substantial capital investment at the initial stages in order to establish specialist facilities for the design, production and delivery of programs. It requires considerable investment in terms of human, physical, financial and technological resources well in advance of the enrolment of students on the courses. It also requires sufficient lead-time to allow for proper planning, preparation and distribution of learning materials, and establishment of learner support services. As far as possible, a Distance Education institution should make maximum use of existing resources and facilities, wherever they are to be found in the country, and should strive to work closely with other institutions and agencies within government and outside. The product of that investment and its economic efficiency can only be measured over several years when the programs will be reaching substantial numbers of students.

There are important technological constraints that hinder Distance Education especially in Africa. In Africa, telephone and other communication infrastructures outside of major cities remains inadequate. Connectivity beyond major capital cities poses a potential problem in creating a national Distance Education strategy. In addition, ISP services are expensive in Africa. In conclusion, access to connectivity remains one of the major challenges in Africa.

Moreover, textbooks and other printed materials certainly would still be part of the curriculum. All the above require funds which many individuals and institutions simply do not have especially in African countries.

Another challenge is the lack of trained professionals to support the implementation of Distance Education. The effective use of Distance Learning technologies demands that faculty be properly trained in using Distance Education as a delivery mode. To date, few African scholars are familiar with teaching in an online environment. This situation poses a major challenge in introducing Distance Education on the continent.

The Association for the Development of Education in Africa [2] summarizes the Distance Education limitations in Africa as following:

- The lack of high level political support for Distance Education by political authorities in Africa
- The lack of recognition of Distance Learning by the public service in its assessment of employee qualifications
- The availability of professionally trained Distance Learning personnel
- The lack of follow-up and support programs
- Limited budgets
- Poor domestic infrastructure.

The main pedagogical challenges, which have been identified during the teaching of "Christian Catechism and Christian Pedagogics" course, are the following:

- Sometimes, not very clearly see the face of the other party in order to form a more correct impression of the reactions taking into account for example the body language or the face expression
- The fact that you have not a visual image of all listeners but only those who are in front of the camera, it creates problems regarding the right communication. A teaching is not based solely on language bus also to other reactions
- If a course is based only on the transmission of knowledge can be carried out in a satisfactory level but if the course has also a practical part (for example a laboratory) there are important issues
- The teaching method followed was that of suggestion and dialogue, but it was not able to immediately discern the causes why some students do not want to participate in the discussion
- It is difficult to put into practice a participative teaching model, to co-decided tasks etc.
- You cannot figure out who is tired or not, happy or unhappy, so that you can modify the used method.

4 Approach

The approach which was used in order to provide the Distance Learning course, was the following:

- The teacher located in University Ecclesiastical Academy of Vella in Ioannina, Greece was preparing each lecture of the course "Christian Catechism and Christian Pedagogics". The language of the course was French.
- In pre-defined time and date the teacher connected with the students located in Orthodox University of Congo in Kinshasa, Congo using the Skype software. The reasons for selecting Skype are the following:
  - It is a well-known, easy to use and stable videoconference software and
  - There is no need for eLearning trained professionals in Congo
- The Synchronous Distance Learning lectures was completed
- The teacher was providing educational material to the students using various Internet software (e.g. e-mail, file transfer, file sharing etc.)

Figures 1 and 2 show characteristic snapshots during the Distance Learning course "Christian Catechism and Christian Pedagogics" between Orthodox University of Congo in Kinshasa, Congo and University Ecclesiastical Academy of Vella in Ioannina, Greece.

Fig. 1. Image 1 during the lesson

Fig. 2. Image 2 during the lesson

5 Results Discussion

Didactic motivations push the person into action. Psychologists distinguish motivation into endogenous and exogenous. The endogenous motivation is the interest that the student derives from the process of learning to which it is subjected by the external environment when it satisfies the natural curiosity and prompts it to explore and discover. Endogenous incentives include motivations that create needs for knowledge, self-expression, self-esteem, cooperation, and social recognition.

Exogenous incentives include the means we use to increase the involvement of the students. Such incentives are usually praises and rewards, as well as any form of recognition of the student's effort to explore and create.

Based on the above-mentioned theoretical context regarding motivation, we believe that the Distance Learning Course creates quite a lot of motivation for the students. They can exchange thoughts, know new worlds, satisfy curiosity about learning and develop research interest. In the case of the Congo, the Distance Learning course satisfies students' learning needs to a great extent, because the third world's socio-economic situation does not provide opportunities for learning, research and self-expression. It is worthy to mention, the significant interest the students expect the lesson to begin, the way they are watching, and how they participate. There was a clear need for learning and immediate response to the requirements of the course. New technologies in this area have provided a lot.

We can summarize the benefits of the used approach in the following points:

- Communication with people who are away is one of the leading technology offers both cognitive and scientific and social development

- The communicative approach gives joy and satisfaction feelings because overcome the limitations of space and time
- A lesson via Synchronous Distance Learning gives you the feeling that you teach in a regular classroom, you feel the interaction. You can see faces, movements, listen proposals, send messages and feel quite the reactions of others.

6 Conclusion - Future Work

Multimedia and hypermedia had and still have a tremendous impact on the evolution of educational software. Internet based systems have been shown to be useful tools for supporting educational communication for teachers and students. In this paper, we analyze and present in detail the lessons learned during Synchronous Distance Learning in university level at Congo. The Synchronous Distance Learning took place between Greece (where the teacher located) and Congo (where the students located). In addition, we present the pedagogical and technical challenges during the above Distance Learning sessions between a European county and an African country.

The main technical challenges which have been identified are the following:

- Communication infrastructures outside of major cities remains inadequate
- Another challenge is the lack of a trained professionals to support the implementation of Distance Education
- The absence of clearly defined national Distance Education policies in most African countries poses another challenge
- Access to connectivity remains one of the major challenges in Africa. The cost of network access is very high comparing for example with Europe and if someone consider also the income of people in Africa the cost of network access is extreme high.

The main pedagogical challenges are the following:

- Sometimes, not very clearly see the face of the other party in order to form a more correct impression of the reactions
- The fact that you have not a visual image of all listeners it creates problem at the right communication
- If a course is based only on the transmission of knowledge can be carried out in a satisfactory level
- The teaching method followed was that of suggestion and dialogue, but it was not able to immediately discern the causes why some students do not want to participate in the discussion
- It is difficult to put into practice a participative teaching model
- You cannot figure out who is tired or not, happy or unhappy.

Our future work includes the continuation of the cooperation between University Ecclesiastical Academy of Vella in Ioannina, Greece and Orthodox University of Congo in Kinshasa, Congo in providing Distance Learning courses using more advance ICT (Information and Communication Technologies) equipment.

References

1. Andrew, B.: A study into the future use of synchronous and asynchronous technologies in the distance ESL education area with an emphasis on collaborative learning. In: Proceedings of ED-MEDIA 97 and ED-TELECOM 97, Calgary, Alberta, Canada, 14–19 June (1997)
2. Association for the Development of Education in Africa (ADEA), Tertiary Distance Learning in Sub-Saharan Africa, ADEA Newsletter, vol. 11, no. 1 (January–March), pp. 1–4 (1999)
3. Barker, B.O., Dickson, M.W.: Distance learning technologies in K-12 schools: past, present, future practice. TechTrends **41**(3), 19–22 (1996)
4. Bork, A.: Technology in education: an historical perspective. In: Muffoletto, R., Knupfer, N. (eds.) Computers in Education: Social, Political, and Historical Perspectives, pp. 71–90. Hampton Press, Cresskill (1993)
5. Bouras, C., Gkamas, A., Tsiatsos, T.: Distributed learning environment using advanced services over the internet. In: Third IASTED International Conference on Internet and Multimedia Systems and Applications, 18–21 October, 1999 - Nassau, Grand Bahamas (1999)
6. Brown, T.: The role of m-learning in the future of e-learning in Africa? In: 21st ICDE World Conference, June 2003, Hong Kong (2003)
7. Chi-hung, N.: Motivational and learning processes of university students in a distance mode of learning: an achievement goal perspective. The Open University of Hong Kong (2000)
8. Darkwa, O., Mazibuko, F.: Creating virtual learning communities in Africa: challenges and prospects. First Monday, [S.l.] (2000). http://firstmonday.org/ojs/index.php/fm/article/view/744. Accessed 05 Sept 2017. ISSN 13960466
9. Fall, B.: Survey of ICT and Education in Africa: Republic of Congo (Congo-Brazzaville) Country Report. InfoDev ICT and Education Series. World Bank, Washington, DC. © World Bank, License: CC BY 3.0 IGO (2007). https://openknowledge.worldbank.org/handle/10986/10672
10. Kinyanjui, P.: Distance education and open learning in Africa: what works or does not work. In: EDI/World Bank Workshop on Teacher Education Through, Distance Learning, Addis Ababa, Ethiopia (1998)
11. Mancillas, A.: Counseling Students' Perceptions of Counseling Effectiveness (2011)
12. Marcelle, G.M.: Strategies for including a gender perspective in African information and communications technologies (ICTs) policy. In: ECA International Conference on African Women and Economic Development, Addis Ababa, 28 April–1 May (1998)
13. Musika, F., Bukaliya, R.: The effectiveness of counseling on students' learning motivation in open and distance education. Int. J. Res. Humanit. Soc. Stud. **2**(7), 85–99 (2015)
14. Ololube, N., Ubogu, A., Elemchuku, D.: ICT and distance education programs in a sub-Saharan African country: a theoretical perspective. JITI J. Inf. Technol. Impact **7**, 181–194 (2007)
15. Yizengaw, T.: Challenges of higher education in Africa and lessons of experience for the Africa-US higher education collaboration initiative: a synthesis report for the Africa-U.S. higher education initiative (2008). http://www.aplu.org/NetCommunity/Document.Doc?id=1183. Accessed 21 July 2013

Case Study: Integrating Computational Thinking into the Introductory Course of Computer Science via the Use of the Programming Language Python

Steka Maria[1(✉)] and Thrasyvoulos Tsiatsos[2]

[1] Aristotle University of Thessaloniki, Thessaloniki, Greece
maria.g.steka@gmail.com
[2] Department of Informatics, Aristotle University of Thessaloniki, Thessaloniki, Greece
tsiatsos@csd.auth.gr

Abstract. The purpose of this study is the integration of computational thinking in the introductory course "Introduction to Computer Science" of Aristotle University of Thessaloniki. After the publication of Jeanette Wing's innovative article, computational thinking became of interest for the educational community and nowadays is considered as important as writing, reading and arithmetic. This research attempts to integrate the computational thinking skill in the introductory course of the Computer Science department. In order for this goal to be achieved and its results to be as good as possible, the related literature was looked into, for the lesson to meet, specifically the part of which is dealing with the computational thinking concepts, the formal and informal criteria that have been set. For the purposes of the research some presentations were developed based on four fundamental concepts of computational thinking, decomposition, pattern recognition, abstraction and algorithms. Also, a part of them was about Python, the programming language which was an accompanying tool of the lesson and consequently of computational thinking. Except from the presentations, two assignments were designed and developed in order to assess the progress of the skill in the participants and also a part of the final test was dedicated to that assessment, too. The results showed that there was no statistically significant difference in the performance of participants between the two tests/assignments conducted.

Keywords: Computational thinking · Python

1 Introduction

The main subject of this project is computational thinking and the way to include it in the course "Introduction to Informatics" that is being taught to first year students of the Computer Science Department of Aristotle University of Thessaloniki. To begin with, there was an excessive research on the bibliography that is related to computational thinking. The research was focus on what computational thinking is, why it is a necessary ability of the 21st century, what concept composes it, what approaches are followed in order to teach it etc. There were many articles taken into consideration

M. E. Auer and T. Tsiatsos (Eds.): IMCL 2017, AISC 725, pp. 531–541, 2018.
https://doi.org/10.1007/978-3-319-75175-7_52

about various ages that happened because computational thinking is still quite vague and the students' theoretical background varies when it comes to this ability. Afterwards, there was the design of the teaching material that was the product of the research's analysis. In order for the design of the material to be successful, the multimedia designing principles were followed, that let the students to learn essentially the meaning of computational thinking in depth, which could be reached during the courses. Moreover, the presentations' structure was based on Bloom's taxonomy and there were a self- assessment exercise in every subcategory, in order for the participants to understand their weaknesses and improve themselves. There was, also, another research considering the evaluation of computational thinking. The main purposes of which was the understanding of the designing of the evaluation process, the importance of the validity of it and the understanding of the existing evaluations that have been practiced and where they may have gone wrong. The result of this research was the evaluation process that was developed and implemented in this project and the results and outcomes come based on it.

2 Theoretical Background

2.1 What is Computational Thinking

As stated in [1, 2] Wing like others, in [3] have described computational thinking as a generic analytical approach for solving problems, designing systems and understanding human behavior. Although computational thinking is based on concepts related to computer use and computer science, it includes practices like problem representation, abstraction, decomposition, simulation, verification and prediction. Many researchers have defined it in their own way since then. Computational thinking, basically covers a wide range of abilities, which they include the formation of the problem, the abstractions and the formulation of the solutions, in ways that can be used by computers. These skills vary from algorithms and data structures to presentations and visualizations. It differs from other ways of thinking that are mentioned in educational bibliography, like critical thinking which has a much disciplined goal. On the other hand Wing's goal in her article is to use computational thinking to basic skills like reading, writing and arithmetic, the definition may be really limited and "tight" in comparison to more traditional definition in computer science, as stated in [6] in order to be acceptable outside of the range of this science. Basically, it is a type of analytical thinking, which has a lot in common with mathematical thinking, considering the ways by which someone can approach the solution of a problem. Additionally, according to Wing "Computational Thinking includes a variety of mental tools which reflect the depth of the computer science, it represents a universal applicable behavior and an ensemble of skills which everyone, not just computer scientists must be willing to learn and use, to think like a computer scientist it means a lot more than just be able to program a computer. It demands to think in multiple levels of abstraction". Taking as a given that computational thinking is going to be used everywhere in the future, it will affect everyone directly or indirectly. So questions arise like when and how it will be included in the educational teaching.

2.2 Similar Studies

For this paper, various researches were studied, as stated in [7–12], they focus in the development of computational thinking courses, mostly having as a subject the computer science. Their studies were important for researching the way of developing computational thinking courses and suitable techniques to teach the skill to the students, make proposals for optimization of the educational results, help avoiding probable obstacles and urge for development of computational thinking courses or the inclusion of it in already existing courses.

3 Methodology

As it was said before these skills are a necessary tool for everyone, regardless of their area of interest. However, the fact that abroad there are trying to include the skills in the educational programs, in the educational system of Greece there are substantial deficiencies. Therefore, because of this luck of background regarding the first year students, it was decided in this transitional and experimenting step to enter computational thinking in the already existing introductive course in order to familiarize the students with the concept.

3.1 Course Structure

The course was available for first year students, in the Computer Science department, once a week as a three hour class. The whole duration of the course was thirteen weeks and took place during the winter semester of 2016–2017. Five weeks out of the thirteen were dedicated to Computational Thinking and to the programming language Python, which was used as a complementary tool in understanding the concept. For the computation thinking part, a special educational material was created based on the research of the bibliography and a relevant book was used, considering the programming language Python, as stated in [14] which was suggested as complementary material to the students. During the course except for the lectures, two exercises where given concerning computational thinking and Python, the first one was given before the CT teaching was started and the second one after the CT teaching was completed, it happened so there would be a way to have results concerning the students' understanding of the subject, there would be further analysis about this later. Also the students were given one final project, a programming one, a part of which was related to the CT, the project's grade were calculated into the final course grade of the students. After the project was announced the students had the chance to express any questions, they may had through a forum that was created in order the students to converse about the project with the supervision of the professor's assistant. Finally, a part of the final exams was dedicated to CT. It must be noted that during the lectures the professor was referring to the definitions and made examples of CT so the students would be more familiarized with it and they would be able to comprehend it better always within the time given and the depth that the concept was analyzed.

3.2 Design of Presentations

Creation: methods and techniques
The design of the educational material was based on the cognitive theory associated with multimedia learning. The relevant principles were followed in the educational design so as to achieve the best possible learning from the participants. In particular, the design principles followed were: multimedia, contiguity, coherence, personalization and segmentation. Cognitive science explores intelligence and behavior, that is, how it represents the information that the brain takes, how it is processed and how it is transformed through the functions of the nervous system. This theory is based on a synthesis of fields, such as psychology, artificial intelligence, neuroscience, etc. and its main concern is the understanding of how to build and represent knowledge in human memory. Learning is an individual process of building meaning through experience rather than stellar memorization. According to Neiser, human cognitive functions can be described in computational terms. This is why the basic model of this theory resembles the processing system of a computer machine, that is learning (external representations) - knowledge processes (internal representations in the short-term memory processed by the brain) - knowledge (cognitive structures established in the long-time memory). There are several approaches to this theory, such as reverse engineering, which, from the low level (learning mechanisms, decision-making), is led to high (logic and design). There is also the Representational-Computational Theory of the mind (where the function of the mind is equated with that of the computer) and the theory of information processing (information input-sensory memory-working memory-long-term memory). This theory began in the '50 s and its founders were Miller, Bruner, Chomsky, Newell, Turing, and others, as stated in [13].

Objectives of the educational process
The objectives of the educational process were for the trainees to be able to define spherically what computational thinking is, based on the modules of it, which were analyzed in the corresponding presentations. Specifically, they should have the ability at the end of the semester to define what is decomposition, abstraction, pattern recognition and algorithms, and also to describe basic characteristics of these. Equally important was the understanding of where and how the modules are used, in order to basically comprehend, what computational thinking is and of what it is comprised. In addition to these they had to learn the basics of the programming language Python, which has been the basic tool of the present study.

Self-assessment and further study
In each presentation there is a self-assessment exercise on each subject that is explained, so that the trainees can understand more and cope with their weaknesses. Also for each subject, material is provided for further study, in order to give the participants the opportunity to deepen more in each piece of the Computational Thinking and Python.

3.3 Content of Presentations

Presentations created for the Computational Thinking regards the following areas of it, decomposition, pattern recognition, abstraction and algorithms. Also a part of them focuses on Python programming language. The selection of the specific themes from the overall scope of the Computational Thinking was due to the fundamental character of the aforementioned concepts, in order for the learners to construct a first knowledge base on the subject under consideration.

Decomposition

According to Polson and Jeffries, as stated in [20], dividing a problem into sub-problems is a generalized strategy, which is often recommended to solve problems. However, there are not many convincing methodologies available on how this heuristic approach can be applied to real problems. In other words, decomposition is a way of thinking about problems, algorithms, objects, processes and systems from the point of view of the parties that compose them. The individual parts can then be understood, solved, developed and evaluated separately. This makes complicated problems easy to solve and large systems easier to design. Decomposition is one of the four cornerstones of the Computer Science. It also plays a very important role for the Computational Thinking, as it is one of the key features in it. If a problem is not decomposed, it is much harder to solve. Dealing with many different stages at once is much more difficult than dividing a problem into smaller segments and solving one segment at a time. Dividing the problem into smaller parts means that every minor problem can be considered in more detail. Similarly, one can more easily understand how a complex system works using decomposition. In the corresponding presentation which is dealing with the issue of decomposition a first explanation of the technique is made and where and how it is used as well.

Pattern Recognition

What is defined as a pattern? According to Watanabe, as stated in [16] the pattern is defined as "the opposite of chaos, as an entity, which is indefinitely defined and to which could be given a name." For example, a design could be a picture of a finger, a handwritten calligraphy word, a human face, or a speech signal Taking into account a pattern, its recognition/classification may consist of one of the following two tasks, as stated in [16], first of the supervised classification, and secondly of unassigned classification. The role of recognition standards in computer applications are equally important with that in the context of Computational Thinking. Standards help in solving problems when one can recognize them into problems to solve, as the same part of the solution can be reused in them. Forming the idea of what one can expect is a way of finding patterns. The more one looks, the more patterns he or she will discover in nature, in computer objects and in processes. Once a pattern is recognized, the other computational skills of the individual can then be exploited to understand the significance of that. In the context of the presentations, a first palpation of the recognition of standards is made, so that the trainees understand their importance and usefulness. Examples of pattern recognition are given in the presentation created for the corresponding module.

Abstraction

Abstraction is seen as a basic concept in computer science. According to the ACM/IEEE report in the Computer Science Curriculum in 2008, it is mentioned as one of the recurring issues that students will have to deal with, during an undergraduate program in computer science. The role of abstraction in Computational Thinking is particularly important, but the teaching of such concepts is quite difficult. Hazzan, as stated in [17] in his article pointed out that simple lessons for these ideas would make them potentially difficult and irrelevant to students. In order to the trainees to understand ideas, such as abstraction, they must first see and feel them, as well as understand their use. Based on this perspective, there are three suggested ways in which they can teach such ideas to trainees. First, teachers should explain the process that characterizes the concept under consideration. Second, the teaching process should incorporate stochastic processes, and thirdly students throughout the teaching should be active, that is, they should do and reflect on the concepts they are learning. When creating the content of the presentations, the above mentioned methods were taken into account and especially in the section of abstraction.

Algorithms

The concept of the algorithm, according to the IEEE ACM curriculum in 2001, is one of the most basic and fundamental concepts of computer science and is also a concept often introduced at the beginning of introductory courses. The concept becomes known to novice students as a logical and systematic method for solving a problem. Solomon argues that introductory courses in computer science have lost focus because they require specific skills in programming languages to the detriment of basic algorithmic comprehension. Algorithmic in-depth understanding can be applied in many areas of study. By focusing on this basic skill, students develop the ability to move between the intuitive understanding of the problem and its formal description. In addition, trainees have the opportunity to improve their capacity to represent the descriptions of a problem and its solutions, as stated in [7]. According to Yinnan, and Chaosheng, as stated in [4], Computational Thinking is a way of thinking related to problem solving. In other words, it expresses the problem-solving process in a programmatic or mechanized way. As a rule, the problem-solving process is divided into five steps: problem presentation, problem analysis, linking, behavioral selection, and reflection testing. When trainees are faced with problems to solve, according to their acquired knowledge, they have the ability to ask questions, propose an algorithm, and finally perform procedures to test the effect of their solution.

Python

The programming language was chosen in order to allow immediate comprehension on the computing principles. According to Hambrusch et al. as stated in [5] the aim is to enable students to write meaningful programs within a short time without focusing on - and struggling with - the details of the language. Python was chosen as a programming tool in the study mentioned, but also in the present because of its interactive environment, the ability to allow beginner programmers to write fast and non-trivial programs, the adoption of it from many scientific communities, and the support that it offers for many specialized libraries. Python can be run efficiently, fact that makes it a good option, not only for small-scale experiments but also for larger data sets as well as for computational problems. In order to serve the purpose for which the programming language is needed in the introductory course "Introduction to Information Technology", Python was chosen. The main reason for this choice is the positive reputation surrounding it, in terms of teaching, basic concepts and programming structures. Finally, the Python development environment, IDLE, helps the tutor to control all the work done in a lab through the output that appears in the corresponding window.

4 Assessment

In order to evaluate the under study research, the evaluations proposed by the NSF, as stated in [18, 19] and other evaluation methods, from studies focusing on corresponding research questions, as stated in [21–24], were examined. However, for reasons to be further analyzed, it was preferable to create a new assessment test that was more responsive to this research needs.

4.1 Assessment in the Present Study

The selection and exploration of previous studies was done in order to give a comprehensive picture of the assessment frameworks that exist on Computational Thinking and as a way of building a more appropriate technique according to the research. While the ages addressed by some of the assessments or concepts examined, do not converge with those of the present study, they are nevertheless a basis for building an assessment that meets the requirements that have been set. Also, as mentioned above due to the incompletely structured nature of the assessment within the Computational Thinking field, there are no clear boundaries and structure that should be followed. Furthermore, because the Computational Thinking is a part of the "Introduction to Information Technology" lessons and not an integrated lesson, the design of the evaluation was not implemented with PACT, as stated in [19] and was also chosen not to use FCS1, as stated in [18] because the Python programming language was important for the lesson. Additionally, evaluations related to visual programming were excluded, due to the nature of the selected language. However, most of Brennan and Resnick's suggestions were followed to build the best possible Computational Thinking evaluation. In more detail, two tests were developed to evaluate the learner's understanding in the Computational Thinking, in particular in the decomposition, abstraction, recognition of standards in both the algorithms and the Python programming language. The first test was given at the beginning of the semester and the second at the end, in order to produce results on the progress of the students in the field of Computational Thinking. There was also a part in the final examination of the course dedicated to Computational Thinking in order to not only to make relative comparison of the two tests but also with the whole lesson. The structure of the work consisted of two parts, the first concerned the Computational Thinking and the second the participants' programming skills. The part about Computational Thinking consisted of a blank template in which the students were asked to complete the process of Computational Thinking that they followed to solve the problem. While the programming piece was about resolving the problem in Python code. The differences between the two tests are that in the first one, there was a solved exercise and a completed model in order to understand the process and the steps to be followed and that the problems assigned to them were not the same. However, the problems were of the same difficulty in order to avoid problems in exporting the results or to minimize as much as possible the questioning of them. Finally, in the design of the completed template the multimedia and proximity principles, as well as the fade out technique were followed, but also in the solved and unsolved problem, there was no intermediate intersection.

5 Results

The results to be analyzed below are the result of the assessment designed for the part of Computational Thinking, in the context of the course "Introduction to Information Technology". The program used to analyze them is IBM SPSS Statistics 22.

5.1 Participants

The students who participated in the research were first-year students of the Department of Informatics of the Aristotle University of Thessaloniki. The total number of students enrolled in the course was four hundred and eight, of which one hundred and fifty-five were active in the class. Of the total number of active learners, the sample that participated in this study was consisted of one hundred and nineteen people. The people who participated, were not selected against the rest on a basis of some specific criteria, they were selected because they participated in the pre- and post-test distributed to all active participants. The students attended consisted of 84.03% of men and 15.97% of women.

5.2 Analysis of the results

The students participated in a pre- and post-test in order to investigate any improvement with regard to their competence in Computational Thinking and in the final examination of the course, as mentioned previously. The results are shown in the corresponding figures.

According to "Fig. 1" it can be noticed that the average value of the first test is 88.25%, the second 90.87% and the final examination 70.02. The average price of work is high in both cases, with a slight difference, which suggests, however, that the students did well in both tests, with a slight improvement in the second. The results in the final exam are lower than those of the tests by about 20%, which means that the participants went better on the part of the Computational Thinking. The maximum grade of the first test is 100% and the minimum 35%, the second 100% and 40% respectively and the final exam 94% and 0%, however 0% is due to the non-attendance of the students in the final examination, so its interpretation is impossible. A paired sample T-test was then applied between the first and the second test in order to determine whether there is a statistically significant difference between the results. The use of the test was justified and it was shown that the mean values of the scores of the participants in the two tests did not show a statistically significant difference. Therefore,

Statistics

		CT 1st Test	CT 2nd Test	Final Exam
N	Valid	118	118	118
	Missing	0	0	0
Mean		88,25	90,87	70,02
Median		95,00	95,00	74,00
Std. Deviation		14,263	12,929	17,721
Variance		203,435	167,172	314,051
Minimum		35	40	0
Maximum		100	100	94
Sum		10413	10723	8262
Percentiles	25	80,00	87,50	65,00
	50	95,00	95,00	74,00
	75	100,00	100,00	80,00

Fig. 1. Tests and exam statistics

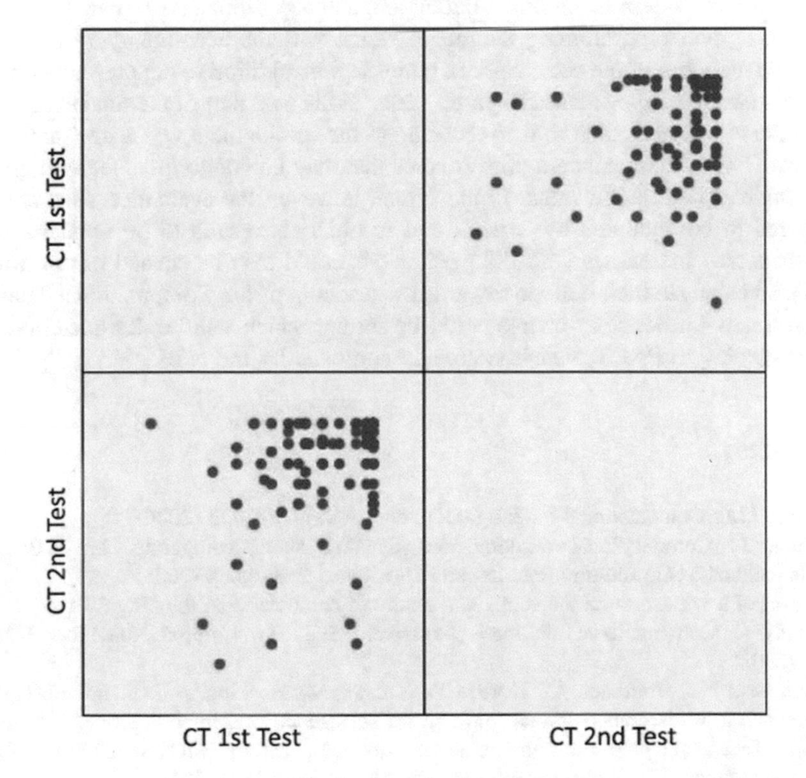

Fig. 2. Spreads diagram

there has been no statistically significant improvement in the Computational Thinking skills, possibly the already high level of students in the university led to this result. A paired sample T-test was also applied between the two tests and the final examination to see if there is a statistically significant difference between the results. Again, it was justified the usage of the test. The mean scores of the participants in the two tests compared to the final test showed a statistically significant difference. Therefore there was a statistically significant improvement in the skills of the Computational Thinking, in relation to the overall performance of the students in the course. In addition, the correlation between the two tasks was also examined. It turned out that there is a positive correlation, very important between the first and the second work. This means that as one of its average values changes, the other changes in the same way. In Fig. 2 their spreads diagram are appeared.

6 Conclusions

The aim of this work was to integrate computational thinking into the introductory course "Introduction to Information Technology" of the first year of Informatics of the Aristotle University of Thessaloniki. This goal came about due to the increasing demand for this skill and its importance in the 21st century. For research purposes, therefore, presentations have been developed on four basic concepts of computational thinking, decay, pattern recognition, deduction, and algorithms. Also, part of them was about the Python programming language, which was the accompanying tool of the lesson and therefore of the computational thinking. In addition to the presentations, two tests were designed to evaluate the participants' skills, and part of the final examination of the course was dedicated to it. According to the results there was a non-statistically significant difference in student performance that may be due to the already high level of the students who participated. In the future, however, the evaluation used could be considered to be validated by experts and in order its results to be unquestionable. Semi-structured interviews with the participants could also be carried out in order to observe and analyze their attitude towards the teaching of the Computational Thinking. Finally, a new introductory course could be set up, which will be designed and evaluated according to PACT, which is already approved by the NSF.

References

1. Wing, J.M.: Computational thinking. Commun. ACM **49**(3), 33 (2006)
2. Wing, J.: Computational thinking and thinking about computing. In: 2008 IEEE International Symposium on Parallel and Distributed Processing (2008)
3. Report of a workshop on the scope and nature of computational thinking (2010)
4. Bundy, A.: Computational thinking is pervasive. J. Sci. Pract. Comput. Noted Rev. **1**(2), 67–69 (2007)
5. Hambrusch, S., Hoffmann, C., Korb, J.T., Haugan, M., Hosking, A.L.: A multidisciplinary approach towards computational thinking for science majors. In: Proceedings of the 40th ACM Technical Symposium on Computer Science Education - SIGCSE 2009 (2009)
6. Hemmendinger, D.: A plea for modesty. ACM Inroads **1**(2), 4 (2010)

7. Qualls, J.A., Grant, M.M., Sherrell, L.B.: CS1 students' understanding of computational thinking concepts. J. Comput. Sci. Coll. **26**(5), 62–71 (2011)
8. Kafura, D., Bart, A.C., Chowdhury, B.: Design and preliminary results from a computational thinking course. In: Proceedings of the 2015 ACM Conference on Innovation and Technology in Computer Science Education - ITiCSE 2015 (2015)
9. Selby, C.C.: Promoting computational thinking with programming. In: Proceedings of the 7th Workshop in Primary and Secondary Computing Education on - WiPSCE 2012 (2012)
10. Kafura, D., Tatar, D.: Initial experience with a computational thinking course for computer science students. In: Proceedings of the 42nd ACM Technical Symposium on Computer Science Education - SIGCSE 2011 (2011)
11. Davies, S.: The effects of emphasizing computational thinking in an introductory programming course. In: 2008 38th Annual Frontiers in Education Conference (2008)
12. Philip, M., Renumol, V.G., Gopeekrishnan, R.: A pragmatic approach to develop computational thinking skills in novices in computing education. In: 2013 IEEE International Conference in MOOC, Innovation and Technology in Education (MITE) (2013)
13. Demetriades, S.N.: Learning Theories and Educational Software. Tziola, p. 302 (2014)
14. Manis, G.: Introduction to programming with Python programming language. SEAB (2015)
15. Bloom, B.S., Engelhart, M.D., Furst, E.J., Hill, W.H., Krathwohl, D.R.: Taxonomy of Educational Objectives, Handbook I: The Cognitive Domain. David McKay Co Inc., New York (1956)
16. Watanabe, S.: Pattern Recognition: Human and Mechanical. Wiley, New York (1985)
17. Hazzan, O.: Reflections on teaching abstraction and other soft ideas. ACM SIGCSE Bull. **40** (2), 40 (2008)
18. Tew, A.E., Guzdial, M.: The FCS1: a language independent assessment of CS1 knowledge. In: Proceedings of the 42nd ACM Technical Symposium on Computer Science Education - SIGCSE 2011 (2011)
19. Bienkowski, M., Snow, E., Rutstein, D., Grover, S.: Assessment design patterns for computational thinking practices in secondary computer science: a first look. SRI International (2015)
20. Polson, P., Jeffries, R.: Instruction in problem-solving skills: an analysis of four approaches. In: Segal, J.W., Chipman, S.F., Glaser, R. (eds.) Thinking and Learning Skills, vol. 1, pp. 417–455. Erlbaum, Hillsdale, NJ (1985)
21. Gouws, L., Bradshaw, K., Wentworth, P.: First year student performance in a test for computational thinking. In: Proceedings of the South African Institute for Computer Scientists and Information Technologists Conference on - SAICSIT 2013 (2013)
22. Park, T.H., Kim, M.C., Chhabra, S., Lee, B., Forte, A.: Reading hierarchies in code. In: Proceedings of the 2016 ACM Conference on Innovation and Technology in Computer Science Education - ITiCSE 2016 (2016)
23. Román-González, M.: Computational thinking test: design guidelines and content validation. In: 7th Annual International Conference on Education and New Learning Technologies, Barcelona, Spain (2015)
24. Brennan, K., Resnick, M.: New frameworks for studying and assessing the development of computational thinking. In: Annual American Educational Research Association meeting, Vancouver, BC, Canada (2012)

Automated Development of Physics Educational Content for Mass Individualized Education

Alexander Chirtsov[1,3], Sergey Sychov[2(✉)], and Alexander Mylnikov[2]

[1] University LETI, Professora Popova 5, St. Peterburg 197022, Russia
alex_chirtsov@mail.ru
[2] ITMO University, 49 Kronverksky Pr., St. Petersburg 197101, Russia
sergey.v.sychev@gmail.com, sasha_mylnikov@mail.ru
[3] Herzen State Pedagogical University of Russia,
6 Kazanskaya (Plekhanova) st., St. Petersburg 191186, Russia

Abstract. The article considers the base structure and mechanisms of developing science education constructors designed with the use of Physical Object-Oriented Models (PhOOM). This approach makes it possible to create a number of customized applications for natural science education. Some models for physics and chemistry courses are mentioned. PhOOM combines model demonstration, test problem design and easy adjustment by the user in one easily manageable product or online service for the purpose of engineering and natural science education. A further development of the approach is a system generating test problems, based on real-time simulation, for advanced plasma physics courses.

Keywords: PhOOM · MOOC · Physics · Generation of tests
Education automation

1 Introduction

The intensive development of computer and information technology is followed by its involvement in almost all areas of human activity. Active computerization in education replaces traditional approaches in a whole range of learning and teaching activities.

From the very beginning of the computerization of education in Russia, it was proposed that the implementation of electronic means must be preceded by a clear understanding of the specific advantages of the use of these means compared to traditional methods (Бутиков Е.И., А.С. 1992). According to these criteria, computer simulations are the most effective methods to support natural science education. They combine advantages of both experimentally and theoretically oriented teaching methods. A number of unique software packages which demonstrate and simulate physical processes in educational purposes were created in Russia by the end of the 90s (Butikov 1992).

The next stage of the development of e-learning tools of this type was a number of electronic constructors of virtual physical systems.

M. E. Auer and T. Tsiatsos (Eds.): IMCL 2017, AISC 725, pp. 542–549, 2018.
https://doi.org/10.1007/978-3-319-75175-7_53

This article contains a brief review of the results for a project having the goal to create easy-to-use virtual electronic constructors of virtual physical systems for in-class usage as well as for distant and mobile education. A further development of the project is software to facilitate chemistry teaching, which is considered too.

2 Physical Object-Oriented Modeling (PhOOM)

At the present time, most of electronic constructors exist in the form of resource-intensive software systems (Interactive Physics: Physics Simulation Software for the Classroom 2016; LabVeiw 2016). Most of them use Object-Oriented Modeling (OOM) (Kook and Novak 1991) - an approach based on the use of highly versatile but very formal concepts requiring special training of users. To use OOM one needs to model and design the algorithm and the working program.

Contrary to this, the set of ideas in the field of methodology, performance algorithms, architecture, and software implementation of electronic designs that simplify physics teaching is called physical object-oriented modeling (PhOOM) (Чирцов et al. 2015). PhOOM is a self-organizing system using physics laws included in the program.

The general algorithm of the program (Fig. 1) is a simple time series that sequentially transfers the focus of activity to all the program's objects that constitute the system. Each of these "elementary" objects requests other objects (capable of interacting with it) about their current state. Sequential iterations produce patterns of objects' own behavior at the current stage of the system's evolution based on the laws of physics firmly established for simple objects. It is not necessary that the user constructing this system has a prior knowledge or prediction about the behavior of the system.

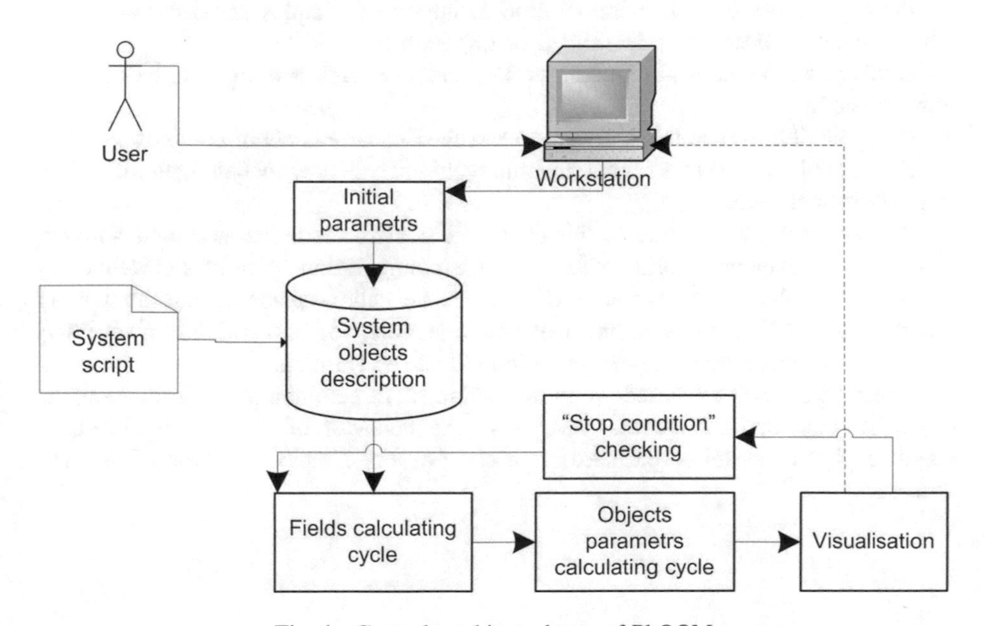

Fig. 1. General working scheme of PhOOM

There is a number of PhOOM models covering almost the entire course of classical physics created with the use of this approach:

"Particles in force field"

"Viewer of configurations of electromagnetic fields of complex sources"

"Rays Generator"

"Diffraction Pattern Generator".

With the methodology presented in (Чирцов 1995), the modern version of the electronic constructor of virtual physical systems does not require additional software installation and allows online use by untrained users. There two options: the first one is to use (with a change of parameters) ready-made computer experiments (predetermined scripts), the second one is when the user develops their own original models (Чирцов 2010). General working scheme of the physical constructor (PhOOM) is presented in Fig. 1.

The constructor design aims for the profound study of physics, having included such software objects as "relativistic particles", "radiation friction force", "field of arbitrary space-time configurations" and "user-customizable interaction laws". These virtual objects allow investigating of some hypothetical problems, that are not possible to be researched experimentally (Figs. 2 and 3). This approach of modeling is coexisting with a cognitive research presented by MIT and advocating Modeling Applied to Problem Solving (MAPS) (Pawl et al. 2009).

The main advantages of electronic constructors of virtual physical systems are:

easy adaptation to a variety of educational software programs and individual plans of teachers;

the ability to demonstrate a gradual approximation of theoretical models to reality (Fig. 3);

the opportunity to demonstrate a time evolution of complex physical systems, in which analytical description is difficult or impossible;

students are encouraged to learn, produce creative work and organize independent mini-research;

the possibility to use this software to test theoretical models of complex systems;

the possibility to visualize and quantitatively analyze new models with difficulties in experimental verification.

Another important feature of this electronic constructor is the automatic software generation of adaptive algorithms (described in configuration script files or defined by the user) that calculate the evolution of systems. This allows using the constructor as a means to automate the development of new e-learning content, which is a necessary component in the organization of mass individualized learning.

These algorithms are based on the following set of common physics equations for objects motion and interaction (Eqs. 1-3). The behavior of elementary objects of classical physics systems (particles) is based on very simple equations of classical dynamics:

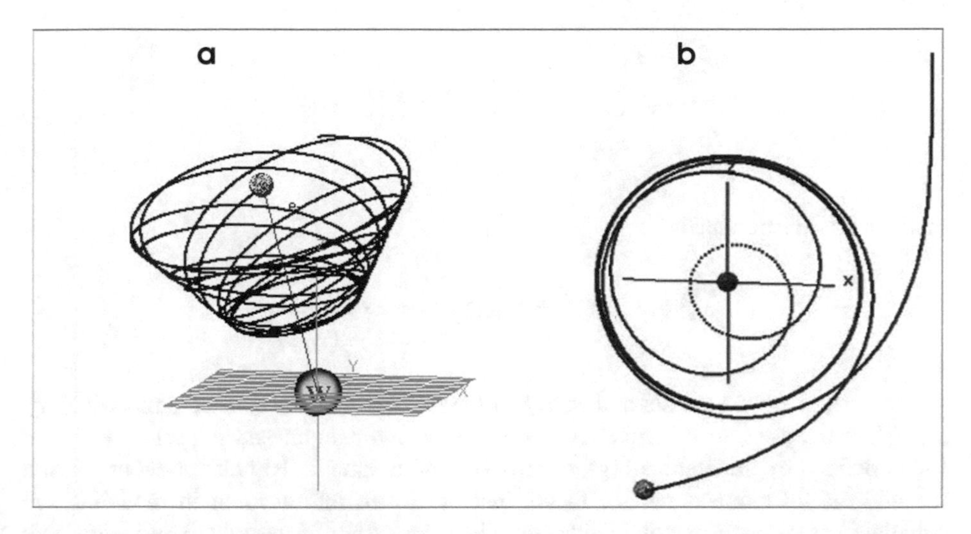

Fig. 2. Examples of using the possibility of overriding the laws of interaction between objects: a - demonstration of a hypothetical system "electron in the stationary magnetic monopole with a positive electric charge"; b - one of the possible forms of movement of nuclear forces corresponding to the Yukawa potential.

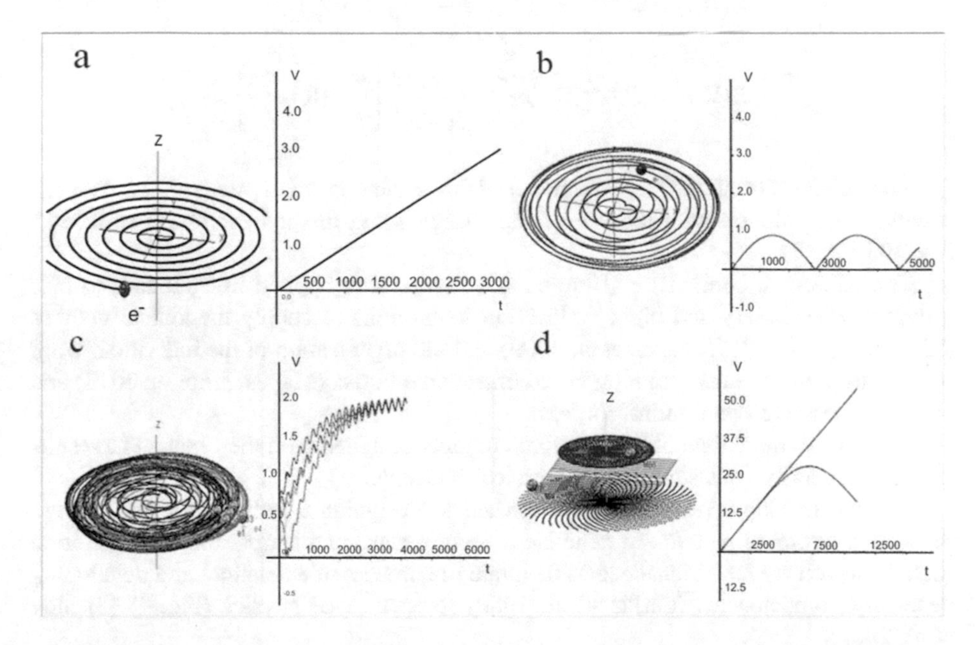

Fig. 3. Different trajectories for a charged particle in an accelerator dependent on the choice of a system physical model: a - classical movement in constant magnetic and alternating electric fields that are tuned to the resonance, b – a violation of resonance case, c - radiation friction effects, d - relativistic effects.

$$\begin{aligned} & m_j \frac{d^2\mathbf{r}_j}{dt^2} = \mathbf{F}_{j\Sigma}, \\ & \mathbf{F}_{j\Sigma} = m_j \mathbf{g}(\mathbf{r}_j, t) + q\mathbf{E}_j(\mathbf{r}_j, t) + \frac{q_j}{c}\left[\frac{d\mathbf{r}_j}{dt}, \mathbf{B}(\mathbf{r}_j, t)\right] - \\ & \eta \frac{d\mathbf{r}_j}{dt} + \sum_{j' \neq j} k(r_j - r_{j'}) \end{aligned} \tag{1}$$

or their relativistic analog:

$$\frac{d\mathbf{p}_j}{dt} = \mathbf{F}_{j\Sigma}(\mathbf{r}, t) + \mathbf{F}_{rad}, \mathbf{p}_j = \frac{m_j \mathbf{v}_j}{\sqrt{1 - (v_j/c)^2}}, \tag{2}$$

where mj и qj are the masses and charge of the particle with the current number j; rj, vj, pj, FjΣ correspond to its radius vector, velocity, momentum and impact of the total force, defined by gravitational (g), electric (E) and magnetic (B) fields, viscous friction (η) and elastic interactions (k). Fields that are taken into account in modeling are calculated as sums of external fields (set when the system is described) and additional contributions of a particle's interaction patterns:

$$\begin{aligned} & \mathbf{g}(\mathbf{R},t) = \mathbf{g}^{(внеш)}(\mathbf{R},t) - \sum_{j' \neq j} G \frac{m_{j'}}{|\mathbf{R} - \mathbf{r}_{j'}|^3} (\mathbf{R} - \mathbf{r}_{j'}), \\ & \mathbf{E}(\mathbf{R},t) = \mathbf{E}^{(внеш)}(\mathbf{R},t) + \sum_{j' \neq j} \frac{q_{j'}}{|\mathbf{R} - \mathbf{r}_{j'}|^3} (\mathbf{R} - \mathbf{r}_{j'}), \\ & \mathbf{B}(\mathbf{R},t) = \mathbf{B}^{(внеш)}(\mathbf{R},t) + \sum_{j' \neq j} \frac{q_{j'}}{|\mathbf{R} - \mathbf{r}_{j'}|^3} \left[\frac{v_{j'}}{c}, (\mathbf{R} - \mathbf{r}_{j'})\right]. \end{aligned} \tag{3}$$

The behavior of all the main elements of this system is strictly determined. We can calculate the solution with the use of the Runge-Kutta method for fourth order differential Eq. (1).

The electronic constructor (Чирцов 2010) has been applied in both face-to-face training in secondary and higher education institutions to supply theoretical courses (Микушев et al. 2015; Марек et al. 2009) and the organization of the individual work of students and in remote training based on seven e-books (Марек, Чирцов 2012) and MOOCs (massive open online courses).

There were more than 500 educational models designed with their help. The results obtained are fully consistent with theoretical concepts.

Further development of this approach led to designing a testing tool that process problem's common patterns to generate a variety tasks and scripts for demonstration models, which are later visualized to facilitate the process of education, and developing necessary competencies (Larkin et al. 1980) for course of physics (Fig. 4) (Sychov et al. 2017).

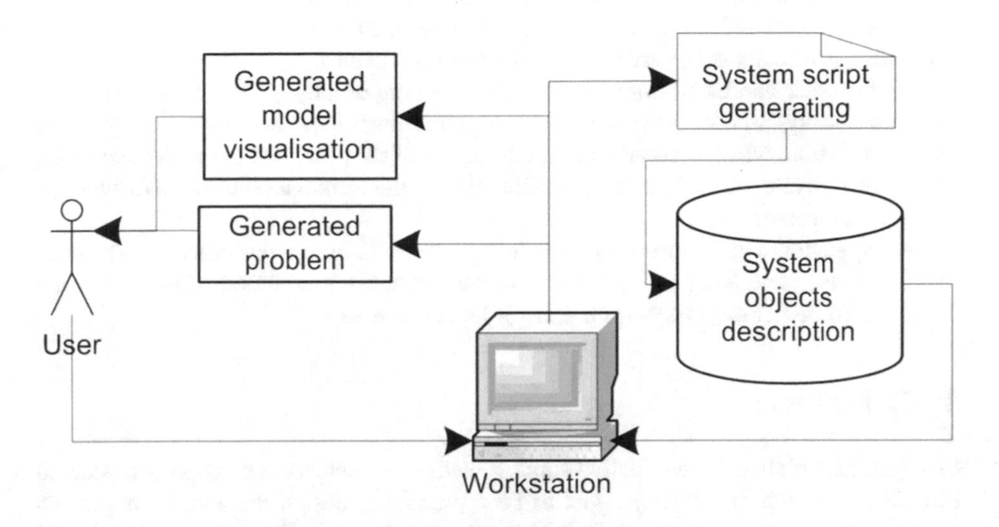

Unlimited variety of tasks in a process of common pattern based task generation allows for a group of students to have a number of tasks corresponding to the number of students in a group though preventing cheating.

3 Phoom Paradigm and Chemical Reactions Modeling

The idea of PhOOM-based electronic constructors to study chemistry was influenced by the constructor "Particles in force fields" and a series of efficient 3D models reflecting the behavior of an ideal gas.

A modern personal portable computer is powerful enough to operate with a PhOOM model of an ideal gas consisting of a few hundreds of particles in real-time as well as with a more resource demanding simplest model of a non-ideal gas in which an ensemble of "classical electric dipoles" that rotate and oscillate is placed in an external electric field (Fig. 4).

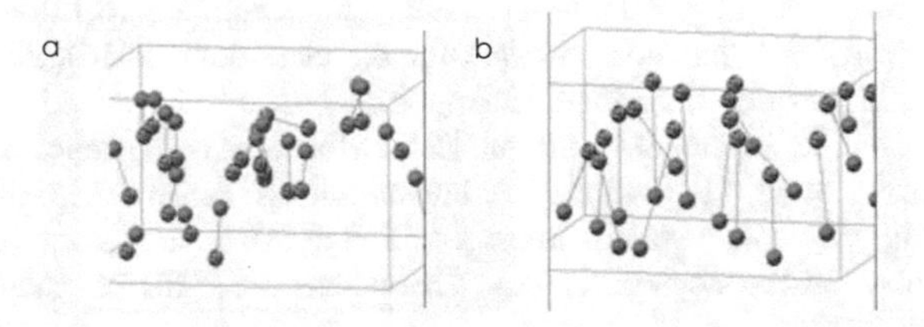

Fig. 4. The algorithm for a series of computer models describing the behavior of the ensemble of "soft classical dipoles" that can oscillate in an external electric field and the corresponding interactive models: b - polarization of solid dielectric, c - the polarization of gas from polar molecules.

In the case of chemical processes, modeling the behavior of the system's elements is essentially probabilistic and multi-variant. The calculation of its quantitative characteristics is actually a non-trivial task of quantum chemistry.

On the other hand, traditions of chemistry teaching usually require formal averaged macroscopically smooth solutions for concentrations instead of the language of microsimulation, which demonstrates the behavior of the system at the molecular scale. This approach allows a substantial simplification of the formulation of tasks to simulate chemical processes.

We applied the first approach for modeling processes in plasma (that is an advanced level) and for generating problems on the base of these calculations and the second approach for generating tasks on a college level chemistry.

4 Conclusions

Computer simulation is an effective support method for science education. Considered PhOOM approach has demonstrated to be a promising design method for numerous applications to support advanced physics courses. Additionally, it can be used as a digital prognostic toll to understand dynamics of intricate physical systems. Computer models developed with the use of PhOOM language can supply an increased number of students with a high-level individualized education.

Common pattern based task generation supported with PhOOM models demonstration proved to be an effective tool for developing necessary skills and competencies for natural science education.

REFERENCES

Pawl, A., Barrantes, A., Pritchard, D.E.: Modeling applied to problem solving. In: AIP Conference Proceedings (2009)

Butikov, E.I.: Planets and satellites. In: Educational software package. Physics Academic Software, American Institute of Physics (1998). 1910th Annual Educational Software Contest winner (1999, Computing in Science and Engineering Magazine), European Academic Software Award Winner (EASA 2004) (1992)

Kook, H.J., Novak Jr., G.S.: Representation of models for expert problem solving in physics. IEEE Trans. Knowl. Data Eng. **3**(1), 48–54 (1991). https://doi.org/10.1109/69.75888

Interactive physics: physics simulation software for the classroom (2016). http://www.design-simulation.com/IP/Index.php. Accessed 15 Aug 2016

Larkin, J.H., Mcdermott, J., Simon, D.P., Simon, H.A.: Models of competence in solving physics problems. Cogn. Sci. **4**(4), 317–345 (1980). https://doi.org/10.1207/s15516709cog0404_1

LabVeiw (2016). http://www.labview.ru. Accessed 15 Aug 2016

Sychov, S., Chirtsov, A.S., Mylnikov, A.: ChemGenerator (2017). chem-generator.com. Accessed 27 Oct 2017

Butikov, E.I., Chirtsov, A.S.: The laws of motion of macroscopic bodies. A package of training and demonstration programs in the course of general physics. In: Model-Oriented Data Analysis, St. Petersburg, p. 27 (1992)

Marek, V.P., Mikushev, V.M., Chirtsov, A.S.: The use of information technologies in creating an innovative educational environment at the physics faculty of the classical university. Int. J. Exp. Educ. **6**, 23–26 (2009)

Marek, V.P., Chirtsov, A.S.: A series of electronic collections of multimedia materials on the course of general physics: original approaches to the creation of multimedia resources and their use. Comput. Tools Educ. **1**, 58–72 (2012)

Mikushev, V.M., Somnov, Y.M., Chirtsov, A.S.: The concept of using MOOC-technologies for remote active individualized training in physics and its approbation. Int. J. Exp. Educ. **12**(3), 359–362 (2015)

Chirtsov, A.S.: Motion of charged particles in the force fields - a package of training programs and a physical designer. In: Paper presented at the Training of International Conference Modern Teaching Technologies, St. Petersburg, 16–17 May 1995

Chirtzov, A.S.: A series of electronic collections of multimedia materials on the course of general physics: new approaches to the creation of electronic designers of virtual physical models with simple remote access. Comput. Tools Educ. **6**, 42–56 (2010)

Chirtsov, A.S., Mikushev, V.M., Tchaikovskaya, O.N.: The use of physical object-oriented modeling in the MEP for mechanics. NTL Publishing House, Tomsk (2015)

Learning to Think and Practice Computationally via a 3D Simulation Game

Nikolaos Pellas(✉) and Spyridon Vosinakis

University of the Aegean, Mytilene, Greece
{npellas, spyrosv}@aegean.gr

Abstract. Various studies have presented controversial results about the way that young students tried to cultivate and practice their computational thinking (CT) skills with Computer science concepts through the game making programming. However, there is still limited evidence addressing how the gameplay of a simulation game (SG) can be associated with the development of computational problem-solving practices. Therefore, the purpose of the present study is threefold: (a) to elaborate a rationale on how a 3D SG can support the development of computational problem-solving practices using OpenSimulator with Scratch4SL, (b) to analyze how the in-game elements should be mapped to assist basic CT skills cultivation and programming concepts to support students in learning how to think and practice computationally, and (c) to summarize the findings from a preliminary mixed methods study following a game playing approach in regard to the learning experience with a total of fifteen (n = 15) junior high school students. The results indicate that students had a greater range of expressing sufficiently alternative and self-explanatory solutions in blended instruction. The instructor's feedback and guidance facilitate them to rationalize decisions taken on the cognitive aspects of computational practices in coding.

Keywords: Computational thinking · Game-based learning · Virtual worlds

1 Introduction

Learning programming as an indispensable part of Computer science (CS) in K-12 education is a cognitively complex and demanding task. It requires a synthesis of cognitive thinking skills, such as problem-solving, logical reasoning and algorithmic thinking that enables students to transfer their thinking solutions into workable plans and algorithms for proposing solutions to real-world problems [1]. Previous research efforts [1, 2] and literature reviews [3–5] came to the conclusion that computational thinking (CT) paves a pathway of recognizing the prerequisites in a broad range of analytical and logical ways of human's thinking on how to solve problems finding the most efficient in order to apply solutions. Subsequently, learning to think and practice computationally includes the pervasive use of fundamental programming constructs (e.g., selection, sequencing, repetition) in a computational problem-solving strategy related to the analysis of designing, planning and debugging a proposed solution that can be applied as a perfect way to evaluate the correctness of a thinking process [6, 7]. Even more, CS curriculums worldwide have been widely intended to provide

M. E. Auer and T. Tsiatsos (Eds.): IMCL 2017, AISC 725, pp. 550–562, 2018.
https://doi.org/10.1007/978-3-319-75175-7_54

game-based learning (GBL) approaches as more appropriate to fulfill students' learning needs and experiences through formal or informal instructional contexts [3], either with exercises for learning programming by designing a game (game making) or with exercises for learning programming by playing a game (game playing) [6, 8]. Two are the most distinctive categories of interactive environments that these approaches are implemented: (a) visual programming environments (VPEs), like Alice, AgentCubes, Scratch etc. [4] or three-dimensional (3D) virtual worlds (VWs), like Second Life (SL) [9, 10], OpenSimulator (OS) [11] or Neverwinter Nights 2 [12].

On the other side, a series of ongoing studies have pointed out reasons causing difficulties in student learning experience by practicing with understanding the use of core programming concepts in relation to CT as a problem-solving process. First, the inadequacy of learning at an initial stage alongside with the development of CT skills by paying attention to syntax or semantics of the first programming language, when CS instructors and students use interactive environments lacking the means to abstract functionality into functions and procedures [3]. This leads to the description of vague abstractions which are specified by expressions of code commands in simple or without purpose projects [2, 9]. Assessing student computational understanding is still unable, due to inability in connecting practically the abstract representations with programming constructs [7]. Second, lack of problem-solving, logical and abstract reasoning skills, which are regarded as essential for spotting and solving problems. Lack of such skills hinders student information on how plan solutions, decompose a problem into smaller subparts and apprehend errors to debug by testing and figuring out possible mistakes before coding to ensure the accuracy of final projects, e.g. games, stories, artifacts [2, 6]. Lastly, with the inefficient process of memorizing and executing continuously the same programming constructs referring to small exercises that are based on school textbooks' core aspects of learning cannot be explicitly investigated how fostering and assessing the transfer of CT skills to future real-world contexts [4]. Such a process is inappropriate for understanding how the cognitive thinking process of solving a problem is reflected in practices for the execution and verification of this process correctness [7, 12].

Although recent studies [2, 13] have argued on how students cultivate and apply CT skills mostly with creative computing or artistic expression tasks in game making programming using various interactive environments, limited research demonstrated how gameplay of a simulation game (SG) can be associated with the development of computational problem-solving and how it can support greatly the implementation of computational practices. Accordingly, a substantial body of literature reviews has come to the statement that there is an overt "gap" concerning, either the creation and use of new interactive environments [3, 4] or the combination of already known "tools" for game playing tasks [8]. Therefore, there is a need to have a better understanding about the impact of digital games for introductory programming on the cultivation of CT skills and on the learning experience from students' computational practices via a SG.

To address the aforementioned challenge, this study's hypothesis is whether the combination of the most significant design features and characteristics of visual programming environments, like the visual palette of Scratch4SL (S4SL) to prevent syntax complexity in coding and the realistic representational fidelity of a 3D VW can support the cultivation of computational practices fostering the transfer of learning outcomes as proposed solutions for a real-world problem in game playing modes. Hence, the purpose of the present study is threefold: (a) to elaborate a rationale on how a 3D SG can support the development of computational problem-solving practices, using OS with S4SL, (b) to analyze how in-game elements should be mapped to basic CT skills and programming concepts, following a game playing approach to supporting students in learning how to think and practice computationally, and (c) to summarize the findings inferring to computational practices related to the use of (simple or nested) sequence, conditionals or iteration methods to solve a computational problem. A preliminary testing about the effectiveness of a 3D SG on students' learning experience was conducted in after-school programming sessions with a total of fifteen (n = 15) junior high school students who participated voluntarily in blended instruction.

2 Background

Wing [13] defines CT as a problem-solving process for conceptualizing, developing abstractions and designing systems that require the use of human's logical and analytical thinking with concepts fundamental to computing. Additionally, relevant literature reviews [3, 5] have also proposed an operational definition for CT covering a wide variety of skills that need to be cultivated. Such skills eventually assist humans to understand how programming can become more effortful and unfold the support of computational tools to think how to solve problems. The literature in GBL focuses on "game making" or "game playing" approaches to facilitate the development of CT skills by utilizing: (i) VPEs and (ii) 3D VWs. The most remarkable features of the former category are [6]: (a) the applicability and visualization of algorithmic control flow (code tracing) using a "drag and drop" process facilitating code blocks organization and documentation, (b) the assembly of coloured blocks that resemble like a jigsaw puzzle through logically and specific commands with constructs (control flow blocks nesting), with a view to eschewing syntax errors and (c) the execution of proposed solutions to a problem that are expressed as solutions created by combining simple or nested code blocks (design patterns).

However, game making using VPEs has become a target of negative criticism from a growing body of literature [3, 6, 12]. Students tend to create game-based applications using trial-and-error modes of design patterns, by copying and pasting the same code of other projects or by using only some programming constructs, rather than patterns emerging from a thinking before the coding process. Thence, the transformation of plans with syntactically correct instructions for execution and assess the consequent results of those instructions inferring to [6, 12, 13]: (a) the presentation of digital artifacts or applications, including stories or digital games that seemed too simple or without purpose. By developing and programming simple games, students cannot adequately articulate CT skills, failing to internalize computational problem-solving practices in a more inferential realistic interpretation through game playing mechanisms, and (b) the interpretation of applications, where proper writing code fragments are executed correctly, but the difficulty in decomposing and formulating a problem is observed. This happens because students struggle to understand if their design patterns can support any proposed solution to a problem, even in abstract manner. Thereupon, code documentation is neither what exactly students would like to present, nor assist their trials to comprehend source code, causing often significant conceptual gaps [2, 6].

Alternatively, through game playing in *Code.org*, students can play computer games, like *Angry Birds* that promote mainly the algorithmic thinking and basic programming skills. Nevertheless, even in this case, previous studies [14, 15] have criticized such an approach because games of this site do not support all possible programming phases and possible biases for the development of CT skills. Another worthy game playing approach is the creation of SGs in 3D VWs, as it be adequately combined the effects of *"flow"* (a state of enjoyment and psychological immersion referring to the optimal experience when users are engaged through in-game challenges to succeed goals of each activity) and *"presence"* (a human's feeling of being somewhere or having an effect by taking part in activities at a different place than truly is his/her location) [16]. Roleplay learning systems, such as SGs that are created in 3D VWs are increasingly applied to foster practical cognitive thinking skills through active/exploratory learning experience facilitating a flow learning experience [9, 10]. The most noticeable characteristics to support GBL in CS courses are as follows: (a) the common and persistent environment to all users can give CS instructors opportunities to evaluate students' computing skills and competences at the same time during the learning process or to provide constructive feedback using a/-synchronous communication tools [11], (b) self-evaluation and reflection upon students' cognitive thinking process are achieved visually or acoustically by integrating behavior in objects or by creating artifacts to link abstract-concept formation to a more concrete game experience for learning gain [9], (c) enhancement of creative computing for constructing 3D artifacts with behavior in SL by avoiding syntax errors of Linden Scripting language (LSL) via S4SL [10]. However, the way in which students try to write syntactically correct the scripting language code of the most known 3D VWs, given its similarity with other general-purpose, like C, is still not well-documented.

Up to date, few notable efforts [4, 6, 8] have referred design frameworks and guidelines that allow users to conceive computational problem-solving strategies required for the cultivation and the connection of problem-solving with the development of programming skills to apply design patterns as solutions to a problem. For example, Lye and Koh [4] have proposed an imperative number of design guidelines and directions towards a constructivist (thinking-doing) problem-solving learning approach. In addition, Kafai and Burke [8] have already recommended the connection of serious gaming opportunities in a simulated world, like *SimCity* with *Scratch 2.0* environment for writing programs.

According to the above-mentioned, the focus of this contribution aims to outline the learning procedure on how students can interact with a 3D SG to examine their computational problem-solving practices based on different design patterns that they propose. OS and S4SL can be considered as a powerful set of tools for bridging the "gap" between problem formulation and solution expression with the intention of imposing how to apply programming constructs in simulated real-life problem-solving contexts. The representational fidelity of OS offers an interactive environment that can support authentic (open-ended) problem-based learning conditions in view of fostering students' cognitive thinking skills. By using S4SL, students can get focus on expressing and applying solutions based on logical reasoning and algorithmic thinking for assessing the correctness of their thinking process into computational practices.

3 Design Decisions and Rationale to Utilize a Simulation Game

SGs support students' problem-solving practices and learning experience reflected on two undisputed keystones of CT inducing the expression [1, 3, 8]: (a) on how the pertinent behaviors are considered from logical and abstract thinking (abstraction) in favor of formulating and testing solutions to a problem by specifying computational rules and concepts and (b) on how to obtain solutions to a problem by performing a sequence of steps that are applied through programming skills as design patterns. It becomes then clear to what extent are these patterns implying the need for simulation to interpret the abstractions (automation) requiring access to computing tools. To this notion, three-goal examples to design are considered as important for helping students to articulate and transfer their thinking solutions from tacit thinking to workable plans and algorithms [1, 6, 15]: (a) *Integration of the learning material within the game interface:* A natural way of formalizing knowledge to an abstract manner in a simulated problem-solving context during gameplay is imperative. Formulating innate thinking into abstract representations using visual metaphors of a 3D VW employs an approach that should infer and predetermine the designer specify algorithmic rules corresponding to multiple movements that are the most appropriate to be done by users. For example, the visual metaphors of OS support introductory programming learning in regard to the conceptualization of algorithmic rules through abstract thinking logic that illustrated in simulated real-world and problem-solving contexts; (b) *Transfer from tacit thinking to concrete thoughts of computational concepts:* In-game activities should allow users to describe the learning situation in which they attend and explicitly link their actions during gameplay through CT skills cultivation. The reflective observation of the concrete experience and implementation of computational problem-solving practices not only assimilate abstract conceptualization without remaining tacit, but also facilitate student understanding on how and why use computational concepts in two aspects: (i) by decomposing abstract representations of the main problem to articulate a natural way in an effort of formalizing tacit knowledge and (ii) by conceptualizing an abstract logic thinking during gameplay to instantiate design patterns for testing and debugging a proposed solution; and (c) *Transform students' thinking knowledge through in-game play settings into formal logic and analysis about a solution in coding:* Student's progress through in-game activities requires the process of concreteness a solution by transforming a natural way of innate thinking to coding. A SG created in OS provides an intuitive-natural modality for user-interaction. Thus, users can articulate and transfer from tacit thinking to concrete their thoughts by practicing with understanding the use of core programming concepts in relation to CT for developing and implementing computational problem-solving practices. S4SL usage eliminates split attention of code syntax and users can make the focus on goals of solutions that are applied as results of computational problem-solving practices (design patterns).

Inevitably, prior works [6, 15] have suggested tools that can foster CT skills cultivation with visual thinking by supporting problem formulation in applications, such as simulations when various evocative spatial metaphors are offered as alternative options for game playing. The in-game use of evocative visual objects can be considered as powerful conceptualization approach. Also, other works [4, 15] have advocated that such characteristics can assist the forging of abstractions which predominately serve as the beginning of a path from problem formulation to solution expression. These tools are expected to support GBL since through spatial reasoning the logical relations satisfy make more clear and understandable the in-game objectives that players need to achieve using inductive and abstract reasoning [14, 15]. Abstract simulations created in 3D VWs can be considered as suitable for helping users to understand the concepts by taking the chance to take advantage from the formation of spatial knowledge representations that support problem-solving learning tasks [6, 11].

Based on all the above and in response to the first purpose of this study, the following proposed instructional guidance in Table 1 is dedicated to blended instruction and it constitutes of: (a) the learning tasks associated with the operational definition of CT as a problem-solving process with specific learning objectives [5], (b) the CT skills that need to be cultivated during in-game process [5], (c) the proposed guidelines (G1-G5) from Pellas and Vosinakis [6] study about the creation of a SG and (d) game activities that can assist students to express and apply computational problem-solving practices:

Table 1. Game activities associated with characteristics and skills of CT

Sessions	1st session	2nd session	3rd session	4th session	5th session	6th session
Operational definition of CT	Formulating problems	Logically organizing and analyzing the data	Representing data through abstraction	Automating solutions through algorithmic thinking	Identifying, analyzing and implementing possible solutions	Generalizing and transferring a problem-solving process to solution
CT skills	Problem-solving	Problem-solving	Abstraction	Algorithmic thinking	Design-based thinking	Pattern recognition
Proposed guidelines	Student motivation (G1)	Student active participation (G2a)	Simulation of an authentic problem (G2b)	System's feedback on user's actions (G3)	Development of computational practices (G4)	Applying design patterns (G5)
In-game activities (Students should be able to…)	Decompose in subparts the main problem	Analyze a cleaning path and describe the RVC 's movements	Designate the RVC's movements in spatially-explicit context	Transform a solution to algorithm and debug it by finding errors	Implement in coding the proposed solutions via Scratch4SL	Examine the effectiveness of the proposed design patterns

4 Method

4.1 Setting and Sampling

This study was conducted in an intensive 2-week period with 6 sessions. The first 4 sessions lasted 4 h in face-to-face sessions and the other 2 sessions lasted 2 h in supplementary online during the Spring trimester 2017. The sample comprised of 7 girls (M_{age}: 13.87, SD: 1.13) and 8 boys (M_{age}: 14.74, SD: 1.15) volunteered to participate and they were from the local schools. All participants were recruited to attend in all after-school sessions and they wanted to learn how to code using interactive environments. Also, all of them were coding novices, but all had a previous experience with Scratch. When the participants were selected, the main researcher contacted their teachers and parents in order to obtain the necessary consent from both the student and the legal guardian for the data collection.

4.2 Instrumentation and Data Analysis

For assessing the experiential dimensions, a mixed-methods study was followed to bring the strengths of research forms in favor of validating the results. At the end of this experiment, quantitative data were gathered through close-ended self-reporting questionnaire responses of participants [17] given the option of writing short comments (Table 2), whilst maintaining anonymity and confidentiality. Their answers analyzed according to the guidelines of user experience studies [18]. Supplementary, qualitative data were collected through open-ended interview questions to understand students' enchantment and engagement using the proposed SG.

Table 2. Short comments on how the SG contributing to LE, LP, and UX

Learning effectiveness (LE)	(a) Roleplay scenario [*n* = 8, 54%]	(b) Exploration and problem description [*n* = 2, 13%]	(c) Learning objectives [*n* = 2, 13%]	(d) Chat or voice communication [*n* = 2, 13%]	(e) Visual feedback [*n* = 1, 7%]
Learning procedure (LP)	(a) OpenSim and Scratch4SL [*n* = 5, 40%]	(b) Instructor's feedback [n = 4, 30%]	(c) Game context [*n* = 2, 10%]	(d) Understanding of user control in the game [*n* = 2, 10%]	(e) In-game visual elements [*n* = 2, 10%]
User experience (UX)	(a) The game setting (RVC, 5 rooms, visual objects, etc.) [*n* = 5, 30%]	(b) In-game problem recognition accuracy [*n* = 3, 20%]	(c) Interactivity with visual objects [*n* = 3, 20%]	(d) The 3D graphical user interface [*n* = 2, 15%]	(e) The anthropomorphic avatar [*n* = 2, 15%]

For assessing the user experience, this study followed the research considerations by Bargas-Avila and Hornbæk [17] who identified several aspects of experiential dimensions that should be utilized. All statements in this work are expressed and rated simply on a 5-point Likert scale [strongly disagree (1) to strongly agree (5)]. The items about the procedure for measuring student learning experience was based on 16 questions, translated into the Greek language and separated in three subparts: learning effectiveness (LE), learning procedure (LP) and user experience (UX). Subparts about students' learning outcomes and experiences concerned with issues that are ubiquitous in respective work; in specific, all identified aspects (aesthetics of interaction engagement, usability, usefulness, visual appeal) related to user experience [17]. Cronbach's alpha (α) of the main questionnaire was 0.835, reflecting on a reasonable internal consistency of the variables to describe students' expectations.

Data were analyzed using: (a) guidelines of usability metrics for evaluating the user experience [18], including each user's response into the top-2-boxes (positive responses) or the bottom-2-boxes (negative responses), (b) probing questions from the instructor provided feedback by posing questions to each participant when s/he seemed to get confused helping them find an adequate direction to propose a solution, and (c) code tracing via S4SL, the instructor evaluated the applicability of algorithmic control flow to identify whether the adoption of selection control-flow blocks and the exploitation of nesting composition among programming constructs were achieved.

4.3 Procedure

The aim of this teaching intervention was the exploitation of a roleplay 3D SG following an instructive guided approach with step-by-step programming exercises and the investigation of its impact on students' learning outcomes depending on computational practices. Having the role of embedded software engineers, students should assist an old woman with special needs, who moves only with her wheelchair and struggles to clean all rooms of her house. In gameplay context, they need to elaborate a solution aimed at creating unique algorithms with logically and precise instructions and finally to propose solutions as design patterns for this problem. In this vein, students first need to navigate, determine movement positions and describe the best cleaning path that an autonomous robot vacuum cleaner (RVC) should follow in sufficient time. Their solutions need to be implemented by integrating behavior from S4SL to OS, where a robot cleaner should move and clean five rooms that are differentiated in spatial geometry layout, in terms of division among house furniture or objects and calculate arithmetically distances without causing hits or damages. House furniture and objects in square floors are seen as evocative spatial metaphors of basic geometric shapes (e.g., triangle, square, hexagon) assisting students to think and practice computationally with an abstract conceptualization approach alongside with a pathfinding for a simulated real-world problem. Abstract spatial representations of geometric shapes were extensively used in this 3D SG, such as a triangle, for example, to prevent hitting a table, players need to determine arithmetic computation between chairs and table distance (e.g., each side's square floor has width and height 5 m respectively) or-/ and calculate degrees of turning correctly (e.g., 90° for square or 45° for equilateral triangle) to traverse the robot a specific cleaning path down from the table (movement on X and Y axis), without hitting the table lamp (impact on crashing to Z axis). This process was becoming more compelling, since it was expected from students to implement these computational strategies through practices via S4SL in order to be presented the shortest path between the present location and the goal location of the robot. The main researcher has to propose the use of at least one programming construct for each room that should be included in the solution, albeit students were free to propose alternative solutions using other constructs.

For learning through gameplay challenges, students need to analyze how to plan a solution for a cleaning path problem, subdivide it into smaller parts, and apprehend hypothetical error situations for retrieving visual feedback by means of OS. They should debug their cognitive thinking process by trying to test and figure out possible misconceptions in computational practices via S4SL. These practices are combined with programming constructs (serial sequence, if/else statement or loop) and instruction/movement commands for executing the design patterns. For example, Fig. 1 depicts a problem on how moving the robot in a cleaning path, by coding a solution, like being square root spiral. When the robot moved under the table (root), the user needs to use the same design patterns with iteration and commands blocks in relation to numbers or variables. It is imperative to take the advantage from the environment's spatial layout comprising all of the rules for performing arithmetic computations for the distance of the robot between the avatar and house furniture. There are also some noteworthy core gameplay mechanics, basic rules, and functions in this SG. First, each player should describe and apply an algorithm that calculates the most efficient route for cleaning. To identify and present the proposed solution, each player should copy and paste the commands and programming constructs via S4SL palette inside the RVC's notecard explaining a step-by-step solution before executing her/his proposed solution. Second, six checkpoints inside each room are allowed for the mapping process which is preceded by the players. The correct pass of the robot above them means no counting on the total time until the final solution finished. Third, whenever the robot is programmed to passing and cleaning all dust gray signs off the floor, for rewarding, it gains energy, giving grades to the player. If gathering the smallest possible number of code blocks for cleaning each room based on resilient planning, execution time and fewer hits on the house furniture or objects, then such a player is declared as the winner.

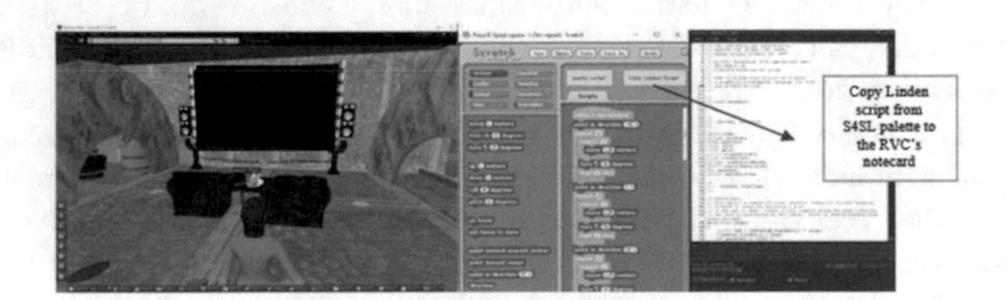

Fig. 1. Illustration of the in-game learning process in room B

5 Results

Regarding the participants' background based on demographics information, almost more than half percent (55%) of them find really important their participation in CS courses with reasoning and learning capabilities to be the implementation of various tasks using programming environments. It seemed that most of them (60%) had previous experience with Scratch. Some of them (20%) answered that they knew about SGs, such as *The Sims* or *Minecraft* and some others (33%) who had utilized them.

The vast majority of participants reported on several points of view about the RVC simulator. In Fig. 2, the top-2-box scores include responses to the two most favorable response options, i.e. ranking percentage based on their answers was e.g., from 87% (13 out of 15 students) about expressing and applying their solutions to 67% (10 out of 15 students) about decomposing in subparts the main problem. Slightly more than half of them (54%) referred that roleplay scenario and problem description contributing to LE (Table 2). A student reported that *"Some facts in the game are really represented well. This helped me not only to rationalize my decisions by applying and explaining my solution but also to know why I used some programming constructs without only proposing zigzag movement as cleaning path"*. Another one said that *"S4SL helped me to apply a proposed solution, as I visually saw the results of the code inside OS"*. In contrast, other users could not easily recognize the interaction between elements inside the house (Visual feedback: 7%) complaining one of them that *"Sometimes I struggled to understand if the robot collided with house furniture or objects, when I was applied for my program"*, albeit in the end their preference than Scratch or Alice was referred. The use of communication tools to succeed the learning objectives were mentioned less from few users (13%), maybe due to the instructor's feedback in face-to-face tasks.

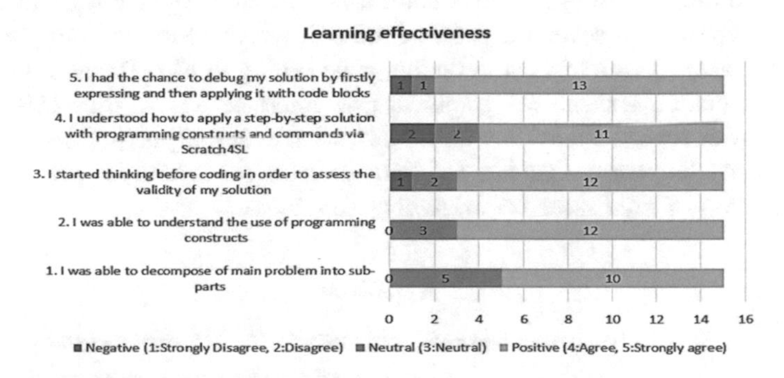

Fig. 2. Horizontal stacked bar chart of top/bottom-2-boxes of user responses about the LE

In terms of LP, again many participants were at the top-2-box scores. Ranking percentage based on their answers was e.g., from 73% (11 out of 15 students) about understanding instructor's feedback to 53% (8 out of 15 students) for the effective communication and successful implementation of design patterns for proposing solutions to each subpart of the main problem (Fig. 3). Others reported on several points of view with regards to the SG that contributed to the LP (Table 2) with the most notable to be the combination of OS with S4SL (40%). After the game context, understanding of in-game user control and visual elements follow with 10% to each. The combination of OS and S4SL was necessary for integrating behavior inside the robot to follow a cleaning path and getting responses of its movement, in an effort of proposing and applying visually solutions through design patterns. The coding phase to visualize a proposed solution was referred by others as an important feature, especially because it enables them to assess their thinking process: *"The S4SL palette enabled me to write correctly the code, while I was previously described and proposed a solution in natural*

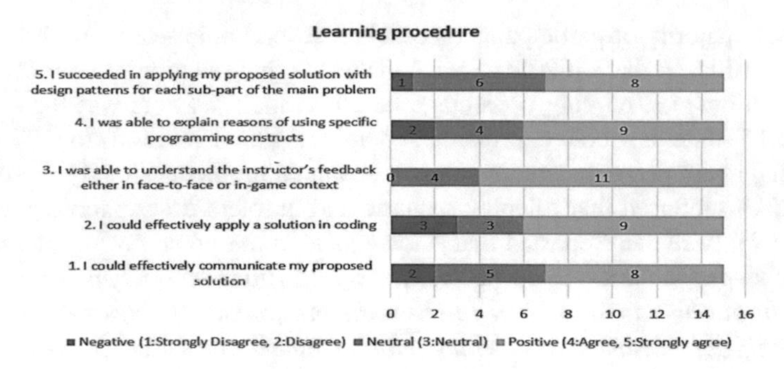

Fig. 3. Horizontal stacked bar chart of top/bottom-2-boxes of user responses about the LP

language". Another one participant referred that *"The instructor guided my practices and he helped me with the code responses in order to be applied my planning"*.

With respect to the UX, most participants were at the top-2-box scores (Fig. 4). For instance, the top-2-box score is 67% (10 out of 15 students) of students who felt engaged with the VRC simulator rating it favorably compared to their counterparts who have opposite opinion according to a bottom-2-score of 13% (2 out of 15 students). Participants reported on several aspects of the SG, which contributed to positive user experience (Table 2) with highest to be the game setting (30%). The anthropomorphic avatar representation and the 3D graphical user interface follow with 15%. A representative answer reported that "*It was a motivating setup of playing in-game tasks"*. Other one said, *"In past, sometimes I did not have the opportunity to present my code and speak of why I used some programming constructs"*.

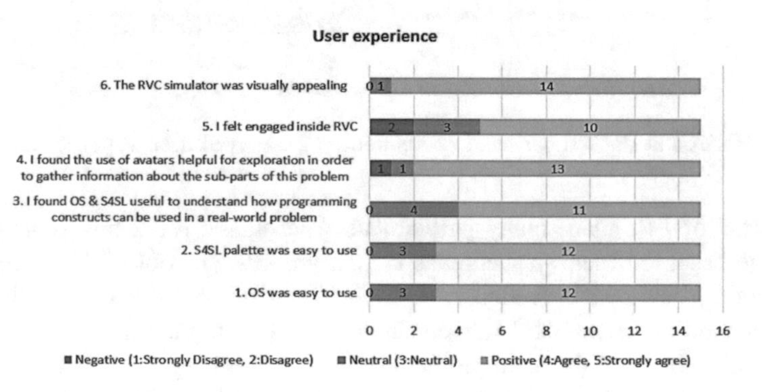

Fig. 4. Horizontal stacked bar chart of top/bottom-2-boxes of user responses about their UX

Negative aspects of the UX were reported about interactivity among visual objects (15%), like *"When the robot stroked a table or a sofa, sometimes I did not recognize the error message, maybe because of the poor quality of graphics"*. Few users struggled to log into OS, said that *"I was observed slow loading times in my entrance"* at the beginning or others did not copy and paste correctly the code into the notecard of RVC.

6 Discussion and Conclusion

The main purpose of this study was to investigate the effectiveness of a 3D SG to introductory programming high school course settings. The RVC simulator provides affordances for instructive guided support through informal blended instruction to CT teaching. Furthermore, it enables the free experimentation and reflection of students in a concrete problem-solving space by exploring and expressing solutions through design patterns. Their answers revealed the positive acceptance about how instruction using S4SL and OS engaged them in innovative and interactive learning situations since they had very satisfactory performance and user experience. Findings of this preliminary testing unveil that a great number of students found this 3D SG interesting, fascinating and relevant to their previous experience with other roleplay SGs, like *The Sims* or *Minecraft*. Without so highly advanced, but with simple design patterns to be nested and presented as final solutions, students appeared not having any difficulties in producing some good computational problem-solving practices. Based on code tracing analysis, the applicability of selection control-flow blocks and the exploitation of nesting composition among programming constructs, for instance, such as mastering if/else conditionals with numbers using S4SL, students were able to propose well-defined solutions and learning outcomes that can be easily visualized in OS. Consistent with Howland's and Good's [12] study findings, a block-based palette is regarded as a reliable tool for high school students to avoid syntax errors in programming and trigger more in problem-solving via 3D roleplay games by expressing and applying more succinct and precise rules with instructions in combination with programming constructs. On the other side, contrary to the results of past efforts [2, 7], students of this study using a 3D SG seemed to have reasonable efforts by answering why they used specific programming constructs and instructions in computational practices, dodging the vague syntax of programming constructs. Such a process can give valuable answers for assessing how students try to think and practice computationally before coding. This can also give evidence of a deeper understanding from the description of a cognitive thinking process to the comprehension and production of coded solutions.

In conclusion, this study's findings may be of interest to instructional designers who want to take in advance a 3D SG and design (in-)formal introductory programming courses in blended instruction to foster students' computational problem-solving practices. Some educational implications that should be underlined using OS with S4SL are as follows: (a) students can think critically and logically so as to organize code blocks design patterns and execute programs for a simulated real-world problem, (b) students were able to understand evocative spatial metaphors inside OS, referring from almost all of them different computational practices in coding, and (c) students' learning experiences and achievements affected positively their overall performance in order to transform easily their thinking solutions into workable plans and algorithms.

The results about computational understanding cannot be easily generalized, due to the limitation of small sample size in a time-intensive after-school program. To widen and generalize a more efficient way to foster computational problem-solving, a quasi-experimental study should conduct with an experimental group that will utilize

OS and S4SL compared to a control group that will use the visual programming environment of Scratch with the same subparts of the main problem illustrated in the RVC simulator.

References

1. Liu, C., Cheng, Y., Huang, C.: The effect of simulation games on the learning of computational problem-solving. Comput. Educ. **57**, 1907–1918 (2011)
2. Mouza, et al.: Development, implementation, and outcomes of an equitable computer science after-school program: findings from middle-school students. J. Res. Technol. Educ. **48**, 84–104 (2016)
3. Grover, S., Pea, R.: Computational thinking in K–12: a review of the state of the field. Educ. Res. **42**, 38–43 (2013)
4. Lye, S., Koh, L.: Review on teaching and learning of computational thinking through programming: what is next for K-12? Comput. Hum. Behav. **41**, 51–61 (2014)
5. Kalelioglu, F., Gülbahar, Y., Kukul, V.: A framework for computational thinking based on a systematic research review. Baltic J. Mod. Comput. **4**, 583–596 (2016)
6. Pellas, N., Vosinakis, S.: How can a simulation game support the development of computational problem-solving strategies? In: EDUCON. IEEE, Athens, pp. 1124–1131 (2017)
7. Brennan, K., Resnick, M.: New frameworks for studying and assessing the development of computational thinking. In: AERA, Vancouver, Canada (2012)
8. Kafai, Y., Burke, Q.: Constructionist gaming: understanding the benefits of making games for learning. Educ. Psychol. **50**, 313–334 (2015)
9. Esteves, M., et al.: Improving teaching and learning of computer programming through the use of the Second Life virtual world. BJET **42**, 624–637 (2011)
10. Girvan, C., Tangney, B., Savage, T.: SLurtles: supporting constructionist learning in Second Life. Comput. Educ. **61**, 115–132 (2013)
11. Rico, M., et al.: Improving the programming experience of high school students by means of virtual worlds. Int. J. Eng. Educ. **27**, 52–60 (2011)
12. Howland, K., Good, J.: Learning to communicate computationally with Flip: a bi-modal programming language for game creation. Comput. Educ. **80**, 224–240 (2015)
13. Wing, J.: Computational thinking. CACM **49**, 33–35 (2006)
14. Theodoropoulos, A., Antoniou, A., Lepouras, G.: How do different cognitive styles affect learning programming? Insights from a game-based approach in Greek schools. ACM Trans. Comput. Educ. **17**, 3 (2016)
15. Román-González, M., Pérez-González, J.-C., Jiménez-Fernández, C.: Which cognitive abilities underlie computational thinking? Criterion validity of the computational thinking test. Comput. Hum. Behav. **72**, 678–691 (2017)
16. Faiola, A., Newlon, C., Pfaff, D., Smyslova, O.: Correlating the effects of flow and telepresence in virtual worlds: enhancing our understanding of user behavior in game-based learning. Comput. Hum. Behav. **29**, 1113–1121 (2013)
17. Bargas-Avila, J.A., Hornbæk, K.: Old wine in new bottles or novel challenges: a critical analysis of empirical studies of user experience. In: SIGCHI. ACM, New York, pp. 2689–2698 (2011)
18. Tullis, T., Albert, W.: Measuring the User Experience: Collecting, Analyzing, and Presenting Usability Metrics. Morgan Kaufmann Publishers Inc., San Francisco (2013)

Using Serious Games for Promoting Blended Learning for People with Intellectual Disabilities and Autism: Literature vs Reality

Stavros Tsikinas(✉), Stelios Xinogalos, Maya Satratzemi, and Lefkothea Kartasidou

University of Macedonia, Thessaloniki, Greece
{s.tsikinas, stelios, maya, lefka}@uom.edu.gr

Abstract. Educating people with intellectual disabilities (ID) or autism spectrum disorder (ASD) is a non-trivial process and differs from the learning methods of typically developed people. Recently, serious games (SGs) have been used to enhance the learning process of these groups and address different skills. On the other hand, blended learning (BL) is applied to formal and informal educational contexts and combines face-to-face and online learning. In this study, we examine if SGs can provide the necessary means for applying BL, especially for people with ID or ASD that could be benefited by personalized learning opportunities. In addition, we examine 43 existing SGs for people with ID or ASD, as well as the perceptions of 93 special education professionals (SEP) and teachers (SET) working in schools and institutions for people with ID or ASD regarding the role of technology and SGs in their education. We concluded that SGs could enhance the learning process of people with ID or ASD in many skills. In addition, the opinions of SEP and SET regarding the importance of technology in the learning process of people with ID or ASD and the familiarity with SGs, indicate that BL could be effectively promoted through SGs.

Keywords: Accessible blended learning · Serious games
Intellectual disabilities and autism

1 Introduction

The learning methods of people with ID or ASD differ from the learning methods of typically developed people, or even of people with learning disabilities [52]. The most common learning methods for people with ID or ASD include role/mimic playing, educational cards, conversation, reading and creative work. However, these learning methods for people with ID or ASD might present issues in generalizing the learning objectives [31] and require enhancements [14]. On the other hand, researchers have indicated specific steps in order to include effectively new technologies in accessible learning [40]. Thus, new technological solutions, such as educational software and SGs have empowered the learning process for people with ID or ASD [10, 27].

SGs are digital games that aim to fulfill additional purposes apart from entertainment [43]. SGs have been used in various disciplines and scientific fields, such as

M. E. Auer and T. Tsiatsos (Eds.): IMCL 2017, AISC 725, pp. 563–574, 2018.
https://doi.org/10.1007/978-3-319-75175-7_55

education and health [43]. The use of SGs in the learning process adds motivation and engagement to the learner [23]. Lately, SGs have been used successfully in the learning process of more inclusive learners, i.e. people with ID or ASD [3, 18]. Due to the effectiveness of SGs, it is our belief that SGs can provide the necessary means for promoting BL for people with ID or ASD. Specifically, SGs can offer great opportunities for implementing all the unique features of BL and applying it effectively: SGs can be applied to informal and formal educational contexts; SGs can combine in-person and online learning; SGs can be designed so as to achieve an important feature of BL, which is to provide students control, to a certain extent, over time, place, path or pace [26]; SGs can incorporate combinations of different learning methods and may include multiple learning tools in order to promote effectively the learning process, exactly in the same manner as in BL [54]. In summary, SGs can be used both at the context of education, as well as tools for informal education at home encompassing various technologies (e.g. online, learning, learning analytics).

The goal of this study is to examine the landscape of SGs for people with ID and ASD, as well as the perceptions of SEP and SET working in schools and institutions for people with ID and ASD regarding the role of technology and SGs in their education. The rest of the paper is organized as follows. In Sect. 2 the goals and the methodology of our study are presented. In Sect. 3 the results are summarized and comparatively analyzed and in Sect. 4 the conclusions of the study are presented.

2 Research Goal and Study Methodology

The main goal of this study is to examine whether SGs can be utilized as a means of promoting BL opportunities for people with ID or ASD, that is if SGs can be used by people with ID or ASD for learning not only in the context of school but at home as well on their own pace. In order to approach this goal, we have to:

- *Objective 1*: Study the landscape of SGs for people with ID or ASD based on: the category of targeted skill; platform (desktop-based, web-based or app-based); teaching approach used; evaluation results.
- *Objective 2*: Study the perceptions of SEP and SET in the field, regarding various factors that could make possible the usage of SGs for supporting BL. The factors studied are: the usage of technology (computers, smartphones, tablets) by people with ID or ASD; the current state regarding the usage of educational software and SGs in the learning process; how effective SGs would be and for which specific adaptive behavior or intellectual functioning skills; what typical teaching approaches should be incorporated in SGs.
- *Main Goal:* Synthesize the results of the state-of-the-art regarding existing SGs for people with ID or ASD along with the perceptions of SEP and SET.

In order to achieve the first objective, we refined a previous work of ours [59] that reviewed 43 SGs for people with ID or ASD. The SGs were grouped based on the category of skill that they address according to the American Association on

Intellectual and Developmental Disabilities (AAIDD) definition of ID [2]. In this study the SGs are organized in two main categories regarding the target group (ID or ASD) and are further grouped based on the targeted skill(s), the type/platform of the game, the teaching approach utilized (where defined) and the evaluation results. The evaluation results indicate if the SG was evaluated for its usability or effectiveness and if the results were positive (+), neutral (±) or negative (−).

In order to study the stakeholders' perceptions (objective 2) an online questionnaire was prepared and distributed to special schools and institutions for people with ID and ASD in Greece. Ninety-three SEP and SET responded and the data collected were analyzed using descriptive statistics.

Finally, in order to reach the main goal, we critically analyze the results of the literature review on SGs along with the results of the questionnaire survey. This analysis aims at investigating whether the SGs developed for people with ID or ASD: utilize the teaching methods proposed by the SEP and SET; address the adaptive behavior or intellectual functioning skills that the SEP and SET believe that such SGs would be more efficient; utilize the platform that better suits people with ID or ASD. This will allow us to draw significant conclusions regarding the inclusion of BL in the learning process of people with ID or ASD using SGs.

3 Results of Literature and Questionnaire

3.1 State-of-the-Art

Tables 1 and 2 present SGs for people with ID and ASD respectively, providing the information described in Sect. 2.

The 14 SGs presented in Table 1 aim to improve adaptive behavior or intellectual functioning skills for people with ID. Most of the SGs presented aim to improve conceptual skills (5) or intellectual functioning skills (6). Regarding the hosting platform, 7 SGs are desktop-based, 2 are app-based, 2 web-based and 2 run on game consoles. One SG [38] runs both on the web and as a desktop application. The majority of the SGs (9) have been positively evaluated for either their usability or their effectiveness. As far as the teaching approach is concerned, in 7 SGs the approach is not presented or defined. However, in most of the rest SGs, it is observed that role-playing (3) and conversation (2) methods are utilized, although other approaches are also identified, such as experimental learning and narration.

The 29 SGs presented in Table 2 aim to improve adaptive behavior or intellectual functioning skills for people with ASD. The majority of the SGs presented aim to improve the social skills of the users (17), addressing social interaction [24], emotion and facial recognition [1] and collaboration [53]. Most of the SGs for people with ASD are desktop-based (21) and fewer are app-based (5). It is observed that there are no web-based SGs. Also, there are studies were a custom hosting system is used [44, 53]. Although the majority of SGs is positively evaluated (18), negative evaluation results were reported for 4 SGs. The teaching approach utilized in most of the SGs for people with ASD is not defined. The teaching approach used in the rest of the SGs is: conversation (4); role-playing (3); and narration (3). In particular, the SGs that aim to improve social skills use mainly social interaction approaches, between the user and an AI agent [4] or between users [53].

Table 1. SGs for people with ID.

Category of skill	Num	Type/platform				Teaching approach (number of SGs)	Evaluation results
		Desktop	Web	App	Console		
Adaptive behavior: conceptual skills							
Money [15, 16, 41]	2	1		1		Experimental learning (1)	Effectiveness + (1) Usability + (1)
Language, literacy [50]	1				1		
Time [48]	1		1			Conversation self-exploration	Effectiveness +
Numbers [6]	1		1			Role-playing	
Adaptive behavior: social skills							
Interpersonal [20, 37]	1	1				Conversation	Usability +
Social responsibility [10]	1	1					Effectiveness +
Adaptive behavior: practical skills							
Daily living [9]	1	1				Narration	Effectiveness -
Work-related [38, 39, 56, 57]	2	2	1			Role-playing (1)	Usability + (1) Usability ± (1)
Healthcare [36, 49]	2	1			1		Effectiveness + (1) Usability/Effect. + (1)
Travel, transportation [7, 8]	1			1			Usability +
Intellectual Functioning							
Cognitive skills [33, 34]	1	1				Role-playing	

In general, it is observed that the existing literature on SGs addresses the importance of using SGs as supplementary tools, in combination with existing typical teaching methods or practices [27]. Therefore, using a SG in combination with the appropriate established learning methods can be considered as BL, because BL encourages combinations of different learning methods [54] utilizing technology.

Table 2. SGs for people with ASD.

Category of skill	Num	Type/platform				Teaching approach (number of SGs)	Evaluation results
		Desktop	Web	App	Console		
Adaptive behavior: conceptual skills							
Language, literacy [46]	1	1					Effectiveness + (1)
Adaptive behavior: conceptual skills							
Interpersonal [1, 4, 12, 19, 21, 24, 25, 27–30, 35, 42, 44, 45, 53, 55, 58, 62]	18	13		3	2	Conversation (4) Role-playing (2) Narration (2)	Usability + (4) Usability – (2) Effectiveness + (4) Effectiveness – (1) Effectiveness ± (1) Effectiveness N/A (1) Usability/Effect. + (1)
Adaptive behavior: practical skills							
Daily living [5]	1				1		Effectiveness +
Safety [61]	1	1					Usability +
Healthcare [13, 22]	2	2					Usability + (1)
Schedules, routines [60]	1			1		Narration	Usability -
Use of Telephone, Internet [51]	1	1				Role-playing	Usability +
Intellectual functioning							
Cognitive skills [11, 17, 32, 47]	4	3		1			Usability + (2) Effectiveness + (2)

3.2 Questionnaire

The respondents comprised of 64 female and 29 male SEP (9) and SET (84) working in different institutions, such as special primary school, special secondary high school, Special Education and Training Workshop, etc. The participants have on average 9.8 years of experience in special education.

Table 3 presents the participants' usage of technology in the learning process. As presented, PC is mainly used in the learning process compared to smartphones or tablets. Furthermore, the use of smartphones and tablets is quite similar, indicating that mobile devices are used equally in the learning process.

Table 3. Usage of technology in the learning process (1 = never, 5 = very much).

	Mean	SD	1	2	3	4	5
PC	3.58	1.15	3.2% (3)	18.3% (17)	20.4% (19)	33.3% (31)	24.7% (23)
Smartphone	2.06	1.22	46.2% (43)	22.6% (21)	12.9% (12)	15.1% (14)	3.2% (3)
Tablet	2	1.26	50.5% (47)	21.5% (20)	12.9% (12)	8.6% (8)	6.5% (6)

The results presented in Table 4, show the use of educational software or SGs in the learning process. It is clear that SEP and SET have preference in using educational software compared to SGs. However, the number of participants that use SGs is encouraging and indicates that SEP and SET are familiar with SGs. The results also indicate that SGs could be included in the learning process of people with ID or ASD and provides further room for exploration.

Table 4. Use of educational software and SGs (1 = minimum degree, 5 = maximum degree).

	Mean	SD	1	2	3	4	5
Educational software	3.1	1.23	9.7% (9)	25.8% (24)	24.7% (23)	24.7% (23)	15.1% (14)
Serious games	2.44	1.35	35.5% (33)	17.2% (16)	24.7% (23)	12.9% (12)	9.7% (9)

Table 5 presents the perceptions of SEP and SET regarding the role of technology in the learning process of people with ID or ASD (1 = very negative, 5 = very positive) and the familiarization of people with ID or ASD with technology (1 = very low, 5 = very high). The results indicate that there is not a significant difference regarding the role of technology between people with ID and people with ASD, which is considered much or very much important. However, SEP and SET believe that people with ASD are more familiar with technology. Therefore, both groups are more or less familiar with technological means and technology is important for their education, and this encourages us to believe that BL could be used successfully.

Table 5. Role of technology in the learning process and familiarization with technology.

Technology	Mean	SD	1	2	3	4	5
ID – Role	4.46	.60	0	0	5.4% (5)	43% (40)	51.6% (48)
ASD – Role	4.37	.69	0	0	11.8% (11)	39.8% (37)	48.4% (45)
ID – Familiarization	3.03	.88	1.1% (1)	28% (26)	43% (40)	22.6% (21)	5.4% (5)
ASD – Familiarization	3.55	1.04	2.2% (2)	16.1% (15)	24.7% (23)	38.7% (36)	18.3% (17)

Figure 1 presents the opinion of SEP and SET on the effectiveness of SGs for people with ID or ASD in specific adaptive behavior and intellectual functioning skills. The participants believe that SGs could effectively improve all the skills presented in the questionnaire for people with ID or ASD. Also, they believe that SGs for people with ID or ASD would be more effective for children or young adults (16–24), apart from work-related skills, where adults (<35) should be the target group. However, as observed in Table 4, the use of SGs in the learning process is limited, so we have to be reserved towards the opinion of the SEP and SET that SGs could be effective.

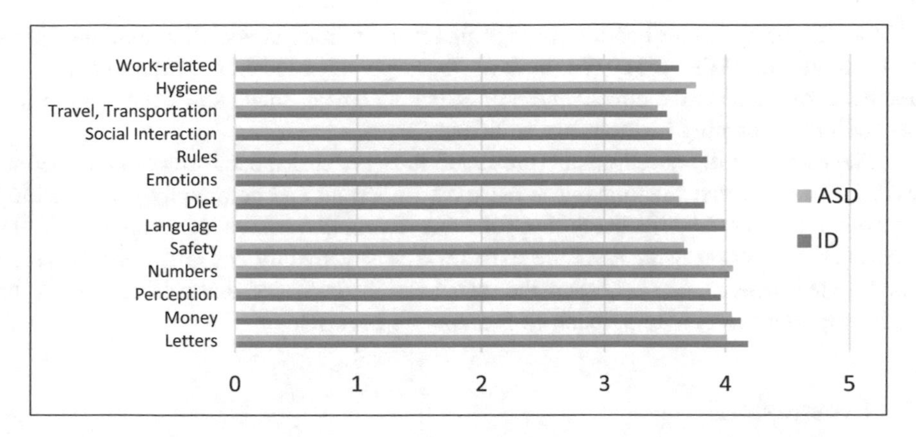

Fig. 1. Effectiveness of SGs on various skills (1 = not effective, 5 = very effective).

Figure 2 presents the importance of incorporating typical teaching approaches in SGs in order for them to be effective. The results indicate that incorporating role-playing is considered important for the effectiveness of SGs for both groups.

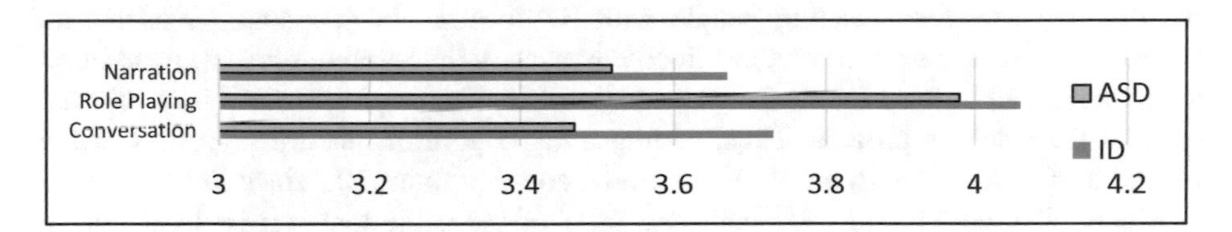

Fig. 2. Importance of teaching approaches in SGs (1 = not at all, 5 = very much).

3.3 Literature vs Reality

The next step of this study is to compare the results of the existing SGs with the perceptions of SEP and SET. Firstly, the most suited electronic device for developing an educational solution for people with ID or ASD is the PC, since it is the main hosting system for the existing SGs (65%) and the most common device that SEP and SET use in their work (58% uses PCs, 18.3% uses smartphones and 15.1% uses tables much or very much).

The results from the literature review indicate that the majority of SGs developed for people with ID address conceptual and cognitive skills. In addition, the SEP and SET believe that SGs addressing conceptual skills would be more effective. In particular, developing SGs for improving the concepts of money, letters and numbers, would be more effective. However, it is observed that SGs for improving social skills (social interaction and emotion control) would not be such effective.

As observed in Fig. 1, SGs for people with ASD would be more effective for conceptual skills. However, the studies of the existing SGs address mainly social skills, such as social interaction, emotion control, facial recognition and collaboration. People with ASD, on the other hand, have significant limitations in social, communicational and emotional skills [63]. We believe that special education professionals' and teachers' opinion on the effectiveness of SGs concerning social skills rely on the fact that common learning methods are sufficient.

The results of the questionnaire presented in Fig. 2 and the existing SGs for people with ID indicate that role-playing is the most important and commonly used teaching approach that could effectively improve the learning process of people with ID. However, for people with ASD the existing SGs use mainly conversational teaching approaches, whereas the results of the questionnaire indicated that role-playing is the most important teaching approach to incorporate in a SG.

4 Conclusions

In the present study, we examined the state-of-the-art of SGs for people with ID or ASD and the results of a close-ended questionnaire. We conducted an analysis of the literature compared to the opinions of SEP and SET. In addition, we aimed to identify whether promoting BL through SGs for people with ID or ASD is possible. Based on the literature and the questionnaire, it is concluded that SGs could effectively enhance the existing learning methods. The implementation of personalized learning models is an effective way for educating people with ID or ASD. In this sense, SGs that are adjusted to the user's profile can be effectively used in the learning process. In addition, such games can be used for the daily evaluation of students' progress and help SEP and SET in the teaching process. Thus, adding a game performance-tracking tool, which exists in SGs examined in our previous study, could promote BL, since in this way the SEP and SET could adjust the path and pace of the game that can be used both at school and home.

Furthermore, the comparison of existing SGs with the results of the questionnaire indicated that SGs for people with ID could be effectively used for promoting conceptual skills, whereas for people with ASD the results are conflicted. Although the existing SGs address social skills, SEP and SET believe that developing SGs to improve conceptual skills would be more effective. In addition, based on the questionnaire, SGs for people with ID that apply a role-playing approach are preferred and considered effective. We have to note, however, that in several SGs the teaching approach utilized is not defined, so the conclusion considering the inclusion of specific teaching approaches cannot be generalized and further research is required.

References

1. Alves, S., Marques, A., Queirós, C., Orvalho, V.: LIFEisGAME prototype: a serious game about emotions for children with autism spectrum disorders. Psychnol. J. **11**(3), 191–211 (2013)
2. American Association of Intellectual Disabilities and Developmental Disabilities (AAIDD): Definition of Intellectual Disability (2013). http://aaidd.org/intellectual-disability/definition#.Vv4m3fmLTUI. Accessed 01 Apr 2016

3. Bartoli, L., Corradi, C., Garzotto, F., Valoriani, M.: Exploring motion-based touchless games for autistic children's learning. In: Proceedings of the 12th International Conference on Interaction Design and Children, pp. 102–111. ACM (2013)
4. Bernardini, S., Porayska-Pomsta, K., Smith, T.J.: ECHOES: an intelligent serious game for fostering social communication in children with autism. Inf. Sci. **264**, 41–60 (2014)
5. Blum-Dimaya, A., Reeve, S.A., Reeve, K.F., Hoch, H.: Teaching children with autism to play a video game using activity schedules and game-embedded simultaneous video modeling. Educ. Treat. Child. **33**(3), 351–370 (2010)
6. Boleracki, M., Farkas, F., Meszely, A., Szikszai, Z., Sik, L.C.: Developing an animal counting game in second life for a young adult with down syndrome. Stud. Health Technol. Inform. **217**, 71–77 (2015)
7. Brown, D.J., McHugh, D., Standen, P., Evett, L., Shopland, N., Battersby, S.: Designing location-based learning experiences for people with intellectual disabilities and additional sensory impairments. Comput. Educ. **56**(1), 11–20 (2011)
8. Brown, D., Standen, P., Saridaki, M., Shopland, N., Roinioti, E., Evett, L. Grantham, S., Smith, P.: Engaging students with intellectual disabilities through games based learning and related technologies. In: Universal Access in Human-Computer Interaction. Applications and Services for Quality of Life, pp. 573–582. Springer, Heidelberg (2013)
9. Burke, K., OBroin, D., McEvoy, J.: Teaching death concepts to adults with intellectual disabilities using computer games. In: European Conference on Games Based Learning, vol. 2, pp. 767–774 (2014)
10. Chang, Y.J., Kang, Y.S., Liu, F.L.: A computer-based interactive game to train persons with cognitive impairments to perform recycling tasks independently. Res. Dev. Disabil. **35**(12), 3672–3677 (2014)
11. Chen, C.: Developing a tablet computer game with visual-spatial concept jigsaw puzzles for autistic children (2013). http://design-cu.jp/iasdr2013/papers/2197-1b.pdf. Accessed 22 Aug 2017
12. Christinaki, E., Vidakis, N., Triantafyllidis, G.: A novel educational game for teaching emotion identification skills to preschoolers with autism diagnosis. Comput. Sci. Inf. Syst. **11**(2), 723–743 (2014)
13. Chukoskie, L., Soomro, A., Townsend, J., Westerfield, M.: 'Looking' better: designing an at-home gaze training system for children with ASD. In: 6th International IEEE/EMBS Conference on Neural Engineering, pp. 1246–1249 (2013)
14. Cote, D.: Problem-based learning software for students with disabilities. Intervent. Sch. Clin. **43**(1), 29–37 (2007)
15. Curatelli, F., Martinengo, C.: Design criteria for educational tools to overcome mathematics learning difficulties. Procedia Comput. Sci. **15**, 92–102 (2012)
16. Curatelli, F., Martinengo, C., Bellotti, F., Berta, R.: Paths for cognitive rehabilitation: from reality to educational software, to serious games, to reality again. In: Games and Learning Alliance, pp. 172–186 (2013)
17. Davis, M., Otero, N., Dautenhahn, K., Nehaniv, C.L., Powell, S.D.: Creating a software to promote understanding about narrative in children with autism: reflecting on the design of feedback and opportunities to reason. In: 6th International IEEE Conference on Development and Learning, pp. 64–69 (2007)
18. Delavarian, M., Bokharaeian, B., Towhidkhah, F., Gharibzadeh, S.: Computer-based working memory training in children with mild intellectual disability. Early Child Dev. Care **185**(1), 66–74 (2015)
19. Deriso, D., Susskind, J., Krieger, L., Bartlett, M.: Emotion mirror: a novel intervention for autism based on real-time expression recognition. In: Computer Vision–ECCV 2012. Workshops and Demonstrations, pp. 671–674 (2012)

20. Fernández-Aranda, F., Jiménez-Murcia, S., Santamaría, J.J., Gunnard, K., Soto, A., Kalapanidas, E., Bults, R.G., Davarakis, C., Ganchev, T., Granero, R., Konstantas, D.: Video games as a complementary therapy tool in mental disorders: PlayMancer, a European multicentre study. J. Ment. Health **12**, 364–374 (2012)
21. Finkelstein, S.L., Nickel, A., Harrison, L., Suma, E.A., Barnes, T.: cMotion: a new game design to teach emotion recognition and programming logic to children using virtual humans. In: VR IEEE Conference, pp. 249–250 (2009)
22. Finkelstein, S., Nickel, A., Lipps, Z., Barnes, T., Wartell, Z., Suma, E.A.: Astrojumper: motivating exercise with an immersive virtual reality exergame. Presence: Teleoperators Virtual Environ. **20**(1), 78–92 (2011)
23. Frank, A.: Balancing three different foci in the design of serious games: engagement, training objective and context. In: DiGRA Conference, September 2007
24. Friedrich, E.V., Sivanathan, A., Lim, T., Suttie, N., Louchart, S., Pillen, S., Pineda, J.A.: An effective neurofeedback intervention to improve social interactions in children with autism spectrum disorder. J. Autism Dev. Disord. **45**(12), 4084–4100 (2015a)
25. Friedrich, E.V., Suttie, N., Sivanathan, A., Lim, T., Louchart, S., Pineda, J.A.: Brain–computer interface game applications for combined neurofeedback and biofeedback treatment for children on the autism spectrum. In: Interaction of BCI with the Underlying Neurological Conditions in Patients: Pros and Cons, vol 7, no. 21, pp 15–21 (2015b)
26. Garrison, D.R., Kanuka, H.: Blended learning: uncovering its transformative potential in higher education. Internet High. Educ. **7**(2), 95–105 (2004)
27. Golan, O., Baron-Cohen, S.: Systemizing empathy: teaching adults with Asperger syndrome or high-functioning autism to recognize complex emotions using interactive multimedia. Dev. Psychopathol. **18**(2), 591–617 (2006)
28. Grynszpan, O., Martin, J.C., Nadel, J.: Multimedia interfaces for users with high functioning autism: an empirical investigation. Int. J. Hum.-Comput. Stud. **66**(8), 628–639 (2008)
29. Hansen, O.B., Abdurihim, A., McCallum, S.: Emotion recognition for mobile devices with a potential use in serious games for autism spectrum disorder. In: Serious Games Development and Applications, pp. 1–14 (2013)
30. Hourcade, J.P., Bullock-Rest, N.E., Hansen, T.E.: Multitouch tablet applications and activities to enhance the social skills of children with autism spectrum disorders. Pers. Ubiquit. Comput. **16**(2), 157–168 (2012)
31. Hughes, C., Rung, L.L., Wehmeyer, M.L., Agran, M., Copeland, S.R., Hwang, B.: Self-prompted communication book use to increase social interaction among high school students. Assoc. Pers. Sev. Handicaps **25**(3), 153–166 (2000)
32. Hulusic, V., Pistoljevic, N.: "LeFCA": Learning framework for children with autism. Procedia Comput. Sci. **15**, 4–16 (2012)
33. Hussaan, A.M., Sehaba, K., Mille, A.: Tailoring serious games with adaptive pedagogical scenarios: a serious game for persons with cognitive disabilities. In: 11th IEEE International Conference on Advanced Learning Technologies, pp. 486–490 (2011)
34. Hussaan, A.M., Sehaba, K., Mille, A.: Helping children with cognitive disabilities through serious games: project CLES. In: The proceedings of the 13th International ACM SIGACCESS Conference on Computers and Accessibility, pp. 251–252 (2011)
35. Jain, S., Tamersoy, B., Zhang, Y., Aggarwal, J.K., Orvalho, V.: An interactive game for teaching facial expressions to children with autism spectrum disorders. In: 5th International Symposium on Communications Control and Signal Processing, pp. 1–4 (2012)
36. Karal, H., Kokoç, M., Ayyıldız, U.: Educational computer games for developing psychomotor ability in children with mild mental impairment. Procedia-Soc. Behav. Sci. **9**, 996–1000 (2010)

37. Kostoulas, T., Mporas, I., Kocsis, O., Ganchev, T., Katsaounos, N., Santamaria, J.J., Jimenez-Murcia, S., Fernandez-Aranda, F., Fakotakis, N.: Affective speech interface in serious games for supporting therapy of mental disorders. Expert Syst. Appl. **39**(12), 11072–11079 (2012)
38. Lányi, C.S., Brown, D.J.: Design of serious games for students with intellectual disability. IHCI **10**, 44–54 (2010)
39. Lányi, C.S., Brown, D.J., Standen, P., Lewis, J., Butkute, V.: Results of user interface evaluation of serious games for students with intellectual disability. Acta Polytech. Hung. **9** (1), 225–245 (2012)
40. Lock, R.H., Kingsley, K.V.: Empower diverse learners with educational technology and digital media. Interv. Sch. Clin. **43**(1), 52–56 (2007)
41. Lopez-Basterretxea, A., Mendez-Zorrilla, A., Garcia-Zapirain, B.: A telemonitoring tool based on serious games addressing money management skills for people with intellectual disability. Int. J. Env. Res. Public Health **11**(3), 2361–2380 (2014)
42. Malinverni, L., Mora-Guiard, J., Padillo, V., Valero, L., Hervás, A., Pares, N.: An inclusive design approach for developing video games for children with Autism Spectrum Disorder. Comput. Hum. Behav. **71**, 535–549 (2017)
43. Michael, D.R., Chen, S.L.: Serious Games: Games that Educate, Train, and Inform. Muska & Lipman/Premier-Trade, USA (2005)
44. Piper, A.M., O'Brien, E., Morris, M.R., Winograd, T.: SIDES: a cooperative tabletop computer game for social skills development. In: Proceedings of the 20th ACM Anniversary Conference on Computer Supported Cooperative Work, pp. 1–10 (2006)
45. Porayska-Pomsta, K., Anderson, K., Bernardini, S., Guldberg, K., Smith, T., Kossivaki, L., Hodgins, S., Lowe, I.: Building an intelligent, authorable serious game for autistic children and their carers. In: Advances in Computer Entertainment, pp. 456–475 (2013)
46. Rahman, M.M., Ferdous, S.M., Ahmed, S.I.: Increasing intelligibility in the speech of the autistic children by an interactive computer game. In: IEEE International Symposium on Multimedia (ISM), pp 383–387 (2010)
47. Retalis, S., Korpa, T., Skaloumpakas, C., Boloudakis, M., Kourakli, M., Altanis, I., Siameri, F., Papadopoulou, P., Lytra, F., Pervanidou, P.: Empowering children with ADHD learning disabilities with the Kinems Kinect learning games. In: European Conference on Games Based Learning, vol 2, pp. 469–477 (2014)
48. Ripamonti, L.A., Maggiorini, D.: Learning in virtual worlds: a new path for supporting cognitive impaired children. In: Foundations of Augmented Cognition. Directing the Future of Adaptive Systems, pp. 462–471 (2011)
49. Salem, Y., Gropack, S.J., Coffin, D., Godwin, E.M.: Effectiveness of a low-cost virtual reality system for children with developmental delay: a preliminary randomised single-blind controlled trial. Physiotherapy **98**(3), 189–195 (2012)
50. Sánchez, J.L.G., Gutiérrez, F.L., Cabrera, M., Zea, N.P.: Design of adaptative video game interfaces: a practical case of use in special education. In: Computer-aided design of user interfaces VI, pp. 71–76 (2009)
51. Sbattella, L., Tedesco, R., Trivilini, A.: Multimodal interaction experience for users with autism in a 3D environment. In: 6th European Conference on Games Based Learning, pp. 442–450 (2012)
52. Shogren, K.A., Wehmeyer, M.L., Davies, D., Stock, S., Palmer, S.B.: Cognitive support technologies for adolescents with disabilities: impact on educator perceptions of capacity and opportunity for self-determination. J. Hum. Dev. Disabil. Soc. Change **21**(1), 67–80 (2013)
53. Silva, G.F.M., Raposo, A., Suplino, M.: PAR: a collaborative game for multitouch tabletop to support social interaction of users with autism. Procedia Comput. Sci. **27**, 84–93 (2014)

54. Singh, H.: Building effective blended learning programs. Educ. Technol.-Saddle Brook Englewood Cliffs NJ- **43**(6), 51–54 (2003)
55. Tanaka, J.W., Wolf, J.M., Klaiman, C., Koenig, K., Cockburn, J., Herlihy, L., Brown, C., Stahl, S., Kaiser, M.D., Schultz, R.T.: Using computerized games to teach face recognition skills to children with autism spectrum disorder: the Let's Face It! program. J. Child Psychol. Psychiatry **51**(8), 944–952 (2010)
56. Torrente, J., Del Blanco, Á., Moreno-Ger, P., Fernández-Manjón, B.: Designing serious games for adult students with cognitive disabilities. In: Neural Information Processing, pp. 603–610 (2012)
57. Torrente, F.J., Blanco Aguado, Á.D., Serrano Laguna, Á., Vallejo Pinto, J.A., Moreno Ger, P., Fernández Manjón, B.: Towards a low cost adaptation of educational games for people with disabilities. ComSIS **11**(1), 369–391 (2014)
58. Tsai, T.W., Lin, M.Y.: An application of interactive game for facial expression of the autisms. In: Edutainment Technologies. Educational Games and Virtual Reality/Augmented Reality Applications, pp. 204–211 (2011)
59. Tsikinas, S., Xinogalos, S., Satratzemi, M.: Review on serious games for people with intellectual disabilities and autism. In: European Conference on Game Based Learning, pp. 696–703 (2016)
60. Yan, F.: A sunny day: Ann and Ron's world an iPad application for children with autism. In: Serious Games Development and Applications, pp. 129–138. Springer (2011)
61. de Urturi, Z.S., Zorrilla, A.M., Zapirain, B.G.: A serious game for android devices to help educate individuals with autism on basic first aid. In: Distributed Computing and Artificial Intelligence, pp 609–616 (2012)
62. Wijnhoven, L.A., Creemers, D.H., Engels, R.C., Granic, I.: The effect of the video game mindlight on anxiety symptoms in children with an Autism Spectrum Disorder. BMC Psychiatry **15**(1), 1–9 (2015)
63. Zakari, H.M., Ma, M., Simmons, D.: A review of serious games for children with autism spectrum disorders (ASD). In: Serious Games Development and Applications, pp. 93–106 (2014)

Design, Implementation and Evaluation of a Computer Science Teacher Training Programme for Learning and Teaching of Python Inside and Outside School

Establishing and Supporting Code Clubs to Learn Computer Programming by Self-contained Examples

Nektarios Moumoutzis[1(✉)], George Boukeas[2], Vassilis Vassilakis[2], Nikos Pappas[1], Chara Xanthaki[2], Ioannis Maragkoudakis[1], Antonios Deligiannakis[1], and Stavros Christodoulakis[1]

[1] Laboratory of Distributed Multimedia Information Systems and Applications, School of Electrical and Computer Engineering, Technical University of Crete, Chania, Greece
{nektar, nikos, imarag, adeli, stavros}@ced.tuc.gr
[2] Secondary Education, Chania, Greece
boukeas@gmail.com, vassilakis1978@gmail.com, chara.xanthaki@gmail.com

Abstract. We present the design, implementation and evaluation of a training programme for Computer Science teachers on the educational use of the Python programming language inside and outside school. The programme used educational resources centered on meaningful, self-contained programming projects. The training programme followed a blended-learning approach thus offering an opportunity to many computer science teachers make their first steps towards the educational use of the Python language within a very promising learner-centered pedagogical framework. Using initially an online course made it possible to reach a much greater number of Computer Science teachers, especially those living in remote areas and through them, have a considerable impact on students through the subsequent establishment of local code clubs. The synchronous interaction with the course facilitators during the monthly sessions, the forming of local groups and the systematic communication through the learning platform used, made it possible to alleviate many of the disadvantages usually linked with online courses as it is evident from the evaluation results.

Keywords: Python programming · Computer science in secondary education Code clubs

1 Introduction

Teaching introductory programming courses has received much attention the last years. This is mainly due to the ubiquitous use of computers, the proliferation of the so called cultures of participation [1], end-user programming [2] and end-user software

M. E. Auer and T. Tsiatsos (Eds.): IMCL 2017, AISC 725, pp. 575–586, 2018.
https://doi.org/10.1007/978-3-319-75175-7_56

engineering [3]. These trends are addressing software tools that provide powerful scripting languages to enable flexible customization and rich interactive content development by end-users. In this respect, knowledge of computer programming concepts is nowadays necessary for most knowledge workers including scientists and engineers. Consequently, many higher education departments have included introductory programming courses in their curricula [4]. Furthermore, many countries extend their curricula in secondary or even primary education to address the development of basic programming skills [5]. The importance of computer programming has received even more attention through computer coding campaigns such as the *The Hour of Code* and *Europe Code Week*. Informal learning opportunities are also offered in many countries following the organizational approach of coding clubs [6].

Following the above trends, Python has recently been introduced in vocational training curricula (professional lyceums) as well as in upper secondary education in Greece. However, many Computer Science teachers are not familiar with Python and all available professional development opportunities (mostly short webinars) are in high demand and overbooked. To address the need of enabling secondary Computer Science teachers in Greece to get familiar with the Python programming language and adopt effective learner-centered pedagogies, a 7-month teacher training programme was designed and implemented with partial funding from the Google CS4HS initiative. This programme is described here along with its evaluation. It was offered in a blended-learning fashion starting with a 3-month distant learning phase to study the Python language through self-contained programming projects and a subsequent 3-month phase with face-to-face collaboration focusing on the establishment of local code clubs. The programme finished with an evaluation phase of one month duration.

The rest of this paper is organized as follows: Sect. 2 presents previous experience in organization of training programmes to enable teachers enrich their teaching practices and use engaging learning scenarios for promoting coding at schools. Section 3 presents very briefly the training portal that we used to support distant training services while Sect. 4 focuses on the training programme philosophy and structure combining distant training with local collaboration thus creating a blended learning framework. Section 5 presents the methodology and the results of the evaluation Sect. 6 concludes and presents future plans.

2 Previous Steps on Promoting Coding in Schools Within a Creative Learning Context

An initial exploration on the potential of programming in education from the perspective of providing opportunities for students in primary and secondary education develop their own coding projects for learning was undertaken during the pSkills project (October 2009 – September 2011) [7]. The target was to build a community of primary and secondary teachers, mainly computer science teachers, to exploit modern educational programming languages in their courses. The approach adopted could be described as a teacher training framework rather than a learning framework. This proved to be rather limiting in exploiting the full potential of the approach. However, the fact that most of the participants were computer science teachers made it possible for them to proceed very quickly with the appropriation of the underlying technologies and enter the classrooms successfully. Other teacher specialties faced significant problems and were unable to do the same.

The pSkills teacher training activities were structured as a three-step process: (a) initial training and community building workshops with over 400 participants in total; (b) pilot workshops, one in each one of the four participating countries with over 40 participants in Greece, Austria, Italy and Estonia; and (c) the pSkills Summer School, a one-week intensive training event with 10 participants. In all these phases, a training portal was used to host all materials and provide communication between the participants before and after the events, thus providing the basis of a blended-learning approach. Though its training activities, the project raised awareness and provided insight and inspiration through indicative learning scenarios targeting courses in primary and secondary education. Furthermore, it offered materials to enable teachers guide their students through an engaging learning process during which they develop their own coding projects.

The focus was on developing digital games using an appropriated educational programming language [8]. This process starts from the inception of a game employing brainstorming. The game design is facilitated by storyboards while testing by evaluation rubrics. Game development was based on Scratch [9] and game distribution on the Scratch community site. The offered materials included: (a) Brainstorming guidelines and selected Scratch projects to provide inspiration; (b) Storyboarding templates in the form of slide presentations; (c) Game skeletons in Scratch along with introductory hands-on tutorials to enable game development; (d) Worksheets for step-by-step development of simple games and ideas for their extension; (e) Rubrics for peer assessment containing criteria regarding playability, usability and user experience qualities; and (f) Links to Scratch community site pages with important information and additional links to related resources on the web.

The ultimate goal of the pSkills project was to foster a systemic change beyond the mainstream focus on the so-called IT literacy towards IT fluency. IT literacy is linked to surface technical skills related to office automation applications (word processing, spreadsheets, presentations etc.) and communication tools (email, web browsers etc.). IT fluency focuses on sufficiently foundational material with "staying power" to promote understanding of computers and their applications and the ability to fully exploit the potential of modern systems and computer applications through programming.

Towards the end of the implementation period of the project, an explosion of interest of computer science teachers on its themes was observed. Over 250 teachers participated in the local workshops held in Crete, Greece. Most of them started to apply the ideas into the classrooms and, more importantly, within the context of creative projects with their students. Following a viral pattern, the use of Scratch was soon spread in other schools and the student projects developed were enough to justify a first attempt to organize a Student's Digital Creativity Fest in 2011. Starting from Crete, this annual event has now reached most of the Greek regions with more than 7000 participating students, around 1000 teachers and more than 400 schools. These developments had a notable contribution in firing a positive change in the national education system: In the new curricula for computer science at secondary education and the supporting textbooks, there is now an explicit focus on game design, game development and the use of appropriate platforms such as Scratch and App Inventor. This observed impact justifies our decision to adopt a bottom-up approach and provides evidence that small changes in everyday learning activities in the classrooms can trigger systemic changes as well.

Following the successful impact of the pSkills project, our focus expanded beyond computer science education to include other domains including non-formal and informal education. The opportunity for this was given within the context of the ALICE project [10] and its decision to include games as a creative language to be adopted by adult trainers along with Music, Digital Storytelling, and Children Narratives. ALICE targeted adult trainers with the aim to enable them design and implement intergenerational creative learning environments. The project designed and piloted a graduate programme for educators in Greece, Italy, Switzerland, UK and Romania. The programme was offered through a learning portal and included face-to-face sessions as well in each one of the five participating countries, thus adopting a blended-learning approach. The training consisted of six learning units and a final project. One of these learning units targeted digital games and a number of participants adopted games as a creative language to work on their individual projects during which they designed and implemented learning activities where adults and children they were invited to learn together in creative ways. The term gaming-literacy was adopted to signify a step towards the accommodation of a new way of thinking, working, collaborating, teaching and learning, as initially introduced by Zimmerman [11].

The gaming literacy learning framework offered the opportunity to the ALICE participants to address topics on three levels: (1) Understanding and evaluating games through critical analysis that promotes the acquisition of the language pertaining to technology, genres, values, stereotypes, production processes of games and their learning value. (2) Critical consumption through reflection on gamers' behavior in order to better exploit free time, foster learning and enrich human relationships. Time spent for video game playing, game preferences, social aspects of game play, type of entertainment and learning offered are issues related to this critical self-reflection. (3) Crafting digital artifacts using modern tools that enable non-technical people to invent their own video games and be engaged in their realization by creating rules, characters, narratives, graphics, audio, and animations.

The work reported in this paper is a third step towards supporting educators on the integration of new technological tools in their teaching practices emphasizing on digital skills and creative learning. This time, the focus is on general purpose programming language, not an educational or creative language as it was the case in the pSkills and the ALICE projects. In particular, the programming language adopted in this case is Python and its use as a first programming language for novices taking into account the parallel developments in the school curricula in Greece and the need to support Geek computer science teachers on their use of the Python language within a pedagogical framework that promotes active learning within engaging learning scenarios in coding clubs. Selecting coding clubs as the pedagogical framework of our approach incorporates the positive prior experience on promoting coding in formal, informal and non-formal learning settings in pSkills and ALICE projects.

3 The Training Portal

The training portal used for setting up the blended learning framework of the training programme is the evolution of MOLE, a multimedia online learning environment [12] that was initially developed to support educational activities in university departments. MOLE integrated tools and services for educational material reuse in an interoperable manner [13]. After its successful adoption in the academic environment, it has been adapted and enhanced under the name *Coursevo* to support professional development and training within a context that enables the establishment and sustainable operation of Communities of Practice (CoP) [14].

Coursevo enables communication between tutors/trainers and trainees, cooperation among trainees and access to coursework information and learning resources. It can combine traditional classroom-based lessons and practical sessions, with self-study and eLearning. Coursevo platform hides the complexity and frees the trainers from tedious system maintenance tasks, since a course or even a full functional learning site can be created in a few steps following the SaaS (Software as a Service) paradigm

Coursevo integrates BigBlueButton (http://bigbluebutton.org/) to enable video teleconferencing. This proved very important for the implementation of the programme: One synchronous teleconference was organized each month to give guidance to participants, present best practices and examples, answer to questions and solve practical problems, especially for the organization of local code clubs. Furthermore, the workgroup support offered by Coursevo was used and appropriately adapted to enable the coordination of work in each group (code club organizers) and facilitate the development and submission of assignments.

4 The Training Programme

The design of the training programme was based on the principles of social constructivism: Initially, participants explored the course material and develop Python skills, as they delved into the programming projects and tackled the assignments. The assignments, meant to induce structure on the learning process, were implemented in groups exploiting the special features of the underlying learning platform. There were no lectures on Python programming. The course facilitators served as peer advisors, guides and coordinators. Following this initial phase of getting familiar with the Python language and the proposed pedagogical methodology, participants were asked to apply the knowledge and skills they have acquired, in workshops or coding clubs for their students, exploiting scenario-based pedagogical approaches. In this context, they eventually composed their own training material and developed strategies for cooperating with other participants from their regional group, learn from and support each other. There was also a strong element of reflection, self- and peer-evaluation at the final phase of the project. As already stated, an important aspect of the training programme was its blended-learning approach to promote collaboration among computer science teachers in many locations parts of Greece including several remote areas.

The program consisted of three phases. In the first phase (3 months) participants studied, explored and evaluated the course material, familiarizing themselves with

Python and the programming projects approach. In the second phase (3 months) participants implemented the course material in coding clubs. The final phase (1 month) involved extensive reflection and evaluation of the course.

The kick off was done via a teleconference that presented the overall structure and objectives of the training programme. Every month there was a live online session with the course facilitators, where participant groups had the opportunity to make presentations or engage in structured discussions. During the course, participants also communicated via discussion forums, online chat rooms and videoconferencing facilities offered by the Coursevo platform. Participants worked together in regional groups and posted their assignments online, each group creating a portfolio that was reviewed by their peers. All results were documented and shared in the form of adaptable learning scenarios (i.e. project-based scenarios and/or lesson plans) that referenced teaching objectives of the Greek CS curricula and were organized in a digital repository that is available to all CS teachers through a Creative Common license for further reuse after the end of the project.

5 Evaluation of the Training Programme

The evaluation of the training programme was based on three complementary elements: (1) a detailed questionnaire for the participating computer science teachers; (2) an initial and a final questionnaire targeting the students participating in code clubs; (3) a self-evaluation report for each one of the established code clubs that was prepared by the organizers, i.e. the computer science teachers participating in the training programme that collaborated in the design and implementation of each code club.

Teacher evaluation of the training

The teachers' questionnaire contained 4 sections: (1) Demographic data; (2) Likert scale questions for the evaluation of the training programme; (3) Likert scale questions for personal and group evaluation; (3) Open-ended questions regarding the objectives of each participant and the degree of satisfaction as well as on the strong and weak points of the training programme. The Table 1 below summarizes the socio-demographic and professional profile of the participating teachers. There was equilibrium in gender while the vast majority of the participants were within the most active age-band for teaching (ages 36–55 sum up to more than 86% of the participants). Their distribution in terms of basic degrees, resonates with the overall distribution in computer science teacher community as a whole. Finally, their level of studies shows a high percentage of participants with postgraduate degrees, which is true for the computer science teacher community as a whole in Greece as well.

By the analysis of the answers to the teacher evaluation questionnaire (N = 80) it is evident that the official goals of the training programme, as listed in Sect. 4, were met (Table 2 below). The training programme provided an opportunity for the participating teachers to make their first steps on the educational use of the Python language within a promising pedagogical framework. Using an online course made it possible to reach a much greater number of teachers (especially those living in remote areas) and, through them, have a considerable impact on students through the establishment of code clubs.

Table 1. Socio-demographic and profesional profile of the participating teachers (N = 80)

Variable	N	(%)
Gender		
Male	40	(50.0)
Female	40	(50.0)
Age band		
<=25	1	(01.3)
26–35	4	(05.0)
36–45	35	(43.8)
46–55	35	(43.8)
>=56	5	(06.3)
Specialization		
University graduate in computer science	61	(76.3)
Technological institution graduate in computer science	17	(21.3)
Other	2	(02.5)
Level of studies		
Bachelor	33	(41.3)
Masters	43	(53.8)
PhD	4	(05.0)

The synchronous interaction with the course facilitators during the monthly sessions, the forming of regional groups and the systematic communication through the forums helped alleviate many of the disadvantages usually linked with online courses as it is evident from the analysis of the results from the teachers' questionnaires and the self-evaluation reports of the code clubs. The three-phase structure of the course enabled participants to gain practical, hands-on experience both while learning to program in Python themselves and while using it to teach programming to their students. The implementation of the course material by the participants ensured that the knowledge and skills acquired can be transferred into the classroom and put to practical use.

The coding club approach adopted for engaging students in programming projects was an important aspect of the training programme. It constituted its pedagogical background that differentiated it from other initiatives for computer science teacher professional development where training is usually decoupled from a certain student learning model. Coding clubs are ideal for implementing social constructivism and related pedagogical ideas that are considered very important for cultivating student initiative, creativity and innovative thinking. Students learn by creating things and teachers have a direct experience of organizing such structures and sustaining them.

Interesting findings were documented regarding issues related to the collaboration between participants in working groups and the establishment of code clubs as summarized in Table 3 below. The participants reported that they generally succeeded in reaching the goals of the programme especially with respect to familiarizing with the basic characteristics of Python and usage of self-contained examples in their teaching.

Table 2. Evaluation of the training programme as a whole (N = 80)

Variable	1	2	3	4	5	Mean
Training progr. success with respect to its initial objectives						
Familiarize with the basic characteristics of Python	0	1	4	44	31	**4.31**
Use self-contained examples in teaching programming	0	1	3	41	35	**4.38**
Enable sustainable local communities of practice	0	3	18	33	26	**4.03**
Evaluation of the learning material						
The learning material can be used without modification	2	5	16	45	12	**3.75**
Autonomous study of learning mat. is easy for novices	4	11	25	25	15	**3.45**
Introduces basic Python concepts in an engaging way	1	2	15	39	23	**4.01**
The content of the learning material is easy to understand	1	3	7	44	25	**4.11**
The learning material is attractive and appealing	0	3	11	40	26	**4.11**
Evaluation of the trainers						
They had adequate knowledge to support trainees	0	0	1	20	59	**4.73**
They effectively transmit their knowledge	0	0	6	28	46	**4.50**
They were supportive in the practical exercises	0	1	10	24	45	**4.41**
They responded to questions	0	0	4	16	60	**4.70**
They were adequaly prepared	0	0	2	18	60	**4.73**
Evaluation of teleconferences						
Presentations were adequate and complementary	0	1	13	40	26	**4.14**
Presentations and discussion addressed my needs	1	9	24	31	15	**3.63**
Teleconferences triggered reflection and discussion	0	2	11	40	27	**4.15**
Teleconferences gave me interesting ideas	0	4	15	41	20	**3.96**
Teleconferences stimulated further study	0	3	24	37	16	**3.83**
Evaluation of the organization and workload of exercises						
The material was adequate for doing the exercises	0	2	4	44	30	**4.23**
There was adequate support for finishing the exercises	0	1	8	43	28	**4.23**
Adequate time was given for doing the exercises	1	13	8	35	23	**3.83**
Evaluation of the training platform (Coursevo)						
The platform is easy to use	0	9	15	40	16	**3.79**
The platform has adequate functionality	0	9	11	46	14	**3.81**
Content navigation is easy and effective	2	13	26	30	9	**3.39**
Support for teleconferences is effective	0	2	6	42	30	**4.25**
Searching for content is adequately supported	5	13	24	29	9	**3.30**
There is adequate support for working in groups	0	6	10	46	18	**3.95**
There is adequate support on technical issues	0	0	10	38	32	**4.28**
Overall evaluation of the training programme						
It had clear objectives	0	6	17	40	17	**3.85**
It had activities that reflect and follow the objectives	0	2	12	43	23	**4.09**
The duration was adequate	0	7	10	48	15	**3.89**
The workload was adequate	3	12	23	31	11	**3.44**
There was adequate guidance by the trainers	0	1	7	33	39	**4.38**
My expectations were fulfilled	1	8	22	34	15	**3.68**

(*continued*)

Table 2. (*continued*)

Variable	1	2	3	4	5	Mean
I can apply what I have learned	0	1	5	42	32	**4.31**
Now I will perform better in my teaching duties	1	3	14	40	22	**3.99**
I will not need support to apply what I have learned	0	10	27	22	21	**3.68**

Their interest and participation was adequately focusing on the different phases of the training programme with a slightly more emphasis on the first phase. This is mainly due to some problems some participants had in combining their professional obligations with the code club creation phase. Positive evaluation of the collaboration of working groups was also reported. However, it was in some cases necessary for the trainers to intervene in working groups to help activities run smoothly. We believe this is important in such blended learning activities: The trainers should proactively and appropriately intervene to avoid problems in working groups. The training platform used offered adequate support for this monitoring taking into account the organization of the training through the use of particular services and the capability to send personal messages to individuals or group messages to members of working groups.

An important aspect of the evaluation is that coding club establishment was seen as a complex and risky task initially while at the end of the programme, the organizers were in many cases surprised by their students asking to continue with the code club and even extend its theme and contents in many other topics exploiting the wide range of uses of the Python programming language. This finding was documented by the self-evaluation forms of the code clubs (prepared by the participating teachers) providing detailed description in free text of their experience. Furthermore, the self-evaluation reports document that the requirement to use and adapt appropriate worksheets to support autonomous learning and personalization during the code club activities, although it was initially confronted with doubts, proved to be extremely effective. Furthermore, the approach to focus on self-contained projects instead of artificial examples and small programming exercises promoted student engagement and contributed to the creation of an atmosphere of meaningful learning.

Student evaluation of the training programme

Apart from teacher evaluation, as already mentioned, student evaluation was also done using an initial and a final questionnaire. The initial questionnaire contained 4 sections: (1) Demographic data; (2) Likert scale questions regarding the information about the code club, reasons for participation and expectations; (3) Likert scale questions about previous knowledge in programming; (4) Likert scale questions regarding attitude towards programming. The final questionnaire contained 5 sections: (1) Demographic data; (2) Likert scale questions regarding the evaluation of the code club; (3) Likert scale questions on the Python programming language and its possible use in secondary education; (4) Likert scale questions on the attitude towards programming; (5) Open-ended questions on the strong and weak points of the code club. By analysing the answers of these questionnaires (465 for the initial questionnaire and 358 for the final one) it is evident that the code clubs had a very positive impact on students in terms of developing programming skills and positive change in their attitude towards

Table 3. Personal and group evaluation of the training programme (N = 80)

Variable	1	2	3	4	5	Mean
Training progr. success with respect to its initial objectives						
I familiarized with the basic characteristics of Python	0	2	11	47	20	**4.06**
I can use self-contained examples in teaching progr.	0	0	12	47	21	**4.11**
I was enabled to participate in a sustainable local CoP	4	6	17	36	17	**3.70**
Evaluation of personal participation						
I was much interested throughout the programme	0	10	20	30	20	**3.75**
I was more interested in Python programming (phase 1)	5	11	20	24	20	**3.54**
I was more interested in code club creation (phase 2)	5	17	27	19	12	**3.20**
I actively participated throughout the programme	0	10	11	38	21	**3.88**
I was more active in Python programming (phase 1)	5	9	22	25	19	**3.55**
I was more active in the code club creation (phase 2)	5	18	23	19	15	**3.26**
I generally did the exercises within the deadlines	2	2	12	34	30	**4.10**
I did all/almost all individual and group exercises	1	8	8	23	40	**4.16**
I assimilated all/almost all the content of the programme	1	8	9	42	20	**3.90**
I participated adequately in the preparation and implementation of the code club of my working group	4	3	11	23	39	**4.13**
Evaluation of my working group						
Adequate collaboration when studying the learning mat.	1	8	20	27	24	**3.81**
Adequate collaboration during code club creation	1	5	12	28	34	**4.11**
Adequate collaboration during group exercises	1	9	9	34	27	**3.96**
Effective distribution of workload among group members	4	6	21	23	26	**3.76**
Fair distribution of workload among group members	7	13	16	24	20	**3.46**

programming which is now seen as an important professional pathway. Table 4 below demonstrates these findings presenting only the change in students' attitudes using the corresponding questions of pre- and post-questionnaires. Due to lack of space we do not present an analysis of the other questions leaving it for a future presentation.

Table 4. Attitudes of students towards programming before and after their participation in the local code clubs (Npre = 465, **Npost = 358**). Percentages are shown for likert scale values.

Variable	1	2	3	4	5	Mean
I believe that:						
Girls and boys are equally competent in coding	09.9	15.3	65.4	03.4	06.0	**2.80**
	07.8	**12.6**	**65.6**	**06.4**	**07.5**	**2.93**
I can collaborate with others when coding	06.5	08.4	24.3	26.7	34.2	**3.74**
	05.0	**07.0**	**21.8**	**29.1**	**37.2**	**3.87**
I can code by myself without help	12.5	23.7	24.9	20.2	18.7	**3.09**
	05.3	**15.4**	**27.9**	**30.4**	**20.9**	**3.46**
It is probable that I follow a computing profession	13.5	17.8	23.0	17.8	27.7	**3.23**
	10.9	**12.3**	**21.8**	**21.2**	**33.8**	**3.55**
Only future computer professionals should code	34.4	26.9	19.6	09.0	10.1	**2.34**
	27.7	**24.9**	**25.7**	**11.7**	**10.1**	**2.52**

6 Conclusions and Future Plans

The paper describes an approach and a concrete pilot experience with respect to teacher training in Python. The proposed training scheme is based on meaningful self-contained programming projects that are undertaken by students in coding clubs. A transfer was observed between knowledge acquired by the teachers and use of this knowledge with their own students. Evaluation was based on questionnaires and self-assessment for teachers and a pre-, post-questionnaire evaluation of the students that participated in the coding clubs established. Focusing on engaging programming projects rather than relying on artificial exercises addressing the syntax and the structure of Python (as it is in many cases the approach in traditional classroom teaching) highlights a wide range of higher-level concepts ranging from functional abstraction and problem-solving strategies to artificial intelligence.

Participating teachers were supported through distant-learning facilities offered by an appropriate learning platform (Coursevo) to study special material on the Python programming language and thus gain confidence in using an alternative, engaging methodology which can serve as a springboard for exposing their students to Computer Science practices and concepts. Participants, working together in regional groups, used these resources (a) to familiarize themselves with the Python programming language, (b) apply these resources in coding clubs employing pedagogically sound learning scenarios and (c) critically evaluate these resources and develop their own, based on the experience they acquire while applying them. The blended-learning approach followed promoted and enabled effective communication between tutors/trainers and trainees, as well as cooperation among trainees and access to courseware and learning resources.

A specific guideline to practitioners that wish to design and deliver effective training to teachers is to understand the importance of combining distant learning facilities with local cooperation of teachers and practical use of the acquired knowledge in organizing learning activities with students. This combination puts the general idea of blended-learning into a context that is highly effective providing motivation for teachers and increasing the positive impact to students.

Future work will address the transfer of the material and the methodologies reported in other domains (e.g. mathematics [15]) that could use coding to make learning playful and more engaging. Furthermore, we plan to explore the integration of visual interfaces to Python capitalizing promising results with respect to the performance gain that those interfaces can offer to novices [16].

Acknowledgments. The work reported in this paper was partially supported by the Google CS4HS programme (https://www.cs4hs.com/).

References

1. Fischer, G.: Understanding, fostering, and supporting cultures of participation. Interactions **18**, 42–53 (2011)
2. Myers, B., Ko, A.J., Burnett, M.: Invited research overview: end-user programming. In: CHI 2006 Extended Abstracts on Human Factors in Computing Systems, pp. 75–80 (2006)

3. Ko, A.J., Abraham, R., Beckwith, L., Blackwell, A., Burnett, M., Erwig, M., Scaffidi, C., Lawrance, J., Lieberman, H., Myers, B.: The state of the art in end-user software engineering. ACM Comput. Surv. (CSUR) **43**(3), 21 (2011)
4. Aleksic, V., Ivanovic, M.: Introductory programming subject in European higher education. Inform. Educ. **15**(2), 163–182 (2016)
5. Balanskat, A., Engelhardt, K.: Computing Our Future: Computer Programming and Coding-Priorities. School Curricula and Initiatives Across Europe. European Schoolnet, Brussels (2014)
6. Smith, N., Sutcliffe, C., Sandvik, L.: Code club: bringing programming to UK primary schools through scratch. In: Proceedings of the 45th ACM Technical Symposium on Computer Science Education, pp. 517–522 (2014)
7. Ovcin, E., Cerato, C., Smith, D., Lameras, P., Moumoutzis, N.: The pSKILLS Experience: Using Modern Educational Programming Languages to Revitalise Computer Science Teaching. International Conference on The Future of Education, Florence, Italy. (2011). http://conference.pixel-online.net/edu_future/common/download/Paper_pdf/ITL70-Ovcin, Cerato.pdf
8. Smith, D., Lameras, P., Moumountzis, N.: Using educational programming languages to enhance teaching in computer science. In: EDGE 2010 Conference on the Use of Technologies in K-12 and Post-Secondary Education, Newfoundland and Labrador, Canada (2010)
9. Peppler, K., Kafai, Y.: Creative Coding: Programming for Personal Expression. In: 8th International Conference on Computer Supported Collaborative Learning (CSCL), Rhodes, Greece (2005). https://goo.gl/pAHtQA. Accessed 11 Sept 2017
10. Moumoutzis, N., Christoulakis, M., Pitsiladis, A., Sifakis, G., Maragkoudakis, G., Christodoulakis, S.: The ALICE experience: a learning framework to promote gaming literacy for educators and its refinement. In: 2014 International Conference on Interactive Mobile Communication Technologies and Learning (IMCL), pp. 257–261 (2014)
11. Zimmerman, E.: Gaming literacy – game design as a model for literacy in the twenty-first century. In: Perron, B., Wolf, M.J.P. (eds.) The Video Game Theory Reader 2. Routledge, New York, London (2009)
12. Pappas, N., Arapi, P., Moumoutzis, N., Mylonakis, M., Christodoulakis, S.: The multimedia open learning environment (MOLE). Never Waste a Crisis! 76 (2011)
13. Mylonakis, M., Arapi, P., Pappas, N., Moumoutzis, N., Christodoulakis, S.: Metadata management and sharing in multimedia open learning environment (MOLE). In: Metadata and Semantic Research, pp. 275–286 (2011)
14. Pappas, N., Arapi, P., Moumoutzis, N., Christodoulakis, S.: Supporting learning communities and communities of practice with Coursevo. In: 2017 IEEE Global Engineering Education Conference (EDUCON), pp. 297–306, Athens, Greece (2017)
15. Lameras, P., Moumoutzis, N.: Towards the gamification of inquiry-based flipped teaching of mathematics: a conceptual analysis and framework. In: 2015 International Conference on Interactive Mobile Communication Technologies and Learning (IMCL), pp. 343–347 (2015)
16. Kyfonidis, C., Moumoutzis, N., Christodoulakis, S.: Block-C: a block-based programming teaching tool to facilitate introductory C programming courses. In: 2017 IEEE Global Engineering Education Conference (EDUCON), Athens, Greece, pp. 570–579 (2017)

Mobivoke: A Mobile System Architecture to Support off School Collaborative Learning Process

Panos K. Papadopoulos[1], Nikolaos C. Zygouris[2(✉)], Maria G. Koziri[2], Thanasis Loukopoulos[1], and Georgios I. Stamoulis[1,3]

[1] Computer Science and Biomedical Informatics Department, University of Thessaly, 2-4 Papasiopoulou street, 35100 Lamia, Greece
{ppapadopoulos, luke}@dib.uth.gr, georges@uth.gr
[2] Computer Science Department, University of Thessaly, 2-4 Papasiopoulou street, 35100 Lamia, Greece
{nzygouris, mkoziri}@uth.gr
[3] Electrical and Computer Engineering Department, University of Thessaly, 37 Glavani street, 38221 Volos, Greece

Abstract. The collaborative learning paradigm offers one of the most solid approaches to increase the participation, interest and knowledge level of pupils (typical achieving and/or learning disabled students) during the educational process. Recent advances in the field have offered a plethora of tools to facilitate collaboration during school time. Nevertheless, the possibility of applying the collaborative learning principles together with personalized exercise/project assignments (whenever deemed necessary) during off school hours is often overlooked. Motivated by the fact that most pupils/students nowadays have access to smart mobile devices, e.g., tablets, in this paper a system architecture (Mobivoke) is proposed that enables the coupling of individual devices into a social group and offers the means to build applications for orchestrating and monitoring the off school learning process in a collaborative manner.

Keywords: Collaborative learning environments · Mobile systems
Mobile applications

1 Introduction

While a plethora of Web based e-class and learning tools exists that can be used during off school hours, they are either specifically tuned towards one activity type, e.g., Hot Potatoes software suite, or completely lack functionalities to support student collaboration. Motivated by the potential of collaborative learning during off school hours, but also by personalized learning demands (particularly for students with learning difficulties), in this paper a mobile system architecture *Mobivoke* is proposed in order to support and facilitate the development of collaborative learning mobile applications.

The vision behind Mobivoke is distinctly broader compared to e-class management and online assignment systems. Namely, the core target of the system is to enable either

M. E. Auer and T. Tsiatsos (Eds.): IMCL 2017, AISC 725, pp. 587–592, 2018.
https://doi.org/10.1007/978-3-319-75175-7_57

direct or indirect control of applications remotely. In other terms, both the tutor and the students will have the ability to invoke and handle applications on the mobile devices of others (given the right permissions) either directly or through notifications. Contrary to desktop sharing which is too intrusive for educational purposes, the Mobivoke system aims at offering a set of communication, synchronization and remote invocation primitives that can be used to develop comprehensive collaborative mobile learning applications.

The rest of the paper is organized as follows. Section 2 summarizes the related work from the particular standpoint of the pedagogical approach targeted (collaborative learning during off school hours). Section 3 presents system functionality, while Sect. 4 includes the results from a preliminary evaluation of system's technical aspects. Finally, Sect. 5 provides the concluding remarks.

2 Related Work

In [3] the authors provided evidence that mobile technologies had become a habitual part of the lives of teachers and learners in UK. They also discuss effective ways of incorporating such technologies in the learning procedure. Early generation of this kind of learning projects tended to propose formally designed activities that were carefully constructed by educators and technologists. The widespread ownership of mobile and wireless devises has increased the amount of learners that are engaged in activities motivated by their personal abilities and/or disabilities [8].

Collaborative learning is a relatively new educational process that became widely used and accepted as a teaching methodology over the last 20 years [2]. The educational theory that leads to the constructive collaborative learning process includes the formation of small groups in which students help each other in the learning development without any particular directions from the teacher [2]. Several works e.g., [7], support that the main elements/characteristics of collaborative learning can be summarized as: (i) the educational procedure is active, (ii) the teacher is usually a facilitator of knowledge, (iii) teaching and learning are shared experiences, (iv) students participate in small group activities and (v) students are stimulated to reflect on their own conclusions. Thus, during the process learners develop their social and team skills. Furthermore, collaborative learning leads to a deeper level of knowledge, critical thinking, understanding the elements of the curriculum and long term storage of the curriculum material [4].

On the other hand, just placing students in groups does not guarantee collaboration [9]. In order to be effective this teaching strategy has to ensure: (i) the positive interdependence (individual success is associated with the group success), (ii) individual contribution (each member's involvement at the learning process admeasures the group's general impact) and (iii) face to face interaction (achieved in groups with a small number of members). It is also supported in [6] that better outcomes are expected when there is diversity concerning school performance among team members. Furthermore, in [1] collaborative learning is advocated for classrooms consisting of both typical achieving students and students with learning disabilities. The most frequent in occurrence and studied learning disability is developmental dyslexia. Dyslexia is diagnosed when a child's literacy skills are significantly lower than those expected from their age group [10] and affects 5 to 15% (depending on a less or more conservative definition) of the student's population [5].

One of the purposes of the present study is to construct a suite of applications in order to provide the teacher with a tool that can obtain the student's interest. The aforementioned suite will prepare among others the students by giving information through their mobile phone and/or tablet for the lesson that has to be delivered according to the curriculum for the next day as shown in Fig. 1. In other words students will be prepared for the lesson and can obtain all the knowledge that is needed in order to discuss their outcomes with their group in their classroom. Another goal is to ensure that students with dyslexia will have more time in order to be prepared for the lesson and through this can gain the benefits of the participation into a group or teammates.

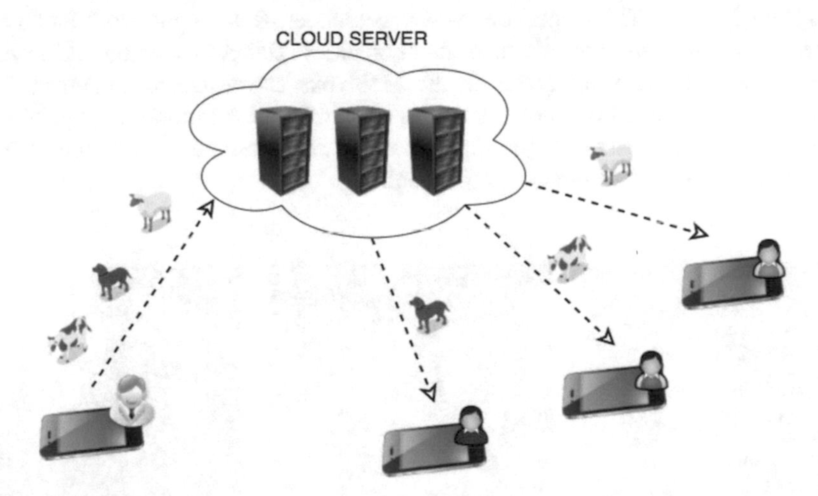

Fig. 1. Example of system's functionality (a teacher distributes personalized learning material for a biology class).

3 System Overview

Mobivoke comprises of two main components. The first one is a native application that is responsible for communication with other devices, as well as application handling in the device it is installed. The second system component is a Web service through which message exchange occurs. Additionally, the Web service is responsible for maintaining user accounts, group participation and performing security checks among others. A user can handle applications installed in other devices by sending suitable messages (ideally through the same application interface). The message payload describes the actions to be carried by the destination device as well as other communication primitives. Such primitives might include whether confirmation is required, whether the action must be synchronized with similar actions to other user devices, e.g., for exam purposes etc. All messages are stored in a database and communication with clients is pull-based, i.e., a native client contacts periodically the service in order to fetch incoming messages. After message parsing suitable notification is created and the end user can accept or reject the involved action.

The Mobivoke system is built using the LAMP stack on the server side and supports Android devices. For an application to be Mobivoke-friendly it must have a specified list of allowable actions following the Android API. A user is able to select the access level allowable to others, e.g., it might specify that one or more actions of an application shouldn't be invoked by anyone.

In order to demonstrate system viability, two applications were built in the ecosystem of Mobivoke. The first was a file sharing application that enables the tutor to distribute course material at different detail levels, e.g., special summaries in the case of pupils with dyslexia. The second one was a messenger. Both applications had a rather simple action interface with two main functionalities related to sending and receiving data and were smoothly integrated in the Mobivoke ecosystem as shown in Fig. 2. Enhancing the collaborative tool ecosystem of Mobivoke with more applications and functionalities is a constant ongoing process that will continue even as the system is currently under pilot deployment in 3 schools.

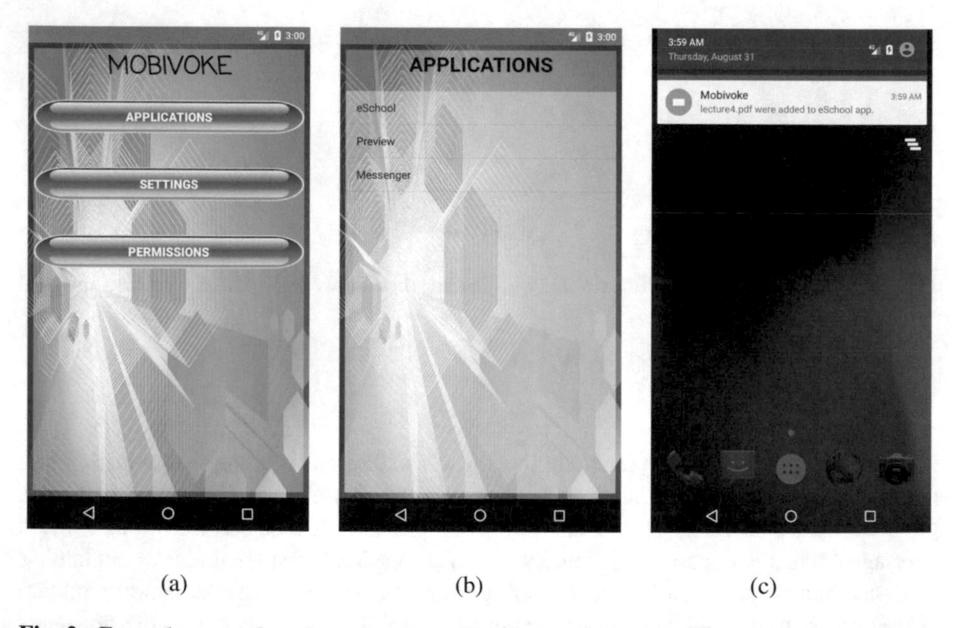

(a) (b) (c)

Fig. 2. Example screenshots from the client side: (a) the Mobivoke initial screen, (b) application list currently available under Mobivoke and (c) notification arrival.

4 Evaluation

4.1 System Performance

Preliminary system performance tests were conducted as follows. A free hosting service (000webhost.com) was used to launch the system's server on a free account. A number of simulated users were created on a laptop with a 2.2 GHz i7 processor. All users were launched simultaneously as separate threads. Each user sent 10 queries to the server accounting for status update. Figure 3 shows the average response time recorded as the time between sending the request and receiving the results. It is worth noting that the trend is linear providing a testament for system's scalability even under the rather adverse scenario upon the performance was tested, whereby a basic hosting service is used and all the clients generate requests from the same machine (thus stretching its load). Lastly, it should be mentioned that even with 100 simultaneous users the delay experienced was in the order of a couple of seconds.

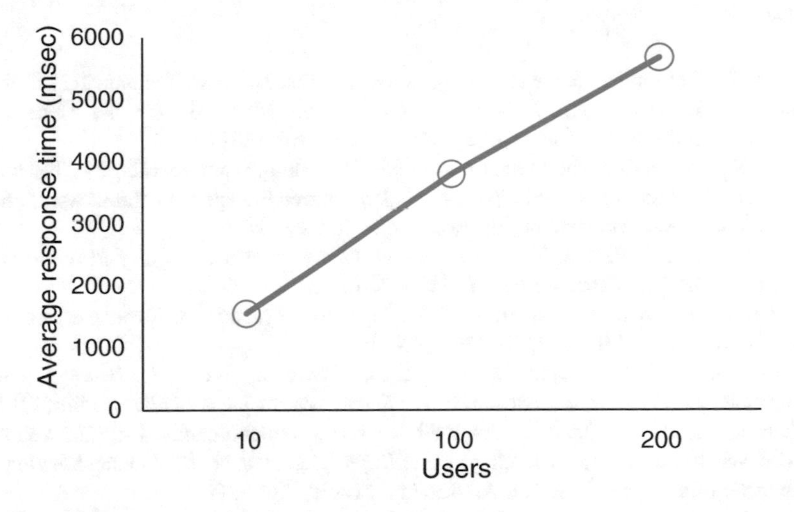

Fig. 3. Query response time.

4.2 Pedagogical Evaluation

Having performed successful system tests, we are in the process of conducting pilot studies on the impact of the Mobivoke system to secondary school teaching process. For this reason both the feedback from teachers and students will be aggregated through questionnaires and statistically analyzed. The findings will be further cross-checked with system use log files in an attempt to identify interesting associations. Finally, aside from evaluating the system as is, effort will be concentrated enhance the Mobivoke ecosystem with new applications following user input.

5 Discussion and Conclusions

In this paper we propose Mobivoke, a system architecture that provides an infrastructure for mobile application cooperation and synchronization. The system potential is demonstrated through a file sharing and a messaging application. Results from evaluating the technical aspects and performance are particularly encouraging, showing good scalability. Evaluation of the system impact to the learning process is planned for an extended version.

Acknowledgments. Panos K. Papadopoulos was supported by scholarship from IKY (State Scholarships Foundation) funded by the Act "Strengthening Human Resources Research Potential via Doctorate Research" of the Operational Program "Human Resources Development Program, Education and Lifelong Learning", 2014–2020 co-financed by the European Social Fund (ESF) and the Greek Government.

References

1. Abbott, C.: Defining assistive technologies-a discussion. J. Assist. Technol. **1**(1), 6–9 (2007)
2. Cohen, E.G., Lotan, R.A.: Designing Groupwork: Strategies for the Heterogeneous Classroom, 3rd edn. Teachers College Press, New York (2014)
3. Facer, K.: Foreword to the literature in mobile technologies and learning. In: Naismith, L., Lonsdale, P., Vavoula, G., Sharples, M. (eds.) Futurelab Report 11. http://www.futurelab.org.uk/resources/documents/lit_reviews/Mobile_Review.pdf
4. Garrison, D.R., Anderson, T., Archer, W.: A theory of critical inquiry in online distance education. Handb. Distance Educ. **1**, 113–127 (2003)
5. Habib, M.: The neurological basis of developmental dyslexia: an overview and working hypothesis. Brain **123**(12), 2373–2399 (2000)
6. Hertz-Lazarowitz, R., Kagan, S., Sharan, S., Slavin, R., Webb, C. (eds.): Learning to Cooperate, Cooperating to Learn. Springer Science & Business Media, Berlin (2013)
7. Kirshner, B., O'Donoghue, J.: Youth-adult research collaborations: Bringing youth voice and development to the research process. Paper presented at the Annual Meeting of the American Educational Research Association, Seattle, April 2001
8. Pettit, J., Kukulska-Hulme, A.: Going with the grain: mobile devices in practice. Australas. J. Educ. Technol. **23**(1) (2007)
9. Soller, A., Lesgold, A., Linton, F., Goodman, B.: What makes peer interaction effective? Modeling effective communication in an intelligent CSCL. In: Proceedings of the 1999 AAAI Fall Symposium: Psychological Models of Communication in Collaborative Systems, pp. 116–123, November 1999
10. Zygouris, N.C., Avramidis, E., Karapetsas, A.V., Stamoulis, G.I.: Differences in dyslexic students before and after a remediation program: a clinical neuropsychological and event related potential study. Appl. Neuropsychol.: Child 1–10 (2017)

Bridging Mindstorms with Arduino for the Exploration of the Internet of Things and Swarm Behaviors in Robotics

Antonis G. Dimitriou[1], Nikolaos Fachantidis[2(✉)], Orestis Milios[1], Petros Perlantidis[3], Arturo Morgado-Estevez[4], Francisco León Zacarías[4], and José Luis Aparicio[4]

[1] School of Electrical and Computer Engineering, Aristotle University of Thessaloniki, Egnatia, 54124 Thessaloniki, Greece
[2] Department of Educational and Social Policy, University of Macedonia, Egnatia 156, 54636 Thessaloniki, Greece
nfachantidis@uom.edu.gr
[3] Department of Informatics and Telecommunications Engineering, University of Western Macedonia, Agios Dimitrios Park, 50100 Kozani, Greece
[4] Department of Engineering in Automation, Electronics, Architecture and Computer Networks, Escuela Superior de Ingeniería de Cádiz, University of Cádiz, c/Chile 1, 11002 Cádiz, Spain

Abstract. In this paper we describe a new testbed that allows for the exploration of rich state-of-the-art STEM content. More specifically, we present the potential of joining two very popular educational/hobbyist platforms in the development of prototype stimulating STEM content. In this context, we present the necessary hardware, needed to provide motion (Mindstorms), low-cost, rich sensing (Arduino) and communication/mobility (wireless transceiver), then explain how to configure it and provide two structured examples in the fields of Internet of Things and Collaborative Robotics. Furthermore, we use this "new" platform as a reconfigurable testbed for the design and testing of a prototype educational-social robot, in the context of project "Stimey".

Keywords: Constructionism · Educational robotics · LEGO robotics Arduino · Internet of Things · Student-led learning

1 Introduction

LEGO Mindstorms represents the most popular robotics platform destined for education. It has originated from the ideas of MIT Media Lab Professor Papert and his learning theory on "Constructionism" [1], involving students drawing their own conclusions through creative experimentation and the making of social objects. The "Mindstorm's" name was inspired by his seminal book [2]. Since 1988, three generations of "bricks" have been commercially available [3]. In January 2013, the latest "EV3" brick was presented. The great success of LEGO Mindstorms is due to the reconfigurability, provided by the famous bricks, combined with a simplified drag-and-drop programming environment. Students use the same set of materials to

M. E. Auer and T. Tsiatsos (Eds.): IMCL 2017, AISC 725, pp. 593–600, 2018.
https://doi.org/10.1007/978-3-319-75175-7_58

create moving vehicles, humanoid robots, industrial machines, musical instruments etc. Teaching is typically carried out by experimentation, instead of a structured plan of lectures. A review of such education activities is presented in [4]. As a drawback, the LEGO kit is designed for robotics-applications only. It includes DC motors, a servo motor and sensors mainly related to motion and obstacle-avoidance, i.e. ultrasonic sensor, infrared sensor (also used for communication), touch sensor, gyro (scopic) sensor and a color sensor. Even though it is primarily intended for the ages 10–18, it has been introduced in undergraduate courses, related to basic programming and robotics [5–8]. The advantage of the Mindstorms kit is the fact that with the same set of bricks, one can create different models. Its disadvantage is its price ($\sim$400€ for the basic education kit) and the limited sensing capabilities. "Sensor-adapters" have been commercialized but at much higher prices compared to their market values for other commercially available microcontroller. Could we exploit LEGO Mindstorms with low-cost electronic-sensors? Imagine how this would increase the potential of the platform, given its great penetration in education.

Arduino is an open-source computer hardware that is based on the work of Barragán [9] on "Wiring", supervised by Massimo Banzi (co-founder of Arduino) and Casey Reas. "Wiring" represents a "teaching language and electronics prototyping system that facilitates and encourages the process of learning, reduces the struggles with electronics design and programming…It is both a programming environment and an electronics prototyping input/output board for exploring the electronics arts and tangible media." [9]. Thanks to its easy programming environment, the "Processing" integrated Development Environment (IDE), the availability of the open-source material and its relatively low-cost, Arduino made it possible for non-technical people to be able to design and construct their own hardware – a property recently named as the "democratization of hardware". The history of Arduino is presented in [10]. The brain of the Arduino is a microcontroller; hence it can process any information (electrical signal as input), either digital or analog and produce the desired (programmed) output. As a result it can exploit every possible electronic sensor that exists in the market and produce any desired action (output visual content, audio, control a motor etc.). Furthermore, it can be paired with transceiver modules, usually in a sandwich-layered hardware architecture ("shields") and thus ensure connectivity and internet access. Thanks to the above properties, it has been exploited in different educational context, including Wireless Sensor Networks and the Internet of Things [11], Physics [12], Robotics [13] etc. In [13], the authors construct specific robotic-bases (with motors and wheels), controlled by the Arduino. Imagine how liberating it would be if we could use the same set of materials to construct different "machines/robots" (moving things), instead of being confined by a specific robotic-base, as in typical "Arduino"-only constructions [13].

In this paper, we present the collaborative learning potentials of bridging LEGO-Mindstorms with Arduino. We wish to exploit the re-usability of the LEGO bricks to create different constructions with the same set of bricks and pair it with the rich, low-cost sensor and communication capabilities of Arduino. By combining the two platforms, the students can use Mindstorms to create any machine and Arduino to interact with the physical world and communicate with other devices. The construction process, but also the exploitation of such a construction in an educational project offers the context for integrating smart devices as collaborative learning tools.

In Sect. 2, we present the necessary hardware, needed to provide motion (Mindstorms), low-cost, rich sensing (Arduino) and communication/mobility (wireless transceiver paired with Arduino), then explain how to configure it. In Sect. 3 two structured examples are given in the fields of IoT and Cooperative Robotics. Furthermore, we use this "new" platform as a reconfigurable testbed for the design and testing of a prototype educational-social robot, in the context of "STIMEY" (Horizon2020 funded project).

2 Bridging Arduino with Mindstorms

Communication between the two platforms is carried out through the "I^2C" interface/protocol. The simplest way to deploy the protocol for the specific connection is to download the "EV3 blocks for I2C Communications" library, developed by "Dexter Industries" [14]. This library introduces a block in the GUI of the EV3 which allows for the exchange of messages between Arduino and EV3. Since a physical hardware connection is essential, one can purchase the 7.5$ "Breadboard Adapter for LEGO Mindstorms" by Dexter Industries, or manually create the adapter, following the schematic from [15]. Either way, the resulting outcome is shown in Fig. 1.

Thanks to the Arduino, wireless connectivity can be easily added, selecting any of the wireless transceiver modules, sold as "shields" (e.g. zigbee, 802.11 WiFi, etc.). We have decided to work with the RFM22 transceiver (see Fig. 3). The reason for this selection is that this module is un-protocolled; therefore it can be programmed by the students, thus allowing for different deployment options and related networking protocols (mesh networks, relays, central management etc.).

An exemplary constructed prototype is shown in Fig. 3. It supports movement, wireless connectivity and interaction with the environment through sensors. The student should program the EV3 and the Arduino separately. For the I2C communication a block of code should be inserted in the EV3 programming environment, as demonstrated in Fig. 2, reading (or writing) data from (to) the Arduino and a block of code in the Arduino sketch (including the "Wire" library) to control I2C messages.

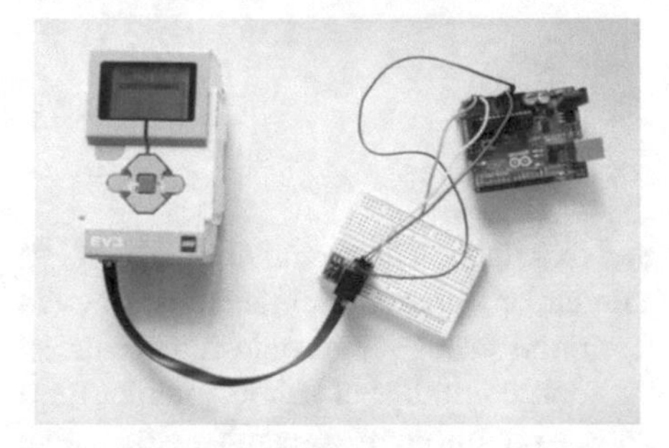

Fig. 1. Schematic for the connection of EV3 with Arduino through I2C.

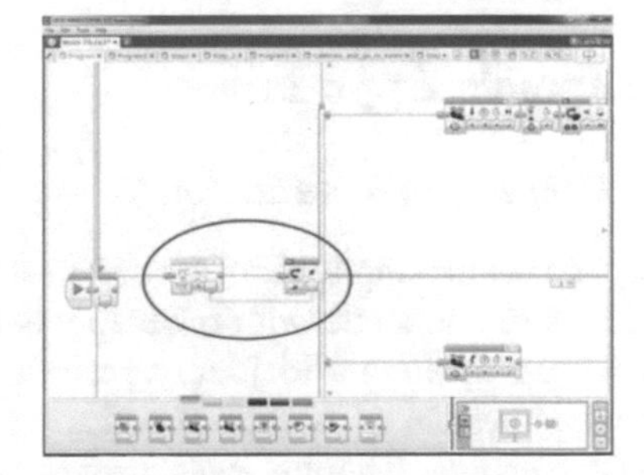

Fig. 2. The I2C block in the above example "reads" from the Arduino and enters the appropriate switch statement.

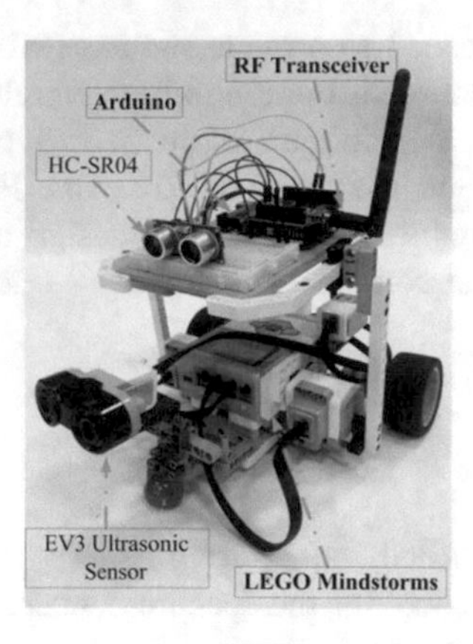

Fig. 3. Constructed prototype with LEGO EV3, Arduino and RF transceiver

Fig. 4. Pair of identical robots meeting each-other, exploiting communication principles.

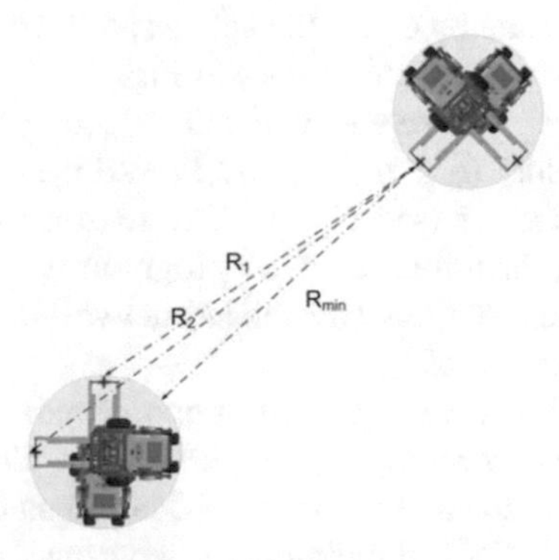

Fig. 5. Basic principle for the "meeting" algorithm

Fig. 6. The IoT setup for Precision Agriculture

3 Example Applications

3.1 Cooperative Localization

In the first course-application, we demonstrate an example on cooperative robots, i.e. "a robot able to cooperate with other robots of similar or different architectures, or even with human operators, to jointly perform a common task". This topic represents an elementary system of "swarm robotics", where a large number of robots interact for a specific outcome. Though swarm robotics spans over different research fields, some basic properties can be addressed through the proposed example. The robots must be able to communicate locally (exchange messages), coordinate their actions, interact with the physical world and produce the desired outcome cooperatively.

3.1.1 The Progressive Inquiry Model

In the context of such a course, we consider application of the "progressive inquiry model". The latter considers the need to connect learning with real-world problems meaningful to learners [16]. Applying the principle of progressive inquiry in a "loop", we consider the following phases, introducing the idea of "swarm robotics":

- Creating the context: The students are shown a video of the potential of having many small robots to cooperate in order to lift a huge load. Apart from that video, they are also shown videos from the animated movie "Big Hero 6", where small "microbots" join together to selectively create any desired large robot. In a sense of "self-reference", students are questioned whether "Mindstorms" is a kind of a "microbot".
- Constructing working theories: The students identify that each robot must be clever enough or at least have a brain. They should also be able to communicate. They should be able to meet. So, they should be able to "see". What are the human senses and how do we map them in a robot?
- Critical evaluation: We want our "microbots" to meet. How can we accomplish that? Is the sensor information adequate to meet? Do we need some kind of processing?
- Searching deepening knowledge: The students discover that electronics make mistakes. Many times, measurements are affected by the environment. When? Why? How?
- Generating subordinate questions: Can we identify an error? Can we correct it?
- Constructing new working theories: Improve the design. Reject meaningless measurements. Average meaningful measurements. Improve. Make the robots meet.

3.1.2 Application of the Model with Mindstorms and Arduino

We assign a mission to a pair of robots randomly positioned (and oriented) inside a room: we ask them to locate each other and meet. To accomplish that, we have constructed two identical robots, shown in Figs. 3, 4. For each robot, we have used a LEGO Mindstorms EV3 education kit, an Arduino Uno, an RFM22 wireless transceiver, an HC-SR04 ultrasonic sensor ($\sim$4$ sensor), the EV3 gyro sensor and the EV3 Ultrasonic Sensor.

The basic idea of the "meeting" algorithm is illustrated in Fig. 5 and is summarized as follows: they rotate around their vertical axis sequentially (not together), while measuring the rotation-angle and the distance between the two ultrasonic HC-SR04 sensors, located at the front of the robot. For the pair of angles, where the distance is minimized, the two robots "look" against each other. They rotate at the angles and start to move forward, until the separation-distance is smaller than a threshold. The EV3 ultrasonic sensor is used to avoid physical obstacles. If such is detected, the robot rotates towards an obstacle-free region and repeats the localization algorithm.

In the above algorithm, there are a few points worth noting. First of all, the HC-SR04 is not an ultrasonic transceiver (hence the low price); it emits 8 ultrasound (40 MHz) pulses through a speaker and "measures" the delay until the pulses "return" to the microphone of the module after being reflected to the surrounding environment. We had to "modify" the operation of the module to measure the time of flight between two HC-SR04 modules (by selectively grounding the ultrasound transmitter of one robot). We want to transmit ultrasound from one module and measure the delay, until the burst reaches the other module. To accomplish that, we need synchronization between the two robots. Exploiting the fact that RF-signal travels 10^6 times faster than the speed of sound, we use the RF transmitter to "synchronize" the two robots, similarly to [15]. When an ultrasonic signal is emitted from "robot 1", it also sends an RF signal to "robot 2" to "turn on" the ultrasonic receiver (then "robot 2" reports the measured signal to "robot 1"). The "constant" error of the sync-process (due to the electronics delays) can be easily measured and subtracted only once at a fixed known distance. The two robots exchange messages among them like "stop", "rotate", "move", etc., in order to accomplish the required task; an essential property of cooperative behavior. STEM content includes physics (sound vs electromagnetic waves), math (sinusoid summation for the calculation of interference pattern due to multipath), engineering (motion, motors, gears), wireless communication (exchange of messages, protocols, synchronization, propagation) and programming (loop, control statements).

3.2 Precision Agriculture

The 2nd example demonstrates the potential of Arduino and exploits ideas from the Internet of Things (IoT) and Precision Agriculture. Imagining a greenhouse of the future, we assume that all plants are monitored by low-cost sensors. Power is drawn in a sustainable manner (from the sun, the ground – thermoelectric, or even the plant itself [17]). Due to the power constraints (and the resulting small range of the transceivers) a drone flies around the greenhouse, to collect the measurements and report the data to the "control center", which then decides on irrigation, fertilization etc.

In order to imitate the above scenario, we place an Arduino with an RFM22 wireless transceiver card at each plant (Fig. 6). The Arduino measures the temperature and the humidity, inside and outside of the soil periodically (saving energy), by deploying two temperature and two soil sensors, while it also measures the light intensity. All 5 sensors cost less than 10$. A robot, also equipped with Arduino and the RFM22 transceiver, moves around the classroom, transmitting beacon RF signals to the surrounding plants. When in the vicinity of a plant, the plant transmits the stored data to the robot, which then wirelessly relays them back to the control center. In the above scenario, the students get familiar with the IoT, energy sustainability, networking etc.

3.3 Testbed for Research

Bridging of the two technologies also represents an excellent testbed for research. Again, the LEGO bricks can be exploited to represent any desired model, while the Arduino will provide access to low cost electronics and communication capabilities. In the context of "STIMEY", a Horizon2020 funded project, we are aiming at the design and construction of a prototype educational, social robot, designed for students. We are using the proposed merging of the two platforms to test aspects of the design, related to sensing, motion and cooperative behaviors.

4 Discussion - Conclusions

In this paper, we demonstrate (hardware and software) how to interconnect two powerful learning platforms and present the advantages of this connection in education. The essence of this work is that "Mindstorms" represent a set of components that can be used to create different machines. The versatility of LEGO-toys demonstrates the wealth of possible outcomes, when putting together the famous bricks. Arduino gives access to low-cost electronics and rich open-source libraries. If Mindstorms represent the equivalent of the mechanical parts of a machine, Arduino represents its senses. By bridging the two technologies one can use the same set of materials to create anything. Mindstorms will bring the construction to life, while Arduino will add information and communication wealth.

The set of presented examples gives only a minor idea of the potential of the technologies. We are currently living what is termed as the "4th Industrial Revolution" (Cyber Physical Systems, Internet of Things, Cloud Computing, Cognitive Computing). Shouldn't the new generation be prepared for the upcoming rapid digitization of the physical world and machine-controlled optimized decision making? Imagine the proposed platform, representing a model of an actual industry of the future (or present), where machines co-operate to control the rate of production, according to sales, where machines control other machines for errors etc.

Acknowledgement. This project has received funding from the European Union's Horizon 2020 research and innovation programme under grant agreement **No. 654109**.

References

1. Harel, I., Papert, S.: Constructionism: Research Reports and Essays 1985–1990 by the Epistemology and Learning Research Group, the Media Lab, Massachusetts Institute of Technology. Ablex Publishing Corporation, Norwood (1991)
2. Papert, S.: Mindstorms: Children, Computers, and Powerful Ideas. Harvester Press, Brighton (1980)
3. History of Lego Robotics. https://www.lego.com/en-us/mindstorms/history. Accessed 27 Oct 2017
4. Danahy, E., Wang, E., Brockman, J., Carberry, A., Shapiro, B., Rogers, C.B.: LEGO-based robotics in higher education: 15 years of student creativity. Int. J. Adv. Robot. Syst. **11**, 27 (2014). https://doi.org/10.5772/58249
5. Klassner, F., Anderson, S.D.: LEGO MindStorms: not just for K-12 anymore. IEEE Robot. Autom. Mag. **10**(2), 12–18 (2003). https://doi.org/10.1109/MRA.2003.1213611
6. Barnes, D.J.: Teaching introductory Java through LEGO MINDSTORMS models. In: Proceedings of the 33rd SIGCSE Technical Symposium on Computer Science Education, Cincinnati, Kentucky (2002). https://doi.org/10.1145/563340.563397
7. Williams, A.B.: The qualitative impact of using LEGO MINDSTORMS robots to teach computer engineering. IEEE Trans. Educ. **46**(1) (2003). https://doi.org/10.1109/te.2002.808260
8. Kim, Y.: Control systems lab using a LEGO mindstorms NXT motor system. IEEE Trans. Educ. **54**(3), 452–461 (2011). https://doi.org/10.1109/TE.2010.2076284

9. Barragán, H.: Wiring: prototyping physical interaction design. M.Sc. thesis, Interaction Design Institute Ivrea (2004)
10. Kushner, D.: The making of Arduino. IEEE Spectrum (2011). https://spectrum.ieee.org/geek-life/hands-on/the-making-of-arduinoAccessed 27 Oct 2017
11. Doukas, C.: Building internet of things with Arduino. CreateSpace Independent Publishing Platform (2012)
12. Zachariadou, K., Yiasemides, K., Trougkakos, N.: A low-cost computer-controlled Arduino-based educational laboratory system for teaching the fundamentals of photovoltaic cells. Eur. J. Phys. **33**(6) (2012)
13. Araújo, A., Portugal, D., Couceiro, M.S., Rocha, R.P.: Integrating Arduino-based educational mobile robots in ROS. In: 2013 13th International Conference on Autonomous Robot Systems (Robotica) (2013). https://doi.org/10.1109/robotica.2013.6623520
14. Connecting the EV3 and the Arduino from Dexter Industries. https://www.dexterindustries.com/howto/connecting-ev3-arduino/. Accessed 27 Oct 2017
15. Micea, M.V., Stancovici, A., Chiciudean, D., Filote, C.: Indoor inter-robot distance measurement in collaborative systems. Adv. Electr. Comput. Eng. **10**(3), 21–26 (2010). https://doi.org/10.4316/aece.2010.03004
16. Hakkarainen, K., Sintonen, M.: The interrogative model of inquiry and computer-supported collaborative learning. Sci. Educ. **11**(1), 25–43 (2002). https://doi.org/10.1023/A:1013076706416
17. Konstantopoulos, C., Koutroulis, E., Mitianoudis, N., Bletsas, A.: Converting a plant to battery with scatter radio and ultra-low cost. IEEE Trans. Instrum. Meas. **65**(2), 388–398 (2016). https://doi.org/10.1109/TIM.2015.2495718

Digital Technology in Sports

Instructional Mirroring Applied in Basketball Shooting Technique

Hippokratis Apostolidis[1], Nikolaos Politopoulos[1](✉), Panagiotis Stylianidis[1], Agisilaos Chaldogeridis[1], Nikolaos Stavropoulos[2], and Thrasyvoulos Tsiatsos[1]

[1] Department of Informatics, Aristotle University of Thessaloniki, Thessaloniki, Greece
{aposti,npolitop,pastylia, achaldog,tsiatsos}@csd.auth.gr

[2] Department of Physical Education and Sports Science, Aristotle University of Thessaloniki, Thessaloniki, Greece
Magic.paok@gmail.com

Abstract. Human posture recognition is a very promising field of computer vision and it is applied in the areas of personal health care, human-computer-interaction, sports etc. In this paper a non-contact view-based approach of human body movement analysis is introduced using Microsoft Kinect V2 sensor. Moreover, there is a developing interdisciplinary scientific field which is trying to integrate modern technology into sports. Computer technology may prove a valuable supportive tool for (a) athletes of all ages to improve their technical skills, and (b) advanced coaching.

The aim of this study is to integrate educational technology and sports in order to support the training of young athletes and boost their performance. This research utilizes Kinect V2 as a 3D depth camera in order to apply body tracking of a young athlete during a shooting attempt. The shooting movement is mirrored to the computer screen with instructions showing the right shooting technique. The application was developed using Microsoft Development Kit 2.0 (SDK) and it was evaluated in a basketball camp. The results of the evaluation were very encouraging concerning the ease of use and the usefulness of this application.

Keywords: Game-based learning · Natura user interfaces · Microsoft Kinect Sports

1 Introduction

Nowadays, playing video games is nearly universal among all ages. Especially kids between 12 and 17 years old, 99% of boys and 94% of girls play them (Lenhart 2008). They are enjoyable and sustainable activities. After 2010, the tendency that emerged was exergames as a profitable market. Exergames or active video games are innovative tools that have the potential to turn a traditional sedentary activity, such as playing video games, in physical exercise (Politopoulos and Katmada 2014). Dance and rhythm

M. E. Auer and T. Tsiatsos (Eds.): IMCL 2017, AISC 725, pp. 603–611, 2018.
https://doi.org/10.1007/978-3-319-75175-7_59

exergames such as Dance Dance Revolution (DDR) are played by frequent gamers and non-gamers play rhythm exergames (Lenhart 2008).

The skills that kids and adolescents acquire during playing exergames can be transferred to real world activities and benefit physical, social, and cognitive development. A video game player becomes an onscreen producer of content, a process highly increased in an exergame when the player uses his body to control the onscreen character's movements. Exergames, through peripheral devices like depth cams, can recognize player's bodily movements as inputs associated and interpret their meanings for game play, transferring real time movement in three dimensional space onto the two-dimensional screen. Because the exergame player is distanced from the character on the screen, s/he must use visual–spatial skills, limbs–eye coordination, and quick reaction time to successfully play the game (Kretschmann 2010). Moreover, exergames allow multiple players to compete or cooperate on a team and as a result, they provide players with both virtual and real social interactions. These social and cognitive impacts of exergames provide additional potential benefits to the physical activity required for game play (Staiano and Calvert 2011).

Fundamental motor skills have been seen as the "building blocks" for lifetime physical activities (Payne and Isaacs 2011). Recent Studies have indicated a positive relationship between the performance of FMS and children's participation physical activity (Fahimi et al. 2013). FMS are activities that have specific motor patterns, which most researchers believe that form the foundation for more advanced and specific sport and non-sport motor activities (Gabbard 2011). FMS are categorized into two groups listed as locomotor skills (e.g., running, jumping, hopping, leaping, galloping, and sliding) and object control skills (e.g., throwing, catching, dribbling, kicking, rolling and striking) (Payne and Isaacs 2011).

The presented research is a first approach to construct a supportive tool functioning as a scaffold to athletes' practice. Many athletes use to stand in front of a mirror and practice on an exercise in order to calibrate themselves and improve their technique. The motivation of this work is to try to take advantage of this real life practice and extend it to a real time virtual mirror which will instruct and support the athlete with corrective signs. Moreover, if a user has an enjoyable experience with a tool and s/he perceives that this product is useful, h/she has a strong motivation to integrate it into his/her daily practice. Thus, we examine the users' perceived usefulness, ease of use and satisfaction of this first attempt of the implementation of such a supportive virtual mirror for sports. This first approach is dealing with basketball shooting technique. In the following section various related works are presented. Then, the evaluation of this work is following. Finally, the conclusions and the future steps are presented.

2 Related Work

Creating tools for sport related skills; developers have to face several heterogeneous problems ranging from psychological questions to technological issues and restrictions. In this section we will discuss the relevant previous work in sport analysis and training based on exergames and then we will focus our attention on different types of feedback delivered in such games, especially visual feedback to players.

Tennis Attack is a game – practicing tool created in order to help tennis players to develop their reaction time. Its design is very simple and based on a previous system that supports athletes to develop their skills of reaction. It uses Microsoft Kinect Sensor V1 as a device to track human movement and it is developed in Unity engine. A player stands in front of a screen and hits the balls that appear on it. Furthermore, there is a specific amount of balls per round and a specific top limit of time that players must react per ball. At the end of every session the player can monitor his reaction time and his improvement (Politopoulos and Katmada 2014). The result of research showed that players can significantly improve their reaction time through practicing.

Free-throw simulator is a virtual reality configurable multimodal platform composed of motion capture facilities and visualization systems. The goal of this system is to assist in the training of basketball free throw skills. The player can simulate through three different kind of shots, free throw, middle range two points shot and three points shot. At the end of the training session the platform gives feedback. It provides the user with information about her performance: success/failure, total score (number of success throws among the total number of trials) and the values of the main parameters at ball release compared to theoretical optimal ones. Evaluation results showed that there was not any significant difference between virtual and real shooting (Covaci et al. 2014).

Vernadakis et al. (2015) investigated the impact of an exergame-based intervention on children's fundamental motor skills. The purpose of this study was to use the Dynamical Systems Theory as a framework to examine whether there is a difference between an exergame-based and a traditional object control (OC) skills training program, in early elementary school children. Analysis of the data illustrated that the OC scores were significantly improved after the intervention. In addition, this research highlighted that kids enjoyed more exercises that used Xbox Kinect's approach than the traditional approaches. Conclusively, findings suggest that the use of Xbox Kinect gaming console as an intervention is a valuable and pleasant approach in order to improve OC skills of elementary school children.

Thus someone could suggest that modern technology applied on sports can combine virtual reality, simulation and other features to an advanced multimodal interaction platform which may improve user training, providing an interesting and amusing intervention mainly to users of young ages. Moreover Xbox Kinect offers an open source highly customizable software development platform and as a gadget is famous and welcomed from young children and teenagers. Therefore our research tried o take advantage of the use of all these technological advances and features in order to present an attractive and effective intervention platform called instructional mirroring. This platform is expected to provide significant support in order to improve young athletes' skills and technique.

3 Research Question

The main focus of this research is to examine young athletes' experience of using a practicing tool. Thus, this article is exploring users' perceives of usefulness, ease of use, ease of learning, and satisfaction after an evaluation activity where they were practicing in shooting technique under instructional mirroring support.

4 Method

This section presents the methodology followed in order to evaluate our research work.

4.1 Participants

In order to test the research hypothesis of this paper 68 young athletes from a basketball academy participated. Athletes were only boys, aged between 10 and 17 years old. All these young athletes were participating in PAOK BC "Junior" basketball program and they were chosen randomly. PAOK is one of the most famous and well known basketball teams of Thessaloniki, Greece and this program offers to young players a combination of games, learning of basketball skills, and scientific analysis of the kids' abilities. Athletes receive individual evaluation certificates with biomechanical and kinematical analyses, which include snapshots of selected basic skills (Fig. 1).

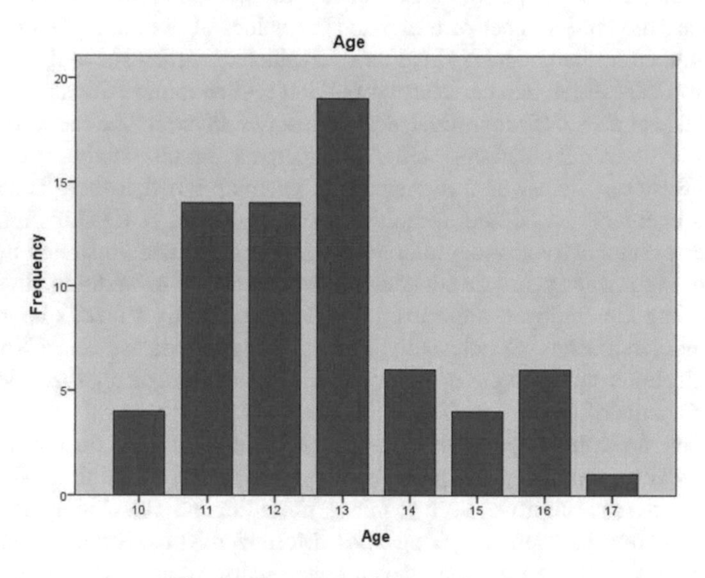

Fig. 1. Frequency histogram for ages

4.2 Instruments

In order to evaluate game's usability, players were asked to complete an online questionnaire, after the activity was concluded. The particular questionnaire was based on Lund's "USE" questionnaire (Lund 2001), which is a short questionnaire designed to measure effectively the most important aspects of a product's usability.

It consists of 30 questions grouped in four dimensions: (a) Usefulness, (b) Ease of use, (c) Ease of Learning, and (d) Satisfaction. The questions' type is a 7-point Likert rating scale with the following anchors: 1 strongly disagree, 2 disagree, 3 slightly disagree, 4 neutral, 5 slightly agree, 6 agree and 7 strongly agree.

4.3 Materials

The Microsoft Kinect V2 sensor is a peripheral device (designed for Xbox One and windows PCs) that functions much like a webcam. However, in addition to providing an RGB image, it also provides a depth map. Meaning for every pixel seen by the sensor, the Kinect measures distance from the sensor. This makes a variety of computer vision problems like background removal, blob detection and more, easier to solve. The Kinect sensor itself only measures color and depth. However, once that information is on a computer, lots more can be done like "skeleton" tracking (i.e. detecting a model of a person and tracking his/her movements) (Table 1).

Table 1. Kinect V2 specifications

	Kinect sensor V2
Color	1920 × 1080 @ 30 fps
Depth	512 × 424 @ 30 fps
Sensor	Time of Flight (ToF)
Range	0.5 – 4.5 m
Angle of view (horizontal/vertical)	70/60°
Body	6 people
Joint	25 Joints/people
Hand state	Open/closed/lasso

5 Application

Microsoft Kinect sensor accompanied with its software development kit (SDK), is a low cost, portable and open source solution applied in human face features recognition, motion capture, skeletal tracking etc. Kinect uses PrimeSense hardware and patent applications to compute body position in two stages ("How does the Kinect work" 2017):

- A depth map is constructed by analyzing patterns of infrared laser light. Kinect combines two computer vision techniques, (a) depth from focus, (b) depth from stereo("How does the Kinect work" 2017; Zhang et al. 2002).
- Body parts are computed and represented using a randomized decision forest algorithm, trained by a database of one million training examples ("How does the Kinect work" 2017).

Kinect technology and its application programming interface (API), offer an easy to use software framework to recognize skeletal joint positions. Thus a human skeleton model is recognized with 25 skeletal joint positions at a frequency up to 30 Hz. The Instructional Mirroring application is developed in C# (Microsoft Visual Studio 2015) using Kinect for Windows SDK 2.0. The presented application called "Instructional Mirroring" (Fig. 2) draws, in real time, the athlete skeletal model with a green-blue line and it puts dots on the recognized joints. The developed software is exploring the technique, the body position and the body movement of a basketball player when he/she is getting ready to shoot and when he/she is shooting the ball to the basket. These two positions are (a) shooting stance (including the user's base, start position of hands and legs and shot line) and (b) the actual shooting movement. During this process when the athlete is getting to the described positions his/her skeletal model is displayed to the computer screen. Besides the user skeletal model, red lines are also displayed showing the technically right bend and direction of every specific skeletal part which is important for establishing a technically correct shot with increased chances of success. Therefore the application recognizes when the user takes a shooting stance displaying instructional red lines for the foot direction and leg, elbow and back bending. Moreover the application becomes aware when the athlete decides to actually shoot the ball displaying instructional red lines for the leg and hand position.

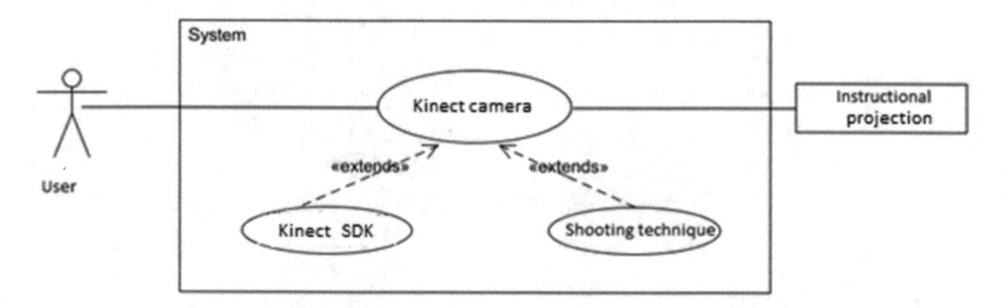

Fig. 2. Use case diagram of instructional mirroring

6 Procedure

Towards the objectives of this research, the authors applied a one-dimensional evaluation process. This process focuses on the investigation of usefulness, ease of use, ease of learning and satisfaction of the game, from players' perspective. Every athlete had to stand in the capture area of Kinect camera. A white tape on the floor marked the

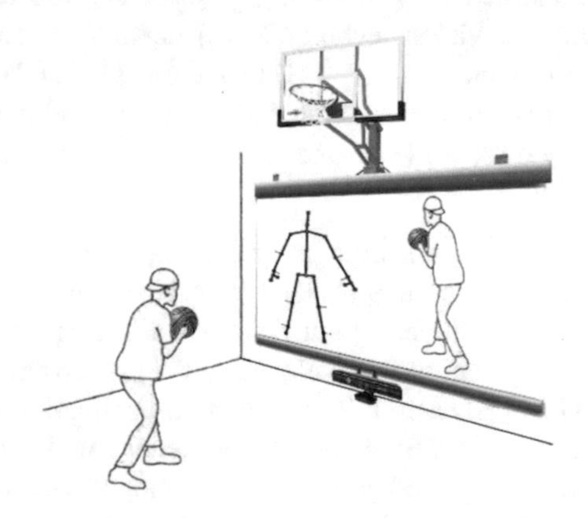

Fig. 3. The activity

landmark position in this capture area. Every young athlete simulated through; (a) free throw shot, (b) jump shot, (c) three-point shot, (d) hook shot and (e) lay up. Under the basketball basket there was a projector screen. The athlete's skeletal model supported with the instructional red lines was projected on this screen (Fig. 3). Therefore during this session the user could calibrate his body and shooting position in order to correct him/her. A coach was monitoring this session and encouraged them to watch themselves on the monitor and make the necessary changes.

7 Data Analysis

7.1 Descriptive Analysis of Responses to Use Questionnaire

The statistical analysis was performed using the SPSS 21 statistical package. Before any statistical test a reliability test was conducted in each one of the four dimensions (usefulness, ease of use, ease of learning and satisfaction of the questionnaire. Each scale had a high level of internal consistency, as determined by Cronbach's alpha greater than 0.700. Descriptive measures of central tendency, such as mean, as well as frequencies, were estimated for each Likert-type item from one to seven. The central tendency and dispersion measures estimated for the students' answers are being presented in Table.

As can be seen, the player' answers are quite satisfactory (Table 2). The game's usability has also been evaluated in total, based on the four aforementioned dimensions, by averaging the responses of students on the individual questions which correspond to each dimension (Boone and Boone 2012). As indicated by the results of the analysis, the players' opinion was positive, since the mean value (M) for the "Usefulness" dimension is 5.53, with standard deviation (SD) = 0.97, whereas M is 5.60, with SD = 1.15, for the "Satisfaction" (Fig. 4) dimension. Moreover, it could be deduced that the players did not encounter any major difficulties in comprehending and using the tool, since the mean value for the "Ease of Use" dimension is equal to 5.57, with SD = 0.83, while M = 5.70, with SD = 1.02, for the dimension "Ease of Learning" (Fig. 5).

Table 2. Mean values and Std. deviation of the four dimensions (Usefulness, Ease of Use, Ease of Learning, and Satisfaction)

	Usefulness	Ease of Use	Ease of Learning	Satisfaction
Mean	5.5276	5.5655	5.6985	5.5945
Std. deviation	.97373	.82887	1.01808	1.15469

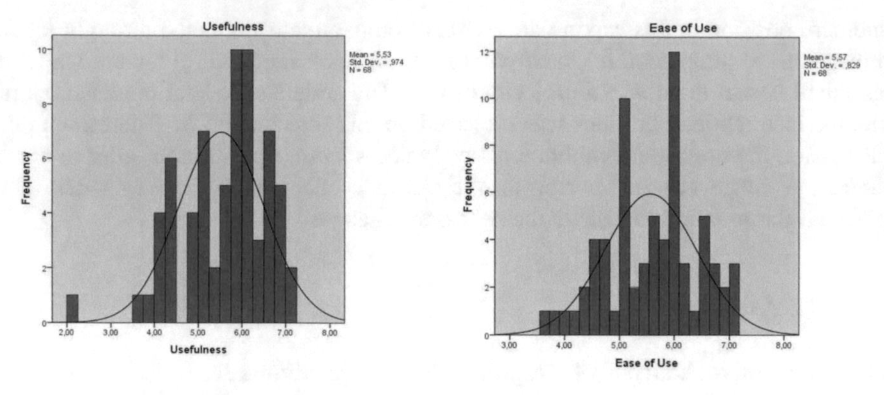

Fig. 4. Frequency diagrams of Usefulness and Easy of Use

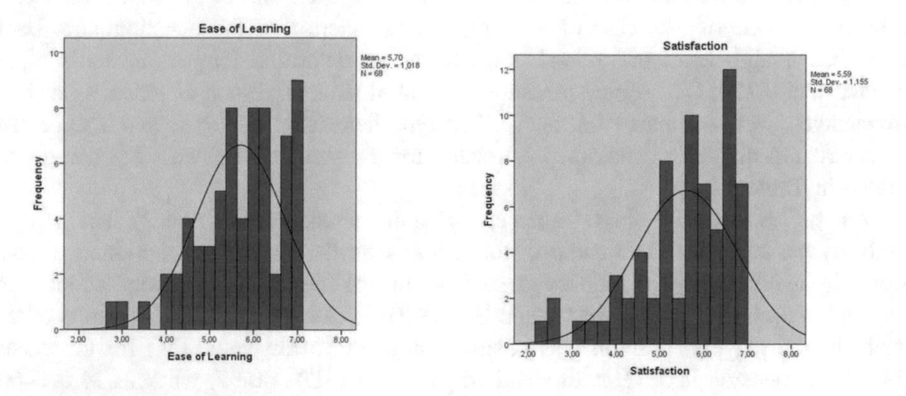

Fig. 5. Frequency diagrams of Ease of Learning and Satisfaction

8 Conclusions

This article presents a first approach of a supportive training system which could be applied in many sports. The first attempt was focused on basketball and the first examined exercise was the shooting technique. The derived evaluation results are very encouraging since the most of a sample of sixty four (64) young athletes found instructional mirroring application quite satisfactory, easy to use and useful. This may be considered a positive indicator on their behavior intension on using this application as a supportive training tool.

9 Further Research

Following up on the results, the next steps of this research are to extend the supportive information given to the training athlete as well as to his/her coach. Thus it is planned to add a second Kinect camera in order to project two user views. The one camera will project the front and the other the side view of the athlete. Moreover it is intended to add more screen indicators i.e. user joint angles. Furthermore it is planned to add more exercises of basketball sport in order to provide a more complete training experience such as control dribble, foot fire dribble, retreat dribble etc. A long term future plan is to extend this kind of training to other sports where the technique improvement is an important factor of successful performance. Finally long term evaluation of instructional mirroring application as a supportive practicing tool will present more safe results of the effectiveness of improvement in athletes' technique.

Acknowledgment. The authors of this article appreciate the PAOK BC "JUNIOR" BASKETBALL PROGRAM for their hospitality and assistance in instructional mirroring analyses of the free throw shot.

References

Lund, A.M.: Measuring usability with the use questionnaire. STC Usability Interface **8**(2) (2001). http://hcibib.org/search:quest=U.lund.2001

Covaci, A., Olivier, A.H., Multon, F.: Third person view and guidance for more natural motor behaviour in immersive basketball playing. In: Proceedings of the 20th ACM Symposium on Virtual Reality Software and Technology, pp. 55–64. ACM, November 2014

Fahimi, M., Aslankhani, M., Shojaee, M., Beni, M., Gholhaki, M.: The effect of four motor programs on motor proficiency in 7–9 years old boys. Middle-East J. Sci. Res. **13**(11), 1526–1532 (2013)

Gabbard, C.: Lifelong Motor Development, 6th edn. Benjamin Cummings, San Francisco (2011)

Boone Jr., H.N., Boone, D.A.: Analyzing likert data. J. Ext. **50**(2) (2012)

How does the Kinect work. https://www.scribd.com/document/237682505/kinect. Accessed 10 June 2017

Kretschmann, R.: Developing competencies by playing digital sports-games. Online Submiss. **7**, 67–75 (2010)

Katmada, K., Politopoulos, N.: The effect of computer games in physical education and health of children and youth. A literature review. In: 2nd Panhellenic Conference "Education in ICTs" Athens, Greece, 22–23 November 2014

Lenhart, A., Kahne, J., Middaugh, E., Macgill, A.R., Evans, C., Vitak, J.: Teens, video games, and civics: teens' gaming experiences are diverse and include significant social interaction and civic engagement. Pew Internet & American Life Project (2008)

Payne, V.G., Isaacs, L.D.: Human Motor Development: A Lifespan Approach, 8th edn. McGraw-Hill, Boston (2011)

Politopoulos, N., Tsiatsos, T., Grouios, G., Ziagkas, E.: Implementation and evaluation of a game using natural user interfaces in order to improve response time. In: 2015 International Conference on Interactive Mobile Communication Technologies and Learning (IMCL), pp. 69–72. IEEE, November 2015

Staiano, A.E., Calvert, S.L.: Exergames for physical education courses: physical, social, and cognitive benefits. Child Dev. Perspect. **5**(2), 93–98 (2011)

Zhang, L., Curless, B., Seitz, M.: Rapid shape acquisition using color structured light and multi-pass dynamic programming. In: Proceedings of the 1st International Symposium on 3D Data Processing, Visualization, and Transmission (3DPVT), Padova, Italy, pp. 24–36, 19–21 June 2002 (2002)

Massive Open Online Course for Basketball Injury Prevention Strategies (BIPS)

Thrasyvoulos Tsiatsos[1], Stella Douka[2], Nikolaos Politopoulos[1], and Panagiotis Stylianidis[1](✉)

[1] Department of Informatics, Aristotle University of Thessaloniki, Thessaloniki, Greece
{tsiatsos,npolitop,pastylia}@csd.auth.gr

[2] Department of Physical Education and Sports Science, Aristotle University of Thessaloniki, Thessaloniki, Greece
sdouka@phed.auth.gr

Abstract. Basketball is one of the most popular sports in Europe and is enjoyed by players of all ages and skill levels. Basketball is a fast game with frequent and aggressive body contacts, so injuries can and do occur. C4BIPS (http://c4bips.csd.auth.gr/) is an Erasmus Plus Sport Small Collaborative Partnership project that focuses on the area of "protecting athletes, especially the youngest, from health and safety hazards by improving training and competition conditions". The main goal of C4BIPS project is to create a Massive Open Online Course for Basketball Injury Prevention Strategies (BIPS).

The aim of this study is to present the project, highlight the importance of BIPS, describe its preliminary results on BIPS understanding by professional athletes, conduct a state of the art about e-learning platforms and summarize C4BIPS future work.

Keywords: MOOCS · Sports injuries · Injury prevention strategies Basketball

1 Introduction

Injuries are an important and largely preventable public health problem in all over the Europe. Main facts could be summarized as follows concerning the negative impact of injuries:

- Taking all age groups together, one out of twelve hospital admissions in the EU relates to an injury.
- Across the board, injuries take a significant share in the total healthcare expenditures in today's society.
- More than 50 million days of hospital care represent about 9% of all days of hospital care
- The number of cases treated in emergency departments outstrips by a factor of six the number of injury patients admitted to hospitals
- The direct medical care costs of all hospital treated injuries (inpatients and outpatients) in the EU is estimated to be at least 78 billion Euro each year

M. E. Auer and T. Tsiatsos (Eds.): IMCL 2017, AISC 725, pp. 612–622, 2018.
https://doi.org/10.1007/978-3-319-75175-7_60

A recent European study (Injuries in the European Union 2013) reports that with 73% of all hospital treated injuries, home, leisure and sports (Fig. 1) is by far the biggest share, which is in contrast to the fact, that home and leisure injury prevention programs appear as far less resourced than programs for road and work-place safety. In general, the tangible and intangible consequences of home, leisure, and sport injuries are also less well covered by insurance systems compared to the compensation schemes for road and work accidents.

	Road traffic	Work-place	School	Sports	Home, leasure	Total of unintentional injuries	Homicide, assault	Suicide, self-harm	Total of all injuries
Fatalites	38 119 16%	[illegible] 2%	1 250 1%	7 000 3%	98 891 42%	150 221 65%	4 704 2%	57 614 25%	232 869 100%
Hospital admissions	668 000 12%	[illegible] 4%	[illegible] 1%	419 000 7%	3 914 000 69%	5 285 000 93%	202 000 4%	213 000 4%	5 700 000 100%
Hospital outpatients	3 524 000 10%	[illegible] 10%	[illegible] 2%	5 644 000 17%	18 951 000 56%	32 465 000 96%	1 231 000 4%	205 000 1%	33 900 000 100%
All hospital patients	4 192 000 11%	[illegible]	[illegible]	6 063 000 14%	22 865 000 59%	37 750 000 95%	1 433 000 4%	415 000 1%	39 600 000 100%

Fig. 1. Comprehensive view on injuries in EU-27 by injury prevention domain [Source: Injuries in the European Union 2013]

1.1 Problem

Basketball is one of the most popular sports in Europe and is enjoyed by players of all ages and skill levels. Basketball is a fast game with frequent and aggressive body contacts, so injuries can and do occur. Drakos et al. (2010) found that there no correlation between injury rate and age, height, weight, or years of NBA experience. This is an essential finding, given that agents and organizations constantly attempt to stratify and predict the injury risk for each player. If there were a correlation between injury rate and player demographics, players at higher risk could be cut from their team.

Both Drakos et al. (2010) and Henry et al. (1982) found that ankle injuries were the most common but that knee injuries accounted for the greatest number of games missed.

Furthermore Starkey (2000) reported on NBA injuries over a 10-year period and similarly reported no correlation between injury rate and player demographics. A 12.4% increase in game-related injuries was noted during the 10-year period, which may be due to an increase in contact in professional basketball. The increase in size and speed of the players, as well as the improvement in diagnostic tools, may also be a factor in the injury increase. Meeuwisse et al. (2003) reported on rates and risks of injury in Canadian intercollegiate competition. Ankle injuries were the most common, but knee injuries resulted in more games missed. Studies of elite basketball in Sweden (Colliander et al. 1986) and high school basketball in Texas (Messina et al. 1999) also found that the ankle was the most commonly injured area, followed by the knee. In conclusion, NBA basketball has evolved to become a highly physical sport with a predictably high rate of injury.

Therefore, the most common types of injuries in basketball are the following:

- Ankle Sprains
- Jammed Fingers
- Knee Injuries
- Deep Thigh Bruising
- Facial Cuts
- Foot Fractures
- Jumper's knee
- Tibia stress fractures
- Achilles tendinopathy and rupture
- Muscle strains

It is estimated that more than 1.6 million injuries are associated with basketball each year (http://www.stopsportsinjuries.org/STOP/Prevent_Injuries/Basketball_Injury_Prevention.aspx 2017).

More injuries are sustained during competition than during training sessions (Fibaeurope.com 2017). In a 16-year review of men's and women's college basketball in the USA, it was found that the rate of injuries in games was two times greater than in practice (Caine et al. 2010). The increased frequency of injuries in games contrary to practice is caused by high intensity level of competition and because of the maximum effort that is expended during games. The athlete is at maximum risk, which might make athletes more vulnerable to injury. In addition it was reported that 3.7 times more serious injuries occurred in games. Opposite to that in professional American basketball, it was reported that male and female players were injured more frequently at practices then at games. Over a 10-year period 56.8% of injuries in male players occurred during a training session. Few studies have investigated the time during a game at which injury occurs, and it was found no significant relationship.

Another study (McGuine et al. 2013), concerning basketball coaches' utilization of ankle injury prevention strategies, showed that these strategies are underutilized. Ankle injuries are the most common high school basketball injury. Little is known regarding the utilization of ankle injury prevention strategies in high school settings. According to McGuine et al. (2013) understanding how prevalent the use of injury prevention strategies is as well as reasons for their disuse is an essential first step to increasing the utilization of these programs in school settings and making it more likely that injury prevention strategies will be successfully implemented in sport settings. Almost half of the coaches who did not utilize an injury prevention program cited a lack of awareness or expertise to utilize these programs. It should also be noted that providing expertise to basketball coaches regarding injury prevention training may not be enough to have them implement these programs for their teams. In one study (Marchi et al. 1999), the authors offered to instruct and provide equipment for a group of coaches in an urban setting and found that nearly two thirds declined to participate because of a lack of time and/or interest. Coaches in this study were provided with the expertise needed to implement the programs but still did not participate. This finding illustrates that working with individual coaches alone may not ensure injury prevention as a priority. Instead, a concerted effort may have to be made at the sport association level down through the school administration to the individual coaches as well as parents and athletes. In this manner, injury prevention may be thought of as an integral component of offering an interscholastic basketball program.

Learning about the formats and components of a Basketball Injury Prevention Strategy (BIPS) preferred by coaches may be key to the successful implementation of these programs. According to McGuine et al. (2013) coaches preferred programs that require minimal equipment and are interesting. This may indicate that sports medicine providers need to emphasize that a minimal level of equipment (at low cost) can be utilized effectively.

2 Related Work

In this section the relevant previous work in sport injuries and training through MOOCS will be discussed and different types of content provided will be highlighted.

A lot of videos about BIPS are uploaded in online video repositories (YouTube, Vimeo, etc.) but there are only a few complete courses that provide coherent learning material. Some of these are the following:

- The Body Matters: Why Exercise Makes You Healthy and How to Stay Uninjured (edX Course by McGill a Canadian post-secondary institution: https://www.edx.org/course/body-matters-why-exercise-makes-you-mcgillx-body101x-0). This course provides the following topics:
  - An understanding of the benefits of physical activity
  - General principles on how to train and how to prevent injuries
  - An understanding of how to recover from injury
- How To Avoid Injury And Illness (Course provided by Athlete Learning Gateway: http://onlinecourse.olympic.org/course/baseview.php?id=14). This course provides the following topics:
  - What puts you at risk of illness and injury?
  - How you can reduce these risks to ensure you are competing at peak fitness.
- Diploma in Football Medicine (Course provided by FIFA: http://fifamedicinediploma.com/). The FIFA Diploma in Football Medicine is a free online course, designed to help clinicians learn how to diagnose and manage common football-related injuries and illnesses. The diploma is presented by FIFA, in collaboration with a number of international experts in football medicine.

3 C4BIPS Project

Injury prevention programs are often designed for health care professionals such as Physiotherapists and Sports Medicine Doctors. Such a strategy is counterproductive because these professionals often attend the athletes after injury. Coaches on the other hand spend a lot more time with the athletes, even in the absence of injuries and they are suitably positioned to implement strategies for injury prevention. In addition coaches start to train athletes from young ages in which there is a lack of health care professionals to implement injury prevention strategies. Furthermore implementing such corrective and preventative strategies from early age can save athletes from the additive effects of unnecessary strains and injuries in their bodies. This project is unique because it specifically targets coaches as the most suitable professionals to implement injury prevention.

The most important reason cited by the coaches for not implementing injury prevention strategies in their arsenal is the lack of knowledge. This is probably because of lack of time to spend on developing the skills normally utilized by health care professional after many years of formal training. Thus, coaches perceive these kind of skills as difficult to acquire and time consuming in their implementation.

Also they are concerned with equipment, cost and time off tactical training for their team. The other innovative aspects of this project concern the mode of delivery of the coaches' training and the overall implementation of injury prevention program. The project will be designed so that the users can increase their skills in a non-threatening, self-paced, continuous professional development environment.

C4BIPS (http://c4bips.csd.auth.gr/) is an Erasmus Plus Sport Small Collaborative Partnership project that focuses on the area of "protecting athletes, especially the youngest, from health and safety hazards by improving training and competition conditions". The main goal of C4BIPS project is to create a Massive Open Online Course for Basketball Injury Prevention Strategies (BIPS). Experts in the field of sports science will create content that will allow coaches to attend the most practical easy to implement strategies for injury prevention. The curriculum will also provide the necessary skills to identify individual risk factors in their athletes by observing their habitual motor strategies during drills and training, and later on implement easy, low cost modifications in their everyday training programs to correct these risk factors. C4BIPS can also serve as a useful resource to constantly update the knowledge of the coaches as well as provide them with the opportunity to share effective modes of implementation with the community of their peers. In this way the project will be constantly renewing and updating itself and always keep the focus and practicality necessary to attract more and more users.

C4BIPS consists of a consortium of various organizations with the participation of people with expertise in appropriate fields such as sport policy and practice (training, competitions, coaching, etc.), with academic expertise as well as their ability to reach out wider audiences. The consortium includes the following organizations:

- Aristotle University of Thessaloniki (AUTH) is the largest university in Greece, with 11 Faculties organized into 41 Schools. Two departments of Aristotle University of Thessaloniki will be engaged in the C4BIPS project (a) The Department of Informatics (http://www.csd.auth.gr) and more especially its Multimedia Laboratory (http://mlab.csd.auth.gr) and (b) The Department of Physical Education and Sports Science (DPESS).
- University of Nicosia Research Foundation (UNRF), Cyprus which manages research activities and is closely linked to the University of Nicosia.
- ASD Margherita Sports and Vita (MSV), which is a basketball academy in Italy.
- Asociación de Baloncestistas Profesionales (ABP), which represents the professional basketball players in Spain.

This project will be a useful addition in Sports Science education programs especially in the Basketball coach specialization. It will then provide the necessary skills for injury prevention to active athletes and future coaches in advance and not after specialization. This will tackle the problems created by lack of interest by the coaches to be trained in such strategies after they start working. So the project takes a proactive approach in creating coaches with endogenous abilities to identify risk factors and implement corrective and preventing measures. These already educated coaches are expected to disseminate knowledge to their peer via seminars and conferences as well as impose a sports association level down approach into coaches' training.

Furthermore, the Europe 2020 strategy stresses the need for transforming educational content for instruction and training in a way that engages, motivates and immerses people to develop personal experiences of constructing their learning experiences. Hence it is widely contested in Europe that transferring training content from traditional contexts will not discern value to the way education and training is moving forward, but rather it will emulate pedagogical modalities already practiced in conventional settings. In congruence to Europe's observation in not ushering innovation on how training content is used, this is an occurring problem in the field of sports training. In response therefore to the significant difficulty of how to design, develop, implement and share highly interactive, process-based content (http://www.openeducationeuropa.eu/en/initiative, 2017), it is important to create such content that permeates personal construction of meaning, activity-based learning meaningful assessment and rapid feedback.

C4BIPS innovates in two ways:

- By introducing and applying multimedia based training content in the Sports field for developing of personal knowledge related with BIPS. Furthermore, the envisaged Massive Open Online Courses (MOOCs) for BIPS will support a holistic educational experience of athletes concerning BIPS. The BIPS course of this project will be based on annotated video material which will be created by an interdisciplinary team of experts (engineers, human computer interaction and distance education faculty, professional basketball athletes, professional basketball coaches and physiotherapists). C4BIPS' MOOC will provide a flexible solution to athletes (and especially the youngest) and coaches to attend educational programs, especially for those who are travelling in professional athletic events and don't possess the time to invest in cultivating their training or education. Hence C4BIPS' goal is to share knowledge on the importance of BIPS to athletes, youngest and coaches and the significance in applying them in everyday training. C4BIPS aims to harness the potential of MOOCS to empower athletes to invest in their personal development by accessing and enrolling in modules that have a focus to improve their training conditions.
- By introducing educational modules for BIPS, which have not been considered or applied in any European MOOC in terms of offering a full-course online experience encompassing associated content, learning design and facilitation of interaction among peers following a structured study guide/syllabus. Educational programs for

athletes have been created statically, using a traditional web-page not foreseeing the importance of providing a BIPS educational modality widely to a massive amount of athletes around Europe who wish to acquire knowledge and skills via flexible learning opportunities provided by a MOOC.

4 Data Collection Method Concerning Athletes Understanding of BIPS

In order to create educational content, C4BIPS' team formulated a research plan to measure athletes' understanding about BIPS and their opinions about injuries prevention. In this section, the data collection method is analyzed.

4.1 Inventory

In order to collect data about athletes understanding of BIPS an online inventory (http://c4bips.csd.auth.gr/first-inventory/) was created by experts in the project consortium.

It consists of 26 questions grouped in four dimensions: (a) Personal Data, (b) Engagement with basketball - Athletic career, (c) Sports related injuries they suffered, and (d) Their opinion about the sports related injuries. Some questions are 6-point Likert-type rating scale with the following anchors: 1 Don't know, 2 not at all, 3 minimally important, 4 not related to sport injuries prevention, 5 very important, 6 extremely important and some of them are YES/NO questions followed up with a count selection, for example: have you ever suffered a neck injury? If YES, how many times?

4.2 Participants

By the time this research was conducted the inventory collection process was open. 70 complete questionnaires have been collected out of which, 64 had accepted the policy agreement, the rest were considered invalid. By the end of this process, it is estimated that more than 400 athletes will be engaged.

5 Statistical Analysis

In this section partial results will be presented due to the fact that the collection process is ongoing, as mentioned previously. The statistical analysis was performed using the SPSS 21 statistical package.

5.1 Personal Data – Engagement with Basketball

92% of the athletes that completed the inventory were men and only 8% were women. 53% of them have already finished their professional career and only 47% were active athletes. 61% of participants have a college degree or higher (Fig. 2).

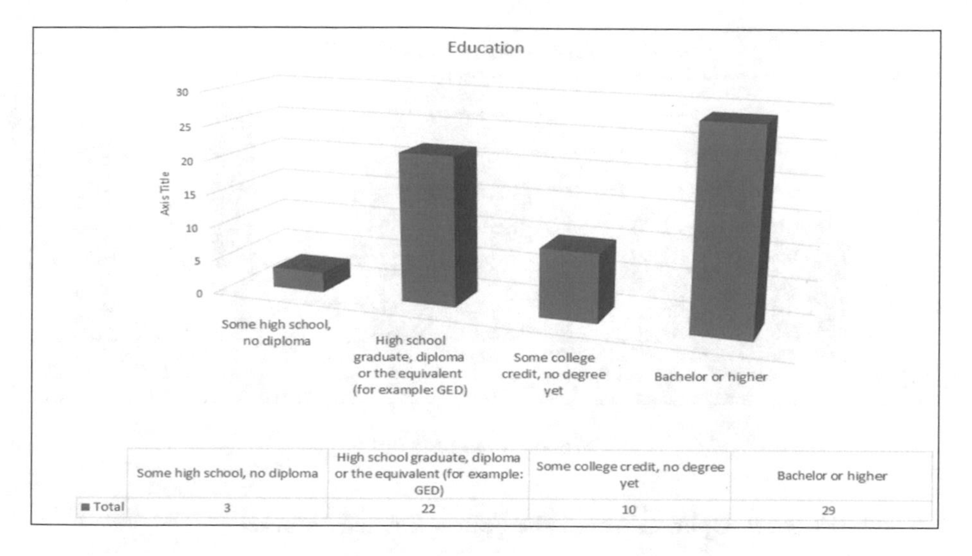

Fig. 2. Level of education

5.2 Sports Related Injuries Suffered

Participants mean value of injuries during their professional career is 6.58 (injuries per athlete). Figure 3 shows the most common areas of injury for a professional athlete. As it was estimated most injuries are located at the lower body (back to foot). In Fig. 4 are shown the most common types of injuries. More than 25% of participants believe that injuries affected their career and more than 40% believe that injuries affected their overall performance.

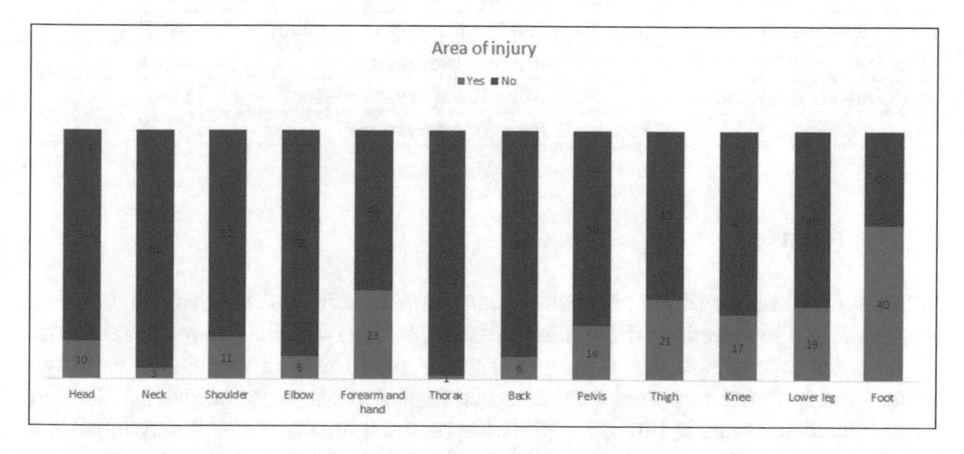

Fig. 3. Area of injury

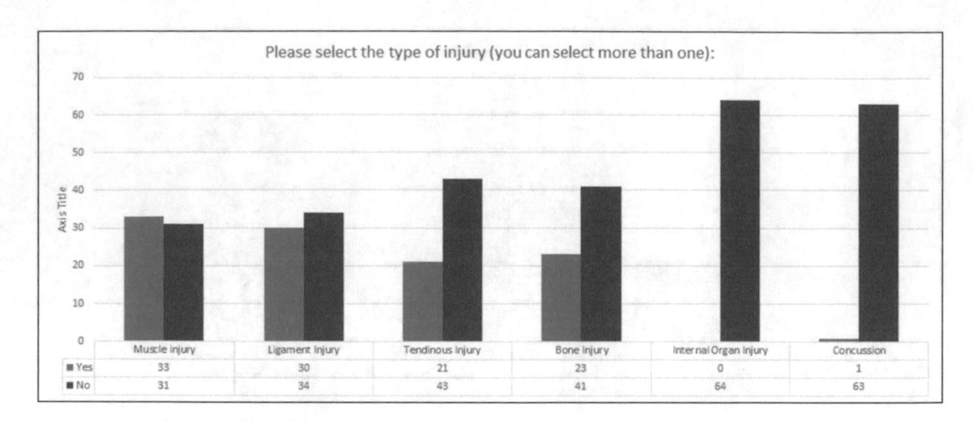

Fig. 4. Type of injury

5.3 Participants Opinion About the Sports Related Injuries

95% of participants believe that injury prevention is very important and higher. Table 1 shows the participants opinion about the importance of BIPS in specific areas.

Table 1. Participants' opinion about the importance of BIPS in specific areas

Question	Very important and higher
How important do you think warm up is for injury prevention?	>97%
How important do you think stretching is for injury prevention?	>77%
How important do you think strengthening is for injury prevention?	>97%
How important do you think eccentric strength is for injury prevention?	>89%
How important do you think proprioception is for injury prevention?	>84%
How important do you think taping is for injury prevention?	>41%
How important do you think motor control is for injury prevention?	>84%
How important do you think bracing is for injury prevention?	>77%

6 Platform

In this section a state of the art about e-learning platforms will be conducted and the result will be the selection of the most suitable platform that fits the project's needs.

In order to evaluate the existing e-learning platforms, an evaluation table was created and importance points were assigned to features that are needed on the platform. Importance scale is 1 to 3, 1 slightly important, 2 important and 3 very important. If a platform qualifies for a feature it was assigned an X on the table. The values are summarized at the bottom of Table 2.

As, can be seen from the evaluation, the platforms that most suits the needs of the project is Moodle (https://moodle.org/). Moodle is a free and open-source software

Table 2. E-learning platforms state of the art

Tools	Importance points (1–3)	E-learning platforms						
		Moodle 3.3.1	Open e-class 3.5.6	Sakai 11	ATutor 2.2.2	Blackboard Learn 9.1	Dokeos CE	Open edX
Communication tools								
Discussion forums	2	X	X	X	X	X	X	X
Discussion management	2	X	X	X	X	X	X	X
File sharing	2	X	X	X	X	X	X	X
Personal messaging	1	X	X	X	X	X	X	X
Collaboration tools								
Group work	3	X	X	X	X	X	X	X
Community networking	3	X	X	X	X	X	X	X
System tools								
PHP/MySQL	3	X	X	–	X	X	X	X
Multilingual package	3	X	X	X	–	X	–	–
Developed plugins	3	X	–	–	–	–	–	–
Open source	3	X	X	X	X	–	X	X
Results								
		25	22	19	19	19	19	19

learning management system written in PHP and distributed under the GNU General Public License. Developed on pedagogical principles, Moodle is used for blended learning, distance education, flipped classroom and other e-learning projects in schools, universities, workplaces and other sectors.

With customizable management features, it is used to create private websites with online courses for educators and trainers to achieve learning goals. Moodle (acronym for modular object-oriented dynamic learning environment) allows for extending and tailoring learning environments using community sourced plugins.

7 Conclusion

The results produced so far are in-line with previous researches about injury types and areas in basketball. Athletes believe that injury prevention is very important and as a result this highlights the importance of gathering and presenting these prevention strategies in an accessible online platform.

By the end of the data collection process and the statistical analysis, a complete curriculum about BIPS will be created. This curriculum will be provided through a MOOC. The online lessons will include annotated videos with multilingual subtitles, reading material, online presentations and online tests. At the end of the project, an evaluation period will occur in order to measure participants' change of opinion and understanding about BIPS.

Acknowledgement. This project has been funded with support from the European Commission. This publication reflects the views only of the author, and the Commission cannot be held responsible for any use which may be made of the information contained therein.

The authors of this research would like to thank C4BIPS team who generously shared their time, experience, and materials for the purposes of this project.

References

Injuries in the European Union: Summary of Injury Statistics, for the Years 2008–2010 (2013). http://ec.europa.eu/health/data_collection/docs/idb_report_2013_en.pdf

Drakos, M.C., Domb, B., Starkey, C., Callahan, L., Allen, A.A.: Injury in the National Basketball Association - a 17-year overview. Sports Health. **2**(4), 284–290 (2010). https://doi.org/10.1177/1941738109357303. http://www.ncbi.nlm.nih.gov/pmc/articles/PMC3445097/

Henry, J.H., Lareau, B., Neigut, D.: The injury rate in professional basketball. Am. J. Sports Med. **10**, 16–18 (1982)

Starkey, C.: Injuries and illnesses in the National Basketball Association: a 10-year perspective. J. Athl. Train. **35**, 161–167 (2000)

Meeuwisse, W.H., Sellmer, R., Hagel, B.E.: Rates and risks of injury during intercollegiate basketball. Am. J. Sports Med. **31**, 379–385 (2003)

Colliander, E., Eriksson, E., Herkel, M., Skold, P.: Injuries in Swedish elite basketball. Orthopedics **9**, 225–227 (1986)

Messina, D.F., Farney, W.C., DeLee, J.C.: The incidence of injury in Texas high school basketball: a prospective study among male and female athletes. Am. J. Sports Med. **27**, 294–299 (1999)

http://www.stopsportsinjuries.org/STOP/Prevent_Injuries/Basketball_Injury_Prevention.aspx

FIBA Europe: Basketball Injuries - Where, When and Why? http://www.fibaeurope.com/cid_VVN9zdHHJOEO8iyoqkT3E3.coid_Q6n2q4t4HuovtdaDGnj0N2.articleMode_on.html

Caine, D.J., Harnmer, P.A., Shiff, M.: Epidemiology of Injuries in Olympic Sports. Blackwell Publishing Ltd, Hoboken (2010)

McGuine, T.A., Hetzel, S., Pennuto, A., Brooks, A.: Basketball coaches' utilization of ankle injury prevention strategies. Sports Health **5**(5), 410–416 (2013). https://doi.org/10.1177/1941738113491072

Marchi, A.G., Di Bello, D., Messi, G., Gazzola, G.: Permanent sequelae in sports injuries: a population based study. Arch. Dis. Child. **814**, 324–328 (1999)

http://www.openeducationeuropa.eu/en/initiative. Accessed 2017

Gamified and Online Activities for Learning to Support Dual Career of Athletes (GOAL)

Thrasyvoulos Tsiatsos[1], Stella Douka[2], Nikolaos Politopoulos[1](✉), Panagiotis Stylianidis[1], Efthymios Ziagkas[2], and Vasiliki Zilidou[2]

[1] Department of Informatics, Aristotle University of Thessaloniki, Thessaloniki, Greece
{tsiatsos, npolitop, pastylia}@csd.auth.gr
[2] Department of Physical Education and Sports Science, Aristotle University of Thessaloniki, Thessaloniki, Greece
{sdouka, eziagkas, vizilidou}@phed.auth.gr

Abstract. Dual careers defined as enabling education or work, promote the attainment of a new career after the sporting career, and protect and safeguard the position of athletes. The success of dual career arrangements is largely dependent on individual, personalised pathways enabled through the use of rich-mediated technologies that will facilitate athlete's effort to adapt to the changing employment needs. GOAL (Gamified and Online Activities for Learning to support dual career of athletes, http://goal.csd.auth.gr/) will effectively start dual careers for athletes embracing their awareness to balance sport training and education and, at a later stage, sport training and employment.

The aim of this study is to present the project, highlight the importance of dual careers, describe the results of the project's needs analysis, conduct a state of the art about e-learning platforms and serious games design and summarize GOAL's future work.

Keywords: MOOCS · Serious games · Dual career · Sports

1 Introduction

Nowadays, elite sport has reached a high level of professionalism (Brackenridge 2004). Athletes' training hours have increased resulting in more than 40 h of work when considering training hours, competition travel time, and study requirements (Amara et al. 2004). The elite sport career entails five to ten years dedicated to sport (Alfermann and Stambulova 2007). Along these lines, balancing studies along with the sport career allows the athlete to better prepare for future employment (Aquilina 2013). A significant body of research has been accumulated on athletic career, examining career development, transitions, and especially athletic retirement (Stambulova et al. 2009). Contemporary research emphasizes the need for "whole career" and "whole person" approach, highlighting that athletes go through several transitions in sport, education, and psycho-social development simultaneously (Wylleman and Lavallee 2004). Dual career research is a response to the call for this holistic approach and it has become a growing area of study (Aquilina and Henry 2010; Burnett 2010)

M. E. Auer and T. Tsiatsos (Eds.): IMCL 2017, AISC 725, pp. 623–634, 2018.
https://doi.org/10.1007/978-3-319-75175-7_61

The combination of education and training often becomes a challenge for athletes. Transitions are taking place often at this stage when athletes are changing homes, sports clubs and have to make new training and sports arrangements. Sports and Physical Education faculties in Europe are focused only on sports training without offering any flexible courses, predominantly through distance learning. This kind of learning may provide to athletes the flexibility in terms of the timing and location of their sporting and academic activities. Distance learning programmes in Europe for supporting the dual career of athletes have not been convincing, in terms of quality, level, accessibility and interactive character.

Europe favours intra-European mobility. The European Qualifications Framework (EQF) ensures international mobility for students in accredited education. It also assesses the standard of degrees from non-accredited institutions and courses. The framework was established by the European Council and is monitored and maintained by an independent national organisation in each MS. In addition, the EU supports the development of more flexible, outcome-driven learning systems to allow the validation of competencies acquired outside of formal education. This effort relies heavily on the widespread use of digital technology in education, to unlock and exploit freely available knowledge (Rethinking Education: Investing in skills for better socio-economic outcomes 2012). All of these policy goals are relevant to sport and for Dual Career. For example, student mobility facilitates competition and training in other member states, and flexible learning systems are supportive of combining sport and education.

Transition to the post-sport career is an inevitable transition for athletes that mixes athletic context with non-athletic context relevant to starting a new career after sports. The average professional athlete's career is over by age 33. For physically-demanding sports, like rythmic gymanstics, it's as young as 18.5 (Fig. 1).

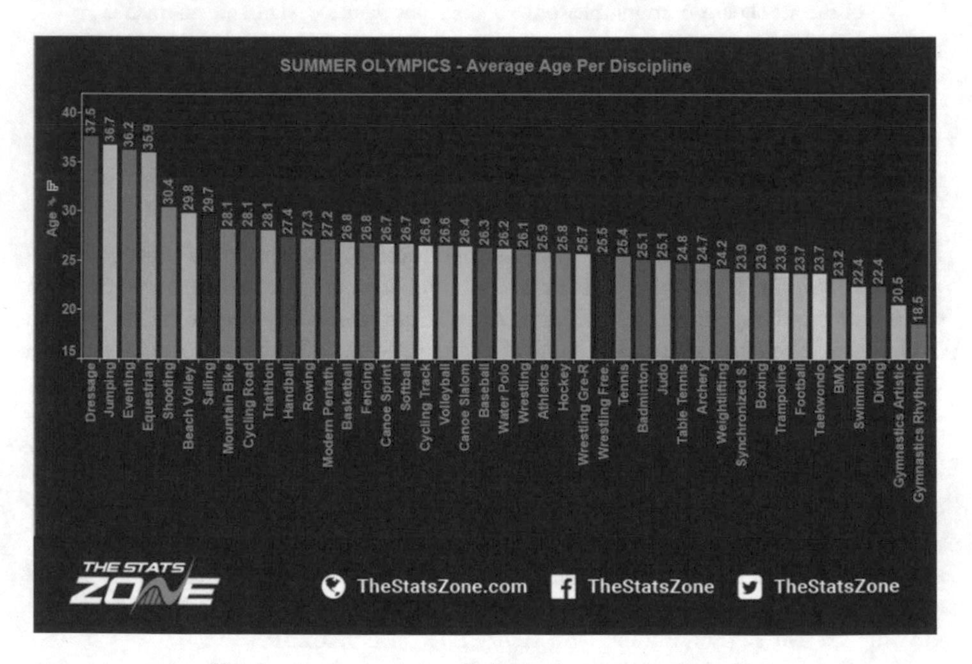

Fig. 1. Average age per discipline in summer Olympics

2 Related Work

In this section the relevant previous work in athletes' dual career programs will be discussed and different types of content provided will be highlighted. Some of these are the following:

- Management training in multicultural events within an international context - Erasmus+ project (http://tasem.inefc.cat/index.php/academic-program/contents),
- Enterprise and Entrepreneurship – University of Kent (https://www.kent.ac.uk/courses/undergraduate/1396/sport-management#!structure),
- Human Resources and Organizational Behavior – University of Kent (https://www.kent.ac.uk/courses/undergraduate/1396/sport-management#!structure),
- International Sport Marketing and Sponsorship - Sheffield Hallam University (https://www.shu.ac.uk/study-here/find-a-course/llm-international-sports-law-in-practice)

All of the aforementioned courses have reading material and some multimedia, none of them uses gamified activities or serious games implementation. That is the gap tha GOAL project wants to fill.

3 GOAL Project

GOAL aims to support active and non-active athletes in the development of their professional endeavours, after the end of their athletic career. By active and non-active, we mean athletes that are already doing a sport and wish to be prepared on achieving a smooth transition between the end of their professional careers and further educational/business endeavours and those athletes who discontinued their professional athletic career and experience challenges with their integration in education, training and the open labour market.

GOAL consists of a consortium of various organizations with the participation of people with expertise in appropriate fields such as sport policy and practice (training, competitions, coaching, etc.), interactive ICT-based tools, sports marketing and entrepreneurship with academic expertise as well as their ability to reach out wider audiences. The consortium includes the following organizations:

- Aristotle University of Thessaloniki (AUTH) is the largest university in Greece, with 11 Faculties organized into 41 Schools. Two departments of Aristotle University of Thessaloniki will be engaged in the GOAL project (a) The Department of Informatics (http://www.csd.auth.gr) and more especially its Multimedia Laboratory (http://mlab.csd.auth.gr) and (b) The Department of Physical Education and Sports Science (DPESS).
- University of Nicosia Research Foundation (UNRF), Cyprus which manages research activities and is closely linked to the University of Nicosia.
- Asociación de Baloncestistas Profesionales (ABP), which represents the professional basketball players in Spain.

- Coventry University (COVUNI), which is expertised in serious games development and gamification.
- YES- European Confederation of Young Entrepreneurs, is the expert partner to cover the project's needs on sports business and entrepreneurship content.
- Asociacion De Jugadores De Futbol Sala (AJFS), which represents the professional futsal players in Spain.
- Portuguese Football Players' Union (Sindicato dos Jogadores Profissionais de Futebol, SJPF), which represents the professional football players in Portugal.
- V4Sport, which is expertised in organization of sports events, volunteer coordination, training and consulting and public relations.

The project is determined to create awareness about the topic on dual career by providing an enabling environment for addressing athletes' dual career incommensurable goals whilst leveraging athletes' skills and competencies (e.g. problem solving, decision-making, communicating, teamwork and leadership) for understanding, applying and sharing best practices as means to help their integration in education, training and open labour market. Furthermore GOAL provides information to support athletes in life coaching, particularly focused on how to cope with different dual career related transitions and topics; and share knowledge on the importance of education and the significance in preparing for the business world during and after their sports careers.

In particular GOAL will identify and test gamified learning and training activities to form best practices for supporting dual careers of athletes using digital technology such as games/gamification in sports. A set of interactive ICT-based tools will be offered to active and non-active athletes for acquiring skills and competencies necessary to consciously discover, plan and determine their future career goals once they complete their competitive sports career. Such skills are critical in developing athlete's continuous professional career development including efforts of coping with transition and change both as individual personalities being part of a wider community (e.g. active citizenship) as well as professionals that will be following a career after sports competition, and thereby preparing them for a new job.

GOAL tackles the need for European Commission to consider taking action in discerning the introduction and implementation of dual career programmes by designing an online educational programme with innovative features. GOAL's approach highlights two innovative aspects:

- GOAL innovates by introducing and applying training content in the Sports field for developing of personal capacities that are critical in developing the athletes both as individual personalities (such as teamwork, goal setting, coaching) as well as professionals that will be following a career after sports competition. Thus GOAL will propose teaching as well as physical training via serious games, and underlying mechanics and dynamics (i.e. goals, challenges, levels, progress bars, scoring mechanisms, dialogues, avatars for coaching/mentoring about how to train a sport). In particular, GOAL designs, develops and implements serious games for sports. These games will encompass rules, goals and coaching and free-play for athletes to be trained on how to conceptualise abstract theoretical aspects of a specific sport (e.g. sport's history, rules) and also how to understand the application of this specific sports activities in practice. These game will integrate content (i.e. questions) for three specific sports football, basketball and futsal The games will be complemented by specific sport activity scenarios instantiated via an activity-based approach for eliciting inquiry, questioning and learning by doing. They will also include all those features and aspects that resemble an instructional process situating learning activities, teaching strategies, feedback and assessment balanced with game mechanics and dynamics;

- Entrepreneurship for athletes is a relatively new conceptualisation specific to sports. As the EC expert group report (Developing the creative and innovative potential of young people through non-formal learning in ways that are relevant to employability 2014), there is a need to integrate non-formal learning via rich mediated technological interventions to sports for supporting athletes' re-integration to employability. To foster this perceived inclusivity, GOAL offers an entrepreneurship game coupled with scenarios relevant to athlete's dual careers by creating awareness on the different employability options and pathways that can be chosen conducive to their own sports expertise and business mode. To situate scenarios to specific athletic business preferences and variety of choices, GOAL carries out user-oriented needs analysis to precisely capture and discern the entrepreneurship needs, challenges and types of businesses most relevant to athletes and integrated into the entrepreneurship game. This is congruent in terms of creating a methodology that is constantly evolving and adjusted to target group's requirements in all phases of the project including needs analysis, games development and implementation ensuring that perceived changes in athlete's behaviour in creating new business ideas are identified, captured, analysed and escalated in an informed and systematic way before the actual games development and module design is initiated.

The added value of European cooperation in GOAL is demonstrated in the following ways:

- GOAL project fills an identified gap and a direct need for developing distance learning programmes and tools that combine educational content with a range of flexible forms of education delivery, in this case comprising serious games and multimedia blended content with personalised feedback provision to student-athletes. The project provides an online virtual learning environment where educational content for skills and knowledge development is blended with capabilities for social interaction and support from peers and experts, providing an opportunity for athletes around Europe to seamlessly develop or sustain their dual careers while being on the move. This virtual environment provides added value to athletes in mobility, facilitating further personal development in an individualised and undisrupted manner and additional psychological support emphasised by the feeling of belonging to a community that shares similar experiences, problems and needs.

- The mix of diverse sports and educational organisations from different European countries collaborating in the GOAL project will allow a multitude of stakeholder views to unravel and maximize the effect and impact of proposed activities.
- Moreover, the online educational content and the created community of practice that will dually support the development of knowledge and skills in sports and business will be co-developed and shared by the participating EU educational and sports organisations and subsequently made available to European sports stakeholders across Europe, avoiding duplication of efforts that would unavoidably occur if the initiative is taken at a national level in different countries.
- GOAL project will essentially develop a network of dual career practitioners by nurturing a virtual community of practice among athletes, sports educators, coaches, sports entrepreneurs and managers of sports activities, clubs and centres. The European character of this network will be supported by the project consortium members and will be further extended through their connections with sports' stakeholders across Europe. Most importantly, support received from the community will be fully integrated in the sport, educational, vocational and lifestyle systems of the athletes rather than remaining isolated outside the sport context and will be based on direct contact with athletes, parents, coaches, performance directors and other stakeholders.

4 Needs Analysis

In order to create educational content, GOAL's' team formulated a research plan to measure athletes', experience in ICT's and distance learning, expectations about starting their professional career and their interest in a number of courses that were proposed. At the end, the results of online surveys will be presented. The statistical analysis was performed using the SPSS 21 statistical package.

4.1 Survey

In order to collect data about matters that were mentioned above, an online survey (http://goal.csd.auth.gr/first-inventory/) was created by experts in the project's consortium.

It consists of 34 questions grouped in six dimensions: (a) Personal Data, (b) Athletic career, (c) Experience in ICT's, (d) Experience in distance learning, (e) Expectations about starting a professional career and (f) Preferred courses. 24 questions are 5-point Likert rating scale with the following anchors: 1 disagree, 2 slightly disagree, 3 neutral, 4 slightly agree, and 5 agree, 1 multiple selection and the rest of them are answering questions.

4.2 Results

Participants Personal Data and Athletic Career
By the end of the survey more than 1400 athletes were approached and 890 of them completed the survey. 853 of those read and agreed to the GOAL's Privacy Policy and answered the questionnaire, thus 37 of them didn't agree to the GOAL's Privacy Policy so they were excluded from the study. Mean age of athletes was 30.8 years (S.D = ±10.91).

Athletes that completed the survey come from 19 different countries. Most of them are from Portugal (N = 458), followed by Spain with 179 athletes, and then Greece with 62 athletes. The rest of them were from Poland, Cyprus, Belgium, Italy, Albania, Austria, Andorra, Belarus, France, Georgia, Germany, Netherlands, Slovenia, Switzerland and United Kingdom (UK). 616 of them were male athletes and the remaining 237 female. Most of them have a bachelor degree or higher (N = 385), 171 athletes have some college credit but no degree yet and only 12 had not completed school.

Participants have been athletes, at means, for 15.8 years (S.D = ± 8.26). 514 of them were still active athletes and the rest 339 are retired athletes. Retired athletes have stopped their athletic career at means 10.0 years ago (S.D. = ±10.52).

Experience in ICT's and Distance Learning
More than 78.55% of the participants believe that they are good at using computers, 77.38% say that they can learn to operate any basic software application, and more than 85% (85.47%) that they are good at using electronic devices such as mobiles phones etc., but only 40.89% believe that they can resolve computer related issues (software issues etc.).

Additionally 88.86% believe that they are good at using Internet and 83.82% that can learn to operate any basic web application (e-mail etc.), but only 55.21% that can resolve issues that may come up with using the Internet.

As it was highlighted before, the participants are familiar operating PC's and although 93.67% believe that Internet is useful for finding information and 71.86% that online courses are useful to their knowledge, only 25.44 had participated in distance learning activities frequently. The reason about that might be that only 20.52% prefer distance learning than traditional learning methods (face to face), 30.36 are neutral. What is important and positive to be mentioned is that only 26.38% don't want to participate in online courses.

Expectations About Starting a Professional Career
Athletes want to start a professional career in sport industry (71.28%), but they are not so font of starting a new career in a different industry (29.55%). As can be seen in Table 1, most of the participants prefer to get professionally involved with the sport that they are/were athletes.

Table 1. Preferrable career orientation after athletic career

Question	Slightly agree and higher
How much you would like to start a professional career as a coach in the sport that you were an athlete?	57.90%
How much you would like to start a career as a coach in a different sport than the one that you were an athlete?	11.49%
How much you would like to start a business career (e.g., management, administration, and marketing) in the sport that you were an athlete?	45.26%
How much you would like to start a business career (e.g., management, administration, marketing) in a different sport that the one that you were an athlete?	23.57%
How much you would like to work for the development to the sport that you were an athlete?	77.37%
How much you would like to work generally for the development of sport?	70%
How much you would like to be involved in the administration of sport associations in the sport that you were an athlete	64.59%

Preferred Course
In Fig. 2 can be seen the ranking of participants' preferred courses.

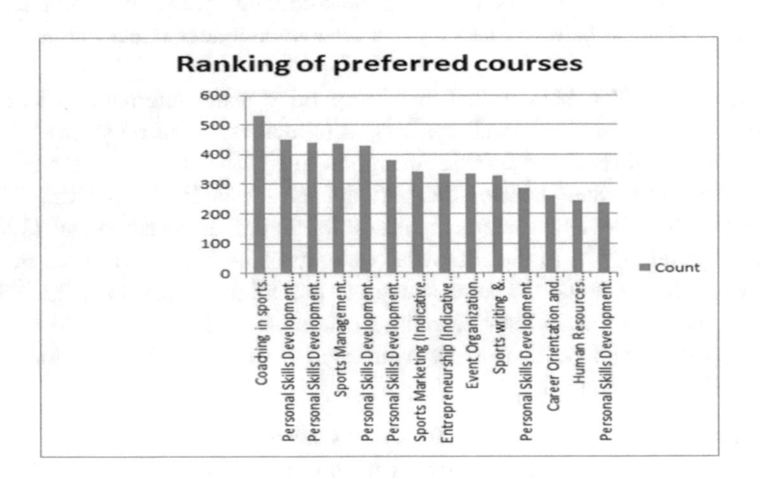

Fig. 2. Ranking of preferred courses

With that in hand, it has been decided that learning material will be developed for the following courses, divided in two cycles (Table 2).

Having settled on the curriculum, the next step was to design the technological solution for content and service delivery.

Table 2. Goal curriculum

Cycle 1	Cycle 2
• Entrepreneurship • Personal skills development - Teamworking skills • Personal skills development - Decision making skills	• Sports management • Sports marketing • Coaching in sports

5 Design of Technological Solution

5.1 MOOC Platform

In this section a state of the art about e-learning platforms will be conducted and the result will be the selection of the most suitable platform that fits the project's needs.

In order to evaluate the existing e-learning platforms, an evaluation table was created and importance points were assigned to features that are needed on the platform. Importance scale is 1 to 3, 1 slightly important, 2 important and 3 very important. If a platform qualifies for a feature it was assigned an X on the table. The values are summarized at the bottom of Table 3.

Table 3. E-learning platforms state of the art

Tools	Importance points (1–4)	E-learning platforms						
		Moodle 3.3.1	Open e-class 3.5.6	Sakai 11	ATutor 2.2.2	Blackboard Learn 9.1	Dokeos CE	Open edX
Communication tools								
Discussion forums	2	X	X	X	X	X	X	X
Discussion management	2	X	X	X	X	X	X	X
File sharing	2	X	X	X	X	X	X	X
Personal messaging	1	X	X	X	X	X	X	X
Collaboration tools								
Group work	3	X	X	X	X	X	X	X
Community networking	3	X	X	X	X	X	X	X
System tools								
PHP/MySQL	3	X	X	–	X	X	X	X
Multilingual package	4	X	X	X	–	X	–	–
Developed plugins	4	X	–	–	–	–	–	–
Open source	4	X	X	X	X	–	X	X
Results								
		28	24	21	20	20	20	20

As, can be seen from the evaluation, the platforms that most suits the needs of the project is Moodle (https://moodle.org/).

5.2 Games

The movement towards the use of serious games as training and learning is proliferated by the perceived ability of such games to create a memorable and engaging learning experience. Various commentators and practitioners alike argue that serious games may develop and reinforce 21st century skills such as collaboration, problem solving and communication. While in the past, practitioners and trainers have been reluctant in using serious games for improving skills and competencies there is an increasing interest, to explore how serious games could be used to improve specific skills and competencies. The overarching assumption made is that serious games are built on sound learning principles encompassing teaching and training approaches that support the design of authentic and situated learning activities in an engaging and immersive way.

Developing serious games based on activity-centred pedagogies that enable trainees to engage actively with questions and problems associated with sport activity and dual careers is an empowering approach with benefits for subject learning as well as for developing a wide range of important high-order intellectual attributes including the notion of 'transferability – that is being able to situate specific skills in different settings and contexts, a competence much needed for active and non-active athletes as means to establish their own business and being competitive European citizens and role models for the society.

The serious games and gamification mechanics of GOAL will be based on scenarios that will bring-up gameplays tied up to athletes' attempts to start their own sport business and playing a sport either individually or collaboratively. Aspects of how to create their idea to practical implementations on how to create and run a business will be introduced in a fun and educative way. The games will target athletes at the end-of-sporting career including those who leave the system earlier than planned with the objective to re-integrate athletes into education and labour market and transform them into highly qualified employees on the European labour market.

In order to develop personal skills and based on the statistical analysis, it was decided to create adventure games and especially interactive movies.

Adventure Games - Interactive Movies

An interactive movie contains pre-filmed full-motion cartoons or live-action sequences, such as dialogues, where the player controls some of the moves of the main character or his/her answers. For example, when in danger, the player decides which move, action, or combination to choose. In these games, the only activity the player has is to choose or guess the move the designers intend him to make. Elements of interactive movies have been adapted for game cut scenes, in the form of Quick Time Events, to keep the player alert.

State of the Art

In order to evaluate the existing game engines suitable to create these types of games, an evaluation table was created and importance points were assigned to features that are needed on the platform. Importance scale is 1 to 3, 1 slightly important, 2 important and 3 very important. If a game engine qualifies for a feature it was assigned an X on the table. The values are summarized at the bottom of the Table 4.

As, can be seen from the evaluation, the game engine that most suits the needs of the project is Unity (https://unity3d.com/). A concept design of these games was created (Fig. 3) to demonstrate the main structural idea behind them. Every player will be

Table 4. Game engines state of the art

Characteristic	Degree of significance	Unity +Fungus	GameMaker	Ren'Py	AGS	eAdventure
Dialogue tools	3	X	X	X	X	X
Graphic tools	3	X	X	X	X	X
Sound tools	2	X	X	X	X	X
Variables	3	X	X	X	X	X
Free export to WEB	3	X	–	–	–	–
Free export to MOBILE	3	X	–	–	–	–
Scenario editing	3	X	X	X	X	X
VLE incorporation	1	–	–	–	–	X
Cross-platform – Free export	3	X	–	–	–	–
Total		23	14	14	14	15

Fig. 3. Interactive movie prototype design

involved in every day scenarios that assess and improve his personal skills, such as communication skills and allow him to self-regulate.

6 Conclusion

The results produced so far are in-line with previous researches about athletes' dual career (Wylleman and Reints 2010). Athletes after the end of their professional career want to be involved in different aspects of the sports' industry, but they prefer to stay in the "safe zone" of the sport that they participate. While many prefer traditional learning methods most of the athletes would like to participate in distance learning due to the flexibility it provides.

The next step, after the needs collection and technological design, is to create the syllabus for each selected course and develop the appropriate learning material. This curriculum will be provided through a MOOC. The online lessons will include serious games, reading material, online presentations and online tests all of them under a unified gamification framework. At the end of this stage, a pilot study will be conducted to measure learning outcome and user satisfaction.

Acknowledgement. This project has been funded with support from the European Commission. This publication reflects the views only of the author, and the Commission cannot be held responsible for any use which may be made of the information contained therein.

The authors of this research would like to thank GOAL project team who generously shared their time, experience, and materials for the purposes of this project.

References

Alfermann, D., Stambulova, N.: Career transitions and career termination. In: Tenenbaum, G., Eklund, R.C. (eds.) Handbook of Sport Psychology, pp. 712–736. Wiley, New York (2007)

Amara, M., Aquilina, D., Henry, I., PMP: Education of young sportspersons (lot 1). European Commission, Brussels (2004)

Aquilina, D.: A study of the relationship between elite athletes' educational development and sporting performance. Int. J. Hist. Sport **30**(4), 374–392 (2013). https://doi.org/10.1080/09523367.2013.765723

Aquilina, D., Henry, I.: Elite athletes and university education in Europe: a review of policy and practice in higher education in the European Union Member States. Int. J. Sport Policy **2**, 25–47 (2010)

Brackenridge, C.: Women and children first? Child abuse and child protection in sport. Sport Soc. **7**(3), 322–337 (2004)

Burnett, C.: Student versus athlete: professional socialisation influx. Afr. J. Phys. Health Educ. Recreat. Dance **1**, 193–203 (2010)

Developing the creative and innovative potential of young people through non-formal learning in ways that are relevant to employability. European Commission (2014). http://ec.europa.eu/assets/eac/youth/news/2014/documents/report-creative-potential_en.pdf. Accessed 12 Sept 2017

Rethinking Education: Investing in skills for better socio-economic outcomes. European Commission, Strasbourg (2012)

Stambulova, N., Alfermann, D., Statler, T., Côté, J.: ISSP position stand: career development and transitions of athletes. Int. J. Sport Exerc. Psychol. **7**, 395–412 (2009)

Wylleman, P., Lavallee, D.: A developmental perspective on transitions faced by athletes. In: Weiss, M.R. (ed.) Developmental Sport and Exercise Psychology: A Lifespan Perspective, pp. 507–527. Fitness Information Technology, Morgantown (2004)

Wylleman, P., Reints, A.: A lifespan perspective on the career of talented and elite athletes: perspectives on high-intensity sports. Scand. J. Med. Sci. Sport **20**, 88–94 (2010). https://doi.org/10.1111/j.1600-0838.2010.01194.x

Designing Exergames for Working Memory Training Using MaKey MaKey

Agisilaos Chaldogeridis(✉), Nikolaos Politopoulos, and Thrasyvoulos Tsiatsos

Department of Informatics, Aristotle University of Thessaloniki, Thessaloniki, Greece
{achaldog, npolitop, tsiatsos}@csd.auth.gr

Abstract. Functioning at a higher cognitive level is becoming an increasing necessity over the years, as our society and especially the professional market demand individuals to perform as highly as they can for longer time periods. Starting from school and moving to a professional career, each person has to develop high cognitive skills in order to complete others. Enhancing and improving cognitive skills through cognitive training, is a promising field that becomes considerably important. A core cognitive function is memory and specifically working memory as its condition affects other cognitive functions. Training working memory through cognitive training programmes appears to be beneficial for people and affect their overall cognitive performance. In this work we describe the design process of a cognitive training computer game for increasing the working memory capacity combining the MaKey MaKey (MM) device as input method and its implementation process in order to evaluate its effects to higher education students. The ulterior goal is to extract general patterns and strategies, for game based cognitive training programmes.

Keywords: Working memory training · Computer-based Cognitive Training Exergames

1 Introduction and Motivation

Unlike previous years, people nowadays have increased needs for higher brain functionality in a demanding education and working environment and daily living likewise. Starting from school and moving all the way up to a professional career, people constantly have to strengthen their brain connectivity and develop advanced cognitive skills like complex problem-solving, novel thinking, emotional intelligence, strategic leadership, flexibility and agility. Furthermore, education-wise, each student has his own needs and preferences and especially for those who face learning and cognitive disabilities, it is important to be able to keep up with the pace of rest of the students. But also, students have to develop higher skills and perform advanced educational progress in order to be effective and competitive for a future working environment.

Apart from that, the well-known ageing-related cognitive decline has to be a top priority issue, more than ever before, given the growing life expectancy globally. Older people should remain mentally sharp, for as long as possible, thus they can be able to

M. E. Auer and T. Tsiatsos (Eds.): IMCL 2017, AISC 725, pp. 635–643, 2018.
https://doi.org/10.1007/978-3-319-75175-7_62

live autonomously and normally. Furthermore, considering individuals who suffer from mental and cognitive declines (e.g. dementia, Alzheimer's, brain injuries etc.), brain training as rehabilitation method is considered to be more than useful and of significant effectiveness.

Brain fitness is based on the fact that the human brain has the so called "neuroplasticity" which is the ability to form new neurons and reorganize their interconnection between them (synapses) through experience. This is against the previous theory which dominated over the last years, that human brain is fixed and it cannot change through years, but that is obsolete already. Over the last years, psychologists developed several batteries for training and stimulating the human brain, by implementing specific cognitive exercises for corresponding cognitive skills, thus there is much interest on how to exploit neuroplasticity for brain enhancement, delay cognitive declines and lead to better living. The promising fact that it is feasible to strengthen brain circuits through education, professions and daily living, as well as through meditation for cognitive rehabilitation makes brain fitness a key-feature in our lives.

This research describes the design of a game for working memory training using alternative input methods, based on the platform MM, trying to build an engaging training environment for higher education students in order to increase their WM capacity for better academic performance. In the following sections, the theoretical background for game based WM training is described, followed by a series of mockups screens, presenting the application's design and lastly the future work.

2 Theoretical Background

2.1 Working Memory

Cognitive skills include a set of mental processes and abilities referred to knowledge, attention, memory, working memory, judgment, evaluation, reasoning, "computation", problem solving, decision making, comprehension and production of language (Chaldogeridis 2015). Working memory is the cognitive system located in the prefrontal cortex of the brain and has a crucial role in reasoning skills development, as it underlies several cognitive abilities, like logical reasoning and problem solving (Klingberg et al. 2002).

Working memory (WM) capacity can be described as a measure of short-term information retention and involves maintenance and retrieval, but also reflects abilities associated with attention, reasoning and problem solving. These critical abilities reflect in real-world conditions like school, work and overall daily living. Apparently, being able to increase the capacity of WM leads to advanced cognitive skills like multi-tasking, processing speed of different types of information, complex problem-solving and high reasoning.

Over the last years, the increasing interest in brain fitness has led to the development of several applications for training cognitive skills and thus WM. Cognitive Training (CT) is general term that concludes a set of different methods for exercising the human brain in order to increase a person's mental abilities and cognitive skills by personalized therapeutic sessions and computerized intervention programmes. CT can be applied by various ways, but, computerized WM training enriched with game-based characteristics is proven to be more effective than any other form, as it can dramatically increase the engagement and immersion level for advanced WM training activities.

There are many research teams that developed many applications for cognitive assessments and therapies, mindfulness apps, virtual reality:

- BrainHQ (https://www.brainhq.com/),
- CogniFit (https://www.cognifit.com/),
- Akili (http://www.akiliinteractive.com/),
- Pear Therapeutics (https://peartherapeutics.com/),
- Click Therapeutics (http://clicktherapeutics.com/),
- Cogniciti (https://www.cogniciti.com/),
- Headspace (https://www.headspace.com/),
- Claritas Mindsciences (https://www.claritasmind.com/),
- Emotiv (https://www.emotiv.com/),
- NeuroSky (http://neurosky.com/),
- MindMaze (https://www.mindmaze.com/),
- Braintrain (www.braintrain.com),
- Lumosity (http://www.lumosity.com/),
- Brain Metrix (www.brainmetrix.com),
- PASAT (http://en.wikipedia.org/wiki/Paced_Auditory_Serial_Addition_Test),
- CogniFit Brain Fitness (www.cognifit.com) and more.

Physical exercise and cognitive skills
Scientific evidence based on neuroimaging approaches over the last decade has demonstrated the efficacy of physical activity improving cognitive health across the human lifespan. Aerobic fitness spares age-related loss of brain tissue during aging, and enhances functional aspects of higher order regions involved in the control of cognition. More active or higher fit individuals are capable of allocating greater attentional resources toward the environment and are able to process information more quickly. These data are suggestive that aerobic fitness enhances cognitive strategies enabling to respond effectively to an imposed challenge with a better yield in task performance.

Exercise helps memory and thinking through both direct and indirect means. The benefits of exercise come directly from its ability to reduce insulin resistance, reduce inflammation, and stimulate the release of growth factors—chemicals in the brain that affect the health of brain cells, the growth of new blood vessels in the brain, and even the abundance and survival of new brain cells.

Indirectly, exercise improves mood and sleep, and reduces stress and anxiety. Problems in these areas frequently cause or contribute to cognitive impairment.

Many studies have suggested that the parts of the brain that control thinking and memory (the prefrontal cortex and medial temporal cortex) have greater volume in people who exercise versus people who don’t.

2.2 Serious Games

Studies showed that by the time a teenager begins his/her professional career, he has already played over 10,000 h (Prensky 2001; European Summary Report 2012), or the equivalent to 5 years of fulltime employment. In recent years, the interest on gaming has led to a rapid growth of the game industry, and in particular commercial entertainment games. Video games allow individuals to reach high levels of motivation and engagement and they have proven to be more successful than schools in attracting interest from young people (Caperton 2007; Boyle et al. 2016).

The movement towards the use of serious games as training and learning is proliferated by the perceived ability of such games to create a memorable and engaging learning experience. Various commentators and practitioners alike argue that serious games may develop and reinforce 21st century skills such as collaboration, problem solving and communication. While in the past, practitioners and trainers have been reluctant in using serious games for improving skills and competencies there is an increasing interest, to explore how serious games could be used to improve specific skills and competencies. The overarching assumption made is that serious games are built on sound learning principles encompassing teaching and training approaches that support the design of authentic and situated learning activities in an engaging and immersive way.

Developing serious games based on activity-centered pedagogies that enable trainees to engage actively with problems associated with cognitive training and specifically working memory training is an empowering approach with benefits for learning as well as for developing a wide range of important high-order intellectual attributes.

Cognitive training applications, and specifically working memory training systems are designed to improve the user's working memory. However, conventional systems are frequently considered tedious or repetitive which deeply affects the user's motivation to learn and consequently the potential for learning transfer (Green and Bavelier 2008). According to Prins et al. (2011) working memory training with game elements significantly improves motivation and training performance. In recent years, developers have created video games, both for commercial and research purposes that promote exercise and healthy lifestyles. These games are called Active Video Games or "exergames".

Exergames

Exergames are innovative tools have the potential to turn a traditional sedentary activity, playing video games, in physical exercise (Politopoulos and Katmada 2014). Exergames have the potential to increase exercise by shifting attention away from aversive aspects and toward motivating features such as competition and three-dimensional (3D) scenery (Politopoulos et al. 2015). Participation in exergaming compared with traditional exercise can lead to greater frequency and intensity (Annesi and Mazas 1997) and enhanced health outcomes (Lieberman 2009) (Fig. 1).

Van Schaik et al. (2008) reported that compared with traditional stationary cycling, older adults preferred cycling with interactive gaming. Although promising, there are limited published data on whether interactive exergaming technologies are reliably associated with enhanced physical and cognitive health outcomes, and more-controlled research on the effects of health games is needed.

Fig. 1. Examples of exergames

3 Game Design

The main idea is to create an interactive video game enhanced with natural user interfaces in order to train working memory. Players would use their bodies as controllers in order to interact with the game. The main idea is to use MM as input method. In order to evaluate the available tools for prototyping tangible user interfaces, an evaluation table was used (Table 1, Beginner's Mind Collective and Shaw 2012).

Table 1. Tangible interface evaluation

	Microcontrollers	Sensor boards	Hacking existing devices	MaKey MaKey
Quick start for beginners	–	X	–	X
Works with any software	–	–	X	X
Nature-based interfaces	X	–	–	X
Programming not required	–	–	X	X
Soldering not required	–	X	–	X

3.1 MaKey MaKey

As, can be seen from the evaluation, the tool that most suits the needs of the project is MM (http://makeymakey.com/). MM was developed by a research team at MIT together with SparkFun Electronics. The toolkit comprises a printed circuit board with an Arduino microcontroller, alligator clips and a USB cable. It uses the HID protocol to communicate with a computer to send key presses, mouse clicks, and mouse movements. There are six inputs (the 4 arrow keys, the spacebar and a mouse click) positioned on the front of the board that alligator clips are clipped onto in order to connect with a computer via the USB cable (see Fig. 2). The other ends of the clips can be attached to any non-insulating object, such as a vegetable or piece of fruit. Thus, instead of using the computer keyboard buttons to interact with the computer, external objects such as bananas are used. The computer thinks MM is just like a keyboard or mouse. An example is to play a digital piano app using bananas as keys rather than keys on the computer keyboard. When they are touched they make a connection to the board, and MM sends the computer a keyboard message.

Fig. 2. MaKey MaKey

The flexibility of MM for creating physical interfaces out of everyday objects, especially including natural materials, means it will facilitate bricolage in the physical world. At the same time, because it is compatible with just about any software you can think of (anything that takes keyboard or mouse input), it can be used for a kind of digital bricolage. In this way, it is believed that MM as a platform opens up a new space for bricolage, combining creative use of found materials in both the physical and digital worlds. Many researchers believe that MM is the interface that bridges tangible interfaces with natural user interfaces.

3.2 Game Prototyping

The main idea is to create a platform game, where the player always starts from the entry point (SP). Then, the player watches which is the next step, by looking at the flashing tile on the grid. When the player starts walking the correct path will be hidden. The goal is to remember and correctly walk the mentioned path. After completing it, the sequence of steps will be increased by one and will flash again on screen (Fig. 3). The player returns to SP and has to repeat again.

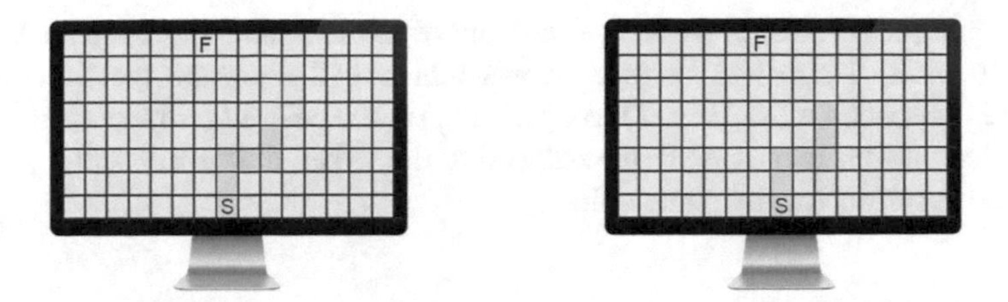

Fig. 3. Two consecutive stages of the game.

If the path is incorrect, the player goes again to the SP and repeats the steps for one last attempt. If the path is incorrect for two consecutive attempts, the path decreases by one step, and the player tries again from the SP.

While the player steps on the correct tile, this will be turned green on screen (Fig. 4(a)), and the player will have a positive feedback. On the other side when he makes a wrong step, the tile will be turned red (Fig. 4(b)).

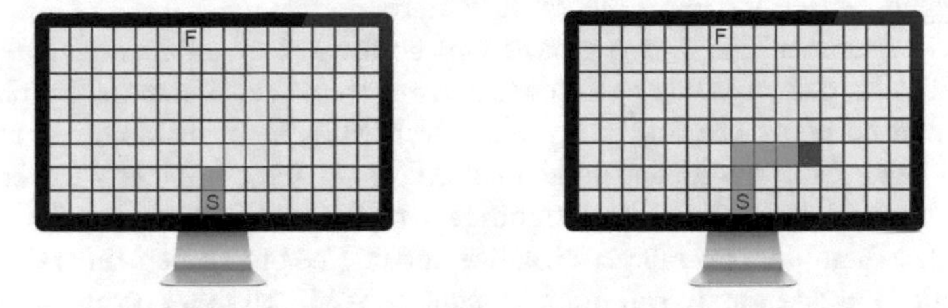

Fig. 4. (a) Feedback for correct step and (b) feedback for wrong step

Fig. 5. Gameplay example

The main idea is that, players stand in front of a screen and interact with MM using their body parts. There will be four arrows which will represent the four directions where the player can move (Fig. 5). By pressing on the desired arrow with her/his foot, the on screen avatar is moving to the respective tile. Also, a fifth key will represent the start/pause function for the game's flow.

4 Future Work

Right after the design process, the implementation procedure is following. A state of the art research will be conducted in order to select the developing platform for the game. Each available platform should be reviewed in order to meet the needs of the design process. The goal is to make use of a complete software that allows the interconnection between an online database for storing critical information (user's personal data, performance indexes etc.) and the main application, and also, supports enriched multimedia features. By completing development process, the software has to be tested and debugged, before the experimental face starts.

For the experiment itself, two groups will be formed by university students. Both groups will take part in weekly sessions, twice a week, for 3 months by playing the game for half an hour. The first group will use MM as input method, and the second will make use of the traditional input method, using the computer's keyboard. The experiment's goal is to detect any significant differences concerning the two input methods in each group's overall performance, the degree of engagement and immersion in the application and the overall improvement of WM. All these potential differences will be detected by pre-post tests before and after the intervention procedure.

Finally, the next step would be to introduce Microsoft's Kinect, or any other similar natural user interface, as an alternate input method, that will detect the user's body position and rotation, allowing user to perform more natural movements and providing advanced immersive experience.

References

Annesi, J.J., Mazas, J.: Effects of virtual reality-enhanced exercise equipment on adherence and exercise-induced feeling states. Percept. Mot. Skills **85**(3 Pt 1), 835–844 (1997)

Ariës, R.J., Groot, W., van den Brink, H.M.: Improving reasoning skills in secondary history education by working memory training. Br. Educ. Res. J. **41**, 210–228 (2015). https://doi.org/10.1002/berj.3142

Chaldogeridis, A.: Cognitive training supported by information and communication technologies. In: 2015 International Conference on Interactive Mobile Communication Technologies and Learning (IMCL), pp. 425–429. IEEE (2015)

Green, C.S., Bavelier, D.: Exercising your brain: a review of human brain plasticity and training-induced learning. Psychol. aging **23**(4), 692–701 (2008)

Prins, P.J., Dovis, S., Ponsioen, A., ten Brink, E., van der Oord, S.: Does computerized working memory training with game elements enhance motivation and training efficacy in children with ADHD? Cyberpsychol. Behav. Soc. Netw. **14**(3), 115–122 (2011)

Prensky, M.: Digital Game-Based Learning. McGraw Hill, New York (2001)

Caperton, I.: Video games and education. In: OECD Background Paper for OECD-ENLACES Expert Meeting (2007)
Boyle, E.A., Hainey, T., Connolly, T.M., Gray, G., Earp, J., Ott, M., Lim, T., Ninaus, M., Ribeiro, C., Pereira, J.: An update to the systematic literature review of empirical evidence of the impacts and outcomes of computer games and serious games. Comput. Educ. **94**, 178–192 (2016). https://doi.org/10.1016/j.compedu.2015.11.003. ISSN 0360-1315
Politopoulos, N., Katmada, A.: The effect of computer games in physical education and health of children and youth. A literature review. In: 2nd Panhellenic Conference "Education in ICTs", Athens, Greece (2014)
Politopoulos, N., Tsiatsos, T., Grouios, G., Ziagkas, E.: Implementation and evaluation of a game using natural user interfaces in order to improve response time. In: 2015 International Conference on Interactive Mobile Communication Technologies and Learning (IMCL), pp. 69–72. IEEE, November 2015
Lieberman, D.A.: Designing serious games for learning and health in informal and formal settings. In: Ritterfeld, U., Cody, M., Vorderer, P. (eds.) Serious Games: Mechanisms and Effects, pp. 117–130. Routledge, New York (2009)
Van Schaik, P., Blake, J., Pernet, F., Spears, I., Fencott, C.: Virtual augmented exercise gaming for older adults. CyberPsychol. Behav. **11**(1), 103–106 (2008)
Beginner's Mind Collective, Shaw, D.: MaKey MaKey: improvising tangible and nature-based user interfaces. In: Proceedings of 6th International Conference on Tangible, Embedded and Embodied Interaction, pp. 367–370. ACM (2012)
Gomez-Pinilla, F., Hillman, C.: The influence of exercise on cognitive abilities. Compr. Physiol. **3**(1), 403–428 (2013). https://doi.org/10.1002/cphy.c110063
Klingberg, T.: The Overflowing Brain: Information Overload and the Limits of Working Memory. Oxford University Press, New York (2009). (Trans. by, N. Betteridge)
Schwaighofer, M., Fischer, F., Bühner, M.: Does working memory training transfer? A meta-analysis including training conditions as moderators. Educ. Psychol. **50**(2), 138–166 (2015). https://doi.org/10.1080/00461520.2015.1036274
Holmes, J., Gathercole, S.E.: Taking working memory training from the laboratory into schools. Educ. Psychol. **34**(4), 440–450 (2014). https://doi.org/10.1080/01443410.2013.797338
Baddeley, A.D.: The episodic buffer: a new component of working memory? Trends Cogn. Sci. **4**(11), 417–423 (2000)

The Effect of a 12 Week Reaction Time Training Using Active Video Game Tennis Attack on Reaction Time and Tennis Performance

Efthymios Ziagkas[1(✉)], Vasiliki Zilidou[1], Nikolaos Politopoulos[2], Stella Douka[1], Thrasyvoulos Tsiatsos[2], and George Grouios[1]

[1] Department of Physical Education and Sport Science, Faculty of Physical Education and Sport Sciences, Aristotle University of Thessaloniki, Thessaloniki, Greece
{eziagkas, sdouka, ggrouios}@phed.auth.gr, vickyzilidou@gmail.com

[2] Department of Informatics, Faculty of Science, Aristotle University of Thessaloniki, Thessaloniki, Greece
{npolitop, tsiatsos}@csd.auth.gr

Abstract. Reaction time is one of human abilities that involves in physical activity almost in all sports and it is important factor to winning when an athlete dominate others. Tennis Attack is an interactive 3D video game, which promises to improve response time of, including but not limited to, athletes using entertainment as its vehicle. The aim of the present study was to identify if Tennis Attack can improve the reaction time of the video game player and to explore if a possible improvement on reaction time affects the performance in tennis. The sample is consisted of 42 male undergraduate students. The sample divided into two groups. The first group (N = 20) was the experimental group and the other (N = 22) was the control group. Both groups attended two-hour tennis lessons once a week. All participants, at baseline and at end line, played Tennis Attack in order to examine their reaction time. The experimental group also played Tennis Attack for half an hour 2 times a week, as intervention. Additionally, both groups, at base line and at end line performed the Hitting Accuracy Tennis Test in order to evaluate their performance on tennis strokes. After the intervention, as regards reaction time, both groups showed an improvement in reaction time. Concerning tennis performance, the experimental group showed significant higher scores at means than the control group. Concerning tennis serve performance score both groups mentioned higher mean scores, although there were no significant differences between two groups. With respect to groundstrokes performance, after intervention, the experimental group mentioned significant better scores.

Keywords: Tennis · Reaction time · Tennis Attack · Performance

M. E. Auer and T. Tsiatsos (Eds.): IMCL 2017, AISC 725, pp. 644–652, 2018.
https://doi.org/10.1007/978-3-319-75175-7_63

1 Introduction

At the present time youths are involved in videogames, watching TV, movies and exploring internet. On the other hand, in order to keep themselves active and physically healthy, are also involved in sports like tennis, volleyball, badminton, football. These sports not only make them physically healthy but would also improve their alertness and concentration. In sports and games, in which movements of a participant are conditioned by signals, by movements of opponents, or by motion of the ball, reaction time is of great importance. The athlete's ability to react to a stimulus is a component of athletics that should not be underestimated. Tennis Attack is an interactive 3D video game, which promises to improve response time of, including but not limited to, athletes using entertainment as its vehicle.

1.1 Reaction Time

Response or Reaction time (RT) is defined as the time interval between the onset of a sudden and unforeseen auditory or visual stimulus to starting a predetermined kinetic response to it (Grouios 1988). Reaction time is one of human abilities that involves in physical activity almost in all sports and it is important factor to winning when an athlete dominate others, it means that he/she should produce faster response (Schmidt and Lee 2005). Environmental data arrives as input into processing stage which takes place in the brain and ultimately lead to motor behaviour which is known as output (Schmidt and Lee 2005). In fact this is decision making process that is match with reaction time process.

The reaction time can be measured in a variety of experimental tasks – detection, discrimination, localization, recognition, remembering, etc. The simple reaction time measures the speed of the nervous influx, while the discrimination reaction time measures the basic time (the speed of the nervous influx) combined with the identification time – the decision time concerning the significance of the stimulus (capable or not to generate a motor response). By subtracting the corresponding reaction time pairs one can measure the necessary time for the mental operation of identification (Grigore et al. 2015).

Reaction time is a decisive factor in many sports and can be developed by regular training (Bompa 1998). Charu (2008), studied the effect of special motor skills on reaction time and find that these skills are effective in improving reaction time. Zwierko (2010), conducted a study on volleyball athletes and non-athletes reaction time, and find that visual stimuli's transmutes faster in athletes so she attributed this to speed of information processing and said that physical activity has improved it (Foroghipour et al. 2013).

1.2 The Active Video Game Tennis Attack

Tennis Attack is an interactive 3D video game, which uses the natural user interface Microsoft Kinect, in order to improve response time of, including but not limited to, athletes. The game aims to improve, using entertainment as its vehicle, the response time of the user to important stimuli of everyday life.

Tennis Attack belongs to the category of serious games and in particular, to the category of ActiveVideoGames or AVG that promote exercise and healthy lifestyle converting a traditional sedentary activity into physical exercise.

Unity game engine, C# scripts and Zigfu Development Kit or ZDK were used in order to develop Tennis Attack. The game can serve as a tool for measuring and improving simple reaction time (one stimulus - one response). It's partially configurable by the players and allows them to control the number of the balls and the rate at which they will appear on screen. Scores are stored in a database through a local server and allows the users to monitor their performance and observe their progress (Politopoulos et al. 2015).

1.3 Reaction Time as a Factor on Tennis Performance

The athlete's ability to react to a stimulus to which they are confronted is a component of athletics that should not be underestimated. Once the body is placed in the proper position, the hands often complete the task of athletic success. Activities such as the tennis return, and stick, glove, or hand saves in such sports as lacrosse, hockey, and soccer require fast moving hands to ensure task success. A tennis match includes intermittent anaerobic exercise bouts of varying intensities and a multitude of rest periods over a long duration, allowing the aerobic energy systems to aid in recovery. The time of contact between the ball and racquet is between 0.003 and 0.006 s, and the racquet and ball must be in optimal orientation to execute the desired stroke (Renstrom 2002). Therefore, high-level tennis requires quick reaction time during a tennis match (Mero et al. 1991).

Van Biesen (2010), through study on tennis players found that reaction time, coordination, eye movement and performance are significantly better than control group. He concluded that tennis training have led to these changes. Delignieres (1994), have studied the effect of exercise on choice reaction time in athlete and non-athlete groups. Results showed that in athlete group more attempt led to better performance but more attempt led to instability in non-athletes. Also, Devranche et al. (2006), found that physical activity reduces reaction time. However many researches confirm the positive effect on reducing reaction time, simultaneously there are different researches results that show other findings. Ando et al. (2009) found that reaction time increases in high physical difference between karate and volleyball athletes reaction time. Draper et al. (2010), found that simple and choice reaction time have significant difference in intensive and slight exercises and intensive exercises have negative effect on choice reaction time but have no significant effect on simple reaction time. Epuran et al. in 2001 mention the importance of quick decision-making for the best action in sports, because the situations are rapidly changing during the competitions. Today, the use of computer technology makes the precision and accuracy of registrations to be assured. The movements associated with device manipulation (buttons, levers, pedals) are known as instrumental movements (Anitei 2007). Concerning reaction time, the duration of resting pauses and the intensity of the stimulus greatly influenced this aspect. At the same time, also the training process influences the values of the reaction time, decreasing them and improving the performance, and after a while, the athletes reach a set limit in their executions. Studies show (Levitt and Gutin 1971) that the subjects, who were exercising sufficiently to produce a heart rate of 115 beats per minute, had the fastest reaction times.

The study of relation of the reaction time to motor skill performance in sport is not new, but a few things are known about how video games affect the reaction time and especially, if a video game can be used in order to improve reaction time in sports. The aim of the present study is two-fold. Firstly, it attempts to identify if Tennis Attack can improve the reaction time of the video game player. Secondly it purports to explore if a possible improvement on reaction time affect the performance in tennis. For the purpose of this study two hypothesis have been raised. The first hypothesis is that playing Tennis Attack will improve players' reaction time and the second hypothesis is that the players who mentioned better reaction time on the game will show better performance in tennis.

2 Methods

2.1 Sample

The sample is consisted of 42 male undergraduate students who participated in tennis course during the spring semester at the Department of Physical Education and Sport Science of Aristotle University of Thessaloniki. Participants' age ranged from 19 to 22 years. All participants were beginner tennis players who hadn't play tennis before.

2.2 Materials

The interactive 3D video game Tennis Attack was used in order to examine the reaction time pre and post intervention. Also Tennis Attack was used as the intervention of the experimental group.

In order to examine tennis performance pre-post intervention, we used the Hitting Accuracy Tennis Test, as has been described by Strecker et al. (2011). All participants were called to execute 20 forehand and 20 backhand groundstrokes from the baseline of the court on predetermined targets. A tennis ball machine was used in order to feed the balls (frequency of 20 balls per minute) (Fig. 1). Additionally, all participants were called to execute 10 serves from each side court (deuce court - advantage court) in order to examine tennis serve performance (Fig. 2). Two experienced tennis coaches documented the valid stokes and mentioned the total score of each participant.

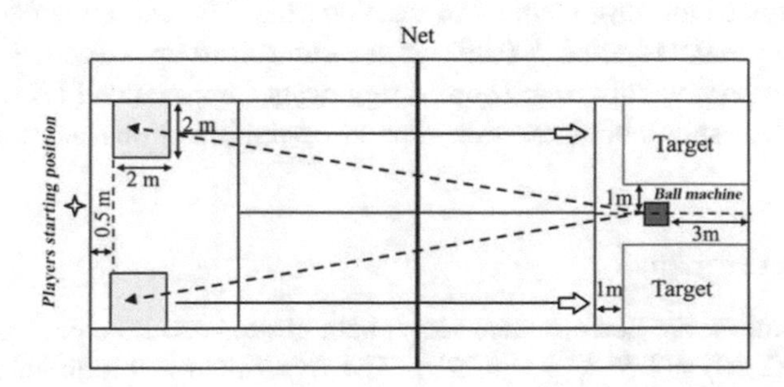

Fig. 1. The hitting accuracy tennis test as described by Strecker et al. (2011).

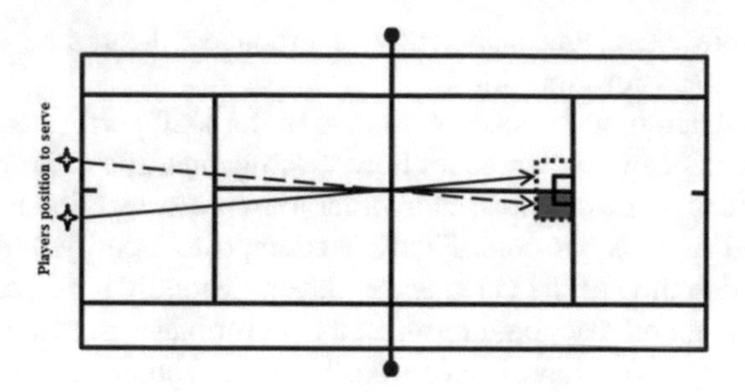

Fig. 2. The tennis serve performance test.

2.3 Procedure

Using a random number table the sample divided into two groups. The first group (N = 20) was the experimental group and the other (N = 22) was the control group. Both groups attended two-hour tennis lessons once a week. All participants, at baseline and at end line, played Tennis Attack in order to examine their reaction time. The experimental group, on top of their casual training, played Tennis Attack for half an hour 2 times a week, as intervention.

Both groups, at base line and at end line performed the Hitting Accuracy Tennis Test (Strecker et al. 2011) in order to evaluate their performance on groundstrokes and a tennis serve accuracy test in order to examine their performance on tennis serve. Two experienced tennis coaches documented the valid stokes and mentioned the total score of each participant. Data processed using SPSS version 24.0. In order to test our hypothesis we used oneWay Anova and the significance was set at $p = 0.05$.

3 Results

3.1 Reaction Time

As regards reaction time, at the base line the experimental group had reaction time at means RTexp = 1.219 s, sd = 0 ± .152 and the control group mention at means RTcon = 1.213 s, sd = ±0.151. No significant difference were found between the experimental and the control group [$F(1.40) = 0.016$, $p = 0.899$]. After the intervention both groups showed an improvement in reaction time. The experimental group shown at means RTexp = 0.844 s, sd = 0.092 an the control group RTcont = 1.045 s, sd = ±0.205. The experimental group showed significant improvement [$F(1.40) = 16.328$, $p = 0.000$]. The results about reaction time at baseline and endline are presented in Table 1.

3.2 Serve Performance

Concerning tennis serve performance score, both groups had no significant differences at baseline [$F(1.40) = 1.532$, $p = 0.223$]. The experimental group mentioned mean score 10.500, sd = 2.013 and the control group mentioned mean score 9.727 sd =

Table 1. Descriptive statistics concerning reaction time

	Group	N	Mean	Std. deviation
Reaction time baseline	Experimental	20	1.219	0.152
	Control	22	1.213	0.151
	Total	42	1.216	0.150
Reaction time endline	Experimental	20	0.844	0.092
	Control	22	1.045	0.205
	Total	42	0.949	0.189

2.028. After the intervention, both groups mentioned higher mean scores, although there were no significant differences between two groups [F(1.40) = 0.394, p = 0.534]. Detailed data about performance on tennis serve are presented in Table 2.

Table 2. Descriptive statistics about the performance on tennis serve.

	Group	N	Mean	Std. deviation
Serve total performance baseline	Experimental	20	10.500	2.013
	Control	22	9.727	2.028
	Total	42	10.095	2.034
Serve total performance endline	Experimental	20	14.100	2.426
	Control	22	14.546	2.176
	Total	42	14.333	2.281

3.3 Groundstrokes Performance

With respect to groundstrokes performance at base line, no significant differences found between two groups F(1.40) = 3.224, p = 0.080. The experimental group mentioned mean score 18.300, sd = ±2.342 and the control group mentioned mean score 17.091, sd = ±2.022. After the intervention experimental group mentioned mean score 31.200, sd = ±1.795 and the control group mentioned mean score 23.590, sd = ±1.869. This difference was significant F(1.40) = 180.363, p = 0.000. Descriptive statistics about groundstrokes performance are presented in Table 3.

Table 3. Descriptive statistics about performance on ground stokes.

	Group	N	Mean	Std. deviation
Ground strokes total performance baseline	Experimental	20	18.300	2.342
	Control	22	17.091	2.022
	Total	42	17.667	2.238
Ground strokes total performance endline	Experimental	20	31.200	1.795
	Control	22	23.590	1.869
	Total	42	27.214	4.251

4 Discussion

The aim of the present study was two-fold. Firstly, it attempted to identify if Tennis Attack can improve the reaction time of the video game player. Secondly it purported to explore if a possible improvement on reaction time affect the performance in tennis. For the purpose of this study two hypotheses had been raised. The first hypothesis is that playing Tennis Attack will improve players' reaction time and the second hypothesis is that the players who mentioned better reaction time on the game will show better performance in tennis.

As regards reaction time, the first hypothesis was accepted as the intervention group who played Tennis Attack two times a week for half an hour, shown significant better results in reaction time on Tennis Attack than those of the control group. This result in online with the aspect of game creator, whom in a recent study, mentioned that this game can be used as a tool for measuring and improving simple reaction time (one stimulus - one response) (Politopoulos et al. 2015).

According to the effect of reaction time on tennis performance, the second hypothesis is half accepted. In tennis serve performance we found no significant differences between the two groups pre-post intervention. Although, both groups showed better performance at the endline. This improvement may occur on tennis serve technique improvement due to the tennis lessons. The reaction time seems not to affect performance in tennis serve. A possible explanation about this finding is the fact that tennis serve is the only shot in tennis where the player depends solely on himself (closed feedback task), and therefore the technique supposes the most important role in the shot (de Subijana and Navarro 2010).

Concerning the performance in Hitting Accuracy Tennis Test our results shown that although at baseline the two groups had no statistical differences, after the intervention the experimental group mentioned significant better results in this test. A possible explanation of this finding is that racquet sports such as tennis, badminton, and squash have been classified as reaction sports (Marion and Suzanne 1989). In tennis specifically, the incredible speed of the ball and the distance it travels between opponents allows a very minimal amount of time to react and execute shots. Tennis player has to give proper and quick response during the game. They have to hit the ball in proper direction (Bhabhor et al. 2013). As the player, the racquet and ball must be in optimal orientation to execute the desired stroke (Renstrom 2002) and high-level tennis requires quick reaction time during a tennis match (Mero et al. 1991) the better results of the experimental group is supposed to occur due to the improvement in reaction time by the training on Tennis Attack.

Further in field research is required in order to examine the effects of active video games like Tennis Attack, not only in reaction time, but possibly in other performance factors such as strategy, predictability etc.

5 Conclusion

Tennis Attack is an appropriate tool that not only can improve reaction time of video game players, but also a useful instrument for athletes and coaches who through the improvement on reaction time aim to improve athletes' sport performance.

References

Ando, S., Yamada, Y., Tanaka, T., Oda, S., Kokubu, M.: Reaction time to peripheral visual stimuli during exercise under normoxia and hyperoxia. Eur. J. Appl. Physiol. **106**, 61–69 (2009)

Anitei, M.: Psihologie experimentala. Polirom, Iasi (2007)

Bhabhor, M.K., Vidja, K., Bhanderi, P., Dodhia, S., Kathrotia, R., Joshi, V.: A comparative study of visual reaction time in table tennis players and healthy controls. Indian J. Physiol. Pharmacol. **57**, 439–442 (2013)

Bompa, T.O.: Antrenman Kuramı ve Yöntemi. Qlknur, K.A., Burcu, T. (Çev). Bagırgan Yayınevi, Ankara (1998)

Charu, S., Rini, K., Arun, P.: Reaction time of a group of physics students. Department of Physics and Electronics, S G T B Khalsa College, University of Delhi, Delhi 110 007, India (2008)

de Subijana, C.L., Navarro, E.: Kinetic energy transfer during the tennis serve. Biol. Sport **27**, 3–11 (2010)

Delignières, D., Brisswalter, J., Legros, P.: Influence of physical exercise on choice reaction time in sport experts: the mediating role of resource allocation. J. Hum. Mov. Stud. **27**, 173–188 (1994)

Devranche, K., Burle, B., Audiffren, M., Hasbroucq, T.: Physical exercise facilitates motor processes in simple reaction time performance. An electromyography analysis. Neurosci. Lett. **396**, 54–56 (2006)

Draper, S., McMorris, T., Parker, J.K.: Effect of acute exercise of differing intensities on simple and choice reaction and movement times. J. Psychol. Sport Exerc. **11**, 536–541 (2010)

Epuran, M., Holdevici, I., Tonita, F.: Psihologia sportului de performanta. Teorie şi practica. FEST, Bucureşti (2001)

Foroghipour, H., Monfared, M.O., Pirmohammadi, M., Saboonchi, R.: Comparison of simple and choice reaction time in tennis and volleyball players. Int. J. Sport Stud. **3**, 74–79 (2013)

Grigore, V., Mitrachea, G., Paunescua, M., Predoiua, R.: The decision time, the simple and the discrimination reaction time in elite Romanian junior tennis players Procedia. Soc. Behav. Sci. **190**, 539–544 (2015)

Grouios, G.: The effect of mental rehearsal on the reaction time of top level sports participants. Unpublished Doctoral thesis. University of Manchester (1988)

Levitt, S., Gutin, B.: Multiple choice reaction time and movement time during physical exertion. Res. Q. **42**, 405–410 (1971)

Marion, J.L.A., Suzanne, L.B.: An analysis of fitness and time motion characteristics of handball. Am. J. Sports Med. **17**, 76–82 (1989)

Mero, A., Jaakkola, L., Komi, P.V.: Relationship between muscle fibre characteristics and physical performance capacity in trained athletic boys. J. Sports Sci. **9**, 161–171 (1991)

Politopoulos, N., Tsiatsos, T., Grouios, G., Ziagkas, E.: Implementation and evaluation of a game using natural user interfaces in order to improve response time. In: IMCL 2015, 19–20 November 2015, Thessaloniki, Greece (2015)

Renstrom, P.A.F.H. (ed.): Tennis. Blackwell, Malden (2002)
Schmidt, R.A., Lee, T.D.: Motor Control and Learning: A Behavioral Emphasis, 4th edn. Human Kinetics, Champaign (2005)
Strecker, E., Foster, E.B., Pascoe, D.D.: Test-retest reliability for hitting accuracy tennis test. J. Strength Cond. Res. **25**, 3501–3505 (2011)
Van Biesen, D., Verellen, J., Meyer, C., Mactavish, J., Van de Vliet, P., Vanlandewijck, Y.: The ability of elite table tennis players with intellectual disabilities to adapt their service/return. Adapt. Phys. Act. Q. **27**, 242–257 (2010)
Zwierko, T., Wiesław, W., Damian, F.: Speed of visual sensor motor processes and conductivity of visual pathway in volleyball players. J. Hum. Kinet. **23**, 21–27 (2010)

Match Fixing in Sport: Usefulness of Co-creation in Developing Online Educational Tools

Andreas Loukovitis(✉) and Vassilis Barkoukis

Department of Physical Education and Sport Science, Aristotle University of Thessaloniki, Thessaloniki, Greece
louko-vitis@hotmail.com, bark@phed.auth.gr

Abstract. Match-fixing is an ongoing threat to the integrity, reputation, and societal welfare dimension of sports. The present study aimed to develop an online educational tool against match-fixing by using the co-creation approach in the context of the "Fix the Fixing" Project. Participants were 18 young athletes from Greece who took part in the co-creation workshops. The analysis of the co-creation workshops revealed several social and psychological consequences related to match-fixing. Also, participants emphasized the need for education with respect to match fixing for those involved in competitive sports. Another significant issue that emerged was the importance of fair play and the usefulness of sharing experiences of match fixing as information source. The findings of the present study provide valuable information about the athletes needs for education material provided by online means regarding the complex global threat of match-fixing applying the usefulness of a co-creation approach in guiding future preventive and educational strategies against match-fixing.

Keywords: Match-fixing · Co-creation · Interviews · Online educational tool

1 Introduction

Match-fixing represents and ongoing threat to the integrity, reputation, and societal welfare dimension of sports. It is an illegal activity with an international dimension across all levels and types of sport. Match-fixing involves the manipulation of an outcome or contingency by competitors, teams, sports agents, support staff, referees and officials and venue staff. A recent report (2014) by Sorbonne University and the International Centre for Sport Security showed that match-fixing is constantly increasing, and approximately 300 to 700 events can be suspected for being fixed every year on a global basis since 2010. The rise of reported instances of match-fixing in sport is staggering. Match-fixing has been called one of the most serious threats to the integrity of sport [1]. Considering the increasing number of match-fixing cases across different levels and types of sport, it becomes increasingly important to safeguard sport from immoral and illegal practices such as match fixing. Doing so requires a comprehensive strategy to combat the complex global threat of match-fixing thorough

M. E. Auer and T. Tsiatsos (Eds.): IMCL 2017, AISC 725, pp. 653–660, 2018.
https://doi.org/10.1007/978-3-319-75175-7_64

understanding of its nature and aspects, as well as the coordination of prevention activities such as education.

Towards this end, the need for a better understanding of the consequences related to match-fixing and information sources about it, as well as the educational needs of individuals with respect to match fixing is apparent. Reflecting on such aspects can further elucidate the key points related to match fixing and inform future anti-match fixing campaigns. So far, existing education efforts have been largely relied on increasing athletes' and stakeholders' awareness about match fixing. However, to date, there has been a relative paucity of related research on the effects of match-fixing and also, there are limited existing studies on the need for education related to match fixing issues. More importantly, there is no evidence on the content and delivery mode of education aiming to tackle match fixing.

Therefore, the development and the promotion of education programs and training materials as an effective response to match fixing, is an area which requires more systematic empirical study. Research evidence shows that there have been some efforts to outline the threat of match-fixing and accordingly, a number of seminars have been organized, in the framework of campaigns against match-fixing. However, the related educational practices are limited and all efforts are delivered face to face. Thus, there is need for a well-designed online educational material related to match-fixing which will be accessible to all individuals involved in competitive sports. Online educational resources provide more flexible access to content and instruction at any time, from any place. Research data have demonstrated the effectiveness of online learning compared to those of face-to-face instruction [2]. Subsequently, an online educational tool against match-fixing will provide everyone with the opportunity to learn about the key areas of this continually rising threat.

1.1 Co-creation as a Means to Develop Educational Practices Against Match-Fixing

To promote training, education and prevention as effective responses to match fixing, they should be targeting match-fixing in an innovative way incorporating different methods such as the co-creation approach. Project "Fix the Fixing", in the context of which the present study was conducted, utilizes mixed methods design (structured questionnaires and focus group interviews), action research (target group members as co-researchers) and sentiment analysis of open public social media data for data collection. This methodology will allow for the co-creation of the training tool, and ensure that the training tool is both scientifically updated and relevant to the needs and experiences of individuals involved in competitive sports.

Research data have shown that co-creation generates positive outcomes [3]. But research on the applicability of co-creation approaches in educational settings is rather limited. This gap in the literature is surprising given the attention this concept has been receiving in services marketing scholarship, particularly in terms of the way customers are viewed as partial employees of the organization [4]. Specifically, Song and Adams demonstrated that the co-creation approach was effective in developing educational material related to customer participation. Since then, several scholars promoted this concept in consumer-based research and education [5]. Similarly, Kristensson et al., supported the concept of co-creation as a meaningful means towards new product development. According to Kristensson et al. co-creation helps in getting with more creative ideas, ideas that are more highly valued by customers, and they are more realistic and easily implemented [6]. In service research, value co-creation is one of the most important foundational premises of Vargo and Lusch's seminal Service Dominant Logic framework [7].

Considering that education can be perceived as a service product, co-creation implemented in the services marketing literature was adopted to derive information that would result into an effective tool to tackle match-fixing. Similar efforts have been endorsed education literature [8], and shown that effective learning can be achieved by the co-creation [9]. Co-creation allows for collaboration and describes a dynamic process between multiple actors [10].

1.2 The Present Study

The present study was set out to adopt a co-creation approach to gain information about athletes' needs about an online educational material for the effective prevention of match-fixing. This approach is bottom-up and lead to educational outputs that have been developed through the active engagement of the intended target groups. Thus, the adoption of a co-creation approach can provide valuable information that can be used in the development of an educational material. This material is expected to be more relevant to the needs of the target groups as the content of the material was derived from them. In this respect, co-creation can ensure greater relevance and impact of the project outputs to the needs of the targeted populations. The study reported here is part of a larger-scale, European-wide research project funded by European Commission (Project Fix the Fixing) and was concerned with research and education against match-fixing. Specifically, "Fix the Fixing" aims to enhance the combat against match-fixing in sport through the utilization of an evidence-based, bottom-up and innovative education approach. Also, "Fix the Fixing" aims at enriching the evidence-base for the scientific study of match-fixing in sport by utilizing a multidisciplinary approach as well as empowering policy-makers and relevant stakeholders to more effectively tackle match-fixing across levels and types of sport. In the present the results of the co-creation workshops performed in Greece are presented.

2 Method

2.1 Sample

The sample of the study consisted of 18 participants (males = 9 and females = 9). The participants derived from the initial sample of a quantitative and qualitative study which conducted in Greek athlete population in the context of the "Fix the Fixing" project. Participants' age ranged from 18 to 24 years. All participants were athletes competing in a range of different sports (i.e., football, tennis, volleyball, basketball) and voluntarily took part in the co-creation workshops.

2.2 Co-creation Workshop Matrix

The themes which were used for the design of the online educational tool against match-fixing were emerged through a semi-structured interview in the context of an interactive co-creation workshop. An interview matrix was prepared including a series of questions which were aiming to extract participants beliefs and opinions about match fixing through an interactive manner. In particular, the first part of the matrix of the co-creation workshops included the outline of athletes' feelings related to match-fixing as well as the identification of the impact that match-fixing has on the lives of the people involved. Participants were asked to describe the feelings that they perceived athletes may experience when fix a match. A key question was about the pressure athletes involved in match-fixing may experience. In the second part of the matrix there were questions about the needed education related to match-fixing issues. Example question was, "What type of education is needed?". In addition, participants were asked to describe the content of a well-designed educational program (e.g., "What should an educational material should include to tackle match fixing?") and participants were further probed about whether it should include legal consequences, psychological consequences, resistance techniques etc. They were also asked to identify the people or the agencies who should deliver this education and in which format education should be delivered. The third part of the matrix was dedicated to sport and sport principles. It aimed to outline the concept of values in sport. Thus, questions were asked to gain insight into the most important values of sport and the possibility to be achieved in modern sport. Also, athletes' beliefs about the role of sport values in match fixing prevention were requested. Finally, participants were encouraged to express their thoughts about the information sources related to match-fixing. An example question was, "Would examples of other athletes help you better understand the concept of match fixing?".

2.3 Procedure

The study's design was in line with the Code of Ethics for Research of the Aristotle University of Thessaloniki, Greece. The study consisted of two phases. Phase one a semi-structured interview was organized in the context of an interactive workshop with 10 participants of the total sample. During phase two, there was the training of the co-researchers and the workshops. Specifically, two of the participants who took part in the previous phase were taught how to conduct the same workshop and served as co-researchers. These co-researchers further interviewed four other athletes (e.g., fellow athletes or same age friends). Therefore, two more co-creation workshops were conducted (1 co-researcher and 4 participants took part in each one). After processing the data from the interactive workshops and co-researchers' workshops, responses were codified and a thematic analysis was performed to identify the themes that would guide the development of the online educational tool.

Participants were purposefully selected for the workshops and were first contacted through email or telephone. They were informed about the aim of the study and were assured they would remain anonymous throughout this study. Furthermore, they were asked if the interviews could be recorded to facilitate further analysis. All participants agreed and signed the informed consent. Interviews took place at the university or at a quiet location of the participant's choice. Co-researchers rewarded for their participation by receiving a certificate of participation to the project. All participants consented to participate in the workshops and audiotape the interviews to facilitate data analysis.

2.4 Data Collection

Data were collected through a qualitative approach. By utilizing the co-creation, we actively involved the 10 participants in the co-creation of the information materials on match fixing. Particularly, for the interactive workshop a semi-structured interview was contacted. During the interview, participants were asked to discuss and present their opinions about the consequences related to match fixing, as well as the importance of values in sport. They discussed about the training needs of those involved in competitive sports with respect to match fixing. Also, questions were asked to gain insight into the information sources. Subsequently, in the second phase, the two co-researchers' workshops provided a more detailed insight for the themes related to match-fixing which emerged from the first co-creation workshops.

2.5 Data Analysis

Interviews (from the phases one and two) were audiotaped and transcribed verbatim. Transcripts were analyzed with the assistance of ATLAS.ti, a qualitative data management program. All transcripts were reviewed and coded by a member of the research team. Codes and emerging themes were discussed continually among the principal investigators and agreed on or revised through a process of consensus. Coding is the process that was used to organize and sort the data. Codes served as a way to label, compile and organize the data in a meaningful way. The data were coded by assigning a word or a phrase to each coding category. The word which was assigned to the item of data was a code, and codes were labels that classified items of information. More specifically, the ideas, concepts and themes from the interview transcripts were coded to fit the categories. Transcripts analyzed to identify thematic and overarching categories underpinning respondents' experiences.

3 Results and Discussion

Using the co-creation approach the present study attempted to develop an online educational tool against match-fixing. As it was stated previously, it is of great importance that the data from all the partner countries of the "Fix the Fixing" project will be used for the co-creation of the online educational tool. In the present study, participants in the co-creation workshops perceived match-fixing as a phenomenon with increasing frequency and identified various social and psychological consequences related to it. In addition, they acknowledged the need for education with respect to match fixing for those involved in competitive sports. Another significant theme that emerged was the importance of fair play.

3.1 Social and Psychological Consequences Related to Match-Fixing

Participants viewed match-fixing as a threat with various social and psychological consequences. Most of them identified financial pressure as the most important social consequence related to match-fixing. Athletes subject to considerable financial temptations and pressure to influence sporting objectives and outcomes in two ways. Firstly, those who are involved in fixing matches use financial rewards as a means to get players to play in a such a way that they achieve a pre-determined result. Examples of such financial gains were receiving a new or improved contract or an increase in prize money. Secondly, in some cases athletes are threatened that if they don't fix a match they will not receive their salary. Furthermore, the athletes reported that they could be tempted to fix a match if in a different case it would have an impact on their career. If an athlete wants to advance his/her career, he/she should accept to be manipulated. As one athlete stated, "Unfortunately, people who are engaged in match-fixing ranging from criminals running illegal betting syndicates to the players themselves, promote the athletes who meet certain conditions and be manipulated".

Regarding psychological consequences, psychological pressure was described as related to match-fixing. This includes negative feelings of guilt, shame, anxiety and fear of getting caught. Participants emphasized that an athlete does always his best and gives maximum effort and consequently he/she feels bad when he is not allowed to do it and he/she is manipulated. The athlete feels guilt and shame in such cases. At psychological level, athletes could be tempted to fix a match when they perceived a high pressure to gain money or fame. But participants acknowledged that it depends on athlete's personality and moral standards. Some athletes are willing to fix a match while others don't because they feel bad. One athlete stated, "I would be afraid that others would learn it and I would also feel guilty about behaving in this way to my teammates". Fear was the worst feeling delineated by participants. Participants believed that all people are afraid, both the conscious and the unconscious.

3.2 The Importance of Fair Play

Participants noted that fair play is a factor that can influence the consequences related to match-fixing. Specifically, the athletes who participated in the workshops talked about the importance of fair play in sport. It was believed that athletes' performance can be improved just by hard training and this is a part of fair play. As one athlete said, "The most important values of sport are the respect for the opponent as well as the recognition of opponent's superiority". Participants reported that in sport there are three possible results: the defeat, the draw and the win. They claimed that it is important for an athlete to accept both the win and the defeat. Fair play was identified as a means to protect sport from threats such as match-fixing. Values like fair play, respect for the opponent, acceptance of the result and dedication to the sport were emerged as playing a role in match fixing prevention.

3.3 Training Needs of Individuals Involved in Competitive Sports with Respect to Match-Fixing

Education was emerged as a theme of great importance with respect to match fixing. Participants mentioned that the education should include the promotion of athletes-models who adopt an appropriate behavior both in the context of sport and in the society. One of the athletes participated in the co-creation workshops said that, "The teaching of meritocracy to those involved in competitive sports helps them to avoid match-fixing". Participants also stated that stakeholders can use techniques which derive from the field of psychology in order to satisfy the educational needs related to match fixing. In addition, psychological values like self-esteem, honesty and purity were seen as necessary to help athletes to avoid match-fixing. As an athlete stated, "Children should realize that money isn't everything but the most important are the ethics. They should be aware of the impact of the behavior on the society". Also, participants mentioned that education should include examples of athletes' cases. The following statement from an athlete highlights the role of examples use in the education related to match-fixing: "Introducing the negative consequences might prevent some athletes from match-fixing".

Regarding the format of the education, it was accepted by all participants that education should be done online since everyone have access to the internet. Online was described as the most preferred delivery mode of messages and information related to match fixing, as reflected in an athlete's statement, "Social media as well as online applications or platforms are the best way to deliver the education on match-fixing. There is no athlete today who does not have access to them". Another participant pointed out that, "The books are outdated and attending a conference is really boring. That's why an on-line training tool is the fastest, easy to use and interesting way".

In conjunction with the above, participants recommended starting the education process early. The athletes participated in the co-creation workshops talked about the importance of starting at an early age to train individuals involved in competitive sports with respect to match fixing. Also, they believed that education must be initiated since someone start to engage with sport and must be continued during the sporting career. Before someone getting involved with sports, it is needed to know the sports values. One athlete stated, "I think a lot has to do with instilling an attitude of wrong and right behaviors with the kids early on and just expecting them to do things for themselves".

4 Conclusions

The findings of the present study highlight athletes' need for online education. Online education was perceived more effective as compared to typical educational approaches as young people have easily access to the internet and they are familiar with this way of working. The results of the co-creation workshop in Greece revealed important information about the content of the such an educational material aiming to tackle match fixing. More specifically, consequences related to match-fixing, the importance of fair play, and the training needs with respect to match-fixing emerged as important themes that should be included in such an educational material. Finally, more evidence is needed by the partners of the project "Fix the Fixing" to ensure that these themes are valid in other European countries as well.

References

1. Holden, J.T., Rodenberg, R.M.: Lone-wolf match-fixing: global policy considerations. Int. J. Sport Policy Polit. **9**(1), 137–151 (2017)
2. Means, B., Toyama, Y., Murphy, R., Bakia, M., Jones, K.: Evaluation of evidence-based practices in online learning: a meta-analysis and review of online learning studies. US Department of Education (2009)
3. Giner, G.R., Rillo, A.P.: Structural equation modeling of co-creation and its influence on the student's satisfaction and loyalty towards university. J. Comput. Appl. Math. **291**, 257–263 (2016)
4. Groth, M.: Customers as good soldiers: examining citizenship behaviors in internet service deliveries. J. Manag. **31**(1), 7–27 (2005)
5. Song, J.H., Adams, C.R.: Differentiation through customer involvement in production or delivery. J. Consum. Mark. **10**(2), 4–12 (1993)
6. Kristensson, P., Magnusson, P.R., Matthing, J.: Users as a hidden resource for creativity: findings from an experimental study on user involvement. Creat. Innov. Manag. **11**(1), 55–61 (2002)
7. Vargo, S.L., Lusch, R.F.: Evolving to a new dominant logic for marketing. J. Mark. **68**(1), 1–17 (2004)
8. Mavondo, F., Tsarenko, Y., Gabbott, M.: International and local student satisfaction: resources and capabilities perspective. J. Mark. High. Educ. **14**(1), 41–60 (2004)
9. Ng, I.C., Forbes, J.: Education as service: The understanding of university experience through the service logic. J. Mark. High. Educ. **19**(1), 38–64 (2009)
10. Woratschek, H., Horbel, C., Popp, B.: Introduction: value co-creation in sport management. Eur. Sport Manag. Q. **14**(1), 1–5 (2014)

The Co-creation of an Online Educational Tool to Raise Awareness About Whistleblowing

Vassilis Barkoukis[1(✉)], Andreas Loukovitis[1], Lambros Lazuras[2], and Haralambos Tsorbatzoudis[1]

[1] Department of Physical Education and Sport Science, Aristotle University of Thessaloniki, Thessaloniki, Greece
{bark, lambo}@phed.auth.gr, louko-vitis@hotmail.com
[2] Department of Psychology, Sociology and Politics, Sheffield Hallam University, Sheffield, UK
l.lazuras@shu.ac.uk

Abstract. Considering the rise of reported instances of harmful irregularities in sport, it is of paramount importance to safeguard sport from unethical and illegal practices through whistleblowing. The present study aimed to develop an online educational tool to raise awareness about whistleblowing by using the co-creation approach, in the context of the "Sport WHISTLE" Project. Participants were 18 young athletes from Greece who took part in the co-creation workshops. Regarding the results, athletes didn't perceive what the whistleblowing is. A significant issue that was emerged was the legitimacy and thc responsibility of whistleblowing. Also, athletes acknowledged that attitudes, social norms and controllability play a key role in whistleblowing. Finally, the reasons for or against whistleblowing were identified. The findings of the present study offer a better insight into the athletes needs for online education regarding the whistleblowing of harmful irregularities in sport applying the usefulness of a co-creation approach in guiding future preventive and educational strategies promoting whistleblowing.

Keywords: Whistleblowing · Co-creation · Interviews
Online educational tool

1 Introduction

According to the European Commission, sport is an integral part of society contributing to financial growth, social welfare and solidarity, as well as personal development and self-fulfillment [1]. Nevertheless, harmful irregularities in sport, such as doping, corruption and match fixing, have a detrimental impact on public attitudes towards sports and European-level action is needed to help member-states and sport governing bodies and authorities to proactively and more effectively tackle those threats. Improving existing policies for good governance, transparency and sport integrity is essential, but it is also important that member-states and international and national sport governing bodies attend to behavioral factors, such as the ability to recognize, resist, and report

M. E. Auer and T. Tsiatsos (Eds.): IMCL 2017, AISC 725, pp. 661–667, 2018.
https://doi.org/10.1007/978-3-319-75175-7_65

harmful sport irregularities, especially among athletes, coaches and other relevant stakeholders directly involved in sport and sport governance.

Education plays an important role regarding the recognition, resistance and reporting of harmful irregularities in sport. More specifically, education can facilitate accurate recognition of harmful irregularities by raising awareness about the different types of irregularities and their defining features. Accordingly, through education, athletes, coaches and other sport stakeholders can learn effective ways to resist direct and indirect, intentional and unintentional involvement in harmful sport irregularities. Importantly, education about whistleblowing can also empower athletes, coaches and relevant sport stakeholders to speak out and report effectively incidents of harmful irregularities using secure and effective means, such as existing whistleblowing platforms (e.g., WADA's "Speak Up" or IOC's "integrity and compliance hotline"). Essentially, if education is implemented across the wider spectrum of recognition, resistance and reporting behaviour, this may lead to important changes in safeguarding sport from unethical and illegal practices such as doping and match fixing.

To achieve this goal, collaborative efforts are needed among both academic partners and sport governing bodies and authorities. Academic partners can provide important input for an evidence-based approach in developing education against harmful irregularities in sport, whereas sport governing bodies can readily apply this knowledge and translate it into initiatives, interventions and policies to safeguard sports. To date, such synergies are scarce, partly because there is a relative paucity of research on sport whistleblowing behaviour, and because there are limited education initiatives related to sport whistleblowing. Therefore, the development and the promotion of educational materials for recognizing, resisting and reporting harmful irregularities in sport is of paramount importance for safeguarding sports and is an area that requires more systematic empirical study. In support of this argument, a large number of scholars argued that education is an effective way to prevent and curb doping use in elite and amateur sport [2, 3], and it is reasonable to believe that this approach can be extended to other domains of harmful sport irregularities such as corruption and match fixing in sports.

1.1 Co-creation as a Means to Develop Educational Practices Related to Whistleblowing

Co-creation is an approach that originates in marketing and business management research. Its use is particularly popular in the field of management since it is the most recent and dynamic approach that is adopted by researchers [4]. Co-creation is an important aspect of open innovation, the process of utilizing knowledge and experience from outside sources (e.g., external to a services offering organization) to develop and deliver novel and improved services [5]. Co-creation is also gaining prominence in other fields, such as education. Bovill showed that co-creating academic curricula through student-lecturer synergies led to greater student engagement with the course of study and improved decision outcomes in terms of academic leadership [6, 7].

The present study recognizes the value of co-creation as a dynamic approach in developing timely and relevant educational resources that are mapped onto the real training needs of education recipients. Our contention is that co-creation can be used as the guiding methodological framework to identify the needs and areas of interest for education and training pertaining to recognizing, resisting, and reporting harmful sport irregularities, such as doping, bad governance and corruption, and match fixing.

1.2 The Present Study

The present study employed the co-creation approach in order to facilitate the understanding of the end-users' needs towards the development of meaningful and reliable online education material for the effective prevention of harmful irregularities in sport through whistleblowing. By using the co-creation approach, the educational outputs can be developed through the active engagement of the intended target groups and thus, the educational tool will be relevant to the needs of the intended target populations. This is important in order to maximize the reach and impact of proactive whistleblowing activities especially against doping and match-fixing. By adopting the co-creation approach, the educational material is expected to better represent and respond to the needs of the target groups. Thus, the aim of the present study was the co-creation of an online educational tool related to whistleblowing of harmful irregularities in sport. The study reported here is part of a larger-scale, European-wide research project funded by European Commission (Project Sport Whistle). In the present study, a preliminary analysis of the co-creation workshops performed in Greece is presented.

2 Method

2.1 Sample

A total of 18 athletes participated in the study with an age range between 18 and 24 years old (males = 9, females = 9). The participants were athletes involved in a range of different sports such as basketball and tennis. The study conforms to AUTh's ethics policies.

2.2 Co-creation Workshop Matrix

Athletes participated in semi-structured interviews in the context of interactive co-creation workshops and presented their experiences from whistleblowing and their views on related issues. A semi-structured interview guide was organized which consisted of questions addressing certain topics. Firstly, the introductory part of the interview included the evaluation of athletes' knowledge about the concept of whistleblowing. The first part of the matrix of the co-creation workshops included discussion about why people choose to become whistleblowers. As whistleblowing is a topic of ongoing ethical debate, participants were asked whether they consider whistleblowing as ethical and why. An example question was, "Do you believe that whistleblowing counts as "morally obligatory"?". Furthermore, a key question was about the protections which are available to the whistleblowers. Also, participants were asked to describe the circumstances under which an athlete is justified to become a whistleblower. The second part of the interview was dedicated to the anticipated emotions on the whistleblowing decision. Specifically, an example question was, "Do you think that possible regret effects are associated with either blowing the whistle or staying silent?". In addition, participants were asked to describe the expected outcomes of whistleblowing by answering in questions such as, "what do you think will happen to a whistleblower as a result of his/her action?". In the third part of the matrix there were questions about the social acceptance of the whistleblowing, as well as the related normative aspects. For example, participants were asked the following question, "What are the requirements in relation to whistleblowers' safety are requested?". In addition, they were asked to identify the persons who should blow the whistle. The fourth part of the matrix questions were asked to gain insight into the factors that would facilitate or hinder the whistleblowing. Participants were asked to describe the factors that they perceived may prevent an athlete from whistleblowing. Finally, participants were encouraged to express their thoughts about why an athlete decides to become a whistleblower or not. Particularly, athletes' beliefs about the motivations for whistle-blowing were asked. An example question was, "What motivates a whistleblower? Why risk to whistleblow?". The semi-structured format of the interview also allowed for spontaneous interaction.

2.3 Procedure

The present study involves three different but interconnected co-creation workshops. More specifically, a semi-structured interview was conducted in the context of the first workshop with 10 participants of the total sample. Then two of the participants who participated in the first workshop were trained as co-researchers and conducted the workshops with 4 other athletes to collect data on the same subject. Thus, two co-creation workshops (1 co-researcher and 4 participants participated in each one) were conducted.

All participants were duly informed about the aims and purposes of the study, provided implied informed consent by voluntary participation, were informed about their participations rights (i.e., voluntary participation, termination of participation at any time without prior notice and negative consequences), and were reassured about the anonymity and confidentiality of their data. Also they gave their consent to audiotape the interviews in order to facilitate further analysis.

2.4 Data Analysis

Atlas.ti computer software was used for data analysis. Interviews were audio-recorded and transcribed verbatim. To organize and sort the data, coding was used. Codes served as a way to label, compile and organize the data in a meaningful way. They also allowed us to summarize and synthesize what is happening in our data. Coding was done by assigning a word or a phrase to each coding category. This was done in all interview transcripts in a systematic way. For example, codes were emerged by asking questions as, "What is this saying?" or "What does it represent?". The word that we assigned to the item of data in answering such questions was a code, and codes are labels that classify items of information. All transcripts were reviewed and coded by a member of the research team. Codes and emerging themes were discussed continually among the authors and agreed on or revised through a process of consensus. More specifically, the ideas, concepts and themes from the interview transcripts were coded to fit the categories. Transcripts analyzed to identify thematic and overarching categories underpinning respondents' experiences.

3 Results and Discussion

Using the co-creation approach the present study attempted to advance an online educational tool to raise awareness about whistleblowing and teach athletes how to report harmful irregularities in sport. It should be noted that the co-creation approach will ensure greater commitment of all the participants in the development of the educational material and that the educational tool will better represent and respond to the needs of the target groups. The results of the thematic analysis revealed a theme about the definition of whistleblowing. Also, significant themes that emerged were the legitimacy and the responsibility of whistleblowing. In addition, participants acknowledged that attitudes, and social norms a key determinants of whistleblowing.

3.1 Definition of Whistleblowing

Participants reported that they didn't know the concept of whistleblowing. They even said that they had never heard about it before. A representative answer is the following, "I don't really know what whistleblowing means, I think this is relevant to financial institutions but I am not sure how can someone become a whistleblower in sport".

3.2 Legitimacy and Responsibility of Whistleblowing

Participants agreed that irregularities in sport such as doping and match-fixing harm the splendor of sport. For this reason, they acknowledged that such cases of irregularities should be reported to the relevant authorities. In fact, what they propose is whistleblowing, although they initially stated that they didn't know what this concept means. Regarding the perceived legitimacy of whistleblowing, athletes who participated in the workshops strongly support that detailed evidence is needed in order to become a whistleblower. Concrete and specific evidence was highlighted as a very important component in case of whistleblowing. Otherwise, as one athlete stated, "Whistleblowing can be considered as defamation". Most of the participants believed that suspicion or doubt is not enough to claim an irregularity and stated that whistleblowing without evidence can seriously harm one's prestige and damage his/her career. One athlete acknowledged the responsibility of a potential whistleblower, but he wondered whether the efforts are recognized.

Furthermore, the protection of a whistleblower's anonymity was described as a sensitive area because of the increased incidence and the risk to which the whistleblower will be exposed due to the absence of such a protection. Participants emphasized that whistleblowers fear retaliation and thus, the reassurance of their anonymity will help them to feel safe. Also, the athletes suggested that it would be a good idea for the facilitation of the whistleblowing if there was an anonymous reporting mechanism (e.g., a hotline or a website). However, one athlete expressed his opposition to whistleblower's anonymity by claiming that, "Maybe by disclosing the identity, a whistleblower is more responsible".

Regarding the justification of whistleblowing, the participants reported that whistleblowers probably experienced injustice in the past and by revealing an irregularity they feel that they help to restore justice. On the other hand, as reported by some participants, in some cases an athlete becomes a whistleblower as an act of revenge or to defame someone else (e.g. his/her opponent). However, all participants agreed that regardless of why someone becomes a whistleblower, the whistleblowing serves the purpose to prevent future injustice and promote clean sports.

3.3 Attitudes to Whistleblowing

Emotions like pleasure and satisfaction were reported to be experienced by whistleblowers. This was reflected in the following quote: "If I had revealed an irregularity, I would be really pleased because I had contributed to clean sport". In addition, the most participants noted that they would be more satisfied if the authorities did something (e.g., by imposing sanctions) about what they had previously denounced. On the contrary, they agreed that if the authorities didn't take any action, they are not sure that they would become whistleblowers the next time. As one athlete stated, "If an irregularity, that a whistleblower had revealed, wouldn't be punished, I would be tempted to be involved in a fixing match to make a profit".

Regarding the outcomes, participants expected that whistleblowing will prevent any further injustice and corruption in sport. The following statement from an athlete highlights the role of whistleblowing in sport integrity: "Whistleblowing can only have positive consequences. Or at least, the benefits of whistleblowing hugely outweigh the costs or the risks". On the other hand, the disturbance that will be caused in case of whistleblowing was emerged as a particularly unwanted outcome. But participants acknowledged that the worst that could be caused by whistleblowing is the defamation of the whistleblower as reflected in an athlete's statement, "If a whistleblower fails to use reasonable evidence to establish truth, he/she may face defamation". Such cases were considered by the most participants to harm the integrity of sport.

3.4 Social Norms

It was believed that everyone who wants to fight corruption in sports should blow the whistle. The whistleblower can be an athlete, a coach or a club manager. Participants highlighted that it is important for an athlete to know what is whistleblowing and how an athlete can become a whistleblower. One participant stated, "I have just heard about whistleblowing today. I believe that if athletes be informed about it, it will be more likely to bring evidence to the surface by becoming whistleblowers".

Recognizing the normative aspects of whistleblowing, the athletes who participated in the workshops talked about the social acceptance of it. They noted that whistleblowing depends on the culture since it plays an important role in the social approval of whistleblowing. As an athlete said, "Whistleblowing is most effective when it operates within an open-door culture where persons are actively encouraged to raise their concerns". Participants reported that in other countries it may be easier to blow the whistle. But, in Greece, whistleblowers are sometimes seen as traitors or "defectors and some even accuse them of solely pursuing personal glory and fame. This may explain why whistleblowing is not as prevalent in Greece, as reported by an athlete. However, participants claimed that people should be encouraged to take initiative with respect to whistleblowing.

4 Conclusions

The results of the analysis supported the need of athletes for an online educational tool. This approach was deemed more effective comparing to standard educational approaches as young people have easily access to the internet and they are familiar with this way of working. Also, the results of the co-creation workshop in Greece showed that the content of such an educational material should include the definition of whistleblowing, themes related to legitimacy and responsibility of whistleblowing, the attitudes, and social norms associated to whistleblowing. More evidence is needed by the partners of the project "Sport WHISTLE" to ensure that these themes are valid in other European countries as well and provide further themes that may be identifiable in Greece too.

References

1. European Commission: The White Paper on Sport, Brussels, COM(2007) 391 final (2007)
2. Backhouse, S.H., Patterson, L., McKenna, J.: Achieving the olympic ideal: preventing doping in sport. Perform. Enhanc. Health **1**(2), 83–85 (2012)
3. Cléret, L.: The role of anti-doping education in delivering WADA's mission. Int. J. Sport Policy Polit. **3**(2), 271–277 (2011)
4. Agrawal, A.K., Kaushik, A.K., Rahman, Z.: Co-creation of social value through integration of stakeholders. Procedia-Soc. Behav. Sci. **189**, 442–448 (2015)
5. Chesbrough, H.W.: Bringing open innovation to services. MIT Sloan Manag. Rev. **52**(2), 85 (2011)
6. Bovill, C.: An investigation of co-created curricula within higher education in the UK, Ireland and the USA. Innov. Educ. Teach. Int. **51**(1), 15–25 (2014)
7. Bovill, C., Cook-Sather, A., Felten, P.: Students as co-creators of teaching approaches, course design, and curricula: implications for academic developers. Int. J. Acad. Dev. **16**(2), 133–145 (2011)

Technology's Role on Physical Activity for Elderly People

Vasiliki Zilidou[1,2(✉)], Stella Douka[2], Efthymios Ziagkas[2], Evangelia Romanopoulou[1], Nikolaos Politopoulos[3], Thrasyvoulos Tsiatsos[3], and Panagiotis Bamidis[1]

[1] Lab of Medical Physics, Medical School, Aristotle University of Thessaloniki, Thessaloniki, Greece
vickyzilidou@gmail.com
[2] Faculty of Physical Education and Sport Sciences, Department of Physical Education and Sport Science, Aristotle University of Thessaloniki, Thessaloniki, Greece
[3] Department of Informatics, Aristotle University of Thessaloniki, Thessaloniki, Greece

Abstract. The purpose of this study is to investigate the effectiveness of two physical exercise interventions in the elderly, one with the traditional way of exercising and the other with the use of new technologies and by comparing the two interventions and evaluating the most beneficial effects in their functioning ability and their quality of life. The exercise protocols included activities to improve aerobic capacity, muscle strength, mobility (flexibility) and coordination skills (balance). Forty-four elderly women aged 60–80 years old agreed to participate, grouped by twenty-two elderly in each intervention. They evaluated at baseline (pre assessment), as well as at the end of the intervention (post assessment). Both interventions took place at Day Care Centers and lasted 10–12 weeks (min 30 sessions, 2–3 times/week, 60'). The results of this study are in line with the reports of similar studies of the last decade pertaining to the positive role of the physical exercise and the exergames to the psychological well-being and quality of life in older adults. Both interventions revealed the improvement of body functions and balance associated with daily activities, improve their physical and mental health and therefore quality of their life.

Keywords: Physical training · Elderly and technologies
Active and healthy aging

1 Introduction

The worldwide phenomenon of aging is strongly affected by the limitation of the birth or the prolongation of life expectancy as well as immigration. During aging, some biological changes occur. The first signs are in appearance. Internally, the fat replaces lean body mass and many people are gain weight, there is loss of bone and muscle, the lungs lose their ability to take air and our respiratory efficiency decreases, the functions of the cardiovascular and renal systems reduce the number of strokes cells decreases, as is the total brain mass, and finally vision and hearing are impaired [1].

M. E. Auer and T. Tsiatsos (Eds.): IMCL 2017, AISC 725, pp. 668–678, 2018.
https://doi.org/10.1007/978-3-319-75175-7_66

According to the World Health Organization (WHO) it is estimated that the world population will be over two billion people aged over 60 by 2050, where 80% will live in developing countries [2]. These indications force many countries to follow new directions promoting active aging. The primary purpose is to raise awareness and to inform the responsible bodies and society about the problems that aging causes in the population. This fact leads to the need of developing methods and ways to address and meet needs existed in health and social security, promoting the autonomy of the elderly, ensuring the mutual obligations between the generations, even for the continuous improvement of a network of effective care and social support for the elderly.

The elderly people experience several changes, such as physical (kinetic, cardiovascular, respiratory, secretory, autonomous, reproductive system, reduced sensory stiffness), changes in cognitive functions (speed processing, memory, and intelligence), changes in the family's environment or feeling marginalization on the part of their family. When these changes appear, they can have positive or negative effects on their life, but also on the way they are getting old. These changes are a consequence of all these attributes acquired in their lives, such as health, education, employment, financial situation [3].

The major problem faced by the elderly is the social isolation and the blockade occurred by the modern socio-economic developments. According to the agency AGE Concern UK, the definition of social exclusion indicates the lack of accessibility to things of everyday life [4]. Physical impotence is the largest cause of social isolation. The participation of the elderly in a variety of social activities promotes their health and contributes to the greatest possible delay in aging. Aging is accompanied by a decrease in strength, energy and good physical condition and is therefore associated with poor quality of life. Improving elderly fitness capabilities (strength, balance, endurance and flexibility) through effective and comprehensive physical activity programs, enables them to maintain their functional abilities. Thus, elderly become able to adopt a more active lifestyle and improve their quality of life, delaying or reversing the decline in physical performance.

Marques et al. [5] and Mangani et al. [6] mentioned the effect of aerobic exercise against resistance in muscle strength and physical performance in the elderly. Their results presented that aerobic exercise improves the physical performance measures, as balance and gait speed enhances muscle strength improvement. Also, randomized controlled trials of combination of resistance and aerobic exercise have reported improvements in both the muscular strength and the physical performance [7].

Lack of balance in the elderly often leads them to fall where about 10% of falls leading to fracture or other serious injury [8]. Programs designed to improve balance and strength show significant reductions of falls in frail elderly. The component of all positive tests included the balance training [9, 10]. Several studies have shown that physical activity can contribute to the improvement of certain cognitive functions [11] and in particular when combined with cognitive exercise [12]. In a recent neuroscience study, presented data on how the preventive action affects cognitive and physical exercise on cognitive impairment and dementia [13, 14].

Many researches' efforts are being made to support technology in order to find solutions about the elderly's physical exercise through games. The "exergaming" design, if properly implemented, significantly increases the motivation of the elderly while training their body with the appropriate physical activity. The effects of exergaming are now well documented as the positive changes in mood, strengthening socialization and overall improvement in quality of life [15]. Serious games for the elderly, are mainly focused on physical or cognitive training, are considered as preventive/therapeutic interventions. In the composition of an important and promising part of serious games, the exergames are serious games that focus on engaging older people in physical activity through games. It has also been shown that serious games promote and enable the socialization of the elderly [9]. New technologies are a source of social support and general and an instrument for integration in society as citizens [16].

Therefore, there is a growing interest to investigate whether the negative effects of aging can be improved, or even delay their development, aiming at an autonomous and independent living for the elderly and at quality in their life. The aim of this study is to investigate the effectiveness of a traditional physical training intervention for elderly people with the presence of an expert and a computerized physical training intervention evaluating the most beneficial effects.

2 Materials and Methods

2.1 Participants

Forty-four (44) elderly women aged 60–80 years old participated in this study. One group of twenty two women with an average age of 67.09 (SD = 5.95) and an average span of education at 6.54 (SD = 2.17) (cf. Table 1) followed the intervention with traditional aerobic exercise while the other group of twenty two women participated in the computerized physical exercise intervention. A total of 44 women were invited to participate with the prerequisite of performing a medical examination in which good functional, psychological and emotional state were ensured. People with heart failure, hypertension and respiratory failure were excluded from the study. Additional criteria involved non participation in similar intervention programs. They were fully informed about the aims of the study and finally signed the informed consent form. The evaluation was conducted at baseline (pre assessment), as well as at the end of the intervention (post assessment).

Table 1. Demographic characteristics of participants across two different groups to test effects of two exercise-based intervention programs

	Web physical training *n* = 22		Physical training *n* = 22	
	Mean	SD	Mean	SD
Age (years)	*67.09*	*5.95*	*70.70*	*6.0*
Education (years)	*6.54*	*2.17*	*5.50*	*3.59*
Weight (kg)	*77.5*	*9.08*	*74.40*	*9.55*
Height (cm)	*156.7*	*5.93*	*154.9*	*5.23*
BMI	*31.42*	*3.37*	*31.13*	*4.75*
Smoker	*3 (13.6%)*		*0*	

2.2 Intervention

Two different training interventions were designed in compliance to the ACSM/AHA recommendations. The exercise protocols included activities to improve aerobic capacity, muscle strength, mobility (flexibility) and coordination skills (balance) (cf. Table 2). The intensity was 50%–60% of the maximum heart rate and 78–90 b/min [9, 17]. The first protocol consisted of computerized physical exercises using the personalized training program webFitForAll while the second protocol follows a traditional physical exercise intervention.

Table 2. Description of the content structure of two different training programs

Type of activity	Computerized physical training	Traditional physical training	Time (min)
Warm-up period	√	√	7
Stretching	√	√	8–10
Strength	√	√	10–15
Flexibility	√	√	8–10
Balance-strength	√	√	15
Cool down period	√	√	5
Use of technology	Yes	No	

Both interventions took place at the Day Care Centres (KAPI) of Municipality of Pella (cf. Fig. 1) and lasted 10–12 weeks (min 30 sessions, 2–3 times/week, 60').

Fig. 1. Day care centers of municipality of Pella

The webFitForAll exergaming platform, supports a large number of physical exercises suitable for the elderly and complies with the majority of design recommendations and practice suggestions. Adopts MS Kinect as the main gaming controller for delivering physical exercises blended with games (exergames) to the elderly. It enables the professional to create and edit the physical exercise protocols (e.g. type of exercises/games, game parameters, duration) and assign them to the elderly [9, 15]. The aerobic exercises navigate the elderly through a Google maps virtual environment (cf. Fig. 2) based on the walking on the spot or cycling on static bicycle speed. The balance exercises delivered through games that require body balance alterations such as Fishing (balance alterations on the vertical axis), or the Arkanoid game (balance alterations on the horizontal axis) and Golf (balance alterations on both axis) (cf. Fig. 3).

Fig. 2. Aerobic is represented through a Google Maps virtual environment

Fig. 3. During the fishing game the users control the boat's position with body movements on the vertical axis.

2.3 Physical Assessment

The tools selected for the assessment of their physical status and their functional capacity were: the Fullerton Senior Fitness Test (Chair stand, 8 Foot Up and Go, Back Scratch, Arm Curl, Chair Sit and Reach, Two Min Step) [18] to assess the overall physical status, the Berg Balance Scale [19] for the balance and risk of falls, the Tinetti Test [20] for walking and risk of falls and the Stork Balance Stand Test [21] for assessing the balance when standing on one leg.

2.4 Quality of Life and Neuropsychological Evaluation

The WHOQOL questionnaire was used, developed by the World Health Organization, which aims at promoting an intercultural Quality of Life assessment system and the use of the wider health sector. It includes 26 questions divided into four thematic sections (WHOQOL Group 2004) where the respective questions consider: (a) physical health, (b) mental health, (c) social relationships and (d) the environment [22]. It should be noted that there were many cases in which not only the aforementioned assessment tools were used, but also self-report questionnaires were collected. For the evaluation of some cognitive functions (depression, functional, physical and mental health, anxiety, sociability, risk of falls), the following were used: GDS, PHQ-9, IADL, SF-12, BECK, Risk of Falls, Friendship Scale.

3 Results

The statistical analysis was conducted via a 2 × 2 Mixed Model ANOVA. The groups of intervention (Web Physical Training and Physical Training) served as between-subjects factor whereas the time (pre-post) as within-subject factor. Results revealed a statistically significant interaction between time and intervention for the Chair Stand score difference [$F(1,42) = 6.522$, $p = 0.014$]. Furthermore, an interaction between time and intervention was also revealed for the Tinetti score [$F(1,42) = 12.47$, $p = 0.001$], the Berg Balance score [$F(1,42) = 23.48$, $p = 0.000$], the PHQ-9 score [$F(1,42) = 6.54$, $p = 0.014$] (cf. Table 3), (Fig. 4).

Table 3. 2 × 2 ANOVA results

Test	Subdomain	F(1,42)	Time/Interv.
Fullerton	Chair stand	6.522	0.014
Tinetti		12.472	0.001
Berg		23.488	0.000
PHQ-9		6.542	0.014

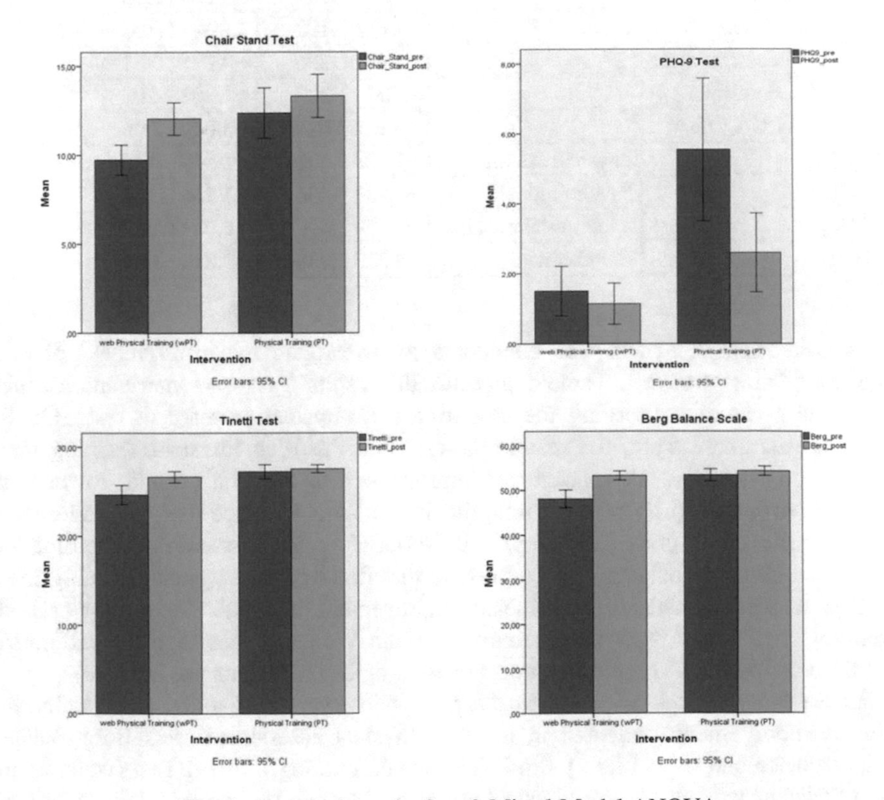

Fig. 4. Results from the 2 × 2 Mixed Model ANOVA

Table 4. Efficacy results in terms of Fullerton, Berg, Stork, Tinetti, GDS, IADL, PHQ-9, SF12, Beck, Friendship Scale, Risk of Falls and WHOQOL. Bold p values denote

Test	Subdomain	Web physical training $n = 22$		Physical training $n = 22$	
		t (21)	p value	t (21)	p value
Fullerton	Chair stand	−6.098	**.000**	−2.546	**.019**
	Arm curl	−5.116	**.000**	−4.134	**.000**
	2-min walk in place	−3.794	**.001**	−4.348	**.000**
	Sit and reach	−1.262	.221	0.596	.558
	Back scratch	−3.223	**.004**	−1.064	.299
	8-Foot-Up-And-Go	4.569	**.000**	3.281	**.004**
Berg balance scale		−6.689	**.000**	−2.160	**.042**
Stork test		−2.863	**.009**	−3.476	**.002**
Tinetti test		−6.079	**.000**	−.924	.366
GDS		3.648	**.002**	0.539	.595
IADL		1.368	.186	1.449	.162
PHQ-9		1.219	.236	3.052	**.006**
SF12pcs		−2.046	.054	−2.064	.052
SF12mcs		−0.340	.737	−2.208	**.038**
Beck anxiety inventory		−1.611	.122	1.694	.105
Friendship scale		−1.071	.296	−0.995	.331
Risk of falls		1.000	.329	−1.308	.205
WHO QOL		−2.162	**.042**	−3.049	**.006**
	Physical health	−1.328	.198	−1.548	.136
	Psychological	−0.957	.350	−1.566	.132
	Social relationships	−1.779	.090	−2.082	**.050**
	Environment	−2.220	**.038**	−3.819	**.001**

A paired-samples t-test was conducted to investigate the effectiveness of each physical training protocol. Table 4 presents the results (post–pre intervention differences and p-value) concerning the cognitive and physical assessment tests. On the physical assessment front, the results indicate that the Web Physical Training intervention evoked statistically significant improvement in the Chair Stand (lower body strength), Arm Curl (upper body strength), 2-min Walk in place (aerobic endurance), Back Scratch (upper body flexibility) and 8-Foot-Up-And-Go (complex coordination, agility and dynamic balance). In contrast to that finding, the Physical Training intervention showed statistically significant improvement in Chair Stand (lower body strength), Arm Curl (upper body strength), 2-min Walk in place (aerobic endurance) and 8-Foot-Up-And-Go (complex coordination, agility and dynamic balance).

In the same notion, Web Physical Training intervention evoked statistically significant improvements reflected in the rest physical assessment tests Berg Balance Scale (balance and risk of falls), Stork (balance on one leg), Tinetti Test (walking and risk of falls) as well as the non-physical assessment test GDS (geriatric depression) and WHO QOL (quality of life).

The participants of the Physical Training intervention showed statistically significant improvements in the Berg Balance Scale (balance and risk of falls), Stork (balance on one leg), PHQ-9 (Patient Health), SF12mcs (Physical and Mental Health) and WHO QOL (quality of life).

4 Discussion

The present study describes the design of two interventions and the conductance of a pilot test with elderly. The overall aim is to investigate whether physical training with or without the use of technology can evoke significant effects in both physical and physiological health, as well as the quality of life of the participants.

In Chair Stand test, which assesses the lower limbs strength and the dynamic balance, statistically significant differences post-pre assessments were revealed in both interventions. Also, statistically significant results were found in the Arm Curl test, evaluated the upper body strength, in the 8-Foot Up and Go test, as well as in the Two Minute Step test, which values the aerobic capacity in both interventions. The upper body flexibility was significantly improved in the Web Physical Training intervention according to the Back Scratch test.

Furthermore, significant results were presented in Berg Balance Scale test, which evaluates the balance and the risk of falling, in the Stork test, assessing the static balance capacity, as well as in WHO QOL (quality of life) in both interventions. The Tinetti test, which evaluates balance in different postures (sitting, standing and lying on one leg, seating and standing), as well as GDS test (geriatric depression) showed statistically significant results only for the participants of the Web Physical Training intervention. On the contrary, significant improvements showed to the participants of the Physical Training intervention as measured by PHQ-9 (Patient Health) and SF12mcs (Mental Health).

The American College of Sports Medicine (ACSM) [23], states that exercise is important for improving and maintaining the health of older people. Aerobic, balance, strength and flexibility are suitable for this population. However, there is no consensus on the frequency, intensity and type of balance exercises for them. Previous studies, showed that the combination of physical activity and brain training with computerized means significantly improved verbal memory after 16 weeks [24]. Moreover, our results are in line with the reports of similar studies of the last decade pertaining to the positive role of the physical exercise and the exergames to the psychological well-being and quality of life in older adults [25, 26]. Both interventions revealed the improvement of body functionalities and balance that are connected to daily activities like carrying luggage or shopping, as well as, more confidence and independence in general [27].

Significant improvements were indicated especially in the Web Physical intervention regarding the depression and quality of life assessment tests before and after intervention. This fact is considered to occur as the aerobic exercises, such as hiking and cycling, were presented through the environment of Google maps. Thus, the participants were able to visit and explore cities or natural places that visited before or heard of in their real life. This fact could provoke the feeling of nostalgia or the challenge to visit and explore new places [9].

The findings showed that there was a consistent positive and simultaneously strong relationship between many of the health indicators, physical state and functional capacity with a total volume of activity and at least moderate activity. In comparison of the results between the two interventions, we observed that the web Physical Training intervention showed statistically significant results in the testing of the strength of the lower limbs, in the balance in different postures and in monitoring and calculating the severity of depression. Physical activity is enjoyable but many older people tend to suffer from loneliness and lack of this. We believe that the use of technology through exergames can offer fun and physical and mental health.

5 Conclusion

The present study investigated the role of technology in physical training programs specialized for the elderly people which is considered to provide benefits to their physical and functional condition in order to deliver autonomous and independent living. Both interventions revealed the improvement of body functions and balance associated with daily activities, improve their physical and mental health and therefore quality of their life. The serious games, which used in the protocol, in addition to the fun, also offered an improvement in their balance and gait, strength of the lower body as they improve their attitude.

References

1. Hooyman, N.R., Kiyak, H.A.: Social Gerontology: A Multidisciplinary Perspective, 9th edn. Pearson, Upper Saddle River (2011)
2. World Health Organization: Active Ageing. A policy framework (Geneva 2002). http://whqlibdoc.who.int/hq/2002/WHO_NMH_NPH_02.8.pdf
3. Kostaridou-Efkleidi, A.: Topics of geropsychology and gerontology. Greek Letters (1999)
4. AGE Concern H.B. http://www.ageconcern.org.uk/AgeConcer/social_inclusion
5. Marques, E.A., Wanderley, F., Machado, L., Sousa, F., Viana, J.L., Moreira-Gonçalves, D., Moreira, P., Mota, J., Carvalho, J.: Effects of resistance and aerobic exercise on physical function, bone mineral density, OPG and RANKL in older women. Exp. Gerontol. **46**(7), 524–532 (2011)
6. Mangani, I., Cesari, M., Kritchevsky, S.B., Maraldi, C., Carter, C.S., Atkinson, H.H., Penninx, B.W., Marchionni, N., Pahor, M.: Physical exercise and comorbidity. Results from the Fitness and Arthritis in Seniors Trial (FAST). Aging Clin. Exp. Res. **18**(5), 374–380 (2006)
7. Marques, E.A., Mota, J., Machado, L., Sousa, F., Coelho, M., Moreira, P., Carvalho, J.: Multicomponent training program with weight-bearing exercises elicits favorable bone density, muscle strength, and balance adaptations in older women. Calcif. Tissue Int. **88**(2), 117–129 (2011)
8. Centers for Disease Control and Prevention (CDC): Self-reported falls and fall-related injuries among persons aged > 65 years - United States, 2006. MMWR Morb. Mortal. Wkly Rep. **57**, 225–229 (2008)

9. Zilidou, V., Konstantinidis, E., Romanopoulou, E., Karagianni, M., Kartsidis, P., Bamidis, P.: Investigating the effectiveness of physical training through exergames: focus on balance and aerobic protocols. In: 1st International Conference on Technology and Innovation in Sports, Health and Wellbeing (TISHW), Vila Real, Portugal (2016)
10. Tinetti, M.E., Kumar, C.: The patient who falls. "It's always a trade-off". JAMA **303**(258–266), 2010 (2010)
11. Erickson, K.I., Voss, M.W., Prakash, R.S., Basak, C., Szabo, A., Chaddock, L., Kim, J.S., Heo, S., Alves, H., White, S.M., Wojcicki, T.R., Mailey, E., Vieira, V.J., Martin, S.A., Pence, B.D., Woods, J.A., McAuley, E., Kramer, A.F.: Exercise training increases size of hippocampus and improves memory. Proc. Natl. Acad. Sci. **108**(7), 3017–3022 (2011)
12. Bamidis, P.D., Fissler, P., Papageorgiou, S.G., Zilidou, V., Konstantinidis, E.I., Billis, A.S., Romanopoulou, E., Karagianni, M., Beratis, I., Tsapanou, A., Tsilikopoulou, G., Grigoriadou, E., Ladas, A., Kyrillidou, A., Tsolaki, A., Frantzidis, C., Sidiropoulos, E., Siountas, A., Matsi, S., Papatriantafyllou, J., Margioti, E., Nika, A., Schlee, W., Elbert, T., Tsolaki, M., Vivas, A.B., Kolassa, I.-T.: Gains in cognition through combined cognitive and physical training: the role of training dosage and severity of neurocognitive disorder. Front. Aging Neurosci. **7** (2015)
13. Bamidis, P.D., Vivas, A.B., Styliadis, C., Frantzidis, C., Klados, M., Schlee, W., Siountas, A., Papageorgiou, S.G.: A review of physical and cognitive interventions in aging. Neurosci. Biobehav. Rev. **44**, 206–220 (2014)
14. Frantzidis, C.A., Ladas, A.-K.I., Vivas, A.B., Tsolaki, M., Bamidis, P.D.: Cognitive and physical training for the elderly: evaluating outcome efficacy by means of neurophysiological synchronization. Int. J. Psychophysiol. **93**(1), 1–11 (2014)
15. Konstantinidis, E.I., Billis, A.S., Mouzakidis, C.A., Zilidou, V.I., Antoniou, P.E., Bamidis, P.D.: Design, implementation, and wide pilot deployment of fitforall: an easy to use exergaming platform improving physical fitness and life quality of senior citizens. IEEE J. Biomed. Health Inform. **20**(1), 189–200 (2016)
16. Czaja, S.J., Charness, N., Fisk, A.D., Hertzog, C., Nair, S.N., Rogers, W.A., Sharit, J.: Factors predicting the use of technology: findings from the Center for Research and Education on Aging and Technology Enhancement (CREATE). Psychol. Aging **21**(2), 333 (2006)
17. Nelson, M.E., Rejeski, W.J., Blair, S.N., Duncan, P.W., Judge, J.O., King, A.C., Macera, C. A., Castaneda-Sceppa, C.: Physical activity and public health in older adults: recommendation from the American College of Sports Medicine and the American Heart Association. Circulation **116**(9), 1094–1105 (2007)
18. Jones, C.J., Rikli, R.E.: Measuring functional fitness of older adults. J. Act. Aging, no. March April, pp. 24–30 (2002)
19. Berg, K., Wood-Dauphine, S., Williams, J.I., Gayton, D.: Measuring balance in the elderly: preliminary development of an instrument (2009). http://dx.doi.org/10.3138/ptc.41.6.304
20. Tinetti, M.E.: Performance-oriented assessment of mobility problems in elderly patients. J. Am. Geriatr. Soc. **34**(2), 119–126 (1986)
21. Johnson, B.L., Nelson, J.K.: Practical Measurements for Evaluation in Physical Education (1969)
22. WHOQOL Group: The World Health Organization's WHOQOL-BREF quality of life assessment: Psychometric properties and results of the international field trial. A report from the WHOQOL Group. Qual. Life Res. **13**, 299–310 (2004)
23. Garber, C.E., Blissmer, B., Deschenes, M.R., et al.: American College Sports of Medicine position stand: quantity and quality of exercise for developing and maintaining cardiorespiratory, musculoskeletal, and neuromotor fitness in apparently healthy adults: guidance for prescribing exercise. Med. Sci. Sports Exer. **43**(7), 1334–1359 (2011)

24. Shah, T., Verdile, G., Sohrabi, H., Campbell, A., Putland, E., Cheetham, C., Dhaliwal, S., Weinborn, M., Maruff, P., Darby, D., Martins, R.: A combination of physical activity and computerized brain training improves verbal memory and increases cerebral glucose metabolism in the elderly. Trans. Res. **4** (2014). https://doi.org/10.1038/tp.2014.122
25. Berger, B.G.: The role of physical activity in the life quality of older adults. Am. Acad. Phys. Educ. Pap. **22**, 42–58 (1988)
26. O'Connor, P.J., Aenchbacher III, L.E., Dishman, R.K.: Physical activity and depression in the elderly. J. Aging Phys. Act. **1**(1), 34–58 (1993)
27. Jones, C.J., Rikli, R.E.: Measuring functional fitness of older adults. J. Act. Aging, no. March April, pp. 24–30 (2002)

Mobile Health Care and Training

Controlling Diabetes with a Mobile Application: Diabetes Friend

Angela Pereira[1](✉), Micaela Esteves[2], Ana Margarida Weber[3], and Mónica Francisco[3]

[1] CiTUR - Tourism Applied Research Centre - ESTM, Polytechnic Institute of Leiria, Leiria, Portugal
angela.pereira@ipleiria.pt
[2] CIIC - Computer Science and Communication Research - ESTG, Polytechnic Institute of Leiria, Leiria, Portugal
micaela.dinis@ipleiria.pt
[3] Polytechnic Institute of Leiria, Leiria, Portugal
{2130308,2130776}@my.ipleiria.pt

Abstract. Researchers have recognized that mobile applications have a great potential to support diabetes self-management. However, the majority of the mobile applications available presents some limitations and are not suitable for patient's needs. Usability is one of the most pointed problems and consequently the over-age of diabetes application usage is low.

In this context, the authors in collaboration with Portuguese healthcare professionals developed a Diabetes Friend system formed by a mobile and a web application. To develop this system, a User-Centered Design approach was used with the aim to create an easier interface and adapts to the end user's needs. This system has the particularly of allowing the healthcare professional to help and monitoring the diabetes patients.

Initial tests have shown that the system has a friendly-interface and the end users are satisfied.

Keywords: Health technologies · Mobile application · Mobile healthcare
Diabetes self-management · Web application

1 Introduction

Diabetes is a chronic disease requiring ongoing medical care, the continuous patient education, and management. It is progressive and leads to macro and microvascular complications', including heart disease, stroke, hypertension, nephropathy, neuropathy and in the end, it can cause death. It is one of the main responsible for the global burden of disease [1]. It can be an auto-immune condition that usually develops in childhood or adolescence known as type 1 diabetes (T1D). Another typology of this disease is type 2 diabetes (T2D) that is more widespread and develops in adulthood when body tissues become resistant to the action of insulin, even in a state of circulating hyper-insulinemia [1]. The T2D represents 90% of the global cases of diabetes. According to International Diabetes Federation [1], it is estimated that 193 million people with diabetes are undiagnosed and are therefore at risk of developing complications.

M. E. Auer and T. Tsiatsos (Eds.): IMCL 2017, AISC 725, pp. 681–690, 2018.
https://doi.org/10.1007/978-3-319-75175-7_67

In Portugal, in 2015, were diagnosed 168 new cases of diabetes per day and it is known as the European Union country with the highest prevalence of this chronic disease [1]. The diabetic patients should have a healthy and balanced diet plan; make regular exercise, monitor the blood glucose levels and apply insulin at the right time, if necessary. In order to achieve this, it is essential to motivate people through innovative solutions that truly meet their needs.

The evolution and widespread use of mobile technologies have boosted the industry to develop applications and services. The mobile devices contributed to this expansion and have become part of people's daily moments, leading to a rapid growth of the development of software applications for these platforms. The mobile health applications (apps) are one example of this rapid growth with over 31,000 health and medical apps available for download [2], once these apps have the possibility to help patients to manage their disease. Researchers [3–5] highlight that diabetes apps, in particular, have great potential to support diabetes self-management. However, the majority of diabetes applications available [5, 10] have some limitations, for example, are not intuitive to use, lack of personalized feedback, serving only as an electronic logbook and have usability problems.

This study seeks to present a computer system development that helps diabetic patients to self-manage their condition with a close range to healthcare professionals. In the next section of this paper, the related work is outlined and the project methodology is described. Finally, we present the system architecture and model, the conclusions and future work.

2 Related Work

Through a simple research on Google Play or Apple store, it is possible to find out many applications related to diabetes self-management. However, only a few of them are open to everyone, while others have only some premium functionality available at an additional cost. In this research, a set of 25 applications was considered, 13 from Google Play and 12 from Apple Store, in English and Portuguese, as well as blood glucose monitorization as a minimum requirement. Studies were also reviewed from the literature.

The functionalities presented in all applications analysed (Table 1) there are timely insulin dosage or medication and registration of blood glucose levels. Some of these applications have indications suitable for diet, physical exercise, weight control, and blood pressure. Other support alert/notifications and integration with social media such as Facebook and Twitter. In related of uploading blood glucose levels and blood pressure, the majority of commercial applications use manual data entry. This issue is considered a limitation, as the manual data entry implies extra steps to transfer data from the measurement device to the phone and is prone to errors.

The applications reported in the literature include self-management tasks such as personalized feedback, blood pressure monitorisation, self-management education, communication and patient monitorisation by clinicians. However, some of these applications are not available in the market.

Table 1. Functionalities found in diabetes mobile applications

Functionalities	Frequency (percentage)
Insulin dosage	25 (100%)
Medication	25 (100%)
Blood glucose	25 (100%)
Blood pressure	6 (23%)
Diet	17 (68%)
Weight control	7 (25%)
Physical exercise	11 (41%)
Alerts/notifications	19 (76%)
Data entry automation	3 (10%)
Education	5 (18%)
Social media	4 (15%)

El-Gayar [4] argues that it would be helpful to the patient to provide an individual analysis and interpretation of results. However, in the majority of the applications, it only provides insulin dosage suggestion, based on the data record, located in the patient mobile device [5, 10]. These applications are limited because they do not consider other issues related to the clinical history of the patient that only the professional healthcare knowns.

On the other hand, diabetes researchers consider that self-management education has a positive impact on clinical outcomes [11]. The commercial applications, 18% have some components of self-management education, such as guideline for diabetes care and food information. Anyway, a personalized education is considered more appropriated to promote changes in patient's behaviour based on their profile, context, education, and goals of the medical treatment [12].

One issue that has not been taken into account by developers is usability. Usability is a quality or attribute that represents the ease which a human-computer interface is used and provides an effective, efficient, and satisfying experience [6, 7].

The more common usability problems referred in studies are limited functionality and interaction, multi-steps task and difficult navigation system [5, 10]. Moreover, the rate of diabetes application usage is low [8]. Fu et al. [5] consider these applications can have some usability problems. One way to overcome the usability issues is to associate in the app development process to the patients and physicians [7].

In this context, the authors developed a computer system with the aim to help diabetes self-management of the patients, with a near physician monitorisation named Diabetes Friend. This computer system is based on a cloud computing that integrates two components, one mobile application for patients and a web application for physicians.

3 Methodology

A well-designed interface can contribute to the success of a software. A key aspect in the development of user interfaces is to know the characteristics of the end user [9]. The User-Centered Design (UCD) is a user interface design process that focuses on

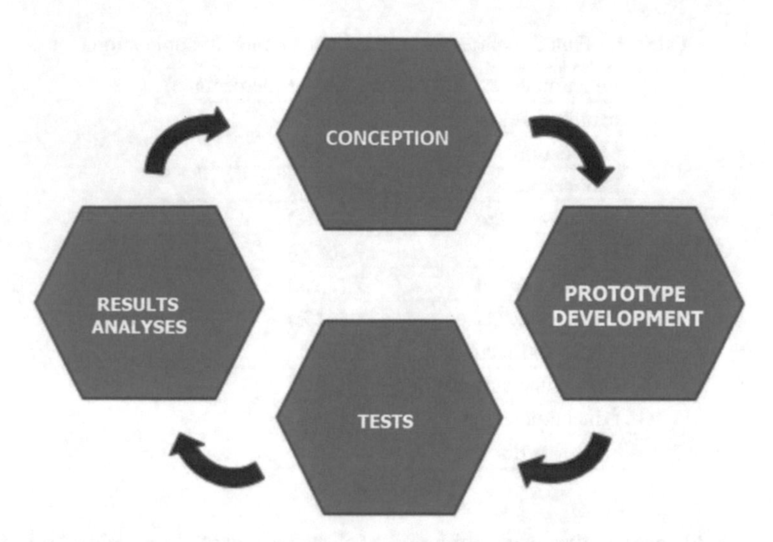

Fig. 1. Cycle methodology

usability goals, user characteristics, environment, tasks and workflow in the design of an interface. In the 1980s, Norman defined and promoted two key user-centered design concepts: users should be the center of the design process and the design process should match user needs to product design elements [13].

In order to develop a software to be convenient and valuable by the end users, it is necessary to understand the correct user requirements. Thus, the Diabetic Friend system should not only include powerful functionalities, but also present user-friendly interfaces. Therefore, this study adopts a UCD process, which is an effective approach to satisfy users' needs and to improve the user interface.

To develop the Diabetes Friend system a cycle methodology (Fig. 1) was used which consist of four phases: conception, prototype development, tests and result analysis [14]. These phases were repeated until users were fully satisfied with the functionalities and interface.

3.1 Conception Phase

The research strategy began to understand how the healthcare mobile applications available, on the market, helped the diabetic patient. With this purpose, the authors have tested different diabetes self-management mobile applications. From this research, several limitations were identified, mentioned above, in related work and some functionalities were considered to be included in the Diabetes Friend system. Patients and physicians were directly involved in the system development process to accomplish a friendly usability interface that will be suitable for users' needs. The main target for this system are patients with type 2 diabetes, which is the most common in Portugal, with age above 50. The authors made structured interviews and applied surveys to the healthcare professional (physicians, nurses) and diabetes patients. Based on the feedback from the patients and the healthcare professionals a set of functionalities were reached.

From the clinical guidelines [15–17] the following features were considered that are important for diabetes self-management: education and personalized feedback; diet and weight management; physical activity; insulin and medication management; other therapeutics (foot, eye care) and psychosocial care; communication and patient monitorisation are done by primary care providers.

All this process leaded to a set of functionalities to be implemented in Diabetes Friend system which are:

Mobile Application Functionalities

- Patient initial configuration – personal profile and clinical history. This functionality is configured by the healthcare professional through the web application that is automatically synchronized with the patient mobile application.
- Daily registration – blood glucose levels, blood pressure and important annotation such as food intake changes, medication, physical exercise and others. To be more precise about the food intake, the patient can take a photo. All this information is synchronized with the healthcare professional application.
- Notifications/Alerts – alert notifications about taking medication (insulin or others), date of medical consultation, assessment and others alerts that are important for the patient health.
- Data analysis – The information that is recorded daily (blood glucose levels, blood pressure, and weight) are represented in a chart and table format. In conjunction with the patient representation values, the chart represents the normal values of the patient's age.
- Chat – allows the patient to communicate with the healthcare professional whenever he/she has a doubt. Also, it is possible to the patients to communicate with each other's.
- Education – presents a set of information about diabetes disease and healthcare that patient must have. The healthcare professional through the web application share with the patient mobile application, a set of specific healthcare information according to their needs.
- Prescription – the physician directions about medication dose are presented in the mobile application through the web application.

Web Application Functionalities

- Patient configuration – allow the healthcare professional insert, edit and query data of the patient.
- Data transfer – the healthcare professional can transfer data to the patient mobile application such as: prescription, education and diet plan, physical exercises. Also, the web application receive the data about patient daily registration.
- Chat – allows the healthcare professional to communicate with an individual patient or send messages to a group.
- Alerts – the healthcare professional receives alerts when a patient has abnormal values related with blood glucose, blood pressure and others.
- Data visualization – the healthcare professional can observe patient data in several formats for analysis and study the disease progress.

3.2 Prototype Development

In order to organize the functionalities, a low-fidelity prototype was created and designed with several mockups for the same functionality, with the purpose to find the most suitable solution. These mockups were evaluated by the end users, who choose the best solution for them. Then a high-fidelity prototype was built with the purpose to assess the interface usability. The mobile application (Fig. 2) was developed for a smartphone and the web application (Fig. 3) was developed for personal computer.

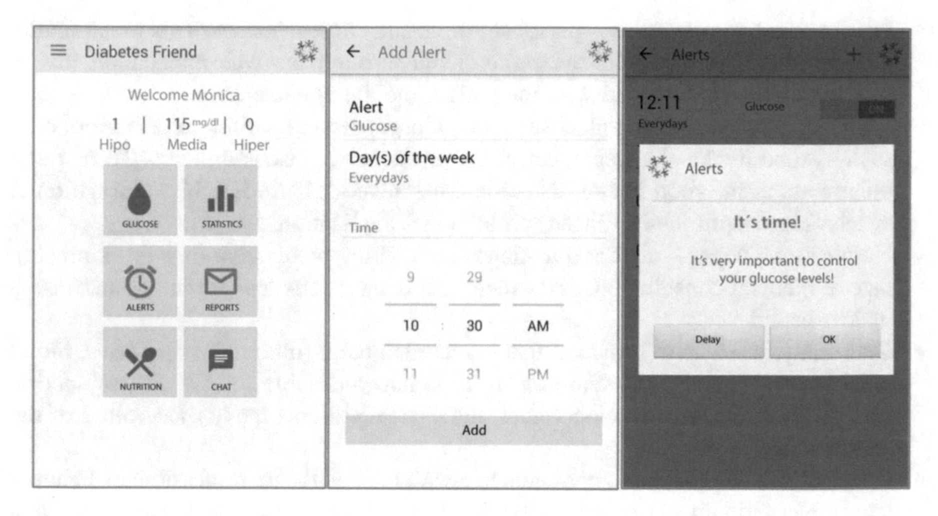

Fig. 2. Patient mobile application: high-fidelity prototype

3.3 Tests and Results

Firstly, Nielsen's heuristics [9] were used. This is a usability engineering method to find the usability problems in a user interface design that can be attended to as part of an iterative design process. A set of four assessors examine the interface and judge its compliance by using usability principles (the "heuristics"). To avoid bias the authors did not participate in any evaluation. At the end of this evaluation, minor errors were found, which were corrected in the implementation phase.

Afterwards, the authors conducted usability tests with end users, healthcare professionals, and diabetes patients, to validate high-level design concepts, discover and fix usability problems and learn more about users and their tasks. The method adopted was user testing, suggested by Nielsen [9], that is one of the most basic and useful tests, which has 3 components:

- Get hold of some representative users.
- Ask the users to perform representative tasks with the design.
- Observe what the users do, where they succeed, and where they have difficulties with the user interface.

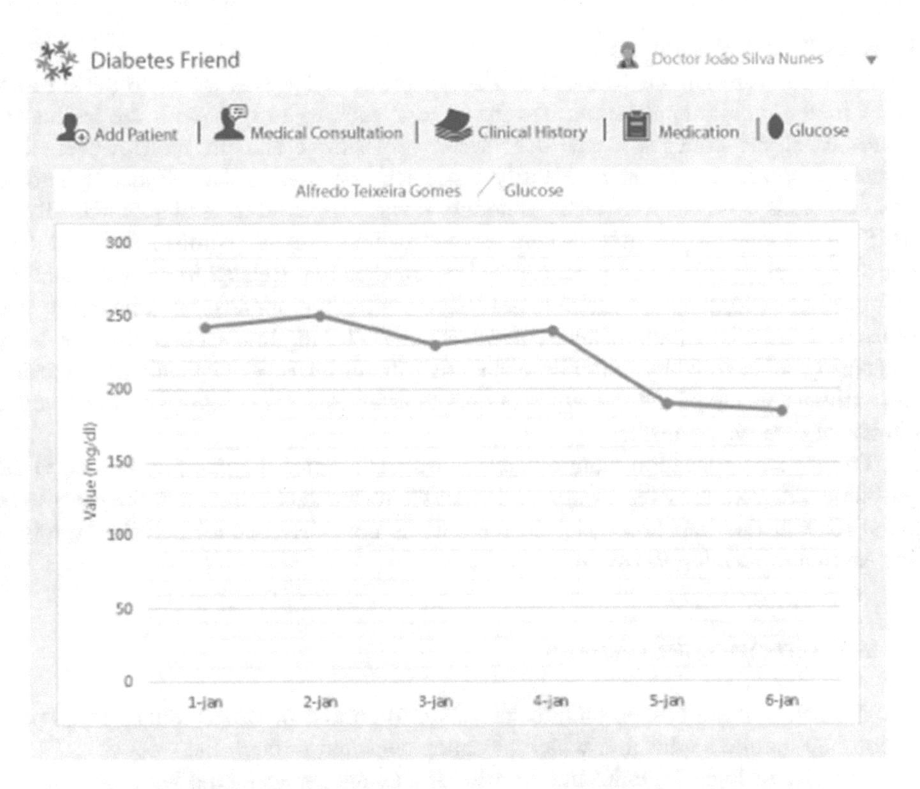

Fig. 3. Healthcare professional web application: high-fidelity prototype

According to Nielsen [9], to identify a design's most important usability problems, it is only needs to test 5 users, rather running a big and expensive study. Thus, the tests were made by four healthcare professionals, two physicians and two nurses, and twelve patients' type 2 diabetes.

Healthcare Professional Test and Analysis

Healthcare professional was asked to perform six tasks that were observed and recorded by an outsider specialist. The purpose of these tasks were to evaluate the efficacy, efficiency and satisfaction. The basic tasks, such as access to patient information, visualization of the daily patient registration, were all accomplished easily by the users.

The tasks with a medium difficult level like, to create a medical prescription and send it to the patient mobile application, were performed quickly in general. However, the users made some mistakes, for example, by sending the prescription to the chat instead of the mobile app. The patient mobile application icon it was not considered very intuitive, so it should be improved. Regarding sending a message to a group of patients, this must be improve as they have some difficulty in selecting the patients.

In general, the healthcare professionals appreciate the application interface and it would be very useful to them.

Patients Test and Analysis

Patient tests were similar to the healthcare professional it was asked to perform six tasks in the mobile application. The basic task such as to chat with the healthcare professional and colleges it was considered very simple and useful. However, they suggest that was important to identify those who are online and offline. Regarding creating alerts it was an easy task. Although, it was suggested to send automatically an SMS to the patient caregiver whenever the patient did not confirm the alert task conclusion, especially when it concerns about medication. Related to diary registration of blood glucose, it was a difficult task once the mobile application supports to two types of data register (automatically and manually). Taking into account that not every measuring device used by Portuguese diabetic patients have automatically data transfer, the authors considered the two types of data transfer. So, it is necessary to improve the choice of data type transfer.

To turn the application more efficient it was redesigned the interface related to the problems referred. New tests were made and the results had shown that the users were satisfied with this improvements. So the authors considered concluded the prototype and started to develop the system.

4 Architecture and Model

The Diabetes Friend system architecture (Fig. 4) consist in two modules, web application and mobile application which are interconnected through the web.

The system logical architecture consist of a Client-server model with three layers: Client, Middleware and Server. The Client consists of two applications, one web application for the healthcare professional and other android mobile application for

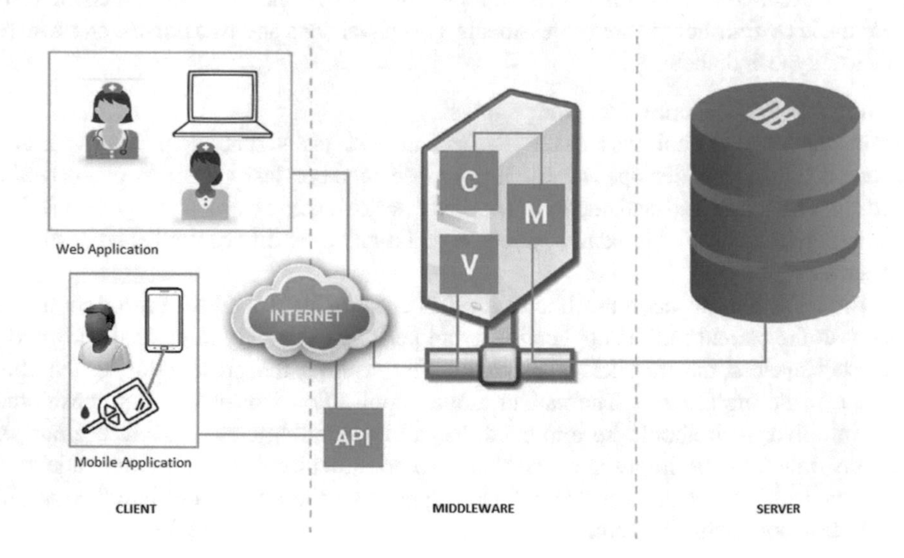

Fig. 4. Diabetes friend system architecture

patients. Firstly, the authors considered only the android operating system because it is more used by patients in Portugal. Depending on the patient acceptability the authors will develop for iOS system.

In the middleware layer also known by business, the data processing is made through MVC model (Model/View/Controller).

In the last layer, Server, the CRUD (create, read, update and delete) operations are made by database (DB).

Currently, the Diabetes Friendly system has been developed. Although the mobile application is finished but the web application still in progress. Thus, the authors only considered to make final test after all the system be done. This test includes the study of clinical effectiveness, which is defined as improvement in glycaemic control.

5 Conclusions and Future Work

Diabetes is a chronic illness that requires constant monitorisation and management of glycaemic fluctuations through medication, diet, and exercise. Patients diagnosed with diabetes face a challenge of incorporating these various monitoring components into their lives daily. The mobile technologies have made diabetes applications available and these have a great potential to support diabetes self-management. Furthermore, some applications are not suitable to patients needs once it has lack of personalized feedback and usability issues, particularly for elderly people. On the other hand, the existence of many diabetes applications in the market starts to be strenuous to find an application in this plethora of options that are suitable for patient's own needs. Along with these lines, the adoption of an application for diabetes self-management only makes sense if the patient feels secure and confident in its use and have a close support of their healthcare.

This article presents the Diabetes Friend system development, with a UCD approach which intends to help the patients and healthcare professional to control the diseases in conjunction. For this purpose, multiple usability measurement methods were combined to attain more diversified results. More specifically, Nielsen's heuristics, as well as, usability tests were applied to assess the interface of this system during the design process.

As future work, the authors intend to study the clinical effectiveness of the Diabetes Friend System. This will be a long-term study carried out in conjunction with Portuguese healthcare professionals.

References

1. Atlas, I.D.: International Diabetes Federation, Brussels (2015). http://www.oedg.at/pdf/1606_IDF_Atlas_2015_UK.pdf. Accessed 10 Aug 2017
2. Payne, H.E., Lister, C., West, J.H., Bernhardt, J.M.: Behavioral functionality of mobile apps in health interventions: a systematic review of the literature. JMIR mHealth uHealth **3**(1), e20 (2015)

3. Chomutare, T., Fernandez-Luque, L., Årsand, E., Hartvigsen, G.: Features of mobile diabetes applications: review of the literature and analysis of current applications compared against evidence-based guidelines. J. Med. Internet Res. **13**(3), e65 (2011)
4. El-Gayar, O., Timsina, P., Nawar, N., Eid, W.: Mobile applications for diabetes self-management: status and potential. J. Diab. Sci. Technol. **7**(1), 247–262 (2013)
5. Fu, H., McMahon, S.K., Gross, C.R., Adam, T.J., Wyman, J.F.: Usability and clinical efficacy of diabetes mobile applications for adults with type 2 diabetes: a systematic review. Diab. Res. Clin. Pract. **131**, 70–81 (2017)
6. ISO 9241-11. International standards for HCI and usability standards related to usability can be categorised as primarily concerned with: development of ISO standards 1998:1–13. http://www.usabilitynet.org/tools/r_international.htm#9241-11. Accessed 10 Aug 2017
7. Esteves, M., Pereira, A.: YSYD-You Stay You Demand: user-centered design approach for mobile hospitality application. In: 2015 International Conference on Interactive Mobile Communication Technologies and Learning (IMCL), pp. 318–322. IEEE (2015)
8. Research2Guidance. Diabetes App Market Report 2014. http://www.reportsnreports.com/reports/276919-diabetes-app-marketreport-2014.html. Accessed 10 Aug 2017
9. Nielsen, J.: Heuristic evaluation. In: Nielsen, J., Mack, R.L. (eds.) Usability Inspection Methods. Wiley, New York (1994)
10. Padhye, N.S., Wang, J.: Pattern of active and inactive sequences of diabetes self-monitoring in mobile phone and paper diary users. In: 2015 37th Annual International Conference of the IEEE Engineering in Medicine and Biology Society (EMBC), pp. 7630–7633. IEEE (2015)
11. Norris, S.L., Lau, J., Smith, S.J., Schmid, C.H., Engelgau, M.M.: Self-management education for adults with type 2 diabetes. Diab. Care **25**(7), 1159–1171 (2002)
12. Clement, S.: Diabetes self-management education. Diab. Care **18**(8), 1204–1214 (1995)
13. Norman, D.: The design of everyday things. Verlag Franz Vahlen GmbH (2016)
14. Hartson, R., Pyla, P.S.: The UX Book: Process and Guidelines for Ensuring a Quality User Experience. Elsevier, Amsterdam (2012)
15. Paulweber, B., Valensi, P., Lindström, J., Lalic, N.M., Greaves, C.J., McKee, M., Kissimova-Skarbek, K., Liatis, S., Cosson, E., Szendroedi, J., Sheppard, K.E.: A European evidence-based guideline for the prevention of type 2 diabetes. Horm. Metab. Res. **42**(S 01), S3–S36 (2010)
16. Sibal, L., Home, P.D.: Management of type 2 diabetes: NICE guidelines. Clin. Med. **9**(4), 353–357 (2009)
17. Funnell, M.M.: Standards of care for diabetes: what's new? Nursing2017 **40**(10), 54–56 (2010)

A Preliminary Study on a Mobile System for Voice Assessment and Vocal Hygiene Training: The Case of Teachers

Eugenia I. Toki[1(✉)], Dionysios Tafiadis[2], Konstantinos Rizos[2], Marina Primikiri[2], Georgios Tatsis[2], Nausica Ziavra[1], and Vassiliki Siafaka[1]

[1] Laboratory of Audiology, Neurootology and Neurosciences, Department of Speech and Language Therapy, School of Health and Welfare, Technological Educational Institute (TEI) of Epirus, Ioannina, Greece
{toki,nziavra,vsiafaka}@ioa.teiep.gr
[2] Department of Speech and Language Therapy, School of Health and Welfare, Technological Educational Institute (TEI) of Epirus, Ioannina, Greece
{d.tafiadis,g.tatsis}@ioa.teiep.gr, konrizos@yahoo.com, MARINAPRIM12@hotmail.com

Abstract. The aim of this study was to present preliminary verification results on a voice assessment mobile system used by teachers. A summary of voice characteristics is presented in terms of means and standard deviation. Significant statistical differences were noted for Voice Handicap Index Total and its dimensions and Reflux Symptoms Index scores between school teachers without voice disorders using the mobile system and an aged matched voice professional population with voice disorders that were traditionally administrated. No significant differences were obtained between teachers without voice disorders using the mobile system and the corresponding population without voice disorders using the above questionnaires traditionally. Finally the results of the study are discussed together with the perspective of teacher education on vocal hygiene routines indicating the accuracy and therefore the verification of the mobile system according to the requirements and design specifications.

Keywords: Mobile system · Voice assessment · Vocal hygiene Teacher

1 Introduction

Preschool and primary school teachers have a vocally demanding profession and are often at higher risk for developing a voice disorder [1]. Anatomic, physical, or functional abnormal changes in the voice mechanism may cause a voice disorder [2] leading to vocal malfunction or inefficiency [3]. Some of the parameters and mechanisms that lead to abnormal voicing and voice disorders are summarized by Johnson [4] and include factors such as smoking, reflux, voice misuse and overuse. Roughly 9% of the general population is affected by a voice disorder [5] with teachers having the second

M. E. Auer and T. Tsiatsos (Eds.): IMCL 2017, AISC 725, pp. 691–699, 2018.
https://doi.org/10.1007/978-3-319-75175-7_68

greatest incidences [6, 7]. The occurrence of vocal symptoms are indicators for developing voice disorders [6, 7].

Teachers seem to have low awareness of the vocal demands in their professional careers [6, 7]. They rely on their voices to be clear and stable all over the day without any influences from various factors [6, 8]. Voice evaluation is an important procedure which usually takes place with traditional and/or mobile procedures.

More than 20% of the Dutch teachers sought medical help or had been treated for a voice problem [9]. Female teachers more frequently reported voice complaints and absence from work due to their voice problems [9]. Among teaching professions, music teachers are roughly four times more likely than classroom teachers to develop voice-related problems [10]. School music teachers perform significant improvement if they receive vocal hygiene and behavior modification instructions helping them to correct their vocal problems [11].

The prevalence of reporting a current voice problem was significantly greater to the prevalence of voice disorders during lifetime between teachers, especially women, and other professions [12–14]. Additionally risk factors for voice disorders for Slovenian teachers' were found: (i) aged over 40 years, (ii) gastroesophageal reflux disease (GERD) and allergies [15].

In contrast to the elaborate literature describing the vocal risk factors, less attention has been paid to vocal rehabilitation [16]. Furthermore, few studies have investigated whether teachers received information about their voice and the use of vocal hygiene [16]. Moreover, according to Van Houtte et al. [16] only 27.8% of teachers received information about vocal hygiene and vocal techniques. They also state that a very small portion of them are familiar with vocal hygiene methods (e.g. increasing fluid intake or avoid speaking loud in a noisy environment). Additionally, there is a strong argue about the effectiveness of preventative strategies even though it is documented by previous studies [16–20].

Vocal care has not been taken in account in any educational teachers' program. The aforementioned probably justifies the fact that younger teachers have greater vocal difficulties and having poorer vocal hygiene techniques than their more experienced peers as Mjaavatn reported [21].

Voice evaluation is an important procedure which usually takes place with traditional [22, 23] or digital methods [24, 25]. In traditional voice evaluation there are well established evidence-based models and multidimensional protocols [26]. According to Tafiadis et al. "*these protocols included laryngeal imaging, aerodynamic, perceptual-acoustic evaluation, and the impact of voice on the quality of life via self-perceived questionnaires*" [26]. In the case of digital based methods on voice evaluation there is less research evidence [24, 25]. Towards this innovative direction it is crucial to verify the accuracy of voice evaluation when using the mobile system. A powerful common method to evaluate the mobile system is the usability testing through the analysis of typical end users interacting with the system [27].

The aim of this study was to present preliminary results verifying a voice assessment mobile system when used by voice professionals - teachers. The ultimate aim is to embed individual voice hygiene feedback.

2 Material and Methods

2.1 Participants

In this study, 130 teachers (54 male and 76 female) were enrolled. All consented teachers, after being informed of the research purposes, used the mobile system. The participants who experienced any laryngeal or respiratory disorders for a period of 2 weeks were excluded from this study. Furthermore, participants who experienced symptoms of gastroesophageal reflux or laryngopharyngeal reflux, history of alcohol or drug addiction, working or living under environmental conditions that may influence voice (noise, exposure to chemicals etc.), or any history of voice abuse were excluded.

This research was approved by the Department of Speech and Language Therapy Research Ethic Committee of TEI of Epirus, and also the Institute of Educational Policy of the Ministry of Education, Research and Religious Affairs. Data were collected from schools settings from the regions of Ioannina and Fthiotida, Greece in 2017.

2.2 Data Collection

The data were collected by an under development mobile system. This system focuses on voice assessment procedures in order to give individualized immediate vocal hygiene feedback. Three digital questionnaires were included to gather the user's digital voice profile; the translated in Greek version of the Voice Evaluation Form (VEF) [28], the standardized Hellenic version of the Voice Handicap Index (VHI) [29] and Hellenic Reflux Symptom Index (RSI) [30–36]. These, were filled out by all participants. The VEF was developed by the American Speech-Language-Hearing Association as a consensus template for voice disorders. The VEF was also used to exclude subjects with history of voice disorders. The VHI is a 30-item questionnaire summarizing a VHI-T from 0 to 120 points. This score is split into three equal domains (0–40 points): Voice Handicap Index-Emotional (VHI-E), Voice Handicap Index-Physical (VHI-P) and Voice Handicap Index Functional (VHI-F). The RSI [30–32] is a 9-item questionnaire summarizing a total score of 45 for most severe symptoms. The RSI [30–32] it's a self-administered nine-item questionnaire designed to assess various vocal symptoms of GERD and symptoms related to laryngopharyngeal reflux [30].

Also two (2) voice samples of each participant were recorded in a room where background noise did not exceed 30 dB (sustained voicing of /a/ and /e/ sounds). Analysis was achieved by 5105 MDVP software, 2.3 edition and the signal was digitized at 50 kHz sample rate for a total of 228 voice recordings.

Acoustic analysis was conducted in order to determine with an objective method whether participant voice characteristics were or not within normal range [37] and whether they are in line with the output of the mobile system. Then, the output of the mobile system was compared with aged matched professional voice user results from the literature [29]. In detail Fig. 1 depicts the methodology for verifying the voice assessment mobile system.

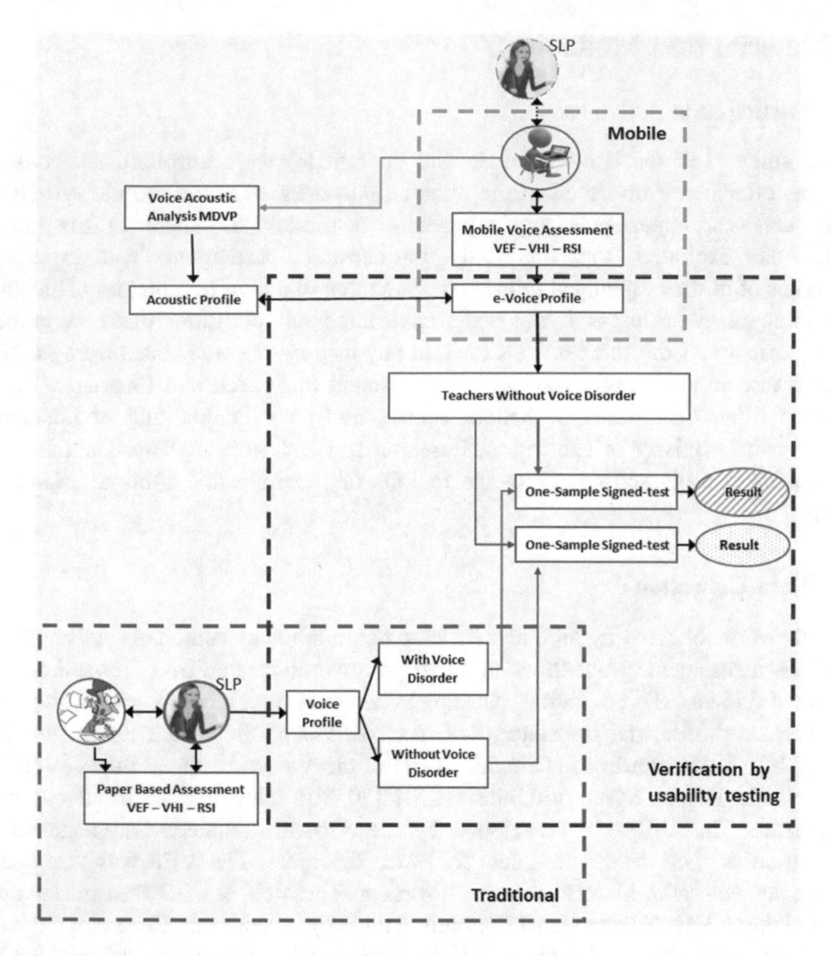

Fig. 1. Methodology for verifying the voice assessment mobile system

2.3 Statistical Analysis

A report on voice characteristics values (F0, Shimmer, and Jitter) from data of primary school teachers (male and female) is presented in terms of mean and standard deviation.

Differences were examined between Greek normative data for VHI (Total, Functional, Physical and Emotional domains) and RSI scores and data of primary school teachers. After controlling for normality with Kolmogorov-Smirnov test ($p < 0.05$) & the related Q-Q plots, a rather high kurtosis was observed, revealing no symmetry in the data distribution. Therefore, these differences were examined using the one sample signed-rank test.

Analysis was conducted using SPSS statistical software.

3 Results and Discussion

Table 1 presents voice acoustic characteristics for teachers that used the mobile system for the sustained voicing of /a/ and /e/ sounds. The normative data that were used for the comparisons were obtained from researches that had age and gender equivalent populations to this study. According to Table 1 voice characteristics of the sample of the study are in normal range [37].

Table 1. Voice characteristics for teachers used the mobile system

N = 114	Sex	N	Mean (SD)	SD error mean	Normative voice data[a]
MF0 /e/	Male	54	149.472 (±44.355)	6.469	118
	Female	76	228.047 (±54.638)	6.675	189
HF0 /e/	Male	54	146.797 (±42.764)	6.237	118
	Female	76	224.096 (±56.145)	6.859	189
Jitter /e/	Male	54	5.597 (±19.218)	2.803	7.700
	Female	76	2.795 (±2.523)	0.308	6.500
Shimmer /e/	Male	54	8.452 (±5.456)	0.795	3.700
	Female	76	7.873 (±4.648)	0.567	2.300
MF0 /a/	Male	54	129.730 (±45.577)	6.648	118
	Female	76	213.046 (±47.507)	5.803	189
HF0 /a/	Male	54	130.805 (±41.280)	6.021	118
	Female	76	211.468 (±48.447)	5.918	189
Jitter /a/	Male	54	1.735 (±1.360)	0.198	7.700
	Female	76	1.497 (±1.250)	0.152	6.500
Shimmer /a/	Male	54	6.528 (±2.539)	0.370	4.700
	Female	76	5.648 (±3.359)	0.410	3.300

Normative data by: [a][37]

In Table 2 VHI and RSI scores and dispersion values are presented for teachers that used the mobile system.

Table 2. VHI and RSI scores for teachers that used the mobile system

N = 114	Mean (SD)	Median (Range)	VHI norms[a] & RSI norms[b]	
			Controls	Pathological
VHI-T	5.73 (10.025)	2.00(78)	7.41	37.00
VHI-F	2.02 (3.207)	1.00(19)	3.21	10.00
VHI-P	2.58 (4.025)	1.00(27)	2.84	18.00
VHI-E	1.12 (3.537)	0.00(32)	1.36	9.00
RSI	2.74 (4.228)	1.50(23)	2.32	24.76

Normative data by: [a][29], [b][30]

A detailed description of the differences between Greek normative data for VHI (Total, Functional, Physical and Emotional domains) and RSI scores, and data of primary school teachers, follows. After controlling for normality of all the above scores, with Kolmogorov-Smirnov test (p = .000) & the related Q-Q plots, no symmetry in the data distribution was revealed. Therefore, these differences were examined using the one sample signed-rank test.

The results of the study stated that there was a significant statistical difference at the $p < 0.05$ level for VHI Total (p = .000), VHI Physical (p = .000), VHI Functional (p = .000) and VHI Emotional (p = .000) of teachers using the mobile system without voice disorders when compared to teachers using the traditional paper version and had voice disorders [16].

A one sample signed-rank test was also conducted for teachers using the mobile system and the traditional paper version that both had no vocal pathologies resulted to no significant statistical difference at the $p < 0.05$ level for VHI Total (p = .102), VHI Physical (p = .448), VHI-Functional (p = .434) and VHI Emotional (p = .471) [16].

The results of the study reported a significant statistical difference at the $p < 0.05$ level for RSI (p = .000) of teachers using the mobile system when compared to a pathological population using the traditional paper version [30–32] and no significant statistical difference at the $p < 0.05$ level for RSI (p = .973) when compared to a normal population using the traditional paper version [30–32].

It is obvious in the current study, that teachers can understand their voice status and VHI using either mobile or paper version. It may also help towards a better awareness of their vocal needs. That is in agreement with Roy et al. [38] research who studied the effects of two treatment approaches (indirect method with proper vocal hygiene and direct method using vocal function exercises) in voice-disordered teachers and they discovered improvement in the exercise group using VHI. Additively, Chan [39] studied the outcomes of a voice hygiene program in kindergarten teachers and a control group. They concluded that training the voice hygiene group resulted in improving their voices, based on acoustic data whereas the control group experienced significantly more vocal fatigue. Furthermore, previous studies [39–41] using acoustic parameters analysis found that voice hygiene education were effective. In those studies, significant changes were found for voice perturbations (decreased jitter change during the working day and shimmer mean at the end of the term), decreased self-reported difficulty during phonation and better perceptual voice quality. Thus these needs for better vocal health can be addressed by a mobile individualized expert voice system.

The results of this study confirm the verification of this mobile system and therefore the system fulfils this part of the specification requirements. Future research may focus on examining the verification concerning other voice professionals and comparisons between them as well as the verification on other parts of the mobile system. Furthermore it can be extended towards objective analysis of voice characteristics embedding individualized hygiene voice advice.

Although, the current mobile system is not yet tested for the hygiene advisory part it is an important step for future research as literature underlines its need [19, 20, 41–50]. Traditional vocal education programs are assumed as valuable in preventing voice disorders among teachers but with conjunction with routinely voice screening [47, 49] and therefore a mobile system with individualized vocal hygiene feedback has promising potentials.

4 Conclusion

The study presents the verification of a mobile voice assessment system pointing towards results with high statistical confidence. When the scores of teachers without voice disorders using the mobile system were compared to teachers using the traditional paper version (i) without voice disorders, results are in line, and (ii) with voice disorders, scores deviate significantly as expected.

Acknowledgements. Our thanks are due to the teachers of the primary schools from the regions of Ioannina and Fthiotida in Greece, who contributed to data collection.

References

1. Chong, E.Y., Chan, A.H.: Subjective health complaints of teachers from primary and secondary schools in Hong Kong. Int. J. Occup. Saf. Ergon. **16**, 23–39 (2010)
2. Coyle, S.M., Weinrich, B.D., Stemple, J.C.: Shifts in relative prevalence of laryngeal pathology in a treatment-seeking population. J. Voice **15**, 424–440 (2001)
3. Stemple, J.C., Glaze, L.E., Gerdeman, B.K.: Clinical Voice Pathology: Theory and Management, pp. 25–37. Cengage Learning, San Diego (2000)
4. Roy, N., Merrill, R.M., Gray, S.D., Smith, E.M.: Voice disorders in the general population: prevalence, risk factors, and occupational impact. Laryngoscope **115**, 1988–1995 (2005)
5. Russell, A., Oates, J., Greenwood, K.M.: Prevalence of voice problems in teachers. J. Voice **12**, 467–479 (1998)
6. Titze, I.R., Lemke, J., Montequin, D.: Populations in the US workforce who rely on voice as a primary tool of trade: a preliminary report. J. Voice **11**, 254–259 (1997)
7. Verdolini, K., Ramig, L.O.: Review: occupational risks for voice problems. Logoped. Phoniatr. Vocol. **26**, 37–46 (2001)
8. Warhurst, S., Madill, C., McCabe, P., Heard, R., Yiu, E.: The vocal clarity of female speech-language pathology students: an exploratory study. J. Voice **26**(1), 63–68 (2012)
9. De Jong, F.I.C.R.S., Kooijman, P.G.C., Thomas, G., Huinck, W.J., Graamans, K., Schutte, H.K.: Epidemiology of voice problems in Dutch teachers. Folia Phoniatr. Logop. **58**, 186–198 (2006)
10. Morrow, S.L., Connor, N.P.: Comparison of voice-use profiles between elementary classroom and music teachers. J. Voice **25**(3), 367–372 (2011)
11. Hackworth, R.S.: The effect of vocal hygiene and behavior modification instruction on the self-reported vocal health habits of public school music teachers. Int. J. Music Educ. **25**(1), 20–30 (2007)
12. Roy, N., Merrill, R.M., Thibeault, S., Parsa, R.A., Gray, S.D., Smith, E.M.: Prevalence of voice disorders in teachers and the general population. J. Speech Lang. Hear. Res. **47**, 281–293 (2004)
13. Van Houtte, E., Claeys, S., Wuyts, F., Van Lierde, K.: The impact of voice disorders among teachers: vocal complaints, treatment-seeking behavior, knowledge of vocal care, and voice-related absenteeism. J. Voice **25**(5), 570–575 (2011)

14. Van Houtte, E., Claeys, S., Wuyts, F., Van Lierde, K.: Voice disorders in teachers: occupational risk factors and psycho-emotional factors. Logop. Phoniatr. Vocol. **37**(3), 107–116 (2012)
15. Bahar, M.S., Kosak, T.S., Boltezar, I.H.: Voice problems among Slovenian physicians compared to the teachers: prevalence and risk factors. Slov. Med. J. **81**(9), 626–633 (2012)
16. Van Houtte, E., Claeys, S., Wuyts, F., Van Lierde, K.: The impact of voice disorders among teachers: vocal complaints, treatment-seeking behavior, knowledge of vocal care, and voice-related absenteeism. J. Voice **25**(5), 570–575 (2011)
17. Pasa, G., Oates, J., Dacakis, G.: The relative effectiveness of vocal hygiene training and vocal function exercises in preventing voice disorders in primary school teachers. Logoped. Phoniatr. Vocol. **32**, 128–140 (2007)
18. Gillivan-Murphy, P., Drinnan, M.J., O'Dwyer, T.P., Ridha, H., Carding, P.: The effectiveness of a voice treatment approach for teachers with self-reported voice problems. J. Voice **20**, 423–431 (2006)
19. Bovo, R., Galceran, M., Petruccelli, J., Hatzopoulos, S.: Vocal problems among teachers: evaluation of a preventive voice program. J. Voice **21**, 705–722 (2007)
20. Duffy, O.M., Hazlett, D.E.: The impact of preventive voice care programs for training teachers: a longitudinal study. J. Voice **18**(1), 63–70 (2004)
21. Mjaavatn, P.E.: Voice difficulties among teachers. Paper presented at the XVIII Congress of the International Association of Logopedics and Phoniatrics, Washington, DC (1980)
22. Dejonckere, P.H., Bradley, P., Clemente, P., Cornut, G., Crevier-Buchman, L., Friedrich, G., Van De Heyning, P., Remacle, M., Woisard, V.: A basic protocol for functional assessment of voice pathology, especially for investigating the efficacy of (phonosurgical) treatments and evaluating new assessment techniques. Eur. Arch. Oto-rhino-laryngol. **258**(2), 77–82 (2001)
23. Schutte, H.K., Seidner, W.: Recommendation by the Union of European Phoniatricians (UEP): standardizing voice area measurement/phonetography. Folia Phoniatr. Logop. **35**, 286–288 (1983). https://doi.org/10.1159/000265703
24. Oliveira, G., Fava, G., Baglione, M., Pimpinella, M.: Mobile digital recording: adequacy of the iRig and iOS device for acoustic and perceptual analysis of normal voice. J. Voice **31**(2), 236–242 (2017)
25. Maryn, Y., Ysenbaert, F., Zarowski, A., Vanspauwen, R.: Mobile communication devices, ambient noise, and acoustic voice measures. J. Voice **31**(2), 248-e11 (2017)
26. Tafiadis, D., Chronopoulos, S.K., Siafaka, V., Drosos, K., Kosma, E.I., Toki, E.I., Ziavra, N.: Comparison of voice handicap index scores between female students of speech therapy and other health professions. J. Voice (2017). https://doi.org/10.1016/j.jvoice.2017.01.013
27. Kushniruk, A.: Evaluation in the design of health information systems: application of approaches emerging from usability engineering. Comput. Biol. Med. **32**(3), 141–149 (2002)
28. American Speech Hearing Association. Voice Evaluation Template. http://www.asha.org/uploadedFiles/slp/healthcare/AATVoiceEvaluation.pdf. Accessed 14 Feb 2015
29. Helidoni, M.E., Murry, T., Moschandreas, J., Lionis, C., Printza, A., Velegrakis, G.A.: Cross-cultural adaptation and validation of the voice handicap index into Greek. J. Voice **24** (2), 221–227 (2010)
30. Spantideas, N., Drosou, E., Bougea, A., Assimakopoulos, D.: Laryngopharyngeal reflux disease in the Greek general population, prevalence and risk factors. BMC Ear Nose Throat Disord. **15**(1), 7 (2015)
31. Spantideas, N., Drosou, E., Bougea, A., Assimakopoulos, D.: Gastroesophageal reflux disease symptoms in the Greek general population: prevalence and risk factors. Clin. Exp. Gastroenterol. **9**, 143 (2016)

32. Spantideas, N., Drosou, E., Karatsis, A., Assimakopoulos, D.: Voice disorders in the general Greek population and in patients with laryngopharyngeal reflux. Prevalence and risk factors. J. Voice **29**(3), 389-e27 (2015)
33. Belafsky, P.C., Postma, G.N., Koufman, J.A.: Validity and reliability of the reflux symptom index (RSI). J. Voice **16**(2), 274–277 (2002)
34. Printza, A., Kyrgidis, A., Oikonomidou, E., Triaridis, S.: Assessing laryngopharyngeal reflux symptoms with the Reflux Symptom Index: validation and prevalence in the Greek population. Otolaryngol.-Head Neck Surg. **145**(6), 974–980 (2011)
35. Lagergren, J., Bergström, R., Lindgren, A., Nyrén, O.: Symptomatic gastroesophageal reflux as a risk factor for esophageal adenocarcinoma. N. Engl. J. Med. **340**(11), 825–831 (1999)
36. Green, J.A., Amaro, R., Barkin, J.S.: Symptomatic gastroesophageal reflux as a risk factor for esophageal adenocarcinoma. Dig. Dis. Sci. **45**(12), 2367–2368 (2000)
37. Colton, R.H., Casper, J.K., Leonard, R.: Understanding Voice Problems. A Physiological Perspective for Diagnosis and Treatment, 4th edn. Lippincott Williams & Wilkins, Baltimore (2014)
38. Roy, N., Gray, S.D., Simon, M., Dove, H., Corbin-Lewis, K., Stemple, J.C.: An evaluation of the effects of two treatment approaches for teachers with voice disorders: a prospective randomized clinical trial. J. Speech Lang. Hear. Res. **44**(286), 96 (2001)
39. Chan, R.W.: Does the voice improve with vocal hygiene education? A study of some instrumental voice measures in a group of kindergarten teachers. J. Voice **8**(279), 91 (1994)
40. Bovo, R., Trevisi, P., Emanuelli, E., Martini, A.: Voice amplification for primary school teachers with voice disorders: a randomized clinical trial. Int. J. Occup. Med. Environ. Health **26**(3), 363–372 (2013)
41. Ilomäki, I., Laukkanen, A.M., Leppänen, K., Vilkman, E.: Effects of voice training and voice hygiene education on acoustic and perceptual speech parameters and self-reported vocal well-being in female teachers. Logop. Phoniatr. Vocol. **33**(2), 83–92 (2008)
42. Böhme, G.: Berufsstimmstorungen [Occupational Voice Disorders]. Munchener Medizinische Wochenschrift **116**, 1721–1726 (1974)
43. Calas, M., Verhuist, J., Lecoq, M., Dalleas, B., Seilhean, M.: Vocal pathology of teachers. Rev. Laryngol. **110**, 397–406 (1989)
44. Cooper, M.: Teacher, save that voice. Grade Teacher March, pp. 71–76 (1970)
45. Gundermann, V.H.: Phoniatrische Bemerkungen zur sogenannten Lehrerkrankheit. Das Deutsche Gesundheitswesen **18**, 69–72 (1963)
46. Labastida, L.: On the subject of 150 phoniatric surveys on primary school teachers. Acta oto-rino-laringol. Ibero-Amer. **12**, 200–203 (1961)
47. Lejska Sapir, S., Keidar, A., Van Velzen, D.: Vocalattritioninteachers: survey findings. Eur. J. Disord. Commun. **28**, 177–185 (1993)
48. Unger, E., Bastian, H.J.: Professional dysphonias. Deutsche Gesundheitswensen **36**, 1461–1464 (1981)
49. Anderson, V.A.: Speech needs and abilities of prospective teachers. Q. J. Speech **26**, 221–225 (1940)
50. Comins, R.: Student teachers recognise their voice needs. Hum. Commun. **August**, 18–19 (1993)

A Preliminary Study on a Mobile System for Voice Assessment and Vocal Hygiene Training in Military Personnel

Eugenia I. Toki[1(✉)], Vassiliki Siafaka[1], Dimitrios Moutselakis[2], Prodromos Ampatziadis[2], Dionysios Tafiadis[2], Georgios Tatsis[2], and Nausica Ziavra[1]

[1] Laboratory of Audiology, Neurootology and Neurosciences, Department of Speech and Language Therapy, School of Health and Welfare, Technological Educational Institute (TEI) of Epirus, Ioannina, Greece
{toki,vsiafaka,nziavra}@ioa.teiep.gr

[2] Department of Speech and Language Therapy, School of Health and Welfare, Technological Educational Institute (TEI) of Epirus, Ioannina, Greece
dimitrismoutselakis@yahoo.gr,
ampatziadispro@gmail.com,
{d.tafiadis,g.tatsis}@ioa.teiep.gr

Abstract. The aim of this study is to verify part of a mobile voice assessment system, during the development phase to determine whether the specified phase requirements are met. Preliminary results serving this goal are reported. Voice quality characteristics are presented tabulating their statistical means and standard deviations. The results show that the voice characteristics of the sample of the study were in normal range. Significant statistical differences were noted for Voice Handicap Index Total and its dimensions and Reflux Symptoms Index between (i) the military population without voice disorders using the mobile system and (ii) an aged matched voice professional population with voice disorders that were traditionally administrated. No significant statistical difference were observed between (i) the military population without voice disorders using the mobile system and (ii) the corresponding population without voice disorders using the above questionnaires traditionally. Results of the study are discussed together with the perspective of vocal hygiene routines indicating the accuracy and therefore the verification of the mobile system according to requirements and design specifications.

Keywords: Mobile system · Verification · Voice assessment
Vocal hygiene · Military personnel

1 Introduction

Individuals working in vocally demanding professions usually suffer from voice disorders [1]. Voice disorders can result from anatomic, physical, or functional abnormal changes in the voice mechanism [2] leading to vocal malfunction or inefficiency [1]. Previous studies have underlined a wide range of voice disorders occurrence [3]. It has

M. E. Auer and T. Tsiatsos (Eds.): IMCL 2017, AISC 725, pp. 700–708, 2018.
https://doi.org/10.1007/978-3-319-75175-7_69

been reported that voice disorders affect approximately 9% of the population [4] with women having a higher occurrence of voice disorder than men [5] and professional voice users being at higher risk for developing a voice disorder [6].

Vocal symptoms are indicators for developing voice disorders [6]. Parameters and mechanisms that lead to abnormal voicing and voice disorders are summarized by Titze et al. [6] and include factors such as smoking, reflux, vocal abuse and/or misuse. A population with voice disorders are professionals voice users such as individuals serving in the military. Military personnel, seem to have low awareness of the vocal demands [7, 8].

Clinically, Vocal hygiene (VH) was always considered a therapeutic tool for behavioral treatment of a voice patient [9, 10]. Vocal hygiene, as an indirect therapy, majorly includes many principles for engaging better vocal habits and the improvement of a person's vocal health. VH addresses both speech (i.e. adapting patients voice to an easy level of loudness and pitch) and non-speech (i.e. minimizing the incidences of daily throat cleaning, raising the hydration daily intake and changing habitual voice misuse- excessive yelling or talking over noise) negative influencing voice factors [9–11]. The basic goal of those programs is to promote communicative efficiency via better vocal well-being [12]. Often vocal hygiene programs are suggested as an isolated sufficient treatment method even the scientific evidence are not so adequate [10].

Bad vocal hygiene habits are still not well defined for general population and professional voice users [10]. Ruotsalainen et al. [13] conducted a Cochrane review and by analyζing 46 researches concluded that dysphonia is associated with decreased quality of life and a higher probability of work loss. They also, concluded that traditional training of populations at risk, such as military employees, lack of evidence to establish their efficacy. They also suggested that long term programs may lead to the opposite results. The above findings ascertain that well-designed studies and protocols (traditional and mobile-online) are necessary and critical for the prevention of developing vocal symptoms in professional voice users.

Current research reports that individuals working in military forces are at risk to develop a voice disorder even after their deployment [14]. This may be due to the high work load under rough and adverse conditions [7]. Additionally, another study by Gurevich-Uvena et al. [8] reported that paradoxical vocal fold motion is recurrent in the military population which was strongly associated with high levels of physical activity and stress. They also reported that this type of voice disorder is correlated with comorbid medical conditions that affect voice (asthma, allergies or GERD).

Voice evaluation is an important procedure which usually takes place with traditional [15, 16] or mobile system procedures [17, 18]. In traditional voice evaluation there are well established evidence-based models and multidimensional protocols [19]. According to Tafiadis et al. "*these protocols included laryngeal imaging, aerodynamic, perceptual-acoustic evaluation, and the impact of voice on the quality of life via self-perceived questionnaires*" [19]. In the case of mobile system procedures for voice evaluation there is little research evidence [17, 18]. Towards this innovative direction it is crucial to verify the accuracy of voice evaluation when using a mobile system. A powerful common method to evaluate the accuracy of fulfilling the specified phase requirements of a mobile system is the usability testing through the analysis of typical end users interacting with the system [20].

The aim of this study was to present preliminary results verifying part of a voice assessment mobile system when used by military personnel. The ultimate aim is to embed individual voice hygiene feedback.

2 Material and Methods

2.1 Participants

One hundred and one males serving at Greek Military Forces were recruited for this study. All participants did not experience any laryngeal or respiratory disorders in the last 2 weeks before enrollment. Also, they did not have any history of any addictions influencing voice (alcohol or drug abuse), consistent symptoms of gastroesophageal reflux disease or laryngopharyngeal reflux, and reported environmental factor that could deteriorate their voices' condition.

2.2 Data Collection

An under development mobile system was used for data collection. This system focuses on voice assessment procedures in order to give individualized immediate vocal hygiene feedback. Three questionnaires were included to gather the user digital voice profile. Precisely, all participants filled out the digital version on a mobile system of the Voice Evaluation Form (VEF) [21], the Hellenic Voice Handicap Index (VHI) [22] and the Hellenic Reflux Symptom Index (RSI) [23–29].

VEF is a consensus template (including over 70 items/questions) that was developed by the American Speech Hearing Association to copy the current and the former background of an examinee's voice status. VEF served as a voice history form for this study.

The Greek standardized version of VHI was also administrated. VHI consisted of 30 questions, which were summarized to a total score (VHI-T) split into three domains: emotional (VHI-E), physical (VHI-P), and functional (VHI-F). Each domain included 10 questions (30 items in total). VHI was included to this study as it has been used in different populations with voice and laryngeal disorders [22, 30–34]. Also, the Greek version of RSI was fulfilled by all participants.

The RSI [23] it's a self-administered nine-item questionnaire designed to assess various symptoms related to laryngopharyngeal reflux. It includes questions for the diagnosis of GERD [23–29]. Each item is rated on a Likert scale from 0 (no problem) to 5 (severe problem), with a total score of 45 for most severe symptoms. In addition, RSI can give information about the prevalence of symptoms but also about their severity. The research was approved by the Department of Speech and Language Therapy Research Ethic Committee. The data collected in military camps or the region of east Macedonia-Greece.

Finally each participant recorded a sustained voicing of /a/ and /e/ sounds in a room where background noise did not exceed 30 dB. Analysis was achieved by 5105 MDVP software, 2.3 edition and the signal was digitized at 50 kHz sample rate for a total of 202 voice recordings.

Acoustic analysis was conducted in order to determine with an objective method whether participant voice characteristics were or not within normal range [34] and whether they are in line with the output of the mobile system. Then, the output of the mobile system was compared with aged matched professional voice user results from the literature [22]. In detail Fig. 1 depicts the methodology for verifying the voice assessment mobile system.

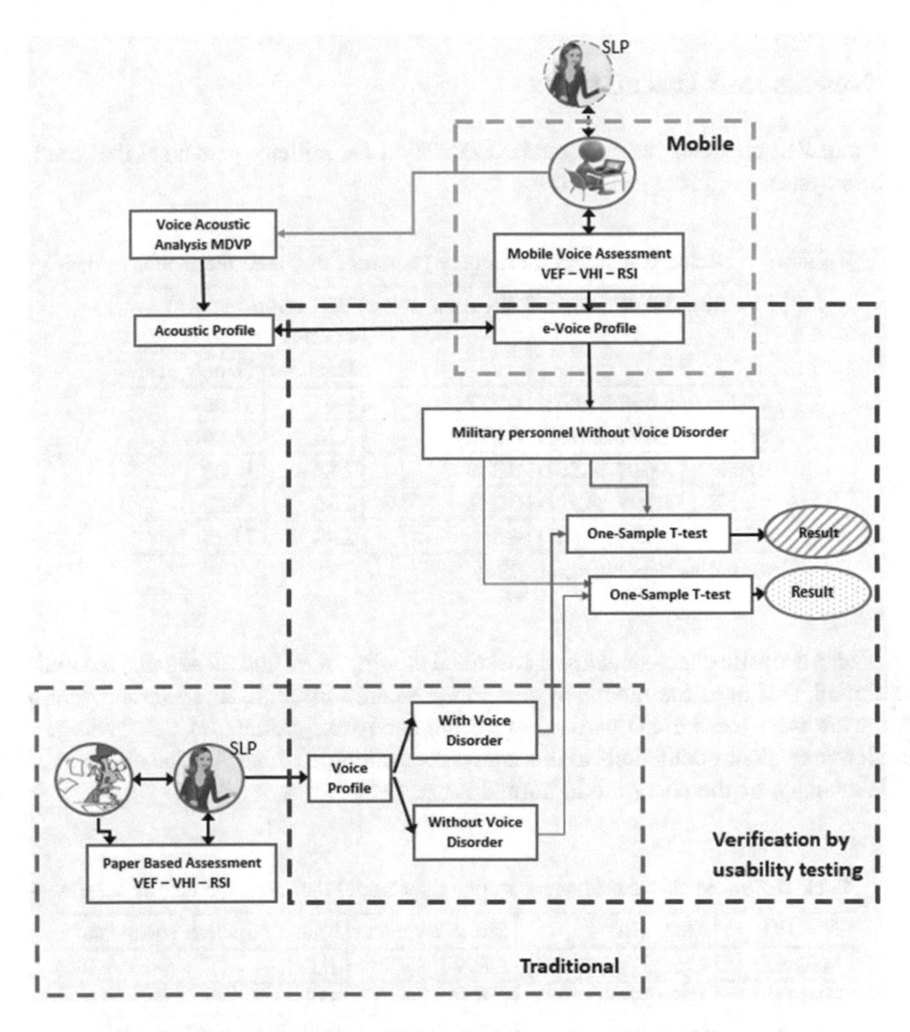

Fig. 1. Methodology for verifying the voice assessment mobile system.

2.3 Statistical Analysis

All normal distributed variables are expressed in mean (M) and standard deviations (SD) on the other hand s all skewed variables are expressed through a median calculation (interquartile range). The one sample t-test was used to calculate the differences between the international norms of voice characteristics (F0, Shimmer, and Jitter) from data of male serving to the military. Likewise, to calculate any differences between Greek normative data for VHI (Total, Functional, Physical and Emotional domains) and RSI scores from data of male serving to the military, the one sample t-test was used. All reported P values were two-tailed. The statistical significance was set at $P < 0.05$. All data were analyzed using SPSS statistical software (version 19.0, Armonk, NY, USA).

3 Results and Discussion

VHI and RSI mean scores and standard deviation for military personnel that used the mobile system are presented in Table 1.

Table 1. VHI and RSI scores for military personnel that used the mobile system

N = 101	Mean (SD)	SD. error mean	VHI norms[a] & RSI norms[b]	
			Controls	Pathological
VHI-T	6.66 (±7.801)	0.777	7.41	37.00
VHI-F	2.85 (±2.791)	0.278	3.21	10.00
VHI-P	2.61 (±3.507)	0.349	2.84	18.00
VHI-E	1.16 (±2.521)	0.251	1.36	9.00
RSI	2.49 (±3.681)	0.366	2.32	24.76

Normative data by: [a][23], [b][24]

Voice acoustic characteristics of sustained voicing of /a/ and /e/ sounds, for military personnel, that used the mobile system are presented in Table 2. The normative data that were used for the comparisons were obtained from researches that had age and gender equivalent populations to this study. According to Table 2 voice characteristics of the sample of the study are in normal range [34].

Table 2. Voice characteristics for military personnel that used the mobile system

N = 101	Mean (SD)	Std. error mean	Male normative voice data[a]
MF0 /a/	125.081 (±26.350)	2.622	118
HF0 /a/	125.169 (±26.357)	2.622	118
Jitter /a/	0.929 (±0.465)	0.046	0.87
Shimmer /a/	4.092 (±2.025)	0.202	4.700
MF0 /e/	132.300 (±29.855)	2.970	118
HF0 /e/	132.437 (±29.952)	2.980	118
Jitter /e/	1.092 (±0.707)	0.704	0.99
Shimmer /e/	2.820 (±1.018)	0.101	3.700

Normative data by: [a][34]

Using the one sample t-test a significant statistical difference at the $p < 0.05$ level was reported for VHI Total (p = .000), VHI Functional (p = .000), VHI Physical (p = .000) and VHI Emotional (p = .000) when military personnel using the mobile system without voice disorders was compared to a population with voice disorders using the traditional paper version [22].

Contradictory, when scores of military personnel without voice disorders using the mobile system were compared to a population without voice disorders using the traditional paper version there was no significant statistical difference at the $p < 0.05$ level for VHI Total (p = .339), VHI Functional (p = .200), VHI Physical (p = .518) and VHI Emotional (p = .424) [20].

A significant statistical difference at the $p < 0.05$ level for RSI (p = .000) was found when scores of military personnel without voice disorders using the mobile system where compared to a population with voice disorders using the traditional paper version [23–25]. No significant statistical difference at the $p < 0.05$ level was reported when scores of military personnel without voice disorders using the mobile system were compared to a population without voice disorders using the traditional paper version [23–25].

Individuals working in the Greek Military Forces even if they didn't have significant difference compared to testing value of VHI total and its domains they had higher scores than other population. These higher scores can be related to the demanding military training that may have an effect on vocal effort and an impact on intensity parameters [7]. Especially for VHI-Physical and VHI-Functional higher scores maybe be also linked to similar finding reported in the literature such to body fluid reduction [35–40] leading to dehydration of the vocal fold mucosa which is triggering higher effort for voice production [7, 40]. Additionally, they may be also linked to the requirement of using high pitched and louder voice to give commands and execute any type of function in accordance to current research [7, 41]. The above findings reinforce the importance of focusing intensively on vocal care for people serving in the military and support them with customized mobile software that promotes vocal health likewise recent literature [17, 18].

The results of this study indicate the accuracy of verification of this mobile system and therefore the voice assessment mobile system fulfils the specification requirements. Future research may focus on different sample of voice populations (other professions, age groups, vocal habits) may embed and investigate in depth features such as hygiene voice feedback on mobile devices. Furthermore it may be extended towards objective analysis of voice characteristics and individualized hygiene voice advice according to user needs and upcoming trends.

4 Conclusion

The study presents the verification of part of developing mobile system on voice assessment pointing towards results with high statistical confidence. When the scores of military personnel without voice disorders using the mobile system were compared to a population using the traditional paper version (i) without voice disorders, results are in line, and (ii) with voice disorders, scores deviate significantly, as expected.

Acknowledgements. We would like to thank the Commander of 4th Army Corps Lieutenant General Georgios Kampas as well as the commander and the staff of XXV Armored Brigade for their valuable help during this research.

References

1. Stemple, J.C., Glaze, L.E., Gerdeman, B.K.: Clinical Voice Pathology: Theory and Management, pp. 25–37. Cengage Learning, San Diego (2000)
2. Coyle, S.M., Weinrich, B.D., Stemple, J.C.: Shifts in relative prevalence of laryngeal pathology in a treatment-seeking population. J. Voice **15**, 424–440 (2001)
3. Roy, N., Merrill, R.M., Gray, S.D., et al.: Voice disorders in the general population: prevalence, risk factors, and occupational impact. Laryngoscope **115**, 1988–1995 (2005)
4. Russell, A., Oates, J., Greenwood, K.M.: Prevalence of voice problems in teachers. J. Voice **12**, 467–479 (1998)
5. Verdolini, K., Ramig, L.O.: Review: occupational risks for voice problems. Logoped. Phoniatr. Vocol. **26**, 37–46 (2001)
6. Titze, I.R., Lemke, J., Montequin, D.: Populations in the US workforce who rely on voice as a primary tool of trade: a preliminary report. J. Voice **11**, 254–259 (1997)
7. Nascimento, C.L., Constantini, A.C., Mourão, L.F.: Vocal effects in military students submitted to an intense recruit training: a pilot study. J. Voice **30**(1), 61–69 (2016)
8. Gurevich-Uvena, J., Parker, J.M., Fitzpatrick, T.M., Makashay, M.J., Perello, M.M., Blair, E.A., Solomon, N.P.: Medical comorbidities for paradoxical vocal fold motion (vocal cord dysfunction) in the military population. J. Voice **24**(6), 728–731 (2010)
9. Thomas, L.B., Stemple, J.: Voice therapy: does science support the art? Commun. Disord. Rev. **1**. 49–77 (2007)
10. Behlau, M., Oliveira, G.: Vocal hygiene for the voice professional. Curr. Opin. Otolaryngol. Head Neck Surg. **17**(3), 149–154 (2009)
11. Carding, P., Horsley, I., Docherty, G.: A study of the effectiveness of voice therapy in the treatment of 45 patients with nonorganic dysphonia. J. Voice **13**, 72–104 (1999)
12. Ilomaki, I., Laukkanen, A.-M., Leppanen, K., Vilkman, E.: Effects of voice training and voice hygiene education on acoustic and perceptual speech parameters and self-reported vocal well-being in female teachers. Logoped. Phoniatr. Vocol. **33**, 83–92 (2008)
13. Ruotsalainen, J.H., Sellman, J., Lehto, L., et al.: Interventions for preventing voice disorders in adults. Cochrane Database Syst. Rev. **17**, CD006372 (2007)
14. Dion, G.R., Miller, C.L., Ramos, R.G., O'Connor, P.D., Howard, N.S.: Characterization of voice disorders in deployed and nondeployed US army soldiers. J. Voice **27**, 57–60 (2013)
15. Dejonckere, P.H., Bradley, P., Clemente, P., Cornut, G., Crevier-Buchman, L., Friedrich, G., Woisard, V.: A basic protocol for functional assessment of voice pathology, especially for investigating the efficacy of (phonosurgical) treatments and evaluating new assessment techniques. Eur. Arch. Oto-Rhino-Laryngology **258**(2), 77–82 (2001)
16. Schutte, H.K., Seidner, W.: Recommendation by the Union of European Phoniatricians (UEP): standardizing voice area measurement/phonetography. Folia Phoniatr. Logop. **35**, 286–288 (1983). https://doi.org/10.1159/000265703
17. Oliveira, G., Fava, G., Baglione, M., Pimpinella, M.: Mobile digital recording: adequacy of the iRig and iOS device for acoustic and perceptual analysis of normal voice. J. Voice **31**(2), 236–242 (2017)
18. Maryn, Y., Ysenbaert, F., Zarowski, A., Vanspauwen, R.: Mobile communication devices, ambient noise, and acoustic voice measures. J. Voice **31**(2), 248-e11 (2017)

19. Tafiadis, D., Chronopoulos, S.K., Siafaka, V., Drosos, K., Kosma, E.I., Toki, E.I., Ziavra, N.: Comparison of voice handicap index scores between female students of speech therapy and other health professions. J. Voice **31**(5), 583–588 (2017). https://doi.org/10.1016/j.jvoice.2017.01.013
20. Kushniruk, A.: Evaluation in the design of health information systems: application of approaches emerging from usability engineering. Comput. Biol. Med. **32**(3), 141–149 (2002)
21. American Speech Hearing Association: Voice Evaluation Template. http://www.asha.org/uploadedFiles/slp/healthcare/AATVoiceEvaluation.pdf. Accessed 14 Feb 2015
22. Helidoni, M.E., Murry, T., Moschandreas, J., Lionis, C., Printza, A., Velegrakis, G.A.: Cross-cultural adaptation and validation of the voice handicap index into Greek. J. Voice **24** (2), 221–227 (2010)
23. Spantideas, N., Drosou, E., Bougea, A., Assimakopoulos, D.: Laryngopharyngeal reflux disease in the Greek general population, prevalence and risk factors. BMC Ear Nose Throat Disord. **15**(1), 7 (2015)
24. Spantideas, N., Drosou, E., Bougea, A., Assimakopoulos, D.: Gastroesophageal reflux disease symptoms in the Greek general population: prevalence and risk factors. Clin. Exp. Gastroenterol. **9**, 143 (2016)
25. Spantideas, N., Drosou, E., Karatsis, A., Assimakopoulos, D.: Voice disorders in the general greek population and in patients with laryngopharyngeal reflux. Prevalence and risk factors. J. Voice **29**(3), 389-e27 (2015)
26. Belafsky, P.C., Postma, G.N., Koufman, J.A.: Validity and reliability of the reflux symptom index (RSI). J. Voice **16**(2), 274–277 (2002)
27. Printza, A., Kyrgidis, A., Oikonomidou, E., Triaridis, S.: Assessing laryngopharyngeal reflux symptoms with the reflux symptom index: validation and prevalence in the Greek population. Otolaryngol.–Head Neck Surg. **145**(6), 974–980 (2011)
28. Lagergren, J., Bergström, R., Lindgren, A., Nyrén, O.: Symptomatic gastroesophageal reflux as a risk factor for esophageal adenocarcinoma. N. Engl. J. Med. **340**(11), 825–831 (1999)
29. Green, J.A., Amaro, R., Barkin, J.S.: Symptomatic gastroesophageal reflux as a risk factor for esophageal adenocarcinoma. Dig. Dis. Sci. **45**(12), 2367–2368 (2000)
30. Sanuki, T., Yumoto, E., Kodama, N., Minoda, R., Kumai, Y.: Long-term voice handicap index after type II thyroplasty using titanium bridges for adductor spasmodic dysphonia. Auris Nasus Larynx **41**(3), 285–289 (2014). https://doi.org/10.1016/j.anl.2013.11.001
31. Gillespie, A.I., Gooding, W., Rosen, C., Gartner-Schmidt, J.: Correlation of VHI-10 to voice laboratory measurements across five common voice disorders. J. Voice **28**(4), 440–448 (2014). https://doi.org/10.1016/j.jvoice.2013.10.023
32. Tafiadis, D., Kosma, E.I., Chronopoulos, S.K., Papadopoulos, A., Drosos, K., Siafaka, V., Toki, E.I., Ziavra, N.: Voice handicap index and interpretation of the cutoff points using receiver operating characteristic curve as screening for young adult female smokers. J. Voice (2017). https://doi.org/10.1016/j.jvoice.2017.03.009
33. da Rocha, L.M., de Lima, B.S., do Amaral, P.L., Behlau, M., de Mattos Souza, L.D.: Risk factors for the incidence of perceived voice disorders in elementary and middle school teachers. J. Voice **31**(2), 258-e7 (2017)
34. Colton, R.H., Casper, J.K., Leonard, R.: Understanding Voice Problems. A Physiological Perspective for Diagnosis and Treatment, 4th edn. Lippincott Williams & Wilkins, Baltimore (2014)
35. Witt, R.E., Taylor, L.N., Regner, M.F., Jiang, J.J.: Effects of surface dehydration on mucosal wave amplitude and frequency in excised canine larynges. Otolaryngol.-Head Neck Surg. **144**(1), 108–113 (2011)

36. McGlinchey, E.L., Talbot, L.S., Chang, K.H., Kaplan, K.A., Dahl, R.E., Harvey, A.G.: The effect of sleep deprivation on vocal expression of emotion in adolescents and adults. Sleep **34**(9), 1233–1241 (2011)
37. Franca, M.C., Simpson, K.O.: Effects of hydration on voice acoustics. Contemp. Issues Commun. Sci. Disord. **36**, 142–148 (2009)
38. Sivasankar, M., Leydon, C.: The role of hydration in vocal fold physiology. Curr. Opin. Otolaryngol. Head Neck Surg. **18**(3), 171 (2010)
39. Verdolini, K., Titze, I.R., Fennell, A.: Dependence of phonatory effort on hydration level. J. Speech Lang. Hear. Res. **37**(5), 1001–1007 (1994)
40. Verdolini, K., Rosen, C.A., Branski, R.C. (eds.): Classification Manual for Voice Disorders-I. Special Interest Division 3, Voice and Voice Disorders. ASHA, Rockville (2006)
41. Goy, H., Fernandes, D.N., Pichora-Fuller, M.K., van Lieshout, P.: Normative voice data for younger and older adults. J. Voice **27**, 545–555 (2013)

Social Networking for Obesity Prevention and Eradication for Algerian Teenagers

Abdelkarim Benatmane, Ramzeddine Bouchachi, Samir Akhrouf, and Yahia Belayadi(✉)

University of Bordj Bou Arreridj, El Anasser, Algeria
ak.benatmane@gmail.com, ramzi_mi@hotmail.com, samir.akrouf@gmail.com, belayadi_yahia@yahoo.fr

Abstract. Obesity can be classified as one of the most dangerous diseases in the world, as the number of obese people keeps on increasing. Obesity is related to many chronic diseases such as cancer, type 2 diabetes and heart disease, also to many factors such as eating habits and inactivity. This last can be caused by people's addiction to Internet and especially social networks. This research introduces a new approach in the daily fight against obesity, by proposing a system that supports a community of overweight people, obese people, their families and friends, physicians, and anyone who might need help and support from this community or who can offer help to members of the community. In addition to the new system proposed a specific purpose social network to support this community was developed in order to help them discuss their problems, express their specific needs, share their experiences and motivate other members by sharing their experience with obesity… etc.

Keywords: Obesity · Social networks · Community · eHealth

1 Introduction

The Internet in general and social networks in particular have been recognized as some of the main contributing factors in people's inactivity, which is the principal factor of obesity. On one hand, users of such technologies spend most of their out-of-work time navigating the Internet and communicating through various social networks. Moreover, people with overweight tend to e-socialize with people with close or similar profiles and or health issues. On the other hand, the same technologies, namely Internet and social networks, can be used as a center-stone medium in obesity prevention and treatment through appropriate awareness, prevention, positive energy, and community support. In this research, the aim is to positively use social networks in the daily fight against obesity. We aim to design and develop a specific-purpose social network to support a community of overweight people, obese people, their families and friends, physicians, and anyone who might need help and support from this community or who can offer help to members of the community.

M. E. Auer and T. Tsiatsos (Eds.): IMCL 2017, AISC 725, pp. 709–719, 2018.
https://doi.org/10.1007/978-3-319-75175-7_70

2 Obesity

Obesity is a state of an individual in where he/she has too much body fat. The World Health Organization (WHO) defines obesity and overweight "as an abnormal or excessive fat accumulation that may impair health" [1]. Obesity is significantly growing as earn epidemic, as the number of the obese people keeps on increasing in the whole world. According to WHO, worldwide obesity has doubled since 1980; in 2014, 600 million adults werc obese among 1.9 billion overweight [1]. Children are not doing better either; globally, 42 million preschool children were overweight in 2013 [2]. Moreover, overweight children are more likely to become obese adults. Researches has shown that consequences related to obesity are serious and such as: type 2 diabetes, cancer, heart disease, high blood pressure [3]. In addition, a study that has been conducted recently shows that increasing population fatness will have important implication for global resource requirements including food intake [4]. Energy intake and expenditure are significantly associated with the obesity epidemic, thus to overcome obesity, individuals who have serious obesity problem must monitor their dietary food and physical activities [5]. WHO considers obesity as a global epidemiology [6], and emphasizes the need of fighting obesity by any means necessary. Hence, the public health is very disconcerting and huge efforts from different populations in multidisciplinary approaches are needed in order to help in obesity prevention [2].

3 Social Networks

Our consideration of social networks is driven by the recent statistics showing that the number of social network users worldwide has been growing over the last five years. In fact, more than 31% of the world population was active on social networks on January 2016 on various platforms as depicted by Fig. 1. Moreover, Fig. 2 illustrates that, in 2018, number of social network users will increase to 2.55 billion, compared to 1.87 billion in 2014.

Fig. 1. Global digital snapshot January 2016 [7]

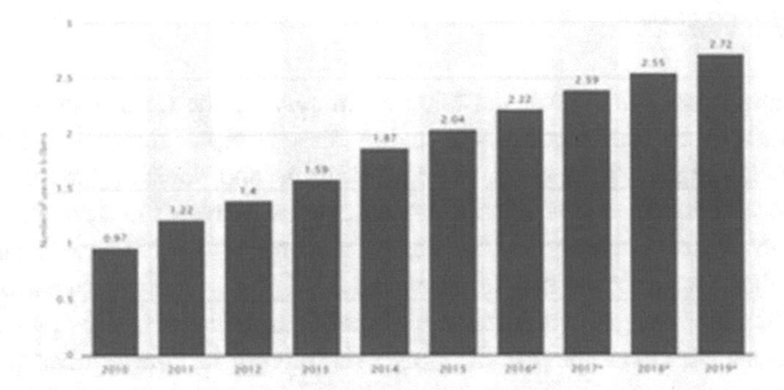

Fig. 2. Number of social network users 2010–2019 (globally) [8] (*: Internet users who use a social network site via any device at least once per month.)

4 Obesity and Social Networks

People are interconnected and so their health is interconnected, Smith and Christakis in their research [9] conclude that the existence of social networks means that people's health is interdependent. Moreover, a study found that overweight adolescents were more likely to have overweight friends than normal-weight friends [10]. Other study found that school friendship are significantly similar in terms of eating behaviors, and bodyweight [11]. Thus, Ashrafian et al. [12] performed a systematic review and meta-analysis to assess the role of social networking in modifying BMI, the study showed that interventions using social networking services produced significant reduction in BMI from baseline for the people who participated in the studies' interventions.

5 Social Networks in Healthcare

The use of social networks in obesity prevention is very limited; however, there have been more researches in where social networks plays a good role helping prevent other healthcare problems such as Diabetes. Thus, a study conducted in [13] showed that social interaction by regular meetings and online interaction can influence self-monitoring outcomes positively. Another systematic study [14] was conducted to evaluate literature published from 2006 to April 2013 regarding social support in adults with diabetes conducted in the USA and Europe, it showed that social support can result in a positive influence on both the ability of the patient to initiate and sustain diabetes management that can potentially result in positive health outcomes. Another study showed that Left Ventricular Assist Device (LVAD) patients who use social media to exchange information or advices or related things benefits in participating in these groups [15].

6 Related Works

Recently, many researchers have participated in solving the obesity epidemic that has spread nowadays. In this section, we review the previous researches that used the following technologies, namely mobile applications and social networks, for weight loss management to overcome obesity. A study has proven the need of engineering approaches in obesity prevention [16]. Hence, researchers in [17] have reviewed many reported studies that concern the involvement of ICT in obesity prevention, and also online weight loss management systems and mobile applications. This paper shows the effectiveness and the usefulness of ICT, as it is playing its role in obesity prevention and healthcare in general. Different methods have been proposed in order to study and overcome obesity, one such method is a GPS-based mobile games. Therefore, a team has developed a collaborative mobile serious game named STARSRACE [18], in which a two players (obese patients) have to collect, starting from locations pre-defined on the map and gain more points. Those starts points are distributed by the therapist depending on the players' profiles. The game was tested and evaluated using an evaluation model by Nokia Research Center [19]. Different obese patients with different ages participated in this test, they showed good impressions about the game attraction, user interfaces and its idea of combating obesity. A related research [20]; a mobile application was designed to encourage making better dietary decision through just-in-time motivation at the point of purchase, by helping users to compare between foods at a glance. Another mobile application was developed in [21] named PmEB, a classic mobile application for self-monitoring real time caloric balance, where a user have to enter his food intake and any physical activities. The collected data will be sent to a central server every 24 h to be analyzed, then suggestions and recommendations will be sent to the user based on values he/she has entered. A different method was used in [22], a team developed a web-based system for obesity prevention. After the user logs in and records his/her health and life style information including his/her body weight, food intake, alcohol consumed physical exercises. First the system detects 10 user's habits which are considered as risk factors of obesity, secondly detects individual factors of weight gain form these habits by calculating the correlation coefficients between the 10 habits and fat mass, and finally presents this information to the user to change his/her habits. Since calorie intake is an important factor to achieve best results, the team developed and compared three different dietary information logging methods: a hierarchical list method, a simple list method, and a photo based method. After the experiments were conducted, it turned out that the simple list method is enough to detect calorie intake. In [23], the authors designed a self-monitoring system based on Hadoop. In this system, the user creates his/her own profile and stores his/her personal information (age, sex, height and weight). This system computes energy gap based on the following formula:

$$\text{Energy Gap} = \text{Energy Intake} - \text{Energy Expenditure} \tag{1}$$

In order to calculate the energy gap, the user must enter his/her physical record details (jogging, walking, physical exercises), and the food item name and the amount consumed, to assess the energy expenditure and energy intake respectively. The system then will generate the appropriate diet plan that will suite the user, and possibly recommend physical exercises. To increase the availability of this system, it was designed to be used on mobile phones, or browser. Although, the system was designed to be used by individuals to monitor their personal obesity level, it can be used in schools where parents (school authorities) can monitor the obesity level of their children (students), or in hospitals where nurses can monitor the level of obesity of their obese patients following recommendations and suggestions received from the doctors.

7 The Proposed System

In this section, we will discuss in details our proposed solution in the daily fight against obesity. Our system embraces social networks and combines offline and online activities to achieve our goal. The goal of this research is to raise people awareness of the problem and the possible solutions, by building a community of specific persons who are linked by obesity.

7.1 Scenario

Our system will provide possible facilities to members of this community to help them reduce their weight. The system will be available for concerned people who are interested. However, normal people also can join and participate in this community and get advices from other members (physician, gym coaches, dieticians, obese, overweight, normal people…) to keep on shape and avoid obesity and its related medical, social, and economical side effects. When a member creates his/her account and logs into the community, he/she must provide correct personal and health information. Whenever his/her information is correct it increases the ease of helping him/her by other members of the community. After that, the user can search and find other people that are mentioned previously to connect with them. During the period of searching, which is one of important phases to get to know better people who can offer a helping hand, our system shall give each expert's (physicians, gym coaches, dieticians…) profile a rank for the best who can give effective advices. Non-experts people also can give good advices, but their advices must be approved by experts before sharing it in the community. When sharing information, the system should also guarantee the privacy of each member of the community, e.g. detailed information that physicians have access to should not be exposed to other members.

Once (X) is a member in this community, he starts to form a network within it. This network should be a group of experts (including physicians (P), gym coaches (G), dieticians (D)…), also regular members (including obese (Y), overweight (Z), normal people (N)) as depicted by Fig. 3. People in this network should exchange health data, medical advices, physical activities programs, experiences with obesity…etc. as illustrated in Fig. 4.

An important thing, that this community could provide, is when one of the overweight or obese members succeeds in losing weight and getting a better shape. At this moment, he/she can help and motivate other members by sharing his/her experience

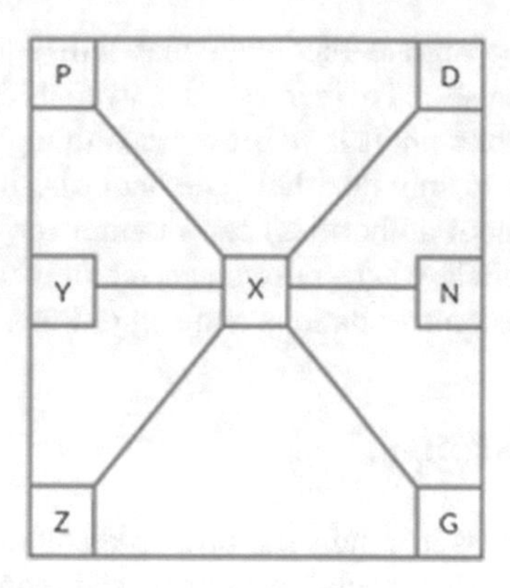

Fig. 3. Sample of a small network within the community

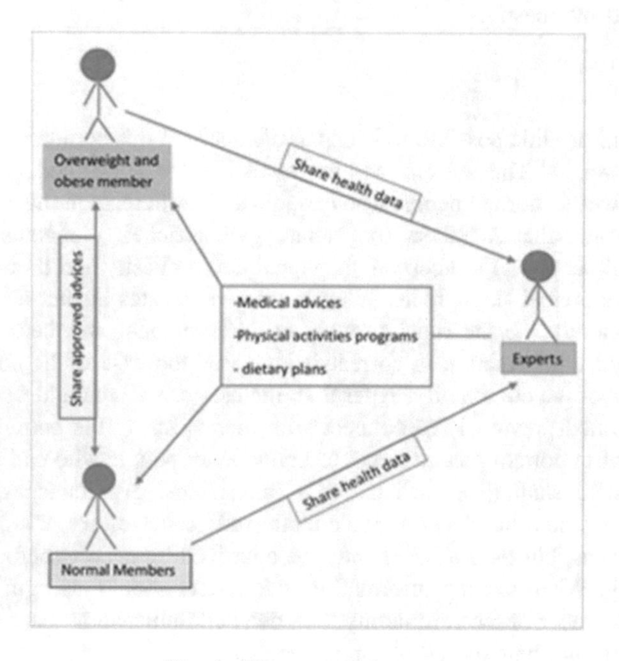

Fig. 4. The proposed system

with obesity, e.g. suggesting to them the best doctor, giving them effective programs approved by experts… etc.

7.2 System Pillars

This community is built around the same pillar principles mentioned [24]. These principles are: awareness dissemination, early prevention, community support, inspiring through positivity, and appropriate share of information.

7.3 System Implementation

For the development of the social network, we developed it in house. To that extent, a web site was developed and hosted on an online server under domain name "www.helpweightonline.com" as shown in Fig. 5. It describes the register/login page. While mobile applications for all major platforms will be developed later.

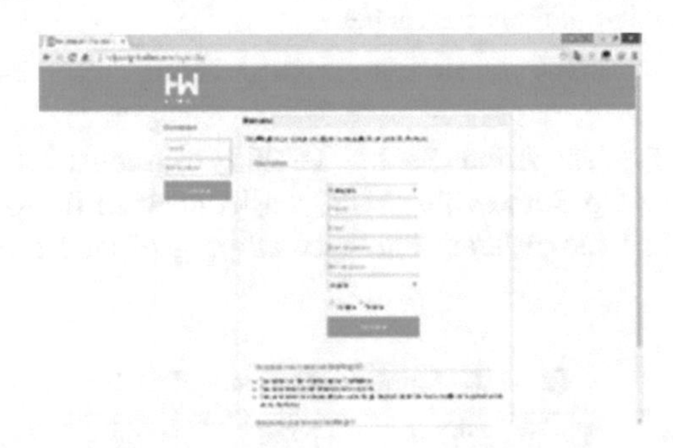

Fig. 5. Help weight register/login page

8 Results

Simulation: Similar studies have been conducted and proven that building a community of obesity expertise and healthcare professionals shall play good role in preventing obesity. Economos et al. in [25] did a study in Massachusetts, USA to test the hypothesis that a community-based environmental change intervention could prevent weight gain in young children (7.6 ± 1.0 years). According to the authors, a number of 1178 children in grades 1 to 3 attending public elementary school participated (see Fig. 6) along with 30 elementary schools. This Community Based Participatory Research study was conducted in three communities (cities) as illustrated in Table 1,

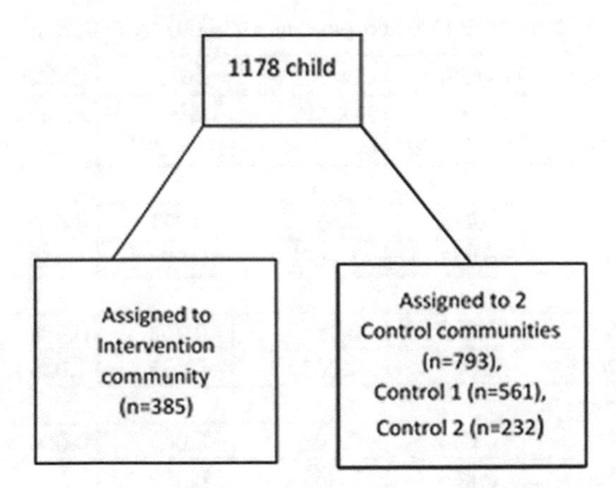

Fig. 6. Number of participated children after filtering

Table 1. Details on the three communities

Community	Intervention	Control 1	Control 2
City	Somerville (Located outside of Boston)	Unknown (Located outside of Boston)	Unknown (Located outside of Boston)
Participating schools	10	15	5
Demographic characteristics	Similar in three communities		

and lasted for over a 3-years period (September 2002 to August 2005). The intervention program called Shape Up Somerville (SUS) was conducted in Somerville city where scholars were keen that the program influence all parts of an elementary school's day (see Table 2).

Table 2. Some components of SUS

Environments	Before school	During school	After school	Home	Community
Some of the activities	Breakfast program; Walking to school bus	Professional development (nutrition and physical activity) for all school staff; School food service	Increase physical activity; Cooking lessons	Parents outreached and education; Parent nutrition forums	Walking/pedestrian trainings; Physical activity guide

A year data (September/October 2003 to May/June 2004) from this study indicated that BMI *z*-score of population at the intervention community has significantly decreased when compared to the other communities (see Table 3).

Table 3. BMI z-score (pre and post intervention) by community and sex

	Intervention (N = 385)		Control 1 (N = 561)		Control (N = 232)	
	Female (n = *190)*	*Male* (n = *195)*	*Female* (n = *298)*	*Male* (n = *263)*	*Female* (n = *117)*	*Male* (n = *115)*
Pre BMI z-score						
Mean years	0.782	0.918	0.617	0.777	0.679	1.132
SD	1.100	1.021	1.060	0.999	1.055	0.903
Post BMI z-score						
Mean years	0.755	0.882	0.615	0.768	0.688	1.113
SD	1.070	1.022	1.065	0.995	1.055	0.926
(Post – Pre) BMI z-score						
Mean years	−0.027	−0.036	−0.002	−0.009	0.009	−0.018
SD	0.356	0.284	0.265	0.289	0.294	0.253

Pre, September/October 2003; Post, May/June 2004; SD, standard deviation.

9 Discussion

This study demonstrates that community-based interventions can play an effective role in the fight against overweight and obesity.

For the purpose of this research, we will compare between the two studies as it follows:

9.1 Community Groups

The intervention in [25] included many groups and individuals (children, parents, teachers, school food service providers, city departments, policy makers, healthcare providers, before- and after-school programs, restaurants, and the media). Moreover, the main goal, which is preventing weight gain in young children, was successfully achieved.

As it was mentioned in previous sections, the goal of this research is to raise the people awareness of the obesity problem and the possible solutions, by building a community of specific persons (including physicians, dieticians, gym coaches, obese people, overweight people, their families and friends).

9.2 Intervention Program

Using a community participatory process, in [25] their intervention was developed to influence almost every part of an early elementary schoolchild's day.

In our case study we do not plan for a "specific intervention program", everyone within the community can participate and share information based on:

- Medical advices and perceptions have to be approved by physicians before they are shared and suggested to other members.
- Advices about physical activities have to be approved by Gym coaches.
- Eating programs have to be approved by dieticians.
- People who have experience with obesity can share their experience but only when an expert approves it, so other members will benefit.

10 Conclusion

The obesity epidemic represents a serious health problem, hence, huge efforts in multidisciplinary approaches are needed to contribute in order to fight and eliminate this disease. Despite the relation between obesity and social networks, the use of this technology in obesity prevention is very little. Thus, more improvement could help in achieving better results. In this research we presented and discussed a study in where social networking can serve in creating culture of health awareness amongst obese people and their community.

References

1. World Health Organization: Media Centre: obesity and overweight, Fact sheet N°311. http://www.who.int/mediacentre/factsheets/fs311/en/. Accessed 10 Jan 2016
2. World Health Organization: 10 facts on obesity. http://www.who.int/features/factfiles/obesity/en/. Accessed 10 Jan 2016
3. Bray, G.A.: Medical consequences of obesity. J. Clin. Endocrinol. Metab. **89**(6), 2583–2589 (2004)
4. Walpole, S.C., Prieto-Merino, D., Edwards, P., Cleland, J., Stevens, G., Roberts, I.: The weight of nations: an estimation of adult human biomass. BMC Public Health **12**(1), 439 (2012)
5. Sazonov, E.S., Schuckers, S.: Monitoring energy intake and energy expenditure in humans ©, pp. 31–35, February 2010
6. World Health Organization (WHO): Controlling the global obesity epidemic (2003). http://www.who.int/nutrition/topics/obesity/en/. Accessed 18 Mar 2016
7. Digital in 2016. http://wearesocial.com/uk/special-reports/digital-in-2016. Accessed 02 May 2016
8. Number of social network users worldwide from 2010 to 2019. http://www.statista.com/statistics/278414/number-of-worldwide-social-network-users/. Accessed 03 May 2016
9. Smith, K.P., Christakis, N.A.: Social networks and health. Annu. Rev. Sociol. **34**(1), 405–429 (2008)
10. Valente, T.W., Fujimoto, K., Chou, C.P., Spruijt-Metz, D.: Adolescent affiliations and adiposity: a social network analysis of friendships and obesity. J. Adolesc. Health **45**(2), 202–204 (2009)
11. Fletcher, A., Bonell, C., Sorhaindo, A.: You are what your friends eat: systematic review of social network analyses of young people's eating behaviours and bodyweight. J. Epidemiol. Community Health **65**(6), 548–555 (2011)
12. Ashrafian, H., Toma, T., Harling, L., Kerr, K., Athanasiou, T., Darzi, A.: Social networking strategies that aim to reduce obesity have achieved significant although modest results. Health Aff. **33**(9), 1641–1647 (2014)
13. Chomutare, T., Tatara, N., Arsand, E., Hartvigsen, G.: Designing a diabetes mobile application with social network support. Stud. Health Technol. Inform. **188**, 58–64 (2013)
14. Kirk, J.K., Ebert, C.N., Gamble, G.P., Ebert, C.E.: Social support strategies in adult patients with diabetes: a review of strategies in the USA and Europe. Expert Rev. Endocrinol. Metab. **8**(4), 379–389 (2013)
15. Boling, B., Hart, A., Okoli, C.T.C., Halcomb, T., El-Mallakh, P.: Use of social media as a virtual community and support group by left ventricular assist device (LVAD) patients. VAD J. **1**(1), 18 (2015)
16. Ershow, A.G., Hill, J.O., Baldwin, J.T.: Novel engineering approaches to obesity, overweight, and energy balance: public health needs and research opportunities. In: Conference of the IEEE Engineering in Medicine and Biology Society Proceedings, vol. 7, pp. 5212–5214 (2004)
17. Yusof, A.F., Iahad, N.A.: Review on online and mobile weight loss management system for overcoming obesity. In: 2012 International Conference on Computer and Information Science, ICCIS 2012 - A Conference of World Engineering, Science and Technology Congress, ESTCON 2012 - Conference Proceedings, vol. 1, pp. 198–203 (2012)
18. Al-Qurishi, M.S., Mostafa, M.A., Alrakhami, M.S., Alamri, A.M.: StarsRace: a mobile collaborative serious game for obesity. In: Proceedings of the IEEE International Conference on Multimedia and Expo (2014)

19. Korhonen, H., Koivisto, E.M.I.: Playability heuristics for mobile multi-player games. In: Proceedings of the 2nd International Conference on Digital Interactive Media in Entertainment and Arts, pp. 28–35 (2007)
20. Intille, S.S., Kukla, C., Farzanfar, R., Bakr, W.: Just-in-time technology to encourage incremental, dietary behavior change. In: AMIA (2003)
21. Lee, G., Tsai, C., Griswold, W.G., Raab, F., Patrick, K.: PmEB: a mobile phone application for monitoring caloric balance. In: CHI 2006 Extended Abstracts on Human Factors in Computing Systems, pp. 1013–1018 (2006)
22. Kato, Y., Suzuki, T., Kobayashi, K., Nakauchi, Y.: Comparison of dietary information logging methods for obesity prevention system. In: 2012 IEEE/SICE International Symposium on System Integration (SII), pp. 615–620 (2012)
23. Govindarajan, R., Madan, S., Shetty, A.: Hadoop based obesity monitoring system (2011)
24. Harous, S., Serhani, M.A., El Menshawy, M., Benharref, A.: Hybrid obesity monitoring model using sensors and community engagement. In: 2017 13th International Wireless Communications and Mobile Computing Conference (IWCMC), pp. 888–893 (2017)
25. Economos, C.D., et al.: A community intervention reduces BMI z-score in children: Shape Up Somerville first year results. Obesity (Silver Spring) **15**(5), 1325–1336 (2007)

A Multi-agent Framework for Medical Diagnosis Driven Smart Data in a Big Data Environment

Zakarya Elaggoune, Ramdane Maamri(✉), and Imane Boussebough

LIRE Laboratory, University of Constantine 2-Abdelhamid Mehri, Constantine, Algeria
{zakarya.elaggoune,ramdane.maamri}@univ-constantine2.dz, iboussebough@gmail.com

Abstract. In the era of big data, recent developments in the field of information and communication technologies are facilitating organizations to innovate and grow. These technological developments and wide adaptation of ubiquitous computing enable numerous opportunities for government and companies to discover useful trends or patterns that are used in health-care decision making. A common problem affecting data quality is the presence of noise and irrelevant information which can lead decision makers to a wrong decision. Intelligent Decision Support System (IDSS) an automated judgment that supports decision making is composed of human and computer interaction to help in decision-making accuracy. Also, multi-agent systems (MAS) are collections of independent intelligent entities that collaborate in the joint resolution of a complex problem. Multi-agent IDSS can be used to solve large-scale convention problem. In this paper, we introduce a multiagent-MapReduce framework based dimension reduction for medical diagnosis that can filter the noise and irrelevant information and keeps only smart data, which can lead to a reduced storage space in one hand and produce a better health-care decision in the other hand.

Keywords: Big data · MapReduce · Multi-agent · Smart data
Intelligent Decision Support Systems · Diagnosis · Health-care

1 Introduction

The current world is the one of data, commonly used applications such as social networks, forums, messaging systems, research articles, online transactions and corporate data produce heterogeneous data that is enormous in volume and generated exponentially. These data can be very effective in developing business strategies and planning effective decisions, but the era of big data and its characteristics put many challenges on the storage and analysis of these data that require intelligent mechanisms and tools to manage these data sets.

In the health-care context, volume, velocity, and variety are treated as the main primary characteristics of big data and all these properties are seriously

M. E. Auer and T. Tsiatsos (Eds.): IMCL 2017, AISC 725, pp. 720–727, 2018.
https://doi.org/10.1007/978-3-319-75175-7_71

considered in the literature [1,2]. Recently two other new characteristics of big data have been introduced - veracity and value - [3]. These last two characteristics are very important factors; in one side big data generates value and ensures the veracity only when applied to make better and faster decisions, on the other side the collection of irrelevant data increases noisy data that affects the operational cost of enterprises. Therefore, the collection of fine-grained, highly relevant, and reduced data from users is another challenge that requires serious attention while designing big data systems.

The aim of this paper is to propose a multiagent-MapReduce framework for medical diagnosis driven smart data (relevant data) that can help hospitals to achieve smart health-care. The framework can introduce some intelligent management systems to support the digital collection, processing, storage, transmission, and sharing of internal citizen information that can be used by an IDSS for intelligent management and supervision of public health diagnosis.

Multi-agent systems can manage complex and distributed computer scenarios. Moreover, the agent-based frameworks have not been taken into account seriously in the existing MapReduce frameworks. Therefore we need to combine the two concepts (MAS and MapReduce), in order to provide more accepted solutions in solving many real-world problems with big datasets.

The rest of this article is organized as follows. In Sect. 2, we review the existing literature. In Sect. 3, we review some preliminary knowledge related to big data and smart data. In Sect. 4 the multi-agent-MapReduce framework is presented, describing in detail the different steps of smart data extraction. Then we discuss the solution, after which we offer some new research directions. Lastly, we conclude our study in Sect. 6.

2 Related Work

In recent years, some research about Decision Support Systems (DSS) have been made. They can be classified into two categories as follow:

- *Conventional Business Intelligence (BI) systems.*
  These systems use the Extract, Transform and Load (ETL) process for data warehousing, they are used by companies that collect only structured data because it is very difficult to handle semi-structured and unstructured data with a data warehouse especially with the arrival of the era of big data [4,5].
- *DSS based MapReduce parallelism mechanism.*
  With the arrival of the big data, many works in the literature have proposed different DSS based on the use of the framework Hadoop and the MapReduce process [6–9], which can undoubtedly remedy the 3Vs - Velocity, Volume and Variety - challenges of big data, but these works concentrate only on these 3Vs and neglects the two others challenges - Value and Veracity, which lead to a higher cost because of the noisy data-sets stocked and the high computational complexity that can lead to a wrong decisions.

Most of The existing research in the literature try to solve only the challenges associated with the 3Vs - Velocity, Volume and Variety and ignore some others challenges.

The central research question in this work will be "How can we extract relevant to improve the quality of decision making in a Big Data environment?" This study intends is to find answers to the following questions:

- The redundant data stored in multiple disks and partitions is a serious challenge for big data processing systems, how to cover this challenge?
- How to adapt the decision-making process to support the extraction of relevant information?
- How to cover the all 5Vs challenges?

3 Preliminary Knowledge

3.1 The "5Vs" of Big Data

There are a set of basic characteristics that big data always show: 'volume' that refers to the large amount of data generated from various data sources in the world, 'variety' that refers to the data collected from multiple sources in different formats, and 'velocity' that refers to the speed at which the data is generated and processed. In addition to the above mentioned 3Vs, other important challenges, inherent in the term big data, are known as 'value' and 'veracity'. Value refers to insights that can be extracted from the data, and the latter challenge (veracity) refers to the uncertainty in the data that may be due to incompleteness, ambiguity and the reliability of the data source, this poses a big challenge of authenticity and trust on the data used.

3.2 Smart Data

The term smart data is utilized to denote the challenge of transforming raw data into data that can be processed later to obtain valuable information [10]. Smart data discovery involves filtering big data holding useful information, becoming a subset of data that is important for companies and researchers [11]. Obtaining a reduced/filtered amount of data may involve a large reduction in data storage costs and it influences decision makers to make the right decision.

Figure 1 illustrates an overview of filtering noisy data and extracting smart data (relevant information). The different steps of the process are listed in the following paragraphs:

Step 1 (Extract information from Data). "Although data and information are different, many people speak of them as if they are synonymous, which is almost never true" explains Michael Wu in his web article [11]. Data is simply a record of the events that took place. It is the raw data that describes what happend, when, where, how, who is involved, etc. Data does give us information, they are not the same. Information is only the non-redundant parts of the data [13], the

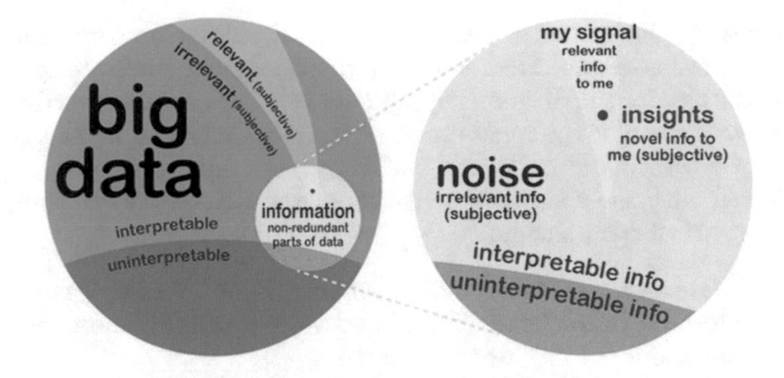

Fig. 1. Extracting smart data from big data [12]

maximum amount of extractable information can be measured through lossless compression algorithms (algorithms that filtering the data redundancy).

Step 2 (Extract relevant information). Information must satisfy some criteria to provide relevant information that are valuable. Authors in [14] define three key attributes for data to be smart, it must be accurate, actionable and agile:

- **Accurate:** Data must be what it says it is with enough precision to drive value. Data quality matters.
- **Actionable:** Data must drive an immediate scalable action in a way that maximizes a business objective like media reached across platforms. Scalable action matters.
- **Agile:** Data must be available in real-time and ready to adapt to the changing business environment. Flexibility matters.

3.3 Big Data Reduction Techniques

Big data reduction methods vary from pure size reduction techniques to compression-based data reduction methods and preprocessing algorithms, deduplication of data at the block level, redundancy elimination and the implementation of network theory concepts (graphic). Dimension reduction techniques are useful for managing the heterogeneity and severity of large data by reducing data from millions of variables to a manageable size [15,16]. These techniques usually work during the data collection phases.

In addition to the aforementioned advantages of data reduction techniques, data reduction is very useful in decision-making because filtering noise (irrelevant data) and keeping relevant data improves the quality of the decision.

4 The Proposed Framework

The proposed solution is intended to provide services for the monitoring and analysis of big data, it can be used for intelligent management and supervision of public health, the system collects data from the different data sources that provide information about citizens and information about the state of water, air and eat in the city for the diagnosis of patients who live in the same city, because the collective disease of the inhabitants of the same city can be caused by the condition of the city: air pollution, water pollution, eat rotten, or any other causes related with the state of the city. The aimed framework is composed of the following set of components:

- *External data centers:* Air Quality Center (AQC), Water Quality Center (WQC), Hospital Data Center (HDC) and any other data center that provide information about the city or the citizens.
- *Apache ActiveMQ:* One of the Apache hadoop ecosystem, it is very useful for communication and events transfer.
- *Hadoop Distributed File System (HDFS):* For data storage.
- *Apache Flume, Apache Sqoop and ODBC/JDBC Connectors:* For data extraction and integration.
- *Apache Hive:* After data pre-processing and filtering of irrelevant data, the data will be stored on Apache Hive for data mining and data analysis, Hive represents a virtual data-warehouse.

Intelligent agents are utilized nowadays in each field of life to take care of complex issues by distributing the work. Agents are software programs that take the autonomous action in different states to achieve design goals. As indicated by [17], responsive, proactive, autonomous and social are essential attributes of agents. In a multi-agent based system, agents work collectively and each agent performs specific tasks according to the role assigned [5].

Figure 2 illustrates an overview of the proposed framework. The different agents and their roles are listed in the following paragraphs.

- **Extractor Agent (EA):** EA builds up a connection with the sources system and extracts data.
- **Driver Agent and Workers Agents (DA and WA):** These agents apply the preprocessing process to the data stored in the HDFS. Its role is summarized in two steps:

• identifying and eliminating contradictions and inconsistencies. Driver and workers agents removes duplicate, missing and redundancy from data.
• identify and remove the irrelevant information. Driver and workers agents filter all the noise and only relevant information that are actionable, accurate and agile will be loaded into the Apache Hive for Analysis. They applie a technique called Automatic Relevance Determination (ARD) and other techniques like Feature Extraction (FE) and Feature Selection (FS) for the dimension reduction [18].

A MapReduce model is implemented, from where the Driver Agent is located in the master node, it is created to execute specific tasks, while the real work is carried out by the workers agents who are located in the slave nodes.

- **Analyzer Agents (AA):** The purpose of these agents is to convert the amount of data stored in the Apache Hive into valuable information by applying a fast and efficient analysis and creating various views and representations of that data. The objective is to carry out machine learning and data mining on behalf of an agent or a user and to report the results to the requesting entity and to all other entities that should be informed.
- **User-Interface Agent (UIA):** The UIA enhances the ability of the system user to use and entirely benefit from the Decision-Support System. It is responsible for all communications between the user of the system and the other agents in order to transmit reports and results to the end user.

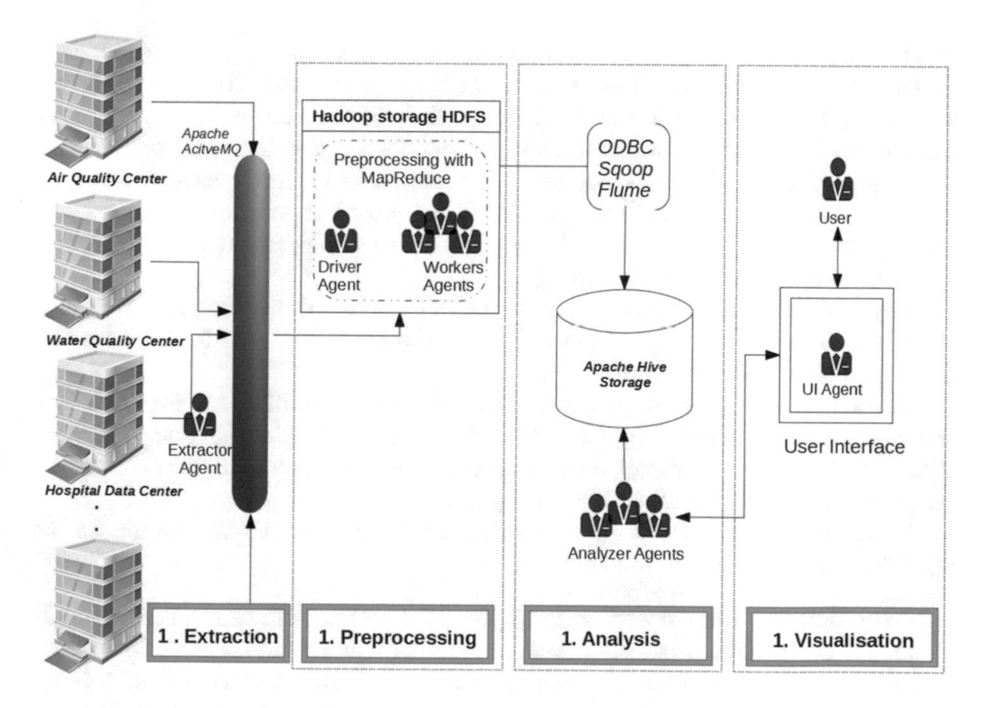

Fig. 2. The proposed framework

5 Discussion

In this paper, we tried to propose a multi-agent framework for public health diagnosis that can deal with huge data-sets. Smart health is an emerging filed, but it is a promising field especially with the era of Internet of Things (IoT) and the arrival of smart cities.

Compared with traditional frameworks, the proposed solution offers additional benefits:

- Firstly, the framework can avoid waste of storage space, which can result in low cost.
- Secondly, adapting decision making process to support the relevant information extraction and filtering the noise leads to a batter quality data and knowledge, therefore a better decision.
- Finally, contrary to other framework that focus only on 3Vs - Volume, Velocity and Veracity challenges, the proposed framework can handle the 3Vs challenges and the two others challenges (Value and Veracity) by transforming big data into smart valuable sata.

6 Conclusion and Future Work

In this paper, we have tackled the problem of noise in big data, which is a crucial step in transforming such raw data into smart data. We have proposed a multi-agent framework for medical diagnosis driven smart data in a big data environment. In the proposed framework, agents work collectively to perform tasks according to the roles assigned to them. The system contains different groups of agents to reduce the extraction time, filter the noise and optimize the performance of decision making. Hadoop is used to ensure fast data loading, fast query processing and efficient storage.

The highly tolerant nature of Hadoops failures, flexibility, extensibility, efficient load balancing and platform independence are also useful features for the development of any distributed process.

The solution may be adapted to different contexts, enabling the user to select the relevant data attributes and apply them in a suitable way into a certain situation. Moreover, the implementation of the agent based scenario for analysis purposes in different fields of life can be done.

Meanwhile, it is worth mentioning that the framework involves some future directions as follows:

1. The redundant data stored on multiple disks and partitions is a serious challenge for big data processing systems, and in this paper we have not shown the true strength of multi-agent systems and their effectiveness in solving complex and distributed problems collectively, so we hope in future work to find a solution for the elimination of the distributed data redundancy that requires intelligent mechanisms that can work in a distributed environment.
2. Noise elimination helps improve decision quality, but data loss can lead to a bad decision, so noise elimination with no data loss is a challenge that needs to be addressed in the future.
3. Finally, the implementation of the framework in a specific way for the smart health field, and in a generic way for general usage must be the next step.

References

1. Sakr, S., Gaber, M. (eds.): Large Scale and Big Data: Processing and Management. Auerbach, Philadelphia (2014)
2. Groves, P., Kayyali, B., Knott, D., Kuiken, S.V.: The big data revolution in healthcare. McKinsey Q. **2**, 3 (2013)
3. James, M., Michael, C., Brad, B., Jacques, B., Richard, D., Charles, R., Angela, H.: Big Data: The Next Frontier for Innovation, Competition, and Productivity. The McKinsey Global Institute, New York (2011)
4. Bologa, A., Bologa, R.: Business intelligence using software agent. Database Syst. J. **2**(4), 31–42 (2011)
5. Talib, R., Hanif, M.K., Fatima, F., Ayesha, S.: A multi-agent framework for data extraction, transformation and loading in data warehouse. Int. J. Adv. Comput. Sci. Appl. **7**(11), 351–354 (2016)
6. Belghache, E., Georg, J., Gleizes, M.: Towards an adaptive multi-agent system for dynamic big data analytics. In: Ubiquitous Intelligence and Computing, Advanced and Trusted Computing, Scalable Computing and Communications, Cloud and Big Data Computing, Internet of People, and Smart World Congress (2016)
7. Twardowski, B., Ryzko, D.: Multi-agent architecture for real-time big data processing. In: International Joint Conferences on Web Intelligence and Intelligent Agent Technologies (2014)
8. El Fazziki, A., Sadiq, A., Ouarzazi, J., Sadgal, M.: A multi-agent framework for a hadoop based air quality decision support system. In: Advenced Information Systems Engineering (2015)
9. Qayumi, K., Norta, A.: Business-intelligence mining of large decentralized multimedia datasets with a distributed multi-agent system. Int. J. Comput. Electr. Autom. Control Inf. Eng. **10**(06), 1160–1169 (2016)
10. Lenk, A., Bonorden, L., Hellmanns, A., Roedder, N., Jaehnichen, S.: Towards a taxonomy of standards in smart data. In: 2015 IEEE International Conference on Big Data (Big Data) (2015)
11. Triguero, I., Maillo, J., Luengo, J., Garcia, S., Herrera, F.: From big data to smart data with the k-nearest neighbours algorithm. In: IEEE International Conference on Smart Data (Smart Data 2016) (2016)
12. Wu, M.: The key to insight discovery: where to look in big data to find insights (2013)
13. Wu, M.: The big data fallacy (2012)
14. Garcia-Gil, D., Luengo, J., Garcia, S., Herrera, F.: Enabling smart data: noise filtering in big data classification. Computer Science (2017). arXiv:1704.01770 [cs.DB]
15. Fu, Y., Jiang, H., Xiao, N.: A scalable inline cluster deduplication framework for big data protection. In: Middleware, pp. 354–373 (2012)
16. Xia, W., Jiang, H., Feng, D., Hua, Y.: SiLo: a similarity-locality based near-exact deduplication scheme with low RAM overhead and high throughput. In: USENIX Annual Technical Conference (2011)
17. Nurse, J., Rahman, S., Creese, S., Goldsmith, M., Lambert, K.: Information quality and trustworthiness: a topical state-of-the-art review. In: International Conference on Computer Applications and Network Security, pp. 492–500. IEEE (2011)
18. Ramirez-Gallego, S., Krawczyk, B., Garcia, S., Wozniak, M., Herrera, F.: A survey on data preprocessing for data stream mining: current status and future directions. Neurocomputing **239**, 39–57 (2017)

Brain Plasticity in Older Adults: Could It Be Better Enhanced by Cognitive Training via an Adaptation of the Virtual Reality Platform FitForAll or via a Commercial Video Game?

Vasiliki Bapka[1(✉)], Irene Bika[1], Charalampos Kavouras[1], Theodore Savvidis[2], Evdokimos Konstantinidis[2], Panagiotis Bamidis[2], Georgia Papantoniou[3], Elvira Masoura[1], and Despina Moraitou[1]

[1] Lab of Psychology, School of Psychology, Aristotle University of Thessaloniki, Thessaloniki, Greece
vasw_mpapka@hotmail.com
[2] Lab of Medical Physics, Medical School of Medicine, Aristotle University of Thessaloniki, Thessaloniki, Greece
[3] Department of Early Childhood Education, University of Ioannina, Ioannina, Greece

Abstract. The aim of the study was to examine the effectiveness of Virtual Reality (VR) and Video Games (VG) as cognitive training tools in community dwelling older adults. A total of 19 older adults aged from 65 to 79 years comprised the sample of the study. According to the intervention program they were submitted to, participants were separated into two groups. The first one was trained using an adaptation of virtual reality platform (FitForAll), developed to train specific cognitive functions (n = 10). Contrary to the above the second group was trained in a go-kart-style video game (Super Mario Kart) on the Wii console (n = 9). There wasn't significant difference between the two groups in terms of age, gender and educational level. In both groups the intervention consisted of 18 sessions of about 40' each, over a six-week period (3 times per week). Specific cognitive functions and state affect were measured before and immediately after training, as well as one month after the intervention. The findings revealed participants' improved performance in executive functions, after being trained in both intervention programs. However, it has been observed that the participants trained using VG had been better at maintaining their improved performance even after the end of the intervention program. VG also seemed to enhance in the long term participants' positive affect.

Keywords: Cognitive aging · Cognitive control · Cognitive rehabilitation
Positive affect

1 Introduction

Nowadays, it is a commonplace that technology has taken over every aspect of life. As a consequence, many technological achievements are also used in the field of cognitive rehabilitation. In this vein, there is an increasing number of studies that focus on how

M. E. Auer and T. Tsiatsos (Eds.): IMCL 2017, AISC 725, pp. 728–742, 2018.
https://doi.org/10.1007/978-3-319-75175-7_72

virtual reality platforms can be effective as cognitive training tool in older adults, focusing mainly on the maintenance but also the enhancement of their cognitive functions (Craik and Bialystok 2006). The age-group of young older adults (65–80 years old) is rather the most interesting in this case, as it experiences age- related cognitive decline, which can occur even under normal aging circumstances (Bischkopf et al. 2002). Hence, because of the fact that life expectancy is increasing, it is important to improve older adults' cognitive functions in order to improve their healthy life as well as their quality of life in general (Jin et al. 2014). Two technological assets most used for this goal are Virtual Reality (VR) platforms and commercial Video Games (VG). Virtual reality is a high level computer interface, which includes real time simulation and interactions through multiple sensory pathways (Burdea 2003; Mirelman et al. 2010). As a cognitive rehabilitation tool it is a rather innovative practice, which uses computer software in order to combine visual, auditory and tactile sensations, as simulation of different aspects of everyday life (Rose et al. 2001). For instance, Optale et al. (2010), after exposing 36 older adults to a combination of a virtual reality program and music sessions, found that there was stabilization or even an improvement of their cognitive functions, spatial-visual processing and executive functions. This kind of software seems to rely on the brain's plasticity- brain's ability to adjust to the current environmental circumstances by reforming its structure (Lövdén et al. 2013). Commercial video games on the other hand, via exposing players to various stimuli can train various cognitive abilities according to the game's category (action, strategy, racing, etc.) (Basak et al. 2008; Zhang and Kaufman 2015; Van Muijden et al. 2012). Both interventions rely on the fact that even aging brain can learn and adjust to new environmental circumstances, which leads us to the conclusion that brain's plasticity is a continuous process that never stops.

Considering the aforementioned theory and findings, the present study's aim was to examine whether VR and VG are effective as cognitive function training tools for community dwelling older adults.

The hypotheses of the study were formulated as follows:

1. VR's intervention developed to recruit specific executive functions (visual sort-term memory, inhibitory control, task switching and attention) was expected to enhance older adults' performance in tests requiring these functions. The improvement was expected to be maintained for over 1 month (follow-up).
2. VG's intervention was expected to improve participants' performance in the same tests. Improvement would be sustained for 1 month (follow-up).
3. However, it was expected that older adults who have participated to the VR's program would perform better to tasks that measure specific executive functions than older adults trained by a commercial VG, as the VR platform used in this study is adjusted to train specific executive functions in a systematic way. Since both types of intervention have a game-like character, it was expected that both of them could increase state positive affect.

2 Method

2.1 Participants

At first, 64 older adults, who attended Open Day Care Center, in Thessaloniki, Greece, were informed about the terms of the study. Only 29 of them were willing to participate. Due to not fulfilling the requirements of the study, 7 participants were excluded, so 22 community dwelling older adults started participating in the study. During the intervention 3 of them left, so the final sample comprised a total of 19 (9 women) community dwelling older adults, who participated voluntarily. Participants were separated into two groups, according to the intervention program to which they were submitted to. The first one consisted of 10 participants aged 65–79 years (M = 70.3, SD = 5.03) who were trained in virtual reality program. The second group consisted of 9 participants aged 65–78 years (M = 71.33, SD = 4.60) and was trained in a go-kart-style racing video game. There was not a significant difference between the two groups as far as gender, $x2(1) = .656$, $p > .05$, age, $x2\ (10) = .921$, $p > .05$ and educational level, $x2(2) = 5.295$, $p > .05$ were concerned (see Table 1). It is worth mentioning that strict statistical criteria were applied regarding to the educational level because of the potential inconsistency of the chi-square test, due to the small number of participants in each group of the sample. So, Monte Carlo and Exact value criteria were conducted and the respective indices were computed. Potential participants who suffered from uncorrected hearing and/or visual loss and any other severe psychiatric, neurological or physical disease were excluded. The Geriatric Depression Scale-15 (GDS-15) was used to examine the presence of depressive symptoms. Participants, who scored higher than 6 in this scale, were excluded from the research's sample (Yesavage et al. 1983; Fountoulakis et al. 1999). Addition exclusionary criterion was the existence of cognitive decline, which was examined using the Mini Mental State Examination (MMSE). A score lower than '23–24' is considered indicative of dementia symptomatology (Folstein et al. 1975; Fountoulakis et al. 2000). In this research, each participant who scored less than '27' was excluded in order to ensure that even mild

Table 1. Participants' distribution according to age, gender and educational level, and screening tests' mean

Intervention	Age	Gender		Education (years)			GDS-15*	MMSE**	MoCA***
	Range	Male	Female	Low	Middle	High	Mean (*SD*)	Mean (*SD*)	Mean (*SD*)
Video game (*n* = 9)	65-78	4	5	8	1	–	0.77 (1.4)	28.9 (1.1)	25.1 (2.8)
Virtual reality (*n* = 10)	65-79	6	4	4	3	3	0.9 (1.1)	28.7 (1.2)	26 (2.2)
Total sample (*N* = 19)	65-79	10	9	12	4	3			

**GDS-15 = Geriatric Depression Scale, **MMSE = Mini Mental State Examination, ***MoCA = Montreal Cognitive Assessment*

cognitive decline was absent. Furthermore, the Montreal Cognitive Assessment (MoCA) was administered (Nasreddine et al. 2004; http://www.mocatest.org/pdf_files/test/MoCA-Test-Greek.pdf), not as a strict screening tool (as the other two aforementioned scales), but in order to help in the formulation of a more accurate assessment of general cognitive ability of the sample.

2.2 Measures

1. *Delis Kaplan Executive Function System (D-KEFS) Color-Word Interference Test (C-WIT;* Delis et al. 2001). The D-KEFS battery is created to examine higher- order cognitive functions supported by the frontal lobe, namely executive functions. It consists of nine stand-alone tests. In the current study, one of them has been used. The psychometric properties of the test have been examined by two of the co-authors of this study in a series of previous studies (see Nazlidou et al. 2014). The C-WIT consists of 4 conditions: the last two measure mainly inhibition and cognitive flexibility, respectively. The first two conditions that measure speed of naming and reading speed of the participants measure basic cognitive abilities and their completion is a prerequisite to proceed with the third and fourth condition which measure executive functioning. Specifically, the third condition, which examined participants' inhibitory control, requested from them to name the color of the ink that words are written, while words may refer to different colors in 180 s. The fourth condition, which examined participants' cognitive flexibility (inhibitory control & task switching), demanded from them to name the color of the ink that words were written and alternately reading words that referred to colors in 180 s. It is worth to mention that in this study we used in both conditions the variable "completion time".
2. *Trail Making Test* (TMT) (Armitage 1946; Vlachou and Kosmidis 2002). The test mainly measures visual attention and task switching. It consists of two tasks in which participants were instructed to connect a set of 25 dots as quickly as possible while trying to be accurate. In Task A, which measures attention, concentration and perceptive/motor ability, participants had to connect the 25 numbered dots in ascending order (e.g. 1, 2, 3, etc.). The reliability indicator of this task varies from .36 to .94 (Vlachou and Kosmidis 2002). In Task B, participants were instructed to connect these 25 dots, which this time contained numbers and letters, alternating between these two categories (e.g. 1-A-2-B-3-C, etc.). Actually, the second task measures cognitive flexibility. As far as the reliability indicator of this task is concerned, it varies from .44 to .87 (Vlachou and Kosmidis 2002). This test seems to be more valid in "rapid visual detection", "cognitive flexibility", "visual-spatial sequence" and "focused mental processing speed" (Vlachou and Kosmidis 2002).
3. *Visual Patterns Test (V.P.T.)* (Della Sala et al. 1997) is a measure of short term visual memory shorn of the spatial components. The test consists of cards, which have patterns like chess and were presented to participants for 3 s. In this duration participants had to memorize them and be able to reproduce them in a plain plexus. There was a gradual increase in the complexity of the patterns and the test was interrupted after 3 consecutive errors in the same level of difficulty. This test has

two versions (Form A/Form B), so that the practice effect is eliminated to the greatest. As far as the test- retest reliability indicator of Form A is concerned, was found to be .75. The test-retest reliability indicator of Form B was found to be .73 (Della Sala et al. 1999).

4. *Positive and Negative Affect Schedule* (PANAS) (Watson et al. 1988; Moraitou and Efklides 2009). This is a self-report questionnaire that consists of a 20-item scale that measures both positive and negative affect. Specifically, participants have to assess in which degree they have felt the emotions that were demonstrated on the test, the past two weeks. Participants had to answer the questions by selecting answers given through a Likert scale from 1 (not at all) to 5 (very much). In addition, PANAS is concerned to be a valid measure of the affect (Clark and Watson 1991; Moraitou and Efklides 2009). Also, its reliability indicator in the Greek adaptation for each of the sub-scales, was found to be: $\alpha = .84$ for the positive affect and $\alpha = .82$ for the negative affect. At last, confirmatory analysis of factors has shown that PANAS actually evaluates two dimensions (Moraitou and Efklides 2009).

2.3 Intervention Programs

1. **Virtual Reality - FitForAll** (Konstantinidis et al. 2016). This virtual reality platform was designed by the Medical Physics Laboratory, A.U.Th. It should be mentioned that the part of this program that was used in the current research was adopted to train specific executive functions. The equipment used for this program was a 32" screen, a "Kinect" sensor and a laptop. Participants were standing in a 2-m distance in front of the screen and the sensor. It was a customized intervention that lasted 18 sessions of 40'each. In every session participants had to complete the four tasks of the program that were formulated as follows:

   - *Apple collection using the right hand.* This task comprises of 3conditions. In the first one, participants have to collect every apple that appears on screen in succession (5 min). This condition examines participants' *attention*. The second one examines inhibition and requires from the participants to collect apples again, avoiding this time the rotten ones that appear on the same time (5 min). In the third condition that examines *switching* and *inhibition*, participants have to collect alternately 2 types of apples (red-green) appeared on screen (5 min).
   - *Fish collection by moving the torso back and forth.* The current task is composed of 2 conditions. In the first one, which examines *attention*, participants have to collect each fish appeared on screen (5 min). The second task examines *inhibition* and requires from the participants to keep collecting fish, while avoiding sharks, which appear among them.
   - *Golf simulation by moving the torso in 4 different directions (right-left-back-forth).* The task is comprised of two sub-tasks. In the first one, participants have to throw the ball into the hole by moving their torso (5 min). This condition is focused on *attention*. The second sub-task is quiet the same as the first one, but here participants have also to avoid the potholes that appear among the regular holes (5 min), testing in this way participants' *inhibitory control.*

- *Break Blocks Game simulation by moving the torso right and left.* This task examines *attention*. Participants move the bar by moving their torso right and left, so that to hit the ball and break the bricks (5 min).

2. **Video Game - Super Mario Kart (Nintendo)** (Nintendo 1992). A Wii console, which consists of a central processing unit, a wireless remote control-steering wheel and a 26" screen, was the equipment used for this intervention. Participants were seated in front of the screen and they were holding the steering wheel. They had to finish the track by driving a car, while trying to beat their opponents and avoid obstacles or attacks from them. It was also a customized intervention that lasted 18 sessions of 40' each. During each session participants had to complete playing 12 tracks.

2.4 Procedure

Firstly, all the participants had to complete an individual-demographics form before the start of the examination process. Secondly, the screening tests, which ensured that the participants met the inclusion criteria, were administered. Afterwards, the battery was administered before, after and one month after the end (follow-up measurement) of the both intervention programs. As it was mentioned before, both intervention programs relied on customized examination.

2.5 Ethical Standards

The authors assert that all procedures contributing to this work comply with the ethical standards of the relevant national and institutional committees on human experimentation and with the Helsinki Declaration of 1975, as revised in 2008. All the participants were informed about the procedure and the aim of the study, and their written informed consent was taken.

2.6 Statistical Analyses

The data were analyzed by using the SPSS (V.22). Mixed Measures ANOVAs were conducted. Specifically, the group type (VR-VG) was defined as the between-subjects factor, and the measurements of every test (before-after-follow-up) as the within-subjects factor. Wherever the results were statistically significant, repeated Measures ANOVAs were conducted, in order to define which intervention program affected the most participants' performance.

3 Results

3.1 D-KEFS Color-Word Interference Test

- **3rd Condition: Inhibitory control.** Mixed measures ANOVA with the Greenhouse-Geisser correction showed that the type of intervention program affected the time participants needed to complete the test, $F(1.4,23.5) = 7.849$, $p = .006$, $\eta 2 = .316$.

Specifically, the Bonferroni correction revealed that among measurements were found statistically significant differences which are due to the type of intervention between the assessment which was carried out before the intervention programs and that one which was carried out in the follow-up measurement after the end of the programs, p = .016. In order to determine which of the two interventions had a greater impact, Repeated Measures ANOVA for each group were conducted subsequently. According to the results VG affected statistically significant the time participants needed among the measurements, $F(1.2,10) = 7.789$, $p = .015$, $\eta 2 = .493$ (Fig. 1). However, this difference was not strong enough so that it can be determined in which of measurements exactly is identified.

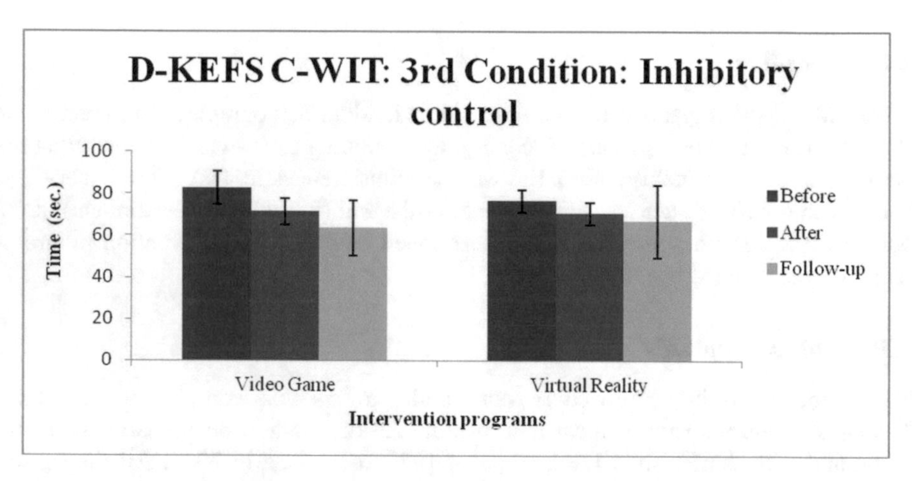

Fig. 1. Participants' performance in the D-KEFS – C-WIT: 3rd condition testing inhibitory control

- **4th Condition: Cognitive Flexibility.** Mixed measures ANOVA with Greenhouse-Geisser correction showed that the time that participants needed to complete the test was affected by the type of the intervention program, $F(1.4, 23.1) = 7.318$, $p = .008$, $\eta 2 = .301$. In particular, Bonferroni correction that conducted afterwards, revealed that due to the type of the intervention program statistically significant differences were found between the before and the follow-up measurement, p = .016. Repeated Measures ANOVA for each group was conducted subsequently and showed that the statistically significant difference were identified in participants' performance, who had been trained by using VG, $F(2,16) = 4.606$, $p = .026$, $\eta 2 = .363$ (Fig. 2). Nevertheless, this difference, as it happened in the previous condition was not strong enough so that it can be determined in which of measurements exactly is identified.

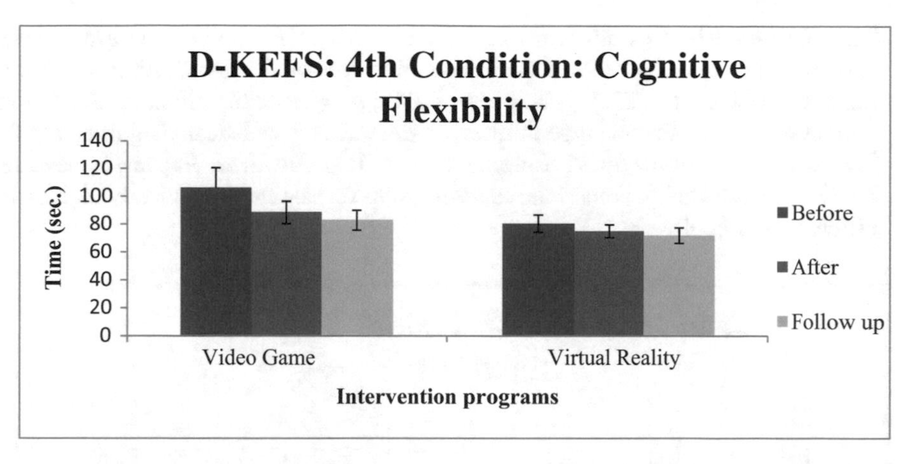

Fig. 2. Participants' performance in the D-KEFS- C-WIT: 4th condition testing cognitive flexibility (inhibition & task switching).

3. **Trail Making Test**.

- **Task A: Attention control.** Mixed measures ANOVA with Greenhouse-Geisser correction revealed a significant effect of the type of intervention program on the participants' performance on this task of TMT, $F(1.4, 23.3) = 4.417$, $p = .036$, $\eta 2 = .206$. More specifically, Bonferroni correction showed a statistically significant difference between the before and the after measurement I–J = 6.600, $p = .033$. Repeated measures ANOVA conducted for each intervention group, showed a statistically significant difference in the VR group, $F(2, 18) = 4.414$, $p = .028$, $\eta 2 = .329$. Specifically, Bonferroni correction revealed a difference between the before and the after, I–J = 6.200, $p = .046$, as well as between the before and the follow-up measurement, I–J = 10.000, $p = .018$ (Fig. 3).

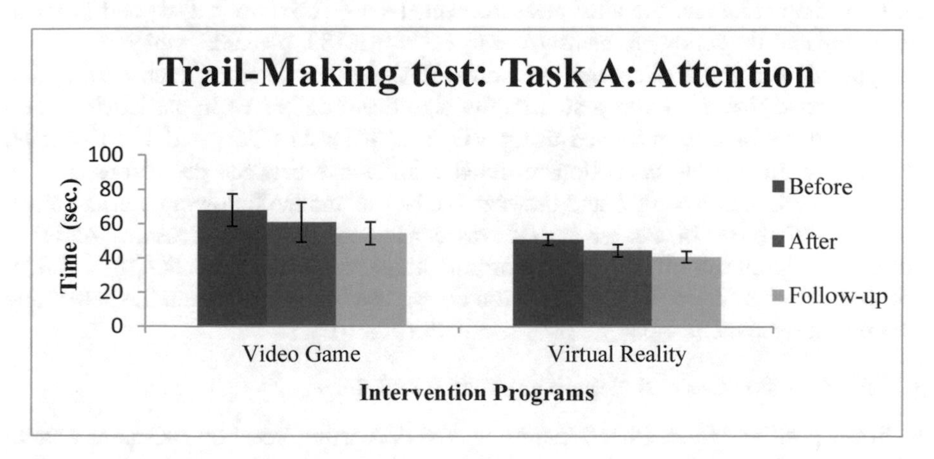

Fig. 3. Participants' performance on Trail-Making test: Task A: total time needed to complete the task examining attention.

- **Task B: Cognitive flexibility.** Mixed measures ANOVA with Greenhouse- Geisser correction revealed the effect of the intervention type on the time participants had to finish the task, F (1.4, 23.5) = 3.885, p = .049, η^2 = .186. In particular, Bonferroni correction showed that the type of intervention affected statistically significantly the before and the follow-up measurement, p = .017. However, Repeated measures ANOVA conducted for each intervention group revealed no statistically significant differences (Fig. 4).

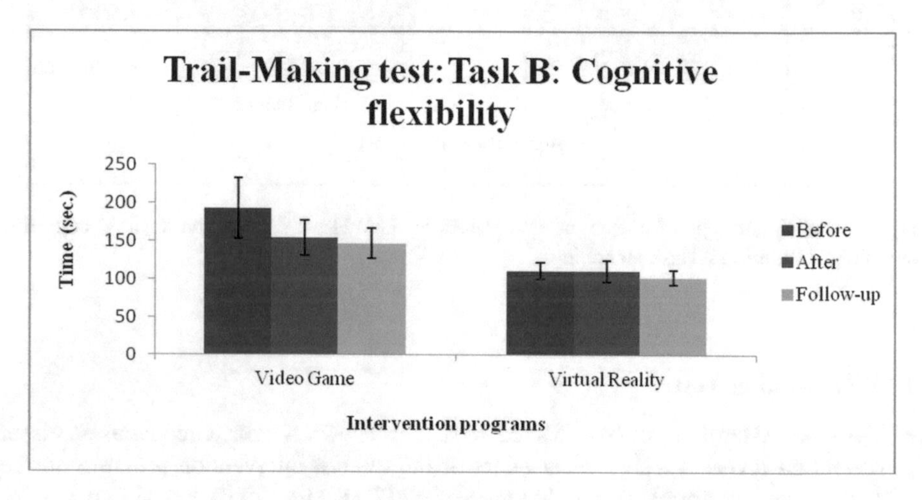

Fig. 4. Participants' performance on Trail Making test: Task B: total time needed to complete the task examining cognitive flexibility (task switching).

4. **Visual Pattern: Visual short-term memory.**

Mixed measures ANOVA showed that participants' short-term visual memory was affected by the type of the intervention program, F(2, 34) = 14.302, p = .000, η^2 = .457. Specifically, Bonferroni correction showed statistically significant difference between the before and the after measurement, I-J = −.2.578, p < .001, and between the before and the follow-up measurement, I–J = −.2.183, p = .005, which are due to the type of intervention. Repeated measures ANOVA conducted for each intervention group showed that there was a statistically significant difference in participants' performance who have been trained using VG, F (2,16) = 11.829, p = .001, η^2 = .597. Particularly, Bonferroni correction revealed a difference between the before and the after, I–J = −2.556, p = .007, and between the before and the follow-up measurement, I–J = −2.667, p = 0.14. As far as VR was concerned, Repeated Measures ANOVA showed a significant difference among the three measurements, F(2,18) = 5.307, p = .015, η^2 = .371. Bonferroni correction showed a difference between the before and the after measurement, I–J = −2.600, p = .008 (Fig. 5).

5. **PANAS - Positive and Negative Affect Schedule.**

- **State positive affect.** Mixed measures ANOVA with Greenhouse-Geisser correction revealed the intervention's type effect on participants' positive affect, F(1.4,

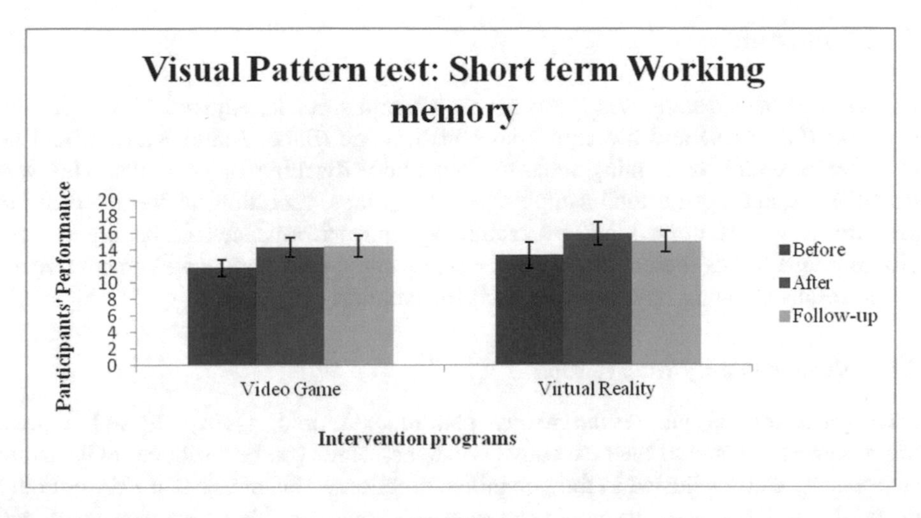

Fig. 5. Participants' performance on the Visual Pattern test, examining the short-term visual memory

24.6) = 6.724, $p = .009$, $\eta^2 = 2.83$. More specifically, Bonferroni correction showed a statistically significant difference between the before and the after measurement, I–J = −4.339, $p = .047$, and the before and the follow-up measurements, I–J = −3.783, $p = .019$. Repeated measures ANOVA with Greenhouse - Geisser conducted for each intervention group, revealed a statistically significant difference among the 3 measurements in the VG group, $F(1.2, 9.4) = 9.622$, $p = .010$, $\eta^2 = .546$. In particular, Bonferroni correction showed differences between the before and the after measurement, I–J = 7.778, $p = .045$ and between the before and the follow-up measurement, I–J = 6.667, $p = .025$. As far as VR program is concerned no statistically significant differences were found (Fig. 6).

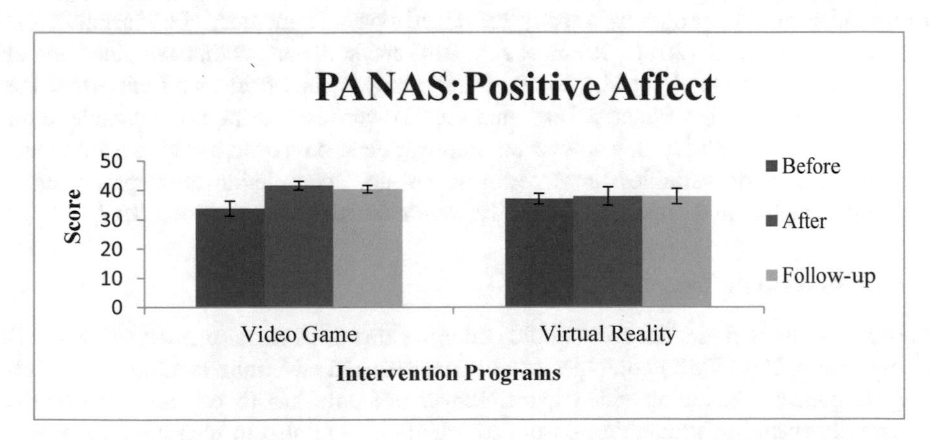

Fig. 6. Participants' perfomance on PANAS, which measures the Positive Affect.

4 Discussion

The purpose of this study was to examine whether a specially adjusted Virtual Reality platform (FitForAll) and a commercial Video Game (Super Mario Kart), would be effective as cognitive training tools to community dwelling older adults. This was tested by separating the total sample into two groups, according to the intervention program they were trained, and by examining their performance in a battery at three different time points (before-after-follow-up measurements). Below will be a reference to the results showing how effective each intervention program is.

4.1 Virtual Reality Intervention

Participants trained via virtual reality platform FitForAll (FFA) showed a great improvement to some of their executive functions, which can be attributed to its nature as it was specially adjusted to train cognitive functions. This means that FFA not only delayed the constantly developing cognitive impairment, but also sustained and improved the existing executive functions. In particular, the elderly have been improved in attention control, which seemed to be maintained even long term. Regarding the existing literature, statistically significant differences in attention control were revealed by a research on elderly people who were diagnosed with Mild Cognitive Impairment (MCI) and were examined cognitively after participating in a VR program (Optale et al. 2010). At the same time, contrary to the results of this study, others studies have shown changes in inhibitory control and in cognitive flexibility, which were perceived in the follow-up measurement. Leading study that showed this pattern of improvement was the study of Tarnanas et al. (2014), in which researchers found a statistically significant improvement in cognitive flexibility, as it was measured by Trail Maiking Test 5 months after completion of a virtual museum navigation intervention in elderly patients diagnosed with MCI. Improved performance was also found in visual short term memory, which however did not manage to be maintained in long term at the same level. Researches that have found improvement in memory, as a result of using VG programs, mostly examined episodic memory. For instance, the research of Man et al. (2011); Jebara et al. (2014) and some others that examined verbal memory, such as Optale et al. (2010). Additionally, it is worth mentioning that the submission of "Design Fluency Test" and the "Tower test" to the same sample, after training at a VR platform, showed an improvement on participants' performance. Particularly, in the variables 'total performance' and 'completion time' participants' inhibitory control and cognitive flexibility were enhanced (Bapka et al. 2017).

4.2 Video Game Intervention

Participants trained via Wii console did better in short-term visual memory. Given that Super Mario Kart (SMK) combines competitiveness and adventure at a high level, is a very cognitive demanding video game. Player not only has to pay attention to the constantly changing stimulation on the surroundings, but also to plan his next movement. As a consequence processing speed, planning and decision making are essential to accomplish this game. Visual spatial processing, which was trained while playing

SMK, improves to a great extend the processes of visual perception and memory. This agrees with the findings of this study as the short term visual memory not only was improved, but also maintained in the follow-up measurement. Many other studies agree with the above, as they have found improvement of short-term visual memory (Basak et al. 2008; Peretz et al. 2011), while others have showed imaging findings of brain's plasticity to the regions responsible for this (short-term visual memory) after playing VG (Basak et al. 2011; Kuhn et al. 2014). Moreover, is very important to mention how playing with SMK improved participants' positive affect after the intervention and one month after. This comes to an agreement with other scientific papers which have found that playing VG is highly associated to positive affect. Lower levels of loneliness have been stated by older adults who played VG instead of watching TV or playing conventional board games (Jung et al. 2009; Kahlbaugh et al. 2011). Also, it seems that older adults who entertained themselves by playing VG had less depression symptoms (Allaire et al. 2013) and better interaction with younger generations (Mahmud et al. 2010). SMK in particular, was specially designed for amusement, so its design contains funny noises, colourful environments, etc., which has a positive effect on the elderly's psychological insight. Last but not least, self-esteem of the players is improved as they accomplish the game's goals, and also they tend to socialize more as this kind of game promotes social interaction with other players.

4.3 Comparison of the Two Intervention Programs

The above helped in reaching to a conclusion for the hypotheses of the current studies. The first hypothesis of the current study was merely confirmed as the results have shown that the participants' performance improved after the intervention using Virtual Reality, but there was not a satisfying maintenance of the improvement as revealed from the follow- up measurement. The second hypothesis of the study was confirmed as the results have shown that participants who trained using a commercial Video Game had an enhanced performance on the battery which was maintained to a large degree even after a month. The last hypothesis though, was not confirmed as the group of older adults who trained by playing Video Game did better in the battery not only in the cognitive part but also in the psychological. Participants who played Video Game came out to be much happier than participants who were trained by using Virtual Reality.

To conclude, it seems that both intervention programs were effective as cognitive training tools, with the difference that they did not trained the same cognitive functions. Specifically, FFA was better in training attention control, while SMK was better in enhancing short-term visual memory and positive affect of the participants. All the above could be interpreted by the theories based on brain's plasticity (Green and Bavalier 2008). Taking for granted that exercising and cognitive training are factors both of paramount importance for the beginning of neuroplastic changes, virtual reality platforms that combines them both, seem to be more effective in the improvement and maintenance of higher level cognitive functions. On the other hand, video games are created for amusement, so they are based on sentimental factors and not on the improvement of cognitive functions. This is the main reason why, in this study, they were better on the improvement of participants' psychological state. Last but not least, older adults did not have to be accustomed to technological means in order to be benefited from its advantages. This can be showed from the fact that neither of the participants in this study had ever played a video game or used a virtual reality platform before, but yet everybody was benefited from the intervention.

4.4 Limitations and Future Research

The restricted nature of the sample, the short duration of both intervention programs (18 sessions) and the small period of time mediated the administrations of the battery (before-after-follow-up measurements) could be considered as limitations in the current study. Moreover, we cannot be sure about the extent the benefits qualified by the intervention programs are adapted to the ederly's everyday life.

Finally, it would be of paramount interest to examine how the intervention programs apply to other groups, such as people with dementia (MCI, Alzheimer's disease, etc.). Also, another topic which could be further examined in the future is the category of the video game that influences the most the cognitive function of older adults. A research comparing different video game categories would be really interesting.

4.5 Conclusion

The hypotheses of the study were partially confirmed as both intervention programs improved participants' executive functions. However, VG seems to be more effective not only in maintaining the results of the cognitive training but also in improving the psychological state of the elderly (positive affect). Despite the fact that the researches conducted in the field of cognitive rehabilitation via technology arise, there is still a lot to be learnt.

References

Allaire, J.C., McLaughlin, A.C., Trujillo, A., Whitlock, L.A., LaPorte, L., Gandy, M.: Successful aging through digital games: socioemotional differences between older adult gamers and non-gamers. Comput. Hum. Behav. **29**(4), 1302–1306 (2013)

Armitage, S.G.: An analysis of certain psychological tests used for the evaluation of brain injury. Psychol. Monogr. **60**(1), i (1946)

Bapka, V., Bika, I., Savvidis, T., Konstantinidis, E., Bamidis, P., Papantoniou, G., Moraitou, D.: Cognitive training in community dwelling older adults via a commercial video game and an adaptation of the virtual reality program FitForAll: comparison of the two intervention programs. Hellenic J. Nucl. Med. **20**(2), 21–29 (2017)

Basak, C., Boot, W.R., Voss, M.W., Kramer, A.F.: Can training in a real-time strategy video game attenuate cognitive decline in older adults? Psychol. Aging **23**(4), 765–777 (2008). https://doi.org/10.1037/a0013494

Basak, C., Voss, M.W., Erickson, K.I., Boot, W.R., Kramer, A.F.: Regional differences in brain volume predict the acquisition of skill in a complex real-time strategy videogame. Brain Cogn. **76**, 407–414 (2011)

Bischkopf, J., Busse, A., Angermeyer, M.C.: Mild cognitive impairment–a review of prevalence, incidence and outcome according to current approaches. Acta Psychiatr. Scand. **106**(6), 403–414 (2002)

Burdea, G.C.: Virtual rehabilitation–benefits and challenges. Methods Inf. Med. **42**(5), 519–523 (2003)

Clark, L., Watson, D.: Tripartite model of anxiety and depression: psychometric evidence and taxonomic implications. J. Abnorm. Psychol. **100**(3), 316–336 (1991)

Craik, F.I.M., Bialystok, E.: Cognition through the lifespan: mechanisms of change. Trends Cogn. Sci. **10**(3), 131–138 (2006)

Della Sala, S., Gray, C., Baddeley, A., Wilson, L.: Visual Patterns Test: A Test of Short-Term Visual Recall. Thames Valley Test Company, Bury St Edmunds (1997)

Della Sala, S., Gray, C., Baddeley, A., Allamano, N., Wilson, L.: Pattern span: a tool for unwelding visuo–spatial memory. Neuropsychologia **37**(10), 1189–1199 (1999)

Delis, D.C., Kaplan, E., Kramer, J.H.: Delis-Kaplan Executive Function System (D-KEFS). Psychological Corporation, San Antonio (2001)

Folstein, M.F., Folstein, S.E., McHugh, P.R.: "Mini-mental state": a practical method for grading the cognitive state of patients for the clinician. J. Psychiatr. Res. **12**(3), 189–198 (1975)

Fountoulakis, K., Tsolaki, M., Iakovides, A., Yesavage, J., O'Hara, R., et al.: The validation of the short term of the Geriatric Depression Scale (GDS) in the Greece. Aging **11**(6), 367–372 (1999)

Fountoulakis, K.N., Tsolaki, M., Chantzi, H., Kazis, A.: Mini mental state examination (MMSE): a validation study in Greece. Am. J. Alzheimer's Dis. **15**(6), 342–345 (2000)

Green, C.S., Bavalier, D.: Exercising your brain: a review of human brain plasticity abd training-induced learning. Psychol. Aging **23**(4), 692 (2008)

Jebara, N., Orriols, E., Zaoui, M., Berthoz, A., Piolino, P.: Effects of enactment in episodic memory: a pilot virtual reality study with young and elderly adults. Front. Aging Neurosci. **6**, 338 (2014)

Jin, K., Simpkins, J.W., Ji, X., Leis, M., Stambler, I.: The critical need to promote research of aging and aging-related diseases to improve health and longevity of the elderly population. Aging Dis. **6**(1), 1–5 (2014)

Jung, Y., Li, K.J., Janissa, N.S., Gladys, W.L.C., Lee, K.M.: Games for a better life: effects of playing Wii games on the well-being of seniors in a long-term care facility. Paper presented at the Proceedings of the Sixth Australasian Conference on Interactive Entertainment, Sydney, Australia (2009)

Kahlbaugh, P.E., Sperandio, A.J., Carlson, A.L., Hauselt, J.: Effects of playing Wii on well-being in the elderly: Physical activity, loneliness, and mood. Act. Adapt. Aging **35**(4), 331–344 (2011)

Konstantinidis, E., Bamparopoulos, G., Bamidis, P.: Moving real exergaming engines on the web: the webFitForAll case study in an active and healthy ageing living lab environment. IEEE J. Biomed. Health Inform. **21**, 859–866 (2016)

Kuhn, S., Gleich, T., Lorenz, R.C., Lindenberger, U., Gallinat, J.: Playing Super Mario induces structural brain plasticity: gray matter changes resulting from training with a commercial video game. Mol. Psychiatry **19**(2), 265–271 (2014)

Lövdén, M., Wenger, E., Mårtensson, J., Lindenberger, U., Bäckman, L.: Structural brain plasticity in adult learning and development. Neurosci. Biobehav. Rev. **37**(9, Part B), 2296–2310 (2013)

Mahmud, A.A., Mubin, O., Shahid, S., Martens, J.-B.: Designing social games for children and older adults: two related case studies. Entertain. Comput. **1**(3–4), 147–156 (2010)

Man, D.W.K., Chung, J.C.C., Lee, G.Y.Y.: Evaluation of a virtual reality-based memory training programme for Hong Kong Chinese older adults with questionable dementia: a pilot study. Int. J. Geriatr. Psychiatry **27**(5), 513–520 (2011)

Mirelman, A., Patritti, B.L., Bonato, P., Deutsch, J.E.: Effects of virtual reality training on gait biomechanics of individuals post-stroke. Gait Posture **31**(4), 433–437 (2010)

Moraitou, D., Efklides, A.: The blank in the mind questionnaire (BIMQ). Eur. J. Psychol. Assess. **25**(2), 115–122 (2009)

Nasreddine, Z.S., Chertkow, H., Phillips, N., Whitehead, V., Collin, I., Cummings, J.L.: The montreal cognitive assessment (MoCA). Neurology **62**(7), A132 (2004)

Nazlidou, E.I., Moraitou, D., Natsopoulos, D., Papantoniou, G.: Social cognition in adults: the role of cognitive control. Hellenic J. Nucl. Med. **18**, 109–121 (2014)

Nintendo, E.A.D.: Super Mario Kart (1992)

Van Muijden, J., Band, G.P., Hommel, B.: Online games training aging brains: limited transfer to cognitive control functions. Front. Hum. Neurosci. **6**, 221 (2012). https://doi.org/10.3389/fnhum.2012.00221

Vlachou, C.H., Kosmidis, M.H.: The Greek trail-making test: preliminary norms for clinical and research application. Psychology **9**(3), 356–362 (2002). (in Greek)

Optale, G., Urgesi, C., Busato, V., Marin, S., Piron, L., Priftis, K., Bordin, A.: Controlling memory impairment in elderly adults using virtual reality memory training: a randomized controlled pilot study. Neurorehabil. Neural Repair **24**(4), 348–357 (2010)

Peretz, C., Korczyn, A.D., Shatil, E., Aharonson, V., Birnboim, S., Giladi, N.: Computer-based, personalized cognitive training versus classical computer games: a randomized double-blind prospective trial of cognitive stimulation. Neuroepidemiology **36**(2), 91–99 (2011)

Rose, F.D., Attree, E.A., Brooks, B.M., Andrews, T.K.: Learning and memory in virtual environments: a role in neurorehabilitation? Questions (and occasional answers) from the University of East London. Presence **10**(4), 345–358 (2001)

Tarnanas, I., Tsolakis, A., Tsolaki, M.: Assessing virtual reality environments as cognitive stimulation method for patients with MCI. In: Technologies of Inclusive Well-Being, vol. 536, pp. 39–74. Springer, Heidelberg (2014)

Watson, D., Clark, L.A., Tellegen, A.: Development and validation of brief measures of positive and negative affect: the PANAS scales. J. Pers. Soc. Psychol. **54**(6), 1063 (1988)

Yesavage, J.A., Brink, T.L., Rose, T.L., Lum, O., Huang, V., Adey, M., Leirer, V.O.: Development and validation of a geriatric depression screening scale: a preliminary report. J. Psychiatr. Res. **17**(1), 37–49 (1983)

Zhang, F., Kaufman, D.: Physical and cognitive impacts of digital games on older adults a meta-analytic review. J. Appl. Gerontol. (2015). https://doi.org/10.1177/0733464814566678

Multimedia Learning in Music Education

Virtual Design Studio Project for Mobile Learning in Byzantine Music

A Cognitive Walkthrough Methodology

Nektarios Paris[1], Dionysios Politis[2(✉)], Rafail Tzimas[2], and Nikolaos Rentakis[3]

[1] Department of Music Science and Art, University of Macedonia, Thessaloniki, Greece
nepa@uom.gr
[2] Department of Informatics, Aristotle University of Thessaloniki, Thessaloniki, Greece
dpolitis@csd.auth.gr, tzimasr@gmail.com
[3] Ecclesiastical Lyceum of Neapoli-Stavroupoli, Thessaloniki, Greece
mail@lyk-ekkl-neapol.thess.sch.gr

Abstract. This research extends the notion of cognitive walkthrough over the uncontested world of Byzantine music liturgical practices for chanting on major feasts. This specific musical system is a Delta musical structure with a "thinner" microtonal scale partition of the well-known Western music scales and therefore new notes are involved. Apart from the tonality distribution, in Byzantine Music Delta notation transitional patterns occur stating qualitative ways for the Delta ascent or descent of the prosodic pitch, i.e. the language dependent part of singing. This paper describes the multimedia environment that has been set up for the melodic melurgic instruction of Byzantine Music chants, in an intermediary to advanced level in terms of music education.

Keywords: Mobile communication · Singing · Byzantine music
Multimedia learning · Cognitive walkthrough

1 The Introduction: Why Promote Mobile Communication in Byzantine Music?

In Mobile Technologies have the advantage over their computer based peers, that communication with the application may take place *in* situ, i.e. on the stage where chanting is performed.

Since, in liturgical terms, the rotating portion of ceremonial corpus for public worship is constantly evolving, the need for retrieving it out of a database like environment emerges as a global trend. Indeed, what should be chanted daily, for the main services of the canonical hours, Vespers and Matins, and to a lesser degree for the Mass itself, results from the coinciding observances of three revolving calendars: the annual, the weekly for each one of the 8 chanting modes, and the perpetual context of the

M. E. Auer and T. Tsiatsos (Eds.): IMCL 2017, AISC 725, pp. 745–757, 2018.
https://doi.org/10.1007/978-3-319-75175-7_73

moveable feasts like Easter, which are based on the annual lunar calendar. Not that important are other Lectionary type rotations of more than one year cycles.

Over the years, as typing has become more affordable, most churches are equipped with the majority of liturgical literature corpus. However, this proves not to be the case with chanting, where the amount of printed music instruction is already reaching immense sizes.

Therefore, the master of a choir has to provide in notes, and generally speaking in semantics, out of a multitude of music books, the correct and exact melodic phrases to be sung each time. These pamphlets or books may be either staged within the church's resources or cropped out of his personal collection.

As international culture interactions increase, this situation becomes more complicated: more languages adhere to what was previously performed within a few major ones, and more variants are constantly added, as the personalization principle is applied, extending its impetus to chanting [1].

Another difficulty with chanting has to do with its music notation. If for the ample use of multilingual texts in the multitude of our linguistic Babel, mobile technology has reached a culmination point only in the last few recent years, for out of mainstream music forms the situation is more difficult.

Lacking formal support, it terms of technology, they have difficulties in using typeset characters with an orientation along a publishing path; also they lack (artist oriented) animation languages.

Of course, these problems are more customarily adhered to the global music and singing community. Although some of these have been sufficiently solved within the mobile communication framework, like how the visual characteristics of notes and the rendering characteristics for special scripting annotations combine (Fig. 1), it is obvious that the promising world of musical multimedia learning has just taken off.

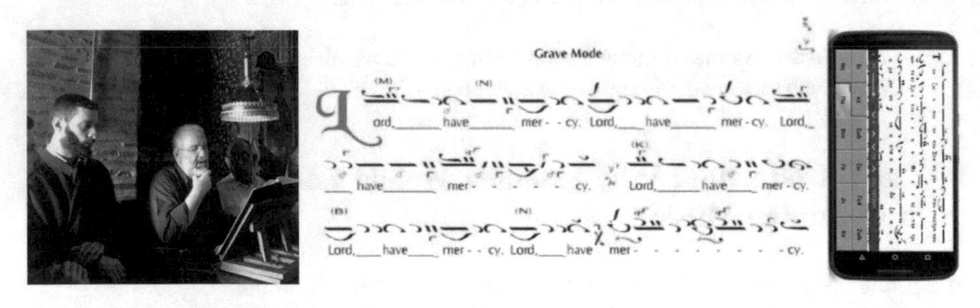

Fig. 1. The interaction scheme for Byzantine Music chanting: left, the choir focuses on the melody sung. The choir master provides the musical texts. Several books are interchangeably used. Center, the musical surfaces involved, as melodically ascribed in computer printed form, St. Anthony's Monastery, Arizona, USA. Right, the portable technology offspring, which enhances the melody with its unison tool, the *Ison*^{TM} vocalizer, providing baseline audio simultaneous to viewing.

2 Diversity and Collaboration of Peers and Suggestions

When we use the term Byzantine Music, we imply the music that was used by the people of the Eastern Roman empire (4th to 15th century), composed to Greek texts as ceremonial festival or church music. Elements were derived from Syrian, Hebrew, and Greek sources [2]. This musical system was not confined only to ecclesiastical music; it was a generalized musical system originating directly from ancient Greek Music and was used as the usual music surface by all the people living in the vast areas of the empire, from Southern Italy and the Balkans up to Russia and Armenia and down to Middle East and Egypt. The nations of Middle East were accustomed to the Greek language and (musical) civilization throughout the conquests of Alexander the Great and the oncoming hellenization.

What gives theoretical background in relating vocal music with technological paraphernalia of the mobile world, was the fact that various instruments of that era were used for training on chanting; the predominant one was the organ, depicted in Fig. 2, with its keyboard-like triggering mechanism [3]. The organ was a musical wind instrument in which sound was produced by one or more sets of pipes, each producing a single pitch by means of a mechanically or electrically controlled wind supply. Several keyboards (manuals) were played with the hands. Projecting knobs (stops) to the sides of the keyboard operated wooden sliders that passed under the mouths of a rank of pipes to stop a particular rank. The pedals of the organ were like another keyboard, played with the feet [2].

Two major characteristics of Byzantine music are its *modal* character and its *homophonic* performance. By the term mode we do not merely imply the ways of ordering the notes of a scale, but a *tropos*, a way, a guideline *of* performance including various side effects like pitch bendings and accompanying prosodic transient phenomena [5].

Byzantine Music is basically modal. Eight modes are used currently: four authentic (Modes A, B, C and D) and their plagal ones (Modes Plagal A, Plagal B, "Grave" and Plagal D). These came from a pool of about 15 modes used in Byzantine era. The ones that were not suitable for the solemnity of ecclesiastical music were diminished and have survived sporadically in Eastern musical traditions. Since Byzantine Music is a direct descendant of Ancient Greek Music, Ancient Greek modes like Dorian, Lydian, Mixolydian, etc. are rehashed. However, although the names of modes are taken from ancient Greek names, there is an ambiguity on the exact correspondence of Ancient Greek modes, Byzantine modes and the re-use of the Ancient Greek mode names in contemporary music forms like Jazz [6].

The Byzantine Music system is a Delta musical system [7]. Having an uninterrupted evolutionary course of about 15 centuries, it was reformed to an analytical system by Patriarchal Music Committees in 1814 and in 1881–85. These committees did not create a new musical system; they gave a more systematic approach to the underlying surface of the Byzantine Musical tradition and founded the theoretical values of contemporary Byzantine Music. Although their approach was focused on

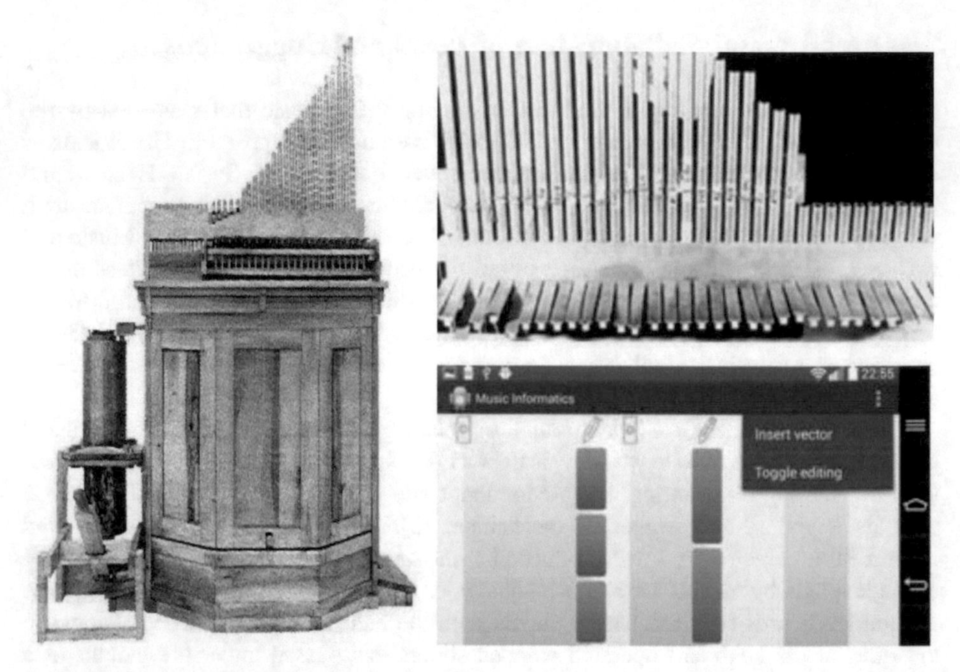

Fig. 2. Left, the 3rd century BC reconstructed hydraulis, by ECCD, the European Cultural Center of Delphi (1999). The precursor of the church organ. Top right, in detail, the keyboard that triggers the pipes. Bottom right, an Android app that can simulate non diatonic, microtal octave distributions [3]. Photograph by G. Ventouris, courtesy of ECCD.

vocal ecclesiastical music, it engulfed the whole structure of Byzantine Music. As a result, the Greek musical heritage, whether ecclesiastical or secular has been transcribed to Byzantine Music notation, its inherent notation [8].

The symbols of this notation came out of the numerous symbols of the earlier neumatic notations, like those exhibited in Fig. 3. These symbols comprise the musical alphabet of Byzantine Music, and as a whole they are described by the term parasimantiki (meaning parasemantics). However, parasimantiki does not describe notes explicitly, but as an increment or decrement from the previous level, as a phonetic transition from the current state to the next one [7]. Consequently, there is an inherent difficulty in using the vast majority of MIDI like resources that have been amply offered to the programming world by computer music industry communities [9].

Therefore, developers cannot take advantage of the plethora of MIDI or CMN oriented hardware and software facilities.

Whereas secular Byzantine Music used instruments, ecclesiastical music was vocal. It is reported however that the organ was used for tuning or ear training, but not for the formal vocal performance [4].

3 Designing Web-Based Training to Maximize Learning

The vast increase in multimedia file circulation over the Internet has created enormous in size repositories for recording, saving and distributing educational content. Especially for Byzantine music, and ardent shift has been tracked.

Some 20 years ago, the files of such content, traceable within various sites and repositories were less than a magnitude of 2 (10^2). Recently, it has been estimated that definitely more than 10,000 (a magnitude of 4) audiovisual files can be more or less downloaded, while some fear that by using mobile devices to record music events on the go, this number will soon surpass the 100,000 (10^6) threshold.

In terms of informal education, the situation could be described as a swift move of the pendulum: 20 years ago Byzantine music students were totally lacking access to audiovisual sources, and therefore they were experiencing a severe constrain in learning resources, while nowadays, the situation has gone all the way around, flooding the learner with a huge amass that he cannot absorb, as far as proper musical learning should be conditioned.

Fig. 3. Byzantine Music Manuscipts. (a) Detail from the 1935 AD microfilm of the so-called "Chartres fragment" with musical notation, beginning of the sticheron "Η σοφια του Θεου", Mode plagal D, early 11th century, Mt. Athos. The original manuscript, Laura Gamma 67, was destroyed during the bombing of Chartres, France, in May 1944. Picture cropped from the Monumenta Musicae Byzantinae site, University of Copenhagen, Denmark. (b) "Doxastarion" of Iakovos Protopsaltes, St. Panteleimon Monastery of Mount Athos, cod. 1013, 1805 AD. Paper, 17.7 × 11 cm, ff. 258.

For the ones designing educational portals the difficult query of proper selection is constantly cropping up. Yes, we are happy that the global community of Byzantine music, using extensively mobile devices records an unprecedented volume of audio-visual happenings, but which ones should be chosen to form an educative sequence?

Even further, the multiple versions of the same class, i.e. many learning objects that claim to be prototypal and accurate renditions of the same "master class" lesson (e.g. Fig 3b) pose the serious problem of which one is the proper one, in educative terms.

The learner does not have in full extent developed a mechanism to sense which is which, i.e. which chant is rich in learning content and which is merely a show off exhibition.

This research aims to create at first level a database with a proper taxonomy. Experts in the field were invited to provide the parasemantics of legendary in their learning appreciation melodies that are used in the usual liturgical and ecclesiastical practice.

The second step, however is more complicated. In Byzantine music and generally speaking in singing, the most difficult issue to resolve is the choice of the proper teacher. Music in higher levels is not taught on a massive scale: from antiquity, the analogy was if not one to one, one to a few. Quotas were never excessive, unless a super talented (in musical perception) learner was involved.

This approach of course is not taking into account the potential for multimedia learning. Before the existence of such an instruction scheme, the master would hear his student, correct his mistakes and counsel him to the endless possibilities of successful singing.

Within the knowledge society, however, such limitations are not acceptable. Multimedia learning has the width, the throughput and the dynamics to surpass these obstacles. Commencing from the usual starting point, in which for objective reasons the learner is deprived of a crafty master, machine learning and multimedia learning offer an alternative, augmented virtual reality environment to cure this handicap. A creditable database with legendary lessons to master provides the learning trajectory for the pool of online learners.

Mobile devices increase this availability. When travelling, when out of the house, when within a university, the learner can ubiquitously access locally or via the Internet a whole list of offered lessons.

A. Cognitive Walkthrough

Cognitive Walkthrough is the basic axis for providing guidance on interactive systems that supply the learning community with professionally accredited learning material. The success of this methodology lies on the fact that it is vivid, provoking the trial-and-error approach by its students, rather than using the classic methodology according to which learners should first read well the instructions manual. This approach guarantees enhanced levels of audiovisual interaction and consequently, fast and sound results.

An example of this methodology is presented by the learning portal that this paper promotes. The lesson explained, in 3 steps, concerns the "service of typika".

(1) *A new approach for mastering this lesson*

Usually this lesson is taught at the end of the second year of studies or the beginning of the third. The IT system primarily provides some liturgical incentive on when and how this service is incorporated into the usual mass (The Eucharist). The text sang comprises of two psalms, 103 (102 in the Septuagint) and 146 (145 in the Septuagint). Typika are chanted interchangeably with the usual antiphons ("Save us, O Son of God, … Alleluia").

In both cases, antiphonal chanting means that both choirs alternatively sing, verse by verse. The decorative, raised initial letter helps each choir to easily trace the text of the antiphony (Fig. 4).

Ἐπιτομὴ Τυπικῶν

Ψαλμὸς ρβ'.

Fig. 4. The service of typika, Mode Plagal D' (Fourth). Abstract from the book "Melurgy I - The Holy Liturgy. As chanted in the Stavropegic and Patriarchical Monastery of Vlatadon", Fr. N. Paris, Thessaloniki 2013, p. 30.

Thus far, the machine learning provided does not differentiate its guidance from the one given by the master to the leaner. From stage 2 and onwards, however, the use of multimedia makes the difference.

(2) *The multimedia lesson*

The learner already has three different pathways to choose.

i. *On the text dictation variations:*

In Fig. 4 the classic lesson (for the last 150 yeas or so) is presented. It is written by Professor N. Paris, it bears minor corrections and is adopted to the contemporary liturgical practice. Some points that make it attractive: it exists in printed and electronic format as well, it uses spacy, ample fonts that make it pleasant to read, while decorated capital first letters support the ecphonetic nature of shifting hence and forth between the two choirs.

It should be noted, that the advanced learner may indulge himself into very important additions that Professor Paris has made: He has designated in full length the beginning and the end of words, by scribing on the first note of the measure the total length, in quarter notes, of the metrical units for the group of syllables.

This is very helpful, indeed, since as seen in Figs. 3 and 4, as all music students know, the symbols of notes are attached on the syllables, not making it easy to discriminate between the beginning and the end off words or series of words that are pronounced as one.

Even further, as Professor Paris explains, music should invest the prosodic content of missal texts and not the opposite; therefore, the usual music interpretation of rhythmic measures may provide wrong tonicity patterns. Denoting the complex measures of 7 or 5 units alongside connection marks, aids the alignment of prosody with the patterns of the rhythm.

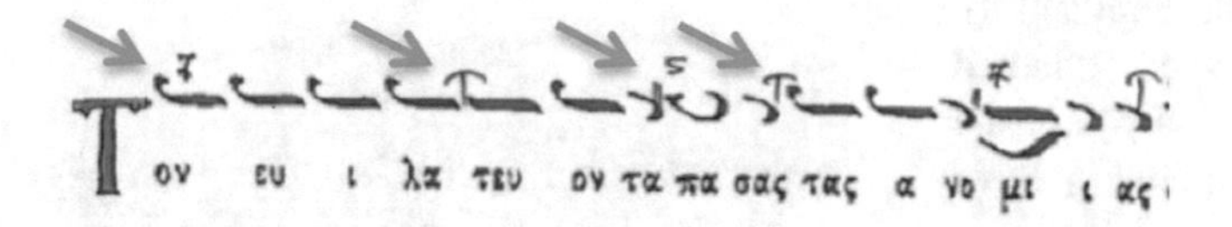

Of course, the sophomore learner is not expected to have such an insightful depth in his understanding or musical interpretation of parasemantics.

ii. *Audio guided, note by note instruction:*

The learner is expected to move with more cautious steps: he sees the text and hears the exact rendition of it. He may hear it many times, to get accustomed with it, and of course, attempt to properly reproduce it (Fig. 5, up).

The novelty of this system is that he has more than one alternatives. He may choose variants of this rendition that have variable depth, as is the previously explained “cantilevered” interpretation of 7, 5, 7 … arrays of note-invested syllables. If he is not that advanced, he may progress with the usual rendition (Fig. 5, up).

iii. *Video guided, spatial oriented instruction*:

Chanting is not only a musical activity. It is linked with the dramatic, symbolic actions that take place as a staged performance. Combined action of the clergy and the lay constituents render music phrases as a major component of the assembly’s actions. However, all parts should be operating at the same rate. The participants, when congested, improvise their steady or consistent rhythm to retain synchronization with the other involved parts (Fig. 5, down). The learner should be aware of such alterations, so usual in exuberant or solemn celebrations.

(3) *Training the (machine) learning system*

As it happens with browser practices, the system keeps records of the apprentice’s wandering through the learning objects. Therefore, it can make an estimate of the user’s strong and weak points.

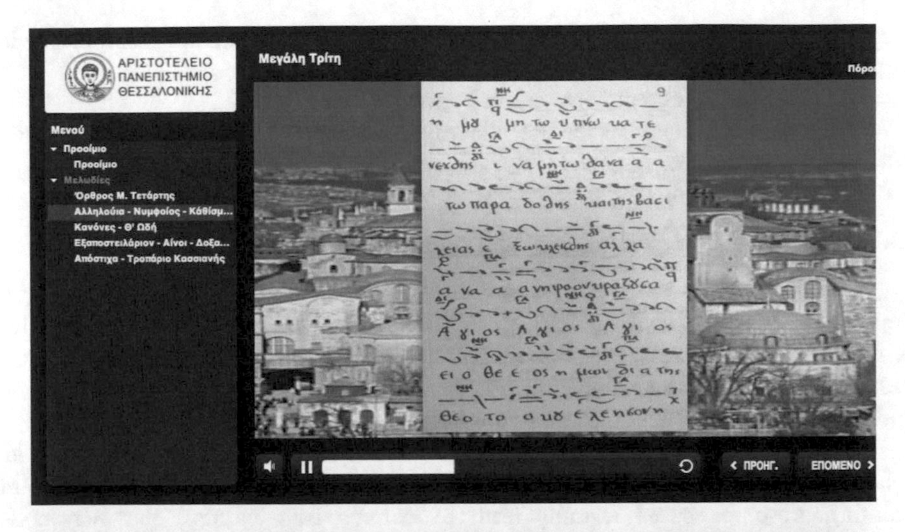

Fig. 5. The mobile version of the learning portal. Up, the learner may choose a guided audio version along with synchronized appearance of the music text. In this still are depicted Good Tuesday chants performed by E. Georgiadis' s choir. Down, the video version of Good Friday, from the St. Therapon church in Thessaloniki. It helps the learner understand how music investment adheres with the rituals and the movements performed by the clergy, the lay assistants, the council, the scouts and the gathering of the congregation.

Even further, a dialogue like phase is involved, and upon the predefined evaluation set by the professor, the interactive system can make an estimate of the level of acquisition of knowledge achieved.

For example, the missal service (Liturgy) comprises of small learning objects, like the "service of typika", which are no big deal for a second year student to master. However, the same missal service incorporates the "cherubic hymn", considered one of the most difficult learning objects to master, for obtaining the diploma, with unprecedented depth and width for its instruction.

The other missal components are in between.

Therefore, the system has the potential to discriminate between a proficient chanter, who chants at the celebrating feast of his church, with the novice who covers the simple liturgy held in week days.

B. Design of Mobile Communication Training Courses to Maximize Learning

All these Courseware Packages sustain the revolution in education offered by in depth management of contemporary VLEs. The pervasive nature of this paradigm shift is not limited to the extravagant facilities that mobile technologies have amply provided. Moreover, it is a change in the mental activity of the learner, when approaching such computer-supported collaborative portals.

Mobile technologies do not simply facilitate ubiquitous podcasting. They provide e-Learning with new tools that shape communities of practice and socio-cultural learning. Resource based learning leads to instructivistic learning, that may solve complex learning problems amassed within the Byzantine Music community [11].

C. Cooperative and Collaborative Learning

It is easily understood as an analytic and reflective approach when visiting the site - currently hosted on the Kalliope portal (http://kalliope.csd.auth.gr).

Musical practice, however, is an entirely different thing. It is a skillful, dramatic interpretation of musical pieces that cannot be mastered in conservatoires. Therefore, the machine learning approach attempts to depict the actual processing that takes place in an organized choir.

The head of the choir instructs constantly his group of fellow chanters on how church services should be performed, according to the written and oral tradition of great, accepted masters (Fig. 6, left). He may even correct his disciples if they are not meeting the performance requirements for a particular situation or activity.

Even further, he takes a leading role in antiphonal chanting, orientating the choir towards the leading in hierarchical terms episcopal authority that confers the orders.

There is an obvious difference when performing within a spacey parochial church in comparison to the narrow floor of a monastery. The level of adaptation for synchronization is not melurgy, since this would lead all the performers, especially the clergy, in unnatural movements (Fig 7).

The performance is dramatic in both melodic and kinaesthetic patterns, but not odd or remarkably uncommon so to induce the feeling of irregularity and extensive pretention.

Melurgy usually adjusts to the kinetic action and not the opposite.

Usually the action of key players involves, even further, chromatic elements as well. They may be indeed colorful annotations, as seen by the interventions of the candelabrum and the two-and-three candles waived by the deacons. In certain feasts it would even include circular movements of the big, illuminated with candles candelabrum.

Fig. 6. Left, the choir master conducting and synchronising the 1st choir around the melurgy. Right, the choir is synchronising its rhythm according to the timing of the kinaesthetic blessing of the congregation by the bishop, who becomes the center of action, in a rotative move from N to S. Priests and deacons move around him, standing always behind him.

Fig. 7. Enhanced synaesthesia - the candelabrum is turned off. Left, the deacons cense the minister and the congregation with the ringing thuribles. Right, the kinaesthetic movement of the bishop's assembly is enhanced by the lights of the three-candle chandelier, the two-candle one and the standing candle held by an altar boy or the church sexton. The minister reverences the 4 major icons placed adversarially on each side of the Altar Gate, starting from right to left, and moving circularly. The 1st choir chants the episcopal euphemism "τὸν Δεσπότη καὶ Ἀρχιερέα ἡμῶν" adjusting its tempo to the minister's moves and not vice versa.

4 Conclusion: Audiovisual Interpretation of Lighting and Movement: Why a Videolesson

New technologies offer a promising approach to Byzantine Music instruction and learning. Machine learning may devise scenarios that are more engageful, more natural in "stakeholder consultation", providing a learning-centric focus [12].

Thus far, due to technology restrictions, the audiovisual broadcasts of Byzantine Music festivities were focusing on the rigid singing order of the typikon. Recently, as the recording potential has being enhanced by portable, easily transferable, not disturbing the attendants' portable devices, the learning process has focused on the acquisition of knowledge and skills through previously unnoticed singing experiences and practices. These ones have a lot to do with the synaesthetic content of performing arts and sciences.

5 Conclusion

New technologies offer a promising approach to Byzantine Music instruction and learning. Machine learning may devise scenarios that are more engageful, more natural in "stakeholder consultation", providing a learning-centric focus [12].

Thus far, due to technology restrictions, the audiovisual broadcasts of Byzantine Music festivities were focusing on the rigid singing order of the typikon. Recently, as the recording potential has being enhanced by portable, easily transferable, not disturbing the attendants' portable devices, the learning process has focused on the acquisition of knowledge and skills through previously unnoticed singing experiences and practices. These ones have a lot to do with the synaesthetic content of performing arts and sciences.

Acknowledgement. The authors would like to express their gratitude to their Eminencies, His All Holiness Bartholomew, Archbishop of Constantinople, New Rome and Ecumenical Patriarch, alongside the Most Reverend Metropolitan Panteleimon of Tyroloa and Serentium, former Abbot, former Rector, Professor Emeritus of the Aristotle University of Thessaloniki, and the Most Reverend Bishop Nikiphoros of Amorion, Abbot of the Holy Patriarchal Stavropegic Vlatadon Monastery, for providing their unreserved support, both morally and materially, apart from their blessing, for the successful completion of this research.

In the sermons recorded in the kyriakon have participated: the Most Reverend Metropolitan Theoliptos of Iconium, the Abbot, Most Reverend Bishop Nikiphoros of Amorion, the Very Reverend Fathers Gabriel, Petros, Georgios and Dorotheos, the Reverend Deacons Petros and Panteleimon, the choir leaders Fr. Nektarios Paris and Symeon Kanakis with their assistants Chrysostomos Vletsis and Philippos Rentakis, and the sextons from the International Boarding Establishment of the Monastery.

The audiovisual material from the Vlatadon Monastery has been recorded by Georgios Varsamis, Nikolaos Rentakis and Dionysios Politis. The audiovisual material from the St. Therapon church has been recorded by Athena Katsanevaki, Constantinos Ginis and Dionysios Politis. It has been released for copyright-free educational use over the Internet by the Very Reverend Archimandrite Fr. Iakovos Athanasiou, Chancellor of the Holy Metropolis of Thessaloniki. The melodic texts used within the learning objects were kindly provided by Professor Patroklos Georgiadis, Aristotle University of Thessaloniki, son of the late Archon Lampadarios in Musicology of the Ecumenical Patriarchate Eleftherios Georgiadis.

The authors gratefully inscribe this research in memoriam of their master in Byzantine Music E. Georgiadis (Constantinople, 1920 - Thessaloniki, 2016).

References

1. Cox, C., Warner, D.: Audio Cultures, Readings in Modern Music. Continuum, London (2007)
2. Columbia Encyclopedia: Electronic Edition (2000)
3. Politis, D., Piskas, G., Tsalicgopoulos, M., Kyriafinis, G.: variPiano™: visualizing musical diversity with a differential tuning mobile interface. Int. J. Interact. Mob. Technol. iJIM **9**(3) (2015)
4. Psachos, K.: The Eight Modes System of Byzantine Ecclesiastical and Folk Music, and that of Harmonic Unison. Polychronakis, Neapolis (1980)
5. Spyridis, H., Politis, D.: Information theory applied to the structural study of Byzantine Ecclesiastical Hymns. ACU-STICA **71**(1), 41–49 (1990)
6. Moysiadis, P., Spyridis, H.: Applied Mathematics on the Science of Music. Zitis, Thessaloniki (1994). (in Greek)
7. Margounakis, D., Politis, D.: Producing music with N-Delta interfaces. In: Proceedings of AXMEDIS 2005, Florence, Italy, 30 December–2 December (2005)
8. Pantelopoulos, P.: 350 Traditional Songs from Greece (in Byzantine Music Notation), Patras, Greece (1996)
9. Paris, N., Politis, D.: Beyond MIDI: theoretical foundations for the voice instrument digital interface for byzantine music. In: Proceedings of 4th Conference on Interdisciplinary Musicology CIM08 MUSICAL STRUCTURE, Dept. Of Musical Studies, Aristotle University of Thessaloniki, 3–6 July (2008)
10. Wentzel, P.: Using mobile technology to enhance students' educational experiences. ECAR Case Study **2**, 1–22 (2005)
11. Gall, M., Breeze, N.: Music composition lessons: the multimodal affordances of technology. Educ. Rev. **57**(4), 4115–4133 (2005)
12. Hampel, R., Hauck, M.: Towards an effective use of audio-conferencing in distance language courses. Lang. Learn. Technol. **8**(1), 66–82 (2004)

Neurophysiology-Based Acoustic Measurements of Singing Contours

An Empathetic Interactive Loudness Approach

Dionysios Politis[1(✉)], Georgios Kyriafinis[2], Jannis Constantinidis[2], and Nektarios Paris[3]

[1] Department of Informatics, Aristotle University of Thessaloniki, Thessaloniki, Greece
dpolitis@csd.auth.gr
[2] 1st ENT Academic Department, AHEPA General Hospital, Aristotle University of Thessaloniki, Thessaloniki, Greece
orlci@med.auth.gr, janconst@otenet.gr
[3] Department of Music Science and Art, University of Macedonia, Thessaloniki, Greece
nepa@uom.gr

Abstract. This research focuses on how singing is perceived when amplifying mechanisms vastly enhance its energy contours. This is especially true in big auditoriums, churches or outdoor performances. In practice, the singing voice is altered in its phonetic characteristics. The same time, the listener's perceptional sensory mechanisms are affected, and in terms of neurophysiological processes, an altered acoustic form is received. Apart from these findings, a new unit for evaluating the equivalent to pure acoustic singing is proposed, when using instrumentation to amplify phonation.

Keywords: Singing voice · Neurophysiological perception · Dynamic activity

1 The Acoustic Background of Singing

Sound is one of the most impressive elements of audiovisual performances with its awareness for formal communication, i.e. by having a meaning to convey in every language, for each audience. Its expressive potential varies from whisper, merely audible, to emphatic in expression very loud concerts or similar forms of entertainment. It may offer with music beauty of form, harmony and expression for emotions. Recently, it impresses audiences as spectacular sound effects or pleasingly to be perceived harmonious musical composition.

How engineers manipulate the expressive elements of musical entertainment, i.e. singing, dancing, performing on stage, distributing to audio channels the sound associated with movement, accompaniment of visual images, and similar combinations, renders the easily distinguishable forms, styles and genres that have hosted the musical revolutions of the 20th century, ongoing lustily to the 21st.

M. E. Auer and T. Tsiatsos (Eds.): IMCL 2017, AISC 725, pp. 758–767, 2018.
https://doi.org/10.1007/978-3-319-75175-7_74

The branch of physical science that relates the sense of hearing with sounds is referred to as Acoustics.

Acoustics is closely related with music, since positive artistic perceptions reveal how the audience embraces the sonic creative skill. In simple terms this means that music and the impression of loudness are interconnected: meaningful interpretation of loudness contours cannot take place without using elements of music, while the laws of harmony cannot be fully exploited without taking into account adequate loudness contours.

In conclusion, ideas or feelings in musical renditions are tuneful, melodious and euphonious when associated with pleasant sounds building public confidence.

2 The Neurophysiological Perception of Loudness

The expression of sonic waves travelling from the sound source to the receiver object has the following vector form:

$$\boldsymbol{P} = \boldsymbol{P}_o \sin 2\pi(f\,t\, - x/\lambda) \tag{1}$$

where

P_o = the pressure intensity at the source (N/m^2)
x = the distance of the preceptor from the source
f = the oscillating frequency of the source
λ = the wavelength.

Sound vibrations travel with a relatively slow velocity through the air or other media that surround us, as seen in (2):

$$V_s = \sqrt{\frac{\gamma RT}{M}} \approx 20.1\sqrt{T} \tag{2}$$

where

γ = C_p/C_v, the ratio of specific heat capacities under steady pressure and volume
R = the universal gas constant
M = the mole of the media for transmission
T = temperature in Kelvin degrees.

The material of the transmission medium designates the properties of sound. We have a different impression of sounds when air is the only medium between our ears and the sonic source and when solids or liquids are involved. So to say, sound is transmitted at 6100 m/s in steel, at 1480 m/s in water and only at 331 m/s in the air at 0 °C. Temperature is another crucial parameter for sound transmission: for every degree in the Celsius scale sonic velocity increases 0.60 m/s.

Thus the speed of sound in the air is approximated by

$$v_s \approx (331 + 0.60\,T^{o})\,m/s \tag{3}$$

with T° the temperature of air in Celsius degrees (C°).

For this research, the properties of sound transmission in the air are encountered, but not in open space; rather the architectural characteristics of the building in which concert takes place are taken into consideration. Audience comprehension is subordinate not only to the purity of the vocal tone but also to the influence of room acoustics, the structure of the roof, refraction due to permanent parts of the construction like pillars and crossbars, the absorbance by moving or immovable materials, the back formation from drapery or the clothes of densely assembled listeners, the capabilities of electroacoustic machinery, as are microphones, amplifiers and loudspeakers, and many other parameters that overall take up and reduce the effectual intensity of sound.

Indeed, the behavior of spreading sound waves in the open air is rather different with the way they function in public buildings, churches and similar constructions. Enclosure of sound sources produces multiple reflections from the enclosure surfaces. Sound waves have a variable behavior due to the fact that given a certain enclosure; the resonant frequencies (i.e. the preferred frequency of vibration having attendant high sound pressures) act as springs for specific frequency zones. At these frequencies the mass of the air bounces on the enclosed air, exerting concentrations and thin air regions [1]. The nature of the fields depends significantly upon the dimensions of the enclosure in relation to the wavelength λ of the transmitted sound (Fig. 1).

Room acoustics have also to do with the kind of music performed. In Table 1 the fundamental frequency (*F0*) ranges of major instruments are demonstrated.

Table 1. Fundamental frequency (*F0*) spectrum for major music instruments

Instrument	F0 frequency range (Hz)
Human voice	70–2000
Piano	30–3500
Violin	200–3000
Flute	260–3000
Pipe organ	16–4000

Human perception however is not limited to the *F0* mental impression. Overtones accompanying the fundamental tone create a complex oscillation, which gives a rich in content awareness through the senses. As a neurophysiological process, including memory, sonic stimuli is categorized more or less to the following bands:

(1) *Sub-bass:* between 20 and 60 Hz

Sounds within this range are rather felt as strong oscillations by the human bones primarily and tissues to a lesser extend than heard by the ears. Such sounds affect the vestibulocochlear sensory pathways of the inner ear, which contribute both to hearing and balance [2].

(2) *Bass:* between 60 and 200 Hz

Voice, instruments and sounds in general that concentrate in this frequency range are denoted The most important components of rhythmic energy are localized within this band, characterized for its deep, resounding, resonant low frequency impression. The well known bass instrument ranges between 80 and 640 Hz (for note pitches). Bass singers and baritones as well are proficient in using this band.

(3) *Bass-medium:* between 200 Hz and 1.5 kHz

The first and stronger harmonics of most instruments can be found within this range.

(4) *High-medium:* between 1.5 and 4 kHz

Important harmonics that shape the special timbre of musical sound are located here. They dignify the character and quality of singing voice as a distinct feature from its pitch and intensity.

(5) *High:* between 4 and 10 kHz

Within this band is clearly expressed and easily understood the tonal behavior of a sound in relation to similar sounds.

(6) *Ultra-high:* between 10 and 20 kHz.

At low frequencies wave effects are not that apparent, as air undergoes uniform compression and expansion. As higher sound frequencies however, λ is comparable to enclosure dimensions and apart from interference diffraction and scattering by the presence of solid obstacles play an important role.

A region around the sound source is unaffected by such phenomena and is called the Direct Field. Complementary to this is the Reverberant Field, in which sound is directed to listeners with multiple reflections. In cases of specially designed rooms, like auditoriums and similar, if the field is omni-directional, the term Diffuse Field is used (Fig. 1).

Sound waves transmit mechanical energy, in the form of with the compressions and expansions movement (Fig. 1). What objectively is hence received by sensory organs is the *sound power* generated by the source. In SI the measurement of energy concentration is defined as its Intensity, i.e. the rate at which energy passes perpendicular through a unit are of space (Watts/m^2).

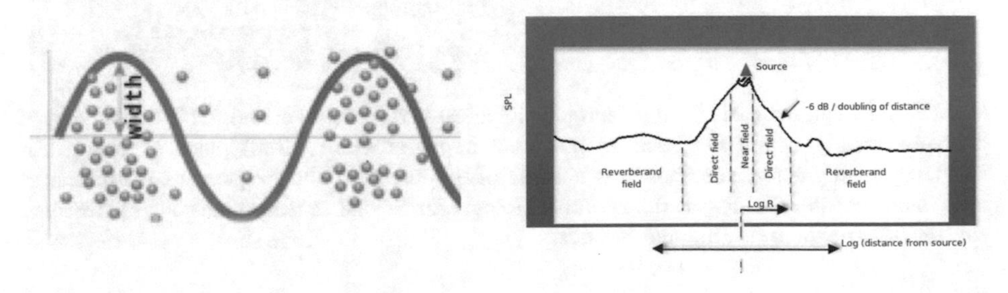

Fig. 1. Left, pulse wavefronts forming atmospheric pressure modules. Right, field regions for sound pressure levels for a source within an enclosure.

Although rational in its SI conception, this way for measuring the energy of a sound proves to be rather constipated. Instead, a reference level intensity I_o is chosen, and the intensity I of a sound wave is expressed in relation to I_o.

One **Bel** corresponds to measured intensity 10:1. Since a logarithm scale is used in most sonic measurements, 2 Bels would encounter a 100:1 relation, a rather considerable extent for intensity estimations. Therefore, for daily use Bel is divided to 10 deciBels (dB).

So, the intensity to be received by our ears is estimated as follows in dB:

$$Sound\, Power\, Level = 10\, log_{10}(I_o/I_o) \tag{4}$$

In terms of neurophysiology, the perception of sonic or musical complexity lies within the inner ear structure (Fig. 1). It comprises an intricate arrangement embedded in the temporal bone, whose diverse organs, like the utricle, saccule, cochlea and three semi-circular canals. While the cochlea is the basic neurotransmitter for the junction of electromechanical vibrations of fluids, membranes and elastic solids with the synapses of the acoustic nerve, the vestibular structures highly influence the sense of balance for the whole body. In acoustic terms this is perceived as irritation within the sub-bass zone (Fig. 2).

When movement is involved along with sonic impressions, the utricle, the larger of the two fluid-filled cavities of the labyrinth within the inner ear, contains hair cells and otoliths that send signals to the brain concerning the orientation of the head. With this organ cerebral sensing takes place when the human body alters its position horizontally. This occurs during physical activity, or abundantly, when people assume a horizontal, retiring position while resting on a supporting surface, like bed. The saccule, the smaller of the two fluid-filled cavities, encloses another region of hair cells

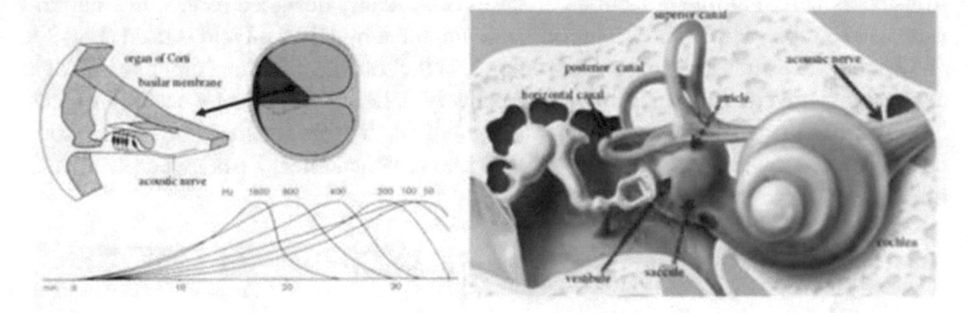

Fig. 2. Inner ear, the basis for the neurophysiological perception of sound. Left: the curving of the spiral cavity within the cochlea, projected in a transverse (axial) plane. The cochlear frequency tuning curves are shown as a result of the threshold and frequency equalization of individual neuron cells. Right: the neurophysiologic arrays and sensory pathways of the inner ear, that contribute to hearing and balance.

and otoliths that send signals interpreting the orientation of the head as vertical acceleration. This is crucial when listening to highly rhythmic music, with strong bass frequencies that vibrate our body, we may perceive a feeling similar to moving within a fast elevator [3].

For neurophysiology, the perceived sound does not have equal response over the hearing spectrum as far its intensity is concerned. The actual listening indication over frequency range is seen in Fig. 3.

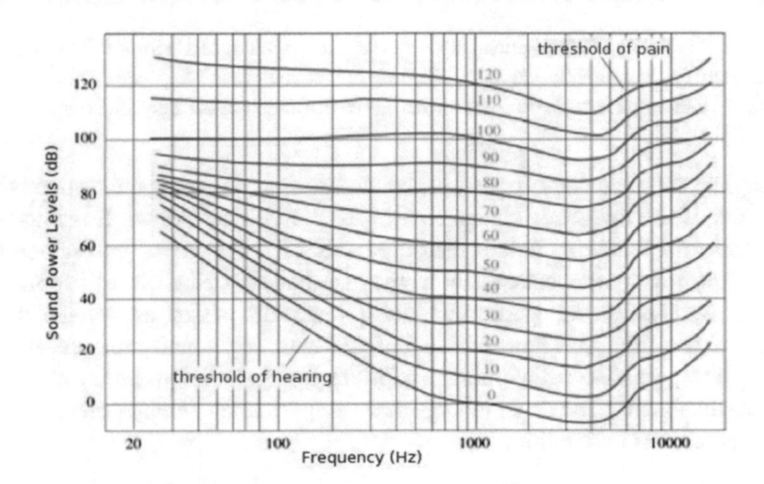

Fig. 3. The human hearing domain depicted in logarithmic curves of equal hearing in neurophysiological terms.

It is a norm of **acoustic efficiency** over the frequency characteristics of incoming signals in cerebral activity terms.

As previously mentioned, the classification of sound sources is meaningful when the produced energy of sound waves is correlated with zones of the audio frequency range 20–20000 Hz.

However, the thresholds for hearing vary considerably from person to person and individual characteristics decline with age as well. A note with pitch of 1.000 Hz is located between notes B_5 (987,77 Hz) and C_6 (1.046,5 Hz); at that region human hearing has exceptional physical characteristics as far as the sideways extent of acceptable sound waves is concerned; indeed it surpasses 120 dB, which in algebraic terms means that if the intensity of sound waves for the threshold of pain is 1 W/m^2, then the threshold of hearing is 10^{-12} W/m^2, making the ear one trillionth times more sensitive (1: 1.000.000.000.000) [4].

In practice, this means that while an orchestra can perform within a conservatoire, it cannot do the same in open space without heavy electromechanical support (Fig. 4).

Fig. 4. Left, AUTh's orchestra rehearsing for an open space concert within the campus, having, centre, heavy electroacoustic support. Right, the same orchestra performing within the University's Hall, in pure acoustic mode with no electromechanical amplification.

In singing, the frequency content of radiated sounds describes the distinctive nature more or less of the frequency characteristics for the voicing source. It is important for this research to be able to obtain measurements of the relative (and if possible the overall) frequency components, either as they are heard in pure acoustic terms, or as it is usual nowadays, as the performed vocal rendition passes as voltage through a microphone, amplifier and loudspeaker before becoming sound pressure at the open field. I.e., how room acoustics mingle with electro acoustics when an electrical low level intensity signal is subjected to large scale amplification and is received as acoustic output of enormous calibration.

3 The Experiment: Loudness Measurements Within an Acoustic Enclave

The experiments described in this section were conducted in three churches of different sizes. Mobile technology for recording the audiovisual happenings was involved, alongside classic multimedia recording devices. When somebody is performing within a confined auditorium enclave, the hearing perception is not the same: the walls, columns, arches, naves along with the extensive wooden fixtures and the cloth based ceremonial investments of any kind (the attendants' ones included as well) dynamically reshape the level of the acoustic perception. Both the performers and the audience are feeling as if they were in a loudspeaker [5].

The architectural shape designates regions of acoustic amplification, acoustic "shadows", or spots with out of control echoing.

Performers lack the presence of monitor loudspeakers and therefore rely on the acoustics of the building to calibrate their level of intervention when handling the microphone.

Fig. 5. The outside and inside 3D architectural characteristics of one of the auditory enclaves where hearing perception sound levels were measured: the church of the Dormition of Our Lady in Alexandria, Imathia district in Central Macedonia. A fairly big construction, typical cross-shaped basilica with dome, with 3 naves, and an inverse Π shaped upper floor construction. Build between 1965–71.

For this research, there were involved:

- A 15 member female choir that was performing on the left central part of the main nave, in front of the solea
- two priests and a deacon, using the 3 standing microphones of the whereabouts of the altar
- four members of the male chanting at the right analogion.

The recording equipment was placed in between the two choirs, under the rear right column of the dome (Fig. 5).

It was equally distanced from both the female choir, which performing in acoustic mode, and the nearest wall mounted 2 way speakers that were connected with the electro mechanic amplifying microphone based sound system [6].

All the performers, acting interchangeably and in a rotating style, according to the *typikon* of the service held, tried to have a comparable level of performance as far as sound loudness was perceived.

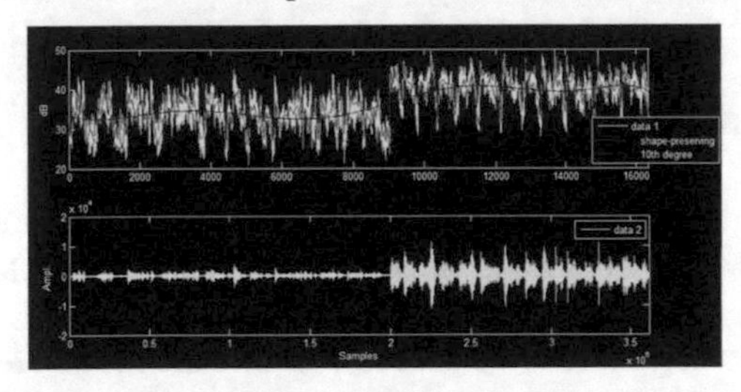

Fig. 6. Solo vs. choral. Energy plots with a 10th degree polynomial fitting and a shape preserving interpolant. In the first part, a member of the female choir is performing a solo recital, while in the second part all 15 members participate.

In Fig. 6 the pure acoustic mode reciting performance of the female choir is evaluated. The energy plot gives an image of what is heard by the attendants situated close to the central nave and under the dome.

Therefore, an indication is given how much more powerful is the $x10^{+}$ prosodic recital in acoustic mode.

In Fig. 7 the comparison stands between a male performer who uses a microphone. The coda part of his singing, as he ends the phrase is denoted by the left arrow. Immediately afterward, the female choir catches up with the onset of their acoustic performance. The onset is demonstrated by the second arrow [7].

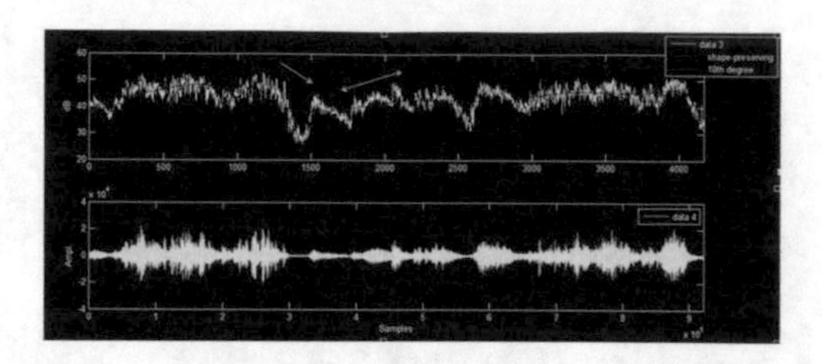

Fig. 7. Energy plots with a10th degree polynomial fitting and a shape preserving interpolant. Left, microphone enhanced solo singing, dwindling by the denoted Coda arrow. Right, acoustic choral singing, achieving comparable energy levels, especially sustained after the initial Onset phase, denoted by the two-headed arrow.

Once a sustainable level of acoustic performance is achieved, the energy reaching the audience's ears under the dome is more or less the same, defining some sort of x10 chanting power, as far as neuro-physiological perceptions involved [8].

It is noteworthy, that although comparable energy levels may be sustained by both male and female performers, the energy distribution over the spectrum is different, giving thus a clearly different temperamental in phonation perception.

This is seen in Fig. 8, where it is obvious that the female choir, commencing to perform after the 7ths, demonstrates a $\sim$100 Hz shift [9].

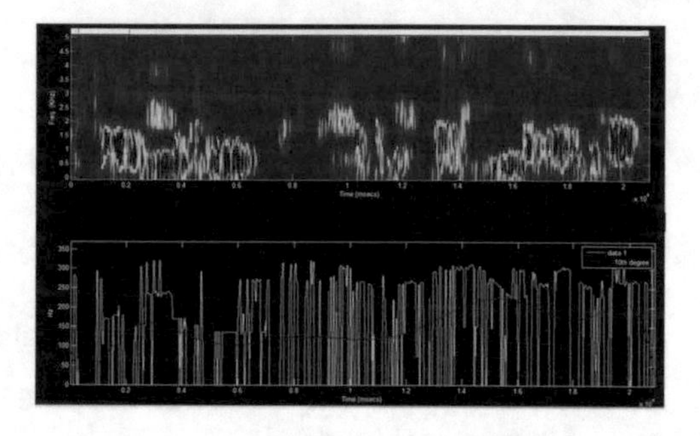

Fig. 8. Energy contours and Fundamental Frequency distributions for a male performer (between 0 and 7 s), chanting with a microphone, and afterwards the 15-member female choir in pure acoustic mode, performing similar in melody verses. A 10th degree polynomial fitting curve gives a first image of the shift performed by the female choir.

4 Conclusion: Towards Chanter Calibrated Metrics

As customers prefer to estimate the speeding performance of their vehicles in "horse power" metrics, rather than kW, the same style measurement is proposed for calibrating the acoustic performance of singers.

It is handy for chanters and sound engineers to be able to consider the actual effect of pure acoustic singing, for single performers or choirs, within the acoustics of a specific architectural arrangement. Even further, it is important to be able to deliver these metrics' when electrical amplification takes place.

In the latter case, the perceptual acoustic signal is estimated with the equivalent of chanters. Each one of them is taken as basis, when his energy contours are estimated at specific points of the acoustic enclave. The position of the source is usually at key liturgical and functional points of the aisle.

Acknowledgement. The authors would like to express their gratitude to the Aristotle University Orchestra (the Organizing Committee of the Orchestra, its Director, Dimitrios Dimopoulos, and its musicians) for providing part of the acoustic material used in this survey.

Another part was recorded in the church of the Dormition of Our Lady in Alexandria, Imathia district of Central Macedonia, under the blessing of His Eminence the Metropolitan of Berea, Naousa and Campania, the Most Reverend Panteleimon.

The remnant of the audio files that were analyzed was recorded in various schools of Thessaloniki, like the Ecclesiastical Lyceum of Neapoli-Stavroupoli and the Experimental Primary School for Intercultural Learning, under the auspices of their Principals, Dr. Aikaterini Galoni and Georgia Panagiotopoulou respectively.

References

1. Hardcastle, W.J., Laver, J., Gibbon, F.: The Handbook of Phonetic Sciences. Wiley, Hoboken (2010)
2. Politis, D., Tsalighopoulos, M., Kyriafinis, G.: Dialectic & reconstructive musicality: stressing the brain-computer interface. In: Proceedings of the 2014 International Conference on Interactive Mobile Communication Technologies and Learning IMCL2014, 27–28 November (2014)
3. Politis, D., Tsalighopoulos, M., Iglezakis, I. (eds.): Digital Tools for Computer Music Production and Distribution. IGI Global, Hearshey (2016)
4. Williamson, V.: You Are the Music: How Music Reveals What it Means to be Human. Icon Books Ltd., London (2014)
5. Meyer, L.: Emotion and Meaning in Music. University of Chicago Press, Chicago (1956)
6. Ouzounian, G.: Visualizing acoustic space. Musiques Contemp. **17**(3), 45–56 (2007)
7. Cook, P.: Music, Cognition and Computerized Sound, An Introduction to Psychoacoustics. MIT Press, Cambridge (1999)
8. Raphael, L., Borden, G., Harris, K.: Speech Science Primer – Physiology, Acoustics and Perception of Speech. Williams & Wilkins, Philadelphia (2007)
9. Barbour, J.M.: Tuning and Temperament: A Historical Survey. Da Capo Press, New York (1972)

Spatial, Kinaesthetic and Bodily Chromaticism

Gesture and Sound Interconnection Within the Framework of Singing Performance

Dimitrios Margounakis[1(✉)], Dionysios Politis[1], Nektarios Paris[2], Symeon Kanakis[2], and Neoklis Lefkopoulos[3]

[1] Department of Informatics, Aristotle University of Thessaloniki, Thessaloniki, Greece
{dmargoun, dpolitis}@csd.auth.gr
[2] Department of Music Science and Art, University of Macedonia, Thessaloniki, Greece
nepa@uom.gr, kamaresfc@gmail.com
[3] Architecture Firm "Domi & Energeia", Thessaloniki, Greece
nlefk@tee.gr

Abstract. This paper extends the notion of chromaticism for the actual performance scenery. In contemporary staged performances, singers not only chant but they also move around, while the theatrical scenery is dynamically recreated with advanced audiovisual components. The lighting components interplay with streaming components, urging for kinaesthetic and bodily communication with audiences. As interaction increases, the notion of "live performance" is enhanced with ubiquitous and mobile communication components. Thus, an augmented sphere of interaction is created, for both in situ and remote participants. The production terms of such performances are contested under the prism of liturgical practices in diachrony and advances in mobile technology in our synchrony.

Keywords: Chromaticism · Spatial stage characteristics
Bodily and kinesthetic postures

1 The Artistic Perception of Acoustic Background in Singing

In recent years, mobile communications alongside cross- referenced mobile media expression provide both researchers and the wide public with enhanced methods and practices of neuroscience. This environment within the global communication sphere is continuously escalating with ample provisions of high quality audiovisual material, recently in 3D video and audio formats, providing a better interpretation on how human and IS interactions can become more realistic in terms of synaesthetic IT-related behaviors [1].

Since antiquity, artistic impression is triggering mental processes for acquiring knowledge through the apprehension of senses. Commencing from Egyptians, Greeks or Chinese, the scientific community has clues how chromatic impressions provided awareness or understanding on the nature of multicolored arrangements (Fig. 1).

M. E. Auer and T. Tsiatsos (Eds.): IMCL 2017, AISC 725, pp. 768–777, 2018.
https://doi.org/10.1007/978-3-319-75175-7_75

Fig. 1. People in ancient civilizations especially enjoyed the impressions of multicoloured creations. Left, Minoan performances, some 15 centuries BC in Crete. (Reconstruction from http://terraeantiqvae.blogia.com). Centre, the Parthenon chromatically reconstructed. Right, for China, even in pre-Christian era times, significant painters had made a name. During the Tang Dynasty (618-907 AD), a considerable wall painting culture made its appearance. Furthermore, the Song Dynasty era culture (960–1279 AD) promoted a notion for more naturalistic image perceptions.

Although visual information is used in diachrony to strategically manipulate audience perception, the exact "nature" of color sensation is not perfectly understood. Perhaps more than any other epoch we have diluted the light constituents to a great detail, but still we lack a coordinated analysis of the relationship between visible impression, gesture, sound and overall performed actions within the context of expressing an idea or meaning. In terms of cognitive neuroscience, performance provides symbolic communication between people assembled at a public event. Since mobile communication aims to enhance the means of connection between people or places, in particular by creating private virtual reality like transmission of information by various Internet based applications, it promulgates audiovisual messaging systems that provided multi-angle recreation of real life, real-time events, in a broadcast mode.

In functional terms, the masters of music have instructed the coordination of position and movement alongside singing performance [2, 3]. Broeckx [4] states that through gesture music can be perceived as kinaesthetic performance, which is empathetically felt with the biggest artistic event of our times, in global terms, the Eurovision song contest.

It is indeed a pageant for mobile computing in many aspects.

This research aims to present a better insight on how the spatiotemporal action of performers is related to the chromaticism of music.

2 The Chromaticism of Singing

The evolutional course of music through centuries has shown an incremental use of chromatic variations by composers and performers for melodies' and music sounds' enrichment. An integrated model for calculating most musical aspects of chromaticism has been presented [5]. The model takes into account both horizontal (melody) and vertical chromaticism (harmony).

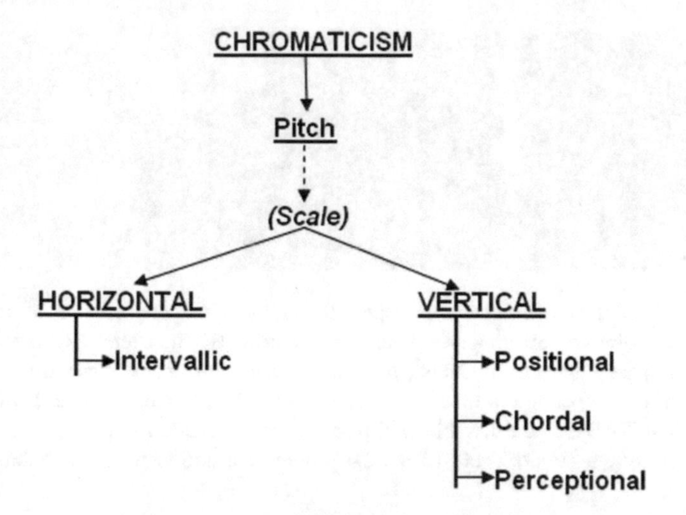

Fig. 2. The factors that affect low-level music chromaticism in a framework

The proposed qualitative and quantitative measures deal with music attributes that relate to the audience's chromatic perception. They namely are: the musical scale, the melodic progress, the chromatic intervals, the rapidity of melody, the direction of melody, music loudness, and harmonic relations. This theoretical framework can lead to semantic music visualizations that reveal music parts of emotional tension. The model, which is shown in Fig. 2, covers the attributes that affect the chromaticism in the melodic and harmonic structure of a musical piece.

When the notion of music chromaticism relates to well- tempered instruments (e.g. piano), it matches well to the western definition of chromaticism: *Chromaticism in music is the use of notes foreign to the mode or diatonic scale upon which a composition is based. Chromaticism is applied in order to intensify or color the melodic line or harmonic texture* [6, 7]. However, when it comes to human singing voice, things are a bit more complicated. Discrete pitch values of piano do not apply on the continuous nature of singing human voice pitch. As a result, the chromatic phenomena of human voice (with fundamental frequency ranging from 70 to 2000 Hz) are of much more capabilities and auditory perceptional effects. Therefore, their chromatic results are usually not straight lines, but curves (Fig. 3).

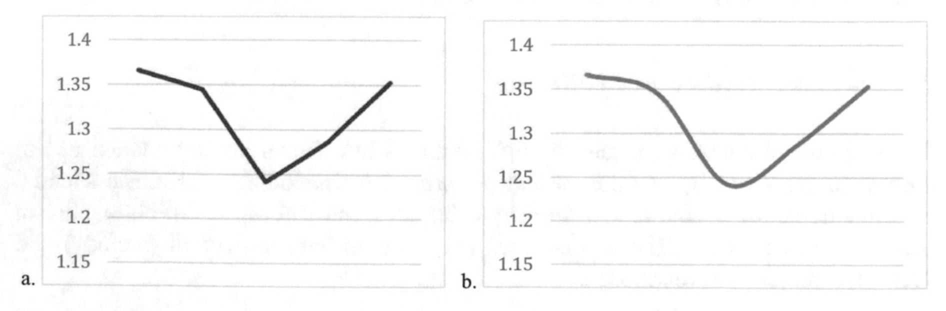

Fig. 3. a. Chromatic indices in Western music notation b. Chromatic indices in human singing voices

The audio performance of singing, which is discussed here, stands as one of the high-level music characteristics, enhancing the music chromaticism and the audience's perception. The second high-level characteristic is visual performance. Both of these characteristics are expressed in live music performances and try to trigger audience's feelings by enhancing the music experience. Moreover, they cannot be described in notation staff (it can be said they even slide it over) (Fig. 4).

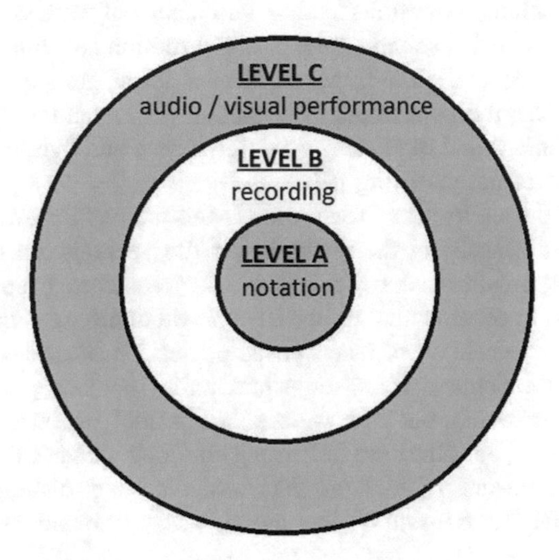

Fig. 4. Levels of music presentation. The outer circles embed more chromaticism compared to the inner circles.

Concerning visual performance, innovative and impressive effects (lights, digital representations, spatial distribution, gestures, movement and dance) are used in several ways: either in accordance to certain music aspects, either in accordance to rhythm, or in a random way.

Digital technology and computing machinery have essential influence to the nature of things to come. They have given increased control to complex arrays for natural movement and the setting of lights, revolutionizing the way with which stage performers handle sound and insinuate motion.

In a lot of cases, it has changed the way with which the stage designers, whether organizers, performers, or technical assistants attest, develop, present and distribute their work in everyday music entertainment habitats [8].

Developments in electronic equipment are not actually something new in the arena of audiovisual presentations. The last couple decades, both engineers and audiences are experiencing a pivotal recreation of audiovisual recreation and provision of new techniques for lighting the performance stage. What is a really new substantial innovation is the increased level of control that Human-Machine Interaction offers between electronic lighting devices, sound dispatching equipment, and computing machinery of all kinds, which collectively produce distinctive audiovisual streams by promoting pioneering ways of interaction [8]. For instance, the capacity to interpret human motions and indicative signs is a sheer, unmitigated increase in the degrees of communication exerted over music production. It reshapes the behavioral communiqué as far as bodily patterns are incorporated in performance semantics.

This way we could say that the space within which the performance is staged becomes an imminent human-machine interface. Everything the performer or the stage technician touches and uses produce elaborate results: in one way or another the whole stage, and even its online "extensions", convey to means of expression. It is recognized by theatre and art people that sound expression and motion are adjusted to the demands of the performing space; this space by it self dictates some, perhaps unwritten-rules. As the audience perceives the performing environment as its natural habitat, audiovisual engineers take advantage and regulate accordingly scene and lighting design to express the collective mood of the gathering [9].

An example of the audio-visual revolution is the famous Eurovision Song Contest. After studying the diachrony of the festival over the years, it can be noticed that the effects of music chromaticism have fallen into a transition: from vocal patterns to audio-visual effects. In other words, in our days if you don't "see" the performed song, you may lose a large percentage of its proposed perception. That was not the case some decades ago, when the richness of the songs laid mainly on their vocal part (polyphony, choir, chromatic alterations), but they were rather "static" on stage.

In large area events, as is the case of Eurovision Song Contest, the floor of the stage is equipped with sensors of various technologies. We distinguish three major approaches, thus far, in translating the mood, aesthetics and artistic content that streamed:

a. Motion sensors - in practice wearable technology, stage based sensors, laser harps, and similar.
b. Remote controlled sensors - video streams, infra red beams, laser lightings, exploited holography, stimulated emissions, LED technology and many more.
c. Internet or mobile telephony network extensions - remote voting, opinion probing, mood sensing via tweets, distant users' reciprocal reaction and similar.

These new technology achievements, together with the performing arts of the participants form a complementary grid of mostly aesthetics to support music in order to produce more intense feelings and enhance the music experience (Fig. 5).

Although the spectacular performance of a contemporary music show seems as a totally new idea, the truth is that elements of this interface can be found from much earlier, e.g. in the performance of Byzantine music during the psaltic ritual. During psaltic ritual, *λόγος* (ecphonetic speech), *μέλος* (chant) and *κίνηση* (movement) are the three pillars that extort what the soul sees, remembers and wishes. Although they are expressed in a simple, modest and uncomplicated way, their interconnection results in a unique effect on singing's perception.

The members of the choir have a specific role, depending on their grade, act musically on the lectern, and contribute to chanting, in order to produce a collective result [10]. In the patriarchal temple, the music choirs consist of the protopsaltes (protocantor, director and leader of the right choir), the lambadarios (leader of the left choir), the domestikos of the first, the domestikos of the second, the first and the second

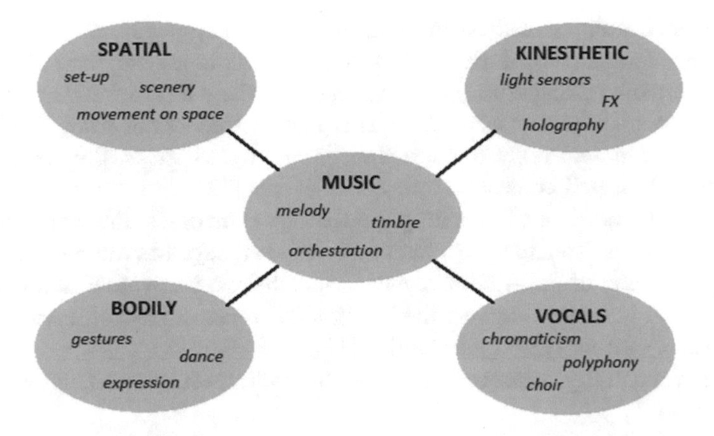

Fig. 5. Interconnection within the framework of music performance

kanonarch of the right choir, the first and the second kanonarch of the left choir, and the other isokrates and regular helpers [11].

3 The Method Used for Detecting Chromatic Spatial Patterns

This research was conducted in a historic site, originating in its symbolism since 51 AD, constructed in its present form, originally in 1351-1371 AD. Having among its donors queen consort Anna Palaiologina (Anna of Savoy), governesses of Thessaloniki, it has been an emblematic monastery within the city walls. The original building has been extended through ages bearing various styles interventions, like the 19th century loggia with colonnade and its westward entrance under a peristyle (Fig. 6).

Fig. 6. a. The chapel on the rock where Apostle Paul preached in 51 AD. b. The original form of the kyriakon of the monastery, in an engraving made by A. Xyngopoulos, 1951. c. The church as seen now with its architectural extensions from SE. d. Ground plan of the church. The black lines denote the original stone and brick structure of the 14th century building. An inversed Π extension has been added since. The loggia in (c.) is a 19th century addition.

Two "mobile" cameras recorded the event alongside two voice recorders, handled independently throughout the staged liturgy, without conspicuously attracting the notice of the audience.

The research was focused on the both the performing stage and the participants. Contradicting current staged performances that seek to excite their audiences by increasing hormonic excitation, the body language, the hand movements, the peal of bells refer to a coordinated array of events that instead of diverting the biological responses of the human body to excitation or aggravated stress, they culminate in a rhythmic unreel of well controlled kinesthetic events [12].

Indeed, well-coordinated chanting performance involves the concretization of notes, signs, neumes. The interpretation of a bodily language has a conveyance with the rhythmic postulation of the Archon chanter. Nonetheless, innovative technology offers new possibilities that may augment the boundaries of musical performance beyond the connotations of current typikon semantics [13].

This human activity, however, is subject to several restrictions (Fig. 7).

Fig. 7. Phases of the Great Entrance: a. The church sextons, with candles, are prepared to lead in procession from the left entrance of the temple. b. The leading priest censes the congregation. c. The leading deacon directs the procession semicircularly around the solea, in front of the temple's Altar Gate. d. The right choir tries to keep synchronization with kinetic timings of the procession.

1. The "translation" system successfully materializing the transformation from one form to the other; for example, a sequence of dancing steps may well and according to the musicological rules be interpreted to instrumental sound.
2. When the human factors of a performance understand how their movements and their activity on stage alters the functional characteristics of the performance.
3. When the whole technological environment functions in such a way that the public comprehends the interaction taking place, without having been previously informed on the performed function.
4. When the audience itself may alter the flow of the presentation; for example the overt movement of the public may redirect the sequence of events thrilling the climax of a concert.

In terms of melurgy the above notions can be concisely presented with the melodic lines of Fig. 8. There are many reasons that dictate how the kinetic and synaesthetic content of a procession may be rendered [13]. The building it self, the length of the aisle, the occasion for which the celebration takes place, the guests themselves, the lighting potential, the sound systems, and similar.

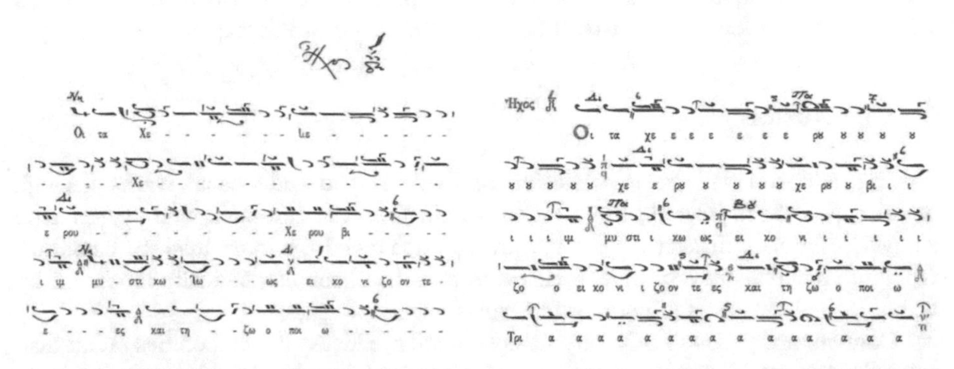

Fig. 8. Classical master classes of Cherubic hymn chants. Left, a melurgy in Fourth mode that can last, depending on the acceleration of the choir leader, from 10 to 12 min. Right, an abbreviated form, in the same mode, that can last from 5 to 7 min. This one is usually performed in the Vlatadon monastery, due to the very short distance of the aisle the procession has to walk in. The difference in tempo is evident.

So to say, the same feast is not celebrated homogenously through out a city or a country, since the constituents of the musical and social activity vary considerably in their capacity to offer a staged performance.

Even the language used is a key factor. A melody like the one depicted in Fig. 8 would have a considerably different rendition, if it were chanted in another language. The differentiation would be slighter for the long version, seen left, while for the short one on the right, the prosodic content of various language would alter its melodic lines to a serious extend. Another element is the mother tongue of the major chanting actors. As they have various ethnic backgrounds, their performance is usually influenced by the pronunciation biases of their mother tongue. In Fig. 7 the clergymen that sing in Greek have come from quite diverse environments: some are Latin oriented (from Spain) and some have strong Slavic influences (being born in Poland, Ukraine, etc.). Their language patterns more or less bias they way they chant. It is obvious that stress and intonation patterns, along with the breath patterns and the harshness for pronouncing consonants shape a particular version of their chanting rendition.

Chanters from northern countries perform usually with considerable variation especially in the bass region, with obvious pressure being laid on the abrasion of labiodental and fricative consonants. The vowels they phonologically produce have minimal contrastive differences, while diphthongs are shaped as single phonetic segments, passing rather quickly (Fig. 8). The nasalization characteristic, so obvious in the articulation of glottal voice fricatives and labiodental nasals in Greek is nearly totally absent. Performers from southern regions have a more lyrical, softer articulation for the phonological production of vowels and diphthongs that come from prolonged tongue and lip movement.

Overall, an important associated contribution comes from the phonetic quality of chanters, where vowel and consonant approaches, that give a distinctive prominence to their singing potential by the shape of their pitch contours. However, a detailed explanation of these features would be out of the scope of this paper.

4 Conclusion

The reproduction and the transmission of lively staged audiovisual works through broadcasts and the Internet, has a lot to benefit from the chromatic and synaesthetic analysis of rituals that have been performed unchanged for more than 10 centuries. Although solemn, they surpass in richness what has been emphatically amplified by technological means as promoting fantastic imagery.

Contemporary models rely only on ephemeral broadcasted commodities rather than tangible artifacts. They offer more strong impressions, but do not enjoy a ubiquitously apprehended music stream that comes with a well digested kinesthetic awareness that triggers senses with sensitive and careful handling.

By defining the axes of chromaticism, audiovisual activities like live broadcasting of staged events may be considerably enhanced, not only for TV watching of liturgical practices but for events taking advantage of the Internet highways. They have a great potential for educational purposes.

Acknowledgement. The authors would like to express their gratitude to their Eminencies, His All Holiness Bartholomew, Archbishop of Constantinople, New Rome and Ecumenical Patriarch, alongside the Most Reverend Metropolitan Panteleimon of Tyroloa and Serentium, former Abbot, former Rector, Professor Emeritus of the Aristotle University of Thessaloniki, and the Most Reverend Bishop Nikiphoros of Amorion, Abbot of the Holy Patriarchal Stavropegic Vlatadon Monastery, for providing their unreserved support, both morally and materially, apart from their blessing, for the successful completion of this research.

In the sermons recorded in the kyriakon have participated: the Most Reverend Metropolitan Theoliptos of Iconium, the Abbot, Most Reverend Bishop Nikiphoros of Amorion, the Very Reverend Fathers Gabriel, Petros, Georgios and Dorotheos, the Very Reverend Deacons Petros and Panteleimon, the choir leaders Fr. Nektarios Paris and Symeon Kanakis with their assistants Chrysostomos Vletsis and Philippos Rentakis, and the sextons from the International Boarding Establishment of the Monastery.

The audiovisual material was recorded by Georgios Varsamis, Nikolaos Rentakis and Dionysios Politis.

References

1. Riedl, M., Young, R.: Narrative planning: balancing plot and character. J. Artif. Intell. Res. **39**, 217–268 (2010). https://doi.org/10.1613/jair.2989. PDF—PostScript
2. Hatten, R.: Musical Meaning in Beethoven: Markedness, Correlation, and Interpretation (Advances in Semiotics). Indiana University Press, Bloomington (2004)
3. Cumming, N.: The Sonic Self: Musical Subjectivity and Signification (Advances in Semiotics). Indiana University Press, Bloomington (2001)

4. Broeckx, J.: Muziek, ratio en affekt, Antwerpen, Metropolis (1981)
5. Margounakis, D., Politis, D., Mokos, K.: Music in colors. In: Politis, D., Tsalighopoulos, M., Iglezakis, I. (eds.) Digital Tools for Computer Music Production and Distribution, pp. 82–115. IGI Global, Hershey (2016)
6. Barsky, V.: Chromaticism. Harwood Academic Publishers, Netherlands (1996)
7. Jacobs, A.: The New Penguin Dictionary of Music. Penguin, USA (1980)
8. Markaki, E., Kokkalidis, I.: Interactive technologies and audiovisual programming for the performing arts: the brave new world of computing reshapes the face of musical entertainment. In: Politis, D., Tsalighopoulos, M., Iglezakis, I. (eds.) Digital Tools for Computer Music Production and Distribution, pp. 137–157. IGI Global, Hershey, PA (2016)
9. Cangeloso, S.: LED Lighting -Illuminate your World with Solid State Technology - A Primer to Lighting the Future. O-Reilly - Maker Press, USA (2012)
10. Kanakis, S.: Οι μουσικοί ρόλοι των ψαλτών των πατριαρχικών χορών μέσα από τις περιγραφές του Άγγελου Βουδούρη. In: Proceedings and perspectives of the Interdisciplinary Research on Psaltiki "Από Χορού και Ομοθυμαδόν" (= all together and in the same mood), 2nd International Interdisciplinary Musicological Conference of the Department of Psaltic Art and Musicology of the Volos Academy for Theological Studies, 9–11 June 2016 (2016)
11. Βουδούρηςς, Α.: Οι Μουσικοί Χοροί της Μεγάλης του Χριστού Εκκλησίας κατά τους κάτω χρόνους. Εν Κωνσταντινουπόλει 1934. Επανέκδοση: Μουσικολογικά Απομνημονεύματα, Ευρωπαϊκό Κέντρο Τέχνης, Αθήναι (1998)
12. Kouroupetroglou, G., Papadakos, C., Kamaris, G., Chryssochoidis, G., Mourjopoulos, J.: Optimal acoustic reverberation evaluation of byzantine chanting in churches. In: Karagounis, K., Kouroupetroglou, G. (eds.) The Psaltic Art as an Autonomous Science: Scientific Branches – Related Scientific Fields – Interdisciplinary Collaborations and Interaction. Academy for Theological Studies of Volos, Department of Psaltic Art and Musicology, Greece (2015)
13. Cox, C., Warner, D.: Audio Cultures: Readings in Modern Music, Continuum, New York (2007)

Mobile Communications Technologies Impact on Radio Frequency Broadcasts

A Cognitive Approach

Dionysios Politis[1], Anastasios Tsirantonakis[1], Veljko Aleksić[2], Panagiotis Nteropoulos[1], and Dimitrios Margounakis[1(✉)]

[1] Department of Informatics, Aristotle University of Thessaloniki, Thessaloniki, Greece
{dpolitis,tsiranto,nteropou,dmargoun}@csd.auth.gr
[2] Faculty of Technical Sciences Čačak, University of Kragujevac, Kragujevac, Serbia
veljko.aleksic@ftn.kg.ac.rs

Abstract. This research examines how regulatory action by policy makers and the states involved may reshape the map of terrestrial television and radio broadcasts, not only in Greece, who serves as a case study, but for many countries who face similar problems. Taking as starting point the recent licensing competition this paper provides suggestions about how deregulating may be the amalgam of technologies that mix free on-line media providers with expensive, state regulated public goods in the frequency domain.

Keywords: TV and radio broadcasts · Emissions over the Internet
Regulating authorities

1 Introduction to the Context of Radio Frequency Broadcasts

Recent advances in Mobile Communications Technologies have reshaped the audio-visual input and output handling of mass communication. Since 2011, in a big country not far away from here, namely Egypt, when the wide global public was informed about a massive, on-going revolution taking place in the central plaza of its capital, the world is witnessing an upheaval of mass communication displacement that collectively submerges television, radio and newspapers by reciprocating an insentient, ubiquitous transfer of power to what is called in ITC terms Mobile Computing, Wireless Internet Communications, Social Media and Streaming Media, to name a few.

All these factors contribute to a "vector" like displacement of media campaigns to Internet based emissions, which alter previously easily identified concepts, used by normal subject statements or positions like "I watch the news", "I am informed about the weather", "I express my reaction" and similar. They extend the range of ordinary knowledge or experience coming from the usual space and time coordinates by interposing patterns of growth involving extravagant media, "prosthetics" and "bionics" that proliferate the availability of information and the accompanying changes in its storage and dissemination.

M. E. Auer and T. Tsiatsos (Eds.): IMCL 2017, AISC 725, pp. 778–788, 2018.
https://doi.org/10.1007/978-3-319-75175-7_76

It is evident, that apart from the ITC central infrastructure credited for this change, there is a steady fast process for distancing the viewer from its television set and having him connected around his Internet portal: viewers watch the news from information sites, blogs, or social media. They continue to watch movies with their TV sets but radiofrequency broadcasts are gradually replaced by video streaming from Netflix™, Hulu™ or Youtube™ emissions. Even further, audiences are given the prospect of evaluating, virtually navigating (and not simply changing channels) and adapting the online emissions by turning audiovisual streams to equivalent objects that can be easily manipulated for storing, retrieving and sending information. It's not strange, after all that virtually TV emissions are readily transformed to mashups or learning objects that heave to massive build ups of communal knowledge.

Therefore, computer networks provide interconnection not only to Web 1.0 tools like e-mails, collaborative educational content mega structures like Wikis (Web 2.0 tools), massive Web 3.0 tools accessible to the general public, like educational portals for formal or informal learning, communication with government agencies, not only for taxation or similar matters, industry portals fro the dissemination of goods, and recently Web 4.0 tools for advanced mobile communication.

It is exactly this approach of Web 4.0 infrastructures that turns the terminal devices, i.e. the mobile computers to advanced, as they are regarded collectively by makers and users, highly adaptive devices that contritely and inventively turn an assortment of electronic equipment to recording activity input channels, professionally alleged to social media means of access, and reproducing gateways for entertainment, information and education.

In brief, audiovisual broadcasts are turned in the perpetual form of adequately recorded forms for video, music, slides or text files that are handled collectively in huge global repositories. Already most of this information is handled, both in its input and output mode, by mobile devices.

2 Problem Formulation: Terms, Trends and Definitions

When using a contemporary mobile device, even in its simple incarnation as IoT device or household equipment, one ubiquitously communicates not with a stand-alone computer but with a smart terminal that immerses him into a very sophisticated global network. Virtually we don't speak for HCI but, rather, for Human - Technology Interaction. New peaks are thus revealed that are characterized by unprecedented maxima of activity or exuberant demand.

Estimating in 1933 the road track since the 19th century for popular commodities, as far as TV and radio were concerned, Ogburn and Gilfillan counted 150 side effects on the character, personal development and social behavior [1, 2]. Many of these are attributed to the Internet, as we experience it nowadays, and have either a positive sign (for instance, reduced discrimination between social classes and economic potential or budgetary assets) or clearly a negative one (e.g. the furious intrusion in cultures under negotiation results to unacceptable assimilation of cultural conclaves into wider societal norms at an alarming speed). Even with recent technological criteria, it was stated by Alexander in 1977 [2] that new paradigms, according to HCI standards were to be deployed in watching TV, orienting this globally attested activity on functional rather than on decorative characteristics, as older audiences may readily recall.

Already, a cultural gap is noticed between users of the first world counties and third world counties: although the former are in a more privileged position, their population curve includes a significant portion of third age users. They are by physical characteristics and old age handicapped in engulfing new technologies, although they posses the financial assets to acquire them. Deploying the full functionality of the latest mobile phone is indeed a rigorous task to perform: both their physical condition, their mentality, which developed in the 2010minus communication era, and their potential to perceive information from the on-line world is adversely attested.

On the other end, societies in the developing counties are characterized by an explosive increase of the young age percentiles of their population curve. Even though they lack financial means, when they afford a smart mobile device, they can exploit its potential to the utmost, as recent examples of indeed heavy use of such devices by refugees has been recorded [3].

As a matter of fact, young people coming in Greece from the third world in some cases did not even posses a TV set, but they are unparalleled in the skillful use of their smartphone as far as on-line navigation is concerned. Indeed, an augmented human intellect process is being revealed in front of our very eyes, with regard to abstract or academic matters. The perplexity of this interaction is depicted as Csikszentmihályi fluxes [2] in Fig. 1.

Already the results of this complex interactivity scheme are seen in practice. Technology is changing the viewing habits of households and individuals as well.

Up to now, houses could not afford to host more than 1 or 2 TV sets. Families had to gather around them scheduling how to view broadcasts on entertainment, information and education. No more than one program could be viewed at a time, and the

Fig. 1. The human technology interaction fluxes perceived as HCI practices as far as Media are concerned [2]. Image cropped from: http://www.clarehooper.net/publications/2013/Interactions2013.pdf

notion of **prime time** was dominant on how TV or radio stations were carrying out their lists of intended events and times. A zonal approach was characterizing how the concentration or consciousness of public was attracted.

The pervasive Internet approach, whether with cable, Wi-Fi or cellular networks, introduced a new arrangement for exchanging information, contacts, professional and social experiences. Viewing is less and less not synonymous to simultaneous reception of broadcasts, and is appropriating, in operational terms, groups or systems of interconnected computers, machines or operations.

TV sets are not expected to disappear under this transformation. Their transition to smart viewing devices, with computer capabilities is evident, though. Direct and continuous flow of information or education is not however anymore synonymous to broadcast. As a result, in order to watch a desired emission one has not necessarily adjust his daily itinerary to the TV schedule and habits; he can alternatively search vast, global, online repositories of stored audiovisual archives of various formats, sizes and orientation. The global community saw in the 1920s the dawn of broadcasting. In 2020 it is experiencing the synaesthetic scattering of broadcast to telecast, webcast, simulcast or cablecast of worldwide broadcasts.

Consequently, information, entertainment and education is not received as prime time synchronization but as prime event repercussion. Viewers are navigating with the appropriate site that hosts the vital pieces of information rather than tuning with specific TV stations, gambling with the probability of occurrence. Contemporary viewers depend more on particular sequences of videos, recordings and hypertext that provide rich alternatives, as contrasted with monolithic, passive and in many cases biased sequences of information.

As seen in Fig. 2, in a survey conducted over 2,368 American adults and published in the Statista™ statistics portal in 2013, audiences are shifting from the weary, uninteresting, tedious character of TV broadcasts to cross-referenced, interactive, with definitely web-based cognizance that is using more than one medium for expression or communication.

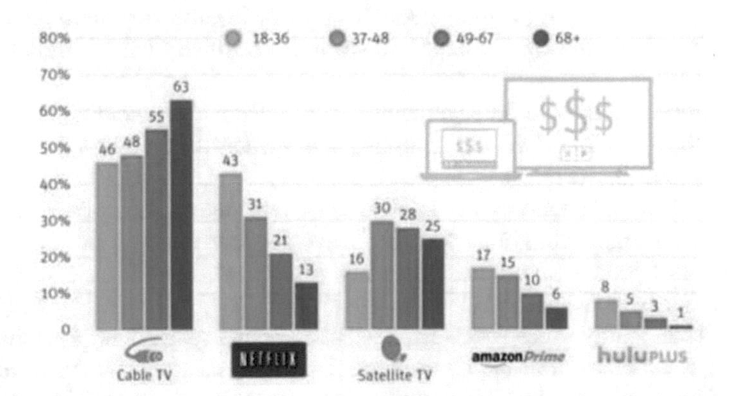

Fig. 2. American young people that subscribe to Pay TV services, as of October 2013. Source: Harris Interactive.

The shift in viewing habits does not have to do merely with the ability of the audience to see or hear something from a specific place. By changing the visual appearance or the particular way under which audiovisual works are exhibited, new prospective for sales emerge, and therefore the economic scenery of TV broadcasts is altered. An agenda for new technological potential means also that online streaming will be attracting more and more financial resources.

Indicative of this trend is the situation on the most influential music industry, the American one. More and more audiophiles are using online streaming services instead of consuming physical media, like CDs or DVDs (Fig. 3).

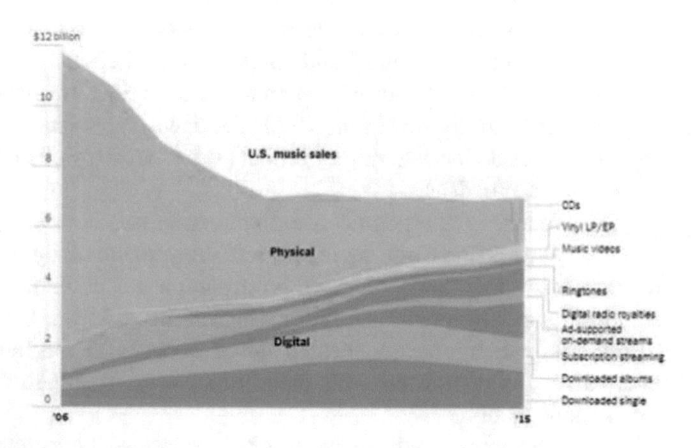

Fig. 3. Music consumption data and trends, as calculated by the recording industry association of America

As global audiences use in indistinguishable manner their Internet connection to receive audiovisual emissions or music they alter the industry's character as well. As recent surveys have shown [4] the personalization factor, the usefulness and pleasure attained with online viewing is already much more positive than traditional viewing (Fig. 4).

Even further, when the Engagement - Disengagement potential is taken into account, or the Attention dynamic of users, it is evident that on-line TV viewing is much more promising (Fig. 5).

When the level of attention that audiences demonstrate has increased, there is an obvious financial implication that does not leave the advertising industry indifferent. Indeed, the most prominent criterion that advertisers had thus far used was audience focus on broadcasted events.

As the Technology Acceptance Model used changes, based on Personalization, Ease of Use, Self Satisfaction and Usefulness, it becomes clear that the Neuromarketing potential of major players, like TV broadcasting corporations and Internet TV channel providers becomes a key issue of gaining increased ratings [5]. The next step of this subtle revolution is the alteration of the economic model, causing drastic reductions of investment and staffing levels: thus far, advertisers have a coarse image of who is

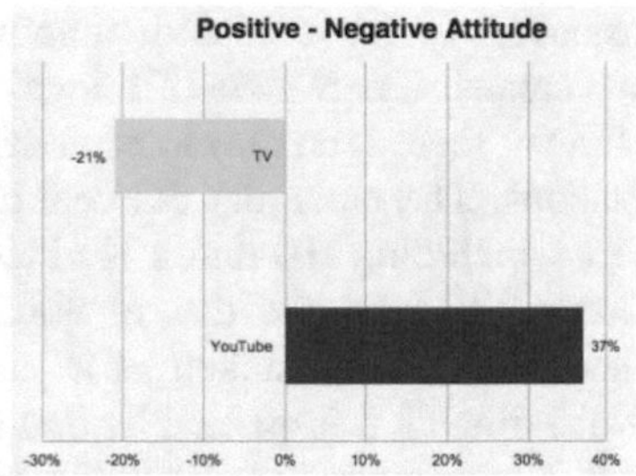

Fig. 4. Attention levels when comparing with EEG data for TV viewing vs. YouTube viewing, as far as the Positive-Negative attitude is concerned. Source: NetValue Inc.

watching a highly projected prime time event. For instance, for a football final it is reasonable to deduce that rather young to middle aged male audiences are apparent to flock in. With the new technological models of audiovisual viewing the audience watching an event is much better identified. Having each one of the viewers' navigation history, or even better, his social network profile, advertising becomes much more targeted and effective [6, 7].

Even further, advertising agents create a culture of repetitive use out of a user's navigation profile, that was not that obvious when the loose-fitting processes of massive target approach was used [8].

3 The Economic and Legal Substrate for Global Viewing

Although the political sense of the wide public was fearing some orchestrated biasing of the broadcasting arena by "big" players of the field [9], it is the change in the field of broadcasting that leverages systematic changes over the predefined means of viewing (Fig. 5).

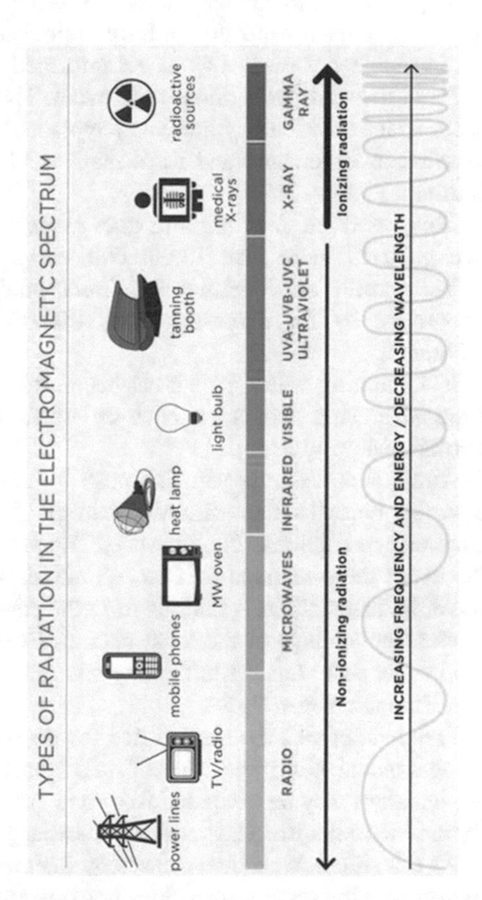

Fig. 5. The radio frequency spectrum distribution. Most of it is used for broadcasting, either terrestrial, or satellite. Recently, mobile communications and IoT get the better of it. Cropped from: shttp://d2mdn1s78c9h86.cloudfront.net/site_21/345-l-artificialEMFs_lrg.jpg

For instance, 4G networks and communication of various gadgets at the 2.4 GHz arenas have created a new state of things.

Even further, there is an unwanted, a negative side effect for Greece's 900 radio and 100 TV stations. The emerging 5G cellular networks are designated to deliver ultra-band services, including HD and 4 K video streaming. Such technology is in practice building cable TV networks that in South Eastern Europe were a rarity. 5G technologies should not be mistaken as a cable TV alternative or cheap substitute. As previously mentioned, it is an enabler that allows cellular facilities to invade the world of machines. It is not merely challenging televised broadcasts; it offers better in quality and far more advanced in Interactivity services.

Under this prism, mobile TV or Internet TV lever a paradigm shift from prime time viewing, that dominated classic radio and TV broadcasts, to prime event relocation of the wide public: viewers do not have to flock around specific places or to stay tuned in certain time zones every day to get informed about stories that make the news.

Even if they miss a prime time event, for several reasons, they may always have access to information anytime and anyplace. This attitude better triggers the power of the mind to understand and form more stable judgments by processes of logic than sentiment.

This paradigm shift is more than evident. Already, popular broadcasts and clips have gathered more than 10.000.000 views in channels hosted on YouTube™ or similar repositories. For American similar products the numbers are really devastating: some have by far surpassed 1.000.000.000 (1 billion) views, setting historical landmarks.

It is a turning point, since it has been estimated that alone in Greece more than 20% of the advertising budget has been shifted from radio frequency broadcasts to on-line, international providers.

And this is not over yet. The push for digital broadcasting via 4G, 5G and WiFi networks, ransacks the existential barrier of FM transmission: that of geographical zones and restrictions. For instance, Thessaloniki's radio broadcasting scenery with 30+ major radio stations will be reshuffled, since the invasion of some 25-30 more robust, in financial terms stations from the country's capital, will drastically reduce the commercial dealings of the local market. Even worse, the new landscape will not be sustainable, since this reshuffle may take place from stronger free market competitors form Europe or even further.

Free market was the major force for the creation of a pluralistic, enterprise driven business model, out of the ashes of state monopolies. Yet, if no provision will be taken, the pendulum may be oscillated from one extreme to the other: healthy entrepreneurial schemes will collapse in favor of broadcasting giants that will operate as monopolies [10].

As this situation becomes biased by international influence or operation, it becomes obvious that the legal quiver does not posses many weapons to regulate this market. Policy makers and politicians in the broader sense, do not have the insight to regulate globalized entrepreneurship tat is not confined within a state's sovereignty or legal grasp.

A digital radio or TV station may be Greek in the sense that it is aiming an audience within this country, but it may not be located for technical or financial reasons within the country's legal sphere.

Thus far, legal systems categorized broadcasting within their domain stations as follows:

- Model A: State monopolies, i.e. legal entities that are not only permitted to broadcast by the state, but also they are state controlled and state-run. Even in its mild version, this system does not guarantee free press administration or equal distances from the major political parties.
- Model B: State controlled enterprises, i.e. public entities that are free market corporations, being however regulated as far as their delivered content to the public is concerned. Legal experts differentiate this scheme with the free press one. There, any entrepreneur may rather unobstructed exercise his right to establish a newspaper, without any preliminary special permit.

This is not the case with a TV station, which, as recent practice has shown in Greece, will be exhaustively scrutinized for its whereabouts before receiving some sort of permit to broadcast. For instance, the state may impose manpower thresholds or other restrictions before auctioning the use of terrestrial broadcast frequencies.

- Model C: Market controlled enterprises. This liberal approach does not put considerable obstacles in regulating state goods, like using the radiofrequency spectrum, but asks only for strict fiscal or tax evading discipline, as it would ask for any company exercising economic activity within the country's financial sovereignty.

In all above cases, however, the state may exercise its right for quality control. Although vague in its notion, a state run Radio and TV Council may give to the impreciseness of quality control a concrete approach [11].

Serbian electronic broadcasting scenery is regulated by the Government Regulatory Agency for Electronic Communications and Postal Services, which includes mobile and internet communications. One of its jurisdiction is providing TV broadcasting licenses (national, regional and local). There are seven TV stations with national coverage license and about 30 stations with regional license. One of the TV stations with national license is Serbian state owned RTS that broadcasts its program on three channels 24/7. Even though it is state owned, RTS has independent editing policy as the board of experts selected on public contest controls the broadcasting content. It's financing is citizen-based, as every household has the legal obligation to pay monthly subscription. It should be noted that the regulatory policy on broadcasting content of private TV stations is very loose, and there are no examples of private TV subscription. Total income from Serbian electronic communication in 2016 was 1,54 billion euro. The income is generated by annual licenses paid by the broadcasters. However, most of it was from mobile operators (59%), while the profit from media content distribution (TV, cable and IPTV broadcast) was only about 150 million euro. There is no public information on TV broadcasters licensing fee algorithm. In total, there are 1.66 million subscribers of media content in Serbia. Most of them are households, so in explicit terms over 90% of population is using some sort of electronic communication regularly. Cable networks provide its service to about 58% of subscribers, 25% uses IPTV and about 16% uses terrestrial digital antennas. Certain TV content is possible to watch on mobile devices by using various applications without monthly subscription agreement. In 2016, there were about 16.000 users of these applications provided by three Serbian operators.

For instance in France, the broadcasting market is controlled by the Government, the Parliament and the Council (Conseil supérieur de l'audiovisuel – CSA). As a result, the have designated that state owned station TF1 should broadcast 24/24, while CanalPlus is forced to a minimum of 18 h at least broadcast on a daily basis. TF1 is obliged to provide at least 1000 h per year of emissions focusing on matters of the younger French generation, while music station M6 is directed to devote as significant portion of its emission time to French composers, alongside composers from abroad. Significant protection is provided to French produced films, broadcasted via TV channels. TV stations should keep a sustainable portion of locally produced media or media that use the French language, instead of having a priori the unlimited capacity to fill in their scheduled emission table with what ever (cheap) film they may amass.

4 Social Intelligence and Collaborative Learning in the Radiofrequency Domain

Although in strict medical terms the perceptive human ability is rather deteriorated within four measurable dimensions (the three spatial ones, such as length, breadth, depth, height, plus time), it seems recently that users can immerse themselves in "bubbles of augmented reality", that constitute an extended multiverse.

Therefore, the term dimension becomes metaphorically representative of concepts that are used to explain a particular occasion, for instance in computer generated environments that are seemingly real.

In numbers, it seems that modern Greeks spend quite a time into the "broadcasting" universe, as seen in Table 1.

Table 1. Broadcasting demographics in Greece, with Internet viewing acting as an alternative

	Media used	Broadcast viewing and listening	Sex	Age	Device
1	Broadcast TV	4.5 h		All ages	
2		6.5 h		>65	
3					
4	Internet use	70.9%		(Ages 13–74)	
5		74.5%	Male		
6		67.3%	Female		
7	Internet: daily use	62.2%			All devices
8	Internet: daily use	141 min			
9	Internet: daily use	187 min		18–24	
10	Internet surfing	35.5%			Smartphone
11	Prime event vs. prime time viewing	65%–35%		24–41	

As expected, audiences flock for many hours daily into the radiofrequencies universe [source: AGP - WebID, March 2015]. This in practice means the following: Older audiences are steadily tuned in their TVs mainly and radio transceivers secondly. They have started using smartphones and mobile devices in general, but fail to take advantage of their broadcasting potential, i.e. their ability to accept analog radio transmissions or WiFi-4G enabled digital connections.

On the contrary, younger audiences although using many hours daily their electronic audiovisual appliances, swiftly move towards cellular or WiFi connections. They use their mobile devices to view the news or hear the news, but they customarily apply a prime event orientation instead of a prime time viewing.

As research companies primarily focus on viewing stats, they fail to discriminate between the prime event paradigm and the prime time viewing their customers want them to appreciate (Fig. 6).

An in class survey was conducted [source: 21 graduate students, November 2016], revealing that young, educated audiences watch indeed the news, they don't disdain or depreciate the concept of news broadcasting, but clearly, they can receive information from interactive, alternative sites that offer, apart from the visual or audio information, collaborative learning in the form of anonymous, public commenting and variety of informal sources, like Facebook, YouTube and similar social networks.

It seems that Internet-based-Protocol-TV (IPTV) is a game changer [11]. It allures the general public by repeating the usual prime time broadcasts in a prime event mode, apart from offering interactive manipulation of the collaborative manner of viewing.

As audiences would prefer to watch the live broadcast of their favorite sports team in a cafe, enjoying the meditative, collective atmospheric reaction, the same way IPTV practices recreate a social media sphere of collaboration in receiving noteworthy information.

As seen in Fig. 6, Prime Event ratings soar after the successful promotion, for various reasons, of the emission media focus on. Some events have successful viewings for quite some time after their launch time, but this trend has not been in depth examined thus far. It is important, though, to estimate the eminent preemptive success of some events. Based on the previous fame they have amassed, they gather audiences that try to get some potentially useful information, and thus create a movement that provides ratings, commercials or indirect advertisement for the main event.

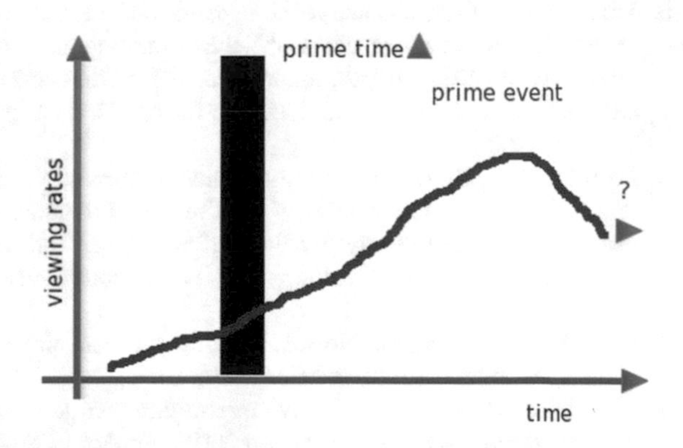

Fig. 6. Prime time viewing of broadcasted events vs. Prime event viewing. The viewing potential of successful Prime event may is quite unpredictable thus far, after some weeks.

The Eurovision song contest is a classic example of such a Prime time vs. Prime event broadcast.

5 Conclusion

New technologies in some countries are dealt with obsolete and general provisions, giving evidence for many ambiguities and legislative gaps.

While most European countries tend to adopt Model B for their Radio and TV whereabouts [11, 12], the fact that contemporary technology forcibly trends to shift to the deregulated Model C culture passes unnoticed.

References

1. Ogburn, W.F., Gilfillan, S.C.: The influence of invention and discovery. Recent Social Trends U.S.: Report of the President's Research Committee on Social Trends, pp. 122–166 (1933). https://archive.org/details/recentsocialtren01unitrich
2. Dix, A.: Human–computer interaction, foundations and new paradigms. J. Vis. Lang. Comput. **42**, 122–134 (2016)
3. Politis, D.: Piracy in musical audio-visual production and distribution: a forensic engineering calculus approach for the economics of deregulation. IJRAE **1**(3), 11 (2015)
4. Norman, D.A.: Emotion and design: attractive things work better. Interact. Mag. **9**(4), 36–42 (2002)
5. CB Insights: The race for AI: Google, Twitter, Intel, Apple in a rush to grab artificial intelligence startups. CB Insights, New York (2016). https://www.cbinsights.com/blog/topacquirers-ai-startups-ma-timeline/
6. Dix, A.: A shifting boundary: the dynamics of internal cognition and the web as external representation. In: Proceedings of the 3rd International Web Science Conference (WebSci 2011). ACM (2011). https://doi.org/10.1145/2527031.2527056
7. Schmidt, A.: Implicit human computer interaction through context. Pers. Technol. **4**(2–3), 191–199 (2000)
8. Moridis, C.N., Klados, M.A., Kokkinakis, I., Terzis, V., Economides, A.A., Karlovasitou, A., Bamidis, P.D., Karabatakis, V.E.: The impact of audio-visual stimulation on alpha brain oscillations: an EEG study. In: Proceedings of the 10th IEEE International Conference on Information Technology and Applications in Biomedicine, Corfu, Greece, 3–5 November 2010 (2010)
9. Czepek, A., Hellwig, M., Nowak, E.: Press Freedom and Pluralism in Europe Concepts and Conditions. Intellect Ltd., Bristol (2009)
10. Kiki, G.: Η ελευθερία των οπτικοακουστικών μέσων (υπό το πρίσμα της Συνταγματικής αναθεώρησης του 2001). Εκδόσεις Σάκκουλα, Αθήνα (2003)
11. Open Society Institute, Television Across Europe, vol. 2 (2005)
12. Tsevas, A.: Διασφάλιση του πλουραλισμού και έλεγχος συγκέντρωσης στα μέσα ενημέρωσης-Ημερίδα. Νομική Βιβλιοθήκη, Αθήνα (2007)

5G Network Infrastructure

Implementation of an SDN-Enabled 5G Experimental Platform for Core and Radio Access Network Support

Kostas Ramantas[1], Elli Kartsakli[1], Mikel Irazabal[1], Angelos Antonopoulos[2](✉), and Christos Verikoukis[2]

[1] Iquadrat Informatica S.L., Barcelona, Spain
{kramantas,ellik,mirazabal}@iquadrat.com
[2] Telecommunications Technological Centre of Catalonia, Barcelona, Spain
{aantonopoulos,cveri}@cttc.es

Abstract. Software Defined Networking (SDN) and Network Function Virtualization (NFV) are very promising technologies for future 5G mobile networks, enabling network operators to enhance the flexibility, efficiency and ease of management of their networks. Although SDN is typically associated with wired networks, it is lately making steady advances in cellular networks. In this paper, we present the architecture of an SDN-enabled 5G experimental platform that employs SDN to support both the core and the radio access network.

Keywords: SDN · 5G · Orchestration · Testbed

1 Introduction

Software Defined Networking (SDN) and Network Function Virtualization (NFV) have been envisioned as promising technologies to increase the efficiency of future networking and enhance the flexibility of network configuration and management [1]. In SDN-enabled networks, the control plane, which handles management operations, is logically centralized and physically decoupled from the data plane, thus enabling high network configurability and programmability. The network programmability and the capability of optimizing the resource allocation and utilization in a centralized way, made possible by the SDN paradigm, are expected to alleviate the burden of the data onslaught expected from data-intensive applications. Hence, SDN, along with network virtualization, are expected to play an important role in future 5G networks, paving the way for the implementation of novel concepts such as end-to-end slicing, thus enabling network operators to micro-segment traffic by application and increase network agility.

While SDN and NFV are pretty well understood in the context of wired networks and data centers, there are several advantages that can be obtained by the application of the SDN principles to the wireless domain as well. For instance, distributed protocols currently employed in cellular networks cannot optimally allocate resources, nor manage handoffs and load balancing between cells. In this context, SDN can enable the implementation of centralized handover and load balancing schemes [2], allowing mobile operators to optimally manage the limited spectrum resources.

M. E. Auer and T. Tsiatsos (Eds.): IMCL 2017, AISC 725, pp. 791–796, 2018.
https://doi.org/10.1007/978-3-319-75175-7_77

Several large-scale SDN testbeds have been designed enabling researchers to run their prototypes and experiments as well as to evaluate the network backhaul of future 5G architectures [3]. Small-scale platforms have also been proposed [4] using single-board computers as SDN switches to lower the deployment cost. Other works address the Radio Access Network (RAN) part of 5G networks, focusing on RAN Virtualization [5, 6], or flexible resource allocation [7]. However, these are largely conceptual and do not consider the associated implementation challenges of real-world RANs. A recent work [8] moves past the theoretical context and focuses on the prototyping challenges of applying SDN principles to solve the Radio Resource Management problem in 5G networks for a number of use cases (interference management, mobile edge computing, RAN sharing).

In this paper, we propose a novel architecture for the implementation of an SDN-enabled 5G experimental platform that integrates both the wired and wireless parts of a cellular network. Our testbed leverages state-of-the-art components from the OpenDaylight [9] and OpenAirInterface [10] open source projects, and employs the FlexRAN controller [8] for centralized SDN control of the RAN. The proposed setup aims to enable the experimentation and validation of novel algorithms and mechanisms in next generation 5G networks.

2 Architecture Description

The proposed testbed architecture, depicted in Fig. 1, considers the application of the SDN paradigm to both the core network and the RAN of a Long Term Evolution (LTE) system. More specifically, the core network and the RAN can be considered as the wired and the wireless part of the network, respectively. In continuation, we describe in detail the proposed testbed architecture and setup, and discuss open issues with respect to the integration of the wired and wireless components.

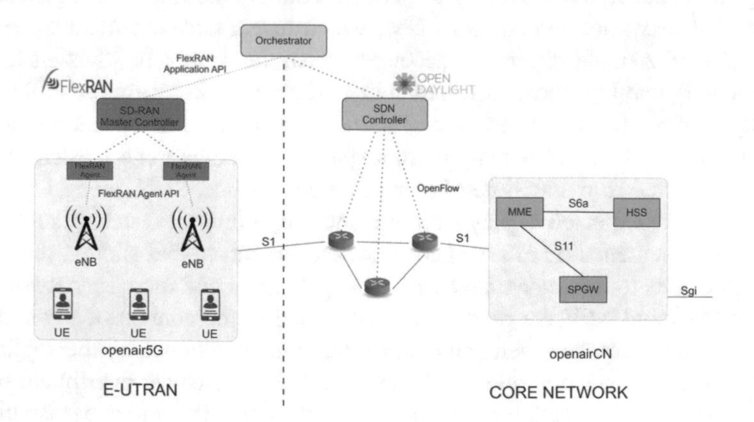

Fig. 1. Testbed architecture

A. *Wired Domain*

A key feature of the SDN technology is the ability to significantly cut on network costs by replacing expensive specialized hardware devices by generic programmable components. The proposed SDN-enabled experimental platform exploits this feature by employing low cost SDN switches to build a transport network, controlled by OpenDaylight. Specifically, the SDN switch has been implemented on an Odroid C2 board that includes a fast 64-bit 1.5 Ghz quad core ARM CPU, 2 GB DDR3 SDRAM, and a Gigabit Ethernet interface (Fig. 2). The packet switching is handled by Open vSwitch, which implements the switching functionality via a kernel module. This module had to be specifically built for the Odroid C2, as it was not included in the Linux distribution provided by the manufacturer. Finally, for the VLAN switch we employed the TL-SG105E 5-Port Gigabit Switch from TP-Link.

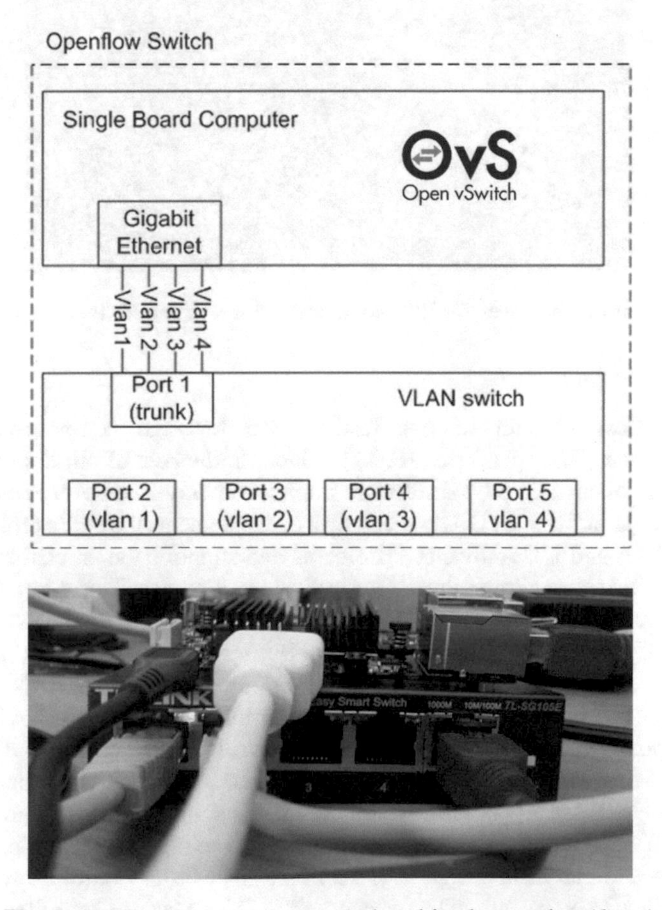

Fig. 2. SDN switch. Architecture (up) and implementation (down)

For the LTE core network, we employed the OpenAirInterface EPC (openairCN) software [11]. OpenairCN is an open source implementation of the 3GPP specifications of EPC [12], i.e., it includes the implementation of the Mobility Management Entity (MME), the Home Subscriber Server (HSS), the Service Gateway (S-GW) and the Packet Gateway (P-GW). The openairCN has been deployed on Ubuntu 16.04 LTS with kernel version 4.7 on top of a hypervisor machine (VMware software). The different EPC components have been inter-connected via virtual interfaces, as depicted in Fig. 3. The first two interfaces (i.e., eth0:11, eth0:21) facilitate the S11 communication between the MME and the S-GW, while the third one (i.e., gtp0) supports GPRS Tunneling. Finally, the S1 communication between EPC and E-UTRAN is realized by the real network interface of the Host machine.

```
Terminal
eth0:11   Link encap:Ethernet  HWaddr 00:0c:29:2b:4d:ea
          inet addr:192.11.56.10  Bcast:192.11.56.255  Mask:255.255.255.0
          UP BROADCAST RUNNING MULTICAST  MTU:1500  Metric:1

eth0:21   Link encap:Ethernet  HWaddr 00:0c:29:2b:4d:ea
          inet addr:192.21.56.10  Bcast:192.21.56.255  Mask:255.255.255.0
          UP BROADCAST RUNNING MULTICAST  MTU:1500  Metric:1

gtp0      Link encap:UNSPEC  HWaddr 00-00-00-00-00-00-00-00-00-00-00-00-00-00-00-00

          inet addr:172.16.0.1  P-t-P:172.16.0.1  Mask:255.240.0.0
          inet6 addr: fe80::964b:41d1:b52f:3be/64 Scope:Link
          UP POINTOPOINT RUNNING NOARP MULTICAST  MTU:1500  Metric:1
          RX packets:0 errors:0 dropped:0 overruns:0 frame:0
          TX packets:0 errors:2 dropped:0 overruns:0 carrier:0
          collisions:0 txqueuelen:1
```

Fig. 3. Virtual interfaces for EPC components for S11 and GTP communication.

B. *Wireless Domain*

For the wireless domain of our testbed, we leverage components from the OpenAirInterface (OAI) platform [10, 11], which, to the best of our knowledge, is the most complete open-source LTE software implementation. The RAN is emulated with the oaisim emulator from the OAI project. Oaisim implements the entire protocol stack of all relevant LTE and LTE-Advanced protocols and supports the emulation of up to 16 UEs and 4 eNBs in a single computer. Our testbed employs a desktop computer equipped with a 6th generation Intel core i5 CPU and 16 GB RAM memory for the RAN emulation.

As it can be observed in Fig. 1, a FlexRAN Master Controller [8] is employed for the SDN-enabled centralized control of the RAN. In our setup, the FlexRAN Master Controller is deployed on Ubuntu 16.04 LTS distribution. FlexRAN is a flexible and programmable platform which separates the RAN control and data planes and supports the design of real-time RAN control applications. In FlexRAN, the RAN plane separation is made by the FlexRAN Agent API, which acts as the southbound API, having the eNB data plane on one side and the FlexRAN control plane on the other side. The control plane follows a hierarchical design with the FlexRAN Master Controller connected to a number of FlexRAN Agents, one for each eNB. The Master Controller controls the underlying infrastructure (eNBs) and orchestrates their operation, facilitating multiple advanced use cases, such as centralized load balancing and handover control.

C. *Client Traffic emulation*

As already mentioned, the proposed testbed implementation emulates the RAN by employing oaisim, which has two modes of operation, namely, full PHY emulation and PHY abstraction [13]. PHY abstraction mode allows us to emulate a large-scale system with multiple UEs and eNBs with low computational complexity and, therefore, low emulation time. In order to run more realistic emulation scenarios, oaisim has the possibility to read scenario descriptions from well-structured (OSD) files which include environment/system, topology, application and emulation configurations. The client traffic is emulated with the OpenAirInterface Traffic Generator (OTG) [14]. OTG is a realistic packet-level traffic generation tool for emerging application scenarios, which is capable of accurately emulating the traffic of different application scenarios, such as Voice over IP (VoIP), online gaming and machine type communication (MTC) among others.

Our 5G platform also supports wired network clients, which are emulated with Mininet [15]. Mininet is a network emulator that runs a collection of end-hosts, switches, routers, and links on a single Linux system. It uses lightweight virtualization (i.e., Linux containers) to emulate a complete network in a single machine. Apparently, Mininet-based networks cannot exceed the available resources (e.g., CPU or bandwidth) on a single server. Mininet hosts can run arbitrary Linux programs and traffic generators. Another powerful feature of Mininet is its ability to interact with real networks, by bridging virtual host interfaces with physical interfaces. Then, virtual hosts can send/receive packets to/from the physical SDN transport switches.

D. *Wired/Wireless orchestration*

The consolidation of the two SDN controllers for the wired and wireless domains would give the potential of centralized core/RAN orchestration, taking advantage of the global network view from both SDN controllers. Our current efforts are focused on the implementation of a centralized orchestrator, which works by consolidating APIs form both OpenDaylight and FlexRAN controllers. This will allow the design of applications which can manage, monitor and control the heterogeneous infrastructure. Such applications can be used to coordinate radio access functions for interference management, mobile edge computing, as well as RAN sharing and virtualization in heterogeneous networks. For example, FlexRAN implements centralized resource scheduling and allocation as well as centralized interference cancelation, which potentially increases the performance in high load and high interference scenarios.

3 Conclusion and Future Work

In this paper, we have presented in detail the architecture and design decisions of a 5G experimental platform that supports SDN control in the wired and wireless domains. The proposed testbed leverages components from major open source projects, such as OpenAirInterface and OpenDaylight. In the future, we plan to deploy a wired/wireless orchestrator and incorporate hardware UEs and eNBs, in a hybrid setup. Taking advantage of centralized network knowledge and analytics, provided by the proposed platform, will allow us to implement traffic engineering across the heterogeneous infrastructure. Moreover, it will enable the implementation and validation of end-to-end network slicing and micro-segmenting traffic of specific applications (network virtualization).

Acknowledgement. This work has been supported by the research projects CellFive (TEC2014-60130-P), 5G-AURA (675806), AGAUR (DI059-2016), 5GSTEPFWD (722429), IoSense (692480) and 5G-PHOS (761989).

References

1. Kim, H., Feamster, N.: Improving network management with software defined networking. IEEE Commun. Mag. **51**(2), 114–119 (2013)
2. Li, L., Mao, Z., Rexford, J.: Toward software-defined cellular networks. In: Proceedings of European Workshop on Software Defined Networking, pp. 7–12 (2012)
3. Huang, T., Yu, F.R., Zhang, C., Liu, J., Zhang, J., Liu, J.: A survey on large-scale software defined networking (SDN) testbeds: approaches and challenges. IEEE Commun. Surv. Tutor. **19**(2), 891–917 (2017)
4. Kim, H., Kim, J., Ko, Y.B.: Developing a cost-effective OpenFlow testbed for small-scale software defined networking. In: 16th International Conference on Advanced Communication Technology, Pyeongchang, pp. 758–761 (2014)
5. Tseliou, G., Adelantado, F., Verikoukis, C.: Scalable RAN virtualization in multi-tenant LTE-A heterogeneous networks. IEEE Trans. Veh. Technol. **65**(8), 6651 (2016)
6. Costanzo, S., et al.: OpeNB: a framework for virtualizing base stations in LTE networks. In: IEEE International Conference on Communications (ICC), pp. 3148–3153 (2014)
7. Tseliou, G., Samdanis, K., Adelantado, F., Pérez, X.C., Verikoukis, C.: A capacity broker architecture and framework for multi-tenant support in LTE-A networks. In: IEEE International Conference on Communication (ICC), pp. 1–6 (2016)
8. Foukas, X., Nikaein, N., Kassem, M., Marina, M., Kontovasilis, K.: FlexRAN: a flexible and programmable platform for software-defined radio access networks. In: International Conference on emerging Networking EXperiments and Technologies (CoNEXT) (2017)
9. OpenDaylight: Open Source SDN Platform. https://www.opendaylight.org/
10. Navid, N., Marina, M.K., Manickam, S., Dawson, A., Knopp, R., Bonnet, C.: OpenAirInterface: a flexible platform for 5G research. SIGCOMM Comput. Commun. Rev. **44**(5), 33–38 (2014)
11. OpenAirInterface Platform. http://www.openairinterface.org/, Repository: https://gitlab.eurecom.fr/oai/
12. 3GPP, Network Architecture. Technical Specification 23.002, 3GPP, December 2011. http://www.3gpp.org/DynaReport/23002.htm
13. Latif, I., Kaltenberger, F., Nikaein, N., Knopp, R.: Large scale system evaluations using PHY abstraction for LTE with OpenAirInterface. In: Workshop on Emulation Tools Methodology and Techniques, Cannes, France, March 2013. http://www.eurecom.fr/en/publication/3925/download/cm-publi-3925
14. Hafsaoui, A., Nikaein, N., Wang, L.: OpenAirInterface traffic generator (OTG): a realistic traffic generation tool for emerging application scenarios. In: 2012 IEEE 20th International Symposium on Modeling, Analysis and Simulation of Computer and Telecommunication Systems, Washington, D.C., pp. 492–494 (2012)
15. Mininet. http://mininet.org/

Quality of Service Provisioning in High-Capacity 5G Fronthaul/Backhaul Networks

John S. Vardakas[1](✉), Elli Kartsakli[1], Sotirios Papaioannou[2], George Kalfas[2], Nikos Pleros[2], Angelos Antonopoulos[3], and Christos Verikoukis[3]

[1] Iquadrat Informatica, Barcelona, Spain
{jvardakas, ellik}@iquadrat.com
[2] Department of Informatics, Aristotle University of Thessaloniki, Thessaloniki, Greece
{sopa, gkalfas, npleros}@csd.auth.gr
[3] Telecommunications Technological Center of Catalonia (CTTC/CERCA), Barcelona, Spain
{aantonopoulos, cveri}@cttc.es

Abstract. Passive Optical Networks (PONs) are an efficient high-performance optical access solution that is able to provide huge bandwidth in a resourceful way through the incorporation of inexpensive passive elements. PONs have been widely used for the provision of high-demand services to a number of wired end-users. However, the PON configuration may be also used for the support of wireless end-users through the integration of the PON with a wireless access system. This converged network can therefore take advantage of the mobility features of the wireless network and the high-bandwidth benefits of the PON. In this article, we highlight the basic features of the existing optical-wireless access systems. Moreover, we discuss the challenges in the medium access control layer of the converged network and viable solutions for the efficient Quality of Service management support.

Keywords: 5G networks · Passive Optical Networks · Fiber-wireless networks Medium access control

1 Introduction

The strong demand for delivering broadband telecommunication services to residential users suggests the utilization of high-capacity technologies in the access domain. To meet these demands, Passive Optical Network (PONs) are introduced as the basic technology for the implementation of Fiber-To-The-Home (FTTH) solutions [1]. PONs provide fiber connections directly to end-users, low operational costs, and tremendously larger bandwidth compared to the traditional copper twisted pair. The installation of a PON also allows the service provider to deliver any current or foreseeable set of broadband services that creates the footing to the migration to 5G systems.

M. E. Auer and T. Tsiatsos (Eds.): IMCL 2017, AISC 725, pp. 797–804, 2018.
https://doi.org/10.1007/978-3-319-75175-7_78

PONs come in different flavours depending on the multiple access scheme that they apply in order to provide service to a high number of users. Current PON configurations are based on the cost-effective Time Division Multiplexing (TDM) technology, while Optical Code Division Multiple Access (OCDMA) [2] and Orthogonal Frequency Division Multiple Access (OFMDA) [3] have been also considered as viable solutions to support current and future bandwidth demands. In addition, more advanced PON-based techniques that are based on the Wavelength Division Multiplexing (WDM) technology are currently being developed and standardized. Recent advances in photonic technology allow the realization of the Ultra-Dense WDM (UDWDM), which is the key technology for the Wavelength-To-The-User (WTTU) concept [4].

PONs are also the key player in the converged optical-wireless access network, where they act as the optical backhaul of a wireless access network front-end, thus forcing all possible FiWi solutions to ensure PON compatibility. Various technologies have been proposed for the wireless fronthaul/backhaul; these solutions include Wireless Fidelity (WiFi), Worldwide interoperability for Microwave Access (WiMAX) and Long-Term Evolution (LTE). Nevertheless, the 5G operational framework is targeted to be the key technology that will enable the roll-out of very dense wireless networks interconnecting over 7 trillion devices and 7 billion people. 5G networks envision to significantly increase the network capacity, mainly through the utilization of the mmWave advantages, such as the small wavelength that allows the application of Multiple-Input Multiple-Output (MIMO) communication schemes that further enhance the efficient exploitation of the spectrum [5]. However, the utilization of the mmWave frequencies may outcome a number of transmission deficiencies, such as blocking-coverage holes (as a result of weaker non-line-of-sight paths), higher path losses (due to the high frequencies) and reduced diversity (as a result of less significant scattering). However, the incorporation of effective mechanisms in the MAC layer can further improve the network capacity, especially when medium transparent schemes are applied that provide support to wireless signals over the optical transport.

In this article, we discuss the challenges that arise from the utilization of PONs as the backhaul solution for 5G networks and study the applicability of Quality of Service (QoS) mechanisms. Specifically, we initially present the technology changes that are necessary to support the high data rates envisioned in 5G through an integrated optical wireless network. Next we present the solutions that should be incorporated both in the physical and network domains, in order to implement a fully converged PON-based 5G network that is able to realize the ambitious targets of 5G networks, even in ultra-dense deployments, and provide QoS differentiation through the application of a packet priority-based mechanism. The proposed method targets to differentiate the network services, by considering different queues for each one of the supported QoS classes, and determines the mean queuing delay in the uplink direction of the network. The results show that the proposed approach is able to reduce the end-to-end delay for delay-sensitive, high-priority QoS classes.

2 5G Integrated PON-Based Network Challenges and Solutions

Mobile network operators are facing significant challenges in order to cope with the ever increasing demands for bandwidth both in the uplink and downlink. Current 4G technology deployments are able to provide uplink and downlink rates that are far from the 5G targets of 50 Mbps and 300 Mbps per user, respectively. Furthermore, current solutions may provide latency up to 10 ms; however, specific low-latency 5G applications require latency values below 1 ms. In addition, the new network must also target to achieve the 1000x capacity increase compared to the current LTE technology, as well as the significant power consumption reduction by a 10x factor. It therefore evident that these ambitious targets cannot be realized by just upgrading or extending the current 4G solutions, since: (i) current frequency bands cannot support the envisioned data rates; (ii) the high cell densification that may increase the total capacity will lead to significantly higher interference levels; (iii) in ultra-dense deployments the connection of every antenna with the backhaul exclusively through fiber will significantly increase the installation cost.

To address the aforementioned challenges, significant research challenges have been considered that are based on newly proposed deployment paradigms like Network Function Virtualization (NFV) [6], and network slicing [7]. The main conclusions of these efforts are the promotion of mmWave bands and massive MIMO antennas, as well as an eventual transition to Ethernet-based fronthaul transport. Still, these solutions may provide inefficient results when applied to high-density urban area environments.

In such a challenging environment, PONs are expected to play a vital role as part of a fiber-wireless network, mainly by utilizing advantages of other technologies both in the physical and network domain, in order to become an efficient 5G network solution. In the physical domain, cost-effective and high-speed transceivers are required, in order to support the emerging 25 Gb/s and 100 Gb/s PON access. In addition, the optical beamforming technology is an attractive solution for offering beamforming to high-bandwidth and ultra-fast massive (e.g. 64×64) MIMO antennas in a cost- and energy efficient way. Furthermore, in order to fully exploit the huge optical bandwidth offered by the WDM technology, Reconfigurable Optical Add Drop Multiplexers (ROADMs) with reduced insertion loss are necessary for the dynamic switching of the wavelengths in and out of the Remote Radio Head (RRH) of the wireless network, over a given time span.

The aforementioned technology in the optical domain should be combined with high-capacity and efficient equipment in the wireless domain. To this end, massive MIMO antennas are essential in order to provide multi-Gbps fronthaul, especially in densely populated areas [8]. The development of such technology requires to dynamically allocating different wireless frequency bands within the mmWave available spectrum to multiple MIMO boards at sub-μsec-scale tunability speeds. These configurations can therefore be combined with the WDM technology in order to simplify high-capacity setups and reduce their energy consumption. However, for such high-capacity MIMO setups, a Digital Signal Processing engine is required for carrying out the channel coding/mapping and MIMO processing tasks.

In Layer 2 new approaches are necessary in order to allocate both optical and wireless bandwidth to the RRHs and end-users. Such solutions should be therefore medium-transparent [9], so as to allow the direct traffic negotiation between the PON's central office and the wireless end-users. This can be implemented through a centralized polling-based synchronization and resource allocation scheme, in order to resolve the current challenges of synchronization and packet delay variation in Ethernet-based fronthaul, and achieve very-high-throughput performance with very low latency values.

In the network domain, a Software Defined Network (SDN) plane can be considered as an effective solution for orchestrating the entire optical-wireless converged network. Such a solution can provide the means for the design, deployment, customization, and optimization of the different network slices and resources of both the optical and the wireless domain. This configuration will therefore allow an advanced and efficient network configuration and management providing the necessary credentials for a holistic network reconfiguration and orchestration over all the available resources across the complete optical-wireless network.

3 System Model

We consider the network architecture of Fig. 1 that supports N rooftop antennas. Each one of the rooftop antennas are connected to the Optical Line Terminal (OLT) in the uplink direction and to a number of small-cell antennas in the downlink direction. Specifically, we consider that in rooftop n ($n = 1,\ldots, N$) a number of B_n small-cell antennas are connected that provide service to $U_{n,bn}$ users ($bn = 1,\ldots, B_n$). Each one of these users has a number of K queues, one for each one of the K QoS classes. Specifically, packets that belong to QoS-class k ($k = 1,\ldots, K$) are stored in the corresponding queue until they are transmitted to the corresponding rooftop antenna.

Each rooftop antenna is able to transmit groups of packets to the OLT, called superframes (see Fig. 1). In order to avoid collisions, each rooftop antenna is able to transmit its superframe during a specific time-interval in each transmission window; we consider that this time-interval has a fixed duration of T_{sp} sec. During T_{sp}, each rooftop transmits the packets of all K QoS-classes, as well as the control packets that are necessary for the coordination of the bandwidth allocation mechanism. In order to support QoS differentiation, we assume that each QoS class allocates a dissimilar percentage of the superframe in such a way so that more packets that belong to higher QoS classes are transmitted compared to the lower QoS classes. Specifically, we assume that QoS-class k transmits up to m_k packets in each transmission window, with $m_1 > m_2 > \ldots > m_K$. Therefore, the total number of packets that are transmitted from each small-cell antenna to the rooftop is:

$$T_{packets} = \sum_{k=1}^{K} m_k \tag{1}$$

The values of m_k are selected based on the mean packet arrival rates of the corresponding QoS class and are assumed to be constant for each transmission window.

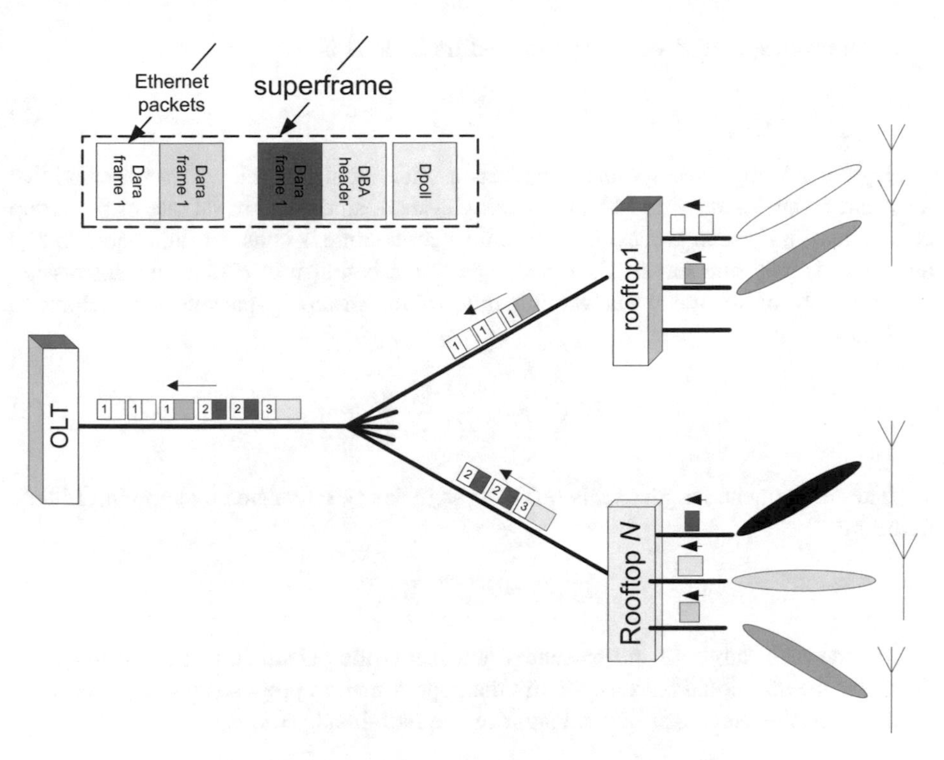

Fig. 1. The converged optical-wireless network.

By considering that each packet is transmitted during a time-slot of duration σ, and that the network applies a safety time interval δ between two consecutive superframes, the transmission window can be calculated by the following equation:

$$T_{TW} = \sum_{n=1}^{N} \sum_{b_n=1}^{B_n} T_{sp}\sigma + (N-1)\delta \tag{2}$$

where $T_{sp} = T_{data} + T_{ctr}$, and T_{ctr} refers to the transmission of the control packets. The transmission window T_W can be considered as the service-time of each one of the queues that are installed in each rooftop antenna, which is constant over time. On the other hand, we assume that the packet arrival rate of each QoS-class for small-cell antenna b_n of rooftop antenna n is Poisson with mean value of $\lambda_{bn,k}$; the total arrival rate of QoS-class packets to the rooftop antenna n is therefore equal to $\lambda_{n,k} = \sum_{b_n=1}^{B_n} \lambda_{b_n,k}$. By considering this value, the packet arrival rate of the group of packets of QoS-class k is

$$\lambda_{n,k}^{group} = \frac{\lambda_{n,k}}{m_k} \tag{3}$$

while the corresponding equivalent offered traffic load is

$$A_{n,k}^{group} = \lambda_{n,k}^{group} \cdot T_{TW} \cdot \sigma \tag{4}$$

By considering these groups of packets as individual entities, we may assume that they can be modeled by an M/D/1 queueing system, since the arrival rate of the group also follows a Poisson process (Eq. (3)), the service time is constant (and equal to T_W) and there is only one server, since one superframe is transmitted in each transmission window. Therefore, the mean waiting time of the group of packets that belong to QoS-class k is:

$$W_{n,k}^{group} = \frac{T_{TW} \cdot \sigma}{2 \cdot (1 - A_{n,k}^{group})} \tag{5}$$

The mean queue length of this M/D/1 system can be calculated by applying Little's theorem:

$$L_{n,k}^{group} = \lambda_{n,k}^{group} \cdot W_{n,k}^{group} \tag{6}$$

As the main target of the presented analysis is the calculation of mean queueing delay for the individual packets, we use the approximation proposed in [10] in order to firstly determine the mean queue length of the individual packets:

$$L_{n,k} \approx m_k \cdot L_{n,k}^{group} + P_{n,k}^{w} \frac{m_k - 1}{2} + (1 - P_{n,k}^{w}) \cdot \frac{w_k - 1}{2} \tag{7}$$

where w_k is the minimum number of QoS-class k packets that are transmitted in each transmission window, and $P_{n,k}^{w}$ is the probability of waiting in the corresponding M/D/ m_k queuing system. Finally, the mean queueing delay for the individual packets can be calculated by applying Little's law:

$$W_{n,k} = \frac{L_{n,k}}{\lambda_{n,k}} \tag{8}$$

4 Evaluation

In this section we provide numerical results from the proposed analysis for the mean queueing delay by considering a network of N = 10 rooftop, each one providing service to 3 small-cell antennas. The network supports K = 5 QoS-classes, which transmit their packets during a superframe duration of 50 time-slots, which are distributed to the 5 QoS classes by considering that (m_1, m_2, m_3, m_4, m_5) = (14, 12, 10, 8, 6). The time-slot duration is assumed to be equal to 25 μsec. In Fig. 2 we present analytical results for the queueing delay. In order to prove that the proposed approach results in lower queueing delay values for the high-QoS-classes, we assume that the packet arrival rate of all classes is the same and equal to the values of the x-axis of Fig. 2. As the results of Fig. 2 reveal, low QoS-class packets experience higher delay values, since a smaller number of packets are serviced in each transmission window compared to the higher classes.

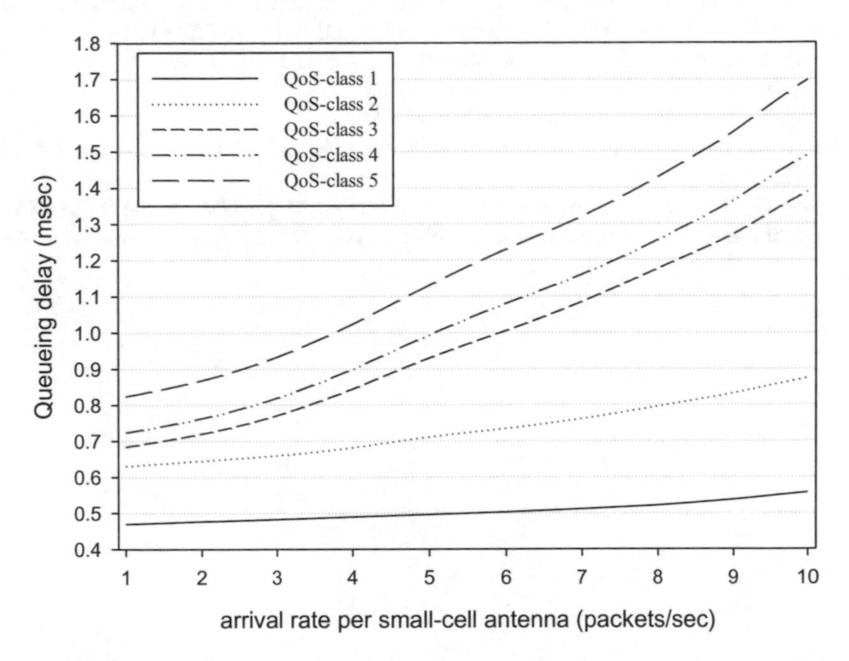

Fig. 2. Numerical results of the queueing delay of each one of the 5 QoS classes.

5 Conclusions

In conclusion, several challenges arise for the full exploitation of an integrated optical and wireless network in order to provide 5G network solutions. These challenges should be addressed by utilizing advances in the area of photonics as well as of network management, in order to architect 5G network that are capable of providing cost-effective and energy efficient solutions, especially in high density urban areas. To this end, we presented an approach for the provision of QoS differentiation in 5G networks and a corresponding analytical model for the determination of the queueing delay. The results showed that the prioritization of the QoS-classes can favor the high-priority classes in terms of queueing delay.

Acknowledgment. This work is funded by the H2020 5GPPP 5G-PHOS research program (grant 761989), MSCA ITN STEP-FWD (grant 722429), MSCA COMANDER (grant 612257) and by the research project IoSense (692480).

References

1. Esmail, M.A., Fathallah, H.: IEEE Commun. Surv. Tutor. **15**(2), 943–958 (2013)
2. Hadi, M., Pakrava, M.R.: J. Lightwave Technol. **35**(14), 2853–2863 (2016)
3. Zhang, W., Zhang, C., Chen, C., Qiu, K.: J. Lightwave Technol. **35**(9), 1524–1530 (2017)
4. Ferreira, R.M., Shahpari, A., Reis, J.D., Teixeira, A.L.: IEEE PTL **29**(11), 909–912 (2017)
5. Bogale, T.E., Le, L.B.: IEEE Veh. Technol. Mag. **11**(1), 64–75 (2016)
6. Wang, C.X., et al.: IEEE Commun. Mag. **52**(2), 122–130 (2014)
7. Peng, M., et al.: IEEE Netw. **29**(2), 6–14 (2015)
8. Wang, R., et al.: IEEE Access **2**, 1187–1195 (2014)
9. Kalfas, G., Pleros, N., Alonso, L., Verikoukis, C.: IEEE/OSA J. Opt. Commun. Netw. **8**(4), 206–220 (2016)
10. Bolch, G., Greiner, S., De Meer, H., Trivedi, K.S.: Queueing Networks and Markov Chains, Modeling and Performance Evaluation with Computer Science Applications. Wiley (2006)

5G Small-Cell Networks Exploiting Optical Technologies with mmWave Massive MIMO and MT-MAC Protocols

Sotirios Papaioannou(✉), George Kalfas, Christos Vagionas, Charoula Mitsolidou, Pavlos Maniotis, Amalia Miliou, and Nikos Pleros

Department of Informatics, Aristotle University of Thessaloniki, Thessaloniki, Greece
{sopa, gkalfas, chvagion, cvmitsol, ppmaniot, amiliou, npleros}@csd.auth.gr

Abstract. Analog optical fronthaul for 5G network architectures is currently being promoted as a bandwidth- and energy-efficient technology that can sustain the data-rate, latency and energy requirements of the emerging 5G era. This paper deals with a new optical fronthaul architecture that can effectively synergize optical transceiver, optical add/drop multiplexer and optical beamforming integrated photonics towards a DSP-assisted analog fronthaul for seamless and medium-transparent 5G small-cell networks. Its main application targets include Urban and Hot-Spot Area networks, promoting the deployment of mmWave massive MIMO Remote Radio Heads (RRHs) that can offer wireless data-rates ranging from 25 to 100 Gb/s and beyond depending on the fronthaul technology employed (e.g. spatial or wavelength division multiplexing transport). Small-cell access and resource allocation is ensured via a Medium-Transparent (MT-) MAC protocol that enables the transparent communication between the central office and the wireless end-users or the access cells via V-band massive MIMO RRHs.

Keywords: 5G networks · Analog optical fronthauling
mmWave massive MIMO · Medium-transparent MAC protocols
Optical beamforming technology

1 Introduction

The advent of 5G mobile communications will enable a wide range of emerging applications and services, supporting massive data volumes and devices, while creating new business models and opportunities. The list of 5G requirements is extensive and highly application-dependent, however an overall 1000-fold increase in wireless capacity, a 10x decrease in power consumption, and sub-ms latency should be achieved [1]. Expansion of the usable spectrum to contain the V-band (mmWave), as well as massive Multiple Input Multiple Output (mMIMO) antennas and Small-Cell

M. E. Auer and T. Tsiatsos (Eds.): IMCL 2017, AISC 725, pp. 805–813, 2018.
https://doi.org/10.1007/978-3-319-75175-7_79

(SC) deployment targeting network densification, have been identified in the literature to be the most promising solutions so as to achieve the demanding capacity and latency 5G requirements[1]. But due to the multiplication of the service access points, necessary to provide mmWave connectivity, in conjunction to the sophisticated hardware, increase greatly the capital and operational expenditures and increases the network's complexity. On the other hand, in an effort to reduce the cost, Cloud Radio Access Network (CRAN) has been introduced, characterized by the concentration of the intelligence and processing power of the RAN at remote centralized Base Band Units (BBU), while installing low-cost and energy-efficient Remote Radio Heads (RRHs) towards the network's edge. This centralized paradigm can fully exploit the potential of mmWave massive MIMO, by enabling the design of advanced coordinated transmission schemes across the network. Besides cost-efficiency, this fully-centralized paradigm enables the use of advanced coordinated multipoint (CoMP) schemes and promotes resource sharing amongst several co-located (virtual) BBUs.

To this day, the Common Public Radio Interface (CPRI)[2] is the only widely adopted framework for communication in the fronthaul (FH), i.e., linking the BBU and RRHs. However, the CPRI is known to add a large bandwidth penalty that can reach up to two orders of magnitude, compared to the original data rate, since it transmits digitized radio waveforms (i.e., I/Q samples), regardless of whether the signal carries data or not. Although this penalty can be tolerated in basic MIMO cases, when extending to massive mmWave multi-Gbps transmission links, this overhead becomes prohibitive. Hence, constantly extending the existing fiber FH infrastructure to achieve RRH densification is not a viable solution [2]. To this end, exploiting also the recent advances in mmWave communication, research efforts are being directed towards a converged fiber-wireless (FiWi) FH approach, which can offer significant advantages and additional degrees of freedom in the network design and base station placement.

The design of a fully converged 5G FiWi network, capable of guaranteeing the required data rates and latency constraints, demands a multidisciplinary holistic approach to seamlessly bring together a wide range of transmission and networking technologies. Multiple challenges arise across the whole architecture: from the design and integration of mMIMO antennas, to beamforming/beam-selection techniques and also to advanced centralized signal processing algorithms and dynamic access allocation schemes for the joint pool of optical/wireless resources.

In this paper, motivated by the upcoming 5G evolution and the foreseen convergence between optical and mmWave access and drawing from recent results in multi-format/bitrate optical communications, optical transceivers and optical beamformers (OBF), we propose a converged analog Radio-over-Fiber (aRoF) 5G FH solution with medium transparent packetized dynamic resource allocation and CoMP beamforming.

---

[1] spectrum.ieee.org/video/telecom/wireless/everything-you-need-to-know-about-5g.

[2] http://www.cpri.info/.

2 Converged FiWi Network Architecture

Taking into account the expected mmWave RAN densification, we have identified two prevalent 5G network scenarios, which can be served by a converged FiWi FH architecture, as depicted in Fig. 1: the Urban-area scenario for high-density metropolitan area RANs and the Hot-spot scenario for highly concentrated population (order of thousands) in very confined geographical areas, such as stadiums, concert halls, etc.

In the Urban-area (left part of Fig. 1), the BBU is connected via a fiber with the mmWave mMIMO RRHs that are in turn wirelessly linked to either mmWave or sub-6 GHz SC that provide the end users with radio access. In compliance with the roadmap of emerging 25 Gb/s and 4×25 Gb/s PON infrastructures [3], the fully-centralized BBU can deliver native wireless data signals via the fiber to the RRH at 25 Gb/s line rate, by exploiting 10 GHz-bandwidth optics, advanced modulation formatting and channel aggregation. The RRHs are capable of receiving the 25 Gb/s incoming optical data streams and transmit them wirelessly at the mmWave band. However, in highly-dense urban areas, e.g. central squares, where the increased number of users results in heavier demands, the proposed FiWi FH infrastructure needs to support scalability of aggregate RoF transmissions to 100 Gb/s and beyond either spatially (e.g. by increasing the number of fibers) or optically (e.g. via wavelength division multiplexing-WDM).

The second proposed FH architecture (right part of Fig. 1) defines a single fiber bus that links the BBU to a series of mMIMO RRHs that surround the Hot-Spot area, offering mmWave Service Access Points directly to the mobile users. Considering that these areas can host up to several thousand users, we can expect that a large number of RRHs will be required to cover the Hot-spot area, surpassing the number of available wavelengths in the fiber network. A wavelength selectivity scheme is necessary to support the tremendous number of users with a significantly smaller number of wavelengths compared to the deployed RRHs. This can be achieved in physical layer by introducing Reconfigurable Add/drop Multiplexers (ROADM) within the RRHs, supported by a wavelength allocation mechanism in MAC layer, such as the FiWi-enabled Medium-Transparent MAC (MT-MAC) protocol, described in Sect. 4.

Figure 2(a) shows the proposed aRoF split-PHY scheme, where most of the FH equipment is shifted to the centralized BBU, leaving the RRH only with the required components. Thus, transceiver optics, analog-to-digital converters and main functions in Digital Signal Processing (DSP), MAC and Software Defined Network (SDN) level are composing the BBU, while the necessary PHY optics and RF components are included in the RRH. Assisted by DSP, the aRoF FH exploits high-rate signals squeezed around a few-GHz Intermediate Frequency (IF) subcarrier [4] to enable signal direction-detection via cost-effective low-bandwidth optics at the RRH site, where a subsequent IF-to-RF up-converter produces the final wireless mmWave carrier signal.

Figure 2(b) depicts the block diagram of the proposed DSP-assisted aRoF FH solution introducing a low-loss high-bandwidth OBF instead of cost-expensive and complex RF circuitry, along with opto-electronic conversion and IF-to-mmWave RF up-conversion for wireless transmission. In this way, the BBU undertakes to carry out all high-demanding DSP processing, enabling mmWave mMIMO antenna implementations by means of OBF-based RRH units.

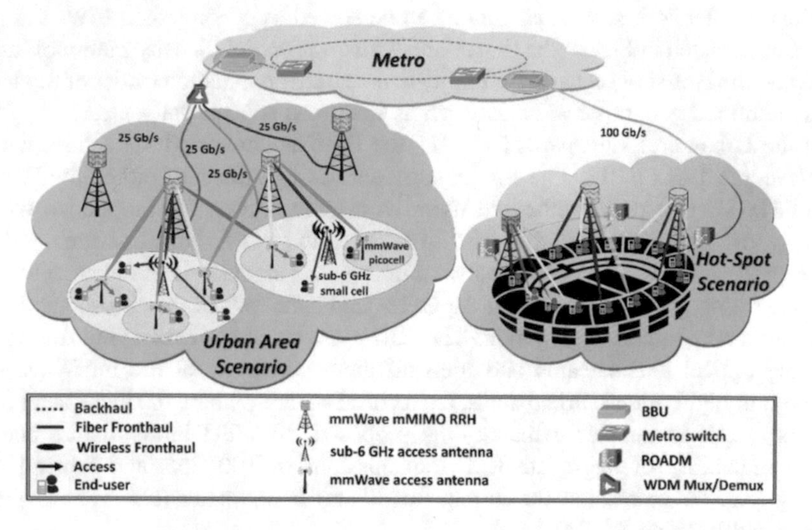

Fig. 1. Network architecture of the converged FiWi network for Urban and Hot-Spot areas.

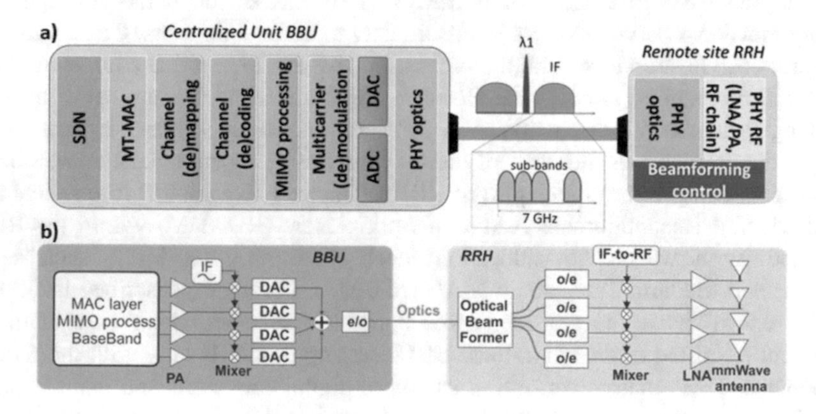

Fig. 2. (a) The proposed split-PHY FH architecture solution. (b) Block diagram of DSP-assisted aRoF FH with OBF network.

3 Optical Technologies for Analog RoF Fronthaul

The proposed aRoF FH scenario needs to exploit and further optimize 3 key optical technologies, towards building a complete technology pool of 5G-capable devices: (i) a linear optical transceiver with low cost modulator-laser and photodiode, (ii) low loss OBF for building a scalable feeder network for mMIMO antennas, (iii) Si photonics-based ROADM supporting WDM technologies and network reconfigurability.

- High-linear External Modulated Lasers (EML)

The envisioned aRoF FH solution can deploy the mature InP optical transceiver technologies to undertake the communication needs between the RRH and the BBU, with large transport capacities up to 100 Gb/s. In order to maintain cost-efficiency, the transceiver will employ a microwave photonic link at the IF and a mixture of advanced modulation formats to benefit from the use of cost-efficient optics. This requirement for analog optical transmission of native wireless signals, places an additional requirement on the linearity of the modulator.

The linearity of the modulator has long been under research investigations, targeting enhanced performance in the digital or analog domain. The Spurious Free Dynamic Range (SFDR) metric has been utilized to characterize the signal noise and the distortion induced in the modulated optical output due to the nonlinearities of the modulator. The widely deployed Mach Zehnder Electro-Optic Modulators are fundamentally limited by their $\cos^2$-shaped transfer function, with typical SFDR values ranging in the order of 90–112 dB-Hz$^{2/3}$, stimulating the emergence of more complex MZI-based structures, e.g. Dual Signal, Cascaded, Resonator-Assisted MZI [5].

In this direction, Electro-Absorption Modulator (EAM) technology has already been shown to achieve high-linear performance and low-cost fabrication. The envisioned aRoF transmitter can thus rely on an Externally Modulator Laser (EML), followed by a highly linear EAM with 10 GHz bandwidth, modulated by the electrical signal emerging at the RRH output. Following this principle, the mmWave wireless signal received by the RRH will be down-converted to an IF and linearly imprinted on the optical carrier of the EML, transferring an advanced modulation format (QPSK, N-QAM). Thus, multiple bits can be loaded on 1 symbol, prior to being fiber-transmitted to the BBU, where a 10 GHz photodiode will convert it back to the electrical domain.

- Optical beamformers (OBF)

On the way to network densification with large MIMO antennas of 64 × 64 patches, beamformer networks would need to utilize multiple phase shifters to control the steering of each antenna unit and in-turn multiple RF chains, that would increase the power consumption, cost and complexity based on an expensive mmWave feeder network. In this context, performing the beamforming directly in the optical domain yields high-bandwidth, low-cost and energy-efficient MIMO antenna structures, emerging as the most suitable candidate for seamlessly interfacing with the optical FH.

OBF can comprise a photonic integrated chip with a cascaded arrangement of 1 × 2 splitters in a tree-like architecture or two-dimensional (2D) phased array matrixes, with optical phase shifters or more appropriately True Time Delay (TTD) units interleaved between the different cascade stages [6, 7]. The optical phase shifters are simpler to achieve, but induce frequency depended phase shifting, distorting the phase front of the input beam, while TTDs support squint free beamsteering. In the first case of tree-architectures [6], by splitting the input data stream in many (e.g. 16) outputs and applying proper time-delay between the input signal constituents, OBF can provide broad instantaneous bandwidth of 7 GHz and large cumulative tunable delays in the order of 1 ns or equivalent time-delay propagation in the air of a few tens of cm. In the second case of the 2D beamsteering networks [7], OBF feature independent tuning elements, dedicated per antenna-unit. Low-loss photonic integrated beamformers and phased array antennas can provide wideband, continuous beamsteering in high-bandwidth and ultra-fast massive MIMO antenna up to 64 × 64 MIMO antenna sites. Such PIC-devices have already been fabricated on silicon-on-insulator or silicon-nitride substrates, yielding low-loss propagation, lightweight and small footprint characteristics, while benefitting also from CMOS compatibility for low cost fabrication.

- Optical Add/Drop Multiplexers (OADM)

OADMs are used in optical long-haul and metro networks to enable wavelength selectivity and re-configurability of the network. In order to achieve this, they typically comprise a wavelength demultiplexer, a wavelength-selective routing device and a wavelength multiplexer. Based on the reconfigurability of the wavelength-selective devices, they are sub-divided in Fixed and Reconfigurable devices, with the latter being the preferred choice, as they facilitate network flexibility, more efficient network planning and bandwidth re-allocation, while interconnecting many ROADMs together can build ROADMs with higher degree of port counts.

The Fixed OADM rely on static filtering devices, e.g. Fiber Bragg gratings with optical circulators, free space grating couplers, and/or planar lightwave circuits (PLCs), while ROADMs mainly rely on Micro-Electro-Mechanical Systems (MEMS), Liquid Crystals on Silicon (LCoS) and Thermo-Optic (TO) switches in PLCs [8]. The two main functionalities of OADMs are the Add and Drop functionalities, with the first describing the insertion of a new wavelength data-channel in a WDM-multiplexed stream of multiple data-channels, and the second describing the selective re-routing of only one channel of a WDM-stream towards a separate port.

OBFs, supporting multi-wavelength signals and, if combined with ring-resonator based ROADM, can eventually introduce WDM multiplexing capabilities to the RoF FH scenario. In a WDM scenario, the ROADM could comprise 4 cascaded Add/Drop rings capable to switch four incoming wavelength (4λ) channels towards 4 distinct OBF networks feeding the antenna elements of the RRH, providing direct compatibility with the current state of the art Ethernet protocol and optical transceiver technology.

As optical network keeps constantly expanding, ROADMs enable adding or removing bandwidths and reconfiguring the logical optical network with reduced CAPEX and OPEX. Silicon Photonics have emerged as an alternative photonic integrated ROADM technology with reduced footprint and energy requirements with faster reconfiguration times, controlled by Software Defined Processes [9], and in this direction, Si-ROADMs can bring similar benefits to aRoF FH.

4 MAC Layer Resource Allocation for Converged 5G FH

The new FiWi FH demands the adoption of novel resource allocation mechanisms that can concurrently address the optical/wireless/time resources in a converged manner. The MT-MAC protocols are a perfect match for the FiWi FH [10, 11] since they administer the wavelength, frequency and time resources directly between the BBU and the mobile terminals.

Figure 3(a) shows a possible MT-MAC operation in the case of urban FH environments, where the MT-MAC protocol creates data window transmission opportunities both in the uplink (UL) as well as the downlink (DL) direction between the cells and the BBU. The inherited MT-MAC's polling mechanism is used to create synchronization amongst all participating entities. The cells only communicate with the BBU through the RRHs, and therefore their transmission/reception lobes are always steered towards the latter, therefore alleviating interference and the need for beamforming/sector training. The MT-MAC initiates communication, by transmitting a Schedule POLL (S-POLL) packet to each cell and receives the cells' replies containing the number of packets in the buffer and their QoS levels. The MT-MAC then creates the transmission schedule and carries it out in two steps: first transmits an RRH Control packet, that carries full instructions set for the RRH regarding the next packet transmissions (containing BF directionality and interim guard spaces). Then the MT-MAC transmits Data POLL (D-POLL) packets directed at each cell with pending UL/DL traffic. The D-POLLs contain the transmission opportunity windows for each cell, including sleep schedules, power levels and channel state. When the cells receive the D-POLL, they begin transmission of UL traffic or await incoming DL transmission from the BBU. At the end of the transmission window, the SCs attach their updated buffer status, so as to facilitate the creation the next transmission window schedule. All DL/UL transmitted packets from all the cells are encapsulated within the MT-MAC frames. A cluster of MT-MAC frames forms the Superframe (SF). The MT-MAC SF is payload agnostic and can thus support encapsulation of many different technologies such as Ethernet.

For the hotspot scenario, the MT-MAC facilitates direct communication between the BBU and the mobile terminals (Fig. 3(b)). Based on the definition of the hotspot scenario, the MT-MAC deals with ultra-high user concentration and thus must ensure (a) dynamic wavelength selectivity to support the large number of RRHs by a single fiber, (b) location discovery and initial access for the mobile users. The first function is supported by transmitting a ROADM configuration packet that informs the ROADM micro-controller on what wavelength to drop at the specific cell and for how long. Then, the MT-MAC transmits an S-POLL packets to each cell that has an assigned wavelength and for every sector, essentially performing a full sector sweep. The mobile nodes that pick-up this transmission, reply with their buffer statuses and QoS demands after a random back-off period. When their replies are received correctly, the MT-MAC creates the net transmission schedule, transmits the RRH Configuration packet and assigns a permanent ID to each mobile client. If two or more nodes choose the same back-off value, their transmissions will collide and the BBU will register noise in the channel. In this event, the BBU will then retransmit POLL packets at that particular sector. When all sectors register no noise, meaning that all mobile clients have been correctly identified, the MT-MAC resumes operation as in the Urban scenario.

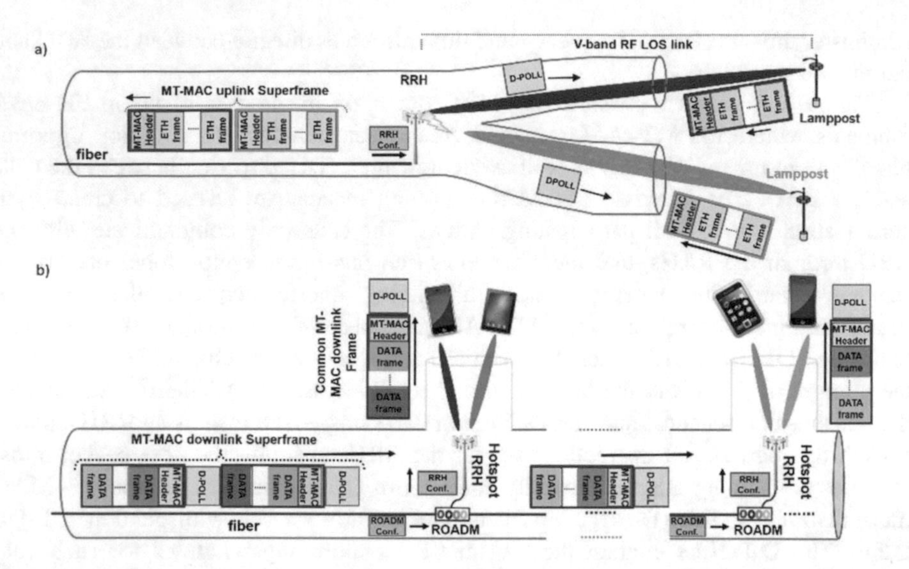

Fig. 3. MT-MAC framework operation for the FiWi FH in UL (a) Urban, (b) Hotspot networks.

5 Conclusions

In this paper, we have presented a new FiWi FH architecture towards fulfilling the emerging requirements in Urban and Hot-Spot Area networks. The envisioned architecture exploits mmWave mMIMO RRHs fiber-connected to BBUs, providing wireless data rates from 25 Gb/s and beyond. In terms of optical integrated technologies, the proposed solution takes advantage of a technology pool of 5G-capable devices, such as cost-effective optical transceivers, OBF and ROADMs. Regarding the MAC layer, a MT-MAC protocol is presented ensuring optical and wireless resource allocation and small-cell access.

References

1. NGMN Alliance: NGMN 5G White Paper. NGMN Board, 17 February 2015. https://www.ngmn.org/uploads/media/NGMN_5G_White_Paper_V1_0.pdf
2. Gao, Z., Dai, L., Mi, D., Wang, Z., Imran, M.A., Shakir, M.Z.: MmWave massive-MIMO-based wireless backhaul for the 5G ultra-dense network. IEEE Wirel. Commun. **22**(5), 13–21 (2015)
3. Effenberger, F.J.: Industrial trends and roadmap of access. J. Lightwave Technol. **35**, 1142–1146 (2017)
4. Kani, J., Terada, J., Suzuki, K.I., Otaka, A.: Solutions for future mobile fronthaul and access-network convergence. J. Lightwave Technol. **35**(3), 527–534 (2017)
5. Dingel, B., Madamopoulos, N., Prescod, A.: Adaptive high linearity intensity modulator for advanced microwave photonic links. In: Optical Communication Technology. InTech (2017)
6. Roeloffzen, C., et al.: Integrated optical beamformers. In: Proceedings of the OFC, Los Angeles, CA, USA, 22–26 March 2015, p. Tu3F.4 (2015)

7. Sun, J., Timurdogan, E., Yaacobi, A., Hosseini, E.S., Watts, M.R.: Large-scale nanophotonic phased array. Nature **493**, 195–199 (2013)
8. Marom, D., et al.: Survey of photonic switching architectures and technologies in support of spatially and spectrally flexible optical networking. J. Opt. Commun. Netw. **9**(1), 1–26 (2017)
9. Nakamura, S., et al.: Optical switches based on silicon photonics for ROADM application. IEEE J. Sel. Top. Quantum Electron. **22**(6), 185–193 (2016)
10. Kalfas, G., Pleros, N.: An agile and medium-transparent MAC protocol for 60 GHz radio-over-fiber local access networks. J. Lightwave Technol. **28**(16), 2315–2326 (2010)
11. Kalfas, G., Vardakas, J., Alonso, L., Verikoukis, C., Pleros, N.: Non-saturation delay analysis of medium transparent MAC protocol for 60 GHz fiber-wireless networks. J. Lightwave Technol. **35**(18), 3945–3955 (2017)

A Two-Tiers Framework for Cloud Data Storage (CDS) Security Based on Agent

Oussama Arki(✉) and Abdelhafid Zitouni

LIRE Labs, Abdelhamid Mehri Constantine 2 University,
Ali Mendjli, 25000 Constantine, Algeria
{oussama.arki,abdelhafid.zitouni}@univ-constantine2.dz

Abstract. Cloud Computing is a new model for delivering resources such as computing and storage on demand through a network. It has many advantages like reducing the cost, however, it remains the problem of security, which is the major brake for the migration towards Cloud Computing.

Cloud Data Storage (CDS) is an important service of Cloud, it provides the way to store and manage data remotely in the Cloud, but data security remains the problem, which makes users worried about the confidentiality and the integrity of their data. In this paper, we propose a framework of security to ensure the CDS, which is based on agents. It contains two layers: Cloud Provider layer and the User layer.

Keywords: Cloud Computing · Cloud Data Storage
Multi-Agent System · Integrity · Confidentiality

1 Introduction

In the last years, the Information Technology (IT) is rapidly developed, and both computing and storage techniques are widely emerging, the main reason is decreasing costs and increasing power of the computer resources, which led to the birth of the Cloud Computing.

Cloud Computing is a new IT model, it provides storage and computation resources as a service on demand, often through a network (typically the Internet). Cloud Data Storage is one of the Cloud services. It allows users to store their data in the Cloud, it also provides the powerful way of managing data. Cloud Data Storage allows to store and to manage the data remotely by reserving a virtual space in the Cloud. Besides the advantages of this service like the reducing of cost and the access to the Data at anytime from anywhere, they remain some concerns for the adoption of this service. The major brake for the adoption of this service is the data security. The security of data in the Cloud is the biggest challenge of Cloud Providers (CP).

To answer the problem of security in CDS, we propose in this paper a framework of security, which contains two layers: the User layer, the Cloud Provider layer. It is based on mobile agents.

M. E. Auer and T. Tsiatsos (Eds.): IMCL 2017, AISC 725, pp. 814–823, 2018.
https://doi.org/10.1007/978-3-319-75175-7_80

This paper is organized as follows: Sect. 2 introduces Cloud Computing and Cloud Storage; Sect. 3 discusses the Information Security and Cloud Storage concerns. Section 4 is about the related works. Section 5 presents our framework and the last section is a conclusion.

2 Cloud Computing and Cloud Storage

This section it is an overview of two terms: the Cloud Computing, and the Cloud Storage.

2.1 Cloud Computing

There are a lot of definitions of Cloud Computing, according to the NIST (National Institute of Standards and Technology): Cloud Computing is a model for enabling ubiquitous, convenient, on-demand network access to a shared pool of configurable computing resources (e.g., networks, servers, storage, applications, and services) that can be rapidly provisioned and released with minimal management effort or service provider interaction [1]. This Cloud model is composed of five essential characteristics, three service models, and four deployment models [2], Fig. 1 shows the NIST Visual Model of Cloud Computing.

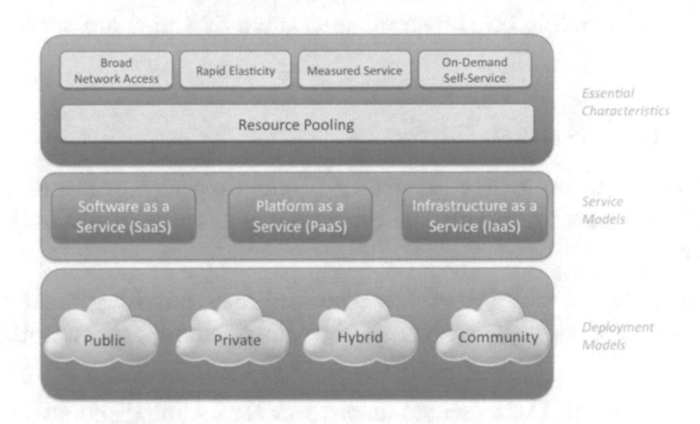

Fig. 1. NIST Visual model of cloud computing definition [3]

2.2 Cloud Data Storage

Cloud storage is a model of networked online storage where data is stored on multiple virtual servers, generally hosted by third parties, rather than being hosted on dedicated servers. Hosting companies operate large data centers; and people who require their data to be hosted buy or lease storage capacity from them and use it for their storage needs. The data center operators, in the background, virtualized the resources according to the requirements of the customer and expose them as storage pools, which the customers can themselves use to store files or data objects [4].

3 Information Security and Cloud Data Storage Concerns

3.1 Information Security

The term Information Security means protecting information and information systems from unauthorized access, use, disclosure, disruption, modification, or destruction in order to provide integrity, confidentiality, and availability [5].

- **Data Confidentiality:** is important for users to store their private or confidential data in the cloud. Authentication and access control strategies are used to ensure data confidentiality. The data confidentiality, authentication, and access control issues in cloud computing could be addressed by increasing the cloud reliability and trustworthiness [6].
- **Data Integrity:** is one of the most critical elements in any information system. Data integrity means protecting data from unauthorized deletion, modification, or fabrication. Managing entitys admittance and rights to specific enterprise resources ensures that valuable data and services are not abused, misappropriated, or stolen [6].
- **Data Availability:** means the following, when accidents such as hard disk damage, IDC fire, and network failures occur, the extent that users data can be used or recovered and how the users verify their data by techniques rather than depending on the credit guarantee by the cloud service provider alone [6].

3.2 Cloud Data Storage Concerns

Cloud storage is an important service of Cloud Computing, however this new paradigm of data hosting and data access services introduces two major security concerns [2]:

Protection of data integrity: Data owners may not fully trust the Cloud server and worry that data stored in the Cloud could be corrupted or even removed.

Data access control: Data owners may worry that some dishonest servers give data access to unauthorized users, such that they can no longer rely on the servers to conduct data access control.

4 Related Works

In the last years, many researchers used agents in their works, to ensure the security in the Cloud.

In [7] Talib et al., proposed a framework of MAS to facilitate security of Cloud Data Storage, the proposed framework has two layers: Customer layer and Cloud Data Storage layer. In the Customer layer, an interface agent is used for the interaction with the customer, the other agents are distributed in the Cloud Data Storage layer, each one of them has a specific task and they communicate to achieve the global goal. The major problem in this work is that the security of the Customer layer is not dealt with, especially the mechanism for the identification of the customer.

In [8] Islam and Habiba proposed an agent-based framework for providing security to Data Storage in Cloud, which contains three levels: Customer level, Cloud Provider level and Data Storage level. In their work, they considered that the degree of data security is depended on data sensitivity, because of that, they classified the sensitivity of data into five categories, for each category they used a different method of authentication, a different algorithm of encoding and a different function of hashing. By using this technique they increased the Cloud performance, but they risk the data of the customer.

In [9] Benabied et al., proposed a MAS framework to make safe Cloud environment, which contains two layers: Cloud Provider layer and the Customer layer. The communication between the two layers is ensured by the use of mobile agents, to ensure the trust at Cloud environment, this framework based on the use of Trust Model, they used two types of agents, one in Customer side and the other in Cloud Provider side, this model ensures the customer trust and guarantees that only the trusted customers can interact with the Cloud Provider. The weakness of this framework is the monitoring of the Cloud Provider side, which is ignored, like the supervising of the virtual machines.

5 Proposed Framework

To protect the data of the user that reside in the Cloud, we proposed a framework of security, which is a multi-agent-system. It contains two-layer cloud provider layer and the user layer. In each layer, a group of agents is distributed and work together to achieve our global goal. We used three methods to guarantee the security of data: a two levels authentication method for the access management, an encryption method based on RSA for the confidentiality and an integrity method based on Message Authentication Code (MAC).

5.1 Framework Architecture

In our work we based on the use of agent, which is widely used to solve problems in distributed and dynamic environment like the Cloud Computing. Figure 2 presents our framework.

As in Fig. 2, to reach the Cloud provider; the user interacts with the Interface Agent, which is an interface between the user and our framework.

For the authentication of the user, the Interface Agent request the Authenticator Agent, which resides in the user layer. The Authenticator Agent verifies the identity of the user and send the response to the interface agent.

If the user has an identity then, he can communicate with the Cloud Provider, to retrieve his data through the Proxy Agent.

The Proxy Agent is an intermediary between the cloud provider and the user. To store the data of the user, it requests the Encryption Agent to encode them using RSA algorithm, after the encoding operation in the user side, the Proxy Agent creates a Mobile Agent to carry the data of the user towards the Cloud Provider.

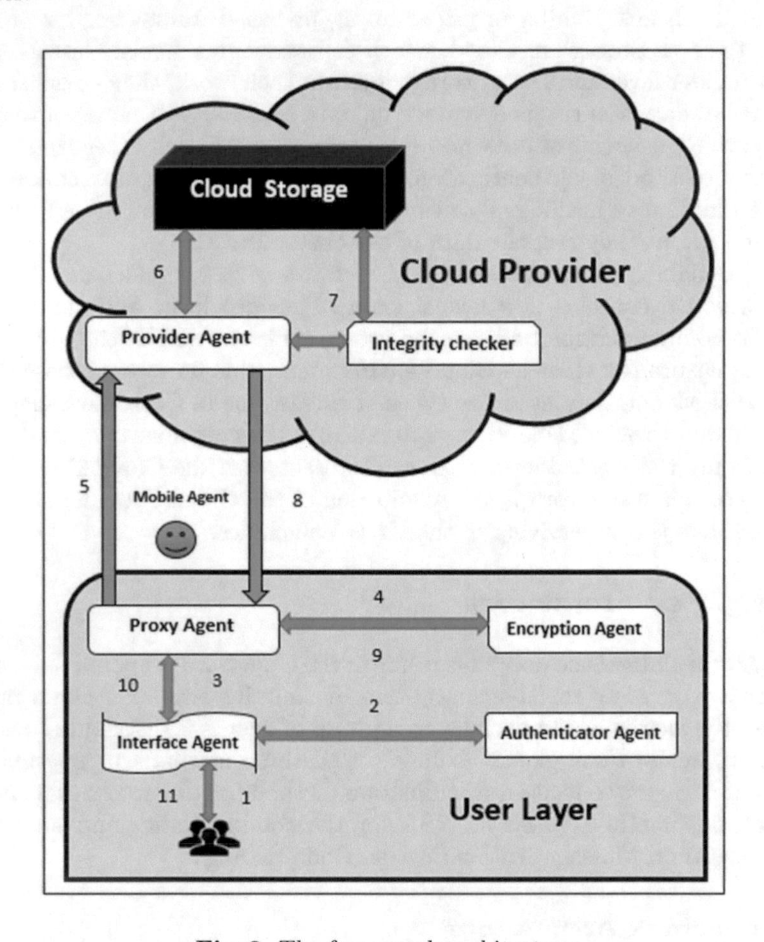

Fig. 2. The framework architecture

In the Cloud Provider layer, we used two agents. One is the Provider Agent, the job of this agent is to execute the request of the user; the other agent is the Integrity Checker Agent, The Provider Agent can request the Integrity Checker Agent, which is used to allow the user to check the integrity of his data, this agent uses an integrity technique, that will be explained in the other section.

The communication between the user and the Cloud provider can be summarized by the sequence diagram as in Fig. 3, where the steps are as follow:

- **Step 01:** the user provides the information to the Interface Agent, to authenticate himself.
- **Step 02:** the Interface Agent analyses the user information, and asks the Authenticator Agent to check the identity of the user.

- **Step 03:** if the response of the Authenticator Agent is good, the Interface Agent connects himself with the Proxy Agent, and the user can request his data, else the access is rejected.
- **Step 04:** if there is a need to encode or decode the data the Proxy Agent can ask the Encryption Agent to do that.
- **Step 05:** the Proxy Agent creates the Mobile Agent, which moves toward the Cloud Provider with the request of the user.
- **Step 06:** the Provider Agent executes the request of the user. It is the same case for storing the data and retrieving them.
- **Step 07:** if there is a need for the integrity check, the Provider Agent can ask the Integrity Checker Agent.
- **Step 08:** the Provider Agent gives the data to the Mobile Agent, which moves towards the user side and turns the data to the Proxy Agent.
- **Step 09:** If there is a need to encode or decode the data, the Proxy Agent can ask the Encryption Agent to do that.
- **Step 10:** After the decoding operation, the Proxy Agent gives the data to the Interface Agent.
- **Step 11:** the interface agent receives the data from the Proxy Agent and displays them to the user.

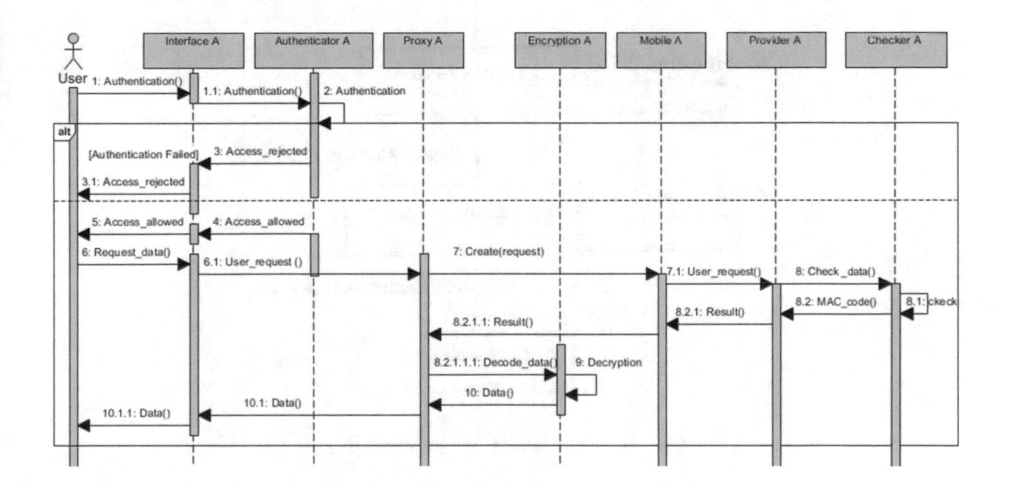

Fig. 3. Interaction between the user and the cloud provider

5.2 Authentication: Two Levels Authentication

The authentication is one of the security principals. It guarantees that only authorized entities can access to the information. There are many methods for the authentication; for our framework, we chose to apply a two levels authentication method. Where the user needs to authenticate himself twice before he can interact with the Cloud Provider. The main reason for this method is the mobile nature of Cloud Storage Services, which allows the users to access their data anywhere at anytime through a mobile device like a smartphone.

In the first step the user authenticate himself, by providing his information of identity (username, password) to the Interface Agent, which asks the Authenticator Agent to check the identity of the user through his data base. If the information given by the user is good, and it reflexs an identity; the Authenticator Agent performs the next step, so it generates a random pin code and sends it to the user. There are two kinds of the second authentication, the user can choose to perform the second authentication by e-mail, so the pin code will be sent to the e-mail of the user, or he can perform it using his mobile phone number, so the code pin will be sent to the user mobile phone. Only if the user provides the right pin code, the system starts the interaction with this user.

If the user fails to perform any step of the authentication, he can try again, until he provides the right information. Figure 4 presents the authentication method.

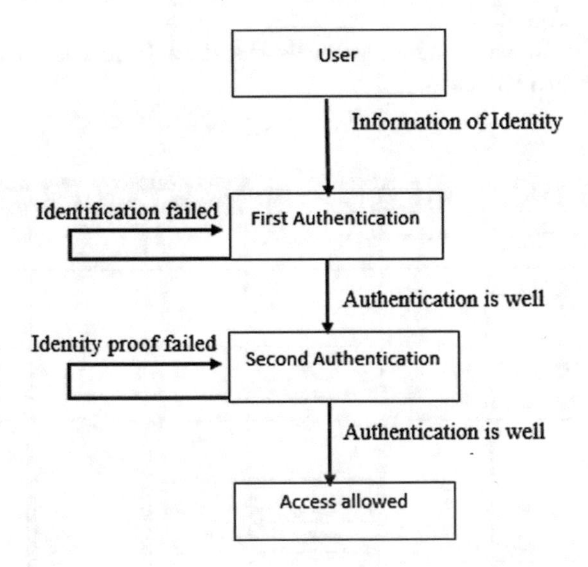

Fig. 4. Two levels authentication

5.3 Data Integrity Check

Integrity, in terms of data security, is nothing; but the guarantee that data can only be accessed or modified by those authorized to do so. In simple word, it is a process of verifying data. Data Integrity is very important among the other Cloud challenges. As Data integrity gives the guarantee that data is of high quality, correct, unmodified [10].

To allow the user to check the integrity of his data stored in the Cloud, we based on the Provable Data Possession (PDP) Scheme based on MAC (Message

Authentication Code), to ensure data integrity of file F stored on Cloud Storage in very simple way. The Encryption Agent computes a MAC of the whole file with a set of secret keys and stores it locally, before outsourcing the file to CP. Whenever a user needs to check the integrity, he requests the CP to recompute the MAC of the file, then compares the re-computed MAC with the previously stored value [10].

There are two types of data integrity check, private check, and public check. In public check the user requests another entity to perform the check. In our framework, we used a private auditing where the user is the responsible for the check of the integrity [11].

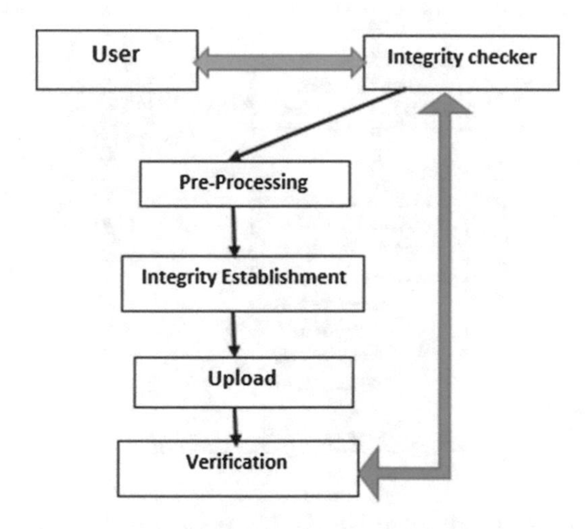

Fig. 5. Integrity check method

As in Fig. 5, the integrity method contains four steps as the following [11]:

- **Step-1 Pre-Processing:** In this step, metadata is generated. For verification, something has to be compared. Metadata is used for this purpose. There is a various technique by which metadata can be generated. In our method of check, we used the MAC technique.
- **Step-2 Integrity Establishment:** In this step, the metadata (which is generated using the MAC proof technique) is stored on the local machine of the user. This metadata will be used to check the Integrity in the next steps.
- **Step-3 Upload:** In this step, only the file of the user is uploaded to the Cloud Storage server.
- **Step-4 Verification:** In this step, the user requests the required metadata for the integrity check of requested file. The User compares the locally stored copy of metadata with the received one, that is computed by the Integrity Checker Agent. If both matches then the file is considered as intact, otherwise it is considered as tampered illegally.

5.4 Encryption Method

Encryption is one of the confidentiality techniques. It guarantees that only who have the keys can reach the real data. In our framework RSA algorithm is used to find out the keys, these keys are used to encode and decode the file [12]. Figure 6 shows the encryption method.

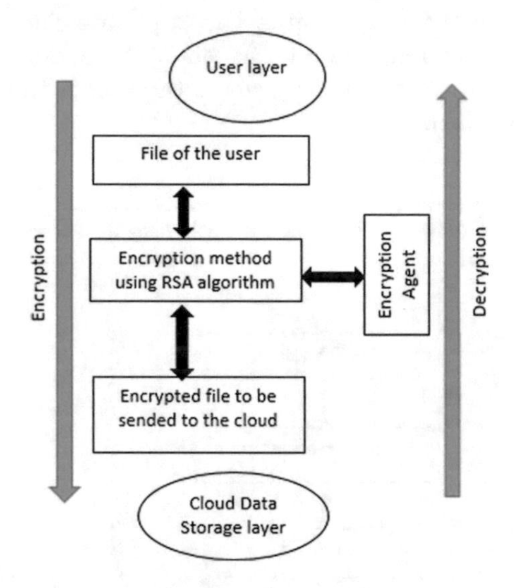

Fig. 6. Encryption method

So the Encryption Agent encodes the file by using the RSA algorithm. After the encoding operation, It returns the encrypted file to the Proxy Agent, which sends it to the Cloud Provider through the Mobile Agent. So the Cloud Provider contains only an encrypted file, which has no means. With the encryption of the user's data at the user side, we guarantee that only the user can recover the real data from the Cloud Storage Provider.

6 Conclusion

Cloud Computing Security is very important for the adoption and the continuity of this model, particularly in Cloud Storage service; because the data security is the biggest challenge of the Cloud Provider. In this paper, we presented our proposed framework, which is a Multi-Agent System of security to ensure the Cloud Data Storage (CDS). It is based on the use of agents especially mobile agent, a two levels authentication method besides the use of RSA algorithm for the encryption and the integrity check based on MAC.

Work is currently going on the framework implantation, where it will be applied to a specific case study. Further research could be realized to improve and to extend the present work, as the use of mobile agents to supervise the virtual machines of the users.

References

1. NIST.: Nist cloud computing standards roadmap. Technical report, National Institute of Standards and Technology (NIST) (2013)
2. Yang, K., Jia, X.: Security for cloud storage systems. Technical report, Springer (2014)
3. CSA.: Security guidance for critical areas of focus in cloud computing v2.1. Technical report, Cloud Security Alliance(CSA) (2009)
4. Balbudhe, P.O., Balbudhe, P.O.: Cloud storage reference model for cloud computing. Int. J. IT Eng. Appl. Sci. Res. (IJIEASR) **2**(3), 83 (2013)
5. Akter, L., Rahman, S.M.M., Hasan, Md.: Information security in cloud computing. Int. J. Inf. Technol. Convergence Serv. (IJITCS) **3**(4), 18 (2013)
6. Sun, Y., Zhang, J., Xiong, Y., Zhu, G.: Data security and privacy in cloud computing. Int. J. Distrib. Sens. Netw. **2014**, 3, 5 (2014)
7. Talib, A.M., Atan, R., Abdullah, R., Murad, M.A.A.: A framework of multi agent system to facilitate security of cloud data storage. In: Annual International Conference on Cloud Computing and Virtualization (2010)
8. Islam, Md.R., Habiba, M.: Agent based framework for providing security to data storage in cloud. IEEE (2012)
9. Benabied, S., Zitouni, A., Djoudi, M.: A cloud security framework based on trust model and mobile agent. IEEE (2015)
10. Giri, M.S., Gaur, B., Tomar, D.: A survey on data integrity techniques in cloud computing. Int. J. Comput. Appl. **122**(2), 27–28 (2015)
11. Desai, C.V., Jethava, G.B.: Survey on data integrity checking techniques in cloud data storage. Int. J. Adv. Res. Comput. Sci. Softw. Eng. (IJARCSSE) **4**(12), 293 (2014)
12. Tikore, S.V., Pradeep, D.K., Prakash, D.B.: Ensuring the data integrity and confidentiality in cloud storage using hash function and TPA. Int. J. Recent Innov. Trends Comput. Commun. (IJRITCC) **3**(5), 2737 (2015)

Network Resilience in Virtualized Architectures

Diomidis S. Michalopoulos[1(✉)], Borislava Gajic[1], Beatriz Gallego-Nicasio Crespo[3], Aravinthan Gopalasingham[2], and Jakob Belschner[4]

[1] Nokia Bell Labs, Munich, Germany
{diomidis.michalopoulos,borislava.gajic}@nokia-bell-labs.com
[2] Nokia Bell Labs, Paris, France
aravinthan.gopalasingham@nokia-bell-labs.com
[3] ATOS, Madrid, Spain
beatriz.gallego-nicasio@atos.net
[4] Deutsche Telekom Technology Innovation, Darmstadt, Germany
jakob.belschner@telekom.de

Abstract. Network resilience represents one of the major requirements of next generation networks. It refers to an increased level of availability, which is of high importance especially for certain critical services. In this work, we argue for resilience as an intrinsic feature that spans multiple network domains, thereby providing a network-wide failsafe operation. Particular focus is put on virtualized architectures envisioned for 5G and beyond. Contrary to traditional architectures where all network functions were hardware-dependent, a virtualized architecture allows a portion of such functions to run in virtualized environment, i.e., in a telco cloud, allowing thus for a wider deployment flexibility. Nonetheless, parts of this architecture such as radio access might still have strong hardware dependency due to, for instance, performance requirements of the physical nature of the network elements. Capitalizing on this architecture, we shed light onto the techniques designed to guarantee resilience at the radio access as well as the telco cloud network domains. Moreover, we highlight the ability of the envisioned architecture to address security-related issues by applying threat monitoring and prevention mechanisms, along with proper reaction approaches that isolate security intrusions to limited zones.

1 Introduction

Next generation networks are expected to be able to cope with stringent requirements in terms of reliability, latency and throughput. Despite their diverse nature, such requirements need to be addressed by using a common network infrastructure, since, otherwise, the cost of deploying separate networks for distinct services would be prohibitive. Such common infrastructure needs thus to provide sufficient flexibility in terms of deployment and operation, such that

M. E. Auer and T. Tsiatsos (Eds.): IMCL 2017, AISC 725, pp. 824–839, 2018.
https://doi.org/10.1007/978-3-319-75175-7_81

diverse services are supported without substantial change on the hardware. The solution towards this end is derived from the concept of *Network Slicing* [1,2], in conjunction with that of Network Function Virtualization *(NFV)* [3]. This results in virtualized network architectures, where the utilization of the available resources is optimized and can be flexibly allocated to the different network slices, for the purpose of the given service.

One of the major challenges associated with virtualized architectures is that of resilience. In this context, the resilience is translated into network robustness to different kinds of unexpected events and problems during the network operation. Resilience is particularly important for industrial applications and mission critical services, which operations have a very low fault tolerance. The problems that might jeopardize the network operation can be related to many aspects e.g. software, virtual or physical infrastructure, the actual implementation, deployment and configuration of the network functions. Thus, the next generation networks need to be built in a resilient way, capable of mitigating problems with critical impact on network operation.

Network resilience comprises a set of approaches, techniques and tools for ensuring the mitigation of network problems. To address this issue in future networks, it is important to adjust such design to the mode of operation in each part of the network. Specifically, it is anticipated that a part of the network functions, particularly those corresponding to high level functionalities and to the higher layers of the RAN protocol stack, are implemented in a virtualized infrastructure. On the other hand, network functions that are part of the lower layers of the Radio Access Network (RAN) often require an implementation in specialized hardware, hence they are implemented in a traditional, non-virtualized infrastructure. As a result, for attaining a sufficient end-to-end resilience level, both non-virtualized hardware and virtualized cloud environments (referred to as telco clouds henceforth) need to demonstrate the required level of robustness. These two domains have potentially different resilience issues and the approaches for achieving resilience might differ accordingly. However, resilience in both domains (i.e., RAN and telco cloud) are important building blocks for achieving overall network service resilience.

The remainder of this paper is structured as follows. A literature overview in resilience and security of existing networks and its connection to future deployments is provided in Sect. 2. Section 3 showcases the directions towards resilience in the RAN domain, while Sect. 4 highlights the basic approaches followed for achieving resilience in the telco cloud. Section 5 illustrates the main security mechanisms devised for providing an advanced security level of next generation networks, while our final concluding remarks are given in Sect. 6.

2 Technological Advances and Challenges in Resilience and Security

In this section we elaborate on the technological progress in the domains of RAN and telco cloud related to resilience, along with security issues associated to the deployment of next generation networks.

2.1 The Challenge of High RAN Reliability

In the RAN domain, the dominant challenge with respect to a reliable operation is managing the highly dynamic radio channel. Providing an Ultra-Reliable Low Latency Communications (URLLC) service requires dedicated approaches to combat e.g. a packet loss due to short-term fading. To this end, a relatively novel approach used to increase the reliability of the RAN domain is multi-connectivity [4,5].

Multi-connectivity is a well-known concept that is used already in the LTE standards, where the main objective is to aggregate two independent radio connections for increasing the overall throughput [6,7]. However, the main concept behind devising multi-connectivity for RAN reliability is to exploit the inherent macro-diversity effect of multiple simultaneous connections, such that the probability that at least one connection is sufficiently strong is increased [8]. As such, multi-connectivity takes a distinct role than the one used in LTE, since the packet flow is now being *duplicated* across multiple links, instead of aggregated. This entails challenges in regard to the system design since special coordination is needed between the links where duplicated data is sent. We shed light onto such design challenges in Sect. 3.

2.2 The Challenge of Resilient Virtualized Networks

Resilient network needs to be able to recover after an unexpected event and to resume its normal operation, without affecting the user experience. This capability is of paramount importance for network reliability and providing a service with satisfying performance especially for critical communication type such as envisioned in URLLC slice.

Advantages of SDN. Software-defined networking (SDN) is growing rapidly in telecommunications due to its capability to efficiently manage end-to-end networks by decoupling control-plane and data-plane. Such scalability and flexibility can bring benefits to network management and maintenance. In general, SDN brings several advantages to mobile network architecture such as high flexibility, programmability, complete control of the network from centralized vantage point, and enables operators to deploy easily new applications, services and fine tune network policies. SDN and NFV are two closely related technologies that are often used together in cloud paradigm to complement and benefit from each other.

The integration of SDN framework in Cloud RAN (C-RAN) can provide several advantages such as dynamic control over fronthaul transport network to allocate available capacity while maintaining overall QoS requirements, realization of centralized SON (e.g., Coordinated scheduling) and configuration and load-balancing between virtual base band units (vBBUs) [9]. Although SDN is a quite matured technology, most of the SDN frameworks have been designed and developed with the major focus on supporting several use cases in fixed and transport networks. However, SDN is an important aspect that can enable dynamic control of radio and networking resources in telco cloud by re-programming/re-configuring VNFs in real-time. Due to the stringent QoS requirements of 5G mobile networks, the SDN framework has to have low latency, resilience and scalability in order to be adapted as a control framework.

Network fault management in telco clouds. In the telco cloud domain, there exist different approaches for increasing the overall resilience. Some of the common techniques for mitigating the network faults in traditional network are self-healing SON solutions [10]. Self-healing SON aims at automatizing the mitigation of outages on the level of individual network cells, including outage detection and root cause analysis. Within such framework different improvements of detection and diagnosis processes can be applied as presented in [11,12]. The introduction of virtualization in network design and deployment brought new challenges in handling the network faults. As the faults can occur on different deployment layers, e.g. physical, virtual, application, the fault management needs to be enhanced in order to master the increased complexity in fault localization and isolation. The work targeting the fault management issues in virtualized environment has been presented in [13] where distributed fault management approach has been chosen.

However, despite the considerable progress in this field, the majority of 5G network architecture proposals did not explicitly or to a large extent target addressing the resilience levels of URLLC. The requirements on resilience has mainly been implicitly addressed by the management and control entities and mechanisms that are designed in a way to promptly react to unexpected events. For instance, as reported in [14], after a violation of QoS requirements is detected on centralized controllers, the problem mitigation is attempted through network reconfigurations. This might involve reconfigurations of network functions parameters, as well as link reconfigurations. In the case that this was not sufficient to overcome the problem, the centralized controllers send a trigger to Management and Orchestration (MANO) blocks, such as the Orchestration entity, in order to perform the action needed for problem mitigation. This might include scale out actions if the resources of network functions are scarce, as well as relocation of existing functions and deployment of new functions.

Although such architecture is capable of reacting to unexpected traffic/network events and mitigate their negative influence to a certain extent, the architecture and mitigation mechanisms are not built under the concept of resilience. In other words, there is no detailed resilience consideration intrinsic to the network design, in the sense that there is no specialized network functions for empowering the resilience or service-specific resilience requirements built in to the network design. Therefore, the aforementioned problem mitigation actions and processes are suboptimal and cannot meet different reliability requirements in an efficient way. In this regard, in Sect. 4 we elaborate on the most prominent challenges and envisioned solution approaches in the context of resilient virtualized networks and the scalability of its control framework.

2.3 The Challenge of Security in Future Networks

In additional to resilience, security is an important factor of the design of next generation systems. In particular, the increased number of connected devices anticipated for future networks poses an certain threats on security, in addition to threats already existing in LTE. Moreover, such threats become even more important in mission critical applications, which are expected to play a vital role in the 5G ecosystem. In this regard, advanced security mechanisms need to be deployed, aiming at preventing and, if this is not possible, minimizing the effects of unexpected events originated deliberately by a human. Such man-made network disruptions either compromise fundamental security properties e.g., integrity, confidentiality, availability in the network or entail other deliberate misuse of the network that can turn into a security threat with major consequences.

Whereas the baseline of the 5G network architecture has been set through different research and standardization organizations and activities, a detailed elaboration on the means for achieving resilience has not been conducted. In Sect. 5 we address this need for a more detailed view on 5G network resilience and provide first insight on the means for achieving the desired level of resilience in 5G networks.

3 RAN Reliability Approaches

Next generation RAN shall allow a higher access reliability level, which is targeted especially for URLLC use cases. From the perspective of the RAN protocol stack design, this is translated into new RAN functionalities that aim at minimizing the radio link outage probability. As discussed in Sect. 2.1, this requirement can be addressed by specially tailored multi-connectivity based solutions that involve data duplication across the radio links [5,8,15].

In the remainder of this section, we highlight the technical features of implementing data duplication in next generation networks. We first discuss the deployment characteristics of data duplication, seen as an extension of LTE's dual connectivity; then, we elaborate on the particular protocol features of its implementation.

3.1 Data Duplication in Heterogeneous Networks (HetNet)

A typical deployment scenario used for the dual-connectivity approach in the LTE standards to increase the throughput is the Heterogeneous Network (HetNet) approach [7], which is also anticipated to provide coverage for data duplication. With this approach, the (UE) connects simultaneously to both a macro cell and a small cell, which usually operate in different frequency bands. It is also assumed that next generation networks will adopt a centralized architecture, where networks functions are split between two RAN units, namely the Central Unit (CU) and the Distributed Unit (DU) [16].

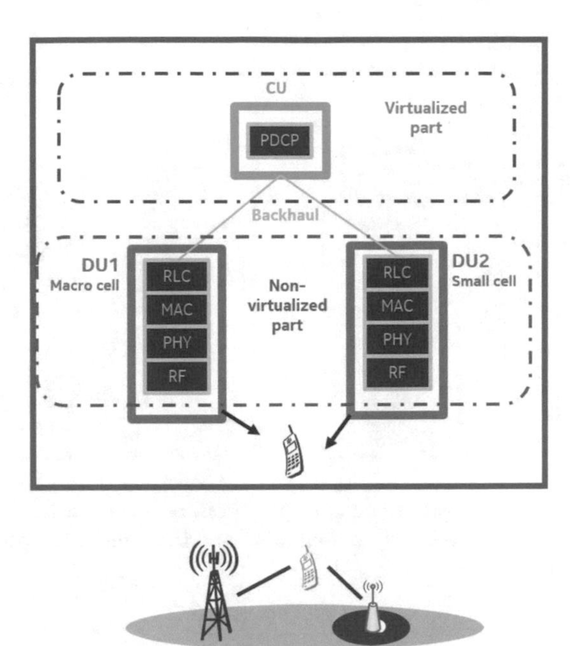

Fig. 1. HetNet deployment under the centralized architecture, where network functions are split between the CU and the DU.

An illustration of the HetNet deployment in the centralized architecture[1] is provided in Fig. 1. As can be seen from Fig. 1, the coverage area of the small cell falls within that of the macro cell. Typically, the location of the small-cell base station is carefully selected so as to fill in coverage gaps from the macro cell. On the basis of the centralized architecture [16], the lower layers of the protocol stack of both the macro- and the small cell take place at the corresponding DUs. Then, the integration of the signal flow to both links involved is carried out at the CU, which contains the higher RAN layers. It is noted that, in the context of NFV, the CU can run in a virtualized implementation, such that it represents part of the telco cloud itself. In such case, the orchestration of the CU resources follows the properties of the telco cloud management, as described in Sects. 2 and 4.

3.2 RAN Protocol View of Data Duplication

The implementation of data duplication requires special coordination of the signal flow at the CU. In particular, the CU needs to take care that duplicate packets are delivered correctly to the UE, and that the overhead of the extra resources needed is minimized. In this regard, we highlight two major points

[1] We adopt "Option 2" from the candidate function split options provided in [16] since it is the most suitable to the use of data duplication.

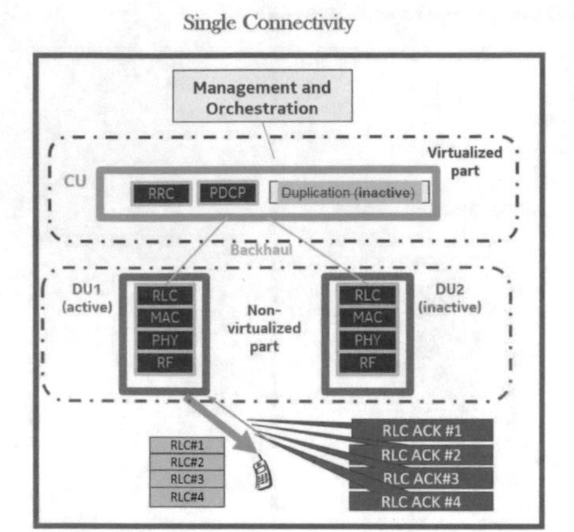

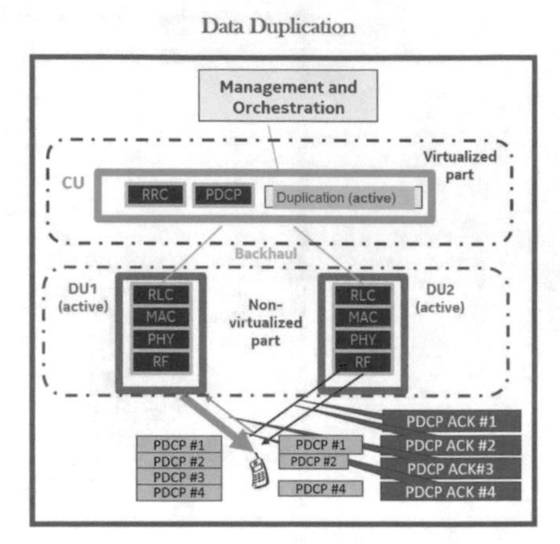

Fig. 2. Single connectivity versus data duplication, as seen via the prism of the new signaling involved within the RAN, as well as to and from the management and orchestration network layer.

where data duplication differs from existing approaches, from the perspective of the underlying technology.

- *Introduction of Packet Data Convergence Protocol (PDCP) acknowledgments.* In LTE standards, the packet acknowledgment feedback (ACK) sent from the receiver to the transmitter in order to indicate whether the transmission was correctly received is carried out in two layers: At the Medium Access Control (MAC) layer by means of Hybrid Automatic Repeat Request (HARQ), and at the Radio Link Convergence (RLC) layer by means of outer Automatic Repeat Request (ARQ). On the contrary, given that the RLC layers of the two involved links do not process the exact same packet sequence (i.e., RLC packet number #2 for DU1 is not necessarily identical to RLC #2 for DU2), in the data duplication case feedback should be sent to the PDCP packet numbering instead. This process is expected to be introduced to 5G systems, and is illustrated in Fig. 2.
- *Management and Orchestration.* The activation of the data duplication process is followed by utilization of additional resources which require special administration and control. In this respect, the data duplication function that resides in the CU (c.f. Fig. 2) is orchestrated by a higher level entity which resides in the MANO layer. Besides the overall orchestration of the virtualized resources, this entity is responsible for deciding whether the data duplication function should be activated, and if so, what is the amount of virtualized resources allocated to it. The RAN Radio Resource Control (RRC) entity is then assigned the task of allocating radio resources between the two involved links, based on the needs of the underlying service.

4 Resilience in Telco Clouds

In order to better address the resilience needs of particular network slice types, e.g. URLLC slices or industrial enterprise slice, the target of this work is in enabling and integrating service-specific resilience and reliability aspects intrinsically in the 5G architecture. Rather than being an afterthought the service-specific resilience needs to be one of the main properties of the 5G network design.

In addition to RAN reliability which has been presented in Sect. 3, in this work we further elaborate on resilience in the telco cloud. In this context, we focus on two main aspects that need to be carefully considered, namely (a) *Network fault management*, taking into account virtualized network functions, and (b) Improving the resilience of individual network elements and functions, with emphasis on the *centralized network controller*.

4.1 Resilient Network Design and Network Fault Management

The main goal of network fault management is to enable the resilience to network failures by monitoring the network state and provide solution to the problems that cause the network performance degradation or failure. As a first step, the detection of changes, potential problems and anomalies in network behavior needs to be performed based on input from monitoring tools. Furthermore, the actual cause of the problem needs to be determined in order to perform the suitable recovery actions. The root-cause analysis enables the localization of the actual problem and consequently its isolation such that the propagation of fault effects and impact to the rest of the network can be minimized. Figure 3 illustrates the main processes and actions involved in the fault management. Such fault management techniques need to be adapted and extended towards the 5G network slicing context. The fault management characteristics and parameters need to be adjusted to the actual service that is supported. This might include e.g., the service-aware design of triggers and thresholds for alarms creation, start of recovery actions, etc.

The virtualization of traditional network elements broke up the tight coupling between hardware and software and introduced additional complexity in handling the faults of network functions. In virtualized networks three layers of deployment can be identified: network function/service logic, virtual infrastructure (e.g. virtual machines, containers) and physical infrastructure (e.g., COTS servers, compute and storage components) as illustrated in Fig. 4. In such an environment there might be different implementation and deployment options for network functions, i.e. there can be many to many relationships between layers of network functions logic, virtual infrastructure and physical infrastructure where the network function resides. Such layered implementation of network function requires enhanced fault management logic which takes into account the actual deployment and interrelations between the layers.

In general, the network fault should be handled at the layer where it occurs, ideally discovered before the major effects take place and/or propagate among

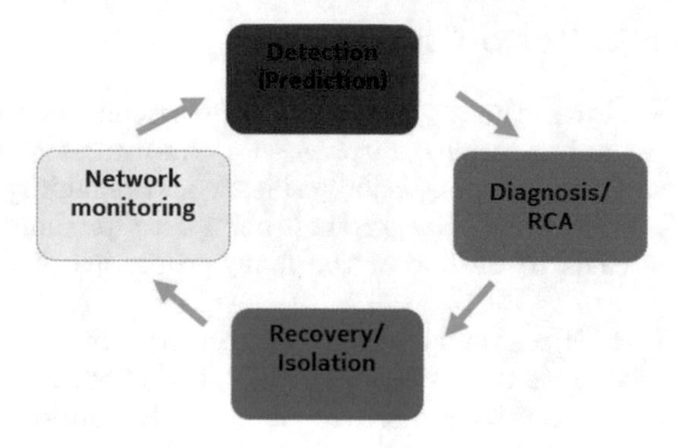

Fig. 3. Main processes and actions involved in network fault management.

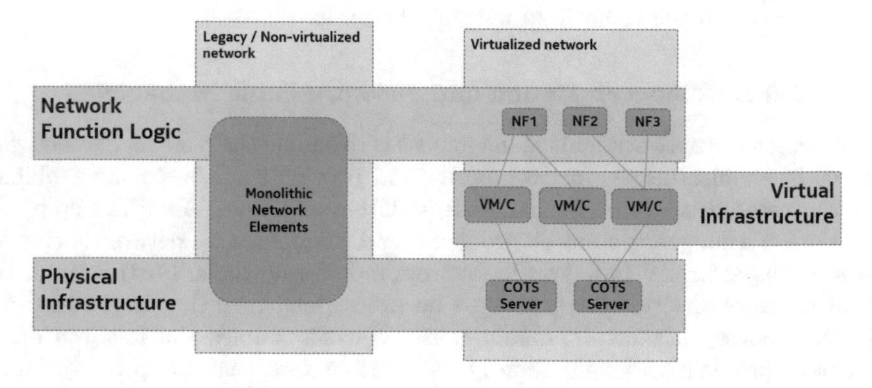

Fig. 4. Layered view of virtualized network: physical, virtual and network function logic layers.

different layers. As the faults can be related to different layers of network function deployment the correlation between fault occurring at different layers is essential for root cause analysis in virtualized networks. Furthermore, the correlation between the resource failures and the impact on the service performance and ultimately on the user satisfaction can create a baseline for better resource provisioning, prioritization and maintenance. However, the correlation is complex task due to many-to-many relations between infrastructure and network functions, service providers, deployments in multi-site and multi-domain data centers etc.

Despite the fact that fault management might be more complex in virtualized networks, the virtualization can be seen as a facilitator for network resilience through much easier and cost-effective redundancy implementation. As the network functions can be implemented on the commodity hardware the network functions can be more easily multiplied and moved across the network. Furthermore, adding redundancy in virtualized environment is more cost-effective as the infrastructure resources of redundant network functions can be more easily re-used. Adding redundancy is especially important for critical network functions or network functions with higher importance/priority. For example, the SDN controllers which have central role in network control might be designed with more redundancy than other network functions, as the outage in network controller might have severe impact on overall network operation. Nevertheless, careful considerations on trade-offs in applying redundancy, e.g. in terms of overprovisioning and resource reservation, needs to be done in order to design efficient and resilient network.

4.2 Resilient and Scalable SDN Control FrameWork

The earliest SDN controller frameworks such as NOX, FOX, Floodlight, Ryu, Beacon considered the architecture to be centralized. Later, with the introduction ONOS and ODL, the control framework can also be deployed in distributed mode avoiding single point of failure and also improving performance, scalability and resilience [17]. The distributed architecture is a key feature of ONOS to support both scaling and fault-tolerance by instantiating and linking multiple instances in the cluster. In such approach, each instance can be an exclusive master for set of switches and failure of any instance leads to the selection of new master for those set of switches by the other instances. Raft consensus [18] algorithm is used for data synchronization and state management between distributed instances in ONOS. ODL has a similar clustering model build with Infinispan NoSQL data-store.

Although the distributed design is intended to improve the control layer resilience, it introduces challenges related to timing, consistency, synchronization and coordination for its adaptability in low latency and time constraint mobile network infrastructure such as telco cloud. In telco cloud, the VNFs corresponding to RAN and Core of particular network slice can be deployed across distributed cloud segments such as Front End Unit, Edge and Central Cloud located in different locations. Moreover, each slice can have different QoS requirements, for example the URLLC slice requires low latency through out its life cycle management starting from deployment to resource allocation. In such scenarios, the control framework needs to have different level of performance and behavior corresponding to different deployment scenarios and use cases.

The current implementation of both ONOS and ODL has its drawbacks by not considering the current and future load in the selection of master control instance for set of devices along with higher data synchronization time i.e., in milliseconds. In summary, as shown in Fig. 5 the successful realization of SDN for telco cloud requires a controller framework that is able to provide scalability and resilience, while satisfying the stringent performance requirement of each use cases. Such framework needs to be load aware and load predictive in selecting master controller instance for each set of devices.

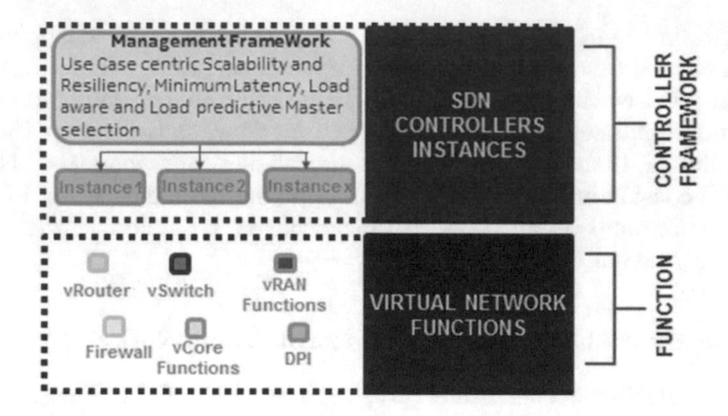

Fig. 5. Use case and load aware scalable and resilient control framework.

5 Implementing Security on Top of Resilience

Security always comes at the extent of high resources consumption and impact on the normal operation of a system. Security mechanisms aim at protecting the system, e.g. by putting additional layers of hardware or software around the ones needed for just providing the functionalities the system was created for. This protection can be implemented in very efficient ways such that the impact on performance is minimized, but the impact per se cannot be avoided (e.g. performance decrease at peak times). In addition, depending on the criticality of the assets to protect, some countermeasures could cause a complete disruption of the network service operation, by completely isolating a portion of the network in order to prevent propagation of attacks or security flaws.

5.1 Security Threats: Prevention, Detection, and Reaction Methods

Any system exposed to the environment and the human interaction is subject to be a target of attacks. Depending on the degree of exposure and the nature of the elements that compose the system, some threats are more likely to occur than others. In the case of IT services based on 5G network infrastructures, there is a wide range of threats that both network tenants and telco operators must be prepared to deal with [19]. Since different technologies are involved, intertwined by multiple software and hardware infrastructure layers, the number of critical assets to protect increases. As a result, the vulnerabilities and weaknesses that can be exploited increase as well. With regard to privacy regulations, data breaches becomes *public enemy no. 1*, and dealing with such threat category makes it necessary to involve not only IT departments within an organization, but Human Resources and Legal Counseling (at least) as well, in order to overcome the so-called *human factor*.

In addition to complexity, some security incidents may have a huge impact on the overall service operation. For instance, in *Man-in-the-Middle* attacks, a user session could be hijacked and used to insert rogue data into a mobile connection to maliciously exhaust network resources. Denial of Service (DoS) attacks are one of the top incident patterns responsible for causing disastrous business downtime, loss of data and application service, with an enormous economic impact, let alone the negative impact in the brand image [20].

An exemplary strategy. With this in mind, it is important to implement a proper strategy which (a) methodologically identifies the threats that may affect the system under analysis and (b) applies the most appropriate mechanisms to address them. In this respect, Fig. 6 depicts an exemplary strategy, which comprises a combination of continuously monitoring of the landscape and active learning.

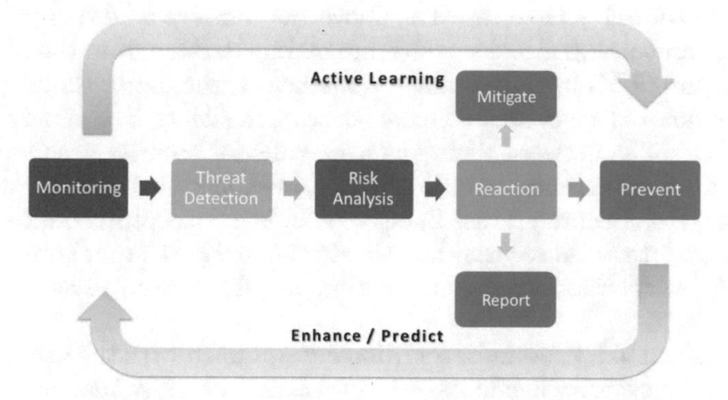

Fig. 6. Security monitoring and active learning process

The importance of the strategy shown in Fig. 6 is explained as follows. On the one hand, looking for known security incident patterns allows for their detection as soon as they occur. It also allows taking the appropriate countermeasures with a minimum delay, minimizing the impact and avoiding propagation. On the other hand, by actively learning from the analysis of anomalous behavior, in contrast to the legitimate or normal behavior, allows to come up with new patterns or evolutions of known ones. Overall, this active learning process is a way to autonomously enhance the knowledge database, adapt to dynamically changing attack vectors and prevent from future security incidents.

Security monitoring. Security monitoring is a conservative mechanism that relies on well proven security directives that permit detecting an (attempt of an) incident with high accuracy. However this is not sufficient nowadays. Advanced attackers put a great deal of efforts in evolving their malicious techniques fast, circumventing any new patch or security obstacle deployed in the system, and making the recently updated detection rules outdated shortly after these are rolled-out. This is the case of Advanced Persistence Threats (APTs), which exemplify the advanced cyber threat due to increasing frequency, sophistication, importance and difficulty in countering in recent years [21].

Threat prevention. Prevention mechanisms aim to overcome this problem since these permit learning from experience and enhance detection rules and reconfigure the security monitoring infrastructure to adapt to new scenarios. However, the main drawback of prevention mechanisms based on machine learning algorithms is the high rate of false positives. The reliability of the alarms raised by such tools is usually not high (especially when the training data is not extensive, rich or varied enough) and thus, the triggered countermeasures must be just preventive rather than reactive. As such, these signals should be used to prepare the system for the worst scenario, which can last for a predefined period of time or until the preventive system identifies that the threat is no longer probable to materialize.

Reaction to security threats. The above methods are used to monitor, detect and possibly preventing attacks, which are mainly derived from the complex and dynamic nature of 5G infrastructures. Nonetheless, the major challenge remains to apply automated responses to cybersecurity incidents in a timely and automated manner. The network slicing concept calls for security architectures that are able to work autonomously within a slice, even in a disconnected way (e.g. at a cloud edge) [19]. Security Trust Zones (STZs) is a concept introduced in [22] to describe an architectural security solution for 5G networks that enhances the so-called AAA (Authorization, Authentication and Accounting) security functions at edge clouds.

The operation of STZs can be additionally equipped with the necessary mechanisms to detect security incidents, take decisions and apply custom countermeasures locally and fast. This *prescriptive security* is based on automating simple and specific threat analysis tasks with sophisticated machine learning and artificial intelligence [23]. In addition, STZs shall have the capabilities to share certain threat intelligence with other zones to avoid propagation and remain self-defending. STZs may cover multi-geographic areas and spread across different network slices, therefore, an inter-slice security management function would be required to govern and orchestrate the overall security response.

5.2 Compromise Between Security and Resilience

Security and Resilience are two related concepts with mutual effect on one another. In particular, a large number of security-related threats can affect to different extent the resilience and functionality of the network fault management. For example, the DoS attack can result in unavailability of machines and network functions running on top of affected machines. Such effect will be detected by the network fault management which will attempt to solve such issue using its restoration capabilities.

If redundant machines, network functions and links are available, the security threat might be mitigated using the existing redundancy. However, this may lead in lowering the current redundancy and consequently resilience level of the network. Depending on the actual service and agreed SLAs with the network tenant as well as the actual severity of the security threat, this might or might not be acceptable. In certain cases, lowering the resilience level for handling the threat might be unacceptable. This is true, for example, in situations where the security threat is assessed to be minor and does not jeopardize the normal network operation, whereas redundancy needs to be kept at the certain level due to risks of software and hardware problems. In such case a security might be compromised for achieving the required resilience/redundancy. On the other hand, as certain security threats might result in severe problems in functionality of individual network functions or the network as a whole, handling such threats might have highest priority, even at the cost of lowering the current redundancy/resilience level. In such case, the resilience might be compromised for security.

In general, measures need to be put in place to guarantee that a certain degree of resilience could pose new threats or attack paths that could be exploited with malicious purposes. Duplicating network resources to ensure availability of a service operation could give attackers another entry point to the system, if such resources are not properly secured. Nevertheless, the solution may not be as straightforward as simply duplicating the security as well, i.e. applying the same security mechanisms to the duplicated network branch. On the contrary, it requires a re-design of the security strategy of the system as a whole, which includes the duplicated network branches and any other plausible resilience mechanism.

6 Conclusions

In this paper the different aspects of network resilience in virtualized architectures were discussed. Specifically, this comprises the RAN reliability challenge of providing URLLC services over a radio link subject to fading, along with the challenge of providing service robustness in the telco cloud. Additionally, the main security challenges of future networks were put forward as an important and related topic. In this context, potential solutions to these challenges were presented. With respect to RAN reliability, multi-connectivity approaches involving data duplication were discussed as a means to reduce the probability of errors. In a telco cloud domain the resilience can be empowered by improved virtualization-aware and service-specific fault management as well as robust and scalable SDN control framework. We also showed that new security threats need enhancements in monitoring, prevention and reaction approaches, taking into account network slicing concept and increasing the level of automation.

An additional topic that was raised is the required compromise between security and resilience, which will mainly represent part of our future work. In addition, our future work plans include proposing a flexible 5G architecture that describes how the concepts and solutions of the individual aspects can jointly form a resilient multi-service network.

Acknowledgment. This work has been performed in the framework of the H2020-ICT-2016-2 project 5G-MoNArch. The authors would like to acknowledge the contributions of their colleagues. This information reflects the consortiums view, but the consortium is not liable for any use that may be made of any of the information contained therein.

References

1. Nokia Networks: Dynamic End-to-End Network Slicing for 5G, White Paper (2016)
2. Rost, P., et al.: Network slicing to enable scalability and flexibility in 5G mobile networks. IEEE Commun. Mag. **55**, 72–79 (2017)
3. ETSI NFV: Network Functions Virtualisation (NFV): Network Operator Perspectives on Industry Progress, White Paper (2014)
4. Ravanshid, A., et al.: Multi-connectivity functional architectures in 5G. In: 2016 IEEE International Conference on Communications Workshops (ICC), Kuala Lumpur, pp. 187–192 (2016)
5. Koudouridis, G.P., Soldati, P., Karlsson, G.: Multiple connectivity and spectrum access utilisation in heterogeneous small cell networks. Int. J. Wireless Inf. Netw. **23**, 1–18 (2016)
6. 3GPP TR 36.808: Evolved Universal Terrestrial Radio Access Network (E-UTRAN); Carrier Aggregation; Base Station (BS) radio transmission and reception (Release 10), July 2013
7. 3GPP TR 36.842: Study on Small Cell Enhancements for E-UTRA and E-UTRAN Higher layer aspects (Release 12), September 2014
8. Michalopoulos, D.S., Viering, I., Du, L.: User-plane multi-connectivity aspects in 5G. In: 23rd International Conference on Telecommunications, ICT 2016, Thessaloniki, Greece (2016)
9. Gopalasingham, A., Roullet, L., Trabelsi, N., Chen, C.S., Hebbar, A., Bizouarn, E.: Generalized software defined network platform for radio access networks. In: IEEE Consumer Communications and Networking Conference (CCNC), January 2016, Las Vegas, USA (2016)
10. Hämäläinen, S., Sanneck, H., Sartori, C.: LTE Self-Organising Networks (SON): Network Management Automation for Operational Efficiency. Wiley, Hoboken (2011)
11. Novaczki, S.: An intelligent anomaly detection and diagnosis assistant for mobile network operators. In: 9th International Conference on the Design of Reliable Communication Networks (DRCN), Budapest, Hungary, June 2013
12. Novaczki, S., Szilagyi, P.: An improved anomaly detection and diagnosis framework for mobile network operators. Demo Presented at the Second International Workshop on Self-organizing Networks, Paris (2012)
13. Miyazawa, M., Hayashi, M., Standler, R.: vNMF: distributed fault detection using clustering approach for network function virtualization. In: IFIP/IEEE International Symposium on Integrated Network Management (IM), Ottawa, Canada (2015)
14. Youusaf, F.Z., et al.: Network slicing with flexible mobility and QoS/QoE support for 5G networks. In: IEEE International Conference on Communications Workshops (ICC Workshops), Paris, France (2017)
15. Martikainen, H., Viering, I., Lobinger, A., Wegmann, B.: Mobility and reliability in LTE-5G dual connectivity scenarios. In: IEEE Vehicular Technology Confernce (VTC) Fall 2017, Toronto, Canada (2017)
16. 3GPP TR 38.801: Technical Specification Group Radio Access Network; Study on new radio access technology: Radio access architecture and interfaces (Release 14), March 2017

17. Sandra, S.H.: Design and deployment of secure, robust, and resilient SDN controllers. In: 2015 1st IEEE Conference on Network Softwarization (NetSoft). IEEE (2015)
18. Ongaro, D., Ousterhout, J.: In search of an understandable consensus algorithm. In: Proceedings of the 2014 USENIX Conference on USENIX Annual Technical Conference (USENIX ATC 2014), pp. 305–320. USENIX Association, Berkeley (2014)
19. Michalopoulos, D.S., Doll, M., Sciancalepore, V., Bega, D., Schneider, P., Rost, P.: Network slicing via flexible function decomposition and flexible network design. In: IEEE Personal, Indoor, and Mobile Radio Communications Conference (PIMRC), Workshop on New Radio Technologies, October 2017
20. Verizon Threat Research Advisory Center: Data Breach Digest. Perspective is Reality, Verizon Cybercrime Case Studies (2017). http://www.verizonenterprise.com/verizon-insights-lab/data-breach-digest/2017/
21. ENISA: ENISA Threat Landscape Report 2016, February 2017. https://www.enisa.europa.eu/publications/enisa-threat-landscape-report-2016
22. Han, B., Wongy, S., Mannweiler, C., Dohler, M., Schotten, H.D.: Security trust zone in 5G networks. In: 24th International Conference on Telecommunications (ICT), May 2017
23. Grigory, A., et al.: Digital Vision for Cybersecurity, Atos Whitepaper, September 2017. https://atos.net/content/dam/uk/white-paper/digital-vision-cyber-security-opinion-paper-new.pdf

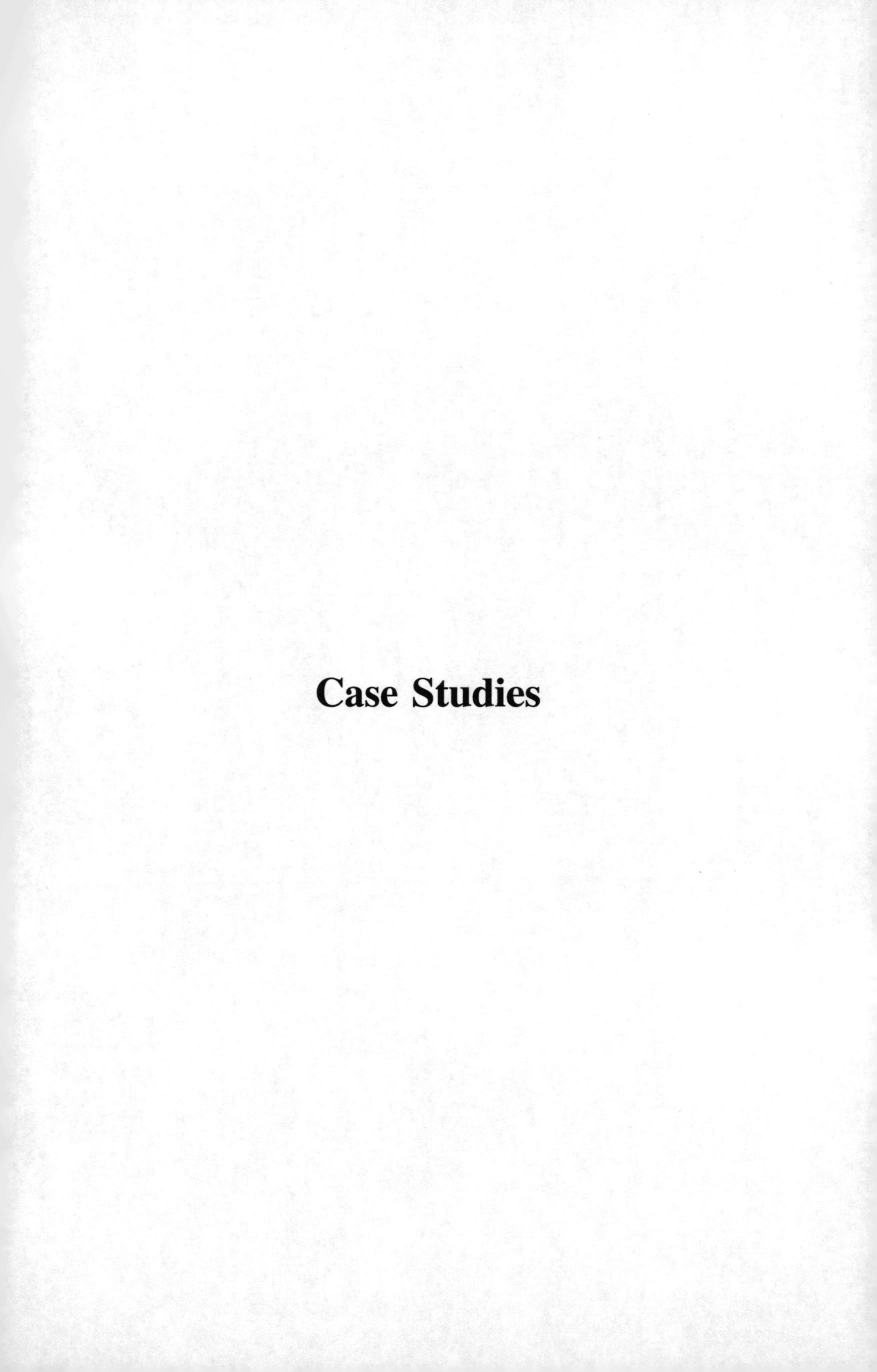

Case Studies

A Cloud Computing Solution for Digital Marketing: A Case Study for Small and Medium-Sized Enterprises

Micaela Esteves[1(✉)], Maria Beatriz Piedade[1], and Angela Pereira[2]

[1] CIIC - Computer Science and Communication Research – ESTG, Polytechnic Institute of Leiria, Leiria, Portugal
{micaela.dinis,beatriz.piedade}@ipleiria.pt
[2] CiTUR - Tourism Applied Research Centre - ESTM, Polytechnic Institute of Leiria, Leiria, Portugal
angela.pereira@ipleiria.pt

Abstract. Nowadays marketing social media has been recognized as a powerful media to promote businesses, communicate and create relations with the customers. However, the management of social media is time consuming and the use of proper systems is crucial for the business's success.

In the present research, the authors conducted a study about how Portuguese Small and Medium-sized Enterprises (SMEs) are using marketing social media. The result shows that these enterprises are interested in social media but do not have an active online presence. One reason for this is the lack of economic conditions for having a marketing department or one person dedicated only to social media.

The authors, in collaboration with SMEs marketers and non-marketers developed a social media system adapted to the needs of these enterprises with the aim to improve their digital presence in social media.

Keywords: Cloud Computing Solution · Digital marketing
Social media strategy · Social media management · Mobile application
Web application

1 Introduction

Nowadays using the internet, social media, mobile applications or other digital communication technology has become part of billions of people's daily lives. Research shows that 87% of adults who use the internet have a presence in a social media. For example, Facebook is the most popular social network with over 1,870 million active users in January 2017 [1].

Killian and McManus [2] refer that "Relationship management has been a key facet of marketing for decades, and social media offer managers a unique tool for building and maintaining relationships". For businesses, social media platforms (including Facebook, Twitter, YouTube, TripAdvisor, LinkedIn, Instagram, Foursquare, Pinterest,

M. E. Auer and T. Tsiatsos (Eds.): IMCL 2017, AISC 725, pp. 843–849, 2018.
https://doi.org/10.1007/978-3-319-75175-7_82

Google+, among others) allow brands to interact with customers in real time, exchange information about services, products and providing feedback.

With the exponential increase of popularity of social media, to keep deeper connections with the customers the brands must have: a cohesive presence with up-to-date and creative contents such as product demonstrations and contests for new advertisement ideas [9]. Also, the brands need to monitor the social media channel constantly, so they can give timely feedback to the customers.

Hilbert [3] considered that posting on social media is "probably one of the most time-consuming, yet important tasks in today's marketing world. Having to make each individual post on each individual network becomes one person's sole job." This brings a great challenge for brands, especially for those who do not have economic conditions to keep a marketing department dedicated only to digital marketing.

The majority of Portuguese Small and Medium-sized Enterprises (SMEs) cannot overcome this challenge due to the lack of economic conditions and appropriate tools. In general, these enterprises do not have experts in Marketing and the communication in digital media is neglected. In this context, the authors made a research concerned to study how the Portuguese SMEs are using marketing social media and developed an application to help the enterprises overcome their digital marketing needs.

This paper aims to present the marketing social media system development which was created in collaboration with SMEs. In the next section of this paper, the related work is presented and then the methodology is described. Finally, we present the system architecture and model, the conclusions and future work.

2 Related Work

Social media management systems, which are used by marketing and communications departments are good options to increase productivity, Return Of Investment (ROI) and marketing impact both internally and externally. Since a good social media management is time consuming the use of a proper system is crucial to the business's success. For instance, there are a lot of businesses with multiple social pages that find it difficult to manage all of them. Imagine how it is if they have to log in and out each individual profile and network to check all of their messages.

Researchers [4–7] argue that Social media management systems have the follow advantages:

- to manage multiple social pages and time management - once business productivity is important to be efficient, having a staff to take care of all social media networks is too expensive for most SMEs;
- to analyse social performance - enable to analyse what kind of posts have more engagement with the customers. Also, which social network has the best results, which demographics are most likely to engage with business content. So, the marketers can plan a better marketing strategy based in this information.
- to create analytics reports - social media managers needs to analyse their marketing strategy results and to show stakeholders across the organization how social media performs.

Actually, there are several software tools in the market that cover a multitude of marketing tasks such as: administer social media accounts, schedule posts, suggest content, boost posts, email marketing, website management, among others. In addition, other software tools are more complete and allow the marketing teams to handle customer relationship management (CRM) and all the functionalities mentioned above through one platform.

The authors search on Google about Social media management systems and found 143 tools [10]. However, only a few of them are open to everyone, while others have only some premium functionalities available with an additional cost. In this research the authors considered the six most popular software tools according to the user reviews and suitability to SMEs. All these tools have a lot of functionalities like analytics, sharing, campaigns, social engagement, advertisements and brand safety. However, the SMEs consider that these tools have a complex interface with too much functionalities and are too technical for non-marketers. On the other hand, the free basic functionalities are not enough, it is necessary to buy the full software package.

In this context, the researchers together with SMEs marketers and non-marketers developed a personalized system named SocialBAM that integrates two components: a web and mobile application.

3 Methodology

The research strategy began by trying to understand how Portuguese SMEs promote their businesses in social media. From the 100 most important Portuguese enterprises was considered fifteen to be analyse. These selected enterprises were limited to the center region, which is a great industrial area. The aim of this study was to understand how these enterprises use social media such as Facebook, Twitter and Instagram to promote their business and interact with the customers through these channels. The results show (Table 1) that all these enterprises have an online presence, at least in one social media. Only 75% of the SMEs have a website, Facebook is the most social media used, for instance, 14 enterprises (93%) have a Facebook presence, but only 7 (50%) of them have activity, followed by LinkedIn and the least used is Pinterest. These results are in line with the users preferences since Facebook is the most used social media in Portugal, although the activity in social networks is low.

Table 1. Social media used by Portuguese SMEs

Social media	Presence	Activity
Facebook	14 (93%)	7 (50%)
Twitter	1 (7%)	1 (100%)
Instagram	7 (47%)	2 (29%)
Google+	2 (13%)	1 (50%)
Linkedin	9 (60%)	4 (44%)
Pinterest	0 (0%)	0 (0%)

Facebook is the main channel used by these enterprises. However they are inactive in relation to the periodicity of publication. In order to understand the reasons for this inactivity a structured interview was applied to the marketers and SMEs owners. From this it was clear that several SMEs don't have economic conditions to maintain a marketing department dedicated to social media. They referred that there is software that allows a Customer Relationship Management (CRM) and manage the Social Media such as Salesforce software, but these types of software are extremely expensive and complex for these enterprises. In this context a Cloud Computing Solution was proposed by SMEs marketers to deliver brand experience across multiple channels. This system should be constituted by two applications, one web-based and another mobile. The main purpose of this system is to support the digital marketing actions and manage the customers activities related to the social media.

3.1 Requirements Identification

The marketers and the non-marketers mentioned that it will be important for their work to have a quick way to monitor and analyse the social media post. Thus, the authors suggested the development of a mobile application that communicates with the web application so the marketers and non-marketers can have a quick access anywhere and at any time.

Based on the feedback from marketers and other people that are responsible for the social communication, a set of functionalities was reached. Also, from software analysis were obtained other functionalities that were validated by the SMEs marketers. All this process led to a set of functionalities to be implemented in system which are:

Web application functionalities:

- User initial configuration – create user, administrator and guest account.
- Manage multiple accounts – add/remove social media accounts (Facebook, Twitter, Instagram, Pinterest, Google+ and others).
- Import clients' database – import the clients' data.
- Plan social media content – create a post with multimedia elements (text, image and video) or select one template and customize it for different platforms.
- Publish social media posts - schedule posts across the different social media.
- Monitoring – respond to posts and comments. With this, easily approve the responses to ensure that the message is in tune with the brand.
- Analyse – get reports and compare results to competitors, influencers and industry benchmarks.

Mobile application functionalities:

- User authentication – This functionality is configured to allow synchronization with the web application.
- Monitorisation – respond to posts and comments. With this, easily approve the responses to ensure that the message is in tune with the brand.
- Notifications/Alerts – alerts related to the posts and fan comments. This item can be personalised.
- Data analysis – get a quick analysis.

In the development of any software application it is crucial to evaluate the usability because it is the only way that it is possible to ensure that the application complies with the learnability, memorability, efficiency and satisfaction [8]. Therefore, before starting the implementation process of this system prototypes were created to evaluate the usability by the end users. Also, a User-Center Design approach was used with the end users integrated in the development team.

All the functionalities mentioned above from the web and mobile application were represented through low-fidelity prototypes. With the aim to find out the best organizations for all functionalities the authors designed several mockups for the same interface. These mockups were evaluated by the end users who chose the best solution for them.

Then a high-fidelity prototype was built with the purpose to assess the interface usability.

4 Architecture and Model

The architecture of the SocialBAM system (Fig. 1) consists of two modules, a web application and a mobile application that are interconnected through the web.

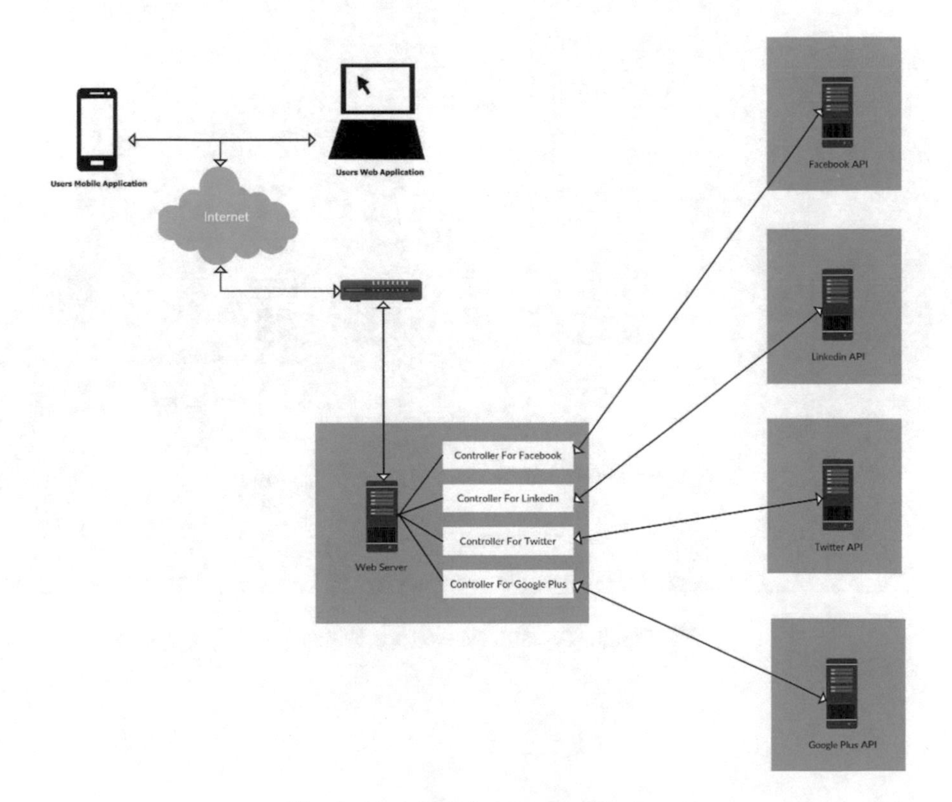

Fig. 1. SocialBAM system architecture

Based on the social media API's the researchers decided to adopt a client-server architecture, in which it is used Representational State Transfer (REST) where clients and servers exchange representations of resources by using a standardized interface and protocol.

The Client consists of two applications, one web application for the marketers to plan and manage the social media communication; another android mobile application that allows a quick access to their work. Firstly, the researchers considered only the android operating system because it was a requirement from SMEs.

At this time, the SocialBAM system is being developed. The web application (Fig. 2) is finished and the mobile application is still in progress.

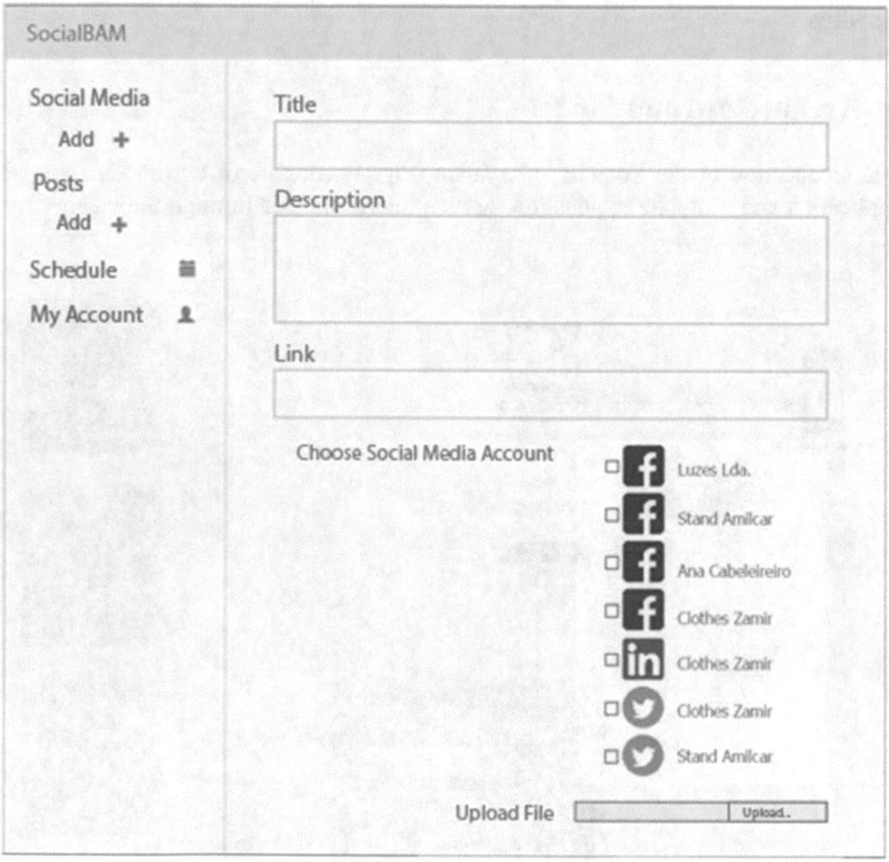

Fig. 2. SocialBAM web application

5 Conclusions

The proliferation of social media applications such as online communities, social networking sites, or blogs gives the people new means of receiving and sharing information. Many opportunities are opening up for brands to communicate with their customers and prospects. However, this is time consuming since the brands need to be constantly feeding the social network in order to be in permanent contact with the customers. The Portuguese SMEs do not have economic conditions to support a Marketing department or to buy social media management systems to help them to improve their marketing strategies.

The developed SocialBAM system is an important resource for SMEs and has been very well accepted by the Portuguese SMEs marketers and non-marketers and also by brand managers although the mobile system is not yet finished. In the future the authors intend to carry out large-scale tests.

References

1. Global social media research summary (2017). http://www.smartinsights.com/social-media-marketing/social-media-strategy/new-global-social-media-research/. Accessed 11 Aug 2017
2. Killian, G., McManus, K.: A marketing communications approach for the digital era: managerial guidelines for social media integration. Bus. Horiz. **58**(5), 539–549 (2015)
3. Hilbert, V.: Marketing software to help small-business productivity (2017). https://www.g2crowd.com/blog/crm-all-in-one/marketing-software-help-small-business-productivity/. Accessed 11 Aug 2017
4. Noe, R.A., Hollenbeck, J.R., Gerhart, B., Wright, P.M.: Gaining a competitive advantage. Irwin: McGraw-Hill, New York (2003)
5. Felix, R., Rauschnabel, P.A., Hinsch, C.: Elements of strategic social media marketing: a holistic framework. J. Bus. Res. **70**, 118–126 (2017)
6. Clark, M., Black, H.G., Judson, K.: Brand community integration and satisfaction with social media sites: a comparative study. J. Res. Interact. Mark. **11**(1), 39–55 (2017)
7. He, W., Wang, F.K., Chen, Y., Zha, S.: An exploratory investigation of social media adoption by small businesses. Inf. Technol. Manage. **18**(2), 149–160 (2017)
8. Esteves, M., Pereira, A.: YSYD-You Stay You Demand: user-centered design approach for mobile hospitality application. In: 2015 International Conference on Interactive Mobile Communication Technologies and Learning (IMCL), pp. 318–322. IEEE (2015)
9. Kaplan, A.M., Haenlein, M.: Users of the world, unite! The challenges and opportunities of social media. Bus. Horiz. **53**(1), 59–68 (2010)
10. Best Social Media Management Software (2017). https://www.g2crowd.com/categories/social-media-mgmt?segment=small-business. Accessed 7 Apr 2017

Teaching Mobile Programming Using Augmented Reality and Collaborative Game Based Learning

Ioannis Kazanidis(✉), Avgoustos Tsinakos, and Chris Lytridis

Advanced Educational Technologies and Mobile Applications Lab,
Eastern Macedonia and Thrace Institute of Technology,
Agios Loukas, Kavala, Greece
{kazanidis, tsinakos, lytridic}@teiemt.gr

Abstract. This paper investigates the potential of using an augmented reality mobile application to teach mobile programming at an undergraduate level and presents the adopted development approach. For the purposes of this study, a mobile application has been developed using open source and free mobile development and augmented reality tools. The application featured a game-based collaborative learning scenario. The application was initially used by a small group of undergraduate students during their "Advanced Mobile Applications" course and the preliminary results were promising.

Keywords: Augmented reality · Mobile learning
Game based collaborative learning · App Inventor

1 Introduction

Rapid developments in Information and Communication Technologies (ICTs) give new perspectives to the learning process. In recent years, there has been great interest in the use of mobile applications in the educational procedure. According to the international literature, mobile applications have been used to all levels of education and various domains. Thereby, innovations in ICTs are gaining more and more research interest because learning can take place through mobile devices without space or time limitations [1], as opposed to every educational method, which takes place either in the classroom or through advanced educational methods which are designed to enhance learning [2]. Collaborative and game based learning are some of those techniques that can exploit mobile applications in order to engage students in learning. Even more, technology advancements and the progress observed in the field of mobile technologies and telecommunications have caused enormous change in learning environments that lead to a conceptualization of "mobile learning," and "blended learning" [3]. One of the new technologies risen to support these learning types is Augmented Reality (AR). AR is quickly gaining momentum in the education sector worldwide, as it has the potential to enable new forms of learning and transform the learning experience by further motivating students with their study.

M. E. Auer and T. Tsiatsos (Eds.): IMCL 2017, AISC 725, pp. 850–859, 2018.
https://doi.org/10.1007/978-3-319-75175-7_83

The aim of the introduction of mobile devices in education is not to replace the traditional school, but the improvement of the educational process [4]. We developed a collaborative game-based application with AR capabilities in order to enhance the educational process of a higher education course titled "Advanced Mobile Applications". At this course, students had to learn, among others, mobile programming using the MIT App Inventor 2, which is a platform that enables the user to create mobile applications without any programming skills using a Lego logic (visual/diagrammatic programming) where different components are chosen and combined in order to produce a program.

This study aims to discover the potential of a mobile collaborative game and AR technology in the instruction of programming through the use of a mobile application developed with MIT App Inventor. The main objective of this paper is a brief description of collaborative learning theory and the technologies that are related to mobile devices, the presentation of the mobile application's design and development procedure and the discovery of the usefulness of a collaborative game and AR technology in teaching mobile programming using MIT App Inventor in a higher education course.

2 Literature Review

2.1 Collaborative Learning

Collaborative learning is defined as a learning system in which small heterogeneous teams of students collaborate in order to accomplish common educational goals. In this scenario, collective success is achieved when a team success. Through this process, knowledge is not transferred from the teacher to the student in the traditional sense, but instead the students acquire knowledge and discover new information through cooperation in common learning activities, which involve common tasks and discussions [5]. Studies have shown that the application of collaborative learning results in a better understanding of the subject matter by the students, the development of critical thinking and social skills, while at the same time being more appreciative of school and themselves [6]. Because of these advantages, collaborative learning has become a widely accepted educational practice in recent years, at all levels of education, from primary schools to universities [5].

2.2 Mobile Learning

The introduction of smart portable devices has created new opportunities in learning, by enabling students to engage in learning activities without being constrained by time or location [7]. Portability, ease of use, availability of content and the possibility of collaborative learning through the mobile devices' multitude of communication options are major advantages of the mobile learning approach [8] and have been the motivation for a significant research interest which focuses on these technologies for a more effective and interesting learning via mobile devices [9]. Moreover, the accessibility features of modern mobile phones make mobile learning possible for students with disabilities [10].

There have already been several studies in the previous years which apply mobile learning in real settings and study its effectiveness, either in the classroom [11], or in fieldwork [12, 13, 14, 15]. for the teaching of a wide variety of subjects. The potential and the effectiveness of mobile learning has been demonstrated and is now being widely used in formal education as well as in informal training scenarios.

It must be mentioned that the introduction of mobile devices in education does not aim to replace the traditional school, but instead it aims at improving the overall educational process [4].

2.3 AR in Education

The continuing development of mobile devices and their increasingly powerful characteristics in terms of processing power and peripheral devices have opened new possibilities not available in earlier devices. One of these possibilities is the use of Augmented Reality in a mobile device and more specifically its use for educational purposes. In this case, it is common to use location awareness in order to allow learning to take place on the location of educational interest (such as in a museum, an archaeological site, etc.), and the educational material that becomes available is directly related to the physical position of the learner e.g. [16, 17]. Augmented reality has been used in a variety of domains in primary and secondary education [18] with positive results. Apart from mobile applications which use AR for educational purposes, there have been authoring tools developed, which allow teachers to create AR educational scenarios. A proven example of such a platform is ARLearn, an authoring tool which supports the creation of training scenarios and serious games based on decision making [19].

3 Approach

The main aim of the adopted approach was to apply collaborative learning in the educational procedure of the higher education course "Advanced Mobile Applications" in order to motivate students and help those producing better results in learning mobile programming using MIT App Inventor.

For this reason we developed a collaborative game for mobile devices which, in combination with augmented reality applications, divides the class into groups of three or four persons that will compete with each other in order to achieve goals. The scenarios of achieving objectives will require interaction with educational materials (e.g. worksheets, course notes) and the performance of each group will be visible to every student. MIT App Inventor was used for the mobile application development and Aurasma Studio [20] for AR support.

3.1 Technologies Used

All the tools used for the application development were open source or free to use. More specifically, the mobile application was developed with the MIT App Inventor (the use of which was one of the main educational objectives of the course). For the AR application the Aurasma platform [21] was chosen since it allows to easily to create AR experiences (auras) using Aurasma Studio. PHP was used as server side scripting language in order to pass and retrieve data to the system database and make appropriate calculations. Finally MySQL was adopted as the database of the application.

MIT App Inventor: This platform enables the user to create mobile applications without any programming skills. It uses a Lego logic where different components are chosen and combined in order to produce a program. It can enable the creation of offline and online applications that may embed pictures, sounds videos and also connection with social networks and e-mail accounts. Via MIT AI Companion creators can check the exact appearance of the application in a smartphone/tablet, also it enables users to download their application and store it in their devices. The biggest advantage of this platform is the fact that the application can be created without programming knowledge and without sacrificing important aspects of the initial thought [22].

Aurasma: Aurasma is an Augmented Reality platform which allows you to see and interact with the world in a new way. With Aurasma, every image, object and even place can have its own aura presented using the technology of Augmented Reality. Auras can be as simple as a video and a link to a web page or as complex as a life like 3D animation. The production of Auras is done either through the, web based, Aurasma Studio or using the Aurasma app available both for Android and iOS, and does not require any programming skills. Every user can make a free account and create his own private or public Auras. The Aurasma app is also working as a "Viewer" of the produced Auras.

3.2 The Mobile Application

The mobile application was designed according to students' needs, using an aesthetic design, with a dark background and colorful buttons.

When a user opens the application, at the top s/he can see the logo of MIT App Inventor and the name of the course. At the bottom of the screen there are three buttons: students, teacher, and exit (Fig. 1a). At this screen the application allows gradual entry of users into a teacher or student role. Figure 1b depicts a part of the code block of this screen and in particular the initialization of the screen and the determination of the actions according to the chosen button.

In case that user selects teacher, some new elements appeared, asking to login. After that, teachers can enroll students, divide them into groups, and monitor their progress and responses.

On the other hand students may login, register in a team and answer collaboratively online quizzes. In particular the application asking student to choose the team s/he wishes to enroll (Fig. 2). There are three available teams. When the student chooses a team, the application checks if student has already inserted his/her name otherwise is asking to do so, as it can be seen on the code of Fig. 3. After this check it calls a .php file in a predefined web server that inserts the user into the application database (Fig. 3).

At the next screen of the application (Fig. 4a) the user may start the collaborative quiz, open Aurasma, show some info regarding the application or close the application. In case the user has already taken the quiz, s/he can check the submitted answers through the Grades button (Fig. 4b) and see the correct results with a short comment.

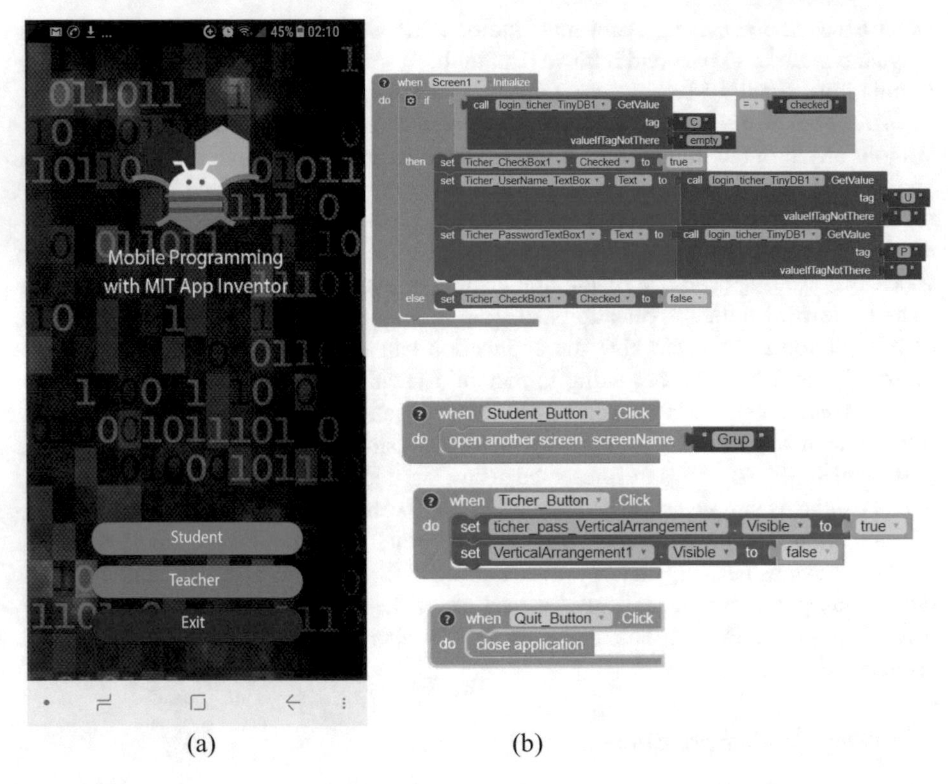

Fig. 1. The first screen of the application with the initial menu and a part of the corresponding code

When a user chooses to answer a quiz, either an image or a video is provided on the screen and s/he asked to answer a question. The type of question can be either short text or multiple choices (Fig. 5). Other students of the team can do the same action. However, this time, the students may see the answers of their fellow team members and upload a new one. The last student of the team that will answer the question is going to finalize the team's answer and send it to the teacher. Students are graded as a team for their answers in the quiz, gaining a score similar to an electronic game.

Furthermore, all students can use the mobile application in order to study supplementary material using AR technology with the aid of Aurasma. In particular, during the course, students have to study specific notes and each week they have to undertake a practical exercise with various tasks. The exercise pages include some figures that explain students what we are expecting to do and which methodology they have to follow. These images are acting as trigger images so that the Aurasma mobile application may provide students, if they point their mobile/tablet on them, with video tutorials and guidelines on the specific task of the exercise. That way, students have access to educational material pointed exactly on the required tasks helping them to fulfill the exercise requirements.

Fig. 2. Asking student to choose a team

Fig. 3. Part of the code that is executed when user chooses team A

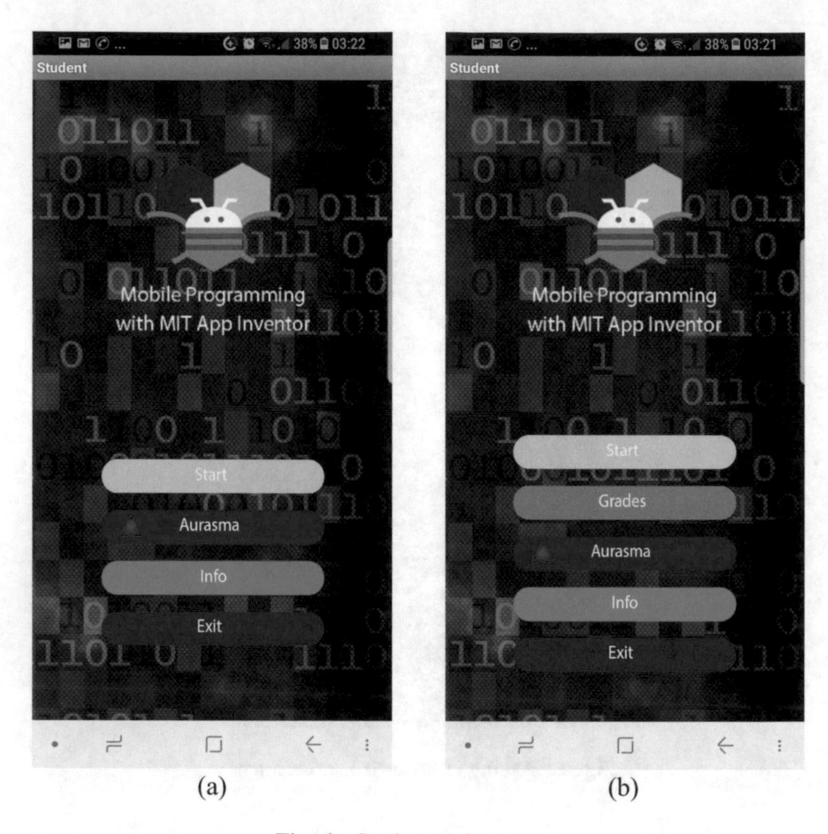

Fig. 4. Student main screens

3.3 Initial Observations

The mobile application was initially tested by a sample of twelve students distinguished into three teams of four students each. The instructors had prepared the educational material which was consisted of quizzes, AR triggered videos and notes all related to the use of MIT App Inventor. The students downloaded the application into their mobile phones and played the game for about thirty minutes answering the questions. After the end of the game a short discussion with the students about their opinion took place.

The general outcome from the discussion was that student were satisfied with the application and considered it as a useful and funny tool for their study. The main application characteristics that students enjoyed more were (a) the way that guidelines and tips on mobile programming assignments were coming to life through the AR technology and (b) the game based collaborative process for answering the provided quiz.

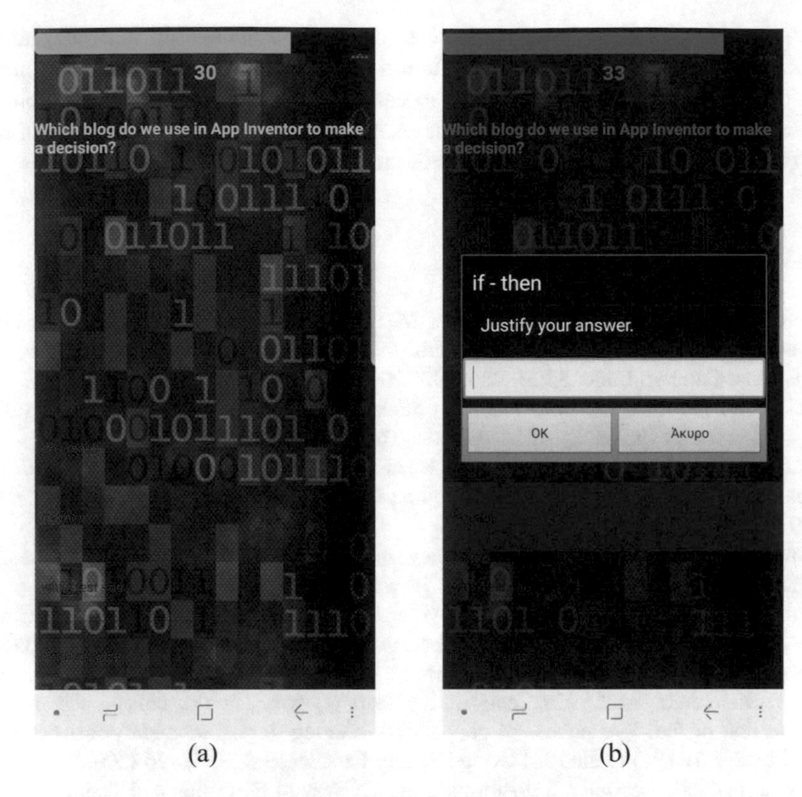

Fig. 5. A multiple choice question (a) and answer justification field (b)

4 Discussion and Conclusion

In conclusion the present study can contribute to the instructional design education by providing evidence of the AR and collaborative game-based applications potential to support teaching and learning of programming. It concludes that mobile applications can be a useful tool both for teachers and students. The results may offer new insights to researchers and provide educators with effective advice and suggestions on how to incorporate this instructional model into their teaching.

The preliminary observations show that this application can be a useful, easy and helpful tool for the instruction of mobile programming using MIT App Inventor. So this application could be a very useful tool also for a blended learning approach since it allows students to collaborate, cooperate and have access to educational content without time or spatial constraints, with tasks being able to be completed both in the classroom and at home.

The first feedback about the application was very positive. However, it remains unclear how this game-based AR application can impact students' motivation and performance in the applied course. Moreover the findings of this study cannot be generalized so easily since we based in a very small sample of students.

Future driven studies should explore in detail the relationship between students' motivational characteristics and exploitation of the developed application into the educational procedure. We are planning to use the developed mobile application at the next semester course "Advanced Mobile Applications" in order to confirm the preliminary results and study on its impact in students' motivation and performance.

References

1. Vavoula, G., Sharples, M., Rudman, P., Meek, J., Lonsdale, P.: Myartspace: design and evaluation of support for learning with multimedia phones between classrooms and museums. Comput. Educ. **53**(2), 286–299 (2009)
2. Makridou-Mpousiou, D., Giouvnakis, T., Samara, C., Tachmatzidou, K.: Θέματα μάθησης και διδακτικής. University of Macedonia, Thessaloniki (2005)
3. Huang, Y.-M., Chiu, P.-S., Liu, T.-C., Chen, T.-S.: The design and implementation of a meaningful learning-based evaluation method for ubiquitous learning. Comput. Educ. **57**(4), 2291–2302 (2011)
4. Molnar, A.: Better understanding the usage of mobile phones for learning purposes. Bull. IEEE Tech. Committee Learn. Technol. **16**(2/3), 18–20 (2014)
5. Johnson, D.W., Johnson, R.T.: What is Cooperative Learning?- An Overview Of Cooperative Learning (2016). Retrieved from Cooperative Learning Institute: http://www.co-operation.org/what-is-cooperative-learning/
6. Jones, K., Jones, J.: Making cooperative learning work in the college classroom: an application of the 'five pillars' of cooperative learning to post-secondary instruction. JET (J. Effective Teach.) Online J. Devoted Teach. Excellence **8**(2), 61–76 (2008)
7. Ally, M.: Mobile Learning: Transforming the Delivery of Education and Training. Collette, Québec (2009)
8. Sharples, M., Arnedillo Sánchez, I., Milrad, M., Vavoula, G.: Mobile Learning: Small devices, Big Issues. Book chapter to appear in Technology Enhanced Learning: Principles and Products, Kaleidoscope Legacy Book. Springer, Berlin (2008)
9. Aamri, A., Suleiman, K.: The use of mobile phones in learning English language by Sultan Qaboos University. Can. J. Sci. Ind. Res. **2**(3), 143–152 (2011)
10. Vavoula, G., Karagiannidis, C.: Review of the state-of-the-art in mobile learning. In: Proceedings of the Third International Conference on Open and Distant Learning, Patras (2005)
11. Born, C.J., Nixon, A.L., Tassava, C.: Closing in on vocabulary acquisition: the use of mobile technologies in a foreign language classroom. In: Models for Interdisciplinary Mobile Learning: Delivering Information to Students, pp. 195–210. IGI Global (2011)
12. Stead, G.: Towards open formats for mobile learning. In: 11th World Conference on Mobile and Contextual Learning, Helsinki, Finland. Citeseer (2012)
13. Tsinakos, A., Ally, M.: Global Mobile Learning Implementations and Trends. China Central Radio and TV University Press, Beijing (2013)
14. Ferry, B.: Using mobile phones to enhance teacher learning in environmental education. In: Herrington, J., Herrington, A., Mantei, J., Olney, I., Ferry, B. (eds.) New Technologies, New Pedagogies: Mobile Learning in Higher Education, pp. 45–55. University of Wollongong, Wollongong (2009)
15. Lytridis, C., Tsinakos, A.: Using a commercial mobile application as a mobile learning platform. In: International Conference on Mobile and Contextual Learning, pp. 29–37. Springer (2014)

16. Cutrí, G., Naccarato, G., Pantano, E.: Mobile cultural heritage: the case study of Locri. In: International Conference on Technologies for E-Learning and Digital Entertainment, pp. 410–420. Springer (2008)
17. Etxeberria, A.I., Asensio, M., Vicent, N., Cuenca, J.M.: Mobile devices: a tool for tourism and learning at archaeological sites. Int. J. Web Based Communities **8**(1), 57–72 (2012)
18. Fotaris, P., Pellas, P., Kazanidis, I., Smith, P.: A systematic review of Augmented Reality game-based applications in primary education. In: Proceeding of the 11th European Conference on Game-Based Learning, (ECGBL 2017), Graz, Austria, 5–6 October 2017, pp. 181–190 (2017)
19. Ternier, S., Klemke, R., Kalz, M., Van Ulzen, P., Specht, M.: ARLearn: augmented reality meets augmented virtuality. J. UCS **18**(15), 2143–2164 (2012)
20. Aurasma (2017). Retrieved from https://www.aurasma.com/
21. Aurasma Studio (2017). Retrieved from https://studio.aurasma.com/landing
22. Karamanoli, P., Tsinakos, A.: A mobile augmented reality application for primary school's history. IOSR J. Res. Method Educ. (IOSR-JRME) **6**(6), 56–65 (2016)

Computational Estimation in the Classroom with Tablets, Interactive Selfie Video and Self-regulated Learning

George Palaigeorgiou(✉), Ioanna Chloptsidou, and Charalampos Lemonidis

University of Western Macedonia, Florina, Greece
{gpalegeo,xlemon}@uowm.gr,
ioanna_chloptsidou@hotmail.com

Abstract. Interactive video combines nonlinear video structuring, dynamic information presentation and several types of interactivity over or next to the video in order to make the video watching experience more participative and constructive. There are no studies that evaluate interactive video's learning effectiveness in the classroom environment. In this study, we focus on assessing an e-learning environment based mainly on interactive videos which enables self-paced learning in the classroom with the use of tablets. The learning goal is to help primary school students to develop computational estimation strategies in math. Fifteen 5th grade students participated in a pilot study in 2 two-hours lessons and they were asked to sit in pairs in front of a tablet, share the provided earphones and follow a learning path concerning computational estimations without any guidance. The main protagonist of the videos was the teacher who went to various locations and presented a series of realistic shopping problems through selfie video. The videos were converted to interactive experiences in the LearnWorlds platform (http://www.learnworlds.com). Data collection was implemented through pre/post test, an attitude questionnaire and a focus group. After the intervention, the students became most flexible in the use of computational estimation strategies and they used a bigger repertoire of strategies. Moreover, students considered this way of learning as more fun and effective and characterized interactive videos as useful, helpful, provocative, informative and participative. The proposed approach managed to promote differentiated, autonomous and authentic learning.

Keywords: Interactive video · Self-paced learning · Self-regulated learning Tablets · Computational estimation · Selfie video

1 Introduction

Video has become the main communication mean for younger students who are video consumers and producers, video senders and receivers. The growing everyday use of mobiles' camera to capture short selfie videos has developed a new media literacy for interacting, developing and understanding video which however is not yet fully exploited for educational purposes. Video-based learning techniques have been

M. E. Auer and T. Tsiatsos (Eds.): IMCL 2017, AISC 725, pp. 860–871, 2018.
https://doi.org/10.1007/978-3-319-75175-7_84

exploited in various settings such as the "flipped" classrooms or MOOCs, and, video has been identified as one of the most diversified and effective virtual learning mediums while it is the most frequently used media in classroom settings from early on [1]. Several studies have shown that video offers a sensory learning environment that supports learners to understand more and recall information better [2] while it can improve the teaching methods.

Although video is a popular digital medium for learning, it is well-known that linear video may become a passive experience and may lead to superficial learning and insufficient viability of the learning effect, what is called the "couch-potato-attitude" [3]. One of the biggest weaknesses of the video is that the students are unable to interact with it [4] and several researchers support that video can reach its full potential only in well-conceptualized learning environments [5]. When learners do not have control over their learning, they are usually less committed and less focused and that detriments the learning results [6].

Exactly for these reason, interactive video - also called "hypervideo"- supports several types of interactivity over or next to the video with the aim of providing a more participatory experience. Users can interact with sensitive regions over the video, answer questions, select how they would like the video story to develop, click on external links, access additional information etc. [7]. The new stream of interactive video tools is easy to use and the interactivity features can be even built on top of common video services such as YouTube or Vimeo (e.g. https://www.hapyak.com, https://edpuzzle.com/, https://www.learnworlds.com/, https://koantic.com/). In a matter of seconds, a video can become interactive without the need for time-consuming editing process.

Till today, interactive video has not been studied as a tool for self-directed learning in the classroom and it has not been explored together with tablets. In this study, we focus on assessing an e-learning environment based mainly on interactive selfie video and which is supposed to enable self-paced learning in classroom with the use of tablets in order to help students to develop computational estimation strategies in math.

2 Learning Through Interactive Video

Most studies adopt a similar definition for the interactive video: "A non-linear, digital video technology that allows students to have their full attention to educational materials and to review each section of video as many times as they wish" [8]. Meixner [9] in her review for hypervideo, approaches interactive video as video-based hypermedia that combines nonlinear video structuring and dynamic information presentations. Meixner [9] explains that the interactive elements over or next to the video, provide access to additional information or allow jumps to other scenes. In an also recent review, Schoeffmann et al. [10] classifies video interaction methods in the following categories: capabilities to annotate, tag or label segments or objects in a video, capabilities to interact together with other users in a synchronized way, to interact with individual objects in the video, to support navigation inside a video, to filter video content and to generate summarized view of the content.

Several researchers have identified better learning results when using non-linear video in opposition to linear video [11]. Wouters et al. [12] supported that there can be functional interactivity on students' actions (e.g. feedback after the student's answer) and cognitive interactivity which involves calls for actions that trigger cognitive and metacognitive processes. For example, a challenge to envisage what will happen next in the video, provokes students to select and organize information and incorporate it into their pre-existing knowledge. Making educational videos with embedded questions can promote students understanding to higher levels of Bloom taxonomy by allowing the learner to think about, review and interpret each segment upon a question [13].

Interestingly, interactive video can become a platform for self-regulating learning [7, 14, 15]. The possibility of controlling the individual speed and of providing links which help avoiding cognitive overload [14], the possibility to seek or overtake a specific portion of the video and the ability to watch a specific portion again if needed [11] provide a useful self-regulated instructional environment. This approach promotes reduced levels of embarrassment or anxiety and allows learners to be comfortable enough to learn new content [16].

3 Tablets and Self-regulated Learning in the Classroom

Studies have shown that tablets can also help students show self-regulated behaviors, take the ownership of the learning process, collaborate and do not receive direct teaching instruction in classroom. For example, tablets seem to increase motivation [17], foster student learning [18], promote personalized learning [19], stimulate face-to-face social interaction between children [20].

Tablets also transform instructors' role since they encourage communication between teachers [18], may improve the quality of pedagogical support [21], bring significant changes in the way teachers approach their professional role as educators and allow a wider range of teaching strategies. Direct real-time feedback from the tablet app moderates the level of students' distraction, since it allows them to flow on to the next task at hand, rather than idling in class and waiting for feedback form their teacher [20]. Hence, tablets do provide substantive opportunities for self-regulated learning. Self-direction is a desirable skill and offers individuals the ability to engage with the materials that interests them, which can lead to greater motivation and success.

4 Computational Estimation and Use of Strategies

The learning domain under consideration for this study was computational estimation strategies in Math. Computational estimation is a widespread process in everyday life of children and adults. We often need to do quick calculations or judgments of numerical magnitudes, without using a computer or paper and pencil, to answer questions such as: Do I have enough money to buy what I need? How much time do we still need to get home [22]? Although, computational estimation is an essential component of number sense and leads to better understanding of rational numbers, students encounter a lot of difficulties in the process of the estimations [23].

Important factors that determine the quality of the estimations and number sense are the flexibility and the variety of strategies that a student can use. Strategic repertoire (which refers to the strategies used) is one of the four dimensions of strategic competence and refers to the various strategies employed by a person to solve a number of problems in a given area [24]. Computational estimation strategies have been investigated in young students [25] and the results show that children use a variety of strategies from a very early stage. In this study, we focus on how students can be supported to develop the strategy of rounding, the front-end strategy, clustering or averaging, special numbers strategy or benchmarking, compatible numbers strategy and prior compensation [22].

5 The Interactive Video Environment

5.1 The Interactive Video Platform

Until today, tablets are used mainly as cognitive tools in the classroom i.e. for capturing ideas, for interacting with simulations or for visualizing processes. However, in this study we will examine whether tablets can become the primary guides for self-regulated learning in the classroom for a whole lesson. In this classroom setting, students are asked to follow a learning path of interactive videos and other learning units by themselves while the instructor becomes a mentor and offers support to students whenever they need to. Students sit in pairs in front of each learning device sharing 2 earphones. We selected the platform LearnWorlds (http://www.learnworlds.com) as the delivery environment since it is tablet friendly, and offers both the opportunity to create lesson paths and embed and edit interactive learning video in these paths.

The learning path, we developed, for familiarizing with computational estimations is mainly consisted of interactive learning videos. The videos have been enriched with the following interactive elements [7]:

- *Pointers*, which are used to control learners' attention and provoke them to think or discuss with their partners.
- *Inductive questions*, which are used for exercising apprehension and for helping students to interpret a hypothesis presented. These questions motivate students to take notes and watch carefully the whole video.
- *Rhetoric questions*, which call students to predict what will happen next in the video. This type of questions helps students externalize their learning misconceptions, and motivates them to be more concentrated in the video in order to infer the answers by themselves.
- *Internal video links*, which allow students to navigate inside the video faster than by clicking randomly on the video bar. Internal links can be presented either at specific time points over the video or can be embedded inside the video play bar and function as content anchors. That way, each video has an internal structure which is clearly visible and accessible for the students.
- *External video links*, which are presented with labels over video and aim to intrigue students to explore further the topic under examination with resources beyond the

ones contained in the learning path. These resources can also contain answers to questions posed inside the video.

- *Inter-path links*, which lead students in different steps of the learning path. These links can be used either to help students remember forgotten issues or to proceed to content of special interest to them.

In the following figure (Fig. 1), at some point of the video, the students are called to decide what to buy. The figure also presents the learning path, content anchors and dynamic labels.

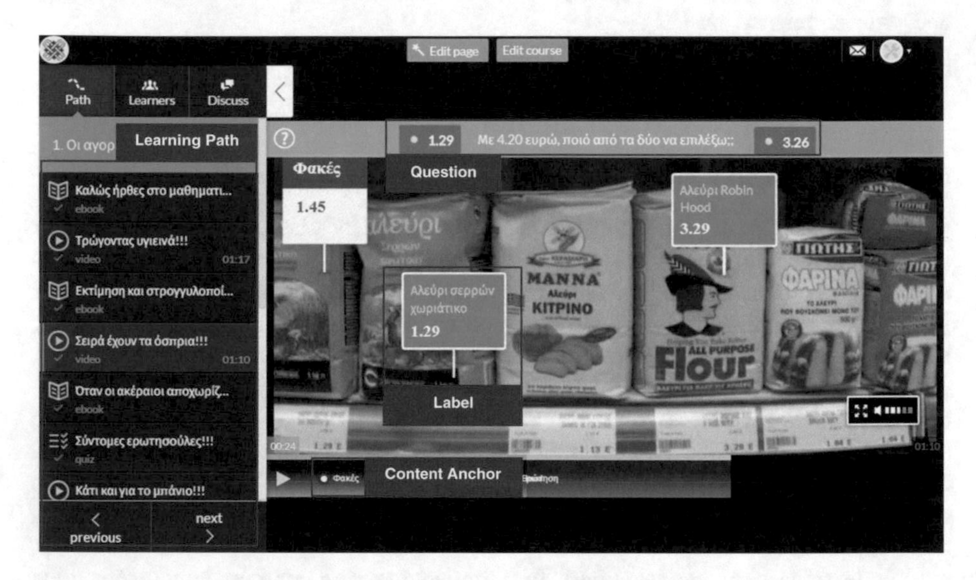

Fig. 1. The interactive video and the learning environment

The goal, when creating the learning material, was to link the computational estimations to original contexts of use in order to raise students' interest and demonstrate the practical value of the learning domain. Four sections were developed for 4 school hours, under the umbrella of a single story. The main protagonist of the video was the teacher who went to various locations and presented a series of realistic shopping problems. The stores selected were close to students' experience and interests, e.g. the first hour was recorded in a super market, the second one in a playhouse, the third one in a sports shop, etc. Each video lasted up to 2 min, although students viewed them for a lot more time because of the required interactions. The format of the videos was similar to a typical selfie video and the videos were recorded with the use of a mobile phone.

The learning environments consisted of interactive videos (17) as well as texts (13) and informal questionnaires that followed the videos (10). There was a mixed content structure and students had for example to watch a video, then read informative material, later take an informal test etc. As time progressed, the text material was reduced and the interactive videos were increased. The researchers took into consideration several design principles regarding the development of the educational videos [27].

6 Methodology

6.1 Participants

A total of 15 students, 7 boys and 8 girls of 5th grade participated in the pilot study. The students came from all three levels of performance (low, medium, high), according to the teacher's comments. All students had access to a tablet in their homes and half of them used it daily (7/15).

6.2 Procedure

The intervention took place in 2, two-hour lessons. The two lessons were conducted in two consecutive days. At the beginning of each lesson, students signed in to the educational environment with pre-ready accounts for their convenience. They formed pairs and shared headsets. Afterwards, they were asked to follow a learning path in the learning environment without any guidance. One instructor was available to help them both with the technical requirements (e.g. navigating in the environment) and the learning content (e.g. conceptual questions).

Pre-tests were given to the students a week before the intervention and post-tests were given a week after. At the end of the second lesson, a questionnaire was given to the students regarding their attitudes towards the learning environment. A focus group followed which lasted about 25 min and was audio-recorded.

6.3 Research Instrument

Data collection was actualized through pre/post tests, an attitude questionnaire and a focus group.

Pre-post tests consisted of 7 demanding calculations (e.g. "which is the sum of the following numbers? 1.26, 4.79, 0.99, 1.37, 2.58"). Students were asked to write down both the calculation results and their thinking strategy so as for the researchers to infer the computational estimation strategy applied.

After the completion of the intervention, students were also requested to answer a questionnaire of 12 five-point Likert-type questions that focused on:

- the perceived learning value of the intervention (e.g. I believe I know more about estimation calculations after this activity, this way of learning suits me better, etc.).
- the perceived learning value of the interactive video (e.g. The video and the questions included led to many discussions with my peers, I would prefer the video not to contain interactions).

Finally, the focus group aimed to draw on students' qualitative reasoning on the issues raised in the questionnaire and also for getting students views on the classroom dynamics, their self-regulation and their cooperation.

7 Results

7.1 Learning Results

Table 1 shows the strategies used by students to solve the 7 tasks (3 problems and 4 proper calculations). It is important to note that for one task, a student could use one or more strategies. In Table 1, strategies were ordered by difficulty from the easiest (i.e. rounding, front-end strategy) to the most difficult (i.e. clustering or averaging, prior compensation). Table 1 shows in the last two columns the variety of strategies used for each task and the total number of strategies used between pre and post tests.

Table 1. Pre-post test usage of computational estimation strategies on different problems. X/Y separates pre and post test values

Tasks	Rounding	Front-end strategy	Compatible numbers	Special numbers strategy	Clustering or averaging	Prior compensation	Variety of strategies	Total number of strategies
T1	7/6	4/4	0/2	0/0	0/0	0/0	2/3	11/12
T2	7/3	1/1	2/9	0/1	0/0	0/0	3/4	10/14
T3	4/4	0/3	0/0	0/0	0/0	0/0	1/2	4/7
T4	12/4	0/3	0/4	0/0	0/3	0/3	1/5	12/17
T5	4/3	7/9	0/0	0/0	0/0	0/0	2/2	11/12
T6	4/10	10/1	0/0	0/0	0/0	0/0	2/2	14/11
T7	7/7	1/1	2/4	0/0	0/0	0/0	3/3	10/12

As can be seen, in the post-test, the strategies repertoire was bigger, both the variety of strategies and the total number of strategies used per task were higher. The variety of strategies in the pre-test was 2 per task while in the posttest 3 strategies per task. The mean of total strategies used in the pre-test were 10.28 strategies per task and in posttest were 12.12. Concerning the quality of the strategies used, we can observe that in the pre-test, students used almost exclusively easiest strategies (rounding and frond-end strategy) while in the post-test they used some more demanding strategies (compatible strategies, special numbers, clustering or averaging and prior compensation). It is interesting that in some tasks, students changed dramatically the strategies exploited as for example in Task 4 and Task 7. Generally, we can say that after the intervention, students became more flexible and could use a bigger repertoire of strategies. These results are particularly encouraging if we take into account that the development of computational estimation strategies requires considerable time.

As can be seen from students' answers in Table 2, they evaluated positively the learning outcome by claiming that they know and can better apply the computational estimation strategies after the two lessons. It is even more exciting that they assessed that the specific way of learning is very pleasant and interesting, and almost everyone agreed that they would like to be applied in other courses as well. It is important to note that the intervention lasted for 4 h and therefore it was not a short technological intervention but a paradigm shift for 4 h and for this reason the answers of the students are of considerable importance.

Table 2. Perceived learning value of the environment

Questions	Min	Max	Mean	SD
I believe that I know more about computational estimations after the use of the environment	2.0	5.0	3.93	1.03
I believe that I cannot make computational estimations more easily after the two lessons	1.0	4.0	2.00	1.07
This way of learning suits me, I learn faster	2.0	5.0	4.33	0.90
I wish there were other courses taught in the same way	3.0	5.0	4.60	0.63

7.2 Attitudes Toward the Learning Environment and the Interactive Video

The interactive video seemed to be a crucial element of the learning environment. The students rated highly the interactions on the video while they answered strongly negatively to the possibility of removing them (Table 3).

Table 3. Students'attitudes towards the learning environment and the interactive video

Questions	Min	Max	Mean	SD
It was easy to use the educational environment	3.0	5.0	4.40	0.74
The lesson with the tablets was very tiring	1.0	5.0	2.00	1.25
With tablets, I learned just as well but in less time than traditional teaching	3.0	5.0	4.60	0.63
Video with questions made me think	3.0	5.0	4.20	0.56
I would prefer the video not to contain interactions	1.0	3.0	1.73	0.70
I loved having to look for the answers	4.0	5.0	4.67	0.49
It was boring just to watch the video. There was a lot of information	1.0	4.0	1.87	1.13
The videos were well-produced and related with the course's subject	3.0	5.0	4.53	0.74

Students also thought the videos were well prepared, even though they were selfie-video in commercial stores. They probably evaluated the combination of the video quality with the scenario and its purpose. Hence, similar learning environments do not seem to require expensive video productions to be effective but a careful combination of a scenario, a problem, and learning interactions even in a selfie format. This finding underlines that the interactive video approach seems affordable and feasible. Students were also excited with the use of the tablets in the classroom and evaluated the learning environment as easy to use.

7.3 Qualitative Data

Students validated the quantitative data on the focus group. As expected, the students were enthused with the use of tablets in the classroom and as a student said *"... tablets are preferable because they are fast, they are newer ... the tablet is ahead, it's technology, ...".*

Students put particular emphasis on the new learning model and described it as playful. They also felt they were involved more with the learning material and that they were the protagonists of the learning process together with their classmates. At the same time, they repeatedly underlined that they considered effective the instructional approach in which the teacher presents real problems in authentic environments through the video. This is a way of connecting the classroom with reality.

> *"We saw it as playing and learning at the same time."*
> *"We liked all the stuff ... we liked watching you shop in the shops and describing your problems while other were seeing you when you were describing the problems ..."*
> *"I liked the tablets and [interactive] video and that the fact that you were inside the video."*
> *"I liked that the prices and the things that you were going to buy were on top of you and you were explaining how to think..."*

When we proposed in the focus group to remove video interactions, 13 of the 15 students disagreed while only two were neutral. The interactive elements were considered useful, helpful, provocative, informative and participative:

> *"I liked the video, the details over the video, I could participate, I could see how I'll have to do them [the computational estimations]."*
> *"I liked the video [with the interactions] too much."*
> *"I liked the questions that stopped the video and asked for my point of view because they helped me to figure out how I should solve the problems."*
> *"... I liked the questions that came up and stopped the video and I had a curiosity to see what it would be shown after each choice ... [and] that if we did not answer correctly to a question, it didn't show us the right answer immediately but motivated us to watch and look better to find the answer."*

The majority of the students felt that the collaborative model was positive. Collaboration is a key component in this approach as students do not directly interact with the teacher but try to solve problems and negotiate concepts and strategies cooperatively with their classmates.

> *"I liked it [the teaching model] because we worked on pairs."*
> *"I was helped by the fact that my classmate was with me because we were discussing how to answer the questions."*
> *"I would like to be even four students [in the team]."*

It is important to note that the pair formation process can be decisive as in this type of learning process, students are not conducting a simple collaborative activity for a short period of time but they learn together for a longer period, they negotiate the pace of their learning with their partner and also negotiate together their understanding. In our case, there was a problem with 3 students who said they did not manage to cooperate in the way they wanted. As one said, *"we like to work on our own, we argued too much because I had to accept his answer; it [the learning environment] was so alive that I wanted to wear the two headphones myself and do it myself."*

8 Conclusions

In this study, we examined whether interactive video together with self-paced learning and tablets may advance the development of computational estimation strategies in the classroom environment. The proposal, which is both technical and pedagogical, requires the transfer of the learning control from the teachers to the students and tries to promote differentiated, autonomous and authentic learning. Its basic premise is that it enhances an important cultural probe in students' life, video and selfie video, so as to become a participative, authentic and joyful learning medium. The pilot study showed that this approach had a learning effect in a complex learning domain such as computational estimations and that students identified the potential of the specific setting. We ought to underline that students were enthused with an application in which they were essentially realizing math calculations for four hours.

Students were also positive towards the selfie videos which were created with a phone camera in authentic contexts for the specific problems and that makes the specific approach affordable and feasible. The classroom dynamics also surpassed the researchers' expectations. We were pleasantly surprised by the ease with which primary school students became independent from the instructor and retained their own pace. The study however was limited and bigger samples of students, more classrooms and different domains may help us understand better the underlying learning mechanisms and validate or not our results.

Although the results for the combination of interactive video with tablets and self-paced learning in the classroom were very positive, the study poses probably more questions than the answers given. There are very few studies in regards to a similar setting, about how to direct an interactive video, about the optimum number of interactions, in regards to the different types of interactions and their instructional value, about the complementary material or the domains that are better suited to this classroom setting. All these questions should be become the pursue of future research with interactive video.

References

1. Feierabend, S., Klingler, W.: Lehrer/-Innen und Medien: Nutzung. Einstellung, Perspektiven (2003)
2. Syed, M.R.: Diminishing the distance in distance education. IEEE Multimed. **8**(3), 18–20 (2001)
3. Ertelt, A., Renkl, A., Spada, H.: Making a difference: exploiting the full potential of instructionally designed on-screen videos. In: Proceedings of the 7th International Conference on Learning Sciences, pp. 154–160. International Society of the Learning Sciences (2006)
4. Laurillard, D.: Teaching as a Design Science. Building Pedagogical Patterns for Learning and Technology. Routledge, New York (2012)
5. Krammer, K., Ratzka, N., Klieme, E., Lipowsky, F., Pauli, C., Reusser, K.: Learning with classroom videos: conception and first results of an online teacher-training program. ZDM **38**(5), 422–432 (2006)

6. Dror, I.E.: Technology enhanced learning: the good, the bad, and the ugly. Pragmat. Cogn. **16**(2), 215–223 (2008)
7. Papadopoulou, A., Palaigeorgiou, G.: Interactive video, tablets and self-paced learning in the classroom: preservice teachers perceptions. International Association for Development of the Information Society (2016)
8. Dimou, A., Tsoumakas, G., Mezaris, V., Kompatsiaris, I., Vlahavas, L. An empirical study of multi-label learning methods for video annotation. In: Seventh International Workshop on Content-Based Multimedia Indexing, CBMI 2009, pp. 19–24. IEEE (2009)
9. Meixner, B.: Hypervideos and Interactive Multimedia Presentations. ACM Comput. Surv. (CSUR) **50**(1), 9 (2017)
10. Schoeffmann, K., Hudelist, M.A., Huber, J.: Video interaction tools: a survey of recent work. ACM Comput. Surv. (CSUR) **48**(1), 14 (2015)
11. Zhang, D., Zhou, L., Briggs, R.O., Nunamaker, J.F.: Instructional video in e-learning: assessing the impact of interactive video on learning effectiveness. Inf. Manag. **43**(1), 15–27 (2006)
12. Wouters, P., Tabbers, H.K., Paas, F.: Interactivity in video-based models. Educ. Psychol. Rev. **19**(3), 327–342 (2007)
13. Kim, J., Glassman, E.L., Monroy-Hernández, A., Morris, M.R.: RIMES: embedding interactive multimedia exercises in lecture videos. In: Proceedings of the 33rd Annual ACM Conference on Human Factors in Computing Systems, pp. 1535–1544. ACM (2015)
14. Chen, Y.T.: A study of learning effects on e-learning with interactive thematic video. J. Educ. Comput. Res. **47**(3), 279–292 (2012)
15. Delen, E., Liew, J., Willson, V.: Effects of interactivity and instructional scaffolding on learning: self-regulation in online video-based environments. Comput. Educ. **78**, 312–320 (2014)
16. Pendell, K., Withers, E., Castek, J., Reder, S.: Tutor-facilitated adult digital literacy learning: insights from a case study. Internet Ref. Serv. Q. **18**(2), 105–125 (2013)
17. Kinash, S., Brand, J., Mathew, T.: Challenging mobile learning discourse through research: student perceptions of blackboard mobile learn and iPads. Australas. J. Educ. Technol. **28**(4), 639–655 (2012)
18. Fernández-López, Á., Rodríguez-Fórtiz, M.J., Rodríguez-Almendros, M.L., Martínez-Segura, M.J.: Mobile learning technology based on iOS devices to support students with special education needs. Comput. Educ. **61**, 77–90 (2013)
19. McClanahan, B., Williams, K., Kennedy, E., Tate, S.: A breakthrough for Josh: how use of an iPad facilitated reading improvement. TechTrends **56**(3), 20–28 (2012)
20. Henderson, S., Yeow, J.: iPad in education: a case study of iPad adoption and use in a primary school. In: 2012 45th Hawaii International Conference on System Science (HICSS), pp. 78–87. IEEE (2012)
21. Murray, O.T., Olcese, N.R.: Teaching and learning with iPads, ready or not? TechTrends **55**(6), 42–48 (2011)
22. Lemonidis, Ch.: Mental Computation and Estimation: Implications for Mathematics Education Research, Teaching and Learning. Routledge, New York (2012)
23. Lemonidis, C., Nolka, E., Nikolantonakis, K.: Students' behaviours in computational estimation correlated with their problem-solving ability. MENON: J. Educ. Res., 1st Thematic Issue, 46–60 (2014)
24. Lemaire, P., Arnaud, L., Lecacheur, M.: Adults' age-related differences in adaptivity of strategy choices: evidence from computational estimation. Psychol. Aging **19**(3), 467–481 (2004)

25. Reys, B.J., Reys, R.W., Penafiel, A.F.: Estimation performance and strategies use of Mexican fifth- and eighth-grade student sample. Educ. Stud. Math. **22**, 353–375 (1991)
26. Mayer, R.E.: Cognitive theory of multimedia learning. In: Mayer, R.E. (ed.) The Cambridge Handbook of Multimedia Learning, pp. 31–48. Cambridge University Press, Cambridge (2005)
27. Palaigeorgiou, G., Despotakis, T.: Known and unknown weaknesses in software animated demonstrations (screencasts): a study in self-paced learning settings. J. Inf. Technol. Educ. **9**, 81–98 (2010)

E-Learning Platform for Disadvantaged Groups

Omar Hasan(✉), Mohamed Taha, Omar Alasali, and Neda Aldaher

Communication Engineering Department,
Princess Sumaya University for Technology, Amman, Jordan
ohasan@psut.edu.jo

Abstract. This paper presents a case study for the design and implementation of a pilot E-Learning laboratory experiments for Communication Engineering students at Princess Sumaya University for Technology (PSUT) in Jordan. Motivated by the Erasmus+ EU project, Improving Higher Education Quality in Jordan using Mobile Technologies for Better Integration of Disadvantaged Groups to Socio-Economic Diversity (mEquity) (This paper is part of mEquity ERASMUS^{+} PROGRAMME, Project N^{o}: 561527-EPP-1-2015-1-BG-EPPKA2-CBHE-JP.). A target group of students with physical and/or socio-economic disabilities has been selected to improve their learning skills. Surveys conducted among teachers and students at the planning phase have shown great interest towards creating such a platform for laboratories. Interestingly, the lab being developed covers experiments taught at the Communication Engineering Lab in PSUT which are not based on available online solutions. Moreover, a statistical assessment for the impact of the pilot experiments in the designed lab on the learning process was conducted using a pre and post questionnaires. The assessment has shown significant improvement after students got acquainted with the provided system.

Keywords: Disadvantaged groups · Disabilities · E-learning
Lab platform · Pre-post surveys · m-Equity · Correlation analysis

1 Introduction

In the last decade, remote lab systems became very popular in engineering curricula. These labs have been implemented using dedicated available boards and physical instruments interfaced using online run engines. This was later improved using virtual instruments [1–3]. An important advantage of using remote learning platforms is in catering for students in our target group, disabled and disadvantaged students. Which can be achieved through the provision of finest communication theory knowledge with access to full laboratory resources via the internet. This will supersede the need for students to attend physically the lab sessions, in turn, eliminates the burden for students with various disabilities or limited economic means. Furthermore, the impact of the designed system on the learning process has been measured statistically using pre and post surveys.

M. E. Auer and T. Tsiatsos (Eds.): IMCL 2017, AISC 725, pp. 872–878, 2018.
https://doi.org/10.1007/978-3-319-75175-7_85

In [4], a remote access to wireless communications systems laboratory was presented. In that work, an online management system has been used to access different antenna elements and evaluate their characteristics. The switching between different antennas was achieved by external antenna-switching controller board (EASCB).

In [3] A Remote instrumentation Engineering Lab based on the VISIR hardware has been presented. A switching matrix was used to alter the interconnections of the hardware.

An implementation of an e-Learning Communications Laboratory utilizing the EMONA board was presented in [5].

Usually the impact of such systems is assessed using statistical analysis. For instance, surveys were conducted by students who used the traditional labs as well as the remote labs in [5]. A more advanced method based on multiple hypothesis testing referred to as Keppel's method was presented for the evaluation of the effectiveness of the remote lab selection in [6]. As for the work in [7], for the analysis of Massive Open Online Courses (MOOCs) technique, pre and post surveys were conducted to measure the skills improvement for the students. All of the previously mentioned work did not target disadvantaged groups.

The current work demonstrates the selection and implementation of hardware and interfacing of the designed lab platform. Moreover, a course management system, where experiments are to be executed and then formally evaluated, will be highlighted. Finally, the contribution of the provided system to the learning process will be evaluated statistically using provided surveys.

This paper is organized as follows, in Sect. 2, the designed platform will be presented. In Sect. 3, the course management platform will be demonstrated. As for Sect. 4, measurement of the impact of the proposed system on student skills progress will be shown. Finally, we conclude in Sect. 5.

2 Platform Design

In this context, the methodology of design and implementation of a real time lab consisting of a series of experiments is provided. This section highlights the selection and implementation of hardware and interfacing of the platform. Each experiment in the current system was built using discrete components (R, L, C, IC's) and interfaced to ELVIS® hardware from National Instruments®. For each experiment, designated Virtual Instruments (VI's) were configured and deployed on a dedicated server at PSUT. The proposed course contains several experiments built on the same board.

Unlike the other labs documented in literature [1–6], this system provides electronic controls to select an experiment to execute, activate and interconnect the appropriate circuits and suitable VI's instead of the conventional jumper wires. For this, a switching mechanism is designed based on the deployment of analogue switches configured according to pre-designed I/O write codes that activate the correct route of signal flow on the board. This solution is considered far simpler, more robust and eliminates human error as opposed to the expensive switching matrix hardware that is usually used. The complete system is shown in Fig. 1.

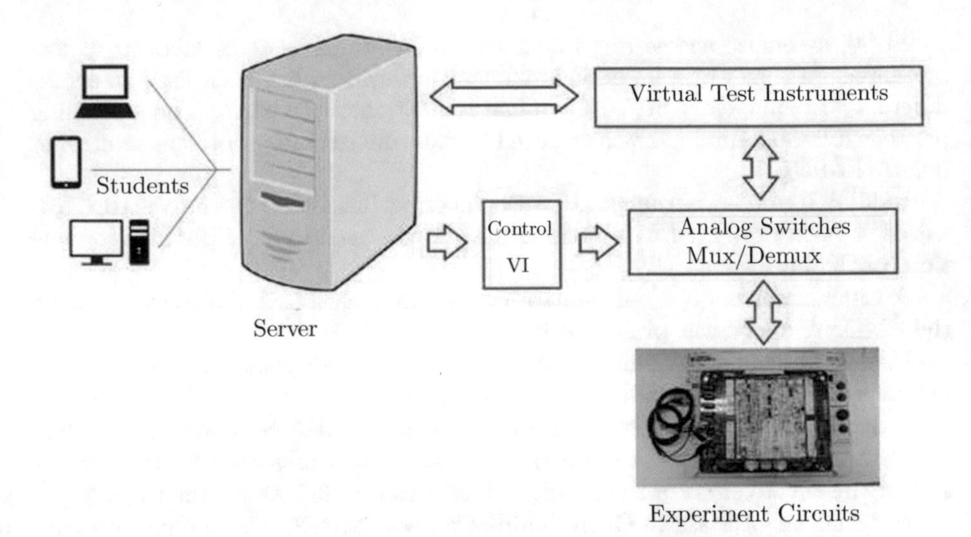

Fig. 1. Complete designed system

When a student selects an experiment the server sends a command via the control VI. The Control VI will in turn activate specific electronic switches to enable the corresponding circuit and suitable virtual instruments such as function generator, oscilloscope, etc. The results of instruments measurements will be directed from the VI's via the server to the appropriate student. The experiment circuit board contains several experiments as shown in Table 1 covering analogue and digital communication concepts. It is worth mentioning that multiple users can access the system simultaneously.

3 Course Management System

The course management system used in our case is Distributed Internet - based Performance Support Environment for Individualized Learning (DIPSEIL) [4], developed by a previous EU-project which enables a multilingual and an open environment system submissions and follows task based learning method as well as feedback to students and performance assessment.

The system is used to host the experiments, technical resources and grading feedback to students. In DIPSEIL, the designed pilot is packaged as a module that contains several experiments also called tasks. Each task covers an aspect of the communication theory to be explained in the experiment and consisting of a description, reference information (resources, presentations, etc.), and task specific training to help the student learn while experimenting.

The conduction of the experiment employs running the system in real time and taking the required measurements. The obtained results will be used to fill an html form. This form found under the field Instructions to Perform was designed to enable the student to carry out the experiment steps and record

Table 1. Circuit board supported experiments

Experiment name	Topics
VCO & PLL	Principles of voltage controlled oscillators and phase locked loops
FM modulation & demodulation	Implementation of FM modulation FSK and FM demodulation loops
Oscillators	Principles of feedback and oscillators with circuit implementation
AM modulation & demodulation	Implementation of AM generation and demodulation
Sampling	Sampling and aliasing

the results online. Also in this section, students will be instructed to plot and analyze the obtained data and then upload the links of their findings through the same form. Once the form is completed, a provided html button can be used to upload the contents of all the results in pdf format. Grading by the instructor and any feedback to the students is also provided for by the DIPSEIL system.

4 System Assessment

For the purpose of assessing our system impact and effectiveness on students pre and post surveys were conducted. The surveys where concerning measurements for usability, relevance, usefulness, easiness, flexibility, future acceptance and sufficiency of technical resources. For measuring the above components, a 4th year class of 19 electrical engineering students were asked to fill in the provided surveys before and after conducting the experiments. A group of them have constrained economic means. The complete survey form is as shown in Fig. 2.

There were 13 respondents, 75% of the sample, filled the surveys. The responses have been collected and analyzed namely through calculating statistical methods that rely on distribution parameters. Also, correlations between different questions were conducted to verify the reliability of the analysis.

Figure 3 shows the overall response for pre and post surveys for each question. As can be seen in the figure, the improvement in acceptance of the lab was 13.19%.

For more specific analysis question 3, 5, 6, and 7 will be further inspected. These questions show significant improvement (greater than 21%).

Question 3 explored improvement of understanding practical aspects of communication systems through remote labs. 61.5% of the students were expecting improvements before conducting the experiment in comparison to 76.9% who felt real improvement after conducting the experiment.

Similarly for question 5, regarding the flexibility of the remote labs, before the experiment, students expectations were 38.5% with a significant improvement to 69.2% (a 30.8% points improvement) this shows that the main objective of the project which is to serve disadvantaged and disabled students has been strongly achieved.

Survey for Remote Laboratory

Please answer the following questions to help improve the experiments and remote lab ("1" means Very Low and "5" means Very High)

1. Does the manual provide sufficient information for easy set up and run of the experiment ?

1 2 3 4 5

2. Do you find it easy to conduct the remote the experiment ?

1 2 3 4 5

3. Does the remote experiment help you understand practical aspects of communication system ?

1 2 3 4 5

4. Do you feel like were operating real equipment ?

1 2 3 4 5

5. Does the flexibility of the remote lab allow you to fit the laboratory work into your schedule?

1 2 3 4 5

6. Based on your experience using the remote lab, do you prefer to use the remote lab in the future?

1 2 3 4 5

7. How do you rate the overall performance of the remote lab?

1 2 3 4 5

8. Do you have any suggestion to improve the remote labs and experiments?

Submit

Fig. 2. Questionnaire form

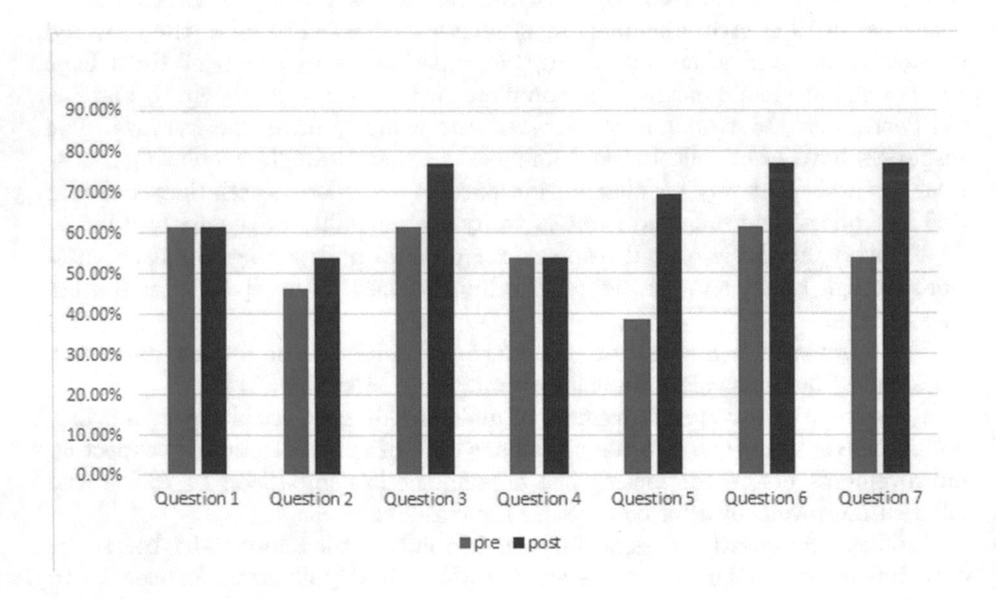

Fig. 3. Responses per question

Question 6, pertaining to the issue of preference to use the remote lab in the future. The improvement in students attitude increased from 61.5% to 76.9% which is an increase of 15.4%. As for question 7 which rates remote labs in overall. Students expectations were 53.9% whereas after conducting the experiment their rating has risen by 23.1% points to 76.2%. This improvement shows that the needs of the disadvantaged and disabled students can be served with distinction. However, for questions 1 and 4 which measure the documentation and manuals as well as measurement of students feel for real hardware. There was no change in documentation effect. As the manuals and theoretical background for both the physical experiment and the remote experiment are very similar. Also, students felt they were not operating real hardware, as exhibited by no change in questionnaire result. This is due to the nature of the remote labs were students do not see nor control hardware manually by hand.

Question 2 showed minor improvement (7.7%) pertaining to the issue of ease of experiment conducting. Since this was a pilot, it was expected to have good to moderate responses from students, overall the responses were encouraging and point toward issues where significant improvement in remote labs design and management can be achieved.

To test the seriousness of the students in conducting the surveys and their answers an overall correlation of questions was 75% which shows that the collected responses were valid for the statistical analysis purposes. Correlations between different questions were conducted to further measure effects of this pilot in general and its impact on students with special needs. Understanding theory was improved because of the ease of the remote experiment as shown by the correlation of 84% between questions 2 and 3. The high correlation of these two questions points to a very good opportunity for the target group to improve their academic performance.

The preference by students for the use of virtual instruments was shown by the acceptable correlation of 64% between questions 4 and 7. Students were highly motivated for using remote labs in the future exhibited by their satisfaction as measured in the correlation of 82% between questions 6 and 7. This correlation also shows how well this methodology serves students with disabilities. Students showed encouragement towards remote labs by their response to lab flexibility and overall performance of remote labs as measured by the correlation (83%) between questions 5 and 7. This also reflects well on our target group. The high correlation of 85% between questions 3 and 6 indicates that remote labs are a preferable teaching method for disadvantaged group. In improving their technical knowledge and ease of access.

5 Conclusion

In this article the design of a remote lab platform for communication engineering highly geared towards serving disabled and disadvantaged students has been proposed. An assessment of the system was performed based on a combination of pre and post surveys conducted by students and the results were provided. Statistical analysis has shown viability of such a system in serving the needs of students and specifically the target group.

Acknowledgment. The authors would like to thank ERASMUS$^+$ PROGRAMME mEquity, Project N^o: 561527-EPP-1-2015-1-BG-EPPKA2-CBHE-JP for funding and support provided to this work.

References

1. Abu-aisheh, A.A., Eppes, T.E., Otoum, O.M., Al-Zoubi, A.Y.: Laboratory collaboration plan in communications engineering. In: IEEE Second Global Engineering Education Conference (EDUCON), Amman, Jordan, 4–6 April 2011, pp. 837–840 (2011)
2. Auer, M.E., Zutin, D.G., Al-Zoubi, A.Y.: Online laboratories for eLearning and eResearch. In: The $5th$ Congress of Scientific Research Outlook and Technology Development in the Arab World (SRO5), Fez, Morocco 25–30 October 2008 (2008)
3. Odeh, S., Anabtawi, M., Alves, G.R., Gustavsson, I., Arafeh, L.: Assessing the remote engineering lab VISIR at Al-Quds University in Palestine. Int. J. Online Eng. **11**(1) (2015)
4. Kafadarova, N., Stoyanova-Petrova, S., Mileva, N., Sahandzhieva, I.: Implementation of remote access to wireless communications systems laboratory. In: Proceedings E-Learning Course, EDULEARN12, pp. 7307–7312
5. Abu-aisheh, A., Eppes, T., Al-Zoubi, A.: Implementation of a remote analog and digital communications laboratory for e-learning. Int. J. Online Eng. **6** (2010)
6. Ursutiu, D., Cotfas, P., Cotfas, D., Stefan, A.: Methods of the quality assurance applied at the remote laboratory selection. In: EDUCON (2010)
7. Castro, M., Tawfik, M., Tovar, E.: Internationalization globalization of engineering. In: Twelfth LACCEI Latin American and Caribbean Conference for Engineering and Technology, (LACCEI 2014) (2014)

Different Uses for Remote Labs in Electrical Engineering Education: Initial Conclusions of an Ongoing Experience

Ana Maria Beltran Pavani[1(✉)], Delberis A. Lima[1], Guilherme P. Temporão[1], and Gustavo R. Alves[2]

[1] Pontifícia Universidade Católica do Rio de Janeiro, Rio de Janeiro, Brazil
{apavani, temporao}@puc-rio.br, delberis@ele.puc-rio.br

[2] Politécnico do Porto, Porto, Portugal
gca@isep.ipp.pt

Abstract. Laboratories are a fundamental part of engineering education due to the very nature of the engineering profession. This is a characteristic of all engineering courses, though it may vary from one curriculum to the other and even in the same curriculum. This paper is dedicated to the analysis of different applications of a remote lab in an Electrical Engineering curriculum. If the types of experiments it offers are concerned, it is classified as an Electric and Electronic Circuits lab; one in a set that also has Digital Electronics, Analog Electronics, Electrical Machines, Control Systems, etc. But it is not a traditional lab – it is a remote lab, with traditional components remotely accessed over the Internet. This work presents the preliminary results of the deployment of VISIR – a remote lab for Electric and Electronic Circuits in some courses. It discusses the different course contexts and how the use of VISIR was adapted to each one. Results of students' opinions are presented and discussed.

Keywords: Remote labs · Electric circuits · General Electricity

1 Introduction

Laboratories are a fundamental part of engineering education due to the very nature of the engineering profession. Feisel and Rosa [1] classified engineering laboratories in three groups: development, research and educational (instructional). The last group is used by students during their engineering courses and this work is focused on them. Electrical Engineering courses rely on different types of labs – Analog and Digital Electronics, Control Systems, Microwave Circuits, Electrical Machines, etc. In many research institutions, students can also have access to research labs, but this fact does not mean that the institution can exclusively rely on them; educational labs are a need and are used by institutions worldwide.

Current ICT – Information and Communication Technology provides many options for engineering education. Some are solely software, like general-purpose products for numerical computation such as MATLAB® and Maple®, both commercial, and SciLab®, which is a free and open software. Other products are specific; CircuitLab®,

M. E. Auer and T. Tsiatsos (Eds.): IMCL 2017, AISC 725, pp. 879–890, 2018.
https://doi.org/10.1007/978-3-319-75175-7_86

Cadence®, LTSpice and Fritzing are based on numerical computation but they solve electric circuits problems. The first two are commercial; the others are free.

The use of ICT is so important that Froyd et al. [2] consider it the fifth major shift in Engineering Education in the last 100 years. The authors even extend the acronym to ICCT - Information, Communication and Computational Technologies.

ICT has also brought a new world of possibilities through the deployment of Remote Labs. The expression Remote Lab in the context of this work is "Remote Access - Real Resource" in the classification presented by Heradio et al. [3].

This paper presents preliminary results on the use of VISIR – Virtual Instrument Systems in Reality, a remote lab for Electric and Electronic Circuits (EEC). It was deployed and started being used at Pontifícia Universidade Católica do Rio de Janeiro (PUC-Rio) in the second semester of 2016 and it has been used for two semesters in different courses. This short experience with VISIR has indicated that its use must carefully be planned to suit the needs and the context of the institution and its courses. This paper is divided in 5 sections and describes the main actions and results. Section 2 addresses the context at PUC-Rio and presents VISIR. Its introduction and use as a learning supporting tool are discussed in Sect. 3 while Sect. 4 shows the main results in different courses. Conclusions and comments are in Sect. 5.

2 The Context at PUC-Rio and VISIR

PUC-Rio is a private and non-profit university in Rio de Janeiro, Brazil. The courses related to Electrical Engineering and Electricity Fundamentals in other engineering curricula have been using different types of ICT tools for about two decades. Blended learning (b-learning) was introduced in the first term of 2014 and since then some courses started being offered in this mode.

VISIR is a remote lab for doing real, physical experiments with EEC. The acronym VISIR stands for Virtual Instrument Systems in Reality.

This section is divided in two subsections: the first describes the context at PUC-Rio and the other focuses on VISIR.

2.1 The Use of ICT Supported Learning at PUC-Rio

The context of ICT supported learning is presented in three steps as follows:

- The Maxwell System (http://www.maxwell.vrac.puc-rio.br/) is a platform that integrates an IR - Institutional Repository as defined by Lynch [4] and an LMS – Learning Management System as defined by Wright et al. [5]. The IR features of the system manage all types of digital contents from ETD – Electronic Theses and Dissertations to interactive courseware of many different types. It hosts both inputs to the learning process as well as its outputs. Since it is an integrated system, students and faculty using the LMS features have seamless access to all types of digital contents. Over and above, contents are items of the collection and not deposited on course folders; courses point to contents that are not replicated, each one has a unique existence on the platform. Maintenance is simple and storage space saved.

- The first integration of the Maxwell System with an external system – focused on simulation that is an important tool in Engineering Education. In order to offer students learning objects that include simulation, the system was integrated with SciLab®. Students, both graduate and undergraduate, and faculty are developing objects that address important topics. When the implementation of VISIR in undergraduate courses began, a set of objects to support EEC was developed. Currently, there are 19 such objects in Portuguese and 10 translations into English; five additional objects are under development. Since SciLab® is a software product, it was installed on a server that is connected to the Maxwell servers.
- The second integration – this was quite different, since the system to be integrated was a Remote Lab; it had not only software but hardware too. This integration had the same objective of the first – VISIR was considered one additional learning resource to be offered students and faculty, and it had to seamlessly be accessed. In order to achieve it, both the IR and the LMS features of the Maxwell System were used. The integration is presented in [6].

The objective of the use of ICT tools is to stimulate students to be more active and independent. Solutions to be integrated must be flexible enough to be expanded and to allow new ones to be added when necessary. Support must be provided to faculty.

PUC-Rio also offers students and faculty other IT solutions; the ones mentioned in this subsection are related to Electrical Engineering only (we are considering Computer Engineering and Control and Automation Engineering as Electrical Engineering sub-types). There is a campus agreement for MATLAB® - students can use it on the computers of the university labs, on-line from the company servers or download it to their own devices. The university also provides LabView® on computers in labs of the Electrical Engineering Department and CircuitLab® on-line from the company servers. There are different types of licenses for these and other products.

2.2 VISIR

VISIR is comprised of two parts – software and hardware. The VISIR software is an open-source released under a GNU General Public License (GPL) [8]. It includes (1) a user interface that handles all the administration, access, and authentication process. This interface may sometimes be invisible to the client depending on how VISIR is integrated into a given Learning Management System (LMS). In the case of PUC-Rio, where VISIR has been integrated into the Maxwell System, only (2) an experiment client interface is visible to the students. The experiment client is a simulated electronics workbench embedded in the HTML code of the user interface. It provides access to a triple-output DC power supply, a digital multimeter (DMM), a two-channel oscilloscope, a function generator and a solderless mounting board (breadboard), as depicted in Fig. 1. The top row of the interface shows the electric and electronic components that are available for a given experiment. The components are physically located on a relay-based matrix that is connected to a PXI® system which contains the 4 instruments previously referred. The matrix and the PXI® system form the hardware part of VISIR. The (3) measurement server is responsible for handling all the remote experiment requests. Every time a client clicks on the “Perform Experiment” button, it reads the circuit description (as built by the client in the experiment client interface), checks if the circuit is valid and then queues the request to (4) the equipment server. This server physically interconnects the components and the instruments on the matrix, applies the stimulus, reads the responses, and then passes this information back to the experiment server to transmit it to the active remote instrument panel. All the operations take a fraction of a second, so VISIR can be described as a batch-mode remote lab, capable of simultaneously serving several clients. In short: the user and experiment interfaces plus the experiment and instrument servers are the software, while the hardware part contains a relay-based matrix and a PXI® system with a DC power supply, a DMM, an oscilloscope, and a function generator.

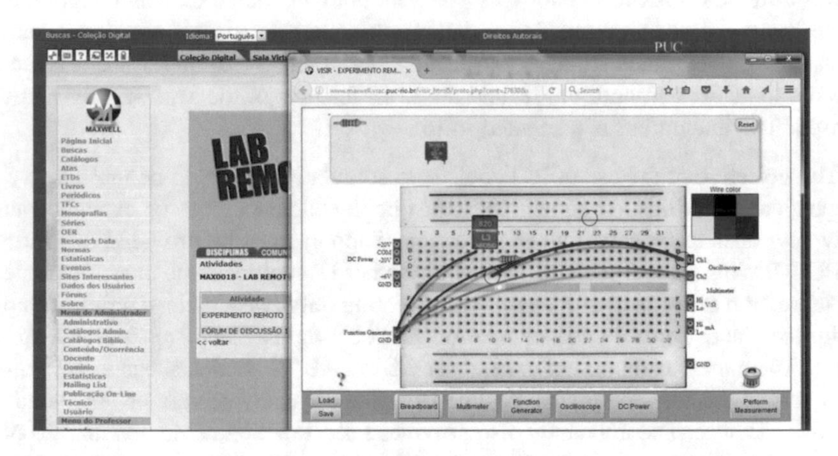

Fig. 1. Experiment client interface (in HTML5) integrated into PUC-Rio Maxwell system.

2.3 Final Remark

The experiments performed in the first semester of 2016 ran from the installation at CUAS – Carinthia University of Applied Sciences in Austria. All other experiments, from the second semester of 2016 on, were performed using the integrated platform at PUC-Rio. This shows the real meaning of the word remote.

3 The Use of VISIR

Different courses in one curriculum and courses in different curricula have distinct characteristics. The distinctions are even bigger when courses of different institutions are considered.

This yields an interesting challenge – examining courses characteristics to find suitable uses of the Remote Lab. They must be adapted to objectives of the course and the context of the institution. This is something that was learned in the second semester of 2016 when VISIR was used for some lab activities of the EEC course. As a result of this experience, it was decided to use it in other courses with different objectives and profiles of students. Two other courses at PUC-Rio were chosen – General Electricity

(GE) and Introduction to Engineering/Electrical Engineering (ItE). At the same time, the VISIR$^+$ Project requires that local partner institutions use the equipment. One of the partners decided to start using; it was Universidade do Estado do Rio de Janeiro (UERJ) and the course was Electric and Magnetic Measurements (EMM). The subsections that follow address the different uses and how their results are being considered to determine how best serve the learning process.

3.1 Learning to Use VISIR

VISIR was the first remote lab to be used at PUC-Rio, so faculty involved on the VISIR$^+$ Project decided to get acquainted with the equipment and its use before the actual installation at the university. CUAS is PUC-Rio's corresponding partner in the project, so its equipment was offered for this first use.

In the first semester of 2016, two experiments were held using the installation in Austria. The first was for one experiment in the EEC course and the second in an extracurricular activity (ECA) aiming at students that had not taken the Circuits course. ECAs are mandatory in the Engineering curricula in Brazil.

First Experiment in EEC. One experiment in the EEC syllabus was selected for adding an extra step, using VISIR, between simulation and real-lab practice. There were 18 students enrolled in the class and 12 responded a three question survey. The three questions were: (1) I found VISIR easy to use; (2) VISIR helped reducing experimental setup errors/setup time; (3) Overall, VISIR had a positive impact in my learning experience. Assigning grades from 1 ("I completely disagree") to 5 ("I completely agree"), the outcomes for the three questions were respectively averages 4.8, 3.5 and 4.6. Further research showed that question 2 had not a high score due to lack of circuit components for that specific lab class.

First Experiment in an ECA. The ECA offering an Electrical Engineering lab activity had 11 students from three engineering curricula: Electrical (6), Mechanical (3) blank;and Computer (2). The students were surveyed and the results were: 6 students graded the activity 5 (highest grade in a 1 to 5 scale), 4 graded 4 and 1 graded 3. When asked if they would recommend their colleagues this ECA, 8 said they strongly would do it and 3 said they would recommend. In addition, this ECA helped the VISIR team to "rehearse" for the use of VISIR in regular courses.

Comment. These two uses of VISIR were of paramount importance in learning how to prepare and run experiments in the following school terms. The insights from these experiences led to the view of the implementation process as presented in [7].

3.2 VISIR and the EEC Course in the Second Semester of 2016

EEC is a mandatory course in three engineering curricula at PUC-Rio: Computer, Controls & Automation, and Electrical. This course has had high rates of students who do not pass it, either they dropout or fail the exams. This is a serious problem since EEC is a prerequisite for many other mandatory courses. Two members of the VISIR team have taught this course and started collecting data about the students' performances in 2011. Since then, many actions have been taken to enhance the results. Quite a few are based on the use of ICT tools – development of learning objects of different natures (animations, videos, simulators, etc.). The switching to b-learning in the first semester of 2014 is one of them. The EEC course has two lab hours per week taught in a traditional way. VISIR was included as an additional activity to offer one more way of experimenting with circuits.

The Experiments Students Must Perform. In the last terms, students have performed 10 experiments: (1) Resistor ladder circuit; (2) Operational amplifier (inverter and non-inverter); (3) Amplifier and adder; (4) Sensor with graphic bar; (5) Super diode; (6) First order circuits (RL and RC); (7) Timer (RC); (8) Second order circuits (RLC); (9) Sinusoidal steady-state; and (10) Frequency response (RL, RC and RLC).

How Students Work at the Lab. Students are required to study before going to the lab. They are assigned a study guide and they have to write an outline of the way the experiments will be performed and the expected results; they are graded for this activity. They also have to simulate experiments (using CircuitLab®) before going to the lab and the results are part of the outline. As mentioned before, they have two hours of lab classes each week but the labs are open from 7 AM to 7 PM every weekday and there are technical staff to support students all the time.

Experiments Using VISIR. The use of VISIR was limited to three experiments: (1) Second order circuits (RLC); (2) Sinusoidal steady-state; and (3) Frequency response (RL, RC and RLC). Students were required to perform one additional task before going to the lab – use VISIR to remotely perform the experiments.

Preparation of the Course for the Use of VISIR. In accordance with the view of the implementation process presented in [7], many items of courseware were developed. They can be found in the series Projeto VISIR+ (https://www.maxwell.vrac.puc-rio.br/series.php?tipBusca=dados&nrseqser=14). Many other supporting learning contents were developed and can be found in Open Access in two other series: Objetos Educacionais em Engenharia Elétrica (https://www.maxwell.vrac.puc-rio.br/series.php?tipBusca=dados&nrseqser=5) and Simulações em Engenharia Elétrica (https://www.maxwell.vrac.puc-rio.br/series.php?tipBusca=dados&nrseqser=12). All other courseware developed for the b-learning mode were available as well.

Grades and Results. A lot of data were collected: the students' GPAs, the numbers of times they had already attended the course, their curricula, etc. The class had 36 students and 3 dropped out; 33 was the final number. The students were handed a survey at the end of the term – 17 (51.51%) filled it but 3 were not identified. This remark is important because it was only possible to match grades with VISIR activities for 14 students. The students were graded by the outlines they presented; 3 had 3 parts (because they included VISIR) and the remaining 7 had only 2. This is something that is commented later. They were also graded by their performances and reports on the experiments. Finally, a test was applied and graded. The final grade of each student was 30% from the average of outlines (1), 30% from the average of experiments and reports (2) and 40% from the test. Table 1 shows the average grades (computed over 10) for all 33 students. When the results of the whole EEC course are considered (lab & theory), the results in terms of grades are shown in Figs. 2 and 3.

Table 1. Average grades for the EEC lab in the second semester of 2016.

Types of grade	Values
Average of all outlines (1)	6.91
Average of all experiments and reports (2)	7.93
Average of all tests	4.55
Average of final grades	6.27

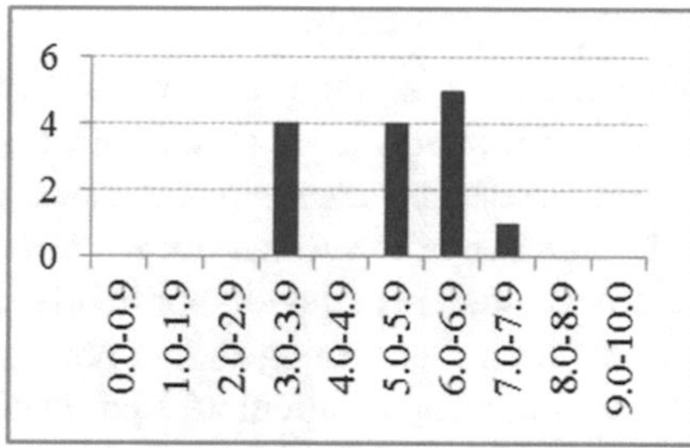

Fig. 2. (Left) numbers of students by final grade – set identified in the survey.

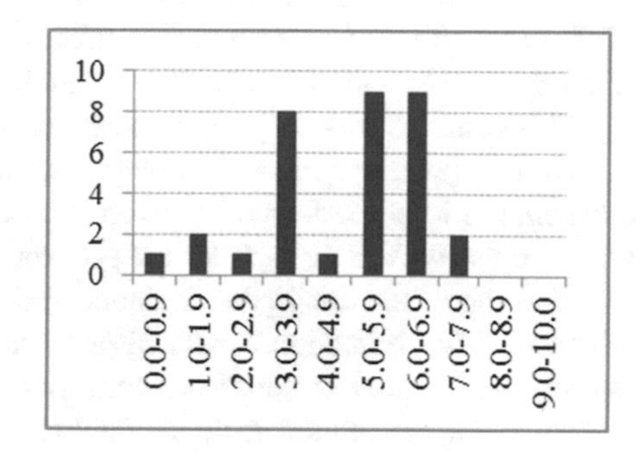

Fig. 3. (Right) numbers of students by final grade – all students in the class.

Gathered data show that students who were identified when replying the questionnaire: (1) represented 42.42% of the class and 50.00% of those who passed the course; and (2) 28.57% were taking the course for second time as compared to 30.30% of the class. The histograms show that the identified students had less unsuccesses. It yields the impression that more dedicated students were concerned in expressing their opinions on the use of VISIR.

The overall average of points in the evaluation of VISIR was 1.99 over a maximum of 5.00. The average number of accesses was 3.96. Some comments were interesting: (1) the weight (in the grade) of experiments outlines requiring different numbers of tasks was the same – a bigger workload was not recognized; (2) the labs are open many hours a day, so the use of a remote equipment does not add that much value; (3) it was difficult to tell the difference between VISIR and a software product; and (4) students cannot perform experiments that are not assembled on the equipment (this can be done in the traditional lab). It is important to remark that experiments used in the course did not explore the potential of the equipment to perform tasks that would not be possible in the traditional lab, as for example, the determination of a Thévènin Equivalent Circuit or the identification of a transfer function of a 1st or a 2nd order circuit; the use of VISIR would allow components to be "hidden" from students.

3.3 Open House to High School Students in the First Semester of 2017

PUC-Rio regularly invites students from public and private schools to visit the University and attend lectures and perform technical activities. In the first semester of

2017, 60 high school students had the opportunity to use an Electrical Engineering lab. It was possible because they used VISIR. For 1 h the students performed basic experiments. The objective was to present results of a real laboratory in a safe way. The activity was successful and the students liked it.

3.4 VISIR and the GE Course in the First Semester of 2017

This course is mandatory to all engineering students who are not required to take EEC. It has a very large number of students and the number of lab classes in the first semester of 2017 was 14 – the total number of students was 273. The number of students who filled the survey was 235 (80.08% - much higher than that of EEC).

This course offers only one hour of lab sessions per week and they are on different days and times. This poses a problem – to keep the experiments in sync; the number of experiments to be performed each term used to be 8. Faculty became very enthusiastic with the use of VISIR in order to: (1) increase the number of experiments each term; (2) offer more lab time to the students; and (3) motivate students who are not involved with electricity related curricula by the use of a new tool. Concerning the number of experiments, it is important to remark that holidays do not impact a remote equipment and, also, experiments can be available for any day since no technical supervision is required; students cannot work in the traditional lab if a technical staff is not present. Three new experiments were added – one divided in 3 steps. This meant a significant enhancement in the number of practical activities.

It is important to remark that GE deals with voltages and currents much higher than VISIR supports. At the same time, basic concepts like Ohm's Law, voltage divider, current divider, linear and nonlinear circuits can be addressed with VISIR with no loss of generality. So VISIR could really add value to the GE lab.

The computation of the results of the survey yielded an average grade of 2.48 out of 5.0; a better result than that of EEC. The average number of accesses was 7.63; almost twice as much as that in EEC. An inspection of the logs of accesses on the Maxwell System showed significant numbers late at night and during weekends. Thus a "mission" of a remote solution seemed to be fulfilled – offer students access to resources when the university is closed.

To illustrate the impact of the VISIR, Fig. 4 presents the grades as functions of numbers of accesses. In addition to that, a trend line was drawn to show that, in the average, the more the students have accessed the VISIR the higher their grades were. As in the EEC course one may think that students who are more committed to their course study more hours per week and access VISIR more times too.

This course is using VISIR in the second semester of 2017.

3.5 VISIR and the ItE Course in the First Semester of 2017

Introduction to Engineering is a Problem Based Learning (PBL) course that is mandatory to all students admitted to one of the engineering curricula. It is taken in the first semester and students are generally grouped in classes the same area of interest. VISIR was used in a class with 17 students distributed in the curricula: (1) Computer

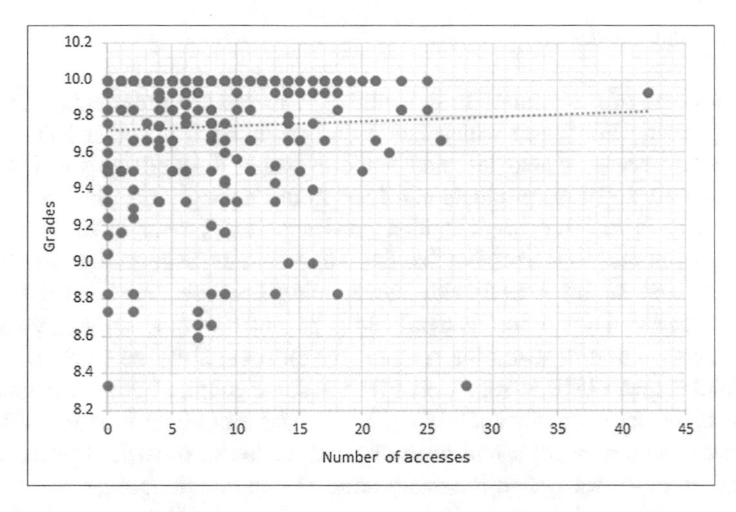

Fig. 4. Grades as functions of accesses.

Engineering – 1; (2) Control and Automation Engineering – 4; (3) Electrical Engineering – 4; and (4) Mechanical Engineering – 9. The focus of their PBL projects was electromechanical systems controlled by microcontrollers. The faculty responsible for the class was in the area of Electrical Engineering.

The computation of the results of the surveys yielded an average grade of 2.50 out of 5.0; also a better result than that of EEC. The number of students who finished the course was 17, the same number of students who filled the survey. Among the 17, 13 wrote comments and 9 said that they liked a remote lab because they could use it any time and any place. One interesting comment was that the student felt comfortable by not risking damaging the equipment. The average number of accesses was 19.35.

3.6 VISIR and the EMM Course in the First Semester of 2017

As mentioned before, the VISIR$^+$ Project requires that participating institutions have local partners. UERJ is one of the partners and was the first to use VISIR. It is a public institution that belongs to the State of Rio de Janeiro. The course that was chosen by UERJ was Electric and Magnetic Measurements which is mandatory for the curriculum of Electrical Engineering. The average number of accesses was 10.93 and the average grade was 3.0. The number of students was 15 and 7 replied the survey. All students wrote comments – 2 students liked the any place any time possibility of use and three had problems with the instruments. One interesting comment was that the student felt comfortable by not risking damaging the equipment, the same as in ItE. This course is using VISIR in the second semester of 2017.

4 Main Outcomes

The first outcome of the VISIR$^+$ Project at PUC-Rio was the introduction of a remote lab in Engineering Education which has been benefitting from other ICT tools. This was important because a real lab was made available through the LMS. Before its deployment, only software products had been in use through this platform.

The second is that the original idea presented in [7] seemed to be right – the inclusion of a remote lab is a project that has many different aspects. It is an educational project that must be integrated with the ongoing learning activities. Students and faculty must feel that a new tool is available to be added to others. Faculty must make sure that course activities using the remote labs must carefully be planned.

The third is that VISIR is very well suited to some courses but not to others. In the EEC course, results were not good when VISIR was employed in many different lab experiments. There was not a real need of a remote lab to provide support to experiments that can be performed in the many hours that the traditional lab is available. At the same time, faculty did not implement the experiments that would only be possible using VISIR. Another point to be examined is the training of the students in the use of the remote lab. In EEC, there was no organized effort in this activity (it as informal) and maybe this contributed to the students' not fully appreciating the equipment.

Nevertheless, there is a huge opportunity for VISIR in EEC in the near future. Students of Electrical, Control and Automation and Computer Engineering enrolled in the university since the first semester of 2017 are now required to attend two EEC courses – EEC I and EEC II – where the traditional lab is not used until the second course. This means that students of EEC I will not have hands-on lab experiments, and VISIR will be used as the sole lab activity. It is also in EEC I that Thévènin Equivalent Circuit is addressed and this is an experiment that perfectly fits VISIR.

In the first semester of 2017, the second with VISIR installed at PUC-Rio, two important things happened. The first is that the VISIR team was more proficient in the use of the remote lab and was able to understand the importance of training students on how to use it. The second was the decision to try VISIR in courses with different characteristics.

The use of VISIR in GE was successful – the number of experiments was increased. The 24/7 availability and the remote access allowed students to experiment for a longer time. Faculty of GE are enhancing the management of VISIR from the experience they got the first time they used it – enhancing the courseware, devoting more time to training students, and allowing students more time when needed to fulfill the tasks. GE has many students, so this yields a big impact, especially because students in this course are not really motivated to courses related to electricity. It is the authors' opinion that the use of VISIR in GE is irreversible in the near future.

ItE yielded results that were considered satisfactory, specially concerning the average number of accesses – 19.35. This class had a very young group of students (first semester of the freshman year) who were involved with PBL activities and did not have traditional circuit lab classes – they easily accepted this new tool since they did not have another lab activity to compare.

EMM is a course with a different profile and taught at another institution. It has 4 weekly hours and traditional lab activities for measurements in high and low levels AC and DC circuits. The experiments the students performed using VISIR were different from the ones of the traditional lab and the most remarkable fact was that they never came to PUC-Rio. This course yielded the highest overall grade in the surveys.

5 Conclusions and Comments

The use of VISIR as one of the steps before traditional laboratory classes of the same experiment did not yield good perceptions among students, since they could compare with the traditional lab where there was a lot more flexibility. At the same time, using VISIR to introduce experiments that are not performed in the traditional way seemed to please students. Faculty were specially satisfied that they could add more experiments to the courses and/or better motivate their students. The use of VISIR in courses other than EEC yielded better results. It is important to observe that the EEC course is in the core of the three engineering curricula, thus students have higher expectations. The grading process was also a problem in EEC.

With the introduction of a new curriculum where EEC will be divided in two courses, there is an opportunity for VISIR in the first course (EEC I), since all traditional lab experiments were left to EEC II. This is expected to add real value to EEC I since students will be able to experiment using concepts that are theoretically taught.

The results discussed in this paper are limited by a very short time frame – one semester in each course. For this reason, it is not possible to state that VISIR helps students achieve better results. Maybe students who access the remote lab more times than the others are more committed to learning regardless of the tools that are used. Two courses (GE and EMM) are currently using VISIR for the second time, so there will be more data to analyze. The VISIR team has no doubt that this study will have to be performed once the equipment has been used for more semesters and different sets of students are exposed to it.

A more extensive and intensive study was performed by Marques et al. [9]. Many of results of this first use at PUC-Rio are in agreement to the results presented in this work. When more data are collected at PUC-Rio, a more detailed comparison on the implementations in two different institutions will be performed to check similarities and differences. Their methodology will be an inspiration for the new study.

A comment is necessary to wrap up: there is a large amount of data already available on the Maxwell System that will be analyzed and presented in a next publication.

Acknowledgements. The VISIR team at PUC-Rio is very grateful to the VISIR team at CUAS for its goodwilled lending of the equipment and competent support in the process of installation.

The VISIR$^+$ Project is funded by the European Commission through grant 561735-EPP-1-2015-1-PT-EPPKA2-CBHE-JP. The PUC-Rio team is grateful for this opportunity of having VISIR and being part of such important project.

References

1. Feise, L.D., Rosa, A.J.: The role of the laboratory in undergraduate engineering education. J. Eng. Educ. **94**(1), 121–130 (2005). https://doi.org/10.1002/j.2168-9830.2005.tb00833.x. Accessed July 2017
2. Froyd, J.E., Wankat, P.C., Smith, K.A.: Five major shifts in 100 years of engineering education. Proc. IEEE **100**(Special Centennial Issue), 1344–1360 (2014). https://doi.org/10.1109/JPROC.2012.2190167. Accessed Sept 2017
3. Heradio, R., de la Torre, L., Galan, D., Cabrerizo, F.J., Herrera-Viedma, E., Dormido, S.: Virtual and remote labs in education: a bibliometric analysis. Comput. Educ. **98**, 14–38 (2016). https://doi.org/10.1016/j.compedu.2016.03.010
4. Lynch, C.: Institutional repositories: essential infrastructure for scholarship in the digital age. ARL Bimonthly Report, 226, United States, February 2003. https://www.cni.org/publications/cliffs-pubs/institutional-repositories-infrastructure-for-scholarship/. Accessed May 2017
5. Wright, C.R., Lopes, V., Montgomerie, T.C., Reju, S.A.: Selecting a learning management system: advice from an academic perspective. EDUCAUSEreview, published 21 April 2014. http://www.educause.edu/ero/article/selecting-learning-management-system-advice-academic-perspective. Accessed 05 Feb 2015
6. Pavani, A.M.B., de Souza Barbosa, W., Calliari, F., de C Pereira, D.B., Lima, V.A.P., Cardoso, G.P.: Integration of an LMS, an IR and a remote lab. In: Porceedings of REV 2017 – 14th International Conference on Remote Engineering and Virtual Instrumentation, United States, pp. 427–442, March 2017
7. Pavani, A.M.B., Lima, D.A., Temporão, G.P., Lima, V.A.P.: Implementação de um Laboratório Remoto: um Porjeto de Múltiplas Facetas". Article presented at COBENG 2016 – Congresso Brasileiro de Educação em Engenharia, Brazil (2016). To appear as a book chapter in 2017 and currently available https://www.maxwell.vrac.puc-rio.br/Busca_etds.php?strSecao=resultado&nrSeq=27615@1
8. Tawfik, M., et al.: Virtual instrument systems in reality (VISIR) for remote wiring and measurement of electronic circuits on breadboard. IEEE Trans. Learn. Technol. **6**(1), 60–72 (2013). https://doi.org/10.1109/tlt.2012.20
9. Marques, M.A., Viegas, M.C., Costa-Lobo, M.C., Fidalgo, A.V., Alves, G.R., Rocha, J.S., Gustavsson, I.: How remote labs impact on course outcomes: various practices using VISIR. IEEE Trans. Educ. **57**(3), 151–159 (2014)

Effectiveness of Integrating Open Educational Resources and Massive Open Online Courses in Student Centred Learning

Vinu Sherimon(✉), Huda Salim Al Shuaily, and Regula Thirupathi

Higher College of Technology, Muscat, Oman
{vinusheri,huda.alshuaily,
regula.thirupathi}@hct.edu.om

Abstract. Open Educational Resources (OER) and Massive Open Online Courses (MOOCS) are successful areas intended at giving students equal access to educational resources and innovative educational opportunities. They facilitate to adapt Student Centred Learning (SCL) in teaching, learning and assessment from many aspects. SCL activities enable the acquisition of new skills by the learners. The great varieties of OER, allows the possibility of differentiated instruction to cater the learner's individual capabilities thereby helping them to depend on themselves. The assessment of the learner is shaped differently with the integration of the resources of OER & MOOCS. This study explores the integration of OER & MOOCS in SCL at Department of Information Technology (IT), Higher College of Technology (HCT), Muscat. The SCL approach is implemented in each course at different level of study through different strategies. In this paper, we present the implementation of SCL in two courses: Introduction to Database and Data Modelling. The first course is offered to Diploma - Year 1 students and the second course is offered to students in Advanced Diploma level. Different surveys were conducted to analyze the effectiveness and it is found that the integration of OER and MOOCS affected positively in the student-learning process.

Keywords: Open Educational Resources · Massive Open Online Courses
Student Centred Learning · Assessment · OER · MOOCS · SCL

1 Introduction

Student-centred learning, also known as learner-centred education, broadly encompasses methods of teaching that shift the focus of instruction from the teacher to the student. Student-centred learning, often referred to as Project-Based Learning (PBL), is a 21st century concept implementing a new curriculum using technology and the student's own abilities to achieve higher standards than the traditional learning styles (Gibson 2016). Stephanie Bell states that PBL is not a supplementary activity to support learning. It is the basis of the curriculum (Bell 2010). The concept of student-centred learning is to bring the classroom and students to life. The teacher is considered a "guide on the side", assisting and guiding students to meet the goals that

M. E. Auer and T. Tsiatsos (Eds.): IMCL 2017, AISC 725, pp. 891–902, 2018.
https://doi.org/10.1007/978-3-319-75175-7_87

have been made by the students and the teacher. This is one of the most important strategies in building a relationship between peers, and in coordinating work within a group.

Higher education institutions around the world have been using the Internet and other digital technologies to develop and distribute teaching and learning for decades. Recently, Open Educational Resources (OER) have gained increased attention for their potential and promise to obviate demographic, economic, and geographic educational boundaries and to promote life-long learning and personalized learning. The term Open Educational Resources (OER) was first introduced at a conference hosted by UNESCO in 2000 and was promoted in the context of providing free access to educational resources on a global scale (McGreal et al. 2013).

A MOOC is not the companionship-based education which is so typical at doctoral level; in fact, it is the opposite of such an education, because the teacher who developed the course is a long way away from the students, even if they do answer e-mails sent to them (Pomerol et al. 2015). The benefits for students include reduced education costs and global access to exclusive institution courses and instructors. However, the benefits for institutions are less clear as there is a financial overhead required to develop and deliver content that is suitable for mass student consumption. This paper describes the effectiveness of implementing OER and MOOCs in the courses of Department of IT, HCT, Muscat. The experience of two courses is presented as a case study.

The rest of the paper is organized as follows: Sect. 2 includes the literature review. Methodology is given in Sect. 3 followed by Implementation in Sect. 4. Discussion & results are included in Sect. 5. Conclusion is presented in Sect. 6 followed by Acknowledgements and References.

2 Literature Review

The future of any country depends on the young generation. It is the responsibility of each citizen to educate their children and prepare them to undertake challenges and to excel in life. Even though teachers have a key role in shaping the future citizens of the country, with the advent of student-centred learning they remain as facilitators transferring the significant role of learning process to students. So, the traditional way of teaching and learning has slowly decreased and every institution around the world has implemented student-centred oriented learning approaches. These approaches emphasize more on the abilities and the learning styles of each student.

To promote interactivity among students, Stuart Glogoff suggested instructional blogging to be a valuable e-learning tool (Glogoff 2005). This encourages students to involve in research activities and discussions with the lecturers. He suggested that Instructional blogging has excellent potentials particularly in virtual classrooms. The idea of flipped classrooms to promote the participation of students was suggested by few researchers (Gilboy et al. 2015). The research provides a template on class activities and assessments to be done before, during and after the class. They implemented it for two undergraduate courses and the results say that students preferred the flipped classroom approach than traditional approaches. Another research investigated the impact of problem based learning (PBL) on undergraduate students as compared with traditional lecture methodologies (Yadav et al. 2011). The result of the research pointed out that the students learning gains were increased twice when compared to the traditional one. A study on student perceptions on active learning was conducted by Angela (Lumpkin et al. 2015). It was assessed by an action research project where students were involved in a variety of in-class and out of class activities and group discussions. The results revealed that the active learning positively influence the learning process of students.

The implication of implementing MOOCs in UK higher education was discussed in a white paper published in 2013 (Yuan et al. 2013). The focus of the paper is to help the decision makers to gain a better understanding about the trends of MOOCs and the impacts of implementing it in higher educational institutions. Even though many institutions have joined this model, the value of the course completion certificates issued by a MOOCs course is yet to be determined. Few challenges of MOOCs are included in the report which includes the sustainability, the quality, completion rates, etc. In 2015, another study analyzed the instructional quality of MOOCs (Margaryan et al. 2015). Seventy-six courses were randomly selected for the analysis. The instructional design quality of the courses was assessed and compared, and the results show that the design quality of majority of the MOOCs courses was found to be low. Another research conducted in 2013 suggested that one of the most flexible ways to offer online quality education is through MOOCs based on OER (Daradoumis et al. 2013). The research proposes a framework that includes software agents to personalize the management of MOOCs courses. A research conducted in 2015 considered MOOCs as a challenge to the traditional educational institutions around the world (Conole 2015). The paper suggested 7Cs of Learning Design framework while designing MOOCs to enhance the experience of the learner and to ensure the quality. The experiences of learner in MOOCs was analyzed in (Veletsianos et al. 2015). The interactions of learners in social networks outside MOOCs were studied and the paper suggests few technological improvements to enhance open learning. Usually the completion rates of MOOCs courses are low compared to traditional courses. In 2016, a research was conducted to analyze the completing rate of MOOCs courses (Pursel et al. 2016). The demographic data, intended behaviours and course interactions of students were analyzed to understand the factors that influence the completion. Similar research was conducted to understand the factors that contribute to learner's performance in MOOCs (Barba et al. 2016).

3 Methodology

Student Centred Learning is part of all courses of Department of IT, HCT. Depending on each course and level, different strategies are employed to implement SCL. In almost all the courses, OER is provided in the eLearning portal of the department. It includes video tutorials, class activities, power point presentations, online quizzes, etc. Self-study topics are identified, and the student is required to study the topic(s) with the help of OER. Another way of implementing SCL is to identify relevant MOOCs which is equivalent to the course outcomes of HCT courses. Few MOOCs are identified, and students are required to enroll and complete the course. In few cases, if the identified

MOOCs is vast, the students are required to complete a module which is relevant. Each course includes separate marks to assess the SCL topics. For example, for OER integrated courses, separate assessment is conducted to assess the self-study topics. If the course has a MOOCs component, then the student is required to produce the course completing certificate to earn the SCL marks.

In this paper, as a case-study, the implementation of SCL in Introduction to Database (ITDB1102) and Data Modelling (ITDB3108) course is presented. The department offers ITDB1102 to Diploma – Year 1 students and ITDB3108 to Advanced Diploma – Database specialization students. In ITDB1102, the eLearning portal is equipped with enough OER and we have integrated a MOOC course in ITDB3108 to achieve student centred learning.

4 Implementation

4.1 Integration of OER in Introduction to Database (ITDB1102) Course

It includes course reference materials, practice exercises, class activities, power point presentations, online videos, sample question papers, etc. All these e-resources are kept in the e-learning portal. Since this is an introductory course, for each topic/sub-topic, simple online videos are included. These videos are self-explanatory. Even in the absence of the lecturer, a student can view a video and learn the concepts. Also, the portal includes videos by HCT faculty, which is more beneficial for the students. Few complex topics are identified, and simple videos are developed to assist the students. This is very helpful for students, particularly during exam period.

Figure 1 display the home page of the course in the e-learning portal. Each chapter is organized into different sections. Figure 2 shows the structure of each chapter. Each chapter includes a power point presentation, study materials, online videos, quizzes, etc. Figures 3, 4 and 5 shows the captured screens from the online video developed by HCT faculty. In this video, an example of mapping entity-relationship diagram

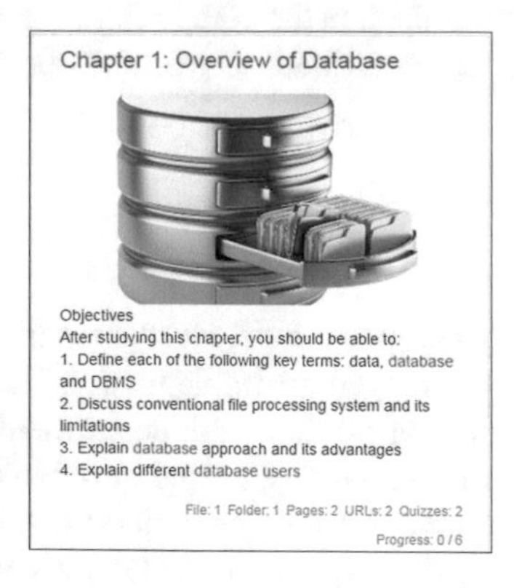

Fig. 1. Course home page in e-learning

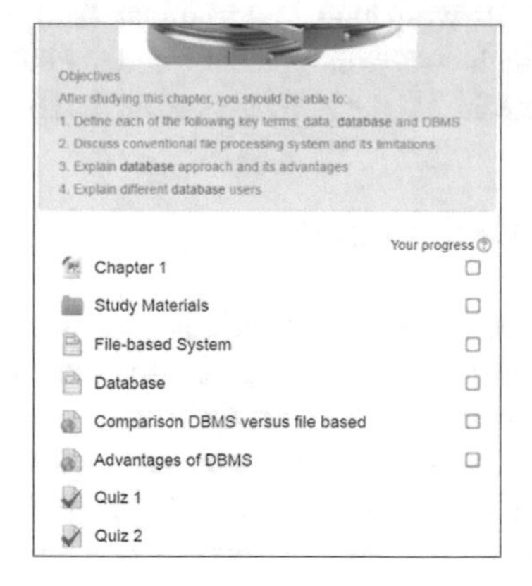

Fig. 2. Course structure

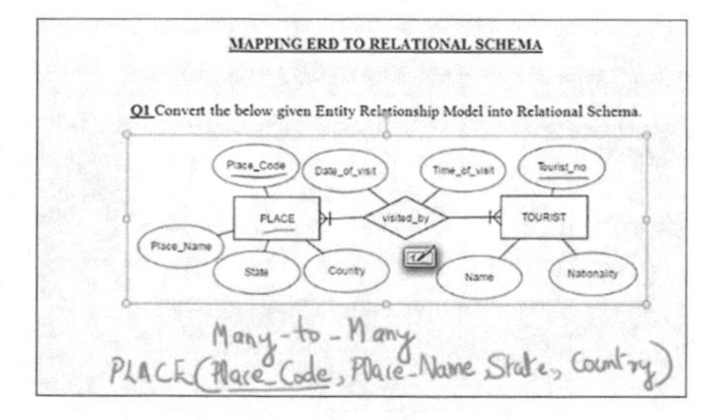

Fig. 3. Screenshot – 1

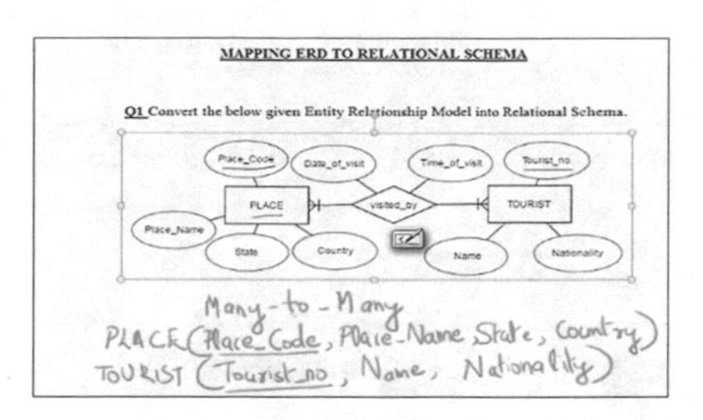

Fig. 4. Screenshot – 2

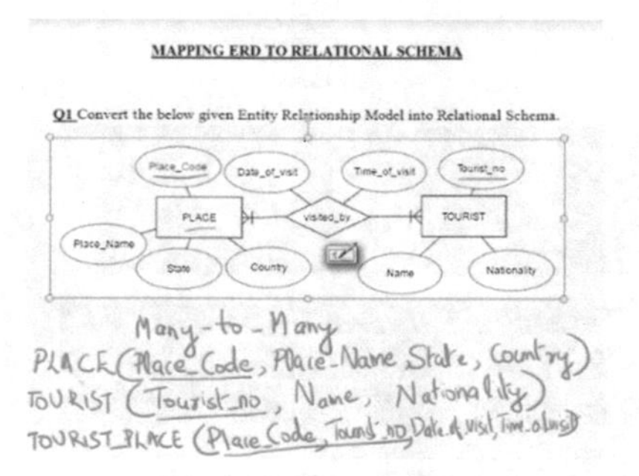

Fig. 5. Screenshot – 3

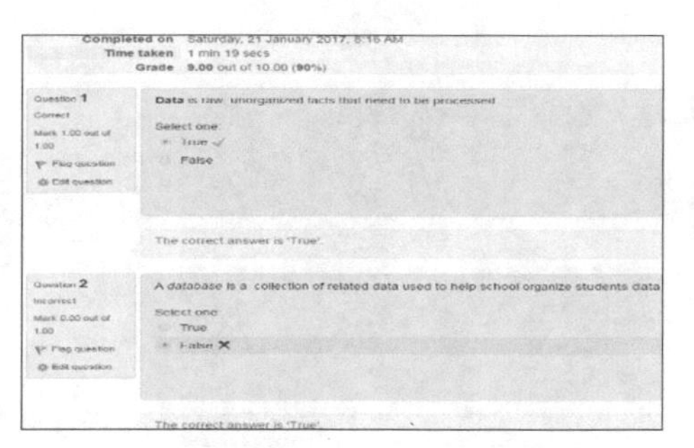

Fig. 6. Sample quiz screen

(ERD) to relational schema is explained. The video includes step-by-step instructions to solve the example.

Online quizzes included in the portal helps students to study themselves. Students can attempt the quiz at any time and once the quiz is submitted, the feedback is given with explanation of correct answers. These quizzes are included as a practice and students can attempt the quiz more number of times. Figure 6 shows a sample quiz screen.

4.2 Implementing MOOC in Data Modelling (ITDB3108) Course

In the e-learning portal, a variety of resources are kept. It includes open textbook, reference materials, practical exercises etc. Apart from these e-resources, an online course offered by Coursera was identified to be integrated as part of the curriculum. Coursera is an educational technology company that offers massive open online courses (MOOCs) [https://en.wikipedia.org/wiki/Coursera]. They offer online courses from different universities in different subjects. To complete the outcomes related to big data modelling, the students were asked to enroll in *Introduction to Big Data* course offered by University of California (Fig. 7). It was a three-week course which includes short videos, slides, hands-on exercises, instructions to install the required software, discussion forums, quizzes, peer-graded assignments, etc. Students need to spend 5–6 h/week to study the topics and to complete quizzes and assignments. Students were

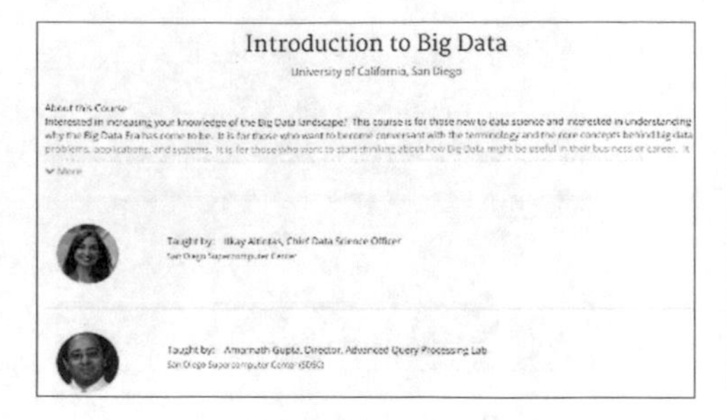

Fig. 7. Course home page in Coursera

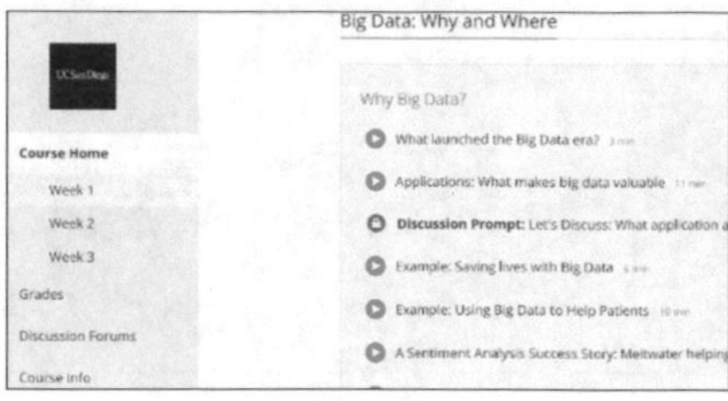

Fig. 8. Sub-topics – week 1

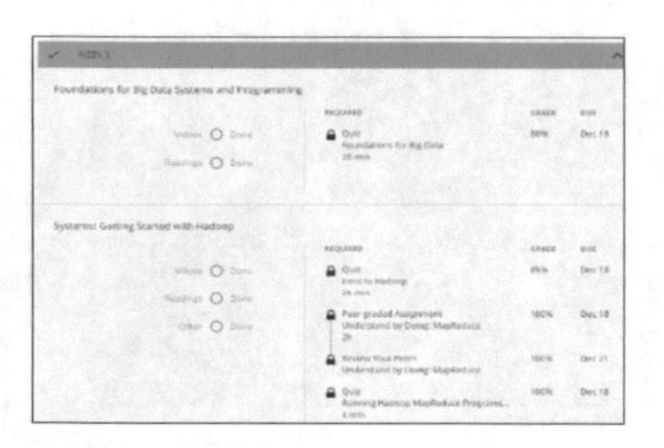

Fig. 9. Course grades

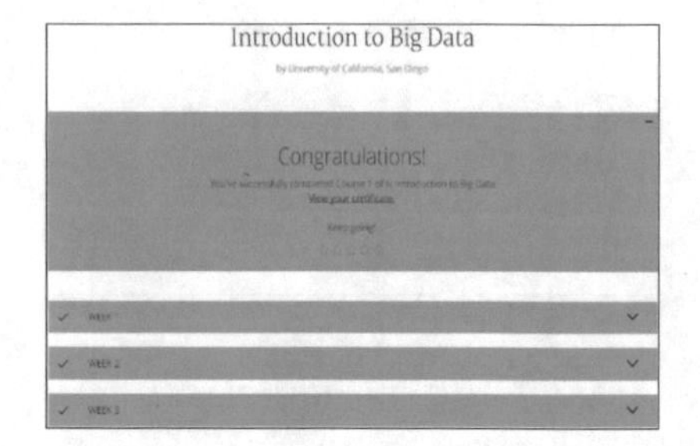

Fig. 10. Course progress

required to submit the assignment on or before the set due date. The due dates helped the students to manage the study schedule. Within the due date, they could re-submit their assignment to improve the grades. At the end, students who earned a passing grade in all the assessments were awarded with a course completion certificate. Figures 8, 9 and 10 shows the concerned pages related to sub-topics, course grades and progress.

5 Discussion and Results

Different surveys are conducted to understand the effectiveness of integrating OER and MOOCs in courses as part of SCL. In this paper, we present two different surveys that is conducted among the staff and students of the department.

OER was part of 81 courses in IT department. So, the first survey was conducted among the course coordinators (81 participants) on the awareness, effectiveness and challenges of integrating OER as part of SCL approaches. Table 1 shows the results of the staff survey. Figure 11 represents the graphical representation of the results.

Regarding the awareness of SCL, 26% of the staff strongly agree and 74% agree to it. 42% of the respondents strongly agree that they possess enough skills to make use of the resources that support SCL and 58% agree to it. The above figures indicate that 100% of the respondents are aware of SCL approaches and possess skills to utilize it.

Table 1. Results – Staff Survey - Effectiveness of integrating OER as part of SCL

Sl. no.	Questions	Strongly agree	Agree	Neutral	Disagree	Strongly disagree
Q1	I am aware of SCL methodologies	21	60	0	0	0
Q2	I possess skills and information to make use of the resources and tools which support SCL	34	47	0	0	0
Q3	My course was completely based on OER	13	21	25	13	9
Q4	I incorporated a MOOCs/Coursera/EdX module in my course	9	17	21	13	21
Q5	Students could learn the self-study topics by themselves	17	51	13	0	0
Q6	There was sufficient time for students to complete the assessments of MOOCs/Coursera/EdX module	4	30	47	0	0
Q7	The resources available in OER/MOOCs/Coursera etc. enhance student skills	30	21	26	4	0
Q8	The introduction of OER encouraged student involvement and participation	21	47	13	0	0
Q9	The marks allocated for assessment based on OER is fair	8	47	26	0	0
Q10	Students performed very well in SCL related activities	9	47	21	4	0
Q11	Time is not enough to cover the course outcomes	9	30	21	21	0
Q12	Class Strength is a major resistance to implement SCL activities	9	21	17	34	0

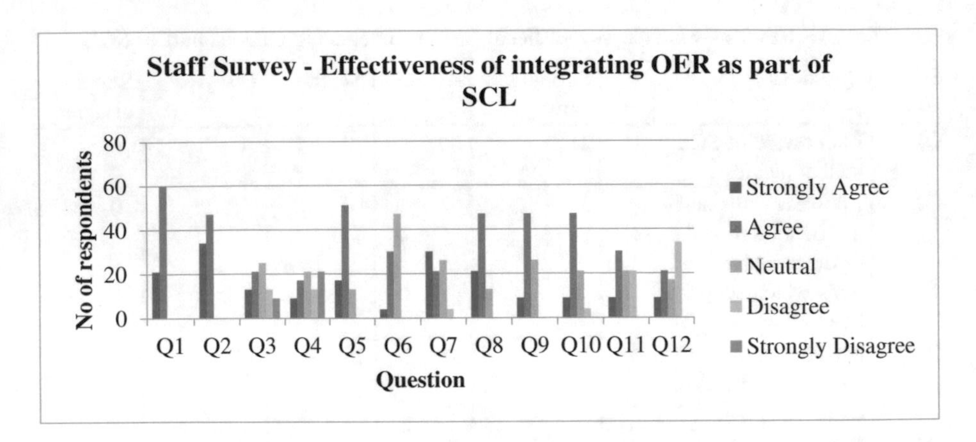

Fig. 11. Effectiveness of integrating OER as part of SCL

Questions from 3 to 10 are based on the effectiveness of OER in SCL. 16% strongly agree and 26% agree that their course was based completely on OER. The percentage of neutral was 31% and 16% disagreed and 11% strongly disagree. As a pilot study, only few sections of some courses in the department were completely based on OER where the students completed the course with no involvement from the lecturers. All the study materials required are uploaded in the e-learning portal. Weekly assessments are also conducted online. 11% strongly agree and 21% agree that they have incorporated a MOOCs based module in their course. 26% were neutral, 16% disagree and 26% strongly disagree. The results show that more than twenty-five percent of the courses in the department have MOOCs based component in their course. It is found that for other courses, a relevant and suitable course that matches the course outcomes and level of the students is difficult to obtain. 21% strongly agree and 63% agree that the students can study the topics by themselves. The remaining 16% were neutral and none disagree. This shows that the students can learn by themselves, except few cases.

5% strongly agree and 37% agree that the students can complete the assessments on time. 58% of the respondents are neutral to it. These results show that more time is required to complete the assessments. Most of the assessments are given one-week time to complete. But since the students study other courses, they are not able to spend more time on a specific course. 37% strongly agree and 26% agree that OER based resources enhances student skills. 32% remain neutral and 5% disagree. 26% strongly agree and 58% agree that students show great enthusiasm in class when OER is introduced. 16% are neutral and none disagree. 10% strongly agree and 58% agree that the marks allocated for OER based assessments are fair. 32% are neutral and none disagree. 11% strongly agree and 58% agree that the students performed very well in SCL related activities. 26% are neutral and 5% disagree. Overall results related to OER shows high interests among students.

Questions 11 and 12 are related to SCL challenges. 11% strongly agree and 37% agree that when SCL is introduced, time is not enough to cover the course outcomes. 26% are neutral and 26% disagree the point. 11% strongly agree and 26% agree that the number of students in the class is one of the challenges in implementing SCL related

Table 2. Results – student survey - effectiveness of implementing MOOCs course as part of SCL

Sl. no.	Questions	Strongly agree	Agree	Undecided	Disagree	Strongly disagree
Q1	The difficulty level of the course is relevant to my (Advanced Diploma) level	12	7	4	2	0
Q2	The course contents are good	5	18	1	1	0
Q3	The explanations of the topics in videos are easy to understand	9	14	0	2	0
Q4	The language in the videos is clear and understandable	10	10	5	0	0
Q5	The Difficulty level of assignment is relevant to me	7	12	3	3	0
Q6	Discussion forum helped me to solve the quizzes and assignments	8	15	1	0	1
Q7	There is sufficient time to submit the assignment	9	13	2	1	0
Q8	All the questions under the Quizzes are easy to solve	5	15	4	1	0
Q9	This course will help in my future career in computer field	12	11	2	0	0
Q10	I am willing to register the next course under big-data to improve my knowledge	11	12	1	1	0
Q11	I am willing to register another course under Coursera as a part of other courses in college	10	14	0	1	0

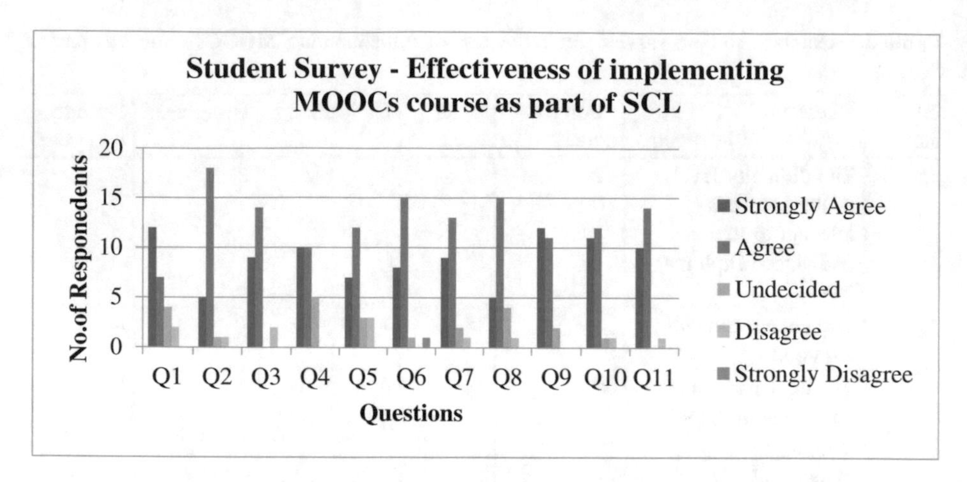

Fig. 12. Effectiveness of implementing MOOCs course as part of SCL

activities. 21% remain neutral, while 42% disagree. The duration of each semester is usually 11–12 weeks including the assessments. Also in Semester I, few classes are cancelled due to Eid and national holidays in Oman and it affects the teaching. So, lecturers are not able to allocate more time to SCL related activities. Also, sometimes there is an imbalance in the number of students among sections. So, if the class strength is more, more time must be devoted for SCL.

The second survey was distributed to all the enrolled students of *Introduction to Big Data*, the MOOC. Table 2 shows the results of the student survey conducted on the effectiveness of implementing MOOCs course as part of SCL. Twenty-five students take part in the survey. Figure 12 represents the graphical representation of the results.

Questions from 1 to 6 are based on the course contents itself. 48% of the students strongly agree and 28% agree that the difficulty level of the course is relevant to their level.16% are neutral while 8% find it not relevant to the level. 20% of the respondents strongly agree and 72% agree that the course contents are good. 4% are neutral and 4% disagree to it. 36% of students strongly agree and 56% of students agree to the point that it is easy to follow the explanation in the video. 8% find it difficult to follow the video topics. Around 80% of the students agree that the language in the video is clear and 20% are neutral.28% of students strongly agree and 48% agree that the difficulty level of assignments as part of the course is relevant. 12% are neutral and 12% disagree. Around 92% of the students agree that the discussion forum helped them a lot to solve the quizzes and assignments. 4% are neutral and 4% strongly disagree it.

Questions 7 and 8 are based on the assessments. 36% of the students strongly agree and 52% agree that the time given to submit the assignment is sufficient.8% is neutral and 4% disagree. Around 80% of the students commented that the quizzes are easy to solve. 16% are neutral and 4% disagree.

Questions 9 to 11 check the overall opinion of the student about the course. 48% of the students strongly believe and 44% agree that the course will help them in the future. 8% are neutral. 44% of the students strongly agree and 48% agree that they are willing to register for the next big data course. 4% are not sure and 4% are not interested. 40%

of the students strongly agree and 56% agree that they are willing to register for another MOOCs course as part of other courses.4% disagree the statement.

The analysis of the above results is as follows: Minority of the class consists of students on academic probation. Such students find the course difficult even though they commented that the contents are good. Most of the tutors in the course are native speakers, so few students find difficulty to follow the explanation in the videos. As the assignments were bit complex, only the best and above average students can solve it easily. The percentage of students who commented that the discussion forum didn't help is negligible. The slow learners in the class find it difficult to complete the assignments on or before the due date. Also, those students who skip some videos are not able to solve the quizzes correctly. Except few minor cases, most the students are happy about the course and are willing to do similar courses in the future.

We compared the results of our survey with the findings of few research reported in the literature section. As indicated in the literature (Barba et al. 2016), the participation is a major factor in predicting the performance of the students. Also, the course was made mandatory to the students as part of their assessments. As the course progressed, the students were motivated too to complete the modules of the course. This motivation also led students to have willingness to enroll in future relevant MOOCs courses. The course was the first MOOC experience for all the enrolled students. So as reported in the literature (Pursel et al. 2016), our research also supports that prior online learning experience has no impact on this course completion. Like the findings of the researchers (Pursel et al. 2016), forum discussions helped the students in the completion of assessments. This is a very positive behavior to have interactions with other fellow students even though forum discussions are non-graded activities. Video tutorials are the main contents in Coursera courses. The quality and the design of these videos had a very positive impact on the students, like the findings of (Veletsianos et al. 2015). We also found that the quality of MOOCs courses can be improved by providing support and guidance to the tutors to make use of modern technologies as indicated by (Conole 2015) and with the help of framework proposed in the study of (Daradoumis et al. 2013).

6 Conclusion

The massive open online courses have considerable potential for growth with high quality products supported by leading universities. The MOOCs have immense social implications for access to higher education in both the advanced and developing worlds. However, they still need to resolve issues including assessment, high dropout rates, and how to maintain viability. The paper presents the author's experiences of integrating Open educational resources and Massive open online courses as part of the curriculum in the courses of Department of IT, HCT, Muscat. The survey results conducted among the staff and students shows that both the approaches are effective, and it accelerates the learning process among the students.

Acknowledgments. We are very grateful to our colleagues and students of Department of IT, Higher College of Technology, Muscat who helped in the survey of the research.

References

Gibson, K.S.: Facilitator Apprehension: Identifying, Describing, and Utilizing Relationships Observed in Educational Technology and the Visual Arts (Doctoral dissertation, Northcentral University) (2016)

Bell, S.: Project-based learning for the 21st century: skills for the future. Clearing House **83**(2), 39–43 (2010)

McGreal, R., Kinuthia, W., Marshall, S.: Open Educational Resources: Innovation, Research and Practice, Commonwealth of Learning & Athabasca University, Vancouver (2013)

Pomerol, J.-C., Epelboin, Y., Thoury, C.: MOOCs and higher education. In: MOOCs. Wiley, Hoboken (2015). https://doi.org/10.1002/9781119081364.ch5

Glogoff, S.: Instructional blogging: Promoting interactivity, student-centred learning, and peer input. Innov.: J. Online Educ. **1**(5), 3 (2005)

Gilboy, M.B., Heinerichs, S., Pazzaglia, G.: Enhancing student engagement using the flipped classroom. J. Nutr. Educ. Behav. **47**(1), 109–114 (2015)

Yadav, A., Subedi, D., Lundeberg, M.A., Bunting, C.F.: Problem-based learning: influence on students' learning in an electrical engineering course. J. Eng. Educ. **100**(2), 253–280 (2011)

Lumpkin, A., Achen, R.M., Dodd, R.K.: Student perceptions of active learning. Coll. Stud. J. **49** (1), 121–133 (2015)

Yuan, L., Powell, S., JISC CETIS: MOOCs and Open Education: Implications for Higher Education (2013)

Margaryan, A., Bianco, M., Littlejohn, A.: Instructional quality of massive open online courses (MOOCs). Comput. Educ. **80**, 77–83 (2015)

Daradoumis, T., Bassi, R., Xhafa, F., Caballé, S.: A review on massive e-learning (MOOC) design, delivery and assessment. In: 2013 Eighth International Conference on P2P, Parallel, Grid, Cloud and Internet Computing (3PGCIC), pp. 208–213. IEEE, October 2013

Conole, G.: MOOCs as disruptive technologies: strategies for enhancing the learner experience and quality of MOOCs. RED, Revista de Educación a Distancia. Número 39 (2013). http://www.um.es/ead/red/39/. 15 de diciembre de 2013

Veletsianos, G., Collier, A., Schneider, E.: Digging deeper into learners' experiences in MOOCs: participation in social networks outside of MOOCs, notetaking and contexts surrounding content consumption. Br. J. Educ. Technol. **46**(3), 570–587 (2015)

Pursel, B.K., Zhang, L., Jablokow, K.W., Choi, G.W., Velegol, D.: Understanding MOOC students: motivations and behaviours indicative of MOOC completion. J. Comput. Assist. Learn. **32**(3), 202–217 (2016)

Barba, P.D., Kennedy, G.E., Ainley, M.D.: The role of students' motivation and participation in predicting performance in a MOOC motivation and participation in MOOCs. J. Comput. Assist. Learn. **32**, 218–231 (2016)

Building a Multiple Linear Regression Model to Predict Students' Marks in a Blended Learning Environment

Vinu Sherimon[1(✉)] and Sherimon Puliprathu Cherian[2]

[1] Higher College of Technology, Muscat, Oman
vinusheri@hct.edu.om
[2] Arab Open University, Muscat, Oman
sherimon@aou.edu.om

Abstract. This research attempts to build a multiple linear regression model to predict the marks of students. As a case study, the course M359 – Relational Database offered to undergraduate students of Arab Open University, Oman is taken. The model is trained using the different assessment marks of students in blended learning mode of the above course. Separate models were built based on the gender. Same datasets were used for training and testing purposes. The open source statistical software gretl was used to build and test the model. The study found that the model generated for male category shows more correlation in the process of prediction than the female category. The findings of the research suggest that it is challenging to build a prediction model for students in blended learning environment.

Keywords: Multiple linear regression · Linear model · Marks prediction Correlation · R-Squared · Gretl · Blended learning · Predictive models

1 Introduction

Systems predicting student performance gains attraction as it helps students to know in prior about the expected marks or grade. A linear regression model is often used in predictions. This paper presents a prediction model for the course M359 – Relational Database offered to undergraduate students of Arab Open University. This course is offered to two categories of students – blended learning and regular mode students. In this study, we have considered only blended learning students. The marks of such students in different assessments were collected. Multiple Linear regression model is used to predict the performance of students. Usually, multiple regression models result in several different model designs, depending on the number of predictor variables. So, it is important to choose the best model available.

The rest of the paper is organized as follows: - Literature Review is described in Sect. 2. Section 3 explains the methodology. Multiple Linear regression model and the proposed model is described here. Section 4 includes the description of model built using gretl, the statistical software. Results are included in Sect. 5 followed by Conclusion and References.

M. E. Auer and T. Tsiatsos (Eds.): IMCL 2017, AISC 725, pp. 903–911, 2018.
https://doi.org/10.1007/978-3-319-75175-7_88

2 Literature Review

In [1] univariate linear regression model is used to predict the final grade by collecting the marks of internal exam. The model helped the students to know the marks required in internal exams to get the desired grade. Authors have implemented an SGPA prediction system to predict the SGPA of engineering students using four classification algorithms among Random Tree, J48, LMT and REP Tree [2]. The research reported that REP Tree algorithm is the best algorithm compared to the other three. In almost all educational institutions, Learning management Systems (LMS) play a vital role in enhancing the knowledge of learner. The learner's utilization of LMS was examined and an educational model is build which predicts the behavior of the learner in [3]. J48 decision tree algorithm and multiple linear regression algorithm was used in the model. The model was primarily aimed to predict the likelihood of getting a passing mark for distance education learners. Using Naïve Bayesian approach, a web-based application was developed by researchers to predict the academic performance of students based on the student's academic history [4]. The research shows that the Bayesian approach provides better accuracy when compared to other methods like neural networks, decision trees, regression, etc. Another study examined the relationship between the marks assigned by the tutors and the marks assigned by the students (self-assessment) [5]. The study revealed that students who performed well underestimated their marks and those who performed less overestimated their marks.

The performance of students in two different courses (a Math course and an IT course) were analyzed in [6] to study the difficulty level of the courses and its impact on student performance. The study proposed that classification and regression tree (CART) enhanced by AdaBoost is the best classification algorithm. A neural network model is proposed by [7] to predict the student's performance. After each lesson, the written comments of the students were collected. Then latent semantic analysis (LSA) techniques were applied to extract semantic information from the comments. In the study proposed in [8], five classification algorithms were compared to analyze and predict the performance of students. The accuracy of the classifier was improved using the bootstrap method. The programming behavior of student was studied to predict the performance of students in basic programming courses in [9].

After going through the literature, it was observed that linear regression model is best suited in our case to build a predictive model.

3 Methodology

3.1 Building Predictive Models – Multiple Linear Regression Model

Multiple Regression model is an extension of simple linear regression model. It is used if the dependent variable is based on more than one predictor variables. So, it is a many-to-one relationship. In Fig. 1, $x_1, x_2, \ldots, x_n$, are predictor variables and y is the dependent variable. The different assumptions that must be true to build a multiple linear regression model are linearity, homoscedasticity, multivariate normality, independent of errors and lack of multi-collinearity. There are certain things to be

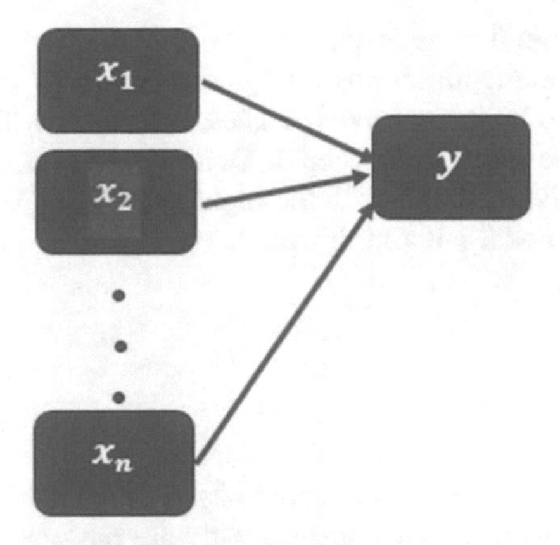

Fig. 1. Multiple linear regression

considered while building models using regression. Sometimes, all the predictor variables are not necessary to generate the best model. When the number of predictor variables increases, the problem of overfitting can occur [11]. So, the best option is to pick up the predictor variables which are relevant and necessary. When the number of predictor variable increases, relationships among them will also increase. This situation is known as multicollinearity, which means that predictor variables are also potentially related to each other [12]. So here the ideal solution is to ensure that the predictor variable(s) are correlated only to the dependent variables, not to themselves.

Consider the multiple regression model given below.

$$y = \beta_0 + \beta_1 x_1 \beta_2 x_2 + \ldots + \beta_p x_p + \in, \tag{1}$$

where β_0 is the intercept; $x_1, x_2, \ldots$ are variables; $\beta_1, \beta_2,..$ are the variable coefficients and $\in$ is the error term.

Now the multiple regression equation is

$$E(y) = \beta_0 + \beta_1 x_1 + \beta_2 x_2 + \ldots + \beta_p x_p \tag{2}$$

The error is assumed to be zero.

The estimated multiple regression equation is

$$\widehat{y} = b_0 + b_1 x_1 + b_2 x_2 + \ldots + b_p x_p, \tag{3}$$

where $\hat{y}$ is the predicted value of the dependent variable and $b_0, b_1,..$ are the estimates of $\beta_1, \beta_2,..$

3.2 Proposed Model

There are different methods to build a multiple regression model. We have used here Backward Elimination, which is the fastest method to build the model. The different steps in Backward elimination method are as follows: -

Step 1 – Select a significance level.
Step 2 – Fit the model with all possible predictions.
Step 3 – Choose the predictor having highest ‘p’ value. If this ‘p’ value is greater than the significance level, go to Step 4. Otherwise the model is ready.
Step 4 – Remove the predictor with the highest ‘p’ value.
Step 5 – Refit the model without this predictor
Step 6 – Repeat Step 3.

4 Experiment

The gretl software is used to describe, and analyze the variable data and to build the model. Gretl is a free open-source statistical software package particularly for econometrics [10]. Gretl stands for Gnu Regression, Econometrics and Time-series Library [10].

4.1 Data Set

The data from the course M359 – Relational Database – Theory & Practice during 2012–2017 were collected. The course is offered to undergraduate students of Arab Open University, Oman. There are three assessments for the course – Tutor Marked Assignment (TMA - 20%), Mid Term Assessment (MTA – 30%) and Final Exam (FE – 50%). The percentage of TMA (Tutor Mark Assignment) and MTA (Mid Term assessment) were used to predict the percentage of marks in Final Exam (FE). The marks are converted to percentages to gain more accuracy.

At the beginning, a sample of 355 student data was collected. The selected students were under the category of open learning. 17 samples were excluded due to missing data on some measures. Out of 338 training data, 203 were females and 135 males. Separate models were built for male students and female students. The complete data was used to train the model.

4.2 Model - Female Dataset

The significance level is selected as 0.05. Next the full model is fitted with both the predictors. The coefficients, standard error, t-ratio and p-value of the model is given below.

Dependent variable: Final				
	coefficient	std. error	t-ratio	P-value
const	8.12947	11.0878	0.7332	0.4643
TMA	0.287832	0.125347	2.296	0.0227**
MTA	0.273261	0.0565809	4.830	2.72e-06***

The multiple linear regression equation is $FE = 8.12947 + 0.287832 \times TMA + 0.273261 \times MTA$.

The relationships between the two predictor variables (TMA & MTA) and the dependent variable (FE) on female data set is plotted in Fig. 2. Both the p-values are less than 0.05, so they are statistically significant.

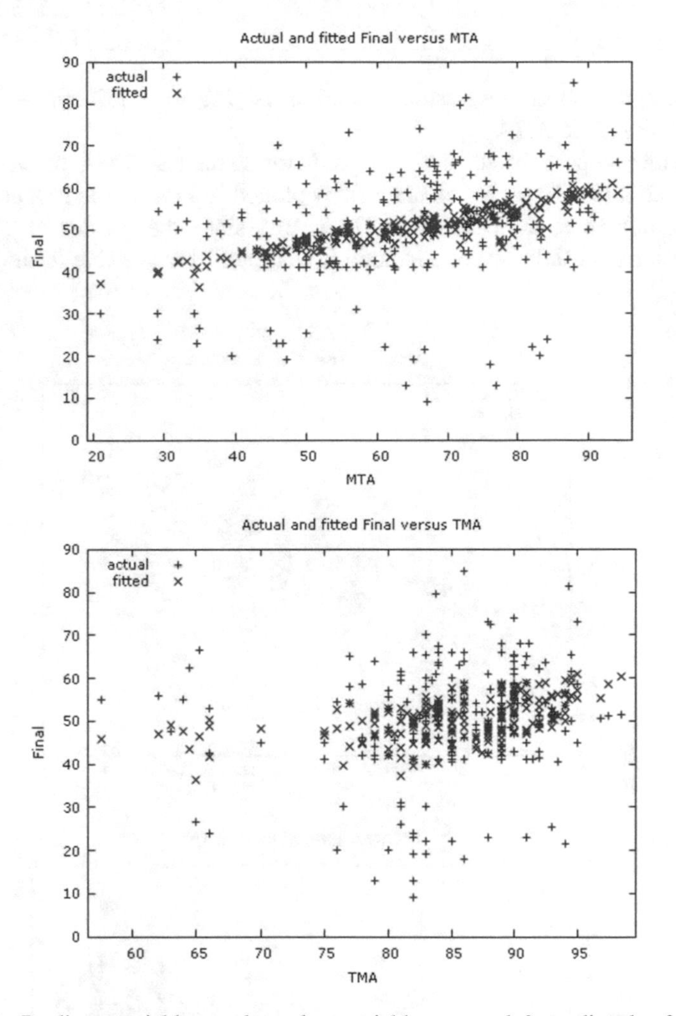

Fig. 2. Predictor variables vs dependent variables – actual & predicted – female

4.3 Model - Male Dataset

The significance level is selected as 0.05. Next the full model is fitted with both the predictors. The coefficients, standard error, t-ratio and p-value of the model is given below.

Dependent variable: Final				
	coefficient	std. error	t-ratio	p-value
const	-1.53698	13.0191	-0.1181	0.9062
TMA	0.279332	0.145254	1.923	0.0566*
MTA	0.532027	0.0652589	8.153	2.42e-013***

The multiple linear regression equation is $FE = -1.53698 + 0.279332 \times TMA + 0.532027 \times MTA$

The relationships between the two predictor variables (TMA & MTA) and the dependent variable (FE) on female data set is plotted in Fig. 3. The p-value of MTA is very less compared to the p-value of TMA. But, since the p-value of TMA is only slightly greater than 0.05, we can include it as a predictor variable in the model.

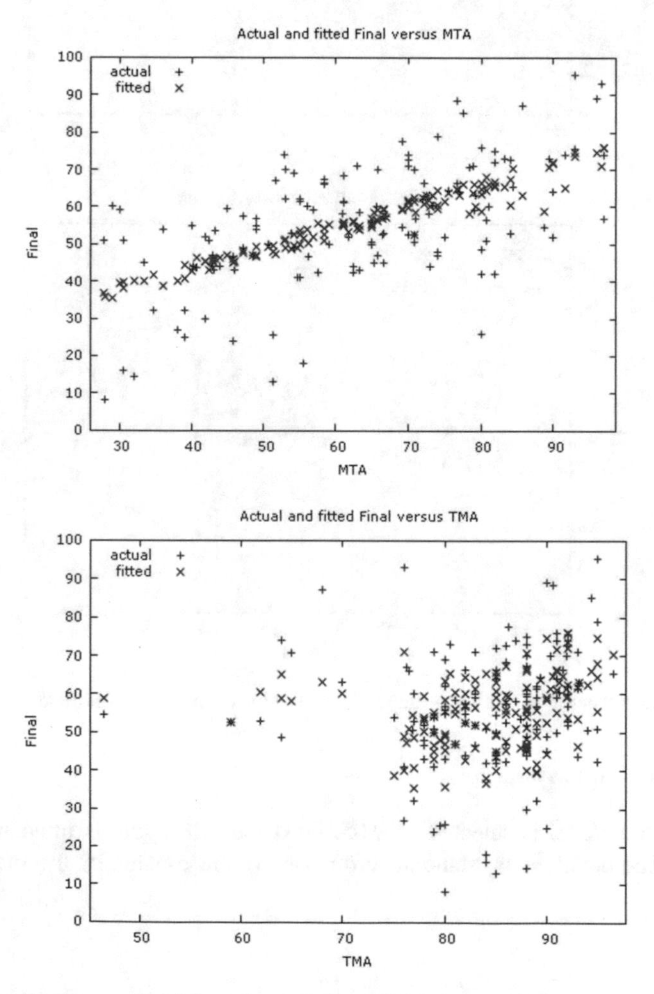

Fig. 3. Predictor variables vs dependent variables – actual & predicted – male

The summary of the models is given in Table 1.

Table 1. Model summary

Model	R	R-squared	Adjusted R-squared	Standard error of regression
Female dataset	0.36	0.131230	0.122542	12.58179
Male dataset	0.59	0.345546	0.335630	13.05753

R-squared is the coefficient of determination, which gives us the measure of how close the data is fitted to the regression line. The value of R is greater for model of male category compared to model built for females. So, we have received a better model for Male Dataset. Still, for both the categories (male & female), the value of R-squared and the adjusted R-squared is a lesser value. It means that the predictive variables don't have a high correlation to the dependent variable. The reason is that the dataset belongs to students of blended learning. Students in a blended learning environment are slow learners compared to full time students. It is found that high performers in MTA doesn't always exhibit high performance in FE. So, there exists not much linear relationship between the dependent variable and the predictor variables.

5 Results

The same data set were used to test the models using the open source software gretl. The prediction results of male students and female students are given in Figs. 4 and 5 respectively. It is observed that the results are not promising. The performance of students in open learning mode varies. Most of the students who performed well in MTA and TMA, didn't perform well in FE. So, in this case, an accurate prediction seems to be difficult.

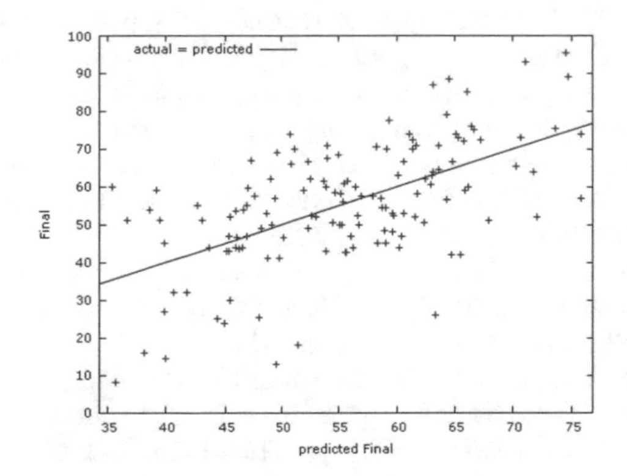

Fig. 4. Prediction results – male

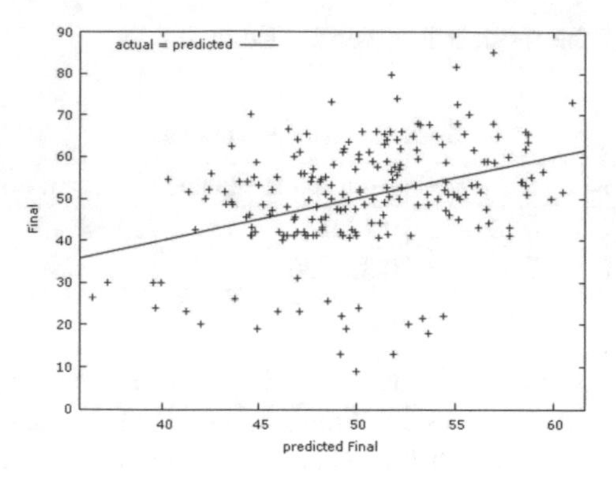

Fig. 5. Prediction results - female

6 Conclusion

This research explains the process of building a multiple regression models to predict student marks. To build the model, two predictor variables, MTA and TMA are considered. Same data set is used for training and testing. Separate models are built for male and female category of students. Even though the predicting results are not that much promising, it is observed that the model built for male students are better in predicting the FE marks when compared to the model of female students. As a future work, we would like to build models using other prediction algorithms to have a comparison with the models mentioned in this research.

References

1. Gadhavi, M., Patel, C.: Student final grade prediction based on linear regression. Indian J. Comput. Sci. Eng. **8**(3), 274–279 (2017)
2. Kaur, P., Singh, W.: Implementation of student SGPA prediction system (SSPS) using optimal selection of classification algorithm. In: 2016 International Conference on Inventive Computation Technologies (ICICT), vol. 2, p. 18, August 2016
3. Comendador, B.E.V., Rabago, L.W., Tanguilig, B.T.: An educational model based on knowledge discovery in databases (KDD) to predict learner's behavior using classification techniques. In: 2016 IEEE International Conference on Signal Processing, Communications and Computing (ICSPCC), p. 16, August 2016
4. Devasia, T., Vinushree, T.P., Hegde, V.: Prediction of students performance using educational data mining. In: 2016 International Conference on Data Mining and Advanced Computing (SAPIENCE), pp. 91–95, March 2016
5. Zvacek, S.M., de Ftima Chouzal, M., Restivo, M.T.: Accuracy of self-assessment among graduate students in mechanical engineering. In: 2015 International Conference on Interactive Collaborative Learning (ICL), pp. 1130–1133, September 2015
6. Kaur, K., Kaur, K.: Analyzing the effect of difficulty level of a course on students performance prediction using data mining. In: 2015 1st International Conference on Next Generation Computing Technologies (NGCT), pp. 756–761, September 2015

7. Sorour, S.E., Mine, T., Goda, K., Hirokawa, S.: Predicting students' grades based on free style comments data by artificial neural network. In: 2014 IEEE Frontiers in Education Conference (FIE) Proceedings, p. 19, October 2014
8. Taruna, S., Pandey, M.: An empirical analysis of classification techniques for predicting academic performance. In: 2014 IEEE International Advance Computing Conference (IACC), pp. 523–528, Febuary 2014
9. Watson, C., Li, F.W.B., Godwin, J.L.: Predicting performance in an introductory programming course by logging and analyzing student programming behavior. In: 2013 IEEE 13th International Conference on Advanced Learning Technologies, pp. 319–323, July 2013
10. http://gretl.sourceforge.net/
11. Babyak, M.A.: What you see may not be what you get: a brief, nontechnical introduction to overfitting in regression-type models. Psychosom. Med. **66**(3), 411–421 (2004)
12. Farrar, D.E., Glauber, R.R.: Multicollinearity in regression analysis: the problem revisited. Rev. Econ. Stat. **49**(1), 92–107 (1967). https://doi.org/10.2307/1937887

Real World Experiences

The Use of Educational Scenarios Using State-of-the-Art IT Technologies Such as Ubiquitous Computing, Mobile Computing and the Internet of Things as an Incentive to Choose a Scientific Career

Kalliopi Magdalinou[1,2(✉)] and Spyros Papadakis[1,3]

[1] Information and Communication Systems,
Open University of Cyprus, Nicosia, Cyprus
kmag388@gmail.com, spapad@cti.gr
[2] Greek Ministry of Education, Research and Relegius Afairs,
Thessaloniki, Greece
[3] Computer Technology Institute & Press "Diophantus", Patras, Greece

Abstract. By definition, technology seeks to exploit scientific knowledge in order to serve today's practical human needs. The promotion of IT-related technologies such as Ubiquitous Computing, Mobile Computing and the Internet of Things (UMI) is a matter of great concern to the scientific community due to their diffusion in all areas of human activities. The educational community recognizes the above technologies as a tool that can be used to improve the quality and effectiveness of the education, while, at the same time, is facing the challenge to highlight ways to utilize the technical characteristics of UMI technologies in order to lead to learning, not just the information accessibility. The presented research contains a proposal for the exploitation of innovative technologies in the field of education and presents the results of its implementation. More specifically, it is a proposal to use UMI technologies in the field of formal Secondary Education based on educational scenarios. Additionally, it is investigated if using UMI while teaching the 15–17-year-old students provides learning- not only ubiquitous access to information- and if it serves as a motivator for choosing a career in science and technology.

Keywords: Ubiquitous Computing/Mobile Computing/Internet of Things (UMI) Computational thinking · Carrier

1 Purpose

The relationship between technology and education has emerged around the 1950s. In particular, developments in computer technology have led researchers to address technology as an active student partner in order to build knowledge [14] and thus improve the quality and effectiveness of the training provided. The advocates of the use of computers and their associated technologies as a way to address educational needs

M. E. Auer and T. Tsiatsos (Eds.): IMCL 2017, AISC 725, pp. 915–923, 2018.
https://doi.org/10.1007/978-3-319-75175-7_89

and problems claim that the aforementioned support student motivation and facilitate collaborative learning, participatory intelligence and problem solving [22].

The justification for this research is based on the following observations:

- The modest level of education in the IT course, which is supported by educators internationally. In particular, the report "Running On Empty : The failure to teach K-12 computer science in the digital age" describes the situation in USA [28]. Similarly, thc report "Shut down or restart: The way forward for computing in UK schools" [23] depicts the problem in UK. In Greece, although there are no specific surveys, the findings are derived from the pupils' results in the PISA competition [21] and the statistics from the Ministry of Education, Research and Religious Affairs on the performance of students in the respective lesson "Development of Applied Applications in a Programming Environment" of the Pan-Hellenic Examinations [18].
- The need – especially as far as the young people are concerned- to acquire knowledge and cultivate skills and attitudes related to UMI technologies that will lead them to understand the real world (physical and digital), to strengthen their digital profiles and, eventually, to be activated to participate in socio-economic processes [3].
- The cultivation of Horizontal Computational Thinking skill. In order to accomplish this goal, educational researchers have proposed integrating Computational Thinking culture into curricula [29]. In fact, many countries (Russia, South Africa, New Zealand, Australia, Great Britain, etc.) aim to cultivate Algorithmic Thinking through programming [12]. According to the research conducted by the Joint Research Center (JRC) of the European Commission's Science and Knowledge Service [1], other approaches suggest the cultivation of Logical Thinking and/or Problem Solving. In Greece, Computational Thinking is not directly involved in the curricula [26].
- The favorable conditions for job creation in the area [4, 13]. Specially, raising students' interest in technology and their self-confidence in using it may affect their decision to pursue a future career in the field.
- The advantages and the reservations that have been made regarding the use of innovative IT technologies in education. The advantages include integration of educational activities in everyday life [11], ubiquitous access to digital resources [17], students' interest in mobile devices [15], customized training for trainees' needs [27], support for collaboration [11], learning in a playful way [25], making inquiry about the educational process [6] and exploiting the potential of mobile cloud computing [10]. Reservations concern the possibility of creating or widening the digital gap between students who have and those who do not have access to innovative technologies [8], pointing out that access to information does not necessarily mean learning [24], the necessity for pedagogical documentation of the educational process [16] and issues concerning the security of data and privacy of communications [19].

It has, therefore, been investigated whether the exposure of students of formal secondary education to UMI technologies during the teaching of "Informatics Applications" in the 1st grade of General Lyceum has an impact to the cultivation of

Computational Thinking, increases the students' interest in technology and affects their self-confidence in the use of technology and their interest for a future career in technology and science.

2 Approach

An educational action research was designed and implemented.

The action research was chosen as the most appropriate type of research since the aim was to intervene in an existing system (school) in order to address a specific problem (level of Computational Thinking) for the benefit of the participants (students and teachers) and to examine the impacts of such an intervene [7]. Design, action, observation and reflection – the four phases proposed for this type of research were implemented (Kemmis and McTaggart 1992). A calendar was kept by the teacher throughout the action.

2.1 Design

The research concerned teaching the course "Informatics Applications" in the First Class of General Lyceum. The curriculum and teaching instructions indirectly aim at cultivating Computational Thinking through the creation of small-scale applications.

The chosen hardware was the Arduino microcontroller and the UDOO NEO and Raspberry Pi single-board computers. These three platforms enable the development of independent interactive objects while at the same time they are credible, have good documentation, and educational targeting.

Two questionnaires were created during the planning of the action. The first one (Pre-test) was answered by the students before the action and the second one (Post-test) after it. The questionnaires traced students' attitudes towards problems and technology, their ability to solve problems, the knowledge of the terms and principles related to the operation of innovative IT technologies as well as students' interest and self-confidence in using technology and their intention to pursue a career in technology and science in the future.

Still, two educational scenarios were designed. The first concerned the introduction of students to the basic terms and principles of the operation of innovative IT technologies and their platforms. The second concerned the teaching of the conditional statement by developing a simple application on the three selected platforms. The conditional statement was chosen as it is a source of great difficulty for novice developers. Many of the difficulties are due to misunderstandings as far as the definition and manipulation of variables are concerned [20]. Elements of Social Constructivism, Activity Theory and Distributed Cognition were exploited. In particular, the Interest-Driven Creator Theory [5] has been adopted to claim that trainees are transformed into creators when a field is of interest to them and provided with appropriate technology support. Indeed, the selected framework includes a recurring cycle of events that trigger, amplify and expand trainees' interest [30]. Additionally, the Computational Thinking Development Framework proposed by Brennan & Resnick was adopted and adapted according to the needs of our research [2]. The suggested activities by the educational scenarios were student-centered and cooperative.

Finally, the calendar structure was selected. It included the fields: Action (Action Description), Start/End (Start Date and Expiration Date), Duration (Required time), Equipment (Required equipment), Tasks (Detailed description of tasks) and Observations.

2.2 Action

The implementation was completed in four phases: completion of the Pre-test (45 min), implementation of the educational scenario "Introductory Scenario - Single board microcontrollers and computers – Internet of Things" (4 h) (see Fig. 1), implementation of the educational scenario "Conditional Statement Scenario – Creating applications" (4 h) (see Figs. 2 and 3), and completion of the Post-test (30 min).

Fig. 1. "Introductory Scenario - Single board microcontrollers and computers – Internet of Things" – Preparing presentation.

Fig. 2. "Conditional Statement Scenario – Creating applications" – The application.

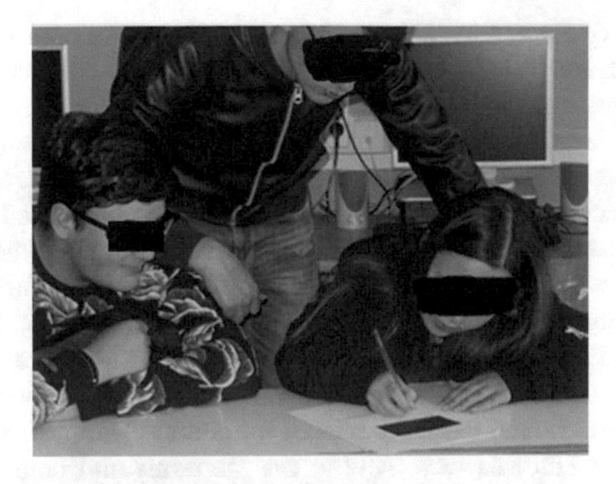

Fig. 3. "Conditional Statement Scenario – Creating applications" - Assessment process.

2.3 Observation

The observation is based on the comparison of the students' answers to Pre-test and Post-test and on the entries of the teacher's calendar.

2.4 Reflection

Reflection includes commenting on results, comparing with expected results, recording research constraints, and proposing new research issues.

One teacher and thirty-eight students participated in the event. The action took place between February 2017 and April 2017.

3 Outcomes

As already mentioned, the two questionnaires and the calendar completed by the teacher were used as criteria for the effectiveness of the action. The findings of the survey were positive and confirmed the expected results.

In particular, from the responses to the two questionnaires the following facts were observed:

- *Students' attitudes towards problems*: Before the action, students recognized the problems as part of their daily routine, but they found it difficult to formulate problems that did not belong to STEM and did not apply some phases (decomposition, pattern recognition). After the action, students said they persist in understanding a problem to a moderate degree, they analyze problems to their components, prefer to solve problems inspired by the real world, and feel modestly confident when solving complex problems.
- *Students' technology attitude (Post-action)*: Students declared a very positive attitude towards technology. They stated that computers offer learning opportunities, recognized the importance of use skills, and said they enjoy dealing with computers.

- *Students' view of the action*: The students expressed their positive experience of the action. They stated that the action increased their interest and self-confidence in using technology and resolved their questions about the applications of technology in real life.
- *Problem solving*: Before the action students were able to solve simple problems but their performance was moderate in solving complex problems and their Algorithmic Thinking was too weak. After the action, the students applied the problem solving procedures to a satisfactory degree, demonstrated good performance in solving complex problems and improved the level of their Algorithmic Thinking.
- *Terms and Principles of UMI Technology*: Prior to the action, a few students were aware of the term Internet of Things without, however, being aware of the precise meaning of the term and the principles of operation of innovative IT technologies. After the action, students were able to use the terms and prepare short technical texts to describe the UMI technology services.
- *Intention for a Future Career*: Prior to the action, students were highly interested in pursuing a career in science and technology. Following the action, the respective percentages have increased further.

From the calendar entries, the following facts were noted:

- *Design phase*: This phase lasted a total of 150 h. The process of creating the Pre-test and the Post-test was interesting. The creating process of the "Introductory Scenario - Single board microcontrollers and computers – Internet of Things" was demanding, especially for a teacher without prior experience in UMI technology. The creation process of the "Conditional Statement Scenario – Creating applications" was very demanding. Compatibility problems between Arduino and Windows 10 were identified, while UDOO NEO had unstable behavior.
- *Action Phase*: No problems were encountered when Pre-test and Post-test were completed. The two educational scenarios were implemented as planned. The students participated in the activities with enthusiasm and achieved the evaluation procedures. Students had difficulty in understanding technical texts in English, cooperating, reading and selecting material from various sources, and programming with Python and Wiring.

4 Conclusions/Recommendations

The questioning of the quality and the results of the provided education in Informatics and, more generally, in the cultivation of students' Computational Thinking, combined with the broad incorporation of innovative IT technologies (Ubiquitous Computing/Mobile Computing/Internet of Objects) in all activities of modern societies and the opportunities offered for scientific career have prompted international organizations and researchers to find ways to increase students' interest in technology.

The action research presented here consists a proposal for the educational use of UMI technologies in the context of formal Secondary Education when teaching Computer Science in 15–17 year-olds. The proposal concerns the mapping of the results of the implementation of two educational scenarios through two specially constructed questionnaires and the calendar entries that were kept by the teacher during the action.

The findings of the research were encouraging as far as knowledge and skills attitudes and at the level of attitudes and future career choices are concerned.

In particular, the ability to implement problem-solving procedures has been positively affected to a satisfactory extent by the intervention. In addition, the ability to express a solution of a problem in algorithms has been positively affected to a great extent. Indeed, students understood difficult concepts such as variables and logical expressions and wrote or modified conditional statements.

On the contrary, student's attitude towards problems was not significantly influenced by the action.

A positive attitude of students towards technology was recorded. The students declared their interest, self-confidence and understanding of the environment increased due to the intervention.

The action has had a positive impact on the understanding and ability to use terms of state-of-the-art IT technologies.

According to students' answers their interest in pursuing a career in science and technology was effected positively by the intervention.

The limitations of research include: (i) limited access to hardware, with the result that students work in groups of 6–7, (ii) the short duration of the action, and (iii) the small sample and the absence of a test group which makes it impossible to draw generalized conclusions.

In the future, it would be useful to deepen on electronic circuits by creating new educational scenarios. Thus, students will acquire the appropriate knowledge to design electronic circuits in order to achieve specific functions or to modify existing plans in order to change their function. This would increase their interest and self-confidence in using technology and could also work as an adjunct to teaching other subjects (Physics).

It would also be interesting to explore the impact of the cross-thematic approach with the use of innovative IT technologies both at the level of school performance and at the level of attitude towards technology.

A research on the contribution of developing intelligent object applications, in the context of computer science courses, aiming the cultivation of students' computational thinking is an interesting proposition. Such an action, however, requires a longer duration of the action as well as and access to additional equipment.

Finally, the implementation of similar actions on a larger scale could yield more, editable by quantitative methods of analysis, data. Furthermore, the use of a test group, so as to have a base of comparison, could be very useful.

The findings of the action research are consistent with in the findings of the Technology Acceptance Model, according to which IT users are positive about new technology when they are convinced of its usefulness and its potentials [9].

Acknowledgment. The UMI-Sci-Ed project has received funding from the European Union's Horizon 2020 research and innovation programme under grant agreement No 710583.

References

1. Bocconi, S., Chioccariello, A., Dettori, G., Ferrari, A., Engelhardt, K.: Developing computational thinking in compulsory education-implications for policy and practice (2016). https://ec.europa.eu/jrc/en/publication/eur-scientific-and-technical-research-reports/developing-computational-thinking-compulsory-education-implications-policy-and-practice. Accessed 22 May 2017
2. Brennan, K., Resnick, M.: New frameworks for studying and assessing the development of computational thinking. American Educational Research Association Meeting, Vancouver, BC, Canada, pp. 1–25 (2012). http://web.media.mit.edu/~kbrennan/files/Brennan_Resnick_AERA2012_CT.pdf. Accessed 13 Feb 2017
3. Brittain, N.: IT education remains mired in uncertainty (2011). http://www.computing.co.uk/ctg/analysis/2110259/education-remains-mired-uncertainty. Accessed 20 Apr 2017
4. Bureau of Labor Statistics. Computer and Information Technology Occupations (2017). https://www.bls.gov/ooh/computer-and-information-technology/. Accessed 07 June 2017
5. Chan, T.W., Looi, C.K., Chang, B.: The IDC theory: creation and the creation loop. In: Kojiri, T., Supnithi, T., Wang, Y., Wu, Y.-T., Ogata, H., Chen, W., Kong, S.C., Qiu, F. (eds.) Workshop Proceedings of the 23rd International Conference on Computers in Education, pp. 814–820. Asia-Pacific Society for Computers in Education, Hangzhou (2015)
6. Cheng, H.C., Liao, W.W.: Establishing an lifelong learning environment using IOT and learning analytics. In: 14th International Conference on Advanced Communication Technology (ICACT), pp. 1178–1183 (2012)
7. Cohen, L., Manion, L.: Research Methods in Education, 4th edn. Routledge, London (1994)
8. Cope, B., Kalantzis, M.: Ubiquitous learning: an agenda for educational transformation. In: Proceedings of the 6th International Conference on Networked Learning, pp. 576–582 (2008). ISBN No 978-1-86220-206-1
9. Davis, F.D.: Perceived usefulness, perceived ease of use, and user acceptance of information technology. MIS Q. **13**, 319–339 (1989)
10. Dinh, H., Lee, C., Niyato, D., Wang, P.: A survey of mobile cloud computing: architecture, applications, and approaches. Wirel. Commun. Mob. Comput. **13**, 1587–1611 (2013)
11. Geddes, S.J.: Mobile Learning in the 21st Century: Benefit for Learners. Knowl. Tree e-journal **30**(3), 214–228 (2004)
12. Grover, S., Pea, R.: Computational thinking in K – 12: a review of the state of the field. Educ. Res. **42**(1), 38–43 (2013). https://doi.org/10.3102/0013189X12463051
13. Horizon 2020. ICT Research & Innovation. https://ec.europa.eu/programmes/horizon2020/en/area/ict-research-innovation. Accessed 07 Jun 2017
14. Kozma, R.B.: Learning with media. Rev. Educ. Res. **61**(2), 179–211 (1991). https://doi.org/10.3102/00346543061002179
15. Kukulska-Hulme, A., Sharples, M., Milrad, M., Arnedillo-Sanchez, I., Vavoula, G.: Innovation in mobile learning: a european perspective. Int. J. Mob. Blended Learn. **1**(1), 13–35 (2009)
16. Lee, M.J.W., Chan, A.: Pervasive, lifestyle-integrated mobile learning for distance learners: an analysis and unexpected results from a podcasting study. Open Learn. **22**(3), 201–218 (2007)
17. Marinagi, C., Skourlas, C., Belsis, P.: Employing ubiquitous computing devices and technologies in the higher education classroom of the future. Proc. – Soc. Behav. Sci. **73**, 487–494 (2013). https://doi.org/10.1016/j.sbspro.2013.02.081

18. Ministry of Education. Research and Religious Affairs. Exams (2016). https://www.minedu.gov.gr/exetaseis-2/baseis-an/21659-16-06-16-statistika-stoixeia-panelladikon-2019. Accessed 18 Apr 2017
19. National Intelligence Council: Disruptive civil technologies – six technologies with potential impacts on US interests out to 2025. Conference report CR 2008-07 (2008). https://www.dni.gov. Accessed 24 May 2017
20. Pane, J., Myers, B.: Usability issues in the design of novice programming systems. School of Computer. Science Technical Reports. Carnegie Mellon University, CMU-CS-96-132 (1996). http://www.cs.cmu.edu/~pane/ftp/CMU-CS-96-132.pdf. Accessed 25 Mar 2017
21. PISA: PISA 2015 Results in Focus. OECD (2016). http://doi.org/10.1787/9789264266490-en
22. Roblyer, M.D.: Integrating Educational Technology into Teaching. Pearson Education Inc., Upper Saddle River (2006)
23. Royal Society: Shut down or restart: the way forward for computing in UK schools (2012). https://royalsociety.org/~/media/Royal_Society_Content/education/policy/computing-in-schools/2012-01-12-Computing-in-Schools.pdf. Accessed 07 Jun 2017
24. Simone, G.C.: Mobile learning : extreme outcomes of everywhere, anytime. In: 12th International Conference Mobile Learning 2016, pp. 139–143 (2016)
25. Smidts, M., Hordijk, R., Huizenga, J.: The world as a learning environment Playful and creative use of GPS and mobile technology in education (2008). http://www.mobieleonderwijsdiensten.nl/attachments/1765201/World_as_learningenvironment.pdf. Accessed 03 May 2017
26. Stavrianos, A., Papadakis, S.: « Εξέλιξη ορισμών της Υπολογιστικής Σκέψης και πολιτικές ενσωμάτωσής της στην Υποχρεωτική Εκπαίδευση στην Ε.Ε. » , 11ο Πανελλήνιο Συνέδριο Καθηγητών Πληροφορικής (2017)
27. Traxler, J.: Current state of mobile learning. International Review on Research in Open and Distance Learning. Athabasca University (2007). http://www.irrodl.org/index.php/irrodl/article/view/346/875. Accessed 05 May 2017
28. Wilson, C., Sudol, L.A., Stephenson, C., Stehlik, C.: Running on Empty: The failure to Teach K-12 Computer Science in the Digital Age. The Association for Computing Machinery and the Computer Science Teachers Association, New York (2010)
29. Wing, J.M.: Computational thinking. Commun. ACM **49**(3), 33–35 (2006)
30. Wong, L.H., Chan, T.W., Chen, Z.H., King, R.B., Wong, S.L.: The IDC theory: interest and the interest loop. In: Kojiri, T., Supnithi, T., Wang, Y., Wu, Y.-T., Ogata, H., Chen, W., Kong, S.C., Qiu, F. (eds.) Workshop Proceedings of the 23rd International Conference on Computers in Education, pp. 804–813. Asia-Pacific Society for Computers in Education, Hangzhou (2015)

Parental Mediation of Tablet Educational Use at Home and at School: Facilitators or Preventers?

George Palaigeorgiou(✉), Kamarina Katerina, Tharrenos Bratitsis, and Stefanos Xefteris

University of Western Macedonia, Florina, Greece
{gpalegeo,bratitsis}@uowm.gr, melenia508@gmail.com, xefteris@gmail.com

Abstract. The digital media age has dramatically transformed how children and parents perceive and react to media. Parental mediation concerns the set of strategies that parents employ in order to maximize the benefits and minimize the risks that the modern digital media induce. The majority of existing research on parental mediation is quantitative in nature, and there is a lack of in-depth understanding of not only the mediation strategies but the overall parents' stance towards the educational usage of tablets both at school and at home. This study presents the results of 54 interviews with Greek parents regarding these attitudes. Our aim was to identify the reasoning behind buying a tablet, the mediation strategies applied, parents' attitudes towards educational apps, their awareness and the sources of information about the educational apps, and whether they were ready and willing to support tablet-based learning at home and in school. Parents responses revealed that they have conflicting views on the educational value of tablets, they are misinformed or uninformed either by ignorance or by their own will and they have several concerns regarding excessive usage, access to unsupervised content and less physical activity. Parents mainly use restrictive mediation practices and, they feel outsmarted by their children. It seems that there is a growing gap between parents, children and the educational use of tablets at home and at school which needs to be addressed.

Keywords: Mobile learning · Parental mediation · Parents' attitudes
Tablets · Tablets in education

1 Introduction

The digital media age has dramatically transformed how we receive and react to media, but parental mediation theory is slow to catch up and largely remains focused on television. Parental mediation studies aim to analyze how parents interact with their children concerning media usage, for example the rules and regulations that they institute, and to identify and propose better strategies and techniques in order to inform the interested stakeholders (parents, educators and policy makers). Parents are facing everyday new challenges, as social interaction opportunities get chaotic, internet

M. E. Auer and T. Tsiatsos (Eds.): IMCL 2017, AISC 725, pp. 924–935, 2018.
https://doi.org/10.1007/978-3-319-75175-7_90

content is difficult to regulate and media device reality is being transformed continuously without the parents being able to familiarize with quickly enough [1]. The available literature on parental mediation of digital media is scarce, especially taking into account the ever increasing adoption and use of mobile technologies by the younger generations [2, 3].

As parents become more concerned and reactive in regulating their children's use of digital devices and applications at home, so has the corresponding academic research gained traction [4]. Both parents and children express concerns about the conflicts that arise at home, about the ways to negotiate and establish rules, as well as to ensure adherence to them [5]. There lies a sensitive dynamic situation with often paradoxical findings. On the one hand, this generation of children is arguably the most watched-over, with parents trying to monitor every aspect of their children's behavior, including online activities [6], with the help of technological tools [7]. On the other hand, parents tend to express that they lack control and they feel quite uncertain what kind of content their children have access to and how it affects their social lives [8].

The media-rich world has changed the parents' habits but also their fears. In studies conducted in the United States, parents expressed fears both about the incoming influences as well as themselves. Parents worry about overly violent or sexual television content [9, 10], as well as advertisement-heavy programs in television [11]. They worry that through the internet their children may become victims of predators or cyber-bullying, as well as have access to sexual content. These fears lead to an overall sense of insecurity and make them feel not ready and totally unprepared to raise children in such a media rich environment [12]. A study by Marais [13] reported on the main three obstacles parents face: (a) lack of awareness of harmful content; (b) high cost of cyber-nanny like software and (c) unfamiliarity with parental control capabilities in mobile OS's. Moreover, they emphatically noted that they feel outsmarted by their children, who are better users of both hardware and software and who find ways to bypass the imposed restrictions [14].

It is well-documented that the parents' media habits and attitudes do influence those of their offspring. The parents' own technological gadgets and their personal attitude towards technology has a significant impact on both the home environment and their children [15, 16]. There are studies showing how parents use technology in the home environment [17–19] while other studies underline the difficulties parents have at managing their own behaviors in the digital landscape [20].

The majority of existing research is quantitative in nature, and there is a lack of in-depth understanding of not only the mediation strategies but also of the parents' stance towards the educational usage of tablet devices both at school and at home. It's important to note that addressing media usage is not a self-evident process since, for example, studies have shown that restrictive strategies, are negatively associated with online risks while they also reduce the offered opportunities for the children. In this study, we try to assess the attitudes parents adopt towards the educational use of tablets for educational purposes. Are parents facilitators or preventers of the tablets' educational usage at home and at school?

2 Literature Review

Parental mediation can be defined as the set of strategies that parents employ in order to maximize the benefits and minimize the risks that the modern digital media induce [1, 21]. This term was coined in the 1980s when (in the United States) deregulation was in effect and children's television standards were low [22].

As with every strategy, the set of parental mediation practices, has its tradeoffs. As parents intervene in the effort to protect their children from "digital risks", research points out that the balance between protection, provision and participation [23] is not perfect; There is an apparent bias towards protection at the expense of the children's participation and needs in the digital media world [24]. The fear of excessive use, cyber-bullying, antisocial behavior, and exposure to inappropriate content, plays a significant role towards swaying the mediation strategies to a more restrictive context [4].

2.1 Mediation Strategies

Although research on mediation strategies has primarily been focused on television [22, 25], some recent studies have shifted the focus to video games [26–28] or internet usage [29, 30]. So, it is crucial to move the study of parental mediation strategies to a broader collection of platforms and account for synergies between media usage patterns [31]. Parental mediation strategies can be broken down into 3 main categories [32]:

- *Restrictive Mediation*: rules are enforced to limit and control children's media usage [33, 34].
- *Active Mediation*: instructive or evaluative conversations are realized in order to explain, discuss and share critical comments in a jargon that children can understand [33, 34].
- *Co-use* (or co-viewing or co-playing): parents engage into shared activities based on common interests which trigger discussions about content and usage [25, 28].

Apart from these three categories, newer research defines and describes two additional models of parental mediation strategies [35, 36]:

- *Participatory learning*: Parents and children learn together about digital media while using them.
- *Distant mediation*: Parents supervise children's media usage from a distance, based on either explicit trust that grants a certain degree of responsibility, or by allowing the children to use digital media independently while keeping an eye from a distance.

2.2 Factors that Influence Mediation Strategies

The way parents perceive digital technologies as entertainment and educational tools, influences the technologies available at home and the accessibility rules for their children. These perceptions are the foundations on which parental mediation strategies are developed, with restrictive strategies being the most popular. In restrictive strategies, parents act as "gatekeepers" [23]. A study in Singapore [35] indicated that parental mediation strategies vary and do not fall into clearly separated activities, but depend upon factors such as children's personality and behavior, gaming activities and preferences as well as the parents' attitude towards parenting practices, lifestyle constraints and use of digital content (in that case, video games). For example, in the same study, gatekeeping activities (restrictive mediation), were used in conjunction with discursive activities, in order to explain the reasoning behind the imposed rules.

Other factors that influence the decisions of mediation strategies have been found to be the country of residence, the socioeconomic status, as well as demographic data [37]. Evidence from a European study [38] indicate that children from more privileged families receive more active mediation than children of poorer families while, for example, parents in the Netherlands are more actively involved in mediation than parents in Estonia who preferably delegate mediation to older children or pose more restrictions. Mediation practices are also influenced by demographics such as the parents' education level [38–40] and digital literacy, with less digitally educated parents preferring more restrictive actions [38]. The child's gender was also found to be of importance, with girls being more actively monitored and restricted than boys, and mothers on the other end being more supportive and communicative than fathers [21, 29, 41].

3 Methodology

3.1 Study Aims

This study aimed at exploring parents' attitudes towards the educational exploitation of tablets at home and at school in Greece. We wanted to identify the reasoning behind buying a tablet, the mediation strategies applied and their effectiveness, parents' attitudes towards educational apps, their awareness and the sources of information about educational apps, the actual educational use of the tablets and whether they were ready to provide their tablets for use in school. We were not focused only to the mediation strategies, in order to identify the whole context of their behavior, whether it was consistent or not, whether parents are ready to promote an effective and efficient educational use of tablets or not.

3.2 Participants

We recruited a total of fifty-four (54) parents. Parents were randomly selected without taking into account specific demographics or characteristics. The only eligibility criterion was the ownership of tablets and having children attending one of the six classes of the primary school. Parents were from 3 different cities of Northern Greece and the majority of their children used tablets daily for at least half an hour (89%).

3.3 Data Collection Process

Parents were randomly selected and were approached when waiting for their children in places with extracurricular activities such as foreign language schools or gyms. They were asked to complete a short questionnaire and afterwards to participate in a semi-formal interview. Several visits in different places were required in order to conduct the 54 interviews.

3.4 Data Collection Instruments

The questionnaire was consisted of six 4-point Likert questions which aimed to quickly identify the perceived educational value of tablets at home and at school (e.g. Do you think that your child's work with the tablet can improve his/her performance at school? Do you think that schools should encourage more the use of tablets during the educational process?).

The semiformal interview was consisted of questions such as:

- Why do you own one?
- Does your child use it for educational purposes?
- Do you download/purchase educational apps? How are you being informed for the relevant apps? How do you decide to get one? Are they useful?
- Which are the benefits, disadvantages and your concerns of tablets' usage for children?
- How do you mediate the tablet usage? Does your child adhere to your rules?
- Would you let your child go to the school with your tablet if the teacher asked for it?

4 Results

4.1 Questionnaire Results

As shown in Table 1, parents' responses exposed their overall negative assessment of tablets' educational value. The majority of the parents stated that tablets hadn't affected in anyway their child's performance at school and that they cannot do that. Moreover,

Table 1. Short questionnaire on parents' attitude towards tablets for education

Question		*AVG*	*SD*
Do you think that your child's use of the tablet can improve his performance at school?		2.20	0.81
Do you think that your child's school performance has improved since using the tablet?		1.94	0.81
Do you consider tablets as useful educational tools?		2.35	0.83
Do you think that schools should encourage more the use of tablets in the educational process?		1.91	0.96
Do you think that the parents' use of new technologies affects the way children interact with new technologies?		2.83	0.50
Which is the most appropriate age for introducing tablets to children?	*Preschool* 6%	*Primary Sch.* 57%	*High Sch.* 37%

they were strongly opposite in letting schools promote the use of tablets in the educational process. Although, all parents had ensured tablet access for their primary school children, 37% of them, believed that the best age to introduce tablets was at high school, and in that way, they underlined their negations with their current status. This short questionnaire worked as an excellent introduction to the interview.

4.2 Interviews with Parents

Parents consider tablets as elaborate toys or buy them for social reasons
Parents mainly consider tablets as elaborate toys and entertainment gadgets. Most of them (22–41%) indicated that they purchased tablets for entertaining reasons for their children. It is interesting that the second reason for buying a tablet was social pressure (10–19%) since they stated that their children were the only ones not having a tablet.

"We had to buy it because they were too envious of their friends who had a tablet."
"He had been persistently asking for a tablet because she saw other friends and classmates to have. Although she is at a young age, and we always told her to wait patiently and when it was the time we will get her one..."

Hence, tablets were bought in many cases without having a specific need in mind. Only in 9 cases (17%), parents said explicitly that they bought it to help children for school assignments while 8 (16%) parents bought tablets for themselves and children started using them later.

Main concerns
Parents' main concerns regarding tablet usage can be organized in the following categories ordered by frequency of reference:

– excessive usage, addiction, nervousness:

"They are dealing with this too many hours, they read, eat, and fall asleep with it. They are incredibly stuck with it."
"...sometimes I see him more nervous when I go to get it. I am very concerned that he becomes more nervous."

– access to unsupervised content:

"I am concerned about various ads on various apps and in general the content of the internet that you cannot control; even on YouTube where you may get the child to see some kid's stuff, he may discovery other content which is inappropriate for him"

– games content and less physical activity:

"I'm often worried about some games so wild with puns and violence."
"I have noticed that when my son goes to play with the other neighbors, instead of taking a ball to play football or something else, all the children together with my son get their tablets and play with them. Instead of playing something else since they are essentially outdoors, they prefer to play with the tablet."

– they feel outsmarted by their children:

"I am concerned that they know how to use tablets better than us, they learn it too easily, I cannot, I have to spend time and think."
"I am surprised that the younger generation in general, from a very early age, from the elementary education not to say in kindergarten, handle it very easily, they know how to download apps. They can do all that faster and much easier than we."

Mediation strategies
When asked about how they mediate the use of tablets on the grounds of their most significant concerns, 38 parents claimed that have found a way of controlling tablets usage, as seen in the Table 2. These findings seemed not to diverge from findings of other studies.

Table 2. How parents mediate the use of tablets at home

Mediation	Parents	Perc. (%)
I set time limits	8	21.05%
I increase my control	8	21.05%
I try to limit the use/forbid the use of tablets	7	18.4%
With fighting	6	15.8%
Through dialogue	5	13.2%
I urge them to go out with friends	4	10.5%
Total	38	100%

The prevalent strategies among most households are restrictive, be it firm time limits, forbidding or even fighting. Dialogue as a mediation technique seems to come at the last place, while some parents prefer an indirect route, that of suggesting their children to increase their socialization, to go outside and play with friends.

"Time is the basic rule I think, because kids are now uncontrolled if you let them use it without restrictions."
"I have set rules in the first-place; they can see specific things on specific sites, which I have already checked."
"In general, I try to watch what she visits, what she does when she has it, and more generally I try to control it as much as I can. I know that as soon as she gets it in her hands, I have to be alerted."
"Now there is no other solution, I take it and I hide it and just forbid its use."
"We face it in a calm way always and through discussion."
"I am suggesting them to do something else to get away from it, for example now that the weather is good I urge them to go out to play football or something else."

Thirty four out of thirty-eight parents have mentioned time limits and stated that this is the main negotiating issue with their children. Interestingly, 17 parents reported that their rules are not applied every time or at all and their application depends on their own mood and their firm monitoring or not of the rules.

Attitudes towards educational use of tablets

Thirty five (65%) parents, in opposition with their initial assessment about tablets' educational value in the short questionnaire, supported that they use tablets for educational reasons and considered them as:

- Complementary instruments for practicing school knowledge (19)
- Sources of encyclopedic knowledge and information (13)
- Motivating learning factors due to their multimedia and game-based nature (7)
- Tools for developing critical thinking and improving perception (3)
- Aids for supporting school curriculum (2)
- Means for getting acquainted with technology (1)

Only 11 parents (20%) have downloaded educational apps in their tablets with focus mainly on mathematics, foreign language learning, and general encyclopedic-memory educational games. In half of these cases, parents decided about the apps to be installed while the children decided for the other half. Only in one case, apps were installed because a teacher asked them to do so. The conflicting three views are depicted in the following quotes:

"We, of course, choose the educational applications. I would not let my children to go and download whatever they want."
"Usually my child downloads an app, most of the times she has seen it from somewhere else, then I check it too."
"The English teacher has selected the apps we have installed on the tablet. We trust the teacher, she obviously knows more things than we as parents."

A reasonable question arises about how parents are being informed about educational apps. Only seven parents (13%) stated that they try to be up to date with the available educational apps by asking other parents, or searching the internet or inquiring teachers. Interestingly, there were parents who explicitly stated that they do not want to learn about educational apps, or do not believe that they need to.

"No, we have not been informed and I do not ask, generally I am not trying to learn from someone."
"I do not think I need any information right now."
"No, I am not getting informed about which apps are appropriate for my child, I would not want any further info since the child does it by himself."

On the other hand, 35% of the parents seem reluctant to identify tablets as important educational tools, stating as reasons either their ignorance of tablets' educational potential or their insistence on using them solely as gaming/entertainment platforms.

"No, we don't use it for educational purposes. I don't even know if there are 'educational purposes' at this age. It is purely used for entertainment."
"We don't want her to associate tablet use with learning, I personally prefer that she associates it with entertainment."

In the same context, some even more conservative objections were raised:

"Educational applications can help, but knowledge acquisition should not begin from there, since they [tablets] serve a more complementary and supplementary purpose [to the main

learning process]. When he learns something new he has to learn it in the traditional way. The tablet and its apps should function as means of revision and practice."

In any case, it is easy to recognize from parents' quotes, that the educational exploitation of tablets is far from being optimal with many parents adopting a non-constructive attitude.

Your tablet in school? No way!
While several parents do use tablets at home for educational purposes, when they were asked about using tablets inside the school environment, they demonstrated strong objections against the "intrusion" of tablets inside schools (see Table 3) as evident in [41]. This objection perseveres even in the case the teachers themselves ask for the tablets in the classroom.

"I personally do not approve school lessons that use tablets. I believe that books and the dialogue between the teacher and the student are much better tools. I don't understand how it can be used in the classroom. Children just have a tablet in front of them and that's it? ...my personal opinion is that books are better."

"I do not think that is the appropriate stimulus for a child at school. There are other tools e. g with a projector and generally other more traditional means to create something just as nice."
"I do not think that tablets have a valid reason of being used at schools. Schools employ teachers. That's a teacher's job, to teach the kids and provide information, while we have the tablet at home in case we need to find more."
"I do not think the tablet is such a good educational tool because teachers are at school that can in word and deed teach children."

These findings are consistent with parents' attitudes towards the educational potential of tablets, as emerged from the closed-type questionnaire. It is apparent that parental mediation can become an impediment for the educational exploitation of tablets, as parents state their preference on what they perceive as "traditional" methods of teaching. They formulate negative attitudes towards the "technological invasion" that threatens to overthrow the human element of the teacher and the familiarity of the school textbook.

Table 3. Parents' reactions toward the use of tablets at school

Answer	*Parents*	*Perc.(%)*
No, I would not give it even if the teacher asked for	30	**55.6%**
Yes, I would give it if the teacher asked	18	33.3%
Yes, if there was no other choice	4	7.4%
Not my own tablet. If the school had its own, I'd have no objection	2	3.7%
Total	54	100%

5 Discussion

Parents have contradictory views on the educational value of tablets, they are misinformed or uninformed either by ignorance or by their own will, they use primitive ways of controlling something that they do not understand so well as their children. In such an occasion, restrictive mediation is an expected stereotype reaction. Parents acknowledge that tablets are helpful in the learning process and they don't object to their children using them, while at the same time they make strong objections against their use in the school environment, underlining their preference to "traditional" teaching methods. Parents consider tablets more as a secondary means of finding information, or of providing practice opportunities. Most parents seem to function as "preventers" or at least as "non-facilitators" of the educational exploitation of tablets.

But this study did not aim to expose or accuse parents but to highlight a growing gap that exists between parents, children education and new media. Why should we expect parents to have a different view on tablets? How should they know every single new educational app or where to learn from? Why should they know how to evaluate, for example, the fit of an app about fractions to their children preferences and knowledge level? This gap, hence, must be perceived as a gap between policy makers, researchers, teachers and parents which is vital to close. Probably the most important link in this chain is teachers. Teachers can and should overcome the excessive engagement with "safe internet" discussions and help parents to exploit the opportunities that are presented with the mobile devices while enabling children to become learners in their own self-paced personalized environment. There is an apparent bias towards protection at the expense of the children's participation and needs in the digital media world.

This study has several limitations such as the sample size, sample representativeness and the results should be contextualized according to the factors mentioned in the literature review that influence parents' mediation such as country, children gender etc. More studies are needed to qualify and describe the gap between parents, children and new media for educational use, in order to propose effective interventions that may reduce it.

References

1. Zaman, B., Nouwen, M., Vanattenhoven, J., et al.: A qualitative inquiry into the contextualized parental mediation practices of young children's digital media use at home. J. Broadcast. Electron. Media **60**, 1–22 (2016)
2. Chaudron, S., Beutel, M., Navarrete, V.D., Dreier, M.: Young children (0-8) and digital technology: a qualitative exploratory study across seven countries (2015)
3. Plowman, L., McPake, J.: Seven myths about young children and technology. Child. Educ. **89**, 27–33 (2013)
4. Nouwen, M., Jafarinaimi, N., Zaman, B.: Parental controls: reimagining technologies for parent-child interaction. Reports of the European Society for Socially Embedded Technologies (2017). https://doi.org/10.18420/ecscw2017-28
5. Ko, M., Choi, S., Yang, S., et al.: FamiLync: facilitating participatory parental mediation of adolescents' smartphone use. In: Proceedings of the ACM International Joint Conference on Pervasive and Ubiquitous Computing (2015)
6. Strausss, W., Howe, N.: Millennials Rising. The Next Great Generation (2000)
7. Herring, S.: Questioning the generational divide: technological exoticism and adult constructions of online youth identity. Youth, identity, and digital media (2008)

8. Blackwell, L., Gardiner, E., Schoenebeck, S.: Managing expectations: technology tensions among parents and teens. In: Proceedings of the 19th ACM Conference on Computer-Supported Cooperative Work & Social Computing (2016)
9. Connell, S.L., Lauricella, A.R., Wartella, E.: Parental co-use of media technology with their young children in the USA. J. Child. Media **9**, 5–21 (2015). https://doi.org/10.1080/17482798.2015.997440
10. Schooler, D., Kim, J., Sorsoli, L.: Setting rules or sitting down: parental mediation of television consumption and adolescent self-esteem, body image, and sexuality. Sex. Res. Soc. Policy **3**, 49–62 (2006)
11. Oates, C., Newman, N., Tziortzi, A.: Parents' beliefs about, and attitudes towards, marketing to children. In: Advertising to Children, pp. 115–136. Palgrave Macmillan, London (2014)
12. Yardi, S., Bruckman, A.: Social and technical challenges in parenting teens' social media use. In: Proceedings of the SIGCHI Conference on Human Factors in Computing Systems, pp. 3237–3246. ACM (2011)
13. Marais, J.: A framework for parental control of mobile devices in South Africa. Nelson Mandela Metropolitan University (2013)
14. Mascheroni, G., Ólafsson, K., Cuman, A., et al.: Mobile internet access and use among European children: initial findings of the Net Children Go Mobile project (2013)
15. Lauricella, A.R., Wartella, E., Rideout, V.J.: Young children's screen time: the complex role of parent and child factors. J. Appl. Dev. Psychol. **36**, 11–17 (2015). https://doi.org/10.1016/j.appdev.2014.12.001
16. Bleakley, A., Piotrowski, J.T., Hennessy, M., Jordan, A.: Predictors of parents' intention to limit children's television viewing. J. Public Health **35**, 525–532 (2013). https://doi.org/10.1093/pubmed/fds104
17. Ammari, T., Kumar, P., Lampe, C., Schoenebeck, S.: Managing children's online identities: how parents decide what to disclose about their children online. In: Proceedings of the 33rd Annual ACM Conference on Human Factors in Computing Systems, pp. 1895–1904. ACM (2015)
18. Ammari, T., Schoenebeck, S.: Networked empowerment on Facebook groups for parents of children with special needs. In: Proceedings of the 33rd Annual ACM, pp. 2805–2814. ACM (2015)
19. Ammari, T., Schoenebeck, S.: Understanding and supporting fathers and fatherhood on social media sites. In: Proceedings of the 33rd Annual ACM, pp. 1905–1914 (2015)
20. Hiniker, A., Sobel, K., Suh, H., et al.: Texting while parenting: how adults use mobile phones while caring for children at the playground. In: Proceedings of the 33rd ACM, pp. 727–736 (2015)
21. Kirwil, L.: Parental mediation of children's internet use in different European countries. J. Child. Media **3**, 394–409 (2009)
22. Mendoza, K.: Surveying parental mediation: connections, challenges and questions for media literacy. J. Media Lit. Educ. **1**(1), 3 (2009)
23. Dias, P., Brito, R., Ribbens, W., Daniela, L.: The role of parents in the engagement of young children with digital technologies: exploring tensions between rights of access and protection, from "Gatekeepers" to "Scaffolders". Global Stud. Child. **6**, 414–427 (2016)
24. Livingstone, S.: Children's digital rights: a priority (2014)
25. Valkenburg, P., Krcmar, M., Peeters, A.: Developing a scale to assess three styles of television mediation: "instructive mediation", "restrictive mediation", and "social coviewing". J. Broadcast. Electron. Media **43**, 52–66 (1999)
26. Nikken, P., Jansz, J.: Parents' interest in videogame ratings and content descriptors in relation to game mediation. Eur. J. Commun. **22**, 315–336 (2007)

27. Nikken, P., Jansz, J.: Developing scales to measure parental mediation of young children's internet use. Learn. Media Technol. **39**, 250–266 (2014)
28. Nikken, P., Jansz, J.: Parental mediation of children's videogame playing: a comparison of the reports by parents and children. Learn. Media Technol. **31**, 181–202 (2006)
29. Eastin, Matthew S., Greenberg, B.S., Hofschire, L.: Parenting the internet. J. Commun. **56**, 486–504 (2006)
30. Livingstone, S., Helsper, E.: Parental mediation of children's internet use. J. Broadcast. Electron. **52**, 581–599 (2008)
31. Rodino-Colocino, M.: Domesticity and new media. In: Berker, T., Hartmann, M., Punie, Y., Ward, K. (eds.) Domestication of Media and Technology. Open University Press, New York (2007). New Media Soc. **9**, 364–371
32. Symons, K., Ponnet, K., Walrave, M., Heirman, W.: A qualitative study into parental mediation of adolescents' internet use. Comput. Hum. Behav. **73**, 423–432 (2017)
33. Gentile, D., Nathanson, A., Rasmussen, E.: Do you see what I see? Parent and child reports of parental monitoring of media. Fam. Relat. **61**, 470–487 (2012)
34. Warren, R.: In words and deeds: parental involvement and mediation of children's television viewing. J. Fam. Commun. **1**, 211–231 (2001)
35. Jiow, H.J., Lim, S.S., Lin, J.: Level up! Refreshing parental mediation theory for our digital media landscape. Commun. Theory **27**, 309–328 (2017)
36. Zaman, B., Nouwen, M., Vanattenhoven, J.: A qualitative inquiry into the contextualized parental mediation practices of young children's digital media use at home. J. Broadcast. Electron. Media **60**, 1–22 (2016)
37. Holloway, D., Green, L., Livingstone, S.: Zero to eight: young children and their internet use (2013)
38. Garmendia, M., Garitaonandia, C.: The effectiveness of parental mediation. In: Children, Risk and Safety on the Internet: Research and Policy Challenges in Comparative Perspective, pp. 231–244 (2012)
39. Shin, W., Huh, J.: Parental mediation of teenagers' video game playing: antecedents and consequences. New Media Soc. (2011)
40. Clark, L.: Parental mediation theory for the digital age. Commun. Theory **21**, 323–343 (2011)
41. Valcke, M., Bonte, S., De Wever, B., Rots, I.: Internet parenting styles and the impact on Inter-net use of primary school children. Comput. Educ. **55**, 454–464 (2010)

Proposal for a Mapping Mechanism Between an E-Learning Platform Users and WebRTC

Cheikhane Seyed(✉), Samuel Ouya, and Jeanne Roux Ngo Bilong

Department Computer Engineering, University Cheikh Anta Diop, Dakar, Senegal
ch.hamod@gmail.com, samuel.ouya@uvs.edu.sn, jeanneroux@yahoo.fr

Abstract. Integrating WebRTC technology into web application platforms is often a challenge for developers because they are forced to implement a solution for mapping users at both the signaling side and the application server side. This paper proposes a plug-in to map the users of an e-learning platform and WEBRTC. It's dedicated to developers to enable them solve the complexity associated with signaling WebRTC, thus offering a convenient user-management system. Indeed, we have implemented an algorithm to merge a user of the e-learning platform and its corresponding signaling side (WebRTC) into a logical structure. This plug-in provides a mechanism that requires a page view of exchanges (videoconferencing, chatting, screen sharing or file transfer) at the level of the web platform. This allows to maintain exchange in real time between the users' side web application and the server WebRTC. To do this, it was necessary to use the WebSocket API to manage bidirectional communication between the web server and the signaling server. This plug-in has been successfully integrated into an e-learning platform based on WebRTC technology.

Keywords: WebRTC technology · User mapping · Web application
Signaling server

1 Introduction

The web is an ever-expanding universe, made up of several components allowing to overflow multimedia streams (voice, video) in an interactive world. This is made possible by the interaction between web technologies and browsers. It is in this light that WebRTC technology not only provides real-time applications, but also facilitates access to these applications because it natively uses web browsers [1].

Several authors [2–4] advocated the use of WebRTC technology in distance learning to provide an easy way to communicate and exchange in real time via web platforms.

These WebRTC applications incorporate a signaling server [5] which manages WebRTC users in the form of a unique identifier regardless of the web application system. Hence the interest of solving the problem of user management complexity via the backend interface. This allows developers to manage users without any control over the signaling mechanism.

M. E. Auer and T. Tsiatsos (Eds.): IMCL 2017, AISC 725, pp. 936–943, 2018.
https://doi.org/10.1007/978-3-319-75175-7_91

The detailed mechanism in this paper proposes a mapping algorithm allowing to combine two physical users (web user and signaling user) in order to obtain a logical user. It integrates a service controller allowing the logical user to simultaneously manage security, web platform management and services offered by WebRTC technology.

This paper is organized as follows: Sect. 2 presents related work on signaling based on the WebRTC and the working environment, Sect. 3 describes the environment of the proposed mechanism, Sect. 4 focuses on the implementation, Sect. 5 presents the results and finally Sect. 6 presents the conclusion and future work.

2 Related Work

2.1 WebRTC Technology

WebRTC is an open source project presented by Google in 2011 [6], which allows real-time communications via a JavaScript API. Point-to-point interactive communications and provides synchronous data exchange [7, 8]. The purpose of WebRTC is to enable communications (audio, video, or written), file transfer, screen sharing, and remote control of the computer. The main WebRTC API components defined by the World Wide Web Consortium (W3C) and Internet Engineering Task Force (IETF) are: [11]:

- MediaStream: allows a browser to access the camera and microphone;
- RTCPeerConnection: Activates audio and video calls;
- RTCDataChannel: allows browsers to send data in a peer-to-peer connection.

2.2 Working Environment

According to the author [9], this university received in 2015 more than ten thousand students wishing to have a degree in trade. These students are young graduates who have no experience in the field of ICT hence the interest of offering them an easy teaching requiring less skills in the field of technology.

Several peers have shown the relevance in terms of security [11] and speed [12] of distance learning systems based on WebRTC technology.

Research carried out by the authors [8, 9] led to the setting up and implementation of e-learning systems via web platforms based on WebRTC technology. These solutions not only provide learners with a way to communicate and exchange in real time through a signaling system, but also facilitate access and use of such a system considering their relatively low level in ICTs.

Difficulties have been reported in the management of users during the implementation of these systems. The latter require both technical and technological skills in the field of real-time communication, in particular the signaling part in WebRTC. This led us to work on the implementation of techniques enabling developers of these systems to concentrate on the implementation of services than on the WebRTC technology backend-related complexities.

Our approach is to design a plugin to make it easier for developers to implement e-learning systems based on WebRTC technology and to provide easy management for users.

3 Description of the Environment of the Proposed Mechanism

3.1 Architecture of the Proposed Mechanism

The proposed architecture (Fig. 1) consists of several components (web server, signaling server, e-learning user, WebRTC user, database server) involved in the design and execution of the mapping mechanism.

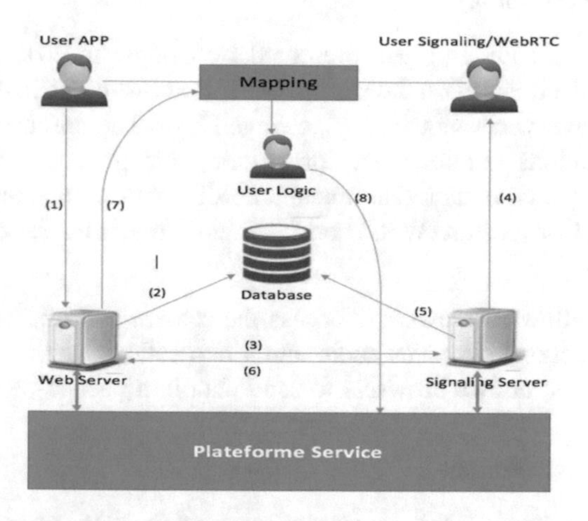

Fig. 1. Architecture of the proposed mechanism

Below, the description of the above architecture:

(1) Creating the platform web user from a sign-up panel.
(2) Adding the user to the database.
(3) Launch offing the signaling server referenced by its IP address and its port through the HTTP protocol.
(4) Generating a new Signaling/WebRTC user from the signaling server.
(5) Insertion by the signaling server of the user generated in the database.
(6) Sending a callback to the web server to confirm the addition of the signaling user in the database.
(7) Launch the API mapping by the web server by passing the information of these two users created in the database.
(8) Transformation of the two users of the system by the API mapping, into a logical user combining the information that allow him to react on all the services of the platform (web services and services signaling).

3.2 Application Fields

We have chosen the e-learning platform designed and developed by the author [9] as the fields of application and the working environment. We have integrated the mechanism in the form of a plugin in order to show the feasibility and usefulness of our approach.

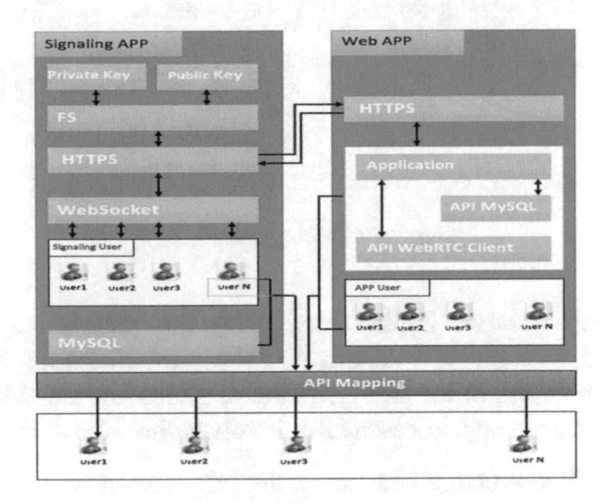

Fig. 2. Component of the study environment

As shown on Fig. 2, the studied system has two platforms (which run on two different servers). The Api mapping intervenes to fulfill the following tasks:

- Retrieve the list of e-learning users from the web platform stored in the database.
- Wait for a user to choose a real-time exchange view on the web platform (video-conference, Chat, screen sharing or file transfer etc.).
- Initiate connection and interaction with the WebRTC server (it uses the HTTP protocol to connect to the signaling server (WebRTC) and the WebSocket protocol to manage bidirectional communication).
- Retrieve the signaling user generated by the WebRTC server with its identifier and merge it with the web user (by launching the page view of the web platform) to make it a logical user.
- Insert this combination of two different users into the database.

This process is repeated for any e-learning user wishing to exchange in real time with another user.

4 Implementation

The implementation of the proposed mechanism consists of two major phases: users mapping and service controller. Figure 3 shows the interaction between the components of these different phases. These components are interfaces that run on two sep-

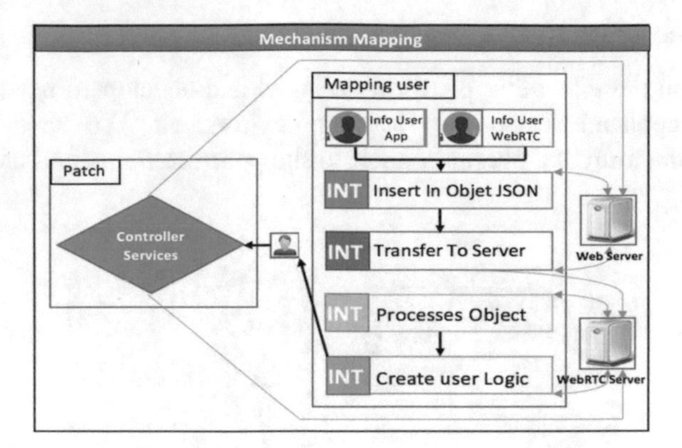

Fig. 3. Mapping interfaces

arate servers (signaling server implemented using the NodeJs language and a web server implemented with the PHP language). These interfaces describe the mapping scenario between the users of the web platform and those of the WEBRTC.

The process of user mapping essentially involves the following points:

1. Creation of a JSON structure containing the information of the connected session (the web user), the address and port of the web server.
2. Transfer of the JSON structure to the signaling server using a WebRTC client method named sendToServer via the Web Socket API.
3. Reception by the signaling server of the JSON object via the WRtcMsg function and division of the object into fields (attributes).
4. The signaling server on its part connects to the database using the MySQLSignaling API and affects the WebRTC user stored in the easyrtcid function the user whose login and password have just been received in a JSON object.
5. Receive a return callback message to inform the successful merge of the information coming from the signaling server (unique identifier) and the web application server (login, password, first name, last name).

5 Results

We implemented this mechanism and integrated it into an e-learning platform (discussed in section B of point III) to show the feasibility of our approach in terms of conversion of a user's identifier at the signaling server. Figure 4 shows the signaling server managing the users in a non-authentic manner. In this figure, the teacher whose

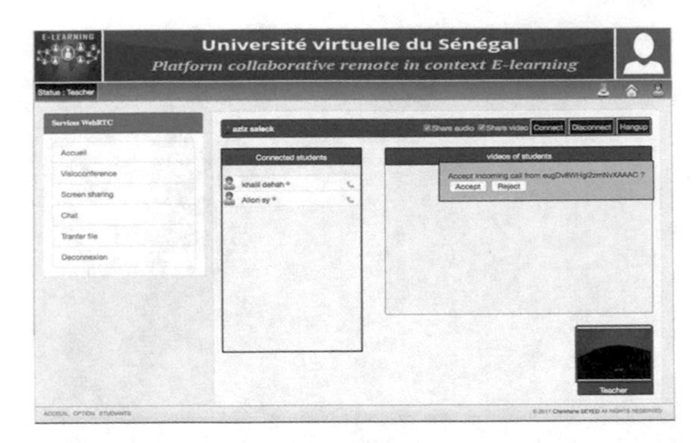

Fig. 4. User not yet mapped

name is aziz salek receives a call from one of his students connected (eugDv6WHg2zm NyXAAAC?) without knowing their identity.

After installing the plugin described on the above mechanism, we show the traces of the results at the signaling server. These traces describe the result of the interfaces implemented, to map the users.

Figure 5 shows the mapping traces on the signaling server between an e-learning user (Teacher) and his WebRTC correspondent.

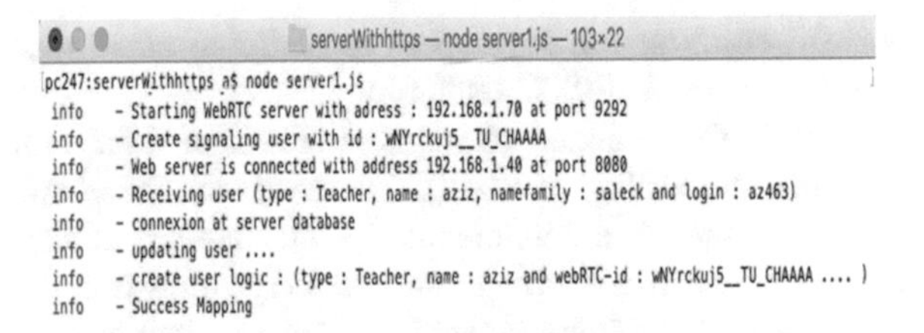

Fig. 5. User mapping (Teacher) with WebRTC correspondent

Figure 6 shows the complete list of e-learning users, containing the two mapped users with their WebRTC correspondents.

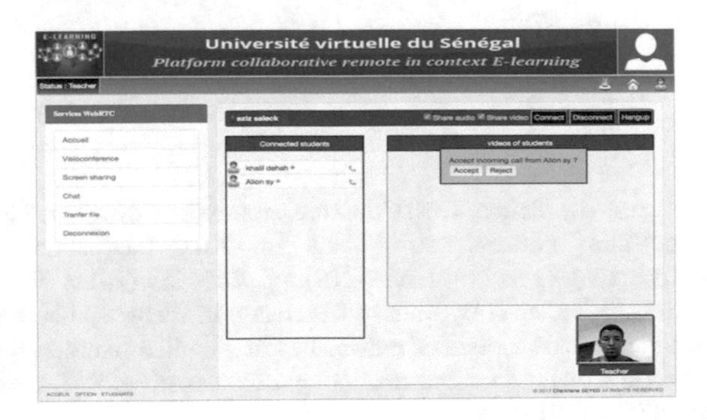

Fig. 6. Authenticated call

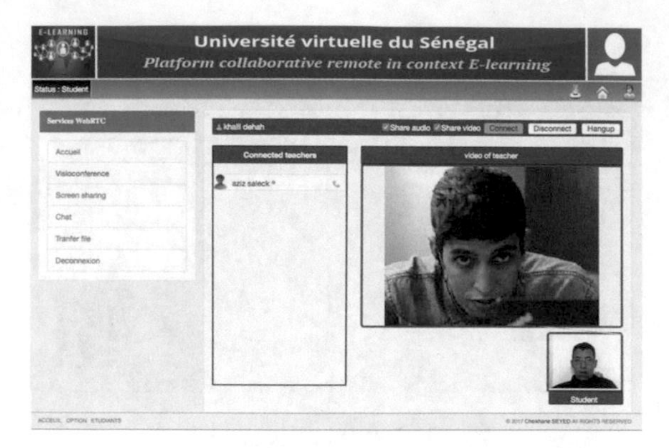

Fig. 7. Videoconferencing between two users mapped

Figure 7 shows that only users who have mapped with their WebRTC correspondent, can access and start the WebRTC services. Both users (aziz with the status of teacher and khalil with student status) launch a videoconference between them.

6 Conclusion

This article provides a solution for developers to facilitate users management for an e-learning platform based on WebRTC technology.

A signaling server is fundamental to manage real-time data communication and exchange between the e-learning users. This server independently manages the WebRTC platform and users. We implemented a mechanism automating user management and mapping between e-learning users and signaling users.

A service controller has been implemented. It allows logical users to be able to load and run WebRTC services.

This work was appreciated by developers of WebRTC technology based solution of the virtual university of Senegal in that it allows to focus more on the implementation of the services than on the complex back end of the WebRTC technology.

For future work, we shall insist on the issue of security, mainly access controls on web platforms integrating WebRTC.

References

1. Wenzel, M., Klinger, A., Meinel, C.: Tele-Board prototyper - distributed 3D modeling in a Web-Based Real-Time collaboration system. In: 2016 International Conference on Collaboration Technologies and Systems (CTS), pp. 446–453 (2016)
2. Ouya, S., Seyed, C., Mbacke, A.B., Mendy, G., Niang, I.: WebRTC platform proposition as a support to the educational system of universities in a limited Internet connection context. In: 2015 5th World Congress on Information and Communication Technologies (WICT), Marrakech, pp. 47–52 (2015)

3. Ouya, S., Sylla, K., Faye, P.M.D., Sow, M.Y., Lishou, C.: Impact of integrating WebRTC in universities' e-learning platforms. In: 2015 5th World Congress on Information and Communication Technologies (WICT), Marrakech, pp. 13–17 (2015)
4. Kokane, A., Singhal, H., Mukherjee, S., Reddy, G.R.M.: Effective e-learning using 3D virtual tutors and WebRTC based multimedia chat. In: 2014 International Conference on Recent Trends in Information Technology, Chennai, pp. 1–6 (2014)
5. Sredojev, B., Samardzija, D., Posarac, D.: WebRTC technology overview and signaling solution design and implementation. In: 2015 38th International Convention on Information and Communication Technology, Electronics and Microelectronics (MIPRO), Opatija, pp. 1006–1009 (2015)
6. Blichar, J., Podhradsky, P.: Liferay as a collaboration and communication platform for m-learning subsystem. In: 2012 Proceedings of ELMAR, WebRTC, 'WebRTC 1.0: Real-time Communication Between Browsers liens' 12–14 September 2012, pp. 175–178 (2012). http://www.webrtc.org/web-apis
7. Elleuch, W.: Models for multimedia conference between browsers based on WebRTC. In: 2013 IEEE 9th International Conference on Wireless and Mobile Computing, Networking and Communications (WiMob), pp. 279–284, 7–9 October 2013
8. Zeidan, A., Lehmann, A., Trick, U.: WebRTC enabled multimedia conferencing and collaboration solution. In: Proceedings of WTC 2014; World Telecommunications Congress 2014. pp. 1–6 13 June 2014
9. Chaczko, Z., Aslanzadeh, S., Klempous, R.: Development of software with cloud computing in 3TZ collaborative team environment. In: International Conference on Broadband and Biomedical Communications (IB2Com), 2011 6th 28 2015 5th World Congress on Information and Communication Technologies (WICT), pp. 318–323, 21-24 November 2011
10. Reis, R., Escudeiro, P., Fonseca, B.: High-Level model for educational collaborative virtual environments development. In: 2012 IEEE 12th International Conference on Advanced Learning Technologies (ICALT), pp. 356–358, 4–6 July 2012
11. Barnes, R.L., Thomson, M.: Browser-to-Browser security assurances for WebRTC. IEEE Internet Comput. **18**(6), 11–17 (2014)
12. Garcia, B., Gortazar, F., Lopez-Fernandez, L., Gallego, M., Paris, M.: WebRTC testing: challenges and practical solutions. IEEE Commun. Stand. Mag. **1**(2), 36–42 (2017)

The Future of Virtual Classroom: Using Existing Features to Move Beyond Traditional Classroom Limitations

Michalis Xenos(✉)

Computer Engineering and Informatics Department,
University of Patras, Patras, Greece
xenos@upatras.gr

Abstract. This paper argues that the true potential of virtual classrooms in education is not fully exploited yet. The features available in most environments that have been incorporated as virtual classrooms are classified into two groups. The first group includes common features, related only to the emulation of a traditional classroom. In this group, the practical differences between traditional and virtual classroom are discussed. In addition, best practices that could aid the professors to make students feel like participating in a typical classroom are presented. The second group comprises of advanced features and practices, which extend the traditional classroom. In this group, examples of successful practices which could not be performed in a traditional classroom are introduced. Finally, a qualitative study with interviews of 21 experts from 15 countries is presented, showing that even these experts are not fully exploiting the advanced features that contemporary virtual classroom environments are offering.

Keywords: eLearning · Virtual classrooms

1 Introduction

Virtual classrooms today are used by educators to replicate a customary practice carried out for centuries, i.e. to teach exactly as they did in a typical classroom. In most cases, this is exactly what learners anticipate, leading virtual classroom usage into a vicious cycle. Although the technology is available for virtual classrooms to move beyond traditional (face-to-face) educational settings and to include practices that cannot be carried out in a traditional classroom, this is not the case and it will probably take some more years to become a widespread practice. This situation is similar in many cases when a modern technology is introduced in a field with established methods and traditions for numerous years.

For example, similar approach has been adopted film industry. While the history of movies began in 1890, all pioneer movies emulated what people (audience and actors) knew from centuries ago: theater. Therefore, the first movies were filmed with stage-bound cameras, the actors did what they knew best before movies, acting on the stage, while the scenes were assumed to follow a linear chronological succession. The

M. E. Auer and T. Tsiatsos (Eds.): IMCL 2017, AISC 725, pp. 944–951, 2018.
https://doi.org/10.1007/978-3-319-75175-7_92

first movie to truly explore some of the potential, which the new medium could offer was the "*Great Train Robbery*" filmed in 1903, where for the first-time location shooting and events happening continuously at identical times but in different places were introduced to the audience [1]. It required a lot of time for all the films to adopt to such practices, which nowadays are common in film making.

This paper argues that while virtual classrooms could move beyond traditional classroom limitations, their usage is still bounded by 'tradition', as in the film industry paradigm. To present this case, the following section includes a brief literature review of virtual classrooms. Section 3 presents experiences and best practices from using virtual classrooms either just to emulate traditional ones, or attempting to move beyond traditional classroom limitations. Section 4 presents a qualitative study with interviews of 21 educators, experts in using virtual classrooms for higher education, which illustrates the ascertainment that even these educators do not fully exploit these features. Finally, Sect. 5 summarizes the main findings of this paper.

2 Virtual Classrooms for eLearning

The use of networked computers to enhance learning was introduced as early as 1980, when Chambers [2] suggested that distance learning experiments should be implemented in a way that could enable in-house learning for some educational materials. The term virtual classroom was introduced in 1986, when Hiltz [3] perceived the use of a computerized conference system as a "*virtual classroom*". The early uses of virtual classrooms focused on practical issues such as sound and video and use of a "*pencil*" for the whiteboard, while the main problems reported were related to limited bandwidth and lack of "*turn-taking*" [4, 5]. As soon as video conference technology evolved and matured, a lot of online synchronous tools for learning have been emerged offering choices for virtual classrooms [6]. Most environments offered features like real-time voice and video, whiteboard, slides presentation, text-based interaction and means for learners' feedback [7].

The use of virtual classrooms was initially driven by necessity, mainly in the context of synchronous distance learning, where a professor had to emulate a typical classroom for distance students. In these early examples, the main goal was to succeed to offer students an experience similar to a face-to-face classroom. In many cases this wasn't succeeded, due to network and equipment limitations that lead into sound and video problems, as well as due to lack of suitable tools (i.e. a discussion administration feature). As new environments started to include more features [6], leaving the sound and video issues in the past, focus was given into the quality and the usability of the environment [8–12]. Using virtual classrooms wasn't only something for distance learners, but also for blended learning, or even as a supplement of on-campus courses [13].

Nowadays, within a virtual classroom, synchronous communication between distance learners may be used to better support personal participation, inducing arousal and motivation [14] and help students to better form a learning community and avoid alienation, which is inversely related to classroom community [15]. Assignments involving collaboration in virtual classroom groups increase the efficiency of the learning process as well as student competencies [16]. In contemporary virtual classroom environments, there is a variety of features available that could be exploited not only to emulate a traditional classroom, but also to move beyond the traditional classroom limitations.

3 Experiences from Using Virtual Classrooms

To present better the author's experiences from twenty years of using virtual classrooms in the tertiary education, the features available in most virtual classroom environments are classified into two groups. The first group (common features) includes features related only to the emulation of a traditional classroom. The second group (advanced features) comprising of features and practices going beyond the traditional classroom. Table 1, includes both categories.

Table 1. Common and advanced features and practices in virtual classrooms

Common features	Advanced features
Video and sound	Retrospective assignments
Chat	Breakout rooms
Students' feedback	Anonymous polling
Whiteboard	Shared whiteboard
Slide presentations	Shared documents and annotating
Discussion administration	Application sharing

To distinguish the advanced features, the following requirements had to be met: (a) being available in most virtual classroom environments that have been incorporated up today, (b) being documented extensively, so most educators could be familiarized with these features, (c) being available only online and not in a traditional classroom, at least not without having to overcome physical and practical limitations. In the following subsections, the practical differences between traditional and virtual classroom are discussed and experiences from the use of these common features are presented, while examples of best practices related to advanced features that could not be performed in a typical traditional classroom are introduced.

3.1 Using Common Features of a Virtual Classroom to Emulate a Traditional One

Nowadays **video and sound** is available for both the professor and the students (or at least sound from all the students). However, this was not the case for the virtual classrooms at the beginning of this century. Video from students increases the sense of community and the best practice is to try to have all students present themselves on video, especially in cases that they haven't met face-to-face. It is a fact that in distance education, having students met at least once is valuable for building a community [17] and in cases that this was not feasible, allowing them to introduce themselves using video and sound is essential.

A **chat feature** can always help overcoming sound problems and, although is not related to a traditional classroom practice, is also included in this list for both historical and practical reasons. Although this is not something occurring very often today, reviewing recordings from 2000 to 2003 reveals that almost one third of the students participating in virtual classrooms faced sound problems during a session [18]. The chat feature, apart from solving sound issues could allow students to better clarify a question, or to allow the professor to collect short responses, especially if the chat supports direct student-to-professor messages, as most contemporary environments do. Experience had revealed that, although action in the chat is a measure of active participation, a single professor is unable to handle both oral and written communication. In this case, a solution is having two educators present (one responsible from collecting chat messages and presenting them orally). When the chat system is expected to be used by students and the number of participants is higher than 30, having more than one professor present is strongly advised.

The **students' feedback feature** allows the professor to effectively monitor participation. Expecting from each student to take a turn and reply to a simple question "is everything OK so far?", or "can everyone hear me?" may take several minutes and distress the normal flow of the lecture, while goes naturally using the feedback mechanism. Most environments offer complex feedback, including emotions, but the best advice is to keep it simple to a "yes" and "no". Normally, the best practice is asking for a confirmation at least every 10 to 15 min, usually following this confirmation with an activity that will further involve students.

The **whiteboard feature**, while in a typical classroom is always a problem for the professor, since extensive use requires to turn the back to audience for a long time, is a major asset in virtual classrooms. The more a professor uses the whiteboard, the more engages the students and the best practice is to frequently allow students to write on whiteboard, or to let them highlight areas they want to discuss further.

The **slide presentations feature** is a valuable feature, only if used with caution. Ideally one should only use slides with complex schemas and images. A virtual classroom based entirely on slide presentation turns out to be a webcast. The best practice is that if something can be sketched in the whiteboard, use the whiteboard instead of a slide. When using a slide is unavoidable, use the virtual laser pointer, add comments, ask students to point, or highlight and do anything possible to engage students into the discussion.

Finally, the **discussion administration feature** facilitates the most challenging task the professor using a virtual classroom must tangle. Controlling the audience, monitoring the 'raised hands' and allowing 'turns' to speak, is something one need to practice for a while, before mastering it in practice. Since it is a quite often phenomenon that some students will 'raise hand' and then cancel it, especially the shy ones, it takes practice from the professor to be able to control the flow of the discussion and do not let students feel left out. The best practice is when the audience is under 10 students to set all microphones on and disable the 'hand raise' feature, while for larger audience using it is required. In some extreme cases of many participants, having an assistant to monitor raised hands from students can be proven extremely helpful (especially when the audience is above 50 to 60 students). In case that after speaking for 10 to 15 min there are no hands raised, the best practice is to take a break and ask something to engage the audience.

3.2 Advanced Features for Going Beyond the Traditional Classroom

Since all the virtual classroom sessions can be recorded and viewed many times, the recording feature can be used for **retrospective assignments**. The best practice, exploiting the recording feature and forcing students to review a session, is to relate assignments to the previous session. Assignments like "*In the 25th minute of the session, a student asked about .... After hearing the discussion, could you offer some more options?*", require from students to review the recorded session and is a valuable educational practice.

The feature of **breakout rooms** is very powerful for engaging students. There are a variety of teamworking practices that could exploit this feature and using it properly could really enhance the educational experience. While in typical classrooms such teamworking is always in terms with physical limitations, in virtual classrooms is something that can be done with ease. The best practice is to engage students using breakout rooms quite frequently during a virtual classroom session (at least once in every session) and to have students report back to the main room their discussion.

Although the technology for **anonymous polling** could also exist in a typical classroom, this requires equipment not commonly available, while it is common in all virtual classroom environments. The use of anonymous polling, engages students and provides the professor with real time feedback. Having one or two review questions every 10 min is the best practice. While the typical student feedback may be 100% "*yes*" in the question "*is everything clear so far*", a couple of review questions may reveal the need to repeat a part of the session, or to start a discussion. In a typical classroom, usually one student will reply correctly to these questions and the rest will silently concur, misleading the professor to think that everything was understood.

The feature of the **shared whiteboard** allows students' participation in activities related to design charts, graphs and similar. Having a few students working together on the whiteboard for a task is a valuable practice. Usually, in the main room the professor could ask for 2 and up to 4 volunteers to work on an exercise, but the best practice is to use a breakout room, allowing a small number of students in each room (depending on the activity 2 and up to 4, or even 5 students). Some virtual classroom environments allow the results of each room whiteboard to be shared back to the main room, but not all. If this feature is not available, usually a working solution is to use a print-screen of the results to report it back to the main room.

The feature of **shared documents and annotating**, allows the professor to have students working together on a document, i.e. reviewing code and annotating as part of a collaborative exercise. Depending on the number of students, this is something that can be done in the main room (usually when up to 10 students attend the virtual classroom), or using breakout rooms. The best practice for larger audiences is to combine breakout rooms with such collaborative exercises.

The feature of **application sharing** is important, not only for the professor sharing an application to demonstrate the use of a software tool, but also for students. In fact, having students share their application to present a problem while the professor comments on that is a powerful and constructive educational experience. In the field of computer science, where students are required to use many software tools, this practice speeds up significantly the process of responding to questions and providing appropriate feedback. It is much easier to view the students' solution and comment on it, rather than having them explain their solution and making assumptions. This is also a very helpful educational practice for all the students participating in the virtual classroom, as long as the discussion is not monopolized on a single student's solution. This practice could be also valuable when a virtual classroom session is used as 'office hours' for responding to students' questions.

4 A Survey on the Exploitation of the Advanced Features

For the six advanced features that could aid the virtual classrooms to go beyond traditional classroom limitations and fully exploit what technology offers, an informal interview was conducted involving 21 educators from the following 15 countries: Austria, Cyprus, Czech Republic, Belgium, Germany, Greece, Finland, Italy, Lithuania, Poland, Portugal, Spain, The Netherlands, Turkey, and the United Kingdom. All the interviewed educators are experts in using virtual classrooms for higher education. The interviews were informal, feeling more like a friendly discussion, trying to minimize note taking and allowing the discussion to include successful experiences from the virtual classroom usage, or anecdotes of failures. The questions asked were the following, starting with "Have you…":

- Q1: … assigned something that would require from students to review the session from the lectures archive?
- Q2: … used breakout rooms to let the students work on a collaborative assignment?
- Q3: … used anonymous online polls during the lecture?
- Q4: … shared the whiteboard to more than one student at the same time?
- Q5: … used a shared document and asked students to annotate?
- Q6: … asked from students to share an application to demonstrate a problem?

The frequency ranges were informally discussed and sometimes the interviewees failed to provide a clear answer or gave answers like "*I don't know if it is five or ten, maybe less than five, maybe closer to ten, but definitely isn't something that fits in my classroom*", so the frequencies are presented as follows:

- F0: Never, or just to test the tool but not in a real classroom.
- F1: A few (one to five) times over all the years, but this never became a customary practice.
- F2: Not frequently, but sometimes and not on every course I teach.
- F3: Frequently (more than once in each course I teach), but this is not a regular process.
- F4: This is a regular process I use in my virtual classroom sessions.

The results for each of the six features are presented in Table 2, where the six rows correspond to the features and the five columns to the usage frequency. Even though most (12 out of 21) educators are from the technology field (teaching STEM courses) and therefore are expected to be familiar with the use of modern virtual classroom features, results showed that in most cases the features were tested and never actually exploited in practice. In fact, considering that the results could be biased towards a positive attitude of the technology–since most of the educators were tech savvy, which made more difficult for them to admit that they didn't use these features–the results indicate that exploitation of the advanced features of virtual classrooms hasn't reached its full potential yet. Some features, like the retrospective assignments start to become part of normal practice, others like the shared whiteboard and documents are included occasionally into some sessions, while other like the anonymous polls, the application sharing and the breakout rooms are just starting to be acknowledged as promising opportunities.

Table 2. Results from the survey

	F0	F1	F2	F3	F4
Q1	6	4	7	2	2
Q2	20		1		
Q3	15	4	1		1
Q4		6	11	3	1
Q5	7	9	5		
Q6	19	2			

5 Conclusions

This study suggests that the use of virtual classrooms hasn't reached its full potential yet and there are features that could be employed to aid towards moving virtual classrooms beyond just emulating traditional ones. The paper presents examples of best practices using such advanced features, based on the author's experience. It also presents practices that could be used to improve teaching, based on the common features of virtual classrooms, which are used mostly to emulate teaching as in traditional classrooms. Nowadays most professors still use virtual classrooms to replicate the practice they are familiar with: teaching in a face-to-face classroom. This is typical when technology is introduced in a practice which exists for many years. The best practices presented in this paper could aid professors to move beyond traditional classroom limitations and fully exploit the entire spectrum of modern virtual classroom features.

References

1. Dirks, T.: Filmsite Movie Review, The Great Train Robbery (1903). http://www.filmsite.org/grea2.html. Accessed 2 June 2017
2. Chambers, J.A., Sprecher, J.W.: Computer assisted instruction: current trends and critical issues. Commun. ACM **23**(6), 332–342 (1980)

3. Hiltz, S.R.: The "virtual classroom": Using computer-mediated communication for university teaching. J. Commun. **36**(2), 95–104 (1986)
4. Dwyer, D., Barbieri, K., Doerr, H.M.: Creating a virtual classroom for interactive education on the Web. Comput. Netw. ISDN Syst. **27**(6), 897–904 (1995)
5. Hiltz, S.R., Wellman, B.: Asynchronous learning networks as a virtual classroom. Commun. ACM **40**(9), 44–49 (1997)
6. Schullo, S., Hilbelink, A., Venable, M., Barron, A.E.: Selecting a virtual classroom system: Elluminate live vs. Macromedia breeze (adobe acrobat connect professional). MERLOT J. Online Learn. Teach. **3**(4), 331–345 (2007)
7. Finkelstein, J.: Learning in Real Time. Jossy-Bass Publishing Company, San Francisco (2006)
8. Xenos, M., Christodoulakis, D.: Software quality: the user's point of view. In: Lee, M., Barta, B.-Z., Juliff, P. (eds.) Software Quality and Productivity. IAICT, pp. 266–272. Springer, Boston, MA (1995). https://doi.org/10.1007/978-0-387-34848-3_41
9. Johnston, J., Killion, J., Oomen, J.: Student satisfaction in the virtual classroom. Internet J. Allied Health Sci. Pract. **3**(2), 6 (2005)
10. Stavrinoudis, D., Xenos, M., Peppas, P., Christodoulakis, D.: Early estimation of users' perception of software quality. Softw. Qual. J. **13**(2), 155–175 (2005)
11. Stefani, A., Vassiliadis, B., Xenos, M.: On the quality assessment of advanced e-learning services. Interact. Technol. Smart Edu. **3**(3), 237–250 (2006)
12. Katsanos, C., Tselios, N., Xenos, M.: Perceived usability evaluation of learning management systems: a first step towards standardization of the system usability scale in Greek. In: 16th Panhellenic Conference on Informatics, PCI 2012, pp. 302–307 (2012)
13. Singh, H.: Building effective blended learning programs. Edu. Technol.-Saddle Brook Then Englewood Cliffs NJ **43**(6), 51–54 (2003)
14. Hrastinski, S.: The potential of synchronous communication to enhance participation in online discussions: a case study of two e-Learning courses. Inf. Manag. **45**(7), 499–506 (2008)
15. Rovai, A.P., Wighting, M.J.: Feelings of alienation and community among higher education students in a virtual classroom. Internet High. Edu. **8**(2), 97–110 (2005)
16. Crişan, A., Enache, R.: Virtual classrooms in collaborative projects and the effectiveness of the learning process. Procedia – Soc. Behav. Sci. **76**, 226–232 (2013)
17. Marsap, A., Narin, M.: The integration of distance learning via internet and face to face learning: why face to face learning is required in distance learning via internet? Procedia – Soc. Behav. Sci. **1**(1), 2871–2878 (2009)
18. Xenos, M., Skodras, A.: Evolving from a Traditional Distance Learning Model to e-Learning. In: 2nd International LeGE-WG Workshop on e-Learning and Grid Technologies: A Fundamental Challenge for Europe, Paris, France, 3rd and 4th March 2003, pp. 121–125 (2003)

Cooperative Learning-Agents for Task Allocation Problem

Farouq Zitouni[1(✉)] and Ramdane Maamri[2]

[1] KASDI Merbah University, Ouargla, Algeria
farouq.zitouni@univ-constantine2.dz
[2] Constantine 2, Abdelhamid MEHRI University, Constantine, Algeria
ramdane.maamri@univ-constantine2.dz

Abstract. In this paper, we will present a working methodology for solving the task allocation problem in a multi-robot system, i.e. assign the tasks being performed to appropriate robots. In fact, the proposed approach combines the advantages of several well-known algorithms (e.g. quantum genetic algorithms, Q-learning machine-learning, etc.), in order to construct a good solution for the task allocation problem.

Besides, the proposed working methodology has been implemented using the Java programming language and the JADE multi-agent platform; also it has been simulated on a real-life scenario, which is the extinction of fires (tasks) in an environment altered by a natural disaster. Finally, the experimental results are promising and show the effectiveness of the proposed methodology.

Keywords: Multi-robot systems · Task allocation · Cooperation
Quantum genetic algorithm · Q-learning

1 Introduction

In the field of cooperative multi-robot systems, the task allocation problem is a very active research track, which has gained a lot of attention from researchers, especially in last years. In fact, when we build and use multi-robot systems we must inevitably answer the following question: which robot, among others, is selected to perform a given task?

The multi-robot task allocation (MRTA) problematic tries to answer the question of how to obtain the assignments, between robots and tasks, that optimize the total cost or utility of a considered system. Generally, if we want to answer this question for most of the encountered situations, even for simple cases, we are faced with problems of the NP-hard class. Therefore, the majority of the approaches proposed in the literature to address this problem are approximate or heuristic, i.e. they product suboptimal solutions. Also, it should be noted that the quality of the proposed algorithms greatly affects the performance of multi-robot systems that must use them to solve the task allocation problem; especially, if we increase the number of robots and tasks. Likewise, it should be mentioned that since 2005 the topic of task allocation has become one of the key subjects in the field of multi-robot systems. In the paper Brian and Maja (Brian and

M. E. Auer and T. Tsiatsos (Eds.): IMCL 2017, AISC 725, pp. 952–968, 2018.
https://doi.org/10.1007/978-3-319-75175-7_93

Maja 2004), we find a good classification of the different task allocation situations, that could be found in multi-robot systems, and it is summarized as follows:

- Single-task robots (ST) versus Multi-task robots (MT): the acronym (ST) means that each robot can perform a single task at a given time. On the other hand, the acronym (MT) means that some robots can perform several tasks at the same time.
- Single-robot tasks (SR) versus Multi-robot tasks (MR): the acronym (SR) means that each task requires only one robot for its accomplishment. On the other hand, the acronym (MR) means that some tasks may require several robots to accomplish them.
- Instantaneous assignment (IA) versus time-extended assignment (TA): the acronym (IA) means that robots are only interested in current tasks without considering future allocations. On the other hand, the acronym (TA) means that robots must prepare a plan for current tasks and future allocations.

Thus, using the taxonomy presented above we can easily distinguish eight types of task allocation, by considering the Cartesian product: $(ST, MT) \times (SR, MR) \times (IA, TA)$. Therefore, the different situations of task allocation that we can face in multi-robot systems can be classified in one of the classes defined by these combinations. In this paper, we deal with task allocation problems that belong to the class $[ST, MR, IA]$, i.e. each robot can perform a single task at the same time, a task can be accomplished by several robots at the same time and the allocations are instantaneous (no plan is provided).

The remainder of the paper is organized as follows: in the next section we present some related work already done to address the task allocation problem; then we expose the section that explains our proposed working methodology for solving the task allocation problem in multi-robot systems; afterwards we present the section that summarizes the simulation of our methodology on a real life scenario and the discussion of experimental results. Finally, we end the work with a conclusion and some perspectives.

2 Related Work

In recent years, several approaches have been proposed in literature to address the task allocation problem. Thereby, the authors of the paper Hang and Liu (2008) have proposed a classification of these approaches into three classes, which are: behavioural, based on market rules and bio-inspired approaches.

First the behavioural approaches, in this category the tasks to be performed are divided into behavioural groups, and the tasks belonging to the same one often have relations between them. Generally, these approaches are robust, fault-tolerant and operate in real time; However, the found solutions to address the task allocation problem are only optimal at the local level. Among the approaches proposed in this category, we can find Alliance Parker (1998), BLE Barry and Maja (2000) et ASyMTRe Fang and Parker (2005).

Second the market-based approaches, in fact most of the approaches that have been proposed recently can be inserted into this category. Typically, these approaches use the Contract Net Protocol Smith (1980) to exchange messages between agents of the considered system. In addition, they try to maximize the income of the system, while minimizing spent costs. Generally, these approaches are scalable, i.e. they are very often used in the field of distributed robots, and produce optimal solutions. However, the robots must cooperate through explicit communications, i.e. very greedy in resource consumption, so if the communication medium breaks down then the performance of the system is degraded significantly Kalra and Martinoli (2006) (Thus, they are well adapted to task allocation problems of small and medium sizes). Among the approaches proposed in this category we can find M+ Botelho and Alami (1999), First Price Auctions Zlot et al. (2002), Dynamic Role Assignment Chaimowicz et al. (2002), MURDOCH Gerkey and Maja (2002), Traderbots and Bernardine (2004) and DEMiR-CF Sanem and Tucker (2006).

Finally, the bio-inspired approaches, in recent years these approaches have gained a great attention from researchers working on task allocation problems. Essentially, their functioning is derived from the behavior of social insects and living-beings. Besides, these approaches have several advantages, among them we can cite the self-organization ability in unknown environments, emergence and adaptability. Moreover, a system that adopts such approaches is considered more robust because there is no central control, i.e. the failure of one individual (or more) does not alter the performance of the whole system. In addition to this, individuals of the system cooperate through implicit communications; so even if the number of individuals increases, then the rate of communication would not increase exponentially. In conclusion, bio-inspired approaches are generally the most suitable to address the MRTA problems, especially in unknown environments. Among the approaches proposed in this category, we can mention the articles Ding et al. (2003); Yang and Wang (2004); Dandan et al. (Dandan et al. 2007); Zhang and Liu (2008); Yu and Liu (2009); Liu and Zhang (2009); Liu and Zhang (2010); Shuhua et al. (2011); Iztok et al. (2014); Nadia et al. (2015).

3 Proposed Working Methodology

It is obvious now that a multi-robot system (MRS) is a distributed society of robots, in which each one can operate alone or cooperate with the other robots to perform some predefined tasks. Generally, this distributed architecture could increase the flexibility and preserve the consistency of the considered system. However, in a partially (or completely) unknown environment it is very often difficult to decide which robots must cooperate to perform a given task. Such a problematic is the focus of our proposed working methodology.

In this paper, we assume that we have an environment that contains a set of tasks, which must be accomplished by cooperative robots (i.e. robots consider the system profit in the first place). In addition, we consider that we have three different types of robots, which are: explorer robots, worker robots and supervisor robots. A brief description of their roles is given as follows:

1. Explorer robots: these robots explore the task environment in order to detect the presence of works to perform and determine, among other things, their requirements.

2. Worker robots: these robots perform the various tasks discovered in the environment, because they offer their requirements.
3. Supervisor robots: these robots play the role of intermediary between explorer robots (tasks) and worker robots, i.e. assign the tasks discovered in the environment to the robots of the considered system.

In fact, the Fig. 1 shows the general architecture of our system and the relationships between its various components (robots). As we can clearly observe, we first have n explorer robots that must scan the environment to look for tasks to perform, and as soon as a task is discovered a message is sent directly to the supervisor robot (here we have a single robot) that must communicate with the m worker robots, and designate which among them must cooperate to accomplish the desired task. In the following sections, we will explain in detail each component of this architecture.

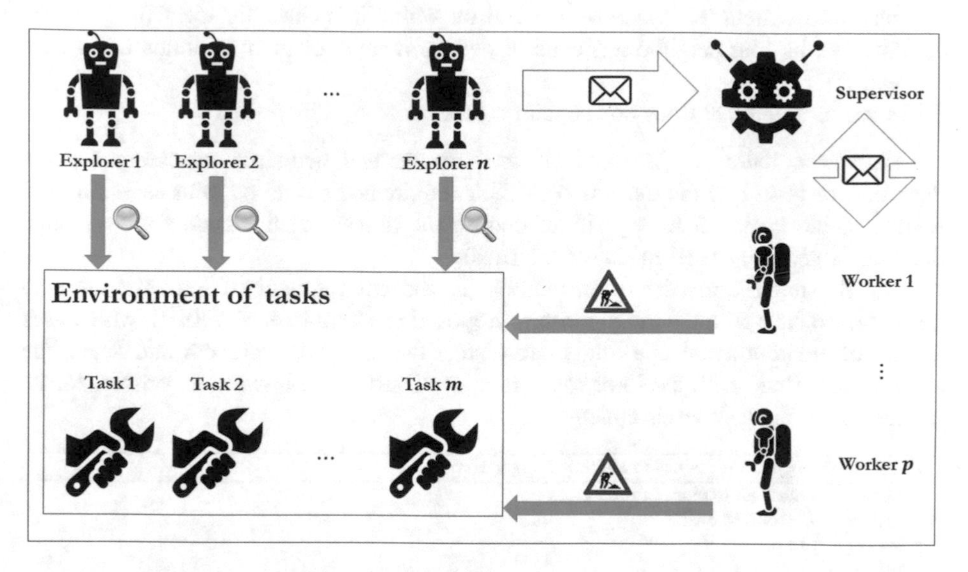

Fig. 1. General architecture of the system

3.1 Task Environment

To simplify things, we have chosen to represent the environment of tasks as a 2D-grid (i.e. it has width and height) of cells. In addition, we assume that each cell (i,j) has four neighboring cells which are: $\{(i-1,j),(i+1,j),(i,j-1),(i,j+1)\}$. In our approach, this environment is denoted by E and a cell of this latter is denoted by x. Moreover, we also give a function V, which for each cell x of E assigns to it a real value, defined as follows:

$$V : x \in E \mapsto V(x) \in \mathbb{R}$$

Therefore, when the task environment is initialized, then all the values of its cells are set to $+\infty$ (i.e. $\forall x \in E : V(x) = +\infty$). Values of cells are used to facilitate the navigation of explorer robots. Also, in this environment we can find tasks (a task occupies a cell) that need skills in order to be performed. In fact, the skill concept is very often mentioned in this paper, and it can be either a sensor (e.g. sensor of heat, light, etc.) or an actuator (e.g. gripper, jack, wheels, etc.). Finally, the tasks in the environment are considered static (i.e. their skills are invariable). In addition, a task has too a position in the environment and an approximate time for its accomplishment.

3.2 Explorer Robots

As mentioned above, these robots explore the task environment to look for a work to perform. Moreover, the environment is discretized into cells, and at any time an explorer robot occupies a cell and perceives its four neighboring cells. Also, several robots can never occupy a given cell at the same time. Briefly, each explorer robot is capable of:

1. Read and write a real value on the cell on which it is currently located.
2. Perceive its four neighboring cells: it detects if none of them contains a task and read their value.
3. Move to a cell that does not contain a task.

Moreover, these robots have all the sensors and actuators necessary for the detection of tasks and the definition of their requirements (this hypothesis is strongly related to the treated field, e.g. if we choose the domain of fire extinction, a smoke detector is generally used to detect the fires).

For the navigation of explorer robots in the environment to look for tasks to perform, we have adopted the algorithm proposed in Simonin et al. (2007), which uses a kind of environmental marking to accelerate the navigation process and detect the tasks faster. Thus, each explorer robot runs cyclically the following algorithm, for its navigation in the task environment:

Algorithm 1: explorer robots navigation algorithm

Input : Given an explorer robot e_i
Output: Discovered tasks

```
1  set V(x_{e_i}) to 0;
2  while (e_i is active) do
3  |   get the list of neighboring cells of the current cell e_i;
4  |   if (there is no a task in neighboring cells) then
5  |   |   if (there exist neighboring cells with +∞ value) then
6  |   |   |   move randomly to a cell marked with +∞ (this cell will be noted x_selected);
7  |   |   else
8  |   |   |   {all the cells are marked};
9  |   |   |   move randomly to one of them (this cell will be noted x_selected);
10 |   |   end
11 |   else
12 |   |   collect all the required information about the task;
13 |   |   send these information to the supervisor robot;
14 |   |   move to a random position;
15 |   |   set V(x_{e_i}) to 0;
16 |   end
17 end
```

In this algorithm, the movements of the explorer robots are based on random displacements with marking, i.e. they explore the task environment and mark it with integer values. However, this is not really a purely random exploration because the values of the marked cells influence the movements of explorer robots: rather they favor the unlabeled cells on those marked ones. As a result, this guided random exploration will considerably accelerate the task detection process. Finally, in order to avoid the repetitive detection of the same task by a given explorer robot, this latter as soon as it discovers a task sends its information to the supervisor robot, and then it chooses a new random position in the task environment and moves to it, after all it restarts the task search process.

3.3 Worker Robots

At first glance, the role of worker robots was mentioned briefly above and consists mainly of the cooperative accomplishment of the discovered tasks. In other words, they help each other to accomplish the different tasks discovered by the explorer robots.

In reality, these worker robots possess skills (sensors or actuators) that they offer to tasks in order to be performed. In other words, tasks need skills to be realized and the worker robots offer them. Thus, a task is said to be feasible if and only if there is a worker robot (or robots) that can satisfy its requirements in terms of skills. In addition to this, a worker robot also has a moving speed (v), an energy level (b), an aging factor (a) and a position in the task environment.

3.4 Supervisor Robot

In fact, the supervisor robot is considered to be the core of our architecture (i.e. its center of gravity). In other words, it has two main roles which are: the allocation of the different discovered tasks to the worker robots and the scheduling of tasks. To do this, this robot on each reception of a message from the explorer robots checks whether it is a new task or not. Thus, if a new task has just been received, then the supervisor robot executes the algorithm represented by the following organigram for its treatment:

As we can see in Fig. 2, the behavior of the supervisor robot consists of the cyclical reading of its mailbox to look for new messages coming from the explorer robots. Then, in case a new message is received, the supervisor robot must check if this message denotes a new task or not (this latter case will simply be ignored). Then, if this message contains a request for a new task, then the supervisor robot creates a virtual agent that must construct a solution for the task allocation problem (the creation of these virtual agents for each new task will allow the parallel processing of multiple tasks simultaneously). Finally, the following figure depicts the working principal of virtual agents (created for each discovered task).

As we can see in Fig. 3, a virtual agent is created by the supervisor robot for each discovered task. Moreover, the working of this agent is very simple. First, it begins by exchanging messages, for a determined time before moving to the next step, with the worker robots. Once the message-exchanging time is over, the virtual agent reads its mailbox (the received messages), collects and organizes all the information needed to make a decision, and finds at the end a solution for the task allocation problem.

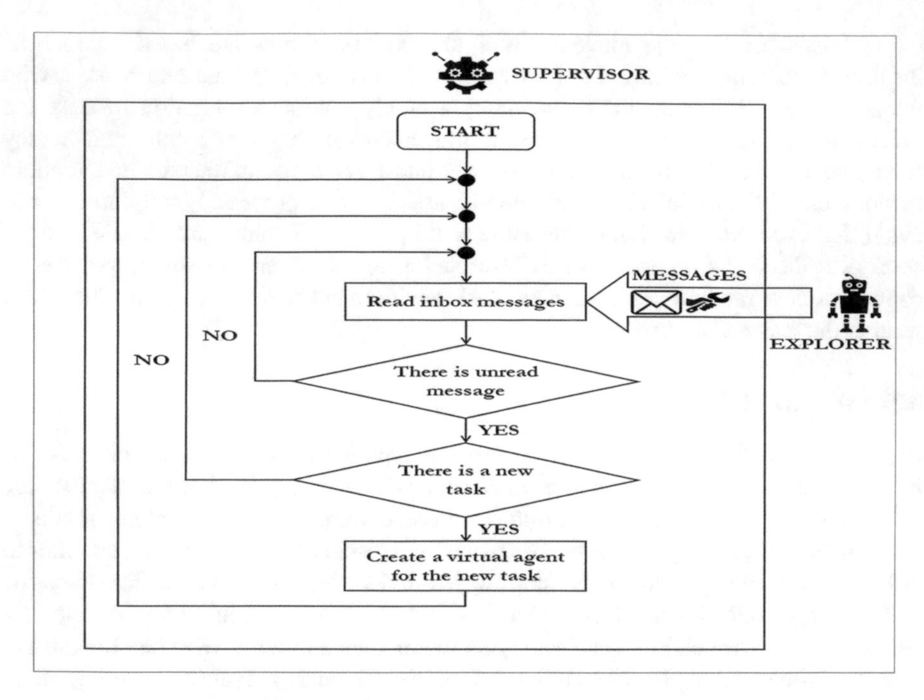

Fig. 2. Working principle of the supervisor robot

Besides, this agent uses a common knowledge base that contains the solutions found during previous experiments. Here, two situations are possible: either the agent finds that a solution is already learned and saved in the common knowledge base, and in this case uses it directly and takes all the necessary measures; either the solution is not in the knowledge base, and in this case it must build a new solution for this situation.

To find a new solution for the task allocation problem, the virtual agent represents all the previously organized information as a graph called *G*, which shows the correspondences between the skills offered by the worker robots and the skills needed by the task being performed, then this graph will be the input of the learning algorithm which will find a solution for the task allocation problem. Finally, the new solution will be saved in the common knowledge base, and the necessary measures will be taken. Now, we will explain how the graph *G* is built and the learning algorithm is used.

3.4.1 The Construction of the Graph *G*

In this section, we give a concrete example to facilitate the understanding of the construction of the graph *G*. First, a quantum genetic algorithm [Rafael (2016)] is used to select the worker robots, which must offer their skills to construct the graph *G* (i.e. offer their skills to perform the task being treated). In fact, the steps of a quantum genetic algorithm are illustrated by the following algorithm:

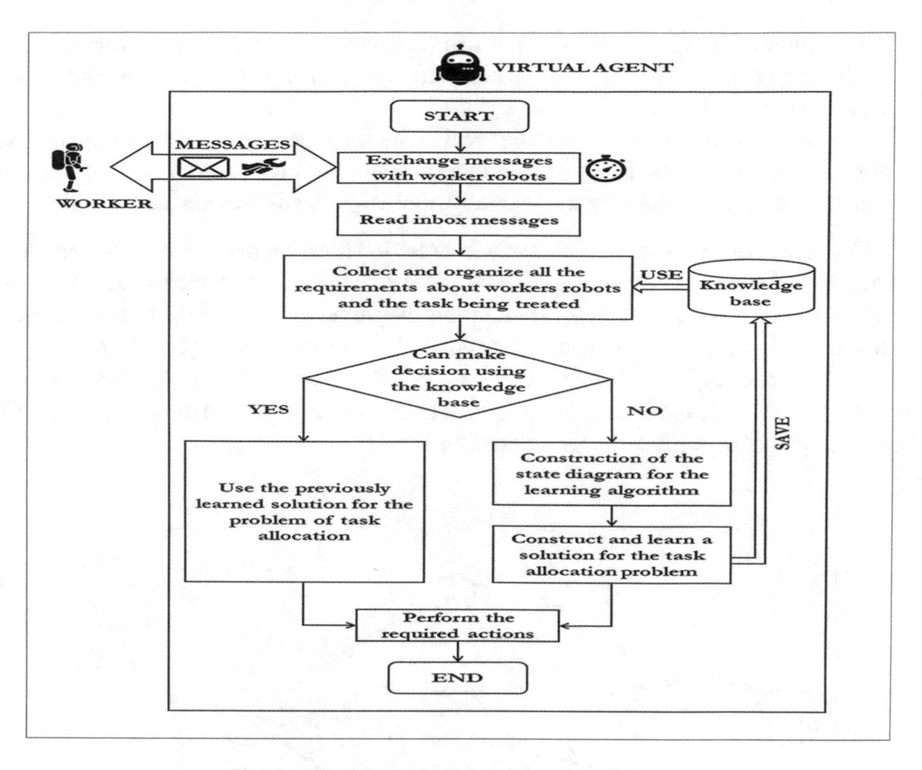

Fig. 3. Working principle of the virtual agents

Algorithm 2: main steps of a quantum genetic algorithm

```
1  Initialize a quantum population Q(0);
2  Make P(0), the measure of every individual Q(0) → P(0);
3  Evaluate P(0);
4  while (not termination condition) do
5  |   t ← t + 1;
6  |   Rotation Q-gate;
7  |   Mutation Q-gate;
8  |   Make a measure Q(t) → P(t);
9  |   Evaluate P(t);
10 end
```

In this algorithm, the chromosome P is a vector of binary values of length n, where n represents the number of worker robots that have answered to the request of the virtual agent. Thus, a nonzero value (i.e. 1) of this vector means that the worker robot occupying this rank contributes to the construction of the graph G, otherwise (i.e. 0) it does not. In addition, we have used the following heuristics for the evaluation of the chromosome P:

1. The evaluation of the chromosome P will be decremented by a certain value n_1, for each worker robot found in this chromosome with a busy state (a busy worker robot can never be selected).
2. The evaluation of the chromosome P will be incremented by a certain value $n_2 \times k$, for each worker robot found in this chromosome with a free state, where k is the number of skills offered by this worker robot to perform the considered task.

Now, we give an illustrative example in order to show the method used to construct the graph G. To do this, we assume that we have a single task to perform named t, with the skill vector $c_t = [1, 1, 0, 1, 0]$. In addition, we assume that we have three worker robots named $\{r_1, r_2, r_3\}$, with the respective skill vectors: $c_{r_1} = [1, 1, 1, 0, 1]$, $c_{r_2} = [0, 1, 1, 1, 0]$ and $c_{r_3} = [1, 1, 0, 0, 0]$. Moreover, the result of the quantum genetic algorithm is $P = [1, 1, 1]$, i.e. the three robots are selected to construct the graph G. Thus, the graph G is given by the following Fig. 4:

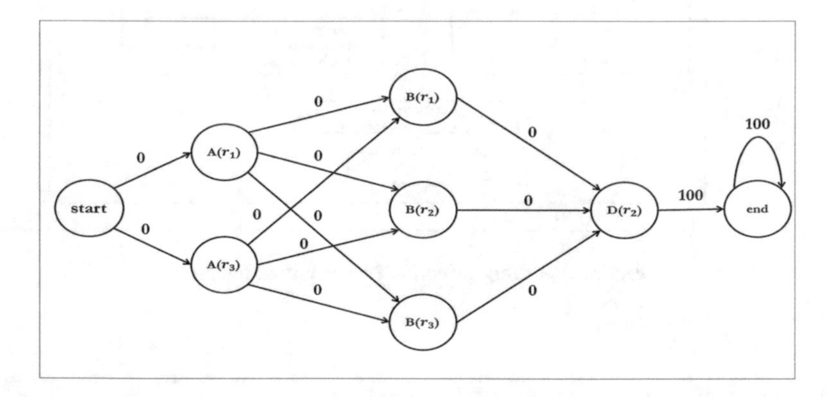

Fig. 4. An example of a graph G

Where the letters A, B and D represent the skills required by the task t and offered by the worker robots $\{r_1, r_2, r_3\}$. Thus, a solution for the task allocation problem is simply a path from the node “start” to the node “end”. However, the best solution is given by the learning algorithm that will be represented in the next section.

3.4.2 The Used Learning Algorithm

Once the graph G is built, it only remains to apply the learning algorithm on it, in order to find a solution for the task allocation problem (assign the worker robots to the considered task). In our approach, we have used the Q-learning reinforcement-learning algorithm [Richard and Andrew (1998)], which is very often used and simple to implement. In fact, this algorithm requires a graph as an input and produces as an output a matrix of assignments, which allocates the skills of the considered task to the different worker robots by considering their skills, too. In addition, the Q-learning algorithm is illustrated by the following code:

Algorithm 3: main steps the Q-Learning algorithm

```
1  Set the γ parameter, and environment rewards in the matrix R;
2  Initialize the matrix Q to zero;
3  foreach (episode) do
4  |  Select a random initial state;
5  |  while (the goal state has not been reached) do
6  |  |  Select one among all possible actions for the current state;
7  |  |  Using this possible action, consider going to the next state;
8  |  |  Get maximum Q value for this next state based on all possible actions;
9  |  |  Compute: Q(state, action) = R(state, action) + γ * Max[Q(nextstate, allactions)];
10 |  |  Set the next state as the current state;
11 |  end
12 end
```

The above algorithm is used by the virtual agent to learn from its previous experiences. In fact, each episode may be considered as a learning session, where the agent explores the graph *G* (represented by the matrix *R*) and receives the rewards until it reaches the final state. Therefore, the purpose of learning is to strengthen the brain of the virtual agent (represented by the matrix *Q*). Finally, the coefficients of the matrix *Q* are updated using the speed (*v*), the energy level (*b*), the aging factor (*a*) and the positions of worker robots with respect to the position of the task being processed, in order to choose the best skills (i.e. if two worker robots offer the same skill to the considered task, the updating of the matrix *Q* will allow us to choose the most relevant). To use the matrix *Q*, the virtual agent simply draws the sequence of states (from the initial state to the final state) by selecting the actions with the highest reward values. Thus, this algorithm is illustrated by the following pseudocode:

Algorithm 4: the algorithm to utilize the matrix Q

```
1 Set current state ← initial state;
2 From current state, find the action with the highest Q value;
3 Set current state ← next state;
4 Repeat Steps 2 and 3 until current state ← goal state;
```

4 Simulation and Discussion of Results

In order to validate the different algorithms and organigrams proposed and explained in the previous sections, then a set of cooperation scenarios between robots were tested and discussed in this section. In fact, these test scenarios consist of an environment altered by a natural disaster, which contains a set of fires that must be extinguished by robots. So, the tasks are the fires.

In each scenario, the proposed working methodology is applied, and the found results are reported and submitted to the evaluation metrics that we have adopted. Besides that, all the test scenarios have two entries, which are: the three types of robots as well as their information and the tasks as well as their information. Finally, the expected outputs after solving each scenario are the best assignments between worker robots and tasks, which respect at best the adopted evaluation metrics.

4.1 Software and Hardware Configuration

To test the proposed working methodology and see how effective it is, for solving the MRTA problem, so we have implemented our own simulator using the Java programming language. In addition, the JADE platform was also exploited to simulate the behavior of the different robots explained above. Therefore, the following figure shows the main graphical interface of our simulator.

As we can clearly see in Fig. 5, the main graphical interface of our simulator is divided into four zones. The first part, is reserved to the initialization of the simulation parameters (e.g. number of robots, number of fires, etc.). The second part, represents the task environment. The third part, shows the different information specific to the tasks (fires) situated in the environment (e.g. position, required skills, etc.). The fourth part shows, the different information specific to the worker robots (e.g. position, speed, level of energy, etc.). For the hardware configuration, all simulations were executed on a DELL laptop intel® Core™ i3 CPU M 380 @ 2.53 GHz, RAM 3.00 Go with a Windows 10–64 bits operating system.

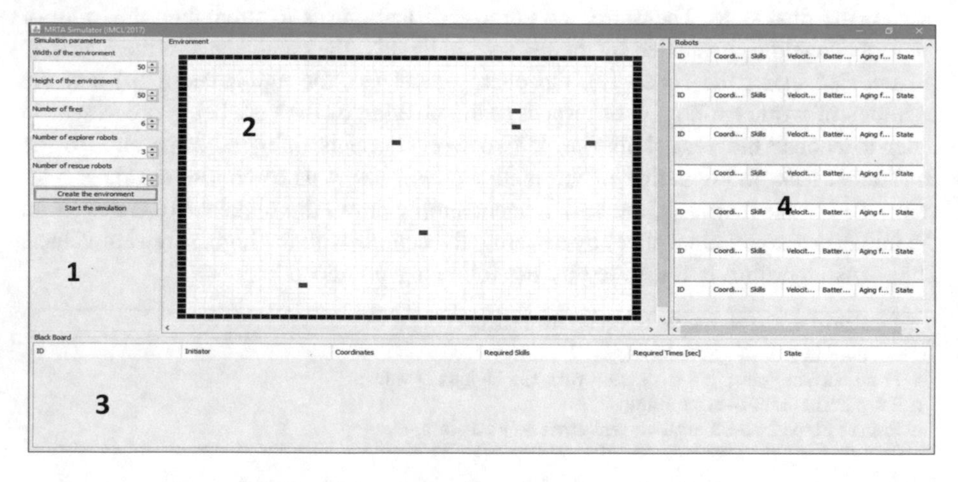

Fig. 5. Main graphical interface of our simulator

4.2 Illustrative Example of Evaluation

Before presenting an illustrative example of simulation in this section, it is imperative to mention that the number of robots and tasks is considered to be one of the major sources of complexity in MRS; thereby as this number is higher as the system is considered more complex. Thus, it was crucial to test qualitatively and quantitatively the proposed working methodology; precisely to measure its ability for solving more or less complex task allocation problems.

In this simulation example, we assume that our environment contains six fires (tasks) that need to be treated as quickly as possible (their positions are generated randomly). Besides, these fires need skills to be extinguished. Also, we assume that we have a single supervisor robot. In addition, we consider five explorer robots (drones) which are equipped with all the sensors and actuators needed to detect fires and determine their needs (e.g. heat sensors, cameras, smoke detector, etc.). Finally, we use too seven worker robots (drones) in this environment that must cooperate to extinguish the various discovered fires. Similarly, these worker robots have skills to extinguish the fires. In fact, this configuration was used to simulate this scenario and to find a solution for the task allocation problem. Therefore, the following tables summarize in detail the different information specific to fires (tasks) and worker robots.

Table 1 shows the different information specific to fires found in the simulated task environment. In fact, the fires are numbered from 000 to 005 (in all we have six fires). Also, each fire is defined by its unique identifier, the identifier of the explorer robot that detected it, its position in the task environment, the skills required for its extinction, the approximate time taken to extinguish it and finally, its state (done or undone).

Table 1. Information specific to fires

ID	Initiator	Coordinates	Required skills	Required times [sec]	State
Fire000	Explorer002	(38.0, 14.0)	[0, 0, 1, 1, 0]	8	Undone
Fire001	Explorer000	(38.0, 6.0)	[0, 1, 0, 0, 0]	6	Undone
Fire002	Explorer002	(21.0, 11.0)	[0, 0, 0, 0, 1]	9	Undone
Fire003	Explorer001	(24.0, 19.0)	[1, 1, 0, 0, 1]	4	Undone
Fire004	Explorer004	(4.0, 19.0)	[1, 0, 1, 1, 0]	4	Undone
Fire005	Explorer004	(38.0, 35.0)	[0, 0, 1, 0, 1]	8	Undone

Table 2 shows the different information specific to worker robots that must extinguish the fires in the simulated task environment. In fact, the robots are numbered from 000 to 006 (in all we have seven worker robots). In addition, each worker robot is defined by its unique identifier, its current position in the environment, the skills that it offers to extinguish the fires, its speed, its battery level, its aging factor and finally its state (free or busy).

Table 2. Information specific to worker robots

ID	Coordinates	Offered skills	Velocity [cm/sec]	Battery level	Aging factor	State
Worker000	(0.0,0.0)	[0, 0, 1, 1, 0]	56.0	87.0	0.97	Free
Worker001	(0.0,1.0)	[0, 0, 0, 0, 1]	81.0	93.0	0.91	Free
Worker002	(0.0,2.0)	[0, 0, 1, 0, 0]	117.0	88.0	0.91	Free
Worker003	(0.0,3.0)	[1, 1, 1, 1, 1]	52.0	85.0	0.95	Free
Worker004	(0.0,4.0)	[0, 0, 1, 1, 0]	52.0	85.0	0.96	Free
Worker005	(0.0,5.0)	[0, 1, 1, 0, 0]	94.0	81.0	0.92	Free
Worker006	(0.0,6.0)	[1, 0, 1, 1, 0]	56.0	96.0	0.97	Free

Now, we explain the meaning of the binary values of skills, i.e. what are the information presented by the vectors of skills specific to worker robots and tasks. First, in this framework we have used five different skills (named *A*, *B*, *C*, *D* and *E*) and by skill we mean either a sensor (e.g. sensor of heat, smoke detector, etc.), or an actuator (e.g. gripper, jack, water pump, etc.). Therefore, if we have a non-zero value in this vector, it means that the skill denoted by the corresponding letter is either offered or requested (i.e. offered by a worker robot or requested by a task).

4.3 Adopted Evaluation Metrics

Before presenting the evaluation metrics that we have adopted in our frameworks, it is strictly important to mention that evaluating the performance of mobile robots and verifying the correctness of their behaviors in different applications remain a very active and open research field [Daniele and Daniele (2009)]. Nevertheless, it should be noted that substantial progress has been made in defining standards for the evaluation processes. However, these standards are often defined by considering strongly the treated field.

In our case, we have chosen to consider two types of time as evaluation metrics, namely: the waiting time for each fire (i.e. the time that elapses between its detection, the designation of worker robots for its extinction and the real beginning of the extinction works) and the total time taken for the extinction of all fires. In fact, we used time as an evaluation metric because this factor is very important in the case of fire extinguishing.

4.4 Discussion of Obtained Results

In this section, we will present the experimental results found by our simulator after executing the framework mentioned above. Thereby, these results are summarized in the following table, which represents the best allocations between worker robots and fires by considering their needs in terms of skills and initial simulation parameters (i.e. a worker robot is assigned to a task if at least it offers a skill to this latter).

As we can clearly see in Table 3, the task having the identifier "Fire000" was assigned to the worker robots having respectively the identifiers "Worker000" and "Worker006" by considering their respective skills "D" and "C". Also, the task with the identifier "Fire002" was assigned to the robot "Worker003" by using its skill "E", and so on…

Now, the following table summarizes the detection time of each fire by the explorer robots (this time represents the order of treatment of fires, i.e. first detected first treated), the time of their management (i.e. the designation of worker robots that must treat them), and finally the approximate time required for their extinction (Table 4).

In the case where two fires are discovered at the same time (have the same priorities), the one with the smallest time of extinction will be treated in the first place. Therefore, the following figure represents the time taken by each task for its accomplishment. Moreover, this time comprises the sum of three different times which are: the time required to choose worker robots that must cooperate to extinguish the fire, i.e. the management time (the red color), the waiting time taken by worker robots to start

Table 3. The solution of the MRTA problem

	Fire000	Fire001	Fire002	Fire003	Fire004	Fire005
Worker000	D	–	–	–	D	–
Worker001	–	–	–	E	–	–
Worker002	–	–	–	–	–	–
Worker003	–	–	E	B	C	E
Worker004	–	–	–	–	–	–
Worker005	–	B	–	–	–	–
Worker006	C	–	–	A	A	C

Table 4. Summary of the different information specific to the order of execution of tasks

Fire	Detection time	Management time	Required times for extinction [sec]	Priority
Fire000	16:07:51	16:07:53	8	4
Fire001	16:07:21	16:07:23	6	2
Fire002	16:07:38	16:07:40	9	3
Fire003	16:08:10	16:08:12	4	5
Fire004	16:07:14	16:07:16	4	1
Fire005	16:07:38	16:07:40	8	3

the extinction of fires, i.e. the effective beginning of works (the yellow color), and finally the approximate time of extinction (the green color). To simplify things, in our case we have ignored the traveling times of worker robots between the locations of the different tasks.

The following table summarizes all the results presented in Fig. 6. Thereby, we can observe that the waiting time for the majority of tasks is zero, except for the fire having the identifier “Fire002”, because this latter was discovered at the same time as the fire

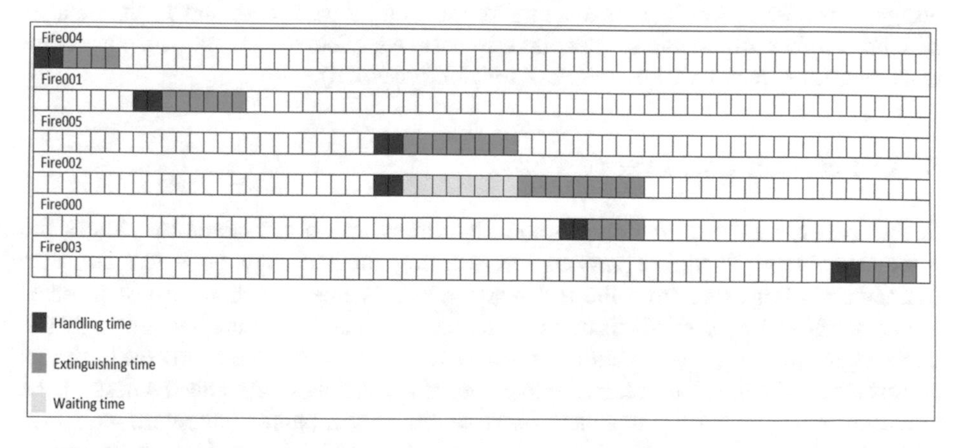

Fig. 6. The different times taken for the accomplishment of tasks (Color figure online)

"Fire005", and the robot "Worker003" was involved in both tasks at the same time; thus the fire having the highest processing time must wait. On the other hand, we can easily calculate the total time taken by all the fires for their extinction which is 59 s (Table 5).

Table 5. Summary of the different information specific to the execution times of tasks

Fire	Handling time [sec]	Extinguishing time [sec]	Waiting time [sec]	Total [sec]
Fire000	2	8	0	10
Fire001	2	6	0	8
Fire002	2	9	8	19
Fire003	2	4	0	6
Fire004	2	4	0	6
Fire005	2	8	0	10
Total [sec]	12	39	8	59

Finally, the found results show the capacity and efficiency of the proposed working methodology for solving the MRTA problem, on a framework of fire extinction in an environment altered by a natural disaster.

4.5 Management of Deadlock Situations

Before concluding this document, it is very important to mention that, like real frameworks of fire extinction, the working methodology that we have developed, to deal with such situations, sometimes does not find solutions to the desired tasks (allocate worker robots to fires). In our paper, such situations are called deadlock situations. Generally, the main cause of the occurrence of such deadlock situations is the lack of skills required by the fires, that is to say there is no worker robots which can offer these skills. Therefore, in order to handle such situations, when we want to designate worker robots to extinguish a given fire, and if we cannot find a solution after a finite number of attempts, then in this case we yield to temptation, and other meta-measures must be taken (e.g. ask for reinforcement).

5 Conclusion and Future Work

In this paper, we have presented some algorithms and organigrams to address the MRTA problem. Then, we applied our working methodology on a test structure (framework), borrowed from the real-world which is the extinction of fires in environments altered by natural disasters. Also, we simulated quantitatively and qualitatively our working methodology on several test frameworks to evaluate its performance. Finally, the obtained experimental results show the effectiveness of the proposed approach for the treatment of these kind of problems. For future work, we will adopt other criteria for better management of fires, for example we can consider:

1. The dynamic of fires and robots in the environment, e.g. the fires can spread in the environment and their needs in terms of skills change over time, the robots operating in the environment can disappear at any time, etc.
2. The priorities of task treatment may change during the various operations.
3. The equitable assignment of worker robots to different tasks (do not overload some robots and let others rest).

Acknowledgment. We would like to thank vividly MS Abida Habiba FARDJALLAH for her help in improving this manuscript.

References

Barry B.W., Maja, J.M.: Broadcast of local eligibility: behavior-based control for strongly cooperative robot teams. In: Proceedings of the Fourth International Conference on Autonomous Agents, pp. 21–22 (2000). https://doi.org/10.1145/336595.336621

Botelho, S.C., Alami, R.: M+: a scheme for multi-robot cooperation through negotiated task allocation and achievement. In: Proceedings 1999 IEEE International Conference on Robotics and Automation (Cat No99CH36288C), vol. 2, pp. 1234–1239 (1999). https://doi.org/10.1109/robot.1999.772530

Brian, P.G., Maja, J.M.: A formal analysis and taxonomy of task allocation in multi-robot systems. Int. J. Robot. Res. **23**(9), 939–954 (2004). https://doi.org/10.1177/0278364904045564

Chaimowicz, L., Campos, M., Kumar, V.: Dynamic role assignment for cooperative robots. In: Proceedings of the IEEE International Conference on Robotics and Automation, vol. 1, pp. 293–298 (2002). https://doi.org/10.1109/robot.2002.1013376

Dandan, Z., Guangming, X., Junzhi, Y., Long, W.: Adaptive task assignment for multiple mobile robots via swarm intelligence approach. Robot. Auton. Syst. **55**(7), 572–588 (2007). https://doi.org/10.1016/j.robot.2007.01.008

Daniele, C., Daniele, N.: Performance evaluation of pure-motion tasks for mobile robots with respect to world models. Auton. Robots **27**, 465–481 (2009). https://doi.org/10.1007/s10514-009-9150-y

Ding, Y., He, Y., Jiang, J.: Multi-robot cooperation method based on the ant algorithm. In: Proceedings of the 2003 IEEE Swarm Intelligence Symposium, pp. 14–18 (2003). https://doi.org/10.1109/sis.2003.1202241

Fang, T., Parker, L.: Asymtre: automated synthesis of multi-robot task solutions through software reconfiguration. In: Proceedings of the 2005 IEEE International Conference on Robotics and Automation, pp. 1501–1508 (2005). doi:https://doi.org/10.1109/robot.2005.1570327

Gerkey, B.P., Maja, J.M.: Sold!: auction methods for multirobot coordination. IEEE Trans. Robot. Autom. **18**(5), 758–768 (2002). https://doi.org/10.1109/TRA.2002.803462

Hang, Y., Liu, S.: Survey of multi-robot task allocation. CAAI Trans. Intell. Syst. **3**(2), 373–376 (2008)

Fister Jr., I., Yang, X.-S., Karin, L., Dusan, F., Janez, B., Iztok, F.: Towards the novel reasoning among particles in PSO by the use of RDF and SPARQL. Sci. World J. **2014**(121782), 10–45 (2014). https://doi.org/10.1155/2014/121782

Kalra, N., Martinoli, A.: Comparative study of market-based and threshold-based task allocation. Proc. Distrib. Auton. Robot. Syst. **7**, 91–101 (2006). https://doi.org/10.1007/4-431-35881-1_10

Liu, S.H., Zhang, Y.: Multi-robot task allocation based on particle swarm and ant colony optimal. J. Northeast Normal Univ. **41**(4), 68–72 (2009)

Liu, S.H., Zhang, Y.: Multi-robot task allocation based on swarm intelligence. J. Jilin Univ. **40** (1), 123–129 (2010)

Bernardine, M., Traderbots, D.: A new paradigm for robust and efficient multirobot coordination in dynamic environments. Ph.D. thesis, Robotics Institute, Carnegie Mellon University (2004)

Nadia, N., de Mendonça, R.M., de Macedo Mourelle, L.: PSO-based distributed algorithm for dynamic task allocation in a robotic swarm. Procedia Comput. Sci. **51**, 326–335 (2015). https://doi.org/10.1016/j.procs.2015.05.250

Parker, L.: Alliance: an architecture for fault tolerant multirobot cooperation. EEE Trans. Robot. Autom. **14**(2), 220–240 (1998). https://doi.org/10.1109/70.681242.60

Rafael, L.B.: Quantum genetic algorithms for computer scientists. Computers **5**(24), 2–31 (2016)

Richard, S.S., Andrew, G.B.: Introduction to Reinforcement Learning. MIT Press, Cambridge (1998)

Sanem, S., Tucker, B.: A distributed multi-robot cooperation framework for real time task achievement. In: Distributed Autonomous Robotic Systems 7, 187–196 (2006). doi:https://doi.org/10.1007/4-431-35881-1_19

Shuhua, L., Tieli, S., Chih-Cheng, H.: Multi-robot task allocation based on swarm intelligence. Multi-Robot Syst. Trends Dev., 393–408 (2011). doi:https://doi.org/10.5772/13106

Simonin, O., Charpillet, F., Thierry, E.: Collective construction of numerical potential fields for the foraging problem. Swarm Intell. **23**, 70 (2007)

Smith, R.G.: The contract net protocol: high-level communication and control in a distributed problem solver. IEEE Trans. Comput. **29**(12), 1104–1113 (1980). https://doi.org/10.1109/tc.1980.1675516

Yu, Z., Liu, S.H.: A quantum-inspired ant colony optimization for robot coalition formation. In: Proceedings of Chinese Control and Decision Conference, pp. 632–637 (2009)

Yang, D., Wang, Z.O.: Improved ant algorithm for assignment problem. J. Tianjin Univ. **37**(4), 373–376 (2004)

Zhang, Y., Liu, S.H.: Large-scale multi-robot task allocation based on ant colony algorithm. In: Proceedings of Chinese Control and Decision Conference, pp. 2057–2062 (2008)

Zlot, R., Stentz, A., Dias, M., Thayer, S.: Multi-robot exploration controlled by a market economy. In: Proceedings 2002 IEEE International Conference on Robotics and Automation (Cat No02CH37292), pp. 3016–3023 (2002). https://doi.org/10.1109/robot.2002.1013690

Inclusive Access to Emergency Services: A Complete System Focused on Hearing-Impaired Citizens

Vaso Constantinou(✉)

Cyprus Interaction Lab, Cyprus University of Technology, 30 Archbishop Kyprianou Str., 3036 Lemesos, Cyprus
va.constantinou@edu.cut.ac.cy

Abstract. In case of emergency, hearing impaired people are not always able to access emergency services and, hence, they do not have equal access to social support and infrastructure. In this work, we undertake the development and evaluation of a system aiming to meet the communication needs of hearing impaired citizens in cases of emergency. The system consists of (i) a mobile application that records and sends the details of an emergency event, and (ii) a central management system that handles these calls from the operation center at the emergency services. The system was completed in four cycles of design, development and evaluation with the involvement of 74 hearing impaired users and three officers from the Cyprus Police (Emergency Response Unit).

Keywords: Technology for hearing impaired people
Inclusive citizen participation · Inclusive design · Action research
Emergency services

1 Introduction

In the European Union 9% of the total population, that is 44 million people, are deaf or hard of hearing [6]. In Cyprus, where this study was conducted, the number of deaf or hard of hearing people is currently approximated at 1000, according to official data from the Ministry of Labor Welfare and Social Insurance. The rights of people with disabilities are registered internationally by the UN Convention on the Rights of Persons with Disabilities and locally, by the constitution and laws of the Republic of Cyprus [13]. Unequal treatment is determined as the impossible or unreasonably difficult access to services and the failure to implement changes, such as the use of specific tools that would facilitate access to services for persons with disabilities. However, although the legislation guarantees the rights of people with disabilities, in practice the implementation of accessibility for all is limited, with an emphasis on physical access. In this work, we emphasize that the concept of access is not limited to physical access, but should also include access to services to all citizens including those with disabilities.

A few studies have focused on the obstacles faced by hearing impaired people when contacting emergency services. These studies have shown that they have

M. E. Auer and T. Tsiatsos (Eds.): IMCL 2017, AISC 725, pp. 969–976, 2018.
https://doi.org/10.1007/978-3-319-75175-7_94

significantly less recorded access to primary care and emergency services [12], which is largely due to infrastructure deficiencies [9]. According to the European Commission, the majority of disabled people have no access to the EU emergency number 112, mainly due to weak infrastructure, equipment and procedures. Seven countries have implemented infrastructures for 112 in order to be accessible by people with hearing disabilities [6]. The solutions implemented to date vary. In some countries, specialized text phones are used and communication is made by exchanging messages, or in some cases, text is translated into voice through a relay service. Another solution implemented in France involves communication by sending a fax using preprinted sheets [6]. Some other solutions allow the exchange of SMS with the emergency services, but have the disadvantage of a possible delay in messaging as supported by Chiu et al. [3] and Meng et al. [8]. Considering that hearing impaired citizens are not always very proficient at using the written language and therefore their ability to use SMS as a communication tool in emergency situations is limited [6], some EU countries have used communication services with predefined SMS messages containing the event location using GPS [6]. As another option, specialized video relay services support the communication between a hearing-impaired person and normal hearing person through an intermediate operator who translates from and to the sign language.

This study undertakes the development and evaluation of a system aiming to meet the communication needs of hearing impaired citizens in cases of emergency. The system consists of (i) a mobile application which records and sends the details of an emergency event, and (ii) a central management system that handles these calls from the end of the emergency services (https://youtu.be/28fGVy41dFY). The implementation was completed in four cycles, for the development of applications and the overall evaluation.

2 Background Work

There are quite a few studies elaborating on the needs of people with hearing impairment and the conditions they face when they need to contact the emergency services. Overall, researchers identify poor accessibility and the need for the development of an effective emergency communication system. Reeves et al. [12], found that deaf or hard of hearing people have significantly less access to primary care and emergency services, facing difficulties at all stages of the process of health services. Similarly, the Northern Virginia Resource Center for Deaf and Hard of Hearing Persons [9] identified weaknesses in emergency response infrastructure and presented an extensive list of recommendations to ensure the reliability and availability of this infrastructure [9].

In terms of empirical work targeting the development and evaluation of relevant applications in this area, a few efforts have been made, while the state of the art is still in its infancy. To name a few examples, the work of Buttussi et al. [2] focused on language barriers between emergency medical responders and hearing-impaired people. The idea was for the emergency medical responders to be able to ask their deaf patients some basic questions such as location and intensity of pain to help them identify the right pathology and proper treatment. The system was developed for PDAs and it allowed health professionals to communicate with deaf patients with the use of pre-proposals relating to the emergency in sign language.

In their study, Zafrulla et al. [16], used the TTY/TDD (Telecommunication Device for the Deaf) technology and developed an emulator for Symbian (Nokia) to allow deaf users to communicate directly with the emergency services using SMS. Also, a group of researchers at the University Trás-os-Montes e Alto Douro [10] have developed SOSPhone, a mobile application that allows emergency calls from users with hearing disability or speaking difficulties. The application uses icons to represent the type of the event and can send an SMS message to the emergency center that contains the selected information, the user profile and the user coordinates if available. However, evaluation data showed that the use of SMS messages may involve risks in the process, because it does not ensure a safe and timely delivery of the message [8]. It also showed a large percentage of failed attempts in some scenarios, without a clear explanation of the reason [10].

The study of Weppner and Lukowicz [15] describes the design, implementation and evaluation of a mobile application that has been introduced as the official app for emergency calls in Germany. The user selects the type of assistance required and the system automatically creates a voice message containing the characteristics of the event. The message is then repeated until a response from the emergency services is recorded. The system was studied on different devices and conditions ranging from silence to people who spoke from a distance of one meter. Although researchers claim that the level and the sound quality were adequate, data on noise levels (dB) and the precision of the result were not reported. Furthermore, the program REACH112 (Responding to All Citizens needing Help) was a 3-year pilot project applied in Sweden, UK, Holland, France and Spain allowing disabled users to communicate directly with emergency services. The pilots have been implemented include various scenarios. In France, the implementation was done by combining (1) operators who are deaf and use sign language in video and messages, and (2) operators who are not deaf and use sound and messages. In the Netherlands, the implementation was done using SMS. In Spain, the implementation was based on video relay services (Video Relay Center) as a link between the deaf caller and the corresponding operator [11]. Also, in their study, Liu et al. [7], evaluated a mobile interface simulator, namely, PeacePHONE which was designed for people with hearing disabilities. Results showed that users appreciated features such as the direct interaction, shortcut keys for reporting emergencies, the ability to receive alerts for emergencies in the event of fire and operation of the outside door bell.

In the field of emergency services and crisis management in general, Díaz et al. [5, 14], analyzed the possibility of involving citizens on crisis management. They proposed a theoretical framework for the contribution of information and improving communication and cooperation between citizens and service agents during these crisis situations. The issue of inclusion of people with disabilities, however, was not considered in this work. Similarly, Aloudat and Michael [1] examined the emergency services implementation issues through the mobile networks of Australia, with a view to implementing an integrated early warning system for mobile devices; yet it is not clear how people with (hearing or other) disabilities could benefit from this warning system.

In the literature examined there is sufficient evidence of the severity of the access problems faced by people with hearing disabilities seeking help from the emergency services. Although a few efforts are recorded there is no universal framework. Fragmentation of solutions, each with its pros and cons, does not help addressing the problem. For example, the implementation using translators or other intermediary tools, such as the pilot implementation of 112 in Spain and France, does not take into account the existence of many different dialects of sign language in the EU, and is in contrast with findings suggesting that deaf users prefer to use communication technologies that do not rely on sign language [4]. Using SMS as a primary means of communication, such as in the pilot implementation of 112 in the Netherlands, may have negative results because many people with disabilities are not good with using the written language [6]. Sending-receiving an SMS may also have significant delays in case of emergency [3, 8]. Overall, our review of the literature regarding inclusive access to emergency services shows that, although a few efforts have been made, there is currently no universal solution, but rather many fragmented ones with proven limitations. In this action research work, we aimed to build a system to address this immediate need.

3 Methodology

The present study was completed in cycles of design, development and evaluation with the involvement of 74 hearing impaired users and three officers from the Cyprus Police (Emergency Response Unit). Data was collected in each cycle and findings lead to improvements of the system, which was subject to investigation in the next cycle, until all requirements were met in a total in four cycles (Fig. 1).

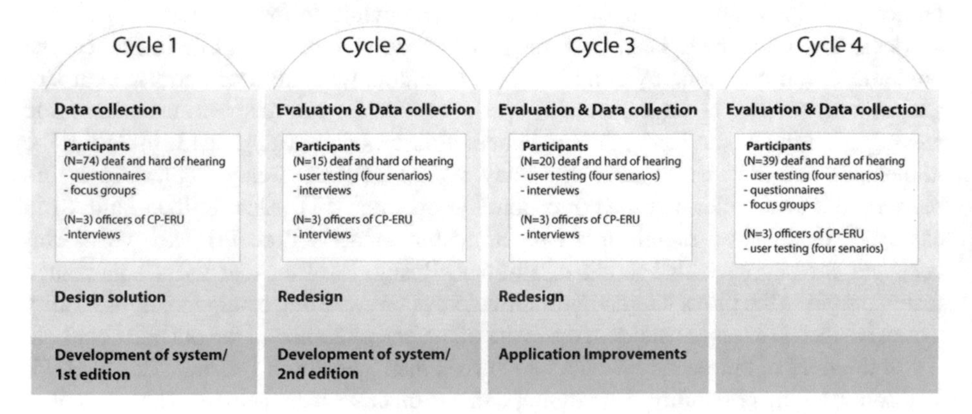

Fig. 1. The cyclic process of design and evaluation

4 Participants

The study involved a total of 74 participants, partially 35 men (47.3%), 61 deaf (82.4%) and 13 (17.6%) hard of hearing. Other participants were three officers from the Cyprus Police - Emergency Response Unit (CP-ERU).

4.1 Procedures

Cycle 1. This cycle involved face-to-face meetings with the deaf participants (N = 74) where data was collected via questionnaires and focus groups. Also, data was collected via interviews with the three officers of CP-ERU. The data collected led us to the identification of requirements, the preparation of the specification and design of a functional prototype of the system.
Cycle 2. This cycle involved a subgroup of 15 deaf participants with range of ages from 21 to 64. The prototype system was examined by each of them completing four scenarios, aiming to present any weaknesses or problems encountered. The scenarios included: (S1) There is an accident on the motorway (police call); (S2) There is a fainted boy (ambulance call); (S3) There is a fire in a rural area (fire call); (S4) You came home and you realize that it has been burgled and your mobile does not have GPS (police call without GPS). Evaluation data fed into the second version of the system that included all the changes proposed, both by users and CP-ERU officers.
Cycle 3. In this cycle, a new subgroup of 20 deaf users (ages 24–61) participated in the evaluation of the application using the same scenarios and data collection procedures as in cycle 2. Another meeting was also held with the CP-ERU officers. The final version of the system is presented in the next session.
Cycle 4. A final round of evaluation was conducted in cycle 4 with 39 deaf participants (ages 22–68) for whom the system was new. Five usability-type measures were gathered: (1) time for completing four scenarios (same scenarios used in cycles 2 and 3; (2) user errors; (3) errors repeated more than once; (4) number of unsuccessful attempts to reach the emergency services, (5) number of unsuccessful attempts to reach the emergency services. Moreover, the participants completed a usability questionnaire and qualitative data was collected via focus groups at the end of the experience. Last, three CP-ERU officers were observed using the system (responding to the 4 scenarios) from the end of the CP emergency services.

5 System

The developed system is based on the below five requirements collected during the first cycle: (i) the system need to be simple to use regardless of age or user skills, particularly language and writing skills, (ii) allow immediate access to emergency services without the need for intermediaries, (iii) allow the necessary data to reach the emergency services so as response to the event is possible (iv) work with simple existing infrastructure available in all EU emergency services and (v) be well received by interested parties/stakeholders (therefore more likely to be adopted).

The system is composed of two parts. First, the system includes a mobile application for deaf users that collects the event data, creates an XML file containing the information and sends it to a specific address on the server. Second, the system includes a data management application at the end of the CP-ERU, recording events into a database (MySQL).

The mobile application was developed using android studio considering the large market share for Android OS. The main screen of the application contains three predefined function buttons for emergency calls: (1) police, (2) ambulance, (3) the fire department, using GPS for tracking the location of the incident. By pressing one of them the application creates an XML that contains all the elements of the event and sends to a specific address. A fourth button was added for emergency calls without GPS, in case the device does not have GPS or the user is located in a space where communication with satellites is not possible (Fig. 2). In this case, the application requires the "Message" field to be completed by the user to record his position.

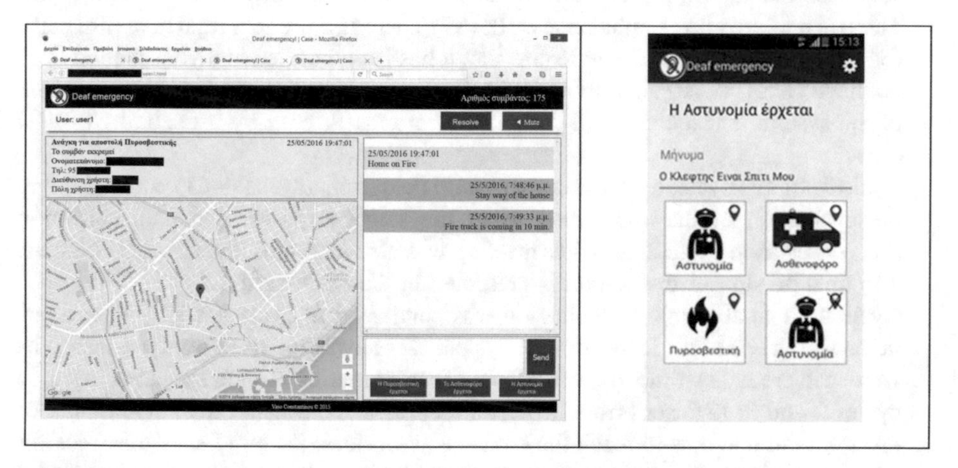

Fig. 2. Data management application with messaging from the operator (left); message received at the end of the user (right)

The data management application makes use of JavaScript for retrieving and displaying data and PHP for XML management and data recording in the database. The application allows for incident data viewing and sending messages to the users. The application checks for new records (every 10 s) and displays the cumulative number of events, listed in three categories: all events, outstanding and resolved.

The emergency services operator can send a predetermined system message to the user such as, "The Police is coming", "The Ambulance is coming" or "The Fire truck is coming", or s/he can type a customized message (Fig. 2). The system allows the recording of all actions taken including messages exchanged between the user and the emergency services.

6 Analysis and Results

Results from the final round of evaluation demonstrated how the system can provide easy and direct access to emergency services, without the need of any intermediate, enabling the inclusion of these citizens in a critical process such as the response to an emergency. The application is simple and easy to use by anyone regardless of age, writing ability or language skills. The system can send the data of the type of assistance required, even when the users are unable to complete the message field. Previous works showed that deaf and hard of hearing users are not in favor of implementations using translators or other intermediaries (e.g., 112 in Spain and France) and prefer to use communication technologies that do not rely on sign language [4]. The present icon-based system gives direct access to emergency services without the use of intermediaries.

The system does not rely on the use of SMS. Because many people with disabilities are not good with using the written language [6], SMS cannot be the primary means of communication and may also result in significant delays in case of emergency [3]. The present icon-based system requires an SMS for the user to record his/her position only when GPS is not available.

Regarding the effectiveness of the system, the final phase of evaluation involved 39 deaf or hard of hearing users completing four emergency scenarios with a success rate 100%. This is a notable progress given that similar studies recorded successful completion rates between 35% and 86% in varied scenarios and were, for the most part, conducted with non-deaf people [10]. During the 4th cycle, we measure the implementation period of four scenarios. The total time ranged from 20 to 70 s with an average completion time of 38.25 s, which included calling for help, typing messages, and reading all the relevant messages sent and received.

Concerning the operation requirements, the system can work with simple existing infrastructure available in all EU countries and does not require additional equipment and specialized staff at the end of the police Emergency Response Unit. The system allows the recording of detailed data of the incidents as well as actions taken for their resolution. Currently, the inability of the deaf to communicate directly with the emergency services leads to no registered statistics on incidents involving this special population in EU and internationally.

Acknowledgment. This project has received funding from the European Union's Horizon 2020 research and innovation programme under grant agreement No. 692058.

References

1. Aloudat, A., Michael, K.: Toward the regulation of ubiquitous mobile government: a case study on location-based emergency services in Australia. Electron. Commer. Res. **11**, 31–74 (2010)
2. Buttussi, F., Chittaro, L., Carchietti, E., Coppo, M.: Using mobile devices to support communication between emergency medical responders and deaf people. In: Proceedings of the 12th International Conference on Human Computer Interaction with Mobile Devices and Services - MobileHCI 2010 (2010)
3. Chiu, H., Liu, C., Hsieh, C., Li, R.: Essential needs and requirements of mobile phones for the deaf. Assistive Technol. **22**, 172–185 (2010)
4. Cromartie, J., Gaffey, B., Seaboldt, M.: Evaluating Communication Technologies for the Deaf and Hard of Hearing (2012)

5. Díaz, P., Aedo, I., Romano, M., Onorati, T.: A framework to integrate large-scale participation in Disaster and Emergency Management. In: Conference: Workshop on Large-Scale Ideation and Deliberation Systems, 6th International Conference on Communities and Technology (2013)
6. EENA European Emergency Number Association.: 112 Accessibility for People with Disabilities (2012)
7. Liu, C., Chiu, H., Hsieh, C., Li, R.: Optimizing the usability of mobile phones for individuals who are deaf. Assistive Technol. **22**, 115–127 (2010)
8. Meng, X., Zerfos, P., Samanta, V., Wong, S., Lu, S.: Analysis of the reliability of a nationwide short message service. In: IEEE INFOCOM 2007 - 26th IEEE International Conference on Computer Communications (2007)
9. Northern Virginia Resource Center for Deaf and Hard of Hearing Persons.: Emergency preparedness and emergency communication access (2004)
10. Paredes, H., Fonseca, B., Cabo, M., Pereira, T., Fernandes, F.: SOSPhone: a mobile application for emergency calls. Univ. Access Inf. Soc. **13**, 277–290 (2013)
11. REACH112 REsponding to All Citizens Needing Help.: Final Project Report (2012)
12. Reeves, D., Kokoruwe, B., Dobbins, J., Newton, V.: Access to Primary Care and Accident and Emergency Services for Deaf People in the North West. A report for the NHS Executive North West Research and Development Directorate (2002)
13. Republic of Cyprus.: A Law to provide for persons with disabilities (2000)
14. Romano, M., Onorati, T., Díaz, P., Aedo, I.: Improving emergency response: citizens performing actions. In: Proceedings of the 11th International ISCRAM Conference (2014)
15. Weppner, J., Lukowicz, P.: Emergency app for people with hearing and speech disabilities: design, implementation and evaluation according to legal requirements in Germany. In: Proceedings of the 8th International Conference on Pervasive Computing Technologies for Healthcare (2014)
16. Zafrulla, Z., Etherton, J., Starner, T.: TTY phone. In: Proceedings of the 10th International ACM SIGACCESS Conference on Computers and Accessibility - Assets 2008 (2008)

Author Index

M. E. Auer and T. Tsiatsos (Eds.): IMCL 2017, AISC 725, pp. 977–980, 2018.
https://doi.org/10.1007/978-3-319-75175-7